MORGANTOWN PUBLIC LIBRARY
373 SPRUCE STREET
MORGANTOWN, WV 26505

$ 30.00

MORGANTOWN PUBLIC LIBRARY
373 SPRUCE STREET
MORGANTOWN, WV 26505

1018443

EXPLORING THE UNKNOWN

NASA SP-4407

EXPLORING THE UNKNOWN

*Selected Documents in the History of the
U.S. Civil Space Program
Volume I: Organizing for Exploration*

John M. Logsdon, Editor
with Linda J. Lear, Jannelle Warren-Findley,
Ray A. Williamson, and Dwayne A. Day

The NASA History Series

National Aeronautics and Space Administration
NASA History Office
Washington, D.C.   1995

MORGANTOWN PUBLIC LIBRARY
373 SPRUCE STREET
MORGANTOWN, WV 26505

Library of Congress Cataloguing-in-Publication Data

Exploring the Unknown: Selected Documents in the History of the U.S. Civil Space Program / John M. Logsdon, editor with Linda J. Lear... [et al.]
p. cm.—(The NASA history series) (NASA SP: 4407)

    Includes bibliographical references and indexes.
    Contents: v. 1. Organizing for exploration
    1. Astronautics—United States--History.   I. Logsdon, John M., 1937– .
II. Lear, Linda J., 1940– .   III. Series.   IV. Series: NASA SP: 4407.
TL789.8.U5E87   1995   95-9066
387.8'0973–dc20   CIP

To the Memory of Eugene M. Emme
(1919–1985)

The First NASA Historian, Without Whose Early Vision
This Collection Would Not Have Been Possible

Contents

Acknowledgments xv

Introduction xvii

Biographies of Volume I Essay Authors xxi

Glossary xxiii

Chapter One

Essay: "Prelude to the Space Age," by Roger D. Launius 1

Documents

I-1 and I-2 Medieval Universe at the Time of Dante; and The Infinite Universe of Thomas Digges 22

I-3   Edward E. Hale, "The Brick Moon," *The Atlantic Monthly*, October 1869, pp. 451-460; November 1869, pp. 603-611; December 1869, pp. 679-688; February 1870, pp. 215-222 23

I-4   Percival Lowell, *Mars* (Boston: Houghton Mifflin Co., 1895), pp. 201-212 55

I-5   K.E. Tsiolkovskiy, "Reactive Flying Machines," in *Collected Works of K.E. Tsiolkovskiy, Volume II—Reactive Flying Machines* (Moscow, 1954) 59

I-6   Hermann Oberth, *Rockets in Planetary Space* (Munich and Berlin, 1923, NASA TT F-9227, December 1964) 84

I-7 and I-8 Robert H. Goddard, *A Method of Reaching Extreme Altitudes*, Smithsonian Miscellaneous Collections, Volume 71, Number 2 (Washington, DC: Smithsonian Institution Press, 1919); and "Topics of the Times," *New York Times*, January 18, 1920, p. 12 86

I-9   Robert H. Goddard, *Liquid-propellant Rocket Development*, Smithsonian Miscellaneous Collections, Volume 95, Number 3 (Washington, DC: Smithsonian Institution Press, 1936) 134

I-10   H.E. Ross, "The B.I.S. Space-ship," Journal of the British *Interplanetary Society*, 5 (January 1939): 4-9 140

I-11   Frank J. Malina and A.M.O. Smith, "Flight Analysis of the Sounding Rocket," *Journal of Aeronautical Sciences*, 5 (1938): 199-202 145

I-12   Theodore von Kármán, "Memorandum on the Possibilities of Long-Range Rocket Projectiles," and H.S. Tsien and F.J. Malina, "A Review and Preliminary Analysis of Long-Range Rocket Projectiles," Jet Propulsion Laboratory, California Institute of Technology, November 20, 1943 153

I-13   "What Are We Waiting For?"; and Dr. Wernher von Braun, "Crossing the Last Frontier," *Collier's*, March 22, 1952, pp. 23-29, 72-73 176

I-14	Dr. Wernher von Braun, "Man on the Moon: The Journey," *Collier's,* October 18, 1952, pp. 52-59 189
I-15	Dr. Fred L. Whipple, "Is There Life on Mars?," *Collier's,* April 30, 1954, p. 21 194
I-16	Dr. Wernher von Braun with Cornelius Ryan, "Can We Get to Mars?," *Collier's,* April 30, 1954, pp. 22-29 195
I-17	IGY Statement by James C. Hagerty, The White House, July 29, 1955 200
I-18	F.C. Durant III, "Report of Meetings of Scientific Advisory Panel on Unidentified Flying Objects Covered by Office of Scientific Intelligence, CIA, January 14-18, 1953," February 16, 1953 201
I-19	"Air Force's 10 Year Study of Unidentified Flying Objects," Department of Defense, Office of Public Information, News Release No. 1083-58, November 5, 1957 207

Chapter Two

Essay: "Origins of U.S. Space Policy: Eisenhower, Open Skies, and Freedom of Space," by R. Cargill Hall 213

Documents

II-1	Louis N. Ridenour, "Pilot Lights of the Apocalypse: A Playlet in One Act," *Fortune,* Vol. 33, January 1946 230
II-2	Douglas Aircraft Co., "Preliminary Design of an Experimental World-Circling Spaceship," Report No. SM-11827, May 2, 1946, pp. i-viii, 1-16, 211-212 236
II-3	J.E. Lipp, et al., "The Utility of a Satellite Vehicle for Reconnaissance," The Rand Corporation, R-217, April 1951, pp. ix, 1-21, 28-39 245
II-4	R.M. Salter, "Engineering Techniques in Relation to Human Travel at Upper Altitudes," *Physics and Medicine of the Upper Atmosphere: A Study of the Aeropause* (Albuquerque: University of New Mexico Press, 1952), pp. 480-487 262
II-5	A.V. Grosse, "Report on the Present Status of the Satellite Problem," August 25, 1953, pp. 2-7 266
II-6	J.E. Lipp and R.M. Salter, "Project Feed Back Summary Report," The Rand Corporation, R-262, Volume II, March 1954, pp. 50-60 269
II-7	Wernher von Braun, "A Minimum Satellite Vehicle: Based on components available from missile developments of the Army Ordnance Corps," September 15, 1954 274
II-8	"On the Utility of an Artificial Unmanned Earth Satellite," *Jet Propulsion,* 25 (February 1955), pp. 71-78 281

II-9	U.S. National Committee for the International Geophysical Year 1957-58, "Summary Minutes of the Eighth Meeting," May 18, 1955	295
II-10	National Security Council, NSC 5520, "Draft Statement of Policy on U.S. Scientific Satellite Program," May 20, 1955	308
II-11	S.F. Singer, "Studies of a Minimum Orbital Unmanned Satellite of the Earth (MOUSE)," *Astronautica Acta,* 1 (1955): 171-184	314
II-12	"Memorandum of Discussion at the 322d Meeting of the National Security Council, Washington, D.C., May 10, 1957," *United Nations and General International Matters,* Vol. XI. *Foreign Relations of the United States, 1955-1957* (Washington, DC: U.S. Government Printing Office, 1988), pp. 748-754	324
II-13	Allen W. Dulles, Director of Central Intelligence, to Donald Quarles, Deputy Secretary of Defense, July 5, 1957	329
II-14	"Announcement of the First Satellite," *Pravda,* October 5, 1957	329
II-15	John Foster Dulles to James C. Hagerty, October 8, 1957	331
II-16	President's Science Advisory Committee, "Introduction to Outer Space," March 26, 1958, pp. 1-2, 6, 13-15	332
II-17	"National Aeronautics and Space Act of 1958," Public Law 85-568, 72 Stat., 426	334
II-18	National Security Council, NSC 5814, "U.S. Policy on Outer Space," June 20, 1958	345
II-19	Nathan F. Twining, Chairman, Joint Chiefs of Staff, Memorandum for the Secretary of Defense, "U.S. Policy on Outer Space (NSC 5814)," August 11, 1958	359
II-20	National Security Council, NSC 5814/1, "Preliminary U.S. Policy on Outer Space," August 18, 1958, pp. 17-19	360
II-21	National Aeronautics and Space Council, "U.S. Policy on Outer Space," January 26, 1960	362
II-22	Cyrus Vance, Deputy Secretary of Defense, Department of Defense Directive Number TS 5105.23, "National Reconnaissance Office," March 27, 1964	373

Chapter Three

Essay: "The Evolution of U.S. Space Policy and Plans," by John M. Logsdon 377

Documents

III-1	Special Committee on Space Technology, "Recommendations to the NASA Regarding A National Civil Space Program," October 28, 1958	394

III-2	Office of Program Planning and Evaluation, "The Long Range Plan of the National Aeronautics and Space Administration," December 16, 1959, pp. 1-3, 9-11, 17-18, 26, 44	403
III-3	President's Science Advisory Committee, "Report of the Ad Hoc Panel on Man-in-Space," December 16, 1960	408
III-4	Richard E. Neustadt, attachment to Memorandum for Senator Kennedy, "Problems of Space Programs," December 20, 1960	413
III-5	"Report to the President-Elect of the Ad Hoc Committee on Space," January 10, 1961	416
III-6	John F. Kennedy, Memorandum for Vice President, April 20, 1961	423
III-7	Robert S. McNamara, Secretary of Defense, Memorandum for the Vice President, "Brief Analysis of Department of Defense Space Program Efforts," April 21, 1961	424
III-8	Lyndon B. Johnson, Vice President, Memorandum for the President, "Evaluation of Space Program," April 28, 1961	427
III-9	Wernher von Braun to the Vice President of the United States, April 29, 1961	429
III-10	"Vice President's Ad Hoc Meeting," May 3, 1961	433
III-11	James E. Webb, NASA Administrator, and Robert S. McNamara, Secretary of Defense, to the Vice President, May 8, 1961, with attached: "Recommendations for Our National Space Program: Changes, Policies, Goals"	439
III-12	John F. Kennedy, Excerpts from "Urgent National Needs," Speech to a Joint Session of Congress, May 25, 1961	453
III-13	Director, Bureau of the Budget, Memorandum for the President, Draft, November 13, 1962, with attached: "Space Activities of the U.S. Government"	454
III-14	James E. Webb, Administrator, NASA, to the President, November 30, 1962	461
III-15	John F. Kennedy, Memorandum for the Vice President, April 9, 1963	467
III-16	Lyndon B. Johnson, Vice President, to the President, May 13, 1963	468
III-17	NASA, *Summary Report: Future Programs Task Group*, January 1965	473
III-18	James E. Webb, Administrator, NASA, to the President, August 26, 1966, with attached: James E. Webb, Administrator, NASA, to Honorable Everett Dirksen, U.S. Senate, August 9, 1966	490
III-19	James E. Webb, [NASA] Administrator, Memorandum to Associate Administrator for Manned Spaceflight, "Termination of the Contract for Procurement of Long Lead Time Items for Vehicles 516 and 517," August 1, 1968	494

III-20	Bureau of the Budget, "National Aeronautics and Space Administration: Highlight Summary," October 30, 1968 495
III-21	Charles Townes, et al., "Report of the Task Force on Space," January 8, 1969 499
III-22	Richard Nixon, Memorandum for the Vice President, the Secretary of Defense, the Acting Administrator, NASA, and the Science Adviser, February 13, 1969 512
III-23	T.O. Paine, Acting Administrator, NASA, Memorandum for the President, "Problems and Opportunities in Manned Space Flight," February 26, 1969 513
III-24	Robert C. Seamans Jr., Secretary of the Air Force, to Honorable Spiro T. Agnew, Vice President, August 4, 1969 519
III-25	Space Task Group, *The Post-Apollo Space Program: Directions for the Future*, September 1969 522
III-26	Robert P. Mayo, Director, Bureau of the Budget, Memorandum for the President, "Space Task Group Report," September 25, 1969 544
III-27	Peter M. Flanigan, Memorandum for the President, December 6, 1969 546
III-28	Caspar W. Weinberger, Deputy Director, Office of Management and Budget, via George P. Shultz, Memorandum for the President, "Future of NASA," August 12, 1971 546
III-29	James C. Fletcher, [NASA] Administrator, Memorandum to Dr. Low, "Meeting with Ed David," August 24, 1971 548
III-30	Klaus P. Heiss and Oskar Morgenstern, Memorandum for Dr. James C. Fletcher, Administrator, NASA, "Factors for a Decision on a New Reusable Space Transportation System," October 28, 1971 549
III-31	James C. Fletcher, "The Space Shuttle," November 22, 1971 555
III-32	George M. Low, Deputy Administrator, NASA, Memorandum for the Record, "Meeting with the President on January 5, 1972," January 12, 1972 558
III-33	Nick MacNeil, Carter-Mondale Transition Planning Group, to Stuart Eizenstat, Al Stern, David Rubenstein, Barry Blechman, and Dick Steadman, "NASA Recommendations," January 31, 1977 559
III-34	Presidential Directive/NSC-37, "National Space Policy," May 11, 1978 574
III-35	Zbigniew Brzezinski, Presidential Directive/NSC-42, "Civil and Further National Space Policy," October 10, 1978 575
III-36	George M. Low, Team Leader, NASA Transition Team, to Mr. Richard Fairbanks, Director, Transition Resources and Development Group, December 19, 1980, with attached: "Report of the Transition Team, National Aeronautics and Space Administration" 579

III-37	Hans Mark and Milton Silveira, "Notes on Long Range Planning," August 1981	587
III-38	National Security Decision Directive Number 42, "National Space Policy," July 4, 1982	590
III-39	National Security Decision Directive 5-83, "Space Station," April 11, 1983	593
III-40	"Revised Talking Points for the Space Station Presentation to the President and the Cabinet Council," November 30, 1983, with attached: "Presentation on Space Station," December 1, 1983	595
III-41	Caspar Weinberger, Secretary of Defense, to James M. Beggs, Administrator, NASA, January 16, 1984	600
III-42	Office of the Press Secretary, "Fact Sheet: Presidential Directive on National Space Policy," February 11, 1988	601

Chapter Four

Essay: "Organizing for Exploration," by Sylvia K. Kraemer 611

Documents

IV-1	J.R. Killian, Jr., "Memorandum on Organizational Alternatives for Space Research and Development," December 30, 1957	628
IV-2	L.A. Minnich, Jr., "Legislative Leadership Meeting, Supplementary Notes," February 4, 1958	631
IV-3	S. Paul Johnston, Memorandum for Dr. J.R. Killian, Jr., "Activities," February 21, 1958, with attached: Memorandum for Dr. J. R. Killian, Jr., "Preliminary Observations on the Organization for the Exploitation of Outer Space," February 21, 1958	632
IV-4	James R. Killian, Jr., Special Assistant for Science and Technology; Percival Brundage, Director, Bureau of the Budget; Nelson A. Rockefeller, Chairman, President's Advisory Committee on Government Organization, Memorandum for the President, "Organization for Civil Space Programs," March 5, 1958, with attached: "Summary of Advantages and Disadvantages of Alternative Organizational Arrangements"	637
IV-5	Maurice H. Stans, Director, Bureau of the Budget, Memorandum for the President, "Responsibility for 'space' programs," May 13, 1958	643
IV-6	W.H. Pickering, Director, Jet Propulsion Laboratory, to Dr. T. Keith Glennan, NASA, March 24, 1959	645
IV-7	T. Keith Glennan, *The Birth of NASA: The Diary of T. Keith Glennan* (Washington, DC: NASA Special Publication-4105, 1993), pp. 1-6	647
IV-8	Anonymous, "Ballad of Charlie McCoffus," n.d.	650

IV-9	*Report to the President on Government Contracting for Research and Development,* Bureau of the Budget, U.S. Senate, Committee on Government Operations, 87th Cong., 2d sess. (Washington, DC: U.S. Government Printing Office, 1962), pp. vii-xiii, 1-24	651
IV-10	Albert F. Siepert to James E. Webb, [NASA] Administrator, "Length of Tours of Certain Military Detailees," February 8, 1963	672
IV-11	U.S. Congress, House, Committee on Science and Astronautics, Subcommittee on NASA Oversight, Staff Study, "Apollo Program Management," 91st Cong., 1st sess. (Washington, DC: U.S. Government Printing Office, July 1969), pp. 59-74	674
IV-12	George M. Low, Deputy Administrator, NASA, Memorandum for the Administrator, "NASA as a Technology Agency," May 25, 1971	685
IV-13	George M. Low, Deputy Administrator, NASA, Memorandum to Addressees, "Space Vehicle Cost Improvement," May 16, 1972	687
IV-14	E.S. Groo, Associate Administrator for Center Operations, NASA, to Center Directors, "Catalog of NASA Center Roles," April 16, 1976	688
IV-15	James C. Fletcher, Administrator, NASA, Memorandum to Bob Frosch, "Problems and Opportunities at NASA," May 9, 1977	711
IV-16	Task Force for the Study of the Mission of NASA, NASA Advisory Council, "Study of the Mission of NASA," October 12, 1983, pp. 1-9	717
IV-17	*Report of the Presidential Commission on the Space Shuttle Challenger Accident,* Vol. I (Washington, DC: U.S. Government Printing Office, June 6, 1986), pp. 164-177	723
IV-18	Samuel C. Phillips, NASA Management Study Group, "Summary Report of the NASA Management Study Group Recommendations," December 30, 1986	730
IV-19	NASA, "The Hubble Space Telescope Optical Systems Failure Report," November 1990, pp. iii-v, 9-1 to 9-4, 10-1 to 10-4	735
IV-20	*Report of the Advisory Committee on the Future of the U.S. Space Program* (Washington, DC: U.S. Government Printing Office, December 1990), pp. 47-48	741

Biographical Appendix ... 745

Index ... 771

The NASA History Series ... 793

Acknowledgments

The idea for creating a reference work that would include documents seminal to the evolution of the civilian space program of the United States came from then-NASA Chief Historian Sylvia K. Kraemer. She recognized that while there were substantial primary resources for future historians and others interested in the early years of the U.S. space programs available in many archives, and particularly in the NASA Historical Reference Collection of the History Office at NASA Headquarters in Washington, D.C., this material was widely scattered and contained a mixture of the significant and the routine. It was her sense that it was important to bring together the "best" of this documentary material in a widely accessible form. This collection, and any long-term value it may have, is first of all the result of that vision. Once Dr. Kraemer left her position as NASA Chief Historian to assume broader responsibilities within the agency, the project was guided with a gentle but firm hand by her successor, Roger D. Launius. Without his subtle prodding and supportive advice, the undertaking might have taken even longer than it has to reach closure.

Jannelle Warren-Findley, an independent intellectual/cultural historian, and Linda J. Lear, an adjunct professor of history at George Washington University, approached the Space Policy Institute of George Washington University's Elliott School of International Affairs with the suggestion that it might serve as the institutional base for a proposal to NASA to undertake the documentary history project. This suggestion found a positive response; the Space Policy Institute was created in 1987 as a center of scholarly research and graduate education regarding space issues, and as a resource for those interested in a knowledgeable but independent perspective on past and current space activities. Having the kind of historical base that would have to be created to carry out the documentary history project would certainly enhance the Institute's capabilities, and so the Space Policy Institute joined with Warren-Findley, Lear, and Ray A. Williamson of the congressional Office of Technology Assessment in preparing a proposal to NASA. Much to our delight, we were awarded the contract for the project in late 1988, and the enterprise got officially under way in May 1989.

The undertaking proved more challenging than anyone had anticipated. The combination of getting ourselves started in the right direction, canvassing and selecting from the immense documentary resources available, commissioning essays to introduce the various sections of the work from external authors and writing several essays ourselves, and dealing with conflicting demands on the time of the four principals in the project has led to a delay in publishing the volume beyond what we anticipated when first undertaking the project. The final pieces of the manuscript for this volume were not delivered to NASA until the end of 1993. By that time, both Jannelle Warren-Findley and Linda Lear had moved on to the next steps in their careers, and Ray Williamson, who had taken a nine-month leave from the Office of Technology Assessment in 1990 to work on the project, had long ago returned to his primary job. This meant that Warren-Findley and Lear did not have the opportunity to make the kinds of contribution to the final product for which they had hoped; nevertheless, without their initiative, the effort would not have been located at George Washington University, and they both made crucial contributions to conceptualizing and organizing the work in its early stages. For that, they deserve high credit. Ray Williamson has been able to stay involved with the project on an occasional basis since returning to the Office of Technology Assessment, and he has made important contributions to several sections of the effort.

In its start-up phase, the project profited from the advice of a distinguished advisory panel that met twice formally; members of the panel were also always available for individual consultation. Included on this panel were: Carroll W. Pursell, Jr., Case Western University (chair); Charlene Bickford, First Congress Project; Herbert Friedman, Naval Research Laboratory; Richard P. Hallion, Air Force Historian; John Hodge, NASA (retired); Sally Gregory Kohlstedt, University of Minnesota; W. Henry Lambright, Syracuse University;

Sharon Thibodeau, National Archives and Records Administration; and John Townsend, NASA (retired). Certainly, none of these individuals bear responsibility for the final content or style of this work, but their advice along the way was invaluable.

We owe thanks to the individuals and organizations that have searched their files for potentially useful materials, and for the staffs at various archives and collections who have helped us locate documents. Without question, first among them is Lee D. Saegesser of the History Office at NASA Headquarters, who has helped compile the NASA Historical Reference Collection that contains many of the documents selected for inclusion in this work. All those who in the future will write on the history of the U.S. space program will owe a debt of thanks to Lee; those who have already worked in this area realize his tireless contributions.

Essential to the project was a system for archiving the documents collected. Charlene Bickford, on the basis of her experience with the First Congress Project, advised on our approach to archiving and to developing document headnotes. The archiving system was developed by graduate student John Morris, who also assisted with document collection. The documentary archive has been nurtured with fervor by another graduate student, Dwayne A. Day; Dwayne has made many major contributions to all aspects of the project. Other students who have worked on the project since its inception include Max Nelson, Jordan Katz, Stewart Money, Michelle Heskett, Robin Auger, and Heather Young. All have been a great help.

Beginning with Linda Lear, a series of individuals has struggled to bring editorial consistency to the essays and headnotes introducing the documents included in this work, and to develop an initial version of the work's index. They have included Erica Angst, Kathie Pett Keel, and, during the final two years of completing this volume, particularly Kimberly Carter. Their contributions have been essential to the lasting quality of the end product. Alita Black also helped set up the indexing system.

There are numerous people at NASA associated with historical study, technical information, and the mechanics of publishing who helped in myriad ways in the preparation of this documentary history. J.D. Hunley, of the NASA History Office, edited and critiqued the text before he departed to take over the History Program at the Dryden Flight Research Center; and his replacement, Stephen J. Garber, prepared the index and helped in the final proofing of the work. Nadine Andreassen of the NASA History Office performed editorial and proofreading work on the project; and the staffs of the NASA Headquarters Library, the Scientific and Technical Information Program, and the NASA Document Services Center provided assistance in locating and preparing for publication the documentary materials in this work. The NASA Headquarters Printing and Design Office developed the layout and handled printing. Specifically, we wish to acknowledge the work of Jane E. Penn, Angela M. La Croix, Patricia M. Talbert, and Jonathan L. Friedman for their editorial and design work. In addition, Michael Crnkovic, Craig A. Larsen, and Larry J. Washington saw the book through the publication process.

Finally, the staff of the Space Policy Institute—Kim Lutz, Paul McDonnell, and Flo Williams—have facilitated the effort throughout.

Introduction

One of the most important developments of the twentieth century has been the movement of humanity into space with machines and people. The underpinnings of that movement—why it took the shape it did; which individuals and organizations were involved; what factors drove a particular choice of scientific objectives and technologies to be used; and the political, economic, managerial, and international contexts in which the events of the space age unfolded—are all important ingredients of this epoch transition from an Earthbound to a spacefaring people. This desire to understand the development of spaceflight in the United States sparked this documentary history.

The extension of human activity into outer space has been accompanied by a high degree of self-awareness of its historical significance. Few large-scale activities have been as extensively chronicled so closely to the time they actually occurred. Many of those who were directly involved were quite conscious that they were making history, and they kept full records of their activities. Because most of the activity in outer space was carried out under government sponsorship, it was accompanied by the documentary record required of public institutions, and there has been a spate of official and privately written histories of most major aspects of space achievement to date. When top leaders considered what course of action to pursue in space, their deliberations and decisions often were carefully put on the record. There is, accordingly, no lack of material for those who aspire to understand the origins and early evolution of U.S. space policies and programs.

This reality forms the rationale for this compilation. Precisely because there is so much historical material available on space matters, the National Aeronautics and Space Administration (NASA) decided in 1988 that it would be extremely useful to have easily available to scholars and the interested public a selective collection of many of the seminal documents related to the evolution of the U.S. civilian space program up to that time. While recognizing that much space activity has taken place under the sponsorship of the Department of Defense and other national security organizations, the U.S. private sector, and in other countries around the world, NASA felt that there would be lasting value in a collection of documentary material primarily focused on the evolution of the U.S. government's civilian space program, most of which has been carried out since 1958 under the agency's auspices. As a result, the NASA History Office contracted with the Space Policy Institute of George Washington University's Elliott School of International Affairs to prepare such a collection, with a 1988 cutoff date for documents to be included. This volume and two additional ones detailing programmatic developments and relations with other organizations that will follow are the result.

The documents collected in this research project were assembled from a diverse number of both public and private sources. A major repository of primary source materials relative to the history of the civil space program is the NASA Historical Reference Collection of the NASA History Office located at the agency's Washington headquarters. Project assistants combed this collection for the "cream" of the wealth of material housed there. Indeed, one purpose of this work from the start was to capture some of the highlights of the holdings at headquarters. Historical materials housed at the other NASA installations, and at institutions of higher learning such as Rice University, Rensselaer Polytechnic Institute, and Virginia Polytechnic University, were also mined for their most significant materials. Other collections from which documents have been drawn include the Eisenhower, Kennedy, Johnson, and Carter Presidential Libraries; the papers of T. Keith Glennan, Thomas O. Paine, James C. Fletcher, George M. Low, and John A. Simpson; and the archives of the National Academy of Sciences, the Rand Corporation, AT&T, the Communications Satellite Corporation, INTELSAT, the Jet Propulsion Laboratory of the California Institute of Technology, and the National Archives and Records Administration.

Copies of more than 2,000 documents in their original form collected during this project (not just the documents selected for inclusion), as well as a data base that provides a guide to their contents, have been deposited in the NASA Historical Reference Collection. Another complete set of project materials is located at the Space Policy Institute at George Washington University. These materials in their original forms are available for use by researchers seeking additional information about the evolution of the U.S. civil space program.

The documents selected for inclusion in this volume are presented in four major sections, each covering a particular aspect of the evolution of U.S. space policies and programs. Those sections address: the antecedents to the U.S. space program; the origins of U.S. space policy in the Eisenhower era; the evolution of U.S. space policies and plans; and the organization of the civilian space effort. A second volume of this work will contain documents arranged in four sections addressing specific relations with other organizations: the NASA/industry/university nexus; civil-military space cooperation; international space cooperation; and NASA, commercialization in space, and communications satellites. A third volume will describe programmatic developments: human spaceflight; space science; Earth observation programs; and space transportation.

Each major section in this volume and the two to follow is introduced by an overview essay, prepared either by a member of the project team or by an individual particularly well-qualified to write on the topic. In the main, these essays are intended to introduce and complement the documents in the section and to place them in a chronological and substantive context. In certain instances the essays go beyond this basic goal to reinterpret specific aspects of the history of the civil space program and to offer historiographical commentary or inquiry about the space program. Each essay contains references to the documents in the section it introduces, and many also contain references to documents in other sections of the collection. These introductory essays were the responsibility of their individual authors, and the views and conclusions contained therein do not necessarily represent the opinions of either George Washington University or NASA.

The documents appended to each chapter were chosen by the essay writer in concert with the editorial team from the more than 2,000 assembled by the research staff for the overall project. The contents of this volume emphasize primary documents or long-out-of-print essays or articles and material from the private recollections of important actors in shaping space affairs. The contents of this volume thus do not comprise in themselves a comprehensive historical account; they must be supplemented by other sources, those both already available and to become available in the future. Indeed, a few of the documents included in this collection are not complete; some portions of them are still subject to security classification. As this collection was being prepared, the U.S. government was involved in declassifying and releasing to the public a number of formerly highly classified documents from the period before 1963. As this declassification process continues, increasingly more information on the early history of NASA and the civil space program will come to light.

The documents included in each section are for the most part arranged chronologically, and each document is assigned its own number in terms of the section in which it is placed. As a result, the first document in the third section of the collection is designated "Document III-1." Each document is accompanied by a headnote setting out its context and providing a background narrative. These headnotes also provide specific information about people and events discussed, as well as bibliographical information about the documents themselves. We have avoided the inclusion of explanatory notes in the documents themselves and have confined such material to the headnotes. The editorial method we adopted for dealing with these documents seeks to preserve spelling, grammar, paragraphing, and use of language as in the original. We have sometimes changed punctuation where it enhances readability. We have used ellipses to note sections of a document not included in this publication, and we have avoided including words and phrases that had been deleted in the original document unless they contribute to an understanding of what was going on in the mind of the writer in making the record. Marginal notations on the original documents are inserted into the text of the documents in brackets, each clearly marked as a marginal comment. When deletions to the original document have been made in the process of declassification, we have noted this with a parenthetical statement in brackets. Except insofar as illustrations and figures are necessary to understanding the text, those items have been omitted from this printed version. Copies of all documents in their original form, however, are available for research by any interested person at the NASA History Office or the Space Policy Institute of George Washington University.

We recognize that there are certain to be quite significant documents left out of this compilation. No two individuals would totally agree on all documents to be included from the more than 2,000 that we collected, and surely we have not been totally successful in locating all relevant records. As a result, this documentary history can raise an immediate question from its users: why were some documents included while others of seemingly equal importance were omitted? There can never be a fully satisfactory answer to this question. Our own criteria for choosing particular documents and omitting others rested on three interrelated factors:

- Is the document the best available, most expressive, most representative reflection of a particular event or development important to the evolution of the space program?

- Is the document not easily accessible except in one or a few locations, or is it included (for example, in published compilations of presidential statements) in reference sources that are widely available and thus not a candidate for inclusion in this collection?

- Is the document protected by copyright, security classification, or some other form of proprietary right and thus unavailable for publication?

Ultimately, as project director I was responsible for the decisions about which documents to include and for the accuracy of the headnotes accompanying them. It has been an occasionally frustrating but consistently exciting experience to be involved with this undertaking; I and my associates hope that those who consult it in the future find our efforts worthwhile.

John M. Logsdon
Director
Space Policy Institute
Elliott School of International Affairs
George Washington University

Biographies of Volume I Essay Authors

R. Cargill Hall is Chief of the Contract Histories Program at the Center for Air Force History in Washington, D.C. He received a B.A. in political science from Whitman College in Walla Walla, Washington, and an M.A. in political science/international relations from San Jose State University. Hall served as a historian for Lockheed Missiles and Space Company (1959-1967) before moving to the California Institute of Technology's Jet Propulsion Laboratory as historian (1967-1977). He joined the U.S. Air Force history program at the headquarters of the Strategic Air Command (1977-1980), subsequently serving as deputy command historian at the headquarters of the Military Airlift Command (1980-1981) and as chief of the Research Division and deputy director of the U.S. Air Force Historical Research Agency (1981-1989), before assuming his present duties. Since the mid-1980s he has assisted other federal history programs focused on aeronautics and astronautics, including those of the National Air and Space Museum and NASA. Hall is the author of *Lunar Impact: A History of Project Ranger* (NASA SP-4210, 1977), editor and contributor to *Lightning Over Bougainville* (Smithsonian Institution Press, 1991), and series editor of the history symposia of the International Academy of Astronautics. He has contributed chapters to and published numerous articles on space law and the history of aeronautics and astronautics in diverse books and journals. He is also contributing editor of *Space Times*, the magazine of the American Astronautical Society, and of *Air & Space Smithsonian*. Hall is a member of the International Academy of Astronautics, the International Institute of Space Law, and he serves on the board of advisors for the Smithsonian Institution Press History of Aviation book series.

Sylvia Katherine Kraemer is a senior executive of NASA's Office of Policy and Plans. Dr. Kraemer joined NASA in 1983 as Director of NASA's History Office. She received her doctorate in the history of ideas from The John Hopkins University in 1969. From 1969 to 1983, she served successively on the faculties of Vassar College, Southern Methodist University, and the University of Maine at Orono. Her many invited lectures and publications include: "Expertise Against Politics: Technology as Ideology on Capitol Hill, 1966-1972" in *Science, Technology, and Human Values* (1983); "The Ideology of Science During the Nixon Years: 1970-76" in *Social Studies of Science* (1984); and "2001 to 1994: Political Environment and the Design of NASA's Space Station System" in *Technology and Culture* (1988), winner of the James Madison Prize of the Society for History in the Federal Government. Her book-length group profile of NASA's Apollo era engineers, *NASA Engineers and the Age of Apollo* (NASA SP-4104), was published in 1992. She co-edited, with Martin J. Collins, *A Spacefaring Nation: Perspectives on American Space History and Policy* (Smithsonian Institution Press, 1991) and *Space: Discovery and Exploration* (Hugh Lauter Levin Associates, Inc., for the Smithsonian Institution, 1993).

Roger D. Launius has been NASA's chief historian at Washington headquarters since 1990. Before that Dr. Launius worked for eight years as a civilian historian with the U.S. Air Force. A graduate of Graceland College, Lamoni, Iowa, he received his Ph.D. from Louisiana State University, Baton Rouge, in 1982. He is the author of articles on the history of aeronautics and space appearing in several journals. He has also published *Joseph Smith III: Pragmatic Prophet* (University of Illinois Press, 1988); *MAC and the Legacy of the Berlin Airlift* (U.S. Air Force, 1989); *Anything, Anywhere, Anytime: An Illustrated History of the Military Airlift Command, 1941-1991* (U.S. Air Force, 1991); *Differing Visions: Dissenters in Mormon History* (University of Illinois Press, 1994); and *NASA: A History of the U.S. Civil Space Program* (Krieger Publishing Co., 1994). He has written or edited six other books. He is currently conducting research for a book-length study of the development of aviation in the American West, covering the period from 1903 to 1945.

John M. Logsdon is Director of both the Center for International Science and Technology Policy and the Space Policy Institute of George Washington University's Elliott School of International Affairs, where he is also Professor of Political Science and International Affairs. He holds a B.S. in physics from Xavier University and a Ph.D. in political science from New York University. He has been at George Washington University since 1970, and he previously taught at The Catholic University of America. Dr. Logsdon's research interests include space policy, the history of the U.S. space program, the structure and process of government decision-making for research and development programs, and international science and technology policy. He is author of *The Decision to Go to the Moon: Project Apollo and the National Interest* (MIT Press, 1970) and has written numerous articles and reports on space policy and science and technology policy. In January 1992 Dr. Logsdon was appointed to Vice President Dan Quayle's Space Policy Advisory Board and served through January 1993. He is a member of the International Academy of Astronautics, of the Board of Advisors of The Planetary Society, of the Board of Directors of the National Space Society, and of the Aeronautics and Space Engineering Board of the National Research Council. In past years he was a member of the National Academy of Sciences's National Academy of Engineering Committee on Space Policy and the National Research Council Committee on a Commercially Developed Space Facility, NASA's Space and Earth Science Advisory Committee and History Advisory Committee, and the Research Advisory Committee of the National Air and Space Museum. He also is a former chair of the Committee on Science and Public Policy of the American Association for the Advancement of Science and of the Education Committee of the International Astronautical Federation. He is a fellow of the American Association for the Advancement of Science and the Explorers Club, as well as an associate fellow of the American Institute of Aeronautics and Astronautics. In addition, he is North American editor for the journal *Space Policy*.

Glossary

ABMA	Army Ballistic Missile Agency
ACDA	Arms Control and Disarmament Agency
ACJP	Air Corps Jet Propulsion
AEC	Atomic Energy Commission
AF	Air Force
Ag	Agriculture
AID	Agency for International Development
AIS	American Interplanetary Society
AMPS	Atmospheric Magnetospheric and Plasmas in Space
AMR	Atlantic Missile Range
AP	Associated Press
APT	Automatic Picture Transmission
ARC	Ames Research Center
ARPA	Advanced Research Projects Agency
ARS	American Rocket Society
AS	Ascent Stage (LEM)
ASAT	Antisatellite
ASEB	Aeronautics and Space Engineering Board
ASP	LEM Ascent Stage and LEM Descent Stage Propulsion
ASTP	Advance Space Technology Program
AT&T	American Telephone & Telegraph
ATS	Applications Technology Satellite
Autour de la Lune	*Around the Moon*
BIS	British Interplanetary Society
BMC	Ballistic Missile Command
BOB	Bureau of the Budget
CAB	Change Analysis Board
CASPER	Committee on Space Research
CCCB	Configuration Change Control Board
CCD	Charge Coupled Device
CIA	Central Intelligence Agency
CIT	California Institute of Technology
CoF	Construction of Facilities
Convair	Consolidated Vaunt Aircraft
CORI	Coaxial Reference Interferometer
CSAGI	Special Committee for the International Geophysical Year
CTS	Canadian Technology Satellite
DCAS	Defense Contract Management Command
DCI	Director of Central Intelligence
De la Terre à la Lune	*From the Earth to the Moon*
De Mirabilibus Mundi	*On the Wonders of the World*
DEW	Distant Early Warning
DFRC	Dryden Flight Research Center
Die Rakete zu den Planetenraumen	*The Rocket in Planetary Space*
DOC	Department of Commerce
DOD	Department of Defense
DOT	Department of Transportation
E pur si muove	*Yet it does move*
EC	European Community
EEO	Equal Opportunity Office
ELINT	Electronic Intelligence
ELV	Expendable Launch Vehicle
EOR	Earth Orbital Rendezvous

EPC	Economics Policy Council
ERDA	Energy Research and Development Agency
EROS	Earth Resources Observatory System
ERTS	Earth Resources Technology Satellites
ESSA	Environmental Science Service Administration
EVA	Extravehicular Activity
FCC	Federal Communications Commission
FY	fiscal year
GALCIT	Guggenheim Aeronautical Laboratory, California Institute of Technology
GAO	Government Accounting Office
GISS	Goddard Institute of Space Studies
GMIC	Guided Missile Intelligence Committee
GNP	Gross National Product
GSAGI	International Council of Scientific Unions Special Committee for the Geophysical Year
GSFC	Goddard Space Flight Center
HEAO	High Energy Astronomy Observatory
HEW	Health Education and Welfare
HRIR	High Resolution Infra Red
HST	Hubble Space Telescope
HTV	Hypersonic Test Vehicle
HUD	Housing and Urban Development
IBM	International Business Machines
ICBM	Intercontinental Ballistic Missile
ICSU	International Council of Scientific Unions
IFF	Identification, Friend or Foe
IG	Inspector General
IGY	International Geophysical Year
INC	Inverse Null Corrector
IR	Infrared
IRBM	Intermediate Range Ballistic Missile
IUGG	International Union of Geodesy and Geophysics
JATO	Jet-Assisted Takeoff
JCS	Joint Chiefs of Staff
JFD	John Foster Dulles
JOP	Jupiter Orbiter Project
JPL	Jet Propulsion Laboratory
JSC	Johnson Space Center
KSC	Kennedy Space Center
LACIE	Large Area Crop Inventory Experiment
LaRC	Langley Research Center
LEM	Lunar Excursion Module
LeRC	Lewis Research Center
LM	Lunar Module
LOC	Launch Operations Center
LOR	Lunar Orbit Rendezvous
LPR	Long Playing Rocket
LRRP	Long-Range Rocket Projectile
LSS	Life Support Subsystem
MDAC-WD	McDonnell Douglas Astronautics Company-Western Division
MFPE	Mission From Planet Earth
MIDAS	Missile Detection Alarm
MIT	Massachusetts Institute of Technology
MLL	Manned Lunar Landing
MOUSE	Minimum Orbital Unmanned Satellite of the Earth

MRB	Material Review Board
MSFC	Marshall Space Flight Center
MTPE	Mission to Planet Earth
NACA	National Advisory Committee for Aeronautics
NAPA	National Academy of Public Administration
NAS	National Academy of Sciences
NASA	National Aeronautics and Space Administration
NASC	National Aeronautics and Space Council
NATO	North Atlantic Treaty Organization
Nauchnoye Obozreniye	*Science Review*
NDRC	National Defense Research Council
NEES	Naval Engineering Experiment Station
NERVA	Nuclear Engine for Rocket Vehicle Application
NIH	National Institutes of Health
NMSG	NASA Management Study Group
NOA	New Obligational Authority
NOAA	National Oceanic and Atmospheric Administration
NOL	Naval Ordnance Laboratory
NOSS	National Oceanic Satellite System
NRC	National Research Council
NRL	National Research Laboratory
NRO	National Reconnaissance Office
NSC	National Security Council
NSDD	National Security Decision Directive
NSF	National Science Foundation
NST	Nuclear and Space Talks
NSTL	National Space Technology Laboratory
NTIA	National Telecommunications and Information Administration
OAO	Orbiting Astronomical Observatories
OAST	Office of Aeronautics and Space Technology
ODM	Office of Defense Mapping
OEO	Office of Economic Opportunity
OFT	Orbiter Flight Test
OGO	Orbiting Geophysical Observatories
OMSF	Office of Manned Space Flight
ONR	Office of Naval Research
OSO	Orbiting Solar Observatory
OSSA	Office of Space Science and Applications
OST	Office of Space Technology
OSTP	Office of Science Technology Policy
OTA	Optical Telescope Assembly
OTP	Office of Technology Policy
P-E	Perkin-Elmer
PAD	Program Approval Document
PMR	Pacific Missile Range
PSAC	President's Science Advisory Committee
R&LO	Reliability and Launch Operations
R&D	Research and Development
R&PM	Research and Program Management
R&T	Research and Technology
RATO	Rocket Assisted Take Off
RCA	Radio Corporation of America
RDT&E	Research, Development, Test, and Evaluation
RIF	Reduction in Force
RMI	Reaction Motors, Inc.
RNC	Reflective Null Corrector

Sat	Saturn
Sidereus Nuncius	Sidereal Messenger
SIG	Senior Interagency Group
Somnium	*Dream*
SPB	Standard Practice Bulletin
SRM	Solid Rocket Motor
STG	Space Task Group
STS	Space Transportation System
T&DA	Training and Data Acquisition
TAN	Task Authorization Notice
TAOS	Thrust Assisted Orbiter Shuttle
TCP	Technological Capabilities Panel
TMIS	Technical Management Information System
TP	Transition Period
TV	Television
USA	United States of America
UCLA	University of California at Los Angeles
UFO	Unidentified Flying Object
ULV	Unmanned Launch Vehicles
UN	United Nations
UNESCO	United Nations Educational Scientific, and Cultural Organization
URSI	International Scientific Radio Union
U.S.	United States
USAF	U.S. Air Force
USC	University of Southern California
USGS	United States Geological Survey
USNC	United States National Committee
USSR	Union of Soviet Socialist Republics
UV	Ultra Violet
Verein fur Raumschiffahrt	Society for Spaceship Travel, or VfR
Voyage dans la Lune	*The Voyage to the Moon*
WAC	Womens Auxiliary Corps, Without Attitude Control
WFC	Wallops Flight Center
WFF	Wallops Flight Facility
XCMS	Experimental Command and Service Module

Chapter One
Prelude to the Space Age
by Roger D. Launius

Curiosity about the universe and other worlds has been one of the few constants in the history of humankind. Prior to the twentieth century, however, there was little opportunity to explore the universe except in fiction and through astronomical observations. These early explorations led to the compilation of a body of knowledge that inspired and in some respects informed the efforts of a body of scientists and engineers who began to think about applying rocket technology to the challenge of spaceflight in the early part of the twentieth century. These individuals were essentially the first spaceflight pioneers, translating centuries of dreams into a reality that matched in some measure the expectations of the public that watched and the governments that supported their efforts. During the period between 1926—when Robert H. Goddard launched his first rocket—and 1957—when the first orbital spacecraft was launched—a dedicated group of spaceflight enthusiasts, scientists, and engineers worked hard to inaugurate the space age.

Early Explorations of the Observable Universe

From ancient times civilizations around the globe have erected great observatories to chart the paths of the Sun, Moon, planets, and stars. Much of this observation became a central part of religion, science, and philosophy, and as a result has informed modern thinking on these subjects. The prehistoric people who built Stonehenge in England apparently used their observations of celestial bodies to chart planting seasons and measure other events, assigning this study a religious significance as well. Astronomers in Babylon about 700 B.C. charted the paths of planets and compiled observations of fixed stars. Later, around 400 B.C., the Babylonians devised the zodiac, the first such mechanism to divide the year into lunar periods and to assign significance to a person's date of birth in foretelling the future. In what became the Americas, the ancient Incan and Aztec cultures built astronomical observatories. North American Indians also observed the supernova of 1054 which created the Crab Nebula—not seen in Europe.[1]

By 150 A.D. the great mathematician, geographer, and astronomer Ptolemy of Alexandria had systematized a large amount of ancient information, and in some cases misinformation, about the universe. Ptolemy based his understanding on classical conceptions of the universe inherited from the Greeks. His great synthesis, *Megiste Syntaxis*, sometimes called the *Almagest*, argued that the Earth was at the center of the universe, and that the Sun, stars, and planets were embedded as jewels in a setting of spheres circling around it. [I-1] The Roman Cicero summarized the Ptolemaic belief this way:

The Universe consists of nine circles, or rather of nine moving globes. The outersphere is that of the heavens, which embraces all the others and under which the stars are fixed. Underneath this, seven globes rotate in the opposite direction from that of the heavens. The first circle carries the star known to

---

1. E.C. Krup, *Echoes of the Ancient Skies: The Astronomy of Lost Civilizations* (New York: Harper and Row, 1983), pp. 27-29; Edward R. Harrison, *Cosmology: The Science of the Universe* (New York: Cambridge University Press, 1981), pp. 10-23, 73; Eugene M. Emme, *A History of Space Flight* (New York: Holt, Rinehart and Winston, 1965), pp. 12-16; Ray A. Williamson, *Living the Sky: The Cosmos of the American Indian* (Norman: University of Oklahoma Press, 1984); Anthony F. Aveni, *Skywatchers of Ancient Mexico* (Austin: University of Texas Press, 1980).

men as Saturn; the second carries Jupiter, benevolent and propitious to humanity; then comes Mars, gleaming red and hateful; below this, occupying the middle region, shines the Sun, the chief, prince and regulator of the other celestial bodies, and the soul of the world which is illuminated and filled by the light of its immense globe. After it, like two companions, come Venus and Mercury. Finally, the lowest orb is occupied by the Moon, which borrows its light from the Sun. Below this last celestial circle there is nothing by what is mortal and perishable except for the minds granted by the gods to the human race. Above the Moon, all things are eternal. Our Earth, placed at the center of the world, and remote from the heavens on all sides, stays motionless and all heavy bodies are impelled towards it by their own weight. The motion of the spheres creates a harmony formed out of their unequal but well-proportioned intervals, combining various bass and treble notes into a melodious concert.[2]

Ptolemy's position accepted the geometric and static harmony of the universe and placed humanity squarely at its center, a view not inconsistent with Christian religious ideals about humanity's special relationship with God.

While this world view was modified to some degree during the following centuries, it was not until the sixteenth century that it began to change appreciably. The Polish astronomer-mathematician Nicolaus Copernicus saw that the irregular flight of some planets could not be explained using the Ptolemaic construct of the universe, so he placed the Earth in orbit around the Sun. [I-2] He was, however, circumspect in his public statements about this finding. Others followed his observation to its logical conclusion, most importantly Galileo Galilei, who used the newly invented telescope to show that the Earth and the planets revolved around the Sun in a dynamic and ever-changing universe. He also learned that the Moon revolved around the Earth and that other planets had satellites as well.[3] "On the seventh day of January in the present year, 1610," Galileo wrote in *Sidereus Nuncius* (*Sidereal Messenger*),

when I was viewing the constellations of the heavens through a telescope, the planet Jupiter presented itself to my view, and as I had prepared for myself a very excellent instrument, I noticed a circumstance which I had never been able to notice before, namely that three little stars, small but very bright, were very near the planet.

He concluded "unhesitatingly, that there are three stars in the heavens moving about Jupiter, as Venus and Mercury around the Sun."[4] Additionally, Galileo's observations of sunspots also led to a view of a dynamic, ever-changing universe.

In 1616 Galileo was brought before the Inquisition in Rome and his ideas were declared heretical because they challenged more than 1,000 years of Christian tradition about humans being at the center of the universe. He received no punishment by agreeing not to teach these ideas, but in 1632 Galileo was brought to trial again for violating his previous compact and was forced into retirement at his home in Florence. He was also compelled to recant his belief that the Earth revolves around the Sun. His legendary response was reported only later, "*E pur si muove*" ("Yet it does move").[5]

Disregarding the narrow position of the church, others took up the cause after Galileo. An Englishman, Isaac Newton, was one of the most important. He formulated the three laws of motion that have been central to the development of space traveling vehicles, and placed both astronomy and physics on a firm scientific foundation. He established physical laws governing all matter that could be both mathematically proven and scientifically

---

2. Quoted in Wernher von Braun and Frederick I. Ordway III, *History of Rocketry and Space Travel* (New York: Thomas Y. Crowell Co., 1975 ed.), p. 9.
3. On this astronomical revolution see Clive Morphet, *Galileo and Copernican Astronomy: A Scientific World View Defined* (London: Butterworths, 1977); Thomas S. Kuhn, *The Copernican Revolution: Planetary Astronomy in the Development of Western Thought* (Cambridge, MA: Harvard University Press, 1957); David C. Knight, *Copernicus: Titan of Modern Astronomy* (New York: Franklin Watts, 1965).
4. Stillman Drake, ed. and trans., *Discoveries and Opinions of Galileo* (Garden City, NY: Doubleday & Co., 1957), pp. 31-34.
5. Lloyd Motz and Jefferson Hane Weaver, *The Story of Physics* (New York: Avon Books, 1989), pp. 37-38.

observed. Newton's observations on motion suggested that the universe was in constant flux and that motion in a predictable pattern was the natural condition. Universal gravitation, Newton argued, accounted for the physical actions of celestial bodies. In particular, he demonstrated that the attraction of the Sun to a planet was directly proportional to the product of the two masses and inversely proportional to the square of the distance separating them. Newton's ideas, so critical to the development of spaceflight, became the established method of explaining the universe during his lifetime and remained so until the twentieth century. During this period the history of astronomy and physics was mainly a story of working out a Newtonian "system of the world."[6]

The Dream of Spaceflight

Not until the twentieth century did technology develop to the extent that actual travel into the observable universe could take place, although many people had posited that it was theoretically possible and longed for the time when humanity could venture beyond Earth. When Galileo first broadcast his findings about the solar system in 1610, he sparked a flood of speculation about lunar flight. Johann Kepler, himself a pathbreaking astronomer, posthumously published a novel, *Somnium (Dream)* (1634), that recounted a dream of a supernatural voyage to the Moon in which the visitors encountered serpentine creatures. He also included much scientific information in the book, speculating on the difficulties of overcoming the Earth's gravitational field, the nature of the elliptical paths of planets, the problems of maintaining life in the vacuum of space, and the geographical features of the Moon.[7]

Other writings sparked by the invention of the telescope and the success of *Somnium* also described fictional trips into space. Cyrano de Bergerac, for example, wrote *Voyage dans la Lune (The Voyage to the Moon)* (1649), describing several attempts by the hero to travel to the Moon. First, he tied a string of bottles filled with dew around himself, so that when the heat of the Sun evaporated the dew he would be drawn upward, but the hero only made it as far as Canada on that attempt. Next, he tried to launch a vehicle from the top of a mountain by means of a spring-loaded catapult, "but because I had not taken my measures aright, I fell with a slosh on the Valley below." Returning to his vehicle, Cyrano's hero found some soldiers mischievously tying firecrackers to it. As they lit the fuse, he jumped into the craft and tier upon tier of explosives ignited like rockets and launched him to the Moon. Thus Cyrano's hero became the first flyer in fiction to reach the Moon by means of rocket thrust, a premonition of Newton's third law of gravity about every action having an equal and opposite reaction. Once on the Moon, the character in this novel had several adventures, and later in the book he also journeyed to the Sun.[8]

Other writers picked up the science fiction format and used it to discuss the possibilities of space travel in the years that followed. For example, Edward Everett Hale, a New England writer and social critic, published in 1869 a short story in the *Atlantic Monthly* entitled "The Brick Moon." [I-3] The first known proposal for an orbital satellite around the Earth, Hale described how a satellite in polar orbit could be used as a navigational aid to ocean-going vessels. "For you see that if, by good luck," he explained,

> there were a ring like Saturn's which stretched around the world, above Greenwich and the meridian of Greenwich,...anyone who wanted to measure his longitude or distance from Greenwich would look out of his window and see how high this ring was above his horizon...So if we only had a

---

6. *Ibid.*, pp. 43-88; Harrison, *Cosmology*, pp. 103-203.
7. Edward Rosen, trans., *Kepler's Somnium: The Dream or Posthumous Work on Lunar Astronomy* (Madison: University of Wisconsin Press, 1967), pp. 17-122; Steven J. Dick, *Plurality of Worlds: The Origins of the Extraterrestrial Life Debate from Democritus to Kant* (Cambridge: Cambridge University Press, 1982), pp. 77-84.
8. Von Braun and Ordway, *History of Rocketry and Space Travel*, p. 12; Emme, *History of Space Flight*, pp. 37-38.

*ring like that... vertical to the plane of the equator, as the brass ring of an artificial globe goes, only far higher in proportion... we could calculate the longitude.*⁹

When the heroes of the story substitute a brick moon for this ring—brick because it could withstand fire—it is to be hurled into orbit 5,000 miles above the Earth. An accident sends the brick moon off prematurely, however, while 37 construction workers and other people were aboard it. In contrast to what is now known about the vacuum of space, these people lived on the outer part of the brick moon, raised food, and enjoyed an almost utopian existence.

Perhaps the most important development in the literary consideration of space travel came following the publication of work by Italian astronomer Giovanni Schiaparelli in 1877 concerning the possibility of canals on Mars. He, and especially others, concluded that the features that he saw on Mars and called canals were the work of intelligent life. This was a startling observation because it meant that science had now validated the speculations of some fiction writers, lending credibility to their claims. Moreover, other scientists sought to explore these ideas, and in the United States Percival Lowell built what became the Lowell Observatory near Flagstaff, Arizona, to study the planets. In 1906 he published *Mars and Its Canals,* which argued that Mars had once been a watery planet and that the topographical features known as canals had been built by intelligent beings. [I-4] Over the course of the next forty years, a steady stream of other works was based upon Lowell's theories about the red planet.¹⁰

While many of these writings were not scientifically valid, that became less true as time passed and more modern science fiction writers such as Jules Verne and H.G. Wells appeared. Both were well aware of the scientific underpinnings of spaceflight, and their speculations reflected reasonably well what was known at the time about its problems and the nature of other worlds. Both Wells and Verne incorporated into their novels a much more sophisticated understanding of the realities of space than had been seen before. Their space vehicles became enclosed capsules powered by electricity, and they possessed some aerodynamic soundness. Most of Wells' and Verne's concepts stood up under some, although not all, scientific scrutiny. For example, in 1865 Verne published *De la Terre à la Lune (From the Earth to the Moon)*. The scientific principles informing this book were very accurate for the period. It described the problems of building a vehicle and launch mechanism to visit the Moon. At the end of the book, Verne's characters were shot into space by a 900-foot-long cannon. Verne picked up the story in a second novel, *Autour de la Lune (Around the Moon)*, describing a lunar orbital flight, but he did not allow his characters actually to land. Wells published *War of the Worlds* in 1897 and *The First Men in the Moon* immediately thereafter. Both used sound scientific principles to describe space travel and encounters with aliens.¹¹

War of the Worlds, furthermore, played upon a theme in space exploration that had been present for many centuries and would continue to appear throughout the twentieth century, humanity's fascination and terror about contact with alien species. Excitement

---

9. Edward E. Hale, "The Brick Moon," *The Atlantic Monthly,* October 1869, pp. 451-460; November 1869, pp. 603-611; December 1869, pp. 679-688.

10. Edward C. Ezell and Linda Neumann Ezell, *On Mars: Exploration of the Red Planet, 1958-1978* (Washington, DC: NASA SP-4212, 1984), pp. 3-5; William Graves Hoyt, *Lowell and Mars* (Tucson: University of Arizona Press, 1976); Frederick I. Ordway III, "The Legacy of Schiaparelli and Lowell," *Journal of the British Interplanetary Society,* January 1986, pp. 18-22.

11. General studies of science fiction can be found in Brian Ash, ed., *The Visual Encyclopedia of Science Fiction* (New York: Harmony Books, 1977); James Gunn, *Alternate Worlds: The Illustrated History of Science Fiction* (Englewood Cliffs, NJ: Prentice-Hall, 1975); Ed Naha, *The Science Fictionary* (New York: Wideview Books, 1980); Franz Rottensteiner, *The Science Fiction Book: An Illustrated History* (New York: Seabury Press, 1975); Jean-Claude Suares, Richard Siegel, and David Owen, *Fantastic Planets* (Danbury, CT: Addison House, 1979); Jules Verne, *De la Terre à la Lune* (Paris: J. Hetzel, 1866); H.G. Wells, *The First Men in the Moon* (London: George Newness, 1901); H.G. Wells, *The Shape of Things to Come* (London: Hutchinson, 1933); H.G. Wells, *The Ultimate Revolution* (New York: Macmillan, 1933); H.G. Wells, *The War of the Worlds* (London: William Heinemann, 1898).

about the prospect that humanity is not alone in the universe, that contact is possible, and that both cultures might be made richer through interaction has been a persistent theme for advocates of the exploration of space. Some science fiction positively expressed this image of contact with aliens—as examples, three novels by C.S. Lewis, *Out of the Silent Planet* (1938), *Perelandra* (1943), and *That Hideous Strength* (1945). At the same time, there has long been a fear that an alien civilization might attack Earth and either enslave or destroy humanity. In *War of the Worlds* the Earth was attacked by invaders from Mars, and eventually only defeated by terrestrial bacteria harmless to humans but deadly to an alien without generations of built-up immunity. These stories, both positive and negative examples of contact, provided some of the inspiration for many scientists and engineers who developed modern rocketry.[12]

The Technology of Rockets

Until the twentieth century, study about the universe and speculation about the nature of spaceflight were not closely related to the technical developments that led to rocket propulsion. A merging of these ideas had to take place before the space age could truly begin. The rocket is a reaction device, based on Newton's Third Law of Motion. Without explicitly understanding that law, humanity had known of the rocket's practicality for centuries. Although it is unclear who first invented rockets, many investigators link the development of the first crude rockets with the discovery of gunpowder. The Chinese, moreover, had been using gunpowder for more than 2,000 years. The first firecrackers appeared during the first two centuries after the beginning of the common era, and the Chinese were using rockets in warfare at least by the time of Genghis Khan (ca. 1155-1227).[13] Not long thereafter the use of rockets in warfare began to spread to the West, and they were in use by the time of the German Albert Magnus, who gave a recipe for making gunpowder and wrapping it "in a skin of paper for flying or for making thunder" in *De Mirabilibus Mundi* (*On the Wonders of the World*). By the time of Konrad Kyser von Eichstadt, who wrote *Bellfortis* in 1405, the use of rockets in military operations was reasonably well known in Europe.[14]

The application of gunpowder rockets was refined through the first part of the nineteenth century. Essentially, the military role of rocketry was as a type of artillery. Sir William Congreve carried rocket technology just about as far as it was to go for another century, developing incendiary barrage missiles for the British military that could be fired from either land or sea. They were used with effect against the United States in the War of 1812; it was probably Congreve's weapons that Francis Scott Key wrote about in the "Star Spangled Banner" while imprisoned on a British warship during the bombardment of Fort McHenry at Baltimore. The military use of the rocket was soon outmoded in the nineteenth century by developments in artillery which became more accurate and more destructive, but new uses for rockets were found in other industries such as whaling and for sea-going shipping where rocket-powered harpoons and rescue lines began to be employed.[15]

---

12. On the issue of contact of two cultures in space see, M. Jane Young, "'Pity the Indians of Outer Space': Native American Views of the Space Program," *Western Folklore* 46 (October 1987): 269-79; Patricia Nelson Limerick, "The Final Frontier?," *Wilson Quarterly* 14 (Summer 1990): 82-83; Ray A. Williamson, "Outer Space as Frontier: Lessons for Today," *Western Folklore* 46 (October 1987): 255-67.

13. Frank H. Winter, "The Genesis of the Rocket in China and its Spread to the East and West," pp. 3-23; Fang-Toh Sun, "Rockets and Rocket Propulsion Devices in Ancient China," pp. 25-40, both in A. Ingemar Skoog, ed., *History of Rocketry and Astronautics: Proceedings of the Twelfth, Thirteenth, and Fourteenth History Symposia of the International Academy of Astronautics* (San Diego: Univelt, Inc., 1990).

14. Von Braun and Ordway, *History of Rocketry and Space Travel*, p. 28.

15. Frank H. Winter, *The First Golden Age of Rocketry: Congreve and Hale Rockets of the Nineteenth Century* (Washington, DC: Smithsonian Institution Press, 1990).

Progenitors of the Space Age

While the technology of rocketry was moving forward on other fronts, some individuals began to see its use for space travel. There were three great pioneering figures in this category. Collectively, they were the progenitors of the modern space age. The earliest was the Russian theoretician Konstantin Eduardovich Tsiolkovskiy. An obscure schoolteacher in a remote part of Tsarist Russia in 1898, he submitted for publication to the Russian journal, *Nauchnoye Obozreniye (Science Review)*, a work based upon years of calculations that laid out many of the principles of modern spaceflight. His article was not published until 1903, but it opened the door to future writings on the subject. In it Tsiolkovskiy described in depth the use of rockets for launching orbital space ships. [I-5] Tsiolkovskiy continued to theorize on the subject of spaceflight until his death, describing in great detail both methods of flight and the technical requirements of space stations. Significantly, he never had the resources—nor perhaps the inclination—to experiment with rockets himself. His theoretical work, however, influenced later rocketeers both in his native land and abroad, and served as the foundation of the Soviet space program.[16]

A second rocketry pioneer was Hermann Oberth, by birth a Transylvanian but by nationality a German. Oberth began studying the nature of spaceflight about the time of World War I and published his classic study, *Die Rakete zu den Planetenraumen (Rockets in Planetary Space)* in 1923. [I-6] It was a thorough discussion of almost every phase of rocket travel. He posited that a rocket could travel in the void of space and that it could move faster than the velocity of its own exhaust gases. He noted that with the proper velocity a rocket could launch a payload into orbit around the Earth, and to accomplish this goal he reviewed several propellant mixtures to increase speed. He also designed a rocket that he believed had the capability to reach the upper atmosphere by using a combination of alcohol and hydrogen as fuel. Oberth followed this up with a long series of publications on rocketry and the prospects of space travel. He became the father of German rocketry.[17] Among his proteges was Wernher von Braun, the senior member of the rocket team that built NASA's Saturn launch vehicle for the trip to the Moon in the 1960s.[18]

Finally, the American Robert H. Goddard pioneered the use of rockets for spaceflight.[19] Motivated by reading science fiction as a boy, Goddard became excited by the possibility of exploring space. In 1901 he wrote a short paper, "The Navigation of Space," that argued that movement could take place by firing several cannons, "arranged like a 'nest' of beakers." He tried unsuccessfully to publish this article in *Popular Science News*.[20] At his high school oration in 1904 he summarized his future life's work: "It is difficult to say what is impossible, for the dream of yesterday is the hope of today and the reality of tomorrow."[21] In 1907 he wrote another paper on the possibility of using radioactive materials to propel a rocket through interplanetary space. He sent this article to several magazines, and all rejected it.[22] Still not dissuaded, as a young physics graduate student he worked on

---

16. Konstantin E. Tsiolkovskiy, *Aerodynamics* (Washington, DC: NASA TT F-236, 1965); Konstantin E. Tsiolkovskiy, *Reactive Flying Machines* (Washington, DC: NASA TT F-237, 1965); Konstantin E. Tsiolkovskiy, *Works on Rocket Technology* (Washington, DC: NASA TT F-243, 1965); Arkady Kosmodemyansky, *Konstantin Tsiolkovskiy* (Moscow: Nauka, 1985).

17. Hermann Oberth, *Ways to Spaceflight* (Washington, D.C.: NASA, TT F-622, 1972); Hermann Oberth, *Rockets into Planetary Space* (Washington, DC: NASA TT F-9227, 1972); H.B. Walters, *Hermann Oberth: Father of Space Travel* (New York: Macmillan, 1962).

18. Interestingly, in 1969 Oberth attended the launch in the United States of Apollo 11 at the request of von Braun, who was then directing a major component of the lunar project.

19. The standard biography of Goddard is Milton Lehman, *This High Man* (New York: Farrar, Straus, 1963), although it is outdated and deserving of replacement.

20. Robert H. Goddard, "Material for an Autobiography," in Esther C. Goddard, ed., and G. Edward Pendray, assoc. ed., *The Papers of Robert H. Goddard* (New York: McGraw-Hill Book Co., 1970), 1:10.

21. Robert H. Goddard, "Of Taking Things for Granted," in *ibid.*, 1:63-66.

22. Robert H. Goddard, "On the Possibility of Navigating Interplanetary Space," in *ibid.*, pp. 81-87; *Scientific American* to R.H. Goddard, October 9, 1907, in *ibid.*, p. 87; William W. Payne to R.H. Goddard, January 15, 1908, in *ibid.*, p. 88.

rocket propulsion and actually received two patents in 1914. One was the first for a rocket using solid and liquid fuel and the other for a multi-stage rocket.[23]

After a stint with the military in World War I, where he worked on solid rocket technology for use in combat, Goddard became a professor of physics at Clark College in Worcester, Massachusetts. There he turned his attention to liquid rocket propulsion, theorizing that liquid oxygen and liquid hydrogen were the best fuels, but learning that oxygen and gasoline were less volatile and therefore more practical. To support his investigations, Goddard applied to the Smithsonian Institution for assistance in 1916 and received a $5,000 grant from its Hodgkins Fund.[24] His research was ultimately published by the Smithsonian as the classic study, *A Method of Reaching Extreme Altitudes*, in 1919. [I-7] In it Goddard argued from a firm theoretical base that rockets could be used to explore the upper atmosphere. Moreover, he suggested that with a velocity of 6.95 miles/second, without air resistance, an object could escape Earth's gravity and head into infinity, or to other celestial bodies.[25] This became known as the Earth's "escape velocity."

It also became the great joke for those who believed spaceflight either impossible or impractical. Some ridiculed Goddard's ideas in the popular press, much to the consternation of the already shy Goddard. Soon after the appearance of his publication, he commented that he had been "interviewed a number of times, and on each occasion have been as uncommunicative as possible."[26] The *New York Times* was especially harsh in its criticisms, referring to him as a dreamer whose ideas had no scientific validity. It also compared his theories to those advanced by novelist Jules Verne, indicating that such musing is "pardonable enough in him as a romancer, but its like is not so easily explained when made by a savant who isn't writing a novel of adventure." [I-8] The *Times* questioned both Goddard's credentials as a scientist and the Smithsonian's rationale for funding his research and publishing his results.[27]

The negative press Goddard received prompted him to be even more secretive and reclusive. It did not, however, stop his work, and he eventually registered 214 patents on various components of rockets. He concentrated on the design of a liquid-fueled rocket, the first such development, and the attendant fuel pumps, motors, and control components. On March 16, 1926 near Auburn, Massachusetts, Goddard launched his first rocket, a liquid oxygen and gasoline vehicle that rose 184 feet in 2.5 seconds.[28] This event heralded the modern age of rocketry. He continued to experiment with rockets and fuels for the next several years. A spectacular launch took place on July 17, 1929, when he flew the first instrumented payload—an aneroid barometer, a thermometer, and a camera—to record the readings. The launch failed; after rising about 90 feet the rocket turned and struck the ground 171 feet away. It caused such a fire that neighbors complained to the state fire marshal and Goddard was enjoined from making further tests in Massachusetts.[29]

This experience, as well as his personal shyness, led him to seek a more remote setting to conduct his experiments. His ability to shroud his research in mystery was greatly enhanced by Charles A. Lindbergh, fresh from his trans-Atlantic solo flight, who helped Goddard obtain a series of grants from the Guggenheim Fund fostering aeronautical activities. This enabled him to obtain a large tract of desolate land near Roswell, New Mexico,

---

23. Robert H. Goddard, "Material for an Autobiography," in *ibid.*, 1:19-20; R.H. Goddard to Josephus Daniels, July 25, 1914, in *ibid.*, 1:126-27.

24. C.D. Walcott to R.H. Goddard, January 5, 1917, in *ibid.*, 1:190; Frederick C. Durant III, "Robert H. Goddard and the Smithsonian Institution," in Frederick C. Durant III and George S. James, eds., *First Steps Toward Space: Proceedings of the First and Second History Symposia of the International Academy of Astronautics* (Washington, DC: Smithsonian Institution Press, 1974), pp. 57-69.

25. Robert H. Goddard, *A Method of Reaching Extreme Altitudes* (Washington, DC: Smithsonian Miscellaneous Collections, Volume 71, Number 2, 1919), p. 54.

26. Robert H. Goddard, "Statement by R.H. Goddard for Newspapers," January 18, 1920, in *ibid.*, 1:409-10.

27. "Topics of the Times," *New York Times*, January 18, 1920, p. 12.

28. Robert H. Goddard, "R.H. Goddard's Diary," March 16-17, 1926, in Goddard and Pendray, *Papers of Goddard*, 2:580-81; Lehman, *This High Man*, pp. 140-44.

29. Lehman, *This High Man*, pp. 156-62.

and to set up an independent laboratory to conduct rocket experiments far away from anyone else. Between 1930 and 1941 Goddard carried out ever more ambitious tests of rocket components in the relative isolation of New Mexico, much of which he summarized in a 1936 study, *Liquid-Propellant Rocket Development*. [I-9] The culmination of this effort was a successful launch of a rocket to an altitude of nearly 9,000 feet in 1937.[30] In late 1941 Goddard entered naval service and spent the duration of World War II developing a Jet-Assisted Takeoff (JATO) rocket to shorten the distance required for heavy aircraft launches. Some of this work led to the development of the throttleable Curtis-Wright XLR25-CW-1 rocket engine that later powered the Bell X-2 and helped overcome the transonic barrier in 1947. Goddard did not live to see this; he died in Baltimore, Maryland, on August 10, 1945.[31]

Goddard accomplished much, but because of his secrecy most people did not know about his achievements during his lifetime. These included the following pioneering activities:

1. Theorizing on the possibilities of jet-powered aircraft, rocket-borne mail and express, passenger travel in space, nuclear-powered rockets; and journeys to the Moon and other planets (1904-1945).
2. First mathematical exploration of the practicality of using rockets to reach high altitudes and achieve escape velocity (1912).
3. First patent on the idea of multi-stage rockets (1914).
4. First experimental proof that a rocket could provide thrust in a vacuum (1915).
5. The basic idea of anti-tank missiles, developed and demonstrated during work for the Army in World War I. This was the prototype for the "Bazooka" infantry weapon (1918).
6. First publication in the United States of the basic mathematical theory underlying rocket propulsion and spaceflight (1919).
7. First development of a rocket motor burning liquid propellants (1920-1926).
8. First development of self-cooling rocket motors, variable-thrust rocket motors, practical rocket landing devices, pumps suitable for liquid fuels, and associated components (1920-1945).
9. First design, construction, and launch of a successful liquid fueled rocket (1926).
10. First development of gyro-stabilization equipment for rockets (1932).
11. First use of deflector vanes in the blast of the rocket motor as a method of stabilizing and guiding rockets (1932).[32]

The U.S. government's recognition of Goddard's work came in 1960 when the Department of Defense and the National Aeronautics and Space Administration (NASA) awarded his estate $1 million for the use of his patents.[33]

Parallel Developments

Concomitant with Goddard's research into liquid fuel rockets, and perhaps more immediately significant because the results were more widely disseminated, were activities in several other quarters. Among the most important of these ventures were those undertaken by the various rocket societies. The largest and most significant was the German organization, the "Verein fur Raumschiffahrt" (Society for Spaceship Travel, or VfR). Although spaceflight aficionados and technicians had organized at other times and in other

---

30. *Ibid.*, 161-312; von Braun and Ordway, *History of Rocketry and Space Travel*, pp. 46-53. Many of these experiments were summarized in Robert H. Goddard, *Liquid-Propellant Rocket Development* (Washington, DC: Smithsonian Miscellaneous Collections, Volume 95, Number 3, 1936).

31. Frank H. Winter, *Rockets into Space* (Cambridge, MA: Harvard University Press, 1990), pp. 33-34.

32. G. Edward Pendray, "Pioneer Rocket Development in the United States," in Eugene M. Emme, ed. *The History of Rocket Technology* (Detroit: Wayne State University Press, 1964), pp. 19-23.

33. See the extensive documentation on this settlement in the "Goddard Patent Infringement" Folders, Biographical Collection, NASA Historical Reference Collection, Washington, D.C.

places, the VfR under the able leadership of Berlin aviator Max Valier emerged soon after its founding on July 5, 1927, as the leading space travel group. It was specifically organized to raise money to test Oberth's rocketry ideas. It was successful in building a base of support in Germany, publishing a magazine and scholarly studies, and of constructing and launching small rockets. One of its strengths from the beginning, however, was the VfR's ability to publicize both its activities and the dream of spaceflight.[34]

The VfR made good on some of those dreams on February 21, 1931, when it launched the LOX-methane liquid-fuel rocket HW-1 near Dessau to an altitude of approximately 2,000 feet. The organization's public relations arm went into high gear after this mission, and emphasized the launch's importance as the first successful European liquid-fuel rocket flight.[35] Wernher von Braun, then a neophyte learning the principles of rocketry from Oberth and Valier, was both enthralled with this flight and impressed with the publicity it engendered. Later, he became the quintessential and movingly eloquent advocate for the dream of spaceflight and a leading architect of its technical development. He began developing both skills while working with the VfR.[36]

There were other national rocketry societies that sprang up during this same period, each contributing to the base of technical knowledge and the popular conception of spaceflight. The American Interplanetary Society (AIS) was one of the more powerful of these institutions. Organized in 1930, within two years the AIS had begun a program of rocket experimentation. On November 12, 1932, the AIS tested its first static test of a LOX-gasoline rocket. It actually launched a rocket on May 14, 1932, attaining an altitude of only 250 feet. But its second and last launch on September 9, 1934, rose over 1,300 feet. Because of the great cost and risk to people involved, after this launch the group concentrated throughout the rest of the 1930s on static firings of engines and published results of its research, the cumulation of which proved significant for later experimentation in rocketry. Almost concomitant with its withdrawal from rocket experimentation, and out of a desire to improve the image of the organization, the AIS changed its name to the American Rocket Society.[37]

That name change may also have been prompted in part by the organization of the British Interplanetary Society (BIS) on October 13, 1933, at Liverpool, England. More oriented toward theoretical studies than rocket experimentation, in the 1930s the BIS became a haven for writers and other intellectuals interested in the idea of spaceflight. By September 1939, at the beginning of World War II, the BIS numbered about 100 members, including several Germans. The BIS periodical, the *Journal of the British Interplanetary Society*, began publication in January 1934, and it quickly became a persistent and powerful voice on behalf of space exploration. The BIS did not undertake field work with rockets (although several members did conduct some crude experiments with potential solid propellants), but in 1938-1939 its members designed a lunar landing vehicle which influenced the Lunar Module used in Project Apollo during the 1960s.[38] [I-10]

---

34. The standard work on the rocket societies is Frank H. Winter, *Prelude to the Space Age: The Rocket Societies, 1924-1940* (Washington, D.C.: Smithsonian Institution Press, 1983). A briefer discussion is available in Winter, *Rockets into Space*, pp. 34-42.

35. Winter, *Rockets into Space*, p. 37.

36. Wernher von Braun, "German Rocketry," in Arthur C. Clarke, ed., *The Coming of the Space Age* (New York: Meredith Press, 1967), pp. 33-55. Von Braun's public relations skills were exceptional throughout his career. Evidence of this can be found in the more than eight linear feet of materials by von Braun held in the Biographical Files of the NASA Historical Reference Collection.

37. Winter, *Prelude to the Space Age*, pp. 73-85; Eugene M. Emme, ed., *Aeronautics and Astronautics: An American Chronology of Science and Technology in the Exploration of Space, 1915-1960* (Washington, DC: National Aeronautics and Space Administration, 1961), p. 31.

38. Winter, *Prelude to the Space Age*, pp. 87-97; *The British Interplanetary Society: Origin and History* (London: British Interplanetary Society Diamond Jubilee Handbook, 1993), pp. 6, 17; H.E. Ross, "The British Interplanetary Society's Astronautical Studies, 1937-1939," in Durant and James, eds., *First Steps Toward Space*, pp. 209-16; H.E. Ross, "The British Interplanetary Society Spaceship," *Journal of the British Interplanetary Society* 5 (January 1939): 4-9.

While both the individual and societal precursors of spaceflight struggled along as best they could, beginning in 1936 the Guggenheim Aeronautical Laboratory, California Institute of Technology (GALCIT), in Pasadena, California, began to pursue its own rocket research program.[39] Frank J. Malina, a young Caltech Ph.D. student at the time, persuaded GALCIT to adopt a research agenda for the design of a high-altitude sounding rocket and enthusiastically began experimentation. Using some of the ideas from the research of Eugen Sänger in Austria and Goddard in New Mexico, Malina and a design team—composed of, among others, H.S. Tsien, a Chinese national who was later deported and became the architect of the ICBM and space launcher programs for the People's Republic of China—began work. Nobel Prize-winning physicist Robert A. Millikan, chair of the Caltech executive council, was especially supportive of this work because he wanted to use sounding rockets for cosmic ray research. Millikan tried to persuade Goddard to join this research project—Malina even visited him at his complex near Roswell, New Mexico, in August 1936—but Goddard refused.[40] In a letter revealing much of Goddard's secretiveness, in September 1936 he wrote disparagingly of Malina to Robert Millikan. Goddard commented that he had tried to help Malina with some of his questions, but "I naturally cannot turn over the results of many years of investigation, still incomplete, for use as a student's thesis."[41]

Beginning in late 1936 Malina and his colleagues started the static testing of rocket engines in the canyons above the Rose Bowl, with mixed results. It was not until November 28, 1936, for example, that the motor ran at all, and then only for 15 seconds. A series of tests thereafter brought incremental improvements; a year later Malina and an associate had learned enough to distill the results into the first scholarly paper on rocketry to come out of GALCIT. [I-11] The test results showed that with proper fuels and motor efficiency a rocket could be constructed with the capability to ascend as high as 1,000 miles.[42]

Because of this research GALCIT's rocketry team obtained funding from outside sources, among them General H.H. (Hap) Arnold, soon to become the Army Air Corps Chief of Staff; he visited GALCIT in the spring of 1938 and was enthusiastic about the work on rockets he saw Malina and co-workers doing. That fall he arranged for additional funding from the National Academy of Sciences to proceed with the project, with the specific goal of research on the possibilities of rocket-assisted takeoff for aircraft. The committee that approved this funding did so with some consternation that it might be money poorly spent. Finally, Jerome C. Hunsaker, head of the Aeronautics Department of the Massachusetts Institute of Technology, told the committee that he would be glad to have Theodore von Kármán, director of GALCIT, "take the Buck Rogers job."[43] GALCIT accepted the task, and beginning in 1939 Malina and his rocket team began working on what became the JATO project. Although Malina always expressed misgivings about working on weaponry, and after World War II accepted employment with the United Nations so he could help prevent such conflict from taking place again, the difficult political climate in 1939 prompted him to support the development of U.S. military capability as a deterrent to

---

39. The history of this organization has been explored in Clayton R. Koppes, *JPL and the American Space Program: A History of the Jet Propulsion Laboratory* (New Haven: Yale University Press, 1982).

40. Malina reflected that while Goddard was pleasant during a visit to his Roswell, New Mexico, facility in August 1936, he left two specific impressions. First, he was bitter toward the press, and therefore exceptionally secretive about his work. Second, "he felt that rockets were his private preserve, so that others working on them took on the aspect of intruders." Malina recalled that Goddard showed him no technical details and declined to participate in the GALCIT program. See Frank J. Malina, "On the GALCIT Rocket Research Project, 1936-1938," in Durant and James, eds., *First Steps Toward Space*, pp. 113-27, quote from p. 117.

41. R.H. Goddard to Robert A. Millikan, September 1, 1936, in Goddard and Pendray, eds., *Papers of Goddard*, 2:1012-13.

42. Frank J. Malina and Apollo M.O. Smith, "Flight Analysis of the Sounding Rocket," *Journal of Aeronautical Sciences* 5 (1938): 199-202; Frank J. Malina, "The Jet Propulsion Laboratory: Its Origins and First Decade of Work," *Spaceflight* 6 (1964): 216-23; Frank J. Malina, "The Rocket Pioneers: Memoirs of the Infant Days of Rocketry at Caltech," *Engineering and Science* 31 (February 1968): 9-13, 30-32.

43. Theodore von Kármán with Lee Edson, *The Wind and Beyond: Theodore von Kármán, Pioneer in Aviation and Pathfinder in Space* (Boston: Little, Brown, 1967), p. 243.

fascism. As a result, Malina and GALCIT engaged throughout the war years in rocketry research for military purposes.[44]

The Rocket and Modern War

Although the work of Goddard, Oberth, and others was pathbreaking, World War II truly altered the course of rocket development. Prior to that conflict technological progress in rocketry had been erratic. The war, however, forced nations to focus attention on the activity and to fund research and development. Such research and development was oriented, of course, toward the advancement of rocket-borne weapons rather than rockets for space exploration and other peaceful purposes. This would remain the case even after the war, as competing nations perceived and supported advances in space technology largely because of their military potential and the national prestige associated with them. The security role of the Department of Defense and the function of NASA as a civilian space agency have been inextricably related ever since.

During World War II virtually every belligerent was involved in developing some type of rocket technology. As an example, the Soviet Union fielded the "Katusha," a solid-fuel rocket six feet in length that carried nearly fifty pounds of explosives and could be fired from either a ground- or truck-mounted launcher. Italy conducted research on solid- and liquid-fuel rockets for small, infantry-carried weapons and torpedoes. Other nations developed various types of hand-held anti-tank and anti-aircraft rockets.[45]

Just before the entry of the United States into World War II, the nation's military began in earnest to acquire a rocket capability, and several efforts were aimed in that direction. One of the most significant was at GALCIT, renamed the Jet Propulsion Laboratory (JPL) in 1943, where von Kármán, Malina, and a group of talented young engineers made important strides based on their research from the latter 1930s. They developed in 1941, for instance, the first JATO solid-fuel rocket system.[46]

In March 1942 the GALCIT team that had developed the JATO system founded Aerojet Engineering Corporation as a vehicle for mass-producing and marketing this new technology; the new company quickly became one of the leading manufacturers of rockets in the United States. Malina recalled that the movement of scientists and engineers into business did not sit well with the military. Within two months of creating Aerojet, von Kármán had brought in two big military production contracts for JATO systems, but the Army Air Forces—successor to the Army Air Corps—canceled its contract even before production began. Von Kármán and Malina flew to Washington, D.C., to protest the decision, and learned that concerns about conflict of interest had prompted the cancellation. "We like you very much, doctor," Colonel Benjamin Chidlaw told von Kármán, "but only in cap and gown to advise us what to do in science. The derby hat of the businessman doesn't befit you." The leaders of Aerojet were able to overcome this problem only because of the dearth of rocket expertise in the United States, but it ceased to be a problem after 1944 when the General Tire and Rubber Company bought a controlling interest in Aerojet and divorced it from its JPL ties.[47]

---

44. Frank J. Malina to parents, October 24, 1938, in "Rocket Research and Development: Excerpts from Letters Written Home by Frank J. Malina between 1936 and 1946," pp. 22-23, Frank J. Malina Folder, Biographical Files, NASA Historical Reference Collection; von Kármán with Edson, *Wind and Beyond*, p. 244; Frank J. Malina, "The U.S. Army Air Corps Jet Propulsion Research Project, GALCIT Project No. 1, 1939-1946: A Memoir," in R. Cargill Hall, ed., *Essays on the History of Rocketry and Astronautics: Proceedings of the Third through the Sixth History Symposia of the International Academy of Astronautics* (San Diego: Univelt, Inc., 1986), pp. 154-60.
45. Frank J. Malina, "A History of Rocket Propulsion up to 1945," in *Jet Propulsion Engines* (Princeton: Princeton University Press, 1959), pp. 18, 22-23.
46. Koppes, *JPL and the American Space Program*, pp. 11-16; von Kármán with Edson, *Wind and Beyond*, pp. 244-56; Theodore von Kármán, "Jet Assisted Take-off," *Interavia*, July 1952, pp. 376-77.
47. Von Kármán with Edson, *Wind and Beyond*, p. 258-60; Malina, "GALCIT Project No. 1," pp. 195-95; Michael H. Gorn, *The Universal Man: Theodore von Kármán's Life in Aeronautics* (Washington, DC: Smithsonian Institution Press, 1992), pp. 90-92; Koppes, *JPL and the American Space Program*, pp. 16-17.

Even as these activities were taking place, in 1943 JPL engineers concluded in a report to the Army Air Forces that "the development of a long-range rocket projectile is within engineering feasibility" and asked for funding to bring it to a reality.[48] [I-12] With some investment financing from the Army, JPL conducted research on engines and other components. Then on January 16, 1945, Malina sent to the Army Ordnance Section a proposal for a liquid-fuel "sounding" rocket that would be able to launch a 25-pound payload to an altitude of 100,000 feet. What emerged from these recommendations was a decision to develop the WAC Corporal, first flown on October 11, 1945; the WAC Corporal became a significant launch vehicle in post-war rocket research.[49]

Less significant, but deserving of attention if only because it was the first U.S. corporation dedicated solely to the development of liquid rocket engines and accessory equipment, Reaction Motors, Inc. (RMI), came into being less than two weeks after the United States entered World War II. Based at Pompton Plains, New Jersey, its founders had been longtime rocket enthusiasts intimately connected with the American Interplanetary Society/American Rocket Society. All were convinced of the military and business potential of the rocket in the expanding world conflict. The company's leadership negotiated a contract with the Navy's Bureau of Aeronautics to develop a 445-Newton (100-pound) thrust regeneratively cooled rocket motor, which was to be employed by the Navy to assist large, heavily laden flying boats during takeoff. By the end of November 1943, RMI was heavily involved with naval research in Annapolis. There, a nitric acid-based rocket program was underway at the Naval Engineering Experiment Station (NEES) where Robert H. Goddard was working on pumps and turbines. Goddard's work was put to good use by RMI, which by early 1944 had succeeded in testing a liquid-fueled engine mounted in a Navy PBM3C flying boat. The company then went on to develop the rocket engine that propelled the first piloted aircraft to fly faster than the speed of sound, the Air Force X-1 in 1947. Thereafter, RMI contributed critical engine components to virtually all U.S. rocket programs.[50]

While the developments in the United States ultimately proved more significant, laying as they did the foundation of much post-war rocket technology, it was in Germany that the most spectacular early successes in developing an operational rocket capability took place. This was probably the case largely because in 1932 the German army hired the charismatic and politically astute Wernher von Braun, then only twenty years old, to work in its military rocket program. While he was the first VfR member to go to work for the German military, he was far from the last.

Von Braun's motivations for this move, with the hindsight of Hitler's rise to power in Germany and the devastation and terror of World War II, have been questioned and criticized. For some he was a visionary who foresaw the potential of human spaceflight, but for others he was little more than an arms merchant who developed brutal weapons of mass destruction. In reality, he seems to have been something of both, all the while never evincing Malina's type of hesitancy about the morality of using scientific and technical knowledge to kill as many people and destroy as many resources as possible. In the 1960s, as the United States was involved in a race with the Soviet Union to see who could land a human on the Moon first, political humorist Tom Lehrer wrote a song about von Braun's pragmatic approach to serving whoever would let him build rockets regardless of their purpose. "Don't say that he's hypocritical, say rather that he's apolitical," Lehrer wrote. "'Once the rockets are up, who cares where they come down? That's not my department,' says Wernher von Braun." Lehrer's biting satire captured the ambivalence of von Braun's atti-

---

48. Theodore von Kármán, "Memorandum on the Possibilities of Long-Range Rocket Projectiles," November 20, 1943, Frank J. Malina Folder, Biographical Files, NASA Historical Reference Collection.

49. Frank J. Malina, "America's First Long-Range-Missile and Space Exploration Program: The ORDCIT Project of the Jet Propulsion Laboratory, 1943-1946, a Memoir," in Hall, ed., *Essays on the History of Rocketry and Astronautics*, pp. 339-83; William R. Corliss, *NASA Sounding Rockets, 1958-1968: A Historical Summary* (Washington, DC: NASA SP-4401, 1971), pp. 17-18.

50. James H. Wyld, "The Liquid Propellant Rocket Engine," *Mechanical Engineering*, June 1947, p. 5; Frederick I. Ordway, III and Frank H. Winter, "Reaction Motors, Inc.: A Corporate History, 1941-1958," Parts I and II, in Roger D. Launius, ed., *History of Rocketry and Astronautics: Proceedings of the Fifteenth and Sixteenth Symposia of the International Academy of Astronautics* (San Diego: Univelt, Inc., 1994), pp. 75-100, 101-27.

tude on moral questions associated with the use of rocket technology.[51]

With military oversight provided by General Walter Dornberger, Germany developed two important aerospace weapons, the V-1 "Buzz Bomb" and the V-2 rocket, the latter built under von Braun's direction. The V-1, first used in June 1944, had one substantial weakness; it was relatively slow, with a top speed of 400 miles per hour. This made it possible for allied pilots and anti-aircraft operators to destroy it. Of the more than 8,000 of these weapons launched, over half were destroyed before reaching their targets. But the "Buzz Bombs" that reached London extracted a toll: several thousand people were killed and wounded.[52]

While the V-1 was essentially an air-breathing cruise missile, the second German weapon was the first true ballistic missile. The brainchild of Wernher von Braun's rocket team operating at a secret laboratory at Peenemünde on the Baltic coast, this rocket was the immediate antecedent of some of those used in the U.S. space program. A liquid propellant missile 46 feet in height and weighing 27,000 pounds at launch, the V-2, called the A-4 by the Germans involved in the project, flew at speeds in excess of 3,500 miles per hour and delivered a 2,200-pound warhead 200 miles away. First successfully flown in October 1942, it was employed against targets in Europe beginning in September 1944, and by the end of the war 1,155 had been fired against England and another 1,675 had been launched against Antwerp and other continental targets. The guidance system for these missiles was imperfect, and many did not reach their targets, but they struck without warning and there was no defense against them. As a result the V-2s had a terror factor far beyond their capabilities.[53]

Germany's astounding success in developing a ballistic missile while the other combatants had not done so was no accident, and it was in no small measure the result of personalities involved in the research. Before 1941 the United States had led the world in rocket technology, chiefly because of Goddard's work. But he failed to gain the significant support of either other scientists or the U.S. government. On the other hand, the energetic Oberth courted his scientific colleagues and those in the German government. For instance, as early as 1929 Oberth had helped kindle the fires of rocketry's promise in Walter Dornberger, later the military commander of the German rocket program. No similar level of salesmanship took place in any other nation. Popular and top-level support was therefore lacking, and Germany was able to capitalize on this with the V-2 development during the war.

Post-War Rocket Technology and Space Science

As World War II was winding down, U.S. military forces brought captured V-1s and V-2s back to the United States for examination. Clearly the technology employed in both of these weapons was worthy of study, and they were the top priority for military intelligence officials sifting through what remained of the impressive array of German military technology. Along with them—as part of a secret military operation called Project Paperclip—came many of the scientists and engineers who had developed these weapons, notably von Braun who intentionally surrendered to the United States in hopes that he could continue his rocketry experiments under U.S. sponsorship. He calculated that his work would be better supported and he would have a freer hand in the United States than in the Soviet Union. The German rocket team was installed at Fort Bliss in El Paso, Texas, and launch facilities for the V-2 test program were set up at the nearby White Sands Proving Ground in New Mexico. Later, in 1950 von Braun's team of over 100 people was moved

---

51. Wayne Biddle, "Science, Morality and the V-2," *New York Times*, October 2, 1992, p. A31; Tom Lehrer, "Wernher von Braun," on the album *That was the Year That Was* (1965).

52. Stanley M. Ulanoff, *Illustrated Guide to U.S. Missiles and Rockets* (Garden City, NY: Doubleday & Co., 1962), pp. 126-27; Kenneth P. Werrell, *The Evolution of the Cruise Missile* (Maxwell Air Force Base: Air University Press, 1987), pp. 40-81.

53. Michael J. Neufeld, "Hitler, the V-2, and the Battle for Priority, 1939-1943," *Journal of Military History* 57 (July 1993): 511-38; Walter Dornberger, *V-2: The Nazi Rocket Weapon* (New York: Viking Press, 1954), p. 97.

to the Redstone Arsenal near Huntsville, Alabama, to concentrate on the development of a new missile for the Army. Meanwhile, in an operation named Project Hermes, the first successful U.S. test firing of the captured V-2s took place at White Sands on April 16, 1946. Between 1946 and 1951, 67 captured V-2s were test-launched on non-orbital flights. The result was a significant expansion of the U.S. knowledge of rocketry.[54]

Although the U.S. Army was using these captured V-2 rockets to learn more about the technology, late in 1945 it offered scientists the opportunity to put experiments on them to study the upper atmosphere. Immediately thereafter the War Department established an Upper Atmosphere Research Panel, and although its name and scope of responsibilities changed periodically during the next several years it continued to coordinate these activities until the birth of NASA in 1958. It prioritized the use of these sounding rockets to study solar and stellar ultraviolet radiation, the aurora, and the nature of the upper atmosphere. As a result, the panel served as the "godfather" of the infant field of space science. Scientific data, while desired, was not the primary purpose of these flights, for Army Ordnance was interested mostly in learning about rocketry to aid in the development of a more advanced generation of weapons.[55]

Throughout the late 1940s and early 1950s rocket technicians conducted ever more demanding test flights and scientists conducted increasingly more complex scientific investigations made possible by the rocket technology. One of the most important series of flights was Project Bumper, which utilized a smaller Army WAC Corporal missile, produced at JPL, as a second stage of a V-2 to obtain data on both high altitudes and the principles of two-stage rockets. The only fully successful launch took place on February 24, 1949, when the V-2/WAC Corporal system reached an altitude of 244 miles and a velocity of 5,150 miles per hour. Much more useful was the Aerobee, a scaled-up version of the WAC Corporal developed by JPL, which could launch at a very economical cost a sizable payload to an altitude of 130 miles. The reliable little booster enjoyed a long career from its first instrumented firing on November 24, 1947, until the January 17, 1985, launch of the 1,037th and last Aerobee. Additionally, the Naval Research Laboratory was involved in sounding rocket research, non-orbital instrument launches, using the Viking launch vehicle built by the Glenn L. Martin Company. Viking 1 was launched from White Sands on May 3, 1949, while the twelfth and last Viking took off on February 4, 1955. The program produced significant scientific information about the upper atmosphere and took impressive high-altitude photographs of Earth. Most important, the Viking pioneered the use of a gimballed engine to control flight and paved the way for later orbiting scientific satellites.[56]

In virtually every instance, rockets developed during the 1950s resulted from the adoption of a basic system built on components that had been tested earlier and mated together into a new booster. For instance, the Scout booster began in 1957 as an attempt by the National Advisory Committee for Aeronautics to build a solid-fuel rocket that could launch a small scientific payload into orbit. To achieve this end, researchers investigated various solid-rocket configurations and finally decided to combine a Jupiter Senior (100,000 pounds of thrust), built by the Aerojet Corporation, with a second stage composed of a Sergeant missile base and two new upper stages descended from the research effort that produced the Vanguard. The Scout's four-stage booster could place a 330-pound satellite into orbit, and it quickly became a workhorse in orbiting small scientific payloads. It was first launched on July 1, 1960, and despite some early deficiencies, by the end of 1968, had

---

54. This effort has been discussed in James McGovern, *Crossbow and Overcast* (New York: William Morrow, 1964); Clarence G. Lasby, *Project Paperclip: German Scientists and the Cold War* (New York: Athenaeum, 1971); Frederick I. Ordway III and Mitchell R. Sharpe, *The Rocket Team* (New York: Crowell, 1979); Linda Hunt, *Secret Agenda: The United States Government, Nazi Scientists, and Project Paperclip, 1945-1990* (New York: St. Martin's Press, 1991). On the rocketry tests at White Sands see Corliss, *NASA Sounding Rockets*, pp. 11-15; Homer E. Newell, *High Altitude Rocket Research* (New York: Academic Press, 1953); David H. DeVorkin, *Science with a Vengeance: How the Military Created the US Space Sciences After World War II* (New York: Springer-Verlag, 1992).

55. John E. Naugle, *First Among Equals: The Selection of NASA Space Science Experiments* (Washington, DC: NASA SP-4215, 1991), pp. 1-4. The scientific research program for space has been discussed in "History of NASA Space Science," by John E. Naugle in this series.

56. DeVorkin, *Science with a Vengeance*, pp. 167-92.

achieved an 85-percent launch success rate.[57]

The Army also developed the Redstone rocket during this same period, a missile capable of sending a small warhead a maximum of 500 miles. Built under the direction of von Braun and his German rocket team in the early 1950s, the Redstone took many features from the V-2, added an engine from the Navaho test missile, and incorporated some of the electronic components from other rocket test programs. The first Redstone was launched from Cape Canaveral, Florida, on August 20, 1953. An additional 36 Redstone launches took place through 1958. This rocket led to the development of the Jupiter C, an intermediate-range ballistic missile that could deliver a nuclear warhead to a target after a non-orbital flight through space. Its capability for this mission was tested on May 16, 1958, when combat-ready troops first fired the rocket. The missile was placed on active service with U.S. units in Germany the next month, and served until 1963. The Redstone later served as the launch vehicle for the first U.S. suborbital launches of astronauts Alan B. Shepard and Gus Grissom in 1961.[58]

The Development of Ballistic Missiles

During this same era all the U.S. armed services worked toward the fielding of intercontinental ballistic missiles (ICBMs) that could deliver warheads to targets half a world away. Competition was keen among the services for a mission in the new "high ground" of space, whose military importance was not lost on the leaders of the world. In April 1946 the Army Air Forces gave the Consolidated Vultee Aircraft (Convair) Division a study contract for an ICBM. This led directly to the development of the Atlas ICBM in the 1950s. At first many engineers believed Atlas to be a high-risk proposition. To limit its weight, Convair Corp. engineers under the direction of Karel J. Bossart, a pre-World War II immigrant from Belgium, designed the booster with a very thin, internally pressurized fuselage instead of massive struts and a thick metal skin. The "steel balloon," as it was sometimes called, employed engineering techniques that ran counter to the conservative engineering approach used by Wernher von Braun and his "Rocket Team" at Huntsville, Alabama. Von Braun, according to Bossart, needlessly designed his boosters like "bridges," to withstand any possible shock. For his part, von Braun thought the Atlas was too flimsy to hold up during launch. The reservations began to melt away, however, when Bossart's team pressurized one of the boosters and dared one of von Braun's engineers to knock a hole in it with a sledge hammer. The blow left the booster unharmed, but the recoil from the hammer nearly clubbed the engineer.[59]

The Titan ICBM effort emerged not long thereafter, and proved to be an enormously important ICBM program and later a civil and military space launch asset. To consolidate efforts, Secretary of Defense Charles E. Wilson issued a decision on November 26, 1956, that effectively took the Army out of the ICBM business and assigned responsibility for land-based systems to the Air Force and sea-launched missiles to the Navy. The Navy immediately stepped up work for the development of the submarine-launched Polaris ICBM, which first successfully operated in January 1960.

The Air Force did the same with land-based ICBMs, and its efforts were already well-developed at the time of the 1956 decision. The Atlas received high priority from the White House and hard-driving management from Brigadier General Bernard A. Schriever,

---

57. Linda Neuman Ezell, *NASA Historical Data Book, Vol II: Programs and Projects, 1958-1968* (Washington, DC: NASA SP-4012, 1988), pp. 61-67; Richard P. Hallion, "The Development of American Launch Vehicles Since 1945," in Paul A. Hanle and Von Del Chamberlain, eds., *Space Science Comes of Age: Perspectives in the History of the Space Sciences* (Washington, DC: Smithsonian Institution Press, 1981), pp. 126-27.

58. Wernher von Braun, "The Redstone, Jupiter, and Juno," in Emme, ed., *History of Rocket Technology*, pp. 107-121.

59. Richard E. Martin, *The Atlas and Centaur "Steel Balloon" Tanks: A Legacy of Karel Bossart* (San Diego: General Dynamics Corp., 1989); Robert L. Perry, "The Atlas, Thor, Titan, and Minuteman," in Emme, ed., *History of Rocket Technology*, pp. 143-55; John L. Sloop, *Liquid Hydrogen as a Propulsion Fuel, 1945-1959* (Washington, DC: NASA SP-4404, 1978), pp. 173-77.

a flamboyant and intense Air Force leader. The first Atlas rocket was test fired on June 11, 1955, and a later generation rocket became operational in 1959. These systems were followed in quick succession by the Titan ICBM and the Thor intermediate-range ballistic missile. By the latter 1950s, therefore, rocket technology had developed sufficiently for the creation of a viable ballistic missile capability. This was a revolutionary development that gave humanity for the first time in its history the ability to attack one continent from another. It effectively shrank the size of the globe, and the United States—which had always before been protected from outside attack by two massive oceans—could no longer rely on natural defensive boundaries or distance from its enemies.[60]

Space and the American Imagination

The development of the United States' rocketry capability, especially with the work on the ICBMs, signaled for the rest of the world that the United States could project military might anywhere in the world. In addition, this military capability could be used for the peaceful projection of a human presence into space. The dreams of Verne and Wells were combined with the pioneering rocketry work of Goddard and Oberth and later developments in technology to create the probability of a dawning space age. Another ingredient entered into this arena, however—imagination, the intangible quality that prompted humans to want to move beyond the atmosphere. There was an especially significant spaceflight "imagination" that came to the fore after World War II and that urged the implementation of an aggressive spaceflight program. It was seen in science fiction books and film, but more importantly, it was fostered by serious and respected scientists, engineers, and politicians. The popular culture became imbued with the romance of spaceflight, and the practical developments in technology reinforced these perceptions that space travel might actually be, for the first time in human history, possible.[61]

The decade following the war brought a change in perceptions, as most Americans went from skepticism about the probabilities of spaceflight to an acceptance of it as a near-term reality. This can be seen in the public opinion polls of the era. For instance, in December 1949 Gallup pollsters found that only 15 percent of Americans believed humans would reach the Moon within 50 years, while 15 percent had no opinion and a whopping 70 percent believed that it would not happen within that time. In October 1957, at the same time as the launching of Sputnik I, only 25 percent believed that it would take longer than 25 years for humanity to reach the Moon, while 41 percent believed firmly that it would happen within 25 years and 34 percent were not sure. An important shift in perceptions took place during that era, and it was largely the result of well-known advances in rocket technology coupled with a public relations campaign based on the real possibility of spaceflight.[62]

Clearly, one of the most important groups that had been consistently enthralled with the promise of spaceflight were the science fiction aficionados and the futurists, many of whom were one and the same. Many science fiction writers were basically hacks writing for a specialized market, but a few broke the boundaries of the genre in the post-war era and contributed significantly to public perceptions of space travel. Perhaps the three most significant authors in this category were Robert A. Heinlein, Isaac Asimov, and Arthur C. Clarke, all of whom took pains to make their science fiction novels and short stories both believable as reality and exciting as works of literature. They found a ready audience in the

---

60. This story is told in Edmund Beard, *Developing the ICBM: A Study in Bureaucratic Politics* (New York: Columbia University Press, 1976); Jacob Neufeld, *Ballistic Missiles in the United States Air Force, 1945-1960* (Washington, DC: Office of Air Force History, 1990).

61. This is the thesis of William Sims Bainbridge, *The Spaceflight Revolution: A Sociological Study* (New York: Wiley, 1976). See also Willy Ley and Chesley Bonestell, *The Conquest of Space* (New York: Viking, 1949).

62. George H. Gallup, *The Gallup Poll: Public Opinion, 1935-1971* (New York: Random House, 1972), 1:875, 1152.

environment of the Cold War, as ever-increasing numbers of Americans could both envision and understand the advance of technology and technocracy, and the merger of bureaucratic and technical expertise in government. Asimov, for one, featured robots in his writings, something more and more Americans could understand as machines of all types took over an ever-increasing part of the workload. Both Asimov and Heinlein played out their stories within the context of complex galactic politics, not unlike those perceived by Americans in the world situation.[63]

Asimov and Clarke also bridged the gap between science fiction and science fact in some very fundamental ways. They each wrote both fiction and popular scientific studies relative to spaceflight, physics, and astronomy. They also identified some interesting potential uses for space technology. For example, in February 1945 Clarke described the use of the German V-2 as a launcher for ionospheric research, even as the war was going on. He specifically suggested that by putting a second stage on a V-2 the rocket could generate enough velocity to launch a small satellite into orbit. "Both of these developments demand nothing in the way of technical resources," he wrote, adding that they "should come within the next five or ten years." He later described the possibility of placing three satellites in geosynchronous orbit 120 degrees apart to "give television and microwave coverage to the entire planet."[64] Later that same year Clarke elaborated on the communications implications of satellites and set in motion the ideas that eventually led to the global communications system first put in place during the 1960s.[65]

Another important way in which the U.S. public became aware that flight into space was a possibility revolved around the rise of films depicting space travel that were firmly rooted in scientific reality. One of the keys in this process was the work of film producer-director George Pal, a master of special effects, who made several space-oriented movies in the 1950s.[66] Especially memorable were two films, *The Day the Earth Stood Still* (1950), directed by Robert Wise, in which the benevolent alien Klaatu warns the Earth to shape up and control its aggressiveness by disarming, and *Forbidden Planet* (1956), about the extinct Krell superintelligent society and the Monster from the Id.[67] These films excited the public with ideas of spaceflight, exploration, and contact with alien civilizations. It is often easy to forget that these sophisticated visions of space travel occurred *before* Sputnik.

Even more important than science fiction literature and film were the public writings and speeches of serious and respected scientists, engineers, and politicians who fostered dreams of spaceflight. Among the most important of these was Wernher von Braun, ensconced in his Army rocket center at Huntsville, Alabama. Von Braun, in addition to being a superbly effective technological entrepreneur within the governmental system, by the early 1950s had learned and was applying daily the skills of public relations on behalf of space travel. His background as a serious rocket engineer, a German emigré, a handsome aristocrat, and a charismatic leader all combined to create a positive impression on the U.S. public. When he managed to seize the powerful print and electronic communication media that the science fiction writers and film makers had been using, no one during the 1950s was a more effective promoter of spaceflight to the public than von Braun.[68]

In 1952 von Braun burst on the broad public stage with a series of articles in *Collier's*

---

63. Sam Moskowitz, "The Growth of Science Fiction from 1900 to the early 1950s," in Frederick I. Ordway III and Randy Lieberman, eds., *Blueprint for Space: Science Fiction to Science Fact* (Washington, DC: Smithsonian Institution Press, 1992), pp. 69-82; Eric Burgess, "Into Space," *Aeronautics*, November 1946, pp. 52-57.

64. Arthur C. Clarke, "V2 for Ionospheric Research?," *Wireless World*, February 1945, p. 58.

65. Arthur C. Clarke, "Extra-Terrestrial Relays: Can Rocket Stations Give World-Wide Radio Coverage?," *Wireless World*, October 1945, pp. 305-308.

66. On Pal's career see, Gail Morgan Hickman, *The Films of George Pal* (South Berwick: A.S. Barnes, 1977); Robert A. Heinlein, "Shooting Destination Moon," *Astounding Science Fiction*, July 1950, p. 6.

67. W.J. Stuart, *Forbidden Planet* (New York: Farrar, Straus & Cudahy, 1956); H. Bates, "Farewell to the Master," *Astounding Science Fiction*, October 1940, p. 58ff.

68. See, as an example of his exceptionally sophisticated spaceflight promoting, Wernher von Braun, *The Mars Project* (Urbana: University of Illinois Press, 1953), based on a German-language series of articles appearing in the magazine *Weltraumfahrt* in 1952.

magazine about the possibilities of spaceflight. The genesis of this series began innocently enough. In 1951 Willy Ley, a former member of the German VfR and himself a skilled promoter of spaceflight, organized a Space Travel Symposium that took place on Columbus Day at the Hayden Planetarium in New York City. Ley wrote to participants that "the time is now ripe to make the public realize that the problem of space travel is to be regarded as a serious branch of science and technology," and he urged them to emphasize that fact in their lectures.[69] By happenstance, two *Collier's* writers attended this meeting. They were most impressed with the ideas presented and suggested to *Collier's* managing editor, Gordon Manning, that his magazine publish several articles promoting the scientific possibility of spaceflight. Recognizing that this idea might have real appeal, Manning asked an assistant editor, Cornelius Ryan, to organize some discussions with Ley and others, among them von Braun. Out of this came a series of important *Collier's* articles over a two-year period, each expertly illustrated with striking images by some of the best illustrators of the era.[70]

The first issue of *Collier's* devoted to space appeared on March 22, 1952. In it readers were asked "What Are We Waiting For?" and were urged to support an aggressive space program. An editorial suggested that spaceflight was possible, not just science fiction, and that it was inevitable that humanity would venture outward. It framed the exploration of space in the context of the Cold War rivalry with the Soviet Union and concluded that "Collier's believes that the time has come for Washington to give priority of attention to the matter of space superiority. The rearmament gap between the East and West has been steadily closing. And nothing, in our opinion, should be left undone that might guarantee the peace of the world. It's as simple as that."[71] [I-13]

Von Braun led off the *Collier's* issue with an impressionistic article describing the overall features of an aggressive spaceflight program. He advocated the orbiting of an artificial satellite to learn more about spaceflight followed by the first orbital flights by humans, development of a reusable spacecraft for travel to and from Earth orbit, the building of a permanently inhabited space station, and finally human exploration of the Moon and planets by spacecraft launched from the space station. [I-14] Willy Ley and several other writers then followed with elaborations on various aspects of spaceflight, ranging from technological viability to space law to biomedicine.[72] The series concluded with a special issue of the magazine devoted to Mars, in which von Braun and others described how to get there and predicted what might be found based on recent scientific data.[73] [I-15, I-16]

The *Collier's* series catapulted von Braun into the public spotlight like none of his previous research activities had been able to do. The magazine was one of the four highest-circulation periodicals in the United States during the early 1950s, with over 3 million copies produced each week. If estimates of readership were indeed four or five people per copy, as the magazine claimed, something on the order of 15 million people were exposed to these spaceflight ideas. *Collier's*, seeing that it had a potential blockbuster, did its part by hyping the series with window ads of the space artwork appearing in the magazine, sending out more than 12,000 press releases, and preparing media kits. It set up interviews on radio and television for von Braun and the other space writers, but especially von Braun, whose natural charisma and enthusiasm for spaceflight translated well through that medium. Von Braun appeared on NBC's "Camel News Caravan" with John Cameron Swayze, on NBC's "Today" show with Dave Garroway, and on CBS's "Gary Moore" program. While *Collier's* was interested in selling magazines with these public appearances, von Braun was

---

69. Willy Ley to Heinz Haber, *et al.*, June 13, 1951, Hayden Planetarium Library, New York, NY.
70. On these articles see Randy Liebermann, "The *Collier's* and Disney Series," in Ordway and Leibermann, eds., *Blueprint for Space*, pp. 135-44.
71. "What Are We Waiting For?," *Collier's*, March 22, 1952, p. 23.
72. "Man Will Conquer Space *Soon*" series, *Collier's*, March 22, 1952, pp. 23-76ff.
73. Wernher von Braun with Cornelius Ryan, "Can We Get to Mars?," *Collier's*, April 30, 1954, pp. 22-28.

interested in selling the idea of space travel to the public.[74]

Following close on the heels of the *Collier's* series, Walt Disney Productions contacted von Braun—through Willy Ley—and asked his assistance in the production of three shows for Disney's weekly television series. The first of these, "Man in Space," premiered on Disney's show on March 9, 1955, with an estimated audience of 42 million. The second show, "Man and the Moon," also aired in 1955 and sported the powerful image of a wheel-like space station as a launching point for a mission to the Moon. The final show, "Mars and Beyond," premiered on December 4, 1957, after the launching of Sputnik I. Von Braun appeared in all three films to explain his concepts for human spaceflight, while Disney's characteristic animation illustrated the basic principles and ideas with wit and humor.[75]

While some scientists and engineers criticized von Braun for his blatant promotion of both spaceflight and himself, the *Collier's* series of articles and especially the three Disney television programs were exceptionally important in changing public attitudes toward spaceflight. Media observers noted the favorable response to the three Disney shows from the public, and recognized that "the thinking of the best scientific minds working on space projects today" went into them, "making the picture[s] more fact than fantasy."[76]

Although an overstatement, some have suggested that the airing of the first Disney space film on March 9, 1955, contributed to President Dwight D. Eisenhower's July 1955 decision to embrace the launching of a scientific satellite as part of the U.S.'s contribution to research during the International Geophysical Year (IGY) in 1957-1958. [I-17] When Disney studio executives wanted to emphasize this possibility, however, von Braun told them, "For God's sake don't put it that this show triggered the presidential announcement." He was apparently concerned that Eisenhower might be embarrassed at the suggestion that a media event influenced his support for the IGY satellite.[77] Regardless of the impetus for Eisenhower's decision, von Braun and the Disney series helped shape the public's perception of spaceflight as something that was no longer fantasy.

Closely tied to the growing public perception of spaceflight as a possibility in the 1950s was the postwar Unidentified Flying Object (UFO) craze that took place in the United States. Between 1947 and 1960 a total of 6,523 UFO sightings were reported in the United States. Many people considered them to be of extraterrestrial origin. The reports slowly began to increase, with 79 reported in 1947, and remained stable until 1951, when 1,501 were recorded. There seems to be a direct tie between public perception of the reality of space travel and these UFO sightings, especially when considering that 701 of the 1957 reports came after the launch of Sputnik I on October 4.[78]

The U.S. Air Force considered the UFO phenomenon significant enough to begin in December 1947 a project to investigate occurrences, especially with a view to learn if "some foreign nation had a form of propulsion possibly nuclear, which is outside our domestic knowledge."[79] Although the researchers recognized the possibility that the UFOs might be extraterrestrial, few thought it was probable and emphasized explanations of the phenomena that were more earthly. For instance, the Scientific Advisory Panel of the Central Intel-

---

74. Liebermann, "The *Collier's* and Disney Series," in Ordway and Liebermann, *Blueprint for Space*, p. 141; Ron Miller, "Days of Future Past," *Omni*, October 1986, pp. 76-81.

75. Liebermann, "The *Collier's* and Disney Series," in Ordway and Liebermann, *Blueprint for Space*, pp. 144-46; David R. Smith, "They're Following Our Script: Walt Disney's Trip to Tomorrowland," *Future*, May 1978, pp. 59-60; Mike Wright, "The Disney-Von Braun Collaboration and Its Influence on Space Exploration," paper presented at conference, "Inner Space, Outer Space: Humanities, Technology, and the Postmodern World," February 12-14, 1993; Willy Ley, *Rockets, Missiles, and Space Travel* (New York: The Viking Press, 1961 ed.), p. 331.

76. *TV Guide*, March 5, 1955, p. 9.

77. Wernher Von Braun to Ward Kimball, August 30, 1955, quoted in Smith, "They're Following Our Script," p. 59. This episode has been discussed in Rip Bulkeley, *The Sputniks Crisis and Early United States Space Policy* (Bloomington: Indiana University Press, 1991), pp. 128-29.

78. Lawrence J. Tacker, *Flying Saucers and the U.S. Air Force* (Princeton: D. van Nostrand Co., 1960), p. 82.

79. Lt. Gen. Nathan F. Twining, Commander Air Material Command, to Commanding General, Army Air Forces, "Flying Discs," September 23, 1947, reprinted in Edward U. Condon, *Final Report of the Scientific Study of Unidentified Flying Objects* (New York: Bantam Books, 1969), p. 895. The letter setting up the study is Maj. Gen. L.C. Craigie to Commanding General, Wright Field, "Flying Discs," December 30, 1947, in Condon, *Final Report*, pp. 896-97.

ligence Agency considered in January 1953 the UFO issue in the United States. [I-18] After a lengthy discussion, members of the panel "concluded that reasonable explanations could be suggested for most sightings." Moreover, concerning one of the central questions this body had about UFOs, it "concluded unanimously that there was no evidence of a direct threat to national security in the objects sighted."[80]

A report released by the Air Force in 1957 reached similar conclusions. It said:

first, there is no evidence that the "unknowns" were inimical or hostile; second, there is no evidence that these "unknowns" were interplanetary space ships; third, there is no evidence that these unknowns represented technological developments or principles outside the range of our present day scientific knowledge; fourth, there is no evidence that these "unknowns" were a threat to the security of the country; and finally there was no physical evidence or material evidence, not even a minute fragment, of a so-called "flying saucer" was ever found.[81]

Even with these studies, however, a fair percentage of Americans still believed that UFOs were probably of extraterrestrial origin.

Explanations of why several thousand people saw something they could not identify and thought was an extraterrestrial spacecraft have ranged far and wide. Humanity has long been intensely interested in supernatural occurrences. The ancient Greeks had their gods who came down from Mount Olympus; people of the Medieval era saw appearances of angels, the Virgin, and devils, as well as fairies and elves. UFO sightings in the 1940s and 1950s—none of which apparently produced any physical evidence—are essentially in the same category.

Humans have always been fascinated and terrified of the unknown. While some feared the stars and planets, others studied them. While some spoke of "the harmony of the spheres," others warned that comets and other stellar phenomena foretold of humanity's destruction. Reports of encounters with extraterrestrials were a response to the duality of fascination and terror of humanity over contact with alien species. Some of the reports were in part a Cold War phenomenon, as Americans longed for the help of a benevolent, wise, and powerful alien race who could chaperon humanity through the possibility of nuclear holocaust *à la* Klaatu from *The Day the Earth Stood Still.* Some reported incidents were negative, harkening back to the terror expressed in Wells' *War of the Worlds.* Some reports reflected American perception of the technological possibilities of space travel. Moreover, if the Earth was on the verge of a space age, what about more advanced civilizations on other worlds? Might they someday journey to Earth? To some the UFOs spoke to the nightmares of humanity. But to others they spoke to some of the sublime dreams of humanity, and they were therefore significant at the time because of what they signaled about public perceptions of what was possible in the emergent space age.[82]

Conclusion

The combination of technological and scientific advance, political competition with the Soviet Union, and changes in popular opinion about spaceflight came together in a very specific way in the 1950s to affect public policy in favor of an aggressive space program. This found tangible expression in 1952 when the International Council of Scientific

---

80. "Report of Meetings of Scientific Advisory Panel on Unidentified Flying Objects Convened by Office of Scientific Intelligence, CIA, January 14-18, 1953," copy in "NACA-UFO, 1948-1958," folder, NASA Historical Reference Collection.

81. "Air Force's 10 Year Study of Unidentified Flying Objects," Department of Defense, Office of Public Information, News Release No. 1083-58, November 5, 1957, copy in "NACA-UFO, 1948-1958," folder, NASA Historical Reference Collection.

82. Carl Sagan, *Broca's Brain: Reflections on the Romance of Science* (New York: Ballantine Books, 1974), pp. 65-70; Philip Klass, *UFOs Explained* (New York: Random House, 1974); Carl Sagan and Thornton Page, eds., *UFOs: A Scientific Debate* (New York: W.W. Norton and Co., 1973).

Unions (ICSU) started planning for an International Polar Year, the third in a series of scientific activities designed to study geophysical phenomena in remote reaches of the planet. The Council agreed that July 1, 1957, to December 31, 1958, would be the period of emphasis in polar research, in part because of a predicted expansion of solar activity; the previous polar years had taken place in 1882-1883 and 1932-1933. Late in 1952 the ICSU expanded the scope of the scientific research effort to include studies that would be conducted using rockets with instrument packages in the upper atmosphere and changed the name to the International Geophysical Year (IGY) to reflect the larger scientific objectives. In October 1954 at a meeting in Rome, Italy, the Council adopted another resolution calling for the launch of artificial satellites during the IGY to help map the Earth's surface. The Soviet Union immediately announced plans to orbit an IGY satellite, virtually assuring that the United States would respond, and this, coupled with the military satellite program, set both the agenda and the stage for most space efforts through 1958. The next year the United States announced Project Vanguard, its own IGY scientific satellite program.[83]

By the end of 1956, less than a year before the launch of Sputnik, the United States was involved in two modest space programs that were moving ahead slowly and staying within strict budgetary constraints. One was a highly visible scientific program as part of the IGY, and the other was a highly classified program to orbit a military reconnaissance satellite. They shared two attributes. They each were separate from the ballistic missile program underway in the Department of Defense, but they shared in the fruits of its research and adapted some of its launch vehicles. They also were oriented toward satisfying a national goal of establishing "freedom of space" for all orbiting satellites. The IGY scientific effort could help establish the precedent of access to space, while a military satellite might excite other nations to press for limiting such access. Because of this goal a military satellite, in which the Eisenhower Administration was most interested, could not under any circumstances precede scientific satellites into orbit. The IGY satellite program, therefore, was a means of securing the larger goal of open access to space. Before it could do so, on October 4, 1957, the Soviet Union launched Sputnik I and began the space age in a way that had not been anticipated by the leaders of the United States.

---

83. A good account of the IGY satellite projects can be found in Bulkeley, *Sputniks Crisis and Early United States Space Policy*, pp. 89-122.

Documents I-1 and I-2

Document title: Medieval universe at the time of Dante, as presented in *The Divine Comedy*, from Edward R. Harrison, *Cosmology: The Science of the Universe* (Cambridge: Cambridge University Press, 1981), p. 77.

Document title: The infinite universe of Thomas Digges, from Edward R. Harrison, *Cosmology: The Science of the Universe* (Cambridge: Cambridge University Press, 1981), p. 79.

Source: Reprinted with the permission of Cambridge University Press.

The ancient conception of the universe as Christianized by Thomas Aquinas in the thirteenth century was carried to its logical conclusion by Dante in his classic work, *The Divine Comedy*. In this representation, I-1, hell became a nether-region inside the Earth's crust, purgatory was the sunlunar region, and the ethereal regions were found to be ideal for the residence of hierarchies of angelic beings. The astronomer and mathematician Thomas Digges modified Dante's medieval conceptions of the universe in his *Description of the Caelestiall Orbes* (1576), I-2, by adopting a Copernican view that placed the Sun in the center of the universe and by eliminating the outermost of the crystalline orbs and dispersing stars throughout an infinite universe beyond.

Document I-1

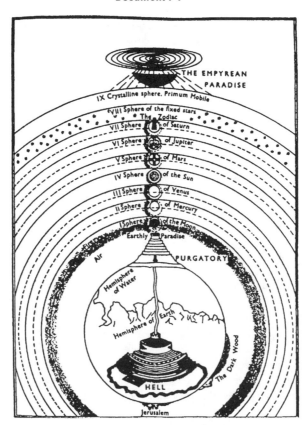

Document I-2

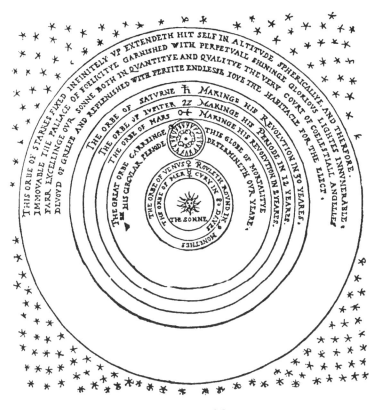

Document I-3

Document title: Edward E. Hale, "The Brick Moon," *The Atlantic Monthly*, October 1869, pp. 451-60, November 1869, pp. 603-11, December 1869, pp. 679-88, February 1870, pp. 215-22. Also published in *His Level Best, and Other Stories* (Boston: James R. Osgood & Co., 1873) pp. 30-124.

Edward Everett Hale was an author and clergyman who was best known for his 1863 short story "The Man Without a Country" (about a member of the Burr Conspiracy being exiled from the United States) He was widely regarded as one of the foremost literary figures of his time and was the primary speaker at Gettysburg in 1863 when Lincoln gave his famous address.

According to Hale, the idea for "The Brick Moon" was inspired by Richard Adams Locke's *Moon Hoax*. Hale further stated that while attending Cambridge University in 1838, the idea came from "an old chart, dreams and plans of college days" and was written while working in a room of his brother's, a professor, at Union College, Schenectady, New York, in 1869. The story was serialized in October, November, and December 1869 in *The Atlantic Monthly*, and a short sequel, "Life in the Brick Moon," appeared in the same magazine in February 1870.

Despite claims by both the Germans and Russians that Oberth and Tsiolkovskiy were the first to discuss Earth satellites, Hale's story is the first account of an artificial Earth satellite. In addition to being the first to mention the concept, Hale also outlined several uses for such an object, navigation being the most important in his view.

The Brick Moon

From the Papers of Captain Frederic Ingham

I.—PREPARATION.

[451] I have no sort of objection now to telling the whole story. The subscribers, of course, have a right to know what became of their money. The astronomers may as well know all about it, before they announce any more asteroids with an enormous movement in declination. And experimenters on the longitude may as well know, so that they may act advisedly in attempting another brick moon or in refusing to do so.

It all began more than thirty years ago, when we were in college; as most good things begin. We were studying in the book which has gray sides and a green back, and is called "Cambridge Astronomy" because it is translated from the French. We came across this business of the longitude, and, as we talked, in the gloom and glamour of the old South Middle dining-hall, we had going the usual number of students' stories about rewards offered by the Board of Longitude for discoveries in that matter,—stories, all of which, so far as I know, are lies. Like all boys, we had tried our hands at perpetual motion. For me, I was sure I could square the circle, if they would give me chalk enough. But as to this business of the longitude, it was reserved for Q, to make the happy hit and to explain it to the rest of us.

I wonder if I can explain it to an unlearned world, which has not studies the book with gray sides and a green cambric back. Let us try.

You know then, dear world that when you look at the North Star, it always appears to you at just the same height above the horizon or what is between you and the horizon: say the Dwight School-house, or the houses in Concord Street; or to me, just now, North College. You know also that, if you were to travel to the North Pole, the North Star would be just over your head. And, if you were to travel to the equator, it would be just on your horizon, if you could see it at all through the red, dusty, hazy mist in the north,—as you could not. If you were just half-way between pole and equator, on the line [452] between us and Canada, the North Star would be half-way up, or 45° from the horizon. So you would know there that you were 45° from the equator. There in Boston, you would find it was 42°20' from the horizon. So you know there that you are 42°20' from the equator. At Seattle again you would find it was 47°40' high, so our friends at Seattle know that they are at 47°40' from the equator. The latitude of a place, in other words, is found very easily by any observation which shows how high the North Star is; if you do not want to measure the North Star, you may take any star when it is just to north of you, and measure its height; wait twelve hours, and if you can find it, measure its height again. Split the difference, and that is the altitude of the pole, or the latitude of you, the observer.

"Of course we know this," says the graduating world. "Do you suppose that is what we borrow your book for, to have you spell out your miserable elementary astronomy?" At which rebuff I should shrink distressed, but that a chorus of voices an octave higher comes up with, "Dear Mr. Ingham, we are ever so much obliged to you; we did not know it at all before, and you make it perfectly clear."

Thank you, my dear, and you, and you. We will not care what the others say. If you do understand it, or do know it, it is more than Mr. Charles Reade knew, or he would not have made his two lovers on the island guess at their latitude, as they did. If they had either of them been educated at a respectable academy for the Middle Classes, they would have fared better.

Now about the longitude.

The latitude, which you have found, measures your distance north or south from the equator or the pole. To find your longitude, you want to find your distance east or west from the meridian of Greenwich. Now if any one would build a good tall tower at Greenwich, straight into the sky,—say a hundred miles into the sky,—of course if you and I were east or west of it, and could see it, we could tell how far east or west we were by measuring the apparent height of the tower above our horizon. If we could see so far, when the lantern with a Drummond's light, "ever so bright," on they very top of the tower, appeared to be on our horizon, we should know we were eight hundred and seventy-three miles away from it. The top of the tower would answer for us as the North Star does when we are measuring the latitude. If we were nearer, our horizon would make a longer angle with the line from the top to our place of vision. If we were farther away, we should need a higher tower.

But nobody will build any such tower at Greenwich, or elsewhere on that meridian, or on any meridian. You see that to be of use to the half the world nearest to it, it would have to be so high that the diameter of the world would seem nothing in proportion. And then, for the other half of the world you would have to erect another tower as high on the other side. It was this difficulty that made Q. suggest the expedient of the Brick Moon.

For you see that if, by good luck, there were a ring like Saturn's which stretched round the world, above greenwich and the meridian of Greenwich, and if it would stay above Greenwich, turning with the world, any one who wanted to measure his longitude or distance from Greenwich would look out of window and see how high this ring was above his horizon. At Greenwich it would be over his head exactly. At New Orleans, which is quarter round the world from Greenwich, it would be just in his horizon. A little west of New Orleans you would begin to look for the other half of the ring on the west instead of the east; and if you went a little west of the Feejee Islands the ring would be over your head again. So if we only had a ring like that, not round the equator of the world,—as Saturn's ring is around Saturn,—but vertical to the plane of the equator, as the brass ring of an artificial globe goes, only far higher in proportion,— [453] "from that ring," said Q., pensively, "we could calculate the longitude."

Failing that, after various propositions, he suggested the Brick Moon. The plan was this: If from the surface of the earth, by a gigantic pea-shooter, you could shoot a pea upward from Greenwich, aimed northward as well as upward; if you drove it so fast and far that when its power of ascent was exhausted, and it began to fall, it should clear the earth, and pass outside the North Pole; if you had given it sufficient power to get it half round the earth without touching, that pea would clear the earth forever. It would continue to rotate above the North Pole, above the Feejee Island place, above the South Pole and Greenwich, forever, with the impulse with which it had first cleared our atmosphere and attraction. If only we could see that pea as it revolved in that convenient orbit, then we could measure the longitude from that, as soon as we knew how high the orbit was, as well as if it were the ring of Saturn.

"But a pea is so small!"

"Yes," said Q., "but we must make a large pea." Then we fell to work on plans for making the pea very large and very light. Large,—that it might be seen far away by storm-tossed navigators: light,—that it might be the easier blown four thousand and odd miles into the air; lest it should fall on the heads of the Greenlanders or the Patagonians; lest they should be injured and the world lose its new moon. But, of course, all this lath-and-plaster had to be given up. For the motion through the air would set fire to this moon just as it does to other aerolites, and all your lath-and-plaster would gather into a few white drops, which no Rosse telescope even could discern. "No," said Q. bravely, "at the least it must be very substantial. It must stand fire well, very well. Iron will not answer. It must be brick; we must have Brick Moon.

Then we had to calculate its size. You can see, on the old moon, an edifice two hundred feet long with any of the fine refractors of our day. But no such refractors as those can be carried by the poor little fishermen whom we wanted to befriend, the bones of whose ships lie white on so many cliffs, their names unreported at any Lloyd's or by any Ross,—themselves the owners and their sons the crew. On the other hand, we did not want our moon two hundred and fifty thousand miles away, as the old moon is, which I will call the Thornbush moon, for distinction. We did not care how near it was, indeed, if it were only far enough away to be seen, in practice, from almost the whole world. There must be a little strip where they could not see it from the surface, unless we threw it infinitely high. "But they need not look from the surface," said Q.; "they might climb to the mast-head. And if they did not see it at all, they would know that they were ninety degrees from the meridian."

This difficulty about what we call "the strip," however, led to an improvement in the plan, which made it better in every way. It was clear that even if "the strip" were quite wide, the moon would have to be a good way off, and, in proportion, hard to see. If, however, we would satisfy ourselves with a moon four thousand miles away, *that* could be seen on the earth's surface for three or four thousand miles on each side; and twice three thousand, or six thousand, is one fourth of the largest circumference of the earth. We did not dare have it nearer than four thousand miles, since even at that distance, it would be eclipsed three hours out of every night; and we wanted it bright and distinct, and not of that lurid, copper, eclipse color. But at four thousand miles' distance the moon could be seen by a belt of observers six or eight thousand miles in diameter. "Start, then, two moon,"—this was my contribution to the plan. "Suppose one over the meridian or Greenwich, and the other over that of New Orleans. Take care that there is a little difference in the radii of their orbits, lest they 'collide' some foul day. Then, in most places, one or other, perhaps [454] two will come in sight. So much the less risk of clouds: and everywhere there may be one, except when it is cloudy. Neither need be more than four thousand miles off; so much the larger and more beautiful will they be. If on the old Thornbush moon old Herschel with his reflector could see a town-house two hundred feet long, on the Brick Moon young Herschel will be able to see a dab of mortar a foot and half long, if he wants to. And people without the reflector, with their opera-glasses, will be able to see sufficiently well." And to this they agreed: that eventually there must be two Brick Moons. Indeed, it were better that there should be four, as each must be below the horizon half the time. That is only as many as Jupiter has. But it was also agreed that we might begin with one.

Why we settled on two hundred feet of diameter I hardly know. I think it was from the statement of dear John Farrar's about the impossibility of there being a state house two hundred feet long not yet discovered, on the sunny side of old Thornbush. That, somehow, made two hundred our fixed point. Besides, a moon of two hundred feet diameter did not seem quite unmanageable. Yet it was evident that a smaller moon would be of no use, unless we meant to have them near the world, when there would be so many that they would be confusing, and eclipsed most of the time. And four thousand miles is a good way off to see a moon even two hundred feet in diameter.

Small though we made them on paper, these two-hundred-foot moons were still too much for us. Of course we meant to build them hollow. But even if hollow there must be some thickness, and the quantity of brick would at best be enormous. Then, to get them up! The pea-shooter, of course, was only an illustration. It was long after that time that Rodman and other guns sent iron balls five or six miles in distance,—say two miles, more or less, in height.

Iron is much heavier than hollow brick, but you can build no gun with a bore of two hundred feet now,—far less could you then. No. Q. again suggested the method of shooting off the moon. It was not to be by any of your sudden explosions. It was to be done as all great things are done,—by the gradual and silent accumulation of power. You all know that a fly-wheel—heavy, very heavy on the circumference, light, very light within it—was made to save up power, from the time when it was produced to the time when it was wanted. Yes? Then, before we began even to build the moon, before we even began to make the brick, we would build two gigantic fly-wheels, the diameter of each should be "ever so great," the circumference heavy beyond all precedent, and thundering strong, so that no temptation might burst it. They should revolve, their edges nearly touching, in opposite directions, for years, if it were necessary, to accumulate power, driven by some waterfall now wasted to the world. One should be a little heavier than the other. When the Brick Moon was finished, and all was ready, it should be gently rolled down a gigantic groove provided for it, till it lighted on the edge of both wheels at the same instant. Of course it would not rest there, not the ten-thousandth part of a second. It would be snapped upward, as a drop of water from a grindstone. Upward and upward; but the heavier wheel would have deflected it a little from the vertical. Upward and northward it would rise, therefore, till it had passed the axis of the world. It would, of course, feel the world's attraction all the time, which would bend its flight gently, but still it would leave the world more and more behind. Upward still, but now southward, till it had traversed more than one hundred and eighty degrees of a circle. Little resistance, indeed, after it had cleared the forty or fifty miles of visible atmosphere. "Now let it fall," said Q., inspired with the vision. "Let it fall, and the sooner the better! The curve it is now on will forever clear the world; [455] and over the meridian of that lonely waterfall,—if only we have rightly adjusted the gigantic flies,—will forever revolve, in its obedient orbit, the Brick Moon, the blessing of all seamen,—as constant in all change as its older sister has been fickle, and the second cynosure of all lovers upon the waves, and of all girls left behind them." "Amen," we cried, and then we sat in silence till the clock struck ten; then shook each other gravely by the hand, and left the South Middle dining-hall.

Of waterfalls there were plenty that we knew.

Fly-wheels could be built of oak and pine, and hooped with iron. Fly-wheels did not discourage us.

But brick? One brick is, say, sixty-four cubic inches only. This moon,—though we made it hollow,—see,—it must take twelve million brick.

The brick alone will cost sixty thousand dollars!

The brick alone would cost sixty thousand dollars. There the scheme of the Brick Moon hung, an airy vision, for seventeen years,—the years that changed us from young men into men. The brick alone, sixty thousand dollars! For, to boys who have still left a few of their college bills unpaid, who cannot think of buying that lovely little Elzevir which Smith has for sale at auction, of which Smith does not dream of the value, sixty thousand dollars seems as intangible as sixty million sestertia. Clarke, second, how much are sixty million sestertia stated in cowries? How much in currency, gold being at 1.37 $1/4$? Right; go up. Stop, I forget myself!

So, to resume, the project of the Brick Moon hung in the ideal, an airy vision, a vision as lovely and as distant as the Brick Moon itself, at this calm moment of midnight when I write, as it poises itself over the shoulder of Orion, in my southern horizon. Stop! I anticipate. Let me keep—as we say in Beadle's Dime Series—to the even current of my story.

Seventeen years passed by, we were no longer boys, though we felt so. For myself, to this hour, I never enter board meeting, committee meeting, or synod, without the queer question, what would happen should any one discover that this bearded man was only a big boy disguised? that the frock-coat and the round hat are none of mine, and that, if I should be spurned from the assembly as an interloper, a judicious public, learning all the facts, would give a verdict, "Served him right." This consideration helps me through many bored meetings which would be else so dismal. What did my old copy say? "Boards are made of wood, they are long and narrow." But we do not get on!

Seventeen years after, I say, or should have said, dear Orcutt entered my room at Haguadavick again. I had not seen him since the Commencement day when we parted at Cambridge. He looked the same, and yet not the same. His smile was the same, his voice, his tender look of sympathy when I spoke to him of a great sorrow, his childlike love of fun. His waistband was different, his pantaloons were different, his smooth chin was buried in a full beard, and he weighed two hundred pounds if he weighed a gramme. O, the good time we had, so like the times of old! Those were happy days for me in Naguadavick. At that moment my double was at work for me at a meeting of the publishing committee of the Sandemanian Review, so I called Orcutt up to my own snuggery, and we talked over old times; talked till tea was ready. Polly came up through the orchard and made tea for us herself there. We talked on and on, till nine, ten at night, and then it was that dear Orcutt asked me if I remembered the Brick Moon. Remember it? of course I did. And without leaving my chair I opened the drawer of my writing-desk, and handed him a portfolio full of working-drawings on which I had engaged myself for my "third"[1] all that winter.

[456] Orcutt was delighted. He turned them over hastily but intelligently, and said: "I am so glad. I could not think you had forgotten. And I have seen Brannan, and Brannan has not forgotten." "Now do you know," said he, "In all this railroading of mine, I have not forgotten. I have learned many things that will help. When I built the great tunnel for the Cattawissa and Opelousas, by which we got rid of the old inclined planes, there was never a stone bigger than a peach-stone within two hundred miles of us. I baked the brick of that tunnel on the line with my own kilns. Ingham, I have made more brick, I believe, than any man living in the world!"

"You are the providential man," said I.

"Am I not, Fred? More than that," said he; "I have succeeded in things the world counts worth more than brick. I have made brick, and I have made money!"

"One of us make money?" asked I, amazed.

"Even so," said dear Orcutt; "one of us has made money." And he proceeded to tell me how. It was not in building tunnels, nor in making brick. No! It was by buying up the original stock of the Cattawissa and Opelousas, at a moment when that stock had hardly a nominal price in the market. There were the first mortgage bonds, and the second mortgage bonds, and the third, and I know not how much floating debt; and, worse than all, the reputation of the road lost, and deservedly lost. Every locomotive it had was asthmatic. Every car it had bore the marks of unprecedented accidents, for which no one was to blame. Rival lines, I know not how many, were cutting each other's throats for its legitimate business. At this juncture, dear George invested all his earnings as a contractor, in the despised original stock,—he actually bought it for $3 \, ^1/_4$ per cent,—good shares that had cost a round hundred to every wretch who had subscribed. Six thousand eight hundred dollars—every cent he had—did George thus invest. Then he went himself to the trustees of the first mortgage, to the trustees of the second, and to the trustees of the third, and told them what he had done.

Now it is personal presence that moves the world. Dear Orcutt has found that out since, if he did not know it before. The trustees who would have sniffed had George written to them, turned round from their desks, and begged him to take a chair, when he came to talk with them. Had he put every penny he was worth into that stock? Then it was worth something which they did not know of, for George Orcutt was no fool about railroads. The man who bridged the Lower Rapidan when a freshet was running was no fool.

"What were his plans?"

George did not tell—no, not to lordly trustees—what his plans were. He had plans, but he kept them to himself. All he told them was that he had plans. On those plans he had staked his all. Now would they or would they not agree to put him in charge of the running of that road, for twelve months, on a nominal salary. The superintendent they had was a rascal. He had proved that by running away. They knew that George was not a rascal. He knew that he could make this road pay expenses, pay bond-holders, and pay a dividend,— a thing no one else had dreamed of for twenty years. Could they do better than try him?

Of course they could not, and they knew they could not. Of course, they sniffed and talked, and waited, and pretended they did not know, and that they must consult, and so forth and so on. But of course they all did try him, on his own terms. He was put in charge of the running of that road.

In one week he showed he should redeem it. In three months he did redeem it!

He advertised boldly the first day: *"Infant children at treble price."*

The novelty attracted instant remark. And it showed many things. First, it showed he was a humane man, who wished to save human life. He would [457] leave these innocents in their cradles, where they belonged.

Second, and chiefly, the world of travellers saw that the Crichton, the Amadis, the

---

1. "Every man," says Dr. Peabody, "should have a vocation and an avocation." To which I add, "A third."

perfect chevalier of the future, had arisen,—a railroad manager caring for the comfort of his passengers!

The first week the number of the C. and O.'s passengers was doubled: in a week or two more freight began to come in, in driblets, on the line which its owners had gone over. As soon as the shops could turn them out, some cars were put on, with arms on which travellers could rest their elbows, with headrests where they could take naps if they were weary. These excited so much curiosity that one was exhibited in the museum at Cattawissa and another at Opelousas. It may not be generally known that the received car of the American roads was devised to secure a premium offered by the Pawtucket and Podunk Company. Their receipts were growing so large that they feared they should forfeit their charter. They advertised, therefore, for a car in which no man could sleep at night or rest by day,—in which the backs should be straight, the heads of passengers unsupported, the feet entangled in the vice, the elbows always knocked by the passing conductor. The pattern was produced which immediately came into use on all the American roads. But on the Cattawissa and Opelousas this time-honored pattern was set aside.

Of course you see the result. Men went hundreds of miles out of their way to ride on the C. and O. The third mortgage was paid off; a reserve fund was piled up for the second; the trustees of the first lived in dread of being paid; and George's stock, which he bought at 3 1/4, rose to 147 before two years had gone by! So was it that, as we sat together in the snuggery, George was worth wellnigh three hundred thousand dollars. Some of his eggs were in the basket where they were laid; some he had taken out and placed in other baskets; some in nests where various hens were brooding over them. Sound eggs they were, wherever placed; and such was the victory of which George had come to tell.

One of us had made money!

On his way he had seen Brannan. Brannan, the pure-minded, right-minded, shifty man of tact, man of brain, man of heart, and man of word, who held New Altona in the hollow of his hand. Brannan had made no money. Not he, nor ever will. But Brannan could do much what he pleased in this world, without money. For whenever Brannan studied the rights and the wrongs of any enterprise, all men knew that what Brannan decided about it was wellnigh the eternal truth; and therefore all men of sense were accustomed to place great confidence in his prophecies. But, more than this, and better, Brannan was an unconscious dog, who believed in the people. So, when he knew what was the right and what was the wrong, he could stand up before two or three thousand people and tell them what was right and what was wrong, and tell them with the same simplicity and freshness with which he would talk to little Horace on his knee. Of the thousands who heard him there would not be one in a hundred who knew that this was eloquence. They were fain to say, as they sat in their shops, talking, that Brannan was not eloquent. Nay, they went so far as to regret that Brannan was not eloquent! If he were only as eloquent as Carker was or as Barker was, how excellent he would be! But when a month after, it was necessary for them to do anything about the thing he had been speaking of, they did what Brannan had told them to do; forgetting, most likely, that he had ever told them, and fancying that these were their own ideas, which, in fact, had, from his liquid, ponderous, transparent, and invisible common sense, distilled unconsciously into their being. I wonder whether Brannan ever knew that he was eloquent. What I knew, and what dear George knew, was, that he was one of the leaders of men!

Courage, my friends, we are steadily advancing to the Brick Moon!

[458] For George had stopped, and seen Brannan; and Brannan had not forgotten. Seventeen years Brannan had remembered, and not a ship had been lost on a lee-shore because her longitude was wrong,—not a baby had wailed its last as it was ground between wrecked spar and cruel rock,—not a swollen corpse unknown had been flung up upon the sand and been buried with a nameless epitaph,—but Brannan had recollected the Brick Moon, and had, in the memory-chamber which rejected nothing, stored away the story of the horror. And now, George was ready to consecrate a round hundred thousand to the building of the Moon; and Brannan was ready, in the thousand ways in which wise men move the people to and fro, to persuade them to give to us a hundred thousand more; and

George had come to ask me if I were not ready to undertake with them the final great effort, of which our old calculations were the embryo. For this I was now to contribute the mathematical certainty and the lore borrowed from naval science, which should blossom and bear fruit when the Brick Moon was snapped like a cherry from the ways on which it was built, was launched into the air by power gathered from a thousand freshets, and, poised at last in its own pre-calculated region of the ether, should begin its course of eternal blessings in one unchanging meridian!

Vision of Beneficence and Wonder! Of course I consented.

O that you were not so eager for the end! O that I might tell you, what now you will never know,—of the great campaign which we then and there inaugurated! How the horrible loss of the Royal Martyr, whose longitude was three degrees awry, startled the whole world, and gave us a point to start from. How I explained to George that he must not subscribe the one hundred thousand dollars in a moment. It must come in bits, when "the cause" needed a stimulus, or the public needed encouragement. How we caught neophyte editors, and explained to them enough to make them think the Moon was wellnigh their own invention and their own thunder. How, beginning in Boston, we sent round to all the men of science, all those of philanthropy, and all those of commerce, three thousand circulars, inviting them to a private meeting at George's parlors at the Revere. How, besides ourselves, and some nice, respectable-looking old gentlemen Brannan had brought over from Podunk with him, paying their fares both ways, there were present only three men,—all adventurers whose projects had failed,—besides the representatives of the press. How, of these representatives, some understood the whole, and some understood nothing. How, the next day, all gave us "first-rate notices." How, a few days after, in the lower Horticultural Hall, we had our first public meeting. How Haliburton brought us fifty people who loved him,—his Bible class, most of them,—to help fill up; how, besides these there were not three persons whom we had not asked personally, or one who could invent an excuse to stay away. How we had hung the walls with intelligible and unintelligible diagrams. How I opened the meeting. Of that meeting, indeed, I must tell something.

First, I spoke. I did not pretend to unfold the scheme. I did not attempt any rhetoric. But I did not make any apologies. I told them simply of the dangers of lee-shores. I told them where they were most dangerous,—when seamen came upon them unawares. I explained to them that, though the costly chronometer, frequently adjusted, made a delusive guide to the voyager who often made a harbor, still the adjustment was treacherous, the instrument beyond the use of the poor, and that, once astray, its error increased forever. I said that we believed we had a method which, if the means were supplied for the experiment, would give the humblest fisherman the very certainty of sunrise and of sunset in his calculations of his place upon the world. And I said that whenever a man knew his place in this world it [459] was always likely all would go well. Then I sat down.

Then dear George spoke,—simply, but very briefly. He said he was a stranger to the Boston people, and that those who knew him at all knew he was not a talking man. He was a civil engineer, and his business was to calculate and to build, and not to talk. But he had come here to say that he had studied this new plan for the longitude from the Top to the Bottom, and that he believed in it through and through. There was his opinion, if that was worth anything to anybody. If that meeting resolved to go forward with the enterprise, or if anybody proposed to, he should offer his services in any capacity, and without any pay, for its success. If he might only work as a bricklayer, he would work as a bricklayer. For he believed, on his soul, that the success of this enterprise promised more for mankind than any enterprise which was ever likely to call for the devotion of his life. "And to the good of mankind," he said, very simply, "my life is devoted." Then he sat down.

Then Brannan got up. Up to this time, excepting that George had dropped this hint about bricklaying, nobody had said a word about the Moon, far less hinted what it was to be made of. So Ben had the whole to open. He did it as if he had been talking to a bright boy of ten years old. He made those people think that he respected them as his equals. But, in fact, he chose every word, as if not one of them knew anything. He explained, as if it were rather more simple to explain than to take for granted. But he explained as if, were they talking, they might be explaining to him. He led them from point to point,—oh! so much more clearly than I have been leading you,—till, as their mouths dropped a little open in their eager interest, and their lids forgot to wink in their gaze upon his face, and so their eyebrows seemed a little lifted in curiosity,—till, I say, each man felt as if he were himself the inventor, who had bridged difficulty after difficulty; as if, indeed, the whole were too simple to be called difficult or complicated. The only wonder was that the Board of Longitude, or the Emperor Napoleon, or the Smithsonian, or somebody, had not sent this little planet on its voyage of blessing long before. Not a syllable that you would have called rhetoric, not a word that you would have thought prepared; and then Brannan sat down.

That was Ben Brannan's way. For my part, I like it better than eloquence.

Then I got up again. We would answer any questions, I said. We represented people who were eager to go forward with this work. (Alas! except Q., all of those represented were on the stage.) We could not go forward without the general assistance of the community. It was not an enterprise which the government could be asked to favor. It was not an enterprise which would yield one penny of profit to any human being. We had therefore, purely on the ground of its benefit to mankind, brought it before an assembly of Boston men and women.

Then there was a pause, and we could hear our watches tick, and our hearts beat. Dear George asked me in a whisper if he should say anything more, but I thought not. The pause became painful, and then Tom Coram, prince of merchants, rose. Had any calculation been made of the probable cost of the experiment of one moon?

I said the calculations were on the table. The brick alone would cost $60,000. Mr. Orcutt had computed that $214,729 would complete two fly-wheels and one moon. This made no allowance for whitewashing the moon, which was not strictly necessary. The fly-wheels and water-power would be equally valuable for the succeeding moons, if any were attempted, and therefore the second moon could be turned off, it was hoped, for $159,732.

Thomas Coram had been standing all the time I spoke, and in an instant he said: "I am no mathematician. But I have had a ship ground to pieces under me on the Laccaqdives because our [460] chronometer was wrong. You need $250,000 to build your first moon. I will be one of twenty men to furnish the money; or I will pay $10,000 tomorrow for this purpose, to any person who may be named as treasurer, to be repaid to me if the moon is not finished this day twenty years."

That was as long a speech as Tom Coram ever made. But it was pointed. The small audience tapped applause.

Orcutt looked at me, and I nodded. "I will be another of the twenty men," cried he. "And I another," said an old bluff Englishman, whom nobody had invited; who proved to be a Mr. Robert Boll, a Sheffield man, who came in from curiosity. He stopped after the meeting; said he should leave the country the next week, and I have never seen him since. But his bill of exchange came all the same.

That was all the public subscribing. Enough more than we had hoped for. We tried to make Coram treasurer, but he refused. We had to make Haliburton treasurer, though we should have liked a man better known than he then was. Then we adjourned. Some nice ladies then came up, and gave, one a dollar, and one five dollars, and one fifty, and so, on,—and some men who have stuck by ever since. I always, in my own mind, call each of those women Damaris, and each of those man Dionysus. But those are not their real names.

How I am wasting time on an old story! Then some of these ladies came the next day and proposed a fair; and out of that, six months after, grew the great Longitude Fair, that you will all remember, if you went to it, I am sure. And the papers the next day gave us first-rate reports; and then, two by two, with our subscription-books, we went at it. But I must not tell the details of that subscription. There were two or three men who subscribed $5,000 each, because they were perfectly certain the amount would never be raised. They wanted, for once, to get the credit of liberality for nothing. There were many men and many women who subscribed from one dollar up to one thousand, not because they cared a straw for the longitude, nor because they believed in the least in the project; but because they believed in Brannan, in Orcutt, in Q., or in me. Love goes far in this world of ours. Some few men subscribed because others had done it: it was the thing to do, and they must not be out of fashion. And three or four, at least, subscribed because each hour of their lives there came up the memory of the day when the news came that the —— was lost, George, or Harry, or John, in the ——, and they knew that George, or Harry, or John might have been at home, had it been easier than it is to read the courses of the stars!

Fair, subscriptions, and Orcutt's reserve,—we counted up $162,000, or nearly so. There would be a little more when all was paid in.

But we could not use a cent, except Orcutt's and our own little subscriptions, till we had got the whole. And at this point it seemed as if the whole world was sick of us, and that we had gathered every penny that was in store for us. The orange was squeezed dry!

II.—HOW WE BUILT IT.

[603] The orange was squeezed dry! And how little any of us knew,—skilful George Orcutt, thoughtful Ben Brannan, loyal Haliburton, ingenious Q., or poor painstaking I,— how little we knew, or any of us, where was another orange, or how we could mix malic acid and tartaric acid, and citric acid and auric acid and sugar and water so as to imitate orange-juice, and fill up the bank-account enough to draw in the conditioned subscriptions, and so begin to build the Moon. How often, as I lay awake at night, have I added up the different subscriptions in some new order, as if that would help the matter: and how steadily they have come out one hundred and sixty-two thousand dollars, or even less, when I must needs, in my sleepiness, forget somebody's name! So Haliburton put into railroad stocks all the money he collected, and the rest of us ground on at our mills, or flew up on our own wings towards Heaven. Thus Orcutt built more tunnels, Q. prepared for more commencements, Haliburton calculated more policies, Ben Brannan created more civilization, and I, as I could, healed the hurt of my people of Naguadavick for the months there were left to me of my stay in that thriving town.

None of us had the wit to see how the problem was to be wrought out further. No. The best things come to us when we have faithfully and well made all the preparation and done our best; but they come in some way that is none of ours. So was it now, that to build the BRICK MOON it was necessary that I should be turned out of Naguadavick ignominiously, and that Jeff. Davis and some seven or eight other bad men should create the Great Rebellion. Hear how it happened.

Dennis Shea, my Double,—otherwise, indeed, called by my name and legally so,— undid me, as my friends supposed, one evening at a public meeting called by poor Isaacs in Naguadavick. Of that transaction I have no occasion here to tell the story. But of that transaction one consequence is that the BRICK MOON now moves in ether. I stop writing, to rest my eye upon it, through a little telescope of Alvan Clark's here, which is always trained near it. It is moving on as placidly as ever.

It came about thus. The morning after poor Dennis, whom I have long since forgiven, made his extraordinary speeches, without any authority from me, in the Town Hall at Naguadavick, I thought, and my wife agreed with me, that we had better both leave town with the children. Auchmuty, our dear friend, thought so too. We left in the ten-thirty Accommodation for Skowhegan, and so came to Township No. 9 in the 3d Range, and there for years we resided. That whole range of townships [604] was set off under a provision admirable in its character, that the first settled minister in each town should receive one hundred acres of land as the "minister's grant," and the first settled schoolmaster eighty. To No. 9, therefore, I came. I constituted a little Sandemanian church. Auchmuty and Delafield came up and installed me, and with these hands I built the cabin in which, with Polly and the little ones, I have since spent many happy nights and days. This is not the place for me to publish a map, which I have by me, of No. 9, nor an account of its many advantages for settlers. Should I ever print my papers called "Stay-at-home Robinsons," it will be easy with them to explain its topography and geography. Suffice it now to say, that, with Alice and Bertha and Polly, I took tramps up and down through the lumbermen's roads, and soon knew the general features of the lay of the land. Nor was it long, of course, before we came out one day upon the curious land-slides, which have more than once averted the flow of the Little Carrotook River, where it has washed the rocks away so far as to let down one section more of the overlying yielding yellow clay.

Think how my eyes flashed, and my wife's, as, struggling through a wilderness of moosewood, we came out one afternoon on this front of yellow clay! Yellow clay, of course, when properly treated by fire, is brick! Here we were surrounded by forests, only waiting to be burned; yonder was clay, only waiting to be baked. Polly looked at me, and I looked at her, and with one voice, we cried out, "The MOON!"

For here was this shouting river at our feet, whose power had been running to waste since the day when the Laurentian hills first heaved themselves above the hot Atlantic; and that day, I am informed by Mr. Agassiz, was the first day in the history of this solid world. Here was water-power enough for forty fly-wheels, were it necessary to send heavenward twenty moons. Here was solid timber enough for a hundred dams, yet only one was necessary to give motion to the fly-wheels. Here was retirement,—freedom from criticism, an escape from the journalists, who would not embarrass us by telling of every cracked brick which had to be rejected from the structure. We had lived in No. 9 now for six weeks, and not an "own correspondent" of them all had yet told what Rev. Mr. Ingham had for dinner.

Of course I wrote to George Orcutt at once of our great discovery, and he came up at once to examine the situation. On the whole, it pleased him. He could not take the site proposed for the dam, because this very clay there made the channel treacherous, and there was danger that the stream would work out a new career. But lower down we found a stony gorge with which George was satisfied; he traced out a line for a railway by which, of their own weight, the brick-cars could run to the centrings; he showed us where, with some excavations, the fly-wheels could be placed exactly above the great mill-wheels, that no power might be wasted, and explained to us how, when the gigantic structure was finished, the BRICK MOON would gently roll down its ways upon the rapid wheels, to be launched instant into the sky!

Shall I ever forget that happy October day of anticipation!

We spent many of those October days in tentative surveys. Alice and Bertha were our chain-men, intelligent and obedient. I drove for George his stakes, or I cut away his brush, or I raised and lowered the shield at which he sighted; and at noon Polly appeared with her baskets, and we would dine *al fresco*, on a pretty point which, not many months after, was wholly covered by the eastern end of the dam. When the fieldwork was finished we retired to the cabin for days, and calculated and drew, and drew and calculated. Estimates for feeding Irishmen, estimates of hay for mules,—George was sure he could work mules better than oxen,—estimates for cement, estimates [605] for the preliminary saw-mills, estimates for rail for the little brickroad, for wheels, for spikes, and for cutting ties; what did we not estimate for—on a basis almost wholly new, you will observe. For here the brick would cost us less than our old conceptions,—our water-power cost us almost nothing,— but our stores and our wages would cost us much more.

These estimates are now to me very curious,—a monument, indeed, to dear George's memory, that in the result they proved so accurate. I would gladly print them here at length, with some illustrative cuts, but that I know the impatience of the public, and its indifference to detail. If we are ever able to print a proper memorial of George, that, perhaps, will be the fitter place for them. Suffice it to say that with the subtractions thus made from the original estimates,—even with the additions forced upon us by working in a wilderness,—George was satisfied that a money charge of $197,327 would build and start THE MOON. As soon as we had determined the site, we marked off eighty acres, which contained all the essential localities, up and down the little Carrotook River,—I engaged George for the first schoolmaster in No. 9, and he took these eighty acres for the schoolmaster's reservation. Alice and Bertha went to school to him the next day, taking lessons in civil engineering; and I wrote to the Brigham trustees to notify them that I had engaged a teacher, and that he had selected his land.

Of course we remembered, still, that we were near forty thousand dollars short of the new estimates, and also that much of our money would not be paid us but on condition that two hundred and fifty thousand were raised. But George said that his own subscription was wholly unhampered: with that we would go to work on the preliminary work of the dam, and on the flies. Then, if the flies would hold together,—and they should hold if mortise and iron could hold them,—they might be at work summers and winters, days and nights, storing up Power for us. This would encourage the subscribers, nay, would encourage us; and all this preliminary work would be out of the way when we were really ready to begin upon the MOON.

Brannan, Haliburton, and Q. readily agreed to this when they were consulted. They were the other trustees under an instrument which we had got St. Leger to draw up. George gave up, as soon as he might, his other appointments; and taught me, meanwhile, where and how I was to rig a little saw-mill, to cut some necessary lumber. I engaged a gang of men to cut the timber for the dam, and to have it ready; and, with the next spring, we were well at work on the dam and on the flies! These needed, of course, the most solid foundation. The least irregularity of their movement might send the MOON awry.

Ah me! would I not gladly tell the history of every bar of iron which was bent into the tires of those flies, and of every log which was mortised into its place in the dam, nay, of every curling mass of foam which played in the eddies beneath, when the dam was finished, and the waste water ran so smoothly over? Alas! that one drop should be wasted of water that might move a world, although a small one! I almost dare say that I remember each and all these,—with such hope and happiness did I lend myself, as I could, each day to the great enterprise; lending to dear George, who was here and there and everywhere, and was this and that and everybody,—lending to him, I say, such a poor help as I could lend, in whatever way. We waked, in the two cabins, in those happy days, just before the sun came up, when the birds were in their loudest clamor of morning joy. Wrapped each in a blanket, George and I stepped out from our doors, each trying to call the other, and often meeting on the grass between. We ran to the river and plunged in,—O, how cold it was!—laughed and screamed like boys, rubbed ourselves aglow, and ran home to [606] build Polly's fire beneath the open chimney which stood beside my cabin. The bread had risen in the night. The water soon boiled above the logs. The children came, laughing, out upon the grass, barefoot, and fearless of the dew. Then Polly appeared with her gridiron and bear-steak, or with her griddle and eggs, and, in fewer minutes than this page has cost me, the breakfast was ready for Alice to carry, dish by dish, to the white-clad table on the piazza. Not Raphael and Adam more enjoyed their watermelons, fox-grapes, and late blueberries! And, in the long croon of the breakfast, lingering at the board, we revenged ourselves for the haste with which it had been prepared.

When we were well at table, a horn from the cabins below sounded the reveille for the drowsier workmen. Soon above the larches rose the blue of their smokes; and when we were at last nodding to the children, to say that they might leave the table, and Polly was folding her napkin as to say she wished we were gone, we would see tall Asaph Langdon, then foreman of the carpenters, sauntering up the valley with a roll of paper, or an adze, or a shingle with some calculations on it,—with something on which he wanted Mr. Orcutt's directions for the day.

An hour of nothings set the carnal machinery of the day agoing. We fed the horses, the cows, the pigs, and the hens. We collected the eggs and cleaned the hen-houses and the barns. We brought in wood enough for the day's fire, and water enough for the day's cooking and cleanliness. These heads describe what I and the children did. Polly's life during that hour was more mysterious. That great first hour of the day is devoted with women to the deepest arcana of the Eleusinian mysteries of the divine science of housekeeping. She who can meet the requisitions of that hour wisely and bravely conquers in the Day's Battle. But what she does in it, let no man try to say! It can be named, but not described, in the comprehensive formula, "Just stepping round."

That hour well given to chores and to digestion, the children went to Mr. Orcutt's open-air school, and I to my rustic study,—a separate cabin, with a rough square table in it, and some book-boxes equally rude. No man entered it, excepting George and me. Here for two hours I worked undisturbed,—how happy the world, had it neither postman nor door-bell!—worked upon my Traces of Sandemanianism in the Sixth and Seventh Centuries, and then was ready to render such service to The Cause and to George as the day might demand. Thus I rode to Lincoln or to Foxcroft to order supplies; I took my gun and lay in wait on Chairback for a bear; I transferred to the hewn lumber the angles or bevels from the careful drawings: as best I could, I filled an apostle's part, and became all things to all these men around me. Happy those days!—and thus the dam was built; in such Arcadian simplicity was reared the mighty wheel; thus grew on each side the towers which were to support the flies; and thus, to our delight not unmixed with wonder, at one day, nor in ten; but in a year or two of happy life,—full of the joy of joys,—the "joy of eventful living."

Yet, for all this, $162,000 was not $197,000, far less was it $250,000; and but for Jeff. Davis and his crew the BRICK MOON would not have been born.

But at last Jeff. Davis was ready. "My preparations being completed," wrote General Beauregard, "I opened fire on Fort Sumter." Little did he know it,—but in that explosion the BRICK MOON also was lifted into the sky!

Little did we know it, when, four weeks after, George came up from the settlements, all excited with the news! The wheels had been turning now for four days, faster of course and faster. George had gone down for money to pay off [607] the men, and he brought us up the news that the Rebellion had begun.

"The last of this happy life," he said; "the last, alas, of our dear MOON." How little he knew and we!

But he paid off the men, and they packed their traps and disappeared, and, before two months were over, were in the lines before the enemy. George packed up, bade us sadly good by, and before a week had offered his service to Governor Fenton in Albany. For us, it took rather longer; but we were soon packed; Polly took the children to her sister's, and I went on to the Department to offer my service there. No sign of life left in No. 9, but the two gigantic Fly-Wheels, moving faster and faster by day and by night, and accumulating Power till it was needed. If only they would hold together till the moment came!

So we all ground through the first slow year of the war. George in his place, I in mine, Brannan in his,—we lifted as we could. But how heavy the weight seemed! It was in the second year, when the second large loan was placed, that Haliburton wrote to me,—I got the letter, I think at Hilton Head,—that he had sold out every penny of our railroad stocks, at the high prices which railroad stocks then bore, and had invested the whole fifty-nine thousand in the new Governments. "I could not call a board meeting," said Haliburton, "for I am here only on leave of absence, and the rest are all away. But the case is clear enough. If the government goes up, the MOON will never go up; and, for one, I do not look beyond the veil." So he wrote to us all, and of course we all approved.

So it was that Jeff. Davis also served. Deep must that man go into the Pit who does not serve, though unconscious. For thus it was that, in the fourth year of the war, when gold was at 290, Haliburton was receiving on his fifty-nine thousand dollars seventeen per cent interest in currency; thus was it that, before the war was over, he had piled up, compounding his interest, more than fifty per cent addition to his capital; thus was it that, as soon as peace came, all his stocks were at a handsome percentage; thus was it that, before I returned from South America, he reported to all the subscribers that the full quarter-million was secured; thus was it that, when I returned after that long cruise of mine in the Florida, I found Polly and the children again at No. 9, George there also, directing a working party of nearly eighty bricklayers and hodmen, the lower centrings wellnigh filled to their diameter, and the BRICK MOON, to the eye, seeming almost half completed.

Here it is that I regret most of all that I cannot print the working-drawings with this paper. If you will cut open the seed-vessel of Spergularia Rubra, or any other carpel that has a free central placenta, and observe how the circular seeds cling around the circular centre, you will have some idea of the arrangement of a transverse horizontal section of the completed MOON. Lay three croquet-balls on the piazza, and call one or two of the children to help you poise seven in one place above the three; then let another child place three more above the seven, and you have the *core* of the MOON completely. If you want a more poetical illustration, it was what Mr. Wordsworth calls a mass

"Of conglobated bubbles undissolved."

Any section through any diameter looked like an immense rose-window, of six circles grouped round a seventh. In truth, each of these sections would reveal the existence of seven chambers in the moon,—each a sphere itself,—whose arches gave solidity to the whole; while yet, of the whole moon, the greater part was air. In all there were thirteen of these moonlets, if I am so to call them; though no one section, of course, would reveal so many. Sustained on each side by their groined arches, the surface of the whole moon was built over them and under them,—simply two domes connected at the bases. The chambers themselves were [608] made lighter by leaving large, round windows or open circles in the parts of their vaults farthest from their points of contact, so that each of them looked not unlike the outer sphere of a Japanese ivory nest of concentric balls. You see the object was to make a moon, which, when left to its own gravity, should be fitly supported or braced within. Dear George was sure that, by this constant repetition of arches, we should with the least weight unite the greatest strength. I believe it still, and experience has proved that there is strength enough.

When I went up to No. 9, on my return from South America, I found the lower centring up, and half full of the working-bees,—who were really Keltic laborers,—all busy in bringing up the lower half-dome of the shell. This lower centring was of wood, in form exactly like a Roman amphitheatre if the seats of it be circular; on this the lower or inverted brick dome was laid. The whole fabric was on one of the terraces which were heaved up in some old geological cataclysm, when some lake gave way, and the Carratook River was born. The level was higher than that of the top of the fly-wheels, which, with an awful velocity now, were circling in their wild career in the ravine below. Three of the lowest moonlets, as I have called them,—separate croquet-balls, if you take my other illustration,—had been completed; their centrings had been taken to pieces and drawn out through the holes, and were now set up again with other new centrings for the second story of cells.

I was received with wonder and delight. I had telegraphed my arrival, but the despatches had never been forwarded from Skowhegan. Of course, we all had a deal to tell; and, for me, there was no end to inquiries which I had to make in turn. I was never tired of exploring the various spheres, and the nameless spaces between them. I was never tired of talking with the laborers. All of us, indeed, became skilful bricklayers; and on a pleasant afternoon you might see Alice and Bertha, and George and me, all laying brick together,— Polly sitting in the shade of some wall which had been built high enough, and reading to us from Jean Ingelow or Monte-Cristo or Jane Austen, while little Clara brought to us our mortar. Happily and lightly went by that summer. Haliburton and his wife made us a visit; Ben Brannan brought up his wife and children; Mrs. Haliburton herself put in the keystone to the central chamber, which had always been named G. on the plans; and at her suggestion, it was named Grace now, because her mother's name was Hannah. Before winter we had passed the diameter of I, J, and K, the three uppermost cells of all; and the surrounding shell was closing in upon them. On the whole, the funds had held out amazingly well. The wages had been rather higher than we meant; but the men had no chances at liquor or dissipation, and had worked faster than we expected; and, with our new brick-machines, we made brick inconceivably fast, while their quality was so good that dear George said there was never so little waste. We celebrated Thanksgiving of that year together,—my family and his family. We had paid off all the laborers; and there were left, of that busy village, only Asaph Langdon and his family, Levi Jordan and Levi Ross, Horace Leonard and Seth Whitman with theirs. "Theirs," I say, but Ross had no family. He was a nice young fellow who was there as Haliburton's representative, to take care of the accounts and the pay-roll; Jordan was the head of the brick-kilns; Leonard, of the carpenters; and Whitman, of the commissariat,—and a good commissary Whitman was.

We celebrated Thanksgiving together! Ah me! what a cheerful, pleasant time we had; how happy the children were together! Polly and I and our bairns were to go to Boston the next day. I was to spend the winter in one final effort to get twenty-five thousand dollars more if I could, with which we might paint the MOON, or put on some ground felspathic granite dust, in a sort of paste, which in its hot flight [609] through the air might fuse into a white enamel. All of us who saw the MOON were so delighted with its success that we felt sure "the friends" would not pause about this trifle. The rest of them were to stay there to watch the winter, and to be ready to begin work the moment the snow had gone. Thanksgiving afternoon,—how well I remember it,—that good fellow, Whitman, came and asked Polly and me to visit his family in their new quarters. They had moved for the winter into cells B and E, so lofty, spacious, and warm, and so much drier than their log cabins. Mrs. Whitman, I remember, was very cheerful and jolly; made my children eat another piece of pie, and stuffed their pockets with raisins; and then with great ceremony and fun we christened room B by the name of Bertha, and E, Ellen, which was Mrs. Whitman's name. And the next day we bade them all good by, little thinking what we said, and with endless promises of what we would send and bring them in the spring.

Here are the scraps of letters from Orcutt, dear fellow, which tell what more there is left to tell:—

"December 10th.

".... After you left we were a little blue, and hung round loose for a day or two. Sunday we missed you especially, but Asaph made a good substitute, and Mrs. Leonard led the singing. The next day we moved the Leonards into L and M, which we christened Leonard and Mary (Mary is for your wife). They are pretty dark, but very dry. Leonard has swung hammocks, as Whitman did.

"Asaph came to me Tuesday and said he thought they had better turn to and put a shed over the unfinished circle, and so take occasion of warm days for dry work there. This we have done, and the occupation is good for us...."

"December 25th.

"I have had no chance to write for a fortnight. The truth is, that the weather has been so open that I let Asaph go down to No. 7 and to Wilder's, and engage five-and-twenty of the best of the men, who, we knew, were hanging round there. We have all been at work most of the time since, with very good success. H is now wholly covered in, and the centring is out. The men have named it Haliburton. I is well advanced. J is as you left it. The work has been good for us all, morally."

"February 11th.

".... We got your mail unexpectedly by some lumbermen on their way to the 9th Range. One of them has cut himself, and takes this down.

"You will be amazed to hear that I and K are both done. We have had splendid weather, and have worked half the time. We had a great jollification when K was closed in,—called it Kilpatrick, for Seth's old general. I wish you could just run up and see us. You must be quick, if you want to put in any of the last licks...."

"March 12th.

"DEAR FRED,—I have but an instant. By all means made your preparations to be here by the end of the month or early in next month. The weather has been faultless, you know. Asaph got in a dozen more men, and we have brought up the surface farther than you could dream. The ways are well forward, and I cannot see why, if the freshet hold off a little, we should not launch her by the 10th or 12th. I do not think it worth while to wait for paint or enamel. Telegraph Brannan that he must be here. You will be amused by our quarters. We, who were the last outsiders, move into A and D tomorrow, for a few weeks. It is much warmer there.

"Ever yours,

"G.O."

I telegraphed Brannan, and in reply he came with his wife and his children to Boston. I told him that he could not possibly get up there, as the roads then were; but Ben said he would go to Skowhegan, and take his chance there. He would, of course, communicate with me as soon as he got there. Accordingly I got a note from him at [610] Skowhegan, saying he had hired a sleigh to go over to No. 9; and in four days more I got this letter:—

"March 27th.

"Dear Fred,—I am most glad I came, and I beg you to bring your wife as soon as possible. The river is very full, the wheels, to which Leonard has added two auxiliaries, are moving as if they could not hold out long, the ways are all but ready, and we thing we must not wait. Start with all hands as soon as you can. I had no difficulty in coming over from Skowhegan. We did it in two days."

This note I sent at once to Haliburton; and we got all the children ready for a winter journey, as the spectacle of the launch of the MOON was one to be remembered their life long. But it was clearly impossible to attempt, at that season, to get the subscribers together. Just as we started, this despatch from Skowhegan was brought me,—the last word I got from them:—

"Stop for nothing. There is a jam below us in the stream, and we fear back-water.

"ORCUTT."

Of course we could not go faster than we could. We missed no connection. At Skowhegan, Haliburton and I took a cutter, leaving the ladies and children to follow at once in larger sleighs. We drove all night, changed horses at Prospect, and kept on all the next day. At No. 7 we had to wait over night. We started early in the morning, and came down the Spoonwood Bill at four in the afternoon, in full sight of our little village.

It was quiet as the grave! Not a smoke, not man, not an adze-blow, nor the tick of a trowel. Only the gigantic fly-wheels were whirling as I saw them last.

There was the lower Coliseum-like centring, somewhat as I first saw it.

But where was the Brick Dome of the MOON?

"Good Heavens! has it fallen on them all? cried I.

Haliburton lashed the beast till he fairly ran down that steep hill. We turned a little point, and came out in front of the centring. There was no MOON there! An empty amphitheatre, with not a brick nor a splinter within!

We were speechless. We left the cutter. We ran up the stairways to the terrace. We ran by the familiar paths into the centring. We came out upon the ways, which we had never seen before. These told the story too well! The ground and crushed surface of the timbers, scorched by the rapidity with which THE MOON had slid down, told that they had done the duty for which they were built.

It was too clear that in some wild rush of the waters the ground had yielded a trifle. We could not find that the foundations had sunk more than six inches, but that was enough. In that fatal six inches' decline of the centring, the MOON had been launched upon the ways just as George had intended that it should be when he was ready. But it had slid, not rolled, down upon these angry fly-wheels, and in an instant, with all our friends, it had been hurled into the sky!

"They have gone up!" said Haliburton; "She has gone up!" said I;—both in one breath. And with a common instinct, we looked up into the blue.

But of course she was not there.

Not a shred of letter or any other tidings could we find in any of the shanties. It was indeed six weeks since George and Fanny and their children had moved into Annie and Diamond,—two unoccupied cells of the MOON,—so much more comfortable had the cells proved than the cabins, for winter life. Returning to No. 7, we found there many of the laborers, who were astonished at what we told them. They had been paid off on the 30th, and told to come up again on the 15th of April, to see the launch. One of them, a man named Rob. Shea, told me that George kept his cousin Peter to help [611] him move back into his house the beginning of the next week.

And that was the last I knew of any of them for more than a year. At first I expected, each hour, to hear that they had fallen somewhere. But time passed by, and of such a fall, where man knows the world's surface, there was no tale. I answered, as best I could, the letters of their friends, by saying I did not know where they were, and had not heard from them. My real thought was, that if this fatal MOON did indeed pass our atmosphere, all in it must have been burned to death in the transit. But this I whispered to no one save to Polly and Annie and Haliburton. In this terrible doubt I remained, till I noticed one day in the "Astronomical Record" the memorandum, which you perhaps remember, of the observation, by Dr. Zitta, of a new asteroid, with an enormous movement in declination.

III.—FULFILMENT.

[679] Looking back upon it now, it seems inconceivable that we said as little to each other as we did, of this horrible catastrophe. That night we did not pretend to sleep. We sat in one of the deserted cabins, now talking fast, now sitting and brooding, without speaking, perhaps, for hours. Riding back the next day to meet the women and children, we still brooded, or we discussed this "if," that "if," and yet others. But after we had once opened it all to them,—and when we had once answered the children's horribly naive questions as best we could,—we very seldom spoke to each other of it again. It was too hateful, all of it, to talk about. I went round to Tom Coram's office one day, and told him all I knew. He saw it was dreadful to me, and, with his eyes full, just squeezed my hand, and never said one word more. We lay awake nights, pondering an wondering, but hardly ever did I to Haliburton or he to me explain our respective notions as they came and went. I believe my general impression was that of which I have spoken, that they were all burned to death on the instant, as the little aerolite fused in its passage through our atmosphere. I believe Haliburton's thought more often was that they were conscious of what had happened, and gasped out their lives in one or two breathless minutes,—so horrible long!—as they shot outside of our atmosphere. But it was all too terrible for words. And that which we could not but think upon, in those dreadful waking nights, we scarcely whispered even to our wives.

Of course I looked and he looked for the miserable thing. But we looked in vain. I returned to the few subscribers the money which I had scraped together towards whitewashing the moon,— "shrouding its guilty face with innocent white" indeed! But we agreed to spend the wretched trifle of the other money, left in the treasury after paying the last bills, for the largest Alvan Clark telescope that we could buy; and we were fortunate in obtaining cheap a second-hand one which came to the hammer when the property of the Shubeal Academy was sold by the mortgagees. But we had, of course, scarce a hint whatever as to where the miserable object was to be found. All we could do was to carry the glass to No. 9, to train it there on the meridian of No. 9, and take turns every night in watching the field, in the hope that this child of sorrow might drift across it in its path of ruin. But, though everything else seemed to drift by, from east to west, nothing came from south to north, as we expected. For a whole month of spring, another of autumn, another of summer, and another of winter, did Haliburton and his wife and Polly and I glue our eyes to that eye-glass, from the twilight of evening to the twilight of morning, and the dead hulk never hove in sight. Wherever else it was, it seemed not to be on that meridian, which was where it ought to be and was made to be! Had ever any dead mass of matter wrought such ruin to its makers, and, of its own stupid inertia, so falsified all the prophecies of its birth! O, the total depravity of things!

It was more than a year after the fatal night,—if it all happened in the night, as I suppose,—that, as I dreamily read through the "Astronomical Record" in the new reading-room of the College Library at Cambridge, I lighted on this scrap:—

"Professor Karl Zitta of Breslau writes to the *Astronomische Nachrichten* to claim the discovery of a new asteroid observed by him on the night of March 31st.

[680]

				(92)					
				App. A.R.		App. Decl.			
Bresl. M T	h.	m.	s.	h. m.	s.	°	'	"	Size.
March 31	12	53	51.9	15 39	52.32	—23	50	26.1	12.9
April 1	1	3	2.1	15 39	52.32	—23	9	1.9	12.9

"He proposes for the asteroid the name of Phoebe. Dr. Zitta states that in the short period which he had for observing Phoebe, for an hour after midnight, her motion in R. A. seemed slight and her motion in declination very rapid."

After this, however, for months, nay even to this moment, nothing more was heard of Dr. Zitta of Breslau.

But, one morning, before I was up, Haliburton came banging at my door on D Street. The mood had taken him, as he returned from some private theatricals at Cambridge, to take the comfort of the new reading-room at night, and thus express in practice his gratitude to the overseers of the college for keeping it open through all the twenty-four hours. Poor Haliburton, he did not sleep well in those times! Well, as he read away on the *Astronomische Nachrichten* itself, what should he find but this in German, which he copied for me, and then, all on foot in the rain and darkness, tramped over with, to South Boston:—

"The most enlightened head professor Dr. Gmelin writes to the director of the Porpol Astronomik at St. Petersburg, to claim the discovery of an asteroid in a very high southern latitude, of a wider inclination of the orbit, as will be noticed, than any asteroid yet observed.

"Planet's apparent α 21^h 20^m $51^s.40$. Planet's apparent ∂—39° 31' 11".9. Comparison star α.

"Dr. Gmelin publishes no separate second observation, but is confident that the declination is diminishing. Dr. Gmelin suggests for the name of this extra-zodiacal planet "Io," as appropriate to its wanderings from the accustomed ways of planetary life, and trusts that the very distinguished Herr Peters, the godfather of so many planets, will relinquish this name, already claimed for the asteroid (85) observed by him, September 15, 1865."

I had run down stairs almost as I was, slippers and dressing-gown being the only claims I had on society. But to me, as to Haliburton, this stuff about "extra-zodiacal wandering" blazed out upon the page, and though there was no evidence that the "most enlightened" Gmelin found anything the next night, yet, if his "diminishing" meant anything, there was, with Zitta's observation,—whoever Zitta might be,—something to start upon. We rushed upon some old bound volumes of the Record and spotted the "enlightened Gmelin." He was chief of a college at Taganrog, where perhaps they had a spyglass. This gave us the parallax of his observation. Breslau, of course, we knew, and so we could place Zitta's, and with these poor data I went to work to construct, if I could, an orbit for this Io-Phoebe mass of brick and mortar. Haliburton, not strong in spherical trigonometry, looked out logarithms for me till breakfast, and, as soon as it would do, went over to Mrs. Bowdoin, to borrow her telescope, ours being left at No. 9.

Mrs. Bowdoin was kind, as she always was, and at noon Haliburton appeared in triumph with the boxes on P. Nolan's job-wagon. We always employ P., in memory of dear old Phil. We got the telescope rigged, and waited for night, only, alas! to be disappointed again. Io had wandered somewhere else, and, with all our sweeping back and forth on the tentative curve I had laid out, Io would not appear. We spent that night in vain.

But we were not going to give it up so. Phoebe might have gone round the world twice before she became Io; might have gone three times, four, five, six,—nay, six hundred,—who knew! Nay, who knew how far off Phoeb-Io [681] was or Io-Phoebe? We sent over for Annie, and she and Polly and George and I went to work again. We calculated in the next week sixty-seven orbits on the supposition of so many different distances from our surface. I laid out on a paper, which we stuck up on the wall opposite, the formula, and then one woman and one man attacked each set of elements, each having the Logarithmic Tables, and so in a week's working-time, the sixty-seven orbits were completed. Sixty-seven possible places for Io-Phoebe to be in on the forthcoming Friday evening. Of these sixty-seven, forty-one were observable above our horizon that night.

She was not in one of the forty-one, nor near it.

But Despair, if Giotto be correct, is the chief of sins. So has he depicted her in the fresco of the Arena in Padua. No sin, that, of ours! After searching all that Friday night, we slept all Saturday (sleeping after sweeping). We all came to the Chapel, Sunday, kept awake there, and taught our Sunday classes special lessons on Perseverance. On Monday we began again, and that week we calculated sixty-seven more orbits. I am sure I do not know why we stopped at sixty-seven. All of these were on the supposition that the revolution of the Brick Moon, or Io-Phoebe, was so fast that it would require either fifteen days to complete its orbit, or sixteen days, or seventeen days, and so on up to eighty-one days. And, with these orbits, on the next Friday we waited for the darkness. As we sat at tea, I asked if I should begin observing at the smallest or at the largest orbit. And there was a great clamor of diverse opinions. But little Bertha said, "Begin in the middle."

"And what is the middle?" said George, chaffing the little girl.

But she was not to be dismayed. She had been in and out all the week, and knew that the first orbit was of fifteen days and the last of eighty-one; and, with true Lincoln School precision, she said, "The mean of the smallest orbit and the largest orbit is forty-eight days."

"Amen!:" said I, as we all laughed. "On forty-eight days we will begin."

Alice ran to the sheets, turned up that number, and read, "R. A. 27° 11'. South declination 34° 49'."

"Convenient places," said George; "good omen, Bertha, my darling! If we find her there, Alice and Bertha and Clara shall all have new dolls."

It was the first word of pleasantry that had been spoken about the horrid thing since Spoonwood Hill!

Night came at last. We trained the glass on the fated spot. I bade Polly take the eye-glass. She did so, shook her head uneasily, screwed the tube northward herself a moment, and then screamed, "It is there! it is there,—a clear disk,—gibbous shape,—and very sharp on the upper edge. Look! look! as big again as Jupiter!"

Polly was right! The Brick Moon was found!

Now we had found it, we never lost it. Zitta and Gmelin, I suppose, had had foggy nights and stormy weather often. But we had some one at the eye-glass all that night, and before morning had very respectable elements, good measurements of angular distance when we got one, and another star in the field of our lowest power. For we could see her even with a good French opera-glass I had, and with a night-glass which I used to carry on the South Atlantic Station. It certainly was an extraordinary illustration of Orcutt's engineering ability, that, flying off as she did, without leave or license, she should have gained so nearly the orbit of our original plan,—nine thousand miles from the earth's centre, five thousand from the surface. He had always stuck to the hope of this, and on his very last tests of the Flies he had said they were almost up to it. But for this accuracy of his, I can hardly suppose we should have found her to this hour, since she had failed, by what cause I then did not know, to take her intended place on the meridian of No. 9. At five thousand miles the MOON appeared as large as the largest satellite of Jupiter [682] appears. And Polly was right in that first observation, when she said she got a good disk with that admirable glass of Mrs. Bowdoin.

The orbit was not on the meridian of No. 9, nor did it remain on any meridian. But it was very nearly South and North.—an enormous motion in declination with a very slight retrograde motion in Right Ascension. At five thousand miles the MOON showed as large as a circle two miles and a third in diameter would have shown on old Thornbush, as we always called her older sister. We longed for an eclipse of Thornbush by B. M., but no such lucky chance is on the cards in any place accessible to us for many years. Of course, with a MOON so near us the terrestrial parallax is enormous.

Now, you know, dear reader, that the gigantic reflector of Lord Rosse, and the exquisite fifteen-inch refractors of the modern observatories, eliminate from the chaotic rubbish-heap of the surface of old Thornbush much smaller objects than such a circle as I have named. If you have read Mr. Locke's amusing Moon Hoax as often as I have, you have those details fresh in your memory. As John Farrar taught us when all this began,—and as I have said already,—if there were a State House in Thornbush two hundred feet long, the first Herschel would have seen it. His magnifying power was 6450; that would have brought this deaf and dumb State House within some forty miles. Go up on Mt. Washington and see white sails eighty miles away, beyond Portland, with your naked eye, and you will find how well he would have seen that State House with his reflector. Lord Rosse's statement is, that with his reflector he can see objects on old Thornbush two hundred and fifty-two feet long. If he can do that he can see on our B. M. objects which are five feet long; and, of course, we were beside ourselves to get control of some instrument which had some approach to such power. Haliburton was for at once building a reflector at No. 9; and perhaps he will do it yet, for Haliburton has been successful in his paper-making and lumbering. But I went to work more promptly.

I remembered, not an apothecary, but an observatory, which had been dormant, as we say of volcanoes, now for ten or a dozen years,—no matter why! The trustees had quarrelled with the director, or the funds had given out, or the director had been shot at the head of his division,—one of those accidents had happened which will happen even in observatories which have fifteen-inch equatorials; and so the equatorial here had been left as useless as a cannon whose metal has been strained or its reputation stained in an experiment. The observatory at Tamworth, dedicated with such enthusiasm,—"another lighthouse in the skies,"—had been, so long as I have said, worthless to the world. To Tamworth, therefore, I travelled. In the neighborhood of the observatory I took lodgings. To the church where worshipped the family which lived in the observatory buildings I repaired; after two Sundays I established acquaintance with John Donald, the head of this family. On the evening of the third, I made acquaintance with his wife in a visit to them. Before three Sundays more he had recommended me to the surviving trustees as his successor as janitor to the buildings. He himself had accepted promotion, and gone, with his household, to keep a store for Haliburton in North Ovid. I sent for Polly and the children, to establish them in the janitor's rooms; and, after writing to her, with trembling eye I waited for the Brick Moon to pass over the field of the fifteen-inch equatorial.

Night came. I was "sole lone!" B. M. came, more than filled the field of vision, of course! but for that I was ready. Heavens! how changed. Red no longer, but green as a meadow in the spring. Still I could see—black on the green—the large twenty-foot circles which I remembered so well, which broke the concave of the dome; and, on the upper edge—were these palm-trees! They were. No, they [683] were hemlocks by their shape, and among them were moving to and fro——flies? Of course, I cannot see flies! But something is moving,—coming, going. One, two, three, ten; there are more than thirty in all!

They are men and women and their children!

Could it be possible! It was possible! Orcutt and Brannan and the rest of them had survived that giddy flight through the ether, and were going and coming on the surface of their own little world, bound to it by its own attraction and living by its own laws!

As I watched, I saw one of them leap from that surface. He passed wholly out of my field of vision, but in a minute, more or less, returned. Why not! Of course, the attraction of his world must be very small, while he retained the same power of muscle he had when he was here. They must be horribly crowded, I thought. No. They had three acres of surface, and there were but thirty-seven of them. Not so much crowded as people are in Roxbury, not nearly so much as in Boston; and, besides, these people are living underground, and have the whole of their surface for their exercise.

I watched their every movement as they approached the edge and as they left it. Often they passed beyond it, so that I could see them no more. Often they sheltered themselves from that tropical sun beneath the trees. Think of living on a world where from the vertical heat of the hottest noon of the equator to the twilight of the poles is a walk of only fifth paces! What atmosphere they had, to temper and diffuse those rays, I could not then conjecture.

I knew that at half past ten they would pass into the inevitable eclipse which struck them every night at this period of their orbit, and must, I thought, be a luxury to them, as recalling old memories of night when they were on this world. As they approached the line of shadow, some fifteen minutes before it was due, I counted on the edge thirty-seven specks arranged evidently in order; and, at one moment, as by one signal, all thirty-seven jumped into the air,—high jumps. Again they did it, and again. Then a low jump; then a high one. I caught the idea in a moment. They were telegraphing to our world, in the hope of an observer. Long leaps and short leaps,—the long and short of Morse's Telegraph Alphabet,—were communicating ideas. My paper and pencil had been of course before me. I jotted down the despatch, whose language I knew perfectly:—

"Show 'I understand' on the Saw-Mill Flat."
"Show 'I understand' on the Saw-Mill Flat."
"Show 'I understand' on the Saw-Mill Flat."

By "I understand" they meant the responsive signal given, in all telegraphy, by an operator who has received and understood a message.

As soon as this exercise had been three times repeated, they proceeded in a solid body—much the most apparent object I had had until now—to Circle No. 3, and then evidently descended into the MOON

The eclipse soon began, but I knew the MOON's path now, and followed the dusky, coppery spot without difficulty. At 1.33 it emerged, and in a very few moments I saw the solid column pass from Circle No. 3 again, deploy on the edge again, and repeat three times the signal:—

"Show 'I understand' on the Saw-Mill Flat."
"Show 'I understand' on the Saw-Mill Flat."
"Show 'I understand' on the Saw-Mill Flat."

It was clear that Orcutt had known that the edge of his little world would be most easy of observation, and that he had guessed that the moments of obscuration and of emersion were the moments when observers would be most careful. After this signal they broke up again, and I could not follow them. With daylight I sent off a despatch to [684] Haliburton, and, grateful and happy in comparison, sank into the first sleep not haunted by horrid dreams, which I had known for years.

Haliburton knew that George Orcutt had taken with him a good Dolland's refractor, which he had bought in London, of a two-inch glass. He knew that this would give Orcutt a very considerable power, if he could only adjust it accurately enough to find No. 9 in the 3d Range. Orcutt had chosen well in selecting the "Saw-Mill Flat," a large meadow, easily distinguished by the peculiar shape of the mill-pond which we had made. Eager though

Haliburton was, to join me, he loyally took moneys, caught the first train to Skowhegan, and, travelling thence, in thirty-six hours more was again descending Spoonwood Hill, for the first time since our futile observations. The snow lay white upon the Flat. With Bob Shea's help, he rapidly unrolled a piece of black cambric twenty yards long, and pinned it to the crust upon the snow; another by its side, and another. Much cambric had he left. They had carried down with them enough for the funerals of two Presidents. Haliburton showed the symbols for "I understand," but he could not resist also displaying · — · —, which are the dots and lines to represent O.K., which, he says is the shortest message of comfort. And not having exhausted the space on the Flat, he and Robert, before night closed in, made a gigantic **O.K.**, fifteen yards from top to bottom, and in marks that were fifteen feet through.

I had telegraphed my great news to Haliburton on Monday night. Tuesday night he was at Skowhegan. Thursday night he was at No. 9. Friday he and Rob. stretched their cambric. Meanwhile, every day I slept. Every night I was glued to the eye-piece. Fifteen minutes before the eclipse every night this weird dance of leaps two hundred feet high, followed by hops of twenty feet high, mingled always in the steady order I have described, spelt out the ghastly message:—

"Show 'I understand' on the Saw-Mill Flat."

And every morning, as the eclipse ended, I saw the column creep along to the horizon, and again, as the duty of opening day, spell out the same:—

"Show 'I understand' on the Saw-Mill Flat."

They had done this twice in every twenty-four hours for nearly two years. For three nights steadily, I read these signals twice each night; only these, and nothing more.

But Friday night all was changed. After "Attention," that dreadful "Show" did not come, but this cheerful signal:—

"Hurrah. All well. Air, food, and friends! what more can man require? Hurrah."

How like George! How like Ben Brannan! How like George's wife! How like them all! And they were all well! Yet poor *I* could not answer. Nay, I could only guess what Haliburton had done. But I have never, I believe, been so grateful since I was born!

After a pause, the united line of leapers resumed their jumps and hops. Long and short spelled out:—

"Your O.K. is twice as large as it need be."

Of the meaning of this, lonely *I* had, of course, no idea.

"I have a power of seven hundred," continued George. How did he get that? He has never told us. But this I can see, that all our analogies deceive us,—of views of the sea from Mt. Washington, or of the Boston State House from Wachusett. For in these views we look through forty or eighty miles of dense terrestrial atmosphere. But Orcutt was looking nearly vertically through an atmosphere which was, most of it, rare indeed, and pure indeed, compared with its lowest stratum.

In the record-book of my observations these despatches are entered as 12 and 13. Of course it was impossible for me to reply. All I could do was to telegraph these in the morning to Skowhegan, sending them to the [685] care of the Moores, that they might forward them. But the next night showed that this had not been necessary.

Friday night George and the others went on for a quarter of an hour. Then they would rest, saying "two," "three," or whatever their next signal time would be. Before morning I had these despatches:—

14. "Write to all hands that we are doing well. Langdon's baby is named Io, and Leonard's is named Phoebe."

How queer that was! What a coincidence! And they had some humor there.

15 was: "Our atmosphere stuck to us. It weighs three tenths of an inch—our weight."

16. "Our rain-fall is regular as the clock. We have made a cistern of Kilpatrick." This meant the spherical chamber of that name.

17. "Write to Darwin that he is all right. We began with lichens and have come as far as palms and hemlocks."

These were the first night's messages. I had scarcely covered the eye-glasses, and adjusted the equatorial for the day, when the bell announced the carriage in which Polly and the children came from the station to relieve me in my solitary service as janitor. I had the joy of showing her the good news. This night's work seemed to fill our cup. For all the day before, when I was awake, I had been haunted by the fear of famine for them. True, I knew that they had stored away in chambers H, I, and J the pork and flour which we had sent up for the workmen through the summer, and the corn and oats for the horses. But this could not last forever.

Now, however, that it proved that in a tropical climate they were forming their own soil, developing their own palms, and eventually even their bread-fruit and bananas, planting their own oats and maize, and developing rice, wheat, and all other cereals, harvesting these six, eight, or ten times—for aught I could see—in one of our years,—why then, there was no danger of famine for them. If, as I thought, they carried up with them heavy drifts of ice and snow in the two chambers which were not covered in when they started, why, they had waters in their firmament quite sufficient for all purposes of thirst and of ablution. And what I had seen of their exercise showed that they were in strength sufficient for the proper development of their little world.

Polly had the messages by heart before an hour was over, and the little girls, of course, knew them sooner than she.

Haliburton, meanwhile, had brought out the Shubael refractor (Alvan Clark), and by night of Friday was in readiness to see what he could see. Shubael of course gave him no such luxury of detail as did my fifteen-inch equatorial. But still he had no difficulty in making out groves of hemlock, and the circular openings. And although he could not make out my thirty-seven flies, still when 10.15 came he saw distinctly the black square crossing from hole Mary to the edge, and begin its Dervish dances. They were on his edge more precisely than on mine. For Orcutt knew nothing of Tamworth, and had thought his best chance was to display for No. 9. So was it that, at the same moment with me, Haliburton also was spelling out Orcutt & Co.'s joyous "Hurrah!"

"Thtephen," lisps Celia, "promith that you will look at yon moon [old Thornbush] at the inthtant I do." So was it with me long and Haliburton.

He was of course informed long before the Moores' messenger came, that, in Orcutt's judgment, twenty feet of length were sufficient for his signals. Orcutt's atmosphere, of course, must be exquisitely clear.

So, on Saturday, Rob. and Haliburton pulled up all their cambric and arranged it on the Flat again, in letters of twenty feet, in this legend:—

RAH. AL WEL.

Haliburton said he could not waste flat or cambric on spelling.

[686] He had had all night since half past ten to consider what next was most important for them to know; and a very difficult question it was, you will observe. They had been gone nearly two years, and much had happened. Which thing was, on the whole, the most interesting and important? He had said we were all well. What then? Did you never find yourself in the same difficulty? When your husband had come home from sea, and kissed you and the children, and wondered at their size, did you never sit silent, and have to think what you should say? Were you never fairly relieved when little Phil said, blustering, "I got three eggs today." The truth is, that silence is very satisfactory intercourse, if we only know all is well. When De Sauty got his original cable going, he had not much to tell after all; only that consols were a quarter per cent higher than they were the day before. "Send me news," lisped he—poor lonely myth!—from Bull's Bay to Valentia,—"send me news; they are mad for news." But how if there be no news worth sending? What do I read in my cable despatch to-day? Only that the Harvard crew pulled at Putney yesterday, which I knew before I opened the paper, and that there had been a riot in Spain, which I also knew. Here is a letter just brought me by the mail from Moreau, Tazewell County, Iowa. It is written by Follansbee, in a good cheerful hand. How glad I am to hear from Follansbee!

Yes; but do I care one straw whether Follansbee planted spring wheat or winter wheat? Not I. All I care for is Follansbee's way of telling it. All these are the remarks by which Haliburton explains the character of the messages he sent in reply to George Orcutt's autographs, which were so thoroughly satisfactory.

Should he say Mr. Borie had left the Navy Department, and Mr. Robeson come in? Should he say the Lords had backed down on the Disendowment Bill? Should he say the telegraph had been landed at Duxbury? Should he say Ingham had removed to Tamworth? What did they care for this? What does anybody ever care for facts? Should he say that the State Constable was enforcing the liquor law on whiskey, but was winking at lager? All this would take him a week, in the most severe condensation,—and for what good? as Haliburton asked. Yet these were the things that the newspapers told, and they told nothing else. There was a nice little poem of Jean Ingelow's in a Transcript Haliburton had with him. He said he was really tempted to spell that out. It was better worth it than all the rest of the newspaper stuff, and would be remembered a thousand years after that was forgotten. "What they wanted," says Haliburton, "was sentiment. That is all that survives and is eternal." So he and Rob. laid out their cambric thus:—

RAH. AL WEL. SO GLAD.

Haliburton hesitated whether he would not add, "Power 5000," to indicate the full power I was using at Tamworth. But he determined not to, and, I think, wisely. The convenience was so great, of receiving the signal at the spot where it could be answered, that for the present he thought it best that they should go on as they did. That night, however, to his dismay, clouds gathered and a grim snow-storm began. He got no observations; and the next day it stormed so heavily that he could not lay his signals out. For me at Tamworth, I had a heavy storm all day, but at midnight it was clear; and as soon as the regular eclipse was past George began with what we saw was an account of the great anaclysm which sent them there. You observe that Orcutt had far greater power of communicating with us than we had with him. He knew this. And it was fortunate he had. For he had, on his little world, much more of interest to tell than we had, on our large one.

18. "It stormed hard. We were all asleep, and knew nothing till morning; the hammocks turned so slowly."

Here was another revelation and relief. I had always supposed that, if [687] they knew anything before they were roasted to death, they had had one wild moment of horror. Instead of this, the gentle slide of the MOON had not wakened them; the flight upward had been as easy as it was rapid, the change from one centre of gravity to another had of course been slow,—and they had actually slept through the whole. After the dancers had rested once, Orcutt continued:—

19. "We cleared E.A. in two seconds, I think. Our outer surface fused and cracked somewhat. So much the better for us."

They moved so fast that the heat of their friction through the air could not propagate itself through the whole brick surface. Indeed, there could have been but little friction after the first five or ten miles. By E.A. he means earth's atmosphere.

His 20th despatch is: "I have no observations of ascent. But by theory our positive ascent ceased in two minutes five seconds, when we fell into our proper orbit, which, as I calculate, is 5,100 miles from your mean surface."

In all this, observe, George dropped no word of regret through these five thousand miles.

His 21st despatch is: "Our rotation on our axis is made once in seven hours, our axis being exactly vertical to the plane of our own orbit. But in each of your daily rotations we get sunned all round."

Of course, they never had lost their identity with us, so far as our rotation and revolution went: our inertia was theirs; all the fatal Fly-Wheels had given them was an additional motion in space of their own.

This was the last despatch before daylight of Sunday morning; and the terrible snow-storm of March, sweeping our hemisphere, cut off our communication with them, both at Tamworth and No. 9, for several days.

But here was ample food for reflection. Our friends were in a world of their own, all thirty-seven of them well, and it seemed they had two more little girls added to their number since they started. They had plenty of vegetables to eat, with prospect of new tropical varieties according to Dr. Darwin. Rob. Shea was sure that they carried up hens; he said he knew Mrs. Whitman had several Middlesexes and Mrs. Leonard two or three Black Spanish fowls, which had been given her by some friends in Foxcroft. Even if they had not yet had time enough for these to develop into Alderneys and venison, they would not be without animal food.

When at last it cleared off, Haliburton had to telegraph: "Repeat from 21"; and this took all his cambric, though he had doubled his stock. Orcutt replied the next night:—

22. "I can see your storms. We have none. When we want to change climate we can walk in less than a minute from midsummer to the depth of winter. But in the inside we have eleven different temperatures, which do not change."

On the whole there is a certain convenience in such an arrangement. With No. 23 he went back to his story:—

"It took us many days, one or two of our months, to adjust ourselves to our new condition. Our greatest grief is that we are not on the meridian. Do you know why?"

Loyal George! He was willing to exile himself and his race from the most of mankind, if only the great purpose of his life could be fulfilled. But his great regret was that it was not fulfilled. He was not on the meridian. I did not know why. But Haliburton, with infinite labor, spelt out on the Flat,

CYC. PROJECT. AD FIN.,

by which he meant, "See article Projectiles in the Cyclopaedia at the end"; and there indeed is the only explanation to be given. When you fire a shot, why does it ever go to the right or left of the plane in which it is projected? Dr. Hutton ascribes it to a whirling motion acquired by the bullet by friction with the gun. Euler thinks it due chiefly to the irregularity of the shape of the ball. In our case the B. M. was regular enough. But on one side, being wholly unprepared for flight, she [688] was heavily stored with pork and corn, while her chambers had in some of them heavy drifts of snow, and some only a few men and women and hens.

Before Orcutt saw Haliburton's advice, he had sent us 24 and 25.

24. "We have established a Sandemanian church, and Brannan preaches. My son Edward and Alice Whitman are to be married this evening."

This despatch unfortunately did not reach Haliburton, though I got it. So, all the happy pair received for our wedding-present was the advice to look in the Cyclopaedia at article Projectiles near the end.

25 was:—

"We shall act 'As You Like It' after the wedding. Dead-head tickets for all of the old set who will come."

Actually, in one week's reunion we had come to joking.

The next night we got 26:—

"Alice says she will not read the Cyclopaedia in the honeymoon, but is much obliged to Mr. Haliburton for his advice."

"How did she ever know it was I?" wrote the matter-of-fact Haliburton to me.

27. "Alice wants to know if Mr. Haliburton will not send here for some rags; says we have plenty, with little need for clothes."

And then despatches began to be more serious again. Brannan and Orcutt had failed in the great scheme for the longitude, to which they had sacrificed their lives,—if, indeed, it were a sacrifice to retire with those they love best to a world of their own. But none the less did they devote themselves, with the rare power of observation they had, to the benefit of our world. Thus, in 28:—

"Your North Pole is an open ocean. It was black, which we think means water, from August 1st to September 29th. Your South Pole is on an island bigger than New Holland. Your Antarctic Continent is a great cluster of islands."

29. "Your Nyanzas are only two of a large group of African lakes. The green of Africa, where there is no water, is wonderful at our distance."

30. "We have not the last numbers of 'Foul Play.' Tell us, in a word or two, how they got home. We can see what we suppose their island was."

31. "We should like to know who proved Right in 'He Knew He was Right.'"

This was a good night's work, as they were then telegraphing. As soon as it cleared, Haliburton displayed,—

BEST HOPES. CARRIER DUCKS.

This was Haliburton's masterpiece. He had no room for more, however, and was obliged to reserve for the next day his answer to No. 31, which was simply,

SHE.

A real equinoctial now parted us for nearly a week, and at the end of that time they were so low in our northern horizon that we could not make out their signals; we and they were obliged to wait till they had passed through two thirds of their month before we could communicate again. I used the time in speeding to No. 9. We got a few carpenters together, and arranged on the Flat two long movable black platforms, which ran in and out on railroad-wheels on tracks, from under green platforms; so that we could display one or both as we chose, and then withdraw them. With this apparatus we could give forty-five signals in a minute, corresponding to the line and dot of the telegraph; and thus could compass some twenty letters in that time, and make out perhaps two hundred and fifty words in an hour. Haliburton thought that, with some improvements, he could send one of Mr. Buchanan's messages up in thirty-seven working-nights.

---

[215]   **IV.—INDEPENDENCE.**

I own to a certain mortification in confessing that after this interregnum, forced upon us by so long a period of non-intercourse, we never resumed precisely the same constancy of communication as that which I have tried to describe at the beginning. The apology for this benumbment, if I may so call it, will suggest itself to the thoughtful reader.

It is indeed astonishing to think that we so readily accept a position when we once understand it. You buy a new house. You are fool enough to take out a staircase that you may put in a bathing-room. This will be done in a fortnight, everybody tells you, and then everybody begins. Plumbers, masons, carpenters, plasterers, skimmers, bell-hangers, speaking-tube men, men who make furnace-pipe, paper-hangers, men who scrape off the old paper, and other men who take off the old paint with alkali, gas men, city water men, and painters begin. To them are joined a considerable number of furnace-men's assistants, stovepipe-men's assistants, mason's assistants, and hodmen who assist the assistants of the masons, the furnace-men, and the pipe-men. For a day or two these all take possession of the house and reduce it to chaos. In the language of Scripture, they enter in and dwell there. Compare, for the details, Matt. xii.45. Then you revisit it at the end of the fortnight, and find it in chaos, with the woman whom you employed to wash the attics the only person on the scene. You ask her where the paper-hanger is; and she says he can do nothing because the plaster is not dry. You ask why the plaster is not dry, and are told it is because the furnace-man has not come. You send for him, and he says he did come, but the stove-pipe man was away. You send for him, and he says he lost a day in coming, but that the mason had not cut the right hole in the chimney. You go and find the mason, and he says they are all fools, and that there is nothing in the house that need take two days to finish.

Then you curse, not the day in which you were born, but the day in which bath-rooms were invented. You say, truly, that your father and mother, from whom you inherit every moral and physical faculty you prize, never had a bath-room till they were past sixty, yet they thrived, and their children. You sneak through back streets, fearful lest your friends shall ask you when your house will be finished. You are sunk in wretchedness, unable even to read you proofs accurately, far less able to attend the primary meetings of the party with which you vote, or to discharge any of the duties of a good citizen. Life is wholly embittered to you.

Yet, six weeks after, you sit before a soft-coal fire, in your new house, with the feeling that you have always lived there. You are not even grateful that you are there. You have forgotten the plumber's name; and if you met in the street that nice carpenter that drove things through, you would just nod to him, and would not think of kissing him or embracing him.

Thus completely have you accepted the situation.

Let me confess that the same experience is that with which, at this writing, I regard the BRICK MOON. It is there in ether. I cannot keep it. I cannot get it down. I cannot well go to it,—though possibly that might be done, as you will see. They are all very happy there,—much happier, as far as I can see, than if they lived in sixth floors in Paris, in lodgings in London, or even in tenement-houses in Phoenix Place, Boston. There are disadvantages attached to their position; but there are also advantages. And what most of all tends to our accepting the situation is, that there is "nothing that we can do about it," as Q. says, but to keep up our correspondence with them, and to express our sympathies.

For them, their responsibilities are reduced, in somewhat the same proportion [216] as the gravitation which binds them down,—I had almost said to earth,—which binds them down to brick, I mean. This decrease of responsibility must make them as lighthearted as the loss of gravitation makes them light-bodied.

On which point I ask for a moment's attention. And as these sheets leave my hand, an illustration turns up, which well serves me. It is the 23rd of October. Yesterday morning all wakeful women in New England were sure there was some one under the bed. This is a certain sign of an earthquake. And when we read the evening newspapers we were made sure there had been an earthquake. What blessings the newspapers are,—and how much information they give us! Well, they said it was not very severe here, but perhaps it was more severe elsewhere; hopes really arising in the editorial mind, that in some Caraccas or Lisbon all churches and the cathedral might have fallen. I did not hope for that. But I did have just the faintest feeling, that *if*—if—if—it should prove that the world had blown up into six or eight pieces, and they had gone off into separate orbits, life would be vastly easier for all of us, on whichever bit we happened to be.

That thing has happened, they say, once. Whenever the big planet between Mars and Jupiter blew up, and divided himself into one hundred and two or more asteroids, the people on each one only knew there had been an earthquake, until they read their morning journals. And then, all that they knew at first was that telegraphic communication had ceased beyond—say two hundred miles. Gradually people and despatches came in, who said that they had parted company with some of the other islands and continents. But, as I say, on each piece the people not only weighed much less, but were much lighter-hearted, had less responsibility.

Now will you imagine the enthusiasm here, at Miss Hale's school, when it should be announced that geography, in future, would be confined to the study of the region east of the Mississippi and west of the Atlantic,—the earth having parted at the seams so named. No more study of Italian, German, French, or Sclavonic,—the people speaking those languages being now in different orbits or other worlds. Imagine also the superior ease of the office-work of the A.B.C.F.M. and kindred societies, the duties of instruction and civilizing, of evangelizing in general, being reduced within so much narrower bounds. For you and me also, who cannot decide what Mr. Gladstone ought to do with the land tenure in Ireland, and who distress ourselves so much about it in conversation, what a satisfaction to know that Great Britain is flung off with one rate of movement, Ireland with another, and the Isle of Man with another, into space, with no more chance of meeting again than there is that you shall have the same hand at whist tonight that you had last night! Even Victoria would sleep easier, and I am sure Mr. Gladstone would.

Thus, I say, were Orcutt's and Brannan's responsibilities so diminished, that after the first I began to see that their contracted position had its decided compensating ameliorations.

In these views, I need not say, the women of our little circle never shared. After we got the new telegraph arrangement in good running-order, I observed that Polly and Annie Haliburton had many private conversations, and the secret came out one morning, when, rising early in the cabins, we men found they had deserted us, and then, going in search of them, found them running the signal boards in and out as rapidly as they could, to tell Mrs. Brannan and the bride Alice Orcutt that flounces were worn an inch and half deeper, and that people trimmed now with harmonizing colors and not with contrasts. I did not say that I believed they wore fig-leaves in B. M., but that was my private impression.

After all, it was hard to laugh at the [217] girls, as these ladies will be called, should they live to be as old as Helen was when she charmed the Trojan senate (that was ninety-three, if Heyne be right in his calculations). It was hard to laugh at them, because this was simple benevolence, and the same benevolence led to a much more practical suggestion, when Polly came to me and told me she had been putting up some baby things for little Io and Phoebe, and some playthings for the older children, and she thought we might "sen dup a bundle."

Of course we could. There were the Flies still moving! or we might go ourselves!

[And here the reader must indulge me in a long parenthesis. I beg him to bear me witness that I never made one before. This parenthesis is on the tense that I am obliged to use in sending to the press these minutes. The reader observes that the last transactions mentioned happen in April and May, 1871. Those to be narrated are the sequence of those already told. Speaking of them in 1870 with the coarse tenses of the English language is very difficult. One needs, for accuracy, a pure future, a second future, a paulo-post future, and a paulum-ante future, none of which does this language have. Failing this, one would be glad of an a-orist,—tense without time,—if the grammarians will not swoon at hearing such language. But the English tongue hath not that either. Doth the learned reader remember that the Hebrew,—language of history and prophecy,—hath only a past and a future tense, but hath no present? Yet that language succeeded tolerably in expressing the present griefs or joys of David and of Solomon. Bear with me, then, O critic! if even in 1870 I use the so-called past tenses in narrating what remaineth of this history up to the summer of 1872. End of the parenthesis.]

On careful consideration, however, no one volunteers to go. To go, if you observe, would require that a man envelope himself thickly in asbestos or some similar non-conducting substance, leap boldly on the rapid Flies, and so be shot through the earth's atmosphere in two seconds and a fraction, carrying with him all the time in a non-conducting receiver the condensed air he needed, and landing quietly on B.M. by a pre-calculated orbit. At the bottom of our hearts I think we were all afraid. Some of us confessed to fear; others said, and said truly, that the population of the Moon was already dense, and that it did not seem reasonable or worth while, on any account, to make it denser. Nor has any movement been renewed for going. But the plan of the bundle of "things" seemed more feasible, as the things would not require oxygen. The only precaution seemed to be that which was necessary for protecting the parcel against combustion as it shot through the earth's atmosphere. We had not asbestos enough. It was at first proposed to pack them all in one of Professor Horsford's safes. But when I telegraphed this plan to Orcutt, he demurred. Their atmosphere was but shallow, and with a little too much force the corner of the safe night knock a very bad hole in the surface of his world. He said if we would send up first a collection of things of no great weight, but of considerable bulk, he would risk that, but he would rather have no compact metals.

I satisfied myself, therefore, with a plan which I still think good. Making the parcel up in heavy old woollen carpets, and cording it with worsted cords, we would case it in a carpet-bag larger than itself, and fill in the interstice with dry sand, as our best nonconductor; cording this tightly again, we would renew the same casing, with more sand; and so continually offer surfaces of sand and woollen, till we had five separate layers between the parcel and the air. Our calculation was that a perceptible time would be necessary for the burning and disintegrating of each sand-bag. If each one, on the average, would stand two fifths of a second, the inner parcel would get [218] through the earth's atmosphere unconsumed. If, on the other hand, they lasted a little longer, the bag, as it fell on B.M., would not be unduly heavy. Of course we could take their night for the experiment, so that we might be sure they should all be in bed and out of the way.

We had very funny and very merry times in selecting things important enough and at the same time bulky and light enough to be safe. Alice and Bertha at once insisted that there must be room for the children's playthings. They wanted to send the most approved of the old ones, and to add some new presents. There was a woolly sheep in particular, and a watering-pot that Rose had given Fanny, about which there was some sentiment; boxes of dominos, packs of cards, magnetic fishes, bows and arrows, checker-boards and croquet sets. Polly and Annie were more considerate. Down to Coleman and Company they sent an order for pins, needles, hooks and eyes, buttons, tapes, and I know not what essentials. India-rubber shoes for the children Mrs. Haliburton insisted on sending. Haliburton himself bought open-eye-shut-eye dolls, though I felt that wax had been, since Icarus's days, the worst article in such an adventure. For the babies he had india-rubber rings: he had tin cows and carved wooden lions for the bigger children, drawing-tools for those older yet, and a box of crochet tools for the ladies. For my part I piled in literature,—a set of my own works, the Legislative Reports of the State of Maine, Jean Ingelow, as I said or intimated, and both volumes of the Earthly Paradise. All these were packed in sand, bagged, and corded,—bagged, sanded, and corded again,—yet again and again,—five times. Then the whole awaited Orcutt's orders and our calculations.

At last the moment came. We had, at Orcutt's order, read the revolutions of the Flies to 7230, which was, as nearly as he knew, the speed on the fatal night. We had soaked the bag for near twelve hours, and, at the moment agreed upon, rolled it on the Flies, and saw it shot into the air. It was so small that it went out of sight too soon for us to see it take fire.

Of course we watched eagerly for signal time. They were all in bed on B.M. when we let fly. But the despatch was a sad disappointment.

107. "Nothing has come through but two croquet balls and a china horse. But we shall send the boys hunting in bushes, and we may find more."

108. "Two Harpers and an Atlantic, badly singed. But we can read all but the parts which were most dry."

109. "We see many small articles revolving round us which may perhaps fall in."

They never did fall in, however. The truth was, that all the bags had burned through. The sand, I suppose, went to its own place, wherever that was. And all the other things in our bundle became little asteroids or aerolites in orbits of their own, except a well-disposed score or two, which persevered far enough to get within the attraction of Brick Moon, and to take to revolving there, not having hit quite square, as the croquet balls did. They had five volumes of the Congressional Globe whirling round like bats within a hundred feet of their heads. Another body, which I am afraid was "The Ingham Papers," flew a little higher, not quite so heavy. Then there was an absurd procession of the woolly sheep, a china cow, an pair of india-rubbers, a lobster Haliburton had chosen to send, a wooden lion, the wax doll, a Salter's balance, the New York Observer, the bow and arrows, a Nuremberg nanny-goat, Rose's watering-pot, and the magnetic fishes, which gravely circled round and round them slowly, and made the petty zodiac of their petty world.

We have never sent another parcel since, but we probably shall at Christmas, gauging the Flies perhaps to one revolution more. The truth is, that although [219] we have never stated to each other in words our difference of opinion or feeling, there is a difference of habit of thought in our little circle as to the position which the B.M. holds. Somewhat similar is the difference to habit of thought in which different statesmen of England regard their colonies.

Is B.M. a part of our world, or is it not? Should its inhabitants be encouraged to maintain their connections with us, or is it better for them to "accept the situation" and gradually wean themselves from us and from our affairs? It would be idle to determine this question in the abstract: it is perhaps idle to decide any question of casuistry in the abstract. But, in practice, there are constantly arising questions which really require some decision of this abstract problem for their solution.

For instance, when that terrible breach occurred in the Sandemanian church, which parted it into the Old School and New School parties, Haliburton thought it very important that Brannan and Orcutt and the church in B.M. under Brannan's ministry should give in their adhesion to our side. Their church would count one more in our registry, and the weight of its influence would not be lost. He therefore spent eight or nine days in telegraphing, from the early proofs, a copy of the address of the chatauque Synod to Brannan, and asked Brannan if he were not willing to have his name signed to it when it was printed. And the only thing which Haliburton takes sorely in the whole experience of the Brick Moon, for the beginning, is that neither Orcutt nor Brannan has ever sent one word of acknowledgment of the despatch. Once, when Haliburton was very low-spirited, I heard him even say that he believed they had never read a word of it, and that he thought he and Rob. Shea had had their labor for their pains in running the signals out and in.

Then he felt quite sure that they would have to establish civil government there. So he made up an excellent collection of books,—De Lolme on the British Constitution; Montesquieu on Laws; Story, Kent, John Adams, and all the authorities here; with ten copies of his own address delivered before the Young Men's Mutual Improvement Society of Podunk, on the "Abnormal Truths of Social Order." He telegraphed to know what night he should send them, and Orcutt replied:—

129. "Go to thunder with your old law-books. We have not had a primary meeting nor a justice court since we have been here, and, D.V., we never will have."

Haliburton says this is as bad as the state of things in Kansas, when, because Frank Pierce would not give them any judges or laws to their mind, they lived a year or so without any. Orcutt added in his next despatch:—

130. "Have not you any new novels? Send up Scribe and the Arabian Nights and Robinson Crusoe and the Three Guardsmen, and Mrs. Whitney's books. We have Thackeray and Miss Austen."

When he read this, Haliburton felt as if they were not only light-footed, but light-headed. And he consulted me quite seriously as to telegraphing to them "Pycroft's Course on Reading." I coaxed him out of that, and he satisfied himself with a serious expostulation with George as to the way in which their young folks would grow up. George replied by telegraphing Brannan's last sermon, 1 Thessalonians iv.11. The sermon had four heads, must have occupied an hour and half in delivery, and took five nights to telegraph. I had another engagement, so that Haliburton had to sit it all out with this eye to Shubael; and he has never entered on that line of discussion again. It was as well, perhaps, that he got enough of it.

The women have never had any misunderstandings. When we had received two or three hundred despatches from B.M., Annie Haliburton came to me and said, in that pretty way of hers, that she thought they had a right to their turn again. She said this lore about the Albert Nyanza and the [220] North Pole was all very well, but, for her part, she wanted to know how they lived, what they did, and what they talked about, whether they took summer journeys, and how and what was the form of society where thirty-seven people lived in such close quarters. This about "the form of society" was merely wool pulled over my eyes. So she said she thought her husband and I had better go off to the Biennial Convention at Assumpink, as she knew we wanted to do, and she and Bridget and Polly and Cordelia would watch for the signals, and would make the replies. She thought they would get on better if we were out of the way.

So we went to the convention, as she called it, which was really not properly a convention, but the Forty-fifth Biennial General Synod, and we left the girls to their own sweet way.

Shall I confess that they kept no record of their own signals, and did not remember very accurately what they were? "I was not going to keep a string of 'says I's' and 'says she's,'" said Polly, boldly. "It shall not be written on my tomb that I have left more annals for people to file or study or bind or dust or catalogue." But they told us that they had begun by asking the "bricks" if they remembered what Maria Theresa said to her ladies-in-waiting.[2] Quicker than any signal had ever been answered, George Orcutt's party replied from the moon, "We hear, and we obey." Then the women-kind had it all to themselves. The brick-women explained at once to our girls that they had sent their men round to the other side to cut ice, and that they were manning the telescope, and running the signals for themselves, and that they could have a nice talk without any bother about the law-books or the magnetic pole. As I say, I do not know what questions Polly and Annie put; but,—to give them their due,—they had put on paper a coherent record of the results arrived at in the answers; though, what were the numbers of the despatches, or in what order they came, I do not know; for the session of the synod kept us at "Assampink for two or three weeks.

Mrs. Brannan was the spokesman. "We tried a good many experiments about day and night. It was very funny at first, not to know when it would be light and when dark, for really the names day and night do not express a great deal for us. Of course the pendulum clocks all went wrong till the men got them overhauled, and I think watches and clocks both will soon go out of fashion. But we have settled down on much the old hours, getting up, without reference to daylight, by our great gong, at your eight o'clock. But when the eclipse season comes, we vary from that for signalling.

"We still make separate families, and Alice's is the seventh. We tried hotel life, and we like it, for there has never been the first quarrel here. You can't quarrel here, where you are never sick, never tired, and need not be ever hungry. But we were satisfied that it was nicer for the children, and for all round, to live separately, and come together at parties, to church, at signal time, and so on. We had something to say then, something to teach, and something to learn.

"Since the carices developed so nicely into flax, we have had one great comfort, which we had lost before, in being able to make and use paper. We have had great fun, and we think the children have made great improvement in writing novels for the Union. The Union to the old Union for Christian work that we had in dear old No. 9. We have two serial novels going on, one called 'Diana of Carrotook,' and the other called 'Ups and Downs'; the first by Levi Ross, and the other by my Blanche. They are really very good, and I wish we could send them to you. But they would not be worth despatching.

"We get up at eight; dress, and fix [221] up at home; a sniff of air, as people choose; breakfast; and then we meet for prayers outside. Where we meet depends on the temperature; for we can choose any temperature we want, from boiling water down, which is convenient. After prayers an hour's talk, lounging, walking, and so on; no flirting, but a favorite time with the young folks.

"Then comes work. Three hours' head-work is the maximum in that line. Of women's work, as in all worlds, there are twenty-four in one of your days, but for my part I like it. Farmers and carpenters have their own laws, as the light serves and the seasons. Dinner is seven hours after breakfast began; always an hour long, as breakfast was. Then every human being sleeps for an hour. Big gong again, and we ride, walk, swim, telegraph, or what not, as the case may be. We have no horses yet, but the Shanghaes are coming up into very good dodos and ostriches, quite big enough for a trot for the children.

"Only two persons of a family take tea at home. The rest always go out to tea without invitation. At 9 P.M. big gong again, and we meet in 'Grace,' which is the prettiest hall, church, concert-room, that you ever saw. We have singing, lectures, theatre, dancing, talk,

---

2. Maria V's husband, Francis, Duke of Tuscany, was hanging about loose one day, and the Empress, who had got a little tired, said to the maids of honor, "Girls, whenever you marry, take care and choose a husband who has something to do outside of the house."

or what the mistress of the night determines, till the curfew sounds at ten, and then we all go home. Evening prayers are in the separate households, and every one is in bed by midnight. The only law on the statute-book is that every one shall sleep nine hours out of every twenty-four.

"Only one thing interrupts this general order. Three taps on the gong means 'telegraph,' and then, I tell you, we are all on hand."

"You cannot think how quickly the days and years go by!"

Of course, however, as I said, this could not last. We could not subdue our world, and be spending all our time in telegraphing our dear B.M. Could it be possible?—perhaps it was possible,—that they there had something else to think of and to do, besides attending to our affairs. Certainly their indifference to Grant's fourth Proclamation, and to Mr. Fish's celebrated protocol in the Tahiti business, looked that way. Could it be that little witch of a Belle Brannan really cared more for their performance of Midsummer Night's Dream, or her father's birthday, than she cared for that pleasant little account I telegraphed up to all the children, of the way we went to muster when we were boys together? Ah well! I ought not to have supposed that all worlds were like this old world. Indeed, I often say this is the queerest world I ever knew. Perhaps theirs is not so queer, and it is I who am the oddity.

Of course it could not last. We just arranged correspondence days, when we would send to them, and they to us. I was meanwhile turned out from my place at Tamworth Observatory. Not but I did my work well, and Polly hers. The observer's room was a miracle of neatness. The children were kept in the basement. Visitors were received with great courtesy; and all the fees were sent to the treasurer; he got three dollars and eleven cents one summer,—that was the year General Grant came there; and that was the largest amount that they ever received from any source but begging. I was not unfaithful to my trust. Nor was it for such infidelity that I was removed. No! But it was discovered that I was a Sandemanian; a Glassite, as in derision I was called. The annual meeting of the trustees came round. There was a large Mechanics' Fair in Tamworth at the time, and an Agricultural Convention. There was no horserace at the convention, but there were two competitive examinations in which running horses competed with each other, and trotting horses competed with each other, and five thousand dollars was given to the best runner and the best trotter. These causes drew all the trustees together. The Rev. Cephas Philpotts presided. His doctrines with [222] regard to free agency were considered much more sound than mine. He took the chair,—in that pretty observatory parlor, which Polly had made so bright with smilax and ivy. Of course I took no chair; I waited, as a janitor should, at the door. Then a brief address. Dr. Philpotts trusted that the observatory might always be administered in the interests of science, of true science; of that science which rightly distinguishes between unlicensed liberty and true freedom; between the unrestrained volition and the freedom of the will. He became eloquent, he became noisy. He sat down. Then three other men spoke, on similar subjects. Then the executive committee which had appointed me was dismissed with thanks. Then a new executive committee was chosen, with Dr. Philpotts at the head. The next day I was discharged. And the next week the Philpotts family moved into the observatory, and their second girl now takes care of the instruments.

I returned to the cure of souls and to healing the hurt of my people. On observation days somebody runs down to No. 9, and by means of Shubael communicates with B.M. We love them, and they love us all the same.

Nor do we grieve for them as we did. Coming home from Pigeon Cove in October, with those nice Wadsworth people, we fell to talking as to the why and wherefore of the summer life we had led. How was it that it was so charming? And why were we a little loath to come back to more comfortable surroundings? "I hate the school," said George Wadsworth. "I hate the making calls," said his mother. "I hate the office hour," said her poor husband; "if there were only a dozen I would not mind, but seventeen hundred thousand in sixty minutes is too many." So that led to asking how many of us there had been at

Pigeon Cove. The children counted up all the six families,—the Haliburtons, the Wadsworths, the Pontefracts, the Midges, the Hayeses, and the Inghams, and the two good-natured girls,—thirty-seven in all,—and the two babies born this summer. "Really," said Mrs. Wadsworth, "I have not spoken to a human being besides these since June; and what is more, Mrs. Ingham, I have not wanted to. We have really lived in a little world of our own."

"World of our own!" Polly fairly jumped from her seat, to Mrs. Wadsworth's wonder. So we had—lived in a world of our own. Polly reads no newspaper since the "Sandemanian" was merged. She has a letter or two tumble in sometimes, but not many; and the truth was that she had been more secluded from General Grant and Mr. Gladstone and the Khedive, and the rest of the important people, than had Brannan or Ross or any of them!

And it had been the happiest summer she had ever known.

Can it be possible that all human sympathies can thrive, and all human powers be exercised, and all human joys increase, if we live with all our might with the thirty or forty people next to us, telegraphing kindly to all other people, to be sure? Can it be possible that our passion for large cities, and large parties, and large theatres, and large churches, develops no faith nor hope nor love which would not find aliment and exercise in a little "world of our own"?

Document I-4

Document title: Percival Lowell, *Mars* (Boston: Houghton Mifflin Co., 1895), pp. 201-12.

Percival Lowell, a Brahmin from Massachusetts, became interested in Mars during the latter part of the nineteenth century. Using personal funds and grants from other sources he built what became the Lowell Observatory near Flagstaff, Arizona, to study the planets. This research led him to publish *Mars and Its Canals* in 1906, which argued that Mars had once been a watery planet and that the topographical features known as canals had been built by intelligent beings. Over the course of the next forty years others used Lowell's observations of Mars as a foundation for their arguments. The idea of intelligent life on Mars stayed in the popular imagination for a long time.

---

[201] To review, now, the chain of reasoning by which we have been led to regard it probable that upon the surface of Mars we see the effects of local intelligence. We find, in the first place, that the broad physical conditions of the planet are not antagonistic to some form of life; secondly, that there is an apparent dearth of water upon the planet's surface, and therefore, if beings of sufficient intelligence inhabited it, they would have to resort to irrigation to support life; thirdly, that there turns out to be a network of markings covering the disk precisely counterparting what a system of irrigation would look like; and, lastly, that there is a set of spots placed where we should expect to find the lands thus artificially fertilized, and behaving as such constructed oases should. All this, of course, may be a set of coincidences, signifying nothing; but the probability points the other way. As to details of explanation, any we may adopt will undoubtedly be found, on closer acquaintance, to vary from the actual Martian state of things; for any Martian life must differ markedly from our own.

[202] The fundamental fact in the matter is the dearth of water. If we keep this in mind we shall see that many of the objections that spontaneously arise answer themselves. The supposed herculean task of constructing such canals disappears at once for, if the canals be dug for irrigation purposes, it is evident that what we see, and call by ellipsis the canal, is not really the canal at all, but the strip of fertilized land bordering it,——the thread of water in the midst of it, the canal itself, being far too small to be perceptible. In the case of an irrigation canal seen at a distance, it is always the strip of verdure, not the canal, that is visible, as we see in looking from afar upon irrigated country on the Earth.

We may, perhaps, in conclusion, consider for a moment how different in its details existence on Mars must be from existence on the Earth. One point out of many bearing on the subject, the simplest and most certain of all, is the effect of mere size of habitat upon the size of the inhabitant; for geometrical conditions alone are most potent factors in the problem of life. Volume and mass determine the force of gravity upon the surface of a planet, and this is more far-reaching in its effects than might at first be thought. Gravity on the surface of Mars is only a little more than one third what it is on the surface of the Earth. This would work in two ways to very different conditions of existence from those to which we are accustomed. To begin with, three times as much work, as for example, in digging a canal, could be done by the same expenditure of muscular force. If we were transported to Mars, we [203] should he pleasingly surprised to find all our manual labor suddenly lightened threefold. But, indirectly, there might result a yet greater gain to our capabilities; for if Nature chose she could afford there to build her inhabitants on three times the scale she does on Earth without their ever finding it out except by interplanetary comparison. Let us see how.

As we all know, a large man is more unwieldy than a small one. An elephant refuses to hop like a flea; not because he considers the act undignified, but simply because he cannot bring it about. If we could, we should all jump straight across the street, instead of painfully paddling through the mud. Our inability to do so depends upon the size of the Earth, not upon what it at first seems to depend on, the size of the street.

To see this, let us consider the very simplest case, that of standing erect. To this every-day feat opposes itself the weight of the body simply, a thing of three dimensions, height, breadth, and thickness, while the ability to accomplish it resides in the cross-section of the [204] muscles of the knee, a thing of only two dimensions, breadth and thickness. Consequently, a person half as large again as another has about twice the supporting capacity of that other, but about three times as much to support. Standing therefore tires him out more quickly. If his size were to go on increasing, he would at last reach a stature at which he would no longer be able to stand at all, but would have to lie down. You shall see the same effect in quite inanimate objects. Take two cylinders of paraffine wax, one made into an ordinary candle, the other into a gigantic facsimile of one, and then stand both upon their bases. To the small one nothing happens. The big one, however, begins to settle, the base actually made viscous by the pressure of the weight above.

Now apply this principle to a possible inhabitant of Mars, and suppose him to be constructed three times as large as a human being in every dimension. If he were on Earth, he would weigh twenty-seven times as much, but on the surface of Mars, since gravity there is only about one third of what it is here, he would weigh but nine times as much. The cross-section of his muscles would be nine times as great. Therefore the ratio of his supporting power to the weight he must support would be the same as ours. Consequently, he would be able to stand with as little fatigue as we. Now [205] consider the work he might be able to do. His muscles, having length, breadth, and thickness, would all be twenty-seven times as effective as ours. He would prove twenty-seven times as strong as we, and could accomplish twenty-seven times as much. But he would further work upon what required, owing to decreased gravity, but one third the effort to overcome. His effective force, therefore, would be eighty-one times as great as man's, whether in digging canals or in other bodily occupation. As gravity on the surface of Mars is really a little more than one third that at the surface of the Earth, the true ratio is not eighty-one, but about fifty; that is, a Martian would be, physically, fifty-fold more efficient than man.

As the reader will observe, there is nothing problematical about this deduction whatever. It expresses an abstract ratio of physical capabilities which must exist between the two planets, quite irrespective of whether there be denizens on either, or how other conditions may further affect their forms. As the reader must also note, the deduction refers to the possibility, not to the probability, of such giants; the calculation being introduced simply to show how different from us any Martians may be, not how different they are.

It must also be remembered that the question of their size has nothing to do with the [206] question of their existence. The arguments for their presence are quite apart from any consideration of avoirdupois. No Herculean labors need to be accounted for; and, if they did, brain is far more potent to the task than brawn.

Something more we may deduce about the characteristics of possible Martians, dependent upon Mars itself, a result of the age of the world they would live in.

A planet may in a very real sense be said to have life of its own, of which what we call life may or may not be a subsequent detail. It is born, has its fiery youth, sobers into middle age, and just before this happens brings forth, if it be going to do so at all, the creatures on its surface which are, in a sense, its offspring. The speed with which it runs through its gamut of change prior to production depends upon its size; for the smaller the body the quicker it cools, and with it loss of heat means beginning of life for its offspring. It cools quicker because, as we saw in a previous chapter, it has relatively less inside for its outside, and it is through its outside that its inside cools. After it has thus become capable of bearing life, the Sun quickens that life and supports it for we know not how long. But its duration is measured at the most by the Sun's life. Now, inasmuch as time and space are not, as some philosophers have from their too mundane standpoint [207] supposed, forms of our intellect, but essential attributes of the universe, the time taken by any process affects the character of the process itself, as does also the size of the body undergoing it. The changes brought about in a large planet by its cooling are not, therefore, the same as those brought about in a small one. Physically, chemically, and, to our present end, organically, the two results are quite diverse. So different, indeed, are they that unless the planet have at least a certain size it will never produce what we call life, meaning our particular chain of changes or closely allied forms of it, at all. As we saw in the case of atmosphere, it will lack even the premise to such conclusion.

Whatever the particular planet's line of development, however, in its own line, it proceeds to greater and greater degrees of evolution, till the process stops, dependent, probably, upon the Sun. The point of development attained is, as regards its capabilities, measured by the planet's own age, since the one follows upon the other.

Now, in the special case of Mars, we have before us the spectacle of a world relatively well on in years, a world much older than the Earth. To so much about his age Mars bears evidence on his face. He shows unmistakable signs of being old. Advancing planetary years have left their mark legible there. His continents are all [208] smoothed down; his oceans have all dried up. *Teres atque rotundus*, he is a steady-going body now. If once he had a chaotic youth, it has long since passed away. Although called after the most turbulent of the gods, he is at the present time, whatever he may have been once, one of the most peaceable of the heavenly host. His name is a sad misnomer; indeed, the ancients seem to have been singularly unfortunate in their choice of planetary cognomens. With Mars so peaceful, Jupiter so young, and Venus bashfully draped in cloud, the planet's names accord but ill with their temperaments.

Mars being thus old himself, we know that evolution on his surface must be similarly advanced. This only informs us of its condition relative to the planet's capabilities. Of its actual state our data are not definite enough to furnish much deduction. But front the fact that our own development has been comparatively a recent thing, and that a long time would be needed to bring even Mars to his present geological condition, we may judge any life he may support to be not only relatively, but really older than our own.

From the little we can see, such appears to be the case. The evidence of handicraft, if such it be, points to a highly intelligent mind behind it. Irrigation, unscientifically conducted, would not give us such truly wonderful mathematical [209] fitness in the several parts to the whole as we there behold. A mind of no mean order would seem to have presided over the system we see,—a mind certainly of considerably more comprehensiveness than that which presides over the various departments of our own public works. Party politics, at all events, have had no part in them; for the system is planet wide. Quite possibly, such Martian folk are possessed of inventions of which we have not dreamed, and with them electrophones and kinetoscopes are things of a bygone past, preserved with veneration in museums as relics of the clumsy contrivances of the simple childhood of the race. Certainly what we see hints at the existence of beings who are in advance of, not behind us, in the journey of life.

Startling as the outcome of these observations may appear at first, in truth there is nothing startling about it whatever. Such possibility has been quite on the cards ever since the existence of Mars itself was recognized by the Chaldean shepherds, or whoever the still more primeval astronomers may have been. Its strangeness is a purely subjective phenomenon, arising from the instinctive reluctance of man to admit the possibility of peers. Such would be comic were it not the inevitable consequence of the constitution of the universe. To be shy of anything resembling himself is part and parcel [210] of man's own individuality. Like the savage who fears nothing so much as a strange man, like Crusoe who grows pale at the sight of footprints not his own, the civilized thinker instinctively turns from the thought of mind other than the one he himself knows. To admit into his conception of the cosmos other finite minds as factors has in it something of the weird. Any hypothesis to explain the facts, no matter how improbable or even palpably absurd it be, is better than this. Snow-caps of solid carbonic acid gas, a planet cracked in a positively mono-maniacal manner, meteors ploughing tracks across its surface with such mathematical precision that they must have been educated to the performance, and so forth and so on, in hypotheses each more astounding than its predecessor, commend themselves to man, if only by such means he may escape the admission of anything approaching his kind. Surely all this is puerile, and should as speedily as possible be outgrown. It is simply an instinct like any other, the projection of the instinct of self-preservation. We ought, therefore, to rise above it, and, where probability points to other things, boldly accept the fact provisionally, as we should the presence of oxygen, or iron, or anything else. Let us not cheat ourselves with words. Conservatism sounds finely, and covers any amount of ignorance and fear.

[211] We must be just as careful not to run to the other extreme, and draw deductions of purely local outgrowth. To talk of Martian beings is not to mean Martian men. Just as the probabilities point to the one, so do they point away from the other. Even on this Earth man is of the nature of an accident. He is the survival of by no means the highest physical organism. He is not even a high form of mammal. Mind has been his making. For aught we can see, some lizard or batrachian might just as well have popped into his place early in the race, and been now the dominant creature of this Earth. Under different physical conditions, he would have been certain to do so. Amid the surroundings that exist on Mars, surroundings so different from our own, we may be practically sure other organisms have been evolved of which we have no cognizance. What manner of beings they may be we lack the data even to conceive.

For answers to such problems we must look to the future. That Mars seems to be inhabited is not the last, but the first word on the subject More important than the mere fact of the existence of living beings there, is the question of what they may be like. Whether we ourselves shall live to learn this cannot, of course, be foretold. One thing, however, we can do, and that speedily: look at things from a standpoint raised above our local point of view; [212] free our minds at least from the shackles that of necessity tether our bodies; recognize the possibility of others in the same light that we do the certainty of ourselves. That we are the sum and substance of the capabilities of the cosmos is something so preposterous as to be exquisitely comic. We pride ourselves upon being men of the world, forgetting that this is but objectionable singularity, unless we are, in some wise, men of more worlds than one. For, after all, we are but a link in a chain. Man is merely this earth's highest production up to date. That he in any sense gauges the possibilities of the universe is humorous. He does not, as we can easily foresee, even gauge those of this planet. He has been steadily bettering from an immemorial past, and will apparently continue to improve through an incalculable future. Still less does he gauge the universe about him. He merely typifies in an imperfect way what is going on elsewhere, and what, to a mathematical certainty, is in some corners of the cosmos indefinitely excelled.

If astronomy teaches anything, it teaches that man is but a detail in the evolution of the universe, and that resemblant though diverse details are inevitably to be expected in the host of orbs around him. He learns that, though he will probably never find his double anywhere, he is destined to discover any number of cousins scattered through space.

Document I-5

Document title: A.A. Blagonravov, Editor in Chief, *Collected Works of K.E. Tsiolkovskiy, Volume II - Reactive Flying Machines,* Translation of "K.E. Tsiolkovskiy, Sobraniye Sochineniy, Tom II - Reaktivnyye Letatal'nyye Apparaty," Izdatel'stvo Akademii Nauk SSSR, Moscow, 1954, NASA TT F-237, 1965, pp. 72-117.

Konstantin Tsiolkovskiy was a school teacher who lived in the small town of Kaluga, Russia. He is regarded by the Russians as the founder of Soviet rocketry, much as Robert Goddard and Hermann Oberth are regarded as the fathers of American and German rocketry in their respective countries. He is responsible for associating the term Sputnik, or "fellow traveller," to artificial satellites. But Tsiolkovskiy's work was almost entirely theoretical and was not widely known or translated outside of Russia, until after his death.

This article, written in 1898 and first published in 1903, established the fundamentals of orbital mechanics and proposed the then-radical use of both liquid oxygen and liquid hydrogen as fuel. It appeared seventeen years before Goddard repeated much of the work in the United States and twenty-three years before he began the first experiments with liquid propellants. It was also the first detailed discussion of a manned space station. The fact that it was not translated until much later meant that its impact on rocket research around the world was minimal.

---

[72]

Exploration of the Universe with Reaction Machines

Heights Reached by Balloons; Their Size and Weight; the Temperature and Density of the Atmosphere

1. So far small unmanned aerostats carrying automatic observation equipment have risen to altitudes of not more than 22 km.

Above this height the difficulties of ascending to higher altitudes by balloon increases rapidly.

Suppose an aerostat is required to climb to an altitude of 27 km carrying a load of 1 kg. The air density at an altitude of 27 km is about $^1/_{50}$ of the density at the surface (760 mm pressure and 0°C). This means that at this altitude a balloon must occupy a volume 50 times greater than on the ground. At sea level, it is filled with, say, at least 2 cubic meters of hydrogen, which at the given altitude will occupy 100 cubic meters. At the same time, the balloon will lift a load of 1 kg, i.e., the automatic instruments, and the balloon itself will weigh about 1 kg. Assuming the diameter of the envelope to be 5.8 m meters, its surface area will be at least 103 square meters. Therefore, every square meter of the material, including the reinforcing mesh sewn into it, should weigh 10 g.

One square meter of ordinary writing paper weighs 100 g, while one square meter of cigarette paper weights 50 g. Thus even cigarette paper would be five times heavier than the material needed for our balloon. Such a material could not be used in the balloon, as an envelope made from it would tear and allow the gas to leak at a rapid rate.

Large balloons may have thicker envelopes. Thus, a balloon [73] with the unprecedentedly large diameter of 58 meters would have an envelope, one square meter of which would weigh about 100 g, i.e., about as much as ordinary writing paper. It would lift a load of 1,000 kg, which is much more than an automatic recorder would weigh.

If we reduce this load to one kilogram, using the same gigantic aerostat, we can make the envelope twice as heavy. In general, such a balloon, while expensive, would be perfectly feasible. At an altitude of 27 km it would occupy a volume of 100,000 cubic meters, and the surface area of its envelope would be 10,300 square meters.

And yet, how miserable these results seem! A mere 27 km. How then could the instruments be raised higher? The aerostat would have to be still larger. But here it should not be forgotten that as the size increases the forces acting on the envelope dominate more and more over the resistance of the material.

Raising instruments beyond the limits of the atmosphere by means of an aerostat is, of course, inconceivable; observations of shooting stars reveal that those limits lie no higher than 200-300 km. In theory, the top of the atmosphere may even be set at 54 km, if we base our calculations on a decrease in air temperature by 5°C per kilometer, which is fairly close to reality, at least with respect to the accessible layers of the atmosphere.*

I present a table of altitudes, temperatures and air densities calculated on this basis. It shows how rapidly the difficulties increase with altitude.

The divisor in the last column indicates the degree of difficulty in constructing a balloon. [74]

Depth of atmosphere in km[†]	Temperature in °C	Air Density
0	0	1
6	- 30	1 : 2
12	- 60	1 : 4.32
18	- 90	1 : 10.6
24	- 120	1 : 30.5
30	- 150	1 : 116
36	- 180	1 : 584
42	- 210	1 : 3900
48	- 240	1 : 28 000
54.5	- 272	0

2. Let us now consider another possible means of reaching high-altitudes—cannon-launched projectiles.

In practice, the initial velocity of a shell does not exceed 1,200 m/sec. Such a projectile, if launched vertically, would rise to an altitude of 73 km, if the ascent took place in a vacuum. In air, however, the height reached would be much lower depending on the shape and mass of the projectile.

If the shape of the projectile were suitable, it might reach a considerable height; but instruments could not be housed inside the projectile, since they would be smashed into fragments on its return to Earth or even while it was still moving through the barrel of the cannon. The danger would be less if the projectile were shot from a [75] tube, but even then it would still be enormous. Suppose, for the sake of simplicity, that the gas pressure on the projectile was uniform, so that the acceleration was W m/sec^2. Then the same acceleration would also be imparted to all the objects inside the projectile, objects forced to share the same motion. As a result, inside the projectile there would develop a relative,

---

[†] Editor's note: According to recent data, in the stratosphere, between 11 and 35 km, the temperature is constant and equal to –56.5°C. In the region between 35 and 50 km a rise in temperature to -30-35°C is observed.

* It is now known that the decrease in temperature continues only as far up as the boundary of the troposphere, i.e., up to 11 km.

apparent gravity,* equal to W/g, where g is acceleration due to gravity at the Earth's surface.

The length of the cannon L may be expressed by the formula

$$L = \frac{V^2}{2(W-g)},$$

whence

$$W = \frac{V^2}{2L} + g,$$

where V is the velocity acquired by the projectile on leaving the muzzle.

It is clear from the formula that W, and hence the increase in relative gravity in the projectile, decreases with increasing barrel length if V is constant, i.e., the longer the cannon the greater the safety of the instruments during the firing of the projectile. But even if the cannon were, in theory, extremely long, which is not feasible in practice, the apparent gravity in the projectile, as the latter accelerates through the barrel, would [76] become so enormous that the sensitive instruments would hardly be able to withstand it. This would make it all the more impossible to dispatch a living organism in the projectile, should this be found necessary.

3. Let us assume that we have succeeded in building a cannon approximately 300 m tall. Suppose it has been erected next to the Eiffel Tower which, of course, is the same height, and let the projectile, under the uniform pressure of the gases, attain a muzzle velocity sufficient to carry it beyond the limits of the atmosphere, e.g., to an altitude of 300 km above the surface. Then the velocity V required for this purpose may be calculated from the formula

$$V = \sqrt{2g \cdot h} \,**,$$

where h is the maximum altitude (we obtain approximately 2,450 m/sec). From the last two formulas, on eliminating V, we obtain

$$\frac{W}{g} = \frac{h}{L} + 1;$$

where W/g expresses the relative or apparent gravity within the projectile. From the formula we find it to be equal to 1001.

Therefore, the weight of all the instruments inside the projectile would increase by more than a thousandfold, i.e., an object weighing one kilogram would experience a pressure of 1,000 kg [77] due to the apparent gravity. There is hardly any physical instrument that can withstand such a pressure. And what a tremendous shock would be experienced by a living organism in a short-barreled cannon and during the ascent to an altitude of more than 300 km!

In order not to lead anyone astray by the words "relative or apparent gravity," let me say that by this I mean a force dependent on the acceleration of a body (for example, a projectile). It also appears during the uniform motion of a body, provided this motion is curvilinear; it is then termed centrifugal force. In general, relative gravity always appears

---

* g force (Editor's note).
**In these calculations air resistance was not taken into account (Editor's note).

whenever a body is acted upon by some mechanical force that disturbs its inertial motion. Relative gravity operates as long as the force engendering it continues to act; once the latter ceases to act, the relative gravity disappears without a trace. If I term this force gravity, it is only because its temporary effect is exactly the same as that of a gravitational force. Just as every material point of a body is subject to gravitation, so relative gravity affects every particle of a body enclosed in a projectile; this is because relative gravity depends on inertia, by which all the material parts of a body are equally affected. Thus, the instruments inside the projectile will become 1,001 times heavier. Even if they could be preserved intact through this terrifying, though momentary (0.24 sec) intensification in relative gravity, there would still be many other obstacles to the use of cannons as a means of reaching the celestial space.

First and foremost, there is the difficulty of building such cannons, even in the future; second, there is the tremendous initial velocity of the projectile. Actually, in the dense lower layers of the atmosphere the projectile would lose much of its velocity owing to air resistance; now this loss in velocity would also considerably reduce the altitude reached by the projectile. Besides, it is difficult to obtain a uniform gas pressure on the projectile, as the latter moves through the cannon barrel, so that the intensification of gravity will be much greater than calculated (1,001). Finally, the safe return of projectile to Earth would be more than doubtful.

Rocket Versus Cannon

4. Thus, the tremendous increase in gravity alone is definitely enough to dispel any notion of using cannons for our purpose.

Instead of cannons or aerostats, I propose the use of reaction [78] machines to explore the atmosphere. By reaction machine I mean a kind of rocket, but a specially designed rocket on a grandiose scale. The idea is not new, but the calculations yield such remarkable results that they simply cannot be ignored.

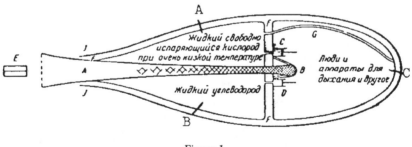

Figure 1

A - Freely evaporating liquid oxygen at very low temperature.
B - Liquid hydrocarbon.
C - Crew, breathing apparatus, etc.

I am far from having investigated every side of the matter, nor have I attempted to solve the practical problems relating to the feasibility of the concept; however, it is already possible to behold, across the veil, such tantalizing and significant glimpses of the distant future as could hardly be dreamed of.

Visualize the following projectile: an elongated metal chamber (the shape of least resistance) equipped with electric light, oxygen, and means of absorbing carbon dioxide, odors, and other animal secretions; a chamber, in short, designed to protect not only various physical instruments but also a human pilot (we shall consider the problem in its broadest terms). The chamber is partly occupied by a large store of substances which, on being mixed, immediately form an explosive mass. This mixture, on exploding in a controlled and [79] fairly uniform manner at a chosen point, flows in the form of hot gases through tubes with flared ends (Fig. 1), shaped like a cornucopia or a trumpet. These tubes are arranged lengthwise along the walls of the chamber. At the narrow end of the tube the explosives are mixed: this is where the dense, burning gases are obtained. After undergoing intense rarefaction and cooling, the gases explode outward into space at a tremendous relative velocity at the other, flared end of the tube. Clearly, under definite conditions, such a projectile will ascend like a rocket.

Automatic instruments are needed to control the motion of the rocket, as I shall call it, and the force of the explosion in accordance with a predetermined schedule.

Schematic View of the Rocket

The two liquid gases are separated by a partition. The place where the gases are mixed and exploded is shown, as is the flared outlet for the intensely rarefied and cooled vapors. The tube is surrounded by a jacket with a rapidly circulating liquid metal. The control surfaces serving to steer the rocket are also visible.

If the resultant of the explosion forces does not pass through the center of inertia of the projectile, the projectile will rotate and, therefore, will not be suitable. Now, it is quite impossible to attain a mathematically precise coincidence of this kind, because the center of inertia is bound to fluctuate owing to the motion of the substances contained by the projectile, in the same way as the direction of the resultant of the gas pressure inside a cannon barrel cannot be mathematically fixed. So long as the projectile is still in the air, it can be guided with control surfaces like a bird, but what can be done in an airless medium where the ether can not provide any appreciable support?

If the resultant is as close as possible to the center of inertia of the projectile, the rotation will be fairly slow. But as soon as it commences, we can shift some mass inside the projectile until the ensuing displacement of the center of inertia causes the projectile to incline in the opposite direction. Thus, by sensing the movements of the projectile and shifting a small mass inside it, we can cause the projectile to swing now in one, now in the other direction, so that the general direction of action of the explosives and the general direction of motion of the projectile do not change.

[80] It may be that manual steering of the projectile will be not only difficult but even infeasible. In this case, it will be necessary to resort to automatic control.

The Earth's gravitational attraction cannot be used as the principal regulating force, since the projectile will be governed only by relative gravity due to the acceleration W, the direction of which will coincide with the relative direction of the outflowing gases or be directly opposite to their resultant pressure. And since this direction varies with the rotation of the projectile, the relative gravity is unsuitable as the basis of a guidance system.

On the other hand, it is possible to use a magnetic needle or the strength of the Sun's rays focused by means of a biconvex lens. Whenever the projectile rotates, the small, bright image of the sun will change its relative position in the projectile, thus causing the expansion of a gas, or creating a pressure or an electric current, and hence the movement of a counterweight to restore the direction of the projectile, so that the light spot again falls on a neutral, insensitive part of the mechanism.

There should be two such automatically actuated masses.

Another basis for the guidance system of the projectile could be a small chamber with two disks rapidly rotating in different planes. The chamber is suspended so that its position or, more exactly, direction is independent of the direction of the projectile. When the projectile rotates, the chamber (if we ignore the friction) retains the same absolute direction (relative to the stars) thanks to its inertia; this property manifests itself to a high degree when the chamber disks rotate rapidly. If fine springs attached to the chamber changed their relative position during the rotation of the projectile, this change could be used to excite a current and produce a shift in the position of the counterweights.

Lastly, rotation of the mouth of the tube might also serve as a means of keeping the projectile on course. The simplest means of steering the rocket would be dual control surfaces mounted externally, close to the mouth of the tube. As for preventing the rotation of the rocket about its longitudinal axis, this can be accomplished by rotating a plate located in the gas flow and aligned with the direction of this flow.*

[81] **Advantages of the Rocket**

5. Before expounding the theory of the rocket or similar reaction-propelled devices, I shall try to interest the reader in the advantages of the rocket as compared with the cannon-launched projectile:

a) Our device, compared with the gigantic cannon, is as light as a feather, relatively cheap, and comparatively easy to realize.

b) The pressure of the explosives, being fairly uniform, creates a uniform acceleration which develops a relative gravity; we can adjust the magnitude of this temporary gravity as desired, i.e., by regulating the force of the explosion, and make it arbitrarily small or many times greater than normal terrestrial gravity. If we assume for simplicity's sake that the force of the explosion diminishes in proportion to the mass of the projectile plus the mass of the remaining, still unexploded fuel, the acceleration of the projectile, and hence the relative gravity, will be constant. Thus, with respect to apparent gravity, a rocket can safely be used to dispatch not only measuring instruments but also human beings, whereas a cannon-launched projectile, even one shot from a colossal cannon as tall as the Eiffel Tower, involves a 1,001-fold increase in relative gravity in ascending to 300 km.

c) Another important advantage of the rocket is that its velocity can be made to increase at a desired rate and in a desired direction; it may be kept constant; or it may uniformly decrease, thus making possible a safe landing. Everything depends on a reliable explosion regulator.

d) At take-off, when the atmosphere is dense and the air resistance at high speeds enormous, the rocket moves comparatively slowly and therefore the losses due to air resistance are low; moreover, the rocket does not become overheated.

The velocity of a rocket increases only very slowly; but later on, as it ascends to more and more rarefied layers of the atmosphere, it can be made to increase more rapidly, until, finally, in airless space the velocity reaches a maximum. Thus, the work done in overcoming air resistance is reduced to a minimum.

[82] **The Rocket in an Atmosphereless, Gravitationless Medium**
The Mass Ratio of the Rocket

6. First let us consider the effect in an atmosphereless, gravitationless medium. As regards the atmosphere, we shall consider only its resistance to the motion of the projectile, disregarding its resistance to the expulsion of exhaust gases. The effect of the atmosphere on the explosion is not altogether clear; on the one hand, it is favorable, since the exploding substances receive some support from the material medium, thus contributing to an increase in the rocket's velocity; on the other hand, the density and pressure of the atmosphere interfere with the expansion of the gases beyond certain known limits, so that these gases do not acquire the velocity they would acquire if expelled in a void. The latter effect is unfavorable, since the increase in the velocity of the rocket is proportional to the

---

*It is noteworthy that here Tsiolkovskiy anticipates the development of modern exhaust control vanes (Editor's note).

velocity of the expelled explosion products.

7. Let us denote by M_1 the mass the projectile together with all it contains, except the supply of explosives; by M_2, the total mass of the explosives; and, lastly, by M, the variable mass of the explosives remaining in the projectile in unexploded form at a given instant.

Thus, the total mass of the rocket at the commencement of the explosion will be: $(M_1 + M_2)$; some time later, however, it will be expressed by the variable $(M_1 + M)$; and finally, when the explosion ends, by the constant M_1.

In order for the rocket to attain its maximum velocity, it must expel the explosion products in a fixed direction relative to the stars. Therefore it must not rotate and, in order for it not to rotate, the resultant of the explosive forces—which passes through their center of pressure—must at the same time pass through the center of inertia of the whole complex of speeding masses. We have already solved the problem of how to accomplish this in practice.

Thus, assuming the optimal expulsion of the gases in a single direction, we obtain the following differential equation based on the law of conservation of momentum:

[83]
$$dV(M_1 + M) = V_1 dM \qquad (8)$$

9. Here dM is an infinitely small mass of explosive material expelled from the mouth of the tube at a constant (relative to the rocket) velocity V_1.

10. I should point out that on the basis of the law of relative motion, given the same conditions, the relative velocity of the exhaust elements is the same throughout the period of the explosion. dV is the increment in the velocity of the rocket together with the remaining unconsumed explosives; this increment, dV, is due to the expulsion of an element dM at the velocity V_1. We shall determine the latter in the proper place.

11. Separating the variables in equation (8) and integrating, we obtain

$$\frac{1}{V_1}\int dV = -\int \frac{dM}{M_1 + M} + C, \qquad (12)$$

or

$$\frac{V}{V_1} = -\ln(M_1 + M) + C. \qquad (13)$$

where C is a constant. When $M = M_2$, i.e., before the explosion, $V = O$; thus we find

$$C = +\ln(M_1 + M_2); \qquad (14)$$

and hence

$$\frac{V}{V_1} = \ln\left(\frac{M_1 + M_2}{M_1 + M}\right). \qquad (15)$$

[84] The velocity of the projectile will be a maximum when $M = 0$, i.e., when the entire fuel supply M_2 has been burned; then, putting $M = 0$ in the preceding equation, we have

$$\frac{V}{V_1} = \ln\left(1 + \frac{M_2}{M_1}\right). \qquad (16)$$

Hence we see that the velocity V of the projectile increases without limit with increase in the amount M_2 of explosives. This means that we can attain different final velocities for different voyages, depending on the store of explosives taken on board. Equation (16) also shows that a definite quantity of explosive is consumed, the velocity of the rocket does not depend on the rate or uniformity of the explosion, so long as the particles of exhaust material move at the same velocity V_1 with respect to the projectile.

17. However, as the quantity M_2 increases, the velocity V of the rocket increases ever more slowly, though without limit. It increases more or less as the logarithm of the increase in the amount of explosives M_2 (if M_2 is large compared with M_1, i.e., if the mass of the explosives is several times larger than the mass of the projectile).

18. Further calculations will be of interest, once we have determined V_1, i.e., the relative and final velocity of the explosion products. Since a gas or vapor leaving the mouth of the tube is much rarefied and cooled (if the tube is sufficiently long) and may even solidify—turn into particles of dust rushing at terrific speeds—it may be assumed that, when an explosion occurs, the entire energy of combustion or chemical combination is transformed into the motion of the combustion products or into kinetic energy. In fact, imagine a given amount of gas expanding in a void, without any restrictions: it will expand in all directions and, consequently, cool until it turns into droplets of liquid or a mist.

This mist will turn into minute crystals, no longer due to expansion but rather to evaporation and radiation into space.

[85] On expanding the gas will release all its manifest and part of its latent energy, which will ultimately be converted into rapid motion of the minute crystals in all directions – since the gas expanded freely. If, however, the gas is forced to expand in a tubular chamber, the tube will orient the motion of the gas molecules in a fixed direction, which is the method we use to propel our rocket.

It would seem that the energy of molecular motion is converted into kinetic motion as long as a substance remains in the gaseous or vapor state. But this is not quite so. In fact, part of the substance may turn to liquid; this involves the release of energy (latent heat of vaporization), which is transmitted to the part remaining as a vapor, thus delaying its transition to the liquid state.

We can observe an effect of this kind in a steam cylinder when steam does work owing to its own expansion and the valve from the boiler to the cylinder is closed. In this case, whatever the temperature of the steam, part of it turns into a mist, i.e., the liquid state, while the rest remains as a vapor and continues to do work, borrowing the latent heat of the condensed and liquefied fraction.

Thus, the molecular energy will continue to be transformed into kinetic energy at least until the liquid state is reached. Once the entire mass turns into droplets, the conversion to kinetic energy will cease almost completely, because the vapors of liquids and solids have only an insignificant pressure when the temperature is low, and their practical utilization is difficult, requiring enormous tubes.

In addition, an insignificant part of the energy is lost, i.e., is not converted into kinetic energy, due to friction against the walls of the tube and the radiation of heat from the heated parts of the tube. However, the tube can be encased in a jacket through which a liquid metal is circulated; this liquid metal will convey heat from the intensely heated end of the tube to the end cooled by the rarefaction of the vapor. Thus, the losses due to radiation and conduction can also be recovered or minimized. In view of the short duration of the explosion, which takes 2 to 5 min at most, the loss due to radiation is negligible, even without any special precautions; the circulation of the liquid metal in the tube jacket is more important for another purpose: the maintenance of a uniformly low tube wall temperature, i.e., the preservation of the mechanical strength of the tube. Despite this it may happen that part of the tube will melt, oxidize, and be carried away by the gases and vapors. To prevent this, the inside walls of the tube could be lined with some special refractory material: carbon, tungsten, etc. Some of the carbon may burn away, but the relatively cool metal tube will suffer little loss of strength.

As for the gaseous product of combustion of carbon—carbon [86] dioxide—this will only intensify the thrust of the rocket. Some kind of crucible material, some mixture of substances, might be used. However, I shall not attempt to solve these and other problems pertaining to reaction-propulsion machines.

In many cases I am limited to guesses or hypotheses. I am not deluding myself and I am perfectly aware that I am not solving the problem in its entirety, that a thousand times more work than I have done must be invested in its solution. My aim is to arouse interest by pointing out its great future significance and the feasibility of a solution....

The liquefaction of hydrogen and oxygen involves no special difficulty. Hydrogen could be replaced by liquid or liquefied hydrocarbons, for example, by acetylene or oil. These liquids must be separated by a partition. Their temperature is very low; therefore it would be expedient to allow them to surround either jackets with circulating liquid metal or the tubes themselves.

Experience will show which is better. Some metals become stronger when cooled; these are the metals that should be employed, copper, for example. I do not recall this clearly, but it seems that experiments on the resistance of iron in liquid air have revealed that its strength at such low temperatures is virtually dozens of times greater. I cannot guarantee the reliability of these experiments, but, in relation to the problem discussed here, they deserve the most diligent attention.* (Why not cool ordinary cannon in the same way before firing, since liquid air is now so easily obtainable.)

[87] Liquid oxygen and liquid hydrogen, when pumped from lands and supplied in a fixed ratio to the narrow inlet of the tube, where they progressively combine, constitute an excellent explosive. The water vapor obtained from the chemical combination of these liquids will expand at a tremendously high temperature in the direction of the wide end or mouth of the tube, until it cools to a liquid racing toward the outlet in the form of an ultrafine mist.

19. Hydrogen and oxygen in the gaseous state release 3825 calories on combining to form 1 kg of water. By the word "calorie" I mean the amount of heat required to raise one kilogram of water through 1°C.

This figure (3825 calories) will be somewhat lower in the present case, since the oxygen and hydrogen are in the liquid rather than in the gaseous state, to which this particular number of calories relates. In fact, the liquids must first be heated and then converted to the gaseous state, which requires some expenditure of energy. In view of the insignificant amount of energy required, as compared with the chemical energy, we shall not reduce this figure (the question has not been completely clarified by science; but we are merely taking oxygen and hydrogen as an example).

Assuming the mechanical equivalent of heat to be 427 kgm, we find that 3825 calories corresponds to 1633 kgm of work; this is enough to raise the explosion products, i.e., one kilogram of matter, to an altitude of 1633 km above the surface of the Earth, that the force of gravity is constant. This work, converted into motion, corresponds to the kinetic

---

*The author mentions a metal, iron, which has proved to be unsuitable with regard to its strength at low temperatures, as already pointed out by R. Lademann in his article "Zum Raketenproblem" (On the Rocket Problem) in the issue of April 28, 1927, of the periodical ZFM.

The author replies in the appendix to the book "Kosmicheskaya raketa, opyinaya podgotovka" (The Space Rocket – Experimental Preparations):

"Concerning the increase in the strength of iron at the temperature of liquid air, in 1903 I merely repeated information that I had read elsewhere, and I shall certainly not insist that it is true, once it has been proved to be untrue. In practice, the explosion tube could not attain such a degree of coldness. It is cooled by oil which, in turn, is cooled by liquid air. It is enough if the tube does not melt or burn, the petroleum does not boil, and the liquid air does not vaporize too quickly. There is no need to reach the temperature of liquid air in the explosion tube." (Editor's note in "Izbrannyye trudy K. E. Tsiolkovskogo" (Selected Works of K. E. Tsiolkovskiy), Book II - "Reaktivnoye dvizheniye" (Reaction propulsion), Moscow, ONTI, 1934.)

energy of a mass of one kilogram moving at a velocity of 5700 m/sec. I know of no group of substances that, on combining chemically, could release such a tremendous amount of energy per unit mass of the resulting product. Moreover, some substances on combining do not form volatile products at all, which is not at all suitable for our purposes. Thus, silicon, on burning in oxygen (Si + O_2 = SiO_2), releases an enormous amount of heat, namely, 3654 calories per unit of mass of the resulting product (SiO_2) but, [88] unfortunately the product is non-volatile.

Having taken liquid oxygen and hydrogen as the most suitable materials for creating an explosion, I gave a somewhat exaggerated figure for their chemical energy per unit mass of product (H_2O), since in a rocket the explosive substances must be in the liquid and not the gaseous state and, moreover, at a very low temperature.

I consider it pertinent to console the reader with the thought that we may expect not only this energy (3825 calories) but an incomparably greater energy in the future, if and when our still embryonic ideas are found to be feasible. In fact, on considering the amount of energy produced by various chemical processes, we find that as a general rule, though naturally with some exceptions, the amount of energy per unit mass of the products of chemical combination depends on the atomic weight of the combining elements: the lower the atomic weight, the greater the heat released during chemical combination. Thus, the formation of sulfur dioxide is accompanied by the release of only 1250 calories, and the formation of cupric oxide by only 546 calories, whereas when carbon dioxide CO_2 is formed, the carbon releases 2204 calories per unit mass of CO_2 hydrogen combining with oxygen, as we have seen, releases even more (3825).

To relate these data to the idea I have just formulated, let me remind you that the atomic weights of the elements named are: hydrogen, 1; oxygen, 15; carbon, 12; sulfur, 32; silicon, 28; copper, 63.

Of course, many exceptions to this rule can be cited, but in general it is valid. In fact, if we imagine a series of points the abscissas of which express the sum (or mathematical product) of the atomic weights of the combining elements, and the ordinates—the corresponding energy of chemical combination, then, on drawing a smooth curve through these points (as close to them as possible), we observe a steady decrease in the ordinates with increase in the abscissas, just as our theory suggests. For this reason, if at some time so-called simple substances prove to be complex and are separated into new elements, the atomic weights of these elements should be smaller than those of the simple substances known to us.

Accordingly, newly discovered elements, upon combining chemically, must release an incomparably larger amount of energy than bodies currently considered simple and having a comparatively large atomic weight.

The very existence of the ether with its almost infinite expansibility and the enormous velocity of its atoms implies that these atoms have an infinitesimally small atomic weight and infinite energy when they combine.

[89] 20. However this may be, for the time being we cannot count on more than 5700 m/sec as the maximum V_1 (see 15 and 19). With time, however, who knows, this figure may increase several times over.

Assuming 5700 m/sec, we can calculate from formula (16) not only the velocity ratio V/V_1 but also the absolute value of the final (maximum) velocity V of the projectile as a function of its $\frac{M_2}{M_1}$ ratio.

21. It is evident from formula (16) that the mass of the rocket together with all passengers and equipment, M_1 may be arbitrarily large without thereby detracting in any way from the velocity V of the projectile, so long as the supply of explosives, M_2, increases in direct proportion to M_1.

Thus, projectiles of any size with any number of passengers may be given any desired velocity. However, as we have seen, an increase in the velocity of the rocket is accompanied by an incomparably more rapid increase in the mass of the explosives. Therefore, though it may be easy to increase the mass of a projectile destined for outer space, it is correspondingly difficult to increase its velocity.

Flight Velocities as a Function of Fuel Consumption

22. From equation (16) we obtain the following table.

[89]

$\dfrac{M_2}{M_1}$	$\dfrac{V}{V_1}$	Velocity m/sec	$\dfrac{M_2}{M_1}$	$\dfrac{V}{V_1}$	Velocity m/sec
0.1	0.095	543	7	2.079	11 800
0.2	0.182	1 037	8	2.197	12 500
0.3	0.262	1 493	9	2.303	13 100
0.4	0.336	1 915	10	2.398	13 650
0.5	0.405	2 308	19	2.996	17 100
1	0.693	3 920	20	3.044	17 330
2	1.098	6 260	30	3.434	19 560
3	1.386	7 880	50	3.932	22 400
4	1.609	9 170	100	4.615	26 280
5	1.792	10 100	193	5.268	30 038
6	1.946	11 100	Infinite	Infinite	Infinite

[90] 23. From this table it is clear that the velocities attainable by reaction propulsion are far from negligible. Thus, when the mass of explosives exceeds 193 times the mass M_1 of the projectile (rocket), the velocity during the final moments of the explosion, when the entire supply of explosives M_2 has been consumed, is equal to the velocity of the Earth around the Sun. It should not be supposed that such an enormous mass of explosives requires a commensurate amount of high-strength material for the vessels in which it is stored. In fact, hydrogen and oxygen in liquid form develop high pressure only if the vessels containing them are closed and if the gases themselves get heated due to the influence surrounding, comparatively warm bodies. [91] In the present case, the liquefied gases must have a free outlet to the tube into which they flow constantly in liquid form and where, on chemically combining, they explode.

The continuous and rapid flow of gases, corresponding to the evaporation of the liquids, cools the latter until their vapors exert hardly any pressure on the surrounding walls. Thus, the vessels containing the explosives need not be made very massive.

24. When the mass of the explosives is equal to the mass of the rocket $\left(\dfrac{M_2}{M_1}=1\right)$, the velocity of the latter is nearly twice as great as would be necessary for a stone or cannon ball, launched by "Selenians" from the surface of the Moon, to leave the Moon forever and become an Earth satellite, a second Moon.

This velocity (3920 m/sec) is nearly enough for bodies launched from the surface of Mars or Mercury to leave these planets forever.

If the mass ratio $\frac{M_2}{M_1}=3$ then, when the entire supply of explosives has been consumed, the projectile will have attained a velocity almost great enough to cause it to revolve like a satellite about the Earth beyond the limits of the atmosphere.

If $\frac{M_2}{M_1}=6$, the velocity of the rocket will be nearly great enough for it to leave Earth forever and revolve around the Sun like an independent planet. If the supply of explosives is big enough, the asteroid belt and even the heavy planets could be reached.

25. It is evident from the table that even if the supply of explosives is small, the final velocity of the projectile will still be adequate for practical purposes. Thus, if the fuel supply accounts for only 0.1 of the rockets weight, the velocity will be 543 m/sec, which is sufficient for the rocket to ascend to 15 km. The table also shows that if the supply is small, after completion of the explosion the velocity will be approximately proportional to the mass of the fuel (M_2); therefore, in this case, the maximum height will be proportional to the square of this mass (M_2). Thus, if the supply of explosives is equal to half the rocket's mass $\left(\frac{M_2}{M_1}\right)=0.5$, the rocket will fly far beyond the limits of the atmosphere.

Efficiency of Rocket During Ascent

26. It is of interest to determine the fraction of the total work done by the explosives, i.e., their chemical energy, that is transmitted to the rocket.

The work done by the explosives may be expressed as $\frac{V_1^2}{2}M_2$; the mechanical work done by a rocket with the velocity V may be expressed in the same unite: $\frac{V^2}{2}M_1$, or, on the basis of formula (16)

$$\frac{V^2}{2}M_1 = \frac{V_1^2}{2}M_1\left\{\ln\left(1+\frac{M_2}{M_1}\right)\right\}^2.$$

On dividing the work done by the rocket by the work done by the explosives, we obtain

$$\frac{M_2}{M_1}\left\{\ln\left(1+\frac{M_2}{M_1}\right)\right\}^2.$$

From this formula we can derive the table of energy utilization by the rocket.
From the table and the formula it is clear that when the amount of explosives is very

small, the utilization (efficiency) is equal to $\frac{M_2}{M_1}$ i.e., the smaller the relative amount of explosives.*

[93]

$\frac{M_2}{M_1}$	Utilization	$\frac{M_2}{M_1}$	Utilization
0.1	0.090	7	0.62
0.2	0.165	8	0.60
0.3	0.223	9	0.59
0.4	0.282	10	0.58
0.5	0.328	19	0.47
1	0.480	20	0.46
2	0.600	30	0.39
3	0.64	50	0.31
4	0.65	100	0.21
5	0.64	193	0.144
6	0.63	Infinity	Zero

[94] Further, as the relative amount of explosives increases, the utilization increases and reaches a maximum (0.65) when $M_2/M_1 = 4$.

Any further increase in the proportion of explosives will gradually but steadily reduce their utilization. When the supply of explosives is infinitely large, the utilization falls to zero, just as when the supply is infinitely small. It is also clear from the table that when M_2/M_1 ranges between 2 and 10 the utilization exceeds one half, i e., more than half the potential energy of the explosives is transmitted to the rocket in the form of kinetic energy. In general, from 1 to 20 it is extremely high and close to 0.5.

Rockets Under the Influence of Gravity. Vertical Ascent

27. We have determined the velocity acquired by the rocket in a gravitationless vacuum as a function of the mass of the rocket, the mass of the explosives and their energy of chemical combination.

We shall now consider the effect of gravity on the vertical motion of the projectile.

When not influenced by gravity, the rocket can acquire dizzy speeds and can utilize a considerable proportion of the explosive energy. This also holds true in a gravitational

---

*In fact, $\ln(1+x) = x - \frac{x^2}{2} + \frac{x^3}{3} - \frac{x^4}{4} ...$

Therefore, approximately, $\frac{M_1}{M_2}\left\{\ln\left(1+\frac{M_2}{M_1}\right)\right\}^2 = \frac{M_1}{M_2} \cdot \frac{M_2^2}{M_1^2} = \frac{M_2}{M_1}$.

environment, so long as the explosion is instantaneous. But this kind of explosion is not suitable for our purposes, since it would result in a lethal shock which could be endured neither by the projectile nor by the equipment and passengers. Clearly, we need a slow explosion; but if the explosion is slow, the useful effect diminishes and may even vanish.

Suppose the explosion is so weak that the resulting acceleration of the rocket is equal to the Earth's acceleration g. Then throughout the explosion the projectile will hang motionlessly in the air without support.

Of course, it will not acquire any velocity, and the utilization of the explosives, in whatever quantity they are present, will be zero. Thus, it is extremely important to analyze the effect of gravity on the projectile.

[95]   **Determining the Acquired Velocity. Examination of the Numerical Values Obtained. Maximum Height.**

When a rocket moves in a gravitationless medium, the time t during which its entire supply of explosives is consumed, is

$$t = \frac{V}{p}, \qquad (28)$$

where V is the velocity of the projectile on completion of the explosion, and p is the constant acceleration imparted to the rocket by the explosives per second.*

In this simple case of uniform acceleration the force of the explosion, i.e., the amount of fuel expended during the explosion per unit of time, will not be constant, but will continually diminish in proportion to the decrease in the mass of the projectile as its supply of explosives is depleted.

29. Knowing p, or the acceleration in a gravitationless medium, we can also determine the apparent (temporary) gravity inside the rocket during its acceleration or during the explosion.

Taking the force of gravity at the Earth's surface as unity, we find the temporary gravity to be p/g, where g is the Earth's acceleration; this formula shows how many times the pressure acting on the base of all the objects in the rocket exceeds the pressure that acts on the same objects when placed on the living room table under normal conditions. It is highly important to know the value of relative gravity in the projectile, since it affects the reliability of the instruments and the health of those setting out to explore the frontiers of space.

[96] 30. Under the influence of a constant or variable gravity of any intensity, the time taken to consume a given supply of explosives will be the same as when there is no gravity at all; it can be expressed by formula (28) or by the following formula:

$$t = \frac{V_2}{p-g}, \qquad (31)$$

where V_2 is velocity of the rocket on completion of the explosion in a gravitational medium with constant acceleration g. Here, of course, it is assumed that p and g are parallel and opposite; p-g expresses the visible acceleration of the projectile (with respect to the Earth), which is the result of two opposite forces: the force of the explosion and the force of gravity.

---

*It is assumed that the mass of the rocket varies in accordance with an exponential law; then the acceleration p due to the thrust will be constant (Editor's note).

32. The action of the force of gravity on the projectile in no way affects the relative gravity inside the projectile, and here the formula p/g still applies. For example, if p = 0, i.e., if there is no explosion, there is no temporary gravity, because p/g = 0. This means that if the explosion ceases and the projectile moves in some direction solely due to its own momentum and the gravitational attraction of the Sun, Earth, and other stars and planets, an observer inside the projectile will himself apparently be completely weightless, and not even the most sensitive spring balance will register when used to weigh any of the objects present inside the rocket along with the observer. On falling or rising inside the rocket under the action of inertia, even at the very surface of the Earth, the observer will not experience the least heaviness unless, of course, the projectile encounters some obstacle in the form of, say, the resistance of the atmosphere, water or solid earth.

33. If p = g, i.e., if the pressure of the exploding gases is equal to the weight of the projectile (p/g = 1), the relative gravity will be equal to terrestrial gravity. If it was stationary at the outset, the projectile will remain stationary throughout the explosion. If, however, the projectile had a certain (upward, lateral, downward) velocity before the explosion, this velocity will remain absolutely unchanged until the entire supply of explosives is consumed; thus the body, that is, the rocket, is balanced and moves as if by inertia in a [97] gravitational medium.

On the basis of formulas (28) and (31) we obtain

$$V = V_2 \left(\frac{p}{p-g} \right). \tag{34}$$

Hence, knowing the velocity V_2 that must be acquired by the projectile after the explosion, we can calculate V, from which, with the aid of formula (16), we can also determine the necessary amount of explosives M_2.

From equations (16) and (34) we obtain:

$$V_2 = -V_1 \left(1 - \frac{g}{p} \right) \cdot \ln \left(\frac{M_2}{M_1} + 1 \right). \tag{35}$$

36. From this, as from the preceding formula, it follows that the velocity acquired by the rocket is smaller in the presence of gravity than in its absence (16). The velocity V_2 may even be zero despite an abundant supply of explosives if (p/g) = 1, i.e., if the acceleration imparted to the projectile by the explosives is equal to the terrestrial acceleration, or if the gas pressure is equal and directly opposite to the effect of gravitational attraction (cf. formulas (34) and (35)).

In this case the rocket will stand motionless for a few minutes without rising and, once the supply of explosives has been consumed, will fall like a stone.

37. The greater the value of p in relation to g, the greater the velocity V_2 acquired by the projectile, given a specific amount of explosives M_2 (formula (35)).
[98] Therefore, if the aim is to climb higher, p must be made as large as possible, i.e., the explosion must be as energetic as possible. This, however, requires, first, a sturdier and more massive projectile and, second, sturdier equipment and instruments inside the projectile, because, according to (32), the relative gravity inside the projectile will be extremely large and, in particular, dangerous to any human observer aboard the rocket.

At any rate, on the basis of formula (35) in the limit

$$V_2 = -V_1 \cdot \ln \left(\frac{M_2}{M_1} + 1 \right),$$

i.e., if p is infinitely large or the explosion instantaneous, the velocity V_2 of the rocket in a gravitational environment will be the same as in a gravitationless environment.

According to formula (30), the explosion time is independent of gravity; it depends solely on the ratio M_2/M_1 and the rate of explosion p.

39. It is important to determine this rate. Suppose in formula (28) V = 11,100 m/sec (22) and p = g = 9.9 m/sec^2, then t = 1133 sec. This means that in a gravitationless medium the rocket would fly for less than 19 minutes with uniform acceleration, even if the amount of explosives were six times the mass of the projectile (22).

In the event of the explosion occurring at the surface of our planet, however, the rocket would stand motionless for the same period of time.

40. In $M_2/M_1 = 1$ then according to the table, V = 3920 m/sec; therefore t = 400 sec or $6\text{-}2/3$ min.

When $M_2/M_1 = 0.1$ V = 543 m/sec, t = 55.4 sec, i.e., less than a minute. In this case the projectile would stand motionless at the Earth's surface for $55\text{-}1/2$ sec.

Hence we can see that an explosion at the surface of a planet, and in general in any medium that is not free of gravity, may be completely ineffective—even if it occurs over a prolonged period of time—if it is of insufficient force; in fact, the projectile then [99] remains stationary and will have no translational velocity unless some has been previously acquired (It will then move over a certain distance at uniform speed). If this motion is upward, the projectile will do some work. If the original velocity is horizontal, the motion will also be horizontal; then no work will be done,* but the projectile could serve the same purpose as a locomotive, steamship or stearable aerostat. The projectile could function in this way only for a few minutes, while the explosion takes place, but even during such a short period of time it could traverse considerable distances, particularly when moving above the atmosphere. However, we do not consider that the rocket is of any practical value for flights through the air.

The time during which a projectile can remain in a gravitational medium proportional to g, i.e., to the force of gravity.

Thus, on the Moon the projectile could stand motionless, without support, for 2 hours if $\frac{M_2}{M_1} = 6$.

41. In formula (35) for an environment with $\frac{g}{p} = 10$ let us put $\frac{M_2}{M_1} = 1$; we than calculate V_2 = 9990 m/sec. Accordingly, the relative gravity will be 10 g, i.e., throughout the explosion time (about 2 min), a person weighing 70 kg will experience gravity ten times as great as on Earth, and, on a spring balance, will weigh 700 kg. This gravity can be safely endured by the traveler only if he observes special precautions: if he is immersed in a special fluid, under special conditions.

On the basis of formula (28) we can also calculate the explosion time, or the duration of the period of intensified gravity; we obtain 113 sec, i.e., less than 2 min. This is very little, and it is amazing that in such a negligible interval of time a projectile could acquire a velocity nearly sufficient to leave the Earth and move around the Sun like a new planet.

We found V_2 = 9990 m/sec, i.e., a velocity only slightly less than the velocity V acquirable in a gravitationless medium under the same explosion conditions (22).

But since during the explosion the projectile also climbs to a [100] certain height, the idea suggests itself that the total work done by the explosives is not less than in a gravitationless medium.

---

*If no allowance is made for the work done in overcoming atmospheric drag (Editor's note).

44. We shall now consider this question.

The acceleration of the projectile in a gravitational medium may be expressed as: $p_1 = p - g$.

At a distance of not more than several hundred versts from the Earth's surface we can assume that g is constant; this does not introduce any appreciable error, and moreover the error will be on the safe side, i.e., the actual figures will be more favorable than those calculated.

The height h reached by the projectile during time t (explosion time) will be

$$h = \frac{1}{2}p_1 t^2 = \frac{p-g}{2} \cdot t^2. \qquad (45)$$

Eliminating t, in accordance with equation (31) we obtain

$$h = \frac{V_2^2}{2(p-g)}, \qquad (46)$$

where V_2 is the velocity of the projectile in a gravitational medium after the entire supply of explosives has been consumed. Now, on eliminating V_2, from (34) and (46) we obtain

$$h = \frac{p-g}{2p^2} \cdot V^2 = \frac{V^2}{2p}\left(1 - \frac{g}{p}\right), \qquad (47)$$

where V. is the velocity acquired by the rocket in a gravitationless medium.

[101] **Efficiency**

The useful work done by the explosives in such a medium may be expressed by:*

$$T = \frac{V^2}{2g}. \qquad (48)$$

On the other hand, depending on the height reached by the projectile and its velocity at the end of the explosion, the work T_1 done in a gravitational medium may be expressed as

$$T_1 = h + \frac{V_2^2}{2g}. \qquad (49)$$

The ratio of $\frac{T_1}{T}$ (T being the ideal value) is thus

$$\frac{T_1}{T} = \frac{2hg + V_2^2}{V^2}. \qquad (50)$$

On eliminating h and V by means of formulas (46) and (34), we find

---

*The calculations in formulas (48) and (49) are for a projectile with a weight equal to unity (Editor's note).

[102]
$$\frac{T_1}{T} = \left(1 - \frac{g}{p}\right), \qquad (51)$$

i.e., the work done in a gravitational medium by a given mass of explosives M2 is less than in a gravitationless medium: this difference $\frac{g}{p}$ is the smaller the higher the exhaust velocity of the gases or the greater the pressure p. For example, under the conditions of note 41 the loss is only $1/10$, while the utilization, according to (51), is 0.9. When p = g, or when the projectile hovers in the air, lacking even a constant velocity, the loss will be complete (1) and the utilization will be zero. The utilization will also be zero if the projectile has a constant horizontal velocity.

52. In note 41 we found V_2 = 9990 m/sec. Applying formula (46) to this case, we find h = 565 km; this means that during the explosion the projectile will travel far beyond the limits of the atmosphere and at the same time acquire a velocity of 9990 m/sec.

Note that this velocity is less than that in a gravitationless medium by 1110 m/sec or exactly $1/10$ of the velocity in a gravitationless medium (22).

Hence it is clear that the loss of velocity obeys the same law as the loss of work (51). Strictly, this also follows from formula (34) which, after transformation, yields

$$V_2 = V\left(1 - \frac{g}{p}\right), \text{ or } V - V_2 = V \cdot \frac{g}{p}.$$

From (51) we find

$$T = T_1 \cdot \left(\frac{p}{p-g}\right), \qquad (56)$$

where T_1 is the work done on the projectile by the explosives in a [103] gravitational medium with an acceleration equal to g.

In order for the projectile to perform the necessary work of climbing, overcoming atmospheric resistance, and acquiring the desired velocity, the total work done most equal T_1.

Having calculated all these forms of work, we find T from formula (56). Knowing T, we can calculate V, i.e., the velocity in a gravitationless medium, from the formula

$$T = M_1 \cdot \frac{V^2}{2g}.$$

Knowing V, we can also calculate the required mass of explosives from formula (16). Thus, we find

$$M_2 = M_1 \left[e^{\sqrt{\frac{T_1 p}{T_2(p-g)}}} - 1 \right].$$

In the calculations, for the sake of brevity, $\left(M_1 \frac{V_1^2}{2g}\right)$ has been replaced by T_2.

Thus, knowing the mass of the projectile M1, together with all it contains apart from the fuel M_2 the mechanical work T_2 done by explosives when their mass is equal to that of the projectile M_1, the work T_1 which must be done by the projectile during its vertical ascent, the acceleration due to the explosion p and gravity g, we can also determine the amount of explosives M2 required to lift the mass M1 of the projectile.

The ratio $\dfrac{T_1}{T_2}$ in the formula will not change if we reduce it by [104] M_1, so that T_1 and T_2 may be construed as the mechanical work T_1 done by a unit mass of the projectile and the mechanical work T2 done by a unit of explosives, respectively.

In general, the gravity g may be construed as the sum of the accelerations due to gravity and the resistance of the medium. But gravity steadily decreases with increasing distance from the Earth's center, so that an increasing fraction of the mechanical work of the explosives is utilized. On the other hand, atmospheric resistance, while very insignificant in comparison with the weight of the projectile, as we shall see, reduces the utilization of the energy of the explosives.

Further, it can be seen that the latter losses, which continue for some time as the projectile races through the atmosphere, are abundantly offset by the gain due to the decrease in gravitational attraction at the considerable distances (500 km) at which the explosion ceases.

Thus, formula (20) can be boldly applied to the vertical flight of a projectile, despite the complications due to the variation in gravity and the resistance of the atmosphere g = 9.8.

Gravitational Field. Vertical Return to Earth

59. First let us consider the process of stopping in a gravitationless medium or a momentary halt in a gravitational medium.

Suppose, for example, that, owing to the force produced by the explosion of some (not all) of the gases, a rocket acquires a velocity of 10,000 m/sec (22). Now in order to stop it, we must give it the same velocity but in the opposite direction. Clearly, in accordance with (22), the remaining amount of explosives must be five times greater than the mass M1 of the projectile. Therefore, on completion of the first part of the explosion (in order to acquire translational velocity) the projectile must have a supply of explosives, the mass of which may be expressed as $5M_1 = M_2$.

60. The total mass including the explosives will be $M_2 + M_1 = 5M_1 + M_1 = 6M_1$. This mass $6M_1$ must have been given a velocity of 10,000 m/sec by the original explosion, and this requires an additional amount of explosives which should also be five times greater (22) than [105] the mass of the projectile plus the mass of the explosives needed to stop the rocket, i.e., $6M_1 \times 5$; thus we obtain $30\ M_1$ which, together with the explosives needed for stopping the rocket, makes $35\ M_1$.

Using the symbol $q = \dfrac{M_2}{M_1}$ to denote the number of times the mass of the explosives exceeds the mass of the projectile, we may express as follows the above reasoning concerning the total mass of explosives $\dfrac{M_3}{M_1}$ needed to acquire and annihilate a given velocity as follows:

$$\frac{M_3}{M_1} = q + (1+q) \cdot q = q(2+q),$$

or, adding and subtracting one from the second part of this equation, we obtain

$$\frac{M_3}{M_1} = 1 + 2q + q^2 - 1 = (1+q)^2 - 1. \qquad (61)$$

whence we find

$$\frac{M_3}{M_1} + 1 = (1+q)^2. \qquad (62)$$

This last expression is easy to remember.

If q is very small, the amount of explosives is approximately 2q (because q^2 will be negligible), i.e., twice as much as needed solely for acquiring a given velocity.

63. On the basis of the above formulas and table (22) we compile the following table:

[106]

V, m/sec	$\frac{M_2}{M_1}$	$\frac{M_3}{M_1}$	V, m/sec	$\frac{M_2}{M_1}$	$\frac{M_3}{M_1}$
543	0.1	0.21	11 800	7	63
1 037	0.2	0.44	12 500	8	80
1 493	0.3	0.69	13 100	9	99
1 915	0.4	0.96	13 650	10	120
2 308	0.5	1.25	17 100	19	399
3 920	1	3	17 330	20	440
6 260	2	8	19 560	30	960
7 880	3	15	22 400	50	2 600
9 170	4	24	26 280	100	10 200
10 100	5	35	30 038	193	37 248
11 100	6	48			

[107] It is evident from this table that if we wanted to acquire and lose a very high velocity an impossibly large supply of explosives would be needed.

From (62) and (16) we have

$$\frac{M_3}{M_1} + 1 = e^{\frac{-2V}{V_1}}, \text{ or } \frac{M_3}{M_1} = e^{\frac{-2V}{V_1}} - 1.$$

Note that the radio $-\frac{2V}{V_1}$ is positive, because the velocities of the projectile and the gases are opposite in direction and therefore differ in sign.

64. If we are in a gravitational medium, then, in the simple case of vertical motion, the process of coming to a halt descending to Earth will be as follows: when, owing to its acquired velocity, the rocket has risen to a certain altitude and stopped there, its earthward fall will begin.

When the projectile reaches the point in its flight where the action of the explosives ceased, it is subjected again to the action of the remainder in the case direction and order. Clearly, when the explosives cease to act and the entire supply is consumed, the rocket will come to a halt at the Earth's surface, whence the flight began. The method of ascent is exactly the same as the method of descent, the only difference being that the velocities are reversed at every point along the path.

Coming to a halt in a gravitational field requires more work and explosives than in a gravitationless medium, and therefore q [in formulas (61) and (62)] must be greater.

Denoting this greater value of q by q_1, on the basis of the foregoing, we find that

$$\frac{q}{q_1} = \frac{T_1}{T} = 1 - \frac{g}{p}, \tag{65}$$

whence

[108]
$$q_1 = q\left(\frac{p}{p-g}\right);$$

substituting q_1 for q in equation (62), we obtain

$$\frac{M_4}{M_1} = (1+q_1)^2 - 1 = \left(1+\frac{pq}{p-g}\right)^2 - 1, \tag{66}$$

here M_4 denotes the amount or mass of explosives needed to ascend from a given point and return to the same point for a rocket coming to a complete stop and traveling in a gravitational medium.

67. On the basis of this last formula we can compile the following table, assuming that p/g = 10, i.e., that the pressure of the explosives is 10 times greater than the weight of the rocket together with the remaining explosives.

Gravitational Field. Oblique Ascent

68. Although a vertical ascent would appear to be more expedient, since the atmosphere is then traversed more rapidly and the projectile rises to a greater height, the work done in rising through the atmosphere is very insignificant compared with the total work done by the explosives and, moreover, given an oblique ascent it is possible to construct a permanent observatory that would travel for an indeterminate length of time around the Earth, like the Moon, beyond the limits of the atmosphere. Furthermore, and most important, in an oblique ascent far more of the explosive energy is utilized than in a vertical ascent.

Let us first consider the special case of horizontal rocket flight [Fig. 2].

[109] **In a Gravitational Field**

V, m/sec	$\dfrac{M_2}{M_3}$	$\dfrac{M_4}{M_1}$
543	0.1	0.235
1 497	0.3	0.778
2 308	0.5	1.420
3 920	1.0	4.457
6 260	2	9.383
7 880	3	17.78
9 170	4	28.64
10 100	5	41.98
11 100	6	57.78
11 800	7	76.05

Denoting by R the resultant of the horizontal acceleration of the rocket, by p the acceleration due to the explosion, and by g the acceleration due to gravity, we have

$$R = \sqrt{p^2 - g^2}. \qquad (70)$$

[110] On the basis of the latter formula,* the kinetic energy acquired by the projectile during time t equals

$$\frac{R}{2} \cdot t^2 \cdot \left(\frac{R}{g}\right) = \frac{R^2}{2g} \cdot t^2 = \frac{p^2 - g^2}{2g} \cdot t^2, \qquad (71)$$

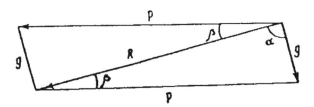

Figure 2

where t is the explosion time. This is also the total useful work done on the rocket. In fact, if we assume the direction of gravity to be constant (which in practice is true only for a short trajectory) the rocket does not climb at all. The work done by the explosives on the rocket in a gravitationless medium is**

---

*Here Tsiolkovskiy calculates the work done by the resultant referred to unit weight of the rocket (Editor's note).
**Tsiolkovskiy calculates the work done by the reaction force referred to unit weight of the rocket (Editor's note).

$$\frac{p}{2}t^2\frac{p}{g} = \frac{p^2 t^2}{2g}. \qquad (72)$$

Dividing the useful work (71) by the total work (72), we obtain [111] the efficiency for horizontal flight.

$$(\frac{p^2-g^2}{2g}\cdot t):(\frac{p^2}{2g}\cdot t) = 1-(\frac{g}{p})^2. \qquad (73)$$

As before, the air resistance has not yet been taken into account.

From this last formula it is evident that the loss of work as compared with a gravitationless medium may be expressed by $(\frac{g}{p})^2$. Hence it follows that this loss is much smaller than during a vertical ascent. Thus, for example, if $\frac{g}{p}=1/10$, the loss will be 1/100 or 1%, whereas in a vertical ascent it would be expressed by $\frac{g}{p}$, i.e., would equal 1/10, that is, 10%.

74. Here is a table in which β is the angle of inclination of the force p to the horizon.

Horizontal Motion

$\frac{p}{g}$	$(\frac{g}{p})^2$	$\frac{g}{p}$	β°
1	1	1	90
2	1 : 4	1 : 2	30
3	1 : 9	1 : 3	19.5
4	1 : 16	1 : 4	14.5
5	1 : 25	1 : 5	11.5
10	1 : 100	1 : 10	5.7
100	1 : 10 000	1 : 100	0.57

[112] **Oblique Ascent. Work done in Lifting the Projectile Referred to the Work in a Gravitationless Medium. Loss of Work.**

75. Now let us solve the general problem—for any angle of inclination of the resultant R. A horizontal trajectory or resultant is undesirable, since a projectile flying horizontally must travel a vastly greater distance through the atmosphere and do a correspondingly greater amount of work in cutting through the air.

Thus, let us keep in mind that a, the angle of inclination of the resultant to the vertical, is greater than a right angle; we have

$$R = \sqrt{p^2+g^2+2pg\cos\gamma}, \qquad (76)$$

where γ = α + β (obtuse angle of parallelogram) in accordance with the sketch.

Further
$$\gamma = \alpha + \beta\,;\ \sin\alpha : \sin\beta : \sin\gamma = p : g : R \qquad (77)$$
and
$$\cos\alpha = \frac{R^2 + g^2 - p^2}{2Rg}. \qquad (78)$$

The kinetic energy is expressed by the formula (71), where R is found from equation (76). The vertical acceleration of the resultant R

$$R_1 = \sin(\alpha - 90°)R = -\cos\alpha R. \qquad (79)$$

[113] Therefore, the work done in lifting the projectile will be

$$\frac{R_1}{2}t^2 = \frac{-\cos\alpha}{2}Rt^2, \qquad (80)$$

where t is the explosion time for the entire supply of explosives. The total work done on the projectile in a gravitational medium [in accordance with (71) and (80)]

$$T_1 = \frac{R^2}{2g}t^2 - \frac{Rt^2\cos\alpha}{2} = \frac{Rt^2}{2}\left(\frac{R}{g} - \cos\alpha\right). \qquad (81)$$

Here ascent of the projectile through unit height in a medium with an acceleration of one g is taken as the unit of work. If $\alpha \rangle\rangle 90°$, in the case of the ascent of the projectile, for example, then $(-\cos\alpha)$ is positive, and vice versa.

In a gravitationless medium the work will be $\dfrac{p^2}{2g}t^2 = T$ in accordance with (72), (let us not forget that the explosion time t is independent of the gravitational forces).

Taking the ratio of these two values of the work, we obtain the efficiency of the explosion as compared with its efficiency in a gravitationless medium, namely:

$$\frac{T_1}{T} = \frac{Rt^2}{2}\left(\frac{R}{q} - \cos\alpha\right) : \left(\frac{p^2}{2g}t^2\right) = \frac{R}{p}\left(\frac{R}{p} - \frac{g}{p}\cos\alpha\right). \qquad (82)$$

Eliminating R in accordance with formula (76), we find

[114]
$$\frac{T_1}{T} = 1 + \frac{g^2}{p^2} + 2\cos\gamma\cdot\frac{g}{p} - \cos\alpha\cdot\frac{g}{p}\sqrt{1 + \frac{g^2}{p^2} + 2\cos\gamma\frac{g}{p}}.$$

Formulas (51) and (73), for example, are merely special cases of this formula, as may be readily ascertained.

84. We shall now find a use for this formula. Assume that a rocket is ascending at an

angle of 14.5° to the horizon; the sine of this angle is 0.25; this means that the atmospheric resistance is four times greater than the value for vertical flight, since it is more or less inversely proportional to the sine of the angle of inclination $(\alpha - 90°)$ of the trajectory to the horizontal.

85. The angle $\alpha = 90 + 14\,^1/_2 = 104\,^1/_2°$; $\cos \alpha = 0.25$; knowing α we can also find β. In fact, from (77) we find

$$\sin \beta = \sin \alpha \frac{g}{p} ;$$

thus, if $\frac{g}{p} = 0.1$;

$$\sin \beta = 0.0968; \beta = 5\,^1/_2°,$$

whence

$$\gamma = 110°, \cos \gamma = 0.342.$$

Now we calculate the efficiency to be 0.966. The loss is 0.034 or about $^1/_{20}$ or, more accurately, 3.4%.

This loss is one-third of the loss in a vertical ascent, not a bad result, especially if we consider that, even in an oblique ascent $(14\,^1/_2°)$, the atmospheric resistance is still less than 1% of the work done in lifting the projectile.

[115] **Oblique Motion**

Degrees				Utilization	Loss
$\alpha-90$	α	β	$\gamma=\alpha+\beta$		
0	90	$5\,^3/_4$	$95\,^2/_3$	0.9900	1 : 100
2	92	$5\,^2/_3$	$97\,^2/_3$	0.9860	1 : 72
5	95	$5\,^2/_3$	$100\,^2/_3$	0.9800	1 : 53
10	100	$5\,^2/_3$	$105\,^2/_3$	0.9731	1 : 37
15	105	$5\,^1/_2$	$110\,^1/_2$	0.9651	1 : 29
20	110	$5\,^1/_3$	$115\,^1/_3$	0.9573	1 : 23.4
30	120	5	125	0.9426	1 : 17.4
40	130	$4\,^1/_3$	$134\,^1/_3$	0.9300	1 : 14.3
45	135	4	139	0.9246	1 : 13.3
90	180	0	180	0.9000	1 : 10

[116] 86. We propose the above table for various approaches: the first column shows the inclination to the horizontal; the last column, the loss of work; β is the deviation of the direction of the pressure exerted by the explosives from the actual line of motion (69).

87. For very small angles of inclination $(\alpha-90°)$ the formula can be much simplified, by replacing the trigonometric values by their arcs and making other simplifications.

We then obtain the following expression for the loss of work:

$$x^2 + \delta x (1-\frac{x^2}{2})+\delta^2 x^2 (x-\frac{\delta}{2}),$$

where δ denotes the angle of inclination ($\alpha - 90°$), expressed as the length of its arc, the radius of white is equal to unity, and x denotes the ratio g/p. On discarding the small quantities of higher orders, we obtain for the loss

$$x^2 + \delta x = (\frac{g}{p})^2 + \delta \frac{g}{p}.$$

Let us put $\delta = 0.02$ N, where 0.02 is the part of a circle corresponding to roughly 1° (1 $^1/_7$) and N is the number of these new degrees. Then the loss of work may be roughly expressed as

$$\frac{g^2}{p^2}+0.02\frac{g}{p}N.$$

From this formula we can readily compile the following table, assuming that

$$\frac{g}{p}=0.1:$$

[117]

N	0	0.5	1	2	3	4	5	6	10
Loss	1/100	1/91	1/83	1/70	1/60	1/55	1/50	1/45	1/33

Hence we see that even at large angles (up to 10°) the discrepancy between this table and the previous, more accurate one is quite small.

We could have considered many other factors too: the work done by gravity, the resistance of the atmosphere; we still have not explained how the explorer could spend a long, even unlimited, time in an environment without even a trace of oxygen. We have not even mentioned the heating of the projectile during its short flight through the atmosphere, nor have we given a general picture of the flight itself and of the extremely interesting phenomena that would (theoretically) accompany it. We have scarcely outlined the magnificent prospects of eventually attaining this still distant goal. Lastly, we could also have considered the subject of rocket trajectories in outer space.

Document I-6

Document title: Hermann Oberth, Rockets in Planetary Space, Translation of "Die Rakete zu den Planeträumen," Verlag von R. Oldenbourg, Munich and Berlin, 1923, NASA TT F-9277, December 1964, Introduction.

Hermann Oberth was born in Transylvania, but considered himself German. In this publication he was the first person to outline many of the fundamentals of spaceflight and to ground them in mathematics and engineering. He proposed that a rocket could be launched to orbit the Earth and that it could travel through the vacuum of space. He also addressed the subject of various rocket fuels and proposed a large rocket that used alcohol and liquid hydrogen as propellant. Oberth's work served as the inspiration for many later pioneers in Germany, particularly Wernher von Braun.

Oberth later refined his proposals in the book *Wege zur Raumschiffahrt* (*Ways to Spaceflight*), which later served as the basis for the spaceship depicted in Fritz Lang's 1929 motion picture *Frau im Mond* (*The Girl in the Moon*). This was the public's first exposure to a realistic spacecraft on the movie screen.

---

[1] **Section 1. Introduction**

1. Given the present state of science and technology, it is possible to build machines which can climb higher than the limits of the atmosphere of the earth.

2. With additional refinement, these machines will be able to attain such velocities that, left to themselves in space, they need not fall back to the earth's surface, and they may even leave the field of gravitation of the earth.

3. These machines can be so constructed that men can be lifted in them, apparently with complete safety.

4. Given certain economical situations, the construction of such machines may even become profitable. Such conditions may prevail within a few decades.

In this work, I intend to prove these four statements. I will first derive some formulas which will give us the necessary theoretical insight into the manner of functioning and the performance capability of these machines. In Part II, I will show that their construction is technically possible, and in Part III I will come to a discussion of the prospects for their invention.

I have strived to be brief. I have been able frequently to simplify the mathematical derivations and formulas by using approximated values, which are easy to use mathematically, for certain quantities. This procedure was used especially when, in the course of a discussion, the facts of a matter could be made more clear. (Incidentally, I have also frequently indicated the actual value of the result, or at least shown how it could be determined from the approximated value, and sometimes I have simply estimated the error.) Technical problems, the solution of which no one doubts, have been covered only briefly. In Part III, I have limited myself to indications, since the subjects treated here still lie rather far off.

It has been my purpose here to cover no more than seemed necessary for an understanding of the invention and for an evaluation of the feasibility, because:

Firstly, it is by no means my intent here to describe a particular model of a machine with all its details, but only to show that machines of this sort are possible. (For example, I need not calculate the exact altitude which a certain rocket might reach if I can show that it is at least possible of meeting the minimum demands placed upon it. Thus I set a constant value c on [2] the exhaust velocity (cf. page 3), even though this value can vary in some cases by as much as 9%, and I discussed the case in which the rocket travels at a velocity of v (cf. page 6), even though the fuel is not consumed most efficiently at this speed. If I estimate the power of the rocket, based upon v and the most unfavorable value of c, and find that the rocket is capable under these circumstances of attaining a required final velocity and altitude, then I have also shown that, in actuality, it can surely attain them.) I believe that the entire picture is clearer if I do not go into too much detail.

Secondly, there are some things which I wish not to reveal (particularly technical solutions which appear favorable), because these are not protected literary property. If my ideas should one day be put into practice, I will naturally want to furnish the exact plans, computations and methods of computation.

Finally, I make no secret of the fact that I consider some of the provisions, in their present form, as by no means being definitive solutions. As I worked out my plans and computations, I naturally had to consider each detail. In so doing, I could at least determine

86   PRELUDE TO THE SPACE AGE

mine that there were no insurmountable difficulties. At the same time, however, it was clear to me that some individual questions could be solved only after the most basic special studies and experimentation lasting perhaps years, at least if the optimum solution were sought.

Document I-7

Document title: Robert H. Goddard, *A Method of Reaching Extreme Altitudes*, Smithsonian Miscellaneous Collections, Volume 71, Number 2 (Washington, DC: Smithsonian Institution Press, 1919). The plates have been omitted from these documents.

Document I-8

Document title: "Topics of the Times," *New York Times*, January 18, 1920, p. 12.

Even before Robert Goddard retreated to New Mexico and began conducting most of his research in seclusion, he rarely published, mostly because of the skepticism and even outright ridicule reflected in the *New York Times* story printed here. His paper, *A Method of Reaching Extreme Altitudes*, was published as part of the Smithsonian Institution's Miscellaneous Collections series, and was a relatively standard scientific publication that would impress colleagues but few others. The first edition was bound in brown paper and numbered 1,750 copies, of which Goddard received 90 complimentary ones. The publication went unnoticed for eight days before suddenly becoming front-page news in several newspapers, including the *Boston American*, the *New York Times*, the *Milwaukee Sentinel*, and the *San Francisco Examiner*. The stories focused exclusively on the most esoteric part of the study, a proposal for traveling to the Moon, which had been played up in an accompanying press release from the Smithsonian. The furor in response to this proposal angered Goddard, particularly since he felt his concept of a rocket itself was being maligned. The controversy did attract the attention of a misinformed editor of the *New York Times*, who derided Goddard's lack of knowledge about ordinary physics. Contrary to the editor's claims, Goddard's speculation on the operation of a rocket in a vacuum was widely accepted at the time, and proved sound in later application.

Document I-7

Preface

The theoretical work herein presented was developed while the writer was at Princeton University in 1912-1913, the basis of the calculations being the assumption that, if nitrocellulose smokeless powder were employed as propellant in a rocket, under such conditions as are here explained, an efficiency of 50 percent might be expected.

Actual experimental investigations were not undertaken until 1915-1916, at which time the tests concerning ordinary rockets, steel chambers and nozzles, and trials *in vacuo*, were performed at Clark University. The original calculations were then repeated, using the data from these experiments, and both the theoretical and experimental results were submitted, in manuscript, to the Smithsonian Institution, in December 1916. This manuscript is here presented in the original form, save for the notes at the end which are now added.

A grant of $5000 from the Hodgkins Fund, Smithsonian Institution, under which work is being done at present, was advanced toward the development of a reloading, or multiple-charge rocket, herein explained in principle, and this work was begun at the Worcester Polytechnic Institute in 1917, and was later undertaken as a war proposition. It was continued, from June 1918 up to very nearly the time of signing of the Armistice, at the Mount Wilson Observatory of the Carnegie Institution of Washington, where most of the experimental results were obtained....

[1] **Outline**

A search for methods of raising recording apparatus beyond the range for sounding balloons (about 20 miles) led the writer to develop a theory of rocket action, in general, taking into account air resistance and gravity. The problem was to determine the minimum initial mass of an ideal rocket necessary, in order that on continuous loss of mass, a final mass of one pound would remain, at any desired altitude.

An approximate method was found necessary, in solving this problem, in order to avoid an unsolved problem in the calculus of variations. The solution that was obtained revealed the fact that surprisingly small initial masses would be necessary (Table VII) *provided the gases were ejected from the rocket at a high velocity*, and also provided that *most of the rocket consisted of propellant material*. The reason for this is, very briefly, that the velocity enters *exponentially* in the expression for the initial mass. Thus if the velocity of the ejected gases be increased fivefold, for example, the initial mass necessary to reach a given height will be *reduced to the fifth root* of that required for the lesser velocity. (A simple calculation shows at once the effectiveness of a rocket apparatus of high efficiency.)

It was obviously desirable to perform certain experiments: First, with the object of finding just how inefficient an ordinary rocket is, and second, to determine to what extent the efficiency could be increased in a rocket of new design. The term "efficiency" here means the ratio of the kinetic energy being calculated from the average velocity of ejection, which was obtained indirectly by observations on the *recoil* of the rocket.

It was found that not only does the powder in an ordinary rocket constitute but a small fraction of the total mass ($1/4$ or $1/5$), but that, furthermore, the efficiency is only 2 percent, the average velocity of ejection being about 1000 ft/sec (Table I). This was true [2] even in the case of the Coston ship rocket, which was found to have a range of a quarter of a mile.

Experiments were next performed with the object of increasing the average velocity of ejection of the gases. Charges of dense smokeless powder were fired in strong steel chambers, these chambers being provided with smooth tapered nozzles, the object of which was to obtain the work of expansion of the gases, much as is done in the de Laval steam turbine. The efficiencies and velocities obtained in this way were remarkably high (Table II), the highest efficiency, or rather "duty," being over 64 percent, and the highest average velocity of ejection being slightly under 8000 ft/sec, which exceeds any velocity hitherto attained by matter in appreciable amounts.

These velocities were proved to be real velocities, and not merely effects due to reaction against the air, by firing the same steel chambers *in vacuo*, and observing the recoil. The velocities obtained in this way were not much different from those obtained in air (Table III).

It will be evident that a heavy steel chamber, such as was used in the above-mentioned experiments, could not compete with the ordinary rocket, even with the high velocities which were obtained. If, however, *successive charges* were fired in the *same chamber*, much as in a rapid-fire gun, *most of the mass of the rocket could consist of propellant*, and the superiority over the ordinary rocket could be increased enormously. Such reloading mechanisms, together with what is termed a "primary and secondary" rocket principle, are the subject of certain United States Patents. Inasmuch as these two features are self-evidently operative, it was not considered necessary to perform experiments concerning them, in order to be certain of the practicability of the general method.

Regarding the heights that could be reached by the above method: an application of the theory to cases which the experiments show must be realizable in practice indicates that a mass of one pound could be elevated to altitudes of 35, 72, and 232 miles, by employing initial masses of from 3.6 to 12.6, from 5.1 to 24.3, and from 9.8 to 89.6 lb, respectively

(Table VII). If a device of the Coston ship-rocket type were used instead, the initial masses would be of the order of magnitude of *those above, raised to the 27th power*. It should be understood that if the mass of the recording instruments alone were one pound, the entire final mass would be 3 or 4 pounds.

[3] Regarding the possibility of recovering apparatus upon its return, calculations show that the times of ascent and descent will be short, and that a small parachute should be sufficient to ensure safe landing.

Calculations indicate, further, that with a rocket of high efficiency, consisting chiefly of propellant material, it should be possible to send small masses even to such great distances as to escape the earth's attraction.

In conclusion, it is believed that not only has a new and valuable method of reaching high altitudes been shown to be *operative in theory*, but that the experiments herein described *settle all the points upon which there could be reasonable doubt*.

The following discussion is divided into three parts: Part I, theory; Part II, experiments; Part III, calculations, based upon the theory and the experimental results.

Importance of the Subject

The greatest altitude at which soundings of the atmosphere have been made by balloons, namely, about 20 miles, is but a small fraction of the height to which the atmosphere is supposed to extend. In fact, the most interesting, and in some ways the most important, part of the atmosphere lies in this unexplored region, a means of exploring which has, up to the present, not seriously been suggested.

A few of the more important matters to be investigated in this region are the following: the density, chemical constitution, and temperature of the atmosphere, as well as the height to which it extends. Other problems are the nature of the aurora, and (with apparatus held by gyroscopes in a fixed direction in space) the nature of the α, β, and γ radioactive rays from matter in the sun as well as the ultraviolet spectrum of this body.

Speculations have been made as to the nature of the upper atmosphere—those by Wegener[1] being, perhaps, the most plausible. By estimating the temperature and percentage composition of the gases present in the atmosphere, Wegener calculates the partial pressures of the constituent gases, and concludes that there are four rather distinct regions or spheres of the atmosphere in which certain gases predominate: the troposphere, in which are the clouds; the stratosphere, predominatingly nitrogen; the hydrogen sphere; and the [4] geocoronium sphere. This highest sphere appears to consist essentially of an element, "geocoronium," a gas undiscovered at the surface of the earth, having a spectrum which is the single aurora line, 557 $\mu\mu$, and being 0.4 as heavy as hydrogen. The existence of such a gas is in agreement with Nicholson's theory of the atom, and its investigation would, of course, be a matter of considerable importance to astronomy and physics as well as to meteorology. It is of interest to note that the greatest altitude attained by sounding balloons extends but one-third through the second region, or stratosphere.

No instruments for obtaining data at these high altitudes are herein discussed, but it will be at once evident that their construction is a problem of small difficulty compared with the attainment of the desired altitudes.

[5] ## Part I. Theory

Method to Be Employed

It is possible to obtain a suggestion as the method that must be employed from the fundamental principles of mechanics, together with a consideration of the conditions of the problem. We are at once limited to an apparatus which reacts against matter, this matter being carried by the apparatus in question. For the entire system we must have: First,

---

1 A. Wegener, *Phys. Zeitscher.* 12, pp. 170-178, 214-222, 1911.

action in accordance with Newton's third law of motion; and, second, energy supplied from some source or sources must be used to give kinetic and potential energy to the apparatus that is being raised; kinetic energy to the matter which, by reaction, produces the desired motion of the apparatus; and also sufficient energy to overcome air resistance.

We are at once limited, since subatomic energy is not available, to a means of propulsion in which jets of gas are employed. This will be evident from the following consideration: First, the matter which, by its being ejected furnishes the necessary reaction, must be taken with the apparatus in reasonably small amounts. Second, energy must be taken with the apparatus in as large amounts as possible. Now, inasmuch as the maximum amount of energy associated with the minimum amount of matter occurs with chemical energy, both the matter and the energy for reaction must be supplied by a substance which, on burning or exploding, liberates a large amount of energy, and permits the ejection of the products that are formed. An ideal substance is evidently smokeless powder, which furnishes a large amount of energy, but does not explode with such violence as to be uncontrollable.

The apparatus must obviously be constructed on the principle of the rocket. An ordinary rocket, however, of reasonably small bulk, can rise to but a very limited altitude. This is due to the fact that the part of the rocket that furnishes the energy is but a rather small fraction of the total mass of the rocket; and also to the fact that only a part of this energy is converted into kinetic energy of the mass which is expelled. It will be expected, then, that the ordinary rocket is an inefficient heat engine. Experiments will be described below which show that this is true to a surprising degree.

[6] By the application of several new principles, an efficiency manyfold greater than that of the ordinary rocket is possible; experimental demonstrations of which will also be described below. Inasmuch as these principles are of some value for military purposes, the writer has protected himself, as well as aerological science in America, by certain United States Patents, of which the following have already been issued: Nos. 1,102,653, 1,103,503, 1,191,299, 1,194,496.

The principles concerning efficiency are essentially three in number. The first concerns thermodynamic efficiency, and is the use of a smooth nozzle, of proper length and taper, through which the gaseous products of combustion are discharged. By this means the work of expansion of the gases is obtained as kinetic energy, and also complete combustion is ensured.

The second principle is embodied in a reloading device, whereby a large mass of explosive material is used, a little at time, in a small, strong, combustion chamber. This enables high chamber pressures to be employed, impossible in an ordinary paper rocket, and also permits most of the mass of the rocket to consist of propellant material.

The third principle consists in the employment of a primary and secondary rocket apparatus, the secondary (a copy in miniature of the primary) being fired when the primary has reached the upper limit of its flight. This is most clearly shown, in principle, in U.S. Patent No. 1,102,653.

By this means the large ratio of propellant material is total mass is kept sensibly the same during the entire flight. This last principle is obviously to avoid damage when the discarded casings reach the ground, each should be fitted with a parachute device, as explained in U.S. Patent No. 1,191,299.

Experiments will be described below which show that, by application of the above principles, it is possible to convert the rocket from a very inefficient heat engine into the most efficient heat engine that ever has been devised.

Statement of the Problem

Before describing the experiments that have been performed, it will be well to deduce the theory of rocket action in general, in order [7] to show the tremendous importance of efficiency in the attainment of very high altitudes. A statement of the problem will therefore be made, which will lead to the differential equation of the motion. An approximate solution of this equation will be made for the initial mass required to raise a mass of one pound to any desired altitude, when said initial mass is a minimum.

A particular form of ideal rocket is chosen for the discussion as being very amenable to theoretical treatment, and at the same time embodying all of the essential points of the practical apparatus. Referring to Fig. 1, a mass H, weighing 1 lb is to be raised as high as possible in a vertical direction[10] by a rocket formed of a cone P, of propellant material, surrounded by a casing K. The material P is expelled downward with a constant velocity c. It is further supposed that the casing K drops away continuously as the propellant material P burns, so that the base of the rocket always remains plane. It will be seen that this approximates to the case of a rocket in which the casing and firing chamber of a primary rocket are discarded after the magazine has been exhausted of cartridges, as well as to the case in which cartridge shells are ejected as fast as the cartridges are fired.

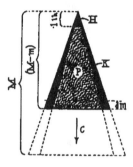

Fig. 1

[8] Let us call

    M = the initial mass of the rocket
    m = the mass that has been ejected up to the time t
    v = the velocity of the rocket, at time t
    c = the velocity of ejection of the mass expelled
    R = the force, in absolute units, due to air resistance
    g = the acceleration of gravity
    dm = the mass expelled during the time dt
    k = the constant fraction of the mass dm that consists of casing K, expelled with zero velocity relative to the remainder of the rocket
    dv = the increment of velocity given the remaining mass of the rocket

The differential equation for this ideal rocket will be the analytical statement of Newton's third law, obtained by equating the momentum at a time t to that at the time $t + dt$, plus the impulse of the forces of air resistance and gravity,

$$(M-m)v = dm(1-k)(v-c) + vk\,dm + (M-m-dm)(v+dv) + [R + g(M-m)]dt$$

If we neglect terms of the second order, this equation reduces to

$$c(1-k)dm = (M-m)dv + [R + g(M-m)]dt \qquad (1)$$

A check upon the correctness of this equation may be had from the analytical expression for the conservation of energy, obtained by equating the heat energy evolved by the burning of the mass of propellant, $dm(1-k)$, to the additional kinetic energy of the system produced by this mass plus the work done against gravity and air resistance during the time dt. The equation thus derived is found to be identical with Eq. (1).

Reduction of Equation to the Simplest Form

In the most general case, it will be found that R and g are most simply expressed when in terms of v and s. In particular, the [9] quantity R, the air resistance of the rocket at time t, depends not only upon the density of the air and the velocity of the rocket, but also upon the cross section S at the time t. The cross section S should obviously be as small as possible; and this condition will be satisfied *at all times*, provided it is the following function of the mass of the rocket $(M-m)$,

$$S = A(M-m)^{2/3} \tag{2}$$

where A is a constant of proportionality. This condition is evidently satisfied by the ideal rocket, Fig. 1. Equation (2) expresses the fact that the shape of the rocket apparatus is at all times similar to the shape at the start; or, expressed differently, S must vary as the square of the linear dimensions, whereas the mass $(M-m)$ varies as the cube. Provision that this condition may approximately be fulfilled is contained in the principle of primary and secondary rockets.

The resistance R may be taken as independent of the length of the rocket by neglecting "skin friction." For velocities exceeding that of sound this is entirely permissible, provided the cross section is greatest at the head of the apparatus, as shown in U.S. Patent No. 1,102,653.

The quantities R, g, and a are evidently expressible most simply in terms of the altitude s, provided the cross section S is also so expressed, giving, in place of Eq. (1),

$$c(1-k)dm = (M-m)dv + \frac{1}{v(s)}[R(s)+g(s)(M-m)]ds \tag{3}$$

Rigorous Solution for Minimum M at Present Impossible

The success of the method depends entirely upon the possibility of using an initial mass M of explosive material that is not impracticably large. It amounts to the same thing, of course, if we say that the mass ejected up to the time t (i.e. m) must be a minimum, conditions for the existence of a minimum being involved in the integration of the equation of motion.

That a minimum mass m exists when a required mass is to be given an assigned upward velocity at a given altitude is evident intuitively from the following consideration: If, at an intermediate altitude, the velocity of ascent be very great, the air resistance R (depending upon the square of the velocity) will also be great. On the other hand, if the velocity of ascent be very small, force will be required to overcome gravity for a long period of time. In both cases the mass necessary to be expelled will be excessively large.

[10] Evidently, then, the velocity of ascent must have some special value at each point of the ascent. In other words, it is necessary to determine an unknown function $f(s)$, defined by

$$v = f(s)$$

such that m is a minimum.

It is possible to put $f(s)$ and $(df(s)/ds)\,ds$ in place of v and dv, in Eq. (3), and to obtain m by integration. But in order that m shall be a minimum δm must be put equal to zero, and the function $f(s)$ determined. The procedure necessary for this determination presents a new and unsolved problem in the calculus of variations.

Solution of the Minimum Problem by an Approximate Method

In order to obtain a solution that will be sufficiently exact to show the possibilities of the method, and will at the same time avoid the difficulties involved in the employment of the rigorous method just described, use may be made of the fact that if we divide the altitude into a large number of parts, let us say n, we may consider the quantities R, g, and also the acceleration, to be *constant over each interval*.

If we denote by a the constant acceleration defined by $v = at$ in any interval, we shall have, in place of the equation of motion (3), a linear equation of the first order in m and t, as follows:

$$\frac{dm}{dt} = \frac{(M-m)(a+g)+R}{c(1-k)}$$

the solution of which, on multiplying and dividing the right number of $(a + g)$, is

$$m = e\left[-\frac{a+g}{c(1-k)}t\right] \cdot \frac{M(a+g)+R}{a+g}\left\{\int e\left[\frac{a+g}{c(1-k)}t\right]\left[\frac{a+g}{c(1-k)}\right]dt + C\right\}$$

$$= e\left[-\frac{a+g}{c(1-k)}t\right] \cdot \frac{M(a+g)+R}{a+g}\left\{e\left[\frac{a+g}{c(1-k)}t\right] + C\right\} \quad (4)$$

where C is an arbitrary constant.

This constant is at once determined as -1 from the fact that m must equal zero when $t = 0$.

We then have

$$m = \left(M + \frac{R}{a+g}\right)\left\{1 - e\left[-\frac{a+g}{c(1-k)}t\right]\right\} \quad (5)$$

This equation applies, of course, to each interval, R, g, and a, being considered constant. We may make a further simplification if, [11] for each interval, we *determine what initial mass M would be required when the final mass in the interval is one pound*. The initial mass at the beginning of the first interval, or what may be called the "*total initial mass*," required to propel the apparatus through the n intervals will then be the *product of the n quantities* obtained in this way.

If we thus place the final mass $(M - m)$, in any interval equal to unity, we have $M = m + 1$ and when this relation is used in Eq. (5), we have for the mass at the beginning of the interval in question

$$M = \frac{R}{a+g}\left\{e\left[\frac{a+g}{c(1-k)}t\right] - 1\right\} + e\frac{a+g}{c(1-k)}t \quad (6)$$

Now the initial mass that would be required to give the one pound mass the same velocity at the end of the interval, if R and g had both been *zero*, is from (6),

$$M = e\frac{at}{c(1-k)} \quad (7)$$

The ratio of Eq. (6) to Eq. (7) is a measure of the additional mass that is required for overcoming the two resistances R and g, and when this ratio is least, we know that M is a minimum for the interval in question. The "total initial mass" required to raise one pound to any desired altitude may thus be had as the product of the minimum M's for each interval obtained in this way.

From Eqs. (6) and (7) we see at once the importance of high efficiency, if the "total initial mass" is to be reduced to a minimum. Consider the exponent of e. The quantities a, g, and t depend upon the particular ascent that is to be made, whereas $c(1 - k)$ depends entirely upon the efficiency of the rocket, c being the velocity of expulsion of the gases, and k the fraction of the entire mass that consists of loading and firing mechanism, and of magazine. In order to see the importance of making $c(1 - k)$ as large as possible, suppose that it were decreased tenfold. Then

$$e^{\frac{a+g}{c(1-k)}t}$$

would be *raised to the 10th power*, in other words, the mass for each interval would be the *original value multiplied by itself ten times.*

[12] **Part II. Experiments**

Efficiency of Ordinary Rocket

The average velocity of ejection of the gases expelled from two sizes of ordinary rocket were determined by a ballistic pendulum. The smaller rockets C, Fig. 1, averaged 120 gm, with a powder charge of 23 gm; and the larger, S, the well-known Coston ship rocket, weighed 640 gm, with a powder charge of 130 gm. Fig. 2, shows the rockets as compared with a yardstick Y.

The ballistic pendulum, was a massive compound pendulum weighing 70.64 kg (155 lb) with a half period of 4.4 sec; large compared with the duration of discharge of the rockets. The efficiencies were obtained from the average velocity of ejection of the gases, found by the usual ballistic-pendulum method, together with the heat value of the powder of the rockets, obtained by a bomb calorimeter for the writer by a Worcester chemist.

The results of these experiments are given in Table I. It will be seen from the table that the efficiency of the ordinary rocket is close to 2 percent;[11] slightly less for the smaller, and slightly more for the larger, rockets, and also that the average velocity of the ejected gases is of the order to 1000 ft/sec. It was found by experiment that a Coston ship rocket, lightened to 510 gm by the removal of the red fire, had a range of a quarter of a mile, the highest point of the trajectory being slightly under 490 ft. A range as large as this is rather remarkable in view of the surprisingly small efficiency of this rocket.

Table I

Type of rocket	Efficiency	Mean efficiency	Velocity corresponding to mean efficiency
Common	2.54%		
Common	1.45		
Common	1.40		
Common	1.95	1.86%	957.6 ft/sec
Coston ship	1.75%		
Coston ship	2.27		
Coston ship	2.62	2.21%	1029.25 ft/sec

[13] *Experiments in Air with Small Steel Chambers*

An apparatus was next constructed, with a view to increasing the efficiency, embodying three radical changes, namely, the use of smokeless powder, of much higher heat value than the black powder employed in ordinary rockets; the use of a strong steel chamber, to permit employment of high pressures; and the use of a tapered nozzle, similar to a steam turbine nozzle, to make available the work of expansion.

Two sizes of chamber were used, one ½-in. diameter, and one 1-in. diameter. The inside and outside diameters of the smaller chamber, Fig. 2a, were, respectively, 1.28 cm and 3.63 cm. The nozzle, polished until very smooth, was of 8° taper, and was adapted to permit the use of two extensions of different lengths. The length of the chamber, as the distance *l* in the figure will be called, could be altered by putting in or removing cylindrical tempered-steel plugs of various lengths, held in place by the breechblock....

Two small chambers were used, practically identical in all respects, one of the soft tool steel, and one of best selected nickel-steel gun-barrel stock, treated to give 100,000 lb tensile strength, for which the writer wishes to express his indebtedness to the Winchester Repeating Arms Company.

Table II

Small

Experiment no.	Chamber	Length of chamber *l* cm	Total mass M gm	Length of nozzle cm	Kind of powder	Mass of powder gm	Mass of wadding and wire gm
1	Soft steel	0.69	3,540.1	Medium	Du Pont	0.7795	0.0345
2	Soft steel	0.69	3,541.9	Medium	Du Pont	0.7060	0.0385
3	Soft steel	1.01	3,538.8	Medium	Du Pont	1.0025	0.0370
4	Soft steel	0.69	3,541.9	Medium	Infallible	0.8247	0.0395
5	Soft steel	1.01	3,538.8	Medium	Infallible	1.2015	0.0380
6	Soft steel	0.69	3,547.9	Short	Du Pont	0.7074	0.0370
7	Soft steel	0.69	3,540.1	Short	Infallible	0.8533	0.0370
8	Soft steel	0.69	3,540.1	Short	Du Pont	0.6825	0.0355
9	Soft steel	1.01	3,645.8	Long	Infallible	1.2397	0.0370
10	Soft steel	1.01	3,645.8	Long	Du Pont	0.9625	0.0365
11	Soft steel	0.69	3,648.93	Long	Du Pont	0.7361	0.0386
12	Soft steel	0.69	3,533.9	Medium	Infallible	0.8985	0.0391
13	Soft steel	0.69	3,645.8	Long	Infallible	0.9068	0.0396
14	Soft steel	0.69	3,533.9	Medium	Du Pont	0.7465	0.0373
29	Nickel steel	0.69	3,553.5	Medium	Infallible	1.0264	0.0445
44	Nickel steel	1.01	6,273.5	Medium	Infallible	1.2731	0.0420
46	Nickel steel	0.69	6,270.5	Medium	Infallible	1.4849	0.0402

Large

51	Chrome-nickel steel	2.28	19,324.0	16.29	Du Pont	8.0522	0.3184
52	Chrome-nickel steel	2.28	19,324.0	16.29	Infallible	9.0259	0.3271

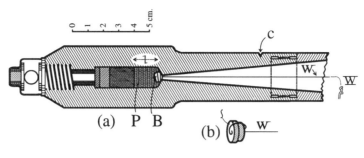

Fig. 2

The charge of powder P, Fig. 2, was fired electrically, by a hot wire in the following way: A fine copper wire w, 0.12-mm diameter, passed through the wadding, Fig. 2b, consisting of two disks of stiff cardboard, and this copper wire joined a short length of platinum or platenoid wire of 0.1-mm diameter f, extending across the inner [14] part of the chamber

Displacement			Length of pendulum	Velocity		Efficiency
d_1	d_2	corrected d				
cm	cm	cm	cm	km/sec	ft/sec	percent
11.55	11.41	11.62	79.15	1.781	5,843	39.01
10.30	10.19	10.35	79.15	1.738	5,703	37.16
15.80	15.70	15.85	79.15	1.907	6,257	44.73
13.60	13.50	13.65	79.15	1.976	6,484	37.13
20.55	20.46	20.59	79.50	2.082	6,832	41.88
9.43	9.38	9.45	79.50	1.585	5,203	30.93
12.59	12.53	12.62	79.50	1.766	5,793	30.12
9.35	9.31	9.37	79.50	1.626	5,336	32.54
20.18	20.10	20.22	79.50	2.045	6,709	40.39
14.20	14.10	14.25	79.50	1.834	6,018	41.38
10.22	10.10	10.28	79.50	1.704	5,592	35.74
13.90	13.83	13.94	79.50	1.850	6,069	33.05
13.85	13.80	13.87	79.50	1.882	6,177	34.24
10.07	10.00	10.10	79.50	1.609	5,279	31.38
17.95	17.85	18.00	79.50	1.969	6,460	37.44
12.58	12.38	12.68	79.50	2.127	6,981	43.73
14.78	14.68	14.93	79.50	2.154	7,064	44.78

chamber

		5.02		2.290	7,515	64.53
		7.08		2.434	7,987	57.25

wadding, in contact with the powder. To the other end of this platinum wire, a short length of the copper wire passed to the side of the wadding, and made electrical contact with the wall of the chamber. A fine steel wire W, 0.24 mm in diameter, served to pull the copper wire w tightly enough to prevent contact of the latter with the nozzle. The wire W was so held that, although it exerted a pull on the wire w, it nevertheless offered no resistance in the direction of motion of the ejected gases.

Two dense smokeless powders were used: Du Pont pistol powder No. 3, a very rapid dense nitrocellulose powder, and Infallible shotgun powder, of the Hercules Powder Company. The heat values in all cases were found by bomb calorimeter.[1] All determinations were made in an atmosphere of carbon dioxide, in order to avoid any heat due to the oxygen of the air. The average heat values were the following:

Powder, in ordinary rocket	545.0 cal/gm
Powder, in Coston ship rocket	528.3
Du Pont Pistol No. 3	972.5
Infallible	1238.5

The ballistic pendulum used in determining the average velocity of ejection, for the small chambers, consisted essentially of a plank B, carrying weights, and supporting the chamber, or gun, C, in a horizontal position. This plank was supported by fine steel wires in such a manner that it remained horizontal during motion. In order to make certain that the plank actually was horizontal in all positions, a test was frequently made by mounting a small vertical mirror on the plank, with its plane perpendicular to the axis of the gun, and observing the image of a horizontal object—as a lead pencil—held several feet away while the pendulum was swinging. Current for firing the charge was led through two drops of mercury to wires on the plank. A record of the displacements was made by a stylus consisting of a steel rod S, pointed and hardened at the lower end. This rod slid freely in a vertical brass sleeve, attached to the under side of the plank, and made a mark upon a smoked-glass strip G. In this way the first backward and forward displacements of the pendulum were recorded, and the elimination of friction was thereby made possible.

The data and results of these experiments are given in Table II, in which d is the displacement corrected for friction.

[15] The velocities and efficiencies were obtained from the usual expression for the velocity in which a ballistic pendulum, with the bob constantly horizontal, is used, namely,

[16]
$$v = \frac{M}{m}\sqrt{2gl(1-\cos\theta)}$$

where M = the total weight of the bob
m = the mass ejected; powder plus wadding
l = the length of the pendulum
θ = the angle through which the pendulum swings
g = the acceleration of gravity

The cosine of θ was corrected for friction by observing the two first displacements d_1 and d_2 and obtaining therefrom

$$d = d_1\sqrt{\frac{d_1}{d_2}}$$

It will be noticed that the highest velocity was obtained with Infallible powder, and

---

1. It was found necessary to use a sample exceeding a certain mass, as otherwise the heat value depended upon the mass of the sample.

was over 7000 ft/sec. The corresponding efficiency was close to 50 percent. In view of the fact that this velocity, corresponding to c in the exponents of Eqs. (6) and (7), is sevenfold greater than for an ordinary rocket, it is easily seen that the employment of a chamber and nozzle such as has just been described must make an enormous reduction in initial mass as compared with that necessary for an ordinary rocket....

[17] *Experiments with Large Chamber*

Inasmuch as all the steel chambers employed in the preceding experiments were of the same internal diameter (1.26 cm), it was considered desirable that at least a few experiments should be performed with a larger chamber, first, in order to be certain that a large chamber is operative; and second, to see if such a chamber is not even more efficient than a small chamber. This latter is to be expected for the reason that heat and frictional losses should increase as the *square* of the linear dimensions of the chamber; and hence increase in a less proportion than the mass of powder that can be used with safety, which will vary as the *cube* of the linear dimensions. Evidence in support of this expectation has already been given. Thus, for ordinary rockets, the larger rocket has the higher efficiency, as evident from Table I.

The large chamber was of nickel-alloy steel (Samson No. 3A), of 115,000 lb tensile strength, for which the writer takes opportunity of thanking the Carpenter Steel Company. This chamber had inside diameter, and diameter of throat, both twice as large as those of the chambers previously used; the thickness of wall of the chamber and the taper of the nozzle were, however, the same. The inside of the nozzle was well polished. Figure 3 shows a section of the chamber; the outer boundary being indicated by dotted lines, P being the powder, and W the wadding. It will be noticed that the wadding is just twice the size of that previously used....

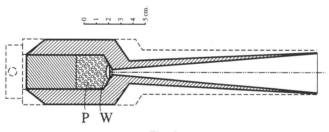

Fig. 3

The chamber was held in the lower end of a [18] 3 ½-ft length of 2-in. pipe P by setscrews. Within this pipe, above the chamber, was fastened a length of 2-in. steel shafting, to increase the mass of the movable system. This system was supported by ½-in. steel pin E.

On firing, the recoil lifted the above system vertically upward against gravity, the extent of this lift, or displacement, being recorded by a thin lead pencil, slidable in a brass sleeve set in the pipe at right angles to the pin E. The point of the pencil was pressed against a vertical cardboard C by the expelled gases will be called the "direct-lift" method; and the theory is given in Appendix A.

Although rebound of the gases from the ground would probably have been negligible, such rebound was eliminated by a short plank D, covered with a piece of heavy sheet iron, and supported at an angle of 45° with the horizontal. This served to deflect the gases to one side.

The results of two experiments, 51 and 52, with this large chamber, are given in Table II. In experiment 51, with Du Pont powder, the powder was packed rather loosely. Any increase in internal diameter was inappreciable, certainly under 0.01 mm. In experiment 52, the Infallible powder was somewhat compressed. After firing, the chamber was found to be slightly bulged for a short distance around the middle of the powder chamber, the inside diameter being increased from 2.6 cm to 2.7 cm, and the outside diameter from 5.08 cm to 5.14 cm. The efficiency (64.53 percent) in experiment 51 and the velocity (7987 ft/sec) in experiment 52 were, respectively, the highest obtained in any of the experiments.

The conclusions to be drawn from these two experiments are, first, that large chambers can be operated, under proper conditions, [19] without involving undue pressures; and second, that large chambers, even with comparatively short nozzles, are more efficient and give higher velocities than small chambers.

It is obvious that large grains of powder should be used in large chambers if dangerous pressures are to be avoided. The bulging in experiment 52 is to be explained by the grains of powder being too small for a chamber of the size under consideration. It is possible, however, that pressures even as great as that developed in experiment 52 could be employed in practice provided the chamber were of "built-up" construction. A similar result might possibly be had if several shots had been fired, of successively increasing amounts of powder. The result of this would have been a hardening of the wall of the chamber by stretching. Such a phenomenon was observed with the soft-steel chamber already described, which was distended by the first few shots of Infallible powder, but thereafter remained unchanged with loads as great as those first used.[12]

Experiments in Vacuo

Introductory

Having obtained average velocities of ejection up to nearly 8000 ft/sec in air, it remained to determine to what extent these represented reaction against the air in the nozzle, or immediately beyond. Although it might be supposed that the reaction due to the air is small, from the fact that the air in the nozzle and immediately beyond is of small mass, it is by no means self-evident that the reaction is zero. For example, when dynamite, lying on an iron plate, is exploded, the particles which constituted the dynamite are moved very rapidly upward, and the reaction to this motion bends the iron plate downward; but reaction of the said particles against the air as they move upward may also play an important role in bending the iron. The experiments now to be described were undertaken with the view of finding to what extend, if any, the "velocity in air" was a fictitious velocity. The experiments were performed with the smaller soft tool-steel and nickel-steel chambers that have already been described.

Method of Supporting the Chamber in Vacuo

For the sake of convenience, the chamber, or gun, should evidently be mounted in a vertical position, so that the expelled gases are shot downward, and the chamber is moved upward by the reaction, either being lifted bodily, or suspended by a spring and set in vibration.
[20] The whole suspended system was therefore designed to be contained in a 3-in. steel pipe, all the essential parts being fastened to a cap, fitting on the top of this pipe. This was done not only for the sake of convenience in handling the heavy chamber, but also from the fact that the only joint that would have to be made airtight for each shot would be at the 3-in. cap.

The means of supporting the chamber from the cap is shown in Plate 6, Fig. 2, and Plate 7, Fig. 1, the apparatus being shown dismantled in Plate 7, Fig. 2. Two $^3/_8$-in. steel rods R, R were threaded tightly by taper (pipe) threads into the cap C. These rods were joined by a yoke, at their lower ends, which served to keep them always parallel. Two collars, or holders, H and H', free to slide along the rods R, R, held the chamber or gun, by three screws in each holder. The inner ends of the screws of the lower holder were made conical, and these fitted into conical depressions c, Fig 2a [page 349], drilled in the side of the gun, so that the lower holder could thus be rigidly attached to the gun. This was made necessary in order that lead sleeves, fitting the gun and resting upon the lower holder H', could be used to increase the mass of the suspended system. Three such sleeves were used, the two largest being molded around thin steel tubes which closely fitted the gun. The rods R, R were lubricated with Vaseline. Two $1/8$-in. steel pins were driven through the rods R, R, just above the yoke Y, in order that the latter could not be driven off by the fall of the heavy chamber and weights when direct lift was employed.

In the experiments in which the chamber and lead sleeves were suspended by a spring, the latter was hooked at its upper end to a screw eye fixed in the cap C. The lower end of the spring was hooked through a small cylinder of fiber. A record of the displacements of the suspended system was made by a stylus S, Plate 6, Fig. 2, in the upper holder H. This stylus was kept pressed against a long narrow strip of smoked glass G by a spring of fine steel wire. This strip of smoked glass was held between two clamps, fastened to a rod, the upper end of which was secured to the cap C, and the lower end to the yoke Y. Except for the largest charges used, it was possible to measure the displacements on both sides of the zero position, and thereby to calculate the decrement and eliminate friction.

When the chamber was suspended by a spring, a deflection as large as a centimeter was unavoidably produced merely by placing the cap C on the 3-in. pipe or removing it, although, in all cases the [21] system would return to within 1 mm (usually much less than this) of the zero position after being displaced. In order to avoid any such displacement as that just mentioned, an eccentric clamp K, Plate 7, Fig. 1, was employed to keep the suspended system rigidly in its zero position during assembling and dismounting the apparatus.

This clamp consisted of an eccentric rod K, free to turn in a hold in the cap C, the lower end being held in a bearing in the yoke Y. Through the upper end of this rod was pinned a small rod K', at right angles to K. The surface of the rod K was smeared with a mixture of beeswax, resin, and Venice turpentine; and the hole in the cap through which K projected was rendered airtight by wax of the same composition.

The suspended system was assembled while the cap C was held by a support touching its under side. When the assembling was complete, the wax was heated by a small alcohol blowtorch until it was soft, then a rubber band was slipped around the rod K' and the outlet pipe E. A trial showed that the cap could now be put in place on the pipe and removed, without the suspended system moving appreciably. After the cap C was in position on the pipe, the rubber band was removed, and the wax heated until the rod K could be turned out of engagement with the holders H, H'. After a shot had been fired, the clamp was again placed in operation until the system had been taken from the 3-in. pipe and the smoked glass removed.

The circuit which carried the electric current to ignite the charge consisted of the insulated wire W, Plate 7, Fig. 1, which passed through a tapered plug of shellacked hard fiber, in the cap C, thence through a glass tube to the yoke Y, to which it was fastened. Below the yoke it was wrapped with insulating tape, except at the lower end where it was shaped to hold the 0.24-mm steel wire, attached to the fine copper wire from the wadding. From the chamber the current passed up the rods R, R and out of the cap, around which was wrapped a heavy bare copper wire V, which together with W, constituted the terminals of the circuit. It should be mentioned, in passing, that a small amount of black powder B, Fig. 2a [page 349], placed over the platinum fuse wire on the wadding, was found necessary as a primer in order to ignite dense smokeless powders *in vacuo*.

In order to make the joint between the cap and the pipe airtight during a determination, the following device was adopted. The outside of the cap C and also a locknut were both turned down to the same diameter. The locknut was made fast to the pipe. These were [22] then painted on the outside with melted wax consisting of equal parts beeswax and resin with a little Venice turpentine.

When a determination was to be made, the cap was screwed into position, a wide rubber band was slipped over the junction between cap and locknut, and the outside of this rubber band was heated with an alcohol blast torch. The result was a joint, for all practical purposes, absolutely airtight, which could, nevertheless, be dismounted at once after pulling off the rubber band.

Theory of the Experiments in Vacuo

The expressions for the velocity of the expelled gases are easily obtained for the two types of motion of the suspended system that were employed, namely, simple harmonic motion produced by a spring, and direct lift.

Simple harmonic motion. Results obtained with simple harmonic motion (slightly damped, of course) were naturally more accurate than with direct lift, as it was impossible in the latter case to eliminate friction. The theory, for simple harmonic motion, in which account is taken of friction is described in Appendix B. The spring was one made to specifications, particularly as regards the magnitude of the force per-centimeter-increase-in-length by the Morgan Spring Company of Worcester, Massachusetts. Care was taken to make certain that in no experiment was the extension of the spring reduced to such a low value as not to lie upon the rectilinear line part of the calibration curve.

Direct lift. The theory of the motion, in this case, has already been given under Appendix A. In this case it might be assumed that a correction could be made for friction by multiplying the displacement s by some particular decrement $\sqrt{d_1/d_2}$ obtained in the experiments with simple harmonic motion, that might reasonably apply. This, as will be shown below, was found to give results in good agreement for the two types of motion, if the direct lift was about 2 cm; but not if it was much larger. It was found that very little frictional resistance was experienced when the mass M was raised by hand, provided the axis of the gun were kept strictly vertical, but a very considerable resistance was experienced if the axis was inclined to one side so that the holders H, H' rubbed against the rods R, R. This sidewise pressure did not take place when the spring was used. It was also found that the trace upon the smoked glass was always slightly sinuous, with direct lift, and [23] straight with the spring. The simple harmonic motion was, therefore, much the more preferable, but could not be used when the powder charges were large.

Means of Eliminating Gaseous Rebound

It should be remembered that the real object of the vacuum experiments is to ascertain what the reaction experienced by the chamber would be, if a given charge of powder were fired in the chamber many miles above the earth's surface. A container is therefore necessary, which, for the purpose at hand, approaches most nearly a container of unlimited capacity. A length of 3-in. pipe, closed at the ends, is evidently unsuitable, because the gas, fired from one end, is sure to rebound from the other end with considerable velocity, and hence to produce a much larger displacement than ought really to be observed. Moreover, any tank of finite size must necessarily produce a finite amount of rebound, from the fact that the whole action is equivalent to liberating suddenly, in the tank, 1 or 2 liters of gas at atmospheric pressure.

There are two possible methods for reducing the velocity of the gas sufficiently to produce a negligible rebound: a *disintegration method*, whereby the stream is broken up into many small streams, sent in all directions (i.e. virtually reconverted into heat); and second, a *friction method*, whereby the individual stream remains moving in one direction, but is gradually slowed down by friction against a solid surface.

As will be shown below, accurate results were obtained by the first method, in what may be called the "cylindrical" tank; and these results were checked satisfactorily by the second method, in what will be called the "circular" tank.

The cylindrical tank was 10 ft 5 in. high and weighed about 500 lb. It consisted of a 6-ft length T, Fig. 4 and Plate 8, Fig. 1, of 12-in. steel pipe, with threaded caps on the ends. Entering the upper cap at a slight angle was the 3-in. pipe P, 4 1/2 ft long, which supported the cap C of Plate 6, Fig. 2, and Plate 7, Fig. 1. The 12-in. pipe was sawn across at the dotted line T_0, so that any device could be placed in the interior of this tank, or removed from it, as desired. The upper section of the tank was lifted off as occasion demanded by a block and tackle. The two ends to be joined were first painted with the wax previously described; and after the tank had been assembled, the joint was painted on the outside with the same wax W, and the entire tank thereafter painted with asphalt varnish.

[24]

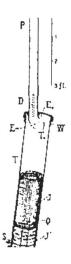

Fig. 4

This tank was used under three conditions:

1. Tank empty, with the elbow E to direct the gas into a swirl such that the gas, while in motion, would not tend to return up the pipe P. In this case some rebound was to be expected from this elbow. This expectation was realized in practice.

2. Tank empty, and elbow cut off along the dotted line E_0. In this case, more rebound was to be expected than in Case 1, which was borne out in practice.

3. Elbow E cut at E_0, and tank half filled with ½-in. square-mesh wire fencing. Two separate devices constructed of this wire fencing were used one above the other. The gas first passed through an Archimedes spiral J, of 2-ft fencing, comprising eight turns, [25] held apart by iron wires bound into the fencing. This construction allowed most of the gas to penetrate the spiral to a considerable distance before being disturbed, and, of course, eliminated regular reflection. This second device J', placed under the first, consisted of a number of 12-in. circular disks of the same fencing, bound to two ¼-in. iron rods Q by iron wires. These disks were spaced 1 in. apart. The three upper disks were single disks, the next lower two were double, with the strands extending in different directions, the next two were triple, and the lowest disk of all, 2 in. from the bottom of the tank, was composed of six individual disks. This lower device necessarily offered large resistance to the passage of the gas; yet strong rebound from any part of it was prevented by the spiral just described. With this third arrangement, small rebound was to be expected, which also was borne out in practice.

This tank was exhausted by way of a stopcock at its lower end, S; and air was also admitted through this same stopcock.

The circular tank, Plate 8, Fig. 2, was 10 ft high and weighed about 200 lb. It consisted of a straight length of 3-in. pipe, carefully fitted, and welded autogenously, to a 4-ft. 3-in. U-pipe. The straight pipe entered the U-pipe on the inner side of the latter, and at as sharp an angle as possible. Another similar U-pipe was bolted to the first by flanges, with ¹⁄₁₆-in. sheetrubber packing between.

In this tank, the gases were shot down the straight pipe, entered the upper U-pipe at a small angle, thus avoiding any considerable rebound, and thence passed around the circular part—not returning up the straight pipe until the velocity had been greatly reduced by friction.

In order to make the time, during which the velocity was being reduced, as long as possible, the pipes were carefully cleaned of scale. They were first pickled, and then cleaned by drawing through them, a number of times; first, a scraper of sheet iron; second, a stiff cylindrical bristle brush, and finally a cloth. All but the most firmly adhering scale was thereby removed. Further, care was taken to cut the hole in the rubber washers, between the flanges, so wide that compression by the flanges would not spread the rubber into the pipe and thereby obstruct the flow of gas.

Notwithstanding all these precautions, evidence was had that the gases became stopped very rapidly. This was to be expected inasmuch as there is solid matter, namely, the wadding and wire, that is [26] ejected with the gas, which accumulates with each successive shot. This solid matter must offer considerable frictional resistance to motion along the U-pipe, and, since the mass of gas is only of the order of a gram, must necessarily act to stop the flow in a very short time. This interval of time was great enough, however, so that this second method afforded a satisfactory check upon the first method.

A possible modification of the above two methods would have been to provide some sort of trapdoor arrangement whereby the gases, after having been reduced in speed in a container as just described, would have been prevented from returning upward into the 3-in. pipe P by this trap, which would be sprung at the instant of firing. In this way gaseous rebound would be entirely eliminated. It was found, however, that results with the two methods already described could be checked sufficiently to make this modification unnecessary.

The tanks were exhausted by a rotary oil pump, No. 1, of the American Rotary Pump Company, supported by a water jet pump. In this way the pressure in the cylindrical tank could be reduced to 1.5 mm of mercury in 25 minutes and to the same pressure in the circular tank, in 10 minutes. The pressures employed in the experiments ranged from 7.5 mm to 0.5 mm.

Methods of Detecting and Measuring Gaseous Rebound

With the two tanks used in the experiments, it was obviously impossible to eliminate gaseous rebound entirely, from the fact that, even if the velocity of the bases is reduced to zero, there still remains the effect of introducing suddenly a certain quantity of gas into the tank. It became necessary, then, to devise some means of detecting, and, if possible, of measuring, the extent of the rebound.

Three devices were employed, one for detecting a *force* of rebound, and two for measuring the magnitude of the *impulse per unit area* produced by the rebounding gas. These latter devices, from the fact that quantitative measurements were possible with them, will be called "impulse meters."

Tissue-paper Detector

The detector for indicating the force of the rebound consisted of a strip of delicate tissue paper I, Plate 6, Fig. 2, and text figure 5a, 0.02 mm thick, with its ends glued to an iron wire W, as shown in Fig. 5a. This iron wire was fastened to the yoke Y, Plate 7, Fig. 1, and held the tissue paper, with its plane horizontal, between the chamber and the wall of the 3-in. pipe P. In many of [27] the experiments, the paper was cut one-third the way across in two places before being used, as shown by the dotted lines b in Fig. 5a. Since the tissue paper has very little mass, the tearing depends upon the magnitude of the force that is momentarily applied, and not upon the force times its duration—i.e. the impulse of the force. The tissue paper will tear, then, if the force produced by the first upward rush of gas, past the chamber into the space in the 3-in. pipe above the chamber, exceeds a certain value. This first upward rush of gas will, of course, produce a greater force than any subsequent rush, as the gas is continually losing velocity. Even though the magnitude of the force that will just tear the tissue paper be not known, it may safely be assumed that if the first upward rush does not tear the paper, the force due to rebound that acts upon the gun must be small compared with the impulse produced by the explosion of the powder.

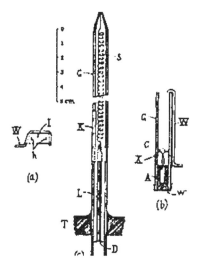

Fig. 5

[28] It should be noted that the tissue paper tells nothing as to whether or not there are a number of successive reflections or rebounds gradually decreasing in magnitude; neither does it give information concerning the *downward* pressure the gases exert upon the chamber tending to decrease the displacement, after they have accumulated in the space between the top of the chamber and the cap C, Plate 6, Fig. 2.

Direct-lift Impulse Meter

A section of the direct-lift impulse meter is shown in Fig. 5b. It is also shown in the photograph Plate 6, Fig. 2, at A. A small cylinder A of aluminum of 1.46 gm mass, hollowed at one end for lightness, was turned down to slide easily in a glass tube G. This tube G was fastened by de Khotinsky cement to an iron wire W, which was in turn fastened to the yoke Y, Plate 7, Fig. 1, so that the glass tube G was held in a vertical position, between the chamber and the wall of the 3-in. pipe—similarly to the tissue paper. Two small wires C, C of spring brass were cemented to the top of the aluminum cylinder, the free ends just touching on opposite sides of the glass tube. The inside of the glass tube was smoked with camphor smoke above the point marked X, so that a record was made of any upward displacement of the aluminum cylinder. The cylinder was prevented from dropping out of the glass tube by a fine steel wire w cemented to the tube and extending across the lower end.

The theory of the direct-lift impulse meter is given in Appendix C. From the theory, we may derive an expression for the ratio Q of the momentum given the gun by the gaseous rebound, to the observed momentum of the suspended system.

There are two disadvantages of this form of impulse meter. First, friction acts unavoidably to reduce the displacement. Second, any jar to which the apparatus is subjected on firing will cause the aluminum cylinder to jump, and thus give a spurious displacement. This latter fact rendered the meter useless for experiments in which direct lift of the chamber took place, as there was always much jar when the heavy chamber fell back, after being displaced upward.

This impulse meter, it will be observed, gave a mean measurement of any successive up-and-down rushes of gas.

[29] Spring Impulse Meter

A section of the spring impulse meter is shown in Fig. 5c. The apparatus consisted of an aluminum disk D, cemented to a lead rod L, of combined mass 5.295 gm, supported by a fine brass spiral spring S. The disk D was of a size sufficient to slide easily in a glass tube G. The upper end of the spring protruded through a small hole in the glass tube, and was fastened at this point by de Khotinsky cement, it thus being easy to make the top of the

[30] Table III

Experiment no.	Type of motion	Length of chamber l cm	Total mass $M + \frac{1}{3} m$ gm	Length of nozzle	Kind of powder	Mass of powder gm	Mass of wadding and wire gm	Mass of black powder gm
15	S.H.M.	0.69	3156.9	Long	Du Pont	0.6747	0.0538	0.007
16	S.H.M.	0.69	3156.9	Long	Du Pont	0.6761	0.0526	0.007
17	S.H.M.	0.69	3156.9	Long	Du Pont	0.6913	0.0508	0.007
18	S.H.M.	0.69	3156.9	Long	Du Pont	0.6929	0.0536	0.007
19	S.H.M.	0.69	3158.9	Long	Du Pont	0.6741	0.0529	0.007
20	S.H.M.	0.69	3156.9	Long	Du Pont	0.7161	0.0516	0.007
21	S.H.M.	0.69	3156.9	Long	Du Pont	0.6495	0.0536	0.007
22	S.H.M.	0.69	3156.9	Long	Du Pont	0.6679	0.0568	0.007
23	S.H.M.	0.69	3156.9	Long	Du Pont	0.6681	0.0537	0.007
24	S.H.M.	0.69	3156.9	Long	Du Pont	0.6693	0.0556	0.007
25	S.H.M.	0.69	2768.1	Medium	Du Pont	0.6998	0.0504	0.007
26	S.H.M.	0.69	2768.1	Medium	Du Pont	0.6715	0.0530	0.007
27	S.H.M.	0.69	2353.8	Short	Du Pont	0.6686	0.0510	0.007
28	S.H.M.	0.69	2353.8	Short	Du Pont	0.6673	0.0510	0.007
30	S.H.M.	0.95	3339.6	Medium	Infallible	0.9186	0.0556	0.010
31	Lift	0.95	2020.7	Medium	Infallible	0.9210	0.0518	0.012
32	Lift	0.95	2020.7	Medium	Infallible	0.9210	0.0601	0.020
33	Lift	0.95	2020.7	Medium	Infallible	0.9210	0.0625	0.020
34	Lift	0.95	2020.7	Medium	Infallible	0.9210	0.0648	0.020
35	S.H.M.	0.95	3339.6	Medium	Infallible	0.9210	0.0614	0.020
36	S.H.M.	0.95	3339.6	Medium	Infallible	0.9210	0.0639	0.020
37	Lift	0.95	2020.7	Medium	Infallible	0.9210	0.0619	0.020
38	Lift	0.95	2135.7	Long	Infallible	0.9210	0.0672	0.020
39	Lift	0.95	2135.7	Long	Infallible	0.9210	0.0608	0.020
40	Lift	0.69	2023.4	Medium	Du Pont	0.6715	0.0576	0.007
41	Lift	0.69	2023.4	Medium	Du Pont	0.6715	0.0599	0.007
42	Lift	0.95	1914.3	Short	Infallible	0.9210	0.0551	0.020
43	Lift	0.95	2020.7	Medium	Infallible	0.9210	0.0641	0.020
45	Lift	1.25	2020.7	Medium	Infallible	1.2581	0.0582	0.020
47	Lift	1.42	3040.5	Medium	Infallible	1.4540	0.0603	0.020
48	Lift	1.42	3040.8	Medium	Infallible	1.3997	0.0607	0.020
49	Lift	1.42	2020.7	Medium	Infallible	1.3997	0.0619	0.020
50	Lift	1.57	3039.0	Medium	Infallible	1.5200	0.0630	0.030

lead rod level with the zero of a paper scale K pasted to the outside of the glass tube. A piece of white paper placed behind the tube G made the motion of the lead rod L very clearly discernible.

This impulse meter was placed in a hole in the upper cap of the 12-in. pipe of the cylindrical tank at D, Fig. 4 and Plate 8, Fig. 1, the same distance from the wall of the 12-in. pipe as the center of the 3-in. pipe. It projected 1 in. through the 12-in. cap which was practically the same as the distance the 3-in. pipe projected. The tube G was kept in position in the cap by being wrapped tightly with insulating tape, the joint being finally painted with the wax already described.

The theory of the spring impulse meter is given in Appendix D, where Q is the ratio already defined in connection with the direct-lift impulse meter. There are two reasons why the ratio Q obtained in the Appendix should be larger than the true percentage at the

Displacement			Tank	Pressure in tank		Paper detector	Rebound impulse to total impulse Q	Velocity		Efficiency
d_1	d_2	Corrected d		Before	After					
cm	cm	cm		mm	mm		percent	km/sec	ft/sec	percent
4.58	4.22	4.97	Cylindrical	5.0	5.5	—	0.000	1.711	5614	36.01
4.78	4.70	4.82	Cylindrical	5.0	10.0	Torn	0.756	1.729	5671	36.75
4.68	4.52	4.70	Cylindrical	4.5	9.0	Torn	0.000	1.671	5481	34.33
4.85	4.55	5.01	Cylindrical	7.0	11.2	Torn	0.000	1.774	5821	38.72
4.66	4.37	4.81	Cylindrical	5.5	10.5	Torn	0.000	1.728	5668	36.71
5.00	4.77	5.12	Cylindrical	7.5	13.0	Not torn	0.000	1.683	5524	34.86
4.73	4.45	4.87	Cylindrical	6.5	10.5	Not torn	0.000	1.780	5840	38.97
4.63	4.34	4.78	Circular	1.5	13.5	Not torn	0.560	1.719	5642	36.37
4.43	4.13	4.59	Cylindrical	1.5	5.5	Not torn	0.000	1.653	5423	33.61
4.68	4.48	4.78	Circular	7.5	22.0	Not torn	0.000	1.719	5642	36.38
4.97	4.31	5.33	Circular	1.5	14.5	Not torn	0.000	1.767	5801	38.46
4.70	3.85	5.19	Cylindrical	1.5	5.0	Not torn	0.000	1.749	5740	37.65
5.05	(#20)	5.17	Circular	1.5	13.0	Not torn	—	1.614	5296	32.05
5.10	(#20)	5.22	Cylindrical	1.5	5.5	Not torn	—	1.630	5347	32.67
7.37	(#20)	7.91	Circular	1.5	21.0	Not torn	—	2.405	7893	55.90
4.60	(#25)	4.94	Cylindrical	1.5	7.5	Torn	—	1.997	6550	39.40
5.87	(#20)	5.90	Circular	1.5	21.0	Torn	—	2.191	7189	46.38
5.30	(#25)	5.69	Circular	4.5	25.5	Torn	—	2.127	6980	43.71
5.50	(#25)	5.90	Cylindrical	2.5	8.0	Not torn	—	2.162	7093	45.15
7.22	(#20)	7.39	Cylindrical	1.5	6.0	Not torn	—	2.336	7665	52.73
7.18	(#20)	7.35	Cylindrical	3.5	9.5	Torn	—	2.319	7610	51.96
5.19	(#25)	5.57	Cylindrical	1.5	7.0	Not torn	—	2.106	6911	42.85
4.83	(#25)	5.18	Cylindrical	1.5	7.0	Not torn	—	2.136	7010	44.09
5.07	(#25)	5.45	Cylindrical	1.5	7.0	Not torn	—	2.202	7227	46.86
2.03	(#25)	2.18	Cylindrical	1.5	4.5	Not torn	—	1.797	5897	39.73
1.95	(#25)	2.09	Cylindrical	0.5	3.0	Not torn	—	1.748	5735	37.93
5.70	(#20)	5.83	Cylindrical	1.5	7.0	Torn	—	2.055	6745	40.82
5.37	(#25)	5.76	Cylindrical	1.0	6.0	Not torn	0.326	2.137	7011	44.14
11.38	(#25)	11.65	Cylindrical	1.5	8.0	Not torn	0.677	2.340	7680	52.93
6.50	(#25)	6.98	Cylindrical	1.5	9.0	Torn	0.690	2.318	7606	51.89
6.03	(#25)	6.47	Cylindrical	1.5	8.5	Not torn	0.730	2.314	7593	51.74
13.00	(#25)	13.96	Cylindrical	1.5	8.5	Not torn	0.735	2.257	7404	49.19
7.28	(#25)	7.82	Cylindrical	1.5	9.5	Not torn	0.790	2.332	7653	52.56

top of the 3-in. pipe. In the first place, friction in the 3-in. pipe will decrease the velocity of the rebounding gas; and, further, the disk D, Fig. 5, is fairly tight-fitting in the glass tube G, whereas there is a considerable space between the gun and the 3-in. pipe, through which the gas may pass and, accumulating above, exert a *downward* pressure on the top of the gun.

One important advantage of the spring impulse meter over that employing direct lift is that the former has very little friction, so that the readings are very reliable. Another advantage is that the displacement of the former will include without any uncertainty the effect of any number of rebounds following one another in rapid succession—i.e. the effect of multiple reflections of the gas, if such reflections are present.

Explanation of Table III

In the vacuum experiments, the soft-steel chamber was used for Du Pont powder, and the nickel-steel chamber for Infallible powder.

The three nozzles called short, medium, and long, were respectively, 9.64, 15.88, and 22.08 cm from the throat to the muzzle.

[31] The length of chamber l, in the third column, is taken as the distance shown in Fig. 2a.

In the cases of simple harmonic motion in which d_2 is not given in the table, the displacements were so large that d_2 was prevented from reaching its full extent by the yoke Y, Plate 7, Fig. 1. Correction for friction was made in these cases by choosing the decrement from some other experiment that would be likely to apply. The number of this experiment is written in parentheses, in the table, in place of d_2. The same procedure is followed in the experiments with direct lift.

Of the experiments in the cylindrical tank, 15 and 16 were performed with the elbow E, Fig;. 4, at the lower end of the 3-in. pipe; experiment 17 was performed with this elbow also in place, with the addition of a sheet-iron sleeve in the pipe, to decrease the curvature *at the elbow*; experiments 18 and 19 were performed with the tank empty; and the remaining experiments were performed with the fencing, already described, in position.

The tissue paper was usually torn at one end, and not torn completely off. It was only torn completely off, with small charges, in the experiments with the cylindrical tank *empty* (experiments 18 and 19). The tissue paper was cut one-third across at each end, as already explained, in experiments 15 to 33, inclusive.

The direct-lift impulse meter was used in experiments 15 to 26, inclusive. In cases in which there was impact of the chamber against the yoke, or pins, at the lower ends of the rods R, R, plate 6, Fig. 2, this impulse meter was useless because of the jar. Only in experiments 16 and 22 was there a measurable displacement, the negligible displacements in the other cases being doubtless due to friction. The spring impulse meter was used only in the last six vacuum experiments.

An inspection of Tables II and III will show that the results, under the same conditions, are in sufficiently close agreement to warrant the comparison of results obtained under various circumstances of firing.

Discussion of Results

1. There is a general tendency for the velocities in vacuo to be larger than those in air, for the same length of chamber I and the same mass of powder.

With Du Pont powder, the medium and short nozzles give greater velocities in vacuo. The long nozzle, however, does not show results very much different from those obtained in air.

[32] There is a large difference, however, with Infallible powder, with all three nozzles. For the medium nozzle a comparison of experiments 4 to 12, inclusive, with 35 and 36 shows that the increase amounts to 22 percent of the velocity in air.

2. The medium nozzle gives, in general, greater velocities than the abort or the long

nozzle with the name length of chamber *l* and approximately the same charges of powder. In all cases, the abort nozzle gives less velocity than the medium or the long nozzle, which is to be expected.

3. The results show no appreciable dependence of the velocities upon the pressure in the tank between 7.5mm and 0.5mm, and it is safe to conclude that the velocities are practically the same from atmospheric pressure down to zero pressure, except as regards the slight increase of velocity with decreasing pressure already mentioned.

4. A comparison of the results when the chamber moved under the influence of the spring with those in which the chamber was merely lifted, shows that the agreement of results obtained by the two methods is good, provided the displacement in the direct-lift experiment is small (compare experiments 40 and 41 with 26). If, on the other hand, the displacement in the direct-lift experiment is large, this method gives considerably lesser velocities than 37, and 43 makes it evident that all the velocities obtained by experiments in which the lift exceeded 4 cm are from 300 to 600 ft/sec too small. This is a very important conclusion, for it means that the highest velocities in vacuo, recorded in Table III, are doubtless considerably less than those which were actually attained.

5. A comparison of the results obtained by means of the circular tank with those obtained by means of the cylindrical tank shows that the velocities range about 100 ft/sec higher for the circular tank—a difference that is so small as to be well within the accidental variations of the experiments.

Concerning the behavior of the cylindrical tank under different conditions, a comparison of experiments shows that the velocities are much the same for all cases. Hence it is safe to conclude that the rebound, at least for small charges, is not excessive even if an empty tank is used, providing it is sufficiently large.

A check of some interest, on the effectiveness of the cylindrical tank, with the retarder J, J' in position inside, was the sound of the shot, which resembled a sharp blow of a hammer on the lower [33] cap of the 12-in. pipe. The impart was most clearly discernible when the hand was on the lowest part of the tank. The sound, in the case of the circular tank, did not appear to come from any particular part. When the tank was grasped during firing, a throb of the entire tank was noticed.

6. Concerning the proportion of the measured reaction that is due to gaseous rebound, the tissue-paper detector, as has already been explained, does not give any information. All that this detector really shows is that the force exerted by the initial upward rush of gas past the chamber is not excessive. The fact that the tissue paper is sometimes torn and sometimes not under identical conditions of firing, shows either that this force differs more or less in various parts of the tank (i.e., the upward rush of gas is not perfectly homogeneous) or that the tissue paper is weakened by each successive shot. This last explanation is the more probable; for fine particles of the wadding rush upward with the gas, as is proved by fine markings on the smoked glass, arid also from the fact that, after a number of shots, the tissue paper is found to be perforated with very small holes.

The gaseous rebound could not be measured accurately with the direct-lift impulse meter. Thus of all the experiments in which this meter could be used, 15 to 26 inclusive, only two, 16 and 22, gave readable displacements; the failure to obtain readable displacements in the other eases being doubtless due to friction, as already mentioned. It will be noticed that the impulse is under 1 percent.

The spring impulse meter used in the last five experiments gave reliable results because of the very slight friction during operation. This impulse meter shows that, if the momentum of the chamber were to be corrected for gaseous rebound, this correction would be much less than 1 percent of the momentum of the chamber. But as has been stated above, the impulse of the rebound at the chamber must be less than that at the impulse meter, from the fact that gases may pass readily behind the chamber and exert a downward pressure, and also because of friction in the 3-in. pipe. The effect of gaseous rebound is therefore negligible, and no account of it has been taken in calculating the velocities and efficiencies.

It now becomes possible to find, from the experimental results, the highest velocity in vacuo upon which dependence may be placed. This is evidently the result of experiment 45 and is 2.34 km/sec or 7680 ft/sec. It is well worth noticing, however, that experiment 50 would have given, without doubt, a velocity even higher, had friction properly been taken into account.

[34] *Discussion of Possible Explanations*

1. The fact that the velocities are higher in vacuo than in air seems explicable only by there being conditions of ignition different in vacuo from those in air; although this may also have been due to the air in the nozzle interfering with the streamlines of the gas, thus producing a jet not strictly unidirectional. It should be remarked that the highest velocity in vacuo recorded, experiment 23, may have been due to unusually good circumstances of ignition; but it may also have been due, in part, to being performed in the circular tank.

2. The fact that the medium nozzle gives in general velocities higher than the long nozzle shows that very likely after travelling the distance from the throat equal approximately to the length of the medium nozzle, the gas is moving so rapidly that it fails to expand fast enough to fill the cross section of the nozzle. A discontinuity in flow is produced at the place where the gas leaves the wall of the nozzle, and this produces eddying and a consequent loss of unidirectional velocity. The efficiency could doubtless be increased by constructing the nozzle in the form of a straight portion, corresponding to a cone of 8° taper, for the length of the medium nozzle with the section beyond this point in the form of a curve concave to the axis of the nozzle.

Conclusions from Experiments

1. The experiments in air and in vacuo prove what was suggested by the photographs of the flash in air, namely, that the phenomenon is really a jet of gas having an extremely high velocity and is not merely an effect of reaction against the air.

2. The velocity attainable depends to a certain extent upon the manner of loading, upon the circumstances of ignition, and upon the form of the nozzle. Hence, in practice, care should be taken to design the cartridge and the nozzle for the density of air at which they are to be used, and to test them in an atmosphere of this particular density.

It is with pleasure that the writer acknowledges the use, as honorary fellow in physics, of the laboratory facilities, and especially the rotary pump, at the Physics Laboratory at Clark University where these experiments were performed.

Significance of the Above Experiments as Regards Constructing a Practical Apparatus

It will be well to dwell at some length upon the significance of the above experiments. In the first place, the lifting power of both [35] powders is remarkable. Experiment 51 shows, for example, that 42 lb can be raised 2 in. by the reaction from less than 0.018 lab of powder. One interesting result is the very high efficiency of the apparatus considered as a heat engine. It exceeds, by a wide margin, the highest efficiency for a heat engine so far attained—the "net efficiency" or duty of the Diesel (internal-combustion) engine being about 40 percent, and that for the best reciprocating steam engine but 21 percent. This high efficiency is, of course, the result of three things: the absence of much heat loss due to the suddenness of the explosion; the almost entire absence of friction; and the high temperature of burning. Owing to these features, it is doubtful if even the most perfect turbine or reciprocating engine could compete successfully with the type of heat engine under consideration.

It is, however, the velocity c in Eqs. (6) and (7) which is of the most interest. The highest velocity obtained in the present experiments is 13 ft/sec under 8000 ft/sec, thus exceeding a mile and a half per second (the "parabolic velocity" at the surface of the moon), and also exceeding anything hitherto attained except with minute quantities of matter by means of electrical discharges in vacuum tubes. Inasmuch as the higher velocities range between seven and eightfold that of the Coston rocket, we should expect a reduction of initial masses to be made possible by employment of the steel chamber, to at least the seventh root of the masses necessary for a chamber like the Coston rocket.

The supposition is, of course, that the mass of propellant material can be made so large in comparison with the mass of the steel chamber, that the latter is comparatively negligible. No attempt was made in the present experiments to reduce the chamber to its minimum weight; in fact, the more massive it was, the more satisfactorily could the ballistic experiments be performed. The minimum weight possible, for the same thickness of wall as in the experiments, was calculated by estimating, first, the volume of a chamber from which all superfluous metal had been removed, as shown by the full lines in Fig. 3, and then calculating the mass of this reduced chamber, from the measured density of the steel. The minimum masses of chamber per gram of powder plus wadding, estimated in this way, were 145, 150, and 120 gm, respectively, for experiments 50, 51, and 52. In the last two cases, a smaller breechblock could doubtless have been used, as evident from Fig. 3; and in the first two cases, the chamber wall, itself, could safely have been reduced in thickness. More important still, a "built-up" construction would much reduce the mass as has already been explained.

[36] It should be mentioned that, for any particular chamber, it will be necessary to determine the maximum possible powder charge to a nicety, from the fact that, as modern rifle practice has demonstrated, one charge of dense smokeless powder may be perfectly safe for any number of shots, whereas a slightly larger amount, or the same amount slightly more compressed (a state in which the powder must exist in the present chamber) will result in very dangerous pressures.

But the whole question of ratio of mass of powder to chamber is without doubt relatively unimportant for the following reason: The photographs of the flash, in experiments 9 and 11, in which the flash was accidentally reflected in the nozzle of the gun, show the nozzle appearing stationary in the photograph, thus demonstrating that the duration of the flash is very small; but this, as already explained, is much longer than the time during which the gases are leaving the nozzle. The time of firing is, therefore, extremely short. This is to be expected, inasmuch as the high pressure in the chamber sets in motion only the small mass of gas and wadding, and hence must exist for a much shorter time than the pressure in a rifle or pistol. For this reason the heat that is developed in the machine gun, due to the hot gases remaining in the barrel for an appreciable time during each shot, as well as that due to the friction of the bullet, will be absent in the type of rapid-fire mechanism under discussion. Hence, a large number of shots, equivalent to a mass of powder greatly exceeding that of the chamber, may be fired in rapid succession without serious heating.[14]

[37] **Part III. Calculations Based on Theory and Experiment**

Application of Approximate Method

As already explained this method consists in employing the equations and

$$M = \frac{R}{a+g}\left\{e^{\left[\frac{a+g}{c(1-k)}t\right]} - 1\right\} + e^{\frac{a+g}{c(1-k)}t} \tag{6}$$

and

$$M = e^{\frac{at}{c(1-k)}} \tag{7}$$

to obtain a minimum M in each interval, where

M = The initial mass, for the interval, when the final mass is one pound, and

R = the air resistance in poundals over the cross section S, at the altitude of the rocket. If we call P the air resistance per unit cross section, we shall have for R, $PS(p/p_o)$ where p is the density at the altitude of the rocket, and to is the density at sea level.
a = the acceleration in feet per second^2, taken conatant throughout the interval,
g = the acceleration of gravity,
t = the time of ascent through the interval, arid
$c(1-k)$ = what will be called the "effective velocity," for the reason that the problem would remain unchanged if the rocket were considered to be composed entirely of propellant material, ejected with the velocity $c(1-k)$. It will be remembered that c actually stands for the true velocity of ejection of the propellant, and k for the fraction of the entire mass that consists of material other than propellant. The effective velocity is taken constant throughout any one calculation.

The altitude is divided into intervals short enough to justify the quantities involved in the above equations being taken as constants. The equations are then used to find the minimum value of M for each interval—the mean values of R and g, in the interval being employed—and the "total initial mass required to raise a final mass of one pound to a desired altitude is then obtained as the product of these M's.

[38] *Values of the Quantities Occurring in the Equations*

The effective velocity $c(1-k)$. The calculation which follows has been carried out with the assumption[15] of a velocity of ejection of 7500 ft/sec and a constant k equal to $1/15$. This velocity is considerably less than those that were actually obtained, both in air and in vacuo. The "effective velocity" will thus be

$$c(1-k) = 7000 \text{ ft/sec}$$

It should be noticed that k could be $1/12$ and yet not necessitate a larger velocity of ejection than 7640 ft/sec, which is also under the highest velocities obtained in the experiments. It is important at this point to remember that the velocities in vacuo would doubtless have been found to be considerably higher than the above value, if friction could have been eliminated in the "direct-lift" method.

The quality R. The mean value of R for any interval is most easily obtained from a graphical representation of P as a function of v, the mean value of P between the beginning and end of the interval being taken. Three curves have been used for this purpose: for velocities ranging from zero to 1000 ft/sec, 1000 to 3000 ft/sec, and from 3000 ft/sec upward. The first curve represented the experimental results of A. Frank[3] obtained with prolate ellipsoids. The second curve represented the experimental results of A. Mallock,[4] whereas the third curve represented an empirical formula by Mallock,[5] which agrees well with experimental results up to 4500 ft/sec—the highest velocity that has been attained by projectiles—and hence may be used for still higher velocities with a fair degree of safety. Mallock's expression, reduced to the absolute fps system and multiplied by $1/4$, the coefficient for projectiles with pointed heads, becomes

$$P = 0.00006432 v^2 \left(\frac{v^1}{a}\right)^{0.375} + 480 \qquad (8)$$

---

3 A. Frank, *Zeitschr. Verein Deutsches Ing.* 55. pp. 593-612, 1906.
4 A. Mallock, *Proc. Roy. Soc.* 79A. pp. 262-273, 1907.
5 A. Mallock, *Proc. Roy. Soc.* 79A. pp. 267, 1907.

where v' = the velocity with which a wave is propagated in the air immediately in front of the projectile; which equals the velocity of the body when that velocity exceeds the velocity of sound in the undisturbed gas

a = the velocity of sound in the undisturbed gas

The constant, 480 poundals, must be added for velocities over 2400 ft/sec owing to the vacuum in the rear of the projectile.

[39] *The quantity p.* The above expression (8), for the resistance, holds only at atmospheric pressure. At high altitudes the pressure, of course, decreases greatly. If we call p the mean density throughout any interval of altitude, and p_o the density at sea level, the right member of (8). On being multiplied by S and p/p_o, will give the air resistance R experienced by the rocket.

A curve representing the relation between density and altitude up to 120,000 ft is shown in Fig. 6. This curve is derived from a table of pressures and temperatures in Arrhenius's Lebrbuch der kosmischen Physik. The ordinates of the curve are the numbers p/p_o.

Beyond 120,000 ft the density is calculated by the empirical rule which assumes the density to become halved at every increase in altitude of 3.5 miles. A comparison was made between two values obtained in this way and those obtained from the very probable pressures deduced by Wegener, in the following way: The mean density between two levels for which Wegener gives pressures was obtained by multiplying the difference in pressure by 13.6, and dividing by the difference in level in centimeters. A comparison showed that the densities used in the present calculations beyond 123,000 ft were from three- to twentyfold larger than those derived from Wegener's data, so that the values used in the present case were doubtless perfectly safe.

Densities beyond 700,000 ft within the geocoronium sphere must be negligible, for not only is the density very small but the resistance to motion is very small—due, according to Wegener, to the properties of geocoroniesio—a conclusion which is supported by the fact that meteors remain, for the most part, invisible above this level.

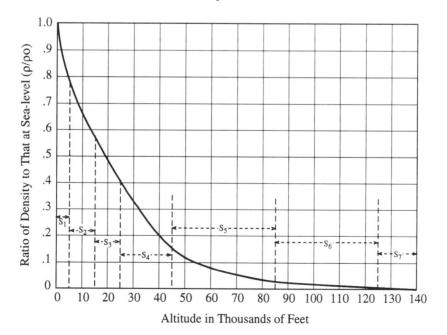

Fig. 6

[40] *Division of the Altitude into Intervals*

In dividing the altitude into intervals the only condition that must be fulfilled is that the densities in any interval shall not differ widely from the mean value in the interval. The least number of intervals which satisfy this condition are given in Table IV. The mean densities in intervals s_1 to s_6 inclusive, were obtained from Fig. 6, on which these intervals are marked. The remaining densities were estimated as already explained.

Table IV

Interval	Length of interval, ft	Height of upper end of interval above sea level, ft	Mean density in terms of p_o	Mean gravity chosen, in terms of gravity at sea level
s_1	5,000	5,000	0.928	1
s_2	10,000	15,000	0.730	1
s_3	10,000	25,000	0.520	1
s_4	20,000	45,000	0.278	1
s_5	40,000	85,000	0.080	1
s_6	40,000	125,000	0.015	1
s_7	75,000	200,000	0.0026	1
s_8	300,000	500,000	0.000025	1
s_9 (a=150)	3,415,000	3,915,000	—	0.839
s_9 (a-50)	8,810,000	9,310,000	—	0.684

Calculation of Minimum Mass for Each Interval

Tables V and VI are calculated for a start, respectively, from sea level and from an altitude of 15,000 ft—i.e., the beginning of s_3. The procedure in each case is, however, identical.

The process of calculation is as follows: At the beginning of any interval we have the velocity already acquired during the previous intervals, let us say vo. This velocity is, of course, zero at the beginning of the first interval. Assume any final velocity at random, v_1 for the interval in question.

[41] The value of *at* may be had from the equation

$$v_1 = v_0 + at \tag{9}$$

and *t* is at once obtained from the relation

$$s = v_0 t + \tfrac{1}{2} a t^2$$

i.e.,

$$t = \frac{s}{v_0 + 1/2\, at} \tag{10}$$

whence, of course, *a* is at once known.

The calculation of

$$\exp\frac{a+g}{c(1-k)}t \quad \text{and} \quad \exp\frac{at}{c(1-k)}$$

calls for no comment; and R is obtained as P, the mean ordinate between v_0 and v_1, from curves as already explained, multiplied by S and p/p_0.

The value of M, the initial mass, for the interval, necessary in order that the final mass in the interval shall be one pound, is then obtained from Eq. (7); and finally, the ratio of Eq. (6) to Eq. (7) is calculated, i.e.,

$$\frac{M}{\exp\left[at/c(1-k)\right]}$$

This is the ratio of the initial mass necessary, including losses due to both R and g, to the mass necessary to give the one pound the same velocity v_1, without overcoming R and g; and the entire calculation must be repeated until a minimum value of this ratio is obtained—when the corresponding mass M will be the minimum mass for the interval in question. Each minimum M is marked in the tables by an asterisk.

This process is carried out for each interval beginning with the first.

It should be noticed that, although P and the density are not really constant in any interval, the result obtained by taking the mean of the quantities must nevertheless give results close to the truth, owing to the fact that P increases during the ascent, whereas the density decreases.

Explanation of Tables V, and VI

It should first be explained why no minimum M has been calculated for the intervals s_7, and s_8. Although the minima for the preceding intervals are clearly defined, a trial will show that a minimum M can occur, for s_7 and s_8, only for extremely high velocities v_1; although for s_7, a secondary minimum occurs for $v_1 = 8000$ ft/sec. Even for $v_1 = 30,000$ ft/sec the minimum has not yet been attained for this interval, although the acceleration required to produce this velocity is 6000 ft/sec². The reason for this state of affairs is evident at once from the fact that the density ratio p/p_0 is very small for s_7, and also from the fact that a occurs in the denominator of the term containing R in Eq. (6), so that the large acceleration counterbalances the increase in R.

Thus, in order that the initial mass for s_7, shall be a minimum, the acceleration must become very large, with consequent severe strains in the rocket apparatus and instruments carried by the rocket, to say nothing of the difficulty of firing with sufficient rapidity to produce such large accelerations. It thus becomes advisable to choose a moderate acceleration in s_7, s_8 and not to assign a velocity v_1, as was done in the preceding intervals. Two accelerations are chosen: 50 ft/sec² and 150 ft/sec², respectively. The interval s_9, also calculated for assigned accelerations, will be explained in detail below. In all cases, when either one of these accelerations is mentioned in connection with s_8, and s_9, this acceleration will be understood as having been taken also in the preceding intervals, beyond s.

In order to see how far the effective velocity $c(1-k)$ may fall short of 7000 ft/sec and still not render the rocket impracticable, a few additional columns for M are calculated.

In the first of the additional columns, M_2, the effective velocity is taken as 3500 ft/sec, namely, half that of the preceding calculations. This allows considerable inefficiency of the apparatus, in a number of ways. For example, the product

$$c(1-k) = 3500$$

may be given by the same proportionality k as before, but with a velocity of ejection of the gases as low as 3750 ft/sec. On the other hand, the velocity of ejection may be as large as before (i.e., 7300 ft/sec); and the proportionality k increased to 0.533; meaning, of course, that the rocket now consists more of mechanism than of propellant.

The second additional calculations M_{R1} are carried out under the assumption that a reloading mechanism is used, with k as in the original calculations ($k = 1/15$), but that the velocity of expulsion of the gases is the mean found by experiment for the Coston ship rockets, namely, 1029.25 ft/sec. In this case the effective velocity is

$$c(1-k) = 1029.25\ (1 - 1/15) = 960.\text{ft/sec}$$

The third additional calculations M_{R2} are carried out for the case of a rocket built up of Coston rockets in bundles (shown in Fig. 7), the lowest bundle of which is fired first and

Table V

Interval		v_1, ft/sec	at	t, sec	a	$\dfrac{at}{c(1-k)}$	$\dfrac{a+g}{c(1-k)}t$	$\exp\dfrac{at}{c(1-k)}$	$\exp\dfrac{(a+g)}{c(1-k)}t$	P, poundals per sq in.	R, PS(p/p_0)	$\dfrac{R}{a+g}$
S_1		500	500	20.0	25	.0716	.1630	1.074	1.176	7.36	6.85	.120
		800	800	12.5	64	.1145	.1720	1.120	1.186	20.0	18.5	.193
	*	1,000	1,000	10.0	100	.143	.1890	1.153	1.207	31.25	29.0	.219
		1,200	1,200	8.34	144	.172	.212	1.185	1.235	61.4	57.0	.323
		1,500	1,500	6.7	226	.215	.2475	1.242	1.276	104.6	98.0	.378
		2,000	2,000	5.0	400	.287	.309	1.332	1.362	202.5	188.0	.436
S_2		1,100	100	9.54	10.47	.0143	.0578	1.014	1.061	153.3	112.1	2.64
	*	1,200	200	9.1	22.0	.0286	.0704	1.034	1.073	166.6	121.6	2.24
		1,400	400	8.33	47.9	.0574	.0954	1.060	1.100	216.0	158.7	1.97
S_3		1,300	100	8.0	12.5	.0143	.0508	1.014	1.052	250.0	130.0	2.925
	*	1,400	200	7.7	25.8	.0286	.0637	1.034	1.066	262.8	136.9	2.37
		1,600	400	7.15	56.4	.0574	.0906	1.06	1.096	294.5	152.6	1.74
S_4		1,500	100	13.8	7.23	.0143	.0775	1.014	1.080	339.0	94.3	2.42
	*	1,600	200	13.33	15.0	.0286	.0898	1.034	1.094	372.0	101.5	2.17
		1,700	300	12.9	23.24	.0429	.1022	1.046	1.107	394.0	109.4	1.975
		1,800	400	12.5	33.25	.0574	.1170	1.060	1.123	424.0	118.0	1.81
S_5		1,700	100	24.25	4.125	.0143	.1258	1.014	1.133	439.0	35.1	.974
	*	1,800	200	23.7	8.45	.0286	.1366	1.034	1.146	480.0	38.4	.951
		2,000	400	22.24	18.0	.0574	.159	1.06	1.173	535.0	42.8	.854
S_6		1,900	100	21.7	4.62	.0143	.1135	1.014	1.12	567.	8.50	.232
	*	2,000	200	21.1	9.50	.0286	.1255	1.034	1.133	603.	9.01	.2175
		2,200	400	20.0	20.0	.0574	.1490	1.06	1.16	669.	10.02	.1923
$S_7(a=150)$		5,160	3,160	21.0	150	.4523	.5452	1.572	1.725	1,878.	4.84	.0264
$(a=50)$		3,393	1,393	27.8	50	.199	.3276	1.218	1.387	1,122.	3.1	.0355
$S_8(a=150)$		10,790	5,630	37.5	150	.804	.976	2.23	2.65	10,600.	0.272	.00146
$(a=50)$		6,833	2,840	55.8	50	.399	.652	1.49	1.92	4,000.	0.0994	.00121
$S_9(a=150)$		33,790	23,000	153.5	150	3.29	3.89	26.9	48.8			
$(a=50)$		30,533	23,700	472.5	50	3.38	4.85	29.13	129.0			

then released; after which the bundle above is fired and then released, and so on. For the Coston ship rocket (having a range of a quarter of a mile, with the charge of red fire removed, as already stated) the ratio of the powder charge to the remaining mass of the rocket is found to be closely $1/4$. Hence the "effective velocity" in this case is only

$$c(1-k) = 1029.25(1-{}^4\!/_5) = 257.3 \text{ ft/sec}$$

The M's in the last two cases are calculated only for the accelerations that make M minima for the first case (effective velocity, 7500 ft/sec). Hence in these cases, the M's are not minima, although only in the last two cases is there probably much discrepancy from the actual minima.

[42]

M, lb	$M/\exp$ at $\dfrac{}{c(1-k)}$	$\exp\dfrac{2(a+g)}{c(1-k)}t$	M_1, lb	$\exp\dfrac{7.28(a+g)}{c(1-k)}t$	M_{R1}, lb	$\exp\dfrac{27.2(a+g)}{c(1-k)}t$	M_{R2}, lb	Time to upper end of interval, sec
1.1972	1.113							
1.2218	1.092							
1.252	1.086	1.458	1.5584	3.94	4.586	167.3	203.90	10.0
1.311	1.106							
1.380	1.112							
1.5195	1.138							
1.222	1.206							
1.237	1.199	1.150	1.4860	1.665	3.155	6.73	20.60	19.1
1.297	1.223							
1.204	1.186							
1.222	1.182	1.137	1.462	1.589	2.974	5.62	16.52	26.8
1.261	1.191							
1.273	1.255							
1.297	1.253	1.198	1.626	1.92	3.91	11.33	33.73	40.13
1.319	1.26							
1.346	1.267							
1.262	1.245							
1.2845	1.242	1.313	1.711	2.694	4.304	40.70	88.45	63.83
1.321	1.246							
1.1478	1.13							
1.162	1.123	1.280	1.3406	2.488	2.810	29.76	36.02	84.93
1.1907	1.124							
1.7442	1.108	2.97	3.022	52.6	53.96	2.63×10^4	2.70×10^4	105.93
1.4007	1.15	1.900	1.9319	10.79	11.13	7.03×10^3	7.28×10^3	112.73
2.6524	1.19	7.02	7.0288	1,192.0	1,193.7	2.88×10^{11}	2.88×10^{11}	143.43
1.9211	1.293	3.680	3.6832	117.4	117.54	4.67×10^7	4.67×10^7	168.53
48.8		2,380.0	2,380.0	$1,906 \times 10^{12}$	$1,906 \times 10^{12}$	5.74×10^{45}	5.74×10^{45}	296.93
129.0		16,700.0	16,700.0	$1,995 \times 10^{15}$	$1,995 \times 10^{15}$	1.25×10^{57}	1.25×10^{57}	641.03

The cross section, *throughout any interval*, is taken as 1 sq in, except for interval s_9. It will be seen from the table that this is justifiable, as the largest mass in intervals s_1 to s_8 does not differ much from one pound.

Fig. 7

[44]   Table VI

Interval	v_1, ft/sec	at	t, sec	a	$\dfrac{at}{c(1-k)}$	$\dfrac{a+g}{c(1-k)}t$	exp $\dfrac{at}{c(1-k)}$	exp $\dfrac{(a+g)}{c(1-k)}t$
s_3	500	500	40.	12.5	0.0715	0.205	1.074	1.29
*	800	800	25.	32.0	0.1147	0.2277	1.120	1.256
	1,000	1,000	20.	50.0	0.142	0.235	1.152	1.263
	1,500	1,500	13.4	112.0	0.2145	0.277	1.24	1.318
s_4	900	100	23.7	4.23	0.0143	0.1227	1.013	1.132
*	1,000	200	22.2	9.00	0.0286	0.1305	1.034	1.137
	1,300	500	19.1	26.2	0.0714	0.1645	1.073	1.177
	1,800	1,000	15.4	65.0	0.1430	0.2136	1.152	1.238
s_5	1,100	100	38.1	2.625	0.0124	0.1888	1.013	1.207
*	1,200	200	36.5	5.47	0.0286	0.1960	1.03	1.215
	1,300	300	34.75	8.64	0.0430	0.202	1.044	1.223
	1,400	400	33.3	12.0	0.0571	0.210	1.058	1.233
	1,500	500	32.1	15.60	0.0715	0.2192	1.073	1.245
	2,000	1,000	26.1	21.40	0.1147	0.268	1.12	1.308
s_6	1,600	300	27.7	10.8	0.0430	0.1690	1.045	1.184
	1,800	500	25.7	19.5	0.0714	0.1890	1.074	1.206
	1,900	600	25.0	24.0	0.0857	0.201	1.091	1.223
*	2,000	700	24.2	28.9	0.1002	0.212	1.105	1.234
	2,100	800	23.6	33.8	0.1142	0.224	1.118	1.249
	2,200	900	22.8	40.0	0.1285	0.237	1.124	1.266

Calculation of Minimum Mass to Raise One Pound to Various Altitudes in the Atmosphere

The "total initial masses" required to raise one pound from sea level to the upper end of intervals s_6, s_7, s_8 are given its Table VII. They are obtained by multiplying together the minimum masses (marked by stars in Table V), from s_1 up to and including the interval in question, and represent, as already explained, the mass in pounds of a rocket which, starting at sea level, would become reduced to one pound at the altitude given.

The highest altitude attained by the one pound mass is not, however, the upper end of the interval in question, but is a very considerable distance higher. This, of course, follows from the fact that the one pound teaches the upper end of each interval with a considerable velocity, and will continue to else after propulsion has ceased until this velocity is reduced to zero, by gravity and air resistance.

If we call v_n the velocity with which the pound mass reaches the upper end of the particular interval where propulsion ceases, b the distance beyond which, the one pound will rise (the cross section still being 1 sq in.), and p the mean air resistance in poundals [47] over the distance b, we have, by the principle of work and energy,

$$h = \frac{v_n^2}{2(g+p)}$$

The values of p are small, owing to small atmospheric density being 1.59 poundals for the b beyond s_6; 0.28 beyond s_7 ($a=50$); and 0.465 beyond s_7 ($a=150$). For s_8 the low density makes this quantity negligible.

The altitudes obtained by adding to the interval the corresponding b are called the "greatest altitude attained" in Table VII.

[46]

P, poundals per sq in.	R, PS(p/p₀)	$\frac{R}{a+g}$	M, lb	M/exp at $\frac{}{c(1-k)}$	exp $\frac{2(a+g)}{c(1-k)}t$	M_2, lb	exp $\frac{7.28(a+g)}{c(1-k)}t$	M_{RI}, lb
11.53	5.97	0.134	1.329	1.236				
30.7	16.00	0.250	1.300	1.162	1.574	1.718	5.225	6.545
46.7	24.3	0.295	1.341	1.165				
165.0	83.3	0.570	1.499	1.207				
95.7	27.7	0.764	1.232	1.216				
108.8	31.4	0.767	1.242	1.200	1.293	1.518	2.581	3.794
165.0	46.25	0.794	1.318	1.227				
305.0	87.90	0.908	1.455	1.263				
150.1	12.0	0.347	1.278	1.261				
170.0	13.55	0.362	1.293	1.255				
195.0	15.65	0.384	1.306	1.250	1.495	1.685	4.32	5.594
218.0	17.49	0.397	1.325	1.252				
243.5	19.45	0.520	1.372	1.280				
417.0	33.4	0.623	1.501	1.340				
343.0	5.16	0.1203	1.206	1.153				
406.0	6.10	0.1186	1.230	1.147				
430.0	6.43	0.1150	1.248	1.142				
460.0	6.90	0.1134	1.260	1.140	1.522	1.581	4.66	5.075
510.0	7.65	0.1165	1.278	1.142				
534.0	8.02	0.1115	1.295	1.151				

Obviously if the start is made at a high elevation, the "total initial mass" required to reach a given height will be less than for a start at less level, due not only to the fact that the apparatus is not raised through so great a height, but also to the fact that the denser part of the atmosphere is avoided. Table VI gives minimum masses M, calculated for a start with zero velocity from the beginning of interval s_3 (i.e., 15,000 ft), the effective velocity being 7000 ft/sec, as in Table V.

It happens that the velocity v_1, for minimum M in the interval s_6 of Table VI is the same as the v_1 for the same interval in Table V. The calculations that have been made for the intervals beyond s_6 apply therefore to the present case, and the only difference between the two cases is that the masses required to reach s_7 will be greater for the start at sea level than for the start at 15,000 ft.

The calculations beginning at 15,000 ft have been carried out in Table VII for all but the lowest "effective velocity;" and it will be observed that the start from a high elevation becomes important only for the lower "effective velocities."

The most striking as well as the most important conclusion to be drawn from Table VII is the small "total initial mass" required to raise one pound to very great altitudes when the "effective velocity" is 7000 ft/sec, the mass for the height of 437 miles (2,510,000 ft), for example, being but 12.33 lb, starting from sea level. Even for an "effective velocity" of 3500 ft/sec, which allows of considerable inefficiency in the rocket apparatus, the mass is sufficiently moderate to render the method perfectly practicable, for in this case an altitude of over 230 miles from sea level, practically the limit of the earth's atmosphere, requires under 90 lb[16]; and an altitude of 118 miles, close under the geocoronium sphere, only 38 lb. For a start at 15,000 ft, the masses are, of course, less, namely, 49.3 lb and 20.9 lb, respectively.[17]

[48] The enormous difference between the total initial masses required for low efficiency rockets compared with those for high, may at first appear surprising; but they should be expected from the exponential nature of Eqs. (6) and (7). Thus if the "effective velocity" is reduced from 7000 ft/sec to half this value, the minimum masses for each interval, neglecting air resistance, will be those for 7000 ft/sec squared; and including air resistance, still greater. Similarly for an effective velocity of 906 ft/sec which is that for reloading

Table VII

Internal	Altitude of upper end of intervals, ft	Greatest altitude attained, ft	Time to reach greatest altitude from sea level, sec	Total initial masses (in lb) Starting	
				$c(1-k)$ = 7,000	$c(1-k)$ = 3,500
s_6	125,00	184,100	144.13	3.665	12.61
s_7 (a = 50) (a = 150)	200,000 200,000	377,500 610,000	217.73 265.93	5.14 6.40	24.36 38.10
s_8 (a = 50) (a = 150)	500,000 500,000	1,228,000 2,310,000	380.53 475.23	9.875 12.33	89.60 267.70
s_9 (a = 50) (a = 150)	9,310,000 3,915,000	∞ ∞	∞ ∞	1,274.0 602.0	1.497×10^6 6.37×10^5

rockets having the same velocity of ejection as Conton ship rockets, the minimum masses will be those for 7000 ft/sec raised [49] to the *7.28th* power; and for bundles for groups of ship rockets, as shown in Fig. 7, the minimum masses will be those for 7000 ft/sec, raised to the *27.2th* power. Even when air resistance is entirely neglected in the calculation for the last case, the masses are of much the same magnitude, as shown in Table VII. The large values of the masses M_{R1}, and M_{R2}, simply express the impossibility of employing rockets of low efficiency. Attention may be called to the particular case under M_{R2} (the groups of slip rockets indicated in Fig. 7) in which one pound is raised to the altitude of 1,228,000 ft (232 miles); the "total initial mass" in this case, even neglecting air resistance entirely, is 2.89×10^{18} lb, or over sixfold greater than the entire mass of the earth.

These large numbers, to be sure, agree with one's first impression as to the probable initial mass of a rocket designed to reach extreme altitudes; but the comparatively small initial masses, possible with high efficiency, are not intuitively evident until one realizes what an enormous reduction is involved in extracting anything at large as the 27th root of a number.

It should be observed that the apparatus is taken as weighing one pound. Strictly speaking, if the recording instruments have a mass of one pound, the entire final mass of the apparatus must be at least 3 or 4 lb. The mass for the recording instruments may be considered as being very small, yet many valuable researches could, of course, be performed with an apparatus weighing no more than this.[18] The entire final apparatus should, if possible, be designed to weigh not over 3 or 4 lb at most, unless the efficiency of the apparatus is so high that the "effective velocity," $c(a-k)$, is at least in the neighborhood of 7000 ft/sec. An examination of Table VII makes very evident the necessity of securing maximum *effectiveness of the apparatus before a rocket for such a purpose as meteorological work, for example, is constructed; in order to make the method as inexpensive as possible.* It should be remarked, however, that the "total initial mass" will really not be increased in as large a proportion as the final mass if the latter is made greater than one pound by virtue of Eq. (2).

Before proceeding further it will be well to consider carefully the question of air resistance as dependent upon the cross section of the rocket during flight. It has already been assumed that the cross section, in the calculation of the minimum M for each

for one pound final mass

from sea level			*Starting from 15,000 ft*		
$c(1-k)$ = 960	$c(1-k)$ = 257.3	$c(1-k)$ = 257.3 R taken = 0	$c(1-k)$ = 7,000	$c(1-k)$ = 3,500	$c(1-k)$ = 960
2,030.0	7.40×10^9	8.63×10^8	2.66	6.95	702.0
2.26×10^4 1.096×10^5	5.46×10^{12} 2.00×10^{15}	6.08×10^{11} 2.28×10^{14}	3.74 4.65	13.38 20.90	7,820 37,800
2.66×10^6 1.318×10^8	2.55×10^{19} 5.77×10^{26}	2.89×10^{18} 6.53×10^{25}	7.19 8.97	49.30 147.30	9.17×10^5 4.51×10^7
5.32×10^{21} 2.49×10^{20}	3.21×10^{76} 3.32×10^{71}	3.63×10^{75} 3.76×10^{70}	926.0 438.0	8.22×10^5 3.51×10^5	1.82×10^{21} 8.59×10^{19}

Interval, was 1 sq in. If we make the apparatus as long, narrow, and compact as possible, the assumption of a cross section of 1 sq [50] in. for an apparatus weighing one pound will not be unreasonable. A glance at Tables V and VI will show that, for "effective velocities" of 7000 ft/sec and 3500 ft/sec, the mass at the beginning of any interval (except s_9) does not greatly exceed one pound—the mass at the end of each interval being one pound—so that the computations are in agreement with this assumption of area of cross section. For the two cases of the adapted Coston rockets, the masses at the beginning of the intervals are much larger; and hence we see that the "total initial masses" in Table VII, large as they are, would have been even larger if a proper value of cross section had been employed.

The important point is, however, that cross-sectional areas of *even less than 1 sq in. should have been used.* The reason for this is obvious when one remembers that in calculating the "total initial masses," when we multiply minimum masses M together, we are also multiplying the cross sections in the same ratio. In other words, we are considering numbers of rockets, each of 1 sq in. cross section, grouped together side by side, into a bundle.

But such an arrangement would have its cross section proportional to its mass and not to the $2/3$ power of its mass, as would be the case if the *shape of the rocket apparatus were at all times similar to the shape at the start* (as in the ideal rocket, Fig. 1). This constant similarity of shape is, as we have seen in Eq. (2), one of the conditions for a minimum initial mass. Hence the "total initial masses" that have been calculated are really larger than the true minima, which would be obtained only by repeating the calculations, assuming a smaller cross section except in the last few intervals, in which the rocket has become so small that the condition of 1 sq in. per pound is approximately satisfied.

Before leaving the subject of air resistance, attention should be called to the fact that the velocities (Table V) do not exceed that for which air resistance has been studied by Mallock until in s_7, for $a = 150$ ft/sec², and in s_8, for $a = 50$ ft/sec²; and furthermore, that the velocities do not become much in excess until the densities have become almost negligible.

Check or Approximate Method of Calculation

A simple calculation, involving only the most elementary formulas instead of Eqs. (6) and (7), will show that the "total initial masses" in Table VII cannot be far from the truth.

Consider, for simplicity, a rocket of the form shown in Fig. 1, and suppose that one-third of the mass of the rocket is fired downward, [51] with a velocity of 7000 ft/sec at the first shot; one-third of the remaining mass, at the second shot; and so on, for successive shots. From the principle of the conservation of momentum it will be evident that the mass that remains is given an additional upward velocity of 35000 ft/sec after each shot.

Thus, after the fourth shot, the mass that remains is $16/81$, of practically $1/5$, of the initial mass, and the velocity is 14,000 ft/sec. This velocity is sufficient, if we neglect air resistance, to raise the part of the rocket that remains to an altitude of 580 miles (by the familiar relation $v^2 = 2gh$). Although the range would be much reduced if air resistance were considered, it should nevertheless be remembered that the values in Table VII are calculated for the condition under which air resistance is a minimum.

The above simple case is not realizable in practice because of the large mass of propellant for each shot compared with the total mass—i.e., provision is not made for the mass of the chamber. The result will be the same, however, if smaller charges are fired in rapid succession, as will be evident from a calculation similar to the above, which is carried out in Appendix E, under the assumption of smaller charges for successive shots.

Recovery of Apparatus on Return

A point of considerable practical importance is the question of finding the apparatus on its return, and of following it during flight, both of which depend in a large measure upon the time of flight.

Concerning the times of ascent, Table VII shows that these are remarkably short. For example, a height of over 250 miles is reached in less than 6 1/2 minutes (s_8; $a= 50$). The reason is, of course, that the rocket under present discussion possesses the advantage of the bullet in attaining a high velocity, with the added advantage of starting gradually from rest. In fact, the motion fulfills closely the ideal conditions for extremely rapid transit—namely, starting from rest with the maximum acceleration possible, and reversing this acceleration, in direction, at the middle of the journey.

The short time of ascent and descent is, of course, highly advantageous as regards following the apparatus during ascent, and recovering it on landing. The path can be followed, by day, by the ejection of smoke at intervals, and at night by flashes. Any distinctive feature, as, for example, a long black streamer, could assist in rendering the instruments visible on the return.

[52] Some means will, of course, be necessary to check the velocity of the returning instruments. It might not appear, at first sight, that a parachute would be operative at a velocity of 10,000 ft/sec or more; but it should be remembered that this velocity will occur in air of very small density, so that the pressure, or force per unit area of the parachute, would not be excessive, notwithstanding the high velocity of the apparatus. The magnitudes of the air resistance will, of course, be much larger than would be indicated from the values of R in Tables V and VI, from the fact that, for motion with the parachute, the cross section will be much larger in proportion to the mass of the rocket than for the cases presented in these tables.

If the parachute is so large that the velocity will be decreased greatly when the denser air is reached, the descent will be so slow that finding of the apparatus will not be so easy as would be the case with a more rapid descent. For this reason, part of the parachute device must be lost automatically when the apparatus has fallen into air of a certain density; or else the parachute must be small enough to facilitate a rapid descent, with additional parachute devices rendered operative as the rocket nears the ground. Such devices are not described in the present paper, but can be of simple and light construction.

The effectiveness of a parachute of even moderate size, operating in a region where the density is small, may be demonstrated by the following concrete example. Suppose that an apparatus weighing 1 lb and having a parachute of 1 sq ft area descends from the altitude 1,228,000 ft (over 200 miles), and does not encounter any atmospheric resistance until it is level with the upper limit of s_6 (125,000 ft). This condition will not, of course, be that which would actually obtain in practice, for a continually increasing resistance will be experienced as the apparatus descends; but if a sufficient braking action can be shown to exist in the present example, the parachute device will *a fortiori* be satisfactory in practice.

The velocity acquired by the apparatus in falling freely under the influence of gravity between the two levels is

$$\sqrt{64 \times 1,103,000} = 8400 \text{ ft/sec}$$

Now the air resistance in poundals per square inch of section at atmospheric pressure for this velocity is, from the plot of Mallock's formula, 360×32 poundals per square inch, making the value of R for the area of the parachute

$$R = 1,653,000 \text{ poundals/in.}^2$$

[53] But the actual resistance is R, multiplied by the relative density at 125,000 ft which is approximately 0.01, giving for the resistance

$$F = 16,530 \text{ poundals/in.}^2$$

A retarding acceleration must therefore act upon the apparatus, of amount given by

$$a = \frac{F}{M} = 16,530 \text{ ft/sec}^2$$

Hence it is safe to say that, long before the apparatus had fallen to the 125,000-ft level, the velocity would have been reduced to, and maintained at, a safe valve, with the employment of even a small parachute. This case, it should be noticed, is entirely different from that of a falling meteor, in that the apparatus under discussion falls from rest, at the highest point reached; whereas the meteor enters the earth's atmosphere with an enormous initial velocity.

If it is considered desirable, for any reason, to dispense with a sufficiently large parachute, the retarding of the apparatus may be accomplished to any degree by having the rocket consist, at its highest point of flight, not merely of instruments plus parachute, but of instruments together with a chamber, and considerable propellant material. Then, after the rocket has descended to some lower level, let us say, to the upper limit of s_6, this propellant material can be ejected, so that the velocity is considerably checked before the apparatus reaches as low an altitude as, say, 5000 ft. For the cases in which the effective velocity $c(1-k)$ is as large as 7000 ft/sec there is little inconvenience in increasing the mass in this way. But for the case in which $c(1-k) = 3500$, this method can hardly be as satisfactory as the parachute method; for if the "final" mass to be elevated is made a number of pounds, let us say n, the "total initial mass" (which is large even for one pound final mass) will be n-fold larger, and the apparatus correspondingly more expensive.

Applications to Daily Observations

Before leaving the subject of the attainment of high altitudes within the earth's atmosphere, it will be well to mention briefly another application of the method herein discussed: namely, to the sending daily of small recording instruments to moderate altitudes, such as 5 or 6 miles. As is already understood, simultaneous daily observations of the vertical gradients of pressure, temperature, and wind velocity, at a large number of stations would doubtless be of great value in weather forecasting. The method herein described [54] is evidently well suited for such a purpose, in that the time of rise and fall would be short, so that the apparatus could easily be found on the return. Thus the expense would be slight, being simply that of a fresh magazine of cartridges for each day.

For this work, as well as for that previously described, the head of the rocket should be prevented from rotating, by means of a gyroscope, such as is explained in U.S. Patent No. 1,102,653.

Calculation of Minimum Mass Required to Raise One Pound to an "Infinite" Altitude

From the fact that the preceding calculation leads us to conclude that such an extreme altitude as 2,310,000 ft (over 437 miles) can be reached by the employment of a moderate mass, provided the efficiency is high, it becomes of interest to speculate as to whether or not a velocity as high as the "parabolic" velocity for the earth could be attained by an apparatus of reasonably small initial mass.

Theoretically, a mass projected from the surface of the earth with a velocity of 6.95 miles/sec would, neglecting air resistance, reach an infinite distance, after an infinite time; or, in short, would never return. Such a projection without air resistance, is, of course, impossible. Moreover, the mass would not reach infinity but would come under the gravitational influence of some other heavenly body.

We may, however, consider the following conceivable case: If a rocket apparatus such as has here been discussed were projected to the upper end of interval s_8, with an acceleration of 50 or 150 ft/sec^2, and this acceleration were maintained to a sufficient distance beyond s_8, until the parabolic velocity were attained, the mass finally remaining would certainly never return.

If we designate as the upper end of s_9 the height at which the velocity of ascent becomes the "parabolic" velocity, it will be evident that this height will be different for the two accelerations chosen, inasmuch as the "parabolic" velocity decreases with increasing distance from the center of the earth.

If we call n = the "parabolic" velocity at a distance H above the surface of the earth
v_1 = the velocity acquired at the upper end of interval s_8,
s_0 = the height of the upper end of s_8 above sea level

we have, taking the radius of the earth as 20,900,000 ft,

$$u = v_1 + at \tag{11}$$
$$H = s_0 + v_1 t + {}^1\!/_2 at^2 \tag{12}$$

[55] and also the equation relating "parabolic" velocity to distance from the center of the earth

$$\frac{36,700}{u} = \sqrt{\frac{20,900,000 + H}{20,900,000}} \tag{13}$$

On putting the values of u and H, from (11) and (12), in (13), we have

$$\sqrt{20,900,000} \times 36,700 = (v_1 + at)\sqrt{21,400,000 + v_1 t + 1/2 at^2} \tag{14}$$

Equation (14) is a biquadratic in t, from which t may easily be obtained (by trial and error). The values of t, for the two accelerations chosen, given in Table V, enable u and the initial masses for s_9 to be at once obtained.

The effect of air resistance in s_9 is negligible, if we accept Wegener's conclusions, above mentioned, concerning the properties of geocoronium. But even if we use the empirical rule of a fall of density to one-half for every 3.5 miles, we shall find the reduction of velocity very small on passing from the upper end of s_8 (500,000 ft) to 1,000,000 ft (beyond which the density is negligible). This is shown in Appendix F.

The "total initial masses," to raise one pound to an "infinite" altitude for the two accelerations chosen, are given in Table VII. It will be observed that they are astonishingly small, provided the efficiency is high. Thus with an "effective velocity" of 7000 ft/sec, and an acceleration of 150 ft/sec^2, the "total initial mass," starting at sea level, is 602 lb, and starting from 15,000 ft is 438 lb.[19] The mass required increases enormously with decreasing efficiency, for, with but half of the former "effective velocity" (3500 ft/sec), the "total initial mass," even for a start from 15,000 ft, is 351,000 lb. The masses would obviously be slightly less if the acceleration exceeded 150 ft/sec^2.

It is of interest to speculate upon the possibility of proving that such extreme altitudes load been reached even if they actually were attained. In general, the proving would be a difficult matter. Thus, even if a mass of flash powder, arranged to be ignited automatically after a long interval of time, were projected vertically upward, the light would at best be very faint, and it would be difficult to foretell, even approximately, the direction in which it would be most likely to appear.

The only reliable procedure would be to send the smallest mass of flash powder possible to the dark surface of the moon when in conjunction (i.e., the "new" moon), in such a way that it would be [56] ignited on impact. The light would then be visible in a powerful telescope. Further, the larger the aperture of the telescope, the greater would be the ease of seeing the flash, from the fact that a telescope enhances the brightness of point sources, and dims a faint background.

An experiment was performed to find the minimum mass of flash powder that should be visible at any particular distance. In order to reproduce, approximately, the conditions that would obtain at the surface of the moon, the flash powder was placed in small capsules C, Plate 9, Fig. 1, held in glass tubes 7; closed by rubber stoppers. The tubes were exhausted to a pressure of from 3 to 10 cm of mercury, and sealed, the stoppers being painted with wax, to preserve the vacuum. Two shellacked wires, passing to the powder, permitted firing of the powder by an automobile spark coil.

It was found that Victor flash powder was slightly superior to a mixture of powdered magnesium and sodium nitrate, in atomic proportions, and much superior to a mixture of powdered magnesium and potassium chlorate, also in atomic proportions.

In the actual test, six samples of Victor flash powder, varying weight from 0.05 gm to 0.0029 gm, were placed in tubes as shown in Plate 9, Fig. 1, and these tubes were fastened in blackened compartments of a box, Plate 9, Fig. 2, and Plate 10, Fig. 1. The ignition system was placed in the back of the same box, as shown in Plate 10, Fig. 2. This system comprised a spark coil, operated by three triple cells of Eveready battery, placed two by two in parallel. The charge was fired on closing the primary switch at the left. The six-point switch at the right served to connect the tubes, in order, to the high-tension side of the coil.

The flashes were observed at a distance of 2.24 miles on a fairly clear night; and it was found that a mass of 0.0029 gm of Victor flash powder was visible, and that 0.015 gm was strikingly visible, all the observations being made with the unaided eye. The minimum mass of flash powder visible at this distance is thus surprisingly small.

From these experiments it is seen that if this flash powder were exploded on the surface of the moon, distant 220,000 miles, and a telescope of 1-ft aperture were used— the exit pupil being not greater than the pupil of the eye (e.g., 2mm)—we should need a mass of flash powder of

2.67 lb to be just visible
13.82 lb *or less* to be strikingly visible

[57] If we consider the final mass of the last "secondary" rocket plus the mass of the flash powder and its container, to be four times the mass of the flash powder alone, we should have, for the final mass of the rocket, four times the above masses. These final masses correspond to the "one pound final mass" which has been mentioned throughout the calculations.

The "total initial masses," or the masses necessary for the start at the earth, are at once obtained from the data given in Table VII. Thus if the start is made from sea level, and the "effective velocity of ejection" is 7000 ft/sec, we need 602 lb for every pound that is to be sent to "infinity."[1]

We arrive, then, at the conclusion that the "total initial masses" necessary would be

6,436 lb or 3.21 tons; flash just visible
33,278 lb or 16.63 tons (or less); flash strikingly visible

A "total initial mass" of 8 or 10 tons would, without doubt, raise sufficient flash powder for clear visibility.[21]

These masses could, of course, be much reduced by the employment of a larger telescope. For example, with an aperture of 2 ft, the masses would be reduced to one-fourth of those just given. The use of such a large telescope would, however, limit considerably the possible number of observers. In all cases, the magnification should be so low that the entire lunar disk is in the field of the telescope.

It should be added that the probability of collision of a small object with meteors of the visible type is negligible, so is indicated in Appendix G.

This plan of sending a mass of flash powder to the surface of the moon, although a matter of much general interest, is not of obvious scientific importance. There are, however, developments of the general method under discussion, which involve a number of

---

1. A simple calculation[20] will show that the total initial mass required to send one pound to the surface of the moon is but slightly less than that required to send the mass to "infinity."

important features not herein mentioned, which could lead to results of much scientific interest. These developments involve many experimental difficulties, to be sure; but they depend upon nothing that is really impossible.

[58] **Summary**

1. An important part of the atmosphere, that extends for many miles beyond the reach of sounding balloons, has up to the present time been considered inaccessible. Data of great value in meteorology and in solar physics could be obtained by recording instruments sent into this region.

2. The rocket, in principle, is ideally suited for reaching high altitudes, in that it carries apparatus without jar, and does not depend upon the presence of air for propulsion. A new form of rocket apparatus, which embodies a number of improvements over the common form, is described in the present paper.

3. A theoretical treatment of the rocket principle shows that, if the velocity of expulsion of the gases were considerably increased and the ratio of propellant material to the entire rocket were also increased, a tremendous increase in range would result, from the fact that these two quantities enter exponentially in the expression for the initial mass of the rocket necessary to raise given mass to a given height.

4. Experiments with ordinary rockets show that the efficiency of such rockets is of the order of 2 percent, and the velocity of ejection of the gases, 1000 ft/sec. For small rockets the values are slightly less.

With a special type of steel chamber and nozzle, an efficiency has been obtained with smokeless powder of over 64 percent (higher than that of any heat engine ever before tested); and a velocity of nearly 8000 ft/sec, which is the highest velocity so far obtained in any way except in electrical discharge work.

5. Experiments were repeated with the same chambers in vacuo, which demonstrated that the high velocity of the ejected gases was a real velocity and not merely an effect of reaction against the air. In fact, experiments performed at pressures such as probably exist at an altitude of 30 miles gave velocities even higher than those obtained in air at atmospheric pressure, the increase in velocity probably being due to a difference in ignition. Results of the experiments indicate also that this velocity could be exceeded, with a modified form of apparatus.

[59] 6. Experiments with a large chamber demonstrated not only that large chambers are operative, but that the velocities and efficiencies are higher than for small chambers.

7. A calculation based upon the theory, involving data that is in part that obtained by experiments, and in part what is considered as realizable in practice, indicates that the initial mass required to raise recording instruments of the order of one pound, even to the extreme upper atmosphere, is moderate. The initial mass necessary is likewise not excessive, even if the effective velocity is reduced by half. Calculations show, however, that any apparatus in which ordinary rockets are used would be impracticable owing to the very large initial masses that would be required.

8. The recovery of the apparatus on its return, need not be a difficult matter, from the fact that the time of ascent even to great altitudes in the atmosphere will be comparatively short, owing to the high speed of the rocket throughout the greater part of its course. The time of descent will also be short; but free fall can be satisfactorily prevented by a suitable parachute. A parachute will be operative for the reason that high velocities and small atmospheric densities are essentially the same as low velocities and ordinary density.

9. Even if a mass of the order of a pound were propelled by the apparatus under consideration until it possessed sufficient velocity to escape the earth's attraction, the initial mass need not be unreasonably large, for an effective velocity of ejection which is without doubt attainable. A method is suggested whereby the passage of a body to such an extreme altitude could be demonstrated.

Conclusion

Although the present paper is not the description of a working mode, it is believed, nevertheless, that the theory and experiments, herein described, together settle all points that could seriously be questioned, and that it remains only to perform certain necessary preliminary experiments before an apparatus can be constructed that will carry recording instruments to any desired altitude.[22]

[60] **Appendix A**

Theory of the Motion with Direct Lift

Let M = the mass of the suspended system, comprising the chamber together with any parts rigidly attached thereto
m_0 = the mass of the expelled charge, comprising wadding and the attached copper wire, the smokeless powder charge (and also, in the experiments in vacuo, the black powder priming charge)
V = the initial upward velocity of the mass M
v = the average downward velocity of the mass M_0
s = the upward displacement of the mass M

We have at once for the initial velocity of the mass M

$$V^2 = 2gs$$

and employing the conservation of momentum, we have for the kinetic energy per gram of mass m_0, expelled,

$$\frac{v^2}{2} = \frac{M^2}{m_0^2} gs$$

Appendix B

Theory of Displacements for Simple Harmonic Motion

In addition to the notation given under Appendix A, the following additional notation must be employed:
Let m_3 = the mass of the spring
F_1 = the force in dynes which produces until extension of the spring
M_1 = the mass in dynes which produces unit extension of the spring
s = the upward displacement of M, resulting from the firing, that would be had if there were no friction

Then, allowing for the mass of the spring, we have, from the theory of simple harmonic motion:

$$Fx = \left(M + \frac{m_3}{3}\right)\left(\frac{2\pi}{P}\right)^2 x$$

where x is any displacement and p is the period of the motion,
[61] But V is the maximum velocity during the motion and hence $V = \omega s$, where s is the maximum displacement and ω is a constant, having the usual significance; also

$$P = \frac{2\pi}{\omega}$$

Hence

$$m_1 g = \left(M + \frac{m_3}{3}\right)\frac{V^2}{s^2}$$

But by the conservation of linear momentum,

$$\left(M + \frac{m_3}{3}\right)V = m_0 v$$

Hence

$$m_1 g = \left(M + \frac{m_3}{3}\right)\left(\frac{m_0 v}{M + m_3 3}\right)^2 \frac{1}{s^2}$$

giving, for the kinetic energy per gram of mass expelled,

$$\frac{v^2}{2} = \frac{(M + m_3/3)(m_{1g})}{2 m_0^2} s^2$$

From this it is possible to obtain the efficiency, by dividing by the heat value of the powder, in ergs; and also the velocity in kilometers per second by multiplying by 2, extracting the square root, and dividing by 10^5.

Correction of the Displacement s for Friction

The displacement s in the preceding calculation is assumed to be the corrected displacement. This is obtained from the upward displacement s_1 and the downward displacement s_2, as

$$s = s_1 \sqrt{\frac{s_1}{s_2}}$$

Appendix C

Theory of Direct-lift Impulse Meter

The theory of the direct-lift impulse-meter is as follows:
Calling I the momentum of the gas that strikes the end of the aluminum cylinder,
- m_c = the mass of the aluminum cylinder
- V_c = the initial upward velocity of the cylinder
- A_c = the area of cross section of the cylinder
- A_g = the area of cross section of the suspended system comprising the gun, lead weight, and holders
- s = the displacement of the aluminum cylinder, as obtained [62] from the trace on the smoked-glass tube

Thus we have, by the principle of the conservation of linear momentum, for the momentum per unit area produced by the gaseous rebound,

$$\frac{I}{A_c} = \frac{m_c V_c}{A_c} = \frac{m_c \sqrt{2gs}}{A_c}$$

Hence the momentum communicated to the suspended system by the gaseous rebound is

$$\frac{m_c A_g \sqrt{2gs}}{A_c}$$

and calling Q the ratio or the momentum given the gun by gaseous rebound to the observed momentum of the suspended system, we have

$$Q = \frac{m_c A_g \sqrt{2gs}}{m_0 A_c v}$$

Appendix D

Theory of Spring Impulse Meter

The theory of the spring impulse meter is as follows: If we use the same notation as in the preceding case, calling, in addition, the mass of the spring m, and the mass required for unit extension of the spring m, we have, by the same theory as that for the gun suspended by a spring,

$$V_c = \frac{\sqrt{m_1 g}}{\sqrt{m_c + 1/3\, m_3}}$$

Hence the momentum per unit area, communicated to the upper cap of the 12-in. pipe, when the chamber is fired, is

$$\frac{I}{A_c} = \frac{(m_c + 1/3\, m_3)V_c}{A_c} = \frac{\sqrt{m_c + 1/3\, m_3}\,\sqrt{m_1 gs}}{A_c}$$

Hence the momentum that would be communicated to the suspended system by the gaseous rebound, provided the system were at the top of the 12-in. pipe, would be

$$\frac{A_g \sqrt{m_c + 1/3 m_3}\,\sqrt{m_1 gs}}{A_c}$$

and the percentage Q of the momentum communicated by the gaseous rebound to the observed momentum of the suspended system is

$$Q = \frac{A_g \sqrt{m_c + 1/3 m_3}\,\sqrt{m_1 gs}}{A_c m_0 v}$$

[63]

Appendix E

Check on Approximate Method of Calculation for Small Charges Fired in Rapid Succession

Consider a rocket weighing 10 lb, having 2 lb of propelling material, fired 2 oz at a time, eight times per second, with a velocity of 6000 ft/sec—much less than the highest velocity attained in the experiments, either in air or in vacuo.

Let us suppose that, for simplicity, the rocket is directed upward and that each shot takes place instantly (a supposition not far from the truth); the velocity remaining constant between successive shots.

After the first shot, the mass, 9 ⅞ lb, has an upward velocity v_0 is at once found by the conservation of momentum. But it is decreased by gravity until, at the end of ⅛ sec, it is reduced to

$$v_0' = v_0 - gt$$

the space passed over during this time being

$$s = v_0 t - \tfrac{1}{2} g t^2$$

We have then, $v_0' = 71.8$ ft/sec, and $s = 9.23$ ft.

At the beginning of the second interval of ⅛ sec, and additional velocity is given the remaining mass, of 76.8 ft/sec, and the final velocity' and space passed over may be found in the same way. By completing the calculations for the remaining intervals we shall have for time just under ½ sec,

$$v' = 293.1 \text{ ft/sec} \qquad s = 91.93 \text{ ft}$$

for time just under 1 sec,

$$v_0' = 603.8 \text{ ft/sec} \qquad s = 335.48 \text{ ft and}$$

for time just under 2 sec,

$$v_0' = 1284.1 \text{ ft/sec} \qquad s = 1315.68 \text{ ft/sec}$$

These figures compare well with those in Table V, for s_1. In the present check, air resistance would doubtless be unimportant until the velocity had reached 1000 ft/sec or so; but the velocity would, even if decreased somewhat by air resistance, compare favorably with that of a projectile fired from a gun.

No more elaborate calculation is necessary to demonstrate the importance of the device, even for military purposes alone; for it combines portability and cheapness (no gun is required for firing it) with a range which compares favorably with the best artillery. Further, all difficulties of the nature of erosion are, of course, avoided.

[64] **Appendix F**

Proof that the Retardation between 500,000 Feet and 1,000,000 Feet is Negligible

The falling off of velocity w, due to air resistance, is given by

$$P \frac{p}{p_0} sh = 1/2 M_0 w_1^2$$

where P – the mean air resistance in poundals per square inch between the altitudes 500,000 and 1,000,000 ft from the previously mentioned velocity curves, the pressure being considered at atmospheric
P = the mean density over this distance
S = the mean area of cross section of the apparatus throughout the distance, taken as 25 sq in, in view of the the average mass M_0 throughout the interval
h = the distance traversed: 500,000 ft

It is thus found that the loss of velocity w is less than 10 ft/sec (for $a = 150$ ft/sec) even when p/p_0 is taken as constant throughout the distance and equal to that at 500,000 ft (i.e., 2.73×10^{-9}).

Appendix G

Probability of Collision with Meteors

The probability of collision with meteors of "visible" size is negligible. This can be shown by deriving an expression for the probability of collision of a sphere with particles moving in directions at random, all having constant velocity, the expression being obtained on the assumption that the speed of the sphere is small compared with the speed of the particles.

If we accept Newton's estimate[7] of the average distance apart of meteors as being 250 miles, we have by considering collision between very small meteors of velocity 30 miles/sec, and a sphere 1 ft in diameter of velocity 1 mile/sec, moving over a distance of 220,000 miles, the probability 23 as 1.23×10-8; which is, of course, practically negligible. The value would be slightly greater if the meteors were considered as having a diameter of several centimeters, rather than being particles[24]; the probability would be less, however, if meteor swarms were avoided.

[65]   **Notes**

10. A step-by-step method of solution similar to that herein employed can evidently be used for oblique projection—other conditions remaining the same.

11. If the efficiency is estimated by the kinetic energy of the rocket itself (from the velocity the average mass of the rocket would acquire, by virtue of the recoil of the gases ejected with the "average velocity" measured), the efficiencies will, of course, be less than the two values given in Table 1, being, respectively, 0.39 and 0.50 percent.

12. Since this manuscript was written, rockets with a single charge, constructed along the general lines here explained, have been considerably further developed.

13. Chambers of considerably reduced weight have since been made and tested for velocities comparable to those here mentioned. For two particular types of loading device, the ratios of weight of chamber to weight of charge (here, 120) were, respectively 63 (also 30 for this case, but at a sacrifice of velocity) and 22; the ratio, for the nozzles, being reducible to comparatively small values. In neither of these cases was any special attempt made to reduce the weight of the chambers.

14. Later experiments support this prediction, and also demonstrate that firing of the charges can take place in rapid succession.

15. The values of c and $(1-k$, here assigned), were chosen as being the largest that could reasonably be expected. Later experiments have shown that lower values are more easily realizable, but it should at the same time be understood that *no special attempt has been made to obtain experimentally the highest values of these quantities.* The numbers chosen may, then, be considered as at least possible limiting values.

It is well to mention, in this connection, that the developments with the multiple-charge rocket have, so far, exceeded original expectations. This is in accord with the fact that the experimental results have, from the start, been more favorable than were expected. Thus an efficiency of 50 percent was at first considered the limit of what could be attained, and 4000 to 5000 ft/sec, the highest possible velocity. Further it was naturally not expected that the velocities obtained in vacuo would actually exceed those in air; nor were chambers as light as those at present used considered producible without considerable experimental difficulty.

16. *Distribution of mass among the secondary rockets for cases of large total initial mass.* For very great altitudes, secondary rockets will be necessary, as already explained, in order to keep the proportion of propellant to total weight sensibly constant. The most extreme cases will require groups of secondary rockets, which groups are discharged in succession.

There are, under any circumstances, two possibilities: either the secondaries may be small, so that each time a secondary rocket, or group of secondaries, is discarded, the total mass is not appreciably changed, as indicated schematically at (*a*), Fig. 8; or a series of as large secondaries as possible may be used (*b*), Fig. 8, in which case the empty casings constitute a considerable fraction of the entire weight at the time the discarding takes place.

[66] Insofar as avoiding difficulties of construction is concerned, the use of a smaller number of larger secondaries is preferable, but they should be long and narrow, as otherwise the air resistance on the nearly empty casings will be greater for the same weight of propellant than would be the case (*a*), were used, in as compact an arrangement as possible. It should be explained, also, that if very small secondaries were employed, the metal of the magazines and casings would become a considerable fraction of the entire weight, as the amount of surface enclosing the propellant would then be a maximum.

---

7. Newton, *Encyc. Brit.*, 9 ed., v. 16.

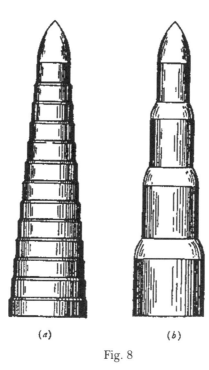

Fig. 8

Possibility of employing case (b). A rough calculation shows at once the possibility of using a comparatively small number of large secondaries [67] (or groups), *provided*, as is, of course, to be expected from dimensional considerations, that the larger any individual rocket, the less, in proportion, need be the ratio of weight of metal to weight of propellant.

Such a calculation can be made by finding the number of secondary rockets, for case *(b)*, that would be required for the same total initial mass, other conditions being the same, as for continuous loss of mass with zero relative velocity, which is practically case *(a)*.

For the latter, Eq. (7), in which R and g are neglected, is evidently sufficient for the purpose, for the reason that the form of the expression, so far as $(1-k)$ is concerned, is the same whether or not R and g are included.

Let us now find what conditions must hold for case (b), in order that the total initial mass shall equal that for case (a). Assume, first, that the casings are discarded successively at the end of n equal intervals of time, no mass being discarded except at these times; the velocity of gas ejection being c, as before. The total initial mass is obtained as the product of the initial masses for each interval, from Eq (7) with $k = 0$, assuming the final mass for each interval is, as before 1 lb, after first multiplying the initial masses by a greater factor than unity, the excess over unit being the weight h of the casings which are discarded at the end of the intervals.

If, in case *(a)*, we divide the time into n equal intervals in the same way, we shall have, as the condition that the total initial masses are the same in the two cases,

$$M = \exp\frac{a(t/n)n}{c(1-k)} = (1+h)^n \exp\frac{a(t/n)n}{c} \tag{15}$$

We obtain, then, on combining (15) with (7),

$$M^k = (1+h)^n$$

from which

$$n = k\frac{\log M}{\log(1+h)} \tag{16}$$

Let us assume, for case *(a)* (many small secondary rockets), as well as for case *(b)* (large secondary rockets), that the ratio of mass of metal to mass of propellant is the minimum reasonable amount that can be expected, which may be put tentatively, at least, as 1:14 and 1:18, respectively.

Two cases will suffice for purpose of illustration: one in which the ratio of initial to final mass is moderately large, e.g., 40, and the other in which the ratio is extreme, e.g., 600.

The number of secondaries (or separate groups) for *(b)*, for these two cases, are, from (16), 5 and 9 respectively, *n* being necessarily an integer.

It is to be understood that the numbers could be made even smaller, although this would necessitate larger total initial masses.

17. If the start were made at a greater elevation than 15,000 ft, for example, at 20,000 or 25,000 ft, the reduction of the "total initial mass" would, of course, be considerably greater. Further, if the rocket were of comparatively small mass, it could be raised to an even greater initial height by balloons.

18. Actually, 300 gm would be sufficient, for many researches.

19. Attention is called to the fact that hydrogen and oxygen, combining in atomic proportions, afford the greatest heat per unit mass of all chemical transformations. For this reason, if the calculations are made under the [68] assumption that hydrogen and oxygen are used (in the liquid or solid state, to avoid weight of the container), giving the same efficiency as that for which Infallible smokeless powder produces respective velocities of, for example, 5500 ft/sec and 7500 ft/sec, the velocities (deducting 218.47 cal/gm as the latent heat plus specific heat, from boiling point to ordinary temperature) would be 9400 ft/sec and 11,900 ft/sec; and the total initial masses for a start from 15,000 ft, respectively, 199 lb and 43.5 lb.

Incidentally, except for difficulties of application, the use of hydrogen and oxygen would have several other evident advantages.

20. This calculation is made under the assumption of stationary centers for the earth and moon.

21. The time of transit for the case under discussion would, of course, be comparatively large. If, however, the final velocity were to exceed by 1000 or 2000 ft/sec the velocity calculated, the time would be reduced to a day or two.

The time can be calculated from solution, by Plana (Memorie della Reale Accademia della Scienze di Torino, Ser. 2, vol. 20, 1863, pp. 1-86), of the analogous problem of the determination of the initial velocity and time of transit of a body, such as a volcanic rock, projected from the moon toward the earth.

22. At the time of signing of the Armistice, the net result of the development of a reloading mechanism had been the demonstration of an operative apparatus that was simple and travelled straight, with the essential parts sufficiently strong and light, using a few cartridges of simple form.

The work remaining, upon which progress has since been made, has been the adaptation of the device for a large proportionate weight of propellant.

23. The probable number of collisions here calculated is the sum of the probable numbers obtained by taking of velocity of the spherical body, and of the meteors, separately equal to zero.

Let v = velocity of the spherical body
  V = velocity of the meteors
  n = the number of meteors per unit volume, which number is, of course, a fraction
    (mutual collisions between meteors being neglected)
  S = the area of cross section of the spherical body

For $v = 0$, the meteors, if any, which strike the sphere during the time t to $t + dt$ will have come from a spherical shell of radii Vt and $V(t + dt)$, neglecting the diameter of the spherical body in comparison with that of the spherical shell. Further, the probable number in any small volume, in this shell, which are so directed as to strike the body, is

$$\frac{S}{4\pi V^2 t^2}$$

being the ratio of the solid angle subtended at the element, by the spherical body, to the whole solid angle 4π. Hence the probable number of collisions N, from all directions, between the time t_1 and t_2 is, evidently,

$$N = nSV(t_2 - t_1)$$

For $V=0$, and expression of the same form is obtained for the probable number of meteors within the space swept out by the spherical body.

[68] The sum of these separate probable numbers is the number 1.23×10^{-8} in Appendix G.

In general, for any values of v and V, the meteors reaching the spherical body at successive instants come from a spherical surface of increasing radius Vt, with moving center distant vt in front of the initial position of the spherical body.

It should be explained that when v differs but little from V, the relative velocity of the body and meteors is small enough to be neglected, for meteors on this expanding spherical surface lying outside a certain cone, the vertex of which coincides with the moving center of the spherical body.

Also if v exceeds V, the only part of the expanding spherical surface which is to be considered is that lying outside the contact circle of the tangent cone, the vertex of which also coincides with the moving center of the spherical body.

Attention is called to the fact that, even if meteor swarms were not avoided, the probable number of collisions would be reduced if the direction of motion were substantially that of the swarm.

24. No difference in the calculation would be necessary if the radius of the sphere were to be increased by the diameter of the meteors, these being then considered as particles.

Document I-8

Topics of the Times

New York Times, January 18, 1920, p. 12 col. 5.

A Severe Stain on Credulity

As a method of sending a missile to the higher, and even to the highest, part of the earth's atmospheric envelope, Professor Goddard's multiple-charge rocket is a practicable, and therefore promising device. Such a rocket, too, might carry self-recording instruments to be released at the limit of its flight, and conceivably parachutes would bring them safely to the ground. It is not obvious, however, that the instruments would return to the point of departure; indeed, it is obvious that they would not, for parachutes drift exactly as balloons do. And the rocket, or what was left of it after the last explosion, would have to be aimed with amazing skill, and in a dead calm, to fall on the spot whence it started.

But that is a slight inconvenience, at least from the scientific standpoint, though it might be serious enough from that of the always innocent bystander a few hundred or thousand yards away from the firing line. It is when one considers the multiple-charge rocket as a traveler to the moon that one begins to doubt and looks again, to see if the dispatch announcing the professor's purposes and hopes says that he is working under the auspices of the Smithsonian Institution. It does say so, and therefore the impulse to do more than doubt the practicability of such a device for such a purpose must be—well, controlled. Still, to be filled with uneasy wonder and to express it will be safe enough, for after the rocket quits our air and really starts on its longer journey, its flight would be neither accelerated nor maintained by the explosion of the charges it then might have left. To claim that it would be is to deny a fundamental law of dynamics, and only **Dr. Einstein** and his chosen dozen, so few and fit, are licensed to do that.

His Plan Is Not Original

That Professor **Goddard**, with his "chair" in Clark College and the countenancing of the Smithsonian Institution, does not know the relation of action to reaction, and of the need to have something better than a vacuum against which to react—to say that would be absurd. Of course he only seems to lack the knowledge ladled out daily in high schools.

But there are such things as intentional mistakes or oversights, and, as it happens, **Jules Verne**, who also knew a thing or two in assorted sciences—and had, besides, a surprising amount of prophetic power—deliberately seemed to make the same mistake that Professor **Goddard** seems to make. For the Frenchman, having got his travelers to go toward the moon into the desperate fix of riding a tiny satellite of the satellite, saved them from circling it forever by means of an explosion, rocket fashion, where an explosion would not have had in the slightest degree the effect of releasing them from their dreadful slavery. That was one of **Verne's** few scientific slips, or else it was a deliberate step aside from scientific accuracy, pardonable enough in him as a romancer, but its like is not so easily explained when made by a savant who isn't writing a novel of adventure.

All the same, if Professor **Goddard's** rocket attains sufficient speed before it passes out of our atmosphere—and if its aiming takes into account all of the many deflective forces that will affect its flight, it may reach the moon. That the rocket could carry enough explosive to make on impact a flash large and bright enough to be seen from the earth by the biggest of our telescopes—that will be believed when it is done.

Document I-9

Document title: Robert H. Goddard, *Liquid-propellant Rocket Development,* **Smithsonian Miscellaneous Collections, Volume 95, Number 3 (Washington, DC: Smithsonian Institution Press, 1936). The plates have been omitted from this document.**

Goddard was conducting tests with liquid-fueled rockets in seclusion in the New Mexican desert when his friend, Charles Lindbergh, began urging him to publish the results. Lindbergh encouraged Goddard to accept an invitation to address an annual convention of the American Association for the Advancement of Science (AAAS). Goddard presented the results of his liquid-propellant rocket experiments at the AAAS convention in St. Louis on December 31, 1935. Goddard was genuinely pleased with the warm reception his work received and was finally convinced by Lindbergh to publish the results. The subsequent paper published by the Smithsonian Institution on March 16, 1936, consisted of ten pages of test and twelve pages of plates illustrating his experiments. It was his first published paper since his previous Smithsonian publication in 1919. The response was overwhelmingly favorable, for Goddard had demonstrated the practicality and scientific basis for rocket research and development.

---

[1] # Liquid-propellant Rocket Development

The following is a report made by the writer to The Daniel and Florence Guggenheim Foundation concerning the rocket development carried out under his direction in Roswell, New Mexico, from July 1930 to July 1932 and from September 1934 to September 1935, supported by this Foundation.

This report is a presentation of the general plan of attack on the problem of developing a sounding rocket, and the results obtained. Further details will be set forth in a later paper, after the main objects of the research have been attained.

Introduction

In a previous paper[1] the author developed a theory of rocket performance and made calculations regarding the heights that might reasonably be expected for a rocket having a high velocity of the ejected gases and a mass at all times small in proportion to the weight of propellant material. It was shown that these conditions would be satisfied by having a tapered nozzle through which the gaseous products of combustion were discharged,[2] by feeding successive portions of propellant material into the rocket combustion chambers,[3] and further by employing a series of rockets, of decreasing size, each fired when the rocket immediately below was employ of fuel.[2] Experimental results with power rockets were also presented in this paper.

Since the above was published, work has been carried on for the purpose of making practical a plan of rocket propulsion set forth in 1914[3] which may be called the liquid-propellant type of rocket. In this rocket, a liquid fuel and combustion-supporting liquid are fed under pressure into a combustion chamber provided with a conical nozzle through which the products of combustion are discharged. [2] The advantages of the liquid-propellant rocket are that the propellant materials possess several times the energy of powders, per unit mass, and that moderate pressures may be employed, thus avoiding the weight of the strong combustion chambers that would be necessary if propulsion took place by successive explosions.

---

1. Smithsonian Misc. Coll., vol. 71, no. 2, 1919.
2. U.S. Patent, "Rocket Apparatus," No. 1,102,653, July 7, 1914.
3. U.S. Patent, "Rocket Apparatus," No. 1,103,503, July 14, 1914.

Experiments with liquid oxygen and various liquid hydrocarbons, including gasoline and liquid propane, as well as ether, were made during the writer's spare time from 1920 to 1922, under a grant by Clark University. Although oxygen and hydrogen, as earlier suggested,[4] possess the greatest heat energy per unit mass, it seems likely that liquid oxygen and liquid methane would afford the greatest heat value of the combinations which could be used without considerable difficulty. The most practical combination, however, appears to be liquid oxygen and gasoline.

In these experiments it was shown that a rocket chamber and nozzle, since termed a "rocket motor," could use liquid oxygen together with a liquid fuel, and could exert a lifting force without danger of explosion and without damage to the chamber and nozzle. These rockets were held by springs in a testing frame, and the liquids were forced into the chamber by the pressure of a noninflammable gas.

The experiments were continued from 1922 to 1930, chiefly under grants from the Smithsonian Institution. Although this work will be made the subject of a later report, it is desirable in the present paper to call attention to some of the results obtained.

On November 1, 1923, a rocket motor operated in the testing frame, using liquid oxygen and gasoline, both supplied by pumps on the rocket.

In December 1925 the simpler plan previously employed of having the liquids fed to the chamber under the pressure of an inert gas in a tank on the rocket was again employed, and the rocket developed by means of the tests was constructed so that it could be operated independently of the testing frame.

The first flight of a liquid-oxygen-gasoline rocket was obtained on March 16, 1926 in Auburn, Massachusetts, and was reported to the Smithsonian Institution May 5, 1926. This rocket is shown in the frame from which it was fired, in Plate 1, Fig. 1. Pressure was produced initially by an outside pressure tank, and after launching by an alcohol heater on the rocket.

It will be seen from the photograph that the combustion chamber and nozzle were located forward of the remainder of the rocket, to which connection was made by two pipes. This plan was of advantage [3] in keeping the flame away from the tanks, but was of no value in producing stabilization. This is evident from the fact that the direction of the propelling force lay along the axis of the rocket, and not in the direction in which it was intended the rocket should travel, the condition therefore being the same as that in which the chamber is at the rear of the rocket. The case is altogether different from pulling an object upward by a force which is constantly vertical, when stability depends merely on having the force applied above the center of gravity.

Plate 1, Fig. 2, shows an assistant igniting the rocket, and Plate 2, Fig. 1, shows the group that witnessed the flight, except for the camera operator. The rocket traveled a distance of 184 feet in 2.5 seconds, as timed by a stopwatch, making the speed along the trajectory about 60 miles per hour.

Other short flights of liquid oxygen-gasoline rockets were made in Auburn, that of July 17, 1929 happening to attract public attention owing to a report from someone who witnessed the flight from a distance and mistook the rocket for a flaming airplane. In this flight the rocket carried a small barometer and a camera, both of which were retrieved intact after the flight (Plate 2, Fig. 2). The combustion chamber was located at the rear of the rocket, which is, incidentally, the best location, inasmuch as no part of the rocket is in the high-velocity stream of ejected gases, and none of the gases are directed at an angle with the rocket axis.

During the college year 1929-1930 tests were carried on at Fort Devens, Massachusetts, on a location which was kindly placed at the disposal of the writer by the War Department. Progress was made, however, with difficulty, chiefly owing to transportation conditions in the winter.

At about this time Col. Charles A. Lindbergh became interested in the work and brought the matter to the attention of the late Daniel Guggenheim. The latter made a grant which permitted the research to be continued under ideal conditions, namely, in

---

4. Smithsonian Misc.Coll., vol 71, no. 2, 1919.

eastern New Mexico; and Clark University at the same time granted the writer leave of absence. An additional grant was made by the Carnegie Institution of Washington to help in getting established.

It was decided that the development should be carried on for two years, at the end of which time a grant making possible two further years' work would be made if an advisory committee, formed at the time the grant was made, should decide that this was justified by the results obtained during the first two years. This advisory committee [4] was as follows: Dr. John C. Merriam, chairman; Dr. C.G. Abbot; Dr. Walter S. Adams; Dr. Wallace W. Atwood; Colonel Henry Breckinridge; Dr. John A. Fleming; Col. Charles A. Lindbergh; Dr. C.F. Marvin; and Dr. Robert A. Millikan.

The Establishment in New Mexico

Although much of the eastern part of New Mexico appeared to be suitable country for flights because of clear air, few storms, moderate winds, and level terrain, it was decided to locate in Roswell, where power and transportation facilities were available.

A shop 30 by 55 feet was erected in September 1930 (Plate 3, Figs. 1, 2), and the 60-foot tower previously used in Auburn at Fort Devens was erected about 15 miles away (Plate 4, Fig. 1). A second tower, 20 feet high (Plate 4, Fig. 2), was built near the shop for static tests, that is, those in which the rocket was prevented from rising by heavy weights, so that the lift and general performance could be studied. These static tests may be thought of as "idling" the rocket motor. A cement gas deflector was constructed under each tower, as may be seen in Plate 4, Figs. 1, 2, whereby the gases from the rocket were directed toward the rear, thus avoiding a cloud of dust which might otherwise hide the rocket during a test.

Static Tests of 1930-1932

Although, as has been stated combustion chambers which operated satisfactorily had been constructed at Clark University, it appeared desirable to conduct a series of thorough tests in which the operating conditions were varied, the lift being recorded as a function of the time. Various modifications in the manner of feeding the liquids under pressure to the combustion chamber were tested, as well as variations in the proportions of the liquids, and in the size and shape of the chambers. The chief conclusions reached were that satisfactory operation of the combustion chambers could be obtained with considerable variation of conditions, and that larger chambers afforded better operation than those of smaller size.

As will be seen from Plate 4, Fig. 2, the supporting frame for the rocket was held down by four steel barrels containing water. Either two of for barrels could be filled, and in the latter case the total weight was about 2000 pounds. This weight was supported by a strong compression spring, which made possible the recording of the lift on a revolving drum (Plate 5, Fig. 1) driven by clockwork.

[5] The combustion chamber finally decided upon for use in flights was $5\,^3/_4$ inches in diameter and weighed 5 pounds. The maximum lift obtained was 289 pounds, and the period of combustion usually exceeded 20 seconds. The shifting force was forced to be very steady, the variation of lift being within 5 percent.

The masses of liquids used during the lifting period were the quantities most difficult to determine. Using the largest likely value of the total mass of liquids ejected and the integral of the lift-time curve obtained mechanically, the velocity of the ejected gases was estimated to be over 5000 feet per second. This gave for the mechanical horsepower of the jet 1030 horsepower, and the horsepower per pound of the combustion chamber, considered as a rocket motor, 206 horsepower. It was found possible to use the chambers repeatedly.

The results of this part of the development were very important, for a rocket to reach great heights can obviously not be made unless a combustion chamber, or rocket motor, can be constructed that is both extremely light and can be used without danger of burning through or exploding.

Flights During the Period 1930-1932

The first flight obtained during this period was on December 30, 1930, with a rocket 11 feet long, weighing 33.5 pounds. The height obtained was 2000 feet, and the maximum speed was about 500 miles per hour. A gas pressure tank was used on the rocket to force the liquid oxygen and the gasoline into the combustion chamber.

In further flights pressure was obtained by gas pressure on the rocket, and also by pumping liquid nitrogen through a vaporizer, the latter means first being employed in a flight on April 19, 1932.

In order to avoid accident, a remote-control system was constructed in September 1931, whereby the operator and observers could be stationed 1000 feet from the tower, and the rocket fired and released at will from this point. This arrangement has proved very satisfactory. Plate 5, Fig. 2, shows the cable being unwound between the tower and the 1000-foot shelter, the latter being seen in the distance, and Plate 6, Fig. 1, shows the control keys being operated at the shelter, which is provided with sandbags on the roof as protection against possible accident. Plate 5, Fig. 2, shows also the level and open nature of the country.

One observer was stationed 3000 feet from the tower, in the rear of the 1000-foot shelter, with a recording telescope (Plate 6, Fig. 2). Two pencils attached to his telescope gave a record of altitude and azimuth respectively, of the rocket, the records being made on a paper [6] strip, moved at a constant speed by clockwork. The sights at the front and rear of the telescope, similar to those on a rifle, were used in following the rocket when the speed was high. In Plate 7, Fig. 1, which shows the clock mechanism in detail, the observer is indicating the altitude trace. This device proved satisfactory except when the trajectory of the rocket was in the plane of the tower and the telescope. For great heights, shortwave radio direction finders, for following the rocket during the decent, will be preferable to telescopes.

During this period a number of flights were made for the purpose of testing the regulation of the nitrogen gas pressure. A beginning on the problem of automatically stabilized vertical flight was also made, and the first flight with gyroscopically controlled vanes was obtained on April 19, 1932 which the same model that employed the first liquid-nitrogen tank. The method of stabilization consisted in forcing vanes into the blast of the rocket[5] by means of gas pressure, this pressure being controlled by a small gyroscope.

As has been found by later tests, the vanes used in the flight of April 19, 1932 were too small to produce sufficiently rapid correction. Nevertheless, the two vanes which, by entering the rocket blast, should have moved the rocket back to the vertical position were found to be warmer than the others after the rocket landed.

This part of the development work, being for the purpose of obtaining satisfactory and reproducible performance of the rocket in the air, was conducted without any special attempt to secure great lightness, and therefore great altitudes.

In May 1932 the result that had been obtained were placed before the advisory committee, which voted to recommend the two additional years of development. Owing to the economic conditions then existing, however, it was found impossible to continue the flights in New Mexico.

A grant from the Smithsonian Institution enabled the writer, who resumed full-time teaching in Clark University in the fall of 1932, to carry out tests that did not require flights, in the physics laboratories of the University during 1932-1933, and a grant was received from the Daniel and Florence Guggenheim Foundation which made possible a more extended program of the same nature in 1933-1934.

Resumption of Flights in New Mexico

A grant made by the Daniel and Florence Guggenheim Foundation in August 1934, together with leave of absence for the writer granted [7] by the Trustees of Clark

---

5. U.S. Patent, "Mechanism for Directing Flight," No. 1, 879, 187, September 1932.

University, made it possible to continue the development on a scale permitting actual flights to be made. This was very desirable, as further laboratory work could not be carried out effectively without flights in which to test performance under practical conditions.

Work was begun in September 1934, the shop being put in running order and the equipment at the tower for the flights being replaced. The system of remote control previously used was further improved and simplified, and a concrete dugout (Plate 7, Fig. 2) was constructed 50 feet from the launching tower in order to make it possible for an observer to watch the launching of the rocket at close range....

Development of Stabilized Flight

It was of the first importance to perfect the means of keeping the rockets in a vertical course automatically, work on which was begun in the preceding series of flights, since a rocket cannot rise vertically to a very great height without a correction being made when it deviates from the vertical course. Such correction is especially important at the time the rocket starts to rise, for a rocket of very great range [8] must be loaded with a maximum amount of propellant and consequently must start with as small an acceleration as possible. At these small initial velocities fixed air vanes, especially those of large size, are worse than useless, as they increase the deviations due to the wind. It should be remarked that fixed air vanes should preferably be small, or dispensed with entirely, it automatic stabilization is employed, to minimize air resistance.

In order to make the construction of the rockets as rapid as possible, combustion chambers were used of the same size as those in the work of 1930-1932, together with the simplest means of supplying pressure, namely, the use of a tank of compressed nitrogen gas on the rocket. The rockets were, at the same time, made as nearly streamline as possible without resorting to special means for forming the jacket or casing.

Pendulum Stabilizer

A pendulum stabilizer was used in the first of the new series of flights to test the directing vanes, for the reason that such a stabilizer could be more easily constructed and repaired than a gyroscope stabilizer, and would require very little adjustment. A pendulum stabilizer could correct the flight for the first few hundred feet, where the acceleration is small, but it would not be satisfactory where the acceleration is large, since the axis of the pendulum extends in a direction which is the resultant of the acceleration of the rocket and the acceleration of gravity, and is therefore inclined from the vertical as soon as the rocket ceases to move in a vertical direction. The pendulum stabilizer, as was expected, gave an indication of operating the vanes for the first few hundred feet, but not thereafter. The rocket rose about 1000 feet continued in a horizontal direction for a time, and finally landed 11,000 feet from the tower, traveling at a velocity of over 700 miles per hour near the end of the period of propulsion, as observed with the recording telescope.

Gyroscope Stabilizer

Inasmuch as control by a small gyroscope is the best as well as the lightest means of operating the directing vanes, the action of the gyroscope being independent of the direction and acceleration of the rocket, a gyroscope having the necessary characteristics was developed after numerous tests.

The gyroscope, shown in Plate 8, Fig. 1, was set to apply controlling force when the axis of the rocket deviated 10° or more from the vertical. In the first flight of the present series of tests with gyroscopic [9] control, on March 28, 1935, the rocket as viewed from the 1000-foot shelter traveled first to the left and then to the right, thereafter describing a smooth and rather flat trajectory. This result was encouraging, as it indicated the presence of an actual stabilizing force of sufficient magnitude to turn the rocket back to a vertical course. The greatest height in this flight was 4800 feet, the horizontal distance 13,000 feet, and the maximum speed 550 miles per hour.

In subsequent flights, with adjustments and improvements in the stabilizing arrangements, the rockets have been stabilized up to the time propulsion ceased, the trajectory being a smooth curve beyond this point. In the rockets so far used, the vanes have moved only during the period of propulsion, but with a continuation of the supply of compressed gas the vanes could evidently act against the slipstream of air as long as the rocket was in motion in air of appreciable density. The oscillations each side of the vertical varied from 10° to 30° and occupied from 1 to 2 seconds. Inasmuch as the rockets started slowly, the first few hundred feet of the flight reminded one of a fish swimming in a vertical direction. The gyroscope and directing vanes were tested carefully before each flight, by inclining and rotating the rocket while it was suspended from the 20-foot tower (Plate 8, Fig. 2). The rocket is shown in the launching tower, ready for a flight, in the close-up (Plate 9, Fig. 1) and also in Plate 9, Fig. 2, which shows the entire tower.

The behavior of the rocket in stabilized flight is shown in Plates 10 and 11, which are enlarged from 16-mm motion-picture films of the flights. The time intervals are 1.0 second for the first 5 seconds, and 0.5 second thereafter. The 60-foot tower from which the rockets rise (Plate 9, Fig. 2) appears small in the first few of each set of motion pictures, since the camera was 1000 feet away, at the shelter shown in Plate 6, Fig. 1. The continually increasing speed of the rockets, with the accompanying steady roar, makes the flights very impressive. In the two flights for which the moving pictures are shown, the rocket left a smoke trail and had a small, intensely white flame issuing from the nozzle, which at times nearly disappeared with no decrease in roar or propelling force. This smoke may be avoided by varying the proportion of the fluids used in the rocket, but is of advantage in following the path of the rocket. The occasional white flashes below the rocket, seen in the photographs are explosions of gasoline vapor in the air.

Plate 10 shows the flight of October 29, 1935, in which the rocket rose 4000 feet, and Plate 11 shows the flight of May 31, 1935, in which the rocket rose 7500 feet. The oscillations from side to side, [10] above mentioned, are evident in the two sets of photographs. These photographs also show the slow rise of the rocket from the launching tower, but do not show the very great increase in speed that takes place a few seconds after leaving the tower, for the reason that the motion picture camera followed the rockets in flight.

A lengthwise quadrant of the rocket casing was painted red in order to show to what extent rotation about the long axis occurred in flight. Such rotation as was observed was always slow, being at the rate of 20 to 60 seconds for one rotation.

As in the flights of 1930-1932 to study rocket performance in the air, no attempt was made in the flights of 1934-1935 to reduce the weight of the rockets, which varied from 58 to 85 pounds. A reduction of weight would be useless before a vertical course of the rocket could be maintained automatically. The speed of 700 miles per hour, although high, was not as much as could be obtained by a light rocket, and the heights, also, were much less than could be obtained by a light rocket of the same power.

It is worth mentioning that inasmuch as the delicate directional apparatus functioned while their rockets were in flight, it would be possible to carry recording instruments on the rocket without damage or changes in adjustment.

Further Development

The next step in the development of the liquid-propellant rocket is the reduction of weight to a minimum. Some progress along this line has already been made. This work, when completed, will be made the subject of a later report.

Conclusion

The chief accomplishments to date are the development of a combustion chamber, or rocket motor, that is extremely light and powerful and can be used repeatedly, and of a means of stabilization that operates automatically while the rocket is in flight.

I wish to express my deep appreciation for the grants from Daniel Guggenheim, The Daniel and Florence Guggenheim Foundation, and the Carnegie Institution of Washington, which have made this work possible, and to President Atwood and the Trustees of Clark University for leave of absence. I wish also to express my indebtedness to Dr. John C. Merriam and the members of the advisory committee, especially to Col. Charles A. Lindbergh for his active interests in the work and to Dr. Charles G. Abbot, Secretary of the Smithsonian Institution, for his help in the early stages of the development and his continued interest.

Document I-10

Document title: H.E. Ross, "The B.I.S. Space-ship," *Journal of the British Interplanetary Society,* 5 (January 1939), 4-9.

The British Interplanetary Society was formed in 1933. This article, by H.E. Ross, one of the society's leaders, outlined the society's most important and well-known contribution to spaceflight, a manned lunar mission. Casual meetings on the subject began in London, leading to the formation of a Technical Committee in February 1937. The committee was split into smaller task groups, including one assigned to conduct extremely crude propellant tests. The result was a solid-propellant spaceship for carrying humans to the Moon and returning them to Earth. Despite the proposal's reliance upon solid propulsion (the committee, ignorant of von Braun's ongoing secret research in Germany, had determined that the pumps and cooling systems required for liquid propulsion were too complex and expensive to develop), it effectively outlined the lunar mission conducted by the United States thirty years later.

[4]

The B.I.S. Space-ship

by H. E. Ross

The B.I.S. space-ship design, as shown on the cover of this issue, is such a radical departure from all previously conceived ideas of a space-ship that a full explanation is called for.

In designing a space-ship the designer has a completely different problem to that involved in the design of any other means of transport. A motor car, railway train, aeroplane or ship consists basically of a vessel and a fuel tank, in the latter being placed the fuel required for a journey or journeys. The shortest space-ship voyage, however, is the journey to the Moon, and with the most optimistic estimates of the fuel energy and motor efficiency the quantity of fuel required will still be such that the fuel tank would require to be much larger than the rest of the ship. Consequently we must revert to the old system of petrol cars, so designing our ship that the cans can be attached outside the ship and thrown away when empty. The last condition does not mean that the cans are cheap—they are actually precision engineering jobs, and horribly expensive—but the cost of the fuel needed to bring them back would be even greater. We find by careful calculation that with the best fuels and motors that we can afford it will require about 1,000 tonnes (metric[1]) of fuel to take a 1 tonne [5] vessel to the moon and back, so our designers' problem has been to design a 1 tonne space-ship with containers for 1,000 tonnes of fuel attached outside and detachable.

The nature of rocket motors has also affected the design considerably. With such motors as aero-engines a larger unit can be made lighter in proportion to its power than a small unit, but in the case of rocket motors quite the reverse is the case; in fact the

---

1. A metric tonne is roughly equivalent to an English ton.

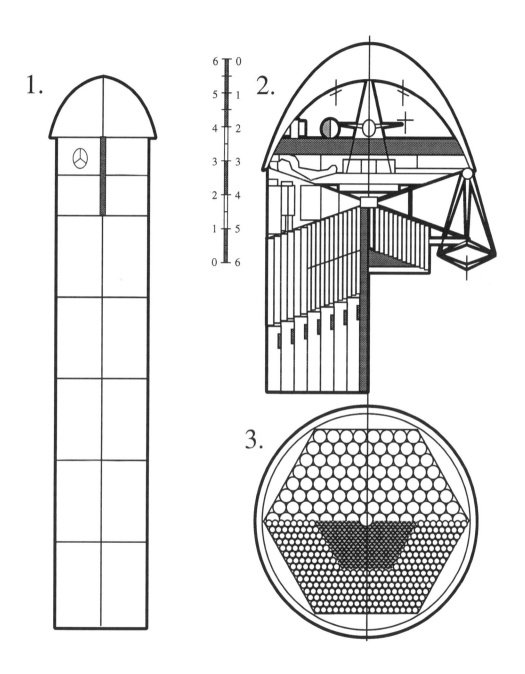

Design for a Lunar Space-ship.

proportionate weight of rocket motors rises so steeply that a motor of more than about 100,000 H.P. is hardly feasible, and as the lifting of the 1,000 tonnes at the start calls for many millions of H.P. this requires a considerable number of small units. Again, since the cost of the motors is less than the cost of the fuel required to bring them back, and as only a few small motors will be required to land the one tonne ship on its return against over a hundred large ones at the start, the motors are jettisoned after use.

For a maximum fuel economy, anything which is to be jettisoned should be jettisoned as soon as possible, and this has led to the cellular space-ship design, with hundreds of small units each comprising a motor and its fuel tank, and each so attached that as soon as it ceases to thrust it falls off. This early detachment of all dead weight has resulted in an enormous increase of efficiency over earlier designs, and has reduced the fuel required for a return voyage to the moon from millions of tonnes to thousands of tonnes.

Owing to the large number of small units, it is possible to start a motor and run it till its load of fuel is exhausted, controlling both thrust and direction by the rate at which fresh tubes are fired. This makes it possible to use solid fuel for the main thrust, with consequent considerable saving in weight, and giving the additional advantages that the strength of the fuel helps to support the parts above and its high density makes the ship very compact. Liquid fuel motors are, however, provide for stages requiring fine control, and also steam jet motors for steering.

Diagram 2 (right) shows the spaceship as it reaches the moon. The approximately hemispherical portion (to the downward pointing cone) is the life container. The portion between the two cones contains the air-lock, air-conditioning plant, heavy stores, batteries and liquid fuel and steam jet motors, etc. Below this are the solid fuel tubes for the return voyage. The whole of the remainder of the vessel (diagrams 1 and 3, consists of the tubes for the outward voyage, which have to be jettisoned by the time of arrival at the moon.

It will be seen that the streamlining is conspicuous by its absence. The form of the ship has been largely dictated by other considerations, and as compared to the terrific power needed to [6] lift the vessel out of the earth's gravitational field the total air resistance is quite negligible (less than 1%), this does not matter greatly. The diameter of the front of the ship is determined as being the smallest reasonable size for the life container. (It should be noted that this design is for a very small space-ship, about the overall size of a large barge. On larger ships this restriction will be somewhat modified). The diameter of the rear of the ship is determined by the firing area required. Too small an area calls for excessive pressures in the motors, and consequently excessively heavy construction. The two diameters being approximately the same has led to the straight-sided form. An increase in central diameter would mean improved streamlining, but this would only decrease the resistance below the velocity of sound, and this is only a small proportion of the whole. On the other hand, the straight-sided form gives the greatest strength, which is of major importance, and also serves to minimize frictional heating. The main body of the space-ship, comprising the motor tubes, is hexagonal in shape; this form giving the closest possible stacking of the tubes.

The form of the nose is intended not so much to reduce the resistance at low velocities, as to split the air at high velocities (several times the velocity of sound), so as to maintain a partial vacuum along the sides. The frontal paraboloidal portion, seen in diagrams 1, 2 and 3, is a reinforced ceramic carapace, capable of withstanding a temperature of 1500°C in air, and by its form the frictional heating is made a maximum on this portion and minimized on the sides. The carapace (which, of course, has no portholes) is detached once the vessel has got away from the earth.

The tubes are stacked in conical layers for greater structural stability, since, apart from the vessel proper—the top portion—the whole strength lies in the tubes, and these are not rigidly fixed together, but simply stacked and held in position by one-way bolts and light webs.

The firing order of the tubes is in rings starting from outside and progressing inwards towards the center. While the motors are firing their thrust holds them in place; when expended, the acceleration of the ship causes them to release from position and they drop off. Those in the inner rings of the bank not yet used do not position themselves for release until their firing thrust carries them a fractional distance up the release bolts. A light metal sheath embraces the outermost ring of tubes; this and the webs are discarded when the whole of the previous bank of motors has been jettisoned.

Diagram 4 gives sections through the vessel at various levels and shows maximum periphery of the carapace. The top half of the diagram represents a section through the large motor tubes [7] stacked in banks A to E; these are used to obtain release from Earth. The lower half of diagram 4 shows the medium and small tubes used for deceleration of the moon (the ship, having been turned end to end, approaches stern first). Fine control for the actual landing is provided by the vertical liquid fuel motors seen within the two cones in diagram 2 and about the hexagon angles in diagram 5. The inner small tubes in diagram 4 are shown in a section through two banks (ref. diagram 2), the lower of these being used for control of deceleration when approaching the moon and the upper bank (ref. diagram 2, right), being used for the return journey.

[8] Adjacent to the top of the liquid fuel motors (diagram 2), are shown four of the tangential tubes. These are necessary in order to provide the crew with artificial gravitation, which is achieved by rotating the ship (approximately 1 revolution in $3 \frac{1}{2}$ seconds). The g value desired is therefor under control of the crew. Not only is this artificial gravitation considered a necessary precaution (the physical affect of long periods of non-gravitation being at present unknown), but in any case haphazard rotation of the vessel would almost certainly take place, making navigational observations impossible. Hence control of rotation is essential. Again, before the moon landing can be attempted it is necessary to stop rotation in order to prevent disaster to the ship when it touches ground.

It is not anticipated that the space-ship can be so accurately manoeuvred that its landing will be without shock. Hydraulic shock absorber arms are therefor incorporated; one of these being shown attached to the frame on the right hand side of diagram 2. These are normally collapsed within the hull, and are extended just prior to landing.

The firing of the motor tubes is carried out by an automatic electrical selector system, but manual control is used for navigational corrections. The ship, being in rotation, is kept thrusting in the correct direction, but this does not prevent "wobble" if firing is not equal on all sides. Manual control of stability is maintained during the first few seconds of ascent, and after that a pendulum conductor automatically controls stability. The main wiring cable to the tubes is led down a central column, provided at each band level with a plug connection which brakes away when its purpose has been served and is then jettisoned.

The hemispherical front of the life-compartment (diagram 2 and 3), is of very light nature; this being made possible on account of the protective carapace above. The segmented carapace (diagram 8), is, of course, discarded after passing out of Earth's atmosphere, and protection of the life-compartment shell is not [9] needed for the ascent from Moon. The return into Earth's atmosphere will be done at low velocities, hence heating of its shell will not be excessive.

Owing to the small scale of the diagrams it has not been possible to show many of the filaments and accessories within the life-compartment, but the following can be noted. Diagram 2 shows one of the seats for the crew of three. These can also be seen pointing radially in diagram 6. The controls for firing are placed on the arms of the chairs, and the chairs themselves move on rails around the life-compartment. The crew recline on these chairs with their heads towards the center of the ship and a circular catwalk is provided for them and around the circumference of the chamber (diagram 2 and 3).

For observation purposes ports are provided in the dome of the life compartment (one shown in diagram 2 and twelve in diagram 7). Under the flange of the carapace, in the rim of the floor of the life-compartment (diagram 1, 2, and 6) are the back-viewing ports; these are covered during thrusting periods. Three forward-viewing ports in the top of the life compartment shell are also provided, see diagram 2 and 7. It should be noted that observation of direction cannot be made during the initial thrusting period in ascent from Earth—it being impossible to see backwards through the tail-blast of the ship—the carapace prevents vision in other directions, and in any case the period is too short to allow of stellar observations. Therefore navigation during this period must be done entirely by means of internal instruments, which consist of an altimeter, speedometer and accelerometer. Another essential is, of course, a chronometer and gyroscope ensures maintenance of direction. A suspended pendulum provides indication of "wobble" and modi-

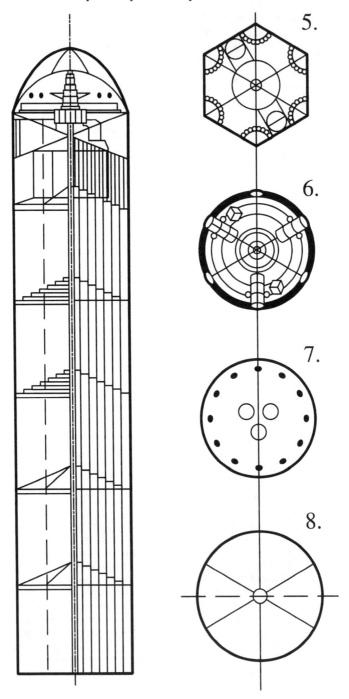

fied sextants and rangefinders are used to determine position. These instruments are placed in convenient juxtaposition to the crew. The cylindrical objects shown just above the catwalk, against the ports (diagram 2) are coelostats. These are synchronized, motor-driven mirror devices something similar to a stroboscope, and it is by means of these that a stationary view of the heavens is provided for navigational observations while the ship is in rotation. The girder structure in the center of the live-compartment is a support for the light shell and also serves to carry navigation instruments. In diagram 1 beneath the carapace and in diagram 6 can be seen the spidered outer and inner doors respectively of the air-lock shown in diagram 5.

A launching device for this ship is necessary on its take-off from Earth, but, being accessory to the ship and somewhat complicated, this will be discussed in a subsequent issue of the Journal.

Document I-11

Document title: Frank J. Malina and A.M.O. Smith, "Flight Analysis of the Sounding Rocket," *Journal of Aeronautical Sciences*, 5 (1938): 199-202.

Frank Malina was a Ph.D. student at Caltech in 1936 when he persuaded the Guggenheim Aeronautical Laboratory, California Institute of Technology (GALCIT), to develop a sounding rocket. He received support in this effort in part because the president of Caltech, Robert A. Millikan, wanted to use rockets for cosmic ray research. In late 1936, a research team began experimenting with rocket engines in the canyons above the Rose Bowl. The tests met with limited success. But by 1938, enough information had been gathered for Malina and a colleague, A.M.O. Smith, to publish GALCIT's first scholarly paper on rocket research. The paper demonstrated that a rocket capable of taking a payload up to 1,000 miles altitude could be developed.

Flight Analysis of the Sounding Rocket

Frank J. Malina and A. M. O. Smith, *California Institute of Technology*

Presented at the Aerodynamics Session, Sixth Annual Meeting, I.Ae.S. January 26, 1938

Introduction

In attempting to reach altitudes above those obtainable by sounding balloons, the rocket motor may be utilized to propel a suitable body. In this analysis a wingless shell of revolution will be considered in vertical flight. It was felt that, before entering into practical experimentation, it was desirable to have a preliminary performance analysis based on simplified assumptions, using the most recent data for air resistance at high speeds. As a matter of fact, this analysis was completed without the knowledge of a similar investigation.[1] However, as this treatment is more general in discussing the influence of the design parameters and more suitable for application to particular cases, the authors believe it is worth while to present the analysis.

The equations of motion for flight *in vacuo* have been included to show the optimum performance and for comparison purposes. After developing similar expressions for flight with air resistance, a series of calculations was carried out using the method of step-by-step integration. The dimensions of the rocket chosen were felt to be feasible for practical

---

1. Ley, Willy, and Schaefer, Herbert, *Les Fusees Volomes Meteorologiques*, L'Aerophile, Vol. 44, No. 10, p. 228-232, October, 1936.

construction. The motor efficiencies for the two cases were chosen to match closely the reported results of R. H. Goddard[2] and Eugene Sänger.[3]

The calculations have not been extended to further cases, as the amount of labor that would be required was not felt justified at the present time.

Assumptions and Notation

Throughout this analysis, the assumption will be made that the rocket motor supplies a thrust of constant magnitude for the period of powered flight. This means that the rate of flow of combustibles and the effective exhaust velocity remain constant. This assumption is of a conservative nature, as theoretical considerations show that the thermal efficiency of the rocket motor and, therefore, the thrust, will increase as the ratio between chamber pressure and exhaust pressure increases.

It has been assumed that the acceleration due to gravity remains constant. This assumption is also conservative.

The following notation has been used for the quantities involved:

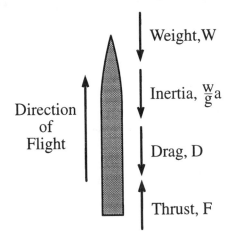

Fig. 1. Forces acting on a rocket in vertical flight.

F = thrust in lbs.
m = mass of exhaust gases flowing per second
c = F/m = effective exhaust velocity in ft./sec.
W_0 = initial weight of rocket, lbs.
W = instantaneous weight of rocket, lbs.
W_{FO} = weight of fuel and oxidizer carried, lbs.
ζ = W_{FO}/W_0, ratio of weight of fuel plus oxidizer to initial weight of rocket
a_0 = initial acceleration, ft./sec.^{2}
a = instantaneous acceleration, ft./sec.^{2}
g = acceleration of gravity, ft./sec.^{2}
V = instantaneous velocity, ft./sec.
V_s = velocity of sound, ft./sec.
h = altitude above sea level, ft.
t = time, sec.
A = largest cross-sectional area of rocket, sq. ft.

2. Africano, Alfred, *Rocket Motor Efficiencies*, Astronautics, No. 34, p. 5, June, 1936.
3. Sänger, Eugen, *Neuere Ergebnisse der Raketenflugtechnik*, Flug, Special Publication No. 1, pp. 6-9, December, 1934.

D = drag due to air resistance, lbs.
σ = air density ratio
ρ_0 = mass density of air at sea level

In Fig. 1 the forces acting on the rocket in vertical flight are shown. Summing the forces:

$$\Sigma \text{ Forces} = 0 = F - W - D - (W/g)a \qquad (1)$$

The thrust developed by the motor is expressed by

$$F = mc \qquad (2)$$

Then from Eqs. (1) and (2):

$$a = (mc - W - D)g/W \qquad (3)$$

[200] If the rate of flow of combustibles is constant during powered flight, one can write:

$$W = W_0 - mgt \qquad (4)$$

At the start of the flight,

$$W = W_0, \, a = a_0, \, V = 0, \, D = 0 \qquad (5)$$

Then Eq. (3) becomes

$$a_0 = mc - W_0 g/W \qquad (6)$$

and

$$m = W_0(a_0 + g)/cg \qquad (7)$$

Eq. (3) can now be evaluated, using Eq. (4) and Eq. (7), and for the acceleration at any instant

$$a = -g + \frac{(a_0 + g)}{1 - \frac{t(a_0 + g)}{c}} - \frac{g}{1 - \frac{t(a_0 + g)}{c}} \frac{D}{W_0} \qquad (8)$$

Flight *in Vacuo*

With no air resistance, the third term of Eq. (8) vanishes so that

$$a = \frac{dV}{dt} = -g + \frac{(a_0 + g)}{1 - \frac{t(a_0 + g)}{c}} \qquad (9)$$

Integrating Eq. (9) one has, for the velocity at any instant:

$$V = \frac{dh}{dt} = -gt - c \log\left[1 - \frac{t(a_0 + g)}{c}\right] + V_0 \qquad (10)$$

Integrating Eq. (10) one has, for the height at any instant:

$$h = -\frac{1}{2}gt^2 + ct + \left(\frac{c^2}{a_0+g} - ct\right)\log\left[1 - \frac{t(a_0+g)}{c}\right] + V_0 t + h_0 \quad (11)$$

The maximum acceleration and maximum velocity will occur at the time that the fuel is exhausted. The time at which thrust ceases is expressed, using Eq. (4), by the relation

$$(1-\zeta)W_0 = W_0 - mgt_p \quad (12)$$

Introducing Eq. (7) into Eq. (12), one obtains, for the duration of powered flight:

$$t_p = \zeta c / (a_0 + g) \quad (13)$$

If, at the start of the flight, $V_0 = 0$ and $h_0 = 0$, then

$$a_{max.} = -g + \frac{a_0 + g}{c - \zeta} \quad (14)$$

$$V_{max.} = -c\left[\frac{g\zeta}{a_0+g} + \log(1-\zeta)\right] \quad (15)$$

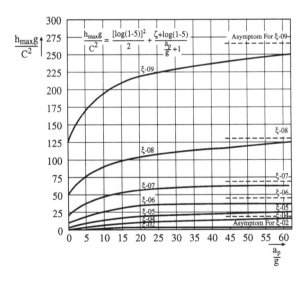

Fig. 2. Variation of $h_{max.}\, g/c^2$ with a_0/g for various ζ for flight *in vacuo*.

The maximum height reached will be the sum of the height at the time the fuel is exhausted and the height resulting from coasting. The height resulting from coasting is given by the expression:

$$h_c = V_{max.}^2 / 2g \quad (16)$$

Adding this to Eq. (11) and evaluating t_p from Eq. (13), the maximum height reached is

$$h_{max.} = \frac{c^2}{g}\left\{\frac{[\log(1-\zeta)]^2}{2} + \frac{\zeta + \log(1-\zeta)}{(a_0/g)+1}\right\} \quad (17)$$

Eq. (17) shows that three parameters determine the rocket performance *in vacuo*. They are a_o, ζ, and c. In Fig. 2 the variation of $h_{max}g/c^2$ is plotted for various values of a_o/g and ζ. The importance of having a large percentage of combustibles is clearly shown. The initial acceleration, a_o, is important until values in the neighborhood of $6g$ are reached.

Flight through a Resisting Medium

Considering flight through the air, the drag of the rocket can be expressed in the form

$$D = p_0 \sigma V^2 C_D A / 2 \quad (18)$$

which, substituted in Eq. (8), gives

$$a = -g + \frac{(a_0 + g)}{1 - \frac{t(a_0+g)}{c}} - \frac{gp_0 \sigma V^2}{2\left[1 - \frac{t(a_0+g)}{c}\right]} \cdot \frac{C_D A}{W_0} \quad (19)$$

This is the fundamental equation for vertical rocket flight. In addition to the performance parameters for flight *in vacuo*, the ratio $C_D A/W_0$ also has important significance in the construction of the sounding rocket. As it appears in a term which reduces the acceleration of the rocket, it should be as small as possible. A rocket [201] of given initial weight should have as small a cross-section as possible and be of a shape that minimizes the drag coefficient.

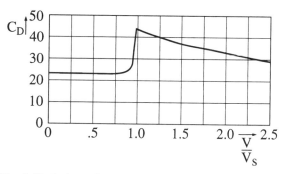

Fig. 3. Variation of the drag coefficient, C_p, with V/V_s.

As the density ratio, σ, and the drag coefficient, C_D, will be subject to great changes during flight, and are difficult to express accurately analytically, two ways of solving Eq. (19) are open. First, approximations can be made for the variation of σ and C_D to make an analytic solution possible, or second, a step-by-step method of integration of any degree of accuracy can be applied. The first method is quite likely to lead to extremely large errors, so that the second method has been chosen.

The variation of σ with height was obtained from references (4) and (5). The variation of C_D with the velocity of flight was taken from reference (6). It has been assumed, due to the lack of information, that the drag of the rocket was identical to the drag of a shell without a jet issuing at its base. The variation of the drag coefficient is reproduced from reference (6) in Fig. 3.

To describe the rocket flight, the following equations were used in the numerical calculations:

$$a_n = -g + \frac{(a_0+g)}{1-\frac{t_n(a_0+g)}{c}} - \frac{g p_0 \sigma_n V^2_{n-1}}{2\left[1-\frac{t_n(a_0+g)}{c}\right]} \cdot \frac{C_D A}{W_0} \tag{20}$$

$$V_n = V_{n-1} + a_{n-1}\Delta t \tag{21}$$

$$h_n = h_{n-1} + V_{n-1}\Delta t + (a_{n-1}/2)(\Delta t)^2 \tag{22}$$

where

Δt = time interval under consideration
n = number of the interval in the r steps of the calculation

The acceleration during coasting is given by

$$a_{c_n} = \frac{F_c}{m} \tag{23}$$

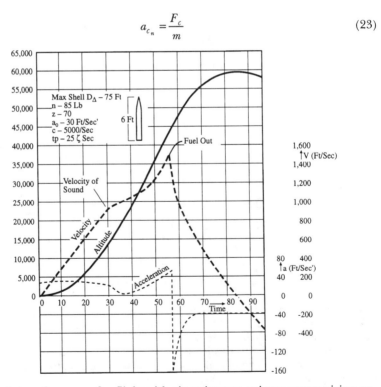

Fig. 4. Rocket performance for flight with air resistance, using a motor giving an effective exhaust velocity of 5000 ft./sec.

where

F_c = weight of the empty rocket plus air resistance or

$$a_{c_n} = -\left[g + \frac{p_0 \sigma_n V^2_{n-1} g}{2(1-\zeta)} \frac{C_D A}{W_0}\right] \quad (24)$$

In the following results to be presented, it was necessary to select dimensions of what may be called a typical sounding rocket. Therefore, the results will apply only to rockets having the same value of the ratio $C_D A/W_o$. For rockets with a different value of the ratio, this analysis serves only as a guide to the performance to be expected.

In Fig. 4 are shown the performance curves of a rocket with $c = 5000$ ft./sec., $\zeta = 0.70$, and $120 = 30$ ft./sec.² The retarding influence of the air is made evident [202] by the decrease in the acceleration as the velocity of sound is approached. The high density of the air at the time the fuel was exhausted prevented the rocket from coasting very high.

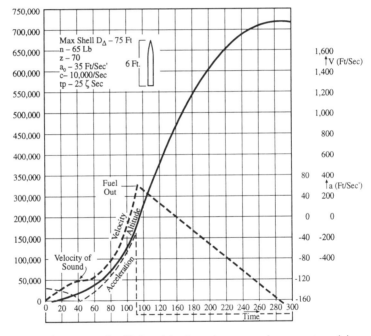

Fig. 5. Rocket performance for flight with air resistance, using a motor giving an effective exhaust velocity of 10,000 ft./sec.

The performance curves shown in Fig. 5 are those for an identical rocket, but with a much more efficient motor which gives an exhaust velocity, c, of 10,000 ft./sec. For the same amount of thrust, the rate of flow of combustible is much smaller, so that the period of powered flight is greatly prolonged. This allows the rocket to get over the hump of the drag curve, and also to travel through less dense air. The velocity at the end of the powered flight will thus be much higher than before, causing the rocket to coast to a much higher altitude.

In Fig. 6 the variation of altitude with the initial acceleration is shown for the two cases. The importance of a high value of the exhaust velocity, c, is clearly evident. This

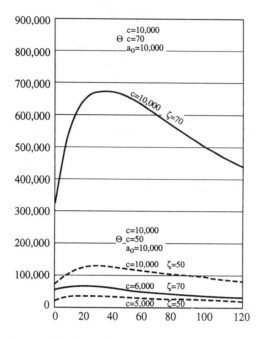

Fig. 6. Effect of a_0 on altitude to be reached for several performance parameter combinations.

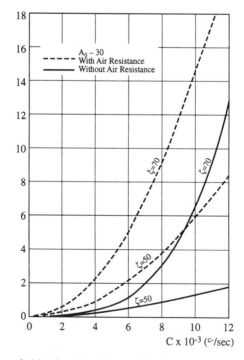

Fig. 7. Variation of altitude with c for $a_0=30$, with and without air resistance.

shows that effort should be directed to develop a motor of high efficiency before flight attempts are made.

This figure also shows that there is a definite initial acceleration corresponding to the maximum possible height. This differs from flight *in vacuo* for which the height reached continually increases with the initial acceleration (see Fig. 2). A high velocity of flight through the dense lower levels of the atmosphere causes the combustibles to be rapidly "eaten up." The advantage to be gained by starting the rocket from a high point is shown in the figure by the calculated height for a rocket started from an initial altitude of 10,000 ft.

The variation of maximum height to be reached with the exhaust velocity, c, for flight *in vacuo* and in air, is shown in Fig. 7. This figure clearly illustrates the amount of height lost due to resistance of the air.

Higher altitudes may be reached by using this step-rocket. A rocket made up of three steps, respectively, of 600, 200, and 100 lbs., the lightest being fired last, with c of 10,000 ft./sec., a_0 of 40 ft./sec.2, and ζ for each step of 0.70, starting from sea level, reaches a calculated altitude of 5,100,000 ft. and a maximum velocity of 11,000 m.p.h.

This analysis definitely shows that, if a rocket motor of high efficiency can be constructed, far greater altitudes can be reached than is possible by any other known means.

Document I-12

Document title: Theodore von Kármán, "Memorandum on the Possibilities of Long-Range Rocket Projectiles," and H.S. Tsien and F.J. Malina, "A Review and Preliminary Analysis of Long-Range Rocket Projectiles," Jet Propulsion Laboratory, California Institute of Technology, November 20, 1943.

Initially, Frank Malina started rocketry research in 1936 with the intention of lofting scientific payloads to high altitudes. But by 1938 GALCIT started receiving money from the National Academy of Sciences, at Army General Henry H. (Hap) Arnold's urging, to develop rockets for assisting heavily-laden aircraft and seaplanes during takeoff. This initial military research later advanced, during World War II, to the study of rockets as weapons of war. In 1943 GALCIT was renamed the Jet Propulsion Laboratory. This report to the Army Air Forces, with a cover letter by Theodore von Kármán, the director of GALCIT and then JPL, concluded that the development of long-range rocket projectiles was feasible and recommended their development at once. By this time, the Germans had already developed and tested the V-2.

H.S. Tsien, the co-author of this secret report, was a Chinese national who was later deported back to China, where he was instrumental in the development of the Chinese ICBM program.

---

[1] Memorandum
on

The Possibilities of Long-Range Rocket Projects

by Th. von Kármán

Recent progress in the field of jet propulsion by the Air Corps Jet Propulsion Research Project, the National Defense Research Committee and the Aerojet Engineering Corporation indicates that the development of a long-range rocket projectile is within engineering feasibility. During the past year reports reached this country crediting the Germans with the possession of extremely large rocket projectiles capable of transmitted to me by the Material Command, Experimental Engineering Division, for Study and

Comment. Comments were submitted in a later dated August 2, 1943.

At the instance of Col W. H. Joiner, A.A.F. Materiel Command Liaison Officer at the California Institute of Technology, two of my associates, Drs. F. J. Malina and H. S. Tsien, prepared a preliminary review and analysis of performance and design of long-rang projectiles which constitutes the substance of this Memorandum. The results of this study show that ranges in excess of 100 miles cannot be realized with propulsive equipment now available in this country. However, with the equipment already developed for super-performance of aircraft, rocket projectiles can be constructed which have a greater range and a much larger explosive load than rocket projectiles currently being used by the Armed Forces. Furthermore, by developing a special type of propulsive [2] equipment of the "athodyd" which utilizes atmospheric air, rangers comparable to those claimed by the Germans might be reached.

It is certain that the solution of the engineering problems connected with such a special jet unit requires considerable time. On the other hand, a large amount of information of immediate usefulness can be accumulated by experimentation with projectiles utilizing aircraft super-performance equipment. The development program should consist of the following coordinated phases:

First, firing tests of a projectile propelled by a restricted burning solid propellant unit produced by the Aerojet Engineering Corporation and accelerated during launching by unrestricted burning solid propellant rockets developed by the NDRC. This projectile would weight approximately 350 lbs and would carry a 50 lb explosive load for a distance of about 10 miles. The firing tests would supply information on problems of launching, stability and control, and for the verification of performance calculations.

Second, the design of a 2000 lb rocket projectile propelled by a liquid propellant jet unit of the type developed by the Air Corps Jet Propulsion Research Project and manufactured by the Aerojet Engineering Corporation. This projectile would carry an explosive load of 200 lbs for approximately 12 miles. This phase should be started as soon as sufficient information has been obtained from Phase I on the design of the projectile shape, stabilization fins and launching technique. At this point the program under Phase I should initiate experiments on the effect of adding wings to the projectile.

Third, it is desirable simultaneously with the first and second phases of projectile development to make a study of the design and [3] characteristics of the "athodyd" type propulsion unit. The "athodyd" or aero-thermodynamic duct jet unit is similar to other thermal jet units that have been developed, with the exception that pressure in the combustion chamber is obtained directly from the dynamic pressure of air resulting from the velocity of flight. The "athodyd" is expected to be more efficient at flight velocities that exceed the velocity of sound. The best means of investigating this type of unit would be to make a ground installation in which tests would be carried out using a compressor unit which is capable of blowing through a duct and combustion chamber system a considerable quantity of air. It appears that such compressor units could be made available in the Los Angeles area. The development of the "athodyd" type unit is not only important of the long range projectile but also has important implications in the general propulsion of aircraft at very high speeds.

Fourth, upon obtaining design information from the first two phases on projectile development and results of the special jet unit development program mentioned under Phase 3, the design and construction of a projectile of 10,000 lbs weight or larger with a range of the order of 75 miles would be undertaken.

It is believed that the projectiles developed in the first two phases would possess immediate military usefulness which would justify the effort expended independently of the general development program. Furthermore, the knowledge that would be obtained on the behavior of wings and control surfaces at supersonic velocities would be most valuable to the designer of high speed aircraft and remotely controlled unmanned missiles. It [4] is understood that missiles such as glide bombs now being developed will be equipped with jet propulsion units. The studies described above will give important information on the possibilities of accelerating such devices up to and beyond sonic velocities. On the other hand, the results collected from the ground launching tests will give important data also for the case of launching rocket propelled devices from aircraft and from surface vessels. In fact, the absence of recoil forces opens up a wide field for application of jet propulsion units. The studies described above will give important information on the possibilities of accelerating such devices up to and beyond sonic velocities. On the other hand, the results collected from the ground launching tests will give important data also for the case of launching rocket propelled devices from aircraft and from surface vessels. In fact, the absence of recoil forces opens up a wide field for application of jet propulsion to large caliber and long range missiles.

[5] # A Review and Preliminary Analysis of Long-Range Rocket Projectiles

by H. S. Tsien and F. J. Malina

I. Consideration of Various Jet Propulsion Methods

The propulsion of missiles or projectiles for military purposes has been the subject of intensive development for many centuries. Perhaps the oldest method that does not utilize muscular energy is that of rocket or jet propulsion. The propulsive force in jet propulsion is obtained from the reaction of a high speed gaseous mass ejected from the body to be propelled. The first jet propelled missiles used black powder for the generation of a high pressure gas. Although the black powder rocket reached a fairly high state of development, it was handicapped by incorrect design features and a propellant of low energy content. Its use as a military weapon was discontinued during the middle of the 19th century.

During the last twenty-five years, jet propulsion has staged a comeback with the assistance of new engineering knowledge and better propellants. Several jet propulsion methods are available, each with its own advantages and limitations. The methods can be divided into two main classes, characterized by independence or dependence on atmospheric air.

These two classes can be further subdivided as follows:

Methods independent of atmospheric air
1. **Solid propellant types**
    a. Unrestricted burning, short duration (0.01 to 2 seconds)
    b. Restricted burning, long duration (5 to 60 seconds)
2. **Liquid propellant types**
    a. Nitric acid type oxidizers (duration limited only by amount of propellant carried) [6]
    b. Liquid oxygen oxidizer (duration limited only by amount of propellant carried)

Methods dependent on atmospheric air
3. **Thermal jet propulsion types**
    a. Air compressor type
    b. Aero-thermodynamic duct type

The salient points of the above types will now be discussed to support the analyses of the long range rocket projectiles in the following parts of this Memorandum.

1. Solid propellant types

a. Unrestricted burning, short duration jet units

The unrestricted burning solid propellant jet unit has been developed to a high degree of perfection by the NDRC. This type uses a smokeless powder (ballistite) grain which has a large burning surface. The jet unit is capable of delivering a high thrust for a short period of time. Units have been tested that deliver as high as 100,000 lb thrust for a very small fraction of a second. The duration is limited by the grain web thickness that can practically be produced and the feasible dimensions of the jet unit.

The fact that this type of jet unit gives a high impulse in a very short time makes it especially suitable for short range missiles such as the Bazooka shell, anti-aircraft rockets, etc. The use of the unrestricted burning jet unit for projectiles whose range exceeds 7 or 8 miles does not appear practical because of the excessive impulse required. [7] The short duration jet unit would become very large in cross section and also the initial acceleration of the projectile would introduce difficult engineering problems. However, as will be pointed out later, the use of the short duration rocket is required in launching a long range projectile.

The general specifications of two short duration jet units developed by the NDRC are listed in Table I.

b. Restricted burning, long duration jet units

The restricted burning solid propellant jet unit was developed by the Air Corps Jet Propulsion Research Project at the California Institute of Technology especially for assisting the take-off of aircraft. The development has been extended to production types by the Aerojet Engineering Corporation. The units utilize an asphalt-base propellant charge which burns on one surface only. The burning takes place in parallel layers perpendicular to the axis of the jet unit. Jet units have been tested that deliver 1000 lb thrust for as long as 45 seconds. Durations of this order of magnitude are near the maximum practically attainable. Thrusts of between 2000 and 3000 lb are believed feasible.

The propellant developed for the restricted burning jet unit is not as effective as the ballistite charges of the unrestricted burning units. At a chamber pressure of 2000 lb per sq in, the former type gives an exhaust velocity of approximately 5500 ft per sec, whereas the latter type gives approximately 6300 ft per sec. On the other hand, [8] the asphalt base propellant is much less sensitive to ambient temperature changes, which is of prime importance in assisted take-off applications and also in projectiles propelled over a large fraction of their flight path.

The specifications of two Aerojet restricted burning solid propellant jet units are listed in Table II, and a drawing is shown in Figure 1.

2. Liquid propellant types

a. Nitric acid type oxidizers

Liquid propellant type jet units that have reached the highest stage of development utilize a propellant consisting of two components — an oxidizer and a fuel. Single liquid compounds exist which contain enough oxygen to sustain combustion; however, their investigation is in preliminary stages. The Germans are reported to have such a propellant.

The ACJP Project, the Navy Bureau of Aeronautics Project, and the Aerojet Engineering Corporation have carried the development of a liquid propellant jet until utilizing nitric acid type oxidizers and fuels spontaneously ignitable with the oxidizers to a high degree of reliability. The jet units have been developed primarily for assisting the take-off of aircraft.

Jet units have been tested by the ACJP Project in which a single motor delivered 6000 lb thrust for 20 seconds, and a motor which delivered 1000 lb thrust for a continuous period exceeding 5 minutes.

[9] At the optimum propellant mixture ratio, and effective exhaust velocity of 6000 ft per sec can be expected at a chamber pressure of 300 psi abs and 6400 ft per sec at 500 psi abs. The chamber pressure attained is controlled by the feed pressure applied to the propellant components. Two methods are available for obtaining feed pressure — compressed gas and pumps. The proper choice of a feed system requires a detailed analysis of the application in mind, and the only general rule that can be safely formulated states that for durations exceeding one minute, the pump system is the lightest.

As mentioned above, effective exhaust velocities of the order of 6400 ft per sec can be reached at chamber pressures around 500 psi abs. However, this velocity is obtained at the price of increasing the feed pressure by 200 psi, which is likely to nullify the improved propellant consumption by the additional weight required in the feed system.

The specifications of the Aerojet production unit 25 ALD-1000 and the Aerojet X40ALJ-6000 unit, which has been designed but not built, together with estimates of a 4000 lb thrust 35 second unit, are listed in Table III. The estimates of the 4000 lb thrust unit are believed to be too conservative and that the duration could be increased by 5 seconds by reducing the empty weight of the unit and increasing the propellant weight carried. At the same time the diameter of the tanks could be reduced from 24 in to 20 in.

b. Liquid oxygen type oxidizer

Work with liquid oxygen in combination with various fuels [10] as a propellant for liquid type jet units has been carried out by Goddard, the American Rocket Society, the Navy Bureau of Aeronautics Project and the ACJP Project. The discussion in connection with nitric acid type oxidizers in the preceding section holds when liquid oxygen is used with the exceptions that will be noted.

Tests at the ACJP Project with the liquid oxygen-gasoline combination have shown that effective exhaust velocities as high as 8500 ft per sec can be obtained at chamber pressures around 500 psi abs as compared to 6400 ft per sec for the nitric acid oxidizer. The design of a liquid oxygen-gasoline jet motor that can operate at the high combustion temperatures attendant with high exhaust velocities has not as yet been satisfactorily accomplished.

The utilization of liquid oxygen in military projectiles is believed to be of doubtful feasibility since it cannot be stored in closed containers because of its very low boiling temperature.

3. Thermal Jet Propulsions Types

a. Air compressor type

The jet units so far discussed made use of a propellant whose oxidizer was carried within the body to be propelled. For that reason the propulsion did not depend on the presence of atmospheric air, and operation could be maintained in empty space.

When flight is to be performed within the lower layers of the atmosphere, it does not seem logical to carry an [11] oxidizer when oxygen is on all sides during the flight. However, it is unfortunate from the point of view of propulsion that the oxygen in air is only available at such a low density and pressure.

Jet propulsion units have been developed that use the oxygen in air to burn a fuel. In general, the thermal jet propulsion unit consists of the following components: an inlet duct to a compressor, a compressor, a combustion chamber, a gas turbine, and an outlet duct or nozzle. In the following section on the aero-thermodynamic duct jet unit it will be shown that under certain conditions the dynamic pressure of air due to the motion of a body can be utilized without the addition of a compressor and a gas turbine.

In the thermal jet propulsion unit the compressor is inserted in order to increase the pressure in the combustion chamber and thus improve the thermodynamic efficiency of the unit at low flight velocities. After the air passes through the compressor, it enters the combustion chamber where a fuel such as gasoline is injected. The fuel burns, and heats the air. The hot air at high pressure then drives the gas turbine which furnishes the power for the compressor. The exhaust from the turbine, being still at a higher pressure than that of the atmosphere, discharges through the nozzle and a net propulsive thrust is imparted to the body.

A number of systems similar to the one described have [12] been developed and thrusts as high as 2700 lb have been obtained from thermal jet units. The thermal jet unit is a complex heavy piece of machinery involving the use of a high speed compressor driven by a gas turbine. It is believed that this type of propulsive unit is not at the present time suitable for the propulsion of projectiles, even though the propellant consumption is 6 to 10 times lower than for jet units utilizing a liquid or solid oxidizer.

b. Aero-thermodynamic duct type (athodyd) (Ramjet)

If flight is to be carried out at velocities near and above the velocity of sound, it may be possible to dispense with the necessity of a compressor in a thermal jet propulsion unit. The Germans are reported to have developed a device of this type referred to as an "athodyd." Pressure in a combustion chamber is obtained directly from the dynamic pressure of air resulting from the velocity of flight. Air is taken in the forward end of the tube, slowed down by means of a diffusor before entering the combustion chamber, and permitted to escape through a nozzle after its temperature has been raised by injecting fuel into the combustion chamber.

There may be possibilities of installing such devices in large projectiles. Propulsion from them is obtained after the projectile has reached a high velocity by some other form of propulsion. In the part of this Memorandum on the reported German projectiles a further discussion of the aero-theromo-dynamic duct will be made.

[13]   **II. Specifications and Performance of Two Projectiles Using Available Jet Units**

1. Specifications

Due to the novelty of the subject of long range rocket projectiles, it seems desirable to work out a program of accelerated development starting with projectiles that can be designed with available jet propulsion units. The problems of launching, stability and other engineering problems can be studied with these projectiles. After an analysis of the available design information concerning both the solid propellant and the liquid propellant units, it is concluded that the following two models of long range rocket projectiles are within immediate possibility:

LRRP-I: (Fig 2) Solid Propellant

Initial weight	350 lb
Thrust	1150 lb
Propellant weight	130 lb
Empty Weight	220 lb
Explosive load	50 lb
Duration of thrust	20 sec
Maximum diameter	10 in
Length	81 in

LRRP-II: (Fig 3) Liquid Propellant

Initial weight	2000 lb
Thrust	4000 lb
Propellant weight	830 lb

Empty Weight	970 lb
Explosive load	200 lb
Duration of thrust	35 sec
Maximum diameter	2 ft
Length	16 ft

2. Equations of Motion of a Rocket Projectile

For the calculation of performance of the projectile, the following assumptions are made:
a. The resistance is always that of a projectile whose [14] axis is tangent to the trajectory.
b. The gravitational acceleration is a constant, invariant with altitude.
c. The density and temperature of the atmosphere are functions of the altitude as given by Table IV according to the Standard Atmosphere.

For a satisfactorily stabilized projectile, the deviation of the axis of the projectile from the tangent to the trajectory must be small. Furthermore, it is known that the increase in air resistance due to small yaw is negligible. Therefore, the first assumption is justified. The second assumption is quite accurate due to the small altitudes involved compared with the radius of the earth.

If M is the mass of the projectile, v the velocity and Θ the inclination of the trajectory at the time instant t, the equations of motion of the projectile are

$$M\frac{dv}{dt} = F - D - Mg \sin\Theta \quad (1)$$

$$Mv\frac{d\Theta}{dt} = -Mg \cos\Theta \quad (2)$$

where g is the gravitational acceleration, F the thrust, and D the air resistance. The first equation expresses the acceleration along the trajectory while the second equation expresses the balance of centrifugal forces (Fig 4). For the time being, the projectile is assumed to be without wings. The effect of the addition of wings will be considered in a later paragraph. The value of F during the [15] powered flight is a constant, neglecting the effect of reduction of atmospheric pressure on the operation of the rocket motor.

The air resistance D can be expressed as

$$D = \frac{1}{2}\zeta V^2 C_D \frac{\pi}{4} d^2 \quad (3)$$

where ζ is the density, C_D the drag coefficient and d the maximum diameter of the projectile. The drag coefficient is a function of Mach number or the ratio of the flight velocity to the velocity of sound at the altitude. Variations in the Reynolds' number or the variation in the kinematical viscosity of air will also influence the drag coefficient, but this effect is not large and will be neglected. The values of C_D for the projectiles concerned are given in the Table V and plotted in Figure 5. These values are obtained by adding an appropriate amount of skin friction to the resistance coefficient of a modern artillery shell. The additional skin friction is necessary in order to account for the length of the rocket projectile and the tail fins.

During the powered flight, those values of drag coefficient are conservative. This is due to the fact that an appreciable fraction of the total resistance of an artillery shell comes from the suction at the base of the shell. This suction is certainly absent during the discharge of gases from the rocket motor. Hence the estimated drag coefficient of the shell is too high for the powered flight.

[16] Let: M_0 = the initial mass of the projectile
v_0 = the launching velocity of the projectile
Θ_0 = the launching angle of the projectile
$\xi = v/v_0$ ratio of the flight velocity to the launching velocity

then Eq (2) can be integrated as

$$\frac{1+\sin\Theta}{1-\sin\Theta} = \frac{1+\sin\Theta_0}{1-\sin\Theta_0} e - \frac{2g}{v_0} \int_0 \frac{dt}{\xi} \quad (4)$$

The equation gives the angle of inclination, if ξ is obtained as a function of time t. To obtain the latter relation, Eq (1) has to be solved. That equation can be written in the more convenient form as

$$\frac{d\xi}{dt} = \frac{g}{V_0} \left[\frac{1}{1-\frac{m}{M_0}t} \left\{ \frac{F}{M_0 g} - \left(\frac{1}{2} \frac{\wp_0 V_0^2 \frac{\pi}{4} d^2}{M_0 g} \right) \sigma C_D \xi^2 \right\} - \sin\Theta \right] \quad (5)$$

where m is the mass discharge of the rocket motor per second, and σ is the ratio of the densities at sea-level and at altitude $\wp/\wp_0$.

At the end of the powered flight, the thrust is zero, and the mass of the projectile is M^1, equal to the sum of the empty mass and the explosive mass. Therefore, Eq (5) reduces to the following form for coasting

$$\frac{d\xi}{dt} = -\frac{g}{V_0} \left[\frac{1}{\frac{M_1}{M_0}} \left(\frac{1}{2} \frac{\wp_0 V_0^2 \frac{\pi}{4} d^2}{M_0 g} \right) \sigma C_D \xi^2 + \sin\Theta \right] \quad (6)$$

Eqs (4), (5), and (6) determine completely the performance of the projectile of the launching conditions are given.

3. Performance without Wings

To obtain the performance, Eqs (4), (5), and (6) have to be integrated numerically; assuming values for v_0 and Θ_0, the main results for the two models for long range rocket projectiles are the following:

[17] LRRP-I:
v_0 = 160 ft/sec
Θ_0 = 66°
Range = 52,700 ft = 9.98 miles
Altitude reached = 18,200 ft
Velocity at end of powered flight = 1,623 ft/sec

LRRP-II:
v_0 = 160 ft/sec
Θ_0 = 82°
Range = 61,600 ft = 11.66 miles
Altitude reached = 29,200 ft
Velocity at end of powered flight = 1,428 ft/sec

The details of the performance are given in Tables VI and VII. This performance calculation is of course conservative in the sense that the launching angles Θ_0 are reasonably chosen but not the optimum. It is interesting to notice that the actual range during coasting is approximately 50% of the theoretical coasting distance without air resistance. This fact will be used in the next section of the Memorandum.

The launching angle is a very high compared with ordinary artillery practice. The reason is that due to the rather small launching velocity of these projectiles, the gravitational pull makes the initial part of the trajectory highly curved. On the other hand, in order to extend the coasting range, the inclination of the projectile at the beginning of coasting should be between 30° and 40°. This condition can only be met by using very large launching angles. The trajectories for the two cases investigated are plotted in Figs 6 and 7.

4. Performance with Wings

All the calculations made above are made under the assumption that the projectile is without wings. The addition of wings produces a life force which is perpendicular to the trajectory, a wing resistance along the trajectory and an aerodynamic moment. The lift force (Fig 8) tends to balance the component of gravitational pull normal to the trajectory. If the forces normal to the trajectory are completely [18] balanced, then the trajectory will be a straight line. In general, the curvanture of the trajectory will be much smaller than that without wings. This effect is beneficial in reaching altitude, as the arc length of the trajectory that the projectile has to travel is smaller and hence the work done for a given resistance is also smaller. However, the addition of wings does increase the drag of the projectile, because of the added skin friction and the induced drag of the wings. Therefore, these two effects tend to cancel each other and in absence of complete data for airfoils at very high speeds, it is reasonable to assume that the maximum altitude reached the distance covered up to the maximum altitude are approximately the same as those for wingless projectiles.

After the maximum altitude is reached, the projectile will glide toward its target. This part of the flight path can be approximated by a straight line with a slope equal to the average value of the ratio between the drag force and the lift force. In subsonic flight, the ratio is quite small due to the efficiency of the wing at lower velocities. A study of available test data on airfoils in supersonic flow shows that this ratio is about 1:4. As an approximation then, the guide will be taken as straight line with a slope equal to 1:4. Then the range estimate of winged projectiles with fundamental designs similar to LRRP-I and LRRP-II is as follows:

LRRP-I-W   Range = 19.7 miles
LRRP-II-W: Range = 28.8 miles

Thus the addition of wings to the projectile is capable of greatly extending the range of the projectile. However, there are several disadvantages which must be considered. First of all, the striking speed of the projectile is greatly reduced due to the extended coasting. Secondly, the addition of wings involves also [19] an increase in structural weight of the projectile and therefore a reduction in payload of a fixed initial gross weight. Finally, the problem of stability and control is greatly complicated, which may require intensive study and research.

[20]   ### III. General Performance Estimate and Related Problems

To study the possible development of the long range rocket projectile, a general performance estimate has to be made. However, the problem is quite complicated and involves many variables. To simplify the problem, a basic model of wingless projectile is assumed and its performance is analyzed. Then by using the results obtained for this basic model, the effect of the variation on propellant consumption is calculated approximately.

The final result will be presented as the ratio of total impulse and initial weight plotted against range for different values of propellant consumption.

1. Performance of a Basic Projectile

Take an improved design of the projectile as follows:

Initial weight	= 10,000 lb
Maximum diameter	= 2.52 ft
Length	= 25.2 ft
Propellant consumption	= 5.03 lb/sec for 1,000 lb thrust
Effective exhaust velocity of rocket monitor	= 6400 ft/sec
Launching velocity	= 160 ft/sec

The drag coefficient C_D for this projectile is assumed to be slightly lower than that for LRRP-I and LRRP-II due to improved design and reduction in skin friction at higher Reynolds numbers. The values of C_D are given in Table VIII and Fig 9.

Previous experience obtained in analyzing the performance of sounding rockets[*] shows that the magnitude of acceleration during the powered flight does not influence the range drastically. Therefore, for the convenience of calculation the acceleration will be assumed to be constant and equal to twice the gravitational acceleration or $2g$. In other words,

[21]
$$v = v_0 + 2gt \tag{7}$$

Then the trajectory during the powered flight can be immediately deduced from Eq (2). The horizontal distance x at the instant t is given by

$$x = \frac{v_0^2}{2g} 4k \left\{ \frac{1}{3}(\xi^{\frac{3}{2}} - 1) - k^2(\sqrt{\xi} - 1) + k^3 (\tan^{-1}\frac{\sqrt{\xi}}{k} - \tan^{-1}\frac{1}{k}) \right\} \tag{8}$$

where

$$k = \cot\left(\frac{\pi}{4} - \frac{\Theta_0}{2}\right) \tag{9}$$

$$\xi = v/v_0 = 1 + \frac{2g}{v_0} t \tag{10}$$

The altitude y at the instant t is given by

$$y = \frac{v_0^2}{2g} \left\{ 2k^2(\xi - 1) - \frac{1}{2}(\xi^2 - 1) - 2k^4 \log \frac{k^2 + \xi}{k^2 + 1} \right\} \tag{11}$$

These formulae determine the velocity and altitude at any instant t, and hence the air resistance D. By denoting $D/M_0 g$ by r:

$$r = \frac{1}{2} \frac{\wp_0 v_0^2 \frac{\pi}{4} d^2}{M_0 g} \sigma C_D \xi^2 = 0.01516 \, \sigma C_D \xi^2 \tag{12}$$

then Eq (1) can be used to calculate the ratio of mass M at the instant t and the initial mass, M_0. The result is

---

[*] F. J. "Malina and A. M. O. Smith, "Flight Analysis of the Sounding Rocket." J. AE. Sc., Vol. 5. pp. 199-202, (1938).

$$\frac{M}{M_0} = \frac{1}{2} \frac{e^{-\frac{\xi}{80}}}{(k^2+\xi)^{\frac{k^2}{40}}} \left\{ 2(k^2+1)^{\frac{k^2}{40}} e^{\frac{1}{80}} - \frac{1}{40}\int_1^{\xi} r(k^2+\xi)^{\frac{k^2}{40}} e^{\frac{\xi}{80}} d\xi \right\} \quad (13)$$

The ratio ζ of propellant discharged to the initial mass is of course $1-\frac{M}{M_0}$. Therefore if c = effective exhaust velocity of the jet motor, the ratio Ω of the total impulse up to the instant t to the initial weight is given by

$$\Omega = (1-\frac{M}{M_0})\frac{c}{32.2} = \frac{\zeta c}{32.2} \text{ (seconds)} \quad (14)$$

The result calculated for $\ominus_0 = 78°$ is given in Table IX.
[22] In Table IX, the values of v, $\sin \ominus$, x, and y are given together with Ω. If the propulsion is stopped at the instant t, the projectile will coast with an initial velocity v and inclination $\ominus$. The distance covered by coasting, neglecting air resistance can then be easily determined. According to the analysis in Part II, the air resistance will reduce this distance by 50%. By applying this reduction factor, the range of the projectile by stopping propulsion at various t can be calculated. Fig 10 shows the impulse ratio Ω plotted against range. This can be taken as an estimate of the performance of a long range rocket projectile. This estimate is of course somewhat conservative as no attempt is made to the vary the launching angel $\ominus_0$ to obtain its optimum value.

2. Performance at other Values of Propellant Consumption

If the propellant consumption is different from the value 5.03 lb/sec per 1000 lb thrust, or the effective exhaust velocity c is different from 6400 ft/sec, then the mass at the end of the powered flight will also be different assuming the same initial weight, acceleration and duration of the thrust. Let Ω, and ζ, be the impulse ratio and fuel weight ratio corresponding to $c = c_1 = 6400$ ft/sec respectively, taken by Eq (14)

$$\Omega_1 = \zeta_1 \frac{6400}{32.2} \quad (15)$$

$$\Omega = \zeta \frac{c}{32.2}$$

Now if the projectile with exhaust velocity c has the same launching angle and acceleration as the basic projectile, the trajectory, the velocity and the inclination at the end of powered flight will be the same. Hence the range obtained is also approximately the same. However, the impulse ratio will be different. First of all the mass at the end of the powered flight is now $M_0(1-\zeta)$ instead of $M_0(1-\zeta_1)$. The thrust towards the end of powered flight is therefore $M_0(1-\zeta)(a+g\sin\ominus)$ instead [23] of $M_0(1-\zeta_1)(a+g\sin\ominus)$ where a is the acceleration along the trajectory. The difference is $M_0(.\zeta_1-\zeta)(a+g\sin\ominus)$. At the initial instant, this difference does not exist as the initial mass is taken as the same. Hence if t is the duration of powered flight the additional impulse necessary is approximately $\frac{1}{2}M_0(\zeta_1-\zeta)(a+g\sin\ominus)t$. But $g = \frac{1}{2}a$ for the basic projectile, and $\sin°\leq 1$, hence the additional impulse is less than $\frac{1}{2}M_0(\zeta_1-\zeta)\frac{3}{2}at = 0.75 M_0(\zeta_1-\zeta)(v_c-v_0)$ where v_c is the velocity at the end of powered flight. v_0 is much smaller than v_c therefore it can be neglected in comparison with v_c. Then the impulse ratio can be written as

$$\Omega = \Omega_1 + 0.75(\zeta_1-\zeta)\frac{v_c}{g} \quad 16$$

Eqs (15) and (16) then given

$$\frac{\Omega}{\Omega_1} = \frac{1+0.75\frac{v_c}{C_1}}{1+0.75\frac{v_c}{C}} \qquad 17$$

This relation can then be used to calculate the impulse weight ratio necessary for a given range at various values of c. The result is plotted in Fig 10. An example of how to use this chart will be given in Section III(4) of this memorandum.

3. Launching of the Projectile

In all the performance analyses carried out in the preceding sections, the launching velocity of the projectile is assumed to be 160 ft/sec. This speed is chosen from the consideration of stability. It is felt that for speeds lower then 160 ft/sec, the tail fins can hardly be expected to give the necessary restoring force when the projectile is disturbed into a yaw. To obtain this launching velocity and to aim the projectile a launcher is necessary. For quick aiming of the projectile and easy transportation, [24] the length of the launcher should be made as short as possible. This means that the projectile should be accelerated as quickly as possible. Quick accelerations can be achieved by using a very large launching thrust. This thrust, being of very short duration, can best be supplied by the unrestricted burning solid propellant rocket.

If $M°$ is the mass of the projectile during the launching run, which can be assumed to be constant, a the constant acceleration, $\ominus_0$ the launching angle and v_0 the launching velocity, then the thrust $F°$ required for launching is

$$F° = M°a + M°g \sin \ominus_0 \qquad (18)$$

But $a = v_0^2/2L$ where L is the length of the launching run.
Hence

$$F° = M°\left[\frac{v_0^2}{2L} + g \sin \ominus_0\right] \qquad (19)$$

The duration T of the launching run is of course given by

$$T = 2\frac{L}{v_0} \qquad (20)$$

Let $L=25$ ft, $v_0 = 160$ ft/sec, the $T = 0.312$ sec, $a = 15.9$ Assuming $M° = 1.2\, M_0$ and $\sin \ominus_0 \approx 1$ then

$$F°/M_0 g = 20.3 \qquad (21)$$

In other words, the launching thrust should be approximately 20 times the weight of the projectile at the beginning of flight. For LRRP-I, the thrust is then 7,000 lb while for LRRP-II, this thrust is 40,000 lb. Thus, the unrestricted burning solid propellant rocket is well suited for the launching purpose.

[25] A preliminary design for the LRRP-1 launcher is shown in Figs 11 and 12.

4. Application of the Analysis

Fig 10 can be used to estimate the range of different projectiles of similar proportions to the basic projectile of 10,000 lb initial weight. For instance, if the effective exhaust velocity c is 6400 ft/sec, then with a propellant weight 62% of the initial weight, Eq (15) gives Ω =123.2. By using the curve in Fig 10 labeled c = 6400 ft/sec, the range is determined as 57.5 miles.

This value of propellant weight is rather high and may be difficult to achieve in a practical construction. To obtain a range in excess of 100 miles, it may be necessary to reduce the propellant consumption or to increase the effective exhaust velocity. With an exhaust velocity c = 9600 ft/sec, a 100 mile range requires Ω = 152 according to Fig 10. The propellant weight is then only 51% of the initial weight. This may well be within the realm of possibility.

The launching thrust for such a projectile is, according to Eq (21) about 200,000 lbs. The time duration of launching run is 0.312 sec and length 25 ft.

5. The Effect of Wings

The addition of wings to the projectile can greatly reduce the guide angle during the coasting flight of the projectile as discussed in Part III. If sufficient wing area is added, the range can be extended by as much as 100%. However, as stated in Part III there are several disadvantages to this practice. The main objections are the reduction in striking speed of the projectile and the increase in structural weight. It then seems that the compromise solution would be the addition of a stub wing to the projectile. The range can then be extended by approximately [26] 50% of that without wings, and at the same time the striking velocity and low structural weight can be maintained.

Another possible solution is to drop the wings at an altitude of about 20,000 ft; after the major portion of coasting flight is completed. This can be accomplished by a time fuse or relay which acts automatically at predetermined time. After dropping the wings, the projectile gains speed rapidly and thus will be able to strike the target with necessary velocity.

6. Stability and Control

For the wingless projectile, the problem of stability is relatively well-known. The experience and knowledge gained in bomb design and in the design of short range rocket projectiles can be immediately utilized. If the projectile is launched with a sufficient velocity for the fins to act, it is believed that the projectile will be inherently stable and the stability problem in connection with optimum fin design can be solved within a reasonably short time by a series of firing tests.

In the case of winged projectiles remote control might especially be required in applications in which a small evasive target is to be attacked. It is understood that both the Army Air Forces and the Navy are investigating control methods and devices and full collaboration with the groups concerned with this problem would be highly desirable.

The problem of stability and control for a winged projectile is believed to be much more difficult due to lack of knowledge and experience on wing design for supersonic speeds. A carefully laid program is necessary for a coordinated investigation of wings by both theoretical analysis and experimental observations.

[27]
IV. Analysis of Information Available on the German Long-Range Rocket Projectile

In this part an attempt is made to reconstruct the German long range rocket projectile on the basis of prisoner of war reports contained in the following British Intelligence reports: A.I. (K) Report No. 184A/1943, A.I. (K) Report No. 227A/1943, and A.I. (K)

Report No. 246B/1943. Upon reconstructing the LRRP an analysis of performance is made along the lines discussed in Part III of this Memorandum.

In Table X the specifications of the LRRP as given by various prisoners of war are listed.

In addition to the data in Table X the following information on the propulsive methods utilized is given:

(i) Projectile propelled by athodyds or rockets or combination of both. When the rocket is nearly burnt out a fuse ignites the burner in the athodyds.

(ii) Around the circumference of the rocket container there are a number of rearward - firing jets, probably six, which function from 10 to 70 seconds according to setting. When they have burned out, the propulsion of the projectile is taken over by the athodyds and the rocket portion falls off in one piece.

(iii) In one flight rockets burned for 18 or 19 seconds and the rocket - container became detached after the projectile had traveled a distance of about 9.3 miles.

(iv) The speed of the rocket gases is about 11.500 ft/sec and the athodyds take over propulsion when the projectile reaches a speed of 3280 ft/sec. with an initial acceleration of 8g.

[28] (v) When the athodyds cease functioning the projectile would have reached a speed of 6500 to 9200 ft/sec and the projectile would have covered half its course.

(vi) The athodyds were said to consume about 125 liters of fuel per second and to have an initial efficiency of 65%, rising to a terminal efficiency of 68 to 70%.

(vii) The pressure in the combustion chambers is between 80 and 100 atmospheres and the maximum temperatures probably of the order of 3,400 to 3,800° C. To prevent overheating of the combustion chambers the nozzles are made to function alternately in two sets of three, so that while one set of three is propelling the projectile, the other set is cooling off.

(viii) The combustion chambers are cooled by means of an air jacket with the intake in front and the venting rearwards. It is claimed that this jacket reduces the efficiency of the athodyds only by some 4%.

(ix) The combustion chambers on the projectile were ellipsoidal. There were six athodyds, each of which was housed in a cylinder, and the six cylinders in turn were filled into a larger cylinder which exactly fitted the rear portion of the projectile.

(x) The fuel reservoir extended down the center of the projectile between the athodyd housings.

A drawing of the projectile made by one of the prisoners of war is reproduced in Fig 13.

The following information is given on the fuel utilized:

(i) The new fuel looks like water, and the specific gravity of its various modifications varies between 0.5 and 0.7; it burns without the addition of oxygen to CO_2 and H_2O, and its heat of combustion is 43,600 Btu per lb, most of which is heat of [29] decomposition.

(ii) The new fuel is slightly yellowish in color, and is translucent. The specific gravity is thought to be 0.92.

(iii) Its general formula is $C_xH_{2x}O_{3x}$ and if a benzene ring is considered as monoplanar the first step in the synthesis is to interlock three such rings mutually at right angles and to substitute oxygen as necessary.

(iv) The lowest calorific content of the fuel is 63,000 Btu per pound.

(v) The rocket attachment is the athodyd—propelled projectile is provided with a normal propellant in solid form, but the athodyds in the main portion of the projectile are fed by the new fuel in liquid form.

From the information above the following specifications will be chosen for the German projectile and then its performance estimated.

Initial weight, lb	132,000
Athodyd propellent weight, lb	33,000
Booster rocket section, lb	55,000
Weight of projectile after booster rocket section dropped, lb	77,000
Explosive load, lb (10% of above)	7,700
Diameter, ft	7.5
Length, ft	20
Initial acceleration, ft/sec²	36

If we assume that the projectile is launched at an initial velocity of 160 ft per sec the launching run required is

$$L = \frac{v_0^2}{2a} = \frac{160^2}{2 \times 36} = 356 \text{ Ft, say 350 Ft}$$

[30] This value checks with the size of launching pit described by the prisoners of war. The initial thrust required can be calculated from the equation Eq (19)

$$F° = M°(\frac{v_0^2}{2L} + g \sin \Theta_0)$$

Let us assume that $\Theta_0 \approx 90°$ so that $\sin \Theta_0 \approx 1$, then

$$F° = \frac{132,000}{32.2}(\frac{160^2}{2 \times 350} + 32.2) = 282,000 \text{ lbs.}$$

If six solid propellant rockets are used as boosters then each rocket must deliver 47,000 lb. The exhaust velocity of the rockets is said to be 11,500 ft per sec, and the rockets act for a period of 10 to 70 seconds. Let us assume that they act for 20 seconds, then the weight of propellant in the rockets will be

$$W = \frac{F°g}{c}t = \frac{282,000}{11,500} \times 32.2 \times 20 = 15,800 \text{ lbs.}$$

If the rockets act for 60 seconds then the propellant weight would be

$$W = 15,000 \times 3 = 47,000 \text{ lbs.}$$

One of the prisoners of war states that the booster rockets, which are dropped when the rockets cease, weights 55,000 lb. This would check with the above calculation for a duration of 60 seconds. However, it is believed that an exhaust velocity of 11,500 ft per sec is excessive and a more probable value is 6,500 ft per sec. On this basis for a 20 second duration

$$W = \frac{282,000}{6,500} \times 32.2 \times 20 = 28,000 \text{ lbs.}$$

If one half of the weight of the booster rockets is in the form of propellant the value of 55,000 lb checks with the $2 \times 28,000 = 56,000$ lb very well. This will be used for later calculations.

[31] It will therefore be assumed that there are six solid propellant rockets each delivering 47,400 lb thrust for 20 seconds. At this point it should be a noted that a solid propellant rocket that delivers 47,400 lb thrust with an exhaust velocity of 6,500 ft per sec would have a diameter of 4.25 ft if the density of the propellant is 100 lb per cu ft and its rate of burning 2 in per sec.

The athodyds are said to take over propulsion when the rocket container drops off. If the athodyds deliver a thrust to give the same ratio of thrust to initial projectile weight as the rockets then their thrust is

$$F = \frac{282{,}000 \times 77{,}000}{132{,}000} = 165{,}000 \text{ lbs.}$$

If there are six athodyds then each athodyd must deliver 27,500 lb thrust.

The athodyds are said to consume about 125 liters of fuel per second. If the density is 0.92 then this corresponds to 250 lb per sec. It is not stated if all six or each one consumes this amount of fuel. If all consume this amount then the effective exhaust velocity based on the fuel above is

$$c = \frac{165{,}000 \times 32.2}{250} = 21{,}300 \text{ Ft/sec.}$$

This value of effective exhaust velocity is believed to be too high. A reasonable estimate shows that c is probably around 12,000 ft/sec. By assuming the same fuel consumption as before the thrust of the athodyds becomes 93,000 lbs. If the total weight of the propellant is 33,000 lbs, the duration would be 33,000/250 = 132 seconds.

Then impulse imparted to the projectile by the athodyds is then 93,000 × 132 = 12,270,000 lb sec. The impulse imparted to the projectile, excluding the booster rockets, after the launching run is then 282,000 × 77,000/132,000 × 10 = 1,643,000 lb sec. The total impulse imparted to [32] the projectile alone after the launching run is then 12,270,000 + 1,643,000 = 13,910,000 lb sec. The impulse weight ratio Ω is then 13,910,000/77,000 = 180.6. By using Fig 10 the ragne is estimated to be 140 to 150 miles. This checks very closely with the information given by the prisoners of war.

From the above analysis, a summary of the data for the German long range rocket projectile is given in Table XI. The velocities are estimated from the thrust data and are, of course, only a rough approximation.

[33]

Table I

Jet Unit	NDRC-CIT 3A	NDRC Budd 4.5"
Ave. Thrust, lb	2,000	6,000
Duration, sec	0.90	0.18
Impulse, lb sec	1,800	1,080
Eff. exhaust velocity, ft/sec	6,300	6,700
Propellant weight, lb	8.5	4.8
Motor weight (full), lb	41.5	30
Motor length, in	40.0	23.25
Motor 0. D., in	3.25	4.5

Table II

Jet Unit	Aerojet X20AS-1000	Aerojet X30AS-1000
Ave. Thrust, lb	1,000	1,000
Duration, sec	20	30
Impulse, lb sec	23,000	34,500
Eff. exhaust velocity, ft/sec	5,500	5,500
Propellant weight, lb	130	195
Motor weight (full), lb	270	385
Motor length, in	56.5	73.5
Motor O.D., in	9.625	9.625

Table III

Jet Unit	Aerojet 25ALD-1000	Aerojet X40ALJ-6000	Estimated Projectile 4000 lb-Thrust 30 sec Unit
Ave. Thrust, lb	1,000	6,000	4,000
Duration, sec	25	40	35
Impulse, lb sec	28,000	240,000	140,000
Eff. exhaust velocity, ft/sec	5,500	5,800	5,800
Propellant and nitrogen wt., lb	173	1,500	830
Jet unit weight (full), lb	420	2,900	1,700
Length of unit, in	69.0	104	144
Diameter of unit, in	—	—	24
Max. width, in	22.5	38	—
Max. height, in	24.0	47	—

Table IV

Altitude Ft.	Temp. °F, abs	Vel. of Sound ft/sec	Pressure Ratio	Density Ratio
0	519.0	1120	1.0000	1.0000
1000	515.4	1116	.9643	.9710
2000	511.8	1112	.9297	.9428
3000	508.4	1109	.8962	.9151
4000	504.8	1105	.8636	.8881
5000	501.2	1101	.8320	.8616
6000	497.6	1097	.8013	.8358
7000	494.0	1093	.7716	.8106
8000	490.6	1089	.7426	.7859
9000	487.0	1085	.7147	.7618
10000	483.4	1081	.6876	.7364
11000	479.8	1077	.6613	.7154
12000	476.2	1073	.6366	.6931
13000	472.6	1069	.6112	.6712
14000	469.1	1065	.5874	.6499
15000	465.5	1061	.5642	..6291
16000	461.9	1057	.5418	.6088
17000	458.3	1053	.5201	.5891
18000	454.7	1048	.4992	.5693
19000	451.3	1044	.4789	.5509
20000	447.7	1040	.4593	.5327
21000	444.1	1036	.4404	.5148
22000	440.5	1032	.4221	.4974
23000	436.9	1028	.4045	.4805
24000	433.5	1023	.3874	.4640
25000	429.9	1019	.3709	.4480
26000	426.3	1015	.3550	.4323
27000	422.7	1011	.3396	.4171
28000	419.1	1007	.3249	.4023
29000	415.5	1002	.3105	.3879
30000	412.1	997.9	.2968	.3740
31000	408.5	993.5	.2836	.3603
32000	404.9	989.1	.2708	.3472
33000	401.3	984.7	.2584	.3343
34000	397.7	980.3	.2466	.3218
35000	394.3	976.1	.2351	.3098
36000	393.0	974.5	.2242	.2962
37000	393.0	974.5	.2137	.2824
38000	393.0	974.5	.2038	.2692
39000	393.0	974.5	.1942	.2566
40000	393.0	974.5	.1851	.2447

[35]

Altitude Ft.	Temp. °F, abs	Vel. of Sound ft/sec	Pressure Ratio	Density Ratio
41000	393.0	974.5	.1766	.2332
42000	393.0	974.5	.1683	.2224
43000	393.0	974.5	.1605	.2120
43000	393.0	974.5	.1530	.2121
45000	393.0	974.5	.1458	.1926
46000	393.0	974.5	.1391	.1837
47000	393.0	974.5	.1325	.1751
48000	393.0	974.5	.1264	.1669
49000	393.0	974.5	.1205	.1591
50000	393.0	974.5	.1149	.1517
51000	393.0	974.5	.1095	.1446
52000	393.0	974.5	.1044	.1379
53000	393.0	974.5	.09953	.1315
54000	393.0	974.5	.09489	.1254
55000	393.0	974.5	.09047	.1195
56000	393.0	974.5	.08625	.1139
57000	393.0	974.5	.08222	.1086
58000	393.0	974.5	.07839	.1036
59000	393.0	974.5	.07474	.09872
60000	393.0	974.5	.07125	.09412
61000	393.0	974.5	.06793	.08974
62000	393.0	974.5	.06476	.08555
63000	393.0	974.5	.06174	.08155
64000	393.0	974.5	.05886	.07775
65000	393.0	974.5	.05612	.07413
66000	393.0	974.5	.05350	.07067
67000	393.0	974.5	.05100	.06737
68000	393.0	974.5	.04862	.06422
69000	393.0	974.5	.04636	.06123
70000	393.0	974.5	.04420	.05838
71000	393.0	974.5	.04345	.05739
72000	393.0	974.5	.04017	.05306
73000	393.0	974.5	.03829	.05058
74000	393.0	974.5	.03651	.04823
75000	393.0	974.5	.03480	.04597
76000	393.0	974.5	.03318	.04383
77000	393.0	974.5	.03163	.04178
78000	393.0	974.5	.03016	.03984
79000	393.0	974.5	.02875	.03798
80000	393.0	974.5	.02741	.03621

[36]

Altitude Ft.	Temp. °F, abs	Vel. of Sound ft/sec	Pressure Ratio	Density Ratio
81000	393.0	974.5	.02613	.03452
82000	393.0	974.5	.02491	.03291
83000	393.0	974.5	.02375	.03160
84000	393.0	974.5	.02265	.02991
85000	393.0	974.5	.02159	.02852
86000	393.0	974.5	.02058	.02719
87000	393.0	974.5	.01962	.02592
88000	393.0	974.5	.01871	.02471
89000	393.0	974.5	.01783	.02356
90000	393.0	974.5	.01700	.02246
91000	393.0	974.5	.01621	.02141
92000	393.0	974.5	.01545	.02041
93000	393.0	974.5	.01473	.01946
94000	393.0	974.5	.01405	.01855
95000	393.0	974.5	.01339	.01769
96000	393.0	974.5	.01277	.01686
97000	393.0	974.5	.01217	.01608
98000	393.0	974.5	.01160	.01533
99000	393.0	974.5	.01106	.01461
100000	393.0	974.5	.01055	.01393
101000	393.0	974.5	.01005	.01328
102000	393.0	974.5	.009585	.01266
103000	393.0	974.5	.009138	.01207
104000	393.0	974.5	.008712	.01151
105000	393.0	974.5	.008306	.01097
106000	393.0	974.5	.007919	.01046
107000	393.0	974.5	.007549	.009972
108000	393.0	974.5	.007197	.009507
109000	393.0	974.5	.006862	.009064
110000	393.0	974.5	.006541	.008641
111000	393.0	974.5	.006236	.008238
112000	393.0	974.5	.005946	.007854
113000	393.0	974.5	.005668	.007488
114000	393.0	974.5	.005404	.007138
115000	393.0	974.5	.005152	.006805
116000	393.0	974.5	.004912	.006488
117000	393.0	974.5	.004683	.006185
118000	393.0	974.5	.004464	.005897
119000	393.0	974.5	.004256	.005622
120000	393.0	974.5	.004058	.005360
121000	393.0	974.5	.003868	.005110
122000	393.0	974.5	.003688	.004871
123000	393.0	974.5	.003516	.004644
124000	393.0	974.5	.003352	.004428
125000	393.0	974.5	.003196	.004221

[37]

Table V

Drag Coefficient C_D for LRRP-I & LRRP-II

Mach Number, v/v_s	Drag Coefficient, C_D
0	0.3636
0.25	0.3568
0.50	0.3317
0.75	0.2974
0.95	0.3889
1.00	0.5354
1.10	0.5444
1.50	0.4910
2.00	0.4274
2.50	0.3944
3.00	0.3684
3.50	0.3434
4.00	0.3224

[38]

Table VI

Performance of LRRP-I

Launching Velocity vo, ft/sec	160
Launching Angle	66°
m/M_{01} sec^{-1}	0.01856
$F/M^0 g$	3.288
$\frac{1}{2}\zeta_0 v_0^2 \frac{\pi}{4} d^2 / M_0 g$	0.05225
Altitude at end of Powered Flight, ft	12,063
Distance at end of Powered Flight, ft	14,340
Velocity at end of Powered Flight, ft/sec	1,623
Inclination of Trajectory at end of Powered Flight	31.2°
Maximum Altitude reached, ft	18,200
Range, ft	52,700
Distance Covered by Coasting, ft	38,360
Distance Covered by Coasting, No Air Resistance, ft	88,700
Ratio of Coasting Distance with and without Air-Resistance	0.433

[39]

Table VII

Performance of LRRP-II

Launching Velocity v_0, ft/sec	160
Launching Angle,	82°
m/M_{01} sec^{-1}	0.01185
$F/M^0 g$	2.000
$\frac{1}{2}\zeta_0 v_0^2 \frac{\pi}{4} d^2 / M_0 g$	0.0478
Altitude at end of Powered Flight, ft	21,606
Distance at end of Powered Flight, ft	17,190
Velocity at end of Powered Flight, ft/sec	1,428
Inclination at end of Powered Flight	37.0°
Maximum Altitude reached, ft	29,200
Range, ft	61,600
Distance covered by Coasting, ft	43,410
Distance covered by Coasting, No Air Resistance, ft	82,100
Ratio of Coasting Distance with and without Air Resistance	0.528

[40]

Table VIII

Drag Coefficient C_D for the Basic Projectile

Mach Number, v/v_s	Drag Coefficient, C_D
0	0.3353
0.25	0.3285
0.50	0.3034
0.75	0.2691
0.95	0.3606
1.00	0.5071
1.10	0.5161
1.50	0.4627
2.00	0.3991
2.50	0.3661
3.00	0.3401
3.50	0.3151
4.00	0.2941

[41]

Table IX

Performance of the Basic Projectile

$v_0 = 160$ ft/sec
$\Theta_0 = 78°$

t, sec	x, ft	y, ft	vc, ft/sec	Ω	$\sin\Theta$	$\cos\Theta$	Range, Miles
39.75	79,470	80,490	2720	118.9	0.580	0.815	43.3
44.72	87,520	86,050	3040	121.4	0.567	0.823	50.7
49.69	95,810	91,770	3360	123.8	0.553	0.833	58.5
54.66	104,700	97,510	3680	126.2	0.540	0.841	67.0
59.63	113,930	103,330	4000	128.5	0.5275	0.849	75.9
64.60	123,580	109,270	4320	130.7	0.5185	0.855	85.5
69.57	133,670	115,210	4640	133.0	0.502	0.865	95.2
74.53	144,270	121,250	4960	135.0	0.490	0.872	105.1

[42]

Table X

Initial Wt. lb	Prop. Wt. lb	Explosive lb	Max. Dia. ft	Length ft	Flight Speed ft/sec	Range miles	Initial Accel ft/sec²	Rocket Container Wt., lb	Rocket Container Dia., ft	Rocket Container Length, ft	Launching Distance ft
220,000	33,000		8.2	16.4 - 19.6	3280	310	36.0				
132,000	33,000					155	257.6 8G	55,000	9.8 - 11.5	13.1	394 (pit)
132,000			7.4		6560 (Ave.)	125	257.6 8G				
		308		19.8		125					

[43] Table XI

Estimated Performance of a German Long Range Rocket Projectile

Items	Magnitude
Initial Weight, 1b	132,000
Rocket Booster Weight, lb	55,000
Projectile Weight, Booster Rejected, 1b	77,000
Rocket Propellant Weight, 1b	28,000
Athodyd Propellant Weight, 1b	33,000
Effective Exhaust Velocity of Rockets, ft/sec	6,500
Effective Exhaust Velocity of Athodyd, ft/sec	12,000
Duration of Rockets, sec	20
Duraction of Athodyds, sec	132
Launching Velocity, ft/sec	160
Velocity at Instant of Rejection of Rockets, ft/sec	800
Maximum Velocity of Projectile, ft/sec	6000
Distance Travelled before Rocket Rejection, miles	1.8
Range, Miles	145

Document I-13

Document title: The Editors of *Collier's*, "What Are We Waiting For?," and Dr. Wernher von Braun, "Crossing the Last Frontier," *Collier's*, March 22, 1952, pp. 23-29, 27-73.

Document I-14

Document title: Dr. Wernher von Braun, "Man on the Moon: The Journey," *Collier's*, October 18, 1952, pp. 52-59.

Document I-15

Document title: Dr. Fred L. Whipple, "Is There Life on Mars?," *Collier's*, April 30, 1954, p. 21.

Document I-16

Document title: Dr. Wernher von Braun with Cornelius Ryan, "Can We Get to Mars?," *Collier's*, April 30, 1954, pp. 22-29.

Collier's was a popular, family-oriented information magazine similar to *Life* and *The Saturday Evening Post*. Such magazines flourished in the post-war period until the advent of television and at its peak, *Collier's* had a circulation of over 3 million. On Columbus Day 1951, a Space Travel Symposium was held at the Hayden Planetarium of the New York, Museum of Natural History. The event had been organized by Willy Ley, a German emigré and author of the 1949 best-selling book, *The Conquest of Space*. Two journalists from *Collier's* were present at the symposium and notified their managing editor, Gordon Manning, about what was discussed there. His interest piqued, Manning sent associate editor Cornelius

Ryan to a conference on space medicine held in San Antonio, Texas, in November 1951. After talking to Wernher von Braun, Fred Whipple, and Joseph Kaplan at the conference, Ryan became enthusiastic about the prospects of space travel. Ryan convinced Manning to hold an internal *Collier's* symposium on the subject. Based on this internal symposium, a series of eight feature articles appeared in the magazine from 1952 to 1954. The articles were authored by noted experts such as von Braun, James Van Allen, Fred Whipple, Fritz Haber, and Joseph Kaplan. The articles were accompanied by illustrations by Chesley Bonestell, who had illustrated Ley's book, as well as by Fred Freeman and Rolf Klep.

These articles were the first to be published in a mainstream publication exposing the American public to the details of space exploration. They later led to a series of Disney animated films on the same subject and contributed to the popular historical image of space exploration.

Document I-13

[23]
What Are We Waiting For?

On the following pages Collier's presents what may be one of the most important scientific symposiums ever published by an national magazine. It is the story of the inevitability of man's conquest of space.

What you will read is not science fiction. It is serious fact. Moreover, it is an urgent warning that the U.S. must immediately embark on a long-range development program to secure for the West "space superiority." If we do not, somebody else will. That somebody else very probably would be the Soviet Union.

The scientists of the Soviet Union, like those of the U.S., have reached the conclusion that it is now possible to establish an artificial satellite or "space station" in which man can live and work far beyond the earth's atmosphere. In the past it has been correctly said that the first nation to do this will control the earth. And it is too much to assume that Moscow's military planners have overlooked the military potentialities of such an instrument.

A ruthless foe established on a space station could actually subjugate the peoples of the world. Sweeping around the earth in a fixed orbit. Like a second moon, this man-made island in the heavens could be used as a platform from which to launch guided missiles. Armed with atomic war heads, radar-controlled projectiles could be aimed at any target on the earth's surface with devastating accuracy.

Furthermore, because of the enormous speeds and relatively small size, it would be almost impossible to intercept them. In other words: whoever is the first to build a station in space can prevent any other nation from doing likewise.

We know that the Soviet Union, like the U.S., has an extensive guided missile and rocket program under way. Recently, however the Soviets, intimated that they were investigating the development of huge rockets capable of leaving the earth's atmosphere. One of their top scientists, Dr. M. K. Tikhonravov, a member of the Red Army's Military Academy of Artillery, let it be known that on the basis of Soviet scientific development such rocket ships could be built and, also, that the creation of a space station was not only feasible but definitely probable. Soviet engineers could even now, he declared, calculate precisely the characteristics of such space vehicles; and be added that Soviet developments in this field equaled, if not exceeded, those of the Western World.

We have already learned, to our sorrow, that Soviet scientists and engineers should never be underestimated. They produced the atomic bomb years earlier than was anticipated. Our air superiority over the Korean battlefields is being challenged by their excellent MIG-15 jet fighters which, at certain altitudes, have proved much faster than ours. And while it is not believed that the Soviet Union has actually begun work on a major project to capture space superiority, U.S. scientists point out that the basic knowledge for such a program has been available for the last 20 years.

What is the U.S. doing, if anything, in this field?

In December, 1948, the late James Forrestal, then Secretary of Defense, spoke of the existence of an "earth satellite vehicle program." But in the opinion of competent military observers this was little more than a preliminary study. And so far as is known today, little further progress has been made. Collier's feels justified in asking; What are we waiting for?

We have the scientists and the engineers. We enjoy industrial superiority. We have the inventive genius. Why therefore, have we not embarked on a major space program equivalent to that which was undertaken in developing the atomic bomb? The issue is virtually the same.

The atomic bomb was enabled the U.S. to buy time since the end of World War II. Speaking in Boston 1949, Winston Churchill put it this way: "Europe would have been communized and London under bombardment sometime ago but for the deterrent of the atomic bomb in the hands of the United States." The same could be said for a space station. In the hands of the West a space station, permanently established beyond the atmosphere, would be the greatest hope for peace the world has ever known. No nation could undertake preparations for war without the certain knowledge that it was being observed by the ever-watching eyes aboard the "sentinel in space." It would be the end of Iron Curtains wherever they might be.

Furthermore, the establishment of a space station would mean the dawning of a new era for mankind. For the first time, exploration of the heavens would be possible, and the great secrets of the universe would be revealed.

When the atomic bomb program—the Manhattan Project—was initiated, nobody really knew whether such a weapon could actually be made. The famous Smyth Report on atomic energy tells us that among the scientists where were many who had grave and fundamental doubts of the success of the undertaking. It was a two-billion-dollar technical gamble.

Such would not be the case with a space program. The claim that huge rocket shops can be built and a space station created still stands unchallenged by any serious scientist. Our engineers can spell out right now (as you will see) the technical specifications for the rocket ship and space station in cut-and-dried figures. And they detail the design features. All they need is time (about 10 years), money and authority.

Even the cost has been estimated: $4,000,000,000. And when one considers that we have spent nearly $54,000,000,000 on rearmament since the Korean war began, the expenditure of $4,000,000,000 to produce an instrument which would guarantee the peace of the world seems negligible.

Collier's became interested in this whole program last October when members of our editorial staff attended the First Annual Symposium on Space Travel, held at New York's Hayden Planetarium. On the basis of their findings. Collier's invited the top scientists in the field of space research to New York for a series of discussions. The magazine symposium on these pages was born of these round table sessions.

The scientists who have worked with us over the last five months on this project and whose views are presented in succeeding pages are:

- **Dr. Wernher von Braun**, Technical Director of the Army Ordnance Guided Missiles Development Group. At forty, he is considered the foremost rocket engineer in the world today. He was brought to this country from Germany by the U.S. government in 1945.
- **Dr. Fred L. Whipple**, Chairman Department of Astronomy, Harvard University. One of the nations outstanding astronomers, he has spent most of his forty-five years studying the behavior of meteorites.
- **Dr. Joseph Kaplan**, Professor of Physics at UCLA. One of the nation's top physicists and a world renowned authority on the upper atmosphere, the forty-nine-year-old scientists was decorated in 1947 for work in connection with B-29 bomber operations.
- **Dr. Heinz Haber**, of the U.S. Air Force's Department of Space Medicine. Author of more than 25 scientific papers since our government brought him to this country from Germany in 1947. Dr. Haber, thirty-eight, is one of a small group of scientists working on the medical aspects of man in space.

- **Willy Ley**, who acted as adviser to Collier's in the preparation of this project. Mr. Ley, forty-six is perhaps the best-known magazine science writer in the U.S. today. Originally a paleontologist, he was one of the founders of the German Rocket Society in 1927 and was Dr. Wernher von Braun's first tutor in rocket research.

Others who made outstanding contributions to this issue include:

- **Oscar Schachter**, Deputy Director of the UN Legal department. A recognized authority on international law, this thirty-six-year-old lawyer has frequently given legal advice on matters pertaining to international scientific questions, which lately have included the problems of space travel.
- **Chesley Bonestell**, whose art has appeared in the pages of Collier's many times before. Famous for his astronomical painting, Mr. Bonestell began his career as an architect, but has spent most of his life painting for magazines and lately for Hollywood.
- Artists **Fred Freeman** and **Rolf Klep**. Both spent many months working in conjunction with the scientists.

For Collier's, associate editor Cornelius Ryan supervised assembly of the material for the symposium. The views expressed by the contributors are necessarily their own and in no way reflect those of the organizations to which they are attached.

Collier's believes that the time has come for Washington to give priority of attention to the matter of space superiority. The rearmament gap between the East and West has been steadily closing. And nothing, in our opinion, should be left undone that might guarantee the peace of the world. It's as simple as that.

THE EDITORS

[25]
Crossing the Last Frontier

By Dr. Wernher von Braun

Technical Director, Army Ordnance Guided Missiles
Development Group, Huntsville, Alabama

Scientists and engineers now how to build a station in Space that would circle the earth 1,075 miles up. The job would take 10 years, and cost twice as much as the atom bomb. If we do it, we can not only preserve the peace but we can take a long step toward uniting mankind.

[26] Within the next 10 or 15 years, the Earth will have a new companion in the skies, a man-made satellite that could be either the greatest force for peace ever devised, or one of the most terrible weapons of war—depending on who makes and controls it. Inhabited by humans, and visible from the ground as a fast-moving star, it will sweep around the earth at an incredible rate of speed in that dark void beyond the atmosphere which is known as "space."

In the opinion of many top experts, this artificial moon—which will be carried into space, piece by piece, by rocket ships—will travel along a celestial route 1,075 miles above the earth, completing a trip around the globe every two hours. Nature will provide the motive power; a neat balance between its speed and the earth's gravitational pull will keep it on course (just as the moon is fixed in its orbit by the same two factors). The speed at which the 250 foot-wide, "wheel"-shaped satellite will move will be an almost unbelievable 4.4 miles per second, or 15,840 miles per hour—20 time the speed of sound. However, this terrific velocity will not be apparent to its occupants. To them, the space station will appear to be a perfectly steady platform.

From this platform, a trip to the moon itself will be just a step, as scientists reckon distance in space.

The choice of the so-called "two-hour" orbit—in preference to a faster one, closer to the earth or a slower one like the 29-day orbit of the moon—has one major advantage:

although far enough up to avoid the hazards of the earth's atmosphere it is close enough to afford a superb observation post.

Technicians in this space station—using specially designed, powerful telescopes attached to large optical screen radarscopes and cameras—will keep under constant inspection every ocean, content, country and city. Even small towns will be clearly visible through optical instruments that will give the watchers in space the same vantage point enjoyed by a man in an observation plane only 5,000 feet off the ground.

Nothing will go unobserved. Within each two-hour period as the earth revolves inside the satellite's orbit one twelfth of the globe's territory will pass into the view of the space station occupants within each 24-hour period the entire surface of the earth will have been visible.

Over North America for example, the space station might pass over the East Coast at say 10:00 am and after having completed a full revolution around the earth would—because the [27] earth itself has turned meanwhile—pass over the West Coast two hours later. In the course of that one revolution it would have been north as far as Nome, Alaska, and south almost to Little America on the Antarctic Continent. At 10:00 am the next day, it would appear once again over the East Coast.

Despite the vast territory thus covered, selected spots on the earth could receive pinpoint examination. For example, troop maneuvers, planes being readied on the flight deck of an aircraft carrier, or bombers forming into groups over an airfield will be clearly discernible. Because of the telescopic eyes and cameras of the space station, it will be almost impossible for any nation to hide warlike preparations for any length of time.

* * *

These things we know from High-altitude photographs and astronomical studies: to the naked eye, the earth, more than 1,000 miles below, will appear an awe-inspiring sight. One the earth's "day" side, the space station's crew will see glaring white patches of overcast reflecting the light of the sun. The continents will stand out in shades of gray and brown bordering the brilliant blue of the seas. North America will look like a great patchwork of brown, gray and green reaching all the way to the snow-covered Rockies. And one polar cap—whichever happens to be enjoying summer at the time—will show as a blinding white, too brilliant to look at with the naked eye.

On the earth's "night" side, the world's cities will be clearly visible as twinkling points of light. Surrounded by the hazy aura of its atmosphere—that great ocean of air in which we live—the earth will be framed by the absolute black of space.

Development of the space station is as inevitable as the rising of the sun; man has already poked his nose into space and he is not likely to pull it back.

On the 14th of September, 1944, a German V-2 rocket, launched from a small island in the Baltic, soared to a peak altitude of 109 miles. Two years later on December 17, 1946, another V-2, fired at the Army Ordnance's White Sands Proving Ground, New Mexico, reached a height of 114 miles—more then five times the highest altitude ever attained by a metrological sounding balloon. And on the 24th of February, 1949, a "two-stage rocket" (small rocket names the "WAC Corporal" fired from the nose of a V-2 acting as carrier or "first stage") soared up to a height of 250 miles—roughly the distance between New York and Washington, but straight up!

These projectiles utilized the same principle of propulsion as the jet airplane. It is based on Isaac Newton's third law of motion, which can be stated in this way: for every action there most be a reaction of equal force, but in the opposite direction. A good example is the firing of a bullet from a rifle. When you pull the trigger the bullet speeds out of the barrel, there is a recoil which slams the rifle butt back against your shoulder. If the rifle were lighter and the explosion of the cartridge more powerful, the gun might go flying over your shoulder for a considerable distance.

This is the way a rocket works. The body of the rocket is like the rifle barrel; the gasses ejected from its tail are like the bullet. And the power of a rocket is measured not in horsepower, but in pounds or tons of recoil—called "thrust." Because it depends on the recoil principle, this method of propulsion does not require air.

There is nothing mysterious about making use of this principle as the first step toward making our space station a reality. On the basis of present engineering knowledge, only a determined effort and the money to back it up are required. And if we don't do it, another nation—possibly less peace-minded—will. If we were to begin in immediately, and could keep going at top speed, the whole program would take about 10 years. The estimated cost would be $4,000,000,000—about twice the cost of developing the atomic bomb, but less then one quarter the price of military materials ordered by the Defense Department during the last half of 1951.

Our first need would be a huge rocket capable of carrying a crew and some 30 or 40 tons of cargo into the "two-hour" orbit. This can be built. To understand how, we again use the modern gun as an example.

A shell swiftly attains a certain speed within the gun barrel, then merely coasts through a curved path toward its target. A long-range rocket also requires its initial speed during a comparatively short time, then is carried by momentum.

For example, the V-2 rocket in a 200-mile flight is under power for only 65 seconds, during which it travels 20 miles. At the end of this 65-second period of propulsion it reaches a cut-off speed of 3,600 miles per hour; it coasts the remaining 180 miles. Logically, therefore, if we want to step up the range of the rocket, we must increase its speed during the period of powered flight. If we could step up its cut-off speed to 8,280 miles per hour, it would travel 1,000 miles.

To make a shell hit its target, the gun barrel has to be elevated and pointed in the proper direction. If the barrel were pointed straight up into the sky the shell would climb to a certain altitude and simply fall back, landing quite close to the gun. Exactly the same thing happens when a rocket is fired vertically. But to make the rocket reach a distant target after its vertical take-off, it must be tilted after it reaches a certain height above the ground. In rockets capable of carrying a crew and cargo, the tilting would be done by swivel-mounted rocket motors, which by blasting sideways, would cause the rocket to veer.

* * *

Employing this method, at a cutoff speed of 17,460 miles per hour, a rocket would coast halfway around the globe before striking ground. And by boosting to just a little higher cut-off speed—4.86 miles per second or 17,500 miles per hour—its coasting path, after the power had been cut off would match the curvature of the earth. The rocket would actually be "falling around the earth," because its speed and the earth's gravitational pull would balance exactly.

It would never fall back to the ground, for it would now be an artificial satellite, circling according to the same laws that govern the moon's path about the earth.

Making it do this would require delicate timing—but when you think of the split-second predictions of the eclipses, you will grant that there can hardly be any branch of natural science more accurate than the one dealing with the motion of heavenly bodies.

Will it be possible to attain this fantastic speed of 17,500 miles per hour necessary to reach our chosen two-hour orbit? This is almost five times as fast as the V-2. Of course, we can replace the V-2's alcohol and liquid oxygen by a more powerful propellants, and even, by improving the design, reduce the rocket's dead weight and thereby boost the speed by some 40 or 50 per cent; but we would still have a long way to go.

The WAC Corporal, starting from the nose of a V-2 and climbing to 250 miles, has shown us what we must do if we want to step up drastically the speed of a rocket. The WAC started its own rocket motor the moment the V-2 carrying it had reached it maximum speed. It thereby added its own speed to that already achieved by the first stage. As mentioned earlier, such a piggyback arrangement is called "two-stage rocket;" and by putting a two-stage rocket on [28] another still larger booster we get a three-stage rocket. A three-stage rocket then, could treble the speed attainable by one rocket stage alone (which would give it enough speed to become a satellite).

In fact, it could do even better. The three-stage rocket may be considered as a rocket with three sets of motors; after the first set has given its utmost and has expired, it is jettisoned—and so is the second set in its turn. The third stage or nose of the rocket continues on its way, relieved of all that excess weight.

* * *

Besides the loss of the first two stages, other factors make the rocket's journey easier the higher it goes. First, the atmosphere is dense, and tends to hinder the passage of the rocket; once past it, the going is faster. Second, the rocket motors operate more efficiently in the rarefied upper layers of the atmosphere. Third, after passing through the densest portion of the atmosphere, the rocket no longer need climb vertically.

Imagine the size of this huge three-stage rocket ship: it stands 265 feet tall approximately the height of a 24-story office building. Its base measures 65 feet in diameter. And the over-all weight of this monster rocket ship is 14,000,000 pounds, or 4,000 tons—about the same weight as a light destroyer.

Its three huge power plants are driven by a combination of nitric acid and hydrazine, the latter being a liquid compound of nitrogen and hydrogen somewhat resembling its better-known cousin ammonia. These propellants are fed into the rocket motors by means of turbopumps.

Fifty-one rocket motors, pushing with a combined thrust of 14,000 tons, power the first stage (tail section). These motors consume a total of 5,250 tons of propellants in the incredibly short time of 84 seconds. Thus, in less than a minute and a half, the rocket loses 75 percent of its total original weight.

The second stage (middle section), mounted on top of the first, has 34 rocket motors with a total thrust of 1,750 tons and burns 700 tons of propellants. It operates for only 124 seconds.

The third and final stage (nose section)—carrying the crew, equipment and payload—has five rocket motors with a combined thrust of 200 tons. This "Body" or cabin stage of the rocket ship carries 90 tons of propellants, including ample reserves for the return trip to earth. In addition, it is capable of carrying a cargo or payload of about 36 tons into our two-hour orbit 1,075 miles above sea level. (Also, in expectation of the turn trip, the nose section will have wings something like an airplane's. They will be used only during the decent, after re-entry the earth's atmosphere.)

Years before the actual take-off, smaller rocket ships, called instrument carriers, will have been sent up to the two-hour orbit. They will circle there, sending back information by the same electronic method already in use with current rockets. Based on the data thus obtained, scientists, astronomers and engineers, along with experts from the armed forces, will plan the complete development of the huge cargo-carrying rocket ship.

The choice of the take-off site poses another problem, because of the vast amount of auxiliary equipment—such as fuel storage tanks and machine shops, and other items like radio, radar, astronomical and meteorological stations—an extensive area is required. Furthermore, it is essential, for reasons which will be explained later that the rocket ship fly over the ocean during the early part of the flight. The tiny U.S. possession known as Johnston Island, in the Pacific, or the Air Force Proving Ground at Cocoa, Florida, are presently considered by the experts to be suitable sites.

At the launching area, the heavy rocket ship is assembled on a great platform. Then the platform is wheeled into place over a tunnel-like "jet deflector" which drains off the fiery gases of the first stage's rocket motors. Finally, with a mighty roar which is heard many miles away, the rocket ship slowly takes off—so slowly, in fact that in the first second is travels less then 15 feet. Gradually, however, it begins to pickup speed, and 20 seconds later it has disappeared into the clouds.

Because of the terrific acceleration which will be experienced one minute later, the crew—located of course, in the nose—will be lying flat in "contour" chairs at take-off, facing up. Throughout the whole of its flight to the two-hour orbit, the rocket is under the control of an automatic gyropilot. The timing of its flight and the various maneuvers which take place have to be so precise that only a machine can be trusted to do the job.

After a short interval, the automatic pilot tilts the rocket into a shallow path, by 84 seconds after take-off when, the fuels of the first stage (tail section) are nearly exhausted, the rocket ship is climbing at a gentle angle of 20.5 degrees.

When it reaches an altitude of 24.9 miles it will have a speed of 1.46 miles per second or 5,256 miles per hour. To enable the upper stages to break away from the tail or first stage the tail's power has to be throttled down to almost zero. The motors of the second stage now begin to operate, and the connection between the now-useless first stage and the rest of the rocket ship is severed. The tail section drops behind, while the two upper stages of the rocket ship forge ahead.

After the separation, a ring-shaped ribbon parachute, made of fine steal wire mesh, is automatically released by the first stage. This chute has a diameter of 27 feet and gradually it slows down the tail section. But under its own momentum, this empty hull continues to climb, reaching a height of 40 miles before slowly descending. It is because the tail section could be irreparably damaged if it struck solid ground (and might be dangerous, besides) that the initial part of the trip must be over sea. After the first stage lands in the water, it is collected and brought back to the launching site.

The same procedure is repeated 124 seconds later. The second stage (middle section) is dropped into the ocean. The rocket ship by this time has attained an altitude of 40 miles and 332 miles from the take-off site. It also has reached a tremendous speed—14,364 miles per hour.

Now the third and last stage—the nose section or cabin equipped space ship proper—proceeds under the power of its own rocket motors. Just 84 seconds after the dropping of the second stage, the rocket ship, now moving at 18,468 miles per hour reaches a height of 63.3 miles above the earth.

At this point we must recall the comparison between the rocket and the coasting rifle shell to understand what occurs. The moment the rocket reaches a speed of 18,468 miles per hour at an altitude of 63.3 miles, the motors are cut off even though the fuel supply is by no means exhausted. The rocket ship continues on an unpowered trajectory until it reaches 1,075 miles above the earth. This is the high point, or "apogee"; in this case it is exactly halfway around the globe from the cut-off place. The rocket ship is now in the two-hour orbit where we intend to build the space station.

* * *

Just one more maneuver has to be performed however. In coasting up from 63.3 miles to 1,075 miles, the rocket ship has been slowed by the earth's gravitational pull to 14,770 miles per hour. This is not sufficient to keep the ship in our chosen orbit. If we do not increase the speed the craft will swing back halfway around the earth to the 63.3 mile altitude. Then it would continue on past the earth until as it curves around to the other side of the globe, it would be back at the same apogee at the 1,075-mile altitude.

The rocket ship would already be a satellite and behave like a second moon in the heavens, swinging on its elliptical path over and over for a long time. One might ask: Why not be satisfied with this? The reason is that part of this particular orbit is in the atmosphere at only 63.3 miles. And while the air resistance there is very low, in time it would cause the rocket ship to fall back to earth.

Our chosen two-hour orbit is one which, at *all* points, is exactly 1,075 miles above the earth. The last maneuver which stabilizes the rocket ship in this orbit, is accomplished by turning on the rocket motors for about 15 seconds. The velocity is thus increased by 1,030 miles per hour bringing the total speed to 15,800 miles per hour. This is the speed necessary for remaining in the orbit permanently. We have reached our goal.

[29] An extraordinary fact about the flight from the earth is this: it has taken only

56 minutes, during which the rocket ship was powered for only five minutes.

From our vantage point, 1,075 miles up, the earth to the rocket ship's crew appears to be rotating once every two hours. This apparent fast spin of the globe is the only indication of the tremendous speed at which the rocket ship is moving. The earth, of course, still requires a full 24 hours to complete one revolution on its axis, but the rocket ship is making 12 revolutions around the earth during the time the earth makes one.

We now begin to unload the 36 tons of cargo which we have carried up with us. But how and where shall we unload the material? There is nothing but the blackness of empty space all around us.

We simply dump it out of the ship. For the cargo, too has become a satellite! So have the crew members. Wearing grotesque-looking pressurized suits and carrying oxygen for breathing they can now leave the rocket ship and float about unsupported.

Just as a man on the ground is not conscious of the fact that he is moving with the earth around the sun at the rate of 66,600 miles per hour, so the men in the space ship are not aware of the fantastic speed with which they are going around the earth. Unlike men on the ground, however, the men in space do not experience any gravitational pull. If one of them, while working, should drift off into space, it will be far less serious than slipping off a scaffold. Drifting off merely means that the man has acquired a very slight speed in an unforeseen direction.

He can stop himself in the same manner in which any speed is increased or stopped in space—by reaction. He must do this, theoretically, by firing a revolver in the direction of his inadvertent movement. But in actual practice the suit will be equipped with small rocket motor. He could also propel himself by squirting some compressed oxygen from a tank on his back. It is highly probable, however, that each crew member will have a safety line securing him to the rocket as he works. The tools he uses will also be secured to him by lines; otherwise they might float away into space.

* * *

The spacemen—for that is what the crew members now are—will begin sorting the equipment brought up. Floating in strange positions among structural units and machinery, their work will proceed in absolute silence, for there is no air to carry sound. Only when two people are working on the same piece of material, both actually touching it, will one be able to hear the noises made by another, because sound is conducted by most materials. They will, however, be able to converse with built-in "walkie-talkie" radio equipment. The cargo moves easily; there is no weight, and no friction. To push it, our crew member need only turn on his rocket motor (if he shoved a heavy piece of equipment without rocket power, he might fly backward!).

Obviously the pay load of our rocket ship—though equivalent to that of two huge Super Constellations—will not be sufficient to begin construction of the huge, three-deck, 250-foot-wide space station. Many more loads will be required. Other rocket ships, all timed to arrive at the same point in a continuous procession as the work progresses, will carry up the reminder of the prefabricated satellite. This will be an expensive proposition. Each rocket trip will cost more than half a million dollars *for propellants alone*. Thus, weight and shipping space limitations will greatly affect the specifications of a space station.

In at least one design, the station consists of 20 sections made of flexible nylon-and-plastic fabric. Each of these sections is an independent unit which later, after assembly into a closed ring, will provide compartmentation similar to that found in submarines. To save shipping space, these sections will be carried to the orbit in a collapsed condition. After the "wheel" has been put together and sealed, it will then be inflated like an automobile tire to slightly less than normal atmospheric pressure. This pressure will not only provide a breathable atmosphere within the ring but will give the whole structure its necessary rigidity. The atmosphere will, of course, have to be renewed as the men inside exhaust it.

On solid earth, most of our daily activities are conditioned by gravity. We put something on a table and it stays there because the earth attracts it, pulling it against the table. When we pour a glass of milk, gravity draws it out of the bottle and we catch falling liquid in a glass. In space, however, everything is weightless. And this includes man.

This odd condition in no way spells danger, at least for a limited period of time. We experience weightlessness for short periods when we jump from a diving board into a pool. To be sure, there are some medical men who are concerned at the prospect of permanent weightlessness—not because of any known danger, but because of the unknown possibilities. Most experts discount these nameless fears.

However, there can be no doubt that permanent weightlessness might often prove inconvenient. What we require, therefore, is a "synthetic" gravity within the space station. And we can produce centrifugal force—which acts as a substitute for gravity—by making the "wheel" slowly spin about its hub (a part of which can be made stationary).

To the space station proper, we attach a tiny rocket motor which can produce enough power to rotate the satellite. Since [72] there is no resistance which would slow the "wheel" down, the rocket motor does not have to function continuously. It will operate only long enough to give the desired rotation. Then it is shut off.

Now, how fast would we like our station to spin? That depends on how much "synthetic gravity" we want. If your 250-foot ring performed one full revolution every 12.3 seconds we would get a synthetic gravity equal to that which we normally experience on the ground. This is known as "one gravity" or, abbreviated, "1 g." For a number of reasons, it may be advantageous not to produce one full "g." Consequently, the ring can spin more slowly: for example, it might make one full revolution every 22 seconds, which would result in a "synthetic gravity" of about one third of normal surface gravity.

The centrifugal force created by the slow spin of the space station forces everything out from the hub. No matter where the crew members sit, stand or walk inside, their heads will always point toward the hub. In other words, the inside wall of the "wheel's" outer rim serves as the floor.

* * *

How about the temperature within the space station? Maybe you, too have heard the old fairy tale that outer space is extremely cold—absolute zero. It's cold, all right but not that cold—and not in the satellite. The ironical fact is that the engineering problem in this respect will be to keep the space station comfortably cool, rather than to heat it up. In outer space, the temperature of any structure depends entirely on its absorption and dissipation of the sun's rays. The space station happens to be in the unfortunate position of receiving not only direct heat from the sun but also reflected heat from the earth.

If we paint the space station white, it will then absorb a minimum of solar heat. Being surrounded by a perfect vacuum, it will be, except for its shape, a sort of thermos bottle which keeps hot what is hot and cold what is cold.

In addition, we can scatter over the surface of the space station a number of black patches which, in turn, can be covered by shutters closely resembling white Venetian blinds. When these blinds are open on the sunny side, the black patches will absorb more heat and warm up the station. When the blinds are open on the shaded side, black patches will absorb more heat and warm up the station. When the blinds are open on the shaded side, the black patches will radiate more heat into space, thereby cooling the station. Operate all these blinds with little electric motors, hook them to a thermostat, and tie the whole system in with the station's air conditioning plant—and there's your temperature control system.

Inflating the space station with air will, as we have indicated, provide a breathable atmosphere for a limited time only. The crew will consume oxygen at a rate of approximately three pounds per man per day. At intervals, therefore, this life-giving oxygen will have to be replenished by supply ships from earth. At the same time, carbon dioxide and toxic or odorous products must be constantly removed from the air-circulation system.

The air must also be dehumidified inasmuch as through breathing and perspiration each crew member will loose more than three pounds of water per day to the air system (just as men do on earth).

This water can be collected in a dehumidifier, from which it can economically be salvaged, purified and reused.

Both the air-conditioning and water recovery units need power. So do the radar systems, radio transmitters, astronomical equipment, electronic cookers and other machinery. As a source for this power we have the sun. On the earth, solar power is reliable in only a few places where clouds rarely obscure the sky, but in space there are no clouds, and the sun is the simplest answer to the station's power needs.

Our power plant will consist of a condensing mirror and boiler. The condensing mirror will be a highly polished sheet metal trough running around the "wheel." The position of the space station can be arranged so that the side to which the mirror is attached will always point toward the sun. The mirror then focuses the sun's rays on a steel pipe which runs the length of the mirrored trough. Liquid mercury is fed under pressure into one end of this pipe and hot mercury vapor is taken out at the other end. This vapor drives a turbogenerator which produces about 500 kilowatts of electricity.

Of course, the mercury vapor has to be used over and over again so after it has done its work in the turbine it is returned to the "boiler" pipe in the mirror. Before this can be done, the vapor has to be condensed back into liquid mercury by cooling. This is achieved by passing the vapor through pipes located behind the mirror in the shade. These pipes dissipate the heat of the vapor into space.

Thus we have within the space station a complete synthetic environment capable of sustaining man in space. Of course, man will face hazards—some of them, like cosmic radiation and possible collision with meteorites, potentially severe. These problems are being studied, however, and they are considered far from insurmountable.

Our "wheel" will not be alone in the two-hour orbit. There will nearly always be one or two rocket ships unloading supplies. They will be parked some distance away to avoid the possibility of damaging the space station by collision or by the blast from the vehicle's rocket motors. To ferry men and materials from rocket ship to space station, small rocket-powered metal craft of limited range, shaped very much like overgrown watermelons, will be used. These "space taxis" will be pressurized and, after boarding them, passengers can remove their space suits.

On approaching the space station, the tiny shuttle-craft will drive directly into an air lock at the top or bottom of the stationary hub. The space taxi will be built to fit exactly into the airlock, sealing the opening like a plug. The occupants can then enter the space station proper without having been exposed to the airlessness of space at any time since leaving the air lock of the rocket ship.

* * *

There will also be a space observatory, a small structure some distance away from the main satellite, housing telescopic cameras for taking long exposure photographs. (The space station itself will carry extremely powerful cameras but its spin, though slow, will permit only slow exposures.) The space observatory will not be manned, for if it were, the movements of an operator would disturb the alignment. Floating outside the structure in space suits, technicians will load a camera with special plates or film, and then withdraw. The camera will be aimed and the shutter snapped by remote radio control from the space station.

Most of the pictures taken of the earth, however, will be by the space station's cameras. The observatory will be used manly to record the outer reaches of the universe, from the neighboring planets to the distant galaxies of stars. This mapping of the heavens will produce results which no observatory on earth could possibly duplicate. And, while the scientists are probing the secrets of the universe with their cameras, they will also be planning another trip through space—this time to examine the moon.

Suppose we take the power plant out of our rocket ship's last stage and attach it to a lightweight skeleton frame of aluminum girders. Then we suspend some large collapsible fuel containers in this structure and fill them with propellants. Finally, we connect some plumbing and wiring and top the whole structure with a cabin for the crew, completely equipped with air and water regeneration systems, and navigation and guidance equipment.

The result will be an oddly shaped vehicle [73] not much larger than the rocket ship's third stage, but capable of carrying a crew of several people to a point beyond the rear side of the moon, then back to the space station. This vehicle will bear little resemblance to the moon rockets depicted in science fiction. There is a very simple reason: conventional streamlining is not necessary in space.

The space station, as mentioned previously has a speed of 15,840 miles per hour. Our round-the-moon ship, to leave the two hour orbit, has to have a speed of 22,100 miles per hour to cover the 238,000 mile distance to the moon. This additional speed is acquired by means of a short rocket blast lasting barely two minutes. This throws the round-the-moon ship into a long arch or ellipse, with its remotest point beyond the moon. The space ship will then coast out this distance, unpowered, like a thrown stone. It will lose speed along the way, due to the steady action of the earth's gravitational pull—which, though weakening with distance extends far out into space.

Roughly five days after departure, the space ship will come almost to a standstill and if we have timed our departure correctly the moon will now pass some 200 miles below us with the earth on its far side. On this one trip we can photograph most of the unknown half of the moon, the half which has never been seen from the earth. Furthermore, we now have an excellent opportunity to view the earth from the farthest point yet, at this distance it appears not unlike a miniature reproduction of itself (from the vicinity of the moon the earth will look about four times as large as the full moon does to earth-bound man).

It is not necessary to turn on the space ship's motors for the return trip. The moon's gravity is too slightly to affect us substantially; like the shell which was fired vertically we simply "fall back" to the space station's orbit. The long five day "fall" causes the space ship to regain its initial speed of 22,100 miles per hour. This is 6,340 miles faster than the speed of the space station, but, as we have fallen back tail first, we simply turn on the motors for just two minutes, which reduces our speed to the correct rate which permits us to re-enter the two-hour orbit.

* * *

Besides its use as a springboard for the exploration of the solar system, and as a watchdog of the peace, the space station will have many other functions. Meteorologists, by observing cloud patterns over large areas of the earth, will be able to predict the resultant weather more easily more accurately and further into the future. Navigators on the seas and in the air will utilize the space station as a "fix" for it will always be recognizable.

But there will also be another possible use for the space station—and a most terrifying one. It can be converted into a terribly effective atomic bomb carrier.

Small winged rocket missiles with atomic war heads could be launched from the station in such a manner that they would strike their targets at supersonic speeds. By simultaneous radar tracking from both missile and target, these atomic-headed rockers could be accurately guided to any spot on the earth.

In view of the station's ability to pass over all inhabited regions on earth, such atom bombing techniques would offer the satellite's builders the most important tactical and strategic advance in military history. Furthermore its observers probably could spot in plenty of time any attempt by an enemy to launch a rocket aimed at colliding with the giant "wheel" and intercept it.

We have discussed how to get from the ground to the two-hour orbit, how to build the space station and how to get a look at the unknown half of the moon by way of a round trip from our station in space. But how do we return to earth?

Unlike the ascent to the orbit, which was controlled by an automatic pilot, the decent is in the hands of an experienced "space pilot."

To leave the two-hour orbit in the third stage, or nose section, of the rocket ship, the pilot slows down the vehicle in the same manner in which the returning around-the-moon ship slowed down. He reduces the speed by 1,070 miles per hour. Unpowered, the rocket ship the swings back toward the earth. After 51 minutes, during which we half circumnavigate the globe, the rocket ship enters the upper layers of the atmosphere. Again, it has fallen tail first; now the pilot turns it so that it enters the atmosphere nose first.

* * *

About 50 miles above the earth, due to our downward gravity powered swing from the space station's orbit our speed had increased to 18,500 miles per hour. At his altitude there is already considerable resistance.

With its wings and control surfaces, the rocket closely resembles an airplane. At first however, the wings do not have to carry the rocket ship. On the contrary, they must prevent it from soaring out of the atmosphere and back into the space station's orbit again.

His eyes glued to the altimeter, the pilot will push his control stick forward and force the ship to stay at an altitude of exactly 50 miles. At this height, the air resistance gradually slows the rocket ship down. Only then can the descent into the denser atmosphere begin, from there on the wings bear more and more of the ship's weight. After covering a distance of about 10,000 miles in the atmosphere, the rocket's speed will still be as high as 13,300 miles per hour. After another 3,000 miles the speed will be down to 5,760 miles per hour. The rocket ship will by now have descended to a height of 29 miles.

The progress of the ship through the upper atmosphere has been so fast that air friction has heated the outer metal skin of body and wings to a temperature of about 1,300 degrees Fahrenheit. The rocket ship has actually turned color, from steel blue to cherry red! This should not cause undo concern however inasmuch as we have heat resistant steels which can easily endure such temperatures. The canopy and windows will be built of double paned glass with a liquid coolant flowing between the panes. And the crew and cargo spaces will be properly heat-insulated and cooled by means of a refrigerator-type air conditioning system. Similar problems have already been solved on a somewhat smaller scale in present-day supersonic airplanes.

At a point 15 miles above the earth the rocket ship finally slows down to the speed of sound—roughly 750 miles per hour. From here on, it spirals down to the ground like a normal airplane. It can land on conventional landing gear, on a runway adjacent to the launching site. The touch-down speed will be approximately 65 miles per hour, which is less than that of today's air liners. And if the pilot should miss the runway a small rocket motor will enable him to circle once more and make a second approach.

After a thorough checkup, the third stage will be ready for another ascent into the orbit. The first and second stages (or tail and middle sections), which were parachuted down to the ocean, have been collected in specially made seagoing dry docks. They were calculated to fall at 189 miles and 906 miles respectively from the launching site. They will be found relatively undamaged, because at a point 150 feet above the water their parachute fall was broken by a set of cordite rockets which were automatically set off by a proximity fuse.

They, too, undergo a through inspection with some replacement of parts damaged by the ditching. Then all three stages are put together again in a towerlike hanger, right on the launching platform, and, after refueling and final check, platform and ship are wheeled out to the launching site—ready for another journey into man's oldest and last frontier: the heavens themselves.

THE END

Document I-14

[52] # Man on the Moon: The Journey

By Dr. Wernher von Braun

Technical Director, Army Ordnance Guided Missiles Development Group,
Redstone Arsenal, Huntsville, Alabama

For five days, the expectation speeds through space on its historic voyage—50 men on three ungainly craft, bound for the great unknown

HERE is how we shall go to the moon. The pioneer expedition, 50 scientists and technicians, will take off from the space station's orbit in three clumsy-looking but highly efficient rocket ships. They won't be streamlined: all travel will be in space, where there is no air to impede motion. Two will be loaded with propellant for the five day, 239,000-mile trip and the return journey. The third, which will not return will carry only enough propellant for a one-way trip; the extra room will be filled with supplies and equipment for the scientists' six-week stay.

On the outward voyage, the rocket ships will hit a top speed of 19,500 miles per hour about 33 minutes after departure. Then the motors will be stopped and the ships will fall the rest of the way to the moon.

[53] Such a trip takes a great deal of planning. For a beginning we must decide what flight path to follow, how to construct the ships and where to land. But the project could be completed within the next 25 years. There are no problems involved to which we don't have the answers—or the ability to find them—right now.

First, where shall we land? We may have a wide choice, once we have had a close look at the moon. We'll get that look on a preliminary survey flight. A small rocket ship taking off from the space station will take us to within 50 miles of the moon to get pictures of its meteor-pitted surface—including the "back" part never visible from the earth.

We'll study the photographs for a suitable site. Several considerations limit our selection. Because the Moon's surface has 14,600,000 square miles—about one thirteenth that of the earth—we won't be able to explore more than a small area in detail, perhaps part of a section 500 miles in diameter. Our scientists want to see as many kinds of lunar features as possible, so we'll pick a spot of particular interest to them. We want radio contact with the earth so that means we'll have to stick to the moon's "face" for radio waves won't reach across space to any point the eye won't reach.

We can't land at the moon's equator because its noonday temperatures reach an unbearable 220-degrees Fahrenheit, more than hot enough to boil water. We can't land where the surface is too rugged because we need a flat place to set down. Yet the site can't be too flat either—grain sized meteors constantly bombard the moon at speeds of several miles a second; we have to set camp in a crevice where we have protection from these bullets.

There's one section of the moon that meets all our requirements, and unless something better turns up on closer inspection that's where we land. It's an area called *Sinus Rolis* or Dewy Bay on the northern branch of a plain known as *Oceanus Procellarum* or Stormy Ocean (so called by early astronomers who thought the moon's plains were great seas). Dr. Fred L. Whipple chairman of Harvard University astronomy department, says *Sinus Rolis* is ideal for our purposes—about 650 miles from the lunar north pole where the daylight temperature averages a reasonably pleasant 40 degrees and the terrain is flat enough to land on, yet irregular enough to hide [54] in. With a satisfactory site located we start detailed planning.

To save fuel and time, we want to take the shortest practical course. The moon moves around the earth in an elliptical path once every 27 $1/3$ days. The space station, our point of departure, circles the earth once every two hours. Every two weeks their paths are such that a rocket ship from the space station will intercept the moon in just five days. The best conditions for the return trip will occur two weeks later, and again two weeks after that with their stay limited to multiples of two weeks our scientists have set themselves a six week limit for the first exploration of the moon—long enough to accomplish some constructive research, but not long enough to require a prohibitive supply of essentials like liquid oxygen, water and food.

Six months before our scheduled take-off, we begin piling up construction materials, supplies and equipment at the space station. This operation is a massive, impressive one, involving huge shuttling cargo rocket ships, scores of hard working handlers, and tremendous amounts of equipment. Twice a day pairs of sleek rocket transports from the earth sweep into the satellite's orbit and swarms of workers unload the 36 tons of cargo each carries. With the arrival of the first shipment of material, work on the first of the three moon-going space craft gets underway picking up intensity as more and more equipment arrives.

The supplies are not stacked inside the space station; they're just left floating in space. They don't have to be secured and here's why: the satellite is traveling around the earth at 15,840 miles an hour; at that speed, it can't be affected by the earth's gravity, so it doesn't fall, and it never slows down because there's no air resistance. The same applies to any other object brought into the orbit at the same speed: to park beside the space station a rocket ship merely adjusts its speed to 15,840 miles per hour: and it, too, becomes a satellite. Crates moved out of its hold are traveling at the same speed in relation to the earth, so they also are weightless satellites.

As the weeks pass and the unloading of cargo ships continues, the construction area covers several littered square miles. Tons of equipment lie about—aluminum girders, collapsed nylon-and-plastic fuel tanks, rocket motor units, turbopumps, bundles of thin aluminum plates are a great many nylon bags containing smaller parts. It's a bewildering scene, but not to the moon-ship builders. All construction parts are color-coded—with blue tipped cross braces fitting into blue sockets red joining members keyed to others of the same color and so forth. Work proceeds swiftly.

In fact, the workers accomplish wonders, considering the obstacles confronting a man forced to struggle with unwieldy objects in space. The men move clumsily, hampered by bulky pressurized suits equipped with such necessities of space-life as air conditioning, oxygen tanks, walkie-talkie radios, and tiny rocket motors for propulsion. The work is laborious, for although objects are weightless they still have inertia. A man who shoves a one-ton girder makes it move all right but he makes himself move too. As his inertia is less than the girder's he shoots backward much farther than he pushes the big piece of metal forward.

The small personal rocket motors help the workers move some of the construction parts; the big stuff is hitched to space taxis, tiny pressurized rocket vehicles used for short trips outside the space station.

As the framework of the new rocket ships takes form; big, folded nylon-and plastic bundles are brought over. They're the personnel cabins; pumped full of air, they become spherical, and plastic astrodomes are fitted to the top of sides of each. Other sacks are pumped full of propellant and balloon into the shapes of globes and cylinders. Soon the three moon-going space ships begin to emerge in their final form. The two round-trip ships resemble and arrangement of hourglasses inside a metal framework; the one-way cargo carrier has much the same framework, but instead of hourglasses it has a central structure which looks like a great silo.

Dimensions of the Rocket Ship

Each ship is 160 feet long (nine feet more than the height of the Statue of Liberty) and about 110 feet wide. Each has at its base a battery of 30 rocket motors, and each is topped by the sphere which houses the crew members, scientists and technicians on five floors. Under the sphere are two long arms set on a circular track which enables them to rotate almost a full 360 degrees. These light booms which fold against the vehicles during take-off and landing to avoid damage, carry two vital pieces of equipment: a radio antenna dish for short-wave communication and a solar mirror generating power.

The solar mirror is a curved sheet of highly polished metal which concentrates the sun's rays on a mercury-filled pipe. The intense heat vaporizes the mercury, and the vapor drives a turbo-generator, producing 35 kilowatts of electric power—enough to run a small factory. Its work done, the vapor cools, returns to its liquid state and starts the cycle all over again.

Under the radio and mirror booms of the passenger ships hang 18 propellant tanks carrying nearly 800,000 gallons of ammonialike hydrazine (our fuel) and oxygen-rich nitric acid (the combustion agent). Four of the 18 tanks are outsized spheres, more than 33 feet in diameter. They are attached to light frames on the outside of the rocket ship's structure. More than half our propellant supply—580,000 gallons—is in these large balls: that's the amount needed for take-off. As soon as it's exhausted, the big tanks will be jetisoned. Four other large tanks carry propellant for the landing; they will be left on the moon.

We also carry a supply of hydrogen peroxide [56] to run the turbopumps which force the propellant into the rocket motors. Besides the 14 cylindrical propellant tanks and the four spherical ones, eight small helium containers are strung throughout the framework. The lighter-than-air helium will be pumped into partly emptied fuel tanks to keep their shape under acceleration and to create pressure for the turbopumps.

The cost of the propellant required for this first trip to the moon, the bulk of it used for the supply ships during the build-up period, is enormous—about $300,000,000, roughly 60 percent of the half-billion-dollar cost of the entire operation. (That doesn't count the $4,000,000,000 cost of erecting the space station, whose main purpose is strategic rather then scientific.)

The cargo ship carries only enough fuel for a one-way trip so it has fewer tanks; four discardable spheres like those on the passenger craft, and four cylindrical containers with 162,000 gallons of propellant for the moon landing.

In one respect, the cargo carrier is the most interesting of the three space vehicles. Its big silo-like storage cabin, 75 feet long and 36 feet wide, was built to serve a double purpose. Once we reach the moon and the big cranes folded against the framework have swung out and unloaded the 285 tons of supplies in a cylinder, the silo will be detached from the rest of the rocket ship. The winch-driven cables slung from the cranes will then raise half of the cylinder, in sections, which it will deposit on trailers drawn by tractors. The tractors will take them to a protective crevice on the moon's surface at the place chosen for our camp. Then the other lengthwise half will be similarly moved—giving us two ready-to-use Quonset huts.

Now that we have our space ships built and have provided ourselves with living quarters for our stay on the moon. A couple of important items remain; we must protect ourselves against two of the principal hazards of space travel, flying meteors and extreme temperatures.

For Protection Against Meteors

To guard against meteors, all vital parts of the three craft—propellant tanks, personnel spheres, cargo cabin—are given a thin covering of sheet metal, set on studs which leave at least one inch space between this outer shield and inside wall. The covering, called meter bumper, will take the full impact of the flying particles (we don't expect to be struck by any meteors much larger than a grain of sand) and will cause them to disintegrate before they can do damage.

For protection against excessive heat, all parts of the three rocket ships are painted white because white absorbs little of the sun's radiation. Then, to guard against cold, small black patches are scattered over the tanks and personnel spheres. The patches are covered by white blinds, automatically controlled by thermostats. When the blinds on the sunny side are open, the spots absorb heat and warm the cabins and tanks when the blinds are closed and all white surface is exposed to the sun permitting little heat to enter. When the blinds on the shaded side are open, the black spots radiate heat and the temperature drops.

Now we're ready to take off from the space station's orbit to the moon.

The bustle of our departure—hurrying space taxis, the nervous last-minute checks by engineers, the loading of late cargo and finally the take-off itself—will be watched by millions. Television cameras on the space station will transmit the scene to receivers all over the world. And people on the earth's dark side will be able to turn from their screens to catch a fleeting glimpse of light—high in the heavens—the combined flash of 90 rocket motors, looking from the earth like the sudden birth of a new short lived star.

Our departure is slow. The big rocket ships rise ponderously, one after the other, green flames streaming from their batteries of rockets, and then they pick up speed. Actually we don't need to gain much speed. The velocity required to get us to our destination is 19,500 miles an hour but we've had a running start, while "resting" in the space station's orbit, we are really streaking through space at 15,840 miles an hour. We need an additional 3,660 miles an hour.

Thirty-three minutes from take-off we have it. Now we cut off our motors; momentum and the moon's gravity will do the rest.

The moon itself is visible to us as we coast through space, but it's so far off one side that it's hard to believe we won't miss it. In the five days of our journey, though, it will travel a great distance and so will we; at the end of that time we shall reach the farthest point, or apogee, of our elliptical course, and the moon shall be right in front of us.

The earth is visible, too—an enormous ball most of it bulking pale black against the deeper black of space but with a wide crescent of day light where the sun strikes it. Within the crescent, the continents enjoying summer stand out as vast green terrain maps surrounded by the brilliant [58] blue of the oceans. Patches of white cloud obscure some of the detail; other white blobs are snow and ice on mountains ranges and polar areas.

Against the blackness of the earth's night side is a gleaming spot—the space station, reflecting the light of the sun.

Two hours and 54 minutes after departure we are 17,750 miles from the earth's surface. Our speed has dropped sharply to 10,500 miles and hour. Five hours and eight minutes en route, the earth is 32,950 miles away, and our speed is 8,000 miles an hour; after 20 hours, we're 132,000 miles from the earth traveling at 4,300 miles an hour.

On this first day, we discard the empty departure tanks. Engineers in protective suits step outside the cabin, stand for a moment in space, then make their way down the girders to the big spheres. They pump any remaining propellant into reserve tanks, disconnect the useless containers, and give them a gentle shove. For a while the tanks drift along beside us; soon they float out of sight. Eventually they will crash on the moon.

There is no hazard for the engineers in this operation. As a precaution they are secured to the ship by safety lines. But they could probably have done as well without them. There is no air in space to blow them away.

That's just one of the peculiarities of space to which we must adapt ourselves. Lacking a natural sequence of night and day, we live by an arbitrary time schedule. Because nothing has weight; cooking and eating are special problems. Kitchen utensils have magnetic strips or clamps so they won't float away. The heating of food is done on electronic ranges. They have many advantages: they're clean, easy to operate, and their short-wave rays don't burn up precious oxygen.

Difficulties of Dining in Space

We have no knives, spoons or forks. All solid food is precut; all liquids are served in plastic bottles and forced directly into the mouth by squeezing. Our mess kits had spring operated covers; our only eating utensils are tonglike devices; if we open the covers carefully, we can grab a mouthful of food without getting it all over the cabin.

From the start of the trip, the ship's crew has been maintaining a round-the-clock schedule, standing eight hour watches. Captains, navigators and radio men spend most of their time checking and rechecking our flight track, ready to start up the rockets for a change in course if an error turns up. Technicians back up this operation with reports from the complex and delicate "electronic brains"—computers, gyroscopes, switchboards and other instruments—on the control deck. Other specialists keep watch over the air-conditioning, temperature, pressure and oxygen systems.

But the busiest crew members are the maintenance engineers and their assistants, tireless men who been bustling back and forth between ships since shortly after the voyage started, anxiously checking propellant tanks, tubing, rocket motors, turbopumps and all other vital equipment. Excessive heat could cause dangerous hairline cracks in the rocket motors; unexpectedly large meteors could smash through the thin bumpers surrounding the propellant tanks; fittings could come loose. The engineers have to be careful.

We are still slowing down. At the start of the fourth day, our speed has dropped to 800 [59] miles an hour, only slightly more than the speed of a conventional jet fighter. Ahead, the harsh surface features of the moon are clearly outlined. Behind, the blue-green ball of the earth appears to be barely a yard in diameter.

Our fleet of unpowered rocket ships is now passing the neutral point between the gravitational fields of the earth and the moon. Our momentum has dropped off to almost nothing—yet we're about to pick up speed. For now we begin falling toward the moon, about 23,600 miles away. With no atmosphere to slow us we'll smash into the moon at 6,000 miles an hour unless we do something about it.

Rotating the Moon Ship

This is what we do: aboard each ship, near its center of gravity is a positioning device consisting of three fly-wheels set at right angles to one another and operated by electric motors. One of the wheels heads in the same direction as our flight path—in other words; along the longitudinal axis of the vehicle, like the rear wheels of a car. Another parallels the latitudinal axis, like steering wheel of an ocean vessel. The third lies along the horizontal axis like the rear steering wheel of a hook and ladder truck. If we start anyone of the wheels spinning it causes our rocket ship to turn slowly in the other direction (pilots know this "torque" effect as increased power causes a plane's propeller to spin more rapidly in one direction, the pilot has to fight his controls to keep the plane from rolling in the other direction).

The captain of our space ship orders the longitudinal flywheel set in motion. Slowly our craft begins to cartwheel; when it has turned a revolution, it stops. We are going toward the moon tail-end-first, a position which will enable us to brake our fall with our rocket motors when the right time comes.

Tension increases aboard the three ships. The landing is tricky—so tricky that it will be done entirely by automatic pilot, to diminish the possibility of human error. Our scientists compute our rate of descent, the spot at which we expect to strike; the speed and direction of the moon (it's traveling at 2,280 miles an hour at right angles to our path). These and other essential statistics are fed into a tape. The tape, based on the same principle as the player-piano roll and the automatic business-machine card, will control the automatic pilot. (Actually, a number of tapes intended to provide for all eventualities will be fixed up long before the flight, but last minute-checks are necessary to see which tape to use and to see whether a manual correction of our course is required before the autopilot takes over.)

Now we lower part of our landing gear—four spiderlike legs, hinged to the square rocket assembly, which have been folded against the framework.

As we near the end of our trip, the gravity of the moon which is still to one side of us, begins to pull us off our elliptical course, and we turn the ship to conform to this change of direction. At an altitude of 550 miles the rocket motors begin firing; we feel the shock of their blasts inside the personnel sphere and suddenly our weight returns. Objects which have been not secured before hand tumble to the floor. The force of the rocket motors is such that we have about one third our normal earth weight.

The final 10 minutes are especially tense. The tape-guided automatic pilots are now in full control. We fall more and more slowly, floating over the landing area like descending helicopters as we approach, the fifth leg of our landing gear—a big telescoping shock absorber which has been housed in the center of the rocket assembly is lowered through the fiery blast of the motors. The long green rocket flames being to slash against the baked lunar surface. Swirling clouds of brown-gray dust are thrown out sideways; they settle immediately instead of hanging in air, as they would on the earth.

The broad round shoe of the telescopic landing leg digs into the soft volcanic ground. If it strikes too hard an electronic mechanism inside it immediately calls on the rocket motors for more power to cushion the blow. For a few seconds, we balance on the single leg then the four outrigger legs slide out to help support the weight of the ship, and are locked into position. The whirring of machinery dies away. There is absolute silence. We have reached the moon.

Now we shall explore it.

Document I-15

[21]

Is There Life on Mars?

By Dr. Fred L. Whipple

Chairman, Department of Astronomy, Harvard University

Astronomers—planning to give the great red planet its closest scrutiny in history this summer— are nearer than ever before to answering the most fascinating question of all.

On July 2nd, the planet Mars, swinging through its lopsided orbit around the sun, will be closer to the earth than any time since 1941. All over the world scientists will train batteries of telescopes and cameras on the big red sphere in history's greatest effort to unravel some of the mystery surrounding this most intriguing of the planets.

Next to Venus, Mars is our closet planetary neighbor. Even so, it will be 40,000,000 miles away as it passes by this summer (compared to 250,000,000 miles at its farthest point from the earth); on the most powerful of telescopes it will look no larger than a coffee saucer. Still it will be close enough to provide astronomers important facts about its size, atmosphere and surface conditions—and the possibility that some kind of life exists there.

We already know a great deal.

Mars's diameter is roughly half the size of the earth, the Martian day is 24 hours, 37 minutes long, but its year is nearly twice as long as ours—67 Martian days. During daylight hours the temperature on Mars shoots into the eighties, but at night a numbing cold grips the planet; the temperature drops suddenly to 95 below zero, Fahrenheit.

There is no evidence of oxygen on Mars's thin blue atmosphere. Moreover, its atmospheric pressure is so low that an earth man couldn't survive without a pressurized suit. If life of any kind does exist on Mars it must be extremely rugged.

Through the telescope, astronomers can clearly see Mars's great reddish deserts, blue-tinted cloud formations and—especially conspicuous—its distinctive polar caps.

The Martian polar caps cover about 4,000,000 square miles in the wintertime—an area roughly half the size of the North American continent. But as they melt in spring strange blue-green areas develop near their retreating edges. Some months later these color patches, now covering great areas of the planet's surface turn brownish, finally in the deep of Martin Winter they're dark chocolate color. Do the seasonal color variations indicate some sort of planet vegetable life? That's one of the riddles we'd like to solve.

There's another big question mark: Mars's so-called canals. Although most modern astronomers have long since discounted the once popular theory that the faint tracings seen by some on Mars are actually a network of waterways (and, therefore perhaps constructed by intelligent beings), we still don't know what they are—or if they exist at all.

The "canals" have had a controversial history. They were first reported in 1877 by an Italian astronomer named Giovanni Schiaparelli who said he had seen delicate lines tracing a gridlike pattern over vast areas of the planet. He called them *canali*—"canals" or "channels."

Since Schaparelli, many astronomers (especially Dr. Percival Lowell, who established an observatory for the primary purpose of studying Mars) have reported observing the delicate vein-like lines. Others, just as keensighted, have spent years studying the Martian face without once seeing the disputed markings.

This year we may get an opportunity to clear up the canal confusion once and for all. An American team sponsored jointly by the National Geographic Society and Lowell Observatory, will photograph Mars from Bloemfontein, South Africa, where Mars will appear almost directly over head nightly during early July. The U.S. team, using new photographic techniques and the latest in fast film emulsions, expects to get the most detailed photographs of the planet yet obtained.

But great as the 1954 Mars observation program promises to be, it's only the curtain raiser for 1956, when Mars will approach to within 35,000,000 miles of the earth. Not for another 15 years, in 1971, will it be so close again.

When all the finding have been evaluated we may be able to make some intelligent guesses as to the possibilities of life on Mars. Chances are that bacteria are the only type of animal life which could exist in a planet's oxygenless atmosphere. There also may be some sort of tough primitive plant life—perhaps lichens or mosses which produce their own oxygen and water. Such plants might explain the changing colors of the Martian seasons.

There's one other possibility.

How can we say with absolute certainty that there isn't a different form of life existing on Mars—a kind of life that we know nothing about? We can't. There's only one way to find out for sure what is on Mars—and that's to go there.

Document I-16

[22]

Can We Get to Mars?

By Dr. Wernher von Braun with Cornelius Ryan

Chief, Guided Missile Development Division,
Redstone Arsenal, Huntsville, Alabama

[23]   *Man's trial-blazing journey to Mars will be a breath-taking experience*
         —with problems to match

The first man who set out for Mars had better make sure they leave everything at home in apple-pie order. They won't get back to earth for more than two and a half years.

The difficulties of a trip to Mars are formidable. The outbound journey, following a huge arc 255,000,000 miles long, will take eight months—even with rocket ships that travel many thousands of miles an hour. For more than a year, the explorers will have to live on the great red planet, waiting for it to swing into a favorable position for the return trip. Another eight months will pass before the 70 members of the pioneer expedition set foot on earth again. All during that time, they will be exposed to a multitude of dangers and strains, some of them impossible to foresee on the basis of today's knowledge.

Will man ever go to Mars? I am sure he will—but it will be a century or more before he's ready. In that time scientists and engineers will learn more about the physical and mental rigors of interplanetary flight—and about the known dangers of life on another planet. Some of that information may become available within the next 25 years or so, through the erection of a space station above the earth (where telescope viewings will not be blurred by the earth's atmosphere) and through the subsequent exploration of the moon, as described in previous issues of Collier's.

Even now science can detail the technical requirements of a Mars expedition down to the last ton of fuel. Our knowledge of the laws governing the solar system—so accurate that astronomers can predict an eclipse of the sun to within a fraction of a second—enables scientists to determine exactly the speed a space ship must have to reach Mars, the course that will intercept the planet's orbit at exactly the right moment, the methods to be used for the landing, take-off and other maneuvering. [24] We know, from these calculations, that we already have chemical rocket fuels adequate for the trip.

Better propellants are almost certain to emerge during the next 100 years. In fact, scientific advances will undoubtedly make obsolete many of the engineering concepts on which this article, and the accompanying illustrations, are based. Nevertheless, it's possible to discuss the problems of a flight to Mars in terms of what is known today. We can assume, for example, that such an expedition will involve about 70 scientists and crew members. A force that size would require a flotilla of 10 massive space ships, each weighing more than 4,000 tons—not only because there's safety in numbers, but because of the tons of fuel, scientific equipment, rations, oxygen, water and the like necessary for the trip and for a stay of about 31 months away from earth.

All that information can be computed scientifically. But science can't apply a slide rule to man; he's the unknown quantity, the weak spot that makes a Mars expedition a project for the far distant, rather than the immediate, future. The 70 explorers will endure hazards and stresses the like of which no men before them have ever known. Some of these hardships must be eased—or at least better understood—before the long voyage becomes practical.

For months at a time, during the actual period of travel, the expedition members will be weightless. Can the human body stand prolonged weightlessness? The crews of rocket ships plying between the ground and earth's space station about 1,000 miles away will soon grow accustomed to the absence of gravity—but they will experience this odd sensation for no more than a few hours at a time. Prolonged weightlessness will be a different story.

Over a period of months in outer space, muscles accustomed to fighting the pull of gravity could shrink from disuse—just as do the muscles of people who are bedridden or encased in plaster casts for a long time. The members of a Mars expedition might be seriously handicapped by such a disability. Faced with a rigorous work schedule on the unexplored planet, they will have to be strong and fit upon arrival.

The problem will have to be solved aboard the space vehicles. Some sort of elaborate spring exercisers may be the answer. Or perhaps synthetic gravity could be produced aboard the rocket ships by designing them to rotate as they coast through space, creating enough centrifugal force to act as a substitute for gravity.

Far worse than the risk of atrophied muscles is the hazard of cosmic rays. An overdose of these deep-penetrating atomic particles, which act like the invisible radiation of an atomic-bomb burst, can cause blindness, cell damage and possibly cancer.

Scientists have measured the intensity of cosmic radiation close to the earth. They have learned that the rays dissipate harmlessly in our atmosphere. They also have deduced that man can safely venture as far as the moon without risking an overdose of radiation. But that's a comparatively brief trip. What will happen to men who are exposed to rays for months on end? There is no material that offers practical protection against cosmic rays—practical, that is for space travel. Space engineers could provide a barrier by making the cabin walls of lead several feet thick—but that would add hundreds of tons to the weight of the space vehicle. A more realistic plan might be to surround the cabin with the fuel tanks, thus providing the added safeguard of a two- or three- foot thickness of liquid.

The best bet would seem to be a reliance on man's ingenuity; by the time an expedition from the earth is ready to take off for Mars, perhaps in the mid-2000s, it is quite likely that researchers will have perfected a drug which will enable men to endure radiation for comparatively long periods. Unmanned rockets, equipped with instruments which send information back to earth, probably will blaze the first trail to our sister planet, helping to clear up many mysteries of the journey.

Small Meteors Could Do Little Damage

Meteors, for example. Many billions of these tiny bullets, most of them about the size of a grain of sand, speed wildly through space at speeds of more than 150,000 miles an hour. For short trips, we can protect space ships from these lightning fast pellets by covering all vital areas—fuel tanks, rocket motors, cargo bins, cabins and the like—with light metal outer shields called meteor bumpers. The tiny meteors will explode against this outer shell, leaving the inner skin of the ship—and the occupants—unharmed.

But in the 16 months of space travel required for a visit to mars, much larger projectiles might be encountered. Scientists know that the density of large meteors is greater near the red planet than it is around the earth. If, by some chance, a rock the size of a baseball should plow through the thin shell of one of the rocketships it could do terrible damage—especially if it struck a large solid object inside. A meteor that size, traveling at terrific speed, could explode with the force of 100 pounds of TNT. In the cabin of a space vehicle, such an explosion would cause tremendous destruction.

Fortunately, meteors that size will be extremely rare, even near Mars.

Dime-sized chunks are more likely to be encountered. They will be a danger, too, although not so bad as the larger rocks. They'll rip through the bumper and skin like machine-gun bullets. If they strike anything solid, they'll explode with some force. If not, they'll leave through the other side of the ship—but even then they may cause trouble. Holes will have to be plugged to maintain cabin pressure. The shock wave created by the meteors' extreme speed may hurt the ship's occupants: there will be a deafening report and a blinding flash; the friction created by their passage through the cabin atmosphere will create enough heat to singe the [25] eyebrows of a man standing close by. And, of course, a person in the direct path of a pebble-sized meteor could be severely injured. A fragile piece of machinery could be destroyed, and it's even possible that the entire rocket ship would have to be abandoned after sustaining one or more hits by space projectiles that size (astronomers estimate that one out of 10 ships on a 16-month voyage might be damaged badly, although even that is unlikely).

If one of the Mars-bound vehicles does suffer serious damage, the incident needn't be disastrous. In a pinch, a disabled space vehicle can be abandoned easily. All of the ships will carry small self propelled craft—space taxis—which are easily built and easily maneuvered. They will be fully pressurized, and will be used for routine trips between the ships of the convoy, as well as for emergencies. If for some reason the space taxis aren't available to the occupants of a damaged ship, they will be able to don pressurized suits and step calmly out into space. Individual rocket guns, manually operated, will enable each of them to make his way to the nearest spaceship in the convoy. Space suited explorers will have no difficulty traveling between ships. There's no air to impede motion, no gravitational pull and no sense of speed. When they leave their ship the men will have to overcome only their own inertia. They'll be traveling through the solar system at more than 70,000 miles an hour, but they will be no more aware of it then we on earth are aware that every molecule of our bodies is moving at a speed of 66,600 miles an hour around the sun.

Science ultimately will solve the problems posed by cosmic rays, meteors and other natural phenomena of space. But man will still face one great hazard: himself.

Man must breathe. He must guard himself against a great variety of illnesses and ailments. He must be entertained. And he must be protected from many psychological hazards, some of them still obscure.

How will science provide a synthetic atmosphere within the space-ship cabins and Martian dwellings for two and a half years? When men are locked into a confined, airtight area for only a few days or weeks oxygen can be replenished, and exhaled carbon dioxide and other impurities extracted, without difficulty. Submarine engineers solved the problem long ago. But a conventional submarine surfaces after a brief submersion and blows out its stale air. High-altitude pressurized aircraft have mechanisms which automatically introduce fresh air and expel contaminated air.

There's no breathable air in space or on Mars; the men who visit the red planet will have to carry with them enough oxygen to last many months.

When Men Live Too Close Together

During that time they will live, work and perform all bodily functions within the cramped confines of a rocket-ship cabin or a pressurized—and probably mobile—Martian dwelling. (I believe the first men to visit Mars will take along inflatable, spherical cabins, perhaps 30 feet across, which can be mounted atop tractor chassis.) Even with plenty of oxygen, the atmosphere in those living quarters is sure to pose a problem.

Within the small cabins, the expedition members will wash, perform personal functions, sweat, cough, cook, create garbage. Every one of those activities will feed poisons into the synthetic air—just as they do within the earth's atmosphere.

No less than 29 toxic agents are generated during the daily routine of the average American household. Some of them are body wastes, others come from cooking. When you fry an egg, the burned fat releases a potent irritant called acrolein. Its effect is negligible on earth because the amount is so small that it's almost instantly dissipated in the air. But that microscopic quantity of acrolein in the personnel quarters of a Mars expedition could prove dangerous; unless there was some way to remove it from the atmosphere it would be circulated again and again through the air-conditioning system.

Besides the poisons resulting from cooking and the like, the engineering equipment—lubricants, hydraulic fluids, plastics, the metals in the vehicles—will give off vapors which could contaminate the atmosphere.

What can be done about this problem? No one has all the answers right now, but there's little doubt that by using chemical filters, and by cooling and washing the air as it passes through the air-conditioning apparatus, the synthetic atmosphere can be made safe to live in.

Besides removing the impurities from the man-made air, it may be necessary to add a few. Man has lived so long with the impurities in the earth's atmosphere that no one knows whether he can exist without them. By the time of the Mars expedition, the scientists may decide to add traces of dust, smoke and oil to the synthetic air—and possibly iodine and salt as well.

I am convinced that we have, or will acquire, the basic knowledge to solve all the physical problems of a flight to Mars. But how about the psychological problem? Can a man retain his sanity while cooped up with many other men in a crowded area, perhaps twice the length of your living room, for more then thirty months?

Share a small room with a dozen people completely cut off from the outside world. In a few weeks the irritations begin to pile up. At the end of [26] a few months, particularly if the occupants of the room are chosen haphazardly, someone is likely to go berserk. Little mannerisms—the way a man cracks his knuckles, blows his nose, the way he grins, talks or gestures—create tension and hatred which could lead to murder.

Imagine yourself in a space ship millions of miles from earth. You see the same people every day. The earth, with all it means to you, is just another bright star in the heavens; you aren't sure you'll ever get back to it. Every noise about the rocket ship suggests a breakdown, every crash a meteor collision. If somebody does crack, you can't call off the expedition and return to earth. You'll have to take him with you.

The psychological problem probably will be at its worst during the two eight-month travel periods. On Mars, there will be plenty to do, plenty to see. To be sure, there will be certain problems on the planet, too. There will be considerable confinement. The scenery is likely to be grindingly monotonous. The threat of danger from some unknown source will hang over the explorers constantly. So will the knowledge that an extremely complicated process, subject to possible breakdown, will be required to get them started on their way back home. Still, Columbus's crew at sea faced much the same problems the explorers will face on Mars: the fifteenth-century sailors felt the psychological tension, but no one went mad.

But Columbus traveled only ten weeks to reach America; certainly his men would never have stood an eight-month voyage. The travelers to Mars will [27] have to, and psychologists undoubtedly will make careful plans to keep up the morale of the voyagers.

The fleet will be in constant radio communication with the earth (there probably will be no television transmission, owing to the great distance). Radio programs will help relieve the boredom, but it's possible that the broadcasts will be censored before transmission; there's no way of telling how a man might react, say, to the news that his home town was the center of a flood disaster. Knowing would do him no good—and it might cause him to crack.

Besides radio broadcasts, each ship will be able to receive (and send) radio pictures. There also will be films which can be circulated among the space ships. Reading matter will probably be carried in the form of microfilms to save space. These activities—plus frequent intership visiting, lectures and crew rotations—will help to relieve the monotony.

There is another possibility, seemingly fantastic but worth mentioning briefly because experimentation already has indicated it may be practical. The nonworking members of a Mars expedition may actually hibernate during part of the long voyage. French doctors have induced a kind of artificial hibernation in certain patients for short periods in connection with operations for which they will need all their strength (Collier's December 11, 1953—Medicine's New Offensive Against Shock, by J.D. Ratcliff). The process involves a lowering of the body temperature, and the subsequent slowing down of all normal physical processes. On a Mars expedition, such a procedure, over a longer [28] period, would solve much of the psychological problem, would cut sharply into the amount of food required for the trip, and would, if successful, leave the expedition members in superb physical condition for the ordeal of exploring the planet.

Certainly if a Mars expedition were planned for the next 10 or 15 years, no one would seriously consider hibernation as a solution for any of the problems of the trip. But we're talking of a voyage to be made 100 years from now; I believe that if the French experiments bear fruit, hibernation may actually be considered at that time.

Finally, there has been one engineering development which may also simplify both the psychological and physical problems of a Mars voyage. Scientists are on the track of a new fuel, useful only in the vacuum of space, which would be so economical that it would make possible far greater speeds for space journeys. It could be used to shorten the travel time, or to lighten the load of each space ship, or both. Obviously, a four- or six-month Mars flight would create far fewer psychological hazards than a trip lasting eight months.

In any case, it seems certain that members of an expedition to Mars will have to be selected with great care. Scientists estimate that only one person in every 6,000 will be qualified, physically, mentally and emotionally, for routine space flight. But can 70 men be found who will have those qualities—and also the scientific background necessary to explore Mars? I'm sure of it.

One day a century or so from now, a fleet of rocket ships will take off for Mars. The trip could be made with 10 ships launched from an orbit 1,000 miles out in space, that girdles our globe at its equator. (It would take tremendous power and vast quantities of fuel to leave directly from the earth. Launching a Mars voyage from an orbit about 1,000 miles out, far from the earth's gravitational pull, will require relatively little fuel.) The Mars-bound vehicles, assembled in the orbit, will look like bulky bundles of girders, with propellant tanks hung on the outside and great passenger cabins perched on top. Three of them will have torpedo-shaped noses and massive wings—dismantled, but strapped to their sides for future use. Those bullet noses will be detached and will serve as landing craft, the only vehicles that will actually land on the neighbor planet. When the 10 ships are 5,700 miles from the earth, they will cut off their rocket motors; from there on, they will coast unpowered toward Mars.

After eight months they will swing into an orbit around Mars, about 600 miles up, and adjust speed to keep from hurtling into space again. The expedition will take this intermediate step, instead of preceding directly to Mars, for two main reasons: first, the ships (except for the three detachable torpedo-shaped noses) will lack the streamlining required for flight in the Martian atmosphere; second, it will be more economical to avoid carrying all the fuel needed for the return to earth (which now comprises the bulk of the cargo) all the way down to Mars and then back up again.

Upon reaching the 600-mile orbit—and after some exploratory probings of Mars's atmosphere with unmanned rockets—the first of the three landing craft will be assembled. The torpedo nose will be unhooked, to become the fuselage of a rocket plane. The wings and set of landing skis will be attached, and the plane launched toward the surface of Mars.

The landing of the first plane will be made on the planet's snow covered polar cap—the only spot where there is any reasonable certainty of finding a smooth surface. Once down, the pioneer landing party will unload its tractors and supplies, inflate its balloon-like living quarters, and start on a 4,000 mile overland journey to the Martian equator, where the expedition's Main base will be set up (it is the most livable part of the planet—well within the area that scientists want most to investigate). At the equator, the advance party will construct a landing strip for the other two rocket planes. (The first landing craft will be abandoned at the pole.)

In all, the expedition will remain on the planet 15 months. That's a long time—but it still will be too short to learn all that science would like to know about Mars.

When, at last, Mars and the earth begin to swing toward each other in the heavens, and it's time to go back, the two ships that landed on the equator will be stripped of their wings and landing gear, set on their tails and, at the proper moment, rocketed back to the 600-mile orbit on the flat leg of the return journey.

What curious information will these first explorers carry back from Mars? Nobody knows—and its extremely doubtful that anyone now living will ever know. All that can be said with certainty today is this: the trip can be made, and will be made...someday.

Document I-17

Document title: Statement by James C. Hagerty, The White House, July 29, 1955.

Source: NASA Historical Reference Collection, NASA History Office, NASA Headquarters, Washington, D.C.

NSC 5520, "Draft Statement of Policy on U.S. Scientific Satellite Program," recommended the creation of a scientific satellite program as part of the International Geophysical Year (IGY) as well as the development of satellites for reconnaissance purposes. Based upon this report, the National Security Council approved the IGY satellite on May 26, 1955. However, it was not until July 28 that a public announcement was made during an oral briefing at the White House. The formal statement was dated July 29. This statement emphasized that the satellite program was intended to be the U.S. contribution to the IGY and that the scientific data was to benefit scientists of all nations.

July 29, 1955

The White House

Statement by James C. Hagerty

On behalf of the President, I am now announcing that the President has approved plans by this country for going ahead with launching of small unmanned earth-circling satellites as part of the United States participation in the International Geophysical Year which takes place between July 1957 and December 1958. This program will for the first time in history enable scientists throughout the world to make sustained observations in the regions beyond the earth's atmosphere.

The President expressed personal gratification that the American program will provide scientists of all nations this important and unique opportunity for the advancement of science.

Documents I-18 and I-19

Document title: F.C. Durant, "Report of Meetings of Scientific Advisory Panel on Unidentified Flying Objects Covered by Office of Scientific Intelligence, CIA, January 14-18, 1953," February 16, 1953.

Document title: "Air Force's 10 Year Study of Unidentified Flying Objects," Department of Defense, Office of Public Information, News Release No. 1083-58, November 5, 1957.

Sources: NASA Historical Reference Collection, NASA History Office, NASA Headquarters, Washington, D.C.

This CIA report on Unidentified Flying Objects (UFOs), which was declassified in December 1974, is frequently cited by UFO conspiracy theorists who claim that the government is covering up knowledge of extraterrestrial visits. Several studies of UFOs were conducted by the U.S. military throughout the 1950s and 1960s, partly out of Cold War concern that UFOs were actually Soviet spycraft, and partly in response to public outcry.

Document I-18

Report of Meetings of the Office of Scientific Intelligence Scientific Advisory Panel on Unidentified Flying Objects

Covered by Office of Scientific Intelligence, CIA

January 14–18, 1953

[1] MEMORANDUM FOR: Assistant Director for Scientific Intelligence

FROM: F. C. Durant

SUBJECT: Report of Meetings of the Office of Scientific Intelligence Scientific Advisory Panel on Unidentified Flying Objects, January 14 - 18, 1953

PURPOSE:

The purpose of this memorandum is to present:

a. A brief history of the meetings of the O/SI Advisory Panel On Unidentified Flying Objects (Part I),

b. An unofficial supplement to the official Panel Report to AD/SI setting forth comments and suggestions of the Panel Members which they believed were inappropriate for inclusion in the formal report (Part II).

Part I: History of Meetings

General

After consideration of the subject of "unidentified flying objects" at the 4 December meeting of the Intelligence Advisory Committee, the following action was agreed:

"The Director of Central Intelligence will:
a. Enlist the services of selected scientists to review and appraise the available evidence in the light of pertinent scientific theories...."

Following the delegation of this action to the Assistant Director for Scientific Intelligence and preliminary investigation, [2] an Advisory Panel of selected scientists was assembled. In cooperation with the Air Technical Intelligence Center, case histories of reported sightings and related material were made available for their study and consideration.

Present at the initial meeting (0930 Wednesday, 14 January) were: Dr. H P. Robertson, Dr. Luis W. Alvares, Dr. Thornton Page, Dr. Samuel A. Goudsmit, Mr. Philip G. Strong, Lt. Col. Frederick C. E. Oder (P&E Division), Mr. David B. Stevenson (W&E Division), and the writer. Panel Member, Dr. Lloyd V. Berkner, was absent until Friday afternoon. Messrs. Oder and Stevenson were present throughout the sessions to familiarize themselves with the subject, represent the substantive interest of their Divisions, and assist in administrative support of the meetings. (A list of personnel concerned with the meetings is given in Tab A.)

Wednesday Morning

The AD/SI opened the meeting, reviewing CIA interest in the subject and action taken. This review included the mention of the O/SI Study Group of August 1952 (Strong, Eng, and Durant) culminating in the briefing of the DCI, the ATIC November 21 briefing, 4 December IAC consideration, visit to ATIC (Chadwell, Robertson and Durant), and O/SI concern over potential dangers to national security indirectly related to these sightings. Mr. Strong enumerated these potential dangers. Following this introduction, Mr. Chadwell turned the meeting over to Dr. Robertson as Chairman of the Panel. Dr. Robertson enumerated the evidence available and requested consideration of specific reports and letters be taken by certain individuals present (Tab B). For example, case histories involving radar or radar and visual sightings were selected for Dr. Alvares while reports of Green Fireball phenomena, nocturnal lights, and suggested programs of investigations were routed to Dr. Page. Following these remarks, the motion pictures of the sightings at Tremonton, Utah (2 July 1952) and Great Falls, Montana (15 August 1950) were shown. The meeting adjourned at 1200.

Wednesday Afternoon

The second meeting of the Panel opened at 1400. Lt. R. S. Neasham, USN, and Mr. Harry Woo of the USN Photo Interpretation Laboratory, Anacostia, presented the results of their analyses of the films mentioned above. This analysis evolved considerable discussion as elaborated upon below. Besides Panel members and CIA personnel, Capt. E. J. Ruppelt, Dr. J. Allen Nyack, Mr. Dewey J. Fournet, Capt. Harry B. Smith (2-e-2), and Dr. Stephen Possony were present.

Following the Photo Interpretation Lab presentation, Mr. E. J. Ruppelt spoke for about 40 minutes on ATIC methods of handling and evaluating reports of sighting and their efforts to improve the quality of reports. The meeting was adjourned at 1715.

Thursday Morning

The third and fourth meetings of the Panel were held Thursday, 15 January, commencing at 0900 with a two-hour break for luncheon. Besides Panel members and CIA personnel, Mr. Ruppelt and Dr. Hynak were present for both sessions. In the morning, Mr. Ruppelt continued his briefing on ATIC collection and analysis procedures. The Project STORK support at Battelle Memorial Institute, Columbus, was described by Dr. Hynek. A number of case histories were discussed in detail and a motion picture film of seagulls was shown. A two hour break for lunch was taken at 1200.

Thursday Afternoon

At 1400 hours Lt. Col. Oder gave a 40-minute briefing of Project TWINKLE the investigatory project conducted by the Air Force Meteorological Research Center at Cambridge, Mass. In this briefing he pointed out the many problems of setting up and manning 24-hour instrumentation watches of patrol cameras searching for sightings of U.F.O.'s.

At 1615 Brig. Gen. William N. Garland joined the meeting with AD/SI. General Garland expressed his support of the Panel's efforts and stated three personal opinions:

a. That greater use of Air Force intelligence officers in the field (for follow-up investigation) appeared desirable, but that they required thorough briefing.

b. That vigorous effort should be made to declassify as many of the reports as possible.

c. That some increase in the ATIC section devoted to U.F.O. analysis was indicated.

This meeting was adjourned at 1700.

Friday Morning

The fifth session of the Panel convened at 0900 with the same personnel present as enumerated for Thursday (with the exception of Brig. Gen. Garland).

From 0900–100 [6] there was general discussion and study of reference material. Also, Dr. Hynek read a prepared paper making certain observations and conclusions. At 1000 Mr. Fournet gave a briefing on his fifteen months experience in Washington as Project Office for U.F.O.'s and his personal conclusions. There was considerable discussion of individual case histories of sightings to which he referred. Following Mr. Fournet's presentation, a number of additional case histories were examined and discussed with Messrs. Fournet, Ruppelt, and Hynek. The meeting adjourned at 1200 for luncheon.

Friday Afternoon

This session opened at 1400. Besides Panel members and CIA personnel, Dr. Hynek was present. Dr. Lloyd V. Berkner, as Panel Member, was present at this meeting for the first time. Progress of the meetings was reviewed by the Panel Chairman and tentative [6] conclusions reached. A general discussion followed and tentative recommendations considered. It was agreed that the Chairman should draft a report of the Panel to AD/SI that evening for review by the Panel the next morning. The meeting adjourned at 1715.

Saturday Morning

At 0945 the Chairman opened the seventh session and submitted a rough draft of the Panel Report to the members. This draft had been reviewed and approved earlier by Dr. Berkner. The next two and one-half hours were consumed in discussion and revision of the draft. At 1100 the AD/SI joined the meeting and reported that he had shown and discussed a copy of the initial rough draft to the Director of Intelligence, USAF, whose reaction was favorable. At 1200 the meeting was adjourned.

Saturday Afternoon

At 1400 the eighth and final meeting of the Panel was opened. Discussions and rewording of certain sentences of the Report occupied the first hour. (A copy of the final report is appended as Tab C.) This was followed by a review of work accomplished by the Panel, and restatement of individual Panel Member's opinions and suggestions on details that were felt inappropriate for inclusion in the formal report. It was agreed that the writer would incorporate these comments in an internal report to the AD/SI. The material below represents this information.

Part VI: Comments and Suggestions of Panel

General

The Panel Members were impressed (as have been others, including O/SI personnel) in the lack of sound data in the great majority of case histories; also, in the lack of speedy follow-up due primarily to the modest size and limited facilities of the ATIC section concerned. Among the case histories of significant sightings discussed in detail were the following:

Bellefontaine, Ohio (1 August 1952); Tremonton, Utah (2 July 1952); Great Falls, Montana (15 August 1950); Yaak, Montana (1 September 1952); Washington, D.C. area (19 July 1952); and Haneda A.F.B., Japan (5 August 1952), Port Huran, Michigan (29 July 1952); and Presque Isle, Maine (10 October 1952).

After review and discussion of these cases (and about 15 others, in less detail), the Panel concluded that reasonable explanations could be suggested for most sightings and "by deduction and scientific method it could be induced (given additional data) that other cases might be explained in a similar manner." The Panel pointed out that because of the brevity of some sightings (e.g. 2-5 seconds) and the inability of the witnesses to express themselves clearly (sometimes) that conclusive explanations could not be expected for every case reported. Furthermore, it was considered that, normally, it would be a great waste of effort to try to solve most of the sightings, unless such action would benefit a training and educational program (see below). The writings of Charles Fort were referenced [8] to show that "strange things in the sky" had been recorded for hundreds of years. It appeared obvious that there was no single explanation for a majority of the things seen. The presence of radar and astronomical specialists on the Panel proved of value at once in their confident recognition of phenomena related to their fields. It was apparent that specialists in such additional fields as psychology, meteorology, aerodynamics, ornithology and military air operations would extend the ability of the Panel to recognize many more categories of little-known phenomena.

On Lack of Danger

The Panel concluded unanimously that there was no evidence of a direct threat to national security in the objects sighted. Instances of "Foo Fighters" were cited. These were unexplained phenomena sighted by aircraft pilots during World War II in both European and Far East theaters of operation wherein "balls of light" would fly near or with the aircraft and maneuver rapidly. They were believed to be electrostatic (similar to St. Elmo's fire) or electromagnetic phenomena or possibly light reflections from ice crystals in the air, but their exact cause or nature was never defined. Both Robertson and Alvarez had been concerned in the investigation of these phenomena, but David T. Griggs (Professor of Geophysics at the University of California at Los Angeles) is believed to have been the most knowledgeable person on this subject. If the term "flying saucers" had been popular in 1943–1945, these objects would [9] have been so labeled. It was interesting that in at least two cases reviewed that the object sighted was categorized by Robertson and Alvarez as probably "Foo Fighters" to date unexplained but not dangerous; they were not happy thus to dismiss the sightings by calling them names. It was their feeling that these phenomena are not beyond the domain of present knowledge of physical science, however.

Air Force Reporting System

It was the Panel's opinion that some of the Air Force concern over U.F.O.'s (notwithstanding Air Defense Command anxiety over fast radar tracks) was probably caused by public pressure. The result today is that the Air Force has instituted a fine channel for receiving reports of nearly anything anyone sees in the sky and fails to understand. This has been particularly encouraged in popular articles on this and other subjects, such as space travel and science fiction. The result is the mass receipt of low-grade reports which tend to overload channels of communication with material quite irrelevant to hostile objects that might some day appear. The Panel agreed generally that this mass of poor-quality reports containing little, if any, scientific data was of no value. Quite the opposite, it was possibly dangerous in having a military service foster public concern in "nocturnal meandering lights." The implication being, since the interested agency was military, that these objects were or might be potential direct threats to national security. Accordingly, the need for deemphasisation made itself apparent. Comments on a possible educational program are enumerated below.

Tab C

SCIENTIFIC ADVISORY PANEL ON UNIDENTIFIED FLYING OBJECTS

14 - 17 January 1953

Members	Field of Competence	Organization
Dr. H. P. Robertson (Chairman)	California Institute of Technology	Physics, weapons systems
Dr. Louis W. Alvares	University of California	Physics, radar
Dr. Lloyd V. Berkner	Associated Universities, Inc.	Geophysics
Dr. Samuel Goudsmit	Brookhaven National Laboratories	Atomic structure, statistical problems
Dr. Thornton Page	Office of Research Operations, Johns Hopkins University	Astronomy, Astrophysics

Associate Members

Dr. J. Allan Hynek	Ohio State University	Astronomy
Mr. Frederick C. Durant	Arthur D. Little, Inc.	Rockets, guided missiles

Interviewees

Brig. Gen. William M. Garland	Commanding General, ATIC	Scientific and technical intelligence
Dr. E. Marshall Chadwell	Assistant Director, O/SI, CIA	Scientific and technical intelligence
Mr. Ralph L. Clark	Deputy Assistant Director, O/SI, CIA	Scientific and technical intelligence

Document I-19

[1]
Air Force's 10 Year Study of Unidentified Flying Objects

November 5, 1957

In response to queries as to results of previous investigation of Unidentified Flying Object reports, the Air Force said today that after 10 years of investigation and evaluation of UFO's, no evidence has been discovered to confirm the existence of so-called "Flying Saucers."

Reporting, investigation, analysis and evaluation procedures have improved considerably since the first sighting of a "flying saucer" was made on 27 June 1947. The study and analysis of reported sightings of UFO's is conducted by a selected scientific group under the supervision of the Air Force.

Dr. J. Allen Hynek, Professor of Astrophysics and Astronomy at Ohio State University, is the Chief Scientific Consultant to the Air Force on the subject of Unidentified Flying Objects.

The selected, qualified scientists, engineers, and other personnel involved in these analyses are completely objective and open minded on the subject of "flying saucers." They apply scientific methods of examination to all cases in reaching their conclusions. The attempted identification of the phenomenon observed is generally derived from human impressions and interpretations and not from scientific devices or measurements. The data in the sightings reported are almost invariably subjective in nature. However, no report is considered unsuitable for study and categorization and no lack of valid evidence of physical matter in the case studies is assumed to be "prima facie" evidence that so-called "flying saucers" or interplanetary vehicles do not exist.

General categories of identification are balloons, aircraft, astronomical, other, insufficient data and unknowns.

Approximately 4,000 balloons are released in the U. S. every day. There are two general types of balloons: weather balloons and upper-air research balloons. Balloons will vary from small types 4 feet in diameter to large types 200 feet in diameter. The majority released at night carry running lights which often contribute to weird or unusual appearances when observed at night. This also holds true when observed near dawn or sunset because of the effect of the slant rays of the sun upon the balloon surfaces. The large balloons, if caught in jet streams, may assume a near horizontal position when partially inflated, and move with speeds of over 200 MPH. Large types may be [2] observed flattened on top. The effect of the latter two conditions can be startling even to experienced pilots.

Many modern aircraft, particularly swept and delta wing types, under adverse weather and sighting conditions are reported as unusual objects and "flying saucers." When observed at high altitudes, reflecting sunlight off their surfaces, or when only their jet exhausts are visible at night, aircraft can have appearances ranging from disc to rocket in shape. Single jet bombers having multi-jet pods under their swept-back wings have been reported as UFOs or "saucers" in "V" formation. Vapor trails will often appear to glow with fiery red or orange streaks when reflecting sunlight. Afterburners are frequently reported as UFOs.

The astronomical category includes bright stars, planets, meteors, comets, and other celestial bodies. When observed through haze, light fog, or moving clouds, the planets Venus, Mars, and Jupiter have often been reported as unconventional, moving objects. Attempts to observe astronomical bodies through hand-held binoculars under adverse sky conditions has been a source of many UFO reports.

The "other" category includes reflections, searchlights, birds, kites, blimps, clouds, sun-dogs, spurious radar indications, hoaxes, firework displays, flares, fireballs, ice crystals, bolides, etc., as examples: Large Canadian geese flying low over a city at night, with street lights reflecting off their bodies; searchlights playing on scattered clouds, appearing as moving disc-like shapes.

The insufficient data category include all sightings where essential or pertinent items of information are missing, making it impossible to form a valid conclusion. These include description of the size, shape or color of the object; direction and altitude; exact time and location; wind weather conditions, etc. This category is not used as a convenient way to get rid of what might be referred to as "unknowns." However, if the data received is insufficient or unrelated, the analysts must then place that particular report in this category. The Air Force needs complete information to reach a valid conclusion. Air Force officials stressed the fact that an observer should send a complete report of a bona fide sighting to the nearest Air Force activity. There the report will be promptly forwarded to the proper office for analysis and evaluation.

A sighting is considered an "unknown" when a report contains all pertinent data necessary to normally suggest at least one valid hypothesis on the cause or explanation of the sighting but when the description of the object and its maneuvers cannot be correlated with any known object or phenomenon. In its Project Blue Book Special Report #14, released in October, 1955, the Air Force showed that evaluated sightings in the "unknown" category had been reduced to 3 percent at that time.

Previously "unknown" sightings had been 9% in 1953 and 1954 and in the previous years "flying saucer" sightings had run as high as 20% "unknowns." Project Blue Book Special Report #14, covered "flying saucer" investigations from June 1947 to May 1955. Latest Air Force statistics show the number of unknowns has since been reduced to less than 2%.

[3] The following table presents the results of the evaluation of all reports received by the Air Force from the time that Project Blue Book, Special Report #14, was completed through June 1957. The table gives the percentage of all the reports received by the Air Force during each time period.

	1955 June thru December	1956	1957 January thru June
Balloons	27.4%	26.0%	26.4%
Aircraft	29.3%	24.6%	28.8%
Astronomical	20.1%	26.3%	24.4%
Other (Hoax, searchlight, birds, etc.)	12.3%	6.8%	6.4%
Insufficient Information	8.8%	14.1%	12.1%
Unknown	2.1%	2.2%	1.9%
TOTAL NUMBER OF SIGHTINGS	273	778	250

Air Force conclusions for the ten years of UFO sightings involving approximately 5,700 reports were: first, there is no evidence that the "unknowns" were inimical or hostile; second, there is no evidence that these "unknowns" were interplanetary space ships; third, there is no evidence that these unknowns represented technological developments or principles outside the range of our present day scientific knowledge; fourth, there is no evidence that these "unknowns" were a threat to the security of the country; and finally there was no physical or material evidence, not even a minute fragment, of a so-called "flying saucer" was ever found.

The Air Force emphasized the belief that if more immediate detailed objective observational data could have been obtained on the "unknowns" these too would have been satisfactorily explained.

A critical examination of the reports revealed that a high percentage of them were submitted by serious people, mystified by what they had seen and motivated by patriotic responsibility.

Reports of UFOs have aroused much interest on this subject throughout the country and a number of civilian clubs, committees and organizations have been formed to study or investigate air phenomena. These private organizations are not governmental agencies and do not reflect official opinion with respect to their theories or beliefs based upon observed phenomena or illusions.

No books, motion pictures, pamphlets, or other informational material on the subject of unidentified flying objects have been cleared, sponsored, or otherwise coordinated by the U. S. Air Force, with the exception of the official press releases issued by Headquarters, USAF, in Washington.

The Air Force, assigned the responsibility for the Air Defense of the United States, will continue to investigate, through the Air Defense Command, all reports of unusual aerial objects over the U.S., including objects that may become labeled Unidentified Flying Objects. The services of qualified scientists and technicians will continue to be utilized to investigate and analyze these reports, and periodic public statements will be made as warranted.

END

[1] **Summary**

(Analysis of Reports of Unidentified Aerial Objects)

Reports of unidentified aerial objects (popularly termed "Flying saucers" or "flying discs") have been received by the U.S. Air Force since mid-1947 from many and diverse sources. Although there was no evidence that the unexplained reports of unidentified objects constituted a threat to the security of the United States, the Air Force determined that all reports of unidentified aerial objects should be investigated and evaluated to determine if "flying saucers" represented technological developments not known to this country.

In order to discover any pertinent trend or pattern inherent in the data, and to evaluate or explain any trend or pattern found, appropriate methods of reducing these data from reports of unidentified aerial objects to a form amenable to scientific appraisal were employed. In general, the original data upon which this study was bases consisted of impressions and interpretations of apparently unexplainable events, and seldom contained reliable measurements of physical attributes. This subjectivity of the data presented a major limitation to the drawing of significant conclusions, but did not invalidate the application of scientific methods of study.

The reports received by the U.S. Air Force on unidentified aerial objects were reduced to IBM punched-card abstracts of data by means of logically developed forms and standardized evaluation procedures. Evaluation of sighting reports, a crucial step in the preparation of the data for statistical treatment, consisted of an appraisal of the reports and the subsequent categorization of the object or objects described in each report. A detailed description of this phase of the study stresses the careful attempt to maintain complete objectivity and consistency.

Analysis of the refined and evaluated data derived from the original reports of sightings consisted of (1) a systematic attempt to ferret out any distinguishing characteristics inherent in the data of any of their segments, (2) a concentrated study of any trend or pattern found, and (3) an attempt to determine the probability that any of the UNKNOWNS represent observations of technological developments not known to this country.

The first step in the analysis of the data revealed the existence of certain apparent similarities between cases of objects definitely identified and those not identified. Statistical methods of testing when applied indicated a low probability that these apparent similarities were significant. An attempt to determine the probability that any of the

UNKNOWNS represented observations of technological developments not known to this country necessitated a thorough re-examination and re-evaluation of the cases of objects not originally identified; this led to the conclusion that this probability was very small.

[2] The special study which resulted in this report (Analysis of Reports of Unidentified Aerial Objects, 5 May 1955) started in 1953. To provide the study group with a complete set of files, the information cut-off date was established as of the end of 1952. It will accordingly be noted that the statistics contained in all charts and tables in this report are terminated with the year 1952. In these charts, 3201 cases have been used.

As the study progressed, a constant program was maintained for the purpose of making comparisons between the current cases received after 1 January 1953, and those being used for the report. This was done in order that any change or significant trend which might arise from current developments could be incorporated in the summary of this report.

The 1953 and 1954 cases show a general and expected trend of increasing percentages in the finally identified categories. They also show decreasing percentages in categories where there was insufficient information and those where the phenomena could not be explained. This trend had been anticipated in the light of improved reporting and investigating procedures.

Official reports on hand at the end of 1954 totaled 4834. Of these, 425 were produced in 1953 and 429 in 1954. These 1953 and 1954 individual reports (a total of 854), were evaluated on the same basis as were those received before the end of 1952. The results are as follows:

Balloons	16 per cent
Aircraft	20 per cent
Astronomical	25 per cent
Other	13 per cent
Insufficient Information	17 per cent
Unknown	9 per cent

As the study of the current cases progressed, it became increasingly obvious that if reporting and investigating procedures could be further improved, the percentages of those cases which contained insufficient information and those remaining unexplained would be greatly reduced. The key to a higher percentage of solutions appeared to be in rapid "on the spot" investigations by trained personnel. On the basis of this, a revised program was established by Air Force Regulation 200-2, Subject: "Unidentified Flying Objects Reporting" (Short Title: UFOB), dated 12 August 1954.

This new program, which had begun to show marked results before January 1955, provided primarily that the 4602d Air Intelligence Service Squadron (Air Defense Command) would carry out all field investigations. This squadron has sufficient units and is so deployed as to be able to arrive "on the spot" within a very short time after a report is received. After treatment by the 4602d Air Intelligence Service Squadron, all information is supplied to the Air Technical Intelligence Center for final evaluation. This cooperative program has resulted, since 1 January 1955, in reducing the insufficient information cases to seven percent and the unknown cases to three percent of the totals.

[3] The period 1 January 1955 to 5 May 1955 accounted for 131 unidentified aerial object reports received. Evaluation percentages of these are as follows:

Balloons	26 per cent
Aircraft	21 per cent
Astronomical	23 per cent
Other	20 per cent
Insufficient Information	7 per cent
Unknown	3 per cent

All available data were included in this study which was prepared by a panel of scientists both in an out of the Air Force. On the basis of this study it is believed that all the unidentified aerial objects could have been explained if more complete observational data had been available. Insofar as the reported aerial objects which still remain unexplained are concerned, there exists little information other than the impressions and interpretations of their observers. As these impressions and interpretations have been replaced by the use of improved methods of investigation and reporting, and by scientific analysis, the number of unexplained cases has decreased rapidly towards the vanishing point.

Therefore, on the basis of this evaluation of the information, it is considered to be highly improbably that reports of unidentified aerial objects examined in this study represent observations of technological developments outside of the range of present-day scientific knowledge. It is emphasized that there has been a complete lack of any valid evidence of physical matter in any case of a reported unidentified aerial object.

Chapter Two

Origins of U.S. Space Policy: Eisenhower, Open Skies, and Freedom of Space

by R. Cargill Hall

During World War II, America's civilian and military leadership embraced scientific research for a multitude of advanced weapons.[1] Indeed, at war's end in 1945, General H.H. Arnold, commander of the Army Air Forces, could confidently assure Secretary of War Robert Patterson that the United States would shortly build long-range ballistic missiles to deliver atomic explosives and "space ships capable of operating outside the atmosphere."[2] Thirteen years later, both of the programs that Arnold forecast were underway. This period, the immediate prelude to the space age, spawned America's civil and military space programs—programs that were in the beginning opposite sides of the same coin. These programs were shaped and initiated at the direction of one U.S. president, Dwight D. Eisenhower. Elements of them would become instrumental in forewarning of surprise attack, monitoring compliance with international treaties, and maintaining a delicate peace between the Soviet Union and the United States. For contemporary reasons of national security, the actions that framed this enterprise and the space policy that President Eisenhower and his advisors created for it were made obscure even to many of those directly involved.

Beginnings of the American Space Program

When in late 1945 General Arnold counseled the secretary of war on prospective weapon developments, he also acted to ensure that the Army Air Forces would in the future be equipped with modern weapons superior to any held by a potential adversary. The Army Air Forces commander set up an independent consultant group, Project RAND, to perform operations research and provide advice. To guide a formative RAND and oversee aeronautical research, he created a new position at Army Air Forces headquarters, that of Deputy Chief of Air Staff for Research and Development. Arnold selected Major General Curtis E. LeMay for this position, a young man with a reputation for accomplishing formidable assignments.[3]

During 1946 and 1947, at a time of demobilization and declining budgets, LeMay directed improvements in research and development. In March 1946, among the first investigations at Project RAND, he asked for an engineering analysis of an Earth satellite

---

1. Daniel J. Kevles, *The Physicists* (New York: Vintage Books, 1979), Chapters 19 and 20.
2. U.S. Army Air Forces, *Third Report of the Commanding General of the Army Air Forces to the Secretary of War*, by General H.H. Arnold, November 12, 1945, p. 68.
3. Project RAND was contracted to the Douglas Aircraft Company in Santa Monica, California. The acronym is thought by some old-timers to represent Research and Development, and by others, Research for America's National Defense. See Bruce L.R. Smith, *The Rand Corporation: Case Study of a Non-profit Advisory Corporation* (Cambridge, MA: Harvard University Press, 1966), pp. 40-47.

vehicle[4] after learning of a similar investigation at the Navy Bureau of Aeronautics.[5] He wanted the RAND evaluation completed swiftly, in time to match the Navy presentation scheduled for the next meeting of the War Department's Aeronautical Board.[6] Representatives of the Army Air Forces and the Navy presented their preliminary findings at a May 15, 1946, meeting of the board's Research and Development Committee. Although RAND engineers ruled out the satellite as a weapons carrier, they claimed for it a number of important military support functions, including meteorological observation of cloud patterns and short-range weather forecasting, strategic reconnaissance, and the relaying of military communications.[7] [II-2] The Navy representatives likewise emphasized using Earth satellites in defense support applications: for fleet communications and as a navigation platform from which to guide missiles and pilotless aircraft.[8] The military members, however, could not agree on a joint satellite program or confirm that these uses of an Earth satellite would justify the anticipated costs of building, launching, and operating such a vehicle.

Studies of automatic Earth satellites continued at RAND and the Navy Bureau of Aeronautics while the post-war armed services jockeyed for position in a sweeping military reorganization. President Truman signed the National Security Act on July 26, 1947, that created the National Military Establishment and separate military departments of the Army, Navy, and Air Force. Beginning in September 1947, the three service secretaries reported to a new cabinet officer, the Secretary of Defense. But the reorganization did not immediately assign to any of the military services responsibility for new weapons. A newly formed Research and Development Board in the Department of Defense postponed any decisions of service jurisdiction over deployment or control of intermediate range and intercontinental ballistic missiles—rockets that would be required to propel human-made satellites into Earth orbit.[9]

The Research and Development Board inherited supervision of the satellite studies in the Defense Department, and assigned them in December 1947 to its Committee on Guided Missiles. This committee, in turn, formed a Technical Evaluation Group composed of civilian scientists to evaluate the Navy and Air Force programs and recommend a preferred course of action. Chaired by Walter MacNair of Bell Laboratories, on March 29, 1948, the group delivered its findings and recommendation. The members judged the technical feasibility of an Earth satellite to be clearly established; they concluded, however, that neither service had as yet established a military or scientific utility commensurate with the vehicle's anticipated costs. Consequently, the group recommended deferring construction of Earth satellites and consolidating all further studies of their use at RAND.[10] Adopted

    4. Curtis E. LeMay with Mackinlay Kantor, *Mission with LeMay: My Story* (Garden City, NY: Doubleday & Co., 1965), pp. 399-400.
    5. R. Cargill Hall, "Earth Satellites, A First Look by the United States Navy," in R. Cargill Hall, ed., *History of Rocketry and Astronautics: Proceedings of the Third through the Sixth History Symposia of the International Academy of Astronautics* (San Diego: Univelt, Inc., 1986), AAS History Series, Vol. 7, Part II, pp. 253-278.
    6. The Aeronautical Board, formed during World War I and eventually made up of ranking military members of the Army and Navy air arms, reviewed aeronautical developments and attempted to reconcile "the viewpoints of the two services for the mutual benefit of aviation." The Earth satellite proposals passed from the Aeronautical Board to the War Department's Joint Research and Development Board (JRDB) in early 1947 and, in late 1947, to the JRDB's successor, the Research and Development Board (RDB). Civilian scientists directed and were well represented on the JRDB and RDB, which evaluated and approved all missile and aeronautical research and development within the military departments, and attempted, often without success, to prevent duplication of effort.
    7. Douglas Aircraft Company, Inc.,"Preliminary Design of an Experimental World-Circling Spaceship," Report No. SM-11827, May 2, 1946, copy in NASA Historical Reference Collection, NASA History Office, NASA Headquarters, Washington, DC.
    8. Research and Development Committee, Aeronautical Board, Case No. 244, Report No. 1, May 15, 1946, pp. 1-2, Archives, Jet Propulsion Laboratory, Pasadena, CA.
    9. Charles S. Maier, introduction to *A Scientist at the White House: The Private Diary of President Eisenhower's Special Assistant for Science and Technology*, by George B. Kistiakowsky (Cambridge, MA: Harvard University Press, 1976), pp. xxxiii-xxxiv, also pp. 95-96; Max Rosenberg, *The Air Force and the National Guided Missile Program, 1944-1950* (Washington, DC: Air Force Historical Division Liaison Office, 1964), pp. 22, 63, 84-85.
    10. "Satellite Vehicle Program," Technical Evaluation Group, Committee on Guided Missiles, RDB, GM 13/7, MEG 24/1, March 29, 1948, NASA Historical Reference Collection.

by the Research and Development Board, these recommendations ended Navy satellite work for a number of years and focused the study of military satellites at RAND's headquarters on the West Coast, in Santa Monica, California.[11]

RAND's Earth satellite work in the late 1940s and early 1950s embraced system and subsystem engineering design, the preparation of equipment specifications, and studies of military uses. [II-5] It attracted a host of uncommonly able individuals, among them James Lipp, Robert Salter, Merton Davies, Amron Katz, Edward Stearns, William Kellogg, Louis Ridenour, Francis Clauser, and Eugene Root. Luminaries from academe, such as Bernard Brodie and Harold Lasswell of Yale University and Ansley Coale of Princeton, participated in special conferences, such as the one held at RAND in 1949 that surveyed the prospective political and psychological effects of Earth satellites.[12] All of these men had a hand in shaping the formative space program. And all of them could agree by the early 1950s that the most valuable, first-priority use of a satellite vehicle involved one strategic application: a platform from which to observe and record activity on the Earth.

Back in November 1945, with nuclear weapons and jet aircraft at hand, General Arnold concluded that the next war would provide the country little opportunity to mobilize, much less rearm or train reserves. [II-1] The United States could not again afford an intelligence failure like the one at Pearl Harbor; it could not again be caught unaware in another surprise attack. In the future, he had cautioned Secretary of War Patterson, "continuous knowledge of potential enemies," including all facets of their "political, social, industrial, scientific and military life" would be necessary "to provide warning of impending danger." Arnold also stated, "the targets of the future may be very large or extremely small—such as sites for launching guided missiles." Identifying them, like advance warning, also required "exact intelligence information."[13]

The extreme secrecy that cloaked events within the Soviet Union promoted the focus on intelligence gathering. When relations between the United States and the U.S.S.R. soured after World War II, little information about contemporary Soviet military capabilities existed in the West. In the absence of hard facts in the late 1940s, U.S. leaders acted on their perception of a "growing intent toward expansion and aggression on the part of the Soviet Union."[14] Shortly after the Soviets detonated an atomic bomb in 1949, the newly formed Board of National Intelligence Estimates in the Central Intelligence Agency (CIA)

11. In 1948 Project Rand reorganized as a non-profit advisory group, The Rand Corporation. In Washington, the Defense Department's Research and Development Board continued fitfully to operate until the fall of 1953 when its functions were subsumed in a new Office of Assistant Secretary of Defense for Research and Development; President Dwight D. Eisenhower appointed its first occupant: Donald A. Quarles.

12. Rand Research Memorandum, RM-120, "Conference on Methods for Studying the Psychological Effects of Unconventional Weapons," January 26-28, 1949; Paul Kecskemeti, RM-567, "The Satellite Rocket Vehicle: Political and Psychological Problems," October 4, 1950, both in Rand Library, Santa Monica, CA; see also R. Cargill Hall, "Early U.S. Satellite Proposals," *Technology and Culture* 4 (Fall 1963): 430-31.

Five months after an atomic bomb fell on Hiroshima, Japan, Louis Ridenour provided the American public a first, sobering assessment of future international atomic warfare conducted with Earth-mines and Earth-orbiting satellites. (In the 1950s, fears of a nuclear/thermonuclear surprise attack would move President Eisenhower to fold Earth satellites into an intelligence system designed to preclude such a catastrophe, and establish policy ensuring that spaceflight operations remained devoted to "peaceful purposes.") See L.N. Ridenour, "Pilot Lights of the Apocalypse," and the editor's introductory comment in *Fortune* 33 (January 1946): 116-17, 219.

Robert Salter contributed one of the first and most prescient surveys of the prospects for manned spaceflight in 1951, although the title he selected for it, doubtless to avoid peer ridicule, belied the subject. See Robert M. Salter, "Engineering Techniques in Relation to Human Travel at Upper Altitudes," *Physics and Medicine of the Upper Atmosphere: A Study of the Aeropause* (Albuquerque: University of New Mexico Press, 1952), pp. 480-487.

13. U.S. Army Air Forces, *Third Report of the Commanding General*, pp. 65-67.

14. Harry R. Borowski, *A Hollow Threat: Strategic Air Power and Containment Before Korea* (Westport, CT: Greenwood Press, 1982), p. 6; see also John Prados, *The Soviet Estimate: U.S. Intelligence Analysis and Russian Military Strength* (New York: The Dial Press, 1982), pp. 6-8, 19. See also the newly declassified CIA Office of Research Estimates and later National Intelligence Estimates at the National Archives, including: Central Intelligence Group, "Soviet Foreign and Military Policy," ORE-1, July 23, 1946; Historical Review Group, CIA, National Archives, Box 1, Folder 1; and Central Intelligence Agency, "The Possibility of Direct Soviet Military Action During 1949," ORE-46-49, May 3, 1949, Historical Review Group, CIA, National Archives, Box 3, Folder 102.

warned of the possibility of a Soviet nuclear surprise attack, albeit a limited one, against the United States. That prospect, underscored by the surprise Korean conflict in June 1950 and the development of thermonuclear devices between 1952 and 1954, haunted the nation's military and civilian leadership.[15]

Among America's leaders in the 1950s, the desire to preclude a nuclear or thermonuclear surprise attack was particularly acute. As Dwight D. Eisenhower's biographer aptly phrased it, they had "Pearl Harbor burned into their souls in a way that younger men, the leaders in the later decades of the Cold War, had not." Certainly this was true of Eisenhower in 1953 when he took the oath of office as president, for the subject completely dominated his thinking about disarmament and relations with the Soviets for the next eight years. Besides seeking ways to prevent a surprise attack, Eisenhower also sought "to lessen, if he could not eliminate, the financial cost and the fear that were the price of the Pearl Harbor mentality."[16] To that end, he could agree entirely with General Arnold's views that continuous knowledge of one's potential adversaries was essential "to provide warning of impending danger." The way to get it, Eisenhower also knew from wartime experience, was through aerial reconnaissance.

To secure hard intelligence about the Soviet Union, the CIA and the Air Force undertook a variety of projects at the beginning of the 1950s. Intelligence officers sifted captured German documents for aerial reconnaissance photographs of the U.S.S.R.; that these photographs dated from the early 1940s suggests the magnitude of the problem facing U.S. planners. The interrogation of German and Japanese prisoners of war returning from forced labor in the Soviet Union between 1949 and 1953 helped shed more light on the status of that country's military and industrial might. The Strategic Air Command began flying aircraft on the periphery of the U.S.S.R. on reconnaissance missions, and obtained considerable information about border installations and defenses. But these missions yielded nothing substantial about the Soviet heartland and the state of its economy, society, or military capabilities and preparations.[17]

Seeking this information, RAND proposed and the Air Force conducted the WS 119L program. Beginning in early January 1956, with the approval of President Eisenhower, Air Force personnel loaded automatic cameras in gondolas suspended beneath large Skyhook weather balloons, and during the next four weeks launched 516 of these vehicles in Western Europe. The balloons, equipped with radio beacons that allowed tracking, drifted on prevailing winds at high altitudes eastward across the Eurasian continent, through Soviet airspace. Under the terms of international law to which the United States was a party, the balloons clearly violated Soviet national sovereignty. Those that succeeded in crossing released their gondolas on parachutes, which were recovered in mid-air by C-119 cargo aircraft near Japan and Alaska.[18] Because the aerial path of the balloons could not be controlled, however, the pictures might as easily be of cloud cover or a Siberian forest as of a factory or an airfield. This program, which produced limited intelligence and strongly worded Soviet protests, was quietly canceled on February 6, 1956, at the president's direction. Although the Air Force would subsequently launch a few more of these balloons that operated at yet higher altitudes, Eisenhower quickly terminated that effort as well. Meanwhile, other, more promising avenues of gathering information had appeared.[19]

---

15. James R. Killian, Jr., *Sputnik, Scientists, and Eisenhower: A Memoir of the First Special Assistant to the President for Science and Technology* (Cambridge, MA: The MIT Press, 1977), pp. 68, 94; Prados, *The Soviet Estimate*, p. 21. U.S. intelligence was caught almost completely unaware of the development of the Soviet hydrogen bomb. See, for example, "Estimate of the Effects of the Soviet Possession of the Atomic Bomb Upon the Security of the United States and Upon the Probabilities of Direct Soviet Military Action," ORE 91-49, April 6, 1950, Historical Review Group, CIA, National Archives, Box 4, Folder 131, p. 11.

16. Stephen E. Ambrose, *Eisenhower: Volume II, The President* (New York: Simon and Schuster, 1984), p. 257. The president's decision in favor of aerial reconnaissance is explained on pp. 258-59.

17. David A. Rosenberg, "The Origins of Overkill: Nuclear Weapons and American Strategy, 1945-1960," *International Security* 7 (Spring 1983): 20-21; Prados, *The Soviet Estimate*, pp. 57-58.

18. In the event aerial retrieval failed, the gondolas were designed to float on the ocean's surface and radiate a signal for twenty-four hours. Although many of the gondolas came down in the Soviet Union, sixty-seven of them actually reached the recovery area; of these, the Air Force retrieved forty-four.

19. Tom D. Crouch, *The Eagle Aloft: Two Centuries of the Balloon in America* (Washington, DC: Smithsonian Institution Press, 1983), pp. 644-49; Ambrose, *Eisenhower, Vol. II*, pp. 309-11; Killian, *Sputnik, Scientists, and*

Research and Initial Development

While the CIA and the Air Force endeavored to gather information about the Soviet Union from any source, the Department of Defense acted on the issue of military roles and missions. On March 21, 1950, Secretary of Defense Louis Johnson assigned the Air Force responsibility for long-range strategic missiles, including ICBMs. A few weeks later, the Research and Development Board vested jurisdiction for military satellites in the same service. With these responsibilities, Air Force leaders directed RAND to complete studies of a military Earth satellite.[20]

The resultant RAND report, issued in April 1951, described a spacecraft fully stabilized on three axes and that employed a television camera to scan the Earth and transmit the images to receiving stations. [II-3] The television coverage thus acquired, RAND reminded the service had to occur when "weather permits ground observation."[21] The RAND report encouraged Air Force leaders to believe that directed, periodic observation of the Soviet Union might soon be conducted from extremely high altitudes. To confirm these findings, on December 19, 1951, Air Force headquarters authorized the firm to subcontract for detailed spacecraft subsystem studies. A few weeks later, in January 1952, the service convened a seminal "Beacon Hill" study group to assay strategic aerial reconnaissance under the auspices of Project Lincoln at the Massachusetts Institute of Technology.[22]

The Beacon Hill study group, which first met between January 7 and February 15, 1952, considered improvements in Air Force aerial intelligence processing, sensors, and vehicles. Chaired by Carl Overhage of Eastman Kodak, the fifteen-member group included Air Force optics specialist Lieutenant Colonel Richard Leghorn (later, the founder of Itek), James Baker of the Harvard Observatory, Edwin Land (the founder of Polaroid), Stuart Miller of Bell Labs, Richard Perkin (co-founder of Perkin-Elmer), scientific consultant Louis Ridenour, Allen Donovan of Cornell Aeronautical Labs, and Edward Purcell of Harvard University. These individuals concluded their deliberations in May and issued a final report in June 1952.

The Beacon Hill report recommended to the Air Force specific improvements in the orientation, emphasis, and priority assigned to strategic intelligence, and solutions to the problems involved in its collection, reduction, and use. The study group also suggested refinements in sensors. The improved sensors, the group advised, could be flown near Soviet territory in advanced high-altitude aircraft, high-altitude balloons (later, WS 119L), sounding rockets, and long-range drones such as the Snark or Navaho air-breathing missiles. Whatever the choice of vehicles, study group participants cautioned the service that actual "intrusion" over Soviet territory and violation of its national sovereignty required approval of political authorities "at the highest level." Space satellites, mentioned only in passing and then only as vehicles of the future in the grip of Newtonian mechanics, were, however, identified as certain intruders that would have to "overfly" the Soviet Union.[23]

Eisenhower, p. 12; Paul E. Worthman recollections, cited by W. W. Rostow in *Open Skies: Eisenhower's Proposal of July 21, 1955* (Austin: University of Texas Press, 1982), pp. 189-94. Project "Moby Dick," the test of WS 119L, was conducted in the United States during 1952-1955 and accounted for numerous UFO sightings—as did later tests of the U-2 and A-12.

20. Enclosure with recommendations for guided missiles to Memo 1620/17, for Secretary of Defense Louis Johnson, from the Joint Chiefs of Staff, March 15, 1950; Memo for the Joint Chiefs of Staff from Louis Johnson, "Department of Defense Guided Missiles Program," approving recommendations, March 21, 1950; Report, Air Research and Development Command, *Space System Development Plan*, WDPP-59-11, January 30, 1959, Tab I, "Background," p. I-1-1, all in NASA Historical Reference Collection.

21. J.E. Lipp, R.M. Salter, Jr., and R.S. Wehner, "The Utility of a Satellite Vehicle for Reconnaissance," The RAND Corporation, R-217, April 1951, p. 80, Rand Library.

22. RCA-Rand, "Progress Report (Project Feed Back)," Report RM-999, January 1, 1953, Rand Library. Background of the Beacon Hill study and related developments in 1951 is contained in Herbert F. York and G. Allen Greb, "Strategic Reconnaissance," *Bulletin of the Atomic Scientists*, April 1977, p. 34.

23. "Beacon Hill Report: Problems of Air Force Intelligence and Reconnaissance," Project Lincoln, Massachusetts Institute of Technology, Boston, MA, June 15, 1951, *passim*, JPL Archives.

Elsewhere around the country, various firms under contract to RAND were designing and evaluating specific satellite equipment, including a television payload (Radio Corporation of America), vehicle guidance and attitude-control devices (North American Aviation), and a nuclear auxiliary electrical power source (Westinghouse Electric Corporation, Bendix Aviation, Allis-Chalmers, and the Vitro Corporation). This effort, known collectively as Project Feed Back, confirmed that automated satellites could be built without exceptional delays and at an affordable cost. Whatever the legal ramifications of overflight in outer space might be, in September 1953, RAND officials recommended that a satellite be built.[24] [II-4] A few months later, they concluded their preliminary work and published a final report.

Issued on March 1, 1954, the Project Feed Back report described a military satellite for observation, mapping, and weather analysis, along with examples of the necessary space hardware and ground support systems. [II-6] The second stage booster-satellite would be placed in a low-altitude, "sun synchronous" polar orbit inclined 83 degrees to the equator. Launched at the proper time of day at this inclination, the satellite would precess in one year through 360 degrees, allowing a television camera to operate in maximum daylight brightness throughout all seasons.[25] RAND engineers estimated this satellite system would produce "30 million pictures in one year of operation," a sum equivalent to all the pictures held in the USAF Photo Records and Services Division acquired from all sources in peace and war over the previous twenty-five years![26] Where the Air Force might find the photo-interpreters needed to evaluate this mountain of information, RAND did not say.

In early 1954, however, the problem that faced U.S. policy-makers was not too much intelligence information about the Soviet Union, but far too little. Attempts to fly around the U.S.S.R. had thus far produced inadequate information; details of Soviet military preparations and capabilities remained as much an enigma as ever. Continued Soviet production of atomic weapons, and the means to deliver them, such as the Bison long-range bomber, combined in August 1953 with the Soviet detonation of a thermonuclear device, particularly disturbed President Eisenhower. Former Supreme Commander of the Allied Expeditionary Force in Western Europe, Eisenhower had helped engineer the destruction of the Axis powers in World War II and knew firsthand the enormous devastation that accompanied modern total war.

Any aerial surprise attack on the United States with nuclear weapons, even a limited one, could lay waste to most of the metropolitan areas on the East and West coasts. Moreover, with government agencies unable to gauge the exact nature and extent of a Soviet military threat, the president found himself at a distinct disadvantage in selecting the appropriate level of military preparedness to combat it. This situation, Eisenhower made clear at a meeting of his National Security Council on February 24, 1954, had to be resolved—and soon. As a first step to counter a possible surprise attack, he had already approved a prior council recommendation to design and construct, with Canadian approval, a Distant Early Warning (DEW) picket line of radars across the North American Arctic, to detect and track any Soviet bombers that might be directed against the two countries.[27]

Civilian scientists appointed to the Science Advisory Committee in the Office of Defense Mobilization, meanwhile, had been examining similar issues under the prodding of

---

24. Perry, *Origins of the USAF Space Program*, pp. 35, 39; and Merton E. Davies and William R. Harris, *RAND's Role in the Evolution of Balloon and Satellite Observation Systems and Related U.S. Space Technology* (Santa Monica, CA: The RAND Corporation, 1988), p. 47.

25. J.E. Lipp & R.M. Salter, "Project Feed Back Summary Report," The RAND Corporation, R-262, Volume II, March, 1954, pp. 109-10, Rand Library.

26. *Ibid.*, pp. 85-86.

27. Stephen E. Ambrose, *Ike's Spies: Eisenhower and the Espionage Establishment* (Garden City, NY: Doubleday & Co., 1981), pp. 253, 267; Rpt., Aerospace Defense Command, *A Chronology of Air Defense, 1914-1972*, ADC Historical Study No. 19, March 1973, p. 33; see also NSC 159/4 and attached statement of policy on "Continental Defense," September 25, 1953, and NSC 5408, "Report to the National Security Council by the National Security Planning Board," February 11, 1954, as reprinted in William Z. Slany, ed., *Foreign Relations of the United States, 1952-1954, Volume II: National Security Affairs, Part 1* (Washington, DC: U.S. Government Printing Office, 1984), pp. 475-89, 609-24.

Trevor Gardner, the "technologically evangelical assistant secretary of the Air Force for research and development." Learning of these studies, the president's special assistant for security affairs, General Robert Cutler, invited key committee members to the White House. Meeting with them on March 27, 1954, Eisenhower discussed his concerns about a surprise attack on the United States and the prospects for avoiding or containing it. "Modern weapons," he warned, "had made it easier for a hostile nation with a closed society to plan an attack in secrecy and thus gain an advantage denied to the nation with an open society." In spite of the Oppenheimer case, he apparently viewed the scientists as honest brokers in a partisan city, and he challenged them to tackle this problem.[28]

They did. Lee A. DuBridge, president of the California Institute of Technology and chair of the Science Advisory Committee, and James R. Killian, Jr., president of the Massachusetts Institute of Technology, formed a special task force to consider three areas of national security: continental defense, strike forces, and intelligence, with supporting studies in communications and technical manpower. Approved by President Eisenhower in the spring, the Surprise Attack Panel, or the Technological Capabilities Panel (TCP) as it was subsequently renamed, chaired by Killian, conducted its work between August 1954 and January 1955. Its membership included most of those who had produced the Beacon Hill Report and represented the best that American science and engineering offered. The panel's extraordinary two-volume report, *Meeting the Threat of Surprise Attack*, was issued on February 14, 1955. By all published accounts, the report affected the course of national security affairs enormously.[29]

The TCP report resulted in a number of significant alterations in U.S. defense preparedness. Among other things, it recommended accelerating procurement of intercontinental ballistic missiles (Atlas, and later Titan and Minuteman ICBMs), constructing land- and sea-based intermediate-range ballistic missiles (later Thor, Jupiter, and Polaris IRBMs), and speeding construction of the DEW line in the Arctic (declared operational in August 1957). The TCP also identified a timetable of changes in the relative military and technical positions of the two superpowers. Even more important, perhaps, were the recommendations to acquire and use strategic pre-hostilities intelligence. The intelligence panel, chaired by Edwin Land, urged construction and deployment of the U-2 aircraft[30] that could, if called upon, overfly the Soviet Union at very high altitudes.[31] Any mention of the U-2, however, was excluded from the report proper. In its section on intelligence applications

28. The description of Gardner, and Eisenhower as quoted, in Killian, *Sputnik, Scientists, and Eisenhower*, p. 68; see also, Prados, *The Soviet Estimate*, p. 60.

29. *Meeting the Threat of Surprise Attack*, Vol I and Vol II, February 14, 1955, JPL Archives; see also Killian, *Sputnik, Scientists, and Eisenhower*, pp. 11-12, 70-82; Herbert F. York and G. Allen Greb, "Military Research and Development: A Postwar History," *Bulletin of the Atomic Scientists*, January 1977, p. 22; also York and Greb, "Strategic Reconnaissance," p. 35. For the next two years, the deliberations of the National Security Council turned frequently to the findings and recommendations contained in this report. See John P. Glennon, ed., *Foreign Relations of the United States, 1955-1957: Volume XIX, National Security Policy* (Washington, DC: U.S. Government Printing Office, 1990), hereafter referred to as *Volume XIX*.

30. Eisenhower approved development of the U-2 during the TCP deliberations, on November 24, 1954, and assigned the project to the CIA instead of the Air Force. Under the guidance of Richard M. Bissell, Jr., CIA Special Assistant to the Director of Central Intelligence, Colonel O. J. Ritland, USAF, and Clarence L. "Kelly" Johnson of the Lockheed Aircraft Corporation, the first U-2 was airborne within eight months, on August 6, 1955. Ambrose, *Ike's Spies*, p. 268; Leonard Mosley, *Dulles: A Biography of Eleanor, Allen, and John Foster Dulles and Their Family Network* (New York: Dial Press, 1978), pp. 365-66.

31. Dwight D. Eisenhower, *Waging Peace, 1956-1961* (Garden City, NY: Doubleday & Co., Inc., 1965), p. 470; Killian, *Sputnik, Scientists, and Eisenhower*, pp. 71-84; Rpt., *A Chronology of Air Defense, 1914-1972*, p. 46. The cleared recommendations of the TCP are reprinted in *Volume XIX*, pp. 46-56.

Throughout the 1950s Eisenhower withheld knowledge of the U-2's existence from all but those few directly involved. The program never appeared as an item in National Security Council deliberations until "it tore its britches" in 1960. Karl G. Harr, Jr., "Eisenhower's Approach to National Security Decision Making," in Kenneth W. Thompson, ed., *The Eisenhower Presidency: Eleven Intimate Perspectives of Dwight D. Eisenhower*, Vol. 3 in *Portraits of American Presidents* (Lanham, MD: University Press of America, 1984), p. 97. The product of the U-2 flights was even more closely held, and Eisenhower refused to refute political charges that an American "bomber gap" and, later, a "missile gap" existed, even though he knew them to be false. The latter issue, artfully exploited by John Kennedy, may well have cost Richard Nixon the 1960 presidential election. Since that time, to avoid an unwanted repetition, candidates have been "briefed" on national security affairs before a presidential campaign begins. All of these events square with the perceptive thesis of Eisenhower governance elucidated by Fred I. Greenstein, *The Hidden-Hand Presidency: Eisenhower as Leader* (New York: Basic Books, Inc., 1982).

of science, the report recommended beginning immediately a program to develop a small scientific satellite that would operate at extreme altitudes above national airspace, intended to establish the principle of "freedom of space" in international law for subsequent military satellites.[32] Although committee members could hope that scientific satellites might set such a precedent, James Killian, who chaired the TCP, viewed RAND's proposed military observation satellite as a "peripheral project" and would refuse to support it until the Soviets launched Sputnik I nearly three years later.

Back in the summer of 1954, shortly after authorizing the surprise-attack study, President Eisenhower approved the formation of an organization devoted exclusively to that subject: the National Indications Center. This center, chaired by the Deputy Director of Central Intelligence and composed of specialists drawn from U.S. intelligence agencies, and the Departments of Defense and State, formed the interagency staff of the National Watch Committee, which consisted of presidential confidants such as the Secretaries of State and Defense, and the Director of Central Intelligence (DCI). Chartered on July 1, 1954, for the express purpose of "preventing strategic surprise," the center drew on information furnished by all national intelligence organizations. Eisenhower, one of the participants recalled vividly, was a man "boresighted on early warning of surprise attack."[33]

The National Indications Center assessed the military, economic, and social demands involved in mounting a surprise attack and issued a weekly "watch report" to the Watch Committee members. Staffers expanded a list of key indicators developed earlier under the direction of James J. Hitchcock in the CIA, and applied it to developments that would presage surprise attack in the nuclear age.[34] That is, presuming rational political leadership, one state intending to attack another would need to prepare carefully, say, by dispersing its industry and population many months in advance, and by deploying its military forces on land and sea just days or hours before "M-Day." Thus, the proper intelligence "indicators" applied against this matrix would yield readily identifiable signals, much like a traffic light: green—normal activity; amber—caution; and red—warning.[35] These strategic warning indicators, eventually linked to "defense conditions" (DEFCON 5 through 1), enabled U.S. leaders to mobilize resources and establish force readiness postures. The military, economic, and technical indicators listed in this matrix successfully predicted the Suez War in 1956, and have been monitored and reported in one form or another to the president and other command authorities ever since. The National Indications Center itself, however, was dissolved in March 1975.[36]

32. *Meeting the Threat of Surprise Attack*, Vol. II, pp. 146-48; Memo for the Record, L. B. Kirkpatrick, "Meeting with the President's Board of Consultants, Saturday, 28 Sep. 1957, 11 a.m to 2 p.m.," Eisenhower Library, Abilene, KS.

33. Interview with James J. Hitchcock, May 23, 1986; Cynthia M. Grabo, "The Watch Committee and the National Indications Center: The Evolution of U.S. Strategic Warning, 1950-1975," *International Journal of Intelligence and CounterIntelligence* 3 (Fall 1989): 369-70; see also Eisenhower letter to Winston Churchill, cited in Killian, *Sputnik, Scientists, and Eisenhower*, p. 88. One has only to peruse the documents in *Volume XIX* to gain an appreciation for Eisenhower's fixation on surprise attack and his dedication to forestalling such an event. See especially [8] at p. 40.

34. A RAND study doubtless figured in these deliberations and actions, though a direct linkage is not established at this time. One year earlier, three months after President Eisenhower's inauguration, Andrew W. Marshall and James F. Digby issued RAND Special Memorandum SM-14, *The Military Value of Advanced Warning of Hostilities and its Implications for Intelligence Indicators*, April 1953 (rev. July 1953). The authors compared intelligence warning of attack to the performance of military forces, and urged attention to short-term indications of Soviet preparations for surprise attack. Copies unquestionably circulated within intelligence circles, including the CIA.

35. The British first developed an indicators list in 1948 to identify actions the Soviets would have to take to occupy Berlin. Hitchcock subsequently altered and expanded the list at the CIA in the late 1940s and early 1950s to identify actions that would warn of a surprise attack against the United States. The best available source in the open literature that describes related RAND activities in the 1940s and 1950s is Davies and Harris, *RAND's Role in the Evolution of Balloon and Satellite Observation Systems and Related U.S. Space Technology*.

36. Grabo, "The Watch Committee and the National Indications Center," p. 384; *Volume XIX* [19]; another survey of this subject in the open literature is Duncan E. MacDonald, "The Requirements for Information and Systems," in F. J. Ossenbeck and P. C. Kroeck, eds., *Open Space and Peace: A Symposium on the Effects of Observation* (Stanford, CA: The Hoover Institution, 1964), pp. 64-83. The NSC Planning Board, also at the president's direction, in November 1954 had established a "net capabilities evaluation subcommittee" that performed a function similar to the National Indications Center for the council. See [1 and 19] in *Volume XIX*.

Establishing National Space Policy

President Eisenhower, to be sure, worried considerably about the danger of a Soviet surprise attack in the mid-1950s, and judged strategic warning absolutely vital to counter or preclude it. Shortly after the TCP submitted its report to the National Security Council, in the spring of 1955 the president's closest advisors determined, if at all possible, to keep outer space a region open to all, where the spacecraft of any state might overfly all states, a region free of military posturing. By adopting a policy that favored a legal regime for outer space analogous to that of the high seas, the United States might make possible the precedent of "freedom of space" with all that implied for overflight. This choice also favored non-aggressive, peaceful spaceflight operations, especially the launch of scientific Earth satellites to explore outer space that civilian scientists now urged as part of the U.S. contribution to the International Geophysical Year (IGY).[37] [II-8, II-11] This program, proposed by the U.S. National Committee for the IGY of the National Academy of Sciences in a March 14, 1955, report, had been approved by the academy and sent to National Science Foundation director Alan T. Waterman for government consideration.[38] [II-9]

By this time, a number of prominent scientists and military leaders actively sought approval for spaceflight missions. A few months after RAND's Feed Back report appeared, the Air Force had acted on its recommendations. On November 29, 1954, the Air Research and Development Command issued System Requirement No. 5, which called for competitive system-design studies of a military satellite. On March 16, 1955, while the National Academy of Sciences was completing its satellite deliberations, the USAF issued General Operational Requirement No. 80 (SA-2c), which approved construction of and provided technical requirements for military observation satellites. At the same time, the service named this observation satellite the WS 117L program. In April, the Naval Research Laboratory submitted to the Defense Department a "Scientific Satellite Program" for the IGY, eventually known as Vanguard, which proposed using as a first-stage booster the Viking sounding rocket. Meanwhile, the Army's Redstone rocket team led by Major General John B. Medaris and Wernher von Braun had for some months urged a small, inert Earth satellite launched with the Jupiter IRBM, called Project Orbiter (later named Explorer). [II-7] These and other events soon to follow made 1955 the most momentous of years for the fledgling U.S. space program.[39]

In May 1955, administration officials agreed that the country should launch scientific Earth satellites as a contribution to the IGY. In early May, Assistant Secretary of Defense for Research and Development Donald Quarles referred the Army and Navy IGY satellite proposals to his Committee on Special Capabilities, and requested a scientific

---

37. In 1952 the International Council of Scientific Unions (ICSU) established a committee to arrange another International Polar Year to study geophysical phenomena in remote areas of the Earth (two previous polar years had been conducted, one in 1882-1883 and another in 1932-1933). Late in 1952 the council expanded the scope of this effort, planned for 1957-1958, to include rocket research in the upper atmosphere and changed the name to the International Geophysical Year. In October 1954 the ICSU, meeting in Rome, Italy, adopted another resolution that called for launching scientific Earth satellites during the IGY. "Editorial Note," in John P. Glennon, ed., *Foreign Relations of the United States, 1955-1957: Volume XI, United Nations and General International Matters* (Washington DC: U.S. Government Printing Office, 1988), [361], pp. 784-85.

38. A few months earlier, in December 1954, the American Rocket Society's Committee on Space Flight completed a similar report on the utility of scientific Earth satellites, including a proposal by John Robinson Pierce of Bell Laboratories for a passive communication satellite that much resembled the later Project Echo, and submitted it to National Science Foundation Director Alan T. Waterman. By the spring of 1955 a number of Earth-satellite proposals had landed on the desks of officials at the National Science Foundation and the Department of Defense. See R. Cargill Hall, "Origins and Development of the Vanguard and Explorer Satellite Programs," *Airpower Historian* 9 (October 1964): 106-108.

39. *Ibid.*, pp. 102-104. Project Orbiter first appeared with the name "A Minimum Satellite Vehicle," the result of an August 3, 1954, meeting between Army officials at the Redstone Arsenal and Navy representatives from the Office of Naval Research. See Dr. Wernher von Braun, "A Minimum Satellite Vehicle: Based on components available from missile developments of the Army Ordnance Corps," September 15, 1954, NASA Historical Reference Collection.

satellite proposal from the Air Force.[40] He instructed committee members to evaluate these proposals and recommend a preferred program. Quarles, who warmly embraced the satellite recommendations of Killian's Technological Capabilities Panel and urged an IGY satellite program, subsequently drafted a policy for the launching of these and other spacecraft and submitted it on May 20 to the National Security Council (NSC). NSC members meeting on May 26 endorsed the Quarles' proposal and accompanying national policy guidance. A scientific satellite program for the IGY would not interfere with development of high-priority ICBM and IRBM weapons. Emphasis would be placed on the peaceful purposes of the endeavor. The scientific satellites would help establish the principle in international law of "freedom of space" and the right of unimpeded overflight that went with it, and these IGY satellites would serve as technical precursors for subsequent U.S. military satellites. "Considerable prestige and psychological benefits," the policy concluded, "will accrue to the nation which first is successful in launching a satellite."[41] The next day, "after sleeping on it," President Eisenhower approved this plan.[42] [II-10]

With the president's decision, the United States had tentatively set out to prosecute two closely associated space programs: instrumented military applications and civilian scientific satellites. Presidential advisors still perceived the more complex military spacecraft to be a long way off, but the IGY scientific satellite program was clearly identified as a stalking horse to establish the precedent of overflight in space for the eventual operation of military reconnaissance satellites. Charged with the WS 117L program, the Air Force earlier in 1955 had selected three firms to compete in a one-year design study of a preferred vehicle. Neither the military nor the scientific satellite program had selected a contractor to conduct the work, and neither shared a national priority.

In Burbank, California, in Kelly Johnson's Lockheed "skunk works," the U-2 project unquestionably claimed the highest of national priorities. With the first of these turbojet-powered gliders nearing completion, Eisenhower learned that the United States could soon overfly parts of Soviet airspace at will.[43] The U-2 had an anticipated operating ceiling in excess of 70,000 feet. No known jet fighter operated at altitudes above 50,000 feet. But however safe piloted aerial overflight, or however attractive this opportunity to acquire intelligence on Soviet military preparations, might be, any unauthorized penetration of another state's airspace represented a clear violation of international law—a violation, that is, unless the leaders concerned agreed to such flights beforehand.

While the U-2 neared its first test flight in Nevada, on July 21, 1955, at a summit conference in Geneva, Eisenhower advised Soviet leaders of just such a plan. The president, in an unannounced addition to a disarmament proposal, directly addressed the subject that most concerned him. The absence of trust and the presence of "terrible weapons" among states, he asserted, provoked in the world "fears and dangers of surprise attack." To eliminate these fears, he urged that the Soviet Union and the United States provide "facili-

40. The Air Force proposal, called "World Series," featured an Atlas first stage and Aerobee-Hi second stage; it was submitted to the Committee on Special Capabilities (Stewart Committee) during the first week of July 1955. Because World Series conflicted with the WS 117L program, Air Force leaders gave it scant support.
    Throughout the Eisenhower presidency until his death in office, Donald A. Quarles would influence greatly the choice of policy and missions for the civilian and military satellite programs, first as Assistant Secretary of Defense for Research and Development (September 1953 to August 1955), then as Secretary of the Air Force (August 1955 to April 1957), and finally as Deputy Secretary of Defense (April 1957 to May 1959).
    41. National Security Council, NSC 5520, "Draft Statement of Policy on U.S. Scientific Satellite Program," May 20, 1955, pp. 1-3. See also Annex B, accompanying Memorandum from Nelson A. Rockefeller to Mr. James S. Lay, Jr., Executive Secretary, "U.S. Scientific Satellite Program," May 17, 1955. These documents reprinted, along with the NSC endorsement, in John P. Glennon, ed., *Foreign Relations of the United States, 1955-1957: Volume XI, United Nations and General International Matters* (Washington DC: U.S. Government Printing Office, 1988), [340/341], pp. 723-33, hereafter referred to as *Volume XI*. Air Force leaders enthusiastically embraced the dictum that IGY satellites would not interfere with the ICBM, IRBM, and military satellite programs; Perry, *Origins of the USAF Space Program*, pp. 43-44.
    42. Eisenhower quoted in Lee Bowen, *An Air Force History of Space Activities, 1945-1959* (USAF Historical Division Liaison Office, August 1964), p. 64. Eisenhower did approve the IGY satellite program in NSC 5520 the next day, on May 27, 1955; see *Volume XI* [341], p. 733.
    43. Ambrose, *Ike's Spies*, p. 271; Clarence "Kelly" Johnson, interview with Morley Safer on CBS "60 Minutes," October 17, 1982; Eisenhower, *Waging Peace*, pp. 544-45.

ties for aerial photography to the other country" and conduct mutually supervised reconnaissance overflights.[44] Before the day ended, the Chair of the Soviet Council of Ministers, Nikolai Bulganin, and First Secretary of the Communist Party Nikita Khrushchev privately rejected the president's plan, known eventually as the "Open Skies" doctrine, as an obvious U.S. attempt to "accumulate target information." "We knew the Soviets wouldn't accept it," Eisenhower later confided in an interview, "but we took a look and thought it was a good move."[45] Though the Soviets might object, they were forewarned.[46] Eleven months later, some five months after he terminated the balloon reconnaissance program, Eisenhower approved the first U-2 overflight of the U.S.S.R.[47]

Back in the United States, late in the evening of July 25, 1955, Eisenhower informed the nation in a radio address of the results of the summit conference. On July 27, Eisenhower met with National Science Foundation Director Waterman, Assistant Secretary of Defense Quarles, and Undersecretary of State Herbert Hoover, Jr., to discuss how best to make known the existence of a U.S. IGY satellite program. A general statement, it was decided, would come from the White House after congressional leaders had been notified. These statements would emphasize the satellite project "as a contribution benefiting science throughout the world," and would not link it in any way "to military missile development." Two days later, on July 29, 1955, the president publicly announced plans for launching "small unmanned, Earth circling satellites as part of the U.S. participation in the International Geophysical Year" scheduled between July 1957 and December 1958. [I-17] His statement avoided any hint at the underlying purpose of the enterprise, and assigned to the National Science Foundation responsibility for directing the project, with "logistic and technical support" to be furnished by the Department of Defense. Donald Quarles' Committee on Special Capabilities in early August selected for the IGY satellite project the Naval Research Laboratory's Vanguard proposal, one that combined modified Viking and Aerobee-Hi sounding rockets for the scientific satellite booster, and placed the U.S. Navy in charge of logistics and technical support.[48]

In June 1956, the Air Force chose Lockheed's Missile Systems Division in Sunnyvale, California, to design and build the military satellites for the WS 117L program. Lockheed's winning proposal featured a large, second-stage booster satellite that could be stabilized in orbit on three axes with a high pointing accuracy. To become known as "Agena," this vehicle would be designed and tested to meet Air Force plans for an operational capability in the third quarter of 1963. While the diminutive Vanguard scientific satellite was projected to weigh tens of pounds and be launched by a modified sounding rocket, the

---

44. "Statement on Disarmament, July 21," *The Department of State Bulletin*, 33, No. 841, August 1, 1955, p. 174; Elie Abel, "Eisenhower Calls Upon Soviet Union to Exchange Arms Blueprints," *New York Times*, July 22, 1955, p. 1; also Prados, *The Soviet Estimate*, pp. 31-32. The term "Open Skies" was coined later by the popular press and applied to Eisenhower's statement on disarmament. The background of this proposal, as advanced by the president's special assistant, Harold Stassen, and debated in the National Security Council, is contained in John P. Glennon, ed., *Foreign Relations of the United States, 1955-1957: Volume XX, Regulation of Armaments; Atomic Energy* (Washington, DC: U.S. Government Printing Office, 1990), see especially [33 through 48]. By 1956-1957, Eisenhower and other key administration leaders would view aerial reconnaissance as an "inspection system" that could serve two critical functions: to forewarn of surprise attack *and* supervise and verify arms-reduction and nuclear-test-ban agreements.

45. Herbert S. Parmet, *Eisenhower and the American Crusades* (New York: The Macmillan Company, 1972), p. 406; see also W. W. Rostow, *Open Skies*, pp. 7-8.

46. Richard Leghorn, then working for Eisenhower's special assistant Harold Stassen, wrote the paper on which the "Open Skies" doctrine was predicated. He also produced the 32-page booklet explaining this disarmament proposal given to those attending the Big Four Geneva Conference. Richard S. Leghorn, "U.S. Can Photograph Russia from the Air Now," *U.S. News & World Report*, August 5, 1955, pp. 70-75; "Editor's Note" at p. 71. Cleared by the White House, this important article explained the administration's rationale for Open Skies and the implications of this plan for arms reduction.

47. Ambrose, *Ike's Spies*, pp. 31-34, 266.

48. Attendees at the July 27 meeting included Eisenhower's staff secretary and defense liaison, Colonel Andrew Goodpaster, U.S. Army. Goodpaster, "Memorandum of Conference with the President, July 27, 1955, 11:45AM." The news release is reprinted in *Volume XX* [342], p. 734; see also for related events and the Quarles' IGY selection process, Constance McL. Green and Milton Lomask, *Vanguard: A History* (Washington DC: NASA SP-4202, 1970), pp. 37-38, 55-56.

proposed Air Force satellite would weigh thousands of pounds and be launched atop an Atlas ICBM.[49]

Among other payloads, Lockheed recommended for development those projects already identified by the Navy and RAND, and added one of its own: an infrared radiometer and telescope to detect the hot exhaust gases emitted by long-range jet bombers and, more important, large rockets as they ascended under power through the atmosphere. This novel aircraft-tracker and missile-detection innovation advanced by Joseph J. Knopow, a young Lockheed engineer, fit nicely into the strategic warning efforts of the day and unquestionably helped tip the scales in Lockheed's favor.[50] The Air Force awarded the firm a contract for this program a few months later, in October 1956.[51]

Thus, a year before Sputnik, the two modest U.S. space programs moved ahead slowly, staying within strict funding limits and avoiding unwanted interference, with development of the nation's long-range ballistic missiles just underway. They shared a lower priority than other high-technology defense department programs. To avoid provoking an international debate over "freedom of space," Eisenhower administration leaders in 1956 restrained government officials from any public discussion of spaceflight.[52] At the Pentagon, after a WS 117L program briefing on November 17, Donald Quarles, now Secretary of the Air Force, instructed Lieutenant General Donald Putt, Deputy Chief of Staff for Research and Development, to cease all efforts toward vehicle construction. He expressly forbade fabrication of a mockup or of the first satellite without his personal permission. A military satellite, the Air Force learned, would under no circumstances precede a scientific satellite into orbit.[53]

In early 1957 President Eisenhower remained undecided whether the United States needed to launch more than six IGY satellites for science. Moreover, Secretary of Defense Charles Wilson remained unimpressed with expensive astronautical ventures of any kind.

49. In the mid 1950s, Convair's James W. Crooks, Jr., constantly reminded audiences at Wright-Patterson AFB and elsewhere that the Atlas could lift the weight of a new Chevrolet, 3,500 lbs., into low-Earth orbit. As events turned out, Atlas with a powered upper stage could lift a good deal more—about 10,000 lbs.—into low-Earth orbit.

50. In time, this payload proposal would be separated and identified as the Missile Detection and Alarm System (MIDAS), then evolve to become the contemporary Defense Support Program (DSP). Today, this remarkable set of military satellites can detect and provide advance warning of a missile attack within moments of a launch at sea or on land.

51. LMSD 1536, *Pied Piper Development Plan*, Vol II, March 1, 1956, Subsystem Plan, A. Airframe, A-Apdx., pp. 3-4; and Vol. I, System Plan, *passim*, Eisenhower Library.

52. Unwitting of the National Security Council deliberations and of the ground rules established for the nation's space program, contemporary American military leaders failed entirely to comprehend the rationale that prompted this restriction on public discussion. See, for example, Maj. Gen John B. Medaris, U.S. Army, with Arthur Gordon, *Countdown for Decision* (New York: Paperback Library, Inc., 1960), pp. 101, 124; and testimony of Lt. Gen James M. Gavin, Deputy Chief of Staff Research and Development, U.S. Army, in U.S. Senate, *Inquiry into Satellite and Missile Programs*, "Hearings before the Senate Preparedness Investigating Subcommittee of the Committee on Armed Services," Part II, 6 January 1958, p. 1474, and Part I, 13 December 1957, p. 509. Air Force General Bernard Schriever, charged with the missile and space efforts of that service in the mid-to-late 1950s, was still fuming in 1985. Recalling a February 1957 speech, he announced that the Air Force was ready to "move forward rapidly into space. I received instruction the next day from the Pentagon that I shouldn't use the word 'space' in any of my future speeches. Now that was February 1957! They [the administration] had the IGY going, you know, which was kind of a scientific boondoggle." Richard H. Kohn, June 1985 interview with Generals Doolittle, Schriever, Phillips, Marsh, and Dr. Getting, in Jacob Neufeld, ed., *USAF Research and Development* (Washington, DC: Office of Air Force History, 1990) p. 105. Regarding priority, GOR No. 80 of March 16, 1955, specified a date of "operational availability" for the military satellites in the mid 1960s, a date that bespoke a low priority and bracketed this system to follow the U-2. Certainly, the first military spaceflights would trail by many months those of the scientific satellites. IGY space program priorities considered in "Memorandum of Discussion at the 283d Meeting of the National Security Council, Washington, May 3, 1956," in *Volume XI* [343], pp. 740-41.

53. *USAF Space Programs, 1945-1962, Volume 1* (USAF Historical Division Liaison Office, October 1962), p. 18. The historian added: "...it was apparent that the possible political repercussions arising from use of a military space vehicle were causing concern." On the West Coast, Schriever complained vigorously. The next year, in 1957, he declared, "I finally got $10 million [for WS 117L] from Don Quarles, who was Secretary of the Air Force, with instructions that we could not use that money in any way except component development. No systems work whatsoever. $10 million!" Schriever comments in *USAF Research and Development*, pp. 105-106. The Quarles' stricture remained in effect for nearly an entire year, and was not lifted until September 1957.

"A 'damn orange' up in the air," he snapped to confidants. In May 1957, as costs to build and launch the original six IGY vehicles soared from an estimated $20 million to $100 million, he told Eisenhower that Earth satellites, whatever their merit, "had too many promoters and no bankers."[54] [II-12] Donald Quarles, named Deputy Secretary of Defense one month earlier, nonetheless supported the U.S. IGY satellite effort while he kept an eye on related developments in the U.S.S.R. At his request near the end of June, CIA Director Allen Dulles assessed recent Soviet hints of an impending satellite launch. "The U.S. [intelligence] community," Dulles advised, "estimates that for prestige and psychological factors, the U.S.S.R. would endeavor to be the first in launching an earth satellite." Moreover, he said, it "probably is capable of launching a satellite in 1957."[55] [II-13] However accurate the CIA assessment might be, advocates of the WS 117L program found themselves unable to secure active support within the administration, and in July the Defense Department imposed sharp spending limits that effectively constrained their work to the "study level."[56]

This state of affairs changed dramatically a few months later, in October-November 1957, after the Soviet Union launched Sputniks I and II. Despite presidential assurances, the Soviet space accomplishments fueled a national debate over U.S. defense and science policies.[57] [II-14, II-15] Having downplayed the space program for purposes of their own, Eisenhower and his advisors underestimated the psychological shock value of the satellites that RAND had identified, the Technological Capabilities Panel had acknowledged, and the National Security Council had underscored just a few years before. What began as an evenly, if slowly paced, research and development effort was soon to receive high priority.[58]

Sputniks I and II, with their "Pearl Harbor" effect on public opinion, introduced into space affairs the issues of national pride and international prestige. The administration now moved quickly to restore confidence at home and prestige abroad. The Defense Department authorized the Army to launch a scientific satellite as a backup to the National Science Foundation-Navy Vanguard Project, and the president created the Advanced Research Projects Agency (ARPA), assigning it temporary responsibility for directing all U.S. space projects. James Killian, recently named Science Advisor to the President, also changed his mind. More funds were made available to the military space program, and in early 1958 the administration approved launching these satellites sooner with Thor IRBM boosters. Secretary of Defense Neil McElroy, who succeeded Charles Wilson in Sputnik's aftermath, ordered ARPA to launch space vehicles to "provide a closer look at the moon."[59]

54. Wilson as quoted by Harr, "Eisenhower's Approach to National Security Decision Making," p. 96, and as quoted in "Memorandum of Discussion at the 322d Meeting of the National Security Council, Washington, May 10, 1957," in *Volume XI* [345], p. 752.

55. Allen W. Dulles, Director of Central Intelligence, to The Honorable Donald Quarles, Deputy Secretary of Defense, July 5, 1957, Eisenhower Library.

56. Quarles subsequently drew congressional fire for also restricting the flow of funds to the high-priority missile program. See "Quarles on the Spot," in *Washington Roundup, Aviation Week*, October 28, 1957, p. 25.

57. In his first news conference after the launch of Sputnik I on October 9, 1957, President Eisenhower let slip his true interest in the event, though it went unnoticed in the excitement of the day. "From what they say they have put one small ball in the air," the President declared, adding, "at this moment you [don't] have to fear the intelligence aspects of this." *Public Papers of the President of the United States: Dwight David Eisenhower, 1957* (Washington DC: U.S. Government Printing Office, 1958), p. 724.

58. Eisenhower's advisors had anticipated the launch of a Soviet satellite before the United States, and the Operations Coordinating Board, established within the structure of the National Security Council by Executive Order 10700, February 25, 1957, had prepared a contingency statement to be handled by the National Academy of Sciences. See Operations Coordinating Board, "Memorandum of Meeting: Working Group on Certain Aspects of NSC 5520 (Earth Satellite), Fourth Meeting held 3:30 P.M., June 17, 1957, Room 357 Executive Office Building," and attachment: "Contingency Statement; Proposed Statement by Dr. Detlev W. Bronk, President of the National Academy of Sciences, in the Event the U.S.S.R. Announces Plans for or the Actual Launching of an Earth Satellite," NASA Historical Reference Collection; Herbert F. York, *Race to Oblivion* (New York: Simon and Schuster, Clarion Book, 1970), pp. 106, 146.

59. Defense Secretary Wilson had announced plans to resign before the launch of Sputnik I. These actions and events are described in National Security Council (NSC) Action No. 1846, January 22, 1958, as cited in National Security Council, NSC 5814/1, "Preliminary U.S. Policy on Outer Space," August 18, 1958, p. 20; Mosely, *Dulles: A Biography of Eleanor, Allen, and John Foster Dulles*, p. 432; Prados, *The Soviet Estimate*, pp. 106-107; DOD News Release No. 288-58, March 27, 1958; see also ARPA Orders No. 1-58 and 2-58, March 27, 1958, all in NASA Historical Reference Collection. The new satellite project is described by Kistiakowsky in *A Scientist at the White House*, p. 378.

There was an undeniable public concern with Soviet leadership in outer space exploration. Eisenhower declared on April 2, 1958, that a unified national space agency had to be established.[60] Few disagreed, certainly not the U.S. scientists who had begun to seriously consider the future of research in space, the prospects for obtaining more federal funds for this activity, and the ways of organizing it within the government.[61] [II-16] During the subsequent dialogue and in legislative action, the nation's political leaders endorsed the president's choice of civilian control of expanded U.S. space activities. Except for national defense space operations, for which the Department of Defense remained responsible, the National Aeronautics and Space Act declared that all non-military aeronautical and space endeavors sponsored by the United States would be directed by a civilian agency guided by eight objectives. First among them was basic scientific research, defined as "the expansion of human knowledge of phenomena in the atmosphere and space...." Signed into law by President Eisenhower on July 29, the act wrote a broad and comprehensive mandate for the peaceful pursuit of new knowledge and accompanying technology in space.[62] [II-17]

The National Aeronautics and Space Administration (NASA), formed with the National Advisory Committee for Aeronautics (NACA) as its nucleus, began operating on October 1, 1958, with the ongoing scientific satellite and planetary exploration projects inherited from the National Science Foundation and ARPA. Air Force and other service leaders, limited exclusively to approved military space missions, still had to translate existing plans into functioning systems. Those military satellite projects already underway and projected at the end of 1958 formed the basic military space program.[63] It encompassed five functional areas and, with one exception, consisted of non-piloted military spaceflight projects (see Table 1).[64] In years to come, the Air Force would for the most part retain responsibility for technically managing and launching military spacecraft. Operational direction of the individual projects frequently was assigned elsewhere.[65]

---

60. Robert Vexler, ed., *Dwight D. Eisenhower, 1880-1969, Chronology, Documents, Bibliographical Aids* (Dobbs Ferry, NY: Oceana Publications, Inc., 1972), p. 42. NASA's enabling act was drafted by the NACA General Counsel Paul G. Dembling in January-February 1958. Endorsed by James Killian and other White House officials, and submitted to Congress by the President on April 2, the act passed essentially as first drawn—with the addition of a National Aeronautics and Space Council perhaps the most notable change. In recent years, however, some scholars have argued that congressional agitation forced the issue of a civil space agency on a reluctant president. See, for example, Derek W. Elliott, "Finding an Appropriate Commitment: Space Policy Development Under Eisenhower and Kennedy, 1954-1963," Ph.D. dissertation, The George Washington University, May 10, 1992.

61. See Chapter Four of this volume for a discussion of the debate over organizing the space agency.

62. National Aeronautics and Space Act of 1958, Sec. 102(a) and 102(c); Frank W. Anderson, Jr., *Orders of Magnitude: A History of NACA and NASA, 1915-1980* (Washington, DC: NASA SP-4401, 1981), p. 17; Maier, in Kistiakowsky, *A Scientist at the White House*, pp. xxxviii-xxxxix. An elucidation of the reasons for and objectives of using and exploring space are contained in a contemporary brochure issued by the President's Science Advisory Committee, "Introduction to Outer Space," March 26, 1958, NASA Historical Reference Collection.

63. Various Air Force officials, it is true, attempting to gain responsibility for directing the nation's space program in 1958, did graft to this basic plan and present to Congress all sorts of exotic space proposals, including manned and unmanned orbital bombardment systems and even lunar military bases from which to attack countries on Earth. Besides flying in the face of stated administration commitments to explore and use outer space for peaceful and defensive purposes only, these proposals gained few adherents other than those who already viewed the Soviet sputniks with unalloyed hysteria.

64. This program plan, it is also true, does not appear in this form in contemporary documents. The proposed manned rocket bomber (ROBO), later called Dyna-Soar (X-20), remained the sole exception to space robotics and in research and development until canceled in the early 1960s. Notwithstanding the variations that marked it afterward, the 1958 plan featured automated spacecraft and reflects the basic American military space program in effect today.

65. Neil McElroy, Secretary of Defense, Memorandum to Chairman of the Joint Chiefs of Staff, "Responsibility for Space Systems," September 18, 1959, in Alice C. Cole, *et. al.*, eds., *The Department of Defense: Documents on Establishment and Organization* (Washington DC: Office of the Secretary of Defense, 1978), p. 325; also DOD Directive No. 5160.32, "Development of Space Systems," March 6, 1961, as reprinted in *Ibid*.

Table 1

Military Space Program Plan
(November 1958)

Functions	Projects
Navigation	Transit navigation satellite system; assigned to the Navy on May 9, 1960
Meteorology	Tiros television (RCA) satellite system assigned to NASA; military system proposed, but held to studies while negotiations for a single civil-military system were underway with NASA and the Department of Commerce (Weather Bureau)
Communication	Courier active (repeater) strategic and tactical communication satellite system; assigned to the Army on September 15, 1960
Missile Detection and Space Defense	Infrared radiometers that detect focused and Space Defense heat sources (Missile Detection and Alarm—MIDAS)
	Detection of nuclear detonations (Vela Hotel)
	Satellite inspector
	ROBO/Dyna-Soar (X-20)
	Radar tracking of Earth satellites (SPASUR/SPADATS)
	Optical tracking of satellites (from IGY Baker-Nunn system)
	Distant Early Warning (DEW) radar net and, by the early 1960s, the Ballistic Missile Early Warning System (BMEWS) radar net
Reconnaissance	Other automated satellites

Making Straight the Way

When NASA opened for business in October 1958, periodic U-2 flights over limited areas of the U.S.S.R. had been underway for two years. The Soviets protested vigorously, albeit privately, through diplomatic channels, and administration leaders knew that improved ground-to-air missiles would soon preclude all such missions.[66] Late in the year, President Eisenhower officially notified the Russians once again that the United States specifically sought to allay fears of surprise attack and create an inspection system to supervise arms-reduction agreements by means of aerial *and* space observation. He did so by

---

66. Eisenhower himself viewed these overflights in Soviet airspace as exceptionally provocative and a grave violation of national sovereignty; before personally approving each mission, he had to be convinced of the overriding need for it.

submitting a third, much more significant Open Skies proposal at an extraordinary "Surprise Attack Conference" sponsored by the United Nations in Geneva.[67]

Making his proposal the more remarkable, Eisenhower authorized his representatives, William C. Foster, later head of the Arms Control and Disarmament Agency, and Harvard chemist George Kistiakowsky, to include a "sanitized" version of the threat-and-warning portions of the surprise-attack indications matrix supplied by the National Indications Center. He thus furnished Soviet officials key indicators with which to assess the military status of states in the North Atlantic Treaty Organization—if they had not already devised similar warning indicators independently. The Soviets once again rejected Open Skies, though the U.S. position on the issue was made plain.[68] Even if the Soviets continued to reject the concept in international conference, might not the precepts of international law now be applied to achieve it?

One year earlier, Sputniks I and II had overflown international boundaries without provoking diplomatic protests. Four days after Sputnik I, in fact, Eisenhower and Deputy Secretary of Defense Donald Quarles discussed the issue. Quarles observed: "...the Russians have...done us a good turn, unintentionally, in establishing the concept of freedom of international space.... The President then looked ahead...and asked about a reconnaissance [satellite] vehicle."[69] The U.S. IGY Explorer and Vanguard satellites that followed the first sputniks into orbit in early 1958 likewise transited the world, and again not a single state objected to these overflights. The civil spacecraft would make straight the way for their military counterparts. Testifying before the U.S. House of Representatives in May 1958, Quarles underscored this point for a member of Congress skeptical that the United States should not object to Soviet reconnaissance satellites. "In a military sense," Quarles said, careful to speak only for the Department of Defense, "it seems to me that objects orbiting in outer space have an international character by the very nature of their position there, and it would be inappropriate for us to take the position that what you could see from there of our area would be improper for them to see.... I just think we cannot establish that kind of position that these [military satellites] are improper or objectionable or offensive. So I would have the view that we would not seek to object to such reconnaissance."[70] This tenuous "freedom of space" principle, the right of unrestricted overflight in outer space, the evidence indicates President Eisenhower purposely sought to exploit and codify when he signed the 1958 Space Act. That signature formally divided U.S. astronautics between civilian science and military applications directed to "peaceful"—that is, scientific—or defensive and nonaggressive purposes.

---

67. The second proposal Eisenhower submitted directly to Nikolai A. Bulganin, Chairman of the Soviet Council of Ministers, on March 2, 1956, eight months after the original proposal in Geneva. In it, Eisenhower agreed to accept on-site inspection teams if the Soviets would accept Open Skies. It, too, was rejected. See Ambrose, *Eisenhower: Volume II*, p. 311.

68. Annex 5 and Annex 6 of "Report of the Conference of Experts for the Study of Possible Measures Which Might be Helpful in Preventing Surprise Attack and for the Preparation of a Report Thereon to Government," United Nations General Assembly, A/4078, S/4145, January 5, 1959; William C. Foster, "Official Report of the United States Delegation to the Conference of Experts for the Study of Possible Measures Which Might be Helpful in Preventing Surprise Attack and for the Preparation of a Report Thereon to Governments," Geneva, Switzerland, November 10-December 18, 1958, p. 10, Eisenhower Library.

69. Quarles and Eisenhower remarks quoted in Walter A. McDougall, *The Heavens and the Earth: A Political History of the Space Age* (New York: Basic Books, Inc., 1985), p. 134; an abridged version, less the reference to military satellites, appears in "Memorandum of a Conference, President's Office, White House, Washington, October 8, 1957, 8:30 a.m.," *Volume XI* [347], pp. 755-56. Walter McDougall and Stephen Ambrose, without access to classified documents, correctly perceived the intent of Eisenhower's satellite decision and the rationale behind it. McDougall, *The Heavens and the Earth*, chapter 5; Ambrose, *Eisenhower: Volume II*, pp. 428, 513-14. Quarles, architect of the nation's space policy, reiterated for administration leaders the importance of the principle "freedom of space" and its implications for military observation satellites at a meeting of the National Security Council on October 10, 1957, in *Volume XI* [348], p. 759.

70. U.S. Congress, House, Select Committee on Astronautics and Space Exploration, *Astronautics and Space Exploration*, 85th Cong., 2d sess. (Washington, DC: U.S. Government Printing Office, 1958), p. 1109.

President Eisenhower amplified his space policy with National Security Council directives in June and August 1958 and January 1960. Anticipating the launch of military satellites, the first directive called for a "political framework which will place the uses of U.S. reconnaissance satellites in a political and psychological context most favorable to the United States." The second directive judged these spacecraft to be of "critical importance to U.S. national security," identified them with the peaceful uses of outer space, and set as an objective the "'opening up' of the Soviet Bloc through improved intelligence and programs of scientific cooperation." The third directive described the military support missions in space that fell within the rubric of peaceful uses, identified offensive space-weapon systems for study, and noted a positive political milestone in international law. The United Nations Ad Hoc Committee on the Peaceful Uses of Outer Space now accepted the "permissibility of the launching and flight of space vehicles...regardless of what territory they passed over during the course of their flight through outer space." But the UN Committee, the directive confided, at the same time stipulated that this principle pertained only to flights involved in the "peaceful uses of outer space."[71] [II-18, II-19, II-20, II-21]

Hewing to the policy of "freedom of space" and the peaceful space activities they defined for it, Eisenhower administration officials would in the months ahead permit only the study of offensive space weapons such as space-based antiballistic missile systems, satellite interceptors, and orbital bombers that could threaten the precedent of free passage.[72] This space policy, endorsed by President Eisenhower's successor, John F. Kennedy, secured two objectives simultaneously and permitted the launch and operation of military reconnaissance spacecraft. First, it reinforced the "sputnik precedent" as an accepted principle among states, officially recognizing free access to and unimpeded passage through outer space for peaceful purposes. [II-22] Second, by limiting military spacefaring to defense-support functions, it avoided a direct confrontation with the Soviet Union over observation of the Earth from space and ensured at least an opportunity to achieve Open Skies at altitudes above the territorial airspace of nation states. Thus, without formal convention, the United States could fashion unilaterally an "inspection system" to forewarn of surprise attack and supervise and verify future arms-reduction and nuclear-test-ban treaties.

But if the IGY scientific satellites had set an international precedent, and if publicly the United States was committed to a visible space program under civilian management, at the end of 1958 the actual launch and operation of military spacecraft had still to test President Eisenhower's policy—and Soviet reaction.

---

71. NSC 5814, "U.S. Policy on Outer Space," June 20, 1958, paragraph 54; NSC 5814/1, "Preliminary U.S. Policy on Outer Space," August 18, 1958, paragraphs 21, 30, and 47; NSC 5918, "U.S. Policy on Outer Space," December 17, 1959, paragraphs 18, 19, and 23.

72. The administration's rationale in opposing anything more than the study of space-based weapons is explained in Kistiakowsky, *A Scientist at the White House*, pp. 229-30, 239-40, and 245-46. A few days after the launch of Sputnik I, having just discussed this rationale with Eisenhower, Deputy Secretary of Defense Donald Quarles surprised and chagrined Air Force leaders who briefed him on the military satellite program and the potential of satellites for offensive applications: "Mr Quarles took very strong and specific exception to the inclusion in the presentation of any thoughts on the use of a satellite as a (nuclear) weapons carrier and stated that the Air Force was out of line in advancing this as a possible application of the satellite. He verbally directed that any such applications not be considered further in Air Force planning. Although both General [Curtis] LeMay and General [Donald] Putt voiced objection to this...on the grounds that we had no assurance that the U.S.S.R. would not explore this potential of satellites and could be expected to do so, Mr. Quarles remained adamant." Colonel F. C. E. Oder, USAF, Director, WS 117L, Memorandum for the Record, "Briefing of Deputy Secretary of Defense Mr. Quarles on WS 117L on 16 October 1957," October 25, 1957, Eisenhower Library.

Amplifying administration policy a year later, on October 20, 1958, ARPA Director Roy Johnson ordered the Air Force to cease using the Weapon System (WS) designation in the military satellite program "to minimize the aggressive international implications of overflight.... It is desired to emphasize the defensive, surprise-prevention aspects of the system. This change...should reduce the effectiveness of possible diplomatic protest against peacetime employment." Roy Johnson, Director, ARPA, to Maj. General Bernard Schriever, Cmdr., Air Force Ballistic Missile Division, Air Research and Development Command, n.s., October 20, 1958, Eisenhower Library. Despite these and subsequent messages that canceled offensive space-based, weapon-research programs, Air Force military leaders at that time seemed unable to grasp—or unwilling to accept—the meaning of President Eisenhower's "peaceful uses of outer space," or the rationale behind it.

Document II-1

Document title: Louis N. Ridenour, "Pilot Lights of the Apocalypse: A Playlet in One Act," *Fortune*, Vol. 33, January 1946.

This brief "playlet" offered the first public account of an intercontinental nuclear war directed from underground command centers and conducted using space-based weapons. Written by Louis Ridenour, a physicist who helped develop radar technology at MIT's Radiation Laboratory during World War II, it appeared in print just five months after the atomic bombing of Hiroshima and Nagasaki.

Concern over a nuclear surprise attack led President Dwight D. Eisenhower to propose "Open Skies" and establish a national policy (with the intent to promote an international precedent) of "freedom of space." That policy and subsequent precedent permitted the United States to employ without contest early warning satellites and other Earth-orbiting observation systems.

Pilot Lights of the Apocalypse

A Playlet in One Act

by Louis N. Ridenour

Louis N. Ridenour wrote "Military Security and the Atomic Bomb" in the November issue of Fortune, *attacking the notion that the U.S. could achieve "security by concealment" of its scientific knowledge of atomic power. Since then the principle underlying his view seems to have been incorporated in Anglo-American policy. Dr. Ridenour, a nuclear physicist and a professor of physics at the University of Pennsylvania, went to M.I.T.'s Radiation Laboratory four years ago and helped develop radar weapons there. He believes that while "security by achievement" may determine any victory, the difference between victor and vanquished in a war fought with atomic power would be merely a few percentiles of obliteration. And he fears that such a war will be inescapable if there is an atomic armaments race, since the slightest error, such as often occurs when men under great tension become trigger-happy, would touch off stupendous destruction.*

Dr. Ridenour has not tried to write a fantasy; he has tried to put into the form of a grim playlet the sober conclusions to which he feels driven by knowledge of physics, weapons, and human behavior. His moral: we should do all that decently can be done to avoid an atomic-armaments race.

What that "all" may be, he does not, as a natural scientist, attempt to define. Further definition is the job of social scientists, statesmen, philosophers, and citizens.

The curtain rises to disclose the operations room of the Western Defense Command, somewhere in the San Francisco area and a hundred feet underground. Two sergeants, RIGHT, are tending a row of teletype machines that connect the room with the world's principal cities. Two others, REAR, sit before a sort of telephone switchboard with key switches, lights, and labels representing the world's major cities. Behind them stands a captain. At a large desk, CENTER, sit a brigadier general and two colonels, all reading teletype messages. The wall, LEFT, has a sturdy barred door, a world map, and a framed motto: "Remember Pearl Harbor."

TIME: Some years after all the industrialized nations have mastered the production and use of atomic power.

BRIGADIER (laying down the message he has been reading): Nothing much tonight, I'd say. We'd better get tidied up a little. Captain Briggs!

CAPTAIN (facing about and standing at attention): Yes, sir.

BRIGADIER: Ready for company?

CAPTAIN: Yes, sir. I think so, sir.

BRIGADIER: See that the men look busy—on their toes and busy.

CAPTAIN: Yes, sir. (A bell rings.) Schwartz, you get the door. (One of the sergeants crosses to the door and opens it. All stand rigidly at attention. A little confused, the sergeant goes through the formality of examining passes. He then admits a group of four: a four-star general, a major, and two civilians.)

GENERAL: Carry on. (The men relax. The General leads the two civilians over to the Brigadier and the Colonels. The Major takes up his station by the door. Nobody pays any attention to him.) Mr. President, this is General Anderson, Watch Officer in charge of the Operations Room.

THE PRESIDENT: How do you do?

BRIGADIER: How do you do, Sir? (They shake hands.)

GENERAL: Colonel Sparks and Colonel Peabody, Deputy Watch Officers on duty.

THE PRESIDENT: Glad to meet you both. (They shake hands.)

GENERAL: Dr. Thompson—General Anderson, Colonel Sparks, and Colonel Peabody. (All nod and smile.) Now, Mr. President, this is the nerve center of our counterattack organization for the western area. The teletype machines you see over there (pointing) are on radio circuits that connect us with our people in all the principal cities of the world, and with the other continental defense commands. The stations, and their statuses, are marked on the map. (He gestures toward the map.) We've just come from the defense center, where the radar plots are kept and the guns and the fighters controlled. That's defense. But this is counterattack. Along that wall (waving toward the rear) is our control board. If you'll step over here, sir, I'll show you how it works.

THE PRESIDENT (moving with the General toward the telephone switchboard against the back wall): Defense and counterattack, eh? Why keep them separate?

GENERAL: Well, the defense has to move quickly, or it's no good at all. They don't have time to think. But counterattack—well, counterattack has to move quickly, too. But we want them to have time to decide what they need to do. You can't tell just from the direction of an attack who launched it. An attack might be staged entirely by mines planted inside our borders, so there wouldn't be any direction connected with it. And then again, we have pretty good information that some other countries besides us have got bombs up above the stratosphere, 800 miles above the earth, going round us in orbits like little moons. We put up 2,000 and we can see about 5,400 on our radar. Any time, somebody can call down that odd 3,400 by radio and send them wherever they want. There's no telling from trajectory which nation controls those bombs. What this all means is that the data these fellows here have to go on is mainly political. Radar doesn't do them any good. What they need is intelligence, and that's what comes in all the time, as complete and up-to-date as we can get it on the teletypes. In the defense center, you saw scientists and technicians. The officers here are political scientists.

THE PRESIDENT: That's very interesting. Maybe you'll give me a job here if I ever need one. I'm a political scientist.

GENERAL (laughing just enough): Yes, sir!

THE PRESIDENT: General, you haven't told me what all these gadgets are for. (He waves toward the switchboard.)

GENERAL: No, sir, I haven't. This is our counterattack control board. You see that every station is marked with the name of a city. And every station has three pilot lights: red, yellow, and green.

THE PRESIDENT: All the green ones are on.

GENERAL: That's right, sir. We have unattended radio transmitters, each with three spares, in stations in every city covered on this board. If one of the transmitters goes on the blink, a spare is automatically switched on. But if all four transmitters in any station are destroyed, well, we lose the signal from that station. When that happens the green light goes out and the yellow light comes on.

THE PRESIDENT: How about the red light?

GENERAL: That comes on instead of the yellow when all our stations in the whole city go off the air. Yellow means partial destruction—red means substantially complete destruction.

THE PRESIDENT: And green means peace.

GENERAL: Yes, sir. But this isn't just a monitoring board. You see this key here?

THE PRESIDENT: Yes.

GENERAL: That sets off our mines. We have them planted in a great many cities, and the radio control circuit can be unlocked from here.

THE PRESIDENT: Is the whole world mined now?

GENERAL: Well, no. We haven't bothered much with Asia. And some countries are so hard to get into that coverage is spotty. Our schedule calls for completion of mine installations in two more years. But we have another card to play. You remember I told you about the satellite bombs—the ones that are circling around, 800 miles up?

THE PRESIDENT: Yes.

GENERAL: Well, this other key here will bring them down on the city shown on the marker—we are looking at Calcutta—one of those satellite bombs every time it is pressed.

THE PRESIDENT: Is one of those bombs earmarked for each particular city?

GENERAL: No, sir. The bomb that happens to be in the most favorable location at the time this key is pressed is the one brought down. It might be any one of the whole 2,000.

THE PRESIDENT: This is all damned clever.

GENERAL: We have Dr. Thompson to thank for most of it. His people worked out all the technical stuff. All the Army has to do is man the installations and watch the intelligence as it comes along.

DR. THOMPSON: Good of you to say that, General. But seriously, Mr. President, as people pointed out soon after the first atomic bomb was dropped, there isn't any other nation with the industrial know-how to do a job like this.

THE PRESIDENT: It's very impressive, I must say. Are the other Defense Commands equipped the same way?

GENERAL: Yes, sir. As a matter of fact, to guard against accidents, each Defense Command has two complete operations rooms like this, either one of which can take full control if the other is destroyed.

THE PRESIDENT: We've kept ahead in the armaments race. Who'd dare attack us when we're set up like this?

DR. THOMPSON: Surely nobody would. I don't think you need to expect any trouble.

THE PRESIDENT: Well, this has all been very interesting. (To the Brigadier) General, have you had any exciting times here you can tell me about?

BRIGADIER: Yes, sir. Every time a meteorite comes down—a shooting star, you know—our radar boys track it, shoot it down, and send us in an alert. We have a few bad moments until we get the spectrographic report. If it's iron and nickel—and it always has been so far—we know God sent it, and relax. Someday it'll be uranium, and then we'll have to push a button. Or plutonium.

THE PRESIDENT: How many shooting stars have you shot?

COLONEL SPARKS (laughing politely): We get an average of twelve a month. In August it's the worst, of course. The Perseids, you know.

THE PRESIDENT (puzzled): Iran. . . ?

BRIGADIER (hastily): No, sir. The Perseid meteors. Named after Perseus. Astronomers are a classical bunch.

THE PRESIDENT (recovering): Oh, sure. (Turning to the Colonels) Gentlemen, how do you like this job?

COLONEL SPARKS: We have a feeling of grave responsibility.

THE PRESIDENT: The fate of the nation is in your hands. But always remember that our nation is the most precious.

BOTH COLONELS (awed): Yes, sir.

GENERAL: Well, Mr. President. We've fallen a little behind our schedule. They'll be waiting for us at the mess.

THE PRESIDENT: All right, General, let's get along. General Anderson, Colonel Sparks, Colonel Peabody, I've enjoyed very much seeing your installation. Keep on your toes. We're all depending on you.

GENERAL AND COLONELS (together): Yes, sir.

(Schwartz goes over, opens the door, and stands stiffly at attention as the visitors file out amid a general chorus of "Goodbye" and "Goodbye, sir." Schwartz closes the door. The Brigadier and Colonels sit at their desks.)

BRIGADIER: Well, that's that. The Old Man gave him a good story; I couldn't have done better myself.

COLONEL SPARKS (still in the clouds): He is depending on us.

BRIGADIER: Don't take it too hard. All we're supposed to do is make the other guy sorry. We can't save any lives or rebuild any cities. Never forget what those buttons do.

COLONEL SPARKS: Just the same, sir, I'm glad I was born an American. We've got the know-how. I'm glad I'm on the side that's ahead in the race.

COLONEL PEABODY (disgusted): Sparks, you talk like a damn high-school kid. For this job, you're supposed to have some good sense and detachment.

(Just then, there is a dull rumble. The floor and the walls of the room shake, and a couple of sizable chunks of concrete fall out of the ceiling. The lights go out, except for the green ones on the control board. Emergency lights, dimmer than the regular ones, come on at once. All the men are on their feet.)

BRIGADIER: Good God! What was that? (Recollecting himself) Peabody, get on the phone to headquarters. Sparks, get out the red-line messages for the last twenty-four hours. Captain, anything from the defense center?

CAPTAIN: My line to them seems to be out, sir.

BRIGADIER: What have you got for status? Anybody showing yellow or red?

CAPTAIN: San Francisco is red, sir.

COLONEL SPARKS (riffling wildly through teletype messages): Oh, Jesus. This must be it. San Francisco! (Screaming) San Francisco gone!

BRIGADIER: Shut up, Sparks. Take it easy. (To Peabody) Can't you get headquarters?

COLONEL PEABODY: My line is dead. I can't get reserve operations, either. Maybe this is the real thing.

COLONEL SPARKS (still half hysterical): We better do something. Remember what it says in the book: counterattack must take action before the enemy's destruction of our centers is complete.

BRIGADIER: First we need an enemy. Who's got the highest negative rating in the latest State Department digest?

COLONEL PEABODY (who has quietly taken the messages from in front of Sparks): Denmark, sir. But it's well below the danger point. All we've got is this: (reading) COPENHAGEN 1635 HOURS 22 JANUARY. WIDESPREAD DISAPPROVAL OF WILLIAMS FOUNTAIN, STATUARY GROUP PRESENTED THE KING DENMARK BY THE U.S., BEING SHOWN BY PEOPLE COPENHAGEN. FOUNTAIN HAS BEEN PELTED VEGETABLES BY HOODLUM GROUPS THREE OCCASIONS. FORMAL PROTEST STATING STATUE INSULTS KING RECEIVED FROM ROYAL ACADEMY ART IN FOLLOWING TERMS QUOTE...and so on. Nothing there, I'd say.

COLONEL SPARKS: Nothing there! San Francisco's in ruins, you damn fool, and we're sitting here like three warts on a pickle. All that over a lousy set of statues. I say let'em have it.

BRIGADIER (to Peabody): Is that the hottest you've got?

COLONEL PEABODY: Yes, sir. I don't think it could have been Denmark. Though that sculptor, Williams, does live in San Francisco.

BRIGADIER: We'd better wait and be sure. Captain, how are your lines now?

COLONEL SPARKS (with rising hysteria): What have we got this stuff for if we don't use it? My God, didn't you hear what the President said? He's depending on us; they're all depending on us. If you haven't got the guts, I have. (Before he can be stopped, he rushes to the control board and shoves a sergeant to the floor. Peabody is after Sparks in a flash. He pulls him around and knocks him to the floor. Sparks's head hits hard, and he lies still.

COLONEL PEABODY: General, he did it! Copenhagen shows red!

SERGEANT (at a teletype): Sir, here's a message from the defense center. They've got their line working again. (He tears it off and brings it to the Brigadier.)

CAPTAIN: Stockholm's gone red, sir.

COLONEL PEABODY: Sure. The Danes thought it was the Swedes. That export-duties row.

BRIGADIER: And the Swedes have got two hot arguments on their hands. They'll take the British, too, just to be sure. The British soak the Russians, and then we're next. (He reads the message he has been holding, and drops into a chair.) My God! Peabody, that was an earthquake. Epicenter right smack in San Francisco.

CAPTAIN: London's gone red, sir. And Edinburgh, and Manchester, and Nottingham, and—

COLONEL PEABODY: Dark ages, here I come. It's a pity the Security Council didn't have time to consider all this.

BRIGADIER: Peabody, you're beginning to sound a little like Sparks. Come to think of it, there was nothing wrong with him but too much patriotism and too little sense. Captain, we probably can't pull this out of the fire, but we've got to try. Send a message on all circuits. (The Captain sits down at a teletype keyboard.)

CAPTAIN: Ready, sir.

BRIGADIER: To all stations: URGE IMMEDIATE WORLDWIDE BROADCAST THIS MESSAGE: DESTRUCTION COPENHAGEN 1910 HOURS THIS DATE INITIATED BY THIS STATION THROUGH GRIEVOUS ERROR. ATTACKS MADE SINCE BASED ON IDEA DESTRUCTION COPENHAGEN WAS ACT OF WAR, WHICH IT WAS NOT REPEAT NOT. URGE ATTACKS BE STOPPED UNTIL SITUATION CAN BE CLARIFIED. THERE IS NO REPEAT NO WAR. END.

COLONEL PEABODY (who has been watching board): The hell there isn't. New York's gone red, and Chicago, and. . . (The room rocks, the lights go out. With a dull, powerful rumble, the roof caves in.)

CURTAIN

Document II-2

Document title: Douglas Aircraft Company, Inc., "Preliminary Design of an Experimental World-Circling Spaceship," Report No. SM-11827, May 2, 1946, pp. i-viii, 1-16, 211-12.

Source: Archives, The Rand Corporation, Santa Monica, California.

The newly formed Rand group, a unit of Douglas Aircraft, was directed by General Curtis LeMay to investigate the possible uses of satellites for the Air Force. LeMay took this action after he learned that the Navy was conducting a similar study. The resulting report, released in May 1946, was the first study completed by Rand and the first comprehensive analysis of the military uses of satellites. It suggested that satellites had broad uses in meteorology, reconnaissance, and communications. But while extensive in scope and providing much new information on the value of satellites, including the possibility of a vehicle that could carry humans, the report was virtually ignored by the Air Force, which was unconvinced as to the utility of satellites and unwilling to support a report that questioned the role of the manned bomber. These excerpts from the report, which contained 236 pages plus several lengthy appendices, give a sense of the broader thinking that guided the engineering analyses that comprised the bulk of the document.

---

[i]   **Summary**

This report presents an engineering analysis of the possibilities of designing a man-made satellite. The questions of power plants, structural weights, multiple stages, optimum design values; trajectories, stability, and landing are considered in detail. The results are used to furnish designs for two proposed vehicles. The first is a four stage rocket using alcohol and liquid oxygen as propellants. The second is a two stage rocket using liquid hydrogen and liquid oxygen as propellants. The latter rocket offers better specific consumption rates, but this is found to be partially offset by the greater structural weight necessitated by the use of hydrogen. It is concluded that modern technology has advanced to a point where it now appears feasible to undertake the design of a satellite vehicle.

[ii]   **Abstract**

In this report, we have undertaken a conservative and realistic engineering appraisal of the possibilities of building a spaceship which will circle the earth as a satellite. The work has been based on our present state of technological advancement and has not included such possible future developments as atomic energy.

If a vehicle can be accelerated to a speed of about 17,000 m.p.h. and aimed properly, it will revolve on a great circle path above the earth's atmosphere as a new satellite. The centrifugal force will just balance the pull of gravity. Such a vehicle will make a complete circuit of the earth in approximately 1-$\frac{1}{2}$ hours. Of all the possible orbits, most of them will not pass over the same ground stations on successive circuits because the earth will turn about $\frac{1}{16}$ of a turn under the orbit during each circuit. The equator is the only such repeating path and consequently is recommended for early attempts at establishing satellites so that a single set of telemetering stations may be used.

Such a vehicle will undoubtedly prove to be of great military value. However, the present study was centered around a vehicle to be used in obtaining much desired scientific information on cosmic rays, gravitation, geophysics, terrestrial magnetism, astronomy, meteorology, and properties of the upper atmosphere. For this purpose, a payload of 500 lbs. and 20 cu ft. was selected as a reasonable estimate of the requirements for scientific apparatus capable of obtaining results sufficiently far-reaching to make the undertaking worthwhile. It was found necessary to establish the orbit at an altitude of about 300 miles to insure sufficiently [iii] low drag so that the vehicle could travel for 10 days or more, without power, before losing satellite speed.

The only type of power plant capable of accelerating a vehicle to a speed of 17,000 m.p.h. on the outer limits of the atmosphere is the rocket. The two most important performance characteristics of a rocket vehicle are the exhaust velocity of the rocket and the ratio of the weight of propellants to the gross weight. Very careful studies were made to establish engineering estimates of the values that can be obtained for these two characteristics.

The study of rocket performance indicated that while liquid hydrogen ranks highest among fuels having large exhaust velocities, its low density, low temperature and wide explosive range cause great trouble in engineering design. On the other hand, alcohol, though having a lower exhaust velocity, has the benefit of extensive development in the German V-2. Consequently it was decided to conduct parallel preliminary design studies of vehicles using liquid hydrogen-liquid oxygen and alcohol-liquid oxygen as propellants.

It has been frequently assumed in the past that structural weight ratios become increasingly favorable as rockets increase in size, and fixed weight items such as radio equipment become insignificant weight items. However, the study of weight ratios indicated that for large sizes the weight of tanks and similar items actually become less favorable. Consequently, there is an optimum middle range of sizes. Improvements in weight ratios over that of the German V-2 are possible only by the slow process of technological development, not by the brute force methods of increase in size. This study showed that an alcohol-oxygen vehicle [iv] could be built whose entire structural weight (including motors, controls, etc.) was about 16% of the gross weight. On the other hand, the difficulties with liquid hydrogen, such as increased tank size, necessitated an entire structural weight of about 25% of the gross weight. These studies also indicated that a maximum acceleration of about 6.5 times that of gravity gave the best overall performance for the vehicles considered. If the acceleration is greater, the increased structural design loads increase the structural weight. If the acceleration is less, rocket thrust is inefficiently used to support the weight of the vehicle without producing the desired acceleration.

Using the above results, it was found that neither hydrogen-oxygen nor alcohol-oxygen is capable of accelerating a single unassisted vehicle to orbital speeds. By the use of a multi-stage rocket, these velocities can be attained by vehicles feasible within the limits of our present knowledge. To illustrate the concept of a multi-stage rocket, first consider a vehicle composed of two parts. The primary vehicle, complete with its rocket motor, tanks, propellants and controls is carried along as the "payload" of a similar vehicle of much greater size. The rocket of the large vehicle is used to accelerate the combination to as great a speed as possible, after which, the large vehicle is discarded and the small vehicle accelerates under its own power, adding its velocity increase to that of the large vehicle. By this means we have obtained an effective decrease in the amount of structural weight that must be accelerated to high speeds. This same idea can be used in designing vehicles with a greater number of stages. A careful analysis of the advantages of staging showed that for a given set of performance requirements, [v] an optimum number of stages exists. If the stages are too few in number, the required velocities can be attained only by the undesirable process of exchanging payload for fuel. If they are too many, the multiplication of tanks, motors, etc. eliminates any possible gain in the effective weight ratio. For the alcohol-oxygen rocket it was found that four stages were best. For the hydrogen-oxygen rocket, preliminary analysis indicated that the best choice for the number of stages was two, but refinements showed the optimum number of stages was three. Unfortunately, insufficient time was available to change the design, so the work on the hydrogen-oxygen was completed using two stages. The characteristics of the vehicles studies are tabulated below....

Vehicle Powered by Alcohol-Oxygen Rockets

Stage	1	2	3	4
Gross Wt. (lbs.)	233,669	53,689	11,829	2,868
Weight less fuel (lbs.)	93,669	21,489	4,729	1,148
Payload (lbs.)	53,689	11,829	2,868	500
Max. Diameter (in.)	157	138	105	90

Vehicle Powered by Hydrogen-Oxygen Rockets

Stage	1	2
Gross Wt. (lbs.)	291,564	15,364
Weight less fuel (lbs.)	84,564	4,464
Payload (lbs.)	15,364	500
Max. Diameter (in.)	248	167

[vi] (had three stages been used for the hydrogen-oxygen rockets, the overall gross weight of this vehicle could have been reduced to about 84,000 lbs. indicating this combination should be given serious consideration in any future study).

In arriving at the above design figures, a detailed study was made of the effects of exhaust velocity, structural weight, gravity, drag, acceleration, flight path inclination, and relative size of stages on the performance of the vehicles so that an optimum design could be achieved or reasonable compromises made.

It was found that the vehicle could best be guided during its accelerated flight by mounting control surfaces in the rocket jets and rotating the entire vehicle so that lateral components of the jet thrust could be used to produce the desired control forces. It is planned to fire the rocket vertically upward for several miles and then gradually curve the flight path over in the direction in which it is desired that the vehicle shall travel. In order to establish the vehicle on an orbit at an altitude of about 300 miles without using excessive amounts of control it was found desirable to allow the vehicle to coast without thrust on an extended elliptic arc just preceding the firing of the rocket of the last stage. As the vehicle approaches the summit of this arc, which is at the final altitude, the rocket of the last stage is fired and the vehicle is accelerated so that it becomes a freely revolving satellite.

It was shown that excessive amounts of rocket propellants are required to make corrections if the orbit is incorrectly established in direction or in velocity. Therefore, considerable attention was devoted to the stability and control problem during the acceleration to orbital [vii] speeds. It was concluded that the orbit could be established with sufficient precision so that the vehicle would not inadvertently re-enter the atmosphere because of an eccentric orbit.

Once the vehicle has been established on its orbit, the questions arise as to what are the possibilities of damage by meteorites, what temperatures will it experience, and can its orientation in space be controlled? Although the probability of being hit by very small meteorites is great, it was found that by using reasonable thickness plating, adequate protection could be obtained against all meteorites up to a size where the frequency of occurrence was very small. The temperatures of the satellite vehicle will range from about 40°F when it is on the side of the earth facing the sun to about -20°F when it is in the earth's shadow. Either small flywheels or small jets of compressed gas appear to offer feasible methods of controlling the vehicle's orientation after the cessation of rocket thrust.

An investigation was made of the possibility of safely landing the vehicle without allowing it to enter the atmosphere at such great speeds that it would be destroyed by the heat of air resistance. It was found that by the use of wings on the small final vehicle, the rate of descent could be controlled so that the heat would be dissipated by radiation at temperatures the structure could safely withstand. These same wings could be used to land the vehicle on the surface of the earth.

An interesting outcome of the study is that the maximum acceleration and temperatures can be kept within limits which can be safely withstood by a human being. Since the vehicle is not likely to be damaged by meteorites and can be safely brought back to earth, there is good reason [viii] to hope that future satellite vehicles will be built to carry human beings.

It has been estimated that to design, construct and launch a satellite vehicle will cost about $150,000,000. Such an undertaking could be accomplished in approximately 5 years time. The launching would probably be made from one of the Pacific islands near the equator. A series of telemetering stations would be established around the equator to obtain the data from the scientific apparatus contained in the vehicle. The first vehicles will probably be allowed to burn up on plunging back into the atmosphere. Later vehicles will be designed so that they can be brought back to earth. Such vehicles can be used either as long range missiles or for carrying human beings....

[1] 1. **Introduction**

Technology and experience have now reached the point where it is possible to design and construct craft which can penetrate the atmosphere and achieve sufficient velocity to become satellites of the earth. This statement is documented in this report, which is a design study for a satellite vehicle judiciously based on German experience with V-2, and which relies for its success only on sound engineering development which can logically be expected as a consequence of intensive application to this effort. The craft which would result from such an undertaking would almost certainly do the job of becoming a satellite, but it would clearly be bulky, expensive, and inefficient in terms of the spaceship we shall be able to design after twenty years of intensive work in this field. In making the decision as to whether or not to undertake construction of such a craft now, it is not inappropriate to view our present situation as similar to that in airplanes prior to the flight of the Wright brothers. We can see no more clearly all the utility and implications of spaceships than the Wright brothers could see fleets of B-29's bombing Japan and air transports circling the globe.

Though the crystal ball is cloudy, two things seem clear:

1. A satellite vehicle with appropriate instrumentation can be expected to be one of the most potent scientific tools of the Twentieth Century.

[2] 2. The achievement of a satellite craft by the United States would inflame the imagination of mankind, and would probably produce repercussions in the world comparable to the explosion of the atomic bomb.

Chapter 2 of this report attempts to indicate briefly some of the concrete results to be derived from a spaceship which circles the world on a stable orbit.

As the first major activity under contract W33-038AC-14105, we have been asked by the Air Forces to explore the possibilities of making a satellite vehicle, and to present a program which would aid in the development of such a vehicle. Our approach to this task is along two related lines:

1. To undertake a design study which will evaluate the possibility of making a satellite vehicle using known methods of engineering and propulsion.

2. To explore the fields of science in an attempt to discover and to stimulate research and development along lines which will ultimately be of benefit in the design of such a satellite vehicle and which will improve its efficiency or decrease its complexity and cost.

This report concerns itself solely with the first line of approach. It is a practical study based on techniques that we now know. The implications of atomic energy are not considered here. This and other possibilities in the fields of science may be the subject of future [3] reports, which will cover the second line of approach.

In the preliminary design study analytical methods have been developed which may be used as a basis for future studies in this new field of astronautical engineering. Among these are the following:

1. Analysis of single- and multi-stage rocket performance and methods for selecting the optimum number of stages for any given application.

2. Dimensional analysis of varying size and gross weight of rockets, deriving laws which are useful in design scaling. These laws are also of assistance in appraisal of the effect of shape and proportions on the design of multi-stage rockets.

3. The effect of acceleration and inclination of the trajectory on structural weight and performance of a satellite rocket.

4. Methods of determining the optimum trajectory for satellite rockets.

5. Variation of rocket performance with altitude and its effect on the proportioning of stages.

6. Preliminary study of effect of atmospheric drag on the rocket and how it affects the choice of stages, acceleration, and trajectory.

7. Analysis of dynamic stability and control throughout the entire trajectory.

[4] 8. Method of safely landing a satellite vehicle.

It cannot be emphasized too strongly that the primary contributions of this report are in methods, and not in the specific figures in this design study. *One point in particular should be highlighted: - the design gross weight, which is of the greatest importance in estimating cost or in comparing any two proposals in this field is the least definitely ascertained single feature in the whole process.* This fact is fundamental in the design of a satellite or spaceship, since the slightest variation in some of the minor details of construction or in propulsive efficiency of the fuel may result in a large change in gross weight. The figures in this report represent a reasonable compromise between the extremes which are possible with the data now in hand. The most important thing is that a satellite vehicle can be made at all in the present state of the art. Even our more conservative engineers agree that it is definitely possible to undertake design and construction now of a vehicle which would become a satellite of the earth.

Another important result of this design study is the conclusion on liquid hydrogen and oxygen as fuel versus liquid oxygen and alcohol (the Germans' fuel). The relative merits of these fuels have occasioned spirited controversy ever since liquid fuel rockets have been under development. In the past, the fact which has clinched the arguments has been the difficulty of handling, storing, and using liquid hydrogen. The present design study has approached this subject from another viewpoint. On the assumption that all these nasty problems can be solved, a design analysis has [5] been made for the structure and performance of rockets using both types of fuels. Because of the low density of liquid hydrogen, the greater tankage weight and volume tends to offset the increase in specific impulse. Early in the design study it was necessary to make a choice of the number of stages for both proposed vehicles. Based on the design information available, a decision was made to use four stages for the alcohol-oxygen rocket and two stages for the hydrogen-oxygen rocket. Of these two designs, the alcohol-oxygen rocket proved to be somewhat smaller in weight and size. However, the problem was later re-examined when more reliable data were available. It was found that, while the choice of four stages for alcohol-oxygen had been wise, the hydrogen-oxygen rocket could have been substantially improved by using three stages. The improvement was sufficient to indicate that the three stage hydrogen-oxygen rocket would have been definitely superior to the four stage alcohol-oxygen rocket. Unfortunately, the work had progressed so far that it was impossible to alter the number of stages for the hydrogen-oxygen rocket.

One of the most important conclusions of this design study is that in order to achieve the required performance it is necessary to have multi-state rockets for either type of fuel. The general characteristics of both types are shown in the following table:

4 Stage Alcohol-Oxygen Rocket

Payload 500#

Stage	1	2	3	4
Gross weight (lbs.)	233,669	53,689	11,829	2,868
Fuel weight (lbs.)	140,000	32,200	7,100	1,720

[6] 2 Stage Hydrogen-Oxygen Rocket

Payload 500#

Stage	1	2
Gross weight (lbs.)	291,564	15,364
Total Fuel wt. (lbs.)	217,900	10,000

The design represents a series of compromises. The payload is chosen to be as small as is consistent with carrying enough experimental equipment to achieve significant results. This is done for the purpose of keeping the gross weight within reasonable limits, since the gross weight increases roughly in proportion to the payload above a certain minimum value. The design altitude was originally chosen as 100 miles, since previous calculations indicated that the atmospheric drag there was not great enough to disturb the orbit of the satellite for a few revolutions, and since for communications purposes it was desirable to keep the satellite below the ionosphere. The more refined drag studies made in the present design study show that these early estimates were in serious error, and indicate that the satellite will have to be established at altitudes of 300 to 400 miles to ensure the completion of multiple revolutions around the earth.

It is interesting that the design analysis shows that the optimum accelerations are well within the limits which the human body can stand. Further, it appears possible to achieve a safe landing with the type of vehicle which is required. Future developments may bring an increase in payload and decrease in gross weight, sufficient to produce a large manned spaceship able to accomplish important things in a scientific [7] and military way.

We turn now from the design study phase to the basic research approach of the scientists. Our consultants have all made suggestions which have been taken into consideration in the preparation of this report. In the future it is our expectation that the services of these scientists will be of the greatest benefit in planning and initiating broad research programs to explore new fundamental approaches to the problem of space travel.

The real white hope for the future of spaceships is, of course, atomic energy. If this intense source of energy can be harnessed for rocket propulsion, then spaceships of moderate size and high performance may become a reality, and conceivably could even serve efficiently as intercontinental transports in the remote future. We are fortunate in having the consulting services of Drs. Alvarez, McMillan, and Ridenour, well known in scientific circles. Alvarez and McMillan were two of the key men at the Los Alamos Laboratory of the Manhattan Project. With the benefit of their advice, we hope to achieve a degree of competence in the fields of application of nuclear energy to propulsion.

Alvarez and Ridenour, who are also radar experts, have made basic analyses of the radio and radar problems associated with a satellite. These are of service in planning the new equipment which seems to be necessary to make the satellite a useful tool.

Kistiakowsky, a specialist in physical chemistry, has made valuable suggestions for the development of new rocket propellants.

[8] Schiff has contributed to our knowledge of the optimum trajectories to be used in launching the vehicle.

More important than the ideas and suggestions received to date is the fact that these consultants, who are among the leaders in U.S. science, have begun to think and work on these problems. It is our earnest hope that under the terms of this new study and research contract with the Army Air Forces we may be able to enlist the active cooperation of an important fraction of the scientific resources of the country to solve problems in the wholly new fields which man's imagination has opened. Of these, space travel is one of the most important and challenging.

[9] **2. The Significance of a Satellite Vehicle**

Attempting in early 1946 to estimate the values to be derived from a development program aimed at the establishment of a satellite circling the earth above the atmosphere is as difficult as it would have been, some years before the Wright brothers flew at Kitty Hawk, to visualize the current uses of aviation in war and in peace. Some of the fields in which important results are to be expected are obvious; others, which may include some of the most important, will certainly be overlooked because of the novelty of the undertaking. The following considerations assume the future development of a satellite with large payload. Only a portion of these may be accomplished by the satellite described in the design study of this report.

The Military Importance of a Satellite – The military importance of establishing vehicles in satellite orbits arises largely from the circumstance that defenses against airborne attack are rapidly improving. Modern radar will detect aircraft at distances up to a few hundred miles, and can give continuous, precise data on their position. Anti-aircraft artillery and anti-aircraft guided missiles are able to engage such vehicles at considerable range, and the proximity fuze increases several fold the effectiveness of anti-aircraft fire. Under these circumstances, a considerable premium is put on high missile velocity, to increase the difficulty of interception.

This being so, we can assume that an air offensive of the future will be carried out largely or altogether by high-speed pilotless missiles. The minimum-energy trajectory for such a space-missile without [10] aerodynamic lift at long range is very flat, intersecting the earth at a shallow angle. This means that small errors in the trajectory of such a missile will produce large errors in the point of impact. It has been suggested that the accuracy can be increased by firing such a missile along the same general course as that being followed by a satellite, and at such a time that the two are close to one another at the center of the trajectory of the missile. Under these circumstances, precise observations of the position of the missile can be made from the satellite, and a final control impulse applied to bring the missile down on its intended target. This scheme, while it involves considerable complexity in instrumentation, seems entirely feasible. Alternatively, the satellite itself can be considered as the missile. After observations of its trajectory, a control impulse can be applied in such direction and amount, and at such a time, that the satellite is brought down on its target.

There is little difference in design and performance between an intercontinental rocket missile and a satellite. Thus a rocket missile with a free space-trajectory of 6,000 miles requires a minimum energy of launching which corresponds to an initial velocity of 4.4 miles per second, while a satellite requires 5.1. Consequently the development of a satellite will be directly applicable to the development of an inter-continental rocket missile.

It should also be remarked that the satellite offers an observation aircraft which cannot be brought down by an enemy who has not mastered similar techniques. In fact, a simple computation from the radar [11] equation shows that such a satellite is virtually undetectable from the ground by means of present-day radar. Perhaps the two most important classes of observation which can be made from such a satellite are the spotting of the points of impact of bombs launched by us, and the observation of weather conditions over enemy territory. As remarked below, short-range weather forecasting anywhere in the vicinity of the orbit of the satellite is extremely simple.

Certainly the full military usefulness of this technique cannot be evaluated today. There are doubtless many important possibilities which will be revealed only as work on the project proceeds.

The Satellite as an Aid to Research – The usefulness of a satellite in scientific research is very great. Typical of the outstanding problems which it can help to attack are the following:

One of the fastest-moving fields of investigation in modern nuclear physics is the study of cosmic rays. Even at the highest altitudes which have been reached with unmanned sounding balloons, a considerable depth of atmosphere has been traversed by the cosmic rays before their observation. On board such a satellite, the primary cosmic rays could be studied without the complications which arise within the atmosphere. From this study may come more important clues to unleashing the energy of the atomic nucleus.

Studies of gravitation with precision hitherto impossible may be made. This is possible because for the first time in history, a satellite would provide an acceleration-free laboratory where the ever present pull of the earth's gravitational field is cancelled by the centrifugal force [12] of the rotating satellite. Such studies might lead to an understanding of the cause of gravitation—which is now the greatest riddle of physics.

The variations in the earth's gravitational field over the face of the earth could be measured from a satellite. This would supply one very fundamental set of data needed by the geologists and geophysicists to understand the causes of mountain-building, etc.

Similarly, the variations in the earth's magnetic field could be measured with a completeness and rapidity hitherto impossible.

The satellite laboratory could undertake comprehensive research at the low pressures of space. The value of this in comparison with pressures now attainable in the laboratory might be great.

For the astronomer, a satellite would provide great assistance. Dr. Shapley, director of the Harvard Observatory, has expressed the view that measurements of the ultra-violet spectrum of the sun and stars would contribute greatly to an understanding of the source of the sun's surface energy, and perhaps would help explain sunspots. He also looks forward to the satellite observatory to provide an explanation for the "light of the night sky."

Astronomical observations made on the surface of the earth are seriously hampered by difficulties of "seeing," which arise because of variations in the refractive index of the column of air through which any terrestrial telescope must view the heavens. These difficulties are greatest in connection with the observation of any celestial body whose image is an actual disk, within which features of structure can be [13] recognized: the moon, the sun, the planets, and certain nebulae. A telescope even of modest size could, at a point outside the earth's atmosphere, make observations on such bodies which would be superior to those now made with the largest terrestrial telescopes. Because there would be no scattering of light by an atmosphere, continuous observation of the solar corona and the solar prominences should also be possible. Astronomical images could, of course, be sent back to the earth from an unmanned satellite by television means.

From a satellite at an altitude of hundreds of miles, circling the earth in a period of about one and one half hours, observations of the cloud patterns on the earth, and of their changes with time, could be made with great ease and convenience. This information should be of extreme value in connection with short-range weather forecasting, and tabulation of such data over a period of time might prove extremely valuable to long-range weather forecasting. A satellite on a North-South orbit could observe the whole surface of the world once a day, and entirely in the daylight.

The properties of the ionosphere could be studied in a new way from such a satellite. Present ionospheric measurements are all made by studying the reflection of radio waves from the ionized upper atmosphere. A satellite would permit these measurements to be extended by studying the transmission properties of the ionosphere at various frequencies, angles of incidence, and times. Reflection measurements could also be made from the top of the ionosphere. Since we now know that disruption of the ionosphere accompanying auroral displays is caused by the impact [14] of a cloud of matter from space, the satellite could determine the nature, and maybe the source of that cloud.

Biologists and medical scientists would want to study life in the acceleration-free environment of the satellite. This is an important pre-requisite to space travel by man, and it may also lead to important new observations in lower forms of life.

The Satellite as a Communications Relay Station – Long-range radio communication, except at extremely low frequencies (of the order of a few kc/sec), is based entirely on the reflection of radio waves from the ionosphere. Since the properties of the earth's ionized layer vary profoundly with the time of day, the season, sunspot activity, and other factors, it is difficult to maintain reliable long-range communication by means of radio. A satellite offers the possibility of establishing a relay station above the earth, through which long-range communications can be maintained independent of any except geometrical factors.

The enormous bandwidths attainable at microwave frequencies enable a very large number of independent channels to be handled with simple equipment, and the only difficulty which the scheme appears to offer is that a low-altitude (300 mile) satellite would remain in the view of a single ground station only for about 2,100 miles of its orbit.

For communications purposes it would be desirable to operate the satellites at an altitude greater than 300 miles. If they could be at such an altitude (approximately 25,000 miles) that their rotational period was the same as that of the earth, not only would the "shadow" effect of the earth be greatly reduced, but also a given relay station could be associated with a given communication terminus on the earth, so that the communication system problem might be very greatly simplified.

[15] An idea of the potential commercial importance of this development may be gained from the fact that the ionosphere is now used as the equivalent of about $10,000,000,000. in long-lines, and is jammed to the limit with transmissions.

[16] **The Satellite as a Forerunner of Interplanetary Travel** – The most fascinating aspect of successfully launching a satellite would be the pulse quickening stimulation it would give to considerations of interplanetary travel. Whose imagination is not fired by the possibility of voyaging out beyond the limits of our earth, traveling to the Moon, to Venus and Mars? Such thoughts when put on paper now seem like idle fancy. But, a man-made satellite, circling our globe beyond the limits of the atmosphere is the first step. The other necessary steps would surely follow in rapid succession. Who would be so bold as to say that this might not come within our time?...

[211] 14. **Possibilities of a Man Carrying Vehicle**

Throughout the present design study of a satellite vehicle, it has been assumed that it would be used primarily as an uninhabited scientific laboratory. Later developments could alter its capabilities for use as an instrument of warfare.

However, it must be confessed that in the back of many minds of the men working on this study there lingered the hope that our impartial engineering analysis would bring forth a vehicle not unsuited to human transportation.

It was of course realized that 500 lbs. and 20 cubic feet were insufficient allotment for a man who was to spend many days in the vehicle. However, these values were sufficient to give assurance that livable accommodation could be provided on some future vehicle.

The first question to be considered in determining the possibility of building a man carrying vehicle is whether prohibitively high accelerations can be avoided during the ascent. The V-2 gave hope that this was possible. Our own studies have likewise shown that the optimal accelerations do not exceed about 6.5g. A man can withstand such acceleration for the periods of time involved (several minutes) if he is properly supported with his trunk lying normal to the directions of the acceleration. In Chapter 8, it will be remembered, the analysis showed that the performance could be improved a small amount by throttling each rocket motor during the letter portion of its burning period in order to reduce the structural loads. Under these conditions, the maximum accelerations could be profitably reduced to about 4 g. All these findings confirm [212] that ascent offers no insurmountable obstacle to the construction of an inhabited satellite vehicle.

Next we consider the safety and welfare of the man after the vehicle has been established on the orbit. Popular fiction writers have devoted considerable thought and ingenuity to means of furnishing him with air, food and water. The most ingenious of these solutions is that of the balanced vivarium in which plants and man completely supply each others needs. Leaving these problems to the inventors, we ask ourselves the engineering questions of whether we can provide livable temperatures and a reasonable protection against meteors. In Chapter 11 we have seen that the answers are tentatively in the affirmative.

Lastly we consider the problem of safely returning the vehicle's inhabitant to the surface of the earth. In Chapter 12, we have seen that, with reasonable area wings, we can control the descent sufficiently to avoid dangerously high temperatures. These same wings are adequate to accomplish the final landing on the earth's surface.

The above thoughts are far from final answers on this problem. However, they do give a note of assurance that the hope of an inhabited satellite is not futile....

Document II-3

Document title: J.E. Lipp, R.M. Salter, Jr., and R.S. Wehner, et.al., "The Utility of a Satellite Vehicle for Reconnaissance," The Rand Corporation, R-217. April 1951, pp. ix, 1-21, 28-39.

Source: National Security Archive, Washington, D.C.

After Rand had recommended advanced study into the uses of satellites for strategic reconnaissance in November 1950, the Air Force authorized Rand to undertake further research. The results of a Rand study on "The Utility of a Satellite Vehicle for Reconnaissance" were presented to the Air Force in April 1951. They demonstrated the viability of the concept and recommended further research. This recommendation eventually led to the much larger "Project Feed Back" [II-7] in 1954. These excerpts from the over-135-page report contain a general discussion in terms of orbits and instruments of the feasibility of satellite reconnaissance.

[ix]
Summary

Utility of an earth-circling space vehicle as a reconnaissance device is considered here in detail. A satellite (initially placed on its orbit by rocket power) which televises ground scenes and weather information to surface receiving stations is investigated. Particular attention is given to the television, communication, and electrical-power-supply problems, since these are the major determining factors in payload utility of a reconnaissance satellite. Some important corollary aspects namely attitude control and equipment reliability, are also discussed.

In order to round out the study, performance and weight estimates of the rocket vehicle required to carry a television payload are included.

The general conclusion of the report is that television satellites are feasible and that they would be useful if built and operated. Various essential lines of research in television, auxiliary power, and reliability are indicated....

[1]
Introduction

The basic feasibility of satellites from the point of view of rocket performance was considered in a previous group of RAND reports, Refs. 3 through 14. That investigation pointed to several important conclusions. First, the engineering of a rocket vehicle of adequate performance for use as a satellite would require but minor development beyond the then-existing technology. Secondly, the payload would have to be small (not more than 2000 lb) to keep the gross weight within reason; hence destructive payloads are not likely to be economically worthwhile for many years to come. Thirdly, returning the vehicle to earth intact would be difficult and should not be attempted in the early versions.

The above factors indicated that the payload would be restricted to instrumentation and communication equipment and prompted the RDB (Technical Evaluation Group) and the Air Force to request that further attention be given to the question of utility. RAND's effort since 1947 on the satellite study has been closely tied to the payload—its description and military usefulness. Most attention has been directed toward reconnaissance, since that is a field in which a satellite may very well show advantages over other types of vehicles.

It now appears fortunate that reconnaissance was selected for the first payload investigation. As will be seen later in the report, pioneer reconnaissance (general location and determination of appropriate targets) and weather reconnaissance are suitable with the resolving power presently available to a satellite television system. These two classes of reconnaissance have also been growing in importance to the Air Force because of the vastness of Russia and the difficulty of gaining information by conventional means.

To explore further the possibility of reconnaissance by means of a satellite, it is necessary to investigate the various constraints imposed in conducting such an observation from a remote, unattended vehicle.

The first step in such an analysis logically considers the movement of the satellite as a vehicle with respect to the targets to be viewed. Consideration must be given to the degrees of freedom at our disposal in the type and position of orbits and to the frequency of the satellite within the orbit. This approach, from a macroscopic standpoint, gives rise to information on how often and under what conditions the satellite can be placed over a given target. This is discussed in Section I, "Satellite Orbits and Ground Coverage."

Naturally following this step is the microscopic inquiry into the feasibility of viewing a target from the satellite. Television has been selected as the only practical way known at present for transmitting back to earth that which can be seen from the vehicle. Thus an evaluation of television-camera-equipment capabilities, along with a discussion of associated problems of transmission of the picture, is presented in Section II, "Reconnaissance by Television."

[2] Moreover, since there is an intimate interdependence between the type of reconnaissance desirable and the most fruitful way of obtaining such reconnaissance, some simultaneous consideration should be given to the presentation of satellite position (orbits), television scanning, and picture quality. This may be found at the conclusion of Section II.

The remaining problems can be classed as attendant ones peculiar to obtaining remote television broadcasts from the satellite. In order to scan the surface of the earth with the television camera, the vehicle must be properly oriented (attitudewise) with respect to the earth's surface. This is covered in Section III, "Orbital Attitude Measurement and Control."

The television and attitude control equipments require electrical energy that must be supplied by an auxiliary powerplant. A discussion of this powerplant, as well as the estimated power and weight requirements it must meet, is given in Section IV, "Auxiliary Powerplant."

Section V, "Reliability of the Satellite," includes an analysis of the anticipated reliability of the television and auxiliary equipment. This is a particularly important problem, since the equipment must operate automatically for a long period in an inaccessible location.

Finally, the characteristics of the vehicle itself necessary to place the television payload in a given orbit around the earth are presented in Section VI.

The several appendixes furnish correlative data and extensions of the remarks concerning some of the more salient features resulting from the study of the technical feasibility of utility of satellites for reconnaissance.

[3] **I. Satellite Orbits and Ground Coverage**

This section presents a general discussion of the pertinent facts about orbits which are essential to the utility of a satellite as a reconnaissance vehicle and of the problems concerning the establishment of a rocket-vehicle satellite on an approximately circular oblique orbit relative to the earth. Since the primary utility aspect considered is reconnaissance, the effect of orbits on scanning (i.e., viewing) angles, as well as some discussion of the limitations imposed by optical and radio transmission requirements, are included.

Orbits Generally

A satellite is defined as an attendant body revolving about a larger one; a moon and a man-made object revolving about the earth are thus satellites. The earth itself is a satellite of the sun. The shape of a satellite orbit, which can be either circular or elliptical, is dependent principally on the initial conditions of velocity, position, and direction of motion.

A circular orbit is of course the most desirable for an artificial satellite. Any marked deviations or eccentricity would cause some portion of the flight path to pass through more dense atmosphere and thus decrease the endurance of the satellite (for the likely range of orbital altitudes).

In order to remain on an orbit, the velocity of a satellite must be such that its centrifugal force is sufficient to overcome the earth's gravitational forces upon the satellite at the orbital altitude. Initial trajectory control is required to be such that the velocity is at least that necessary for a circular orbit[298] at the design altitude, and the path angle is within $1/2°$.[(7)] These limits are attainable with present control equipment.

RAND's previous studies were devoted primarily to equatorial orbits, which are still of prime interest for preliminary, experimental satellite flights. However, it is obvious that a reconnaissance satellite must be placed on an oblique orbit[299] to view targets of military interest most efficiently.

[4] **Review of Orbital Features**

Figure 1 illustrates the orbit of a satellite placed on a circular path. Such a path, *if unperturbed*, would maintain a fixed orientation in space (in this case, as in all others to be discussed here, the centers of the satellite's path reference frame coincide with that of the earth's). Thus the satellite reference frame would move around the sun with the earth but would not be affected by the earth's own rotation. Further, the position of the satellite orbit relative to the sunny side of the earth would change with the earth's seasons.[300] Figure 2 depicts this relative change of orbital position for a hypothetical satellite whose orbit is undisturbed by external influences.

Orbital Regression and Resultant Periods

As pointed out in Ref. 14, the orbit is affected by the presence of other astronomical bodies, such as the sun and the moon, and by the shape of the earth. The effect of the sun and the moon on a satellite orbit is nearly identical with their effect upon free-water surfaces of the earth (tides) and results in approximately a 3-ft orbital variation.

The oblate shape of the earth, however, exerts a much larger influence on the satellite rocket. The earth's polar diameter is about 25 mi less than its equatorial diameter. Although a polar orbit will have a vertical variation of approximately 1 mi (an orbit around the equator will have a negligible variation), this effect of the earth's shape is not of direct concern. The interesting and important effect is a corollary of the polar [5] perturbation, namely, a significant regression of the nodes[301] when the satellite orbit is oblique. This regression is similar to the precession of a gyroscope caused by externally applied torques.

Further, the regression period[302] of the satellite orbit will vary, depending on the orbital altitude and obliquity. After the method of Ref. 14, the orbital regression periods relative to the sunny side of the earth and to celestial space are plotted as functions of altitude and orbital angle in Fig. 3. For useful reconnaissance orbits, 45° to 60° obliquity and 350 to 500 mi altitude, the change in period relative to the earth is not great.

---

298. Velocities less than that required for a circular orbit obviously prevent the vehicle from establishing the prescribed orbit; hence the satellite will either fall to earth or assume an elliptical orbit which will cause marked altitude variations. Velocities greater than required yield less disastrous, but also undesirable, elliptical paths.

299. In this report an orbit will be designated by the degrees of an angle between it and the equator. An alternative but equivalent description is the maximum latitude to which the orbit is tangent. Thus a 0° orbit is equatorial, a 90° orbit is polar, and a 56° orbit is 56° oblique to the equator and tangent at 56° latitude.

300. An exception here is an orbit around the equator where this seasonal change is irrelevant.

301. Regression of the nodes may be visualized as a westerly rotation of the line of intersection (nodal line) between the satellite's orbital plane and the earth's equatorial plane (see Figs. 1, 4, and 5).

302. Regression period, as used here, is the time required for the intersection line (see footnote above) to make one complete revolution relative either to the sunny side of the earth or to celestial space, as applicable (see Fig. 4).

For illustrative purposes only, a 56° oblique orbit, approximately the latitude of Moscow, will be studied for most of the balance of this discussion. Figure 4 depicts the nodal regression for a vehicle on such a path. It may be seen in this illustration that the position of the orbit relative to the sunny side of the earth changes not with the earth's seasons, but much more rapidly; for this particular orbit, the period relative to the earth is 70 days rather than a year.

Under these conditions, the satellite can see a given target in the daytime only during alternate 35-day intervals regardless of whether the satellite circles the earth once a day or a thousand times; Fig. 5 amplifies this point. Thus a single satellite cannot [6] give a continuous record of daytime viewing of a particular target, but only during alternate 35-day periods. If continuous chronological daytime coverage is desired for longer periods, a minimum of two vehicles would be required. Further, if contrast requirements exclude twilight intervals, then three satellites operating on 8-hr shifts, with paths as shown in Fig. 6, are necessary.

Altitude, Velocity, and Duration

So far, discussion has been centered on the path of the satellite in its orbit. Its speed and altitude will now be considered. Figure 7 gives a plot of the required satellite velocity as a function of altitude above the earth for a nearly circular orbit. Since this velocity is independent of the earth's rotation, a satellite launched eastward gains by the component of the earth's peripheral speed in that direction. Figure 7 also shows the number of satellite revolutions per day as affected by orbital altitude.

The duration of an orbiting vehicle depends on the amount of atmosphere tending to slow it down. This in turn means that the higher the altitude, the longer the satellite [7] can stay up.

Figure 8, taken from Ref. 3, gives anticipated duration as a function of altitude. At a 100-mi altitude the vehicle will be pulled to earth in less than one revolution because of the atmospheric drag. At 350 mi the duration is about 2 years. At 500 mi the satellite will stay up around 50 years; at 600 mi, several centuries. From this standpoint alone, it is desirable to use as high an altitude as possible. Also, the range of line-of-sight[303] radio transmission increases with altitude. Counterbalancing these factors is the greater size of the satellite required to put a given payload on an orbit at higher altitudes (e.g., 10 to 20 per cent higher gross weight is required to increase altitude from 350 to 500 mi; see Fig. 40, page 77). Another deterrent factor is the increased size and weight of camera equipment necessary to scan the earth from higher altitudes, which requires higher resolving power for an equivalent picture. Therefore, the desirable altitude will represent a compromise between these opposing features but will probably lie between 350 and 500 mi. For purposes of consistency, a 350-mi altitude will be used in the remainder of this report, except where altitude is considered as a variable.

Effect of Orbital Altitude on Ground Coverage and Related Problems

At orbital altitudes of 350 to 500 mi, the satellite circles the earth fifteen to fourteen times a day (see Fig. 7). The satellite tracks cross the equator at intervals of 24° to [8] 25° longitude or, roughly, there are 1700 mi (measured east-west at the equator) between tracks for the 350-mi altitude.

At 56° latitude, for example, this interval is about 800 mi; near the tangent latitude the tracks recross each other several times. Figure 9 indicates the tracks for a satellite at an orbital altitude of 350 mi and at an orbital angle of 56°. Also shown is the average daytime coverage during the daylight "season" with a 400-mi optical scan to either side of the satel-

---

303. Only line-of-sight transmission can be used because high-frequency waves are necessary for television equipment. Also, long radio wavelengths will be adversely affected by the ionosphere; for instance, reflection by the Heaviside layer will prevent such wavelengths from reaching the earth rather than to increase their range.

lite (800-mi optical-scanning band); the light-green area shows targets covered once a day; medium green, those covered twice; and dark green, those covered three or more times. White areas (below the tangent latitude) are those viewed less than once a day; as indicated in the figure, for the assumed satellite orbit, coverage in any one day is not complete below 30° N. latitude.

The 800-mi optical-scanning band at 350 mi altitude represents approximately a 94° included scanning angle, i.e., a 47° scan to either side of the vertical. The included angle of the horizon is 135°, but the value of pictures taken beyond 45° on either side of vertical is questionable. This point is shown schematically in Fig. 10, which also gives a plot of horizon angle as a function of altitude. A discussion of the effects of scanning angle, as well as those of the orbital inclination, upon the minimum resolvable surface dimension is presented in Appendix I.

Proper initial selection of the orbital altitude would enable the satellite to make an integral number of revolutions for one revolution of the earth relative to the orbital plane (not necessarily per 24-hr day, since the orbital plane regresses[304]). Integral [9] numbers of satellite revolutions every other (24.3-hr) day, every third day, etc., are also possible. Such orbital conditions, however, cannot be made accurately enough with present control equipment to afford the same trace on the earth's surface day after day. Thus a drift can be expected so that the satellite will come within a few miles of its track on the previous day (or the previous alternate day, etc.). The significant fact is that by adjusting an orbital period so that it is nearly integral on alternate days, one can obtain, the following day, a picture in the center of the camera scan of a target which was on the periphery the day before (see Fig. 9), except, of course, near the tangent latitude, in which region still greater amounts of overlap are obtained.

As mentioned earlier, one factor indicating the desirability of a 500-mi altitude is the need to receive the satellite's television broadcasts by stations sited either in friendly territories or on ships. Figure 11 shows the area of reconnaissance interest which would be covered by transmission ranges of 1396 and 1743 mi with 5 stations and 2000 mi [10] with 4 stations (see Fig. 22, page 30, for range as a function of altitude and elevation angle). Transmission must be "line-of-sight" because of the required radiation frequencies. It is estimated that the maximum range for acceptable transmission[305] from a 350-mi altitude is about 1400 mi. At this range, 5 stations would be required to pick up Asiatic observations, but about 15 per cent of the USSR, a significant portion near 105°E longitude, would be left out. Increasing the satellite's altitude to 500 mi affords (on the same basis) a range of approximately 1750 mi. With this range and the same 5 stations, the unobserved area is reduced to a small amount.

With a 2000-mi range (not shown), the unobserved area would be eliminated. However, by accepting a small unobserved area near 95°E longitude, 4 sea-borne stations could be employed. At this latter range, the orbital altitude required for equivalent clarity of the transmission exceeds 600 mi (see Fig. 22, page 30); it may be possible to attain a 2000-mi range from a 500-mi altitude, although some uncertainty and signal distortion would occur in the 100- to 250-mi extremity.

The possibility of eliminating so-called unobservable areas by using delayed broadcasting becomes apparent. It is well to note, however, that the number of frames to be filed would cause the transmitting device to be so bulky and complex that this method does not appear to warrant further investigation at the present time.

[11] The effect of different altitudes upon target viewing, as well as upon television camera resolution and contrast, is discussed further in the next section.

---

304. The period for a 350-mi altitude, 56° orbit is 24.31 hr, which is termed a day throughout the remainder of this discussion.

305. It is assumed that a minimum elevation angle (above the horizon) of 5° be employed for completely acceptable signal reception.

Summary

To summarize briefly, the orbiting characteristics are critically dependent on the altitude; a substantially circular orbit is most desirable. Although equatorial orbits are desirable for test purposes, oblique orbits are necessary for meaningful reconnaissance. For example, a 350-mi altitude, 56° orbit, in combination with an 87-day regression period of the orbital plane relative to celestial space and to the seasonal motion of the earth around the sun, will afford daylight views of a specific target during alternate approximately 35-day intervals (one-half the 70-day regression period relative to the earth). Completely target-system coverage, from the eastern to western limits of Russian-controlled territory, will reduce the unproductive interval by about one-half.

[12] ### II. Reconnaissance by Television

In the first section, the macroscopic aspects of satellite reconnaissance have been discussed, namely, the placement of the vehicle in appropriate orbits for bringing targets of military significance under scrutiny. The means of viewing and transmitting these scenes to ground stations will now be weighed. At the present time, it is felt desirable to consider only remote transmission of picture information by high-frequency radio waves. Other possible alternatives, such as using a conventional aerial photographic camera and returning the satellite to earth on command, appear to involve difficulties that would make early versions of the satellite impractical.

Two systems, television and photographic facsimile transmission, are available for consideration for photographing and sending on reconnaissance data. The latter system uses a camera film to record temporarily scene information; this film is then scanned electronically and the impulses transmitted as in the standard "wirephoto" system. A reusable film must be employed because, otherwise, roughly $3/4$ ton of camera film would be required per month's operation. Since we know of no re-usable film (or other less bulky storage strip) under development, the photographic facsimile system will be ruled out for the present; future requirements, such as those for delayed picture transmission, may cause reconsideration of this system.

The use of television emerges, then, as of prime import in viewing and sending to ground stations reconnaissance information for recording and for evaluation. The ability of such a system to accommodate reconnaissance requisites will be considered in detail, both for viewing weather and for observing ground targets. Each of these latter types of reconnaissance has its own peculiar needs, which will be discussed first in this section.

The effect of reconnaissance requirements on camera equipment is considered next. It will be demonstrated that daytime viewing is possible, but nighttime light levels are too low for practical televising. The discussion of daytime viewing is then expanded to include specific numbers of the minimum resolvable ground dimensions as functions of scene contrast, frame speed, and the number of lines per inch resolution of the camera. A correlation between the ground area to be covered and the frame speed, and the need for an optical-scanning system, are determined. The above investigation is of a general nature and would apply to any "camera," whether it uses film, is a television tube, or is the human eye.

Logically following the above discussion, the television camera tubes are examined in relation to the foregoing optical parameters. The commercial Image Orthicon and the Vidicon tubes are shown to be within the realm of possibility for satellite viewing.

A discussion is then presented of the television camera system in context with the reconnaissance requirements and of the various combinations of characteristics that could be employed to produce an over-all optical-scanning system for use in both weather [13] and terrestrial reconnaissance. Also included are actual photographs of a simulated ground scene by a commercial Image Orthicon camera. It is shown that even by present commercial television standards, useful scene information can be obtained.

The transmission of the televised scenes, the necessary television mechanisms, the effects of signal wavelength, the position of the satellite relative to ground stations, and the possibility of enemy interception and jamming are included in the next subsection. Following this is an analysis of the reception and presentation of the televised signal as would be done by the ground monitoring stations.

Weight estimates and power requirements for the satellite television camera-transmitter system are then presented. For a more complete analysis of the television system's design considerations, see Appendix II.

Reconnaissance Requirements

To obtain any useful information from altitudes of 350 to 500 mi appears at the outset to be an extremely difficult operation. It is the purpose here to examine the constraints imposed on the television system in conducting reconnaissance of a worthwhile nature. Two types of observation will be considered: weather and terrestrial.

Weather Reconnaissance

Reference 1 offers a far more complete analysis of the requirements for weather reconnaissance than can be given here. However, in this report it is desirable to discuss briefly these requisites for the purpose of continuity. Information in Ref. 1 reveals that details of cloud structure as small as several hundred feet in dimension may possess meteorological significance. For weather observations, resolutions as poor as 500 to 1000 ft can be utilized, although a better minimum resolvable dimension would be 200 ft. This latter resolution is ample to determine a major portion of the characteristics necessary to predict weather. At this resolution, orientation and structure of clouds, direction of winds, and presence of fronts can be seen.

To explore deeply into the problem, the prevailing contrasts of weather scenes must be examined. For weather reconnaissance, the contrast is a function of the albedos[306] of various types of clouds and of the background. An albedo of 0.8, commonly given for average cloud formations (see Fig. 49, page 94), is used with the albedos of the various surface backgrounds to determine the degree of contrast available. Figure 12 shows graphically that contrasts of 50 per cent or more are produced by virtually all ground-surface background conditions except that of fresh snow; also shown in Fig. 12 is a similar graph for smooth sea surfaces and various solar elevations.

An additional feature of weather reconnaissance is the need to encompass the entire area in question with a daily observational coverage.

At the risk of being premature in describing the television system, a few illustrative remarks will be made here. An optical-scanning system viewing a band on the earth's surface of 800 mi width and taking frames, or pictures, at the rate of ten per second [14] will be assumed. A standard Image Orthicon television camera with appropriate optics at the 10 frame/sec speed and for the pertinent contrasts prevailing in weather scenes will resolve a dimension of 200 ft. Therefore the conclusion is that such a system could be completely adequate and useful for meteorological observational purposes.

Terrestrial Reconnaissance

The requirements for viewing targets—of military significance—on the ground are now considered. Taking the cue from the above discussion, it is apparent that a 200-ft resolution can be easily attained at prevailing weather contrast levels (which are nearly all at contrasts above 20 per cent) and that a complete daily area coverage can be expected with this system. However, that contrasts of less than 20 per cent do exist on military targets and that a 200-ft resolution will not be completely adequate will be discussed subsequently.

---

306. Albedo is the ratio of the amount of light reflected from a prefectly diffuse surface to the total light falling upon it.

Figure 13 depicts contrasts that may be expected from various military targets against representative backgrounds. Year-round observation from the satellite will yield a number of pictures of a given target during the different seasons, possibly with informative [15] results. For example, an asphalt airstrip and its adjacent ground cover may have no albedo difference, hence no contrast, during the spring-to-fall period; during early winter, however, a thin layer of snow on the adjacent ground cover, but melted or removed from the airstrip, may result in contrasts as high as 85 percent. Furthermore, continued observations throughout seasonal ground-cover variations will tend to reduce the effectiveness of camouflage. It is readily apparent that the conditions for taking such pictures are dependent on the number of clear days during the period the satellite passes over a specified military target. However, if a continuous chronological record is broadcast from the satellite for a year, it is reasonable to expect that each target will be seen at some time on a clear day.

The second criterion for terrestrial reconnaissance is, of course, the allowable minimum resolvable surface dimension. The ultimate choice of the figure for this dimension will remain with intelligence personnel skilled at interpreting information. The 200-ft resolution is probably adequate for ferreting out major airfields and for noting the presence of large highway or railroad right-of-ways (even though lateral dimensions may be [16] considerably less than 200 ft). Large factory buildings will be seen, although their exact shape may be indeterminable. Square buildings of 200 ft on a side will tend to be confused with round fuel-storage tanks of similar size.

A 50-ft resolvable dimension will afford considerable improvement in detailed information. The structure of urban areas can be determined. Large aircraft can be identified, as can gun emplacements, revetments, etc.

Assessment of bomb damage will probably require even better resolving power (perhaps as low as 10 ft) and may well be beyond the scope of the satellite system.

From the above discussion, it is seen that the previously assumed camera and scanning system is, on the basis of minimum resolvable surface dimension, useful and adequate for pioneer terrestrial reconnaissance. However, such a system is inadequate for reconnaissance concerned with detailed target identification. The ways in which detailed reconnaissance can be achieved are discussed later in this section, but it can be stated briefly that either a fundamental improvement in the television camera tube or a reduction of the observable area on the ground—so that complete coverage is not made every day but every 10 days or so—must be made.

Summary of Reconnaissance Requirements for the Television System

It has been shown that minimum resolvable dimensions of 50 to 500 ft are acceptable, depending on the type of observations made. Thus the television camera must be capable of resolving dimensions on the ground—or near the ground for weather—of the same order of magnitude measured in feet as $1/10$ to 1 times that of the optical range measured in miles. This resolving power, $0.001°$ to $0.01°$, implies a small angle of view, as will be demonstrated later.

Optical scanning over a reasonably wide swath on the earth's surface will require (in conjunction with the small field of view) a large number of frames in a given time interval—of the order of 10 to 30 frames/sec.

Contrast levels of 20 per cent or higher will normally be needed.

Nighttime and Color Television

So far, discussion has been predicated on the conditions that would prevail in taking black-and-white pictures in the daytime. For black-and-white shots at night, the same scene contrasts would be expected, *but the overall scene brightness would be considerably reduced.*

The use of television for transmitting scenes viewed by a satellite at night, while physically possible (see Appendix II), is considered impractical. A camera system sufficiently flexible to accommodate both daytime and nighttime viewing not only would be complex, but also would require an f/0.6 optical system, which in turn would require a 30-in. aperture for a 20-in. focal length (as compared with a 2-in. aperture for daytime viewing). The total size and weight of such a device as presently conceived would be prohibitive.

Although color television has lower resolution than black-and-white video, the use of color television might result in more effective photo interpretation. For example, a [17] black airstrip surrounded by green grass can readily be identified in color even though its black-and-white contrast may be zero. It is doubtful, however, that color television could counteract camouflage because the TV camera does not see more of the infrared spectrum than does the human eye. Size, weight, and complexity of such a system do not warrant its further investigation at this time.

Hence the remainder of the section will be devoted to black-and-white television, with viewing done only in daylight.

Optical System

Resolving Power and Contrast

Resolvable detail in photographs made by satellite television (or any other type of camera) is dependent on brightness, scene contrast, exposure time, and geometrical factors. It is also a function of the inherent resolution of the camera itself. A television camera is characterized by the number of television lines per inch (equal to roughly twice the number of optical lines per inch) and this parameter is an index of the tube's resolution.[307]

In Table 12, on page 102, may be found an enumeration of the minimum surface dimensions, δ, resolvable by day for various contrasts, TV lines per inch camera resolution, and frame frequencies and for the various required optical parameters of focal length and aperture size. Along with δ, the relative power, P, required for picture transmission is also listed.

A digest of Table 12 is given in Table 1, below, which shows what can be accomplished with an f/10, 2-in. camera aperture, 20-in. focal length camera, operating at a frequency of 10 frames/sec.

[18] It is expected that the Image Orthicon camera tube (see page 20) will give resolutions of the order of 1000 TV lines/in., which means that with the above optical system, a 200-ft minimum resolvable surface dimension can be anticipated for contrasts as low as 20 per cent.

By changing the camera optics to restrict the field of view and by increasing frame frequency, it may be noted from Table 12, page 102, that considerable improvement in ∂ can be wrought. Values as low as $\delta = 40$ ft (at 25 per cent contrast) are obtained with the same TV tube resolution of 1000 TV lines/in.

The Optical-scanning System

The f/10, 20-in. focal length optical system will view, in a single frame, a ground-projected square area of 17.5 mi on a side directly under the satellite. (Figure 15, page 22, shows an equivalent southwest sector of Los Angeles which would be taken by one frame.) A 47° viewing angle will cover a ground-surface width of 800 mi from an altitude of 350 mi. Because of the curvature and obliquity of the surface of the earth, as shown in Fig. 10, page 9, the ground area seen by the optical system at 47° (the angle measured from the vertical) is nearly doubled, the transverse ground dimension being about 35 mi. Consequently, 39 to 40 frames are needed to view the 800-mi band in one transverse sweep.

During this same time, the satellite is moving forward 17.5 mi at a speed of approximately 5 mi/sec, which allows about 3.5 sec/transverse sweep and, for 39 to 40 frames,

---

307. Resolution by a photographic camera is commonly defined by the minimum spacing of lines that can just be discerned in a photograph by the camera. In television the index is based on the distance from one of the lines to the center of the intervening space between the lines (this distance being called one television "line"). Thus one optical line is equivalent, approximately, to two television lines. It should be noted that a single index of this type is inadequate to describe fully the quality of a camera, and it is assumed in this report that the television cameras have good characteristics with respect to sensitivity as a function of the various sizes of the objects viewed.

checks generally with the previously assumed 10 frames/sec.

The motion of the optical scan must be such that the area under observation is "stopped" relative to the photocathode, which requires indexing between successive frames in the transverse sweep, and the fore-and-aft motion to compensate for the satellite's forward motion.

Transverse and longitudinal camera positions relative to the satellite structure are shown in Fig. 14 as functions of time per frame. Also shown is a proposed scanning system. It may be possible to synthesize this complex motion by an appropriately designed, continuously rotating prism (not shown).

Satellite attitude control of yaw, pitch, and roll, relative to "stopping" the picture, is discussed in Section III. Further discussions of the scanning angle, of the orbital inclination, of the frame frequency, and of the resolvable surface dimensions are presented in Appendixes I and II.

The Television Camera

The task of televising a ground scene from a satellite differs from the ordinary video pickup problem in three principal ways: (1) as just indicated, a high-resolution, scanning, optical system is required, (2) the equipment must operate over a relatively [19] long period of time from a remote, unattended station, and (3) each frame is a completely different picture. This latter subject is discussed further under "Reliability of the Satellite," Section V, page 63. It is probable that presently available television-tube resolutions are adequate for preliminary reconnaissance of either weather or terrain; however, it is anticipated that the normal trend in television research will yield higher resolutions by the time a satellite requires such a system.

[20] ## Limiting Resolution of Pickup Tubes

The basic elements of modern television camera tubes are (1) a photosensitive target, upon which the viewed scene is projected and reproduced as a pattern of static electric charges, and (2) an electron beam which scans the charge pattern on the target, reading and erasing it and transforming it into a time-varying electrical signal. The scanning beam is usually made to cover the target in a series of horizontal lines or in two interlaced series of lines, and the beam moves at such speed that the entire picture, or frame, is scanned in a small fraction of a second. Present commercial television practice employs 525 scanning lines/frames at a rate of 30 frames/sec, and the pickup-tube resolution is therefore limited to 525 TV lines/frame (or slightly more than 250 optical lines). A more fundamental limitation on the resolution of a pickup tube than the number of scanning lines is the finite size of the cross section of the scanning beam or the finite size of the elements composing the target, whichever is larger. It is of interest to note that the scanning-beam sizes in electron microscopes is an order of magnitude smaller (10^{-3} min spot size).

The resolution of the best available photoemissive pickup tubes (Image Orthicons) is limited by target structure. This tube uses a thin two-sided target, upon one side of which the charge pattern representing the scene televised is deposited by secondary emission. Photoelectrons from the primary cathode, or photo-cathode, of the tube focus upon the target and impinge upon it under conditions which result in a high secondary emission ratio; each incident photoelectron ejects several secondary electrons from the target face, the charge pattern on the target being correspondingly more intense than that on the photocathode. These secondary electrons are collected by a grid of very fine wire mounted close to the target on the photocathode side. The grid effectively breaks up the otherwise continuous target surface into a mosaic of elements of size corresponding to its mesh spacing. In commercial Image Orthicons, the grid contains slightly more than 500 mesh spacings/linear in., and the limiting resolution is therefore about 500 optical lines/in., or 1000 TV lines, of target surface. Experimental Image Orthicons have been made with fine grid meshes, corresponding to limiting resolutions better than 1500 TV lines/in.

Resolving power of present photoconductive pickup tubes (Vidicons) is limited by the cross-sectional size of the scanning beam. The photoconductive process is inherently more sensitive than photoemission and no preliminary amplification of the target charge pattern by secondary emission is required; no collecting grid is involved and the Vidicon target is essentially continuous. The smallest resolvable target element is therefore determined approximately by the half-power width of the scanning beam (at the target). In a recently developed Vidicon, this beamwidth is about 0.00125 inclusive, corresponding to a limiting tube resolution of about 1600 TV lines to an *inch*. However, this does not mean that the present Vidicons have a higher resolution than do Image Orthicons. On the contrary, the present target size of the Vidicon is considerably less than 1 in., and the number of TV lines to a *frame* is less than in the Image Orthicon. Major difficulty would be experienced in attempting to increase the Vidicon target sizes to about an inch (as in the Image Orthicon) because of the increasing electrical capacitance of [21] the target. Nevertheless, this does not preclude the possibility of using several Vidicons to replace one Image Orthicon, with the attendant reduction of reliability.

The possibility of a really significant improvement in the limiting resolution of an Image-Orthicon-type pickup tube is regarded as remote, because of the great difficulties inherent in constructing and mounting collecting screens composed of conductors much smaller than about one-thousandth of an inch in diameter. Significant improvement seems more likely in the case of photoconductive tubes, since much narrower scanning beams are theoretically possible by improvement of the optical design of the electron gun and of the focusing and scanning fields. But a limit will soon be reached at which further reduction in beam spot size results in no further improvement in resolution and at which resolution will be limited by the finite conductivity of the target. This follows from the fact that the thickness and conductivity of the target must be such as to allow for dissipation of the charge pattern by conduction through the target in a period not much greater than the frame time; if the conductivity is such that it permits this desired charge motion, it will also (assuming isotropic target material) allow charges to move laterally over the target face so that even an initial point charge will be spread over a circle of diffusion, the diameter of which will ultimately determine the limiting resolution regardless of the spot size of the scanning beam.

It appears probable therefore that 1500 TV line/in. is a reasonable maximum value for the limiting resolution of pickup tubes for some time to come, and that a practical value for unattended operation of present tubes in a satellite vehicle might well be considerably less than this, say about 1000 TV lines/in....

[28] **Transmission of the Television Pictures**

On the basis that the satellite's television camera system can collect valuable information, it is necessary to transmit the pictures from the satellite to surface receiving stations and to record and portray the pictures in useful form. The range at which satellite signals can be received by ground stations will be discussed first, since this [29] range affects the disposition of the stations and ultimately determines the completeness of coverage of enemy territory by direct broadcast to receiving points in friendly territory. Possible locations of receiving stations was discussed in Section I.

Next, consideration will be given to the tracking system and to the effects of switching the television broadcast reception from one station to the succeeding one. The following subsection is devoted to the antenna gain and to the power required, since these are intimately related to the system employed to track the satellite. Logically following this there is a discussion of the proper choice of wavelength. Finally, the possibilities of interception and jamming of the television signal will be considered.

Before continuing further, however, it is felt desirable at this point to outline the over-all television system. Figure 21 is a block diagram of the proposed system. Component parts will be (and have been) described as they appear in the discussion.

Range of Transmission

There are quite good reasons for not attempting to track or communicate with the satellite from surface stations when its angular elevation above the horizon is less than [30] 5°.[10] Consequently, this value of the elevation angle has been considered as determining the maximum range over which completely acceptable television transmission should be required. Because of geographic limitations, however, it may be both desirable and necessary to transmit at ranges greater than those indicated by the 5° limitation. An immediately apparent expedient is to consider the use of some portion of the additional range potentially available by transmitting and receiving when the satellite is at an angular elevation of less than 5°. Figure 22 shows maximum radio ranges as functions of altitude at angular elevations of 0°, 2°, and 5°.

[31] Reliable transmission of radio waves several centimeters long can be expected at a 5°-elevation angle; at angles of 1° or 2°, some dispersion will be prevalent.[10] The principal effect will be loss in resolution, but even this may be desirable in place of no picture at all.

The Tracking System

It is evident that if power requirements are considered, it is necessary for the television transmitter to have a directional antenna which can be oriented toward the receiving station. On the basis of orbital computations, a receiving station will know the approximate location of the satellite at any given time.

A station with an appropriately sized receiving antenna will be able to track a 350-mi-altitude satellite for about 3000 mi. The vehicle traverses the distance in approximately 11 min at an average angular tracking rate of 15°/min (the rate is faster at the zenith, being of the order of 34°/min, or 0.6°/sec), this implies that the tracking system must be carefully keyed in with the satellite's system and, further (within the limits of reasonable satellite power consumption), that the ground station's antenna should be as small as possible. The diameters of the satellite's antenna and of the ground station's receiving antenna are assumed to be 1 ft and 16 ft, respectively. Thus the size of the ground station's antenna would be small enough to be amenable to reasonable engineering in mounting, etc.

It is proposed that, for reasons of stability as well as of reliability, the satellite's television camera and transmitter system be turned on, warmed up, and adjusted at the start of the flight, and that it be left on continuously thereafter. The satellite will therefore always be ready to televise on demand of the appropriate ground station.

Of the many possible methods by which antenna-tracking could be accomplished, the optimum would be that which minimizes the complexity, weight, and power requirements of the space-borne equipment. On this basis, the most attractive system yet considered is one in which a tracking receiver in the satellite operates on the continuous-wave signal of a ground beacon to direct the satellite antenna toward the ground station, and— once the space-borne tracking is accomplished—the ground station's receiving antenna is directed to follow the satellite by means of an auxiliary tracking receiver operating on the television signal. The space-borne tracker would operate on a microwave frequency different from and considerably lower than that used for television transmission, but would work through the satellites's television transmitting antenna. The ground beacon would work into a directional antenna separate from that used for television reception, but would be slaved to the latter so as to follow the satellite when the ground tracker takes over. The general nature of the operations of the tracking system is described in the following paragraphs.

The 1-ft-diameter satellite antenna is mounted in gimbals in such a manner that it is free to rotate about a vertical axis and so that the antenna axis can assume any angle with the downward vertical up to a maximum of about 60° (the direction of a ray from the satellite in a 350-mi orbit to a ground station at which it subtends to a minimum angle of 11°; see Fig. 23). The dish-shaped antenna is provided with two feeds: one is fixed on the axis for television transmission at about 10,000 Mc; and the other [32] is offset from the axis and rotates about the axis so that a conical scan is provided for the tracking receiver at some suitable lower frequency, say 3000 Mc. In the search phase, the axis of the dish is maintained at 60° from the downward vertical, the antenna assembly rotates about the vertical axis at a rate of the order of 3 rps, and the nutating feed simultaneously executes a conical scan at a rate of the order of 30 rps. When the ground station appears above the horizon with respect to the satellite, the signals from the ground station's beacon will be received by the conically scanning tracking receiver, which will then operate to disengage the slow search rotation and to maintain the antenna axis in the direction of the beacon, using the conventional servo technique. When the satellite reaches the opposite side of its transit with respect to the ground station, the ground beacon shuts off and the tracker ceases to operate. The satellite's antenna is then rotated about its axis once at the same angle with the vertical as was used in transmitting to the last receiving station. If the next station tracker is not engaged, then the antenna is returned to the primary angle of 60° by stages (probably two revolutions).

This procedure is illustrated in Fig. 24. In the extreme case (provided the next station is not over the horizon), a loss of less than 1 sec in reception may be anticipated. Usually this interval will be about $1/6$ sec and will not cause difficulty in continuity of picture-area coverage since successive disengagements from one station to the next (on alternate days, for example) can be made at different points in the orbit. The 3000-Mc tracking frequency was chosen, since an included angle ($\Delta\phi$) of $27 \, 1/2°$ will illuminate the satellite from a wide range of succeeding ground stations.

Because of the high, and continuously varying, radial velocity of the satellite with respect to the ground station, the 3000-Mc signal of the ground beacon may suffer a Doppler shift of as much as ± 150 kc when it is received at the satellite. The satellite's tracking receiver must therefore have an effective bandwidth in excess of 300 kc if the complexities of automatic-frequency search-and-control circuitry are to be avoided. This wide bandwidth implies that a directional beacon antenna of rather high gain [33] will be necessary if the beacon output power is to be reasonable. Fortunately, it is expected that the establishment of the orbit of the satellite can be made sufficiently precise so that the azimuth angle at which the satellite will appear above the horizon with respect to a given ground station, on a given orbital revolution, may be predicted to within one or two degrees. Hence a beacon antenna having a power gain of about 1000 will yield a broad enough beam to illuminate the satellite when it appears above the horizon.

The design of the ground station's tracker is virtually unrestricted by considerations of circuit complexity and power consumption and could take any of several forms. It could, for example, include a conically scanning tracking receiver similar to that used in the satellite. Such a tracker could use the television transmitter in the satellite as a beacon. The ground station's 16-ft-diameter receiving antenna would have a single feed connected through a power divider to two receivers: one of about 3-Mc bandwidth for television reception and the other of about 400-kc bandwidth for tracking. Both receivers would operate on a television frequency of 10,000 Mc. The feed would be offset from the dish so that a conical scan at a rate of the order of 30 cps would be provided. The search phase would consist in aiming the axis of the dish in the direction of the satellite's scheduled appearance and in oscillating the conically scanning feed back and forth through the axis of the dish in such a manner that the axis of the scan [34] would describe an arc, parallel to and above the horizon, centered on the direction of the satellite. This oscillatory search scan would occur at a rate much slower than the conical scan, say of the order of 3 cps. When the satellite appears, the satellite's tracker first will contact the ground station's beacon, thus aligning the satellite's transmitting antenna with the ground station. The oscillatory search motion of the ground station would then be stopped and, with the axis of the conical scan fixed with respect to the antenna's axis, the entire antenna assembly would be driven by the usual servo system to follow the satellite.

Antenna Gain and Power Required

Appendix II develops the relation of antenna sizes (d for the satellite and D for the ground station), transmitted power, P, transmission wavelength, λ, and all the other factors constraining the signal transmission (signal-to-noise ratio, range, etc.). If these latter factors are considered as constant, K, then the following relation may be stated:

$$P = K \frac{\lambda^2}{d^2 D^2} = \frac{125 \lambda^2}{d^2 D^2}, \qquad (1)$$

where P is in watts, λ is in centimeters, and d and D are in feet.

The power supplied to the transmitter, E, is not directly proportional to the output power, P, although it is desirable to reduce E to an absolute minimum, not much can be gained by reducing P below 4 watts. Using 4.4 watts for P and an assumed wave-length of 3 cm ($v = 10{,}000$ Mc) yields $d^2 D^2 = 256$. Choosing $d = 1$ ft yields $D = 16$ ft. These figures are purely arbitrary; they are based on engineering judgment and may be considerably different from those used in the ultimate system. Wavelength, λ, is discussed later.

It is assumed in the above formula that the two antennas are highly directional, with half-power beamwidths of 0.80° and 13° for the ground and the satellite antennas, respectively. Should less directional antennas be employed, considerably greater transmitter power would be required.

Since the antenna beamwidths are small compared with the total of the solid angles over which communication will be required, means must be provided for aligning the axes of the two antennas shortly after the satellite appears above the horizon at a given ground station and for maintaining that alignment as the satellite passes by in its orbit. This has been described previously.

Choice of Transmission Wavelength

While the optimum frequency for television transmission between the satellite and the surface receiving stations will undoubtedly lie in the centimeter wavelength band, its precise value will be determined as a compromise of many factors. One prime consideration, however, is that of minimizing the required power output of the satellite transmitter, which, other things being equal, may be accomplished by maximizing the product of the satellite's transmitting antenna gain and the transmission efficiency of the circuit. A steerable aperture antenna (such as a conventional paraboloid) is required [35] for transmission from a satellite in an oblique orbit, and its gain, for a given aperture area, will be inversely proportional to the square of the transmission wavelength, so that, from this point of view, the frequency should be as high as possible. The transmission efficiency at very high frequencies will be largely determined by atmospheric absorption (water vapor and oxygen) and by losses caused by scattering due to condensed cloud and rain droplets, the total atmospheric losses increasing rapidly with the decrease in wavelength in the high microwave region. (See, for example, Fig. 50, page 109.) The optimum wavelength will depend on the maximum antenna size, on the minimum satellite elevation angle at which transmission is required, and on the least favorable meteorological condition, which is likely to be encountered at the ground station. For example, with a 1-ft-diameter transmitting antenna on the satellite, the optimum frequency for transmission at a minimum elevation angle of 5° to a ground station located in a region in which moderate rain is falling at the rate of 15mm/hr may be shown to be about 15,000 Mc (2-cm wavelength); the optimum wavelength for transmission under the same conditions, but through a tropical downpour, would be about 5000 Mc (6-cm wavelength). Many other considerations enter into the choice of frequency, among which are system losses, ground-receiving, antenna-tracking accuracy, efficiency and reliability of transmitting tubes, etc., the ultimate optimum probably being greater than 5000 Mc and less than 15,000 Mc (10,000 Mc has, of course, been employed in this study).

Transmission of the picture from the satellite to a surface receiving station, as well as the method of presentation or assembly of individual scenes into a meaningful whole, presents problems regarding deterioration of the clarity of the televised picture, which are discussed more fully in subsequent parts of this section. Transmission should be done with a large enough frequency bandwidth so that this part of the over-all system is equivalent to a considerably higher resolution than that component limiting the resolving power, namely, the TV tube.

Enemy Interception and Jamming

Detection and Tracking by Radar. The microwave-radar cross section of the satellite is estimated as averaging less than about 1 m^2. Detection of so small a target in rapid motion and at slant ranges, which vary from a maximum of about 1700 mi on the horizon to a minimum of 350 mi at the zenith, can be shown to be well beyond the capabilities of the most powerful American radars, either now existent, under development, or being proposed.

It is conceivable that the satellite might be detected and tracked by a radar designed expressly for the purpose, one that employs narrow-band techniques at low vhf frequencies at which relatively high average power is available and at which the satellite might behave as a resonant scatterer of a much higher radar cross section. But the frequencies in question (20 to 40 Mc) are subject to severe ionospheric attenuation and refraction effects. The antenna of such a radar would be enormous, with an aperture area measured in thousands of square meters. The difficulties involved in searching for and following a rapidly moving object with such equipment are obvious. Further, [36] even if the system could be made to work, its accuracy and information rate would probably be too low to be useful.

Detection and Tracking by Passive Techniques. The power density of the television signal transmitted from the satellite will be from 10^{-15} to 10^{-18} watts/m^2 at points on the earth's surface illuminated by the main beam of the satellite's transmitting antenna and will be of the order of 10^{-15} to 10^{-18} watts/m^2 at surface points outside the beam.

There is little doubt that an enemy equipped with suitable interceptor receivers *could* detect and track the satellite by means of its television signal and from a site *sufficiently close* to a friendly ground station's receiving station (within about 40 to 200 mi) to be illuminated by the main beam of the satellite's transmitting antenna. The equipment required would be relatively conventional, based on any of a variety of direction-finders and passive radar techniques. The tracking would be in direction only, with crude range information supplied by triangulation from the data obtained at two or more sites. As a primary difficulty would lie in the first acquisition of the satellite's signal, the enemy, unaided by intelligence information, would have to search through a wide band of frequencies and a solid angle of nearly 2π for a source of radiation which would be above the horizon at a given site for a period of only a few minutes per day.

Detection of the satellite's signal by the enemy from sites not illuminated by the main beam from the satellite would be very difficult. While it could be done, so doing would require the use of narrow-band search receivers worked into very large antennas. The probability of making an interception under these conditions would be of the order of 1000 times less than the already low value applied to the more favorable case previously discussed.

Interception of the Satellite's Transmission (Monitoring). Television transmission from satellite to surface would require high-gain tracking antennas at both ends of the circuit. The enemy could receive the message from the satellite only if he had comparable tracking equipment[308] and then only if he managed to acquire the satellite's tracker before

---

308. It would not be necessary for the enemy actually to receive the signal so long as he was able to acquire the satellite's antenna by sending in the 300-Mc tracking signal. However, this type of interception (in effect, jamming) could be overcome by requiring a pulsed tracking signal, similar to an IFF system.

a friendly receiving station did. Successful interception would require that the enemy know almost every detail of the system and its operation.

Interference and Other Countermeasures. The television link can be relatively easily jammed by an enemy who knows the approximate locations of the ground receiving station and the frequency of transmission and who is able to get a jammer within line-of-sight range of a ground station. Even though the ground station's receiving antenna is highly directional (peak gain probably in excess of 20,000) and tracks the satellite, so that the jamming signal will be discriminated against by a factor ranging from a minimum of 1000 (for 30 db peak-side lobes) to an average of more than 20,000, the jammer can take tremendous advantage of the pulse transmission. For example, an air-borne pulse jammer of 10- to 100-kw peak output worked into an antenna [37] of modest gain (100 to 10) carried by an aircraft at 20,000 ft to within -200 mi of the receiving station could prevent reception of a usable picture. Such jammer powers (peak pulse, at a duty cycle of about 1 per cent) and antenna sizes are comparable with, or modest compared with, those of ordinary airborne radars, and spot-frequency jamming is therefore quite feasible.

If the enemy can be denied access to within line-of-sight range of the ground stations, the television system will be relatively invulnerable to interference by the enemy. Counter-measures applied at the satellite-end of the circuit presume possession by the enemy of adequate search and tracking facilities (the difficulties of which were previously discussed) and can be directed only against the satellite's tracking receiver.

Reception and Presentation of the Television Signal

Reception

A description has already been given of the ground station's receiving antenna and tracking system. Consideration is now devoted to the assimilation of the TV pictures after they have arrived at ground level.

Concurrently to read and interpret information on a single television screen at the rate of 10 completely different frames per second is obviously impossible. Furthermore, each ground station receives only a piece of the target system under scrutiny. Thus it appears necessary to record the transmitted data with as little loss in resolution as possible and to forward it to a central evaluation center.

At a first glance, it would seem that a prodigious amount of film would be required to record all the pertinent television frames transmitted. However, analysis reveals that 2.9 hr/day, at most, are spent over USSR and her satellites, China included. At 10 frames/sec, 2.9 hr are equivalent to 1.0×10^5 frames/day. It has been shown previously that one satellite will observe a given area only on alternate 35-day periods in daylight. Thus, for the first 35 days' operation, 3.5×10^4 frames would be recorded. This is 265,000 ft of 35 mm camera film, or about that used in filming several feature-length movies.

[38] It is believed that during the first 30 to 40 days' operation a fairly comprehensive picture of the USSR would be obtained, and subsequent operations would be concentrated on specific target systems or areas, perhaps with the narrow-scanning-width lens system previously described.

The equipment required at any forward receiving station is not complex. The receiving antenna has already been discussed. An ordinary television receiver will probably suffice for monitoring purposes (to see if the picture quality is satisfactory). Its viewing scope, however, must have a high-persistence screen which will project about one frame out of a hundred.

For recording, a second television receiver is needed. Its scope must be as large as possible and its electric beam spot size must be reduced to a minimum; in short, the whole set must be tailored to the criterion of putting the image on the screen with as little loss in resolution as possible.

The image will then be reduced by camera optics to the appropriate film size; 35 mm may be adequate, but if a significant amount of detail is lost, then 70 mm can be employed. The film does not have to be very "fast," but should be of a fine-grain variety.[309] The camera will be similar to a movie camera but will operate at about one-half the frequency.

Each forward station will be furnished with a time schedule for operating the cameras computed on the basis of the satellite's orbit. Such a schedule will vary from day to day, as mentioned in the discussion on orbits. Some sort of time coding will be included with each frame; a feed-back from the tracking-antenna control will also be fed into this coding, but this is only a crude location device to show up any gross errors in evaluation.

Presentation

The *central evaluation station* will receive the composite films from the forward stations and assemble the story into an integrated whole. Standard photogrammetric techniques call for synchronizing one or more sets of films, together with overlays, etc., [39] to aid in interpretation of results. In such a device, the frames are projected in a mosaic form and compose as the scenes appear on the earth. Also projected could be a master overlay made up of geographical coordinates and, later, after a number of films are taken, of that area of the ground already filmed. The over-all area can be enlarged to any extent necessary for rapid determination of the worth of the films being evaluated. For instance, if a large area is covered by clouds, then just those frames having glimpses of the ground could be separated for subsequent addition to the master mosaic of the USSR.

The cloud pictures would be placed on a larger-scaled photomap so that daily weather maps could be made and preserved.

The entire presentation system should be simple, rapid, reliable, and amenable to standard evaluation techniques.

Summary

To summarize, a 350-mi altitude satellite, having an f/10, 2-in. aperture, 20-in. focal length, Image Orthicon TV camera of 1000 TV lines/in. with a speed of 10 frames/sec, would be capable of resolving scenes of contrast greater than 20 per cent to about 200 ft. Transmitting and receiving antennas for the described system will require careful analysis and design, but their accomplishment does not present any serious research problems. Presentation of the viewed scenes by photographic and photogrammetric methods appears within the limits of known, practiced techniques.

Such a system, employing presently available equipment, is considered satisfactory for both weather and pioneer terrestrial reconnaissance. However, in order to obtain acceptably detailed target evaluation of bomb-damage assessment, the minimum resolvable surface dimension will have to be improved; several possible methods are suggested.

For example, by keeping the frame speed constant but optically reducing the field of view and thereby reducing the scanned bandwidth on the ground, acceptable values for most terrestrial reconnaissance can be attained with present television tubes. This results in not having a daily coverage of the entire target area.

Other means of improvement of the resolvable surface dimensions are an increase in the inherent tube resolution (an increase of about 50 per cent is visualized at this time) and an increase in the frame frequency to 30/sec (about 45 per cent improvement of resolvable surface dimension).

The estimated over-all power, weight, and space requirements of the electronic transmitting system are 350 watts, 300 lb, and 2.25 ft^3, respectively....

---

309. Such expedients as, for example, using blue sensitive film with a blue cathode-ray screen can be used to bring out certain details in the viewed scene.

Document II-4

Document title: R.M. Salter, "Engineering Techniques in Relation to Human Travel at Upper Altitudes," *Physics and Medicine of the Upper Atmosphere: A Study of the Aeropause* (Albuquerque: University of New Mexico Press, 1952), pp. 480-487.

Few scientists and engineers outside of the narrow field of rocketry took human space travel seriously until the mid- to late 1950s. Many felt that it was little more than science fiction fantasy unworthy of serious study; they frowned upon their peers who devoted time and effort to such a frivolous topic. Robert Salter's article was most likely the first serious treatment in American academic and engineering circles of the problems of human spaceflight.

---

[480] Introduction

The subject now under consideration is the current and predicted status of engineering techniques related to the travel of man in the upper atmosphere. In other words, we are to discuss the "how" and "when" of manned space flight. The "why" of human participation in such a venture should also be examined since it is not immediately obvious that we cannot always substitute electronic equipment in place of a pilot.

In order to correlate properly the various data available it is necessary to distinguish between physical limitations, those imposed by actual physical laws (or absolute limits), and purely engineering constraints. Quite often an operation is said to be impractical when actually it is only infeasible on the basis of current engineering techniques. On the other hand, physical considerations usually furnish a clear-cut indication that the particular problem in question either can or cannot be solved. This is not a completely rigorous limitation, since some phases of physics, notably nuclear physics, are currently in a rather fluid state so that our ideas may change in the future.

Thus on the basis of the laws of motion, etc., as we now know them, the various allowable regimes of operation in the aeropause can be enumerated. It can be said that, without the employment of a rather unique release of nuclear energy, certain modes and areas of space travel must be excluded. For example, long-duration flights at 50 mi altitude would be excluded. On the other hand, it now appears that a large portion of space travel of interest can be accomplished with present-day types of propulsion and energy sources! The day for successful interplanetary travel awaits only the decision of man to provide the [481] prodigious and concerted effort required. It might be mentioned in passing that nuclear fuels are not necessary to such an operation and that the basic techniques required have been known for centuries. It is pertinent to note, for instance, that a two-stage rocket was successfully tried in 1855—nearly a hundred years ago.

Regimes of Flight in the Aeropause

Some of the first questions to be answered are, "how high," "how fast," and "how long" can flight be sustained in the upper atmosphere? Emphasis must be given the last item if a pilot is carried in the vehicle. Obviously, other than for the purposes of physiological experimentation or establishing a record, one would not conceive of a manned sounding rocket. Here, there is not time or need to supplant electronic equipment for making observations. However, in cases where flight duration is of sufficient length that electronic reliability is a problem, where computer operations (such as having adequate "memory" included) are too complex, and, in particular, where judgment in unforeseen circumstances is needed—then the participation of man will be required. It may be seen that the first two requisites are limited only by prevailing engineering development, while the last is clearly a basic constraint.

The employment of pilots in supersonic rocket planes and in balloons is an example of present approaches to the problem, and has been covered in previous papers. In the case of air-borne vehicles (those using forward motion to derive lift from the atmosphere) we must consider duration of flights as well as altitude. It is convenient to subdivide this class of vehicles into those using rocket engines and those using air-breathing power plants. This latter type is represented by the various jet propelled aircraft and missiles. In order to fly at very high altitudes it is necessary for such a vehicle to operate at supersonic speeds, not only to provide sufficient lift but also for adequate thrust. At an altitude of 20 mi (32.2 km), for example, the required Mach number for a ramjet is over 5 and the resultant incoming air has a stagnation temperature of the order of 2000° F. Since energy must be imparted to this air at higher temperatures it may be seen that a present engineering limitation on suitable fuels and materials is approached. This is particularly true with the use of nuclear heating. Thus the air-breathing vehicle is limited in altitude. As for duration, a nuclear ramjet might be capable of cruising for indefinite periods around the earth and comprises a very interesting possibility for travel in the near aeropause.

[482]   **Physics and Medicine of the Upper Atmosphere**

With rocket vehicles, higher altitudes can be attained. At 50 mi (80.6 km) and at a near-satellite speed of 4 mi/sec a vehicle can support its weight in gliding. However, in the region of 20 to 100 miles (32.2 to 160.9 km), flights with propulsion that can be envisioned at the present time will be of short duration (an hour or so) and less than a revolution about the earth. In this altitude region, the justification for incorporating a pilot is doubtful.

The gliding trajectory mentioned above naturally leads us to a satellite. The feasibility of establishing a vehicle in a stable orbit around the earth has been theoretically demonstrated both here and abroad. Placing the vehicle in the direction of the instaneous tangent to the earth at the proper speed will result in a circular orbit. At 150 mi (241.4 km) altitude nearly 5 mi/sec is needed; at 500 mi (804.6 km) this speed is a little slower at about $4\frac{1}{2}$ mi/sec. At 150 mi the duration would be of a day or so, drag slowing the vehicle down. At 500 mi, it is estimated that several decades would elapse before a satellite would come down of its own accord.

It is apparent that automatic operation of complicated scientific observational equipment is a tenuous proposition for long periods of time. The temptation for employing a human observer on a one-way basis is ever present, and it is probable that a number of volunteers willing to devote their lives to science could be obtained. However, it is physically possible to bring a satellite back without a great additional source of power.

This is not easy and would require considerable development in control equipment. In launching a satellite, a long, coasting (elliptical) trajectory is indicated, with a small additional kick provided to pull it into orbit. This same kick in reverse will put the vehicle back into the original ballistic flight path, but the vehicle might burn on the way down. By using a carefully selected and maintained gliding trajectory it is believed possible to enter the atmosphere without disastrous skin temperatures and high landing speeds. In fact, terminal speeds slightly over sonic are indicated, at which point parts of the vehicle could be landed with parachutes.

The main problem, then, of the returnable satellite is that it requires a very accurate control during the descent phase—automatic programmed control at the least, and possibly the continuously computed variety.

The satellite, especially a returnable one, is important as a step in the direction of interplanetary space travel. The principle of orbiting [483] about a given planet will undoubtedly be incorporated in future vehicle operations. For example, computations have shown that the establishment of an elliptical orbital path about the earth and moon will require not appreciably more energy than that of a low-altitude satellite—in fact, the energy required is less than that needed to establish a circular orbit at 6400 mi (10,299.9 km) height.

To escape from the earth's gravitational field, a velocity of 40 percent more than the low-altitude satellite or about 7 mi/sec is required. By properly timing the trajectory it should be possible to arrive at a near planet in a reasonable length of time. Some additional control thrust will probably be required to allow the vehicle to be captured by this planet as a satellite. At this point the vehicles can elect to land, or return to earth by orbital escape at the proper time.

Motivating Techniques Required

It is apparent from the foregoing that, as more complicated operations are visualized, an effectively greater thrust impulse is necessary for a given vehicle. Staging is one expedient, and an important one, for effecting space travel. However, if it requires a million pounds take-off weight to put 10,000 pounds on a distant planet, it will take a billion pounds initial weight for a round trip (to a planet of similar gravitational pull and return). This is the weight of ten battleships. This condition prevails if all the fuel must be carried with the vehicle.

Instead of multiplying stages the vehicle might carry a pilot plant for making its own fuel for the return voyage. This would, say, double the pay load and thus only double the gross weight.

The launching of a one-way space vehicle is felt to be possible with highly developed structural techniques and chemical fuels. What gains, then, can be made with the use of nuclear fuels?

Two possibilities exist with nuclear energy, using the fission particles directly, and degrading the energy into heat for increasing the momentum of a secondary fuel. At first sight the direct use of fission fragments and neutrons seems attractive. However, at the vehicle speeds where most of the thrust force is expended, many orders of magnitude greater energy release is required for the direct employment of fission particles to attain a given thrust than is needed for sufficient heating of a secondary fuel. This is because the mass of the particles is so small. One can consider the "dilution" of the particle momentum with inert particles of greater mass. The energy required is proportional to mV^2, [484] while the momentum change for thrust is proportional to $m\bar{V}^2$ if the velocity of the particle is greatly different from that of the vehicle,[298] the low-mass, high-energy propellant is inefficient. (In the above discussion it has been assumed that it is possible to direct the fission particles rearward, which, in itself, is unrealistic considering the tremendous amount of heat generated in an absorbing chamber designed for such purposes.)

It is, of course, well known that such orders of energy production are probably physically realizable, but to accommodate such an energy release in a vehicle is beyond the scope of present engineering thinking.

On the other hand, the use of a nuclear reactor to heat a secondary propellant does not impose a large strain on the imagination. The impulse for a given energy release is roughly proportional to the square roots of the temperatures and to the inverse of the atomic weights of the propellant components. Thus the ability to use hydrogen or methane alone, and/or higher temperatures, indicates possibilities for nuclear propulsion. However, it is not immediately apparent that an improvement over a chemical-fuel system will result or, if so, that the amount of improvement will be significant.

Engineering Limitations

We have explored, in a qualitative fashion, the allowable regimes of upper atmosphere flight from the standpoint of that which is physically conceivable. How much then is realizable on the basis of engineering? In other words, if say, a satellite vehicle can be

---

298. The optimum velocity of the particle is approximately twice that of the vehicle. For particle speeds significantly greater than vehicle speeds the propulsive efficiency is proportional to Vv/Vp.

theoretically predicated, do we have the practical engineering know-how to implement such a venture?

The tentative answer to this question is yes. As a result of war-instigated research we have V2 rockets and microwave radio techniques. Even in 1945 a satellite vehicle could have been built in a "quick and dirty" fashion by staging, and by such crude expedients as clustering existing rocket motors together to form a single power plant. Since then power plants with improved fuel combinations and of a larger size, better materials for certain applications such as titanium metal, and exploratory research in the near upper atmosphere have tended to guarantee even greater success for such a program.

[485] The fact that a satellite has not been practical in a strict military sense has retarded its development in favor of guided missiles. On the other hand, the continual improvements of techniques in propulsion, aerodynamics, structures, and electronics have been brought about by missile programs.

It is on the subject of electronics that we shall dwell for the moment. Approximately half of the effort in the V2 development was expended in the simple radio control that established this vehicle on its relatively short-range ballistic trajectory. With this in mind, it is not hard to extrapolate to the difficulties involved in guiding a long-range missile over a complicated trajectory or in placing a satellite in its orbit. As was mentioned previously, this was one of the considerations in favor of having a pilot in a rocket plane for supersonic flight research. Ever-increasing needs for more complex control equipment call for better electronics. This, in turn, requires special consideration of the electron tubes themselves, which, for a particular application, went through the following interesting stages of evolution.

Initial forms of radio-detection devices included the crystal detector and the triode. Subsequent tube development resulted in more and more complex elements within the tube and in multipurpose tubes. With the recent upsurge of electronic application in microwave radio and in digital computers (or electronic brains) it has become apparent that many simple and reliable diodes and triodes are to be preferred for such circuits. Further exploitation in this direction has centered interest on the semiconductor or transistor, and the equivalents to both diodes and triodes have been made in this form. What is a transistor? It is just an improvement of the old cat-whisker-crystal affair used in the early radio sets. This cycle aptly illustrates an underlying precept of development. As Boss Kettering says, "the ultimate solution to a problem is usually the simplest." Another way of stating this is, "If it's complicated, it's wrong."

The transistor consists of a simple blob of material the size of a lead-pencil eraser (or smaller) with several wires leading to it. Its power requirements may be as small as one-hundredth that of an equivalent electron tube. The reliability of the transistor also probably will be considerably better. In short, it has been heralded as the forerunner of the coming electronic age, where many of man's more menial tasks will be replaced by computer controls.

In the actual construction of an upper-atmosphere vehicle many unforseen problems undoubtedly will arise. Usually, if a wide gap [486] exists between the physical and engineering limits on a device, then a large number of possible solutions exist. In the early days of aircraft, for example, many weird configurations evolved. As research in piston-engine types progressed it became apparent that monoplanes with thin wings were needed. Eventually a physical limit of 500 mph was approached since the additional power required resulted in a larger engine-cooling drag that offset the benefits of increased engine size. Thus, considerable work and careful design were required to reach this limiting speed.

By analogy, it may be said that we are still in the Curtiss biplane stage of rocketry, and such considerations as accessibility of instruments for ground testing frequently predominate. A good example of such freedom may be found in a recent news release, "Newsmen who recently attended the firing of a rocket asked why the rocket was exactly 32 inches in diameter. After a long technical explanation involving the ratio of length to diameter and effects of air drag on a large projectile, the engineer concluded, "Besides, it so happened that the metal plate we were able to obtain made a cylinder exactly 32 inches across."

Summary

Past experience has shown that most advances in the field of human endeavor are not made as a result of some completely new and different concept, but rather by skillful day-to-day improvement of existing technology. This does not mean that intelligence is not required. On the contrary, the human mind is quite capable of solving multi-parameter problems, an operation which is usually termed ingenuity. Very often a design created on the back of an old envelope is perfectly suitable (it can also be completely wrong).

The continuing development of computing machinery has resulted in powerful tools for rapid, simultaneous solution of problems of many variables. However, it should be emphasized that the machines themselves do not possess intelligence. It is quite embarrassing, for example, to find that solutions are insensitive to a given parameter and upon subsequent investigation find that this factor did not belong in the problem in the first place.

There is, and will be, no substitute for sound and thoughtful planning and direction of research. The middle road between the no-stone-left-unturned school and the advocates of the "brilliant hunch" type of investigation will afford the most fruitful course of action.

[487] Just how long before space travel is accomplished cannot be predicted accurately since a very large weighing factor must be assessed to man's own incentives and decisions. If it became necessary to our very existence to conduct interplanetary flights tomorrow, the development period required would be materially shortened-probably within our lifetime. Without such an impetus it may be many generations before such a program is attempted. Let us recall that 50 years ago most of the mechanics of a complete rocket vehicle were known.

To recapitulate, most of the components comprising an upper-atmosphere vehicle probably will be refinements of existing rocket devices. Rather than having an appreciable increase in rocket-plane altitude, the next step probably will be a satellite, with a returnable version used for manned flights. Considerable improvement in electronics from the standpoint of reliability, weight, and power consumption is indicated. The transistor may pave the way toward this end. Many of the more complex operations in the development of rocket vehicles, as well as within the vehicle itself, will be implemented by self-sustaining computers. In the final analysis, though, man himself, with his ability to use judgment and his physical limitations, will provide the key to space flight.

Document II-5

Document title: A.V. Grosse, The Research Institute of Temple University, to Donald A. Quarles, Assistant Secretary of Defense for Research and Development, "Report on the Present Status of the Satellite Problem," August 25, 1953, pp. 2-7.

Source: NASA Historical Reference Collection, NASA History Office, NASA Headquarters, Washington, D.C.

In 1952, President Truman requested Aristid V. Grosse, a physicist at Temple University who had worked on the Manhattan Project, to study the "satellite problem." Major General Kenneth D. Nichols, formerly deputy for Lieutenant General Leslie R. Groves on the Manhattan Project, arranged Grosse's meetings with space scientists at Huntsville, particularly Wernher von Braun. The report was finished after Truman left office; it was delivered to Donald Quarles, the new Assistant Secretary of Defense for R&D under President Eisenhower, on September 24, 1953. Quarles later became a major advocate of the use of space for military purposes. Another copy of the report was sent to Dr. John R. Dunning, dean of the School of Engineering, Columbia University. Dunning discussed it with President Eisenhower, and the report contributed to the initiation of Project Vanguard. Grosse's report represents the first time that the potential propaganda consequences of a Soviet first launch of a satellite were reported directly to top levels of the government.

---

[2] **The Present Status of the Satellite Problem**
A satellite is a man-made or artificial moon which will rotate around the earth beyond the furthermost extent of its atmosphere, for many years or indefinitely. After it has once reached its orbital velocity the centrifugal force of its motion is held in exact balance by the gravitational attraction of the earth; thus the satellite once on its orbit around the earth *does not require* any additional power to keep it there. Usually altitudes of 300 to 1000 miles above the earth's surface are considered.

As an example, at an altitude of 346 miles above sea level the time necessary for the satellite to travel once around the earth, i.e. its period of revolution, will be exactly 96 minutes or 15 revolutions per day. Its orbital velocity will then be 4.71 miles per second. Similarly, a satellite at an altitude of 1037 miles above sea level will have a period of revolution of exactly 120 minutes or 12 revolutions per day and a velocity of 4.37 miles per second.

The satellite could be made to travel over the surface of the earth in a wavy line so that in the course of a few successive days most of the North American continent, Europe, Africa and Asia, could be observed from it.

It could be made visible to the naked eye, under clear atmospheric conditions, at dawn and at dusk as a bright fast moving star.

The technical problem of creating a satellite should logically be divided into the two following steps:
1, The unmanned satellite and
2. The manned satellite.

[3] The accomplishment of the first step, in the opinion of even the most skeptical engineers, is possible with the present know-how and engineering knowledge. Since it is not manned by human beings it would not require any essentially new research and development.

A satellite of about 30 feet in length would require the stepping up of the German V-2 rocket by a take-off weight factor of 6-7. This would require essentially the addition of a third large stage to the present well known two stage rockets such as the WAC Corporal mounted on a V-2, which reached an altitude of 250 miles at the White Sands Proving Grounds in February 1949. A design for such a large stage was already on the drawing boards of Dr. von Braun and his associates in Peenemünde, Germany, in 1945. This German project "A-9 + A-10" was designed for transatlantic bombing of the United States. The A-9 stage was a slightly enlarged V-2 (take-off weight 16.3 metric tons vs. 12.8 tons of the V-2) whereas the A-10 stage had a take-off weight of 69 metric tons. Such a three stage rocket would use conventional fuels giving a specific thrust of 220-240 seconds (for example, liquid oxygen + ethyl alcohol 75%, water 25% = 239 seconds, red fuming nitric acid + aniline = 221 seconds). Conventional combustion chambers, pumps, tanks, ignition devices, etc., could be used.

Research scientists have recently demonstrated that much larger specific thrusts can be realized. For example, a liquid fluorine-liquid hydrogen rocket motor can generate a thrust of about 380 seconds. This would permit the use of a much smaller rocket to achieve satellite velocity. However, the engineers feel that this advantage is offset by the necessity of doing a lot of additional research and development in order to bring the high thrust rocket motors and their accessories to [4] the same stage of reliability and smoothness of operation as the conventional rockets. All of this new development would thus cause a loss in time. This would be unwise because it is felt by all engineers that the present rocket fuels and motors will be able to do the satellite job.

The second step or a satellite manned by human beings is decidedly a much more difficult problem. Ultimately, if solved, it would mean the beginning of man's conquest of interstellar space and would have infinite possibilities for the human race. The solution of this problem, however, involves overcoming all the obstacles in the way of man's existence in the vacuum of outer space. It means the overcoming of the absence of a gravitational field on the functions of the human body and the effects of cosmic radiation on it. Although all of these problems have a possibility of ultimate solution, it would require at least a 20-fold expense of human effort, money and time, as compared to Step 1, coupled with an inestimable amount of human ingenuity and invention.

It is felt that the accomplishment of the first step would help solve many of the problems of the second. This writer feels that probably after the successful launching of the first unmanned satellite, a number of such unmanned satellites will be in existence at various altitudes above the surface of the earth, for various purposes. It is thus felt that at this time the main effort should be directed toward solving the unmanned satellite problem.

The value of an unmanned satellite would fall into the following categories.

a) **Scientific** — with proper electronic and telemetering equipment and devices it would enable us to obtain valuable scientific information regarding the various physical conditions existing in outer space. [5] The satellite would need a concentrated source of energy, which should be *light in weight* and should produce *power for a number of years*. It is considered that such a power plant could be produced by using alpha-active radioactive substances of an average life of a few years in concentrated form, if the appropriate resources of the Atomic Energy Commission could be mobilized.

b) **Military** — again, with the equipment referred to above coupled with televising devices, a satellite station could be a valuable observation post.

c) **Psychological** — with appropriate signaling or broadcasting devices such a satellite could develop into a highly effective sky messenger of the free world.

In the opinion of this writer the last item, i.e., the psychological effect, would be considered of utmost value by the members of the Soviet Politbureau. They would recognize that in the case of atomic and hydrogen bombs the people of the belligerent countries would be subjected to their effects only after the die of World War III is already cast. On the other hand, the satellite would have the enormous advantage of influencing the minds of millions of people the world over during the so-called period of "cold war" or during the peace years *preceding* a possible World War III. In the countries of Asia, where the star gazer since time immemorial has been influencing his countrymen, the spectacle of a man-made satellite would make a profound impression on the minds of the people. The Soviet Union has demonstrated that it has been able to develop the atomic bomb and recently to follow that up with the accomplishment of a thermonuclear reaction on August 12, 1953, [6] as confirmed by the Chairman of the U.S. Atomic Energy Commission, Admiral L. Strauss, much faster than had been generally expected by our scientists and engineers. The building of an unmanned satellite would be a feat of much smaller magnitude than the construction of an atomic bomb since all the basic information was available to the Germans at the end of World War II and is since known both to this country and to the Soviet Union. Furthermore, the industrial plant necessary for the construction of a satellite is much simpler and is now being developed for the guided missiles programs in both countries.

In the Soviet Union the construction of a satellite would amount to only a fraction of the cost in this country, a) because of the use of cheap or slave labor; b) no necessity for great safety precautions, and c) no need for tracking the satellite in the early stages of its flight.

Since the Soviet Union has been following us in the atomic and hydrogen bomb developments, it should not be excluded that the Politbureau might like to take the *lead* in the development of a satellite. They may also decide to dispense with a lot of the complicated instrumentation that we would consider necessary to put into our satellite to accomplish the main purpose, namely, of putting a visible satellite into the heavens first. If the Soviet Union should accomplish this ahead of us it would be a serious blow to the technical and engineering prestige of America the world over. It would be used by Soviet propaganda for all it is worth. Of course, the probable reaction of the American people to a

Soviet satellite circling about 300 miles above Washington, New York, Chicago and Los Angeles, would have to be considered.

At the present time our engineering efforts in this field are limited in scope and distributed over various government agencies. It [7] is recommended as a first step in solving the satellite problem that a small but effective committee be set up composed of our top engineers and scientists in the rocket field, with representatives of the Defense and State Departments. This Committee should report to the top levels of our government and should have for its use and evaluation, all data available to our government and industry on this subject. It should report in detail as to what steps should be taken to launch a satellite successfully into outer space and to estimate the cost and time required for such a development. It is felt that if such a committee were in existence and a definite decision taken by our government regarding the construction of a satellite, that it would fire the enthusiasm and imagination of our engineers and scientists and effectively increase our success in the whole field of rockets and guided missiles.

Document II-6

Document title: J.E. Lipp and R.M. Salter, "Project Feed Back Summary Report," The Rand Corporation, R-262, Volume II, March, 1954, pp. 50-60. The figures have been omitted from this document.

Source: Archives, The Rand Corporation, Santa Monica, California.

In November 1950, Rand recommended to Air Force headquarters that it pursue further advanced research on satellite reconnaissance. Two Rand reports, including "The Utility of a Satellite Vehicle for Reconnaissance" [II-3], were completed in April 1951 and were enthusiastically received by the Air Force. Some members of the Air Force recognized the valuable role that satellites could play in providing strategic reconnaissance of areas not reachable by other means. As a result, the Air Research and Development Command authorized Rand to make specific recommendations on a satellite reconnaissance system. Project Feed Back involved hundreds of scientists and engineers from Rand and a host of subcontractors. Its results were presented to the Air Force on March 1, 1954, and became the basis for the first military satellite program. Many of its specific proposals, such as the use of television transmission of reconnaissance images, were not adopted until many years later. The section of the report dealing with "television payload equipment" is still classified. Volume I of this report has not been cleared for public release. In this excerpt from Volume II, the report discusses the scanning technique that was the basis for obtaining useful data from Earth orbit. The figures have been omitted.

---

[50] **Scanning**

The scanning problem arises for an obvious reason. The limited size and resolving power of the Image Orthicon result in each picture's being able to contain only a finite number of bits of information. Elsewhere in this report it is shown that in order to keep the time between successive views of a particular ground area to a reasonable value, the television optical system must cover a strip extending for 200 mi on each side of the flight line. If this area were to be covered by a single picture, about 1 in. on a side, the scale would then be 1:25,000,000; if the spot size of the scanning beam in the camera tube could be kept down to 0.001 in, the image projected on the ground by the optical system would be 2100 ft in diameter. Anything much smaller than a mile in its principal dimension would be difficult to detect.

At the scale of 1:500,000, one picture is about 8 mi on a side. A strip 400 mi wide will require fifty pictures to cover it. This number of pictures must be transmitted in the time it takes the satellite to move forward 8 mi (1.68 sec), requiring a frame rate of about thirty per second, which is present commercial practice. At this scale, a spot 0.001 in. in diameter will cover a circle on the ground approximately 40 ft in diameter. Two television lines are equivalent to one optical line of resolution, and an object, to have a high probability of detection, must be covered by about two optical resolution lines. This gives, in the present case, a limiting object size of somewhere between 150 ft and 200 ft, approximately the size of bombing aircraft—hence the gain realized by the complication of the addition of a scanning system.

Because the Image Orthicon is an integrating device, it requires a finite exposure time during which the image must remain fixed on the photocathode. If it were not for this, the scanning problem would be reduced to the simple one of two cameras viewing the ground by reflection in continuously rotating mirrors. Two cameras would be required to eliminate the "dead time," i.e., the time during which the mirror into which each camera is looking would be [51] returned to its initial position to start a new sweep. But scanning in the direction of the line of flight is affected by the vehicle's motion. From the standpoint of reliability and long life, intermittent mechanisms which have been proposed for the projection of motion pictures from continuously moving film are applicable. A few of those suggested in Ref. 24 may be difficult to fabricate, but they can be used successfully here, because many of the restrictions imposed by their application to theater projectors do not occur (e.g., the f number and back focal length, in particular, present no problems).

The most promising arrangement that has been investigated is the one proposed by RCA in their study of the problem. It consists of a number of mirror pairs mounted on the periphery of a continuously rotating wheel; each mirror pair deflects a ray through a fixed angle in a plane perpendicular to their line of intersection, independently of any rotation of the mirror pair about any line parallel to their line of intersection. In Fig. 29, M_1 and M_2 are two plane mirrors perpendicular to the plane of the paper, and I is their line of intersection. If the two mirrors rotate slightly, as a unit, the change of deviation of the ray produced by reflection in the first mirror is of the same magnitude and opposite sense as that produced by the second mirror, leaving the deviation produced by the pair of mirrors unchanged. The deviation of the pair depends, therefore, only on the angle between them. RCA's device is shown in Figs. 30 and 31. For a scale of 1:500,000, an altitude of 300 mi, and a strip 200-mi wide on each side of the line of flight, the rotating drum will have eighteen pairs of mirrors, each pair equally spaced around the periphery.

This drum, whose axis is parallel to the direction of motion of the vehicle, rotates at a speed of 38.5 revolutions per minute in a clockwise direction as shown. Since there are eighteen equally spaced pairs of mirrors on the drum, the mirror pairs are spaced 20 deg apart. Both television cameras are spaced 90 deg apart as in Fig. 31. This means that at the instant that camera "A" is viewing a ground scene through a mirror pair, camera "B" is viewing the transition point between two successive mirror pairs. The sequence of ground scenes scanned is shown in Fig. 32.

Lateral image immobilization is achieved during the entire time that a pair, or part of a pair, of mirrors is in line with the optical axis of the camera. Except [52] for a short interval during which the image is recorded, a composite picture of two successive segments is seen because of vignetting effects. The situation perhaps can be explained better by describing the direction of view of one of the cameras. As the drum rotates, a scene, completely stationary except for the image motion caused by the forward motion of the vehicle (the lateral scanning introduces no image motion), can be observed for a period of time depending on the mirror size. The next scene will then be picked up and will start to blend with the first scene, both remaining completely immobilized. At some point, only the second scene will be observed; then the entire cycle will be repeated for adjacent fields of view.

Such a sequence is demonstrated by Fig. 33. The ordinate in this diagram is comparable to the intensity of illumination on the photocathode due to the fields of view indicated by the numbers above each peak, which can be made to correspond to the numbered fields shown in Fig. 31. RCA's proposal to avoid the resulting confusion of images is to "pulse" the image section of the Image [53] Orthicon. That is, the accelerating potential will be applied to the photoelectrons liberated by the optical image on the photocathode only during that part of exposure on, say, field 3, when the light from fields 1 and 5 is less than some minimum value, say 5 percent of full aperture. It should be noted that because the output of both cameras is to be transmitted over the same carrier wave and received on the same device, it is important that they be accurately interdigitated timewise. The curves in Fig. 33 represent the case where the dimensions of the mirror pairs are such that the full-aperture condition obtains only instantaneously. As the size of the mirrors is increased, the full-aperture exposure time increases and the exposure curves, shown in Fig. 34, develop [54] flat tops and bottoms. From the point of view of the most efficient time use, the optimum is reached when the exposure time is one-fourth of the frame frequency for both cameras; the exposure curve has the form shown in Fig. 34 for the camera on one side of the mirror drum.

Starting with field 2 (Fig. 34), full aperture is reached at the abscissa value of 1.5 and is maintained until 2.5. During this time the accelerating potential is applied and a charge image is built up on the target plate in the Image Orthicon. From 2.5 to 3.5 the image-stage voltage is shut off and the scanning beam discharges the image on the target plate. In this same interval (2.5 to 3.5) the camera on the other side of the mirror drum is being exposed. At 3.5 a new exposure is started in the first camera at the same time that the picture in the second camera is being scanned and transmitted. In this way there is always one and only one picture being exposed. There is no "dead time" for the transmitter.

[55] The price that must be paid for this more efficient use of transmitter and exposure time is, of course, weight and bulk—the scanning drum must be larger. Variation of the drum radius with the percentage of the exposure time occurring at full aperture is shown in Fig. 35.

To get the resolving power being discussed, any motion of the image, during the exposure, that can be predicted must be eliminated. Such an image motion is one that is due to the high forward velocity of the vehicle—roughly 25,000 ft/sec. At an exposure time 0.001 sec, the image of the photocathode projected on the ground by the camera lens will move 25 ft, a barely tolerable amount. If the scanning mirror dimensions are chosen so that one-quarter of the frame time is available for exposure, the exposure time at 25 frames/sec will be 0.01 sec and the image motion during this time will be 250 ft, requiring some sort of image-motion compensation.

Here again the type of mechanism devised for the projection of motion pictures from continuously moving film can be used, but there is such a small motion, in terms of percentage of frame height (at most 1000 ft out of 8 mi), [56] to be corrected that, in RCA's opinion, it can be handled (electrically) by scanning in the image stage of the camera tube. All that is required is the addition of a coil above the image stage and one below the image stage, the plane of the coils being parallel to the axis of the tube. These coils can be energized by a "saw toothed" oscillator whose frequency is equal to the frame frequency. As the optical image moves on the photocathode, the photoelectrons liberated by a (moving) point in the image can be brought to a focus at a fixed spot on the target plate where the charge accumulates. RCA workers say that a motion of 5 percent of the frame height is easily corrected in this way, whereas 10 percent is possible to correct, but very difficult. Since the motion in this instance is around 2 percent, it should not be difficult to correct.

The problem of obtaining reconnaissance data is essentially that of typifying various ground-target scenes with patterns of bits varying in intensity. The number of bits in a given period of time determines the bandwidth, or the information rate, of the system. Here, information rates of perhaps three times those of standard television systems have been considered, i.e., bandwidths of about 8 Mc. It is obvious that all components in the television system should be compatible with regard to bandwidth.

A bandwidth corresponding to the above frame rate in tube resolution is about $6\frac{1}{2}$ Mc. It is expected that a slightly higher bandwidth may be employed in the surrounding circuitry of the television camera tube so that no unnecessary degradation of signal will be introduced. Bandwidths of the order of 9 Mc have been employed in the simulation television setup for the photographs used in Vol. I of this report. However, the use of these bandwidths is not standard studio practice because the standard-tube studio television is limited by FCC regulations to about 3 1/2 Mc. Otto Schade, of RCA, has used bandwidths up to 20 Mc in some experimental television equipment, particularly for circuits surrounding the 4 1/2-in. Image Orthicon camera tube.

The next component encountered by the television signal is the magnetic-tape recording system. Magnetic-tape recorders for the purpose of recording video signals have already been investigated and brought to a primitive stage of development.

A magnetic-tape recorder (Fig. 36) will be similar in many respects to the home audio-tape recorder except that it will handle much more information in a given length of time. Two reels, one for feeding the tape and one for winding it, are needed; also, the tape passes over a capstan and several other pulleys. Heads for recording information magnetically on the tape are provided, both [57] for recording and for playing back the information into recording heads, and for taking the information in playback.

An RCA video recording system was exhibited recently. It consists in using either a single track for the video signal, the black-and-white system, or a color system having three tracks on the tape. Tape speed is 30 ft/sec.

Bing Crosby Enterprises have a system using a somewhat slower tape speed. In their device, the black-and-white television is recorded with a number of tracks on the tape.

[58] Both systems are designed for standard studio bandwidths and will have to be increased by a factor of two or three in order to be compatible with the bandwidth proposed for the Feed Back system. Personnel of both RCA and Bing Crosby Enterprises have expressed the opinion that within the development period allotted for Feed Back, such a recording system can be developed.

A suitable tape is one having a cellulose acetate plastic base of 0.0017-in. thickness, similar to the one developed by Minnesota Mining and Manufacturing Company. Lubricating methods developed by them are believed to be adequate. The magnetic surface of the tape is an iron oxide coating of 0.0005-in. thickness, which is impregnated on the plastic base. It is believed that the tape cannot be run continuously over the capstans for a year's period. Even if the tape itself can be made to withstand this length of service, it is probable that the magnetic heads will be worn down because tape has characteristics not too different from those of crocus cloth. Any system assumed for the present report allows for intermittent operation and includes motors for starting and stopping the reels every time a recording or a playback is made. In fact, it is probable that the system will be started in one direction for recording and played back in the opposite direction. Discussion of the programming of the record playback magnetic-tape storage may be found under "Communication Link," page 85.

Next, in its progress through the television equipment, the signal encounters a modulator and transmitter unit. These components must have at least the 8-Mc bandwidth postulated for the other units in a television chain. Engineering of the equipment will be fairly straightforward.

The transmitter in the vehicle will be a frequency-modulated oscillator operating in the X-band and having a power output of about 10 watts. Center frequency of the transmission will be controlled by reference to a very stable high-Q resonant cavity. However, there is some difficulty in obtaining an output transmitting tube capable of transmitting at the megacycle frequency required and also having a year's life capability at reasonable power requirements.

In earlier work it was assumed that 10,000 Mc would be used for the transmission signal. However, RCA believes that 7500 Mc is a more appropriate figure, and this frequency will give a greater capability in transmission through heavy rainstorms. A frequency that is too low will require more power input in the transmitter; therefore, the 7500-Mc frequency is a compromise. RCA has recommended that the output stage be a frequency-modulated magnetron with a 20-watt output for reasonably low power consumption.

However, at present, magnetrons have not been developed to have a year's life, the longest, perhaps, being a month. It is probable (but not certain) that the life length of the [59] magnetron can be improved. Also, it is possible that traveling-wave tubes will be developed to a state of refinement that will allow them to be considered for use as the Feed Back transmitting tubes.

For the example selected here, which discussed informally with RCA, two klystrons developing a total of 5 watts have been used. (A larger antenna compensates for the reduction in output from the 20 watts stated above.) These tubes now have a reliability compatible with Feed Back requirements (over 10,000-hr lifetime in one reported instance). Previously, use of the klystron was not felt possible because power requirements of this tube are quite high. However, in putting together the various parts of the over-all Feed Back system, it became apparent that the 400-watt input required by the klystrons (compared with a tenth as much power required by the magnetron) was not dominant in the total payload power requirement.

Payload power requirements are already in the realm of several kilowatts, so that once a reactor is selected for the auxiliary powerplant, a $1/2$ kw more power can be obtained for about 25 lb additional radiator weight.

A transmitting antenna with a diameter of about $3\ 1/2$ ft is needed for the 5-watt klystron systems. A 20 watt magnetron, on the other hand, requires only a 1-ft-diameter antenna. By placing the antennas in the locations shown on the vehicle drawing (see Fig. 1), it should be possible to enclose, within the vehicle, antennas several feet in diameter, despite their attendant complexity and weight for this particular component.

RCA has proposed an antenna system consisting of two separate paraboloid dishes: one dish receives the 3000-Mc signal and thus is able to track the ground station by means of a conical scan, and the other, the transmitting dish, is slaved to follow the receiving dish by means of a servomechanism.

Rotating parts, such as the antennas, and also the mirror wheel for the optical system, are assumed to be counterbalanced by devices of comparable moment of inertia rotating in the opposite direction.

The antenna system is to be mounted just below the throat of the secondstage rocket motor, so that upon separation of stages it will be exposed to the atmosphere and will be allowed ample freedom to scan not only directly below the vehicle, but to the horizon as well.

Approximately three video stages of amplification will be necessary between the camera equipment and the output tube. It is estimated that about 300 watts will be required to operate the transmitter circuits, exclusive of the tube requirements. The temperature of the compartment which houses the electronic [60] equipment must be regulated to within about 10° C of a desired value, and this eliminates the need for an automatic-frequency control circuit.

A tracking command receiver will also be included in the television system. It will be a simple superheterodyne type with a bandwidth sufficient to accommodate the doppler shifts due to vehicle velocity plus an information bandwidth a few kilocycles wide, which is sufficient to permit transfer of all needed command information for the most extensive case in a period of less that 1 min.

The purpose of this receiver is to receive command information from the ground, particularly to set up the scanning, recording, playback, and transmitting operation for successive passes of the vehicle. Commands will be transmitted in the form of Baudot types of symbols and will be recorded on the rotating drum of the programmer, in accordance with the present sequence arrangement, which is capable of erasing and changing all of the drum information in a period of 1 min or less. At the conclusion of each command cycle, the program drum will be played back to the ground through the data transmitter and will be checked for accuracy against the transmitted commands.

Also included in the operation is the programmer just mentioned. The programmer will probably operate in a manner very similar to that of a timer on an automatic washing machine; i.e., it will consist in a linear sequence of operations. Because successive programs differ only in variations of the length of time (including 0) that operations can take place, the required programmer is inherently simple. More comments on the ground-to-vehicle link and programming will be found under "Communication Link," page 85.

Results of RCA's investigations up to the present time are given in their various progress reports....

Document II-7

Document title: Wernher von Braun, "A Minimum Satellite Vehicle: Based on components available from missile developments of the Army Ordnance Corps," September 15, 1954.

Source: NASA Historical Reference Collection, NASA History Office, NASA Headquarters, Washington, D.C.

On June 23, 1954, Frederick C. Durant III, former president of the American Rocket Society and then current president of the International Astronautical Federation, called Wernher von Braun at the Redstone Arsenal and invited him to a meeting two days later in Washington, D.C., at the Office of Naval Research. At this meeting, plans were discussed for developing a satellite program using already existing rocket components. Further meetings followed at which the Army gave tentative approval, provided that the cost was not too great and the plan did not interfere with missile development. Von Braun's secret report, submitted to the Army in September, summarized what he had said at the earlier meetings. This proposal became the Army's candidate for an IGY satellite project, which was later abandoned in favor of Project Vanguard.

---

[1]   1. **Summary**.

a. The realization of a relatively inexpensive Minimum Satellite Vehicle with a payload of 5 lb. is possible with components available from weapons development of the Army Ordnance Corps. Such components have reached an advanced development stage and are expected to attain a sufficient degree of reliability by 1956, to warrant their use in a satellite vehicle.

b. In view of the launching and tracking problems of a satellite vehicle it is suggested to establish a joint Army-Navy-Air Force Minimum Satellite Vehicle Project. Office of Naval Research, endeavoring to establish a Minimum Satellite Vehicle Project, has expressed definite interest in the proposal laid down in this memorandum.

c. While the feasibility of a Minimum Satellite Vehicle based on existing Ordnance Corps hardware may be considered as firmly established, further theoretical investigations, particularly on the tracking and "lifetime" aspects, are necessary. It is suggested to authorize such studies to the tune of approximately $100,000 for the present fiscal year. Office of Naval Research representatives have indicated their willingness to financially support such studies.

2. **Introduction**.

a. A satellite vehicle circling the earth would be of enormous value to science, especially upper atmosphere, meteorological and radiological research. Up to now any satellite project has been considered an extremely expensive undertaking, and a matter of more than 10 years even if an all-out effort were made. This memorandum proposes to show that, by limiting the payload of a first Minimum Satellite Vehicle to approximately 5 lb, such a project is feasible with presently available missile hardware, and that despite its payload limitations, such a Minimum Satellite Vehicle Project would be a worthwhile initial step.

b. The Office of Naval Research, Washington, D.C., has expressed its desire to support a Minimum Satellite Vehicle Project. On 25 June 1954 a meeting was held at ONR during which the desirability of action toward a Minimum Satellite Vehicle was discussed. Various proposals on how such a project could materialize in the near future were discussed. At the end of the meeting all participants agreed that the most promising approach to a Minimum Satellite Vehicle was a 4-stage missile, using a REDSTONE missile as first stage, and a cluster of LOKI rockets for the three upper stages.

c. The justification of the artificial satellite was summarized as follows:

(1) Upper air research. Due to the stay time of a Satellite Vehicle in outer space (several days or even months) a tremendous wealth of information, particularly about primary solar radiation effects on weather and radio communications [2] can be obtained with *one* firing. Considering the large number of balloon and rocket ascents presently conducted to gather such information, an instrumented satellite vehicle would be a very worthwhile investment.

(2) A rocket capable of carrying the weight of such data collection equipment into an orbit is still many years off. If payload requirements are drastically reduced, orbital speed can be reached with a proper combination of existing equipment. Such an experiment could serve to solve the problems of tracking and orbital stability and would be a logical first step.

(3) The establishment of a man-made satellite, no matter how humble, would be a scientific achievement of tremendous impact. Since it is a project that could be realized within a few years with rocket and guided missile experience available *now*, it is only logical to assume that other countries could do the same. It would be a blow to U. S. prestige if we did not do it first.

d. On 3 August 1954 Commander Wright and Lt. Commander Hoover of the Office of Naval Research visited Redstone Arsenal and discussed the possibilities of Army participation in a joint Army-Navy-Air Force satellite project with Brig. Gen. H. N. Toftoy and Dr. W. von Braun.

3. Description of the Minimum Satellite Vehicle.

The proposed Minimum Satellite Vehicle is a 4-stage rocket consisting of the REDSTONE Missile as 1st stage (or booster), and a three stage missile of clusters of the solid propellant rocket LOKI IIA with the following staging:

2nd Stage:	Cluster of 24 LOKI IIA rockets
3rd Stage:	Cluster of 6 LOKI IIA rockets
4th Stage:	(attaining satellite speed): 1 LOKI IIA rocket with 5 lb payload.

a. Description of REDSTONE Missile, used as 1st stage (or booster).

(1) REDSTONE Missiles #27 and #29 for re-entry studies.

The present REDSTONE R&D program provides that missiles #27 and #29 will be used, among other test purposes, for the study of the re-entry problem of a ballistic 400 to 600 N. Mi.-missile. These two missiles are normal REDSTONE missiles with a few minor modifications. The [3] usual 6900-lb payload (simulated by a concrete filling in the nose section during R & D launchings) is utilized for additional propellants. Thus the burning time is increased from 110 sec to 132.4 sec. The additional propellants are accommodated in an extra-long tank section. For this purpose the length of the cylindrical midsection of the missile is extended by 62 in. (This constitutes a minor change since it does not involve any changes in tooling.) The standard steel warhead hull is replaced by a lightweight aluminum hull for the guidance equipment. A special "re-entry nose" of about 24 in. base diameter and 500 lb. weight, carried in the missile's top, is separated from the missile after cut-off. Both missile and "re-entry nose" attain an altitude of approximately 265 statute miles. Fall into the atmosphere from this altitude duplicates the conditions during re-entry of a ballistic missile of a range from 400 to 600 N. Mi. The "nose cones" of missiles #27 and # 29 will carry telemeter equipment.

(2) REDSTONE as 1st Stage of Minimum Satellite Vehicle.
In the proposed Minimum Satellite Vehicle the 500-lb. "re-entry nose" is replaced by a three stage LOKI cluster of similar total weight. Thus, after successful firing of missile #27 and #29 a proven design of a booster is available that can be directly applied for the Minimum Satellite Vehicle.

b. Description of LOKI Cluster.

LOKI is an unguided, anti-aircraft, solid rocket developed under the auspices of Redstone Arsenal for the Ordnance Corps. LOKI I is presently in production in quantity (59,000 rounds on order). LOKI IIA, an advanced version, is scheduled to soon replace LOKI I production.

The LOKI clusters for the upper stages of the proposed Minimum Satellite Vehicle are obtained by bundling LOKI IIA rockets. With a payload of 5 lb, a particularly favorable staging combination consists of

24 LOKI's as first stage of the cluster (Satellite Vehicle's second stage)
6 LOKI's in the second stage of the cluster (Satellite Vehicle's third stage)
1 LOKI's in the third stage of the cluster (Satellite Vehicle's fourth and final stage).

This combination has been arrived at by comparison of more than 50 different payload and bundling combinations.

[4] Two arrangements of the total LOKI cluster (Satellite Vehicle's 2nd, 3rd and 4th stage) are shown.... It can be seen that both designs provide for a telescoped arrangement of the LOKI's. The total LOKI cluster is carried in a zero launcher. The zero launcher can be rotated about a king-pin mounted on ball bearings.

c. Description of Operation.

(1) The proposed Minimum Satellite Vehicle can be launched with standard REDSTONE launching equipment according to established REDSTONE procedure. The modified nose station contains standard REDSTONE guidance and control equipment. In order to obtain the necessary high accuracy in aiming of the LOKI cluster, the standard air-bearing type stabilized platform of the REDSTONE will be used as guidance head. The LOKI cluster and its zero launcher is brought to rotation at 1800 rpm prior to launching of the entire vehicle by means of an electromotor driven from ground power supply. The rotational speed is maintained during the REDSTONE boost phase from on-missile power supply. (The possible disturbing influence of gyroscopic effects on the REDSTONE missile control during the titling program have been studied and were found negligible.)

(2) After REDSTONE cut-off (at about 55 miles altitude) the nose station is separated from the booster. The nose section's orientation in space is now controlled by the standard REDSTONE spatial attitude control system, which consists of 8 small compressed-air jet nozzles. The tilting program in the guidance head, which determines the reference axes for the desired attitude, continues to run after nose section separation. Due to the absence of corodynamic forces at the altitudes involved, the tilt program, with the aid of the compressed air nozzles, now rotate the nose section into such an attitude, that, by arrival at the apex of its trajectory, it will be parallel to the surface of the earth. The gyroscopic effect of the rotating LOKI cluster will thereby be utilized in such manner that the correcting moments created by the air jets will cause the nose section to "process" into the desired attitude. The summit of the nose section's trajectory is reached at an altitude of approximately 186 statute miles.

(3) The first LOKI stage is ignited at the summit point, which is determined accurately by the REDSTONE guidance system in order to avoid a vertical component of the

trajectory. Each LOKI stage has a burning time of 1.9 sec. The launching of the two successive stages is initiated with time switches set at intervals of about 2.5 sec.

[5] The velocities reached by the stages are:

+	REDSTONE nose section at summit	4298 ft/sec
+	rotation of earth at equator	1519 ft/sec
		5817 ft/sec
+	2nd stage (24 LOKIs)	6283 ft/sec
+	3rd stage (6 LOKIs)	7152 ft/sec
+	4th stage (1 LOKI plus 5 lb payload "Satellite")	6972 ft/sec
+	Total satellite velocity	26224 ft/sec

(4) The satellite thus reaches a velocity *exceeding* the circular velocity of 25368 ft/sec at 186 miles altitude by 856 ft/sec. The satellite therefore enters an *elliptical* orbit around the earth with the following characteristics:

	perigee	apogee
For 0° deviation from horizontal flight path of 4th stage	186 mi (300 km)	819 mi (1318 km)
For 1.6° deviation from horizontal flight path of 4th stage	155 mi (250 km)	850 mi (1368 km)

(5) Suppose, now, that the 5-lb payload of the 4th (satellite)-stage has the shape of a sphere of 20 in. diameter. The lifetime of such a spherical satellite has been determined on the basis of data of the upper atmosphere as adopted by the Upper Atmosphere Research Panel. For the two orbits listed in the foregoing paragraph the lifetimes were found to be 360 days (for perigee at 186 mi altitude) and 90 days (for the perigee at 155 mi altitude). Within these lifetimes the elliptical orbits gradually change to circular orbits at (approximately) perigeal altitude due to aero-dynamic drag. The circular orbit then rapidly converts into a spiral path toward the earth, and the satellite is finally destroyed by aerodynamic heating like a meteor.

(6) The Minimum Satellite Vehicle can be tracked by optical means. For details see chapters 4.e. and 4.f.

4. Discussion of Main Problems.

a. **REDSTONE Missile**.

The REDSTONE missile is being developed as a ground support weapon for a payload of 6900 lbs. As of this date, 4 missiles have been launched, [6] 3 of which were successful.... The R&D program provides that missiles #27 and #29 will be equipped with enlarged tanks and used for re-entry studies. Launching of those two missiles is scheduled for Spring of 1956. This particular version of the REDSTONE is suited for the booster for the proposed Minimum Satellite Vehicle without changes. It is obvious that a high degree of reliability is a prerequisite for the REDSTONE's application for a satellite project. Thus the R&D program of the REDSTONE as a weapon will in no way be affected by the proposal presented herein. Its successful completion is rather a prerequisite for the Minimum Satellite Vehicle.

b. **Cluster LOKIs**.

LOKI is an anti-aircraft solid propellant rocket developed for the Ordnance Corps. LOKI I is presently in production (Production cost less than $100.00 per missile).... The highest performance is obtained by the improved LOKI IIA type, which is scheduled for production soon. Performance data of LOKI IIA are summarized in the following table:

Weight of rocket hardware	5 lb
Weight of propellant	17.3 lb
Total weight	22.3 lb
Specific impulse	219 sec
Combustion chamber head pressure	1320 psi
Thrust	2000 lb
Burning time	1.9 sec

The Air Research and Development Command has initiated a program for the development of a "Hypersonic Test Vehicle" (HTV). This vehicle is a two-stage solid rocket, using a cluster of 7 LOKIs in the first, and a cluster of 4 LOKIs in the second stage.... The first four rounds have been contracted with Aerophysics Development Corporation, Pacific Palisades, California, and will be fired in late 1954, starting about eight weeks from this writing. Firings will be at White Sands Proving Ground.

In view of Redstone Arsenal's great concern with the problem of re-entry of ballistic missiles of extended ranges, supporting research funds for FY 1955 and FY 1956 have been requested for the purpose of expanding this Air Force-sponsored program. Such a joint Army-Air Force program is expected to furnish, at minimum expense, vitally needed data on heat transfer, and behavior of structures and material, at Mach numbers from 8 to 13.

[7] The Hypersonic Test Vehicle may be considered a natural forerunner of the three-stage LOKI cluster for the proposed Minimum Satellite Vehicle. But here again, the Satellite Vehicle Project could never delay the development of the Hypersonic Test Vehicle, since success with the HTV is a prerequisite for the former.

c. **Accuracy of launching LOKI cluster from REDSTONE nose**.

An important problem is the accuracy of aiming the REDSTONE nose section with its LOKI cluster into the horizontal direction at the summit of the first stage trajectory. This becomes evident by comparing the lifetime of the two orbits described in chapter 3.c.(5). The standard spatial attitude control of the REDSTONE missile nose necessitates at least two blasts of the control jets for the correction of an angular deviation of any of the three missile axes: one blast to turn the missile to the zero position and another one to stop it in this position. In reality the zero position can be approached only after repeated application of control jets, because of oscillations around the zero position (poor damping). This spatial attitude control method, therefore, keeps the missile axis continuously and slowly oscillating around the desired zero position. While this is entirely adequate for the standard REDSTONE trajectory, it constitutes a severe handicap for the launching of the LOKI's at the apex of the first-stage trajectory.

The proposed pre-rotation of the zero launcher with its LOKI cluster at 1800 rpm has the effect that launcher and LOKI cluster are gyro-stabilized as long as no external forces are applied. But their joint longitudinal axis can be precessed into any desired direction by applying a force perpendicular to the desired angular direction of movement. The turning of the axis stops, if the applied force stops. It is evident that this method avoids those undesirable oscillations around the desired firing direction of the LOKIs and results in a higher accuracy of the aiming of the cluster. The pre-rotation of the LOKI cluster launcher has the additional and important advantage, that it minimizes the error introduced by inaccuracies in the alignment of the LOKI bundles, of their thrust axis, ignition delays, differences in burning times, and possible torques caused in the rockets leaving the zero launcher. A preliminary analysis indicates that by using the pre-rotation method the total error angle built up during the burning time of the three-stage LOKI cluster may be kept well under 1 degree, which, according to the figures listed in chapter 3.c.(5) is adequate for an extended orbital lifetime of the uppermost stage. The exact magnitude of such error angles will be known after a number of Hypersonic Test Vehicles have been launched.

d. **A stepping stone: Very high altitude firing of the Minimum Satellite Vehicle.**

As an intermediate step toward a Minimum Satellite Vehicle it is suggested to launch a missile, consisting of a REDSTONE as a booster, and a LOKI cluster as upper stages, from the REDSTONE launching site at [8] Patrick Air Force Base, in a nearly vertical trajectory (approximately 2 degrees). This missile would consist of a REDSTONE (with enlarged tanks) and a cluster of 24-6-1 LOKIs, and would reach an altitude of 6400 miles (almost two earth radii!). The purpose of this firing would be to test the combination of REDSTONE booster and spinning LOKI cluster during the REDSTONE flight phase, the technique of pre-rotation of the LOKI launcher, and the launching of the LOKI bundle. Such a near-vertical firing would permit accurate tracking from take-off all the way to, and including, the uppermost stage of the LOKI cluster, and the deviations of the latter's trajectory from the flight path tangent at REDSTONE cut-off. Additional data could be telemetered if desired. The velocities reached by the various stages would be

1st stage (REDSTONE)	7900 ft/sec
2nd stage	14150 ft/sec
3rd stage	21300 ft/sec
4th stage	28300 ft/sec

Such an experiment would involve no dangers from the standpoint of flight safety. The REDSTONE booster would drop into the ocean at a range of about 110 N. Mi. (the same range as anticipated for re-entry missiles #27 and #29). The upper stages would never reach the earth surface again but burn up like meteors upon re-entering the atmosphere. Since LOKI rockets are made of aluminum, re-entry temperatures are sufficient to melt and even vaporize the falling rockets.

e. **Selection of orbit for the Minimum Satellite Vehicle.**

The selection of a suitable orbit for the Minimum Satellite Vehicle is closely linked to the problem of tracking and the number of tracking stations required. As a result of the combined effects of orbital motion and earth's rotation, any orbit inclined to the equatorial plane leads over a vast portion of the earth's surface and the Minimum Satellite Vehicle may thus be visible from many regions on earth. From the "propaganda" angle, this may be desirable, but it may also entail less desirable political problems. From the scientific aspect the main disadvantage of an inclined orbit lies in the fact that it is impossible to set up a sufficient number of tracking stations to "keep an eye" on the Minimum Satellite Vehicle. It must be kept in mind that the Minimum Satellite Vehicle is of limited lifetime because it is still affected by drag in the uppermost layers of the atmosphere. This means that the coordinates of the elliptical path change slightly with each revolution. Such slight changes, on the other hand, offer an ideal opportunity to determine the atmospheric density at altitudes from 150 to 800 miles.

Effective tracking requires that the Satellite passes repeatedly over the same station or stations. Therefore, an orbit in the equatorial [9] plane is most advantageous. In order to reduce the logistic problem of the firing preparations and launching, the Minimum Satellite Vehicle could be launched from the Navy's experimental guided missile ship USS "Norton Sound." In addition to the reasons mentioned before, the equatorial plane offers the great advantage that a velocity gain of 1519 ft/sec is obtained due to the rotation of the earth, if the satellite vehicle is launched in an eastern direction.

f. **Visibility of Minimum Satellite Vehicle in the orbit.**

Another prerequisite for successful tracking is sufficient light for optical tracking instruments. The simplest method to provide sufficient illumination and contrast would be to observe the Satellite shortly before dawn or shortly after sunset. The Satellite, illuminated by sunlight, would then be visible as a fast-moving star against a relatively dark sky. A study of this possibility indicates that, in order to obtain a brightness equivalent to that of a star of first magnitude (e.g. "Capella"), the object would have to be at least 20 feet in diameter. In principle, it appears quite feasible to provide a reflecting surface of this size even within the payload limitation of 5 lb. A particularly simple solution would be a balloon, carried aloft collapsed in the fourth stage's nose, and inflated with Helium after the orbit has been attained. However, it is likely that such a balloon would soon be punctured by cosmic dust particles and become ineffective as a light reflector. Therefore, some kite-like structure may be better suited.

Unfortunately, the conditions for visibility by a tracking station are severely restricted by the (unpredictable) deviations from the desired orbit caused by lack of accurate cut-off velocity control and possible cut-off tangent dispersions of the uppermost stage, as well as uncertainty of upper atmosphere density. Tracking at dusk and dawn will, therefore, be a rather haphazardous endeavor and should be supplemented by an active light source in the uppermost stage. Such possibilities have been discussed with Mr. E.P. Martz, Jr., Chief Optical Systems Section, Flight Determination Laboratory, White Sands Proving Ground. Mr. Martz has suggested to equip the Minimum Satellite Vehicle's uppermost stage with a gaseous discharge tube actuated by solar storage cells utilizing the solar radiation during the sunlit portion of flight and re-emitting as a flashing light during the night. He believes that use of solar storage batteries such as developed by Bell Telephone and by Wright Field are promising. It appears probable that such a light source, adequate for instrument tracking for a period exceeding one month, can be built within the payload limitation of 5 lb. Further studies will be required to determine whether the light flashes can be made bright enough for visibility with the naked eye.

Further methods discussed to improve optical visibility of the satellite vehicle include painting of the uppermost stage with fluorescent paint (brightness could be doubled because ultraviolet light is [10] converted into visible light) and luminescent paint (will absorb sunlight during the day and omit light during the night). There is also a faint possibility for the successful use of chemical smoke trails, such as used in tracking of solid rockets. (It has to be further investigated whether this method is suited to a Satellite.) Chemical flares or shaped charges may also be feasible. Another possibility would utilize solar or artificially induced fluorescence of sodium, mercury or other metallic vapors. (The fluorescence of such vapors is greatly increased by the high ultra-violet radiation from the sun in the vacuum of outer space.) Finally, use of fluorescence of solid mediums has been discussed. (The solid mediums would be activated by small electric current and additionally by solar radiation and radioactive substances.)

For ground tracking stations, normal meteor tracking cameras appear to be best suited. Such equipment is available at White Sands Proving Ground.

5. Acknowledgements.

This report has been based on detail studies prepared by:

Dr. William Bollay	Aerophysics Development Corporation,
Mr. J. B. Kendrick	Pacific Palisades, California.
Mr. E. P. Martz, Jr.	Chief Optical Systems Section, Flight Determination Laboratory, White Sands Proving Ground, New Mexico.
Dr. Wernher von Braun	Guided Missile Development Division,
Mr. Gerhard Heller	Ordnance Missile Laboratories,

Mr. Hans R. Palaero   Redstone Arsenal, Huntsville, Alabama.
Mr. Fritz K. Pauli

6. References.

a. "Observing the weather from a satellite vehicle", lecture presented at the American Museum - Hayden Planetarium, May 4, 1954 by Harry Wexler, Chief, Scientific Services Division, U.S. Department of Commerce, Weather Bureau.

b. "Minimum Orbital Unmanned Satellite of the Earth (MOUSE)" for Astrophysical Research. Lecture presented at the American Museum - Hayden Planetarium, May 4, 1954 by Dr. Fred S. Singer, Professor, Dept. of Physics, University of Maryland.

Document II-8

Document title: "On the Utility of an Artificial Unmanned Earth Satellite: A Proposal to the National Science Foundation, Prepared by the ARS Space Flight Committee, November 24, 1954," *Jet Propulsion*, 25 (February 1955): 71-78. [Copyright American Rocket Society (now American Institute of Aeronautics and Astronautics), 1955. Used with permission.]

In 1934, the American Interplanetary Society, one of the earliest U.S. advocates of spaceflight, had changed its name to the American Rocket Society (ARS) to improve its technical legitimacy. In 1953, the ARS invited Alan T. Waterman, director of the National Science Foundation (NSF), to attend a meeting of the society's Space Flight Committee. Soon after, the committee issued a confidential report calling for the NSF to study "the utility of an unmanned satellite vehicle to science, commerce and industry, and national defense." This report was followed in 1954 by a formal proposal, "On the Utility of an Artificial Unmanned Earth Satellite." It was partly because of advocacy groups such as the ARS that satellites were put on the government's scientific agenda.

---

[71]   **Introduction**

This is a proposal to the National Science Foundation that the Foundation sponsor a study of the utility of an unmanned, earth-satellite vehicle. The proposal is made by the American Rocket Society in the normal exercise of its functions.[1] The role of the Society in this matter is made clear by the following policy statement adopted by the Board of Directors: "The American Rocket Society should act as a 'catalyst' and should promote interest and sound public and professional thinking on the subject of space flight. It should not attempt to evaluate the merits of individual proposals or undertake work on the subject of its own accord. It should, however, encourage such activity on the part of other organizations."

It is apparent, then, that the Society cannot undertake to make the study. It can, however, serve the National Science Foundation in a number of ways, and believes it is doing so in bringing this subject to the Foundation's attention. Should the Foundation elect to sponsor the study the Society could assist by encouraging scientists and engineers both inside and outside the Society to participate. The Society would be willing to perform any other service within its functions and abilities to assist the National Science Foundation in implementing this proposal.

---

1. The American Rocket Society is a professional engineering and scientific organization devoted to the encouragement of research and development of jet and rocket propulsion devices and their application to problems of transportation and communication. It is actively concerned with various technical aspects of space flight, and at the present time is also interested in military applications of the reaction principle.

Background

The proposal was prepared by a Space Flight Committee appointed by the President of the Society. The Committee decided that the study of the utility of an unmanned, earth satellite would be one of the most important steps that could be taken immediately to advance the cause of space flight, and that this step would also increase the country's scientific knowledge and, indirectly, promote its defense.

Why an unmanned, earth satellite? Although many satellite proposals have been put forward, the small, unmanned satellite is the only one for which feasibility can now be shown. This opinion is held by many responsible engineers and scientists involved in rocket and guided-missile work and in upper-atmosphere research. At any rate, most of these people agree that the unmanned earth satellite would be the first step toward more ambitious undertakings. It is felt generally that, although the satellite vehicle has yet to be built, the components, i.e., power plants, airframes, stabilization systems, etc., are either available or under development in conjunction with the nation's guided-missile effort. Furthermore, the country now has nine years of experience in the techniques of instrumenting high-altitude sounding rockets, which techniques would have application to the earth satellite.

Why study utility? Although many claims have been made for the utility of a satellite vehicle and many uses have been proposed, the subject has not been thoroughly investigated by a responsible organization and, at present, does not rest upon a firm foundation. On the other hand, enough is known about possible useful purposes so that most cases are readily amenable to study, and, if such a study were made, reasonably positive conclusions could be drawn. Because of recent advances in guided missiles, the cost of producing a small, unmanned satellite is probably not the mammoth sum that was at one time considered necessary. Nevertheless, the creation of even a small satellite is still a major undertaking and will require a sizable amount of money. It is important that there be justification for the expenditure of this money. The Society feels that to create a satellite merely for the purpose of saying it has been done would not justify the cost. Rather, the satellite should serve useful purposes—purposes which can command the respect of the officials who sponsor it, the scientists and engineers who produce it, and the community who pays for it. The Society feels, therefore, that the study of utility is one of the most important tasks to be accomplished prior to creation of a satellite.

It was apparent to the Committee in its early deliberations that the subject of utility could not be entirely divorced from feasibility, and that some concept of feasibility would have to be assumed. This was done not to be restrictive, but to provide a frame of reference from which those considering utility could proceed. It was assumed that it would be feasible to establish a small payload in an orbit, the difficulty increasing with the size of the payload, and that means could be provided for communicating information from the satellite to the surface of the earth. With this concept in mind, various fields of utility were suggested as follows:

Astronomy and Astrophysics. A satellite could overcome some of the limitations on observations made through the earth's atmosphere.

Biology. Of early importance would be the effects of outer space radiations on living cells.

Communication. A satellite might provide a broad-band transoceanic communication link. A future possibility is that of obtaining continental coverage when the satellite is used as a relay station for radio or television broadcasts.

Geodesy (including Navigation and Mapping). The size and shape of the earth, the intensity of its gravitational field, and other geodetic constants might be determined more accurately. Practical benefits to navigation at sea and mapping over large distances would ensue.

Geophysics (including Meteorology). The study of incoming radiation and its effect upon the earth's atmosphere might lead eventually to better methods of long-range weather prediction.

Experiments Arising from Unusual Environment. The characteristics of the environment (weightlessness, high vacuum, temperature extremes, etc.) will suggest experiments that could not be performed elsewhere.

This list is by no means complete—it is probable that the study would reveal other fields of equal or greater utility.

In order to provide a preliminary sampling of opinion, the committee asked a number of scientists (chosen at random from those known to Committee members) to give brief summaries of their opinions on the utility of an unmanned, satellite vehicle. These papers are presented as appendixes to this proposal and include the following: (A) "Astronomical [72] Observation from a Satellite," by Ira S. Bowen; (B) "Biological Experimentation with an Unmanned Temporary Satellite," by Hermann J. Schaefer; (C) "The Satellite Vehicle and Physics of the Earth's Upper Atmosphere," by Homer E. Newell, Jr.; (D) "Comments Concerning Meteorological Interests in an Orbiting, Unmanned Space Vehicle," by Eugene Bollay; (E) "The Geodetic Significance of an Artificial Satellite," by John O'Keefe; (F) "Orbital Radio Relays," by John R. Pierce.

Recommendation

In view of the facts cited, it is proposed that the National Science Foundation sponsor a study of the utility of an artificial, unmanned earth satellite.

ANDREW G. HALEY, *President*
MILTON W. ROSEN, *Chairman,*
Space Flight Committee

COMMITTEE MEMBERSHIP: Harry J. Archer; William J. Barr; B. L. Dorman; Andrew G. Haley; Kenneth H. Jacobs; Chester M. McCloskey; Keith K. McDaniel; William P. Munger; James R. Patton, Jr.; Richard W. Porter; Darrell C. Romick; Milton W. Rosen; Michael J. Samek; Howard S. Seifert; Willis Sprattling, Jr.; Kurt R. Stehling; and Ivan E. Tuhy.

Appendix A
Astronomical Observations from a Satellite

IRA S. BOWEN
Mount Wilson and Palomar Observatories

The following comments are an expansion of a conversation held by H. S. Seifert with Dr. Ira S. Bowen, Director of the Mount Wilson and Palomar Observatories. The ideas herein originated with Dr. Bowen and have been reviewed by him for accuracy.

Astronomical observations through the Earth's atmosphere are at present limited by three factors: (a) The resolution of detail is degraded at least tenfold by atmospheric turbulence (poor seeing). (b) Exposure time, and hence limiting star magnitude, is curtailed by fogging due to light scattered in the atmosphere. (c) Certain radiations, i.e., regions of the spectrum, are completely absorbed in the atmosphere. Thus, if optical equipment equivalent to that now available at the surface could be placed outside the atmosphere, much additional information in the form of planetary detail, new, remote, or faint objects, and short wave-length spectra would be obtained. This information would be of great interest and value to astronomers and to the sciences generally. The ideal situation would be to place the 200-in. telescope and its accessories on a firm platform such as the moon.

Since optical equipment projected into an orbit on a man-made satellite will be riding on a small and relatively unsteady base, certain practical limits and difficulties will be found, as follows:

Angular Resolution. The best optical resolution which the atmosphere will permit, on days of optimum seeing, which occur only a few times yearly, is of the order of $1/4$ to $1/2$ sec of arc. The 200-in. telescope would permit a theoretical resolving power of 0.025 sec of arc, and a 20-in. to 40-in. telescope would permit a resolving power of 0.25 to 0.125 sec of arc, if free from atmospheric effects and geometric distortions. Thus in order to make use of the transparency of space and secure more detail than can be seen from the ground, an automatic satellite orienting and guiding system would be needed which was stable to an accuracy lying between 0.10 and 0.01 sec of arc.

Limiting Magnitude. Because of night sky light scattered by the atmosphere, objects fainter than a certain limiting magnitude cannot be distinguished from the general background fog. In the case of the 200-in. telescope, exposures longer than half an hour are, for this reason, not useful. In order to record objects of the same faintness, as can be done with the 200-in., a telescope of reasonable size for transport on a satellite, say 30 in., would require an exposure time of 10 to 24 hours. Since the orbital period is of the order of $1\,^1/_4$ hours, of which less than half is spent in the Earth's shadow, a mechanism would be required for shielding the telescope and camera during the sunlit periods while maintaining precise orientation.

Short Wave-Length Spectra. By the use of sounding rockets equipped with sun-following servos, it has been possible to photograph solar spectra down to 1200 A with low resolution. The long exposures required for high-resolution solar and stellar spectra in this wave-length region cannot be obtained during the few minutes or even seconds of high-altitude flight time typical of sounding rockets. Adequately high resolution could be obtained from a spectrograph using light collected with a 12-in. mirror for detailed spectra of the brighter stars, with exposures of several hours. Since the physical dimensions of the equipment are not large and the orientation tolerances are less strict than for telescopic images, a spectrograph would probably be simpler than a telescope to get into proper working order.

Telemetering of Data. If one assumes that a photographic plate cannot be recovered from a satellite, certain problems arise. Any data collected must be capable of being translated into a radio or optical signal and relayed to the ground. The photographic plate has the fundamental advantage that photons are registered simultaneously in all resolvable parts of the spectrum or image. Thus shortening required exposure time. While electronic photon counters exist which equal or excel the sensitivity of the photographic plate, they can collect energy from only one part of the spectrum or image at a time, thus increasing the required exposure time.

A possible technique might be worked out in which plates are exposed and developed automatically (after the manner of the Polaroid-Land Camera), after which the fixed image could be scanned and transmitted sequentially when convenient by a photoelectronic system. Thus the stored data might be held and relayed by a transponder activated from the earth's surface only when the satellite was within radio range of a particular ground station, thus eliminating the need for a dozen or more ground telemetering stations spaced around the equator.

Appendix B
Biological Experimentation with an Unmanned Temporary Satellite

HERMANN J. SCHAEFER
U.S. Naval School of Aviation Medicine

If humans are to fly in the regions at the upper end of and outside the atmosphere, an "artificial environment" has to be provided which maintains full or near sea-level values of the various physical conditions. Whereas this task can be handled, though with considerable technical expenditure, for most of the factors involved, two novel phenomena develop in vehicles moving outside the atmosphere which cannot be compensated very easily. These are the heavy components of the primary cosmic radiation and the state of weightlessness. The technical means of restoring normalcy with regard to these conditions, though theoretically available, imply prohibitive measures with regard to weight and power. The only way out is to study the effects of both influences on the human organism with the aim in mind that a tolerance can be established which does not impose too severe limitations upon the flight of humans in extra-atmospheric regions.

In the discussion of the usefulness of a small artificial satellite for the conquest of space, the question has been put [73] whether biological experiments could be performed in such a vehicle for gaining information as to the two aforementioned problems. The artificial satellite, in this discussion, is conceived as an unmanned and rather small vehicle which will stay in the orbit a limited time only, a few weeks at the most. A further limitation is that preferably all experiments are to be carried out by preset servomechanisms and telemetered recording. This would by-pass the complex and difficult task of recovery.

It should be pointed out from the very beginning that the last-mentioned condition imposes serious restrictions on any biological cosmic-ray experimentation to the degree where it seems questionable whether useful experiments can be designed at all. Exposure to the heavy components of the primary cosmic radiation belongs to the category of so-called long-dosage long-term irradiations which are characterized by slowly developing, initially inconspicuous, but insidious tissue damage. The peculiar feature about this exposure hazard is that a total ionization dosage which nominally remains well within the permissible limits of the official international definition is actually administered in an extremely uneven distribution. As a consequence of it, a small number of cells of the exposed tissue receive ionization dosages up to 10^5 times larger than the average total ionization dosage. It is already well established experimentally that such "heavy nuclei hits" produce severe local radiation injury in the cellular structure of living tissue. No information, however, is available on how many of such hits can be administered to the mammalian organism before a general reaction, i.e., manifest radiation injury, develops.

A conclusive answer to the latter question requires exposure of test animals over an extended period of time. Rocket or balloon flights are entirely inadequate for this purpose, the former because of the short duration, and the latter because of the too low altitude. The study of the radiation effects of the genuinely primary cosmic radiation in regions entirely outside the atmosphere, with exposure times of many days, is of fundamental value for both basic research in radiobiology and the development of high altitude flight. Experiments carried out in an artificial satellite would contribute greatly toward a solution of this problem and would be incomparably more effective than any balloon or rocket flight can ever be.

There are a multitude of radiation effects on living tissue which lend themselves to experiments of the type under consideration. Local radiation injury from heavy nuclei hits has been demonstrated recently very conspicuously in skin tissue of mice. This reaction, however, requires that the animals be recovered alive. Other reactions exist which are of similar sensitivity, but could be tested even if the animal were killed while recovering it as long as the latter is not heavily disfigured or burned. An essential prerequisite for all these reactions is a meticulous autoptic and microscopic examination. That means that recovery of the animals is indispensable. It is suggested, therefore, that the problem of recovery be included in the project from the onset if investigation of biological cosmic-ray effects is contemplated at all.

The discussion of the biological effects of the primary cosmic radiation would be incomplete without mentioning a question which actually concerns the physicist. This question pertains to the fragmentary knowledge of the frequency and the mass spectrum of super-heavy nuclei. Heretofore, iron (Fe) was considered the heaviest regularly occurring component of the heavy spectrum. However, heavier nuclei have been recorded on rare occasions, but no statistics have been established thus far with regard to frequency and mass spectrum. Recordings over extended periods of time in the regions clear of the atmosphere would rapidly accumulate data on these giant nuclei. Such measurements, in contradistinction to actual animal experimentation, could be carried out exclusively by telemetering. As a matter of fact, J. Van Allen has already developed the tools for this type of measurement. His pulse ionization chamber has proved a sensitive and reliable instrument for analyzing the heavy nuclei spectrum and has been repeatedly and successfully used for heavy nuclei recordings at extreme altitudes with rocket balloon tandems. A modification of his method for use in an artificial satellite should be comparatively easy.

The other novel environmental condition to be encountered in orbital travel outside the atmosphere is weightlessness. Present-day knowledge as to its physiological effects is scarce. The state of weightlessness or, as it is also called, the gravity-free state, can be produced artificially in a freely falling elevator car, in the warheads of rockets in unpowered flights outside the atmosphere, and in the powered flights of high-speed aircraft by steering along a free trajectory. The rocket method is most effective and can produce the state of weightlessness for a period of a few minutes. It cannot be used for humans at the present state of development. Second in effectiveness is the trajectory flight of high-speed aircraft. Its present limit stays at a longest duration of about 30 sec. It can and has been applied to humans. No physiological disturbances or incapacitation effects have been so far reported. The same statement holds for animal experiments with mice and monkeys in rockets. Considering the short exposure time of both vehicles, these findings cannot be considered as conclusive evidence that such disturbances will not develop eventually.

In an animal capsule carried by an artificial satellite this problem could be more thoroughly studied and the success of the experiment would not necessarily depend upon the recovery of the animals. A few telemetering channels, reserved for relaying breath and pulse frequencies, would provide interesting information. Telemetering of the locomotion of the animals could also be valuable.

It must be admitted, of course, that the more severe effects of weightlessness of long duration on humans are likely to develop along the psychophysiological line originating from disorientation. Animal experimentation, therefore, can be of limited value only, though some disorientation studies have been successfully performed with mice in a rocket flight.

Evaluating the entire situation on a comparative basis, it must be said that the cosmic-ray problem certainly bears the higher practical importance as well as scientific interest. It should weigh heavily in any debate on the justification of the costs involved in a satellite project, and this all the more since it is closely related and can be combined with research problems concerning the pure physics of cosmic radiation on which paramount importance rests with regard to gaining knowledge of the nuclear forces.

Appendix C
The Satellite Vehicle and Physics of the Earth's Upper Atmosphere

HOMER E. NEWELL, JR.
Naval Research Laboratory

The purpose of the present note is to consider the usefulness of the satellite vehicle for scientific research and to point to a few important experiments which might be done with such a vehicle.

The gas particles, ions, and radiations of the earth's atmosphere act, react, and interact to produce phenomena which are still not fully understood, such as the ionosphere, the aurora, and fluctuations in the earth's magnetic field. In an effort to explain these things a host of researches have been undertaken throughout the past half century; and the effort continues to grow.

Fundamental to the research of the atmosphere has been and is the question of energy. The complex and confusing happenings in the atmosphere are simply a manifestation of [74] an influx of energy from outer space. A detailed knowledge of the nature and magnitudes of the energies concerned would go a long way toward solving some of the important problems. But it is right here that the observer on the ground runs into basic difficulties. The incoming energy in which he has the greatest interest is that which is absorbed at high altitude. The experimenter is, therefore, prevented from observing it in its original form. This leaves many a theorist to speculate on one of the most important ingredients of his theory.

A space station at sufficiently great altitude, say a thousand kilometers or more, would enable the physicist to monitor the energy influx into the earth's atmosphere. A primary carrier of such energy would be electromagnetic radiation from the sun, measurements of which would be of as much interest to the solar physicist as to the geophysicist. Energy is also brought in by particles such as cosmic rays, meteors, and micrometeorites. Because of their extremely high particle energies, the cosmic rays have an important place in current nuclear research. They produce a small but important ionization in the lower atmosphere, but probably have a negligible effect upon the upper atmosphere. Lower energy particles from the sun are believed to cause the aurora, and, in fact, protons have been observed moving downward in auroral displays. It also appears as though such particle radiations may play a significant part in the formation of the ionosphere, particularly the F-region. At the moment the question is wide open and is an important one. Finally, for the sake of completeness, perhaps one ought to mention stellar light, although to the upper-air physicist the corresponding energies are entirely negligible in comparison with those found in solar light.

At the present time rockets are being devoted to an intensive study of the various radiations listed in the preceding paragraph. The rockets permit the experimenter not only to observe and measure the radiations, but also to determine the altitudes at which the different radiations have their effect. The rockets are, however, one-shot affairs, and furnish only a matter of minutes in which to observe. They are not convenient for making a large number of measurements over an extended period of time. It is here that the satellite would be of considerable value. Equipment carried in such a vehicle could be used to measure a specific radiation over long intervals of time.

Arrays of geiger counters could be used to monitor the influx of cosmic rays. By using counters with varying amounts of material to be penetrated by the rays, it would be possible on the low energy side to count the particles in a number of low energy bands. These would perhaps be the simplest cosmic ray experiments to be done. With a little more complication, proportional counters and low efficiency geiger counters could be employed to determine the charges on the particles observed. If the satellite station were made to encircle the earth in a geomagnetic meridian plane, the earth's magnetic field could be used as a rigidity spectrometer, just as is done now in balloon and rocket-borne experiments. By comparing observations made at different geomagnetic latitudes, the low energy end of the cosmic ray spectrum could be studied in detail.

Counters with very thin windows could be used to study auroral particles, which are of several orders of magnitude lower energy than what are commonly termed cosmic rays. As a matter of fact, it may well be that incoming particles will be found to fill out a continuous spectrum of energies between the kilo-electron volts now associated with auroral particles and the billions of electron volts found in the cosmic rays. If so, it will be of considerable interest to observe and study these particles.

Photon counters with various fillings and windows, and photomultiplier tubes, are now used in rockets to study different bands of solar electromagnetic radiation from the visible wave lengths down into the x-ray regions. Present techniques would be adaptable to observations from a satellite station. Such observations would permit a study of the fluctuations over extended periods of time in the different wave-length bands. These fluctuations are connected with solar activity and give rise to distinct effects in the earth's atmosphere. Variations in the near ultra-violet, for example, cause changes in the distribution of ozone in the atmosphere and perhaps have an effect upon weather. Variations in the far ultra-violet and x-rays have a pronounced effect upon the ionosphere, and much more data are required to understand these effects.

The magnitude of the influence of meteors, especially of micrometeors, upon the earth's atmosphere is not yet fully known. Of late there is some tendency to ascribe considerable significance to the role of the micrometeors in the ionization of the higher ionospheric layers. At present, radar and radio techniques are the primary ones in the study of these particles, but methods are being worked out for rocket studies. For example, a very thin sheet of metal used as a resistive element in a circuit would be one means of detecting micrometeors. The particle, upon striking the sheet, would produce a tiny puncture, giving rise to a small but observable pulse. Such a sheet could be used similarly as one plate of a condenser. These rocket techniques, when developed, could also be used in a satellite observatory.

The foregoing experiments are of interest to the upper-air and solar physicist. It seems clear that known techniques, such as those suggested, could be used to carry out the various experiments. Familiar telemetering methods, such as those now used in rocket studies, could be adapted. But, there will be, of course, peculiar experimental difficulties to overcome. A number of these difficulties are already quite plain, although there may be some that are not now apparent.

One of the foremost problems is that of power. It has already been pointed out that a chief advantage of a satellite platform would be the possibility of making continuing observations over long periods of time, such as throughout a given year. The longer the period which could be covered the more satisfying would be the experiment. But to make the various measurements would require energy, which, presumably, would be stored in batteries. Associated with each 50 watt-hours of such stored energy would be something like one pound of battery mass. Including the operation of transmitting measured data, the satellite experiment would probably use energy at the rate of at least 20 watts. Assuming that the mass limitations in an early satellite vehicle would be on the order of 100 lb. this would preclude continuous operation over anything even approaching a year. To surmount this obstacle one would probably resort to periodic operation, to low current components like transistors, and to the use of the sun's light for the recharging of batteries. In this last connection, one may note the recent announcement by the Bell Telephone Laboratories of a solar device, consisting of thin silicon strips with an even thinner covering of boron, which could produce about 50 watts per square yard of exposed surface. The equipment could be turned on and off periodically by some low power timing device, or by radio means from a ground recording station. The latter method might be the preferable one, since it would permit turning on the equipment whenever ground-based observations showed the existence of unusual solar, ionospheric, or cosmic ray activity.

A second problem would be that of temperature. If the satellite were to present the same side always toward the sun, that side would become intolerably hot. By having that station rotate, however, enough of the energy absorbed at any spot on the satellite's surface could be reradiated into outer space to keep the temperature of the station and its equipment at an admissible level. This procedure would, however, introduce some difficulties into the basic experiment. Some omnidirectional arrangement of sensing elements would be required in order to make the experiment independent of the station's orientation. For the high energy cosmic ray [75] experiments, this could be done with crossed counters. In the case of low energy particles and solar ultra-violet light and x-rays, requiring counters with special windows, banks of counters might be the answer. A number of counters connected in parallel could encircle the satellite so as to present a window on every side.

The weightlessness of everything in the satellite might present some vexing problems. Thus, gassing of the batteries used is no cause for concern on the earth, where the bubbles simply rise in the liquid and pass off harmlessly. But in a nonrotating satellite station, the gas bubbles would not rise.

Remaining right where they form, they would cause the electrolyte to foam and fizz, like a bottle of soda. In a rotating satellite the centrifugal force field due to the station's rotation might provide the answer to such problems as this.

The satellite would be giving off gases continuously from the surfaces exposed to space. Also, in the near-vacuum surrounding the station, the metal structure of the satellite would evaporate slowly. Such effects would make it extremely difficult, if not impossible, to measure the original material content of the space in the neighborhood of the vehicle. Hence, although of great interest, such measurements are not suggested at the present time for a satellite experiment.

Appendix D
Comments Concerning Meteorological Interests in an Orbiting Unmanned Space Vehicle

EUGENE BOLLAY
North American Weather Consultants

In view of the fact that unmanned, orbiting space vehicles appear to be feasible with our present engineering knowledge, it would seem appropriate to comment briefly on measurements one might desire to make of meteorological interest in space.

Our present concepts in meteorology revolve largely around the solar balance and the resulting fluid dynamic consequences. Such consequences are produced by heating a gaseous mixture, such as the earth's atmosphere, unevenly under various roughness conditions. A large scientific effort has been made in connection with these hydrodynamic considerations, in contrast to studying the initiating impulse—the solar radiation phenomena.

It would seem that a space-observing platform would be ideally suited for collecting direct solar measurements as well as indirect solar relationships such as magnetic storm activity, etc.

Another item to analyze from a space station is a census of meteoric dust. The recent correlation by Dr. Bowen in Australia of meteoric dust and rainfall deserves rather intense and careful research.

The saying that one does not see the forest because of the trees may be rather appropriate to this problem. On earth we are engulfed in numerous meteorological details which mask or have masked rather completely the initiating circumstances which may become evident from observations in space. Connection of the theories of the General Circulation of the Atmosphere to direct solar influences is still lacking. Information from space may provide data for the solution of this challenging problem.

Appendix E
The Geodetic Significance of an Artificial Satellite

JOHN O'KEEFE
Army Map Service

I. Purpose and Scope
This report is intended to indicate the extent of the usefulness of a small artificial satellite, weighing only a few pounds, in finding out more about the size and shape of the earth, the intensity of its gravitational field, and certain other related constants.

II. Illumination
Most of the possible applications of the satellite would depend on detecting its presence either by visible light or by radar. The latter application falls outside the scope of the present paper, which would be chiefly concerned with the application of visible light. The first question is, then, how the satellite could be illuminated with sufficient brilliance to be observed.

A. Sunlight
Assuming that the satellite were to consist of a hollow aluminum sphere, 8 ft in diameter, such as could be produced by inflating a foil, it would have a surface brightness somewhat greater than that of the moon. The ratio of the total brightness would then be the ratio of the apparent surface area. At a distance of 250 miles, an 8-ft sphere would have a surface area of 1.1×10^{-11} square radians. The moon has a surface area of 1.31×10^{-4} square radians or 1.2×10^7 times as great. The full moon is of -12 magnitude; the above ratio in brightness corresponds to approximately 17.7 magnitudes. The higher reflectivity of

aluminum as compared with lunar surface is compensated for by the fact that the moon has been taken as full while the satellite will be in a partial phase. Thus the object would have a brilliance of a star of magnitude 5.7, barely visible to the naked eye at night in a clear sky. It would not be visible by day even with a large telescope. Also, it would not be visible even in a large telescope when in the shadow of the earth, since its surface brightness would then be less than that of the fully eclipsed moon. It would be visible only between the end of twilight and the time when it passed into the earth's shadow. The end of astronomical twilight would almost coincide with the entrance into the earth's shadow; the interval might be as little as 15'. From the end of nautical twilight to the entrance into the earth's shadow there would be an interval of about 2^m. The interval could be very considerably increased by setting up an orbit which would parallel the line between night and day over the earth.

B. Intrinsic Illumination

1. Evidently the illumination could be somewhat prolonged by making use of a fluorescent coating which could store up solar radiation, and so continue to shine for some time after the sun's light was cut off. This process would not give any more light than direct illumination; but it might produce a longer storage.

2. There is also the possibility of radiant paint. The order of brightness to be expected here is perhaps less than for a fluorescent coating, and considerably less than that for direct illumination.

3. No method of installing a lamp appears to be promising, from the point of view of fuel required to maintain the lamp for more than a few hours.

C. Illumination of a Retrodirective Reflector by a Searchlight

1. There exist some 60-in. searchlights, which yield 800,000,000 candles on the axis. At 250 miles, or 400 kilometers, this corresponds to

$$\frac{8 \times 10^8}{(4 \times 10^5)^2} = 0.5 \times 10^{-2} \text{ lumens}/(\text{meter})^2$$

2. During World War II, the army developed some glass trihedral reflectors. These have the property of returning light over the same path as that along which the light came to them, with a spread of about 12 in. The effective area of the [76] ordinary trihedral reflectors is such that one would receive about 2.5×10^5 lumens, and throw it back in a solid angle of approximately 1.16×10^{-9} square radians, thus yielding 2.16×10^3 candles on the axis of the returning beam, or, on the ground:

$$\frac{2.16 \times 10^3}{(4 \times 10^5)^2} = 1.35 \times 10^{-8} \text{ lumens/sq meter}$$

According to the fundamental measurements of Fabry, a first-magnitude star yields 8.3×10^{-7} lumens/sq meter; hence we should require ten searchlights and six trihedral reflectors to arrive at this amount. To take care of the various aspects of the missile, it might be best to have 72 trihedral reflectors distributed six each on the faces of a regular dodecahedron in order to have adequate returned light in all aspects. There would be a loss of one to three magnitudes from atmospheric absorption in each direction, depending upon the altitude; thus the satellite would appear as an object between the third and seventh magnitudes.

III. Conclusions on Observability

Balancing the disadvantages of irregular motion against the advantages of approximately known position and speed, it appears likely that objects of the 14th magnitude could be observed. On the other hand, we have seen that an object of the 4th magnitude could be produced; there is thus a margin of 10 magnitudes, or a safety factor of 10,000. If the satellite is imagined to be at 1000 miles instead of 250 miles, the brightness is reduced by a factor of 4^4 or 256; there is still a safety factor of 40. However, since the searchlights will cause the sky to appear very bright in the direction in which they are pointed, this margin of safety may turn out to be insufficient. We conclude that at 250 miles the satellite should be an easy object; at 1000 miles it may be a difficult object.

IV. Applications

In studying the applications to geodesy of such a satellite, it will undoubtedly be necessary to proceed by successive approximations, since the geodetic data now available are not adequate to permit the calculation of an accurate orbit. For example, it is believed that the present values for the latitudes and longitudes of points in Europe may be inconsistent with the American system of latitudes and longitudes by several hundred feet. For an object at a distance of 500 miles, this would imply discrepancies of the order of one minute of arc between European and American observations, under certain circumstances. Again, the International figure of the earth may be in error by as much as one part in 20,000; this would lead to a discrepancy of 10 sec per revolution between the position calculated from observations of the linear speed and the actual position; for a 2-hr orbit this would amount to 2 min per day. The problem is thus actually a problem of an overwhelming flood of basic information. It follows that it would be difficult to solve for one element at a time; it is necessary to solve simultaneously for several different elements. The following attempt to sort out the results does not correspond, therefore, to the chronological order in which results would be obtained, except roughly.

A. Determination of Relative Positions Between Continents

1. Simultaneous observations on a satellite missile from two independent triangulation systems would seem to fix their relative positions by a modification of the method now being employed for flare triangulation. A missile at a height of 1000 miles would be easily visible from points 2000 miles away. If the missile were set to follow a track around the earth's equator, the countries in which it would be easily visible would be chiefly those between latitude 30° North and latitude 30° South. In the case of a missile fired at some angle to the equator, this difficulty would disappear; on the other hand, there would be a problem of keeping track of the missile.

2. Assuming an angular accuracy of 5 sec in position, and a precision of $0^s.01$ in the timing, the accuracy of positioning between independent continental systems would be of the order of 125 ft in each coordinate. Thus it is evident that a correction through a missile would be advantageous.

B. Calculation of g

From observations on the satellite made from a single country, it should be possible to obtain the absolute value of the acceleration of gravity, averaged over a large extent of terrain. Ordinary pendulum gravity measurements do not give absolute values of gravity; instead, the pendulum or the gravity meter is brought to a standard point whose gravity is assumed; and comparisons are made by differential methods. From a freely moving missile, on the other hand, the acceleration of gravity could be directly measured in absolute terms. Best of all, the gravity so measured would represent the average value over a considerable area. At present, such a value, which is required for large-scale geodetic studies, is attainable only by laboriously measuring the values at many points, and then averaging; even when this has been done, there is a danger that the mountainous areas are underrepresented, so that systematic errors are created.

C. Calculation of the Earth's Semimajor Axis a

It appears to be possible to calculate the earth's semimajor axis from considerations relating to the earth's linear surface velocity. It may at first sight seem surprising that it is not possible to state with extreme accuracy the speed with which we move around the earth's axis. The angular speed is, of course, very well known. By definition, the earth turns on its axis one full revolution in one sidereal day. The speed of the observer in meters per second due to this rotation could be calculated with great precision if we knew exactly how far he is from the earth's axis. This quantity, however, depends on the earth's semimajor axis a and its flattening f. Of these, f if known with a precision which is adequate for the purpose under discussion; the chief uncertainty arises from the value of a.

V. Conclusions

It appears that the setting up of a satellite capable of being observed by theodolites and the like is an engineering possibility and that it would yield results of high geodetic value. If it could be accurately observed, most of the principal problems of geodesy could be attacked successfully.

Appendix F
Orbital Radio Relays

JOHN R. PIERCE

1. Introduction

Following the announcement last year that the American Telephone and Telegraph Company and the British Post Office have jointly undertaken the construction of a 36-channel two-way submarine telephone cable across the Atlantic at a cost of 35 million dollars, it is natural, at least for a person who is a complete amateur in such matters, to speculate about further developments in trans-oceanic communication, even into the far future.

Would a channel 30 times as wide, which would accommodate 1080 phone conversations or one television signal, be worth 30 × 35 million dollars—that is, a billion dollars? Will someone spend this much trying to make a broad-band channel to Europe? The idea is of course absurd. At the present, there is no commercial demand which would justify such a channel. By the time there is, surely some technical solution to the problem will be sought which does not involve multiplying the cost of the present cable in proportion to the bandwidth.

It is conceivable that such a solution could come about [77] through further development in the field of cables; but a very difficult step must be taken to multiply the channel capacity by 30 or more. In the meantime, other means for obtaining a broad-band channel to Europe have been considered, including routes largely across land rather than across water.

A route from Labrador or Baffin Island to Greenland, around the coast of Greenland, thence to Iceland and via the Faroe Islands to Scotland traverses much nasty country by land and still leaves gaps of several hundred miles by sea. These gaps might conceivably be spanned by radio, using very high power. Perhaps it may be possible to make undersea television cables which would span gaps of a few hundred miles before such cables can be made to span thousands of miles. Even granting the success of a difficult radio link or a broad-band cable, both terrain and climate make this indirect route difficult and unappealing.

A route from Alaska across the Bering Strait to Siberia, and thence overland to Europe is conceivable, but it is difficult and indirect and it has other disadvantages which need not be mentioned.

Radio relay along a continual chain of planes crossing the Atlantic has been proposed. While this is certainly technically feasible, in good weather at least, it seems strange either as a long-range or a short-range solution.

Another "solution" has been proposed to the problem of trans-oceanic communication; that is, relaying by means of a satellite revolving about the earth above the atmosphere. I do not believe that many engineers doubt that it will eventually be possible to put a satellite up and into place, nor to supply it with small amounts of power for long periods and to exercise some sort of radio control over it. However, there is no unclassified information to tell us how long it will be before we could put up a satellite or what it might cost to do so, and there may not even be classified information on the subject. Thus, while I am here considering some aspect of trans-oceanic communication via a satellite, I have nothing at all to say about the over-all feasibility of such communication, which must depend on the feasibility of the satellite itself.

Fortunately, there is a good deal else to be said about the matter. For instance, I have spoken of *trans-oceanic* communication only, and I have a reason for this. We now have trans-continental television circuits. The announced cost of the American Telephone and Telegraph Company's trans-continental TD2 microwave system was 40 million dollars. This is only 5 million dollars more than the 35 million for the 36-channel transatlantic cable; and yet the TD2 system provides a number of television channels in both directions, as well as many telephone channels. Perhaps even more important in an overland system, it provides facilities for dropping and adding channels along the route. Without such flexibility, an overland system would be almost useless.

Some types of satellite relay systems would provide communication only between selected points. These would lack the flexibility required for overland service. Further, there is little reason to believe that a satellite relay could compete with present microwave radio relay or coaxial cable in cost. Present facilities are very satisfactory, so that there is little incentive to replace them with some difficult alternative system, even if it could do the same job. Thus, satellite radio relay seems attractive only for spanning oceans.

Two different sorts of satellite radio repeaters suggest themselves. One consists of enough spheres in relatively near orbits so that one of them is always in sight at the transmitting and receiving locations. The sphere isotropically scatters the transmitted signal, so one has merely to point the transmitter and receiver antennas at it to complete the path. Another system uses a plane mirror or an active repeater with a 24-hr orbital period, located directly above the equator at a radius of around 26,000 miles or an altitude of about 22,000 miles. Such a satellite would be visible to within $9°$ of the poles, that is, in all inhabited latitudes. If it were not for the perturbations of the orbit by the moon and the sun, it could stay fixed relative to the surface of the earth, and large fixed antennas could be used on earth. However, it appears that perturbation of the orbit would be large enough to necessitate steerable antennas on earth and orientation of the satellite antennas or the reflector by remote control.

Even disregarding problems concerned with the making and placing of the satellites, would such satellite relay systems or any satellite relay system be feasible in other respects? To decide this we must consider two sorts of problems: problems of microwave communication, and problems lying in the field of celestial mechanics, concerned with the orbit and orientation of a satellite.

(In his full paper, to appear in a future issue of *Jet Propulsion,* Dr. Pierce proceeds to develop mathematically the power requirements for several types of relays, and he analyzes briefly a few of the orientation and orbit problems.)

2. Summary and Discussion

The best we can do is try to state some sort of conclusions concerning the sorts of systems which have been described.... All of these are for a 5-mc video channel provided by an 8-digit binary pulse code modulation system and a wave of 10 cm. The diameter of the antennas on earth is assumed to be 250 ft.

The great advantages of the passive repeaters over active repeaters are potential channel capacity and flexibility. Once in place, passive repeaters could be used to provide an almost unlimited number of two-way channels between various points at various wave lengths. They would also allow for modifications and improvements in the ground equipment without changes in the repeater.

Spheres, which reflect isotropically, are the most flexible of passive repeaters, because they allow transmission between any two points in sight of them. Moreover, with spheres there is no problem of the angular orientation of the repeaters.

For a 24-hour "fixed" repeater and a 1000-ft sphere, the power required is 10 megawatts, and this seems excessive. However, suppose 10 spheres, each 100 ft in diam, circled the earth above the equator at a fairly low altitude. At low altitudes, one or more would always be in sight. The path length would be only about a tenth that for the 24-hr orbit, and the power required would be around 100 kw, which seems quite feasible.

A plane mirror returns much more power than does a sphere of the same diameter. A 100-ft mirror at an altitude of 22,000 miles would call for a transmitter power of about 20 kw, which again is by no means unreasonable.

The great problem in connection with a plane mirror is that of position and orientation. If it were not for the perturbation caused by the moon and sun, the position and orientation of the mirror could be preserved automatically by a proper [78] disposition of masses attached to the mirror and by the use of damping.

However, as perturbations of position and orientation will be too large to be tolerated, the orientation of the mirror would have to be adjusted by moving masses through radio control. The power required would be small, perhaps less than that required for an active repeater. The advantage of channel capacity would be preserved.

The plane mirror suffers a considerable limitation compared with the sphere, however. If it really hung fixed in the sky, it would provide communication between any point in sight of its face and another particular corresponding point. However, because perturbation by the sun and moon will cause it to wander about in the sky so that the orientation of the mirror must be adjusted to maintain a path between two particular points, a plane mirror can actually be used only to provide channels (and a large number of channels on different frequencies) between two particular points.

The chief disadvantage, then, of the passive repeater is that the power required on the ground is large—though probably attainable.

The attractive feature of an active repeater is the small power required and the small antennas needed at the repeater, as well as the small power required at the ground. The small antennas would have a comparatively small directivity. This, coupled with the fact that for a given angular or positional shift, the beam from a radiator shifts only half as far as the beam from a reflector, makes the orientation problem considerably easier in the case of an active repeater. However, there still is an orientation problem, in contrast to the case of a sphere used as a passive repeater.

The chief disadvantage of the active repeater, aside from disadvantages of power supply and life, is that it provides only the number and sort of channels that are built into it. Once it is in place, its channel capacity cannot be substantially increased by anything done on the ground, although some gain might be made by an increase in transmitter power and receiver sensitivity and by a modification of the nature of the signal.

In conclusion, one can say that, disregarding the feasibility of constructing and placing satellites, it seems reasonably possible to achieve broad-band trans-oceanic communication using satellite repeaters with any one of three general types of repeater: spheres at low altitudes, or a plane reflector or an active repeater in a 24-hr orbit (at an altitude of around 22,000 miles).

At this point, some information from astronomers about orbits and from rocket men about constructing and placing satellites would be decidedly welcome.

Document II-9

Document title: U.S. National Committee for the International Geophysical Year 1957-58. "Summary Minutes of the Eighth Meeting," Washington, D.C., May 18, 1955.

Source: Archives, National Academy of Sciences, Washington, D.C.

The idea for holding an International Geophysical Year (IGY) arose from an informal meeting of scientists at the Maryland home of physicist James Van Allen in 1950. The intention was to coordinate high-altitude research conducted around the world. Supporters took the idea to the International Council of Scientific Unions, where it was supported by sixty-seven nations. In October 1954, Lloyd Berkner, one of the scientists at the original meeting, and ten of his associates discussed the problems and rewards of launching a satellite as part of the IGY and agreed unanimously to recommend it to the Special Committee for the International Geophysical Year (CSAGI). On October 4, the CSAGI issued a statement calling for governments to try to launch Earth satellites. The American Long Playing Rocket Proposal followed from that recommendation. The U.S. National Committee for the International Geophysical Year gave formal approval to the project at its May 18, 1955, meeting. The minutes of that meeting have as attachments background on the U.S. satellite proposal.

---

[1] 1. **Attendance**.
1.1 Members: Joseph Kaplan (Chairman), A. H. Shapley (Vice-Chairman), L. H. Adams, Wallace W. Atwood, Jr., Lloyd V. Berkner, Earl G. Droessler, J. Wallace Joyce, John P. Marble, E. B. Roberts, Walter M. Rudolph, Paul A. Siple, H. K. Stephenson, Merle A. Tuve, E. H. Vestine (alternate).
1.2 USNC-IGY Secretariat: Hugh Odishaw, Executive Secretary, R. C. Peavey, Administrative Officer.

2. **General Business**.
2.1 Dr. Kaplan announced that Senate hearings on the IGY principal budget were scheduled for 3:00 P.M., today.
2.2 J. Wallace Joyce was introduced to the Committee as Head of the NSF Office for the IGY. Earl G. Droessler was introduced as representing the Office of the Assistant Secretary of Defense (R&D).
2.3 L. V. Berkner noted that Dr. Briggs was in the hospital with a broken leg. It was agreed that a letter would be sent to Dr. Briggs in the name of the Committee expressing hope for a rapid recovery.

3. **Discussion on LPR Program**.
3.1 Dr. Kaplan opened a discussion on the Long Playing Rocket Project (LPR). He stated that this USNC Meeting had been called to review and consider formal approval of the LPR policy, program, and budget, which had been outlined by the USNC Executive Committee. He emphasized that the discussion on LPR must be considered private and confidential within the Committee until high policy decisions had been reached and a public announcement had been made by the Executive Branch of the Government.
3.2 Mr. Odishaw reviewed the history of events leading to the formulation of an LPR Program and Budget as an extension of the conventional Rocket Program proposed for the International Geophysical Year. He read the resolutions passed by the International Union of Geodesy and Geophysics (IUGG) [2] on September 20, 1954; the International Scientific Radio Union (URSI), on September 24, 1954; and the International Council of Scientific Unions Special Committee for the International Geophysical Year (CSAGI) on October 4, 1954 (Appendix I to these Minutes).

Mr. Odishaw then discussed the formation of a special LPR Committee consisting of members of the USNC Technical Panel on Rocketry and the USNC Executive Committee, to consider the technical feasibility of the CSAGI proposal and to suggest experiments which should be performed (Appendix II to these Minutes.) He noted that results of this study (Appendix III to these Minutes) and the proposed program and policy position had been approved unanimously by the USNC Executive Committee at its Meeting March 8-10, 1955, which authorized the Chairman to transmit on March 14, 1955, a policy statement on the LPR Project to the President of the National Academy of Sciences and the Director of the National Science Foundation (Appendix IV to these Minutes).

3.3 Detailed discussion ensued on the technical and scientific objectives as well as financial and political aspects of the LPR Project. It was noted that the vehicle, fuel, and launching system would probably involve unavoidable security problems, but it was explicitly understood that the USNC intended for the bird to be freely available for inspection by other nations participating in the LPR and to be tracked in flight by other nations.

3.4 Following this discussion, the USNC gave formal approval to the resolution adopted by the USNC Executive Committee and the policy statement on the LPR Project transmitted to the President, NAS-NRC and the Director, NSF on March 14, 1955, with one member dissenting.

4. **LPR Budget**.

4.1 Mr. Odishaw reported that on May 5, 1955, the USNC Executive Committee had given instructions for the preparation and transmission of a program and budget for the LPR Project to the Director, NSF, for consideration by the National Science Board at its forthcoming meeting on May 20, 1955. He then reviewed in detail estimated costs, totaling $9,734,500, which includes the cost of ten rocket vehicle systems for instrumented birds.

4.2 After discussion, the USNC gave formal approval to the LPR PROGRAM AND BUDGET DOCUMENT drawn up as of May 6, 1955, for transmittal to the Director of the National Science Foundation, with one member abstaining from the vote (Appendix V to these Minutes).

5. **Other Business**.

5.1 Mr. Odishaw reported that detailed program and budget estimates for scientific projects had been received from the USNC Technical Panels and that copies of all project forms would be mailed to the USNC for review and approval as to acceptance in the U. S. Program for the IGY. The USNC was advised that the Supplemental Budget would be presented to the National Science Board for its consideration on May 20, 1955.

5.2 After approval of these procedures discussed under item 5.1 above, the meeting adjourned.

[3]
Appendix I

International Scientific Resolutions
On The Earth Circling Satellite Vehicles

1. **International Union of Geodesy & Geophysics (IUGC)**, September 20, 1954:

"In view of the great importance of observations over extended periods of time of extraterrestrial radiations and geophysical phenomena in the upper atmosphere and the advanced state of present rocket techniques, it is recommended that consideration be given to the launching of small satellite vehicles, their scientific instrumentation and the new problems associated with satellite experiments such as power supply, telemetering, and orientation of the vehicle."

2. **International Scientific Radio Union (URSI)**, September 24, 1954:

"URSI recognizes the extreme importance of continuous observations, from above the E-region of extraterrestrial radiations, especially during the forthcoming AGI.

"URSI therefore draws attention to the fact that an extension of present isolated rocket observations by means of instrumented earth satellite vehicles would allow the continuous monitoring of solar ultraviolet and X-radiation intensity and its effects on the ionosphere, particularly during solar flares thereby greatly enhancing our scientific knowledge of the outer atmosphere."

3. **International Council of Scientific Unions Special Committee for the International Geophysical Year (CSAGI)**, October 4, 1954:

"In view of the great importance of observations during extended periods of time of extraterrestrial radiations and geophysical phenomena in the upper atmosphere, and in view of the advanced state of present rocket techniques, CSAGI recommends that thought be given to the launching of small satellite vehicles, to their scientific instrumentation, and to the new problems associated with satellite experiments, such as power supply, telemetering, and orientation of the vehicle."

[4]
Appendix II

Excerpts
MINUTES
of the
First Meeting
Technical Panel on Rocketry
USNC for the IGY
10:00 - 17:00, January 22, 1955, National Science Foundation
Washington, D. C.

2. **Organization of Panel**

2.1 Dr. Van Allen as convener called the meeting to order and Dr. Spilhaus of the Executive Committee of the U. S. National Committee for the IGY introduced as first order of business a resolution passed by the Executive Committee to be taken up by this panel. A motion was introduced, seconded and carried to the effect that the following part of the meeting on LPR be a closed session, be private, though unclassified, and its record be available to the participants of this closed session only. A record of the Closed Session will be found as Attachment A to these minutes.

[5]
January 22, 1955
Attachment A
REPORT
on the
Closed Session
of the
First Meeting
Technical Panel on Rocketry
U.S.N.C. for the I.G.Y.

1. **Resolution**

1.1 Dr. Spilhaus reported on a resolution passed by the Executive Committee of the U.S. National Committee for the IGY requesting this Panel to perform a study and report on the technical feasibility of the construction of an extended rocket, from here on called LPR, to be launched in connection with scientific activities during the International Geophysical Year.

1.2 This report presents the resolved actions taken on account of the ensuing discussion among the participants of the closed session.

1.3 Participants: N.C. Gerson, B. Haurwitz, J. Kaplan, H.E. Newell, Jr., H. Odishaw, G.F. Schilling, S.F. Singer, A.F. Spilhaus, J.A. Van Allen, F.L. Whipple.

1.4 All contents of this report are classified as private, pending further determination of a suitable security classification, and discussions or communications would be limited to members of the Panel and only those others indicated in 2.2. A motion covering this point was proposed by Dr. Kaplan, seconded, and unanimously passed. Copies of this report, Attachment A to the Minutes of the First Meeting of the Technical Panel on Rocketry of the USNC for the IGY, will be available to participants of this closed session only.

2. **Discussion**

2.1 The ensuing discussion included, in addition to technical details, such topics as the expected reaction of public opinion, the liaison with Government agencies, and the availability of funds.

[6] 2.2 It was finally resolved that a study group to be called LPR Committee would be set up under the chairmanship of Dr. Whipple, consisting of the members of this Panel and the following added consultants: W. Pickering, California Institute of Technology; M.W. Rosen, Naval Research Laboratory; J.W. Townsend, Jr., Naval Research Laboratory.

2.3 This LPR Committee will meet at Pasadena, California, on the evening of February 3, 1955, and possibly on subsequent days, the participants to be informed of exact time and locality by Dr. Van Allen.

2.4 It was resolved that the LPR Committee draft a report on the following topics and/or sub-topics concerning LPR: technical feasibility, budget, geophysical possibilities; controls, motor, manpower, timing, cost estimates, desired orbit; and possibly other pertinent subjects.

2.5 It was resolved that the whole report would either be classified and an unclassified abstract be extracted, or the report would be in two parts, one part carrying a security classification.

2.6 It was resolved that the LPR Committee would send the report to Dr. Spilhaus to be presented to the members of the Executive Committee of the USNC for the IGY, or directly to respective Government agencies such as the National Security Council, upon the discretion of the Chairman of the USNC for the IGY.

2.7 Dr. Haurwitz indicated that due to prior commitments he would be unable to attend the planned meeting on February 3, 1955. He offered to write a letter to Dr. Whipple prior to the planned meeting, containing his contribution for inclusion in the report.

[7]
Appendix III

**National Academy Of Sciences - National Research Council
United States National Committee
For The International Geophysical Year 1957-58
Unclassified Excerpts**

Prepared by: G.F. Schilling, Program Officer, USNC-IGY
Verified by: H.E. Newell, Jr., Exec. Vice Chairman, USNC
Technical Panel on Rocketry
Approved by: Hugh Odishaw, Executive Secretary, USNC - IGY
Date: August 2, 1955

MINUTES
of the Special Meeting
LPR Committee
Technical Panel on Rocketry
USNC for the IGY
09:00 - 13:00, 9 March 1955, IGY Conference Room
Washington, D. C.

1. **Attendance**
1.1 Members Panel on Rocketry: F.L. Whipple (Chairman), W. Berning, W.G. Dow, N.C. Gerson, J. Kaplan, H.E. Newell, Jr., S.F. Singer, W. Stroud, P.H. Wyckoff
Absent: B. Haurwitz, J.A. Van Allen
1.2 IGY Secretariat: H. Odishaw, G.F. Schilling
1.3 LPR Technical Subcommittee: M.W. Rosen, J.W. Townsend
Absent: W.H. Pickering
1.4 Invited Participants: E.L. Eaton, A.F. Spilhaus, T.B. Walker

3. **Business Session**
3.1 Dr. Whipple, Chairman of the Technical Panel on Rocketry of the USNC for the IGY convened the meeting and called upon Mr. Rosen to present a Report of the LPR Technical Subcommittee, prepared by W.H. Pickering, M.W. Rosen, and J.W. Townsend.
[8] 3.2 Mr. Rosen read and submitted a written report (Attachment A to these Minutes), summarizing three conclusions resolved by the Subcommittee.
3.3 The report was accepted by the Chairman of the Panel and Mr. Rosen was called upon to amplify and detail the content of the report.

4. **Detailed Report by M. W. Rosen**
4.1 Mr. Rosen discussed the following three possible approaches to placing a small payload in an orbit around the earth. He emphasized that all three approaches are feasible with present-day knowledge and facilities, and are presented in the order of difficulty and amount of additional development required.
I. Technique Number One:
This technique suggests the use of a one-stage large rocket plus the release of a number of small rockets, launched at or near the top of the flight path of the large rocket. Three existing large rockets are qualified to be used for the first stage. The guidance can be made accurate to one degree of arc.
II. Technique Number Two:
This technique suggests the use of a two-stage rocket plus one or two more stages of small rockets. The guidance problem of the second stage is more difficult here, than with Technique Number One, but it is technically feasible. The technique gives the possibility of greater payload, i.e., instrumented satellite.
III. Technique Number Three:
This technique may represent the most long-term approach, but offers the greatest payload. The basic suggestion is to start out with the biggest power plant presently in development and to build a test vehicle around it. This would involve a development program for the first stage. Before evaluating this program, a preliminary study of two to three months would be necessary.

5. **Discussion on Size Categories**
5.1 Dr. Whipple started a discussion on desirable size categories. It was apparent that an object of the order of magnitude of one pound would not be useful. An object of the order of magnitude of ten pounds would be observable from ground. Into any object of 30 pounds or more some sort of power could be put, thus making it an instrumented satellite. It was agreed that at the present time the use of nuclear or solar power supplies was doubtful and not technically feasible, therefore batteries would be needed.

[9] 5.2 10 pds Observable Object: Dr. Whipple outlined technical and scientific aspects concerning a 10 pd object. He suggested that it should be about 20 inches in diameter, painted white or have a reflecting surface. It could be observed visually from ground at twilight or dawn, representing a star of a 6th magnitude brightness (60% reflection). Dr. Whipple stated firmly that such an object could be found optically with binoculars and telemeter cameras (Askania system), and discussed the technique applicable to find the object once it was in an orbit.

Such an object would permit determination of atmospheric densities and would preferably be placed in an equatorial orbit. An ideal orbit would have a perigee of about 250 miles and an apogee of about 500 miles. If the object were much bigger in diameter than 20 inches, it would spiral back to earth too fast and make determinations of density more difficult and inaccurate. [Paragraph omitted]

Among further scientific results obtainable, the following projects were mentioned: determinations of inter-continental distances to an accuracy of probably 100 feet, if a time accuracy of about 0.01 seconds (equivalent to approximately 200 feet moving path) can be achieved. The mass and density distribution in the earth's crust, e.g., mountain ranges, might be calculated. [Paragraph omitted]

[11] 5.6 Dr. Singer discussed geophysical and astrophysical applications of an instrumented rocket. He went into details concerning the desirability of measuring various geophysical parameters and a general discussion ensued.

5.7 Dr. Newell reported on an engineering study performed at NRL on a 30 to 50 pd object. He presented a theoretical design of an instrumented 15 inch sphere in an equatorial orbit, fitted out with available instruments. He detailed as follows:

Optical tracking: same as Dr. Whipple.
Solar batteries: not practical for next 3 years.

Experiments for an active satellite (the following lists the actually designed instrumentation).

(1) Interplanetary Hydrogen Density (Lyman-alpha)–1 pd, 0.2 watts.
(2) Dual Micrometeor Detector–0.5 pds, 4 watts.
(3) Extraterrestrial and Ionospheric Electric Currents–8 pds,
    5 watts, 0.1 watthour per measurement.
(4) Magnetic Aspect Indicator (for orientation)
    –0.25 pds, 2 watts.
[Sentence omitted]
(6) Telemetering System–3 .8 pds, 71 watts–batteries 20 pds.
(7) Structure–15 pds.

7. Final Discussion

7.1 It was generally agreed upon that instrumentation costs for any type of object would be negligible in comparison with other costs. This means that a number of small objects could be built relatively inexpensively and more types than one would be possible.

7.2 For technical reasons the majority of the Panel members seemed to favor Technique Number Two applied to a 30 pds observable object.

7.3 The Panel designated Dr. Whipple and Dr. Newell to present an unclassified summary of the findings of this meeting to the Executive Committee of the U.S. National Committee for IGY.

[12]
Appendix III

Attachment A
Report on LPR Subcommittee on Vehicle Capabilities

On Friday, February 11, a committee consisting of Dr. W.H. Pickering, Director, Jet Propulsion Laboratory, California Institute of Technology; Mr. M.W. Rosen, Head, Rocket Development Branch, Naval Research Laboratory; and Mr. J.W. Townsend, Assistant Head, Rocket Sonde Branch, Naval Research Laboratory, met in Washington to consider the feasibility of placing a small payload in an orbit around the earth. All three members of the committee were familiar with several proposals for a small satellite, but not necessarily with all such proposals that may exist. In view of the very brief time available for reaching some conclusions, the committee decided that it could not underwrite any specific proposal and that it could not formulate any specific proposal for a satellite.

The Committee reached the following three conclusions:

1. With regard to propulsive power required for reaching satellite velocity, it is feasible to attain the required velocity for payloads up to 10 pounds using combinations of existing rocket vehicles with a moderate amount of additional development work.

2. A more difficult problem is that of the control or guidance necessary to produce an orbit. This problem can be solved using existing control and guidance components. The precision required for producing an orbit is less stringent than that necessary to meet the specifications for at least several existing guided missile projects. An appreciable amount of development work would be required.

3. The creation of a satellite for payloads up to 10 pounds can be realized within two to three years, provided that sufficient funds and manpower are applied....

[13]
Appendix IV

National Academy Of Sciences
National Research Council of the United States Of America

United States National Committee
International Geophysical Year 1957-58
March 14, 1955

Identical Letter was addressed to:
Director, National Science Foundation

Dr. Detlev W. Bronk, President
National Academy of Sciences
2101 Constitution Avenue
Washington 25, D. C.

Dear Dr. Bronk:

A small, approximately fifty-pound, earth-circling satellite which could be freely inspected before launching and tracked in flight by international agencies would be in accord with recommendations of the Comite Special Annee Geophysique Internationale 1957-58 (CSAGI) at its Rome meeting in 1954, and would yield new geophysical data of considerable interest. If such vehicles could be constructed and launched within the spirit of the International Geophysical Year, the Executive Committees of the U.S. National Committee for the International Geophysical Year, on basis of studies and reports by its rocket panel, recommends that the U.S. Government include such vehicles in its rocket program, and provide the U.S. National Committee with opportunity to install the orbiting vehicles for such flights.

A resolution adopted by the Executive Committee of the USNC is presented herewith:

The Executive Committee of the U.S. National Committee for the International Geophysical Year notes that the following resolution was adopted by the CSAGI in Rome, Italy, during 1954: "In view of the great importance of observations over extended periods of time of extraterrestrial radiations and geophysical phenomena in the upper atmosphere, and in view of the advanced state of present rocket techniques, it is recommended that thought be given to the launching of small satellite vehicles, to their scientific instrumentation, and to the new problems associated with satellite experiments, such as power supply, telemetering, and orientation of the vehicle." The Executive Committee of the U.S. National Committee for the International Geophysical [14] Year, basing its opinion on the study of its expert panel on rocketry, feels that a small artificial satellite for geophysical purposes is feasible during the International Geophysical Year if action is initiated promptly and that the realization of such a satellite would give promise of yielding original results of geophysical interest.

I am also submitting the above information and resolution to the Director of the National Science Foundation.

Sincerely yours,
Joseph Kaplan
Chairman

[16]
Appendix V

**National Academy Of Sciences
National Research Council of the United States Of America
United States National Committee
International Geophysical Year 1957-58
May 6, 1955**

The Honorable
Alan T. Waterman, Director
National Science Foundation
1520 - H Street, N.W.
Washington 25, D. C.

Dear Dr. Waterman:

I am writing to present to you the budget and related recommendations of the USNC Executive Committee on the proposed USNC-IGY Project LPR (Long Play Rocket), constituting also an amplification of my letter of March 14, 1955, to the President of the National Academy of Sciences and you on this subject. While this represents an official action by the Executive Committee, final approval by the full USNC is also necessary: a special meeting, as indicated below, will be called as soon as you have advised me on the existence of a favorable Government policy.

The Executive Committee at its Seventh Meeting, May 5, 1955, acted favorably upon the enclosed budget estimate totalling $9,734,500 (see Attachment A). This estimate includes not only provisions for (i) approximately ten "birds" and five observation stations, including the necessary scientific instrumentation, related equipment, and minimum civilian scientific staff but also provisions for (ii) approximately ten vehicles and their associated flight instrumentation. Cost estimates for (i) are $2,234,500; for (ii) $7,500,000; and the total comes to $9,734,500.

Although in our earlier discussions, with which you are familiar, the Executive Committee had considered that the USNC-IGY program need only include item (i), the Executive Committee now believes that item (ii) ought also to be part of the IGY budget. There are several reasons for this conclusion: first, it would be more fully in the spirit of the IGY for the USNC-IGY to sponsor and provide support for the total equipment and instrument needs (analogous to the present rocket program), clearly establishing the basic civilian character of the endeavor; second and somewhat related to the preceding clause, there appears to be no fundamental classification problem involved in the vehicles: without reference to the history of rocketry (in particular, German V-2 developments), such rockets as the Viking and Aerobee, combinations of which are capable of doing the job, [17] are commercially available and involve no security classification considerations, an important factor in terms not only of the philosophy of the IGY but, we believe, in terms of the international relations of the United States; and, third, the inclusion of (i) and (ii) in the USNC-IGY program provides a simplicity in the demarcation of USNC-IGY and DOD responsibilities, the latter, as in the rocket program, having to do with logistics and operational support.

These and related topics are the subject of the narrative accompanying the enclosed budget (see Attachment A); the views of the Executive Committee on the budget itself can be summarized briefly by quoting the Minutes of the Seventh Meeting: "With the understanding that the totals for vehicles for launching of approximately 10 birds at a cost not to exceed $7.5 million for procurement, construction, and necessary system design and development, and $2,234,500 for procurement, construction, and design relating to birds and observing equipment, the USNC Executive Committee recommends to the National Science Foundation that the entire LPR program be funded from IGY funds, provided that procurement, logistics and launching be coordinated by the appropriate DOD agency."

I should like at this time to dwell briefly on the urgency of this matter: namely, unless funds are available within a very brief interval after July 1, 1955, it will be virtually impossible to be sure that the LPR program can be conducted during the IGY. The critical shortage of time can not be over-emphasized. You will recall that my letter of March 14, 1955, contained the phrase "if action is initiated promptly." The enclosed copy of a newspaper article (see Attachment C) may be of interest to you, showing as it does the interest of other nations in such a program.

The Executive Committee assumes that a favorable Government position, emanating from the highest levels within the Executive Branch, will have been reached in the very near future on this program so that the following steps can be taken: (i) approval by the full U.S. National Committee, (ii) submission to the National Science Board at its May meeting, and (iii) submission to the Bureau of the Budget and the Congress during the present session in the hope that funds can be granted by July 1, 1955, or earlier if possible. Aspects of the critical timetable confronting us are described in Attachment B.

I know that you have been pursuing this matter diligently, and the Executive Committee is appreciative of your interest and assistance. I hope to hear from you soon in order to call a special meeting of the USNC, and I and my colleagues shall be pleased to appear before the National Science Board, the Bureau of the Budget, and the Congress on this subject as we have in the past in connection with other aspects of the IGY program.

Sincerely yours,
Joseph Kaplan
Chairman

Attachments

[18]
Appendix V

Attachment A

USNC-IGY LPR (Long Play Rocket) Program
Program and Estimated Budget

1. International Background

The desirability of launching small instrumented vehicles for geophysical research during the International Geophysical Year (1957–58) was the subject of discussion at three international meetings last year:

(i) In August and September, 1954, the International Scientific Radio Union (URSI) considered this matter and endorsed the following resolution:

"Study of Solar Radiation in the Upper Atmosphere.

URSI recognizes the extreme importance of continuous observations, from above the E region of extraterrestrial radiations, especially during the forthcoming AGI.

URSI therefore draws attention to the fact that an extension of present isolated rocket observations by means of instrumented earth satellite vehicles would allow the continuous monitoring of solar ultraviolet and X radiation intensity and its effects on the ionosphere, particularly during solar flares thereby greatly enhancing our scientific knowledge of the outer atmosphere."

(ii) The International Union of Geodesy and Geophysics submitted to the Special Committee for the International Geophysical Year (CSAGI) the following recommendation of its International Association of Terrestrial Magnetism and Electricity (September 20, 1954):

"In view of the great importance of observations over extended periods of time of extraterrestrial radiations and geophysical phenomena in the upper atmosphere and the advanced state of present rocket techniques, it is recommended that consideration be given to the launching of small satellite vehicles, their scientific instrumentation and the new problem associated with satellite experiments such as power supply, telemetering, and orientation of the vehicle."

(iii) The CSAGI at its final plenary session on October 4, 1954, adopted the following resolution:

"In view of the great importance of observations during the extended periods of time of extraterrestrial radiations and geophysical phenomena in the upper atmosphere, and in view of the advanced state of present [20] rocket techniques, CSAGI recommends that thought be given to the launching of small satellite vehicles, to their scientific instrumentation, and to the new problems associated with satellite experiments, such as power supply, telemetering, and orientation of the vehicle."

An indication of international interest is apparent in the news article enclosed herewith as Attachment C.

2. National Background

In view of the above International recommendations and in view of the advanced state of U.S. rocketry developments, the Executive Committee of the U.S. National Committee for the IGY considered the possibility of constructing, launching, and observing an instrumented satellite. A special group for this purpose was established within the USNC Technical Panel on Rocketry.

On the basis of studies made by the above group, the Executive Committee decided that an instrumented satellite program was of scientific importance and was feasible. The Executive Committee summarized its findings as follows (March 8 and 10, 1955):

"A small, approximately fifty-pound, earth-circling satellite which could be freely inspected before launching and tracked in flight by international agencies would be in accord with recommendations of the Comite Special Annee Geophysique Internationale 1957-58 (CSAGI) at its Rome meeting 1954, and would yield new geophysical data of considerable interest. If such vehicles could be constructed and launched within the spirit of the International Geophysical Year, the Executive Committee of the U.S. National Committee for the International Geophysical Year, on basis of studies and reports by its rocket panel, recommends that the U.S. Government include such vehicles in its rocket program, and provide the U.S. National Committee with opportunity to install the orbiting vehicles for such flights."

The Executive Committee adopted the following resolution, March 8 and 10, 1955:

"The Executive Committee of the U.S. National Committee for the International Geophysical Year notes that the following resolution was adopted by the CSAGI in Rome, Italy during 1954: 'In view of the great importance of observations over extended periods of time of extraterrestrial radiations and [21] geophysical phenomena in the upper atmosphere, and in view of the advanced state of present rocket techniques, it is recommended that thought be given to the launching of small satellite vehicles, to their scientific instrumentation, and to the new problems associated with satellite experiments, such as power supply, telemetering, and orientation of the vehicle.' The Executive Committee of the U. S. National Committee for the International Geophysical Year, basing its opinion on the study of its expert panel on rocketry, feels that a small artificial satellite for geophysical purposes is feasible during the International Geophysical Year *if action is initiated promptly* [emphasis added] and that the realization of such a satellite would give promise of yielding original results of geophysical interest."

The Executive Committee authorized the Chairman of the U.S. National Committee to transmit the above findings and resolution to the President of the National Academy of Sciences and the Director of the National Science Foundation. This was done on March 14, 1955.

3. USNC-IGY LPR Program

The Executive Committee of the USNC-IGY proposes a minimum satellite program during the IGY consisting of approximately ten instrumented birds, with the expectation that at least five of the birds will be successfully launched into their orbits, circulating about the earth for a period of approximately two weeks, at a height of about 250 miles, traveling about the equator. Five ground stations will be established for observations and measurement purposes, one each in the Equatorial Pacific, South America, the Atlantic Ocean, Africa, and the Philippines.

The instrumented satellites will permit the performance of a number of important experiments. The simplest and most direct will be the use of the satellite for precise geodetic measurements and the determination of upper air densities. Instrumentation now planned will permit the following investigations: Measurement of solar radiation, measurement of particle radiation such as micrometeorites and those responsible for the aurora, and determination of current flows in the ionosphere associated with magnet storms and radio black-outs. Such a vehicle would also permit the determination of hydrogen in inter-planetary space.

To achieve the objectives of the program, the USNC Executive Committee recommends the following definitions of responsibility:

(i) That the program be an integral part of the USNC-IGY program for both scientific and international reasons and that the execution of the program be under the direction of the USNC and its appropriate Committees and Panels.

[22] (ii) That the USNC-IGY budget include the (a) procurement of instrumented

birds, (b) the procurement of the rocket vehicle systems, (c) ground stations, their scientific instrumentation, and immediately associated supplies, and (d) provisions for the employment (and travel) of 25 scientists; and that the USNC-IGY LPR budget be presented to the Government and Congress under the auspices of the National Science Foundation.

(iii) That the U.S. Government, through the Department of Defense and under the scientific direction of the USNC, assume overall responsibility for the task, establish responsibility within the Department for the task, set up the appropriate working groups and task forces, and provide (a) scientific and technical personnel for vehicles, launching, and instrumentation of vehicles and birds, (b) all technical and field facilities, including laboratories (domestic and field), related structures, and quarters for personnel (including maintenance and subsistence equipment, supplies and services), (c) logistics support, (d) operational support, including transportation and vehicles, domestic and field, and (e) all other types of support, equipment and services essential for the success of the program but not noted in items (a) through (d) and not included in the USNC-IGY LPR budget.

(iv) That cooperation of scientists from other nations be invited with respect to (a) the instrumented bird and (b) the observation program, including provisions for participation in observations at our field stations as well as issuance of data permitting ease of observation from other than our field stations.

4. Budget Estimate and Time Schedule

The attached budget estimate sheet presents current estimates of that portion of the proposed program that appropriately belongs in the IGY budget. If the program is to be effected during the IGY, funds must be available early in Fiscal Year 1956; Attachment B describes aspects of the timetable.

May 6, 1955

[24]
USNC-IGY LPR (Long Play Rocket) Program
Budget Estimates
(May 6, 1955)

Salaries ....$ 215,000
Two professionals at each of 5 stations for one year at an average salary of
$9,500, ten man-years: $95,000; three professionals at each of 5 stations
for one year at an average salary of $8,000, fifteen man-years: $120,000.

Travel 217,500
Round-trip travel for each of 5 men at each of 5 stations for an average of
five shoots, 125 round trips at an average cost of $1,400: $175,000. Per
diem for a period of 20 days for each of the 125 trips at an average daily
cost of $12: $30,000. Domestic travel, including per diem, $12,500.

Transportation of things 222,000
Transportation of supplies and materials, $21,000; transportation of equipment
and facilities, $201,000.

Supplies and materials 140,000
Photographic supplies at $2,000 per station: $10,000; electronic parts and
components at $20,000 per station: $100,000; electrical supplies at $2,000
per station: $10,000; laboratory, office, and miscellaneous supplies at
$4,000 per station: $20,000.

Equipment and facilities 8,940,000
Design of instrumented bird, $100,000; construction of approximately ten
instrumented birds, at approximately $25,000 each: $250,000; five
telemetering stations at $50,000 each: $250,000; three C-W triplet-
antenna Tracking Stations for each of three stations at $200,000 each:
$600,000; two theodolite optical tracking stations for each of two stations
at $100,000 each: $200,000; five sets of manual optical tracking gear for the
five stations at $8,000 per station: $40,000; approximately 10 rocket vehicle
systems, including procurement of rockets and associated control equipment and
related devices and the necessary design and development to synthesize the
vehicle system, at approximately $750,000 each: $7,500,000.

Total 9,734,500

[25]
Attachment B

Factors Affecting USNC-IGY LPR Schedule

The resolution of the USNC Executive Committee, March 8-10, 1955 (see Section 2 of Attachment A) pointed out that the LPR program is feasible during the IGY "*if action is initiated promptly*" [emphasis added]. Funds must be available promptly—no later than early in Fiscal Year 1957; *preferably sooner*—if the objectives of the program are to be achieved during the IGY. There are important reasons for stressing time: the data provided by the LPR program would have considerable added value primarily because this data could be usefully correlated with large bodies of indirect data gathered during the IGY from all parts of the world. It is of interest also to note that at least one other nation has announced plans for a similar program (see Attachment C) under the direction of an extremely able physicist.

It appears that at least the following steps are involved in the carrying of this proposal to the funding stage:

1. Government Policy
    1.1 Departments and Independent Agencies
    1.2 National Security Council
2. U.S. National Committee Approval
3. National Science Board Approval
4. Bureau of the Budget Approval
5. Presentation to the Congress

In view of the limited time between now and the end of the current session of Congress, it is evident that prompt action must characterize Governmental considerations. It is understood that vigorous action with respect to step (1) has been undertaken by the National Science Foundation, following the March 14, 1955, letter from the USNC Chairman: as of May 6, 1955, however a Government position had not been defined. Once this position has been defined, involving basically three considerations [(i) general favorable positions of such agencies as the Department of State, (ii) assumption of responsibility by DOD as outlined in item (iii) of Section (3) of Attachment A but without requiring, at this time, a decision as to how DOD will establish its task group, and (iii) approval by the NSC], a special meeting of the USNC will be called for approval of the Executive Committee recommendations. The next step will require prompt consideration by the National Science Board in order to permit submission of the budget to the Bureau of the Budget and Congress.

May 6, 1955

[25]
Attachment C

**Interplanetary Commission Created
Russians Planning Space Laboratory for Research Beyond Earth's Gravity**

LONDON, April 16 (AP). —Russia announced tonight her top scientists are working on a space laboratory which would revolve around the earth as a satellite.

"A permanent Interdepartmental Commission for Interplanetary Communications has been created in the Astronomic Council of the USSR Academy of Sciences," said Moscow Radio.

"This Commission is coordinating work on problems of mastering cosmic space."

Peter Kapitsa, 61, one of Russia's best known atomic scientists, was among those appointed to the Commission.

Anatoly Karpenko, secretary of the Commission, was quoted as saying:

"One of the first tasks of the Commission lies in organizing work for the creation of an automatic laboratory of scientific research in cosmic space.

"With the ship of such a laboratory—which could, over a long period, revolve around the earth as a satellite, beyond the limit of the atmosphere—it will be possible to carry out observations of phenomena inaccessible under ordinary terrestrial conditions."

With this automatic equipment, Karpenko said, Soviet biologists will be able to obtain information about conditions of life in the absence of gravity.

"Astro-physicists will be able to observe the ultraviolet and roentgen spectra of the radiation of the sun and the stars and, with the help of these observations, to obtain additional data concerning the processes taking place on these bodies.

"Radio physicists will study more completely processes in the ionosphere and will determine the most advantageous conditions for the establishment of radio communications with future space ships."

He said Russia's cosmic laboratory will enable her scientists to penetrate deeper into the secrets of the universe and "will represent the first stage in the solution of the problem of interplanetary communication."

—Washington Post
April 17, 1955

Document II-10

Document title: National Security Council, NSC 5520, "Draft Statement of Policy on U.S. Scientific Satellite Program," May 20, 1955.

Source: Presidential Papers, Dwight D. Eisenhower Library, Abilene, Kansas.

This is the first statement of national policy for outer space. There were three satellite programs under consideration in early 1955. Two—Project Orbiter and WS 117L—were aimed at military and intelligence goals. The other was the IGY satellite being advocated by the National Academy of Sciences and the National Science Foundation. This policy statement emphasizes the political benefits of having the first U.S. satellite launched under international scientific auspices.

---

[1] 1. The U. S. is believed to have the technical capability to establish successfully a small scientific satellite of the earth in the fairly near future. Recent studies by the Department of Defense have indicated that a small scientific satellite weighing 5 to 10 pounds can

be launched into an orbit about the earth using adaptations of existing rocket components. If a decision to embark on such a program is made promptly, the U. S. will probably be able to establish and track such a satellite within the period 1957-58.

2. The report of the Technological Capabilities Panel of the President's Science Advisory Committee recommended [phrase excised during declassification review] an immediate program leading to a very small satellite in orbit around the earth, and that re-examination should be made of the principles or practices of international law with regard to "Freedom of Space" from the standpoint of recent advances in weapon technology.

3. On April 16, 1955, the Soviet Government announced that a permanent high-level, interdepartmental commission for interplanetary communications has been created in the [2] Astronomic Council of the USSR Academy of Sciences. A group of Russia's top scientists is now believed to be working on a satellite program. In September 1954 the Soviet Academy of Sciences announced the establishment of the Tsiolkovsky Gold Medal which would be awarded every three years for outstanding work in the field of interplanetary communications.

4. Some substantial benefits may be derived from establishing small scientific satellites. By careful observation and the analysis of actual orbital decay patterns, much information will be gained about air drag at extreme altitudes and about the fine details of the shape of and the gravitational field of the earth. Such satellites promise to provide direct and continuous determination of the total ion content of the ionosphere. These significant findings will find ready application in defense communication and missile research. When large instrumented satellites are established, a number of other kinds of scientific data may be acquired. The attached Technical Annex (Annex A) contains a further enumeration of scientific benefits.

5. [Paragraph excised during declassification review]

[3] 6. Considerable prestige and psychological benefits will accrue to the nation which first is successful in launching a satellite. The inference of such a demonstration of advanced technology and its unmistakable relationship to inter-continental ballistic missile technology might have important repercussions on the political determination of free world countries to resist Communist threats, especially if the USSR were to be the first to establish a satellite. Furthermore, a small scientific satellite will provide a test of the principle of "Freedom of Space." The implications of this principle are being studied within the Executive Branch. However, preliminary studies indicate that there is no obstacle under international law to the launching of such a satellite.

7. It should be emphasized that a satellite would constitute no active military offensive threat to any country over which it might pass. Although a large satellite might conceivably serve to launch a guided missile at a ground target, it will always be a poor choice for the purpose. A bomb could not be dropped from a satellite on a target below, because anything dropped from a satellite would simply continue alongside in the orbit.

[4] 8. The U. S. is actively collaborating in many scientific programs for the International Geophysical Year (IGY), July 1957 through December 1958. The U. S. National Committee of the IGY has requested U. S. Government support for the establishment of a scientific satellite during the Geophysical Year. The IGY affords an excellent opportunity to mesh a scientific satellite program with the cooperative world-wide geophysical observational program. The U. S. can simultaneously exploit its probable technological capability for launching a small scientific satellite to multiply and enhance the over-all benefits of the International Geophysical Year, to gain scientific prestige, [phrase excised during declassification review] The U. S. should emphasize the peaceful purposes of the launching of such a satellite, although care must be taken as the project advances not to prejudice U. S. freedom of action (1) to proceed outside the IGY should difficulties arise in the IGY procedure, [sentence excised during declassification review]

9. The Department of Defense believes that, if preliminary design studies and initial critical component development are initiated promptly, sufficient assurance of success in

establishing a small scientific satellite during [5] the IGY will be obtained before the end of this calendar year to warrant a response, perhaps qualified, to an IGY request. The satellite itself and much information as to its orbit would be public information. The means of launching would be classified.

10. A program for a small scientific satellite could be developed from existing missile programs already underway within the Department of Defense. Funds of the order of $20 million are estimated to be required to give reasonable assurance that a small scientific satellite can be established during 1957-58 (See Financial Appendix).

[6] Courses of Action

11. Initiate a program in the Department of Defense to develop the capability of launching a small scientific satellite by 1958, with the understanding that this program will not prejudice continued research [phrase excised during declassification review] or materially delay other major Defense programs.

12. Endeavor to launch a small scientific satellite under international auspices, such as the International Geophysical Year, in order to emphasize its peaceful purposes, provided such international auspices are arranged in a manner which:

    a. Preserves U.S. freedom of action in the field of satellites and related programs.

    b. Does not delay or otherwise impede the U.S. satellite program and related research and development programs.

    c. Protects the security of U.S. classified information regarding such matters as the means of launching a scientific satellite.

    d. Does not involve actions which imply a requirement for prior consent by any nation over which the satellite might pass in its orbit, and thereby does not jeopardize the concept of "Freedom of Space."

[7] Financial Appendix

1. Funds of the order of $20 million are estimated to be required to assure a small scientific satellite during the period of the IGY. This figure allows for design and production of adequate vehicles and for scientific instrumentation and observation costs. It also includes preliminary back-up studies of an alternate system without vehicle procurement. The ultimate cost of a scientific satellite program will be conditioned by (1) size and complexity of the satellite, (2) longevity of each satellite, and (3) duration of the scientific observation program. Experience has shown that preliminary budget estimates on new major experimental and design programs may not anticipate many important developmental difficulties, and may therefore be considerably less than final costs.

2. The estimate of funds required is based on:

satellite vehicle	$10-$15 million
instrumentation for tracking	$2.5 million
logistics for launching and tracking	$2.5 million
TOTAL	$15-$20 million

3. These estimates do not include funding for military research and development already part of other missile programs. They include costs for observations that might properly be undertaken by Department of Defense agencies as part of the Department of Defense mission. They do not include costs of other observations that may be proposed by other agencies. They will provide a minimum satellite for which two vehicle systems now under study offer good promise, "Orbiter" and "Viking." They also include exploratory studies for a back-up program based upon the "Atlas" missile and "Aerobee" research rocket development.

[8]
Annex A
Technical Annex

Scientific Values

1. The scientific information that may be expected from a satellite is dependent upon the size of the vehicle and whether it can be instrumented.

2. From a small, inert, trackable satellite, it is reasonable to expect that the following scientific values may be derived:

a. Analysis of currently available information on the upper atmosphere shows a need for additional basic information to support the development of manned craft and missiles for use at high altitudes. More accurate data on air density, pressure and temperature are required. From the analysis of actual orbital "decay" patterns, the air drag at high altitudes can be determined to a greater accuracy than by techniques now available.

b. Electronic tracking would probably permit direct and continuous determination of the total ion content of the ionosphere by comparison of simultaneous electronic and visual observations.

c. Anti-missile missile research will be aided by the experience gained in finding and tracking artificial satellites. It is expected that the satellite will approximate the speed and altitude of an intercontinental ballistic missile.

d. It is probable that a small scientific satellite would yield measurements of high geodetic value. More precise determinations of relative position between continents, the value of the gravitational constant averaged over long distances and the earth's semimajor axis can probably be made by observations of a small scientific satellite.

e. The observation of an uninstrumented satellite in an orbital plane inclined to the equator can permit the determination of the rotation of the orbital plane in space about the earth's polar axis, commonly called the "regression of the nodes." This perturbation is caused by the oblateness of the earth. Its evaluation will have considerable significance in precisely forecasting satellite orbits.

[9] **Military Values**

3. In addition to the scientific values listed above, some of which are clearly relevant to missile and anti-missile research and development programs of the Department of Defense, it may be noted that military communications programs will be enhanced by improvements in knowledge of the ionosphere and by improved knowledge of the rate of earth rotation. To this list must also be added the direct values of experience in organization, operation and logistics accruing to military missile forces detailed to execute a scientific satellite firing program. It is expected that the satellite will approximate the speed and altitude of an intercontinental ballistic missile.

Orbit and Tracking Considerations

4. If a perigee approximating 200 miles and an apogee approximately 1,000 miles are used to fix the desired orbit, the satellite will pass completely around the earth in approximately 90 minutes. If an orbit over the earth's poles or an orbit inclined to the equator is selected, the satellite will pass successively farther west of the launching point on each revolution around the earth. This means that an individual tracking station set up for inclined orbits will not be in an observing position for every revolution. The optimum location for tracking polar orbits is at or near the poles. On the other hand, an equatorial orbit will place each observing station in position to observe every circuit of the satellite. Artificial satellites in a low roughly circular orbit will appear optically similar to a 5.6 magnitude star moving at a high angular rate. Optical observations in broad daylight will be impracticable and observations when the satellite is in the earth's shadow will also be impracticable unless the satellite is illuminated. This means that experiments depending on passive optical tracking of a satellite cannot be conducted except during 50 minutes at

dawn and 50 minutes at dusk. An inclined orbit would thus materially reduce the usable data per station for experiments based on passive optical observations. The usefulness of the satellite and the selection of the desirable orbit is, therefore, closely related to the degree to which the satellite can be acquired and tracked by electronic techniques as well as optical.

5. An inclined orbit utilizing Patrick Air Force Base at Cocoa, Florida, as a launching point has the following advantages over an equatorial orbit:

a. Eliminated necessity to mount tropical expedition to establish launching and tracking sites.

[10] b. Permits observation from Navy Air Missile Test Center, Point Mugu, California; Naval Ordnance Test Station, Inyokern, California; White Sands Proving Ground, New Mexico; British-Australian Guided Missile Range, Woomera, Australia; and a large number of the free world's astronomical observatories.

c. Utilizes the full length (5000 miles) of Long Range Proving Ground for observations of the critical first part of the first orbit.

d. Permits an accumulation of geophysical data over a larger area of the earth's surface.

6. Disadvantages of an inclined orbit when compared to an equatorial orbit are:

a. Inclined orbit provides fewer opportunities to observe from a single base. This is especially critical for small uninstrumented satellites not observable by ordinary radar.

b. Inclined orbit from Patrick Air Force Base reaching a maximum latitude of 35° would result in the satellite passing on different circuits over virtually all of the world between 35°N latitude and 35°S latitude. This might increase substantially the amount of diplomatic negotiations necessary to implement the program.

Hazards to Human Life

7. The launching of a scientific satellite does not appear to threaten in any serious way the safety of air transportation at normal altitudes, nor the safety of personnel and property on the ground. All of the scientific satellites discussed above would be launched from locations where the initial flight of the booster system would be over water. At the end of this stage the booster rocket, which is the largest and potentially most lethal part of the satellite, would separate and fall into the water. Normal precautions taken in launching ordinary guided missiles would suffice to assure adequate safety of the launch and booster phases. The orbiting vehicle in all cases of both instrumented and uninstrumented satellites would be designed with the objective in mind that the entire device would disintegrate and to a large extent vaporize under the heat of re-entry into the earth's atmosphere. This vehicle would, therefore, create negligible hazards after re-entering the atmosphere.

[11]
Annex B
The White House
Washington
Copy May 17, 1955
Memorandum For Mr. James S. Lay, Jr.
Executive Secretary
National Security Council

Subject: U. S. Scientific Satellite Program

1. I should like to register my enthusiastic support of the proposal of the Department of Defense (RD-CGS 202/4) which you sent to me under cover of your memorandum of May 13, 1955.

2. I am impressed by the psychological as well as by the [phrase excised during declassification review] advantages of having the first successful endeavor in this field re-

sult from the initiative of the United States, and by the costly consequences of allowing the Russian initiative to outrun ours through an achievement that will symbolize scientific and technological advancement to peoples everywhere. The stake of prestige that is involved makes this a race that we cannot afford to lose.

3. Because of the basically new questions of ionosphere jurisdiction that are involved, and because the announced Soviet program in interplanetary communications makes it certain that a vigorous propaganda will be employed to exploit all possible derogatory implications of any American success that may be achieved, it is highly important that the U. S. effort be initiated under auspices that are least vulnerable to effective criticism. The extraordinary opportunities for exploitation of superstitions on the one hand and of imputed military hazards on the other that are inherent in a scientific "breakthrough" of such novelty make it imperative to enlist many voices speaking for numbers of nations to allay the potentially boundless fears that may be stirred up, even though they are quite unwarranted.

[12] I agree, therefore, with the suggested procedure of having our Government announce that it is ready to support the project through the U.S. National Committee of the International Geophysical Year. It is important for the following reasons that the U.S. proposal be made public at the time when it is submitted to the IGY:

A. The International Geophysical Year was established by the International Union of Scientific Societies which in turn is affiliated with UNESCO—part of the United Nations structure.

B. I am informed that the IGY in its Rome meeting last year endorsed the launching of a satellite as a desirable scientific step.

C. Since Russia is represented in this organization it would be in a position to know immediately of any U.S. offer made by the Government through the U.S. National Committee to launch a satellite.

D. If the U.S. offer was not made public the Soviets might take immediate action and do one of two things:

    1) Announce it has already launched a satellite.
    2) Make an offer to launch one themselves.

thus reducing the psychological significance and prestige values of the U.S. proposal.

4. The announcement of the U.S. offer might be made by Ambassador Lodge to the United Nations. Although the IGY is affiliated with the United Nations, for public reassurance the Ambassador might state that the United States would welcome some form of direct U.N. sponsorship for the project since its intent was to contribute to the world body of scientific knowledge through study of the satellite in flight. Needless to say, the offer of sharing knowledge would not be extended to the method of launching.

5. The fact that Russia was represented upon the International Geophysical Year which endorsed a satellite launching project can be used to good effect by us in the event that there should be a concerted Communist effort to brand the project as evil or threatening. We should, alternatively, be ready to meet a Soviet statement that it, too, is preparing to launch a satellite upon a shorter time-table or even, at some date, an announcement, true or false, that it has launched one.

[13] 6. Since a U.S. success in being the first to launch a small uninstrumented satellite could be quickly discounted if the Soviets were to follow it with an initial success in the launching of a satellite of more sophisticated type, I believe that the exploratory work on the latter type recommended in paragraph 11 C of the Department of Defense memorandum should be pursued vigorously in the United States concurrently with the program recommended for immediate implementation.

Nelson A. Rockefeller
Special Assistant

Document II-11

Document title: S.F. Singer, "Studies of a Minimum Orbital Unmanned Satellite of the Earth (MOUSE)," *Astronautica Acta*, 1 (1955): 171-84.

Source: NASA Historical Reference Collection, NASA History Office, NASA Headquarters, Washington, D.C.

S. Fred Singer, a physicist at the University of Maryland, proposed a Minimum Orbital Unmanned Satellite of the Earth (MOUSE) at the fourth Congress of the International Astronautics Federation in Zurich, Switzerland, in the summer of 1953. Singer's paper was based on a study prepared two years earlier by members of the British Interplanetary Society who had based their proposal on the use of a V-2 rocket. The Upper Atmosphere Rocket Research Panel at White Sands discussed Singer's plan in April 1954. In May, Singer presented his MOUSE proposal at the Hayden Planetarium's fourth Space Travel Symposium.

Singer had been present at the spring 1950 meeting at James Van Allen's home where the prospect of an International Geophysical Year was discussed, and he became a vocal proponent of building a satellite for the IGY. MOUSE was the first satellite proposal widely discussed in non-governmental engineering and scientific circles.

---

[171]

Studies of a Minimum Orbital Unmanned Satellite of the Earth (MOUSE)[1]

Part I. Geophysical and Astrophysical Applications[2]

by S.F. Singer, College Park/Md.[3], ARS
(*Received August 29, 1955*)

Editor's Note. The announcement by President Eisenhower on July 29, 1955 about the launching of minimum satellites by the United States during the International Geophysical Year 1957-58 has made the present article quite topical.

While it is too early to speculate about the details of the U.S. satellite program, the announced dimensions, payloads and applications resemble very much those of a MOUSE satellite. It will be remembered that in 1954 the International Scientific Radio Union (URSI) in the Hague and the International Union of Geodesy and Geophysics (UGGI) in Rome both endorsed resolutions proposed by Professor Singer to apply artificial satellites to geophysical and astrophysical research.

Abstract. A MOUSE would provide a far-reaching extension of present high altitude rockets in the study of the upper atmosphere and extraterrestrial radiations. Lifetimes of even a few days and payloads as low as 50 pounds would be adequate to allow continuous observations of the *solar ultraviolet* and *X-radiations* which have a profound influence on the ionosphere and therefore on radio communications. The cause of *magnetic storms* and

---

1. Presented in this form at the 25th Anniversary Spring Meeting of the American Rocket Society, Baltimore, Maryland, April 20, 1955. The substance of this paper was first presented at the Fourth Congress of the I.A.F., Zurich, 1953.
2. Part II "Orbits and Lifetimes of Minimum Satellites" was presented as a paper at the New York meeting of the American Rocket Society, Dec. 1954.
3. Associate Professor, Department of Physics, College Park, Maryland, U.S.A.

aurorae could be established with more certainty. Observations of *cosmic rays* would help clear up the question of their origin. Various other astrophysical phenomena, such as *micrometeorites*, could be brought under direct observation. Measurement of the *earth's albedo* (reflected sunlight) would give a measure of total world cloud coverage which could be used to predict long term climatic changes. Radio transmissions from MOUSE would send back all data and allow at the same time a study of the *ionosphere*. The change in the orbit and the lifetime would give information on drag and therefore *upper atmosphere densities*, while observation of a luminous trail of sodium emitted from the satellite would allow studies of *winds, temperature,* and *turbulence* in the outermost layers of the earth's atmosphere.

[172] The technical problems connected with the launching, control and instrumentation of the MOUSE satellite are well within the range of present techniques. It is likely that even smaller satellites will be constructed first to carry out portions of the research program described above....

Introduction

It is the purpose of the present paper to present a strong justification for the establishment of a minimum artificial satellite of the earth in terms of the advances it would lead to in our knowledge of the earth's outer atmosphere, of extraterrestrial radiations and their influences on the earth. It is my belief that only after a justification has been clearly stated and the problems delineated which the [173] satellite would solve for us, does it become possible to deal with the technical problems in an intelligent manner; for example, the questions of optimum altitude of the satellite or of the precision of the orbit or of the necessary lifetime for a satellite cannot really be answered unless the purpose of the satellite is kept clearly in mind.

We will discuss here only the geophysical and astrophysical applications of a *minimum* satellite; i.e., a satellite weighing no more than perhaps fifty pounds, containing a radio transmitter and simple instruments to measure properties of the earth's atmosphere and of the extraterrestrial radiations. This does not mean that a larger satellite carrying more elaborate equipment, such as television cameras, spectrographs, or telescopes with pointing controls, would not be more useful. However, the larger satellite vehicle seems far removed from the standpoint of feasibility. To talk about its obvious usefulness would not add to the very real task of defining the usefulness of a satellite small enough so that it can be constructed and launched within the framework of available techniques. Hence we shall resist the temptation to discuss more elaborate instrumentations and consider only the very simplest types of observations which could be performed by instruments placed in a minimum vehicle above the earth's atmosphere.

Instruments vs. Propulsion

The main tasks would seem to be to decide on what is important to measure, to choose the fields in which crude observations could add appreciable knowledge to our store of information about outer space, and finally to design instruments which can without great refinement yield worthwhile and important data.

One can then investigate how the requirements of such a satellite research program affect the propulsion and guidance necessary to place the vehicle in its orbit. It is obvious that in order to be practical, this cannot be a one-way channel, but rather the propulsion engineer may say to the astrophysicist: "This is as much as we can do, so many pounds of payload, now see what you can do with that." It is necessary, therefore, to place oneself somewhere in the middle, keeping both ears open, one towards propulsion and the other towards the scientific instrumentation and to allow continuously for modifications on both sides in order to produce an end product which will be both *useful* and *feasible*.

This indeed is the heart of the compromise. It is relatively easy to produce a satellite which may be nothing more than a tiny metal slug, but it is hard to justify it on the basis of geophysical usefulness if it cannot be easily observed. On the other hand, an ambitious satellite with elaborate instruments may exceed the limitations set down by the rocket engineer. So, in order to be practical, and this paper will be concerned with a practical approach, a satellite proposal has to be both *feasible* and *worthwhile*.

We will arrange this paper into two parts: The first part will deal with a detailed discussion of the most useful investigations, and their scientific and economic implications. The second part will deal with some of the technical questions pertaining to a satellite, discussed in the light of the above investigations. These technical questions relate primarily to the weight of the satellite, to its physical dimensions, to the structural materials (in particular the skin), to the method of data recovery, to the choice of orbit, to the optimum launching altitude, to the necessary launching accuracy in speed and angle and the resulting precision of the orbit, and to the "lifetime" of the satellite.[4]

[174] **Basic Reasons for Upper Atmosphere Investigations**

In order to study the largest part of extraterrestrial radiations, either electromagnetic or corpuscular, it is necessary to be above the appreciable[5] atmosphere since the radiation on encountering the atmosphere is modified or absorbed. Certain of the radiations, e.g., ultraviolet from the sun, are of great importance for the behavior of the atmosphere, but in every case a study of the incoming radiations reveals much about processes in outer space which could not be determined from sea level or even balloon observations. If we examine the transmission of the atmosphere to electromagnetic radiations [1], we find only two major "windows", one in the visible region from 2900 A to about 7000 A, the other in the radio region. Beyond 7000 A in the infrared there are many absorption bands due to the presence of water vapor, carbon dioxide, and ozone. Occasionally there are "windows" in the atmosphere through which one can see small portions of the infrared spectrum of the sun [2]. A particular prominent window is in the neighborhood of ten microns. It is only when one reaches wave lengths of the order of millimeters, in the microwave region, that the atmosphere again becomes transparent. But when the wave length increases up to a few meters the waves are again prevented from coming to sea level, this time not by absorption, but by reflection from the ionized layers in the upper atmosphere [1]. Going from the optical window towards shorter wave lengths one finds ozone and oxygen to be effective absorbers of ultraviolet radiation. It is only recently that rocket flights above 70 miles have given direct evidence of solar radiation in the far ultraviolet and in the X-ray region.

The situation is even worse with regard to corpuscular radiation. Even the highly energetic ($> 10^9$ ev) cosmic ray primary particles (made up of protons and nuclei of heavier atoms) cannot penetrate far into the atmosphere without undergoing collisions with air nuclei. Only the cosmic ray secondaries produced in these collisions can reach the lower atmosphere [3]. Auroral particles, [175] which are responsible for the northern lights, may contain protons of energy 100 times lower than the lowest cosmic ray energies; they are easily stopped in the upper atmosphere as they give up their energy to excite the auroral glow [4]; strangely enough particles of energy intermediate to cosmic rays and auroral particles are often absent [5]. Particles of even lower velocity (about 3000 km/sec) are extremely difficult to detect if they arrive singly. If they are charged, they will be turned away by the earth's magnetic field, long before they come close to the atmosphere. If, however, they arrive in sufficiently large numbers, in the form of corpuscular streams, instead of singly, then their reaction on the magnetic field is noticeable and may even lead to measurable variations of the earth's field [6].

A third category includes material bodies: interplanetary dust particles, micrometeorites, meteors and meteorites. The last two categories can, of course, be detected from

---

4. This novel concept relates to the duration of the satellite orbit and is discussed in Part II. Astronautics, Acta, Vol. I, Fasc. 4.

5. Where "appreciable" denotes an altitude appropriate to the type of radiation under study, e.g., ~ 25 miles for cosmic rays, ~ 65 miles for solar ultraviolet.

the ground, the meteors by their luminous and ionization trails. If they are extremely small (of the order of a few microns), they may not produce these trails. They could, however, be observed in impacts with detectors placed above the atmosphere [7].

Satellite Observations

Since 1946 upper atmosphere experiments have been carried on in V-2, Aerobee, and Viking high altitude sounding rockets and have furnished a great deal of scientific knowledge about the high atmosphere and solar and cosmic radiations [8]. In a critical study of the rocket program one is left with the feeling that much could be gained in certain fields by more frequent or even continuous observations as against occasional rocket measurements at one location of the earth. Out of such a study emerges a consistent research program for a satellite, consistent in the sense that it will supplement the information which has been [176] derived from high altitude rocket experiments by expanding the time and geographical scale of certain observations. Rocket experiments, because of their greater payloads, can delve more closely into detailed investigations of upper atmosphere phenomena and radiations from outer space.

Electromagnetic Radiations

Probably the most important subject for study from a platform above the atmosphere is the ultraviolet radiation from the sun. The interesting radiations extend all the way into the soft X-ray region with wave lengths of only a few Angstrom units. The portion of the radiation extending from 2900 A to 2200 A is stopped by the ozone (O_3) of the middle atmosphere (about 30 miles). Wavelengths shorter than 2200 A must be observed at altitudes of over 65 miles since the absorption of the residual oxygen in the atmosphere is strong enough to eliminate all traces of this radiation at lower altitudes [1].

A characteristic feature of the solar radiation in the UV region is its great variability; although the sun appears to be emitting steadily in the visible, in the ultraviolet it behaves very much as a variable star. Since the ultraviolet radiation has such profound effects on the earth's upper atmosphere (it produces the radio reflecting layers of the ionosphere and the ozone layer) and since it initiates many photochemical reactions in the upper atmosphere, it is of the utmost importance to keep track of the ultraviolet radiation. Of particular interest in the Lyman-alpha line of hydrogen at 1216 A where a large part of the ultraviolet energy of the sun is concentrated. It is suspected that the intensity of the line can vary considerably depending on the amount of solar activity [9]. During periods of solar activity large amounts of energy seem to be released in the solar atmosphere and can be observed as sudden brightenings ("flares") on the solar surface in the vicinity of sun spots. These brightenings cover only a minute fraction of the solar disc and are detectable only in light filtered in the red line of atomic hydrogen ($H\alpha$ line 6563 A). During these solar flares gaseous material is ejected from the sun into interplanetary space and large amounts of electromagnetic radiation are also emitted. We cannot observe at sea level anything but the visible. There is good evidence, however, for the increased emission of UV through observations of the ionosphere. Large solar flares can produce so-called radio fadeouts which are indicated by the disappearance of reflected radio signals. The fadeouts are caused by excess ionization in the lower D layer of the ionosphere, this excess ionization being produced by a large increase in the UV emission from various levels of the sun's atmosphere. The outermost level, the tenuous corona, emits X-radiations which are similarly enhanced during solar flares. The exact amounts are not known and their variability is quite unknown. Probably, therefore, the most important application of a satellite would be to the study of the Lyman-alpha radiation and the X-radiation of the sun, and of their intensity variation with time during different periods of solar activity. A study of the relationship, during solar flares, of the large increases among the different regions of the solar spectrum may give us valuable information about the manner in which the energy is transferred from the surface of the sun into the sun's outer atmosphere 10,000 miles up. It is suspected that the corona is heated either by magneto-hydrodynamic shock waves or by particle streams, thus causing the high temperatures which are deduced from coronal emission lines observed in the visible. From the travel time of the disturbance we may be able to learn more about the actual mechanism by which the energy is transported in the solar atmosphere.

[177] The economic implications of these studies are quite considerable, particularly if correlated directly with ionospheric observations. The study of the ionosphere has become a vast undertaking carried on by the laboratories of many governments on an international basis in order to derive fundamental information about the radio reflecting layers. The knowledge derived can be applied in a practical way to give predictions necessary for effective radio communication. At the present time this study is handicapped by our very imperfect knowledge about the solar radiations which produce these layers. Although there can be a wide range of argument about which types of satellite observations would be the most useful, it is safe to say that the studies of the solar radiation described above would rank very high.

The instruments for observing solar ultraviolet and X-radiation can be photon geiger-counters or photo sensitive surfaces with appropriate filters, similar to techniques which are now being used in high altitude rockets [8]. It is essential, however, to be able to point the instruments roughly towards the sun, to have them reasonably omnidirectional in case of misorientation, and to be able to observe the sun over as large a fraction of the orbit as possible.

Aside from the UV region, measurements of electromagnetic radiation can also be carried on in the infrared. Here the techniques become more difficult, also the results become less important from the fundamental point of view of solar physics and from the applied point of view of atmospheric effects. Furthermore, the variability in the infrared should be low; therefore, occasional rather than continuous observation may provide the necessary answers. Finally, it is possible to get a great deal of information on the solar infrared from balloon observations above the appreciable water-vapor and carbon dioxide of the lower atmosphere.

Up to now no hard X-rays or gamma rays coming from the sun have been measured. The problem may be one of low intensity, but it is certainly advisable to explore this region in conventional high altitude rockets before discussing applications for a satellite instrumentation. Of particular interest would be the measurement of the 2.2 Mev gamma ray which arises from the radiative nuclear capture of neutrons by protons and would indicate the presence of high energy neutrons on the sun. This is an experiment which could probably be done in a balloon because high energy gamma rays are quite penetrating; thus it is no longer necessary to observe at extremely high altitudes.

Corpuscular Radiation

During periods of great solar activity, one can observe with chronographs prominences of luminous gas being shot out from the solar surface. These observations indicate the great violence of solar processes. Recent radio observations of the sun have shown the existence of streams of charged particles which are shot out through the solar atmosphere during solar flares [9]. But for a half century now it has been hypothesized that the sun can emit clouds of ionized gas with velocities high enough to leave its surface and travel through interplanetary space past the earth. While it has not been possible to observe these gas clouds directly in their travel from the sun, their effects upon the earth are unmistakable. About a day following a strong solar flare (as manifested by visual observations on the solar disk and by ultraviolet enhancements leading to radio fadeouts), one observes a sudden increase of the earth's magnetic field, the so-called "sudden commencement." This is followed a few hours later by a slow decrease of the field which may last for several days. These "magnetic storms" are world-wide and are believed to be produced by the electromagnetic effects [178] of these streams of charged particles as they enter the earth's magnetic field. During these periods one also observes a large enhancement of the aurora in the northern and southern hemispheres. These auroral displays in the upper atmosphere are thought to be due to high speed corpuscles, possibly protons, which come from the sun during these periods of great activity. The regions of the sun which are responsible for the magnetic storms, the so-called *M*-regions, show great persistence; twenty-seven days later, one synodic rotation period of the sun, one may again observe a magnetic storm, an enhancement of the aurora, and associated cosmic ray effects. It is thought, therefore, that the active regions of the sun continue to emit a stream of particles which sweeps interplanetary space, very much like a stream of water from a rotating garden hose [1].

The nature of the solar streams, and the exact mechanism by which they cause magnetic storms, aurorae, and cosmic ray effects, are not well understood. A satellite could contribute to this study in two ways: (1) By intercepting the particles which cause the aurora, we would determine their nature and their intensity, their time variations and their geographical distribution. A satellite traversing an orbit over both poles would also contain the world-wide distribution of auroral particles in the northern and southern hemispheres. This would give an important clue to their origin. (2) By studying the magnetic field above the conducting layers of the ionosphere one would obtain a better picture of the primary effects of the magnetic-storm producing beam since the magnetic effect observed at sea level is distorted by the ionosphere [7]. The instrumentation for these measurements is again well proven from high altitude rocket experiments. The auroral particle measurements could be done with thin-walled geiger-counters whereas the magnetic storm measurements could be done by means of total field magnetometers such as have been used in Aerobee rockets.

The study of magnetic storms and aurorae is of considerable practical importance again from the point of view of radio communication. Magnetic storms have sometimes profound and very violent effects also on long distance telephone communication; the electric fields induced by the strong variations in the earth's magnetic field can easily burn out long distance cables and raise havoc with wire communication.

Cosmic Rays

Cosmic rays are corpuscular radiations of extremely high energies. The primary cosmic rays consist mainly of protons but also of helium nuclei and to a smaller extent of the nuclei of heavier elements. They arrive at the top of the atmosphere with almost the speed of light and with energies ranging from a few billion electron volts (Bev) up to a billion times as much. They constitute the highest energy phenomenon known in nature; but because of the small number of cosmic rays which are received here, the energy they bring in is about equal to the energy of starlight. The effects of cosmic rays on the earth and the earth's atmosphere are, therefore, probably negligible but they constitute one of the most important fields of study in modern physics and provide a challenging problem to the astrophysicist as well as to the nuclear physicist. The nuclear physicist studies cosmic rays because they represent nuclear particles of energy vastly greater than can be produced by even the largest accelerators. The cosmic rays in colliding with the atoms of the upper air produce nuclear reactions which cannot be duplicated in laboratory studies. The nuclear physicist, therefore, views cosmic rays essentially as a tool which nature has provided to help him in the study of high energy physics and with which he hopes to solve the problems of [179] the ultimate constitution of the nucleus and the ultimate nature of the "elementary" particles. Already the study of cosmic rays from that point-of-view has led to the discovery of many new types of elementary particles, the so-called mesons. There is reason to believe that their systematic study will lead eventually to a better understanding of the nature of nuclear forces [3].

The astrophysicist treats cosmic rays essentially as a phenomenon and as an indicator of processes which go on in the galaxy and in the solar system. He is mainly concerned about the origin of the cosmic radiation and about the manner in which they acquire the high energies; he asks about the processes which exist in the universe which can produce such tremendous energies. There is little doubt that these processes are electromagnetic in nature and that, therefore, a study of the origin of cosmic rays will lead to a better understanding of the electromagnetic conditions not only in the vicinity of the earth and in the solar system but also in our galaxy. This knowledge of magnetic fields in the galaxy can have a very profound influence on theories of the origin of galactic systems and on cosmology in general.

One of the most fruitful ways of studying the cosmic radiation is to investigate the distribution-in-energy of the primary rays. This has been accomplished in rocket experiments by observing the cosmic ray flux at different latitudes. The method makes use of the earth's magnetic field, which varies with latitude, and uses this field as an energy analyzer for cosmic radiations. It has led to the rather surprising finding that in the cosmic radiation there is at times an absence of low energy cosmic rays; i.e., below about 0.5 Bev there are very few cosmic rays compared to the number above this energy [5]. The mechanism which either keeps low energy cosmic rays from coming to the earth, or perhaps prevents their ever being produced, is not understood, and if cleared up will probably shed a great deal of light on the origin of cosmic rays themselves. Experimentally this absence of low energy cosmic rays manifests itself as follows: While the cosmic ray flux increases by a factor of ten in going from the equator to geomagnetic latitude 56°, there is no further increase observed between 56° and 90°. If the low energy cosmic radiation were present, the increase between 56° and 90° might be almost another factor of ten.

The most promising method, therefore, for using a satellite for cosmic ray studies would be to investigate the energy spectrum on a continuous basis by allowing the satellite to travel between 0° and 90° latitude to measure the intensity variation of cosmic rays as a function of latitude. We would like to discover, for example, whether the "knee" at 56° is fixed with time or whether its position changes as a function of the solar cycle, whether there are increases in intensity above 56° possibly correlated with phenomena on the sun, and so on.

From cosmic ray studies of the last few years we know that the cosmic ray intensity is not constant. Among the more pronounced effects there are two which seem to be especially suited to satellite observations, because of their large size. They are the cosmic ray increases which sometimes accompany certain bright solar flares, and the cosmic ray decreases which often occur in connection with magnetic storms [3].

The cosmic ray increases associated occasionally with solar flares manifest themselves in a rapid rise of the cosmic ray intensity about ten to thirty minutes after the solar flare. It seems fairly certain that the increases are due to cosmic ray particles, accelerated either on the sun or in the immediate vicinity of the sun, which travel towards the earth and are then deflected by the earth's magnetic field. This deflection causes the particles to be incident at certain locations, the so-called impact zones, with relation to the sun-earth line [10]. From our present [180] sea level observations it seems fairly certain that these cosmic ray increases are due to additional particles of low energy, i.e., not exceeding about 10 Bev. What is quite unknown, however, is the reason why only a few solar flares cause these large increases, four in the last 15 years. Satellite observations could establish whether increases occurred in the primary cosmic rays but were confined to such low energies that no effects could be detected at sea level.

The decreases of the cosmic ray intensity lasting a day or more and associated with magnetic storms are among the most puzzling phenomena. Recent observations that these decreases occur even at the pole establish that we are dealing here with a real decrease in the cosmic ray intensity in the vicinity of the earth, rather than a deflection away from the earth by the ring current which is thought to encircle the earth during periods of magnetic storms [4]. The question as to what produces this decrease in cosmic ray intensity is not at all settled. It is thought likely that the cosmic storms are produced by the corpuscular streams from the sun which are also responsible for magnetic storms. The cosmic ray decreases show the same 27-day recurrence, clearly associated with the 27-day synodic rotation period of the sun. One of the missing links for an interpretation of the phenomenon is again an observation of the primary spectrum during periods of cosmic ray storm decreases [7].

Techniques for observing the cosmic radiation are well developed from work in conventional rockets. Geigercounters of conventional and special design could be used to measure the flux and even the composition of the primary radiation.

Micrometeorites

We cannot study atomic particles of extremely low velocities; i.e., below the velocities of auroral particles, unless they occur in large streams and produce electromagnetic effects as is the case with the corpuscular streams from the sun. We can, however, study low velocity particles of higher mass, i.e., micrometeorites or interplanetary dust particles (dimension of the order of 1 micron). Depending on their orbits with respect to the earth they may enter the atmosphere with velocities up to about seventy kilometers per second. The larger ones produce, of course, the bright flashes and ionization trails associated with meteors but very small ones may escape detection entirely. No direct observations have been made of micrometeorites except for some exploratory rocket experiments in which their impacts have been observed either by condenser microphones or by the pitting of polished plates [7].

Observations of the zodiacal light and of the F-corona, the outer dust corona of the sun, have given some ideas of the density of interplanetary dust between the sun and the earth. Micrometeorites would perhaps have the same dimensions as dust particles but travel with rather high speeds into the atmosphere. The number of observations of the interplanetary dust are not sufficient as yet to establish any significant density variations with the solar cycle or even variations of a shorter time period. A particularly interesting point to investigate would be the effect of solar corpuscular beams on the dust density; it will be possible thereby to evaluate the "sweeping out" effect of a corpuscular beam. This study of the fluctuations in their intensity could be performed in a satellite by counting particle impacts; it would have considerable value in clearing up the origin of the interplanetary dust particles. The measurement of an intensity variation vs. latitude would give information on their momenta and electric charges.

[181] ## Observations of the Earth's Upper Atmosphere

Earth's Albedo

One of the main questions which concerns meteorologists is the heat input to the earth from the sun. The heat balance of the earth can be described roughly as follows: The earth intercepts from the sun an amount of energy equal to the solar constant at the earth's orbit times the cross-sectional area of the earth. The solar constant has the value of 2 calories/cm^2/min; it is believed that variations in the solar ultraviolet emission do not affect its value appreciably although UV radiation does have profound effects on the upper atmosphere. Of the total amount of energy intercepted a certain fraction is reflected in the visible. This reflection is of the order of 35%; it is mainly due to clouds, which have a very high albedo. The albedo of the land surface is of the order of 15% although snow and ice on the land surface will greatly increase the albedo. The largest portion of the earth's surface, the oceans, have an albedo of only 4% in the visible. The net energy, i.e. incident minus reflected, is used to heat the earth's surface and atmosphere. This energy influx is balanced by the heat loss from the earth's surface and the atmosphere; they radiate according to the classical radiation laws. Since their temperature is very low, of the order of 300° K, the radiation occurs mainly in the far infrared, around ten microns. The energy is radiated isotropically into interplanetary space and is lost from the earth. The infrared loss tends to vary slowly because of the large heat capacity of the earth. The great unknown in the heat balance considerations is represented by the amount of reflected sunlight which depends so critically in the day-to-day cloud coverage of the earth. The satellite furnishes a very direct method for measuring the visual albedo of the earth and supplying thereby the vital missing link in the heat balance computations. It should, therefore, be possible to plot more detailed heat flux data for the earth, which in turn could lead to the possibility of predicting long range climate for various latitude belts of the earth and for various seasons. The practical importance of this possibility can hardly be overestimated.

The actual measurement of the earth's albedo is technically a very simple matter. A photocell which views the earth continuously would provide us with the necessary information.

Ionosphere

The radio signal from a satellite which is used to transmit the data from the various instruments can itself be used to yield important information about the ionosphere. As the satellite moves with respect to a fixed station, the total number of electrons between it and the receiving station decreases to a minimum and then increases. This change in index of refraction introduces an easily measured frequency shift. While this ionosphere frequency shift is always superimposed upon a Doppler shift, they can be separated and evaluated independently. The satellite transmitter, therefore, gives us a valuable tool for investigating the ionosphere in a manner which supplements the usual ionospheric investigations with reflected radio signals.

Upper Atmosphere Densities

The measurement of drag deceleration seems to be the only promising method for determining the densities in the very high atmosphere where the molecular mean free path is very much larger than any instrument or any vehicle which can be sent up. Clearly a vertically falling body of any appreciable mass will [182] not experience a measurable deceleration. It is only when the body travels in an orbit in which it can spend a long time in the upper atmosphere, that the product, deceleration X time, leads to an appreciable change in velocity; even though the deceleration is very small, the time interval is long enough to allow the velocity change to be measured.

In the case of a satellite the velocity change will lead to a change in the elements of the orbit; it is, therefore, possible by measuring the change in the elements of the orbit to deduce upper atmosphere densities. A detailed study of the effects of upper atmosphere drag leads to the following results: An initially elliptical orbit which has a perigee sufficiently low in the atmosphere to experience drag, will after a certain number of orbits gradually approach a circular orbit. The rate at which the eccentricity decreases depends not only on the elements of the orbit but critically on the area and mass of the satellite.[6]

---

6. In the elliptical orbit the energy loss occurs mainly at the perigee, the point of closest approach to the earth's surface. It is therefore possible to apply an approximation method in which the energy loss and velocity loss is concentrated at the perigee. This method can be used to predict with good accuracy the lifetime of a satellite after its initial orbit is determined and if its area and mass are known. After the orbit has become more or less circular, the energy loss will occur continuously and the circle will shrink in altitude until finally the energy loss per orbit becomes an appreciable fraction of the total energy. The perturbation method which has been used is then no longer applicable; the satellite rapidly loses altitude and intercepts the earth's surface. (For further details cf. Part II.)

Sodium Trail

It has been remarked earlier that the solar radiation produces reactions in the upper atmosphere which lead to the emission of light. Among the prominent emission lines of the night air glow are those of the "forbidden" oxygen transitions (5577 A and 6300 A) and also the yellow *D*-lines of sodium (5893 A). During twilight, while the lower atmosphere is dark, the sun illuminates the upper atmosphere. Under these conditions the few sodium atoms of the upper atmosphere, because of resonance radiation, exhibit the characteristic yellow sodium line very strongly, the so-called "twilight flash" [1]. Since the sodium atoms are not localized, the emission is observed as a diffuse yellow glow. It has been suggested that if the concentration of sodium were enhanced, the sodium light emission would be similarly increased. Therefore, a novel application of a satellite would be to exhaust sodium vapor into the upper atmosphere so as to produce a defined trail of sodium atoms which would in turn exhibit a defined trail of the sodium emission light. This would lead to rather spectacular results since it should be possible to observe this sodium trail visually from the ground during twilight conditions. From the research point of view the sodium trail offers great advantages. Since the sodium atoms are subject to collisions with other gas atoms, they will soon cool down and share their temperature. From the spreading of the trail, therefore, we would be able to learn about the temperature and turbulence in the outermost layers of the earth's atmosphere and from the distortion of the trail we would be able to deduce the existence of winds in these rarefied regions. The sodium trail certainly promises to be one of the most exciting applications of satellite geophysical research.

Technical Questions

We may now turn our attention to the characteristics of the satellite and its orbit, which are required to make possible the investigations which we have [183] outlined. With the simple instrumentation which these investigations demand, the weight of the satellite can be kept well below fifty pounds. The scientific data would be telemetered to the ground by a radio transmitter which carries superimposed on it a number of telemetering channels; each is assigned to a definite instrument which detects and transmits information about the phenomenon it is sensitive to. The largest portion of the weight will be the power supply and transmitter. Once a radio frequency channel has been established, each additional telemetering channel does not consume very much extra power or weight. The individual instrumentations probably weigh only on the order of ounces. The physical dimensions of the satellite can be similarly small, probably within a cylinder of about one foot diameter and one foot height. With proper precautions the temperature problems in the satellite are not critical; it is only necessary to establish good heat conductivity to prevent hot spots. The average temperature would be of the order of room temperature.

The various experiments outlined earlier strongly suggest an orbit [11] which will go over the poles of the earth rather than an equatorial orbit. Since, however, an equatorial orbit is easier to establish from the propulsion point of view, one probably should not insist too strongly on a polar orbit, at least to begin with, except to point out that it would allow the continuous observation of the sun and, therefore, the continuous production of electric power by means of silicon solar batteries. A polar orbit will also allow a study of the energy spectrum of the cosmic rays and the investigation of auroral particles in the auroral zone. It would further allow scanning of the complete earth's surface in order to obtain the cloud albedo. There are, therefore, many advantages in the choice of a polar orbit rather than an equatorial one; it is hoped that the additional propulsion which a polar orbit demands will not be too difficult to procure. In a polar orbit it would be most economical to store the telemetered information and release it only over the poles, either one or both, since this would demand a minimum of telemetering receiving stations. It is to be kept in mind that the orbit will stay more or less fixed in space as the earth turns underneath it. An orbit, therefore, [184] which is perpendicular to the earth-sun line, will offer the greatest advantage from the point of view of solar observation.

A question equally important as the propulsion problem is the degree of guidance necessary to achieve a desired orbit. The optimum launching altitude and the errors allowable in launching speed and angle are intimately tied up with each other and with the physical properties of the satellite. Together they determine the lifetime of the satellite. This is a matter of detailed considerations and is discussed in Part II. It is to be noted finally that optical visibility and precision of the orbit are of minor importance for a satellite whose main application is geophysical or astrophysical research. It is merely necessary to have it above the atmosphere for a sufficiently long period of time, which may mean only a few days. Astronomical perturbations can be neglected for such short lifetimes. It is seen, therefore, that the guidance and control problem, as well as the propulsion problem for this type of satellite is extremely simple [11] in comparison to satellites which are meant to fulfill more ambitious functions. It is this feature mainly which gives hope for the early accomplishment of a minimum instrumented satellite.

References

1. As a general reference see S. K. Mitra, The Upper Atmosphere, Calcutta: The Asiatic Society, 1952; also R.M. Goody, The Physics of the Stratosphere. Cambridge: University Press, 1954.

2. G. P. Kuiper (ed.), The Earth as a Planet. University of Chicago Press, 1954.

3. J. G. Wilson (ed.), Progress in Cosmic Ray Physics. New York: Interscience Publishers, 1952.

4. A. B. Meinel, Astrophysic. J. 113, 50 (1951).

5. J. A. Van Allen and S. F. Singer, Nature 170, 62 (1952).

6. S. Chapman and J. Bartels, Geomagnetism. Oxford: University Press, 1940; see also S. K. Mitra [1].

7. R.L.F. Boyd, M. J. Seaton and H. S. M. Massey (ed.), Rocket Exploration of the Upper Atmosphere. London: Pergamon Press Ltd., 1954.

8. H. E. Newell, High Altitude Rocket Research. New York: Academic Press Inc., 1953; also S. F. Singer in: Vistas in Astronomy. London: Pergamon Press Ltd., 1955.

9. G. P. Kuiper (ed.), The Sun. University of Chicago Press, 1953.

10. J. Firor, Physic. Rev. 94, 1017 (1954).

11. S. F. Singer, J. Brit. Interplan. Soc. 11, 61 (1952); ibid. 13, 74 (1954); Sky and Telesc. 14, 15 (1954); J. Astronautics 2, No. 3 (1955).

Document II-12

Document title: "Memorandum of Discussion at the 322d Meeting of the National Security Council, Washington, D.C., May 10, 1957," *United Nations and General International Matters,* **Vol. XI.** *Foreign Relations of the United States, 1955-1957* **(Washington, DC: U.S. Government Printing Office, 1988), pp. 748-54. The original document is located in the Whitman File, NSC Series, Eisenhower Library. Top Secret; Eyes Only. Prepared by S. Everett Gleason on May 11, 1957.**

The National Security Council (NSC) had originally approved $20 million for the IGY satellite program, Project Vanguard. But the cost estimate for the program began rising almost immediately until it reached $110 million in April 1957. On May 3, 1957, a four-page "Memorandum for the President" from Percival Brundage, Director of the Bureau of the Budget, had raised the issue of Project Vanguard's cost overruns. This problem prompted President Eisenhower to ask that the scientific satellite program be discussed at the May 10, 1957, meeting of the NSC, including Soviet progress toward developing a space vehicle. At this meeting, the president decided that, despite the cost increases, the United States had no choice but to go ahead with the program.

---

[748] In the course of his briefing [on the scientific satellite program], Mr. [Robert] Cutler [special assistant to the president for national security affairs] explained that another hike in the costs of this program had induced the President to schedule the matter for discussion by the National Security Council. Mr. Cutler said that there would be a presentation by Assistant Secretary of Defense [William M.] Holaday and other officials of the Research and Engineering Division of the Department of Defense. Dr. Detlev Bronk, President of the National Academy of Sciences, and Dr. Alan Waterman, Director of the National Science Foundation, were likewise present, and would comment on the report by the Department of Defense....

After Mr. Cutler had finished his briefing and had noted that the costs of the program had increased from the original estimate (May 1955) of $15-20 million to the estimate of April 1957, of $110 million, he turned to call upon Secretary Holaday to present the Defense Department report. The President, however, interrupted with a vigorous complaint to Mr. Cutler that before he slid over some very important facts it would be well to recall that the original [749] program, calling for six satellites, was primarily a safety program designed to assure that at least one of these six satellites could be successfully orbited. There was no intention necessarily to launch six satellites. Another problem which disturbed the President was the very costly instrumentation currently being provided for the six satellites. Such costly instrumentation had not been envisaged when NSC 5520 had originally been approved by the President. The President therefore stressed that the element of national prestige, so strongly emphasized in NSC 5520, depended on getting a satellite into its orbit, and not on the instrumentation of the scientific satellite.

Mr. Cutler explained that he had not intentionally passed over these problems, and that they would be dealt with in the presentations by the Defense Department which were now to follow. Mr. Cutler then called on Secretary Holaday, who in turn stated that Dr. [John P.] Hagen [director of the Vanguard program] would make the first report on the nature and performance of the earth satellite program and the schedule of test launchings....

Dr. Hagen was followed by Assistant Secretary Holaday, who confined himself to an analysis of the cost aspects of the program to launch an earth satellite, with particular emphasis on the reasons which had led to the marked increases in the estimated costs of completing the program. He concluded his remarks with certain recommendations as to ways and means of funding the remainder of the program.

At the conclusion of Secretary Holaday's remarks, Mr. Cutler called on Dr. Bronk for a statement of the scientific aspects and importance of the earth satellite program. Dr. Bronk said that he would divide his brief report into three main parts. He dealt first with what he described as the immediate practical values to be derived from the successful orbiting of a scientific satellite. Among these, he stressed...information on the determinants of weather; and lastly, the influence of outer space on communications. He commented on the intense anticipation with which scientists were waiting for the receipt of this kind of scientific information.

Dr. Bronk stated that the second aspect of his analysis would be concerned with what might be described as the spiritual aspects of the program. If a satellite were successfully orbited, it would constitute the movement of man into an entirely new area of the universe into which he had never moved before. This was, accordingly, a challenging adventure, and if it were successfully concluded would mark a whole new chapter—indeed, a new epoch—in science and history.

[750] Finally, Dr. Bronk said he would touch on the international aspects of the earth satellite program. These aspects, he said, were of very great concern to our scientists. The fact that our earth satellite program was being carried out in connection with the

International Geophysical Year and in association with scientific groups from many foreign countries, would bring our scientists into a relationship with the scientists of other countries which could be very significant. We are taking the lead, but we are associated with a variety of other nations.

Mr. Cutler then called on Dr. Waterman, who said he would confine himself to discussing the matter of responsibility for funding the earth satellite program, as between the National Science Foundation and the Department of Defense. The gist of Dr. Waterman's remarks was that if it proved necessary to go to the Congress for a supplemental appropriation in order to complete the program set forth in NSC 5520, the Department of Defense was in a much better position, and had a much clearer obligation, to do so than did the National Science Foundation. On the other hand, Dr. Waterman expressed the earnest hope that some way might be found to provide for the costs of completing this program without going up to the Congress with a request for supplemental appropriations.

The President said that two thoughts had come to his mind at once as he had listened to this series of reports and comments. In the first place, there was no particular reason to assume that the latest estimate of the costs of completing the program ($110 million) would prove firmer than the earlier estimates. Indeed, it was quite possible that the costs of completing the program would go to $150 million, or even higher. His second impression, said the President, was that everybody wanted to duck responsibility for finding the money to fund the program.

Mr. Cutler then requested the Director of Central Intelligence to report on what we knew about the Soviet program to launch an earth satellite, and on the world-wide effects of a U.S. decision to abandon its own earth satellite program at this time.

Mr. Dulles indicated that the Soviets had not followed through on their promise to provide the organizers of the International Geophysical Year with the appropriate details of their program,...With respect to the effect of a U.S. abandonment of our program, Mr. Dulles pointed out that the program had been widely advertised and warmly welcomed throughout the world of science. If the Soviets succeeded in orbiting a scientific satellite and the United States did not even try to, the USSR would have achieved a propaganda weapon which they could use to boast about the superiority of Soviet scientists. In the premises, the Soviets would also emphasize the propaganda theme that our abandonment [751] of this peaceful scientific program meant that we were devoting the resources of our scientists to warlike preparations instead of peaceful programs.

Mr. Cutler then invited comments from Secretary [of Defense Charles E.] Wilson. Secretary Wilson replied that when the earth satellite program was first broached in the spring of 1955, it had been clearly and publicly stated that any of the scientific information resulting from the successful launching of an earth satellite would be made available freely to the whole world. Accordingly, our earth satellite program partook of the character of a pure research product rather than of the character of directed research which the Department of Defense could appropriately describe as vital to U.S. national security. Of course, continued Secretary Wilson, we in the Defense Department do have some defense interest in the satellite program. Nevertheless, it was not the kind of program which Defense could properly underwrite and for which it could properly provide money, as it had done lately, out of the DOD emergency funds for research and development. Indeed, Congress had already criticized the Defense Department for allocating money out of its emergency funds to tide over the earth satellite program, and Secretary Wilson said he could not really blame Congressional critics for their attitude. He complained that he was already having enough trouble in providing money out of his emergency funds for research projects which were truly vital to national defense.

The Director of the Budget pointed out to Secretary Wilson that the Department of Defense Emergency Fund ran out each year and had to be renewed each year.

When Mr. Cutler inquired of Secretary [Christian A.] Herter the views of the Department of State, Secretary Herter replied that he felt much as did Mr. Allen Dulles. The State Department favored completing the earth satellite program because of the prestige it would confer on the United States. He could not speak authoritatively of the problem of funding the program, which he said did present a rather frightening picture. Asked for his opinion, Admiral [Lewis L.] Strauss, [Chair of the Atomic Energy Commission], replied that he concurred in the views of Secretary Herter.

The President then commented that there was one lesson to be learned from the experience with the earth satellite program: In the future let us avoid any bragging until we know we have succeeded in accomplishing our objectives. The President then said that he would like to be informed as to how much the increased costs of the earth satellite program derived from increased costs of more elaborate instrumentation. Secondly, he wished to inquire whether the [752] launching of an earth satellite could be rendered easier if the satellite did not contain so much instrumentation as currently planned.

In replying to the President, Secretary Holaday pointed out that the diameter of the earth satellite had been reduced from thirty inches to twenty inches, although he admitted that the instrumentation had become a little "gold-plated", or at least "chromium-plated", as it had developed. Secretary Holaday also admitted that at the start of the earth satellite program we had not realized fully the requirements of the velocity. Likewise, more observation stations were now going to be established than had originally been thought necessary. Such items as these helped to explain the increasing costs of the program.

The President responded by pointing out that although Secretary Holaday had said that the 30-inch sphere had now been reduced to a 20-inch sphere, this was still larger than the "size of the basketball" which had been mentioned when NSC 5520 had first been considered by the Council. The President confessed that he was much annoyed by this tendency to "gold-plate" the satellite in terms of instrumentation before we had proved the basic feasibility of orbiting any kind of earth satellite. Secretary Wilson added the comment that irrespective of the merit of the earth satellite program, this program had too many promoters and no bankers.

Mr. Cutler alluded to a suggestion that if we succeeded in orbiting one of the test vehicles which would have no scientific instrumentation, it might be possible to abandon the rest of the program for launching the fully-instrumented scientific satellite. The trouble with this reasoning, according to Mr. Cutler, was that the six instrumented satellites were already in the pipeline. Accordingly, if we abandoned the attempt to launch these satellites, we wouldn't save very much money and we would miss achieving our objectives.

Secretary Humphrey inquired what was expected to happen if and when we succeeded in orbiting an earth satellite. Would we not then initiate another tremendous program to launch additional satellites and secure additional information about outer space. Secretary Wilson commented that this was the likely eventuality, and that this was the American way of doing everything—bigger and better.

The President observed that it was quite conceivable that the information we achieved from the successful launching of an earth satellite would be so great as to merit a continuing program thereafter. The trouble was that our original "basketball" satellite program had grown bigger, better, and more costly, at the same time that everybody wished to duck financial responsibility for its completion.

Secretary Wilson said that there was another significant factor to account for the increasing costs of programs such as this. Whenever you put a time limit on a new and large scientific program, you [753] immediately encountered financial troubles. The costs were bound to rise if the objective had to be achieved when a specific and relatively short time interval was set.

The President observed that in any event he did not see how the United States could back out of the earth satellite program at this time. We should, however, keep it on no more elaborate a basis than at the present time. Beyond this there was the problem of how to finance the completion of the program. In this respect the President suggested that in view of the fact that we have run out of money, there was no other recourse than for Defense and the National Science Foundation jointly to appear before the Congressional committees, tell them the story, and ask for supplemental funds. Secretary Wilson agreed with the President that we could not now abandon the program, and the President informed Secretary Wilson, Mr. Brundage and Dr. Waterman that they should make arrangements to go before the Congressional committees with a request for funds to finance the program on its present basis. Before doing so, however, the President said he wished the scientists who had been concerned with this program to take another hard look at it to see if there were any ways by which the costs could be cut or minimized. The President said he was not hopeful in this respect, but that it was worth a try. Thereafter the whole truth should be presented to the committees of Congress.

Mr. Cutler said he assumed that the President wished Defense and NSF to make their joint presentation to the same committees of Congress which had been dealing with the earth satellite program in the past. Mr. Cutler also suggested that the President would wish an immediate report to the National Security Council as soon as the Defense Department has succeeded in orbiting a test vehicle.

Mr. Brundage pointed out that the President's decisions would also involve the use of $5.8 million more of the emergency funds of the Department of Defense. The President agreed, and again called for a report by the Defense Department scientists were a little more restricted in their hopes and ambitions for the earth satellite program. Secretary Wilson commented that at least such a review by the scientists might help to prevent a further elaboration of the earth satellite program.

The National Security Council:[1]

a. Discussed the subject, in the light of a presentation by the Department of Defense and comments by the Director, National Science Foundation, the President, National Academy of Sciences, and the Director of Central Intelligence.

b. Noted the President's directive that the U.S. scientific satellite program under NSC 5520 should be continued on no more elaborate basis than at present and under the following conditions:

(1) The necessary arrangements should be made with the Congressional committees which previously dealt with this program, for joint presentations by the Department of Defense and the National Science Foundation, as to:
 (a) The additional funds to be made available from the Defense Department Emergency Fund to continue the program through August 1, 1957; and
 (b) The additional funds which must be appropriated in Fiscal Year 1958 to the Department of Defense in order to complete the program at a total cost not to exceed $110 million.

(2) Prior to the joint presentations under b-(1)-(b) above, the scientists working on this program should again scrutinize it carefully to determine whether the estimated additional funds required can be reduced by restricting the program in ways which will not jeopardize the current objectives under NSC 5520.

(3) In addition to the report required under NSC Action No. 1656-b, the Department of Defense should submit a report to the Council immediately if one of the test vehicles is successfully orbited as a satellite.

Note: The action in b above, as approved by the President, subsequently transmitted to the Secretary of Defense, the Director, Bureau of the Budget, and the Director, National Science Foundation, for implementation....

S. Everett Gleason

1. Paragraphs a-b and Note constitute NSC Action No. 1713. (Department of State, S/S-NSC (Miscellaneous) Files: Lot 66 D 95)

Document II-13

Document title: Allen W. Dulles, Director of Central Intelligence, to The Honorable Donald Quarles, Deputy Secretary of Defense, July 5, 1957.

Source: NASA Historical Reference Collection, NASA History Office, NASA Headquarters, Washington, D.C.

By 1957, the Central Intelligence Agency was aware that the Soviet Union had an active ballistic missile program and was preparing to launch a satellite. But the exact date of the launch was still uncertain. This memorandum from Director of Central Intelligence Allen Dulles to Deputy Secretary of Defense Donald Quarles indicates that American intelligence knew a Soviet space launch was imminent, but, as of early July 1957, was still unsure of the exact date of the launch.

Dear Mr. Secretary:

Thank you for your memorandum of 20 June 1957, transmitting the letter to you from Mr. V. A. Nekrassoff.

Information concerning the timing of the launching of the Soviet's first earth-orbiting satellite is sketchy, and our people here do not believe that the evidence is sufficient as yet for a probability statement on when the Soviets may launch their first satellite.

However, data has been recently received that Alexander Nesmsyanov, President of the Soviet Academy of Sciences, stated that, "soon, literally in the next few months, the earth will get its second satellite." Other information, not so precise, indicates that the USSR probably is capable of launching a satellite in 1957, and may be making preparations to do so on IGY World Days or Special World Days. The U.S. community estimates that for prestige and psychological factors, the USSR would endeavor to be first in launching an earth satellite.

Mr. Nekrassoff's postulation of 17 September 1957, presents an interesting consideration when we note that the public releases on Vanguard project set the first launching of the U.S. satellite in 1958, and the date 17 September 1957, would permit the Russians to attain the objective of a first launching. Further, the Russians like to be dramatic and could well choose the birthday of Tsiolkovsky to accomplish such an operation, especially since this is the one hundredth anniversary of his birth. On the other hand, no IGY World Day has been established in September 1957.

Sincerely,
Allen W. Dulles
Director

Document II-14

Document title: "Announcement of the First Satellite," from *Pravda*, October 5, 1957, F.J. Krieger, *Behind the Sputniks* (Washington, DC: Public Affairs Press, 1958), pp. 311-12.

On October 4, 1957, the Soviet Union launched the first Earth-orbiting satellite to support the scientific research effort undertaken by several nations during the 1957-1958 International Geophysical Year. The Soviets called the satellite "Sputnik" or "fellow traveler" and reported the achievement in a tersely worded press release issued by the official news agency, Tass, printed in the October 5, 1957, issue of *Pravda*. The United States had also been working on a scientific satellite program, Project Vanguard, but it had not yet launched a satellite.

[311] For several years scientific research and experimental design work have been conducted in the Soviet Union on the creation of artificial satellites of the earth.

As already reported in the press, the first launching of the satellites in the USSR were planned for realization in accordance with the scientific research program of the International Geophysical Year.

As a result of very intensive work by scientific research institutes and design bureaus the first artificial satellite in the world has been created. On October 4, 1957, this first satellite was successfully launched in the USSR. According to preliminary data, the carrier rocket has imparted to the satellite the required orbital velocity of about 8000 meters per second. At the present time the satellite is describing elliptical trajectories around the earth, and its flight can be observed in the rays of the rising and setting sun with the aid of very simple optical instruments (binoculars, telescopes, etc.).

According to calculations which now are being supplemented by direct observations, the satellite will travel at altitudes up to 900 kilometers above the surface of the earth; the time for a complete revolution of the satellite will be one hour and thirty-five minutes; the angle of inclination of its orbit to the equatorial plane is 65 degrees. On October 5 the satellite will pass over the Moscow area twice—at 1:46 a.m. and at 6:42 a.m. Moscow time. Reports about the subsequent movement of the first artificial satellite launched in the USSR on October 4 will be issued regularly by broadcasting stations.

The satellite has a spherical shape 58 centimeters in diameter and weighs 83.6 kilograms. It is equipped with two radio transmitters continuously emitting signals at frequencies of 20.005 and 40.002 megacycles per second (wave lengths of about 15 and 7.5 meters, respectively). The power of the transmitters ensures reliable reception of the signals by a broad range of radio amateurs. The signals have the form of telegraph pulses of about 0.3 second's duration with a [312] pause of the same duration. The signal of one frequency is sent during the pause in the signal of the other frequency.

Scientific stations located at various points in the Soviet Union are tracking the satellite and determining the elements of its trajectory. Since the density of the rarefied upper layers of the atmosphere is not accurately known, there are no data at present for the precise determination of the satellite's lifetime and of the point of its entry into the dense layers of the atmosphere. Calculations have shown that owing to the tremendous velocity of the satellite, at the end of its existence it will burn up on reaching the dense layers of the atmosphere at an altitude of several tens of kilometers.

As early as the end of the nineteenth century the possibility of realizing cosmic flights by means of rockets was first scientifically substantiated in Russia by the works of the outstanding Russian scientist K[onstantin] E. Tsiolkovskii [Tsiolkovskiy].

The successful launching of the first man-made earth satellite makes a most important contribution to the treasure-house of world science and culture. The scientific experiment accomplished at such a great height is of tremendous importance for learning the properties of cosmic space and for studying the earth as a planet of our solar system.

During the International Geophysical Year the Soviet Union proposes launching several more artificial earth satellites. These subsequent satellites will be larger and heavier and they will be used to carry out programs of scientific research.

Artificial earth satellites will pave the way to interplanetary travel and, apparently, our contemporaries will witness how the freed and conscientious labor of the people of the new socialist society makes the most daring dreams of mankind a reality.

Document II-15

Document title: John Foster Dulles to James C. Hagerty, October 8, 1957, with attached: "Draft Statements on the Soviet Satellite," October 5, 1957.

Source: John Foster Dulles Papers, Dwight D. Eisenhower Library, Abilene, Kansas.

The Eisenhower administration had anticipated the imminent launch of the first Soviet satellite, and had given some thought to potential public reaction to such an event. But when the launch occurred on October 4, 1957, the administration was surprised by the amount of public concern. Four days after the event, Secretary of State John Foster Dulles sent White House Press Secretary James Hagerty his suggestions for the text of a press release that would place the Sputnik launch in its proper context and reassure the public. Although Dulles' comments did not result in a press release, they did form the basis for much of the administration's "official" comment about the Soviet achievement as well as the core of President Eisenhower's comments at a press conference on October 9th. This document does not contain the draft statement prepared by Allen Dulles, Director of the Central Intelligence Agency and brother of the Secretary of State, which is mentioned in the cover letter.

---

[1] Draft by JFD
10/8/57

The launching by the Soviet Union of the first earth satellite is an event of considerable technical and scientific importance. However, that importance should not be exaggerated. What has happened involves no basic discovery and the value of a satellite to mankind will for a long time be highly problematical.

That the Soviet Union was first in this project is due to the high priority which the Soviet Union gives to scientific training and to the fact that since 1945 the Soviet Union has particularly emphasized developments in the fields of missiles and of outer space. The Germans had made a major advance in this field and the results of their effort were largely taken over by the Russians when they took over the German assets, human and material, at Peenemünde, the principal German base for research and experiment in the use of outer space. This encouraged the Soviets to concentrate upon developments in this field with a use of [2] resources and effort not possible in time of peace to societies where the people are free to engage in pursuits of their own choosing and where public monies are limited by representatives of the people. Despotic societies which can command the activities and resources of all their people can often produce spectacular accomplishments. These, however, do not prove that freedom is not the best way.

While the United States has not given the same priority to outer space developments as has the Soviet Union, it has not neglected this field. It already has a capability to utilize outer space for missiles and it is expected to launch an earth satellite during the present geophysical year in accordance with a program which has been under orderly development over the past two years.

The United States welcomes the peaceful achievement of the Soviet scientists. It hopes that the acclaim which has resulted from [3] their effort will encourage the Soviet Union to seek development along peaceful lines and seek to enrich the spiritual and material welfare of their people.

What is happening with reference to outer space makes more than ever important the proposal made by the United States and the other free world members of the Disarmament Subcommittee. I recall my White House statement of August 28 which emphasized the proposal of the Western Powers at London to establish a study group to the end that "outer space shall be used only for peaceful, not military, purposes."

Document II-16

Document title: President's Science Advisory Committee, "Introduction to Outer Space," March 26, 1958, pp. 1-2, 6, 13-15.

Source: NASA Historical Reference Collection, NASA History Office, NASA Headquarters, Washington, D.C.

An initial assignment for the President's Science Advisory Committee, which was formed in the aftermath of the launches of Sputnik I and II, was to assess the appropriate direction and pace for the U.S. space program. The committee focused heavily on the scientific aspects of the space program. With the president's endorsement, on March 26, 1958, it released a report outlining the importance of space activities, but recommended a cautiously measured pace.

Statement by the President

In connection with a study of space science and technology made at my request, the President's Science Advisory Committee, of which Dr. James R. Killian is Chairman, has prepared a brief "Introduction to Outer Space" for the nontechnical reader.

This is not science fiction. This is a sober, realistic presentation prepared by leading scientists. I have found this statement so informative and interesting that I wish to share it with all the people of America, and indeed with all the people of the earth. I hope that it can be widely disseminated by all news media for it clarifies many aspects of space and space technology in a way which can be helpful to all people as the United States proceeds with its peaceful program in space science and exploration. Every person has the opportunity to share through understanding in the adventures which lie ahead.

This statement of the Science Advisory Committee makes clear the opportunities which a developing space technology can provide to extend man's knowledge of the earth, the solar system, and the universe. These opportunities reinforce my conviction that we and other nations have a great responsibility to promote the peaceful use of space and to utilize the new knowledge obtainable from space science and technology for the benefit of all mankind.

Dwight D. Eisenhower

[1] # Introduction to Outer Space

What are the principal reasons for undertaking a national space program? What can we expect to gain from space science and exploration? What are the scientific laws and facts and the technological means which it would be helpful to know and understand in reaching sound policy decisions for a United States space program and its management by the Federal Government? This statement seeks to provide brief and introductory answers to these questions.

It is useful to distinguish among four factors which give importance, urgency, and inevitability to the advancement of space technology. The first of these factors is the compelling urge of man to explore and to discover, the thrust of curiosity that leads men to try to go where no one has gone before. Most of the surface of the earth has now been explored and men now turn to the exploration of outer space as their next objective.

Second, there is the defense objective for the development of space technology. We wish to be sure that space is not used to endanger our security. If space is to be used for military purposes, we must be prepared to use space to defend ourselves.

Third, there is the factor of national prestige. To be strong and bold in space technology will enhance the prestige of the United States among the peoples of the world and create added confidence in our scientific, technological, industrial, and military strength.

Fourth, space technology affords new opportunities for scientific observation and experiment [2] which will add to our knowledge and understanding of the earth, the solar system, and the universe.

The determination of what our space program should be must take into consideration all four of these objectives. While this statement deals mainly with the use of space for scientific inquiry, we fully recognize the importance of the other three objectives.

In fact it has been the military quest for ultra long-range rockets that has provided man with new machinery so powerful that it can readily put satellites in orbit and, before long, send instruments out to explore the moon and nearby planets. In this way, what was at first a purely military enterprise has opened up an exciting era of exploration that few men, even a decade ago, dreamed would come in this century....

[6] **Will the Results Justify the Costs?**

Since the rocket power plants for space exploration are already in existence or being developed for military need, the cost of additional scientific research, using these rockets, need not be exorbitant. Still, the cost will not be small, either. This raises an important question that scientists and the general public (who will pay the bill) both must face: Since there are still so many unanswered scientific questions and problems all around us on earth, why should we start asking new questions and seeking out new problems in space? How can the results possibly justify the cost?

Scientific research, of course, has never been amenable to rigorous cost accounting in advance. Nor, for that matter, has exploration of any sort. But if we have learned one lesson, it is that research and exploration have a remarkable way of paying off—quite apart from the fact that they demonstrate that man is alive and insatiably curious. And we all feel richer for knowing what explorers and scientists have learned about the universe in which we live.

It is in these terms that we must measure the value of launching satellites and sending rockets into space....

[13] the scientific opportunities are so numerous and so inviting that scientists from many countries will certainly want to participate. Perhaps the International Geophysical Year will suggest a model for the international exploration of space in the years and decades to come.

The timetable...suggests the approximate order in which some of the scientific and technical objectives mentioned in this review may be attained.

The timetable is not broken down into years, since there is yet too much uncertainty about the scale of the effort that will be made. The timetable simply lists various types of space investigations and goals under three broad headings: Early, Later, Still Later....

[14] EARLY
1. Physics
2. Geophysics
3. Meteorology
4. Minimal Moon Contact
5. Experimental Communications
6. Space Physiology

LATER
1. Astronomy
2. Extensive Communications
3. Biology
4. Scientific Lunar Investigation
5. Minimal Planetary Contact
6. Human Flight in Orbit

STILL LATER
1. Automated Lunar Exploration
2. Automated Planetary Exploration
3. Human Lunar Exploration and Return

AND MUCH LATER STILL
Human Planetary Exploration

[15] In conclusion, we venture two observations. Research in outer space affords new opportunities in science, but it does not diminish the importance of science on earth. Many of the secrets of the universe will be fathomed in laboratories on earth, and the progress of our science and technology and the welfare of the Nation require that our regular scientific programs go forward without loss of pace, in fact at an increased pace. It would not be in the national interest to exploit space science at the cost of weakening our efforts in other scientific endeavors. This need not happen if we plan our national program for space science and technology as part of a balanced national effort in all science and technology.

Our second observation is prompted by technical considerations. For the present, the rocketry and other equipment used in space technology must usually be employed at the very limit of its capacity. This means that failures of equipment and uncertainties of schedule are to be expected. It therefore appears wise to be cautious and modest in our predictions and pronouncements about future space activities—and quietly bold in our execution....

Document II-17

Document title: "National Aeronautics and Space Act of 1958," Public Law 85-568, 72 Stat., 426. Signed by the president on July 29, 1958.

Source: Record Group 255, National Archives and Records Administration, Washington, D.C.

After the launch of Sputnik and the publicity surrounding it, the Eisenhower administration moved quickly to create an American civilian space agency. The National Advisory Committee for Aeronautics (NACA) was too small for the task, however; the White House decided that a new agency, with NACA as its core, but also including rocket and space engineers involved in various defense programs, was needed. On March 5, 1958, President Eisenhower approved a final memorandum ordering the Bureau of Budget to draft a space bill immediately. It was ready three weeks later and sent to Congress on April 2. Senator Lyndon Johnson had a great deal of influence on the form of the final bill, which was passed after lengthy congressional deliberations. In particular, Congress added to the administration bill a requirement for a National Aeronautics and Space Council as a presidential-level policy coordinating board.

[1] **AN ACT**

To provide for research into problems of flight within and outside the earth's atmosphere, and for other purposes.

Be it enacted by the Senate and House of Representatives of the United States of America in Congress assembled,

Title I—Short Title, Declaration of Policy, and Definitions
Short Title

Sec. 101. This act may be cited as the "National Aeronautics and Space Act of 1958."

Declaration of Policy and Purpose

Sec. 102. (a) The Congress hereby declares that it is the policy of the United States that activities in space should be devoted to peaceful purposes for the benefit of all mankind.

(b) The Congress declares that the general welfare and security of the United States require that adequate provision be made for aeronautical and space activities. The Congress further declares that such activities shall be the responsibility of, and shall be directed by, a civilian agency exercising control over aeronautical and space activities sponsored by the United States, except that activities peculiar to or primarily associated with the development of weapons systems, military operations, or the defense of the United States (including the research and development necessary to make effective provision for the defense of the United States) shall be the responsibility of, and shall be directed by, the Department of Defense; and that determination as to which such agency has responsibility for and direction of any such activity shall be made by the President in conformity with section 201 (e).

(c) The aeronautical and space activities of the United States shall be conducted so as to contribute materially to one or more of the following objectives:

(1) The expansion of human knowledge of phenomena in the atmosphere and space;

(2) The improvement of the usefulness, performance, speed, safety, and efficiency of aeronautical and space vehicles;

(3) The development and operation of vehicles capable of carrying instruments, equipment, supplies and living organisms through space;

(4) The establishment of long-range studies of the potential benefits to be gained from, the opportunities for, and the problems involved in the utilization of aeronautical and space activities for peaceful and scientific purposes.

(5) The preservation of the role of the United States as a leader in aeronautical and space science and technology and in the application thereof to the conduct of peaceful activities within and outside the atmosphere.

(6) The making available to agencies directly concerned with national defenses of discoveries that have military value or significance, and the furnishing by such agencies, to the civilian agency established to direct and control nonmilitary aeronautical and space activities, of information as to discoveries which have value or significance to that agency;

[2] (7) Cooperation by the United States with other nations and groups of nations in work done pursuant to this Act and in the peaceful application of the results, thereof; and

(8) The most effective utilization of the scientific and engineering resources of the United States, with close cooperation among all interested agencies of the United States in order to avoid unnecessary duplication of effort, facilities, and equipment.

(d) It is the purpose of this Act to carry out and effectuate the policies declared in subsections (a), (b), and (c).

Definitions

Sec. 103. As used in this Act—

(1) the term "aeronautical and space activities" means (A) research into, and the solution of, problems of flight within and outside the earth's atmosphere, (B) the development, construction, testing, and operation for research purposes of aeronautical and space vehicles, and (C) such other activities as may be required for the exploration of space; and

(2) the term "aeronautical and space vehicles" means aircraft, missiles, satellites, and other space vehicles, manned and unmanned, together with related equipment, devices,

components, and parts.

Title II—Coordination of Aeronautical and Space Activities
National Aeronautics and Space Council

Sec. 201. (a) There is hereby established the National Aeronautics and Space Council (hereinafter called the "Council") which shall be composed of—
 (1) the President (who shall preside over meetings of the Council);
 (2) the Secretary of State;
 (3) the Secretary of Defense
 (4) the Administrator of the National Aeronautics and Space Administration;
 (5) the Chairman of the Atomic Energy Commission;
 (6) not more than one additional member appointed by the President from the departments and agencies of the Federal Government; and
 (7) not more than three other members appointed by the President, solely on the basis of established records of distinguished achievement from among individuals in private life who are eminent in science, engineering, technology, education, administration, or public affairs.
 (b) Each member of the Council from a department or agency of the Federal Government may designate another officer of his department or agency to serve on the Council as his alternate in his unavoidable absence.
 (c) Each member of the Council appointed or designated under paragraphs (6) and (7) of subsection (a), and each alternate member designated under subsection (b), shall be appointed or designated to serve as such by and with the advice and consent of the Senate, unless at the time of such appointment or designation he holds an office in the Federal Government to which he was appointed by and with the advice and consent of the Senate.
[3] (d) It shall be the function of the Council to advise the President with respect to the performance of the duties prescribed in subsection (e) of this section.
 (e) In conformity with the provisions of section 102 of this Act, it shall be the duty of the President to—
 (1) survey all significant aeronautical and space activities, including the policies, plans, programs, and accomplishments of all agencies of the United States engaged in such activities;
 (2) develop a comprehensive program of aeronautical and space activities to be conducted by agencies of the United States;
 (3) designate and fix responsibility for the direction of major aeronautical and space activities;
 (4) provide for effective cooperation between the National Aeronautics and Space Administration and the Department of Defense in all such activities, and specify which of such activities may be carried on concurrently by both such agencies notwithstanding the assignment of primary responsibility therefor to one or the other of such agencies; and
 (5) resolve differences arising among departments and agencies of the United States with respect to aeronautical and space activities under this Act, including differences as to whether a particular project is an aeronautical and space activity.
 (f) The Council may employ a staff to be headed by a civilian executive secretary who shall be appointed by the President by and with the advice and the consent of the Senate and shall receive compensation at the rate of $20,000 a year. The executive secretary, subject to the direction of the Council, is authorized to appoint and fix the compensation of such personnel, including not more than three persons who may be appointed without regard to the civil service laws or the Classification Act of 1949 and compensated at the rate of not more that $19,000 a year, as may be necessary to perform such duties as may be prescribed by the Council in connection with the performance of its functions. Each appointment under this subsection shall be subject to the same security requirements as those established for personnel of the National Aeronautics and Space Administration

appointed under section 203 (b) (2) of this Act.

(g) Members of the Council appointed from private life under subsection (a) (7) may be compensated at a rate not to exceed $100 per diem, and may be paid travel expenses and per diem in lieu of subsistence in accordance with the provisions of section 5 of the Administrative Expenses Act of 1946 (5 U.S.C. 73b-2) relating to persons serving without compensation.

National Aeronautics And Space Administration

Sec. 202. (a) There is hereby established the National Aeronautics and Space Administration (hereinafter called the "Administration"). The Administration shall be headed by an Administrator, who shall be appointed from civilian life by the President by and with the advice and consent of the Senate, and shall receive compensation at the rate of $22,500 per annum. Under the supervision and direction of the President, the Administrator shall be responsible for the exercise of all powers and the discharge of all duties of the Administration, and shall have authority and control over all personnel and activities, thereof.

(b) There shall be in the Administration a Deputy Administrator, who shall be appointed from civilian life by the President by and with the advice and consent of the Senate, shall receive compensation of $21,500 per annum, and shall perform such duties and exercise such powers as the Administrator may prescribe. The Deputy Administrator shall act for, and exercise [4] the powers of, the Administrator during his absence or disability.

(c) The Administrator and the Deputy Administrator shall not engage in any other business, vocation, or employment while serving as such.

Functions of the Administration

Sec. 203. (a) The Administration, in order to carry out the purpose of this Act, shall—
   (1) plan, direct, and conduct aeronautical and space activities;
   (2) arrange for participation by the scientific community in planning scientific measurements and observations to be made through use of aeronautical and space vehicles, and conduct or arrange for the conduct of such measurements and observations; and
   (3) provide for the widest practicable and appropriate dissemination of information concerning its activities and the results thereof.

(b) In the performance of its functions the Administration is authorized—
   (1) to make, promulgate, issue, rescind, and amend rules and regulations governing the manner of its operations and the exercise of the powers vested in it by law;
   (2) to appoint and fix the compensation of such officers and employees as may be necessary to carry out such functions. Such officers and employees shall be appointed in accordance with the civil-service laws and their compensation fixed in accordance with the Classification Act of 1949, except that (A) to the extent the Administrator deems such action necessary to the discharge of his responsibilities, he may appoint and fix the compensation (up to a limit of $19,000 a year, or up to a limit of $21,000 a year for a maximum of ten positions) of not more than two hundred and sixty of the scientific, engineering and administrative personnel of the Administration without regard to such laws, and (B) to the extent the Administrator deems such action necessary to recruit specially qualified scientific and engineering talent, he may establish the entrance grade for scientific and engineering personnel without previous service in the Federal Government at a level up to two grades higher than the grade provided for such personnel under the General Schedule established by the Classification Act of 1949, and fix their compensation accordingly;
   (3) to acquire (by purchase, lease, condemnation, or otherwise), construct, improve, repair, operate, and maintain laboratories, research and testing sites and facilities, aeronautical and space vehicles, quarters and related accommodations for employees and dependents of employees of the Administration, and such other real and personal property (including patents), or any interest therein, as the Administration deems necessary within and outside the continental United States; to lease to others such real and personal prop-

erty; to sell and otherwise dispose of real and personal property (including patents and rights thereunder) in accordance with the provisions of the Federal Property and Administrative Service Act of 1949, as amended (40 U.S.C. 471 et seq.); and to provide by contract or otherwise for cafeterias and other necessary facilities for the welfare of employees of the Administration at its installations and purchase and maintain equipment therefor;

[5]   (4) to accept unconditional gifts or donations of services, money, or property, real, personal, or mixed, tangible or intangible;

(5) without regard to section 3648 of the Revised Statutes, as amended (31 U.S.C. 529), to enter into and perform such contracts, leases, cooperative agreements, or other transactions as may be necessary in the conduct of its work and on such terms as it may deem appropriate, with any agency or instrumentality of the United States, or with any State, Territory, or possession, or with any political subdivision thereof, or with any person, firm, association, corporation, educational institution. To the maximum extent practicable and consistent with the accomplishment of the purpose of this Act, such contracts, leases, agreements, and other transactions shall be allocated by the Administrator in a manner which will enable small-business concerns to participate equitably and proportionately in the conduct of the work of the Administration;

(6) to use, with their consent, the services, equipment, personnel, and facilities of Federal and other agencies with or without reimbursement, and on a similar basis to cooperate with other public and private agencies and instrumentalities in the use of services, equipment and facilities. Each department and agency of the Federal Government shall cooperate fully with the Administration in making its services, equipment, personnel, and facilities available to the Administration, and any such department or agency is authorized, notwithstanding any other provision of law, to transfer or to receive from the Administration, without reimbursement, aeronautical and space vehicles, and supplies and equipment other than administrative supplies and equipment;

(7) to appoint such advisory committees as may be appropriate for purposes of consultation and advice to the Administration in the performance of its functions;

(8) to establish within the Administration such offices and procedures as may be appropriate to provide for the greatest possible coordination of its activities under this Act with related scientific and other activities being carried on by other public and private agencies and organizations;

(9) to obtain services as authorized by section 15 of the Act of August 2, 1946 (5 U.S.C. 55a), at rates not to exceed $100 per diem for individuals;

(10) when determined by the Administrator to be necessary, and subject to such security investigations as he may determine to be appropriate, to employ aliens without regard to statutory provisions prohibiting payment of compensation to aliens;

(11) to employ retired commissioned officers of the armed forces of the United States and compensate them at the rate established for the positions occupied by them within the Administration, subject only to the limitations in pay set forth in section 212 of the Act of June 30, 1932 as amended (5 U.S.C. 59a);

(12) with the approval of the President, to enter into cooperative agreements under which members or the Army, Navy, Air Force, and Marine Corps may be detailed by the appropriate Secretary for services in the performance of functions under this Act to the same extent as that to which they might by lawfully assigned in the Department of Defense; and

(13) (A) to consider, ascertain, adjust, determine, settle, and pay, on behalf of the United States, in full satisfaction thereof, any claim for $5,000 or less against the United States for bodily injury, death, or damage to or loss of real or personal property [6] resulting from the conduct of the Administration's functions as specified in subsection (a) of this section, where such claim is presented to the Administration in writing within two years after the accident or incident out of which the claim arises; and

(B) if the Administration considers that a claim in excess of $5,000 is meritorious and would otherwise be covered by this paragraph, to report the facts and circumstances thereof to the Congress for its consideration.

Civilian-Military Liaison Committee

Sec. 204 (a) There shall be a Civilian-Military Liaison Committee consisting of—
(1) a Chairman, who shall be the head thereof and who shall be appointed by the President, shall serve at the pleasure of the President, and shall receive compensation (in the manner provided in subsection (d)) at the rate of $20,000 per annum;
(2) one or more representatives from the Department of Defense, and one or more representatives from each of the Departments of the Army, Navy, and Air Force, to be assigned by the Secretary of Defense to serve on the Committee without additional compensation; and
(3) representatives from the Administration, to be assigned by the Administrator to serve on the Committee without additional compensation, equal in number to the number of representatives assigned to serve on the Committee under paragraph (2).
(b) The Administration and the Department of Defense, through the Liaison Committee, shall advise and consult with each other on all matters within their respective jurisdictions relating to aeronautical and space activities and shall keep each other fully and currently informed with respect to such activities.
(c) If the Secretary of Defense concludes that any request, action, proposed action, or failure to act on the part of the Administrator is adverse to the responsibilities of the Department of Defense, or the Administrator concludes that any request, action, or proposed action, or failure to act on the part of the Department of Defense is adverse to the responsibilities of the Administration, and the Administrator and the Secretary of Defense are unable to reach an agreement with respect thereto, either the Administrator or the Secretary of Defense may refer the matter to the President for his decision (which shall be final) as provided in section 201 (e).
(d) Notwithstanding the provisions of any other law, any active or retired officer of the Army, Navy, or Air Force may serve as Chairman of the Liaison Committee without prejudice to his active or retired status as such officer. The compensation received by any such officer for his service as Chairman of the Liaison Committee shall be equal to the amount (if any) by which the compensation fixed by subsection (a) (1) for such Chairman exceeds his pay and allowances (including special and incentive pays) as an active officer, or his retired pay.

International Cooperation

Sec. 205. The Administration, under the foreign policy guidance of the President, may engage in a program of international cooperation in work done pursuant to the Act, and in the peaceful application of the results thereof, pursuant to agreements made by the President with the advice and consent of the Senate.

[7] ### Reports to the Congress

Sec. 206. (a) The Administration shall submit to the President for transmittal to the Congress, semiannually and at such other times as it deems desirable, a report of its activities and accomplishments.
(b) The President shall transmit to the Congress in January of each year a report, which shall include (1) a comprehensive description of the programmed activities and the accomplishments of all agencies of the United States in the field of aeronautics and space activities during the preceding calendar year, and (2) an evaluation of such activities and accomplishments in terms of the attainment of, or the failure to attain, the objectives described in section 102 (c) of this Act.
(c) Any report made under this section shall contain such recommendations for additional legislation as the Administrator of the President may consider necessary or desirable for the attainment of the objectives described in section 102 (c) of this Act.
(d) No information which has been classified for reasons of national security shall be

included in any report made under this section, unless such information has been declassified by, or pursuant to authorization given by, the President.

Title III—Miscellaneous
National Advisory Committee for Aeronautics

Sec. 301. (a) The National Advisory Committee for Aeronautics, on the effective date of this section, shall cease to exist. On such date all functions, powers, duties, and obligations, and all real and personal property, personnel (other than members of the Committee), funds, and records of that organization, shall be transferred to the Administration.

(b) Section 2302 of title 10 of the United States Code is amended by striking out "or the Executive Secretary of the National Advisory Committee for Aeronautics." and inserting in lieu thereof "or the Administrator of the National Aeronautics and Space Administration."; and section 2303 of such title 10 is amended by striking out "National Advisory Committee for Aeronautics" and inserting in lieu thereof "The National Aeronautics and Space Administration."

(c) The first section of the Act of August 26, 1950 (5 U.S.C. 22-1), is amended by striking out "the Director, National Advisory Committee for Aeronautics" and inserting in lieu thereof "the Administrator of the National Aeronautics and Space Administration", and by striking out "or National Advisory Committee for Aeronautics" and inserting in lieu thereof "or National Aeronautics and Space Administration."

(d) The Unitary Wind Tunnel Plan Act of 1949 (50 U.S.C. 511-515) is amended (1) by striking out "The National Advisory Committee for Aeronautics (hereinafter referred to as the `Committee')" and inserting in lieu thereof "The Administrator of the National Aeronautics and Space Administration (hereinafter referred to as the `Administrator')"; (2) by striking out "Committee" or "Committee's" wherever they appear and inserting in lieu thereof "Administrator" and "Administrator's", respectively; and (3) by striking out "its" wherever it appears and inserting in lieu thereof "his."

(e) This section shall take effect ninety days after the date of the enactment of this Act, or on any earlier date on which the Administrator shall determine, and announce by proclamation published in the Federal Register, that the Administration has been organized and is prepared to discharge the duties and exercise the powers conferred upon it by this Act.

[8] **Transfer of Related Functions**

Sec. 302. (a) Subject to the provisions of this section, the President, for a period of four years after the date of enactment of this Act, may transfer to the Administration any functions (including powers, duties, activities, facilities, and parts of functions) of any other department or agency of the United States, or of any officer or organizational entity thereof, which relate primarily to the functions, powers, and duties of the Administration as prescribed by section 203 of this Act. In connection with any such transfer, the President may, under this section or other applicable authority, provide for appropriate transfers of records, property, civilian personnel, and funds.

(b) Whenever any such transfer is made before January 1, 1959, the President shall transmit to the Speaker of the House of Representatives and the President pro tempore of the Senate a full and complete report concerning the nature and effect of such transfer.

(c) After December 31, 1958, no transfer shall be made under this section until (1) a full and complete report concerning the nature and effect of such proposed transfer has been transmitted by the President to the Congress, and (2) the first period of sixty calendar days of regular session of the Congress following the date of receipt of such report by the Congress has expired without the adoption by the Congress of a concurrent resolution stating that the Congress does not favor such transfer.

Access to Information

Sec. 303. Information obtained or developed by the Administrator in the performance of his functions under the Act shall be made available for public inspection, except (A) information authorized or required by Federal statute to be withheld, and (B) information classified to protect the national security: *Provided,* That nothing in this Act shall authorize the withholding of information by the Administrator from the duly authorized committees of the Congress.

Security

Sec. 304. (a) The Administrator shall establish such security requirements, restrictions, and safeguards as he deems necessary in the interest of the national security. The Administrator may arrange with the Civil Service Commission for the conduct of such security or other personnel investigations of the Administration's officers, employees, and consultants, and its contractors and subcontractors and their officers and employees, actual or prospective, as he deems appropriate; and if any such investigation develops any data reflecting that the individual who is the subject thereof is of questionable loyalty the matter shall be referred to the Federal Bureau of Investigation for the conduct of a full field investigation, the results of which shall be furnished to the Administrator.

(b) The Atomic Energy Commission may authorize any of its employees, or employees of any contractor, prospective contractor, licensee, or prospective licensee of the Atomic Energy Commission or any other person authorized to have access to Restricted Data by the Atomic Energy Commission under subsection 145 b. of the Atomic Energy Act of 1954 (42 U.S.C. 2165 (b)), to permit any member, officer, or employee of the Council, or the Administrator, or any officer, employee, member of an advisory committee, contractor, subcontractor, or officer or employee of a contractor or subcontractor of the Administration, to have access to Restricted Data relating to aeronautical and space activities which is required in the performance of his duties and so certified by the Council or the Administrator, as the case may be, [9] but only if (1) the Council or Administrator or designee thereof has determined, in accordance with the established personnel security procedures and standards of the Council or Administration, that permitting such individual to have access to such Restricted Data will not endanger the common defense and security, and (2) the Council or Administrator or designee thereof finds that the established personnel and other security procedures and standards of the Council or Administration are adequate and in reasonable conformity to the standards established by the Atomic Energy Commission under section 145 of the Atomic Energy Act of 1954 (42 U.S.C. 2165). Any individual granted access to such Restricted Data pursuant to this subsection may exchange such Data with any individual who (A) is an officer or employee of the Department of Defense, or any department or agency thereof, or a member of the armed forces, or a contractor or subcontractor, and (B) has been authorized to have access to Restricted Data under the provisions of section 143 of the Atomic Energy Act of 1954 (42 U.S.C. 2163).

(c) Chapter 37 of title 18 of the United States Code (entitled Espionage and Censorship) is amended by—

(1) adding at the end thereof the following new section:
"799. Violation of regulations of National Aeronautics and Space Administration.

Whoever willfully shall violate, attempt to violate, or conspire to violate any regulation or order promulgated by the Administrator of the National Aeronautics and Space Administration for the protection or security of any laboratory, station, base or other facility, or part thereof, or any aircraft, missile, spacecraft, or similar vehicle, or part thereof, or other property or equipment in the custody of the Administration, or any real or personal property or equipment in the custody of any contractor under any contract with the Administration or any subcontractor of any such contractor, shall be fined not more than

$5,000, or imprisoned not more that one year, or both."

(2) adding at the end of the sectional analysis thereof the following new item: "799. Violation of regulations of National Aeronautics and Space Administration."

(d) Section 1114 of title 18 of the United States Code is amended by inserting immediately before "while engaged in the performance of his official duties" the following: "or any officer or employee of the National Aeronautics and Space Administration directed to guard and protect property of the United States under the administration and control of the National Aeronautics and Space Administration."

(e) The Administrator may direct such of the officers and employees of the Administration as he deems necessary in the public interest to carry firearms while in the conduct of their official duties. The Administrator may also authorize such of those employees of the contractors and subcontractors of the Administration engaged in the protection of property owned by the United States and located at facilities owned by or contracted to the United States as he deems necessary in the public interest, to carry firearms while in the conduct of their official duties.

Property Rights In Inventions

Sec. 305. (a) Whenever any invention is made in the performance of any work under any contract of the Administration, and the Administrator determines that—

(1) the person who made the invention was employed or assigned to perform research, development, or exploration work and the invention is related to the work he was employed or [10] assigned to perform, or that it was within the scope of his employment duties, whether or not it was made during working hours, or with a contribution by the Government of the use of Government facilities, equipment, materials, allocated funds, information proprietary to the Government, or services of Government employees during working hours; or

(2) the person who made the invention was not employed or assigned to perform research, development, or exploration work, but the invention is nevertheless related to the contract, or to the work or duties he was employed or assigned to perform, and was made during working hours, or with a contribution from the Government of the sort referred to in clause (1), such invention shall be the exclusive property of the United States, and if such invention is patentable a patent therefor shall be issued to the United States upon application made by the Administrator, unless the Administrator waives all or any part of the rights of the United States to such invention in conformity with the provisions of subsection (f) of this section.

(b) Each contract entered into by the Administrator with any party for the performance of any work shall contain effective provisions under which such party shall furnish promptly to the Administrator a written report containing full and complete technical information concerning any invention, discovery, improvement, or innovation which may be made in the performance of any such work.

(c) No patent may be issued to any applicant other than the Administrator for any invention which appears to the Commissioner of Patents to have significant utility in the conduct of aeronautical and space activities unless the applicant files with the Commissioner, with the application or within thirty days after request therefor by the Commissioner, a written statement executed under oath setting forth the full facts concerning the circumstances under which such invention was made and stating the relationship (if any) of such invention to the performance of any work under any contract of the Administration. Copies of each such statement and the application to which it relates shall be transmitted forthwith by the Commissioner to the Administrator.

(d) Upon any application as to which any such statement has been transmitted to the Administrator, the Commissioner may, if the invention is patentable, issue a patent to the applicant unless the Administrator, within ninety days after receipt of such application and statement, requests that such patent be issued to him on behalf of the United States. If, within such time, the Administrator files such a request with the Commissioner, the Commissioner shall transmit notice thereof to the applicant, and shall issue such patent to the Administrator unless the applicant within thirty days after receipt of such notice requests a hearing before a Board of Patent Interferences on the question whether the Administrator is entitled under this section to receive such patent. The Board may hear and determine, in accordance with rules and procedures established for interference cases, the question so presented, and its determination shall be subject to appeal by the applicant or by the Administrator to the Court of Customs and Patent Appeals in accordance with procedures governing appeals from decisions of the Board of Patent Interferences in other proceedings.

(e) Whenever any patent has been issued to any applicant in conformity with subsection (d), and the Administrator thereafter has reason to believe that the statement filed by the applicant in connection therewith contained any false representation of any material fact, the Administrator within five years after the date of issuance of such patent may file with the Commissioner a request for the [11] transfer to the Administrator of title to such patent on the records of the Commissioner. Notice of any such request shall be transmitted by the Commissioner to the owner of record of such patent, and title to such patent shall be so transferred to the Administrator unless within thirty days after receipt of such notice such owner of record requests a hearing before a Board of Patent Interferences on the question whether any such false representation was contained in such statement. Such question shall be heard and determined, and determination thereof shall be subject to review, in the manner prescribed by subsection (d) for questions arising thereunder. No request made by the Administrator under this subsection for the transfer of title to any patent, and no prosecution for the violation of any criminal statute, shall be barred by any failure of the Administrator to make a request under subsection (d) for the issuance of such patent to him, or by any notice previously given by the Administrator stating that he had no objection to the issuance of such patent to the applicant therefor.

(f) Under such regulations in conformity with this subsection as the Administrator shall prescribe, he may waive all or any part of the rights of the United States under this section with respect to any invention or class of inventions made or which may be made by any person or class of persons in the performance of any work required by any contract of the Administration if the Administrator determines that the interests of the United States will be served thereby. Any such waiver may be made upon such terms and under such conditions as the Administrator shall determine to be required for the protection of the interests of the United States. Each such waiver made with respect to any invention shall be subject to the reservation by the Administrator of an irrevocable, nonexclusive, nontransferable, royalty-free license for the practice of such invention throughout the world by or on behalf of the United States or any foreign government pursuant to any treaty or agreement with the United States. Each proposal for any waiver under this subsection shall be referred to an Inventions and Contributions Board which shall be established by the Administrator within the Administration. Such Board shall accord to each interested party an opportunity for hearing, and shall transmit to the Administrator its findings of fact with respect to such proposal and its recommendations for action to be taken with respect thereto.

(g) The Administrator shall determine, and promulgate regulations specifying, the terms and conditions upon which licenses will be granted by the Administrator for the practice by any person (other than an agency of the United States) of any invention for which the Administrator holds a patent on behalf of the United States.

(h) The Administrator is authorized to take all suitable and necessary steps to protect any invention or discovery to which he has title, and to require that contractors or persons who retain title to inventions or discoveries under this section protect the inventions or discoveries to which the Administration has or may acquire a license or use.

(i) The Administration shall be considered a defense agency of the United States for the purpose of chapter 17 of title 35 of the United States Code.

(j) As used in this section—
    (1) term "person" means any individual, partnership, corporation, association,

institution, or other entity;

(2) the term "contract" means any actual or proposed contract, agreement, understanding, or other arrangement, and includes any assignment, substitution of parties, or subcontract executed or entered into thereunder; and

[12] (3) the term "made", when used in relation to any invention, means the conception or first actual reduction to practice of such invention.

Contributions Awards

Sec. 306. (a) Subject to the provisions of this section, the Administrator is authorized, upon his own initiative or upon application of any person, to make a monetary award, in such amount and upon such terms as he shall determine to be warranted, to any person (as defined by section 305) for any scientific or technical contribution to the Administration which is determined by the Administrator to have significant value in the conduct of aeronautical and space activities. Each application made for any such award shall be referred to the Inventions and Contributions Board established under section 305 of this Act. Such Board shall accord to each such applicant an opportunity for hearing upon such application, and shall transmit to the Administrator its recommendation as to the terms of the award, if any, to be made to such applicant for such contribution. In determining the terms and conditions of any award the Administrator shall take into account-

(1) the value of the contribution to the United States;

(2) the aggregate amount of any such sums which have been expended by the applicant for the development of such contribution;

(3) the amount of any compensation (other than salary received for services rendered as an officer of employee of the Government) previously received by the applicant for or on account of the use of such contribution by the United States; and

(4) such other factors as the Administrator shall determine to be material.

(b) If more than one applicant under subsection (a) claims an interest in the same contribution, the Administrator shall ascertain and determine the respective interests of such applicants, and shall apportion any award to be made with respect to such contribution among such applicants in such proportions as he shall determine to be equitable. No award may be made under subsection (a) with respect to any contribution—

(1) unless the applicant surrenders, by such means as the Administrator shall determine to be effective, all claims which such applicant may have to receive any compensation (other than the award made under this section) for the use of such contribution or any element thereof at any time by or on behalf of the United States, or by or on behalf of any foreign government pursuant to any treaty or agreement with the United States, within the United States or at any other place;

(2) in any amount exceeding $100,000, unless the Administrator has transmitted to the appropriate committees of the Congress a full and complete report concerning the amount and terms of, and the basis for, such proposed award, and thirty calendar days of regular session of the Congress have expired after receipt of such report by such committees.

[13] ## Appropriations

Sec. 307. (a) There are hereby authorized to be appropriated such sums as may be necessary to carry out this Act except that nothing in this Act shall authorize the appropriation of any amount for (1) the acquisition or condemnation of any real property, or (2) any other item of a capital nature (such as plant or facility acquisition, construction, or expansion) which exceeds $250,000. Sums appropriated pursuant to this subsection for the construction of facilities, or for research and development activities, shall remain available until expended.

(b) Any funds appropriated for the construction of facilities may be used for emergency repairs of existing facilities when such existing facilities are made inoperative by major breakdown, accident, or other circumstances and such repairs are deemed by the Administrator to be of greater urgency than the construction of new facilities.

(Sam Rayburn)
Speaker of the House of Representatives

(Richard Nixon)
Vice President of the United States and President of the Senate.

Document II-18

Document title: National Security Council, NSC 5814, "U.S. Policy on Outer Space," June 20, 1958.

Source: Presidential Papers, Dwight D. Eisenhower Library, Abilene, Kansas.

Even before a new civilian space agency began operation, the Eisenhower administration developed an initial post-Sputnik statement of national space policy. This document was prepared under National Security Council auspices; the discussions included the National Advisory Committee for Aeronautics, which had been chosen by the White House as the core of a new civilian space agency. (The National Aeronautics and Space Council established by the 1958 Space Act had not yet begun to function.) As the document was sent to the National Security Council for discussion on June 20, 1958, disagreements remained on some aspects of the statement of U.S. objectives in space (paragraph 43) and of the policy guidance for U.S. space activities (paragraphs 50 and 56.f).

---

[1]   **Introductory Note**

This statement of U.S. Policy on Outer Space is designated Preliminary because man's understanding of the full implications of outer space is only in its preliminary stages. As man develops a fuller understanding of the new dimension of outer space, it is probable that the long-term results of exploration and exploitation will basically affect international and national political and social institutions.

Perhaps the starkest facts which confront the United States in the immediate and foreseeable future are (1) the USSR has surpassed the United States and the Free World in scientific and technological accomplishments in outer space, which have captured the imagination and admiration of the world; (2) the USSR, if it maintains its present superiority in the exploitation of outer space, will be able to use that superiority as a means of undermining the prestige and leadership of the United States; and (3) the USSR, if it should be the first to achieve a significantly superior military capability in outer space, could create an imbalance of power in favor of the Sino-Soviet Bloc and pose a direct military threat to U.S. security.

The security of the United States requires that we meet these challenges with resourcefulness and vigor.

[2] **General Considerations**
Introduction

Significance of Outer Space to U.S. Security

1. More than by any other imaginative concept, the mind of man is aroused by the thought of exploring the mysteries of outer space.

2. Through such exploration, man hopes to broaden his horizons, add to his knowledge, improve his way of living on earth. Already, man is sure that through further exploration he can obtain certain scientific and military values. It is reasonable for man to believe that there must be, beyond these areas, different and great values still to be discovered.

3. The technical ability to explore outer space has deep psychological implications over and above the stimulation provided by the opportunity to explore the unknown. With its hint of the possibility of the discovery of fundamental truths concerning man, the earth, the solar system, and the universe, space exploration has an appeal to deep insights within man which transcend his earthbound concerns. The manner in which outer space is explored and the uses to which it is put thus take on an unusual and peculiar significance.

4. The beginning stages of man's conquest of space have been focused on technology and have been characterized by national competition. The result has been a tendency to equate achievement in outer space with leadership in science, military capability, industrial technology, and with leadership in general.

5. The initial and subsequent successes by the USSR in launching large earth satellites have profoundly affected the belief of peoples, both in the United States and abroad, in the superiority of U.S. leadership in science and military capability. This psychological reaction of sophisticated and unsophisticated peoples everywhere affects U.S. relations with its allies, with the Communist Bloc, and with neutral and uncommitted nations.

[3] 6. In this situation of national competition and initial successes by the USSR, further demonstrations by the USSR of continuing leadership in outer space capabilities might, in the absence of comparable U.S. achievements in this field,[554] dangerously impair the confidence of these peoples in U.S. over-all leadership. To be strong and bold in space technology will enhance the prestige of the United States among the peoples of the world and create added confidence in U.S. scientific, technological, industrial and military strength.

7. The novel nature of space exploitation offers opportunities for international cooperation in its peaceful aspects. It is likely that certain nations may be willing to enter into cooperative arrangements with the United States. The willingness of the Soviets to cooperate remains to be determined. The fact that the results of cooperation in certain fields, even though entered into for peaceful purposes, could have military application, may condition the extent of such cooperation is those fields.

Problem of Defining Space

8. Many names for the various regions of the earth's atmosphere and the divisions of space have developed over the years. The boundaries of these regions and divisions cannot be precisely defined in physical terms, and authorities differ widely on terminology and meaning.

9. The term **"air space"** has been used to denote the layer of atmosphere surrounding the earth in which military and civilian air vehicles operate. Although national policies and international agreements have dealt extensively with air space and expressly assert the sovereignty of each nation over its air space, the upper limit of air space has not been defined.

---

554. Communist China has announced, furthermore, an intention of proceeding to launch its own earth satellite in the near future. Such a development, which could only result from USSR assistance, would tend to enhance the prestige of the Chinese Communist regime throughout Asia and among the less-developed countries, and could further undermine the reputation of the West for technological leadership unless the accomplishment were matched by a Free World ally.

[4] [Entire page omitted due to classification]

[5] 12. Because the question of rights in "outer space" will undoubtedly arise at the UN General Assembly in September 1958, perhaps in international discussions on post-IGY activities, and perhaps in other international negotiations, it would appear desirable for the United States to develop a common understanding of the term "outer space" as related to particular objects and activities therein.

13. For the purposes of this policy statement, space is divided into to two regions: "air space" and "outer space." "Outer space" is considered as contiguous to "air space", with the lower limit of "outer space" being the upper limit of "air space."

Use of Outer Space

General

14. Outer space can be used:

a. By vehicles or other objects that achieve their primary purpose in outer space; such as

(1) Vehicles or objects that remain in an area directly over a nation's own territory, such as sounding rockets;

(2) Vehicles or objects that orbit the earth;

(3) Vehicles that traverse outer space enroute to the moon, other planets or the sun;

b. For the transmission of electromagnetic energy for such purposes as communications, radar measurement and electronic countermeasures;

c. By [phrase omitted during declassification review] vehicles which traverse outer space, but which achieve their primary purpose upon their return to air space or earth.

15. There are many uses of outer space for peaceful purposes, such as exploration, pure adventure, increase of scientific knowledge, and development and applications of technology. Any use of outer space, however, whatever the purpose it is intended to serve, may have some degree of military or other non-peaceful application. Therefore, U.S. policies relating to international arrangements on uses of outer space for peaceful purposes will have to take into account possible non-peaceful applications in determining the net advantage to U.S. security.

Science and Technology

16. Outer space technology affords new and unique opportunities for scientific observations and experiments which will add greatly to our knowledge and understanding of the earth, the solar system and the universe. These opportunities exist in many fields, including among others:

a. **Geophysics**: Three-dimensional mapping of the earth's gravity and magnetic field.

b. **Physics**: Cosmic ray measurements above the earth's dense atmosphere and experiments in the theory of relativity.

c. **Meteorology**: World-wide cloud-cover mapping for improved forecasting of weather and measurements of incoming and outgoing heat energy which will allow a better understanding of weather.

d. **Biology**: Possible living organisms in space and the effects on man of prolonged exposure to radiation and weightlessness.

e. **Psychological** response of man to a space environment.

f. **Astronomy**: The universe as seen from beyond the earth's atmosphere and measurement of stellar radiation.

g. **Lunar investigations** including the moon's gravity, mass, magnetic field, atmosphere, surface, core, and original state.

h. **Nature of the Planets**.

The foregoing studies would be conducted by means of sounding rockets, earth satellites, lunar vehicles, and interplanetary vehicles.

17. Outer space activity and scientific research would have both military and non-military applications. Examples are satellites as navigational aids; and satellites as relay stations to receive and relay television or radio signals and improve world-wide communications.

18. It is not possible to foresee all applications of outer space activity which may be developed, but our ability to achieve and maintain leadership in such applications will largely depend on the breadth of the scientific research which is undertaken and supported.

[7] **Military**

19. The effective use of outer space by the United States and the Free World will enhance their military capability. Military uses of outer space (some of which may have peaceful applications) may be divided into the following three general categories:

    a. **Now Planned or in Immediate Prospect**

(1) **Ballistic Missiles**. A family of IRBM's and ICBM's is now in the latter stages of development. Components of these missiles can be used to develop other space vehicles, for both military and scientific use.

(2) **Anti-ICBM's** which are now being developed.

(3) **Military Reconnaissance**. (See "Reconnaissance Satellites" section, paragraphs 20-23)

    b. **Feasible in the Near Future**

(1) **Satellites of Weather Observation**.

(2) **Military Communications Satellites**.

(3) **Satellites for Electronic Countermeasures (Jamming)**.

(4) **Satellites as Aids for Navigation**, tracked from the earth's surface visually or by radio.

    c. **Future Possibilities**

(1) **Manned Maintenance and Resupply Outer Space Vehicles**.

(2) **Manned Defensive Outer Space Vehicles**, which might capture, destroy or neutralize an enemy outer space vehicle.

(3) **Bombardment Satellites** (Manned or Unmanned). It is conceivable that, in the future, satellites carrying weapons ready for firing on signal might be used for attacking targets on the earth.

(4) **Manned Lunar Stations**, such as military communications relay sites or reconnaissance stations. Conceivably, launching of missiles to the earth from lunar sites would be possible.

[8] **Reconnaissance Satellites**

20. Reconnaissance satellites are of critical importance to U.S. national security. Those now planned are designed:

(a) [major section omitted during declassification review] Reconnaissance satellites would also have a high potential use as a means of implementing the "open skies" proposal or policing a system of international armaments control.

21. As envisaged in U.S. plans, the instrumentation of reconnaissance satellites would consist primarily of [remainder of page omitted during declassification review] . . .

[9] 23. Some political implications of the use of reconnaissance satellites may be adverse. Therefore, studies must be urgently undertaken in order to determine the most favorable political framework in which such satellites would operate.

Manned Exploration of Outer Space

24. In addition to satisfying man's urge to explore new regions, manned exploration of outer space is of importance to our national security because:

(a) Although present studies in outer space can be carried on satisfactorily by using only unmanned vehicles, the time will undoubtedly come when man's judgment and resourcefulness will be required fully to exploit the potentialities of outer space.

(b) To the layman, manned exploration will represent the true conquest of outer space. No unmanned experiment can substitute for manned exploration in its psychological effect on the peoples of the world.

(c) Discovery and exploration may be required to establish a foundation for the rejection of USSR claims to exclusive sovereignty of other planets which may be visited by nationals of the USSR.

25. The first step in manned outer space travel could be undertaken using rockets and components now under study and development. Travel by man to the moon and beyond will probably require the development of new basic vehicles and equipment.

[10] **Other Implications of Outer Space Activities**
International Cooperation and Control

General

26. International cooperation in certain outer space activities appears highly desirable from a scientific, political and psychological standpoint and may appear desirable in selected instances with U.S. allies from the military standpoint. International cooperation agreements in which the United States participates could have the effect of (a) enhancing the position of the United States as a leader in advocating the uses of outer space for peaceful purposes and international cooperation in science, (b) conserving U.S. resources, (c) speeding up outer space achievements by the pooling of talents, (d) "opening up" the Soviet Bloc, and (e) introducing a degree of order and authority in the necessary international regulations governing certain outer space activities.

27. Various types of international cooperation may be possible through existing international scientific organizations, the United Nations, multilateral and bilateral arrangements with the Free World nations and NATO, and U.S.-Soviet bilateral arrangements. International cooperation by the United States in outer space activities might, as consistent with U.S. security interests, include (a) the collection and exchange of information on outer space; (b) the exchange of scientific instrumentation; (c) contacts among scientists; (d) participation of foreign scientists in U.S. space projects; (e) planning and coordination of certain programs or specific projects to be carried out on a fully international basis (some of which might be: a large instrumented scientific satellite, communication satellites, and meteorological satellites); (f) establishment of regulations governing certain outer space activities; (g) provision and launching of scientific satellites in support of international planning of a program of satellite observations.

28. Under present conditions, the extent of international cooperation, particularly in fields having important military applications such as propulsion and guidance mechanisms, will have to take into account security considerations (see paragraphs 7 and 15).

[11] **U.S. Position**

29. In January 1957 the United States initiated international discussion of the control of outer space by proposing in the UN General Assembly that the testing of outer space vehicles should be carried out and inspected under international auspices. This proposal was based on a policy decision[555] to seek to assure that the sending of objects into outer space should be exclusively for peaceful and scientific purposes and that, under effective control, the production of such objects designed for military purposes should be

---

555. With reference to the relation of the use of outer space to an armaments control system, the Annex to NSC Action No. 1553 (November 21, 1956), which remains in effect, provides: "5. It is the purpose of the United States, as part of an armaments control system, to seek to assure that the sending of objects into outer space shall be exclusively for peaceful and scientific purposes and that under effective control the production of objects designed for travel in or projection through outer space for military purposes shall be prohibited.

Therefore, the United States to propose that, contingent upon the establishment of effective inspection to verify the fulfillment of the commitment, all states agree to provide for international inspection of and participation in tests of outer space objects."

prohibited as part of an armaments control system. It was thought, at the then state of the art, that a control of testing would have precluded development until more comprehensive controls could be agreed upon. The U.S. proposal was altered with the passage of time and, as presented on August 29, 1957 as the Four Power Proposal in London, calls for technical studies of the "design of an inspection system which would make it possible to assure that the sending of objects through outer space will be exclusively for peaceful and scientific purposes." In his letter of January 13, 1958 to Bulganin, the President proposed, as part of a five-point program relating to control of armaments and armed forces, that "we agree that outer space be used only for peaceful purposes" and inquired "can we not stop the productions of such weapons which would use or more accurately misuse, outer space...?" In his later letter to Bulganin, dated February 15, 1958, the President proposed "wholly eliminating the newest types of weapons which use outer space for human destruction."

[12] 30. a. The most recent statement of basic policy relating to the regulation and reduction of armed forces and armaments appears in paragraph 40 of NSC 5810/1 (May 5, 1958).

b. Further consideration of U.S. policy concerning the scope of control and inspection required to assure that outer space could be used only for peaceful purposes, as well as the relationship of any such control arrangement to other aspects of an arms agreement, is deferred pending the recommendation of the Special NSC Committee established to make preparations for a possible Summit Meeting (NSC Action No. 1893). It is understood that the Special NSC Committee will also consider possible interim and more limited arrangements, and take into account the technical feasibility of assuring that outer space can be used only for peaceful purposes.

USSR Position

31. The USSR has proposed an agenda item for the next UN General Assembly meeting calling for the banning of the use of "cosmic space" for military purposes, the elimination of foreign bases on the territories of other countries, and international cooperation in the study of "cosmic space." The Soviets envisage an international agency with the following functions: development and supervision of an international program for launching intercontinental and space rockets to study "cosmic space"; continuation on a permanent basis of the IGY "cosmic space" research; world-wide collection, exchange and dissemination of "cosmic" research information; and coordination of and assistance to national research programs.

United Nations Role

32. The Soviet position makes certain that outer space questions, probably including peaceful uses, control, and organization, will be discussed in the UN General Assembly in September, 1958. The rapid pace of outer space achievements in past months has aroused great interest among all UN members concerning the role of the United Nations in the various aspects of outer space. The maintenance of our posture as the leading exponent of the use of outer space for peaceful purposes requires that the United States take in the General Assembly an imaginative and positive position.

[13] **Legal Problems of Air Space and Outer Space**

33. **Numerous legal problems** will be posed by the development of activities in space. Many of these cannot be settled until we gain more experience and basic information, because the only foundation for a sound rule of law is a body of ascertained fact. It is altogether likely that some issues in the field of space law which will be practical questions in the future are not even identified today. This is not to say that there is an entire lack of international law applicable to activities in space at the present time. For example, Article 51 of the Charter of the United Nations recognizes the inherent right of individual or collective self-defense against armed attack. Clearly this right is available against any space activities employed in such an attack.

34. **International Geophysical Year.** From the arrangements and announcements made in connection with the International Geophysical Year, there may be a general implied consent that scientific satellites be launched and orbited during the IGY. Such implied consent does not necessarily mean, however, that assent has been given to the launching and orbiting or other types of satellites and missiles, or that assent with respect to scientific satellites extends beyond the IGY. It remains to be determined what rules will apply to subsequent satellites; what limitations will govern the types and purposes of satellites in the future. The United States, as well as other countries, has not yet taken positions on these questions and, here again, the answer will depend not only upon what others are likely to do but also upon what activities the United States wishes to be free to engage in.

35. **A problem of jurisdiction in space** on which the United States reserves its position at present is whether celestial bodies in space beyond the earth are susceptible to appropriation by national control or sovereignty.

36. **The problem of legal definitions** is unsolved. As indicated above, there is as yet insufficient basis for legally deciding that air space extends so far and no farther; that outer space begins at a given point above the earth. Because, for some time to come, at least, activities in outer space will be closely connected with activities on the earth and in air space, many legal problems with respect to space activities may well be resolved without the necessity of determining or agreeing upon a line of demarcation between air space and outer space. If, by analogy to the Antarctic proposal of the United States, international agreement can be reached upon permissible activities in space and the rules and regulations to be followed with respect thereto, problems of sovereignty may be avoided or at least deferred.

[14] 37. **Problems of liability** for injury or damage caused by activities in space or by re-entry will also arise. No nations has as yet taken a position as to whether due care against negligence should be the standard or whether liability should be absolute. Here again future experience, and the development of agreement among the nations, will be necessary. Absolute liability as respects objects landing on United States will have to be weighed against absolute liability for U.S. objects landing on other nations.

38. **Problems of national and international regulation** over activities in space will also arise. There is already the need to assign telecommunication wavelengths to communications with satellites and space objects. Other types of regulations having serious security implications will have to be worked out for the identification of space objects and for some type of traffic control to prevent congestion and interference.

39. **Generally speaking, rules will have to be evolved gradually** and pragmatically from experience. While the nations engaging in space activities will play an important role in this field, it will have to be recognized from the nature of the subject that all nations have a legitimate interest in it. The field is not suitable for abstract *a priori* codification.

[15] **Comparison of USSR and U.S.**
Capabilities in Outer Space Activities

40. Conclusive evidence shows that the Soviets are conducting a well-planned outer space program at high priority. The table below attempts to estimate the U.S. and USSR timetable for accomplishment of specific outer space flight activities.

a. Soviet space flight capabilities estimated in the table reflect the earliest possible time periods in which each specific event could be successfully accomplished.

(1) The space flight program is in competition with many other programs, particularly the missile program. The USSR probably cannot successfully accomplish all of the estimated space flight activities within the time periods specified. The USSR will not permit its space flight program to interfere with achieving an early operational capability for ICBM's (which enjoy the highest priority).

(2) The USSR is believed to have the intention to pursue both an active space flight program designed to put man into outer space for military and/or scientific purposes,

and further scientific research utilizing earth satellites, lunar rockets, and probes of Mars and Venus; but it cannot be determined, at this time, whether the basic scientific program or the "man in space" program enjoys the higher priority and will, therefore, be pursued first.

b. U.S. space flight capabilities indicated in the table reflect the earliest possible time periods in which each specific event could be successfully accomplished. Not all of the indicated activities could be successfully accomplished with the time period specified. It must also be recognized that the accomplishment of some of the activities listed would impinge upon space activities already programmed, or upon other military programs.

41. If the USSR high-priority outer space program continues, the USSR will maintain its lead at least for the next few years, as shown in the following table.

[16]
**Earliest Possible Time Periods of Various
Soviet and U.S. Accomplishments in Outer Space**

(NOTE: Generally, Soviet vehicles will be of substantially greater orbital payloads than U.S. vehicles. It should be noted, however, that the comparative capabilities of the United States and the USSR should not be measured by orbital payloads alone.

The United States is estimated to be considerably ahead of the USSR in miniaturization of missile and satellite components, and therefore the effectiveness of U.S. satellites on a "per pound in orbit" basis is estimated to be greater than that of the USSR.)

		Soviet[a]	U.S.[b]
1.	Scientific Earth Satellites (IGY Commitment)	1957-58	1958
2.	Reconnaissance Satellites[c]	1958-59	1959-61
3.	Recoverable Aeromedical Satellites	1958-59	1959
4.	Exploratory Lunar Probes or Lunar Satellites	1958-59	1958-59
5.	"Soft" Lunar Landing	1959-60	Early 1960
6.	Communications Satellites	—	1959-60
7.	Manned Recoverable Satellites		
	a. Capsule-type Satellites	1959-60[d]	1960-63
	b. Glide-type Vehicles	1960-61	1960-63
8.	Mars Probe	Aug. 1958[e]	Oct. 1960
9.	Venus Probe	June 1959[e]	Jan. 1961
10.	25,000-pound Satellite — manned	1961-62	After 1965
11.	Manned Circumlunar Flight	1961-62	1962-64
12.	Manned Lunar Landing	After 1965	1968

---

[a] Estimate by the Guided Missile Intelligence Committee (GMIC) of the IAC as if June 3, 1958.

[b] **Source**: Department of Defense, June 4, 1958.

[c] **Defense Comment**: (See Annex B for test reconnaissance satellites.) The United States plans to launch a reconnaissance satellite of approximately 3,000 pounds in late 1959. During the same time period the USSR is estimated to be capable of launching a 45,000 pound reconnaissance satellite.

[d] The Joint Staff member of GMIC reserves his position on the date 1959.

[e] The Soviets most likely would attempt probes when Venus and Mars are in their most favorable conjunction with the earth for such an undertaking.

Level of Effort

42. a. Because of the highly speculative nature of future activities in outer space, decisions as to the priority and extent of U.S. outer space programs will obviously be a judgment based on limited knowledge. Some activities in outer space would be expedited by the allocation of additional financial resources; others would not, being dependent on research progress. The potentially great importance to U.S. national interests of outer space activities, however, requires taking risks in allocating resources to research and development activities, the success or ultimate utility of which cannot be definitely foreseen.

b. The level of material and scientific effort to be expended on outer space activities must nevertheless be related to other national security programs to ensure that a proper balance is maintained between anticipated scientific, military and psychological gains from outer space programs and the possible loss resulting from reductions in resources allocated to other programs.

Objectives[556]

43. The fullest[557] development and exploitation of U.S. outer space capabilities as needed to achieve U.S. scientific, military and political purposes as follows:

a. A technological capability to meet the requirements of **b**, **c** and **d** below.

b. A degree of competence and a level of achievement in outer space basic and applied research and exploration which is at least on a par with that of any other nation.

c. Applications of outer space technology, research and exploration to achieve a military capability in outer space sufficient to assure the over-all superiority of U.S. [outer space][558] offensive and defensive systems relative to those of the USSR.

d. Applications of outer space technology, research, and exploration for non-military purposes, which are at least on a par with any other nation.

e. World recognition of the United States as, at least, the equal of any other nation in over-all outer space activity and as the leading advocate of the peaceful exploitation of outer space.

[43. The establishment of the United States as the recognized leader in the over-all development and exploitation of outer space for scientific, military and political purposes.][559]

44. As consistent with U.S. security, achievement of international cooperation in the uses of and activities related to outer space: for peaceful purposes, and with selected allies for military purposes.

[19] 45. As consistent with U.S. security, the achievement of suitable international agreements relating to the uses of outer space for peaceful purposes that will assure orderly outer space programs.

46. Utilization of the potentials of outer space to assist in "opening up" the Soviet Bloc through improved intelligence and programs of scientific cooperation.

Policy Guidance[560]

Priority and Scope of Outer Space Effort

47. With a priority and scope sufficient to enable the United States at the earliest practicable time to achieve its scientific, military and political objectives as stated in para-

---

556. See paragraphs 29 and 30 for statement of the status of policy on the regulation and reduction of armed forces and armaments in relation to outer space.
557. Budget proposes to delete "the fullest" and all of the paragraph after "purposes".
558. Defense-JCS proposal.
559. ODM-NACA-USIA alternative paragraph 43.
560. See paragraphs 29 and 30 for statement of the status of policy on the regulation and reduction of armed forces and armaments in relation to outer space.

graph 43, develop and expand selected U.S. activities related to outer space in:

    a. Research and technology required to exploit the military and non-military potentials of outer space.

    b. Outer space exploration required to determine such military and non-military potentials.

    c. Applications of outer space research, technology, and exploration to develop outer space capabilities (in addition to those capabilities which now have the highest national priority[561]) required to achieve such objectives.

48. In addition to undertaking necessary immediate and short-range activities related to outer space, develop plans for outer space activities or the longer range (through at least a ten-year period).

49. Study on a continuing basis the implications which U.S. and foreign exploitation of outer space may hold for international and national political and social institutions. Critically examine such exploitation for possible consequences on activities and on life on earth (e.g., outer space activities which affect weather, health for other factors relating to activities and life on earth).

[50. In the absence of a safeguarded international agreement for the control of armaments and armed forces, place primary emphasis on activities related to outer space necessary to maintain the over-all deterrent capability of the United States and the Free World.][562]

[21] **Psychological Exploitation**

51. In the near future, while the USSR has a superior capability in space technology, judiciously select (without prejudicing activities under paragraph 47) projects for implementation which, while having scientific or military value, are designed to achieve a favorable world-wide psychological impact.

52. Identify, to the greatest extent possible, the interests and aspirations of other Free World nations in outer space with U.S.-sponsored activities and accomplishments.

53. Develop information and other programs that will exploit fully U.S. outer space activities on a continuing basis; especially, during the period while the USSR has superior over-all outer space capabilities, those designed to counter the psychological impact of Soviet outer space activities and to present U.S. outer space progress in the most favorable comparative light.

Reconnaissance Satellites[563]

54. In anticipation of the availability of reconnaissance satellites, seek urgently a political framework which will place the uses of U.S. reconnaissance satellites in a political and psychological context most favorable to the United States.

55. At the earliest technologically practicable date, use reconnaissance satellites to enhance to the maximum extent the U.S. intelligence effort.

International Cooperation in Outer Space Activities

56. Consistent with the objectives in paragraphs 43 and 44, and as a means of maintaining the U.S. position as the leading advocate of the use of outer space for peaceful purposes, be prepared to propose that the United States join with other nations, including the USSR, in cooperative efforts relating to outer space. Specifically:

    a. Encourage a continuation and expansion of the type of cooperation which exists in the IGY programs, through non-governmental international scientific [22] organizations such as the International Council of Scientific Unions; including cooperation in the design of experiments and instrumentation, exchange of information on instrumenta-

---

561. See NSC Action No. 1846.
562. State-Defense-Treasury-JCS proposal. Other Planning Board representatives believe the subject is adequately covered in paragraphs 43 and 47, and would therefore delete paragraph 50.
563. The priority and scope of operational capabilities of reconnaissance satellites are established in NSC Action No. 1846, January 22, 1958.

tion, scientific data and telemetry, exchange of instruments, and in the use of scientific satellites and other scientific vehicles in support of international planning for exploration of outer space.

    b. Recognize UN interests in outer space cooperation, but do not encourage precipitous UN action to establish permanent organizational arrangements. To this end consider: (1) establishment of an ad hoc UN planning committee to formulate recommendations to facilitate international cooperation and appropriate UN organizational arrangements; and (2) in the interim, participation in those joint projects for cooperation and exchange of information for which UN auspices are desirable.

    c. Invite scientists of foreign countries, including the Soviet Bloc in general on a reciprocal basis, to participate in selected U.S. programs for the scientific exploration of space.

    d. Propose scientific bilateral arrangements with other nations (including the USSR) for cooperative ventures related to outer space, provided that the combined existing competence might achieve meaningful scientific and technical advance.

    e. Propose to groups of nations and international organizations independent outer space projects which would be appropriate for multilateral participation.

    f. Assist selected Free World nations willing and able to undertake useful activities related to outer space, [as necessary to assure that the over-all Free World position in outer space developments is at least on a par with that of the Sino-Soviet Bloc].[564]

Limited International Arrangements to Regulate Outer Space Activities
    57. Propose international agreements concerning appropriate means for maintaining a full and current public record of satellite orbits and emission frequencies.

[23] **International Outer Space Law**
    58. [Paragraph excised during declassification review]
    59. Reserve the U.S. position on legal issues of outer space, but undertake on an urgent basis a study of the legal issues that will arise from national and international outer space activities in the near future.

Interim Position in International Negotiations
    60. In negotiations with other nations or organizations dealing with outer space (pending the results of the study referred to in paragraph 58), seek to achieve common agreement to relate such negotiations to the traversing or operating of man-made objects in outer space, rather than to defined regions in outer space.

Security Classification
    61. In considering whether U.S. outer space information and material requires classification under Executive Order No. 10501,[565] take special account of the lead achieved by the USSR in outer space activities and the advantages, including more rapid progress, which could accrue to the United States through liberalizing the general availability and use of such information and material.

Administration of Outer Space Programs
    62. Provide through appropriate legislation for the conduct of U.S. outer space activities under the direction of a civilian agency, except in so far as such activities may be peculiar to or primarily associated with weapons systems or military operations, in the case of which activities the Department of Defense shall be responsible.

---

    564. Budget proposes deletion.
    565. Executive Order No. 10501 ("Safeguarding Official Information in the Interests of the Defense of the United States"), Section 3 provides in part that: "Unnecessary classification and over-classification [of information or material] shall be scrupulously avoided."

[24]

Annex A
The Soviet Space Program

1. **Objectives and Scope of Program.** Conclusive evidence shows that the Soviets are conducting a well-planned space flight program at high priority. This program is apparently aimed at placing both instrumented and manned vehicles into space. Certain successes have been exhibited already in the instrumented vehicles category (including the orbiting of three earth satellites, one containing a dog) and we believe they are fully capable of achieving manned space flight within the next few years.

2. **General.** Evidence of Soviet interest in space flight dates back to a publication in 1903 of a paper, "Investigation of Universal Space by Means of Rocket Flight," by the eminent Russian scientist Tsiolkovsky. This highly scientific treatise for the first time mathematically established the fundamentals of rocket dynamics and included a proposal for an artificial earth satellite. Reactive motion (rockets) was seriously engaged again in the latter '20s and in the '30s. In April 1955, the Interagency Commission for Interplanetary Communications was formed under the Academy of Sciences to establish an automatic laboratory for scientific research in cosmic space as a first step in solving the problems of interplanetary travel. Since early 1955 several hundred articles on space research, earth satellites and space flight have been published in the USSR. Many of the articles have been written by high-caliber Soviet scientists and most deal with the theoretical principles of space flight.

3. **Capabilities.** The Soviet Union dramatically demonstrated its interest and current capability in space flight with the launching of two earth satellites in October and November 1957, and a third in mid-May 1958. The complex facilities and skills needed to operate the large rocket vehicles required for the launching of a satellite or space vehicle are apparently available within the Soviet military. Thus, although the first space flights were doubtless undertaken for the furtherance of scientific knowledge and for whatever psychological and political advantage would accrue, the Soviet military department, by intimate participation of its hardware and personnel, is in a position to utilize immediately such knowledge for the enhancement of the Soviet military position and objectives. The realization of even more advanced space projects particularly those involving manned flight, must be preceded by a vast amount of systematic [25] and well-coordinated scientific and technological work directed toward the development of practical space vehicles, the determination of basic operational requirements and limitations, and the creation of an environment and equipment capable of sustaining human life in outer space. Such a program embraces virtually all fields of science and engineering and the following fields were particularly examined for evidence of Soviet technical capability: guided missiles, re-entry vehicles, propulsion, electronics, space medicine, astrobiology, internal power supplies, and celestial mechanics. While firm association of these areas with a space program varied considerably, it is noted that the state of Soviet art in all sciences required in a space program was such that no scientific barriers of magnitude were detected. Four areas critical to a space program have apparently received considerable attention by the USSR, e.g., development of large rocket-engine propulsion systems, space medicine, cosmic biology and celestial mechanics. We believe the depth and advancement of their research and development makes them world leaders in these areas. In particular their work in space medicine and cosmic biology are strong indicators of their serious intent to put man in space at an early date.

4. **Time Scales.**

a. The following milestones are considered at least partially affiliated with a space program and indicate historically the long-term interest of Soviet Union in this endeavor:

1903	Initial treatise on space flight
1923	Soviet Institute on Theoretical Astronomy founded
1929	First significant rocket studies conducted, "Group for the Investigation of Reactive Motion" founded
1934	Government-sponsored rocket research program established

1940	Flight of first Soviet rocket-powered aircraft
1946-47	Rocket-propelled intercontinental bomber program organized
1953-55	Systematic investigation of moon flight problems undertaken
1955 (Apr.)	Interagency Commission for Interplanetary Communications established
1955-58	Over 500 Soviet articles published dealing with space research, earth satellites and manned space flight.
1957 (Oct.-Nov.)	First artificial earth satellites orbited.

b. **Future Capabilities**. Soviet space flight capabilities estimated in this section are the earliest possible time periods in which each specific event could be successfully accomplished. It is recognized that the space flight program is in competition with many other programs, particularly the missile program, and that the USSR probably cannot successfully accomplish all of the estimated space flight activities within the time periods specified. We believe the USSR has the intention to pursue an active flight program designed to put man into space for military and/or scientific purposes. We also believe they have a definite intention to pursue further scientific research utilizing earth satellites, lunar rockets, and probes of Mars and Venus. We cannot, at this time, determine whether the basic scientific program or the "man in space" program enjoys the higher priority and will, therefore, be pursued first. Whichever approach is adopted will probably result in some slippage in the capability dates indicated for the other program. We believe the Soviet ICBM program still enjoys the highest priority and that the USSR will not permit its space flight program to interfere with achieving an early operational ICBM capability.

(1) **Unmanned Earth Satellites**.

(a) Based on current estimates of Soviet ICBM capabilities, it is estimated that the USSR could orbit scientific satellites weighing on the order of 5,000 pounds within the next several months. The USSR could probably continue to place into orbit more and perhaps larger satellites throughout the period of this estimate. As additional scientific data is obtained, the USSR could refine or develop new scientific instrumentation to be placed into satellites.

(b) It is believed that the USSR could place into orbit and recover aeromedical specimens from satellites early in the period of this estimate. Early recovery of a biological specimen from orbiting satellites is essential and could advance Soviet knowledge of recovery techniques and provide indications of adverse effects of a space environment for man.

(c) The USSR could probably orbit surveillance satellites capable of low resolution (approximately 100-200 feet) at any time within the next year to obtain weather data and perhaps [27] some additional data of military intelligence value such as fleet movements. More sophisticated surveillance satellites, involving improved photographic or TV reconnaissance, infrared photography and/or ELINT, could be developed within a year or two following an initial success. These latter satellites containing this more advanced instrumentation could be capable of providing more diverse scientific and military information. Should they elect to do so, the USSR could also develop a communications relay satellite within the period of this estimate.

(2) **Lunar Rockets**. The USSR has had the capability of launching a lunar probe toward the vicinity of the moon since the fall of 1957 as far as propulsion and guidance requirements are concerned. A Soviet program of lunar probes could commence with experimental rockets followed by rocket landings on the moon with increasingly heavy loads containing scientific and telemetering equipment. Placing a satellite into orbit around the moon requires the use of a retro-rocket and more accurate guidance. It is believed that the USSR could achieve a lunar satellite in late 1958-1959 and have a lunar soft landing about six months thereafter.

(3) **Manned Earth Satellites**. Sufficient scientific data could probably be attained and recovery techniques perfected to permit the USSR to launch a manned satellite into or-

bital flight and recovery by about 1959-1960.[566] A manned capsule-type satellite as well as a manned glide-type vehicle appear to be feasible techniques and within Soviet capabilities. However, it is believed that the first Soviet orbital recovery attempt will probably be with the manned capsule.

(4) **Planetary Probes.** Planetary probe vehicles could utilize existing Soviet ICBM propulsion units for the first stage and presently available guidance components. It is believed that the USSR could launch probes towards Mars and Venus with a good chance of success. The first launchings toward Mars could occur in August 1958, when Mars will be in the most favorable [28] position relative to the earth. More sophisticated probes could occur in October 1960, when Mars will again be in a favorable position relative to the earth. Probes toward Venus could probably occur in June 1959, and more sophisticated probe vehicles could be launched in January 1961.

(5) **Manned Curcumlunar Flights.** Contingent upon their success with manned earth satellites and the development of a new, large booster engine, and concurrent advances in scientific experimentations with lunar rockets, the USSR could achieve a capability for manned circumlunar flight with reasonable chance for success in about 1961-1962.

(6) **Manned Lunar Landings.** It is not believed that the USSR will have a capability for manned lunar landings until some time after 1965.

(7) **Space Platforms.** There is insufficient information on the problems as well as the utility of constructing a platform in space to determine the Soviet capability. It is believed, however, that they are capable of placing a very large satellite (about 25,000 pounds) into orbit in 1961-1962 and that this vehicle could serve some of the scientific functions of a large space platform without the difficulties of joining and constructing such a platform in space.

Annex B
Tentative Schedule of U.S. Vehicle Launchings[567]

(as of June 30, 1958)

1. IGY (Five Vanguard vehicles)
Firing rate about one per month ending late in 1958 (or early 1959).
Payload 22 lbs.
2. Lunar Probes (Three Thor-Vanguard vehicles)
First launching - September 1958 - Payload 25 lbs. plus retro-rockets.
Second launching - October 1958 - Payload 26 lbs. plus retro-rockets.
Third launching - November 1958 - Payload 26 lbs. plus retro-rockets.
3. Satellites and Lunar Probes (Three Juno II, One Juno I)
(Inflatable Sphere) - Fall 1958 - Payload 12-foot reflecting sphere.
(Lunar Probe) - November 1958 - Payload 15 lbs.
(Lunar Probe) - January 1959 - Payload 15 lbs.
(Cosmic ray satellite) - February 1959 - Payload 60 lbs.
4. a. ARGUS Project (Two Juno I)
Both units to be launched as earth satellites in August 1958 - Payload 26 lbs.
 b. ARGUS Project (Six NOTS fly-up satellites)
Satellites are air-launched, have approximately 3-pound payload of instruments designed to detect Argus effect. Satellites are to be launched into polar orbits. Three flights in July 1958 are planned for purpose of testing system. If these first three are successful, three more will be launched in August as part of Argus Project.
[30] 5. Advanced Reconnaissance Satellite Development Program, W/S 117L

---

566. The Joint Staff member reserves his position on the date 1959.
567. Launchings shown are those needed to implement presently-planned programs. These programs are under review and are not to be regarded as final.

a. **Tests**

(1) Thor-boosted program - Up to nineteen vehicles - Test firings for second stage and instrumentation capability.

First launching about November 1958 - Firing rate nearly one per month until completion. Payload 400 lbs. to 135-mile orbit.

(2) Atlas-boosted program - five vehicles - Visual Reconnaissance Components Test.

First launching about July 1959, with firing rate of about one every other month to completion of program. Payload 2600 lbs. to 300-mile orbit. Launchings from Cape Canaveral. Satellites would not pass over the USSR.

(3) Atlas-boosted program - one vehicle - Visual Reconnaissance Components Test. Launching about March 1960 (the earliest date an Atlas 117L launching stand would be available at Cooke). Payload 2600 lbs. to 300-mile orbit.

(4) Atlas-boosted program - four vehicles - Visual Reconnaissance Test.

First launching about May 1960, with firing rate of one every other month to completion November 1960. Payload 2600 lbs. to 300-mile orbit. Launchings from Cooke.

(5) Atlas-boosted program - Three vehicles - Ferret Reconnaissance Test.

First launching about August 1960, with firing rate of one every other month to completion December 1960. Payload 2600 lbs. to 300-mile orbit. Launchings from Cooke.

[Section omitted during declassification review]

(6) X-15 Manned Research Aircraft - Three aircraft - (USAF, NACA, USN). First flight scheduled in first half of CY 1959. Maximum altitude capability, 100-125 miles. Maximum speed about 4500 miles per hour.[568]

Document II-19

Document title: Nathan F. Twining, Chairman, Joint Chiefs of Staff, Memorandum for the Secretary of Defense, "U.S. Policy on Outer Space (NSC 5814)," August 11, 1958.

Source: Presidential Papers, Dwight D. Eisenhower Library, Abilene, Kansas.

While the draft of NSC 5814 was being discussed, Chairman of the Joint Chiefs Nathan F. Twining was asked to review the document, and he recommended changes. His August 11, 1958, memorandum on the subject argued that primary emphasis in U.S. space policy should be given to military activities related to U.S. security.

[1] ## Memorandum for the Secretary of Defense

Subject: U.S. Policy on Outer Space (NSC 5814).

1. The Joint Chiefs of Staff have reviewed the subject draft revision of NSC 5814 prepared by the NSC Planning Board in accordance with NSC Action 1940-c for consideration by the National Security Council at its meeting on 14 August 1958. The Joint Chiefs of Staff consider that while the draft revision is in literal consonance with NSC Action 1940-c, it does not wholly reflect the substance of the recommendation made to the National Security Council by the Deputy Secretary of Defense. As a result, the draft statement of policy does not reflect a proper balance between military and non-military interests in outer space. Paragraph 43 of the draft revision states the following as a U.S. objective: "Development and exploitation of U.S. outer space capabilities as needed to achieve U.S.

---

568. This aircraft was not planned as an orbital vehicle and approved programs do not include modification of system to allow orbiting. Various problems related to re-entry of orbiting or space vehicles can be effectively studied with this aircraft.

scientific, military, and political purposes, and to establish the United States as a recognized leader in this field." The primary product of leadership in the non-military aspect of outer space is world-wide prestige, whereas that of the military aspect is a factor of survival of the United States. Since U.S. resources that can be devoted to outer space activities overall are limited, it is appropriate to indicate the relative priority between military and non-military activities.

2. The Joint Chiefs of Staff wish to emphasize the "preliminary" nature of the draft statement of policy and the need for flexibility in the execution of its provisions.

3. Specific comments on the diverse views contained in the draft revisions follow:

a. *Paragraph 50, page 12.* (Changes to read as follows, indicated in the usual manner):

[50. In the absence of a safeguarded international agreement for the control of armaments and armed forces, *place primary emphasis on activities related to outer space necessary to* maintain the over-all deterrent capability of the United States and the Free World.]

[2] **REASON**: In the absence of a safeguarded international agreement for the control of armaments and armed forces, policy guidance, particularly as it affects future military research and development programs is needed. Such guidance does not appear elsewhere in this paper nor in other U.S. policy papers.

4. Subject to the foregoing,

a. The majority view of the draft revision of policy is acceptable from a military point of view, and

b. The Joint Chiefs of Staff recommend that you concur in the adoption of the draft revision of NSC 5814, "U.S. Policy on Outer Space."

For the Joint Chiefs of Staff:
N.F. TWINING,
Chairman,
Joint Chiefs of Staff.

Document II-20

Document title: National Security Council, NSC 5814/1, "Preliminary U.S. Policy on Outer Space," August 18, 1958, pp. 17-19.

Source: Presidential Papers, Dwight D. Eisenhower Library, Abilene, Kansas.

This statement of U.S. policy, a revision of NSC 5814, was adopted by the National Security Council at its August 14, 1958, meeting and approved by President Eisenhower on August 18, 1958. Coordinating the implementation of the policy was the Operations Coordinating Board, a unit within the Executive Office of the President closely linked to the National Security Council; this was the president's means of making sure that various executive agencies were being responsive to administration policy. At the time this policy statement was approved, the new National Aeronautics and Space Administration had been created, but had not begun operations.

There are a number of language changes between this document and the draft NSC 5814 [II-18], which are relatively insignificant. However, paragraph 44 is a revision of paragraph 43 in the draft document that takes a less expansive view of U.S. purposes in space. Also, paragraph 50 of the earlier document was deleted on the basis of objections from the military community. The major changes are reprinted here.

[17] **Level of Effort**

43. a. Because of the highly speculative nature of future activities in outer space, decisions as to the priority and extent of U.S. outer space programs will obviously be a judgment based on limited knowledge. Some activities in outer space would be expedited by the allocation of additional financial resources; others would not, being dependent on research progress. The potentially great importance to U.S. national interests of outer space activities, however, requires taking risks in allocating resources to research and development activities, the success or ultimate utility of which cannot be definitely foreseen.

b. The level of material and scientific effort to be expended on outer space activities must nevertheless be related to other national security programs to ensure that a proper balance is maintained between anticipated scientific, military and psychological gains from outer space programs and the possible loss resulting from reductions in resources allocated to other programs.

[18] **Objectives**[554]

44. Development and exploitation of U.S. outer space capabilities as needed to achieve U.S. scientific, military, and political purposes, and to establish the U.S. as a recognized leader in this field.

45. As consistent with U.S. security, achievement of international cooperation in the uses of and activities related to outer space; for peaceful purposes, and with selected allies for military purposes.

46. As consistent with U.S. security, the achievement of suitable international agreements relating to the uses of outer space for peaceful purposes that will assure orderly development and regulation of national and international outer space programs.

47. Utilization of the potentials of outer space to assist in "opening up" the Soviet Bloc through improved intelligence and programs of scientific cooperation.

Policy Guidance[555]

Priority and Scope of Outer Space Effort[556]

48. With a priority and scope sufficient to enable the U.S. at the earliest practicable time to achieve its scientific, military and political objectives as stated in paragraph 44, develop and expand selected U.S. activities related to outer space in:

[19] a. Basic and applied research, and exploration required to determine the military and non-military potentials of outer space.

b. Research and technology required to exploit such military and non-military potentials.

c. Application of such outer space research, technology, and exploration to develop outer space capabilities required to achieve such objectives.

49. In addition to undertaking necessary immediate and short-range activities related to outer space, develop plans for outer space activities for the longer range (through at least a ten-year period).

50. Study on a continuing basis the implications which U.S. and foreign exploitation of outer space may hold for international and national political and social institutions. Critically examine such exploitation for possible consequences on activities and life on earth (e.g., outer space activities which affect weather, health, or other factors relating to activities and life on earth)....

---

[554] See paragraphs 30 and 31 for statement of the status of policy on the regulation and reduction of armed forces and armaments in relation to outer space.

[555] See paragraphs 30 and 31 for statement of the status of policy on the regulation and reduction of armed forces and armaments in relation to outer space.

[556] Nothing in this paper shall be construed as affecting priorities established under NSC Action No. 1846 or future priorities approved by the President.

Document II-21

Document title: National Aeronautics and Space Council, "U.S. Policy on Outer Space," January 26, 1960.

Source: Presidential Papers, Dwight D. Eisenhower Library, Abilene, Kansas.

This final Eisenhower administration statement on U.S. space policy was adopted at what was considered a joint meeting of the National Aeronautics and Space Council and the National Security Council on January 12, 1960, and approved by the president on January 26, 1960. It replaced the policy statements contained in National Security Council documents NSC 5520 [II-10] and NSC 5814/1 [II-18, II-20]. Although the draft of this policy statement had been circulated bearing a National Security Council designation (NSC 5918), the approved statement was issued as a National Aeronautics and Space Council document. Because of changes from the draft policy statement that had been circulated in December, this final policy document did not contain a page 14.

---

[1] General Considerations

Scope of Policy

1. This policy is concerned with U.S. interests in scientific, civil, military, and political activities related to outer space. It deals with sounding rockets, earth satellites, and outer space vehicles, their relationship to the exploration and use of outer space, and their political and psychological significance. Although the relation between outer space technology and ballistic missile technology is recognized, U.S. policy on ballistic missiles is not covered in this policy. Anti-missile defense systems also are not covered except to the extent that space vehicles may be used in connection with such systems.

Significance of Outer Space to U.S. Security

2. Outer space presents a new and imposing challenge. Although the full potentialities and significance of outer space remain largely to be explored, it is already clear that there are important scientific, civil, military, and political implications for the national security, including the psychological impact of outer space activities which is of broad significance to national prestige.

3. Outer space generally has been viewed as an area of intense competition which has been characterized to date by comparison of Soviet and U.S. activities. The successes of the Soviet Union in placing the first earth satellite in orbit, in launching the first space probe to reach escape velocity, in achieving the first "hard" landing on the moon and in obtaining the first pictures of the back side of the moon have resulted in substantial and enduring gains in Soviet prestige. The U.S. has launched a greater number of earth satellites and has also launched a space probe which has achieved escape velocity. These U.S. activities have resulted in a number of scientifically significant "firsts." However, the space vehicles launched by the Soviet Union have been substantially heavier than those in the U.S., and weight has been a major point of comparison internationally. In addition, the Soviets have benefited from their ability to conceal any failures from public scrutiny.

4. From the political and psychological standpoint the most significant factor of Soviet space accomplishments is that they have produced new credibility for Soviet statements and claims. Where once the Soviet Union was not generally believed, even its boldest propaganda claims are now apt to be accepted at face value, not only abroad but in the United States. The Soviets have used this credibility for the following purposes:

    a. To claim general superiority for the Soviet system on the grounds that the Sputniks and Luniks demonstrate the ability [2] of the system to produce great results in an extremely short period of time.

b. To claim that the world balance has shifted in favor of Communism.
c. To claim that Communism is the wave of the future.
d. To create a new image of the Soviet Union as a technologically powerful, scientifically sophisticated nation that is equal to the U.S. in most respects, superior in others, and with a far more brilliant future.
e. To create a new military image of the vast manpower of the Communist nations now backed by weaponry that is as scientifically advanced as that of the West, superior in the missile field, and superior in quantity in all fields.

5. Soviet efforts already have achieved a considerable degree of success, and may be expected to show further gains with each notable space accomplishment, and particularly each major "first."

6. Significant advances have been made in restoring U.S. prestige overseas, and in increasing awareness of the scope and magnitude of the U.S. outer space effort. Although most opinion still considers the U.S. as probably leading in general scientific and technical accomplishments, the USSR is viewed in most quarters as leading in space science and technology. There is evidence that a considerable portion of world leadership and the world public expects the United States to "catch up" with the Soviet Union, and further expects this to be demonstrated by U.S. ability to equal Soviet space payloads and to match or surpass Soviet accomplishments. Failure to satisfy such expectations may give rise to the belief that the United States is "second best," thus transferring to the Soviets additional increments of prestige and credibility now enjoyed by the United States.

7. To the layman, manned space flight and exploration will represent the true conquest of outer space and hence the ultimate goal of space activities. No unmanned experiment can substitute for manned space exploration in its psychological effect on the peoples of the world. There is no reason to believe that the Soviets, after getting an earlier start, are placing as much emphasis on their manned space flight program as is the U.S.

[3] 8. The scientific value of space exploration and the prestige accruing therefrom have been demonstrated. The scientific uses of space are a potent factor in the derivation of fundamental information of use in most fields of knowledge. Further, the greater the breadth and precision of the knowledge of the space environment, the greater the ability to exploit its potentials.

9. Among several foreseeable civil applications of earth satellites, two at present offer unique capabilities which are promising in fields of significance to the national economy: communications and meteorology. Other civil potentials are also likely to be identified.

10. [Paragraph excised during declassification review]
11. [Paragraph excised during declassification review]
12. Outer space activities present new opportunities and problems in the conduct of the relations of the U.S. with its allies, neutral states, and the Soviet Bloc; and the establishment of sound international relationships in this new field is of fundamental significance to the national security. Of importance in seeking such relationships is the fact that all nations have an interest in the purposes for which outer space is explored and used and in the achievement of an orderly basis for the conduct of space activities. Moreover, many nations are capable of participating directly in various aspects of outer space activities, and international participation in such applications of space vehicles as those involved in scientific research, weather forecasting, and communications may [4] be essential to full realization of the potentialities of such activities. In addition, an improvement of the international position of the U.S. may be effected through U.S. leadership in extending internationally the benefits of the peaceful purposes of outer space. The fact that the results of the arrangements in certain fields, even though entered into for peaceful purposes, could have military implications, may condition the extent of such arrangements in those fields.

Use of Outer Space

General

13. As further knowledge of outer space is obtained, the advantages to be accrued will become more apparent. At the present time, space activities are directed toward tech-

nological development and scientific exploration; however, it is anticipated that systems will be put into operation, beginning in the near future, that will more directly contribute to national security and well-being and be of international benefit.

14. Present and planned outer space activities will require the use of the following classes of vehicles:

    a. **Sounding Rockets**[1] - Vehicles that are launched vertically or in a ballistic trajectory to heights well outside the earth's atmosphere and return to earth.

    b. **Earth Satellites** - Manned and unmanned vehicles that orbit the earth.

    c. **Space Probes and Interplanetary Space Vehicles** -Manned and unmanned vehicles that escape the earth environment to traverse interplanetary space.

15. It is not possible to foresee all the uses of outer space, but the ability to identify and develop such uses will be significantly influenced by the breadth of the exploratory scientific research which is undertaken.

Scientific Research and Exploration

16. Space technology affords new and unique opportunities for immediate and long-range scientific observation, experimentation, and [5] exploration which will add to our knowledge and understanding of the earth, the solar system, and the universe. Immediate opportunities exist in many areas, including among others:

    a. **Atmosphere** - Study of the structure and composition of the earth's outer atmosphere.

    b. **Ionosphere** - Measurement of the electron density of the earth's outer ionosphere and its temporal and spatial variations.

    c. **Energetic Particles** - Measurement of cosmic ray intensity, radiation belts, and auroral particles and their variations with time and space in the vicinity of the earth and moon.

    d. **Electric and Magnetic Fields** - Measurement of the magnitude and variations of the earth's magnetic field and the associated ionospheric electric currents.

    e. **Gravitational Fields** - Study of the detailed motion of existing and special satellites with the object of determining a more detailed picture of the earth's and moon's gravitational field.

    f. **Astronomy** - Preliminary investigation of the moon; and measurement of spectra, especially in the ultraviolet and X-ray regions, including the brightness and positions of interesting regions of the sky.

    g. **Bio Sciences** - Investigation of the effects of outer space on living organisms, especially those which have most application to the manned exploration of space.

    h. **Geodesy** - Measurement of the size and shape of the earth, and location of land masses and water.

17. Future possibilities for scientific research and exploration include: continuation on a more sophisticated basis of the measurements of atmospheres, ionospheres, electric and magnetic fields, and expansion of such measurements to Mars and Venus and ultimately throughout the solar system; astronomical observations from points beyond the earth's atmosphere; manned and unmanned exploration of the moon and the planets; advance experiments designed to test certain predictions of the theory of relativity and other theories relating to the fundamental nature of the universe; investigation of the occurrence of biological phenomena in outer space.

[6] **Operational Applications of Space Technology**

18. All applications of the technology of outer space that now show promise of early operational utility for military or civilian purposes are based on the earth satellite. These applications ultimately will have to meet one of several criteria if they are to survive in either the defense program or the civilian economy. They will have to make possible the

---

1. Sounding rockets have also been defined as those vertically launched rockets that do not penetrate outer space beyond one earth radius, approximately 4000 statute miles.

more efficient operation of an existing activity or create a new and desirable activity. It is expected that benefits will be gained from these applications, but the full extent of their military, economic, political and social implications has yet to be determined. Military applications are designed to enhance military capabilities by fulfilling stated requirements of the Military Services and are currently being developed for use as operational systems. The applications that are expected to be available earliest are as follows:[2]

    a. **Meteorology** - Satellite systems to provide weather data on a global scale, making use of such techniques as television, optics, infrared detectors and radar. Information on cloud cover, storm locations, precipitation, wind direction, heat balance and water vapor would permit improved weather forecasting, including storm warnings, useful in a variety of civil activities such as agricultural, industrial and transportation activities, and would provide weather information to meet military operational needs.

    b. **Communications** - Satellite systems to improve and extend existing world-wide communications. For the Military Services, such systems would provide more effective global military communications for purposes of command, control, and support of military forces. Civil applications will benefit through more prompt service, increased message capacity, and greater reliability. Direct world-wide transmission of voice and video signals is envisaged.

    c. **Navigation** - Satellite systems to provide global all-weather capability, for land, sea and air vehicles, which will permit accurate determination of position; in the case of the military, secure operations would be possible.

[Paragraph excised during declassification review]

[7] **Manned Space Flight and Exploration**

20. It is expected that manned space flight will add significantly to the effectiveness of many of the scientific, military and civil applications indicated in the foregoing paragraphs. There are a number of important reasons why manned space activities, including the initial step of placing a man in orbit, are being carried out. Primary among these are:

    a. To the layman, manned space flight and exploration will represent the true conquest of outer space. No unmanned experiment can substitute for manned exploration in its psychological effect on the peoples of the world.

    b. Man's judgment, decision-making capability, and resourcefulness will ultimately be needed in many instances to ensure the full exploitation of space technology.

Moreover, manned space flight is required for scientific studies in which man himself is the principal subject of the experiment, because there is no substitute for the conduct in outer space of essential psychological and biological studies of man.

[Footnote excised during declassification review]

[8]   **International Principles, Procedures and Arrangements**

21. National policies and international agreements have dealt extensively with "air space" and expressly assert national sovereignty over this region; however, the upper limit of air space has not been defined. The term "outer space" also has no accepted definition, and the consequences of adopting a definition cannot now be fully anticipated. Although an avowedly arbitrary definition might prove useful for specific purposes, most of the currently foreseeable legal problems of outer space may be resolved without a precise line of demarcation between air space and outer space.

22. The U.S. has advanced and a number of states have accepted the view that outer space is not wholly without law inasmuch as the United Nations Charter and the Statute of the International Court of Justice are not spatially limited. Furthermore, the principles and procedures developed in the past to govern the use of air space and also the sea may provide useful analogies. However, many problems of outer space will be unique in character.

    2. Order of listing does not indicate anticipated order of availability.

23. An initial problem, in which all states have an interest, involves the permissibility of various activities in outer space. With respect to this problem, the report of the United Nations *Ad Hoc* Committee on the Peaceful Uses of Outer Space expresses the following view which the U.S. has supported:

"During the International Geophysical Year 1957-58 and subsequently, countries through the world proceeded on the premise of the permissibility of the launching and flight of the space vehicles which were launched, regardless of what territory they passed over during the course of their flight through outer space. The Committee, bearing in mind that its terms of reference refer exclusively to the peaceful uses of outer space, believes that, with this practice, there may have been initiated the recognition or establishment of a generally accepted rule to the effect that, in principle, outer space is, on conditions of equality, freely available for exploration and use by all in accordance with existing or future international law or agreements."

In this connection, it should be noted that definitions of "peaceful" or "non-interfering" uses of outer space have not been advanced by the United States or other states.

24. Although the U.S. has not to date recognized any upper limit to it sovereignty, a principle of freedom of outer space, such as that expressed by the United Nations *Ad Hoc* Committee, suggests [9] that at least insofar as peaceful exploration and use of outer space are concerned, the right of states to exclude persons and objects may not obtain. However, the full implications of a principle of freedom of outer space, in contrast with a principle of national sovereignty over outer space, remain to be assessed fully.

25. It is possible that certain military applications of space vehicles may be accepted as peaceful or acquiesced in as not-interfering. On the other hand, it may be anticipated that states will not willingly acquiesce in unrestricted use of outer space for activities which may jeopardize or interfere with their national interests.

26. There is frequent and sharpening concern on the part of world opinion over the military implications of unchecked competition in outer space between the U.S. and the Soviet Union, and there is an accompanying interest in international agreements, controls or restrictions to limit the dangers felt to stem from such competition. With regard to the armaments control aspects of outer space, the United States first proposed in 1957, in connection with international consideration of an armaments control system, that a multilateral technical committee be set up to attempt to design an inspection system to ensure that the sending of objects through outer space will be exclusively for peaceful purposes. Furthermore, the United States has offered, if there is a general agreement to proceed with this study without awaiting the conclusion of negotiations on other substantive disarmament proposals. There has not, to date, been multilateral agreement to proceed with such a study, and U.S. policy has not been determined concerning either the scope of control and inspection required to ensure that outer space could be used only for peaceful purposes or the relationship of any such control arrangement to other aspects of an arms agreement.[3]

27. Exploration and use of celestial bodies require separate consideration. Neither the U.S. nor any other state has yet taken a position regarding the questions of whether a celestial body is capable of appropriation to national sovereignty and if so what acts would suffice to found a claim thereto. It is clear that serious problems would arise if a state claimed, on one ground or another, exclusive rights over all or part of a celestial body. At an appropriate time some form of international arrangement may prove useful.

28. Other problems in which all states have an interest arise from the operation of space vehicles. The following problems appear [10] amenable to early treatment with a view to seeking internationally a basis for orderly accomplishment of space vehicle operations: (a) identification and registration of space vehicles; (b) liability for injury or damage caused by space vehicles; (c) reservation of radio frequencies for space vehicles and the related problem of termination of transmission; (d) avoidance of interference between space vehicles and aircraft; and (e) the re-entry and landing of space vehicles, through

---

3. Basic national security policy with respect to disarmament is stated in paragraph 52 of NSC 5906/1.

accident or design, on the territory of other states.

29. Although only a few states may be capable of mounting comprehensive outer space efforts, many states are capable of participating in the conduct of outer space activities, and active international cooperation in selected activities offers scientific, economic, and political opportunities. Continuation and extension of such cooperation in the peaceful uses of outer space through a variety of governmental and non-governmental arrangements should further enhance the position of the United States as the leading advocate of the exploration and use of outer space for the benefit of all. Where space vehicles are employed for military applications, some degree of international cooperation may also prove useful. Any international arrangements for cooperation in outer space activities may require determination of the net advantage to U.S. security.

30. The role most appropriately undertaken by the United Nations with respect to the foregoing matters appears to lie in performing two principal functions: (a) facilitating international cooperation in the exploration and use of outer space, and (b) providing a forum for consultation and agreement respecting international problems arising from outer space activities. Future developments may make it desirable for additional functions to be performed by or under the auspices of the United Nations.

Objectives

31. Carry out energetically a program for the exploration and use of outer space by the U.S., based upon sound scientific and technological progress, designed: (a) to achieve that enhancement of scientific knowledge, military strength, economic capabilities, and political position which may be derived through the advantageous application of space technology and through appropriate international cooperation in related matters, and (b) to obtain the advantages which come from successful achievements in space.

[11]   Policy Guidance

Priority, Scope and Level of Effort

32. While relating the resources and effort to be expended on outer space activities to other programs to ensure that the anticipated gains from such activities are properly related to possible gains from other programs which may be competitive for manpower, facilities, funds or other resources, commit and effectively apply adequate resources with a priority sufficient to enable the U.S. as soon as reasonably practicable to achieve the objectives stated in paragraph 31.

33. In addition to undertaking necessary immediate and short-range activities related to outer space, develop goals and supporting plans for outer space activities for the longer range, through at least a ten-year period.

34. Study on a continuing basis the implications and possible consequences which United States and foreign exploitation of outer space may hold for international and national political and social institutions. Critically examine such exploitation for possible consequences on activities and on life on earth (e.g., the use of nuclear energy for auxiliary or main power sources or for other applications in outer space which may affect health, or other outer space activities which may affect weather or other factors relating to activities and life on earth).

35. Periodically evaluate and compare the space activities of the U.S. and USSR with a view to determining, insofar as possible, the goals and relative rate of progress of each country's program.

Psychological Exploitation

36. To minimize the psychological advantages which the USSR has acquired as a result of space accomplishments, select from among those current or projected U.S. space activities of intrinsic military, scientific or technological value, one or more projects which offer promise of obtaining a demonstrably effective advantage over the Soviets and, so far as is consistent with solid achievements in the overall space program, stress these projects

in present and future programming.

37. Identify, to the greatest extent possible, the interests and aspirations of other Free World nations in outer space with U.S.-sponsored activities and accomplishments.

[12] 38. Develop information programs that will exploit fully U.S. outer space activities on a continuing basis; especially develop programs to counter overseas the psychological impact of Soviet outer space activities and to present U.S. outer space progress in the most favorable light.

[Paragraphs 39 and 40 excised during declassification review]

Manned Space Flight

41. Starting with the recovery from orbit of a manned satellite, proceed as soon as reasonably practicable with manned space flight and exploration.

International Principles, Procedures and Arrangements

42. Continue to support the principle that, insofar as peaceful exploration and use of outer space are concerned, outer space is freely available for exploration and use by all, and in this connection: (a) consider as a possible U.S. position the right of transit through outer space of orbital space vehicles or objects not equipped to inflict injury or damage; (b) where the U.S. contemplates military applications of space vehicles and significant adverse international reaction in anticipated, seek to develop measures designed to minimize or counteract such reaction; and (c) consider the usefulness of international arrangements respecting celestial bodies.

43. Taking into account, among other factors, the relationship of outer space capabilities to the present and future security position of the United States:

  a. Study the scope of control and character of safeguards required in an international system designed to assure that outer space be used for peaceful purposes only; include in this study an assessment of the technical feasibility of a positive enforcement system and an examination of the possibility of multilateral or international control of all outer space activities.

  b. Study the relationship between any international [13] arrangement to assure that outer space be used for peaceful purposes only and other aspects of the regulation and reduction of armed forces and armaments.

  c. In connection with the prosecution of studies enumerated in 43 (a) and (b), give full consideration to the requirements of U.S. security interests.

44. In the interest of establishing an international basis for orderly accomplishment of space flight operations, explore the desirability of and, where so indicated, seek international agreement on such problems as: (a) Some form of identification and registration of space vehicles which is to the net advantage to national security; (b) liability for injury or damage caused by space vehicles; (c) reservation of radio frequencies for space vehicles and the related problem of termination of transmission; (d) avoidance of interference between space vehicles and aircraft; and (e) the re-entry and landing of space vehicles, through accident or design, on the territory of other nations.

45. Seek to increase international cooperation in selected activities relating to the peaceful exploration and use of outer space by such means as: (a) Arrangements within the framework of the international scientific community including the Committee on Space Research (COSPAR) of the International Council of Scientific Unions, and (b) bilateral and multilateral arrangements between the U.S. and other countries including the Soviet Union. International arrangements for cooperation in outer space activities should consider the net advantage to U.S. security.

46. Support the United Nations in facilitating international cooperation in the exploration and use of outer space and in serving as a forum for consultation and agreement respecting international problems arising from outer space activities.

47. Develop means and take appropriate measures to ensure that the U.S. leads the USSR in making the scientific and technological information from its outer space program available to the world at large.

Security Classification

48. In implementing security classification regulations, take special account of the lead achieved by the USSR in outer space activities and the advantages to the U.S. which result from the maximum availability and use of scientific and technological information and material.

[15] ANNEX A

The Soviet Space Program

1. **Soviet Objectives**: The USSR has announced that the objective of its space program is the attainment of manned interplanetary travel. At present, the program appears to be directed toward the acquisition of scientific and technological data which would be applicable to Soviet space activities, their ICBM program, and basic scientific research. While the space program was undoubtedly initiated to serve scientific purposes, one of the primary underlying motivations which continues to give it impetus is the promise of substantial world-wide political and psychological gains for the USSR. Military considerations may have little bearing on the decision to develop certain types of space vehicles, although the successful development of these vehicles may result in military applications. Thus, it can be concluded that the Soviet space program has four major objectives. These objectives will have varying priorities as the program itself progresses and as new political and military requirements develop:

   a. Manned space travel
   b. Scientific research
   c. Propaganda
   d. Military applications

Of the above, it appears now that flight test priority has been on the scientific and propaganda objectives rather that on man-in-space or military applications.

2. **Background**: Russian interest in space flight dates back to 1903 when a scientific paper was published entitled: "Investigation of Universal Space by Means of Rocket Flight," by eminent Russian scientist Tsiolkovsky. Several other Russian actions took place during the succeeding years to the present which have been identified as at least partially associated with a space program. These have included the founding of the Soviet Institute of Theoretical Astronomy in 1923, establishment in 1934 of a government-sponsored rocket research program, flights of animals in vertical rockets since the early 1950's, and systematic investigations of moon flight problems starting in 1953. The establishment in early 1955 of the Interagency Commission for Interplanetary Communications was indicative of the Soviet realization that theory and capability for space flight were both feasible and that accomplishment of a long cherished ambition was within sight.

[16] 3. **Priority**: The Soviets have demonstrated that they are conducting a well-planned space flight program. The importance attached to this program is illustrated by the high quality of the scientists assigned to its direction, by the broad range of facilities and specialists engaged in its implementation, and by the wealth of theoretical and applied research being conducted in its support. However, the numbers of space vehicles actually launched over the past few years have not been as numerous as had been expected and it is apparent that their actual flight program is proceeding at a fairly deliberate pace. While there is no direct evidence on the priority of the overall Soviet space program vis-a-vis the military missile program, it is believed that any interference between the two would be resolved in favor of the missile program. To date, however, there is no indication that the space program has interfered with the missile program.

4. **Capabilities**: The Soviet Union dramatically demonstrated its interest and capability in space flight with the orbiting of two earth satellites in the fall of 1957, and a third in 1958. These were followed by the launching of three lunar associated vehicles in 1959.

Evidence indicates that the Soviet space program has been built on the foundation of military rocketry and guidance systems, with military and other facilities probably engaged dually in supporting tests of military ballistic missiles and space experiments. Thus, although these first space flights were doubtlessly undertaken for the furtherance of scientific knowledge and for whatever psychological and political advantage would accrue, the Soviet military, by intimate participation of its hardware, personnel, and facilities, has been in a position to utilize immediately such knowledge for the enhancement of the Soviet military position and objectives. The realization of more advanced space projects, particularly those involving manned flight, must be preceded by a vast amount of scientific and technological work directed towards the development of useable space vehicles, the determination of basic operational requirements and limitations, and the creation of an environment and equipment capable of sustaining human life in outer space. Since such a program embraces virtually all fields of science and engineering, the following areas were particularly examined for evidence of Soviet technical capability: guided missiles (including vertical rocket launchings), re-entry vehicles and techniques, propulsion, guidance, communications, space medicine, internal power supplies, and celestial mechanics. While firm association of these fields with a space program varied considerably, it is noted that the state of Soviet art in all the sciences required in a space program is such that no scientific or technical barriers of magnitude have been noted. Four areas deemed critical to a space program have apparently received considerable attention by the USSR; e.g., development of large rocket-engine propulsion systems, vertical rocket flights with animals (including recovery devices), space medicine, and celestial mechanics. There are indications that Soviet advanced thinking and study [16] in astro-biology have been de-emphasized in favor of providing an artificial environment within a vehicle suitable for manned space flight.

5. **Future Capabilities**:

a. There is no firm evidence of Soviet future plans for the exploration of outer space with either unmanned or manned vehicles. It is believed they will continue and expand their scientific research with further unmanned earth satellites, lunar probes (including satellites and soft landings), solar and planetary probes. Manned experiments will probably be conducted in earth satellites, circumlunar flights and soft landings on the moon. It is expected that all manned flights into outer space will be preceded by similar tests with animals, unless for political purposes the Soviets attempt a high risk program. Man-in-space programs are confronted by many problems or hazards, the most immediate of which are recovery and life support over extended periods. While data which might lead to solutions or better understandings of both can be obtained from instrumented packages which are orbited and recovered, accomplishment of the same test animals would provide data of more direct application to subsequent attempts with man.

b. The dates estimated for specific Soviet accomplishments in space represent the earliest possible time periods in which each specific event could be accomplished. It is recognized that the various facets of the space flight program are in competition not only among themselves, but with other priority programs, and that the USSR probably cannot undertake all the space flight activities described below at the priority required to meet the time periods specified. At this time it cannot be determined which specific space flight activities enjoy the higher priorities and will be pursued first.

c. No attempt has been made to estimate manned space missions beyond the earth-moon realm. The time periods in which the successful development of sub-systems essential to planetary flight activities can be brought to fruition and integrated into a complete space flight system cannot be foreseen.

d. Similarly, considerations of military applications have been limited to earth orbiting types of space vehicles. Missions beyond this realm are considered only in the scientific or exploratory sense because we believe they cannot be successfully accomplished in the time period considered.

6. An estimate of a possible Soviet space development program is as follows:

[18] **Possible Soviet Space Development Program**

Space Program Objectives   First Possible
                                                                            Capability Rate

These dates represent the earliest possible time period in which each specific event could be accomplished. However, competition between the space program and the military missile program as well as within the space program itself makes it unlikely that all of these objectives will be achieved within the specified time periods.

Unmanned Earth Satellites
5000-10,000 pounds, low orbit satellites 1959
Recoverable (including biological) satellites 1959
Military Satellites: The dates shown are the earliest in which feasibility demonstrations could begin. Generally, militarily useful vehicles would be available 2-3 years after the feasibility demonstration.
[Remainder of paragraph excised during declassification review]

Unmanned Lunar Rockets
Biological Probe 1959
Satellite of the Moon 1959
Soft Landings 1960
Lunar Landing, Return and Earth Recovery 1963-1964

Planetary Probes
Mars .... about Oct. 1960
Venus .... about Jan. 1961

Manned Vertical or Short Down-Range Flight 1959

Manned Earth Satellites - The specified time periods for manned....accomplishments are predicated on the Soviets having previously successfully accomplished a number of similar unmanned ventures.
Capsule-type Vehicles [4] Mid-1960-mid-1961
Glide-type Vehicles .... 1 to 2 years
                                                                                                      after above
Maneuverable (minimum: conventional propulsion) 1963
Maneuverable (nuclear propulsion) .... about 1970
Space Platform (minimum, non-ecological, feasibility demonstration) 1965
Space Platform (long-lived) .... about 1970

Manned Lunar Flights
Circumlunar 1964-1965
Satellites (temporary) 1965-1966
Landings .... about 1970

---

4. Recovery would probably be attempted after the first few orbits but life could probably be sustained for about a week.

[19] ANNEX B

Estimated Funding* Requirements
Summary

	Fiscal Year				
	1960	1961	1962	1963	1964
NASA	524.0	802.0	1031.0	1171.0	1350.0
AEC	46.7	41.5	66.0	60.6	55.2
DEFENSE	438.8	480.7	747.5	750.0	728.0
	1054.5	1324.2	1844.5	1981.6	2133.2

*Figures are in millions of dollars.
More detailed agency estimates given on following pages.

[20]

Estimated Funding* Requirements
of
National Aeronautics and Space Administration

	Fiscal Year				
	1960	1961	1962	1963	1964
RESEARCH & DEVELOPMENT					
Launching Vehicle Dev.	57	140	163	230	375
Space Propulsion Technol.	39	51	118	120	90
Vehicle Systems Technol.	13	30	47	49	50
Manned Space Flight	87	108	120	135	180
Scientific Investig. in Space	82	95	140	145	150
Satellite Applications	11	27	36	60	75
Aeronautical and Space Res.	28	61	70	70	70
Space Flight Operations	16	33	42	50	55
SUB TOTAL - Research & Dev.	333	545	736	859	1045
CONSTRUCTION & EQUIPMENT	100	89	120	137	130
SALARIES & EXPENSES	91	168	175	175	175
TOTAL FUNDS REQUIRED	524	802	1031	1171	1350

*Figures are in millions of dollars.
These figures represent a level of effort which corresponds to an efficient and steadily growing capability. The rate of progress could be improved by an increased funding level, primarily by improving the certainty of the timely completion of the many essential engineering developments.

[21] **Estimated Funding* Requirements
of
Atomic Energy Commission**

PROJECT	1960	1961	Fiscal Year 1962	1963	1964
ROVER - Nuclear rocket	31.5	24.9	33	32.6	26.2
SNAP - Nuclear auxiliary power system	15.2	16.6	33	28	29
Total	46.7	41.5	66.0	60.6	55.2

*Figures are in millions of dollars

Document II-22

Document title: Cyrus Vance, Deputy Secretary of Defense, Department of Defense Directive Number TS 5105.23, "National Reconnaissance Office," March 27, 1964.

Source: Space Policy Institute, George Washington University, Washington, D.C.

In February 1958, President Eisenhower directed the CIA to develop a reconnaissance satellite. He did so because the Air Force's WS 117L was unlikely to be available soon because it relied on the still-untested Atlas rocket. The CIA began the development of a reconnaissance satellite for launch atop Thor IRBMs. The result was that the United States had two reconnaissance satellites in simultaneous development. In August 1960, the National Reconnaissance Office (NRO) was created to manage the development and operation of reconnaissance satellites. The existence of this office was highly classified and was formally established in DoD Directive Number TS 5105.23. The existence of the NRO was not officially acknowledged until September 1992. This document, dated March 27, 1964, is the earliest declassified charter for the NRO. Neither the earlier version of DoD Directive 5105.23 mentioned in this document nor the presidential directive—possibly a National Security Council memorandum—has been publicly released.

[1] **I. General**

Pursuant to the authority vested in the Secretary of Defense and the provisions of the National Security Act of 1947, as amended, including the Department of Defense Reorganization Act of 1958, a National Reconnaissance Office is hereby established as an operating agency of the Department of Defense, under the direction and supervision of the Secretary of Defense.

II. Organization and Responsibility

The National Reconnaissance Office will be organized separately within the Department of Defense under a Director, National Reconnaissance Office, appointed by the Secretary of Defense. The Director will be responsible for consolidation of all Department of Defense satellite and air vehicle overflight projects for intelligence into a single program, defined as the National Reconnaissance Program, and for the complete management and conduct of this Program in accordance with policy guidance and decisions of the Secretary of Defense.

III. Relationships

A. In carrying out his responsibilities for the National Reconnaissance Program, the Director, National Reconnaissance Office shall:
[2] 1. Keep the Director of Defense Research and Engineering and the Assistant Secretary of Defense (Comptroller) personally informed on a regular basis on the status of projects of the National Reconnaissance Program.
2. Similarly inform other Department of Defense personnel as he may determine necessary in the course of carrying out specific project matters.
3. Establish appropriate interfaces between the National Reconnaissance Office and the United States Intelligence Board, the Joint Chiefs of Staff, the Defense Intelligence Agency, and the National Security Agency.
4. Where appropriate, make use of qualified personnel of services and agencies of the Department of Defense as full time members of the National Reconnaissance Office.

B. Officials of the Office of the Secretary of Defense, military departments, and other DoD agencies shall provide support within their respective fields of responsibility, to the Director, National Reconnaissance Office as may be necessary for the Director to carry out his assigned responsibilities and functions. The Director, National Reconnaissance Office will be given support as required from normal staff elements of the military departments and agencies concerned, although these staff elements will not participate in these project matters except as he specifically requests.

IV. Authorities

A. The Director, National Reconnaissance Office, in connection with his assigned responsibilities for the [3] National Reconnaissance Office and the National Reconnaissance Program, is hereby specifically delegated authority to:
1. Organize, staff, and supervise the National Reconnaissance Office.
2. Establish, manage and conduct the National Reconnaissance Program.
3. Assist the Secretary of Defense in the supervision of aircraft and satellite reconnaissance photographic projects, and be his direct representative on these matters both within and outside the Department of Defense.
4. Review all Department of Defense budget requests and expenditures for any items falling within the definition of the National Reconnaissance Program, including studies and preliminary research and development of components and techniques to support such existing or future projects.

B. Other authorities specifically delegated to the Director, National Reconnaissance Office by the Secretary of Defense will be referenced in numbered enclosures to this directive.

V. Project Assignments

All projects falling within the definition of the National Reconnaissance Program are assigned to that program and will be managed as outlined herein unless specific exception is made by the Director, National Reconnaissance Office. Announcements of any such exceptions will be made by numbered enclosures to this directive.

[4]
VI. Security

A. The Director, National Reconnaissance Office will establish the security procedures to be followed for all matters of the National Reconnaissance Program, to protect all elements of the National Reconnaissance Office.
B. All communications pertaining to matters under the National Reconnaissance Program will be subject to special systems of security control under the cognizance of the

Director, Defense Intelligence Agency, except in those instances specifically exempted by either Director, National Reconnaissance Office or the Secretary of Defense.

C. With the single exception of this directive, no mention will be made of the following titles or their abbreviations in any document which is not controlled under the special security control system(s) referred to in B. above: National Reconnaissance Program; National Reconnaissance Office. Where absolutely necessary to refer to the National Reconnaissance Program in communications not controlled under the prescribed special security systems, such reference will be made by use of the terminology: "Matters under the purview of DoD TS-5105.23."

VII. Effective Date

This Directive is effective upon publication.

VIII. Cancellation

Reference (a) is hereby cancelled.

Cyrus Vance
Deputy Secretary of Defense

Chapter Three

The Evolution of U.S. Space Policy and Plans

by John M. Logsdon

The July 1958 National Aeronautics and Space Act [II-17] and the statement of "Preliminary U.S. Policy for Outer Space" adopted by the Eisenhower administration in August 1958 [II-20] provided the framework within which the new space agency, the National Aeronautics and Space Administration (NASA), planned an initial set of programs and projects. In the subsequent thirty years, the interactions among formal statements of national space policy, various presidential decisions on specific space undertakings, and project proposals emanating from internal NASA planning have resulted in an evolving U.S. civilian space program that has been able to meet the Space Act mandate of making the United States "a leader" in space. This essay provides an overview of those interactions as they are manifested in key policy documents included in this work.[1]

The Early Years: Space Policy and Planning, 1959-1960

NASA's initial plan of activity was based on preexisting programs inherited from the Department of Defense and NASA's predecessor organization, the National Advisory Committee for Aeronautics (NACA). [III-1] Within a few months after it began operations, the agency had shaped this inheritance into a short-term program, and its new leaders, Administrator T. Keith Glennan, who had come to NASA from his position as president of the Case Institute of Technology in Cleveland, and Deputy Administrator Hugh L. Dryden, who had been the NACA Director, set about the task of developing their own plans for the new space agency.[2]

By the end of 1959, just over a year after NASA began operations, the space agency had prepared a formal long-range plan. [III-2] The plan noted that NASA's activities during the 1960s "should make feasible the manned exploration of the moon and nearby planets, and this exploration may thus be taken as a long-term goal of NASA activities." The plan called for the "first launching in a program leading to manned circumlunar flight and to a permanent near-earth space station" in the 1965-1967 period. It also called for the first human flight to the Moon sometime "beyond 1970."[3] Although the first NASA long-range plan featured a balanced program of science, applications, and human spaceflight, from the start proposals for the future of the piloted portion of its activities excited the public, and NASA, therefore, became the focal point of future thinking.

---

1. This brief essay cannot purport to be a comprehensive history of the evolution of U.S. space policy; that would require at least book-length treatment. (Such a book does not now exist.) Rather, this essay attempts to put the documents selected for this section of the work in their historical context, so that they can be understood in terms of where they fit in the overall development of U.S. space policy and plans.

2. On Glennan's career, see J.D. Hunley, ed., *The Birth of NASA: The Diary of T. Keith Glennan* (Washington, DC: NASA SP-4105, 1993).

3. Office of Program Planning and Evaluation, "The Long Range Plan of the National Aeronautics and Space Administration," December 16, 1959, NASA Historical Reference Collection, NASA History Office, NASA Headquarters, Washington, DC.

Making human flights to the Moon and planets the stated long-range goal of the NASA program was, to say the least, controversial within the Eisenhower administration. NASA planners since early 1959 had been investigating the appropriate focus for NASA's human spaceflight program, once the initial Project Mercury had demonstrated the ability of the human body to withstand the rigors of launch, weightlessness, and then reentry. In particular, a group chaired by Harry J. Goett, Director of the new Goddard Space Flight Center in Greenbelt, Maryland, concluded by mid-1959 that the appropriate goal for NASA's post-Mercury human spaceflight program was to send humans to the Moon, not just for extended stays in Earth orbit.[4] Even after the lunar landing goal had been included in NASA's long-range plan, it continued to be debated within the top councils of NASA, but by early 1960 a decision was made to proceed with a lunar expedition as a major element in NASA's future planning.

The fact that NASA was contemplating sending people to the Moon did not escape the notice of those advising President Dwight D. Eisenhower on overall science and technology issues. By the end of 1960, an Ad Hoc Panel on Man-in-Space of the President's Science Advisory Committee (PSAC), chaired by chemist Donald F. Hornig of Brown University, had completed its own investigation of NASA's post-Mercury planning. [III-3] The panel concluded that "at the present time. . .man-in-space cannot be justified on purely scientific grounds. . . . On the other hand, it may be argued that much of the motivation and drive for the scientific exploration of space is derived from the dream of man's getting into space himself." The group also estimated that landing humans on the Moon would cost $26-38 billion above the $8-9 billion cost of Earth-orbiting and circumlunar flights, and could not be accomplished until after 1975.[5] The report of the PSAC panel was presented to President Eisenhower at a December 20, 1960, meeting of the National Security Council; Eisenhower's reaction was that "he couldn't care less whether a man ever reached the moon."[6]

At the end of the Eisenhower administration, then, NASA had in place a long-range plan that anticipated a budget gradually increasing during the 1960s to a $2.5 billion level in 1960 dollars, but without White House approval for its centerpiece activities—a post-Mercury program of human spaceflight aimed at an eventual lunar landing and the development of the large boosters required for such a program. Dwight Eisenhower recognized that the United States was in a space race with the Soviet Union, but he was not interested in winning that race at any cost. His attitude is best captured by the policy guidance provided in a January 1960 comprehensive statement of U.S. space policy. [II-21] That statement directed planners to, in orde

To minimize the psychological advantages which the USSR has acquired as a result of space accomplishments, select from among those current or projected U.S. space activities of intrinsic military, scientific or technological value, one or more projects which offer promise of obtaining a demonstrably effective advantage over the Soviets and, so far as is consistent with solid achievements in the over-all space program, stress these projects in present and future programming.[7]

To President Eisenhower, a race to be first on the Moon did not meet the requirement of "intrinsic value," and he was unwilling to approve future human spaceflight efforts that were steps in the direction of such an undertaking. This left NASA in a state of high uncertainty about its future prospects—uncertainty that was resolved within a few months by the new president, John F. Kennedy.

---

4. The work of the Goett Committee is discussed in John M. Logsdon, *The Decision to Go to the Moon: Project Apollo and the National Interest* (Cambridge, MA: MIT Press, 1970), pp. 56-57.
5. The President's Science Advisory Committee, "Report of the Ad Hoc Panel on Man-in-Space," December 16, 1960, pp. 6, 9, NASA Historical Reference Collection.
6. Hunley, ed., *The Birth of NASA*, pp. 292-93.
7. National Aeronautics and Space Council, "U.S. Policy on Outer Space," January 26, 1960, p. 11, paragraph 36, NASA Historical Reference Collection.

The Decision to Go to the Moon[8]

However, this uncertainty was not reduced by the first indication of the posture the new administration might take with respect to the civilian space program. [III-4] After his narrow victory over Richard M. Nixon in November 1960, Kennedy had formed a "transition team" to assess the national space effort. That team was headed by Jerome B. Wiesner, a Massachusetts Institute of Technology physicist who was slated to become Kennedy's science adviser. The "Wiesner Report" [III-5], released on January 10, 1961, was very critical of the quality and technical competence of NASA management and of the emphasis that had been placed on human spaceflight. It called the Mercury program "marginal" because of the limited power of its Atlas booster and criticized the priority given to the Mercury program for strengthening "the popular belief that man in space is the most important aim of our non-military space effort."[9] As the new president took office on January 20, the future course for NASA remained unclear.

Events of the next few months, however, made most of the recommendations of the Wiesner Report moot. President Kennedy announced on January 31 that James E. Webb, a politically skilled and aggressive lawyer-administrator, would become NASA Administrator, and that other senior members of NASA management would remain. Webb took office on February 14, and within six weeks met with the president with a request for a significant acceleration of the NASA program, with an emphasis on larger boosters and a new spacecraft for human spaceflight. Kennedy deferred a decision on Webb's request until the fall of 1961 deliberations on his next budget.

Then, on April 12, the Soviet Union launched the first human, Yuri Gagarin, into orbit. World and domestic reaction to the achievement was universally positive, and within a few days President Kennedy decided that the United States had to not only accept the challenge to a space race put forth by the Soviet Union, but also to enter the race with an intent to win. On April 20, he asked Vice President Lyndon B. Johnson to conduct an "overall survey of where we stand in space." [III-6] In particular, Kennedy asked: "Do we have a chance of beating the Soviets by putting a laboratory in space, or by a trip around the moon, or by a rocket to go to the moon and back with a man? Is there any other space program that promises dramatic results in which we could win?" The president asked for a report on Johnson's findings "at the earliest possible moment."[10]

The review was carried out under the auspices of the National Aeronautics and Space Council, which Kennedy had decided the vice president should chair. Even before the review began, in reaction to presidential guidance at an April 14 White House meeting, NASA had been examining the feasibility and costs of an accelerated civilian space effort. The Department of Defense provided its initial input on April 21. [III-7] Vice President Johnson asked a number of individuals inside and outside government for their views, including rocket engineer and space exploration visionary Wernher von Braun. [III-9] By April 28 the vice president could report to Kennedy [III-8] that "dramatic accomplishments in space are being increasingly identified as a major indicator of world leadership." He added that "if we do not make a strong effort now, the time will soon be reached when the margin of control over space and over men's minds through space accomplishments will have swung so far on the Russian side that we will not be able to catch up, let alone assume leadership," and that "manned exploration of the moon...is not only an achievement with great propaganda value, but it is essential as an objective whether or not we are first in its accomplishment—and we may be able to be first."[11]

---

8. In addition to the specific sources cited below, this account is based on Logsdon, *Decision to Go to the Moon*.

9. "Report to the President-Elect of the Ad Hoc Committee on Space," January 10, 1961, NASA Historical Reference Collection. Quotes are from p. 9 and p. 16.

10. John F. Kennedy, Memorandum for Vice President, April 20, 1961, Presidential Files, John F. Kennedy Presidential Library, Boston, MA.

11. Lyndon B. Johnson, Vice President, Memorandum for the President, "Evaluation of Space Program," April 28, 1961, p. 2, Presidential Papers, Kennedy Presidential Library.

In a meeting on May 3, the vice president brought together a diverse group that included Robert Kerr (D-OK), Chairman of the Senate Committee on Aeronautical and Space Sciences, and the ranking minority member on the committee, Styles Bridges (R-NH), together with others who had been involved in the space review. The primary purpose of the meeting was to ensure that the Senate would support an accelerated space program, but the minutes of the proceedings capture well the state of the debate at that point. [III-10] Of all the participants, NASA Administrator Webb seemed most hesitant to move quickly ahead with a set of ambitious recommendations to the president. Webb apparently wanted to make sure that the White House and Congress would commit the multi-year support needed to carry out those recommendations before he would advocate them to the president.

Webb's hesitation was quickly overcome. On Friday, May 5, the United States launched its first human, Navy Lt. Cdr. Alan B. Shepard, on a suborbital flight, to great acclaim. On the same day, Vice President Johnson learned that he would be leaving the next week for an inspection tour of U.S. military capabilities in Southeast Asia, and requested NASA and the Department of Defense to put together a recommendation to the president before he left. After an intense weekend of work, a report, delivered with a cover letter from James Webb and Secretary of Defense Robert S. McNamara, was ready by Monday morning, May 8. [III-11] The Vice President approved the report without change and transmitted it to the president; he in turn accepted the report's recommendations at a May 10 White House meeting.

The report called for a fundamental reversal of a space policy principle that had been established under President Eisenhower—that competition with the Soviet Union would be based only on projects that had other elements of "intrinsic merit." Rather, it argued that

Major successes, such as orbiting a man as the Soviets have just done, lend national prestige even though the scientific, commercial or military value of the undertaking may by ordinary standards be marginal or economically unjustified.

This nation needs to make a positive decision to pursue space projects aimed at enhancing national prestige. Our attainments are a major element in the international competition between the Soviet system and our own. The non-military, non-commercial, non-scientific but "civilian" projects such as lunar and planetary exploration are, in this sense, part of the battle along the fluid front of the cold war.[12]

The report called for an across-the-board acceleration of U.S. efforts in space and strong central planning for an integrated national space program. It noted that "we are uncertain of Soviet intentions, plans or status" with respect to sending humans to the Moon, but, because "it is man, not merely machines, in space that captures the imagination of the world," the United States should commit itself to a lunar landing program, even though it was not sure it could beat the Soviets to the Moon, because "it is better for us to get there second than not at all."[13]

President John F. Kennedy announced his decision to go to the Moon in what was billed as an unprecedented second "State of the Union" address to a joint session of Congress on May 25, 1961. [III-12] His speech had gone through a number of drafts, with one issue being whether the president should announce 1967, the fiftieth anniversary of the Russian Revolution and a date widely thought to be timely for a Soviet spectacular in space, as the intended date for the first lunar landing. His advisers convinced him that he

---

12. James E. Webb, Administrator, NASA, and Robert S. McNamara, Secretary of Defense, to the Vice President, May 8, 1961, with attached: "Recommendations for Our National Space Program: Changes, Policies, Goals," p. 8, NASA Historical Reference Collection.
13. *Ibid.*, pp. 13-14.

should allow some margin for unexpected delays, and so he called for the lunar landing "before this decade is out." He told Congress and the nation that "now it is time to take longer strides—time for a great new American enterprise—time for this nation to take a clearly leading role in space achievement." Kennedy added in his own hand to the prepared text the words—"which in many ways may hold the key to our future on earth."[14]

Congress quickly accepted the president's call for a more than half-billion dollar supplement to NASA's budget, and with that acceptance gave initial support to the policy of seeking across-the-board leadership—preeminence—in space. Project Apollo, NASA's lunar landing program, became the dominant feature of the U.S. quest for space leadership. Its impacts on the evolution of the U.S. space program were to be pervasive in the decades to come.

Reviewing the Apollo Commitment

President John Kennedy chose to go to the Moon as a response to the political situation in early 1961, not primarily in terms of a long-range vision for the U.S. space program. While Kennedy personally may have come to see space as a particularly important sphere of future-oriented activities, almost from the day that the president announced his intent to set the lunar landing goal, others inside and outside government questioned the wisdom of this commitment. This questioning became more vocal in 1962 and 1963, as the United States forced the Soviet Union to withdraw nuclear missiles from Cuba in October 1962 and as the two nuclear superpowers agreed on a limited test ban treaty and appeared headed toward less tension in their geopolitical rivalry. By September 1963, President Kennedy was ready to go before the United Nations and suggest an end to the space race and the conversion of Apollo into a cooperative U.S.-Russian program.[15]

These public manifestations of possible instability in the U.S. commitment to Apollo and to a preeminent space program were accompanied by major White House reviews in 1962 and 1963. The 1962 review [III-13] was precipitated by the very rapid buildup of the NASA budget, increases in the estimated cost for Apollo, and pressure to *accelerate* the planned date for the initial lunar landing by the individual NASA had chosen to head the Apollo program, D. Brainard Holmes. The controversy over whether Apollo should be carried out on an all-out, "crash" basis or in relative balance with other elements of a program aimed at U.S. space superiority went to President Kennedy for resolution in November 1962. In a follow-up letter summarizing the arguments he had made during his meeting with the president on the question [III-14], NASA Administrator Webb argued that "the objective of our national space program is to become pre-eminent in all important aspects of this endeavor and to conduct the program in such a manner that our emerging scientific, technological, and operational competence is clearly evident." He told the president that "the manned lunar landing program, although of highest national priority, will not by itself create the pre-eminent position we seek."[16]

Based on this reasoning, Webb recommended against providing additional funds for Apollo and moving up the planned date for the first lunar landing. Kennedy accepted Webb's perspective, and soon after Brainard Holmes left NASA. Perhaps more fundamental, the president's acceptance seemed to indicate that across-the-board preeminence was indeed his guiding policy objective for the United States in space.

The 1963 space program review, by contrast, appears to have been stimulated by increasing external criticism of the priority being given to the space program rather than other areas of science and technology, and was focused on those aspects of the program

---

14. President John F. Kennedy, Excerpts from "Urgent National Needs," Speech to a Joint Session of Congress, May 25, 1961, Presidential Files, Kennedy Presidential Library.

15. Representative Thomas Pelly, a Seattle Republican, stood up in the well of the House of Representatives three weeks later and offered an amendment to prohibit the use of government funds to finance a joint expedition. In spite of Kennedy's insistence that his U.N. proposal merely carried out the mandate for international cooperation in NASA's enabling legislation, the amendment passed. See "Major Legislation—Appropriations," *Congressional Quarterly Almanac 1963* (Washington, DC: Congressional Quarterly, 1964), p. 170.

16. James E. Webb, Administrator, NASA, to the President, November 30, 1962, NASA Historical Reference Collection.

not linked to Apollo. In April President Kennedy asked the vice president and the Space Council to conduct a comprehensive review so that he could "obtain a clearer understanding of a number of factual and policy issues relating to the National Space Program which seem to rise repeatedly in public and other contexts."[17] [III-15]

Leading the criticisms of the space program (including Apollo) were many in the scientific and educational communities.[18] For example, Vannevar Bush, who had been a primary architect of the relationship between government and science since World War II, wrote to James Webb in April 1963 with a comprehensive critique of the space effort. Bush argued "that the [space] program, as it has been built up, is not sound," that it was "more expensive than the country can now afford," and that "its results, while interesting, are secondary to our national welfare."[19]

The vice president transmitted the results of the space program review to President Kennedy on May 13, 1963. [III-16] He noted that the central difference between the Eisenhower and Kennedy administration's space programs was that "the plan of the previous Administration represented an effort for a second place runner and the program of the present Administration is designed to make this country the assured leader before the end of the decade." He argued that "our space progr[a]m has an overriding urgency that cannot be calculated solely in terms of industrial, scientific, or military development. The future of society is at stake." Johnson also told the president that all members of the Space Council concurred with the views contained in his report.[20] The president apparently agreed, for the speech he had prepared for delivery in Dallas on November 22, 1963, reaffirmed the administration's strong support for the leadership-oriented space effort he had initiated 2 1/2 years earlier.

Post-Apollo Planning During the Johnson Presidency

There was no comprehensive, presidentially approved statement of national space policy while John Kennedy or Lyndon Johnson were president, as there had been under Dwight Eisenhower; although the staff of the National Aeronautics and Space Council drafted such a document, it never received presidential sanction.[21] After the 1962 and 1963 reviews and the assassination of President Kennedy, any chance of a policy reversal that would downgrade the objective of making the United States first to the Moon disappeared, and the planning focus shifted to what objectives the country should pursue in space after Apollo. On January 30, 1964, President Johnson asked NASA for "a statement of possible objectives beyond those already approved."[22]

James Webb was very skeptical of having NASA come forward with a proposal for its future and then seeking political support for it; rather, he preferred that NASA wait for a "consensus" (which he defined as agreement among politically powerful actors) to form on future objectives for space. Then NASA could develop programs to achieve those objectives.[23] Thus the NASA response to Lyndon Johnson's 1964 request, which was transmitted to the president in February 1965, was not a long-range plan. [III-17] Rather, it described

---

17. John F. Kennedy, Memorandum for the Vice President, April 9, 1963, Presidential Papers, Kennedy Presidential Library.
18. For a sample of the criticisms of Apollo that emerged in the 1962-1964 period, see Amitai Etzioni, *The Moondoggle: Domestic and International Implications of the Space Race* (Garden City, NY: Doubleday & Co., 1964).
19. Vannevar Bush to James E. Webb, Administrator, NASA, April 11, 1963, p. 2, Presidential Papers, Kennedy Library.
20. Lyndon B. Johnson, Vice President, to the President, May 13, 1963, with attached report, Presidential Papers, Kennedy Library.
21. For a discussion of attempts to develop such a policy statement, see *The National Aeronautics and Space Council During the Tenure of Lyndon B. Johnson as Vice President and During His Administration as President* (January 1961-January 1969), a history prepared by the National Aeronautics and Space Council staff. Copy in Lyndon Baines Johnson Library.
22. Lyndon Johnson to James E. Webb, January 30, 1964, NASA Historical Reference Collection.
23. For Webb's views on long-range planning, see Arnold S. Levine, *Managing NASA in the Apollo Era* (Washington, DC: NASA SP-4102, 1982), particularly chapter 9 and Webb's foreword.

"a number of long-range missions that deserve serious attention."[24] The report was in many ways a catalogue of the ways that the capabilities developed during the Apollo buildup could be employed and a "wish list" of future mission possibilities, with no priority indicated among them.

James Webb's hoped-for consensus on future space objectives did not emerge before Webb left NASA in November 1968 near the end of the Johnson administration. While the White House and the majority of Congress seemed willing to sustain the commitment to Apollo, no major new programs were approved, and the NASA budget began to decline after peaking at $5.25 billion in fiscal year 1965. This situation was deeply troubling to Webb. In an August 1966 letter to President Johnson, he pointed out that NASA was already in the process of "liquidation of some of the capabilities we have built up." [III-18] Webb had received Bureau of the Budget guidelines for the next fiscal year that he believed meant that "important options which we have been holding open will be foreclosed." Recognizing the international and domestic pressures of the president, Webb said he had struggled "to try to put myself in your place and to see this from your point of view" but could not "avoid a strong feeling that this is not in the best interests of the country." Webb told the president of his problems with Congress; in order to avoid a $1 billion cut in the NASA budget proposed by Senator William Proxmire (D-WI), Webb had to seek the support of the Republican leader in the Senate, Everett Dirksen (R-IL). Senior Democrats, including Richard Russell (D-GA), chair of the Senate Space Committee, had voted for the cut, which Webb believed would have led to "catastrophic emasculation" of the NASA program.[25]

Through the remaining years of the Johnson presidency, Webb was not able to convince the president to articulate future objectives for the post-Apollo civilian space program; other issues occupied Lyndon Johnson's attention. Meanwhile, as he had said would be necessary, Webb began the process of dismantling the capabilities developed to send Americans to the Moon; in August 1968, he ordered the first steps in shutting down the production line for the giant Saturn booster developed for the lunar mission. [III-19] Decisions on the character of the post-Apollo space program would have to be made by Lyndon Johnson's successor, Richard M. Nixon. [III-20]

Post-Apollo Planning During the Nixon Presidency

When he was sworn in as president on January 20, 1969, Richard Nixon had available—as had John Kennedy—a report from a high-level transition task force on space. [III-21] That task force was chaired by Nobel Prize-winning physicist Charles Townes of the University of California. It recommended proceeding with lunar exploration after Apollo, a program to utilize Apollo hardware and capabilities, an increase in automated solar system exploration and in general better balance between the human spaceflight and automated elements of the NASA program, and more attention to the applications of space and space technologies to useful purposes. The report recommended against commitments to a large space station, low-cost boosters, or human trips to the planets. The task group felt that the program it recommended could be carried out for an annual NASA budget of approximately $4 billion.

President Nixon and his advisers, however, recognized a need for early decisions on the post-Apollo space program. On February 13, 1969, Nixon asked his vice president, Spiro T. Agnew, to chair a Space Task Group (STG) to provide "definitive recommendation on the direction which the U.S. space program should take in the post-Apollo

---

24. NASA, *Summary Report: Future Programs Task Group*, January 1965, p. ii, NASA Historical Reference Collection.

25. James E. Webb, Administrator, NASA, to the President, August 26, 1966, with attached: James E. Webb, Administrator, NASA, to Honorable Everett Dirksen, U.S. Senate, August 9, 1966, NASA Historical Reference Collection.

period." [III-22] Rather than use the National Aeronautics and Space Council, which was chaired by Agnew, as the basis for the review, the president named as the only other members of the STG the acting administrator of NASA, the secretary of defense, and the science adviser. He also asked the science adviser to act as "staff officer" for the group's review.[26]

NASA was not comfortable entrusting its future to the deliberations of the STG. The Nixon administration had not yet selected a NASA administrator, and the man who had been deputy to James Webb and had become acting administrator upon Webb's retirement in November 1968, Thomas O. Paine, was managing the agency. (Paine was selected as the Nixon choice to remain as administrator in March 1969.) Paine was a much different sort of individual than James Webb. He was optimistic, bullish in word and action, and a newcomer to Washington's political ways. Unlike Webb, Paine preferred that NASA have the initiative in outlining future space goals.

Paine decided to try to preempt the work of the STG and attempt to get an early commitment to what NASA saw as its major post-Apollo program, a large space station. The need for an orbital outpost had been part of NASA's planning from the earliest years, but the decision to go to the Moon, particularly using a lunar rendezvous approach, had bypassed this step in space development. As NASA began during 1967 and 1968 to focus attention on its priorities for the next large new program after Apollo, a space station rose to the top of its list. On February 26, Paine sent a lengthy memorandum on "Problems and Opportunities in Manned Space Flight" to the president. [III-23] He suggested that "the case that a space station should be a major future U.S. goal is now strong enough to justify at least a general statement on your part" to this effect.[27]

The White House did not accept Paine's arguments, and he was told that all decisions related to future programs would await the recommendations of the STG. Those recommendations took shape over the summer of 1969, with the U.S. space program at the peak of its prestige and accomplishment as Apollo 11 returned from humanity's first foray on another celestial surface.

Thomas Paine found in Vice President Agnew an ally in calling for a fast-paced, ambitious post-Apollo program. At an early STG meeting, Agnew had asked for an "Apollo for the seventies." In interviews at the Kennedy Space Center following the July 16 launch of the Apollo 11 mission, Agnew said that it was his "individual feeling that we should articulate a simple, ambitious, optimistic goal of a manned flight to Mars by the end of the century."[28]

NASA was prepared to give Agnew what he asked for. The agency's planning for its input to the STG had been built on an "integrated plan" for future human spaceflight that had been developed by George Mueller, the NASA Associate Administrator for Manned Space Flight. That plan focused on activities in the space between the Earth and the Moon. After Agnew's statements, Paine asked Wernher von Braun (who had been thinking about Mars exploration for many years) to add an early Mars expedition to the NASA plan as developed by Mueller. The revised NASA plan was briefed to the STG on August 4, 1969. It included an initial twelve-person expedition to Mars leaving Earth in November 1981, with six members of the expedition spending thirty to sixty days on the Martian surface in late 1982.

The possibility that the STG might actually recommend the program that NASA was proposing was worrisome to Secretary of the Air Force Robert C. Seamans, who represented Secretary of Defense Melvin Laird during the group's deliberations. Seamans had been a senior official in NASA from 1960 to 1967, and he was concerned that the STG would put forward a program that was not politically acceptable. Seamans made his views known to the vice president at the August 4 STG meeting and in a letter dated the same

26. Richard Nixon, Memorandum for the Vice President, the Secretary of Defense, the Acting Administrator, NASA, and the Science Adviser, February 13, 1969, NASA Historical Reference Collection.
27. Thomas O. Paine, Acting Administrator, NASA, Memorandum for the President, "Problems and Opportunities in Manned Space Flight," February 26, 1969, NASA Historical Reference Collection.
28. *New York Times*, July 17, 1969, p. 1.

day. [III-24] Seamans recommended that the NASA program for the 1970s concentrate on using its capabilities for "solution of the problems directly affecting men here on earth" and that the development of a space station or a new, reusable space transportation system (which had emerged during STG deliberations as an attractive option for the future), much less human missions to Mars, not be approved until their feasibility and desirability were more firmly established.[29]

Faced with differing views (Director of the Bureau of the Budget Robert P. Mayo participated only as an observer in the STG, but had made it clear that from a budgetary perspective he was opposed to an ambitious post-Apollo space program), the STG decided on August 4 to present several future program options to President Nixon, rather than attempt to reach consensus on a program that all could support. Over the next month, a report was prepared that outlined three options, each incorporating a space station and a Mars mission, but on different schedules and budget profiles. Another option, at a low budget level, involved terminating the human spaceflight program. As the time came for submission of the report to the president, senior White House aides, particularly Assistant to the President John Erlichman, demanded that it be modified so that the president not receive recommendations that he could not possibly accept, such as a 1982 Mars landing or ending the human spaceflight program so soon after Apollo 11.[30] Changes were hurriedly made, and the report was presented to President Nixon on September 15, 1969.

The STG report [III-25] recommended a "basic goal of a balanced manned and unmanned space program conducted for the benefit of all mankind," with, "as a focus for the development of new capability...the long-range option or goal of manned planetary exploration with a manned Mars mission before the end of this century as the first target." Beyond such general statements, however, the report recommended no commitment to any specific project or schedule of accomplishments. It left to President Richard Nixon the job of setting the future course in space for the United States.

The Nixon Space Policy

Almost six months passed before Nixon issued a formal statement of his views on space in response to the STG report. In that statement, issued on March 7, 1970, the president signaled a significant downgrading in the priority of post-Apollo space efforts; in effect, he rejected even the least ambitious of the options that the STG had recommended. In his statement, Nixon noted the need "to define new goals which make sense for the Seventies." He argued that

many critical problems here on this planet make high priority demands on our attention and our resources. By no means should we allow our space program to stagnate. But—with the entire future and the entire universe before us—we should not try to do everything at once. Our approach to space must be bold—but it must also be balanced.

The president rejected another Apollo-like undertaking, saying that "space activities will be part of our lives for the rest of time," and thus there was no need to plan them "as a series of separate leaps, each requiring a massive concentration of energy and will and accomplished on a crash timetable." Instead, "space expenditures must take their place within a rigorous system of national priorities."[31]

The six-month delay in a presidential response to the STG report was caused by a vigorous battle between NASA and White House political, budgetary, and technical advisers over the content of that response and over the level of the NASA budget for fiscal year 1971—the first post-Apollo 11, post-STG budget. While NASA believed it had in the STG

---

29. Robert C. Seamans Jr., Secretary of the Air Force, to Honorable Spiro T. Agnew, Vice President, August 4, 1969, NASA Historical Reference Collection.

30. For Erlichman's account of this intervention, see *Witness to Power: The Nixon Years* (New York: Simon and Schuster, 1982), pp. 144-45.

31. *Public Papers of the Presidents of the United States: Richard Nixon, 1970* (Washington, DC: U.S. Government Printing Office, 1971), pp. 250-53.

report a mandate for a continuing program of capability development and high visibility achievement, the White House viewed the space program as a place for lowered priority and budget cuts. Nixon's advisers saw few political benefits for the president in continuing to fund a large civilian space program. [III-26, III-27] By the time final budget decisions were announced in January 1970, the NASA budget had been reduced by $402 million from its level of the preceding year, production of the Saturn V heavy lift booster was terminated, and no commitment was made to develop either a space station or a reusable space transportation system, the space shuttle. Clearly, NASA was once again facing an uncertain future.

The Decision to Develop the Space Shuttle

Frustrated by the unwillingness of the Nixon administration to support the kind of space program he thought was in the country's interest and attracted by a lucrative private sector job offer, Thomas Paine announced in July 1970 that he was resigning as of September 15. Paine's deputy, George Low, became acting administrator. Low was a highly respected career NASA engineer who had taken over the Apollo spacecraft program after the 1967 Apollo capsule fire and had come to Washington to be deputy administrator in September 1969. It took until April 1971 for Paine's permanent successor to be named; he was James C. Fletcher, president of the University of Utah. Low stayed on as Fletcher's deputy, and it was the Fletcher-Low team that guided NASA through the critical 1971 decisions that shaped the agency for years to come.

NASA had hoped, as it fought its losing budget battles in the fall of 1969 and the early months of 1970, to come back to the White House at the end of 1970 and get approval for developing both a space station and a space shuttle. But in the months leading up to discussions over NASA's fiscal year 1972 budget, it became clear that there was no enthusiasm in the White House for going ahead with a space station, and that only the proposed reusable transportation system had any chance of approval. But that approval did not come in the fiscal year 1972 decisions, and Fletcher and Low believed that NASA had to get a go-ahead for the shuttle in 1971 if NASA were to maintain its identity as a large development organization with human spaceflight as its central activity. The choice of whether or not to approve the space shuttle thus became a *de facto* policy decision on the kind of civilian space policy and program the United States would pursue during the 1970s and beyond.[32]

At the White House, one individual decided that cuts in the NASA budget were going too far. He was Caspar (Cap) Weinberger, deputy director of the renamed Bureau of the Budget, now the Office of Management and Budget (OMB). In an August 12, 1971, memorandum to the president, Weinberger argued that "there is real merit to the future of NASA, and to its proposed programs." [III-28] OMB was considering, as a means of further cutting the NASA budget, not approving the start of space shuttle development and cancelling the last two Apollo missions, Apollo 16 and 17. Weinberger suggested that such cuts

would be confirming in some respects, a belief that I fear is gaining credence at home and abroad: That our best years are behind us, that we are turning inward, reducing our defense commitments, and voluntarily starting to give up our super-power status, and our desire to maintain world superiority.

America should be able to afford something besides increased welfare, programs to repair our cities, or Appalachian relief and the like.

---

32. For a more detailed discussion of the space shuttle decision, see John M. Logsdon, "The Space Shuttle Decision: Technology and Political Choice," *Journal of Contemporary Business* 7 (1978): 13-30; John M. Logsdon, "The Decision to Develop the Space Shuttle," *Space Policy* 2 (May 1986): 103-19; John M. Logsdon, "The Space Shuttle Program: A Policy Failure," *Science* 232 (May 30, 1986): 1099-1105.

When Weinberger argued that programs such as the space shuttle should be funded, his views found a responsive ear. In a handwritten note on Weinberger's memo, President Richard Nixon indicated "I agree with Cap."[33] OMB Director George Shultz was informed that "the president approved Mr. Weinberger's plan to find enough reductions in other programs to pay for continuing NASA at generally the 3.3-3.4 billion dollar level, or about 400 to 500 million dollars more than the present planning target."[34]

Neither NASA nor the OMB staff knew of this exchange. An often heated struggle over the agency's budget outlook and approval of shuttle development continued through the summer and fall of 1971. [III-29] In May 1971 OMB had told NASA that its budget would be further reduced and then stay level at approximately $3.0 billion/year for the rest of the Nixon presidency. This was the final blow to NASA's hope of getting approval for the approximately $10 billion needed to develop a fully reusable shuttle; between June and December 1971 the agency and its contractors examined many alternatives for a less ambitious development program. The OMB staff, fundamentally convinced that the shuttle was not a desirable program, no matter how cheaply it could be developed, resisted the various concepts NASA put forward.

One line of argument that NASA developed during the shuttle debate was that the program would be cost-effective—i.e., that the savings over the use of existing expendable launch vehicles in launch costs, payload design, and the ability to repair satellites would more than pay for the costs of developing a shuttle. This was the first time NASA had attempted an economic justification for a major program; the approach had been forced on the space agency by OMB. As the decision process regarding the shuttle reached its conclusion in the last months of 1971, NASA's contractor for the cost-effectiveness analysis, Mathematica, Inc., widely circulated a memorandum [III-30] that argued that "*A reusable space transportation system is economically feasible. . . .*"[35]

While an economic argument was one part of NASA's case for shuttle approval, other factors were more important to the agency's arguments. James Fletcher summarized them in a November 1971 memorandum to the White House: [III-31]

1. The U.S. cannot forego manned space flight.

2. The space shuttle is the only meaningful new manned space flight program that can be accomplished on a modest budget.

3. The space shuttle is a necessary next step for the practical use of space. . . .

4. The cost and complexity of today's shuttle is one-half of what it was six months ago.

5. Starting the shuttle now will have a significant positive effect on aerospace employment. Not starting would be a serious blow to both the morale and health of the Aerospace Industry.[36]

After intense debate between NASA and OMB during December, a decision to approve the shuttle program was made over the New Year weekend. The perspective that Weinberger had put forward in August, NASA's arguments in the November memorandum, and the desire to start a new aerospace program that would avoid unemployment in critical states in the 1972 election year were ultimately decisive. NASA was informed of the decision on January 3, 1972. Fletcher and Low, surprised at the go-ahead, made hasty

---

33. Caspar W. Weinberger, Deputy Director, Office of Management and Budget, via George P. Schultz, Memorandum for the President, "Future of NASA," August 12, 1971, Richard M. Nixon Project, National Archives and Records Administration, Washington, DC.

34. Jon M. Huntsman, The White House, Memorandum for George P. Shultz, "The Future of NASA," September 13, 1971. Original memorandum in the files of the Nixon Project, National Archives. Huntsman was one of those in the White House responsible for making sure that presidential decisions were communicated to appropriate officials. Copies of this memorandum were sent to Nixon advisors H.R. Haldeman and Alexander Butterfield and to Caspar Weinberger in OMB.

35. Klaus P. Heiss and Oskar Morgenstern, Memorandum for Dr. James C. Fletcher, Administrator, NASA, "Factors for a Decision on a New Reusable Space Transportation System," October 28, 1971, p. 1, NASA Historical Reference Collection.

36. James C. Fletcher, Administrator, Memorandum for Jonathan Rose, Special Assistant to the President, November 22, 1971, with attached: "The Space Shuttle," NASA Historical Reference Collection.

preparations to fly to California for a January 5 meeting with President Nixon, who was at his western White House in San Clemente, after which shuttle approval would be made public. At that meeting, the president asked NASA to stress the view that the shuttle made economic sense, but "even if it were not a good investment, we would have to do it anyway, because space flight is here to stay."[37] [III-32]

Although it was not specifically part of the set of decisions reached at this time, NASA had justified the costs of developing the shuttle on its use to replace existing expendable space launch vehicles, particularly the Delta, Atlas, and Titan. NASA had also modified the shuttle design during 1970 and 1971 to meet the requirements of the Department of Defense, and the anticipation was that the shuttle would launch all DoD payloads as well as those of the civilian sector. [III-29, III-30, III-31, III-32] The decision to proceed with the space shuttle under these assumptions was the central space choice of the 1970s. It was not, except by default, a policy decision regarding U.S. objectives in space, however. For the rest of the 1970s, the United States would carry out those space missions that could be afforded within a fixed NASA budget after shuttle development costs had been paid. Once the shuttle entered operations, U.S. space objectives would be largely defined in terms of those missions enabled by shuttle capabilities. This was certainly a different approach to space policy than that of the preceding decade.

Space Policy Under President Jimmy Carter

During the brief presidency of Gerald Ford, no major space policy decisions or initiatives were taken, although Ford was generally sympathetic toward the program and as he left the White House approved the start of two major missions. They became the Galileo probe to Jupiter and the Hubble Space Telescope.

President-elect Jimmy Carter did not create a blue-ribbon transition group for space, as had John Kennedy and Richard Nixon. The Carter space transition document was the product of a single individual, who took a generally skeptical tone toward NASA and its programs. [III-33] The document noted:

1. . . . NASA directs our R&D resources towards centralized big technology, maintaining the defense R&D orientation of the aerospace industry.
2. The Shuttle has become the end, rather than the means, because NASA space policy has been shaped by the Office of (Manned) Space Flight. The Offices of Space Science, Applications, and Aeronautics Technology get the funds that are left over.[38]

The Carter administration returned to a practice last followed under President Eisenhower: the development, through an inter-agency process coordinated at the White House level, of formal statements of national space policy. The first of these statements [III-34] was issued on May 11, 1978; it dealt with both national security and civilian uses of space, and large portions of the statement remain classified. A June 20, 1978, White House press release announcing the results of the initial Carter policy review noted that "the major concerns that prompted this review arose from growing interaction among our various space activities" and that the May policy statement resulting from the review did not "deal in detail with the long-term objectives of our defense, commercial, and civil programs." The White House release indicated that its next step would be a comprehensive review of civilian space policy.[39]

---

37. George M. Low, Deputy Administrator, NASA, Memorandum for the Record, "Meeting with the President on January 5, 1972," January 12, 1972, NASA Historical Reference Collection.
38. Nick MacNeil, Carter-Mondale Transition Planning Group, to Stuart Eizenstat, Al Stern, David Rubenstein, Barry Blechman, and Dick Steadman, "NASA Recommendations," January 31, 1977, Jimmy Carter Presidential Library, Atlanta, GA.
39. "United States Space Activities," Announcement of Administration Review, June 20, 1978, in *Public Papers of the Presidents of the United States: Jimmy Carter, 1978* (Washington, DC: U.S. Government Printing Office, 1979), pp. 1135-37.

The results of that policy review were incorporated in Presidential Directive/NSC-42, "Civil and Further National Space Policy," dated October 10, 1978. [III-35] This directive took a measured approach to future U.S. goals in space:

First: Space activities will be pursued because they can be uniquely or more efficiently accomplished in space. Our space policy will become more evolutionary rather than centering around a single, massive engineering feat. Pluralistic objectives and needs of our society will set the course for future space objectives.
Second: Our space policy will reflect a balanced strategy of applications, science and technology development. . . .
Third: It is neither feasible nor necessary at this point to commit the US to a high-challenge, highly-visible space engineering initiative comparable to Apollo.[40]

With this set of guidelines, it was clear that the space program during the Carter administration would be one seeking efficiencies and payoffs from existing capabilities. After considering cancellation of the shuttle program and being dissuaded because of the need for the shuttle to launch satellites critical to his arms control initiatives, Jimmy Carter gave highest priority to completing shuttle development.[41]

Space Policy Under President Ronald Reagan

Unlike his predecessor, Ronald Reagan did assemble a prestigious panel, largely composed of veterans in the space field, as part of his transition effort. The group was headed by George M. Low, who had left NASA in 1976 to become president of Rensselaer Polytechnic Institute. Not surprisingly, the team's report [III-36] was bullish on the space program. It noted:

The year 1980 finds NASA in an untenable position. . . . This unhealthy state of affairs can only be rectified by a conscious decision. Continuation of the prior administration's low level of interest and lack of clear direction would result in an unconscionable waste of human and financial resources.

"NASA and the space program are without a clear purpose and direction," said the transition team. In his cover letter transmitting the report, Low said that the transition team members had asked him to "emphasize our view that NASA and its civil space program represent an opportunity for positive accomplishment by the Reagan administration...NASA can be many things in the future—the best in American accomplishment and inspiration for all citizens."[42]

The Reagan administration selected experienced individuals as the new leaders of NASA. Chosen as administrator was James E. Beggs, an aerospace industry executive who had worked under James Webb in the late 1960s; the designee as deputy administrator was Hans Mark, who had been director of NASA's Ames Research Center before coming to Washington as under secretary and then secretary of the Air Force during the Carter administration. The approach to the space agency that the new pair of NASA managers would take was foreshadowed in a paper prepared by Mark and a senior engineering associate, Milton Silveira, that was widely circulated among the top people in NASA soon after the new leaders assumed control; Beggs accepted the paper as a framework for NASA

---

40. Presidential Directive/NSC-37, "National Space Policy," May 11, 1978, NASA Historical Reference Collection.
41. See Hans Mark, *The Space Shuttle: A Personal Journey* (Durham, NC: Duke University Press, 1987), chapter IX, for an account of the Carter administration decisions on the space shuttle.
42. George M. Low, Team Leader, NASA Transition Team, to Richard Fairbanks, Director, Transition Resources and Development Group, December 19, 1980, with attached: "Report of the Transition Team, National Aeronautics and Space Administration," George M. Low Papers, Archives and Special Collections, Rensselaer Polytechnic Institute, Troy, NY.

planning.⁴³ That approach gave top priority to making the space shuttle operational and then utilizing it frequently while getting approval for a major new development project, the space station. [III-37]

Like the Carter administration before it, the Reagan White House carried out an early, comprehensive review of national space policy. The results of that review were incorporated in a classified national security decision directive issued July 4, 1982. [III-38] The directive provided "the broad framework and the basis for the commitments necessary for the conduct of U.S. space programs." It gave particular emphasis to the role of the space shuttle, which was to be "a major factor in the future evolution of United States space programs."⁴⁴ The directive also transferred White House responsibility for reviewing space policy from the Office of Science and Technology Policy, where it had been vested during the Carter administration, to the National Security Council, and created a Senior Interagency Group (SIG) for Space, chaired by the president's assistant for national security, to oversee the Reagan-era space policy process.

During the following six years, SIG (Space) was the focal point for a series of debates, policy statements, and directives on various aspects of U.S. space efforts. Issues that stimulated these debates included a desire to foster the commercial uses of space, the decision to begin a space station program, controversy over the pricing policy for the space shuttle, and actions required to recover from the January 1986 *Challenger* accident. In 1988, President Reagan approved a revised statement of national space policy that incorporated the results of these individual decisions and directives. [III-42] Reflecting a theme that had been present in U.S. space policy since the beginning, the directive noted that "a fundamental objective guiding United States space activities has been, and continues to be, space leadership."⁴⁵

The Space Station Decision[46]

At his Senate confirmation hearing in June 1981, James Beggs was asked his view on what should be the next major U.S. undertaking in space. He replied that "it seems to me that the next step is a space station."⁴⁷ Between the second half of 1981 and the end of 1983, NASA carried out an intense, and ultimately successful, campaign to gain presidential approval to develop a large, permanently occupied space station as the "next logical step" in space development. Like the space shuttle before it, developing and operating a space station promised to influence the U.S. space program for years to come.

NASA spent most of 1982 laying the foundation for station approval by conducting internal and contractor studies, with a particular focus on identifying the missions that a station might perform. Beggs and Mark pursued a two-pronged strategy for gaining station approval. One path was to work with other government agencies and external constituencies to build a broad coalition in support of the station; the other was to convince President Ronald Reagan that it was in the U.S. interest to go ahead with the program.⁴⁸

The forum for developing an interagency consensus on the station was the National Security Council's SIG (Space). At a March 30, 1983, meeting, SIG (Space) approved terms of reference for a study that would provide the basis for a presidential decision on whether to proceed with the program. To give added weight to the study, a national security

---

43. Mark, *Space Station*, p. 128.
44. National Security Decision Directive Number 42, "National Space Policy," July 4, 1982 (partially declassified June 14, 1990).
45. Office of the Press Secretary, "Fact Sheet: Presidential Directive on National Space Policy," February 11, 1988, NASA Historical Reference Collection.
46. For more details on the space station decision, see Howard E. McCurdy, *The Space Station Decision: Incremental Politics and Technological Choice* (Baltimore, MD: Johns Hopkins University Press, 1990); Mark, *Space Station*.
47. McCurdy, *The Space Station Decision*, p. 40.
48. See Mark, *Space Station*, chapter XIV, for a discussion of this strategy.

decision directive signed by the president and incorporating these terms of reference was drafted. After being briefed on the station program, Reagan approved the directive on April 11. [III-39] The directive identified five policy issues to be studied:

How will a manned Space Station contribute to the maintenance of U.S. space leadership and to the other goals contained in our National Space Policy?. . . How will a manned Space Station best fulfill national and international requirements versus other means of satisfying them?... What are the national security implications of a manned Space Station?. . . What are the foreign policy implications of a manned Space Station?. . .What is the overall economic and social impact of a manned Space Station?

These questions were to be answered with respect to four possible future scenarios:

– Space Shuttle and Unmanned Satellites
– Space Shuttle and Unmanned Platforms
– Space Shuttle and an Evolutionary/Incrementally Developed Space Station
– Space Shuttle and a Fully Functional Space Station

The directive called for study results to be available "not later than September 1983."[49]

In the course of the next several months, NASA discovered that getting a positive recommendation on the station from SIG (Space) was not going to be possible. First of all, the effort got bogged down as the NASA-led team considered the multiple options of the study directive. The process of developing a shorter policy paper containing recommendations to which all SIG (Space) members could agree became stalemated in August; there was significant opposition from the national security members of the group to going ahead with the station. In particular, Secretary of Defense Caspar Weinberger argued against the station project. [III-41] Without SIG (Space) agreement, it seemed, there would be no recommendation to Ronald Reagan to approve the space station.

Over the next few months, however, NASA was able to find an alternative path to get the issue of whether or not to go ahead with the station before the president. It had been James Beggs' position all along that President Reagan would approve the station program, given the opportunity; this had been the second prong of the NASA strategy. NASA's allies in the White House succeeded in getting the station question on the agenda of a December 1, 1983, meeting of the Cabinet Council on Commerce and Trade, one of the organizations that the Reagan administration had created for policy development; the national security community did not have a controlling position among the council's membership.

The NASA presentation to the meeting, which was attended by the president, asked for a decision to proceed with the space station program. [III-40] Primary emphasis in the presentation was given to the station's contribution to U.S. leadership around the world, a theme that Beggs knew was close to Ronald Reagan's heart. The presentation also emphasized the commercial potential of station-based activities, and underlined the fact that the Soviet Union already had a small space station and was expected to develop a larger facility. In concluding the presentation, James Beggs told the president and others in the Cabinet Room that "the time to start a space station is now."[50]

President Reagan approved the station program in an Oval Office meeting a few days later. On January 25, 1984, in his annual State of the Union message, Reagan told Congress and the nations that

America has always been greatest when we dared to be great. We can reach for greatness again. We can follow our dreams to distant stars, living and working in space for peaceful, economic, and scientific

---

49. National Security Decision Directive 5-83, "Space Station," April 11, 1983, National Security Archive, Washington, DC.
50. "Revised Talking Points for the Space Station Presentation to the President and the Cabinet Council," November 30, 1983, with attached: "Presentation on Space Station," December 1, 1983, NASA Historical Reference Collection.

gain. Tonight I am directing NASA to develop a permanently manned space station and to do it within a decade. [51]

Looking Toward the Future

The space station was frequently justified by James Beggs and others in NASA as "the next logical step." When asked "step toward what" NASA most often pointed out the many missions that had been proposed for a permanently occupied orbiting laboratory. During the 1982-1983 debate over station approval, the agency resisted pressure from Presidential Science Adviser George A. Keyworth II to identify the station with the ambitious goal of preparing for human journeys to Mars. The memory of the negative response to the 1969 Space Task Group recommendation for Mars exploration was still strongly in the minds of many at NASA, and Beggs judged that the time was not propitious for linking station approval to such a visionary objective.

Pressure also came from Congress for NASA to articulate its long-term vision of the future in space. In 1984, Congress passed a bill requiring the president to name a National Commission on Space to develop a future space agenda for the United States. The White House in March 1985 chose Thomas Paine as chairman of the commission, who, since leaving NASA fifteen years earlier, had been a tireless spokesman for an expansive view of what should be done in space. The fourteen other commissioners were a diverse group, ranging from Apollo 11 astronaut Neil Armstrong and test pilot Chuck Yeager to the U.S. Ambassador to the United Nations, Jeanne Kirkpatrick.

The commission took most of a year to prepare its report; in addition to its own deliberations, the group solicited public input in hearings throughout the United States. The commission report, *Pioneering the Space Frontier*, was published in a lavishly illustrated, glossy format in May 1986; a summary videotape was also prepared.

The National Commission on Space recommended "a pioneering mission for 21st-century America"—"to lead the exploration and development of the space frontier, advancing science, technology, and enterprise, and building institutions and systems that make accessible vast new resources and support human settlements beyond Earth orbit, from the highlands of the Moon to the plains of Mars."

The report also contained a "Declaration for Space" that included a rationale for exploring and settling the solar system and outlined a long-range space program for the United States.[52]

The United States in 1986 was not in a particularly receptive mood for such bold proposals; the tragic *Challenger* accident in January 1986 had focused attention on the problems with the U.S. space program, not its prospects. But as the year ended, NASA once again began to focus on its long-range objectives. James Fletcher, who had returned for a second tour of duty as NASA administrator in the wake of the shuttle tragedy, asked former astronaut Sally K. Ride to chair a task force to develop options for NASA's future. The group's report, *Leadership and America's Future in Space*, was presented to Fletcher in August 1987.

The Ride report identified four "leadership initiatives" that NASA might choose to pursue, individually or in combination:

1. **Mission to Planet Earth:** *a program that would use the perspective afforded from space to study and characterize our home planet on a global scale.*
2. **Exploration of the Solar System:** *a program to retain U.S. leadership in exploration of the outer solar system, and regain U.S. leadership in the exploration of comets, asteroids, and Mars.*

---

51. Quoted in McCurdy, *Space Station Decision*, p. 190.
52. The Report of the National Commission on Space, *Pioneering the Space Frontier* (New York: Bantam Books, 1986), excerpts.

*3. **Outpost on the Moon**: a program that would build on and extend the legacy of the Apollo program, returning Americans to the Moon to continue exploration, to establish a permanent scientific outpost, and to begin prospecting the Moon's resources.*

*4. **Humans to Mars**: a program to send astronauts on a series of round trips to land on the surface of Mars, leading to eventual establishment of a permanent base.*

In its conclusion, the report referred to the central vision statement of the National Commission on Space, quoted above, and recommended that "the United States needs to define a course of action to make this vision a reality."[53]

Conclusion

Many influences shaped U.S. space policy and the U.S. space program in the three decades between 1958 and 1988. Throughout, leadership in space has been a consistent policy objective, and human exploration of space a constant theme. As a response to the needs of the time, the United States sent twelve people to the surface of the Moon between 1969 and 1972, but this first instance of human exploration of another celestial body did not lead to a sustained program of human exploration. That still lay in the future in 1988; the final Reagan administration statement of space policy set as a long-range goal "to expand human presence and activity beyond Earth orbit into the solar system."[54] While much happened in the early years of the space program, much remains.

---

53. Dr. Sally K. Ride, *Leadership and America's Future in Space: A Report to the Administrator* (Washington, DC: National Aeronautics and Space Administration, August 1987), pp. 21, 58.

54. Office of the Press Secretary, "Fact Sheet: Presidential Directive on National Space Policy," February 11, 1988, NASA Historical Reference Collection.

Document III-1

Document title: Special Committee on Space Technology, "Recommendations to the NASA Regarding A National Civil Space Program," October 28, 1958.

Source: NASA Historical Reference Collection, History Office, NASA Headquarters, Washington, D.C.

By the end of 1957 the National Advisory Committee for Aeronautics (NACA) was heavily involved in space-related research, which constituted forty to fifty percent of its total effort. Sensing that NACA might be the obvious choice for taking the lead in the American space effort after Sputnik, on January 12, 1958, General James Doolittle, chairman of NACA, created a Special Committee on Space Technology. While NACA Director Hugh Dryden addressed the institutional issues involved in transforming NACA into NASA, the Committee on Space Technology was charged with addressing specific areas of space technology deserving early attention. NASA was formally established on October 1, 1958, and the committee issued its final report at the end of that month. The following document reprints the recommendations to NASA on "A National Civil Space Program" offered by the Special Committee on Space Technology on October 28, 1958.

---

[1] **Summary**

The major objectives of a civil space research program are scientific research in the physical and life sciences, advancement of space flight technology, development of manned space flight capability, and exploitation of space flight for human benefit. Inherent in the achievement of these objectives is the development and unification of new scientific concepts of unforeseeably broad import.

Space Research - Instruments mounted in space vehicles can observe and measure "geophysical" and environmental phenomena in the solar system, the results of cosmic processes in outer space, and atmospheric phenomena, as well as the influence of space environment on materials and living organisms. A vigorous, coordinated attack upon the problems of maintaining the performance capabilities of man in the space environment is prerequisite to sophisticated space exploration.

Development - Flight vehicles and simulators should be used for space research and also for developmental testing and evaluation aimed at improved space flight and observational capabilities. Major developmental recommendations include sustained support of a comprehensive instrumentation development program, establishment of versatile dynamic flight simulators, and provision of a coordinated series of vehicles for testing components and sub-systems.

Ground Facilities - Properly diversified space flight operations are impossible without adequate ground facilities. To this end serious study aimed toward providing an equatorial launching capability is recommended. A complete ground instrumentation system consisting of computing centers, communication network, and facilities for tracking and control of and communication (including telemetry) with space vehicles is required. At least part of the system must be capable of real time computation and communication. A competent satellite communications relay system would be most valuable in this regard, and it is recommended that NASA take the lead in determining the specifications of such a system. A coordinated national attack upon the problems of recovery is recommended.

Flight Program - The first recovery vehicles will probably be ballistic, but the control and safety advantages of lifting re-entry vehicles warrant their development. [2] A million-pound-plus booster can be achieved about three years sooner by clustering engines than by developing a new single-barrel engine, but the cluster would not have the growth potential of the larger engine. Further growth potential requires the development of the single-barrel engine. Both developments are needed.

Strong research effort on novel propulsion systems for vacuum operations is urged, and development of high-energy-propellant systems for upper stages should receive full support.

Three generations of space vehicles are immediately available. The first is based on Vanguard-Jupiter C, the Second on IRBM boosters, and the third on ICBM boosters. The performance capabilities of various combinations of existing boosters and upper stages should be evaluated, and intensive development concentrated on those promising greatest usefulness in different general categories of payload.

Introduction

Scientifically, we are at the beginning of a new era. More than two centuries between Newton and Einstein were occupied by the observations, experiments and thought that produced the background necessary for modern science. New scientific knowledge indicates that we are already working in a similar period preceding another long step forward in scientific theory. The information obtained from direct observation, in space, of environment and of cosmological processes will probably be essential to, and will certainly assist in, the formulation of new unifying theories. We can no more predict the results of this work than Galileo could have predicted the industrial revolution that resulted from Newtonian mechanics.

The observation of the nature and effects of the space environment are necessarily paced by the development of space flight capabilities. This report presents suggestions regarding research policies and procedures that should aid in the establishment and improvement of capabilities for space flight and space research.

In preparing this report, the Special Committee on Space Technology has been assisted by the Technical committees of the NACA and the ad hoc Working Groups of the Special Committee. The membership of the Working Groups is listed in an appendix to this report.

The reports of the Working Groups are primarily program-oriented, and while they are not referenced specifically, they have furnished the basis for the preparation of this report. These will be presented to the NASA as separate Working Group reports, independent of this report.

[3] ### Objectives

A national civil space research program to explore, study, and conquer the newly accessible realm beyond the atmosphere will have the following general objectives:

1. Scientific research and exploration in the physical and the life sciences.

Submerged as he always has been beneath the "dirty window" of the atmosphere, man has necessarily inferred the nature of the physical universe from local observations and glimpses of what lies beyond his essentially two-dimensional earth-bound habitat. Little of the radiation and few of the solid particles from outer space reach the earth's surface, yet practically all aspects of man's earthly environment are determined ultimately by extraterrestrial factors. The radiation that does reach the surface is so distorted by passage through the atmosphere that only incomplete observations can be made on the nature of other celestial bodies and the contents of interstellar space.

With the information derived from experiments and directed observations in the actual space environment, man will achieve a better understanding of the universe and of nature phenomena and life on the earth.

An excellent start toward determination of the near-space environment has already been made in connection with the IGY, and the pattern of international cooperation that has developed with this program indicates that mutual understanding and respect among the nations of the earth may be generated by concerted attack upon scientific problems. Inasmuch as national scientific excellence is, to a great extent, now evaluated by the people of the earth in terms of success in the exploration of space, it behooves the United States to achieve and maintain an unselfish leadership in this field.

2. Advancement of the technology of space flight.

[4] Propulsion systems have been developed having the demonstrated capability of putting small instrumented packages into orbit about the earth. However, the reliability of the total vehicle and control system needs improvement in order to conduct much of the desired space program. Larger power plants, and new higher-energy fuels and the equipment to produce them must be developed. If orbits about the earth are to be expanded into practical interplanetary trajectories, new propulsion systems having very low fuel consumption and modest thrust will be required in order that the trajectory can be controlled to perform the mission.

A good start has been made on the development of instrumentation for observing the environment in space. Instrumentation for controlling and navigating the vehicle and for communicating with the earth will require extensive development. Because of the server weight restrictions, all instrumentation must be severely miniaturized. Ground-based communications systems must be expanded to provide for the control of and communication with vehicles on lunar or planetary missions, and for properly controlled re-entry and recovery.

Novel structural problems are posed by space vehicles. Heavy loads of steady acceleration, shock and vibration occur during boost, while weightlessness during unpowered space flight makes possible the use of nonconventional mechanical design principles. For vehicles which must re-enter the earth's atmosphere, problems of structural integrity under high re-entry heating rates, larger thermal gradients, and thermal shock are very important. All of these requirements must be met with an absolute minimum of structural weight.

Extensive human engineering developments are required in order for manned space flight to be successful. Because of the rigorous but largely unknown space environment, these developments will depend critically upon the information obtained in the early probing flights.

A successful National Space Program, therefor requires continuing improvement and development in the pertinent fields of technology.

3. Manned space flight.

Instruments for the collection and transmission of data on the space environment have been designed and put into orbit about the earth. However, man has the capability of correlating unlike events and unexpected observations, a capacity for overall evaluation of situations, and the background knowledge and experience to apply judgment that cannot be provided by instruments; and in many other ways the intellectual functions of man are a necessary complement to the observing and recording functions of complicated instrument systems. Furthermore, man is capable of voice communication for sending detailed descriptions and receiving information whereby the concerted judgment of others may be brought to bear on unforeseen problems that may arise during flight.

[5] Although it is believed that a manned satellite is not necessary for the collection of environmental data in the vicinity of the earth, exploration of the solar system in a sophisticated way will require a human crew.

4. Exploration of space for human benefit.

The practical exploitation of satellites and space vehicles for civil purposes and for human benefit may be as important as–or even more important than–the immediate military uses for space flight. Perhaps the most important example is the use of satellite vehicles for active or passive communications relay. This could extend what are effectively line-of-sight communication links for thousands of miles between points on the ground, with very great bandwidths and none of the capriciousness now characterizing long-range HF communications.

Many indirect benefits will also be derived from the technological developments that will make space flight practical. The necessarily high technological standards required for space flight will certainly accelerate improvement in transportation, communications and other contributions to human welfare.

The unpredictable long-term benefits of space-accelerated scientific and technological advancement will almost certainly far exceed the foreseeable benefits.

Aside from the intentional omission of military and political objectives, the foregoing objectives appear to be in consonance with those mentioned in "Introduction to Outer Space," by the President's Science Advisory Committee (Killian Committee), and with the objectives stated in the National Aeronautics and Space Act of 1958, which is the enabling legislation for the National Aeronautics and Space Administration.

Basic Scientific Research

Space Research

Geophysical observations from satellites and non-orbiting space probes enable the gravitational and magnetic fields in the vicinity of the earth to be mapped to altitudes limited only by the capabilities of the flight vehicle. The interactions among these fields and the particles and radiations approaching the earth from the sun and other [6] space can be studied, and related to the composition and behavior of the gaseous envelope of the earth from troposphere to exosphere. Satellite observations of large-scale cloud movements and other atmospheric phenomena can do much to put meteorology on a more sound scientific basis. As propulsion and guidance systems are improved, "geodetic" and "geophysical" studies can be extended to the moon and other planets.

Telescopes and spectroscopes mounted on earth satellites can utilize the complete radiation spectrum from vacuum ultraviolet to radio frequencies to observe the sun, the planets, stars, and interstellar space. Direct measurements of the space environment should include the nature, direction and intensity of electromagnetic and corpuscular radiation, and the nature and distribution of meteorites. The mass density in space can be measured, and large-scale magneto-hydrodynamic phenomena in and beyond the ionosphere can be studied. These observations and direct measurements will offer tremendous improvements in understanding of cosmic processes.

In addition to scientific observations and environmental measurements, satellite experiments will enable evaluation of the effect of the space environment on all types of material and biological specimens and hardware components. Re-entry phenomena can be studied, and here for the first time, it is possible to investigate the effects of extended periods of weightlessness on instrumentation and living subjects.

Experiments with man and other living organisms, both plant and animal, during extended periods in the space environment may offer new insight to human physiology and psychology and into life processes generally.

Upper Atmosphere Research

Upper atmosphere experiments, utilizing both rocket-propelled and balloon-supported vehicles, can, at reasonable cost, give direct information on both the vertical and time-wise variations of various atmospheric parameters and cosmic radiations. Heat-transfer, ablation, vehicle-control dynamics, and pilot-vehicle interactions can be studied under approximately re-entry conditions. Limited-time biological studies and human physiological and psychological studies under almost space conditions, and with limited periods of weightlessness, can also be investigated.

Ground-Based Supporting Research

In addition to direct study of the space environment, much ground-based research must be conducted as a basis for the space flight program. This will include such factors as radiation effects [7] on materials, instruments, and living organisms, and means of radiation protection. Other physical phenomena pertinent to space flight and re-entry include radio propagation; and behavior, in a space-type environment, of materials, transducers power supplies, and so forth for instrument components; hypersonic gasdynamics, both continuum and noncontinuum; and magnetogasdynamics.

Human factors pertinent to space flight present a real challenge. Those amenable to ground-based study include, among others, acceleration and vibration tolerance and protection, and the influence of new physiological and psychological factors (other than weightlessness) on the performance capabilities of the crew members. A major cooperative effort between the NASA, and the Department of Defense, and other groups concerned with aeromedical and space flight problems is necessary.

Research Techniques and Equipment Development

Vehicle Instrumentation

Vehicle instrumentation presents formidable development problems because of the conflicting requirements of minimum weight, adequate resistance to the accelerations and vibrations of launching and ability to operate correctly for extended periods of time under the conditions of space flight. For scientific observations, a complete range of instrumentation will be required for observing the external environment and recording or telemetering the data. Other special instrumentation will be required to observe experiments conducted within the vehicle.

Navigation and guidance equipment, and instruments for attitude sensing and control for the communication, are required for operation of the vehicle, particularly on extended flights into space. An integrated display of information on the internal environment and the vehicle operation will be required for manned flights. Improved auxiliary power sources will be needed for all types of vehicle-borne instruments.

It is recommended that the NASA organize and give consistent support to a comprehensive program of instrumentation development, comprising not only instruments useful in the development, flight testing, and operation of space vehicles, but also the instruments needed for a broad program of environmental and other experimental research. Special attention should be paid to the novel design possibilities offered by operating such instruments in free fall and in vacuo.

[8] *Ground Simulation of Environment and Operational Problems*

The development and testing of a space vehicle, its components and, for a manned vehicle, its crew require ground simulation of the environment operating problems that will be encountered. The completeness of the simulation may well determine the success or failure of the mission. This will be a continuously changing problem as new information is obtained on the environment and as the operational ranges and durations increase.

Wind tunnels and jets of various types, ballistic rangers and structural test facilities, can simulate, to a reasonable extent aerodynamic effects encountered during launching and re-entry. Vacuum chambers with assorted loading devices and radiation sources will be useful for both instrumental structural tests.

The capacity of a human crew to participate in the operation of a space vehicle is still an unknown quantity. As fast as such capabilities are demonstrated they should be utilized to the extent profitable in operation of the vehicle. Therefore, flight simulators should be designed and built in which the flight dynamics and internal environment of space vehicles can simulated as closely as possible. Such facilities would be used for pilot evaluation and training and for evaluation of the dynamic characteristics of the vehicle-pilot combination.

Flight Testing Techniques

To aid in the advanced development of space vehicles and sub-systems, and to complement the ground-based simulators, it is recommended that the NASA use reliable high-performance rocket-propelled test vehicles which would be standardized for as many tests as possible. In order to minimize the development cost of such vehicles, they should presumably be based on military developments in the missile field.

Two other techniques are recommended for larger-scale tests and for systems development and testing. One of these is a large, high-altitude, balloon-supported laboratory in which most conditions of space environment could be simulated. This balloon-supported laboratory would not only allow a substantial amount of research on the equipment needed by the space crew and on the effects of space environment needed by the space crew and on the effect of space environment on the capsule and its inhabitants, but could also be valuable for basic environmental studies.

[9] The other is a nonorbiting rocket-propelled research vehicle capable of carrying at least two men, or an actual man-carrying satellite capsule. This vehicle should be capable of a number of minutes of free coast well above significant atmospheric influences. Such a vehicle should be used for development and final flight-testing of actual space flight controls and operational instrumentation. In addition, flight crew could be trained and evaluated under longer periods of weightlessness than are possible within the atmosphere.

With the establishment of artificial earth satellites, space flight has become a reality, albeit on only a very limited scale. For more extended space missions, the long-time effects of the space environment on the vehicle and its contents must be known and designed for. This can best be studied in earth satellite vehicles. Strong technological support should be provided for all phases of vehicular development. Specifically, a substantial fraction of space flight missions should be allocated to such technological projects as components tests, materials tests, engine-restart tests, solar power supply systems, et cetera.

Ground Facilities for Space Flight Operations

Range Capabilities and Requirements

In view of the plans to expand the NASA Wallops Island facility for technique development and relatively small probe and satellite launchings, and with the Atlantic and Pacific Missile Ranges capable of substantial further development, there is no present need for another major nonequatoral launching complex. It may be desirable, however, for the NASA to establish permanent field stations at both the Atlantic and Pacific Missile Ranges.

On the other hand, the unique properties of an equatorial orbit lead to a distinct need for an equatorial launching site. These are:
1. Narrow track over the earth's surface.
2. Best departure point for interplanetary operations.
3. Capability for all other orbits.
4. Minimum requirement for ground stations and communication system.

These considerations bringing the Committee to the conclusion that the NASA should establish a study, survey and planning group [10] aimed toward early provision of an equatorial launching capability, including necessary logistic support, for the United States. Fixed-base and ship-based launchings should be considered by the group before reaching a final decision.

Ground-Based Instrumentation Systems

The ground-based instrumentation needs of the civilian space program encompass such things as:

1. Communication with and transmission of commands to vehicles both near the earth and in interplanetary space.
2. Active and passive tracking of space vehicles.
3. Reception of telemetry signals from space.
4. Calculation of real-time search ephemeris data.
5. Calculation of final orbits for scientific analysis.

The instrumentation necessary can thus be listed as:

1. A network of stations suitably located for tracking of a communication with vehicles in interplanetary space. These stations must be tied together with reasonably rapid communication links. The stations will consist of very large antennas, sensitive receiving equipment, and high-power transmitting equipment.
2. A network of radio receiving stations to obtain orbital information from active satellites. These stations may be, in part at least, the same as those in the preceding paragraph.
3. A network of optical stations to make very precise optical observations on some satellites, and a supplementary set of optical observing stations, probably similar to the present Moonwatch teams, for rough orbital data.
4. A set of telemetry receiving stations which will be in part, but not necessarily completely, at the other radio sites.
5. A special network of stations for re-entry experiments.
6. Computing facilities to calculate and publish search ephemeris data.
7. Computing facilities to generate orbital data of sufficient accuracy to satisfy scientific needs.

[11] This complete instrumentation network should be coordinated with similar activities of the Department of Defense, but the special requirements of the civilian space program are such as to require the NASA to establish and operate some of the stations. The technical requirements of the space communication channels, telemetry, et cetera, should likewise be coordinated with the Department of Defense.

In view of the radio frequency requirements of the space program for communication with space vehicles, it is recommended that NASA take the necessary steps to insure that frequency assignments for this purpose are available.

Overseas stations of the NASA could be operated by local technical groups, universities, et cetera, and this phase of the problem should be actively pursued by NASA, for reasons both of efficient and economical operation and of international cooperation.

It is not recommended that the NASA offer to support the continued operation of the present IGY tracking system for an interim period after the expiration of the present IGY support. It is recommended, however, that a study be made of possible radio tracking systems to replace or supplement the present Minitrack stations. It is believed that a permanent radio tracking system should be capable of receiving signals at higher frequencies and from larger numbers of satellites, should probably offer greater angular coverage, and may require a different geographical plan. Special attention needs to be given to the reception of signals of broader bandwidth to take care of future satellites which may have a relatively large quantity of information to transmit back to earth.

Real-Time Communication

Certain projects will require real-time computation of orbits and communication of the data to other ground stations at large earth distances. A capability for communication with the satellite essentially all the time may also be desirable, particularly for manned flights. It appears, however, that such a situation may not be completely feasible, either technically or economically, in the near future, and therefor the communication system which can be provided may prove to be one of the limiting factors in the design of the experiment. Hard wire, which is considered to be the only currently available communication system whose reliability approaches 100 percent, extends only from Hawaii to Italy by commercial cable. All radio systems of substantial range are less reliable, except for line-of-sight operations such as communication satellites might provide. Since many agencies are concerned with this matter, and many important design decisions must be taken to yield the most [12] generally useful satellite communications relay system, NASA should take the initiative in coordinating the various requirements and settling on a preferred system at the earliest possible date. Furthermore, projects requiring real-time communication should formulate a rather complete communications plan early in the project-planning stage.

Recovery

The requirements of recovery of instrumented and manned satellites from orbital flight pose problems involving equipment, communication, and operation which are of very great magnitude. The escape maneuver during both the launch and recovery phases will require recovery capability over large areas of the Atlantic Ocean, the Pacific Ocean, and possibly the United States Zone of the Interior.

It appears that a coordinated national effort is required to cope with this problem.

It is recommended, therefor, that NASA establish a working group on recovery systems which will summarize the experience obtained to date, will define the problems to be solved, and propose operational techniques and equipment which should be developed.

One possible solution would be for the Atlantic, Pacific, and White Sands Missile Ranges to establish coordinated operational groups for these three areas, making maximum use of existing organization and facilities, for all national space programs requiring recovery techniques.

Space Surveillance

It is not considered necessary for NASA to set up the ground equipment and to maintain current ephemerides of all passive satellites, although, of course, ephemerides will be required for all satellites during the course for their experiments and for all satellites intended for recovery.

It is considered important that some kind of control be applied to limit the life of any satellite radio transmitter to a reasonable duration of experiment, in order to prevent cluttering up useful parts of the radio spectrum. However, no non-military need is anticipated, at this time, for a "vacuum cleaner" to remove from orbit the satellites that have outlived their usefulness.

[13] **Flight Program**

Re-entry Vehicles

Types of and uses for non-satellite probes and instrumented satellites have already been commented upon. Manned satellites, however, must be capable of safely re-entering the earth's atmosphere and being recovered. As a result of study of a number of suggested satellite vehicles for manned flight, it is concluded that:

1. The ballistic (pure drag) type vehicle can probably be put in operation soonest because:

    a. The booster problem is simplest by virtue of the low weight of this satellite vehicle.

    b. The aerodynamic heating problem is well understood.

    c. The development of the vehicle appears to be straight-forward.

2. The high-drag, high-lift vehicle study should be carried on concurrently because:

a. The ability to steer during re-entry eases the recovery problem, since it reduces the accuracy required of the retrograde rocket timing and impulse, and allows the vehicle to be flown to or near the ground or sea recovery stations.

b. The danger of excessive accidental decelerations due to malfunction in either the boost phase or re-entry phase of flight is greatly diminished.

3. The low-drag, high-lift vehicle looks less attractive for application to manned space flight for the near future. The advantages of better range control and greater maneuverability after re-entry may eventually make this vehicle more desirable.

Propulsion

There has been much discussion of the relative merits of developing a large booster engine or of clustering small ones. Both of these developments are required.

[14] Schedule studies clearly indicate that a booster of one million pounds thrust or more could be available about three years earlier of it were based on the clustering of existing rocket engines. This would lead to a fourth generation of space vehicles (with Vanguard Jupiter C being the first; IRBM-boosted space vehicles being the second; ICBM-boosted vehicles the third generation.) Progress in the rocket engine field offers a high degree of confidence that multiple-barrel boosters of one to one and a half million pounds total thrust could be ready for flight test in two to three years. Fifth-generation boosters based on the one million pounds-plus thrust, single-barrel engine (whether using one such engine or several) would offer orbital payloads up to 100,000 pounds, and would be available three years later.

It is strongly recommended that a study be made to assess the advisability of developing recoverable first-stage boosters. Recovery techniques should be optimistic from a system point of view.

Strong research effort on novel propulsion systems for vacuum operations is urged, and development of high-energy-propellant systems for upper stages should receive full support.

Vehicles for Early Experiments

In the preceding section several generations of space vehicle boosters are identified in general terms. The first generation, already in being, is capable of putting into orbit payloads of approximately 30 pounds. Such a vehicle enables the observation of a relatively small number of space environmental factors, or the conduct of simple experiments in the space environmental factors, or the conduct of simple experiments in the space environment. The second generation, with payload capabilities up to roughly 300 pounds, enables more sophisticated or larger numbers of experiments and environmental observations. The third-generation vehicles should make possible payloads of 3,000 pounds or more. Heavy or bulky observing instruments with provision for long-time attitude control and data transmission can be carried, and minimal manned space flights should be possible.

In each of these generations a number of boosters and upper stages are either available or under development. Proper combinations of these should make possible a wide spectrum of payloads and performances. Furthermore, it is likely that early generation vehicles will continue to be used even after later generation vehicles are available. Therefor the NASA should make a thorough study of the capabilities of existing stages to determine whether there are any serious gaps in the spectrum, and to select particular combinations of further development and use in these early experiments. [15] With properly selective effort going into the early generations, a more vigorous development program for later generations of boosters and vehicles should be possible.

Conclusion

Scientific advances of the broadest import can result from substantially improved understanding of cosmic processes and their influence upon the environment, and therefor the inhabitants, of the earth. The acquisition of such understanding depends critically upon the establishment of observational vantage points outside the insulation of the earth's atmosphere. The discussions and suggestions regarding research policies, procedures and programs presented in this report are intended to further the rapid and efficient development of the requisite space flight capabilities. All of these suggestions include recommendations, either stated or implicit, for cooperation or close coordination within related work by other civil and military agencies. More detailed discussions and program recommendations in particular fields are treated by Working Group reports....

Document III-2

Document title: Office of Program Planning and Evaluation, "The Long Range Plan of the National Aeronautics and Space Administration," December 16, 1959, pp 1-3, 9-11, 17-18, 26, 44.

Source: NASA Historical Reference Collection, History Office, NASA Headquarters, Washington, D.C.

This initial ten-year plan for NASA was developed during the agency's first year of operation. Because it contained both target dates for various accomplishments and budget estimates for the decade, it received a "Secret" security classification, and was later declassified.

[1]
Introduction

The long-term national objectives of the United States in aeronautical and space activities are stated in general terms in the enabling legislation establishing NASA. It is the responsibility of NASA to interpret the legislative language in more specific terms and to assure that the program so generated provides an efficient means of achieving the following objectives expressed in PL 85-568, Sec. 102(c) as:

"The aeronautical and space activities of the United States shall be conducted so as to contribute materially to one or more of the following objectives:

(1) The expansion of human knowledge of phenomena in the atmosphere and space;

(2) The improvement of the usefulness, performance, speed, safety, and efficiency of aeronautical and space vehicles;

(3) The development and operation of vehicles capable of carrying instruments, equipment, supplies, and living organisms through space;

(4) The establishment of long-range studies of the potential benefits to be gained from, the opportunities for, and the problems involved in the utilization of aeronautical and space activities for peaceful and scientific purposes;

(5) The preservation of the role of the United States as a leader in aeronautical and space science and technology and in the application thereof to the conduct of peaceful activities within and outside the atmosphere;

(6) The making available to agencies directly concerned with national defense of discoveries that have military value or significance, and the furnishing by such agencies, to the civilian [2] agency established to direct and control non-military aeronautical and space activities, of information as to discoveries which have value or significance to that agency;

(7) Cooperation by the United States with other nations and groups of nations in

work done pursuant to this Act and in the peaceful application of the results thereof; and

(8) The most effective utilization of the scientific and engineering resources of the United States, with close cooperation among all interested agencies of the United States, in order to avoid unnecessary duplication of effort, facilities, and equipment."

In operational terms, these objectives are instructions to explore and to utilize both the atmosphere and the regions outside the earth's atmosphere for peaceful and scientific purposes, while at the same time providing research support to the Department of Defense. These objectives can be attained only by means of a broad and soundly conceived program of research, development and operations in space. In the long run, such activities should make feasible the manned exploration of the moon and the nearby planets, and this exploration may thus be taken as a long-term goal of NASA activities. To assure steady and rapid progress toward these objectives, a NASA Long Range Plan has been developed and it is presented in this document.

In interpreting the Plan, it must be remembered that the implications for the national economy reach far beyond the specific program goals. For example, the space science activities cover the frontiers of almost all the major areas of the physical sciences, and these activities thus provide support of the physical sciences in specific applications in the fields of electronics, materials, propulsion, etc., will contribute, directly or indirectly, to all subsequent military weapons developments and to many unforeseen civilian applications. Reciprocally, the NASA program is provided with [3] support, direct or indirect, from all the related research and development activities outside NASA.

The Plan is presented at a level of effort which corresponds to an efficient and steadily growing capability. The rate of progress could be improved by an increased funding level, primarily by improving the certainty of the timely completion of the many essential engineering developments. On the other hand, a significantly lower scale of funding could be accommodated only by arbitrarily limiting the activities to a narrow line and by greatly reducing the rate of approach to the long-term goals.

[9]
Table I
NASA Mission Target Dates

Calendar Year	
1960	First launching of a Meteorological Satellite.
	First launching of a Passive Reflector Communications Satellite.
	First launching of a Scout vehicle.
	First launching of a Thor-Delta vehicle.
	First launching of an Atlas-Agena-B vehicle (by the Department of Defense).
	First suborbital flight of an astronaut.
1961	First launching of a lunar impact vehicle.
	First launching of an Atlas-Centaur vehicle.
1961–1962	Attainment of manned space flight, Project Mercury.
1962	First launching to the vicinity of Venus and/or Mars.
1963	First launching of two stage Saturn vehicle.
1963–1964	First launching of unmanned vehicle for controlled landing on the moon.
	First launching Orbiting Astronomical and Radio Astronomy Observatory.
1964	First launching of unmanned lunar circumnavigation and return to earth vehicle.
	First reconnaissance of Mars and/or Venus by an unmanned vehicle.
1965–1967	First launching in a program leading to manned circumlunar flight and to permanent near-earth space station.
Beyond 1970	Manned flight to the moon.

[10]
Table II
Current Funds & Anticipated Funding Requirements

Fiscal Year	1960*	1961	1962	1963	1964	1965	1966	1967	1968	1969
Research & Development							*Extrapolated*			
Launching vehicle Development	57	140	163	230	375	325	295	235	210	210
Space Propulsion Technology	39	51	118	120	90	75	95	95	95	95
Vehicle Systems Technology	13	30	47	49	50	50	50	50	50	50
Manned Space Flight	87	108	120	135	180	260	260	340	360	360
Scientific Investig. in Space	82	95	140	145	150	165	215	230	300	300
Satellite Applications	11	27	36	60	75	80	75	70	65	65
Aeronautical & Space Research	28	61	70	70	70	70	70	70	70	70
Space Flight Operations	16	33	42	50	55	60	60	60	60	60
Total Research & Development	333	545	736	859	1045	1085	1120	1150	1210	1210
Construction & Equipment	100	90	113	137	130	125	110	105	95	95
Salaries & Expenses	91	168	175	175	175	175	175	175	175	175
Advanced Projects					70	100	120	120	120	
Total Funds Required	524	802	1024	1171	1350	1455	1505	1550	1600	1600

*Includes 1959 Supplemental and 1960 Supplemental Request

[11]
Figure I
Current & Anticipated Funding Requirements

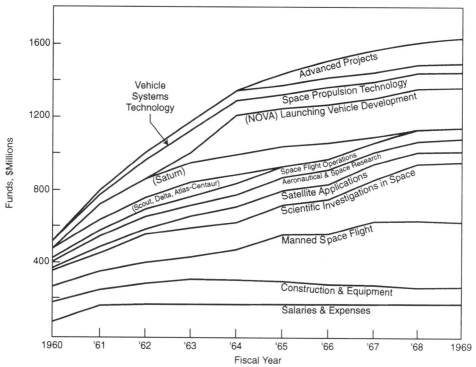

[17]
Table IV
Performance of NASA Launching Vehicles

Vehicle	1st Stage Thrust 1,000 Lbs.	Mission		
		Low Earth Orbit	Moon Probe	Planet Probe
		Spacecraft Wt. Lbs.		
In Use:				
Redstone	80	Used in Project Mercury Develop.		
Atlas	360	Project Mercury Capsule		
Juno II	150	100	20	
Thor-Able	150	200	80	
Atlas-Able	360		370	
Under Development				
Scout	100	200–240		
Thor-Delta	150	400–500	60	
Thor-Agena B[1]	150	1,200–1,500	350[2]	200[2]
Atlas-Agena B[1]	360	4,500–5,500	750–1,000	350–500
Atlas-Centaur	360	8,000–9,000	2,300–2,700	1,500–1,900
Saturn (initially)	1,500	28,500	9,000	7,000

1. DOD Development
2. With additional stage

[18]
Table V
Launching Vehicle Development

Fiscal Year	60	61	62	63	64	65	66	67	68	69
Scout										
Flights	4	2	2							
Funds, $M	2.8									
Thor-Delta										
Flights	1 / 1	1 / 2 / 1	1	5[a]						
Funds, $M	13.3	12.5	3							
Atlas-Vega										
Funds, $M	4.0									
Atlas Centaur										
Decision Points		Δ¹								
Flights		1	5	4	5	5				
Funds, $M	37.0	47.0	55	65	50					
Saturn										
Configuration Analysis	■									
Decision Points	Δ²									
Flights			2	2	3	4	4	4		
Funds, $M	(70)[b]	81[c]	105	115	115	115	85	25		
Nova										
Feasibility Studies	■■■■■■■									
Decision Points				Δ³						
Flights									1	2
Funds, $M				50	210	210	210	210	210	210
Total R&D Funds, $M[d]	57.1	140.5	163	230	375	325	295	235	210	210

(a) Beginning in 1962 Thor-Delta replaced by Thor-Agena B
(b) Funded by Department of Defense
(c) Total FY 1961 funding for Saturn $140M—includes $46M for S&E and 13M for C&E not shown
(d) Vehicle Procurement beyond development phase shown on this table is funded by the using project.

1. Decide time of replacement of Atlas-Agena with Atlas-Centaur
2. Select upper stages for the Saturn vehicle
3. Determine configuration of the Nova vehicle

[26]

Table VII
Vehicle Systems Technology

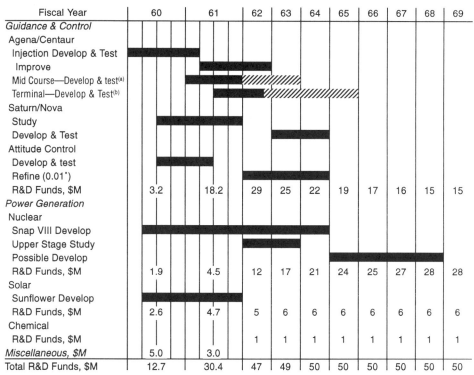

(a) Initial flight use in Mid-1961, refined and improved by 1963. (b) Initial flight use in 1963, refined and improved by 1965.

[44]

Figure II
Distribution of Aeronautical & Space Research Effort by Problem Area

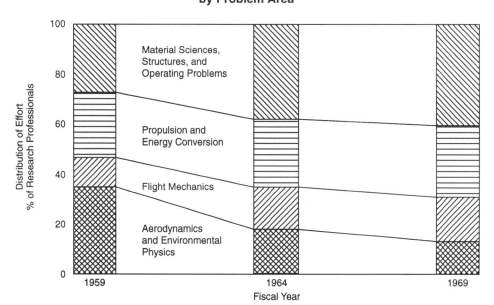

Document III-3

Document title: President's Science Advisory Committee, "Report of the Ad Hoc Panel on Man-in-Space," December 16, 1960.

Source: NASA Historical Reference Collection, History Office, NASA Headquarters, Washington, D.C.

When NASA submitted its 1962 fiscal budget request to the Bureau of the Budget in May 1960, President Eisenhower learned for the first time of the agency's plans for a lunar landing program. He asked Presidential Science Advisor George Kistiatowsky to study "the goals, the missions and the costs" of the manned spaceflight program that NASA had in mind. The study was chaired by Brown University chemistry professor Donald Hornig and was presented to the president at a December 20, 1960, meeting. Eisenhower has been quoted as saying at this time that he was not willing to "hock his jewels" (referring to the decision by Spanish monarchs Ferdinand and Isabella to finance the initial expedition of Christopher Columbus) to send people to the Moon. The handwritten figures included in this report have been omitted.

---

[1] # Report of the Ad Hoc Panel on Man-in-Space

1. Introduction

We have been plunged into a race for the conquest of outer space. As a reason for this undertaking some look to the new and exciting scientific discoveries which are certain to be made. Others feel the challenge to transport man beyond frontiers he scarcely dared dream about until now. But at present the most impelling reason for our effort has been the international political situation which demands that we demonstrate our technological capabilities if we are to maintain our position of leadership. For all of these reasons we have embarked on a complex and costly adventure. It is the purpose of this report to clarify the goals, the missions and the costs of this effort in the foreseeable future, particularly with regard to the man-in-space program.*

This report has been made possible by the complete cooperation of the National Aeronautics and Space Administration. Officials of the NASA presented a very impressive description of their detailed plans for development, utilization and costs of the Saturn vehicle. They also provided technical information on possible follow-on vehicles, advanced propulsion techniques, and possible development and funding schedules, As far as we can tell, the NASA program is well thought through, and we believe that the mission, schedules and costs are as realistic as possible at this time. We had to project their plans beyond 1970, and such projections must be seen as only crude estimates.

[2] ## 2. The Man-in-Space Program

The initial American attempt to launch a manned capsule into orbital flight, Project Mercury, is already well advanced. It is a somewhat marginal effort, limited by the thrust of the Atlas booster. It has as its goal the launching of a one man capsule into orbit around the earth and its successful return to earth. The fact that the thrust of any available American booster is barely sufficient for the purpose means that it is difficult to achieve a high probability of a successful flight while also providing adequate safety for the Astronaut. Achieving reliability on both accounts will strain our capabilities. A difficult decision will soon be necessary as to when or whether a manned flight should be launched. The chief justification for pushing Project Mercury on the present time scale lies in the political desire either to be the first nation to send a man into orbit, or at least to be a close second.

---

*No attempt has been made to include manned space programs initiates, or to be initiated, by the DOD.

The marginal capability cannot be changed substantially until the Saturn booster becomes available. The NASA program for utilizing Saturn involves the development of the so-called Apollo spacecraft. The Saturn rocket which is being developed now (C-1) should be capable of launching a spacecraft of about 19,000 lbs into a low earth orbit. The proposed Apollo spacecraft weight of 15,000 lbs is well within this limit and would enable orbital qualification flights of the Apollo spacecraft (some manned) about 1966-1968. Such a manned flight would occur after about 25 Saturn C-1's have been tested and much depends on whether a demonstrated reliability can be attained in this rather small number of tests. The Apollo spacecraft, as presently envisioned, would carry three men who would exercise control from within the spacecraft and be able to return to earth within a fairly well defined area. The chief purpose of the early Apollo missions would be to gain experience in manned flight, to learn more of the problems encountered by crews under such new conditions and to aid in the development of a spacecraft for more ambitious missions.

The full capabilities of the Saturn booster cannot be utilized until a large hydrogen-oxygen second stage has been developed. The C-2 Saturn, utilizing the new high-performance stage, is expected to enter the test phase about 1965 and may be available for manned flight (No. 17) in 1968 or 1969. There is again a question as to whether 16 flights will be enough to demonstrate sufficient reliability for its use in manned missions.

The Saturn C-2 is expected to lift about 40,000 lbs into low earth orbit and it is planned to utilize this capability to send up an "orbiting laboratory" capable of staying aloft for two weeks or more. It is our opinion that an [3] orbiting laboratory of this size could produce considerably more scientific information if it were wholly instrumented rather than manned. Alternatively, we believe that the valid scientific missions to be performed by a manned laboratory of this size could be accomplished using a much smaller unmanned instrumented spacecraft which would in turn require a smaller booster system. The large manned orbiting laboratory might be of value as a life sciences laboratory to acquire physiological and psychological data on humans, to study life support mechanisms, to perform biological studies, and to carry out engineering tests under gravity free conditions. In short, its major mission appears to be the preparation for further steps in the manned exploration of space.

To take such steps, the Apollo spacecraft may be launched into successively more elliptical orbits which carry it further and further from the earth, culminating about 1970 in a manned flight around the moon and back to the earth. The Apollo program in itself does not reach what might be considered to be the next major goal in manned space flight, i.e. manned landing on the moon. It does, however, appear to represent a logical approach to that goal in that it will develop space craft and crews for space flight and will enable us to gain experience in navigation and successful return from increasingly difficult trips. In the meantime it should be possible to obtain far more detailed information about the moon by unmanned spacecraft and lunar landing craft than the crew of the circumlunar flight could gain.

None of the boosters now planned for development are capable of landing on the moon with sufficient auxiliary equipment to return the crew safely to earth. To achieve this goal, a new program much larger than Saturn will be needed. It is likely to take one of three forms:

1. An all-chemical liquid-fueled rocket, the Nova, might be developed to take the trip directly. It would require a booster with about 6 times the thrust of the Saturn and utilyzing either kerosene or hydrogen-oxygen. The upper stage of the Nova would require hydrogen-oxygen and at least one stage would probably be an existing stage from the Saturn development program.

2. If a suitable nuclear upper stage could be developed, the Nova vehicle could conceivably become a combination chemical-nuclear system. This system would still require the development of a first stage chemical booster with thrust of the same order of magnitude as that described for the all chemical system. If the nuclear development should be as successful as its proponents hope, it might open the way for future developments beyond

the [4] possibilities envisioned for chemical rockets. However, a sound decision on the promise of nuclear rockets cannot be made until about 1963.

3. Rendezvous techniques, utilizing either Saturn C-2 vehicles or some type of advanced Saturn vehicles, could be employed to lift into an earth orbit the hardware and fuel necessary to perform the manned lunar landing mission. In this system, a series of vehicles would be launched into a temporary earth orbit where they would rendezvous to enable fueling of the spacecraft and, if necessary, assembly of the component parts of the spacecraft. This spacecraft would then be used to transport the manned payload to the moon and thence back to Earth. These techniques will require considerable development, and are at present only in a preliminary study phase.

It is clear that any of the routes to land a man on the moon require a development much more ambitious than the present Saturn program. Not only must much bigger boosters probably be developed, but rockets and guidance mechanisms for the safe landing and then for return from moon to earth by means of additional rockets must be developed and tested. Nevertheless, it must be pointed out that this new, major step is implicit in undertaking the proposed manned Saturn program, for the first really big achievement of the man-in-space program would be the lunar landing.

The succeeding step, manned flight to the vicinity of Venus or Mars represents a problem an order of magnitude greater than that involved in the manned lunar landing. Not only does it appear to be insoluble in terms of chemical rockets, thus requiring the development of suitable nuclear rockets or nuclear-powered electric propulsion devices, but it also poses serious problems in terms of life support and radiation shielding for journeys requiring times ranging from many months to years.

3. Unmanned Programs Related to Man-in-Space

A great part of the unmanned program for the scientific exploration of space is a necessary prerequisite to manned flight. The programs which are now planned fall in the following general categories:

1. The general scientific exploration of space. This will take place in a continuing series of flights. This program has been moving along well and has been marked by solid scientific achievement; it could probably be carried on to a high state of advancement using launch vehicles no larger than Centaur (an Atlas with hydrogen-oxygen upper stage).

[5]   2. A rough landing on the moon, with television recording of the impact and with a surviving seismometer to make measurements on the lunar surface, may be made in 1962 or 1963 using an Atlas-Agena-B vehicle.

3. The Centaur rocket, which should make its first flight in 1961, will make it possible to fly instruments past Venus and Mars, making close-up scientific observations for a short time, in 1962 or 1964. It may even be possible to land a 10 lbs instrument capsule through their atmospheres.

4. The Centaur should also make it possible to soft land 190 lbs of scientific gear on the surface of the moon (1964-1966) and to make surface observations from a very close orbit about the moon, including photography comparable to satellite photography of the earth (Samos).

5. The Saturn C-2 will be the first vehicle which can carry an adequately instrumented spacecraft, weighing perhaps 325 lbs, into an orbit about Venus or Mars, and to land a 225 lbs capsule through their atmosphere, giving us direct atmospheric and surface measurements for the first time in about 1967 or 1968. It may then be possible to obtain definite evidence regarding life on Mars. Although such studies can be started with the Saturn C-1 in 1965 or 1966, they really require the C-2 to give reasonable instrument weights.

6. A roving automatic vehicle equipped with television and other sensing instruments to make observations on the surface of the moon can first be landed with the C-2 in about 1967, and is included in present NASA plans.

7. It should also be possible to soft land so object on the moon which is large enough to send a capsule back to earth with a few pound sample of the surface of the moon. This also requires the C-2 and could be tried beginning in about 1968.

8. No booster smaller than the C-2 can carry scientific instruments to the vicinity of Mercury or Jupiter. This, too, should be possible around 1968 to 1970.

9. For unmanned scientific investigations with roving vehicles on the planets, or for more ambitious instrumented missions out of the plane of the ecliptic, even the Saturn C-2 does not provide sufficient payload-carrying capability.

[6] **4. Relation between Manned and Unmanned Space Exploration**

Certainly among the major reasons for attending the manned exploration of space are emotional compulsions and national aspirations. These are not subjects which can be discussed on technical grounds. However, it can be asked whether the presence of a man adds to the variety or quality of the observations which can be made from unmanned vehicles, in short whether there is a scientific justification to include man in space vehicles.

It is said that an astronaut's judgment, decision-making capability and resourcefulness can increase the probability of successful accomplishment of a space mission and expand the variety and quality of observations performed. On the other hand, man's senses can be satisfactorily duplicated at remote locations by the use of available instrumentation and advances in the state of the art are continually increasing the ability to transmit information back to a central receiving point. With such an instrumented system, the decisions requiring man's mental capabilities can be performed by many men in a normal environment and with the aid of elaborate computational aids, where necessary.

The following considerations seem pertinent:

1. Information from unmanned flights is a necessary prerequisite to manned flight.

2. The degree of reliability that can be accepted in the entire mechanism is very much less for unmanned than for manned vehicles. As the systems become more complex this may make a decisive difference in what one dares to undertake at any given time.

3. From a purely scientific point of view it should be noted that unmanned flights to a given objective can be undertaken much earlier. Hence repeated observations, changes of objectives and the learning by experience are more feasible.

It seems, therefore, to us at the present time that man-in-space cannot be justified on purely scientific grounds, although more thought may show that there are situations for which this is not true. On the other hand, it may be argued that much of the motivation and drive for the scientific exploration of space is derived from the dream of man's getting into space himself.

[7] **5. Cost of the NASA Man-in-Space Program**

The NASA man-in-space program, exclusive of the Mercury Project, revolves around the use of the Saturn and Nova vehicles. Development of the Saturn is far enough along that its characteristics are fairly well known, and the costs of its development and use can be predicted with reasonable accuracy. The Nova, required for direct manned operations on the moon, is based on the use of the 1.5 million lbs thrust engine, six of which would probably power the first stage. The character of the vehicle as a whole cannot be clearly determined until the characteristics of this engine are understood. However, the present tentative designs of the Nova configuration are probably adequate to support the very rough cost analysis presented here.

This analysis is based on the rule-of-thumb principle, generally supported by past experience, that the cost of a program of this nature, including development, flight test and use, should be approximately proportional to the dry weight of the booster vehicle and payload on which the program is based. The dry weight of the Nova vehicle is about six times that of the Saturn vehicle, and accordingly a factor of six should be applied to the costs of the two programs. It is pointed out, however, by the NASA that there is some reason to believe that a somewhat smaller factor might be appropriate. There is a good deal of basic engineering that will carry over from Saturn to Nova, and certain of the Nova stages may already have been developed for Saturn. Such considerations are doubtless valid, but they could not justify the use of a factor smaller than four. In the analysis that follows two values of the multiplicative factor are used: four, representing the lower bound on what might be achieved, and six, representing a reasonably conservative estimate.

It is further assumed that the time span required for the development and exploitation of the capabilities of the Nova are the same as that for Saturn. It is assumed, however, that the Nova development follows that of Saturn by seven years. Thus, by 1968 Nova is in a state of development corresponding to that of Saturn in 1961.

With these assumptions in mind, the method of arriving at the yearly costs given in the figure can be stated.

1. The known and estimated costs for the development and use of Saturn are plotted on the curve so labeled in Figure 1. The costs following 1970 are not NASA estimates, but are predicated on the likelihood of some continuing use for this vehicle.

[8] 2. The "Saturn" curve is now displaced to the right by seven years, and the ordinate multiplied by the factors four and six. This produces the solid sections of the curves labeled "Nova" in Figure 1. The dashed left-hand tails of the Nova curves represent pure estimate and have only reasonableness to recommend them.

3. The "Saturn" and "Nova" curves have been added year by year to produce the composite curves of Figure 2. These are taken to represent rough bounds on the cost of the NASA man-in-space program.

4. The integrated areas under the curves represent the total expenditures for the period 1961 through 1975. As indicated on the figures, the total Saturn program costs 8 billion (1961) dollars up to 1975. The Nova program over the same time period comes to 25.5 billion on the lower estimate and 38 billion on the higher estimate. (It will be noted that these totals are not four and six times, respectively, the total Saturn cost. This is because the Nova costs were integrated only out to 1975, when the first manned lunar landing might be achieved. The Saturn costs, on the other hand, were integrated over the entire estimated program.) Figure 2 gives the total composite expenditure to 1975 as 33.5 billion for the lower estimate and 46 billion for the higher.

The cut-off at 1975 is arbitrary and might be misleading. During the five or ten year period preceding this date new developments will be under way to implement new programs for the post 1975 era. It does not seem possible at this time to estimate the incremental costs associated with these programs.

Present indications suggest that alternative methods, described elsewhere in this report, of accomplishing the manned lunar landing mission, could not be expected to alter substantially the over-all cost of mission as analyzed here on the basis of Nova.

In the event that additional flight testing is required to achieve adequate reliability in these programs, it seems likely that the program would be stretched out in time. Thus probably the annual expenditures would not change appreciably, although the integrated expenditure would increase accordingly.

6. Conclusions

1. The first major goal of the man-in-space program is to orbit a man about the earth. It will cost about 350 million dollars.

[9] 2. The next goal, of an intermediate nature, is the manned circum-navigation of the moon. It will cost about 8 billion dollars.

3. The second major goal, landing on the moon, can only be achieved about 1975 after an additional national expenditure in the vicinity of 26 to 38 billion dollars.

4. The Saturn program is a necessary intermediate step toward manned lunar landing but must be followed by a much bigger development before manned lunar landing is possible.

5. The unmanned program is a necessary prerequisite to a manned program. Even if there were no manned program, the unmanned program might yield as much scientific knowledge and on this basis would be justified in its own right.

6. Even if there were no man-in-space program, Saturn C-2 is still a minimum vehicle for close-up instrumented study of Venus and Mars, for unmanned trips to more distant planets, and for putting roving vehicles on the surface of the moon.

7. Manned trips to the vicinity of Venus or Mars are not yet foreseeable....

Document III-4

Document title: Richard E. Neustadt, "Problems of Space Programs," December 20, 1960, attached to Memorandum for Senator Kennedy, "Memo on Space Problems for you to use with Lyndon Johnson," December 23, 1960.

Source: Pre-Presidential Papers, John F. Kennedy Presidential Library, Boston, Massachusetts.

Eisenhower made several recommendations concerning space before leaving office. Some of these, such as the elimination of the Civilian-Military Liaison Committee, were followed by the Kennedy administration. Others, such as the elimination of the National Aeronautics and Space Council (NASC), were not. Eisenhower recommended that the council be abolished, but Richard E. Neustadt, who had worked on the Democratic Party Platform Committee and was serving as consultant to the president-elect, recommended that the vice president be named chairman of the NASC in a memo to Kennedy on December 20, 1960. Neustadt also was the first to bring the Saturn rocket program to Kennedy's attention and to note that it was needed only if the United States intended "to put a man on the moon and get him back before or soon after the Russians do."

[1] December 20, 1960

Problems of Space Programs

The "space" programs both civil and military, raise problems of great difficulty. Superficially these are problems of budget and organization. Essentially they are problems of policy direction.

The following approximate figures include the growth and projected magnitude of the space programs:

New Obligational Authority in millions

	1957 & prior	1958	1959	1960	1961 Approx	1962 Approx	1965 Projected
NASA	--	117	305	524	965	1,110	2,000
AEC (nuclear power for space use)	--	20	33	52	45	53	100
Defense (identifiable space programs)	95	92	511	543	740	825	2,000
Total	95	229	849	1,119	1,750	1,988	4,100

Organizationally, there are two Government space programs: (1) a civilian space program which is the responsibility of NASA (and of AEC with respect to reactor development), and (2) a military space program consisting of activities considered by the Pentagon to be specifically required for defense; these are the responsibility of the Department of Defense.

The existence of two programs has resulted in a certain amount of actual duplication on communications satellites, manned space flight programs, and supporting research and development. The tendency toward duplication has to be watched carefully; there is always danger that it will get out of hand. The Civilian-Military Liaison Committee, established by law, has become inoperative, as a result of experience which showed it to [2] be ineffective in coordination operations. At present there is an administrative established "Aeronautics and Space Coordinating Board" consisting of representatives of NASA and Defense through which operational coordination is being sought.

The Problem of "National Prestige"

Since Sputnik we have been in a race to be "first" in physical achievements of a dramatic sort–the sort which has high visibility and thus makes an impression on mass opinion, especially abroad. This has got us into the business of pressing achievements for the sake of their psychological effect, regardless of concrete scientific or military utility. The dollar costs are high and are bound to grow much higher. The big booster program (see below) is a classic case and demonstrates NASA's expenditures into the future. The dollar costs represent diversion of resources–money, manpower, facilities, scientific skills–from other parts of our national effort.

This is the heart of the problem.

The problem is that we need more funds for research and development of new weapon systems, more funds for science generally, more funds for economic development abroad, more funds for welfare purposes at home. Money spent to serve no concrete purpose save the psychological effect of being "first" is money we could well use for these other needs.

The problem is made sharper by the fact that on the kind of "firsts" which have had most dramatic mass appeal, the Russians may be well ahead of us. We have reason to think that, taken as a whole, our scientific programs of inquiry and exploration in space are more advanced [3] than Moscow's and have yielded more real scientific returns. But we have not yet found means of making our progress drastically apparent to the man-in-the-street around the world.

Two questions arise:

1. If we are behind and are likely to stay behind in the race for "Sputnik-type firsts" should we get out of the race and divert the resources now tied up in it to other uses which have tangible military, scientific or welfare value?

2. By what means, if any, can we make our underlying scientific "firsts" dramatic and appealing, especially aboard? How can we render visible a different sort of "race" which we are more likely to "win"?

Virtually every aspect of the NASA budget now and in the next several fiscal years will be affected by answers to these questions. Admittedly they are very hard questions to answer with anything like a simple yes or no, but reasonably clear answers are needed for the sake of budgetary guidelines in fiscal 1962 and after.

The Big Booster Program

Close to half the NASA budget for 1962 is bound up, in one way or another with this program.

The program has two parts:

First is the so-called Saturn, which, with luck, might become operational in about two years. This is a "bailing wire" devise intended to give us big booster capability for the short-run by combining and adapting devices designed for other purposes. The Saturn is a forced draft operation and an expensive one. Booster capability is needed in this form only in order to put a man on the moon and get him back before or soon after the Russians do. [4] Saturn, in short, is a prestige item. It will *not* affect and is distinct from getting a man in space, *per se*.

Second is the single-engine big booster which is under development for eventual use in a space vehicle designed to transport men and heavy equipment. This is the progenitor of the engines which eventually will be necessary for the "space ship" of the future. Regardless of the Russians, the United States may have reason to go forward with this development. It is a development which stimulates the imagination of young Americans who will be voting before too long. It is a development which *may*, in time, have military and economic uses not foreseeable. Finally, it is part of the whole forward push in technology. We have certainly learned from earlier experience that these forward steps cannot be stopped. But we have also learned that under present circumstances only Government support will get them taken.

The single-engine development is proceeding slowly, in second place for funds and other resources behind the Saturn project. It is necessary to consider: (1) "Prestige" apart, can Government afford, *in the next years*, the diversion of resources needed to bring either or both these boosters to fruition? (2) In the longer run, what proportion of Government resources, for what span of years, should go into developing the technology of space travel?

Civil, Military Duplications of Effort

The decision was made in 1958 to organize governmental space efforts as a civilian enterprise except for programs integrally related to the missions of the military services. The exception, of course, is very significant. Both NASA and the Defense Department will, inevitably, conduct research, development and operations in the space field. The [5] two sets of programs are not always easy to distinguish. For reasons of practicality and economy, NASA uses military facilities for much of its experimentation and testing. Defense, in turn, relies on NASA for some research and development.

The operating relationships between NASA and Defense are bound to be complex, but it does not necessarily follow that they need be inefficient. Nor does it follow that the two programs should be duplicative except where duplication serves a constructive purpose.

Unfortunately, there are many signs of inefficiency in the relationship and many indications of duplicative effort which may not meet the test of useful duplication. There are dual programs in communications, in manned space flight, in vehicles development and in applied research. The relative utility and need for these programs, both on the side of NASA and on the side of Defense, calls not so much for technical as for policy evaluation.

The National Aeronautical and Space Council was originally envisaged as a Cabinet-type advisory committee to help the President with policy evaluation, and to help him also in securing effective coordination of operative relationships. But the Council has not functioned in the past year. Meanwhile, a NASA-Defense committee has been established. Experience to date suggests that this may be a promising development in securing coordination at the working level. It is unlikely to resolve the problems of securing policy advice.

An opportunity now exists to revitalize the National Aeronautical and Space Council under the Chairmanship of the Vice President. Legislation will be required to put him in the chair. It might be timely to simplify the Council's title and to reconsider its statutory membership. If the council is to function effectively in the future, as it has not done in the past, it might be well to keep its membership relatively small and to have it operate selectively on high priority policy issues of the sort mentioned above.

R.E.N.

Document III-5

Document title: "Report to the President-Elect of the Ad Hoc Committee on Space," January 10, 1961.

Source: NASA Historical Reference Collection, History Office, NASA Headquarters, Washington, D.C.

John F. Kennedy was the first president-elect to set up high-level "transition teams" to advise him on issues that he would face upon assuming the presidency. His transition team on space was chaired by Massachusetts Institute of Technology professor Jerome B. Wiesner, a member of President Eisenhower's President's Science Advisory Committee (and thus familiar with discussions inside the Eisenhower administration on space policy and programs). Wiesner had advised Kennedy on science and technology issues during the Presidential campaign and would become the new president's science adviser. The report reflected the widespread skepticism within the scientific elite of the country over the value, and even the feasibility, of human spaceflight.

---

[1]   **I. Introduction**

Activities in space now comprise six major categories:

1. Ballistic missiles.
2. Scientific observations from satellites.
3. The exploration of the solar system with instruments carried in deep space probes.
4. Military space systems.
5. Man in orbit and in space.
6. Non-military applications of space technology.

We rely on the first member of the list, ballistic missiles, for a large part of the retaliatory response to the Russian missile threat.

It is generally assumed by the American citizen that our vast expenditures of money and technical talent in the national space program are primarily designed to meet the overriding needs of our military security. The fact is, however, that the sense of excitement and creativity has moved away from the missile field to the other components of the list, and that missiles, long before they are in condition for us to depend upon them, are slowly being delegated to the category of routine management. Before we proceed in this report to discuss and support the important activities in the other five categories we wish to emphasize the hazard of failing to complete and deploy on time our intercontinental deterrent missiles.

[2] In addition to the need to develop ballistic missiles to provide for our military security, there are five principal motivations for desiring a vital, effective space program. It is important to distinguish among them when attempting to evaluate our national space effort.

First, there is the factor of national prestige. Space exploration and exploits have captured the imagination of the peoples of the world. During the next few years the prestige of the United States will in part be determined by the leadership we demonstrate in space activities. It is within this context that we must consider man in space. Given time, a desire, considerable innovation, and sufficient effort and money, man can eventually explore our solar system. Given his enormous curiosity about the universe in which he lives and his compelling urge to go where no one has ever been before, this will be done.

Second, we believe that some space developments, in addition to missiles, can contribute much to our national security—both in terms of military systems and of arms-limitation inspection and control systems.

Third, the development of space vehicles affords new opportunities for scientific observation and experiment—adding to our knowledge and understanding of the earth, the solar system, and the universe. In the three years since serious space exploration was initiated the United States has been the outstanding contributor to space science. We should make every effort to continue and to improve this position.

Fourth, there are a number of important practical non-military applications for space technology—among them, satellite communications and broadcasting; satellite navigation and geodesy; meteorological reconnaissance; and satellite mapping—which can make important contributions to our civilian efforts and to our economy.

Finally, space activities, particularly in the fields of communications and in the exploration of our solar system, offer exciting possibilities for international cooperation with all the nations of the world. The very ambitious [3] and long-range space projects would prosper if they could be carried out in an atmosphere of cooperation as projects of all mankind instead of in the present atmosphere of national competition.

The ad hoc panel has made a hasty review of the national space program, keeping in mind the objective—to provide a survey of the program and to identify personnel, technical, or administrative problems which require the prompt attention of the Kennedy administration. We have identified a number of major problems in each of these categories, and they will be discussed in this report. It is obvious that there has been inadequate time to examine all facets of the program or to permit full consideration of the possible answers to many of the questions raised.

Because of the overriding necessity to provide more efficient and effective leadership for the program, the group has devoted a major portion of its time to this aspect of the space program. We will, however, indicate important scientific and technical problems which should be thoroughly examined as soon as possible. We have concluded that it is important to reassess thoroughly national objectives in the space effort—particularly in regard to man in space; space, science and exploration; and the non-military applications of space, in order to assure a proper division of effort among these activities. Space activities are so unbelievably expensive and people working in this field are so imaginative that the space program could easily grow to cost many more billions of dollars per year.

While we are now compelled to criticize our space program and its management, we must first give adequate recognition to the dedication and talent which brought about very real progress in space during the last few years. Our scientific accomplishments to date are impressive, but unfortunately, against the background of Soviet accomplishments with large boosters, they have not been impressive enough.

Our review of the United States' space program has disclosed a number of organizational and management deficiencies as well as problems of staffing and direction which should receive prompt attention from the new administration. These include serious problems within NASA, within the military establishment, and at the [4] executive and other policy-making levels of government. These matters are discussed in the sections which follow.

II. The Ballistic Missile Program

The nation's ballistic missile program is lagging. The development of the missiles and of the associated control systems, the base construction, and missile procurement must all be accelerated if we are to have the secure missile deterrent force soon that the country has been led to expect.

While additional funds will undoubtedly be required to accomplish this, we believe that re-establishing an effective, efficient, technically competent arrangement for the program is the overriding necessity.

Though the missile program is not ordinarily regarded as part of the space program, it is important to recognize that for the near future the achievement of an adequate deterrent force is much more important for the nation's security than are most of the space objectives, and that at least part of the difficulty in the management and execution of the program stems from the distraction within the Defense Department and in industry caused by vast new space projects. However, we have no alternative but to press forward, with space developments.

III. Organization and Management

There is an urgent need to establish more effective management and coordination of the United States space effort. The new administration has promised to move our country into a position of preeminence in the broad range of military, cultural, scientific and civilian applications of satellite and other space vehicles. This cannot be done without major improvements in the planning and direction of the program. Neither NASA as presently operated nor the fractionated military space program nor the long-dormant space council have been adequate to meet the challenge that the Soviet thrust into space has posed to our military security and to our position of leadership in the world.

[5] In addition to the difficulties and delays which the program has endured because of the lack of sufficient planning and direction, it has also been handicapped because too few of the country's outstanding scientists and engineers have been deeply committed to the development and research programs in the space field. In changing the management structure and in selecting the administrators for the effort, the need to make space activities attractive to a larger group of competent scientists and engineers should be a guiding principle.

The new administration has announced that it plans to use the National Aeronautics and Space Council for coordinating government space activities, or advising the President on policy on plans and on the implementation of programs. We believe that the space council can fulfill this role only if it is technically well-informed and, moreover, seriously accepts the responsibility for directing the conduct of a coherent national space effort. Particular care should be taken to insure the selection of a very competent and experienced staff to assist the Council

Not only must we provide more vigor, competence and integration in the space field, but we must also relate our space requirements to other vital programs which support our national policy. We refer particularly to the missile needs, already mentioned, and to the continuing need for development and research in the field of aeronautics.

Each of the military services has begun to create its own independent space program. This presents the problem of overlapping programs and duplication of the work of NASA. If the responsibility of all military space developments were to be assigned to one agency or military service within the Department of Defense, the Secretary of Defense would then be able to maintain control of the scope and direction of the program and the Space Council would have the responsibility for settling conflicts of interest between NASA and the Department of Defense.

With its present organizational structure and with the lack of strong technical and scientific personalities in the top echelons, it is highly unlikely [6] that NASA space activities can be greatly improved by vitalization of the Space Council.

We are also concerned by the NASA preoccupation with the development of an in-house research establishment. We feel that too large a fraction of the NASA program, particularly in the scientific fields, is being channeled into NASA-operated facilities. NASA's staff has had to expand much too rapidly and without adequate selectivity, so that many inexperienced people have been placed in positions of major responsibility. This has, in turn, made NASA less willing than would a more mature and competent organization to solicit and accept the advice of competent non-government scientists. This situation appears to be improving at the present time.

One important responsibility of NASA given little attention now in the organization, is that of providing for basic research and advanced development in the field of aeronautics. There is a general belief in the aviation industry that the national preoccupation with space developments has all but halted any advance in the theory and technology of aerodynamic flight. There is ample evidence to support the contention that the Russians and possibly the British, are surpassing us in this field and consequently in the development of supersonic commercial aircraft. We should make a substantial effort to correct this situation, possibly by getting some of NASA's aeronautical and aerodynamic experts back into the field of advanced aircraft research and development. Possibly, after careful investigation, the Space Council would prefer to stimulate this work by non-governmental arrangements, or by placing it entirely in another agency.

We believe that the work of NASA would be facilitated and the task of recruiting staff made possible if an outstanding expert was placed in charge of the direction and management of each of the following important areas of work:

    a. Propulsion and vehicle design and development
    b. The space sciences
[7] c. Non-military exploitation of space technology
    d. Aeronautical sciences and aircraft development

IV. The Booster Program

The inability of our rockets to lift large payloads into space is the key to the serious limitations of our space program. It is the reason for the current Russian advantage in undertaking manned space flight and a variety of ambitious unmanned missions. As a consequence, the rapid development of boosters with a greater weight-lifting capacity is a matter of national urgency.

Payload weight is currently limited by our dependence on modified military rockets as the primary boosters (THOR, JUPITER, ATLAS). Current plans call for the first substantial increase in payload with the addition of the CENTAUR upper stage to the ATLAS in 1962, followed by a second big step with the SATURN booster in 1965.

It is likely that a variety of new booster programs will be proposed in the near future, particularly for military projects. There are no fundamental differences in civilian and military requirements which are foreseeable now. If the national effort is to be focused and the very large expenditures are not to be distributed among an excessive number of booster programs, it is important that we maintain and strengthen the concept of a National Booster Program.

A number of problems may well arise in the National Booster Program. The present MERCURY program, based on the ATLAS, is marginal and if the ATLAS proves inadequate for the job it may be necessary to push alternatives vigorously. The first possibility appears to be the TITAN, although it has not yet demonstrated the reliability which is required. We should study the desirability of carrying out a TITAN-boosted MERCURY program in the event ATLAS should prove to be inadequate.

The CENTAUR rocket involves an entirely new technology and is still untested. If difficulties develop in this program within the next three or four months we must act promptly to initiate an alternate.

[8] Development of the SATURN-booster—a cluster of eight ATLAS engines—should continue to be prosecuted vigorously. However, it would be dangerous to rely on SATURN alone for the solution to our problems, either in the long or short term, for two reasons:

    (a) It is intrinsically so complex that there is a real question whether it can be made to function reliably.

    (b) It represents a maximum elaboration of present technology and provides no route to further development.

Therefore, the development of a very large single engine should proceed as fast as possible so that it may be a back-up for the SATURN cluster and a base for future larger vehicle development. The present F-1 (1.5 million lb. thrust) engine development should be studied to be sure it is progressing fast enough and has enough promise of success to fill this role. If the technological step in going from the present 180,000 lb. thrust engines to 1.5 million lbs. is so big as to make success marginal, a parallel development of a somewhat smaller engine should be started.

The nuclear rocket program (ROVER) presents an area in which some major decisions will have to be taken by the new administration. In principle the nuclear rocket can eventually carry heavier payloads much farther than any chemical rocket. Nevertheless, the technology is so new and the extrapolation from reactors developed now to sizes which would be useful in large rockets is so great that it is not clear how soon they will make an important contribution to the space program. The use of nuclear rockets will raise serious international political problems since the possibility that a reactor could reenter and fall on foreign territory cannot be ignored. A major technical and management review of the ROVER program seems urgent.

Above all we must encourage entirely new ideas which might lead to real breakthroughs. One such idea is the ORION proposal to utilize a large number of small nuclear bombs for rocket propulsion. This proposal should receive careful [9] study with a realization of the international problems associated with such a venture.

[Most of page 9, all of pages 10 and 11, and 1/3 of page 12 excised during declassification review]

[12] **VI. Science in Space and Space Exploration**

In the three years since space exploration began, experiments with satellites and deep space probes have provided a wealth of new scientific results of great significance. In spite of the limitations in our capability of lifting heavy payloads, we now hold a position of leadership in space science. American scientists have discovered the great belt of radiation, trapped within the earth's magnetic field. American scientists have revealed the existence of a system of electric currents that circle our planet. Our space vehicles have probed the interplanetary space to distances of tens of millions of miles from the earth. They have shown that the earth is not moving through an empty space but through an exceedingly thin magnetized plasma. They have intercepted streams of fast-moving plasma ejected from the sun which, upon reaching our planet, produce magnetic storms, trigger off auroral displays and disrupt radio communications.

From these and other experiments, there is gradually emerging an entirely novel picture of the conditions of space around our planet and of the sun-earth relations. One of the important tasks of space science in the next few years will be a full exploration of the new field revealed by the early experiments. There is little doubt that such exploration will lead to further important discoveries.

[13] Another scientific field, where space science promises an early and major breakthrough is that of astronomy. Until a few years ago, visible light from celestial objects, reaching our telescopes through the atmospheric planet, had been the only source of astronomical information available to man. The only other portion of the spectrum capable of penetrating the atmosphere and the ionosphere is that corresponding to short-wave radio signals. In recent years, the development of radio telescopes has made it possible to detect these signals. Radio astronomy has enormously advanced our knowledge of the universe. By means of radio telescopes we can now "see" not only the stars, but also the great masses of gas between the stars; we can detect the high-energy electrons produced by cosmic accelerators located thousands of millions of light years away from the earth.

We are entitled to expect a similar and even perhaps a more spectacular advance the day that we shall have telescopes installed aboard satellites circling the earth above the atmosphere and the ionosphere. These instruments will be capable of detecting the whole of the electro-magnetic spectrum—from long-wave radio signals to gamma-rays.

A third major task of space science in the years to come will be the exploration of the moon and the planets. Scientists are planning to fly instruments to the vicinity of these celestial objects, and eventually to land them upon their surface. From the data supplied by these instruments they expect to obtain information of decisive importance concerning the origin and the evolution of the solar system. Moreover, there is the distinct possibility that planetary exploration may lead to the discovery of extra-terrestrial forms of life. This clearly would be one of the greatest human achievements of all times.

Our present leadership in space science is due to a large extent to the early participation of some of our ablest scientists in our space program—primarily as part of the International Geophysical Year—and to the fact that these scientists were in a position to influence this program. Another important [14] factor was our initial advantage in instrumentation, which helped to offset our disadvantage in propulsion.

We must not delude ourselves into thinking that it will be easy for the U.S.A. to maintain in the future a prominent position in space science. The USSR has a number of competent scientists. It will be easier for them to catch up with us in instrument development than for our engineers to catch up with the Russians in the technique of propulsion. Thus we must push forward in space science as effectively and as forcefully as we can.

Our scientific program in space appears to be basically sound. However, to insure its success, the following requirements must be met.

1. In the planning of our space activities, scientific objectives must be assigned a prominent place.

2. Our space agency must insure a wide participation in its program by scientists from universities and industrial laboratories, where our greatest scientific strength lies.

3. It must provide adequate financial support for the development and construction of scientific payloads.

4. It must exert the greatest wisdom and foresight in the selection of the scientific missions and of the scientists assigned to carry them out.

5. It must initiate immediately a research program in advanced instrumentation, so that we may be ready to exploit fully the capability of flying heavier and more complex payloads that we shall possess several years from now. Problems of automation, processing and transmission of information must be tackled by competent and imaginative research teams.

NASA has not fulfilled all of the above requirements satisfactorily. We believe, as previously stated, that the main obstacle here has been the lack of a strong scientific personality in the top echelons of its organization.

[15] **VII. Man in Space**

We are rapidly approaching the time when the state of technology will make it possible for man to go out into space. It is sure that, as soon as this possibility exists, man will be compelled to make use of it, by the same motives that have compelled him to travel to the poles and to climb the highest mountains of the earth. There are also dimly perceived military and scientific missions in space which may prove to be very important.

Thus, manned exploration of space will certainly come to pass and we believe that the United States must play a vigorous role in this venture. However, in order to achieve an effective and sound program in this field, a number of facts must be clearly understood.

1. Because of our lag in the development of large boosters, it is very unlikely that we shall be first in placing a man into orbit around the earth.

2. While the successful orbiting of a man about the earth is not an end unto itself, it will provide a necessary stepping stone toward the establishment of a space station and for the eventual manned exploration of the moon and the planets. The ultimate goal of this kind of endeavor would, of course, be an actual landing of man on the moon or a planet, followed by his return to the earth. It is not possible to accomplish such a mission with any vehicles that are presently under development.

3. Some day, it may be possible for men in space to accomplish important scientific or technical tasks. For the time being, however, it appears that space exploration must rely on unmanned vehicles. Therefore, a crash program aimed at placing a man into an orbit at the earliest possible time cannot be justified solely on scientific or technical grounds. Indeed, it may hinder the development of our scientific and technical program, even the

[16] future manned space program by diverting manpower, vehicles and funds.

4. The acquisition of new knowledge and the enrichment of human life through technological advances are solid, durable, and worthwhile goals of space activities. There is general lack of appreciation of this simple truism, both at home and abroad. Indeed, by having placed highest national priority on the MERCURY program we have strengthened the popular belief that man in space is the most important aim of our non-military space effort. The manner in which this program has been publicized in our press has further crystallized such belief. It exaggerates the value of that aspect of space activity where we are less likely to achieve success, and discounts those aspects in which we have already achieved great success and will probably reap further successes in the future.

5. A failure in our first attempt to place a man into orbit, resulting in the death of an astronaut, would create a situation of serious national embarrassment. An even more serious situation would result if we fail to safely recover a man from orbit.

On the basis of these facts we would like to submit the following recommendations:

1. By allowing the present MERCURY program to continue unchanged for more than a very few months, the new Administration would effectively endorse this program and take the blame for its possible failures. A thorough and impartial appraisal of the MERCURY program should be urgently made. The objectives of the various phases of this program (including the proposed physiological tests) should be critically examined. The margins of safety should be realistically estimated. If our present man-in-space program [17] appears unsound, we must be prepared to modify it drastically or even to cancel it. It is important that a decision on these matters be reached at the earliest possible date.

2. Whatever we decide to actually do about the man-in-space program, we should stop advertising MERCURY as our major objective in space activities. Indeed, we should make an effort to diminish the significance of this program to its proper proportion before the public, both at home and abroad. We should find effective means to make people appreciate the cultural, public service and military importance of space activities other than space travel.

VIII. Non-military Applications of Space Technology—
An Industry-Government Space Program

As the technical feasibility and reliability of man-made satellites was demonstrated, many possible civilian uses for satellites emerged. With no government support, various groups in private industry have examined the field for areas of study and development and a few substantial projects are already under way.

Industrial and governmental communications satellites appear practical and economically sound. Communication satellites will provide high quality and inexpensive telephone and general communication service between most parts of the earth. A by-product of a communication satellite will almost surely be an international television relay system linking all the nations of the world. On a longer time scale it should be feasible to provide radio and television broadcasting service via satellite-mounted transmitters. Such systems would give the quality broadcast reception now only available in and near urban areas to most of the inhabitants of the earth.

Satellites containing reliable beacons can be used to provide improved means of navigation for aircraft and ships at sea and can greatly advance the field of geodetics.

[18] Proper use of the information gathered by meteorological satellites should greatly increase our understanding of meteorology. With more knowledge of meteorology and with world-wide data frequently available from the satellites, longer-range and more reliable weather predictions should be possible. These projects, dreams a decade ago, bridge areas of technical specialty in which this nation is unexcelled. The United States has the most advanced communication system in the world, with a vast scientific and technological base supporting the communications industry. We are preeminent in the development of our electronic skills in radio, television, telephone and telegraphy. This entire industrial-scientific base is available to apply its art through satellite systems to the civilian needs of the world.

The exploitation of a new area of industrial opportunity for civilian use is normally left by our government to private enterprise. However, in the case of these important space systems, the development investment required is so large that it is beyond the financial resources of even our largest private industry. Furthermore, the use of commercial space satellites will require physical support of government installations as well as financial support.

All of the civilian satellite projects listed here will have direct or indirect military usefulness as well. Furthermore, communication and navigation systems of the type envisaged would be extremely useful in implementing an inspection system which might accompany a disarmament agreement. For these reasons projects of the type proposed might well be undertaken in cooperation with the military services.

We recommend that a vigorous program to exploit the potentialities of practical space systems. The government, through NASA or the Department of Defense, should make available the required physical facilities as well as any extraordinary financial support required to make the undertakings successful.

Organizational machinery is needed within the executive branch of the government to carry out this civilian space program.

[19] **Summary of Recommendations**

1. Make the Space Council an effective agency for managing the national space program.
2. Establish a single responsibility within the military establishments for managing the military portion of the space program.
3. Provide a vigorous, imaginative, and technically competent top management for NASA, including:
    (a) Administrator and deputy administrator
    (b)
        i. A technical director for propulsion and vehicles
        ii. A technical director for the scientific program
        iii. A technical director for the non-military space applications
        iv. A technical director for aerodynamic and aircraft programs.
4. Review the national space program and redefine the objectives in view of the experience gained during the past two years. Particular attention should be given the booster program, manned space technology, military uses of space to the civilian activities of the country.
5. Establish the organizational machinery within the government to administer an industry-government civilian space program.

Document III-6

Document title: John F. Kennedy, Memorandum for Vice President, April 20, 1961.

Source: Presidential Files, John F. Kennedy Presidential Library, Boston, Massachusetts.

This memorandum led directly to the Apollo program. By posing the question "Is there any…space program which promises dramatic results in which we could win?," President Kennedy set in motion a review that concluded that only an effort to send Americans to the Moon met the criteria Kennedy had laid out. This memorandum followed a week of discussion within the White House on how best to respond to the challenge to U.S. interests posed by the April 12, 1961, orbital flight of Yuri Gagarin.

[1] April 20, 1961

MEMORANDUM FOR
VICE PRESIDENT

In accordance with our conversation I would like for you as Chairman of the Space Council to be in charge of making an overall survey of where we stand in space.

1. Do we have a chance of beating the Soviets by putting a laboratory in space, or by a trip around the moon, or by a rocket to land on the moon, or by a rocket to go to the moon and back with a man. Is there any other space program which promises dramatic results in which we could win?

2. How much additional would it cost?

3. Are we working 24 hours a day on existing programs. If not, why not? If not, will you make recommendations to me as to how work can be speeded up.

4. In building large boosters should we put our emphasis on nuclear, chemical or liquid fuel, or a combination of these three?

5. Are we making maximum effort? Are we achieving necessary results?

I have asked Jim Webb, Dr. Weisner, Secretary McNamara and other responsible officials to cooperate with you fully. I would appreciate a report on this at the earliest possible moment.

John F. Kennedy

Document III-7

Document title: Robert S. McNamara, Secretary of Defense, Memorandum for the Vice President, "Brief Analysis of Department of Defense Space Program Efforts," April 21, 1961, without attachment, "Resume of Existing Programs."

Source: Lyndon Baines Johnson Library, Austin, Texas.

This document was the initial Department of Defense response to the review requested by President Kennedy in his April 20 memorandum. Some of the language in this response also appeared in the May 8 recommendations that formed the basis for Apollo and other elements of an accelerated space effort.

---

[1] It is the purpose of the memorandum to outline views with respect to major space programs. This document cannot be adequately supported by detailed analysis at this time. A more complete review is currently under way which will result in a first report on 28 April. That report will include an appraisal in considerable detail of our posture with respect to the Soviets. It will also comment on the Gardner Report and views expressed elsewhere concerning the conduct of our space programs and their proper objectives.

A. General:

1. Programs in space must be undertaken for a variety of reasons. They may be aimed at gaining scientific knowledge. Some, in the future, will be of commercial value. Several current programs are of potential military value for functions such as early warning.

2. All large scale space programs require the mobilization of resources on a national scale. They require the development and successful application of the most advanced technologies. Dramatic achievements in space, therefore, symbolize the technological power and organizing capacity of a nation.

3. It is for reasons such as these that major achievements in space contribute to national prestige. This is true even though the scientific, commercial or military value of the undertaking may, by ordinary standards, be marginal or economically unjustified.

4. What the Soviets do and what they are likely to do are therefore matters of great importance from the viewpoint of national prestige. Our attainments constitute a major element in the international competition between the Soviet system and our own. While the future military value of advanced space capabilities cannot be predicted very well, it, nevertheless, [2] is important to insure that the basic technological building blocks are created in an orderly and timely manner. These building blocks, moreover, must give us capabilities which match the Soviets in all areas of international competition.

5. Because of their national importance and their national scope, it is essential that our space efforts be well planned. It is essential that they be well managed. It is particularly undesirable in this connection to undertake crash programs needed to compensate for inadequate planning. It is likewise undesirable to spread our engineering resources too thinly. It is doubtless necessary to sponsor parallel efforts in the design stage, but it is essential to avoid duplication in the advanced development, procurement and deployment of operational equipment.

The comments in the following paragraphs are based upon these and similar assumptions. They deal with two major areas: launch vehicles and payload recovery.*

B. Launch Vehicles:

It is important from a national standpoint that the launch vehicle "gap" presently separating Soviet and U.S. capabilities be closed in an orderly but timely way. It is also important that our capabilities in the 1965-1970 period continue to grow so that similar important gaps are avoided. There will come a point, of course, at which a superior capability on the part of the Soviets or ourselves will be of little importance since either capability will suffice, but that point will not be reached for many years.

1. **The Current "Gap":**

1.1 The ATLAS-CENTAUR development should continue. If it is successful, it will enable us to boost 8,500 lbs. into a 300-mile orbit which still does not match the Soviet's capability of placing 10,000-14,000 lbs. into a 300-mile orbit.

1.2 CENTAUR like other developments is not assured of success. A substantial delay or major development shortcomings would be serious.

[3] 1.3 It is important, however, that our national launch vehicle program focus on a very small number of devices. A major element in the success of the Soviet program is the orderly, focussed way in which they have placed continued emphasis on the repetitive use of a single booster and a very small family of upper stages.

1.4 It would seem desirable, however, to inaugurate one or two back up programs for ATLAS-CENTAUR. An example would be an advanced upper stage for use with TITAN-II. Another might be a high boost segmented solid rocket booster for use with existing AGENA stages or possibly with an advanced AGENT. There is even the possibility of developing an upper stage using different propellants. It is not possible to decide at this time, but the results of current studies will make it possible to recommend action which fits into an improved over-all plan.

2. **Follow-on Efforts:**

2.1 The SATURN C-1 consisting of a cluster of eight chambers will give a total thrust of about 1.5 million pounds. It is unlikely to become operationally useful for missions such as manned orbital flight until after 1966. Should it prove inadequate or subject to excessive delays, a serious gap in boost capability would develop even if the early undertakings listed above were wholly successful.

2.2 The SATURN, depending as it does upon the clustering of very complex engines, may present very serious reliability problems. It seems almost certain that it will not

---

*Attachment A summarizes the Department of Defense space projects and shows the budgetary changes that were made as a result of the detailed review of FY-62 budget estimates. No further changes in funding levels for FY-62 are recommended for these programs.

be fully usable in its present form for the DYNASOAR mission. For such reasons it appears likely that a suitable parallel effort should be undertaken to insure that present planning and programming provides adequate insurance against the development of a launch vehicle gap in the 1965-1970 period.

2.3 The F-1 engine capable of developing 1.5 million pounds thrust in a single chamber should be more vigorously pushed. Detailed design studies for a suitable first stage utilizing this engine should be undertaken at once. After assessment and analysis, long lead time procurement and development efforts should be begun which give us the opportunity to accelerate and back this route to a 1 1/2 million pound booster if SATURN's progress warrants such a decision.

2.4 Other possibilities also present themselves. Upon further investigation it may be desirable to augment the development of large segmented [4] solid rocket boosters capable of 50-70 million pounds/second thrust. Other proposals such as the development of a high pressure hydrogen-oxygen cluster booster should also be investigated.

2.5 It is important to make sure that the number of such programs is adequate but not excessive. It is essential that engineering resources be focused and not spread too thin. It is vital that hard decisions be made at the critical decision points. Such decisions will include the total termination of unprofitable ventures and the deployment of fiscal and human resources on the highest priority and most promising undertakings.

2.6 It is difficult to estimate at this point the magnitude of the efforts which might be initiated to supplement the SATURN development. It is not unlikely, however, that as much as 50-100 million dollars of additional funds might be utilized in FY-62. These funds represent people and facilities. It is mandatory that policies be followed which utilize and strengthen existing organizations. Our national posture may be worsened rather than improved if added expenditures result in the still greater dispersal of scientific engineering and managerial talent among a variety of organizations too small or too over-loaded to do a fully adequate job.

C. Payload Recovery:

The Soviets have developed a recovery system which enable them to recover large payloads with a comparatively high degree of landing accuracy. This capability is essential to the success of manned experiments in space and is important to the success of many other space missions.

The U.S., however, has developed only the DISCOVERER type of recovery system. It is complex and has not proved to be very reliable.

The MERCURY system using parachutes is not very accurate and requires search operations of an enormous sea area.

The DYNASOAR system will not be testable for many years. Hundreds of millions of dollars will be required before maneuverable entry from orbit can be demonstrated.

It seems most desirable, therefore, to undertake the development of a controlled reentry and recovery system emphasizing simplicity, modest accuracy and high reliability. It is not likely that a recovery system per se will prove desirable. Rather, the development of a standardized space vehicle equipped with such a recovery system and incorporating standardized propulsion [5] and control components is likely to be most attractive. If it is large enough, such a vehicle can be used to carry a great variety of payloads with comparatively minor and largely internal modifications. This aspect of our planning for the future has not been addressed in much detail. It is difficult, therefore, to estimate the amount of additional funds which may be required to begin developments in this direction. It would appear, however, that the amounts involved in FY-62 could be comparatively modest.

Robert S. McNamara

Document III-8

Document title: **Lyndon B. Johnson, Vice President, Memorandum for the President, "Evaluation of Space Program," April 28, 1961.**

Source: **NASA Historical Reference Collection, History Office, NASA Headquarters, Washington, D.C.**

This memorandum, prepared by Edward C. Welsh, the new executive secretary of the National Aeronautics and Space Council, and signed by Vice President Johnson, was the first report to President Kennedy on the results of the review he had ordered on April 20. The report identified a lunar landing by 1966 or 1967 as the first dramatic space project in which the United States could beat the Soviet Union. The vice president identified "leadership" as the appropriate goal of U.S. efforts in space.

[1]
Memorandum for the President

April 28, 1961
Subject: Evaluation of Space Program.

Reference is to your April 20 memorandum asking certain questions regarding this country's space program.

A detailed survey has not been completed in this time period. The examination will continue. However, what we have obtained so far from knowledgeable and responsible persons makes this summary reply possible.

Among those who have participated in our deliberations have been the Secretary and Deputy Secretary of Defense; General Schriever (AF); Admiral Hayward (Navy); Dr. von Braun (NASA); the Administrator, Deputy Administrator, and other top officials of NASA; the Special Assistant to the President on Science and Technology; representatives of the Director of the Bureau of the Budget; and three outstanding non-Government citizens of the general public: Mr. George Brown (Brown & Root, Houston, Texas); Mr. Donald Cook (American Electric Power Service, New York, N.Y.); and Mr. Frank Stanton (Columbia Broadcasting System, New York, N.Y.).

The following general conclusions can be reported:

a. Largely due to their concentrated efforts and their earlier emphasis upon the development of large rocket engines, the Soviets are ahead of the United States in world prestige attained through impressive technological accomplishments in space.

b. The U.S. has greater resources than the USSR for attaining space leadership but has failed to make the necessary hard decisions and to marshal those resources to achieve such leadership.

[2] c. This country should be realistic and recognize that other nations, regardless of their appreciation of our idealistic values, will tend to align themselves with the country which they believe will be the world leader–the winner in the long run. Dramatic accomplishments in space are being increasingly identified as a major indicator of world leadership.

d. The U.S. can, if it will, firm up its objectives and employ its resources with a reasonable chance of attaining world leadership in space during this decade. This will be difficult but can be made probable even recognizing the head start of the Soviets and the likelihood that they will continue to move forward with impressive successes. In certain areas, such as communications, navigation, weather, and mapping, the U.S. can and should exploit its existing advance position.

e. If we do not make the strong effort now, the time will soon be reached when the margin of control over space and over men's minds through space accomplishments will have swung so far on the Russian side that we will not be able to catch up, let alone assume leadership.

f. Even in those areas in which the Soviets already have the capability to be first and are likely to improve upon such capability, the United States should make aggressive efforts as the technological gains as well as the international rewards are essential steps in eventually gaining leadership. The danger of long lags or outright omissions by this country is substantial in view of the possibility of great technological breakthroughs obtained from space exploration.

g. Manned exploration of the moon, for example, is not only an achievement with great propaganda value, but it is essential as an objective whether or not we are first in its accomplishment–and we may be able to be first. We cannot leapfrog such accomplishments, as they are essential sources of knowledge and experience for even greater successes in space. We cannot expect the Russians to transfer the benefits of their experiences or the advantages of their capabilities to us. We must do these things ourselves.

[3] h. The American public should be given the facts as to how we stand in the space race, told of our determination to lead in that race, and advised of the importance of such leadership to our future.

i. More resources and more effort need to be put into our space program as soon as possible. We should move forward with a bold program, while at the same time taking every practical precaution for the safety of the persons actively participating in space flights.

****

As for the specific questions posed in your memorandum, the following brief answers develop from the studies made during the past few days. These conclusions are subject to expansion and more detailed examination as our survey continues.

Q.1- Do we have a chance of beating the Soviets by putting a laboratory in space, or by a trip around the moon, or by a rocket to land on the moon, or by a rocket to go to the moon and back with a man. Is there any other space program which promises dramatic results in which we could win?

A.1- The Soviets now have a rocket capability for putting a multi-manned laboratory into space and have already crash-landed a rocket on the moon. They also have the booster capability of making a soft landing on the moon with a payload of instruments, although we do not know how much preparation they have made for such a project. As for a manned trip around the moon or a safe landing and return by a man to the moon, neither the U.S. nor the USSR has such capability at this time, so far as we know. The Russians have had more experience with large boosters and with flights of dogs and man. Hence they might be conceded a time advantage in circumnavigation of the moon and also in a manned trip to the moon. However, with a strong effort, the United States could conceivably be first in those two accomplishments by 1966 or 1967.

[4] There are a number of programs which the United States could pursue immediately and which promise significant world-wide advantage over the Soviets. Among these are communications satellites, and navigation and mapping satellites. These are all areas in which we have already developed some competence. We have such programs and believe that the Soviets do not. Moreover, they are programs which could be made operational and effective within reasonably short periods of time and could, if properly programmed with the interests of other nations, make useful strides toward world leadership.

Q.2- How much additional would it cost?

A.2- To start upon an accelerated program with the aforementioned objectives clearly in mind, NASA has submitted an analysis indicating that about $500 million would be needed for FY 1962 over and above the amount currently requested of the Congress. A program based upon NASA's analysis would, over a ten-year period, average approximately $1 billion a year above the current estimates of the existing NASA program.

While the Department of Defense plans to make a more detailed submission to me within a few days, the Secretary has taken the position that there is a need for a strong effort to develop a large solid-propellant booster and that his Department is interested in undertaking such a project. It was understood that this would be programmed in accord with the existing arrangement for close cooperation with NASA, which Agency is undertaking some research in this field. He estimated they would need to employ approximately $50 million during FY 1962 for this work but that this could be financed through management of funds already requested in the FY 1962 budget. Future defense budgets would include requests for additional funding for this purpose; a preliminary estimate indicates that about $500 million would be needed in total.

[5] Q.3- Are we working 24 hours a day on existing programs? If not, why not? If not, will you make recommendations to me as to how work can be speeded up?

A.3- There is not a 24-hour-a-day work schedule on existing NASA space programs except for selected areas in Project Mercury, the Saturn C-1 booster, the Centaur engines and the final launching phases of most flight missions. They advise that their schedules have been geared to the availability of facilities and financial resources, and that hence their overtime and 3-shift arrangements exist only in those activities in which there are particular bottlenecks or which are holding up operations in other parts of the programs. For example, they have a 3-shift 7-day-week operation in certain work at Cape Canaveral; the contractor for Project Mercury has averaged a 54-hour week and employs two or three shifts in some areas; Saturn C-1 at Huntsville is working around the clock during critical test periods while the remaining work on this project averages a 47-hour week; the Centaur hydrogen engine is on a 3-shift basis in some portions of the contractor's plants.

This work can be speeded up through firm decisions to go ahead faster if accompanied by additional funds needed for the acceleration.

Q.4- In building large boosters should we put our emphasis on nuclear, chemical or liquid fuel, or a combination of these three?

A.4- It was the consensus that liquid, solid and nuclear boosters should all be accelerated. This conclusion is based not only upon the necessity for back-up methods, but also because of the advantages of the different types of boosters for different missions. A program of such emphasis would meet both so-called civilian needs and defense requirements.

[6] Q.5- Are we making maximum effort? Are we achieving necessary results?

A.5- We are neither making maximum effort nor achieving results necessary if this country is to reach a position of leadership.

Lyndon B. Johnson

Document III-9

Document title: Wernher von Braun to the Vice President of the United States, April 29, 1961, no pagination.

Source: NASA Historical Reference Collection, History Office, NASA Headquarters, Washington, D.C.

Of all those consulted during the presidentially mandated space review, no one had been thinking longer about the future in space than Wernher von Braun. Even when he had led the development of the V-2 rocket for Germany during World War II, von Braun and his associates had been planning future space journeys. After coming to the United States after World War II, von Braun was a major contributor to popularizing the idea of human spaceflight. As he stressed in his letter, von Braun had been asked to participate in the review as an individual, not as the director of NASA's Marshall Space Flight Center. Von Braun told the vice president in his letter that the United States had "an excellent chance" of beating the Russians to a lunar landing.

This is an attempt to answer some of the questions about our national space program raised by The President in his memorandum to you dated April 20, 1961. I should like to emphasize that the following comments are strictly my own and do not necessarily reflect the official position of the National Aeronautics and Space Administration in which I have the honor to serve.

Question 1. Do we have a chance of beating the Soviets by putting a laboratory in space, or by a trip around the moon, or by a rocket to land on the moon, or by a rocket to go to the moon and back with a man? Is there any other space program which promises dramatic results in which we could win?

Answer: With their recent Venus shot, the Soviets demonstrated that they have a rocket at their disposal which can place 14,000 pounds of payload in orbit. When one considers that our own one-man Mercury space capsule weighs only 3900 pounds, it becomes readily apparent that the Soviet carrier rocket should be capable of

- launching *several* astronauts into orbit simultaneously. (Such an enlarged multi-man capsule could be considered and could serve as a small "laboratory in space".)

- soft-landing a substantial payload on the moon. My estimate of the maximum soft-landed net payload weight the Soviet rocket is capable of is about 1400 pounds (one-tenth of its low orbit payload). This weight capability is *not* sufficient to include a rocket for the *return flight* to earth of a man landed on the moon. But it is entirely adequate for a powerful radio transmitter which would relay lunar data back to earth and which would be *abandoned* on the lunar surface after completion of this mission. A similar mission is planned for our "Ranger" project, which uses an Atlas-Agena B boost rocket. The "semi-hard" landed portion of the Ranger package weighs 293 pounds. Launching is scheduled for January 1962.

The existing Soviet rocket could furthermore hurl a 4000 to 5000 pound capsule *around* the moon with ensuing re-entry into the earth atmosphere. This weight allowance must be considered marginal for a one-man round-the-moon voyage. Specifically, it would not suffice to provide the capsule and its occupant with a "safe abort and return" capability, a feature which under NASA ground rules for pilot safety is considered mandatory for all manned space flight missions. One should not overlook the possibility, however, that the Soviets may substantially facilitate their task by simply waiving this requirement.

A rocket about ten times as powerful as the Soviet Venus launch rocket is required to land a man on the moon and bring him back to earth. Development of such a super rocket can be circumvented by orbital rendezvous and refueling of smaller rockets, but the development of this technique by the Soviets would not be hidden from our eyes and would undoubtedly require several years (possibly as long or even longer than the development of a large direct-flight super rocket).

Summing up, it is my belief that

a) we do *not* have a good chance of beating the Soviets to a manned *"laboratory in space."* The Russians could place it in orbit this year while we could establish a (somewhat heavier) laboratory only after the availability of a reliable Saturn C-1 which is in 1964.

b) we *have* a sporting chance of beating the Soviets to a soft-landing of a radio *transmitter station on the moon*. It is hard to say whether this objective is on their program, but as far as the launch rocket is concerned, they could do it at any time. We plan to do it with the Atlas-Agena B-boosted Ranger #3 in early 1962.

[3] c) we have a sporting chance of sending a 3-man crew *around the moon* ahead of the Soviets (1965/66). However, the Soviets could conduct a round-the-moon voyage earlier if they are ready to waive certain emergency safety features and limit the voyage to one man. My estimate is that they could perform this simplified task in 1962 or 1963.

d) we have an excellent chance of beating the Soviets to the *first landing of a crew on the moon* (including return capability, of course). The reason is that a performance jump by a factor 10 over their present rockets is necessary to accomplish this feat. While today we do not have such a rocket, it is unlikely that the Soviets have it. Therefore, we would not have to enter the race toward this obvious next goal in space exploration against hopeless odds favoring the Soviets. With an all-out crash program I think we could accomplish this objective in 1967/68.

Question 2. How much additional would it cost?

Answer: I think I should not attempt to answer this question before the exact objectives and the time plan for an accelerated United States space program have been determined. However, I can say with some degree of certainty that the necessary funding increase to meet objective d) above would be well over $1 Billion for FY 62, and that the required increases for subsequent fiscal years may run twice as high or more.

Question 3. Are we working 24 hours a day on existing programs? If not, why not? If not, will you make recommendations to me as to how work can be speeded up.

Answer: We are *not* working 24 hours a day on existing programs. At present, work on NASA's Saturn project proceeds on a basic one-shift basis, with overtime and multiple shift operations approved in critical "bottleneck" areas.

During the months of January, February and March 1961, NASA's George C. Marshall Space Flight Center, which has systems management for the entire Saturn vehicle and develops the large first stage as an in-house project, has worked an average of 46 hours a week. This includes all administrative and clerical activities. In the areas critical for the Saturn project (design activities, assembly, inspecting, testing), average working time for the same period was 47.7 hours a week, with individual peaks up to 54 hours per week.

Experience indicates that in Research & Development work longer hours are not conducive to progress because of hazards introduced by fatigue. In the aforementioned critical areas, a second shift would greatly alleviate the tight scheduling situation. However, additional funds and personnel spaces are required to hire a second shift, and neither are available at this time. *In this area, help would be most effective.*

Introduction of a *third* shift *cannot* be recommended for Research & Development work. Industry-wide experience indicates that a two-shift operation with moderate but not excessive overtime produces the best results.

In industrial plants engaged in the Saturn program the situation is approximately the same. Moderately increased funding to permit greater use of premium paid overtime, prudently applied to real "bottleneck" areas, can definitely speed up the program.

Question 4. In building large boosters should we put our emphasis on nuclear, chemical or liquid fuel, or a combination of these three?

Answer: It is the consensus of opinion among most rocket men and reactor experts that the future of the nuclear rocket lies in deep-space operations (upper stages of chemically-boosted rockets or nuclear space vehicles departing from an orbit around the earth) rather than in launchings (under nuclear power) from the ground. In addition, there can be little doubt that the basic technology of nuclear rockets is still in its early infancy. The nuclear rocket should therefore be looked upon as a promising means to extend and expand the scope of our space operations in the years beyond 1967 or 1968. It should not be considered as a serious contender in the big booster problem of 1961.

The foregoing comment refers to the simplest and most straightforward type of nuclear rocket, viz. the "heat transfer" or "blow-down" type, whereby liquid hydrogen is evaporated and superheated in a very hot nuclear reactor and subsequently expanded through a nozzle.

There is also a fundamentally different type of nuclear rocket propulsion system in the works which is usually referred to as "ion rocket" or "ion propulsion." Here, the nuclear energy is first converted into electrical power which is then used to expel "ionized" (i.e., electrically charged) particles into the vacuum of outer space at extremely high speeds. The resulting reaction force is the ion rocket's "thrust." It is in the very nature of nuclear ion propulsion systems that they cannot be used in the atmosphere. While very efficient in propellant economy, they are capable only of very small thrust forces. Therefore they do not qualify as "boosters" at all. The future of nuclear ion propulsion lies in its application for low-thrust, high-economy cruise power for interplanetary voyages.

As to "chemical or liquid fuel" The President's question undoubtedly refers to a comparison between "solid" and "liquid" rocket fuels, both of which involve chemical reactions.

At the present time, our most powerful rocket boosters (Atlas, first stage of Titan, first stage of Saturn) are all liquid fuel rockets and all available evidence indicates that the

Soviets are also using liquid fuels for their ICBM's and space launchings. The largest solid fuel rockets in existence today (Nike Zeus booster, first stage Minuteman, first stage Polaris) are substantially smaller and less powerful. There is no question in my mind that, when it comes to building very powerful booster rocket systems, the body of experience available today with liquid fuel systems greatly exceeds that with solid fuel rockets.

There can be no question that larger and more powerful solid fuel rockets can be built and I do not believe that major breakthroughs are required to do so. On the other hand it should not be overlooked that a casing filled with solid propellant and a nozzle attached to it, while entirely capable of producing thrust, is not yet a rocket ship. And although the reliability record of solid fuel rocket *propulsion units*, thanks to their simplicity, is impressive and better than that of liquid propulsion units, this does not apply to *complete rocket systems*, including guidance systems, control elements, stage separation, etc.

Another important point is that booster performance should not be measured in terms of thrust force alone, but in terms of total impulse; i.e., the product of thrust force and operating time. For a number of reasons it is advantageous not to extend the burning time of solid fuel rockets beyond about 60 seconds, whereas most liquid fuel boosters have burning time of 120 seconds and more. Thus, a 3-million pound thrust solid rocket of 60 seconds burning time is actually not more powerful than a 1½-million pound thrust liquid booster of 120 seconds burning time.

[Paragraph excised during declassification review]

My recommendation is to substantially increase the level of effort and funding in the field of solid fuel rockets (by 30 or 50 million dollars for FY 62) with the immediate objectives of

- demonstration of the feasibility of very large segmented solid fuel rockets. (Handling and shipping of multi-million pound solid fuel rockets become unmanageable unless the rockets consist of smaller individual segments which can be assembled in building block fashion at the launching site.)

- development of simple inspection methods to make certain that such huge solid fuel rockets are free of dangerous cracks or voids

- determination of the most suitable operational methods to ship, handle, assemble, check and launch very large solid fuel rockets. This would involve a series of paper studies to answer questions such as

a. Are clusters of smaller solid rockets, or huge, single poured-in-launch-site solid fuel rockets, possibly superior to segmented rockets? This question must be analyzed not just from the propulsion angle, but from the operational point of view for the total space transportation system and its attendant ground support equipment.

b. Launch pad safety and range safety criteria (How is the total operation at Cape Canaveral affected by the presence of loaded multi-million pound solid fuel boosters?)

c. Land vs. off-shore vs. sea launchings of large solid fuel rockets.

d. Requirements for manned launchings (How to shut the booster off in case of trouble to permit safe mission abort and crew capsule recovery? If this is difficult, what other safety procedures should be provided?)

Question 5. Are we making maximum effort? Are we achieving necessary results?

Answer: No, I do *not* think we are making maximum effort.

In my opinion, the most effective steps to improve our national stature in the space field, and to speed things up would be to

- identify a few (the fewer the better) goals in our space program as objectives of highest national priority. (For example: Let's land a man on the moon in 1967 or 1968.)

- identify those elements of our present space program that would qualify as immediate contributions to this objective. (For example, soft landings of suitable instrumentation on the moon to determine the environmental conditions man will find there.)

- put all other elements of our national space program on the "back burner."

- add another more powerful liquid fuel booster to our national launch vehicle program. The design parameters of this booster should allow a certain flexibility for desired program reorientation as more experience is gathered.

Example: Develop in addition to what is being done today, a first-stage liquid fuel booster of twice the total impulse of Saturn's first stage, designed to be used in clusters if needed. With this booster we could

    a. double Saturn's presently envisioned payload. This additional payload capability would be very helpful for soft instrument landings on the moon, for circumlunar flights and for the final objective of a manned landing on the moon (if a few years from now the route via orbital re-fueling should turn out to be the more promising one.)

    b. assemble a much larger unit by strapping three or four boosters together into a cluster. This approach would be taken should, a few years hence, orbital rendezvous and refueling run into difficulties and the "direct route" for the manned lunar landing thus appears more promising.

[Paragraph excised during declassification review]

Summing up, I should like to say that in the space race we are competing with a determined opponent whose peacetime economy is on a wartime footing. Most of our procedures are designed for orderly, peacetime conditions. I do not believe that we can win this race unless we take at least some measures which thus far have been considered acceptable only in times of a national emergency.

Yours respectfully,
Wernher von Braun

Document III-10

Document title: "Vice President's Ad Hoc Meeting," May 3, 1961.

Source: NASA Historical Reference Collection, History Office, NASA Headquarters, Washington, D.C.

As the space review progressed in early May, Vice President Lyndon Johnson called together many of those participating in the review to meet with Senator Robert Kerr (D-OK), who was the new chairman of the Senate Committee on Aeronautical and Space Science, and Senator Styles Bridges (R-NH), the committee's ranking minority member. The point of the meeting was to let these key senators know what was being discussed within the executive branch and to solicit their support for an acceleration of the space program. Evident in the notes of the meeting is some tension between Vice President Johnson, who was pushing for a strong recommendation to the president, and NASA Administrator James Webb, who was not yet sure that a program of the magnitude that the vice president wanted was technically or politically feasible.

[1]
Vice President's Ad Hoc Meeting

May 3 - Room 210 - With Senators

    Vice President: This meeting is to get the benefit of remarks and ideas from the Senators and others for Mr. Webb, the Department of Defense, the President, and the Space Council members. We haven't gone far enough or fast enough. We need a new look, and to know how much it will cost. Mr. Webb has a comprehensive program. It would add $500 million more or less to the current budget. The Department of Defense also has need for about $100 million more. Everyone is requested to give suggestions and recommendations. We are grateful for the participation of the senior Senators. This is not a partisan matter. We are all Americans doing the best job we can for America. In taking a new look, we will call first on Senator Kerr.

Senator Kerr: After all that has gone on, and considering particularly things that have happened recently, we can see that the budget of the outgoing Administration was not adequate to do what the President and others have in mind. The establishment of the Space Council is the most progressive thing that has yet been done. One of the first things they must do is to look at requirements, costs, and come up with a budget, either on an annual or project basis, that will fix the agency of responsibility. We have some [2] great men leading the program.

****

We need to agree on objectives, the timing of those matters, get a decision from the President on what he needs in the budget, and after these preliminary steps, the matter comes to us in the Congress. We need some cold-blooded decisions, but the Senate can be counted on in the end to face up to whatever is required. Senator Bridges is an indispensable man in the matter of getting started and getting the right answer from the Senate.

Senator Bridges: Concurred with the remarks of Senator Kerr. The Vice President will remember the trials and frustrations of 1958 that came with the development of the Space Act. Now we face a new situation. What are our short and long range objectives? We have been attempting to maintain a balance between the military necessity and the scientific desires. A coordination between them has been maintained. It certainly is necessary to attain the highest possible scientific use and to maintain the glory of the United States and its prestige, but basic to the whole matter is the security of the United States. These things have to be in tune. It is a tremendous challenge. The Space Committee of the Senate will cooperate, but it wants to be informed, it wants the truth, and it will carry its share of the load and more.

[3] Vice President: I am asking Mr. Webb to review the high points of his short and long range objectives and his budget needs. He has a paper covering ten or eleven points, and this seems to be a good way to cover the ground.

Mr. Webb: When the amendment to the Space Act was recently passed, the Vice President asked for our views on the subject just announced. These were put together very fast. They start from a basic estimate that we can arrange for a manned lunar landing in 1967 or 1968. Before that can be done, science must have found plenty of new answers or ideas, and much more must be found and understood before we can either put three men in an orbit round the moon or make a soft landing. Before we get to that objective, we have some other things we can do along that path. Some new shots will advance meteorology. We will have communications satellites and a system set up which will serve both military and commercial needs.

In order to answer the first question asked by the President, "Do we have a chance of beating the Soviets?", I have assigned the best 25 scientists on each one of the five projects the task of analyzing the possibilities and probabilities to cover each part of [4] the question. There is a great deal that must be done before the Vice President will be in a position to make recommendations and the President be ready to go to the Congress and ask for the large sums which will be necessary, so we've got to be very careful now. The magnitudes are something like this–$1.7 billion for next year, $3 billion one year later, and $4.4 billion the following year.

The Vice President: Do you feel that you will not be prepared to give me answers for a month? You should be making your recommendations as early as you can. You were desperate for $308 million two weeks ago. You didn't get all of that. Is it going to take you a month to make the decisions necessary to arrive at good targets? I am not trying to rush you. But you must not wait a month or Congress will have gone home.

Webb: There are some overall policy guidances with which we ought to be provided to make a proper start in our estimates.

In the last two weeks we have actually had a new invention. It amounts to a combination of solid and liquid propellants in boosters. If this process is feasible, we can make a national decision –

Vice President: I thought you said you would have your answers in a month. What you are now talking about will take longer than a month, won't it?

[5] Webb: If our top men decide the psychological and military necessities, and that these things require a lunar shot in 1967, then the budget for the first year would be $1.7 billion for NASA in 1962.

Senator Kerr: Does this mean that you will want to adjust your figures and your justification for the recent appropriation increases requested?

Webb: Although there would not be a big change in spending, there would be a considerable problem having to do with long range commitments. There is a real question as to whether a request for $33 billion for these objectives is proper at this time.

Vice President: We want to keep clear here our need for recommendations on next year and on a ten-year total. It seems that your best "horseback" guess was a need for $509 million. (The Vice President then read from paper titled "Major Items in Accelerated Program Requiring Additional Funds for Fiscal 1962." The amount needed for each of the eleven points was read. Total $509 million.)

[6] Now my question is, do you want to change that $509 million figure?

Webb: Our new idea on boosters will force us to change some of these figures, just how much I do not know, because DOD has now decided that they need big boosters and they may take over some or all of this one project cost.

Dryden: Speaking as a technician, I would recommend that we try for the moon as soon as possible, but national policy has other significant guidance factors and we're fishing for some of that guidance in order to know what kind of support we would get for some of our ideas.

Vice President: I don't want anyone here to feel that we are putting him on the spot. We'll wait a month if necessary for people to get guts enough to make solid recommendations. Our purpose today is to have these important Senators get the benefit of consultation and for us to have the benefit of consulting them and we want to consult everyone who can make a contribution.

Dryden: In terms of the overall national interest and objectives, we find ourselves in a pretty narrow part of that, i.e., the space program.

[7] When it comes to its relationship to the USSR, someone else has to tell us how we fit into the overall scheme of things and what we should do to carry our part of the burden.

Webb: We can only do our share – we can't carry the burden of all of those who have responsibility.

Vice President: What we want to know is what would you do if you were President?

Webb: Since we don't know enough about his other problems, we can't judge what we would do if we were President. I think I would be for a moon shot in 1967.

Vice President: In other words, you take the same position that you recommended two weeks ago [4/19/61]. Will your figures be way off?

Dryden: It will take a month to work out our program and its justification to a point which will satisfy the Bureau of the Budget, and until then we can't really tell the Senators much about figures.

Senator Kerr: If the objectives discussed here have the President's backing, NASA and Senator Kerr will share the burden of that load. If the President [8] has not been convinced, Kerr backs out.

Vice President: The President will stand behind the recommendations from the Space Council which will be based on recommendations from Webb and the other members.

Webb: We need a national commitment to defined objectives. If Congress would give us a commitment for big increments to our program in future years, we could do a better job of planning. It would be a national disgrace if we were to start now a big program and then have to stop because of lack of appropriation in future years.

Vice President: You can't get any more than a one-year appropriation but Congress will give you annual appropriations as they do in the case of a big dam. Let's not try to do the impossible. On the other hand, let's hope that you can go on year after year persuading reasonable men and explain the changes as they take place. We got $126 million when we asked for $308 million, but the President explained his reasons and he is ready to listen to new recommendations with an open mind.

Webb: The President did give us our needs for a big booster and he paved the way for more. We need more experience on life in space with men and [9] with animals. We will have to reschedule the use of our $509 million plan. The $182 million we have asked for would actually level out lower if our big booster hopes work out, but we will need more now as we see greatly increased costs coming up later. There are four major booster developments in our program which were not in the budget which went to the President. We need $509 million now, instead of the $182 million we asked the President for. If our new idea of the marriage of the types of boosters is feasible, we will have some new costs and one of those will be $105 million for launching pads. We need new types of space crafts for meteorological work and for communications work.

Senator Bridges: Based on intelligence resources, how does your program compare with the achievements which are to be expected from the USSR?

Webb: They will be ahead until 1967 or 1968.

Senator Bridges: Should we be planning to do less than we know they are doing?

Webb: We can't help it. We can't do more during that period. They have been using an 800,000 lb. thrust and are ready to make the next jump in size, which may be double or more. On the other hand, we are about to jump our size ten times, and there are lots of unsolved problems involved. [10]

Senator Bridges: Who decided on the publicity which has been given this shot that was cancelled the other day? [Shepard]

Webb: The plan for this shot was in existence when we took over. One of the first things we did was to cancel out some arrangements which had been made between the astronauts and commercial publicists. Although there was a half million dollars involved, we came to some pretty firm conclusions and in spite of the important commercial interests, we fixed some limitations. Although there are reports of 500 newsmen down at Cape Canaveral, they are not in the way. They have been eased out of the blockhouse and any other area where they have in the past interfered. This whole problem needs more attention, but with things like a House Committee investigating me and others pressing me every minute, I haven't had time.

Dryden: It's true that the basic decisions were made by the last Administration and these are some of the reasons Cape Canaveral can't be cut off from public view. Photographs can be taken with long lenses and misleading information provided, based on those photographs, so it was decided to brief the press on scheduled events weeks in advance and cut down the scuttlebutt. There are all kind of activities at the base which will give intelligent people benchmarks and ways to judge what is going on.

[11] The recent shot is a case in point. NASA has never announced a date but the press watches Navy ships go to stations and astronauts' wives shifting to other quarters, and they come to conclusions which we are forced to agree to by maintaining silence.

Senator Bridges: Did we ever think of moving?

Dryden: Cape Canaveral belongs to Defense. We have test operations at Wallops Island for small shots. We have considered other locations, but moving in remote areas multiplies the costs, and they are prohibitive.

Senator Bridges: If we fail in this coming shot with publicity as tremendous as it has been, the results will be tragic.

Rubel: (Defense) Defense is working very closely with NASA. The plans which are being developed are the best which can be expected from several points of view. There are many elements in the Department of Defense which are involved, such as the R&D in DOD, Aerospace Corporation (AF), Englewood, California, and the military staff of the Air Force. They all have notions. We all have to put our heads together, and we're almost sure of our position now. [12]

Vice President: In your last report you promised more complete information by April 28th. I hope we can have that soon, and that it will tell us what the Department of Defense will need and will do.

Rubel: It should be ready by Monday, and very little will be omitted from the things we hoped to provide. About $100 million will be needed for 1962. This is the same as we estimated on April 20th but we now have the problem of justifying it for the BOB.

Vice President: (He gave a summary of previous executive meetings for the benefit of the Senators.) Boiled down, it appears that the President should be asked to expand the total program from an estimated $22 billion to be spent in ten years to $33 billion in ten years. As a part of this, $509 million would be added for FY 1962. The Eisenhower budget provided for $2.015 billion for all purposes. The Kennedy add-ons amount to $308 million, a total of $2.323 billion, with a new request for about $600 million, The new total for FY 1962 will be about $3 billion. Now quickly we must get our figures necessary on the recommendations we wish to make on a national effort and then let the BOB decide what part of it is justifiable in the light of overall national needs. [13]

Brown (AEC): He was in total agreement with NASA, and hoped that the $509 million estimate would include the AEC tie-in agreements on their targets with NASA, but he learned that this portion of the $182 million requested of the President had been turned down without prejudice to reconsideration later. He considered it most important to get target dates fixed for these major efforts. The present budget does not provide AEC with what it needs for any new targets. For that purpose we would have to put back in what AEC needed in the $308 million request. Their needs are not reflected in the $509 million request.

Webb: The figures we have show the principal use by AEC of funds after FY 1970. The AEC budget would need to be reconsidered only if we want to go ahead all across the board. Frankly, in order to get some of the other things done on new schedules, perhaps we can't handle the AEC request.

Vice President: We may have to come to that but let's come to that later. This meeting is not for the purpose of making such decisions. The last decision by the President involved a bite which he could take, chew, and swallow at that time. He might now be ready for the next bite, and it's up to us to be prepared with the facts and the estimate. [14]

Webb: It doesn't appear that we can do everything at the same time and to the extent that we find a conflict of objectives, these must be resolved in the national interest.

Vice President: We won't be able to set national objectives until you are sure and you make recommendations. The President won't be able to act for more than a month if that's when you will be ready to advise him. If the program that is ultimately devised is astronomical in cost, we can be sure that it will have to be tailored.

Hansen (BOB): If we all work together in the development of space plans and projects, we can approach them with confidence. The BOB can be depended upon to stay with you and accept and approve budget plans that are justified, in the overall national interest.

Senator Bridges: While the things you are talking about may all be true, if we were to have a war, things would just have to be different. We can start on the basis that we are here to discuss them, but we must realize that war would mean a complete reorganization of the national plans and space programs. [15]

Cook: The agencies here represented know what they are doing and they know that they have to have the right people doing the proper jobs. And like all good soldiers, having made recommendations, we have to take the decision and work with it all together. First a program should be laid out in terms of objectives, then the policy determination should be made, the funds provided and schedules fixed and adhered to.

Vice President: In order to catch up on what has been going on recently, it is suggested that everyone here obtain copies of the testimony which Dr. Welsh in hearings before the Space Committees of the House and the Senate on the amendment to the NASA Act. In reading this testimony along with the changes, one can get a good understanding of the new responsibilities of the new Space Council and also obtain an insight on Dr. Welsh's plans for the Space Council staff and staff activities.

Stanton (CBS): Setting aside for a moment comments on publicity about the imminent space shot and the public image –

With ignorance as to policy needs and further need for briefing on scientific and military requirements, it's difficult for a businessman to [16] set aside his desire to go ahead and get things done. There are times, however, when the scientists want to go deeper into the problem before they start concrete action, and it creates the need for critical examination of the whole problem in the national interest.

Vice President: Do the NASA scientists want to go into this matter to a point that would increase the $509 million estimate?

Dryden: Yes. This, however, is principally in field centers where they usually want more than we in the national office consider reasonable in the interest of program direction.

Stanton: Our interest in the future depends upon being first in science. If a moon shot is judged to be expedient, we should press on. We don't have to be concerned about national support if wise men have decided upon the action necessary in the national interest.

Dryden: To do a moon shot will tax everyone involved to the limit and all along the way there will have to be trial, error, and correction.

Stanton: To get national support, we must give now consideration to this matter of the "goldfish bowl." This was the highlight of the President's [17] talk to the American Newspaper Publishers Association. We can't go on doing as we have in the past. We must be particularly careful about publicizing failures that have the effect of dropping national support. Every good laboratory has its failures, but this isn't generally understood by the public, and there are times when it is far better that we keep quiet.

Hansen: We know now that the USSR is going to be ahead of us for a time. Therefore anything that we do they will attempt to ridicule. We must give our people the best information we have because the absence of complete information is worse when you have some failures and your people are not prepared for them. It is far better to be frank if you want to get full support. We need some aggressive leadership to sell the people that we are going about this in the right way.

Webb: This is really a new frontier. We are up against some hostile voices in the unknown. We really need 50 shots to try things out in an orderly way, but we have to boil this all down to a lot fewer and if we announce each of them, a lot will be riding on every test. Mr. Stanton can help by keeping the focus on our accomplishments, but we certainly ought to avoid broadcasting our objectives. [18]

Stanton: It would be most desirable to put all of the information in the right frame in order properly to get public support and keep public opinion moving in the direction you want it to follow.

(Senators and members of the Committee staffs left at this point.)...

[19] Webb: This whole issue is made complicated by a great many different factors. We mustn't forget that there are many foreign people involved. We have networks of stations all over the free world and they all have to know what we're doing and when we're doing it. Then there are the scientists. It's hard to control the minds of such men. As a matter of fact, we have to remember that we are fighting for men's minds.

Farley: (State) State is somewhat concerned both about the public reaction at home, and foreign reaction. There must be some way that we could cut down publicity and make it all look like a bleacher stunt. There are many things we can do to build up our dignity. We're developing a program to meet the USSR all around the world. By 1967 or 1968 we must get out of second place. If in the meantime we can do some thing that is concrete like the establishment of a communications network by means of satellites and show the free world a program from which they will benefit, we will advance U.S. interests all along the line. We ought to set our goals as soon as possible for the communications network and the supply of weather information from our satellites just as soon as possible. It appears that the communications network is the best short range target.

George Brown: Experience proves that when one pinpoints a long range program, you get things done all along the way. The products that come out at short [20] range are automatic. We shouldn't forecast each of our steps for the USSR. When we do, we can be sure that they will find some way to cover or blank out our desired publicity.

Webb: The previous Administration asked for bids on the communications satellite. Seven were received and we are taking two weeks to evaluate them. We understand that Bell is going ahead with its plans whether or not it gets the award. FCC has a big part of the control in this project. It started its hearings on the 1st of May and promised to give a ruling within 4 to 6 weeks. There are certain military requirements and therefore some reservations about the use of these communications satellites. The expense of this project is somewhat in doubt and the House is calling for hearings before we are ready to talk. Nevertheless we ought to be ready to run an experiment across the Atlantic by mid-1962. We have the Tiros now and the Nimbus to come in 1962. If the AT&T is successful, they have stated that they would have an operational prototype ready by Christmas and be ready to shoot early next year.

Cook: Isn't it better for the public to know what it has to face? Isn't it possible that spoon-feeding and disclosures of something less than full frankness will result in casting doubts on the leadership of the program? [21]

Vice President: Serious consideration is being given to having the President in a message put this whole matter right out on the table and explain it, but before he does this we will see that all of the pros and cons are taken into account.

Each of the men at this conference is requested to prepare a page or a page and a half paper on what our nation space effort should be. Let's define the goals. Your judgment and what you will have to say will not be charged against you. If you will each use your brains [and others theirs] we will add the ideas all together and get a good product. In other words, let's have a short paper by Monday to tell me what you think. Our objectives should be, what we should do [I don't need it from Webb. I have his.]

I assume that Webb will be burning the midnight oil to firm up the outlines of two weeks ago and get the BOB to agree. The President has read the interim memo and asked questions about it. He is keeping in very close touch with changing events. Everyone involved should get details to the BOB and get their justifications confirmed because June 30th is coming and we've got to have it all done. We have to try all sorts of things and in doing that, we'll get something like we want on each of them.

Should we give some thought to having the President enlist peoples' support in our program honestly explaining that there will [22] be limited information? Can we put emphasis on the announcements of our success and in doing so, call it "running gear" and then say that we need $500 million more to use that "gear?" No one is going to be pinned down when he expresses his best judgment. We must hope for the best, but be ready to accept the inevitable. So far NASA has gotten everything it has asked for. I want them to plan and dream big enough to get us out ahead. (Story of electricity for rural Texas). I want to know what the national effort should be in your judgment. By working together, we will achieve the national goal.

Document III-11

Document title: James E. Webb, NASA Administrator, and Robert S. McNamara, Secretary of Defense, to the Vice President, May 8, 1961, with attached: "Recommendations for Our National Space Program: Changes, Policies, Goals."

Source: NASA Historical Reference Collection, History Office, NASA Headquarters, Washington, D.C.

This memorandum is the charter for Project Apollo and for an across-the-board acceleration of U.S. space efforts. It was the hurried product of a weekend of work following the successful suborbital flight of the first U.S. astronaut on Friday, May 5, 1961. The urgency was caused by the vice president's desire to get recommendations to the president before he left on a rapidly arranged inspection tour to Southeast Asia. NASA, the Department of Defense, and the Bureau of the Budget staffs and senior officials met on Saturday and Sunday at the Pentagon to put together the memorandum, which the vice president approved without change and delivered to the president on Monday, May 8. On that same day, Alan Shepard came to Washington for a parade down Pennsylvania Avenue and a White House ceremony with President Kennedy.

---

8 May 1961

Dear Mr. Vice President:

Attached to this letter is a report entitled "Recommendations for Our National Space Program: Changes, Policies, Goals", dated 8 May 1961. This document represents our joint thinking. We recommend that, if you concur with its contents and recommendations, it be transmitted to the President for his information revised and expended objectives which it contains.

> Very respectfully,
> James E. Webb
> Administrator
> National Aeronautics and
> Space Administration
> Robert S. McNamara
> Secretary of Defense...

[1] **Introduction**

It is the purpose of this report (1) to describe changes to our national space efforts requiring additional appropriations for FY 1962; (2) to outline the thinking of the Secretary of Defense and the Administrator of NASA concerning U.S. status, prospects, and policies for space; and (3) to depict the chief goals which in our opinion should become part of Integrated National Space Plan. These matters are covered in Sections I, II, III, respectively.

Three appendices (Tabs A through C) support these sections. Tab A highlights the Soviet space program. The bulk of this Tab (Attachment A) is separated from this report since it bears a special security classification. Tab B includes a description of major U.S. space projects and elements. Tab C provides financial summaries of the present programs, the proposed add-ons, and future costs of the program.

The first joint report contains the results of extensive studies and reappraisals. It is a first and not our last report and does not, of course, represent a complete or final word about our space undertakings.

[2] I. **RECOMMENDATIONS FOR FY-1962 ADD-ONS**

Our recommendations for additional FY 1962 NOA for our space efforts are listed below. They total $626 Million of which all but $77 million is for NASA. Certain of these additions will accelerate projects which need to be accomplished more quickly if national space goals are to be reached on time. Other additions augment projects or programs to afford greater likelihood of success or to acquire concurrently data needed to implement long range goals for which we would otherwise have had to wait. Some of the additions, especially those for rocket engine, booster, and upper stage developments, support parallel programs to insure that the failure or delay in a single launch vehicle development will not place our long range goals in jeopardy. It is our belief that it is feasible to accomplish those objectives of acceleration, augmentation and greater certainty of success through the application of the funds specified. The general objectives of acceleration, augmentation and greater certainty of success through the application of the funds specified. The general objectives and their implementation in each particular case are amply supported by the investigations and assessments of qualified scientific and technical people intimately familiar with our space undertakings in detail and in the large.

These FY 1962 additions represent a vital first stage funding toward implementing longer range goals, including the objective of manned lunar exploration in the latter part of this decade, specified in Section III. They will, we believe, prove to be consistent with the objectives and policies set forth in Section II.

a. Spacecraft for Manned Lunar Landing and Return

To achieve the goal of landing a man on the moon and returning him to earth in the latter part of the current decade requires immediate initiation of an accelerated program of spacecraft development.

The program designated Project Apollo includes initial flights of a multi-manned orbiting laboratory to qualify the [3] spacecraft, and manned flights around the moon before attempting the difficult lunar landing.

The additional funds required will be used to extend tests with the Mercury spacecraft to learn more about the behavior of man during longer duration flights in space, to study the biomedical problems encountered in outer space, to investigate the problems of reentering the earth's atmosphere from lunar return speeds, to initiate the development of the multi-manned spacecraft for the mission, and for ground tracking and other facilities.

	APPROVED BUDGET	RECOMM. BUDGET	ADD'AL
FUNDS:	29.5M	240.0M	210.5M

b. Launch Vehicle Development

1. *For the manned lunar landing*

The advanced goal of manned landing on the moon requires the development of a launch vehicle (Nova) with a thrust of about six times greater then that of the largest vehicle now under development (Saturn). The funds requested are to accelerate the development of 1 1/2 million pounds thrust liquid fueled rocket engine now under development; for design, engineering, and component development of the Nova vehicle; and to initiate the construction of necessary facilities required in support of the vehicle development and test.

	APPROVED BUDGET	RECOMM. BUDGET	ADD'AL
FUNDS:	42.7M	155.2M	112.5M

2. *Solid propellant parallel approach*

To assure a high degree of success in achieving the manned lunar landing goal a parallel approach to the development of a first stage for the large launch vehicle (NOVA) must be undertaken. In addition to the use of liquid fuels for this purpose we must also undertake the immediate development of large solid rocket launch vehicles. When developments in both liquid fueled and solid fueled rockets have [4] progressed to a stage where one or the other can be shown to be the superior approach, it will be pursued as the principal launch vehicle development. Certain elements of the solid rocket development are also believed to have future military importance. The DoD will be responsible for this development, being fully responsive to NASA requirements.

	APPROVED BUDGET	RECOMM. BUDGET	ADD'AL
FUNDS (DoD):	-0-	62M	62M

c. Development of An Upper Stage for Titan II

The Atlas-Agena combination is the most powerful launch vehicle available to the

U.S. until the Atlas-Centaur becomes operational in 1963. The Atlas-Agena cannot place more then 5,000 pounds in the 300-miles orbit. The Atlas-Centaur, if successful, will be capable of launching as much as 8,500 pounds.

Because of an urgent need by both DoD and NASA of a vehicle with the performance of Atlas-Centaur, a back-up for this launch vehicle is considered essential. Therefore it is proposed to initiate development of an upper stage for Titan II which will then provide a strong back-up for the Atlas-Centaur. The Titan II upper stage development will be terminated if the timely success of Atlas-Centaur becomes apparent.

	APPROVED BUDGET	RECOMM. BUDGET	ADD'AL
FUNDS (DoD):	-0-	15M	15M

d. Unmanned Lunar Exploration

Before attempting a manned lunar landing, it is essential to learn more about the phenomena that exist in space near the moon and about the nature of the characteristics of the moon's surface. The programs now underway, designated Ranger and Surveyor, are designed to provide this information. With additional funds, the number of fights in these programs will be increased and the program will be accelerated to provide timely information and greater assurance.

[5]

	APPROVED BUDGET	RECOMM. BUDGET	ADD'AL
FUNDS:	71.67M	134.67M	63.0M

e. Scientific Experiments on Space Environment

Knowledge of the space environment through which man must travel to the moon is now still very meager. The nature and characteristics of the radiation emanating from the sun and from outer space must be thoroughly studied and understood. Man's ability to survive in outer space and on the surface of the moon depends on this knowledge. The additional funds will augment and expedite current programs that will provide this information.

	APPROVED BUDGET	RECOMM. BUDGET	ADD'AL
FUNDS:	72.2M	87.2M	15.0M

f. Satellite Communications Systems

With the launching of the Echo and Courier communications satellites, the United States achieved a position of leadership in the use of satellites for worldwide communications. Studies by many qualified organizations have shown the potential economic, political and military value of these systems. The funding requested will accelerate the developments, lead to a much earlier availability of an operational system, and maintain the United States position of leadership

	APPROVED BUDGET	RECOMM. BUDGET	ADD'AL
FUNDS:	44.6M	94.6M	50.0M

g. Meteorological Satellites for Worldwide Weather Prediction[*]

The outstanding success demonstrated by the U.S. Tiros weather satellite in worldwide weather prediction will enable the U.S. to exploit this technology for the benefit of

---

[*]The Department of Commerce is considering the request of an additional 53.5M in FY 62 to initiate an operational meteorological satellite system.

all mankind. [6] Developments are now under way which will lead to operational systems. To assure success of an operational system at the earliest time it is necessary to augment the funding of this program by the amount shown below.

	APPROVED BUDGET	RECOMM. BUDGET	ADD'AL
FUNDS:	28.2M	50.2M	22.0M

h. Nuclear Rocket Development*

Nuclear rocket development (Rover) must be carried out on an accelerated basis because of its great potential for even more difficult missions than landing a man on the moon. This program faces many difficult technological problems which will require substantial support over a number of years for their solution. The funding will provide for augmented research and development and essential facilities for the conduct of the program.

	APPROVED BUDGET	RECOMM. BUDGET	ADD'AL
FUNDS:	17.5M	40.5M	23.0M

i. Supporting Research and Technology

The successful accomplishment of the accelerated program goals requires immediate expansion of basic researches and advances in technology in many fields. For the conduct and management of this major effort some additional increase in the NASA staff will also be required. Requirements for the additional research, advancement of technology and support of personnel and plant are shown below.

	APPROVED BUDGET	RECOMM. BUDGET	ADD'AL
FUNDS:	393.5M	446.5M	53.0M

[7] ## II. NATIONAL SPACE POLICY

The recommendations made in the preceding Section imply the existence of national space goals and objectives toward which these and other projects are aimed. Major goals are summarized in Section III. Such goals must be formulated in the context of a national policy with respect to undertakings in space. It is the purpose of this Section to highlight our thinking concerning the direction that such national policy needs to take and to present a backdrop against which more specific goals, objectives and detailed policies should, in our opinion, be formulated.

a. Categories of Space Projects

Projects in space may be undertaken for any one of four principal reasons. They may be aimed at gaining scientific knowledge. Some, in the future, will be of commercial or chiefly civilian value. Several current programs are of potential military value for functions such as reconnaissance and early warning. Finally, some space projects may be undertaken chiefly for reasons of national prestige.

The U.S. is not behind in the first three categories. Scientifically and militarily we are ahead. We consider our potential in the commercial/civilian area to be superior. The Soviets lead in space spectaculars which bestow great prestige. They lead in launch vehicles needed for such missions. These bestow a lead in capabilities which may some day become important from a military point of view. For these reasons it is important that we take steps to insure that the current and future disparity between U.S. Soviet launch

---

*The Atomic Energy Commission has requested an additional 7M in FY 62 to support the reactor portion of the Nuclear Rocket Development Program.

capabilities be removed in an orderly but timely way. Many other factors however, are of equal importance.

b. Space Projects for Prestige

All large scale space projects require the mobilization of resources on a national scale. They require the development and successful application of the most advanced technologies. They call for skillful management, centralized control and unflagging pursuit of long range [8] goals. Dramatic achievements in space, therefore, symbolize the technological power and organizing capacity of a nation.

It is for reasons such as these that major achievements in space contribute to national prestige. Major successes, such as orbiting a man as the Soviets have just done, lend national prestige even though the scientific, commercial or military value of the undertaking may by ordinary standards be marginal or economically unjustified.

This nation needs to make a positive decision to pursue space projects aimed at enhancing national prestige. Our attainments are a major element in the international competition between the Soviet system and our own. The non-military, non-commercial, non-scientific but "civilian" projects such as lunar and planetary exploration are, in this sense, part of the battle along the fluid front of the cold war. Such undertakings may affect our military strength only indirectly if at all, but they have an increasing effect upon our national posture.

c. Planning

It is vital to establish specific missions aimed mainly at national prestige. Such planning must be aimed at both the near-term and at the long range future. Near-term objective alone will not suffice. The management mechanisms established to implement long range plans must be capable of sustained centralized direction and control. An immediate task is to specify long-range goals, to describe the missions to be accomplished, to define improved management mechanisms, to select the launch vehicles, the spacecraft, and the essential building blocks needed to meet mission goals. The long-term task is to manage national resources from the national level to make sure our goals are met.

It is absolutely vital that national planning be sufficiently detailed to define the building blocks in an orderly and integrated way. It is absolutely vital that national management be equal to the task of focusing resources, particularly scientific and engineering manpower [9] resources, on the essential building blocks. It is particularly vital that we do not continue to make the error of spreading ourselves too thin and expect to solve our problems through the mere appropriation and expenditure of additional funds.

d. Feasibility

It is technically feasible to match and surpass the Soviets in all areas of national space competition whether scientific, commercial, military or in the area of national prestige. Certain steps need to be undertaken right away. Those requiring additional appropriations in FY 1962 were described in Section I.

Additional important actions have also been defined. They include the necessity to specify standardized "work horse" building block combinations to support our space efforts for the long pull. It is particularly important to define building block combinations of boosters and upper stages for each major class of payload and mission. Conversely, it is important to avoid wherever possible the creation of complex and costly launch vehicles and other equipments optimized for and largely limited in application to a single project or mission. This principle has been recognized and applied to portions of our launch-vehicle developments. It has not been applied to all. It must govern all our efforts in the future. Major sub-elements including guidance systems, control systems, power supplies, telemetry, recovery and other basic system elements must also be standardized and used repetitively to the maximum possible degree.

After fully adequate study we must specify the minimum family of launch vehicle systems that will enable us to accomplish both near term (such as communication satellites) and long range missions (such as lunar or planetary exploration) which will comprise our national space goals.

If properly conceived and designed the vehicles will be used for many years in dozens and perhaps hundreds of costly missions. Their development and procurement will involve the expenditure of billions of [10] dollars over a period of a decade or more. It is essential that we set out [our] foot on the correct path kneading to the future. The decisions made in the beginning will define that path. Once embarked upon it, we cannot turn back or turn aside without losing time which can never be regained.

e. **Background Information Bearing on the Problem**

These words would not be written and this report would not be called for if we were satisfied with our status and our prospects for the future in space. We are not satisfied. We are behind in important ways and it is not clear that we are catching up. In reading the balance of this report, it is important to have in mind some of the highlights of space history, of our posture today, and of our prospects for the future. It is important to realize that more money is not the only answer. While considerable additional effort will be called for, a principal problem is how better to harness, not merely how further to expand, the human and physical resources already at hand.

It is important to note that the recent Soviet attainments are the result of a program planned and executed at the national level over a long period of time. The decisions which led to the current successes were, for the most part, made many years ago. Many of them, in fact, must have been made in the early part of the 1950-1960 decade.

That decade has witnessed a great expansion in U.S. government-sponsored research and development, especially for large-scale defense programs. Enormous strides have been made, particularly in our space efforts and in the development of related ballistic missile technology on a "crash" basis. We have, however, incurred certain liabilities in the process. We have over-encouraged the development of entrepreneurs and the proliferation of new enterprises. As a result, key personnel have been thinly spread. The turnover rate in U.S. defense and space industry has had the effect of removing many key scientific engineering personnel from their jobs before the completion of the projects for which they were employed. Strong concentrations of technical talent needed for the best [11] work on difficult tasks have been seriously weakened. Engineering costs have doubled in the past ten years.

These and other trends have a strong adverse effect on our capacity to do a good job in space. The inflation of costs has an obvious impact and they are still rising at the rate of about seven per cent per year. This fact alone affects forward planning. It has often led to project stretch-outs, and may again in future years. The spreading out of technological personnel among a great many organizations has greatly slowed down the evolution of design and development skills at the working level throughout the country. Precisely the opposite is true in the USSR, where the turnover rate is very low and where skilled cadres of development personnel remain in existence for a great many years.

It is not suggested that we apply Soviet type restrictions and controls upon the exercise of personal liberty and freedom of choice. It is suggested, however, that our American system can be and must be better utilized in the future than in the past.

Our space efforts, like many of our military weapons developments, have suffered because of our tendency to "improve" and to embellish our designs. We have allowed ourselves to strive for the optimum solution to nearly every problem project-by-project. We have often tried to "integrate" very complex system elements at minimum weight and with very little margin for safety or for error. Many have come to think that such techniques are the natural and obvious way to get jobs done. They are not, they will not succeed and they must be changed.

We must address ourselves to these problems more effectively in the future than in the past. We must create mechanisms to lay out and to insist upon achievement, not mere improvement. We must stress performance, not embellishment. We must insist from the top down, that, as the Russians say, "the better is the enemy of the good."

[12] f. **Summary**

Clearly, then, the future of our efforts in space is going to depend on much more than this year's appropriations or tomorrow's new idea. It is going to depend in large measure upon the extent to which this country is able to *establish and to direct* an "Integrated National Space Program."

True, it will be necessary to support our space efforts at a higher funding level than recommended before. Such support will have to be backed by the Administration, by the Congress, and by the American people. If, however, the application of more money leads to still further cost increases and still further thinning out of technical manpower and technical supervision, it is likely that the Russians will be ahead of us ten years from now just as they are today.

It will be necessary, therefore, to find a way to formulate and apply plans and policies aimed at insuring the success of an Integrated National Space Program. Top level scientific and policy direction must be forthcoming from the top management echelon. The mere statement of broad objectives will not be good enough. Periodic budget reviews and their intensification in the spring of each year will not suffice. It will be necessary to impose policy and management actions which will alter many of the trends of the past ten years, particularly in the management of research and engineering resources on a national scale. It will be necessary to impose actions which may involve painful cancellations and redirections.

These and other policies, too, must be supported by the Administration, by the Congress, and by the American people to insure success for the long pull ahead.

Our joint efforts are addressed to the creation of management tools to deal with these and other problems we will face in the years ahead.

[13] III. **MAJOR NATIONAL SPACE GOALS**

It is the purpose of this section to outline some of the principal goals, both long range and short range, toward which our national space efforts should, in our opinion, be directed. It is not the intent to specify all of the goals or even all of the major goals of importance to a National Space Plan. We wish to stress five principal objectives which in our opinion have not been adequately formulated or accepted in the past and which we believe should be accepted as a basis for specific project undertakings in the years ahead.

a. **Manned Lunar Exploration**

We recommend that our National Space Plan include the objective of manned lunar exploration before the end of this decade. It is our belief that manned exploration to the vicinity of and on the surface of the moon represents a major area in which international competition for achievement in space will be conducted. The orbiting of machines is not the same as the orbiting or landing of man. It is man, not merely machines, in space that captures the imagination of the world.

The establishment of this major objective has many implications. It will cost a great deal of money. It will require large efforts for a long time. It requires parallel and supporting undertakings which are also costly and complex. Thus, for example, the RANGER and SURVEYOR Projects and the technology associated with them must be undertaken and must succeed to provide the data, the techniques and the experience without which manned lunar exploration cannot be undertaken.

The Soviets have announced lunar landing as a major objective of their program. They may have begun to plan for such an effort years ago. They may have undertaken important first steps which we have not begun.

It may be argued, therefore, that we undertake such an objective with several strikes against us. We cannot avoid announcing not only our general goals but many of our specific plans, and our successes [14] and our failures along the way. Our cards are and will be face up—their's are face down.

Despite these considerations we recommend proceeding toward this objective. We are uncertain of Soviet intentions, plans or status. Their plans, whatever they may be, are not more certain of success than ours. Just as we accelerated our ICBM program we have accelerated and are passing the Soviets in important areas in space technology. If we set our sights on this difficult objective we may surpass them here as well. Accepting the goal gives us a chance. Finally, even if the Soviets get there first, as they may, and as some think they will, it is better for us to get there second than not at all. In any event we will have mastered the technology. If we fail to accept this challenge it may be interpreted as a lack of national vigor and capacity to respond.

b. **Worldwide Operational Satellite Communication Capability**

It is our belief that advances in technology will make it possible to set up an operational satellite-based telecommunications capability within a few years. It is too early to be sure what kind of capability we should create.

We *are* certain, however, that at least one of several current possibilities is very likely to prove successful and practical. We are also confident that an operational communication satellite capability can have far reaching applications and implications for the U.S.

Many commercial enterprises in this country have displayed great interest in this subject. It is virtually certain that communication satellites will have commercial utility in future years. The Department of Defense is keenly interested in view of its large use of commercial capacity and is also undertaking the development (Project ADVENT) of a satellite aimed principally at fulfilling military requirements.

Finally and perhaps most important of all, communication satellites uniquely provide a way of relaying information from one point [15] to another over great distances. A successful global satellite-based communication system may be looked upon somewhat as an overhead cable thousands of miles in the air to which users all over the world may make connections by means of invisible radio beams.

Accordingly, we have recommended that the NASA Communication Satellite effort be expanded in FY 1962. Within the next few years this country which is already the world's leader in communications of all kinds will be able to deploy a worldwide satellite-based communication capability.

c. **Worldwide Operational Satellite Weather Prediction System**

The TIROS I meteorological satellite operated for several months, transmitting pictures of cloud cover around the world. TIROS II, launched last November, is still transmitting cloud pictures and infra-red data. The Weather Bureau, the NASA, and other interested governmental agencies are closely coordinating their interests with respect to TIROS and follow-on meteorological satellites. A worldwide system of such satellites will be of great value to people in every country, to public and private interests in the U.S., and to our military forces, particularly those at sea and in the air. The addition of $22 million in FY 1962 will accelerate the early attainment of preliminary operational capability.

d. **Scientific Investigation**

Fundamental to and underlying all progress in the exploration and application of space is the knowledge to be gained from the space sciences. It is essential that the national space sciences program be broad and comprehensive both in content and in participation by the scientific community of the world.

Before man can explore in the vast and hostile regions of space more knowledge is required on the effects of hard vacuum, weightlessness, and radiation.

The broad program that is recommended includes the following objectives:

[16] — To understand the nature of the sun and its influence on the earth.
— To investigate the solar system, its nature, and its history.
— To search for life in the solar system.
— To study cosmology, the history and nature of the universe.

Researchers in government laboratories, universities, industry, and other scientific organizations must participate in the space sciences effort and the space science program should support not only the existing talent but also the development of new talent. This requires support of universities in the education of the young scientists who are inspired by the challenge of space and who will strengthen the program with their vitality and new ideas.

For the fiscal year 1962 we recommend adding fifteen million dollars to the program to assure success of these objectives.

e. **Large Scale Boosters for Potential Military Use**

Space technology is in its infancy. The first U.S. satellite was launched only $3\text{-}1/2$ years ago. Vast resources are presently devoted to this field in the Communist world and in our own. The future potentialities and capabilities that space technology will afford cannot be foreseen. Their military potential and implications are largely unknown. It is certain, however, that without the capacity to place large payloads reliably into orbit, our nation will not be able to exploit whatever military potential unfolds in space.

We believe it is important, therefore, to insure that large scale boosters are made available. They should, of course, form part of a family of launch vehicles applicable to many missions and thoroughly integrated with NASA developments and with characteristics suitable for NASA needs.

We have agreed jointly in recommending the addition of $62 million for the DoD in FY 1962 to undertake the development of large scale solid propellant rocket motors.

[17] TAB A: **THE SOVIET PROGRAM AND CAPABILITIES**

Attachment A depicts the Soviet space program in considerable detail. It is separated from this report for security reasons, but it should be read by the key policy makers concerned with the U.S. space program. It should be read not only because of what it tells about the Soviet program in space, but for what it reveals concerning the caliber of the competition we are up against in this important arena of the cold war.

Attachment A reveals a set of Soviet undertakings and achievements of enviable simplicity and unmatched success. With the possible exception of the first two launchings, a single ICBM booster has been used on all Soviet space shots. A small family of intermediate and final stages has been combined to form a total of only three different configurations. The Soviets have placed only 14 space craft into orbit around the earth, the moon, or elsewhere in the planetary system.

The Soviets have employed a single launch complex for all ICBM and space launchings. All shots have been made in the same direction whether for ICBM testing, earth-orbit missions (including man in space), or lunar and deep space missions. These features of the Soviet program evidence long-range planning which in all probability began early in the 1950-1960 decade.

The U.S. did not undertake a corresponding planning effort at the national level until much later. Space was not recognized as an area of importance to be planned for and pursued until after Sputnik I. Great efforts were then made to utilize the tools at hand to enter the space arena. Early projects were undertaken on a crash basis. New governmental and industrial organizations were formed and expanded under pressure and in haste. For several years it was necessary to "make do" with too-small boosters and launch vehicles employing devices developed and deployed with unusual urgency. However dedicated or effective such efforts were, they were not the product of a deliberate effort adhering to preplanning schedules and objectives.

[18] The U.S. space program in the past three years reflects this situation in many ways. We have been forced to design with inadequate margins for error or deficiencies in thrust. We have been forced to develop elaborate and often unreliable new ways to cram complex equipments in a very small space. Our results have, despite many excellent achievements, been disappointing in many important ways. Nearly half of our attempted launchings failed to achieve orbit. Certain programs achieved success, real success, on fewer than a third of all attempts. To a large degree, though not entirely, of course, these disappointments are symptoms of the lack of adequate national planning and guidance for the long pull.

It is possible, of course, that the Soviet program is not actually the result of careful planning toward long range goals. It may only appear that way in retrospect. It is possible, too, that Soviet management and decision making is not as excellent as it appears to date. Perhaps the poverty of their resources forced concentration on a brute force approach which paid off not as the result of initiative and forethought but through the force of circumstances which left no other choice. Perhaps luck played an important part at an early stage and the Soviets were wise enough and swift enough to exploit it far beyond any initial long range plan.

These are conjectures. The evidence points dramatically to the existence of long range planning and competent and flexible technical decision making and managerial direction. It is prudent to suppose that the next decade will be marked with Soviet achievements in space which will be well planned, well directed, and executed with deliberateness and skill.

Attachment A indicates the manner in which we can estimate key milestones of the future of the Soviet program by extrapolating our present knowledge. Doubtless, the Soviet plan is not irrevocably fixed and includes options and decision points which will enable them to pace their achievements in relation to ours and to Soviet national objectives....

[27]   **SPACE ACTIVITIES OF THE UNITED STATES GOVERNMENT**
Tab C **New Obligational Authority/Program Basis—in millions**

Historical Summary and Proposed 1962 Add-ons

	NASA[1]	Defense[2]	AEC[3]	NSF[4]	WB[5]	Total
1955	$ 56.9	$ 3.0	-	-	-	$ 59.9
1956	72.7	30.3	$ 7.0	$7.3	-	117.3
1957	78.2	71.0	21.3	8.4	-	178.9
1958	117.3	205.6	21.3	3.3	-	347.5
1959	338.9	489.5	34.3	-	-	862.7
1960	523.6	560.9	43.3	.1	-	1,127.9
1961	964.0	751.7	62.7	.6	-	1,779.0
1962 Budget	1,109.6	846.9	55.1	1.6	$2.2	$2,015.4
Budget amendments	125.7	159.0	23.5	-	-	308.2
Total present 1962 Budget	1,235.3	1,005.9	78.6	1.6	2.2	2,323.6
Proposed 62 Add-ons	549.0	77.0	7.0[6]	-	53.0[7]	686.0
Total 1962 Proposals	$1,784.3	$1,082.9	$85.6	$1.6	$55.2	$3,009.6

    1. National Aeronautics and Space Administration amounts are totals for all activities of NASA and include totals for NACA prior to establishment of NASA.
    2. Department of Defense amounts are based on identifiable Defense funding for space and space-related effort and do *not* include substantial amounts for (1) construction and operation of the national missile ranges with regard to space programs, (2) the cost of developing missiles such as Thor and Atlas which are also used in space programs, or (3) supporting research and development (such as bio-medical research) which is more or less mutually applicable to programs other than "space."
    3. Atomic Energy Commission amounts are those identifiable with ROVER nuclear rocket and SNAP atomic power source projects.
    4. National Science Foundation amounts are those identifiable with VANGUARD and with the NSF space telescope project.
    5. Weather Bureau amounts are those identifiable with the meteorological satellite program.
    6. AEC add-on for ROVER corresponding to the $23 million included in proposed NASA add-ons.
    7. Add-on for initiation of operational NIMBUS system under consideration by Department of Commerce.

[28] **NATIONAL AERONAUTICS AND SPACE ADMINISTRATION
FY 1962 BUDGET AND PROPOSED ADD-ONS**
(In Millions)

	Original 1962 Budget	1962 Budget Amendment	Present 1962 Budget	Proposed 1962 Add-ons	Total 1962 Proposed
I. Major Vehicle & Propulsion Dev. Projects					
A. Centaur	$ 49.8	$ 25.6	$ 75.4	-	$ 75.4
B. Saturn	212.8	73.0[1]	285.8	-	285.8
C. Rover (NASA only)	13.5	4.0	17.5	$ 23.0[4]	40.5
D. F-1 Engine and Nova Vehicle	33.4	9.3	42.7	112.5	155.2
E. Large Solid Boosters	2.0	-	2.0	-	2.0
F. Other Projects	6.6	-	6.6	-	6.6
II. Major Space Flight Programs					
A. Mercury	74.2	-	74.2	-	74.2
B. Other Manned Space Flight	29.5	-	29.5	210.5	240.0
C. Meteorology	28.2	-	28.2	22.0	50.2
D. Communications	34.6	10.0	44.6	50.0	94.6
E. Scientific Satellites & Sounding Rockets	72.2	-	72.2	15.0	87.2
F. Unmanned Lunar Exploration	71.7	-	71.7	63.0	134.7
G. Unmanned Planetary Exploration	32.2	-	32.2	-	32.2
III. Supporting Development and Operations					
A. NASA Development Centers & Launch Operations	164.0	3.0[2]	167.0	28.0[5]	195.0
B. Tracking and Data Acquisition	60.2	-1.0	59.2	-	59.2
C. Other Supporting Development	57.7	-	57.7	25.0	82.7
IV. Aeronautical and Space Research	142.7	1.3[3]	144.0	-	144.0
V. Program Direction	24.3	.5	24.8	-	24.8
Total NASA	$1,109.6	$125.7	$1,235.3	$549.0	$1,784.3

[1] Total of $78 million identified with Saturn C-2 includes $5 million of supporting costs on line III-A.
[2] Consists of $5 million for support identified with Saturn C-2 less $2 million adjustment in other supporting costs.
[3] Consists of $2 million identified with the supersonic transport less $.7 million adjustment on a vivarium construction project.
[4] Would require corresponding add-on of $7 million for AEC.
[5] Part of this amount may be applied to IV and V.

[29]

DEPARTMENT OF DEFENSE
SPACE AND RELATED PROGRAMS[1]
FY 1962 BUDGET AND PROPOSED ADD-ONS

	Original 1962 Budget	1962 Budget Amendment	Present 1962 Budget	Proposed 1962 Add-ons	Total 1962 Proposed
DISCOVERER	$ 24.9	$ 30.0	$ 54.9	-	$ 54.9
SAMOS	282.2	-	282.2	-	282.2
MIDAS	147.6	60.0	207.6	-	207.6
TRANSIT	22.4	-	22.4	-	22.4
ADVENT	57.0	15.0	72.0	-	72.0
SAINT	12.0	14.0	26.0	-	26.0
SPACETRACK	34.0	-	34.0	-	34.0
SPASUR	4.3	-	4.3	-	4.3
BLUE SCOUT	5.0	10.0	15.0	-	15.0
WESTFORD	4.3	-	4.3	-	4.3
X-15	7.0	-	7.0	-	7.0
DYNASOAR (Step I)	76.5	30.0	106.5	-	106.5
Component development/ Applied Research/Other	169.7	-	169.7	-	169.7
Large Solid Booster	-	-	-	62.0	62.0
CENTAUR BACKUP	-	-	-	15.0	15.0
TOTAL DOD Space and Related Programs	$846.9	$159.0	$1,005.9	$77.0	$1,082.9

1. Covers identifiable DOD funding for space and space-related effort; does *not* include substantial amounts for (1) construction and operation of the national missile ranges with regard to space programs, (2) supporting research and development (such as bio-medical research) which is more or less mutually applicable to programs other than "space," and (3) the cost of developing missiles such as ATLAS and THOR which are also used in space programs.

[30] **ESTIMATED PROJECTIONS OF N.A.S.A. and DEFENSE SPACE PROGRAMS (Excludes A.E.C., Weather Bureau, and N.S.F.)**

New Obligational Authority/Program Basis—In Millions

	1962	1963	1964	1965	1966
BASE PROJECTIONS (Includes only (1) continuing programs and (2) major projects currently underway or approved for initiation in the amended 1962 Budget):					
N.A.S.A.	1235	1390	1275	1215	1070
Defense	1006	1300	1360	1475	1675
Total	2241	2690	2635	2690	2745
OTHER CURRENT PLANS (Projects which would be initiated in 1963 or later Budgets under current agency plans):					
N.A.S.A.	—	799	1051	1520	1595
Defense	************** NOT AVAILABLE **************				
Total	—	799	1051	1520	1595
TOTAL PROJECTIONS OF CURRENT AGENCY PROGRAMS:					
N.A.S.A.	1235	2189	2326	2735	2665
Defense	1006	1300	1360	1475	1675
Total	2241	3489	3686	4210	4340
INCREASED FUNDING REQUIREMENTS RESULTING FROM PROPOSED ADD-ONS:					
N.A.S.A.	549	785	1917	1917	1959
Defense	77	150	160	125	100
Total	626	935	2077	2042	2059
TOTAL PROJECTIONS OF PROPOSED PROGRAMS:					
N.A.S.A.	1784	2974	4243	4652	4624
Defense	1083	1450	1520	1600	1775
Total	2867	4424	5763	6252	6399

Document III-12

Document title: **John F. Kennedy, Excerpts from "Urgent National Needs," Speech to a Joint Session of Congress, May 25, 1961.**

Source: **NASA Historical Reference Collection, History Office, NASA Headquarters, Washington, D.C.**

This is the section of President Kennedy's "reading text" of his address to a Joint Session of Congress in which he called for sending Americans to the Moon "before this decade is out." President Kennedy in his own hand modified the prepared text of his remarks. The text as written, modified, and ultimately delivered varies considerably. Kennedy ad-libbed three additional paragraphs near the end of his speech.

Handwritten additions to the text are contained in brackets. Portions of the text that Kennedy crossed out are contained in parentheses.

---

[63] IX. **Space**

Finally, if we are to win the battle for men's minds, [64] the dramatic achievements in space which occurred in recent weeks should have made clear to us all [as did the Sputnik in 1957] the impact of this new frontier of human adventure. Since early in my term, our efforts in space have been under review. With the advice of the Vice President [who is Chairman of the National Space Council] we have examined where we are strong and where we are not, where we may succeed and where we may not. Now it is time to take longer strides — time for a great new American enterprise — time for this nation to take a clearly leading role in space achievement [which in many ways may hold the key to our future on earth].

[65] I believe we possess all the resources and all the talents necessary. But the facts of the matter are that we have never made the national decisions or marshalled the national resources required for such leadership. We have never specified long range goals on an urgent time schedule, or managed our resources and our time so as to insure their fulfillment.

Recognizing the head start obtained by the Soviets with their large rocket engines, which gives them many months of lead-time, [66] and recognizing the likelihood that they will exploit this lead for some time to come in still more impressive successes, we nevertheless are required to make new efforts. For while we cannot guarantee that we shall one day be first, we can guarantee that any failure to make this effort will find us last. We take an additional risk by making it in full view of the world — but as shown by the feat of astronaut Shepard, this very risk enhances our stature when we are successful. But this is not merely a race. [67] Space is open to us now; and our eagerness to share its meaning is not governed by the efforts of others. We go into space because whatever mankind must undertake, *free* men must fully share.

I therefore ask the Congress, above and beyond the increases I have earlier requested for space activities, to provide the funds which are needed to meet the following national goals:

First, I believe that this nation should commit itself to achieving the goal, before this decade is out, of landing a man on the moon and returning him safely to earth. [68] No single space project in this period will be more (exciting or) impressive [to mankind as it makes its judgement of whether the world is free] or more important for the long-range exploration of space; and none will be so difficult or expensive to accomplish. (Including necessary supporting research, this objective will require an additional $531 million this year and still higher sums in the future.) We propose to accelerate development of the appropriate lunar space craft. We propose to develop alternate liquid and solid fuel boosters much larger than any now being developed, until certain which is superior. [69] We propose additional funds for other engine development and for unmanned explorations —explorations which are particularly important for one purpose which this nation will never overlook: the survival of the man who first makes this daring flight. But in a very real sense, it will not be one man going to the moon — it will be an entire nation. For all of us must work to put him there.

Second, an additional $23 million, together with $7 million already available, will accelerate development of the ROVER nuclear rocket. [70] This (is a technological enterprise in which we are well on the way to striking progress, and which) gives promise of some day providing a means for even more exciting and ambitious exploration of space, perhaps beyond the moon, perhaps to the very ends of the solar system itself.

Third, an additional $50 million will make the most of our present leadership by accelerating the use of space satellites for world-wide communications. When we have put into space a system that will enable people in remote areas of the earth to exchange messages, hold conversations, [71] and eventually see television programs, we will have achieved a success as beneficial as it will be striking.

Fourth, an additional $75 million — of which $53 million is for the Weather Bureau —will help give us at the earliest possible time a satellite system for world-wide weather observation. (Such a system will be of inestimable commercial and scientific value; and the information it provides will be made freely available to all the nations of the world.)

Let it be clear that I am asking the Congress and the country to accept a firm commitment to a new course of action — [72] a course which will last for many years and carry very heavy costs [531 million dollars this year] — an estimated $7-9 billion additional over the next five years. If we were to go only halfway, or reduce our sights in the face of difficulty, it would be better not to go at all. [this is the choice and finally you and the American public must decide for itself.]

Let me stress also that more money alone will not do the job. This decision demands a major national commitment of scientific and technical manpower, material and facilities, and the possibility of their diversion from other important activities where they are already thinly spread. It means a degree of dedication, [73] organization and discipline which have not always characterized our research and development efforts. It means we cannot afford undue work stoppages, inflated costs of material or talent, wasteful interagency rivalries, or a high turnover of key personnel.

New objectives and new money cannot solve these problems. They could, in fact, aggravate them further—unless every scientist, every engineer, every serviceman, every technician, contractor, and civil servant involved gives his personal pledge that this nation will move forward, with the full speed of freedom, in the exciting adventure of space.

Document III-13

Document title: Director, Bureau of the Budget, Memorandum for the President, Draft, November 13, 1962, with attached: "Space Activities of the U.S. Government."

Source: NASA Historical Reference Collection, History Office, NASA Headquarters, Washington, D.C.

This memorandum summarized the results of a special review of the space program carried out by the Bureau of the Budget, NASA, and the Department of Defense in the second half of 1962. The assessment was in response to President's Kennedy's request for "an especially critical review" of the total national space efforts. Other factors justifying the review included NASA's decision to adopt the lunar orbital rendezvous approach to the lunar mission and a subsequent upward revision in the budget estimates for Apollo; a suggestion by Brainard Holmes, the individual in charge of the Apollo program, that the target date for the first landing attempt be moved up from late 1967 to late 1966; and the lack of evidence that the Soviet Union was itself carrying out a lunar landing program.

[1] # Memorandum for the President

This memorandum is to report the status and results to date of the special review of space programs which we have been conducting with the National Aeronautics and Space Administration and the Department of Defense, and to present two policy questions on which your guidance is needed at this time:

    1. The pace at which the manned lunar landing program should proceed, in view of the budgetary implications and other considerations; and

    2. The approach that should be taken to other space programs in the 1964 budget, i.e., should they as a matter of policy be exempted from or subjected to the restrictive budgetary ground rules applicable in 1964 to other programs of the Government.

Decisions on specific programs, and the final amounts to be included in the 1964 budget can wait. However, advance decisions on the above two policy questions are essential to guide the preparation of refined estimates and specific recommendations, especially in the case of NASA.

The special space review

A special space review was begun last summer in response to your request that the space programs of all agencies be given an especially critical review and be presented to you as a whole. As a part of the 1964 budget preview process we arranged to have the tentative 5-year space programs of NASA, Defense, the Atomic Energy Commission, and the Weather Bureau, as they stood last August, laid out on a comparable basis in considerable detail for consideration and [2] review. Subsequently the agencies have made some significant revisions in the programs and cost estimates—notably an upward revision in the cost estimates for the NASA manned lunar landing program—and the agencies and the Bureau of the Budget have developed a variety of higher and lower alternative programs, have reviewed the more important individual programs, and have given special consideration to areas where the programs and interests of the agencies overlap.

The 1964 budget estimates of the agencies now under consideration reflect many of the results of the special review, and serve as a useful basis for the consideration of the various policy alternatives outlined below. A more detailed table is attached as an appendix.

Current Agency Estimates
New obligational authority - in billions

	1962	1963	1964	1965	1966	1967
Manned lunar landing program	$1.3	$2.7	$4.6	$3.4	$2.6	$1.8
All other NASA	.5	1.0	1.6	2.6	3.4	4.2
Total, NASA	1.8	3.7	6.2	6.0	6.0	6.0
Department of Defense	1.1	1.6	1.6	1.6*	1.6*	1.6*
AEC and Weather Bureau	.2	.3	.4	.4	.5	.5
Total NOA, all space programs	3.1	5.6	8.2	8.0	8.1	8.1
Total expenditures, all space programs	2.3	3.9	6.5	8.0	8.1	8.1

*Illustrative amounts; current estimates not yet projected by DOD

Manned lunar landing program

The question of the pace and budget level of the manned lunar landing program revolves around (1) the acceptability of both the schedule and funding [3] requirements of the program currently proposed by NASA; (2) the desirability, cost, and practical feasibility of measures that might be taken to accelerate the program, which have been set forth in a letter by Mr. Webb in reply to your question on the possibility of acceleration; and (3) the merits of lower alternatives which would delay the program to some degree but would ease the burden on the 1964 budget. There are three recent significant developments relating to the manned lunar landing program. One is that a firm decision has been reached to proceed with the "lunar orbit rendezvous" approach. As you know, Dr. Wiesner and his advisory committees have had strong reservations with respect to this approach and advocated further studies and reconsideration of other alternatives. After the latest round of studies and discussions, however, Dr. Wiesner has now agreed that while it might have been better to have concentrated on the earth orbit rendezvous or a 2-man direct ascent approach from the start, in the present circumstances the NASA decision to proceed with the LOR approach is appropriate and offers the best possibility for accomplishing the mission at the earliest practicable date. It is, however, desirable to continue the studies of the 2-man direct mode.

A second development is that NASA's latest estimates, based on the details of the LOR approach as they have now been worked out, indicate that substantially higher amounts would be required in 1963 and 1964 to keep the entire program on an optimum schedule–over $400 million in 1963 above the amounts now appropriated and about $550 million in 1964 above the initial LOR estimates last August. These revised estimates, reflected in the [4] alternatives below, accentuate the budgetary problem, and illustrate once again the tendency for repeated increases in estimated costs of large and complex development projects, while there are reasons to believe that the present estimates are much firmer than previous ones, we cannot with any confidence say that there will not be still further increases in this, without doubt, the largest and most complex single development project the nation has ever undertaken. Third, our understanding of the latest intelligence estimates is that there is no evidence yet that the Russians are actually developing either a larger booster of the size required for a manned lunar landing attempt or rendezvous techniques of the sort that would be required to assemble a manned lunar landing vehicle in earth orbit using their available boosters. While not conclusive, this suggests that extreme measures to advance somewhat our own target dates may not be necessary to preserve a good possibility that we will be first.

The range of possible alternatives is as follows. As indicated in the explanations, all of the alternatives are not equally feasible and have not been worked out in the same detail. In all of the alternatives the "schedule" is to be understood as the target date established for program planning and estimating purposes, not as a forecast of when the first manned lunar landing attempts would actually be made. Experience has shown that on a realistic basis slippage of as much as a year must be anticipated.

[5] **Manned Lunar Landing Program**

	MLL target date	1963	1964	NOA in billions 1965	1966	1967
Alternative 1	late 1967	$2.7	$4.6	$3.4	$2.6	$1.8
Alternative 2	mid-1967	3.1	4.6	3.2	2.4	1.8
Alternative 3	late 1966	3.6	5.4	3.9	3.0	1.0
Alternative 4	late 1968	2.7	3.7	3.5	2.7	2.1

Alternative 1. Assumes no 1963 supplemental and a late 1967 target date, which is regarded as the earliest feasible without a 1963 supplemental. It is included in NASA's current 1964 budget estimates as the alternative preferred by NASA on the basis of current policy guidance, recognizing the practical problems involved in getting timely approval of a 1963 supplemental authorization and appropriation. This alternative involves a sharp peaking of fund requirements in 1964, because the normal funding curves for all of the principal subprojects Gemini, Apollo, Advanced Saturn, etc.—have to peak in the same year—in order to meet the assumed schedule. (There is some doubt whether the requirements in 1965 will drop as much as present estimates indicate.)

Alternative 2. Assumes a 1963 supplemental of about $425 million with approval to proceed immediately on a deficiency basis in anticipation of the supplemental, and a mid-1967 target date. This is the "optimum" schedule referred to above. This alternative, which might accelerate the schedule by about 6 months, would require a strong presidential endorsement and the concurrence of congressional leaders and the appropriations committees with the decision to proceed on a deficiency basis. Because of the practical problems, it is not strongly advocated by Mr. Webb as the appropriate course for the administration to take.

[6] Alternative 3. Assumes a 1963 supplemental of $900 million, approval to proceed on a deficiency basis in 1963, and a decision to proceed on an all-out "crash" basis. NASA estimates that these measures of maximum acceleration might advance the date of a first attempt by as much as one year. This alternative would also require strong Presidential endorsement and congressional concurrence. It would create enormous additional management problems, and in NASA's view and ours would not appear to offer enough assurance of actually advancing the date of a successful attempt to be worth the cost and other problems involved.

Alternative 4. This is an estimate of the minimum amount ($3.7 billion) that could be provided in 1964 and still accommodate a program based on a target date about one year later than alternative 1. A new detailed program would have to be worked out under these dollar and schedule assumptions, and there would be considerable dislocations in activities now underway in 1963. This alternative is significant as indicating probably about the lowest 1964 estimate under which the first actual manned lunar landing might still be expected to occur during this decade, after a realistic allowance for slippage. As such it could be regarded as being in accord with the announced administration policy of achieving a manned lunar landing before the end of this decade. It would also represent an approach to the manned lunar landing program more closely corresponding to the restrictive approach we are taking with respect to other parts of the 1964 budget.

I agree with Mr. Webb that alternative 1, the NASA recommendation, is probably the most appropriate choice at this time to press forward to achieve a manned lunar landing ahead of the Russians. While it will be criticized [7] in some quarters as representing slightly less than a maximum effort, I believe that practical as well as budgetory considerations make a more accelerated program, like alternatives 2 or 3, inadvisable....

Other NASA programs

The special attention give to the manned lunar landing program has sometimes obscured the other program objectives being pursued by NASA. Perhaps the most important are the programs for scientific investigations in space, in which the United States has from the start been the recognized world leader, which have important intrinsic value, which have been the focus of significant programs of international cooperation, and which in some cases, for example if the Mariner spacecraft succeeds in its voyage to Venus, can provide spectacular achievements with some of the same popular appeal as manned space flights. Less costly, but most important, are the programs directed at developing practical applications of space technology, chiefly in the meteorological and communications fields. Finally, there is the continuing research and development effort required to lay the technical foundation for and begin the development of engines and other components, space vehicles, and techniques for future manned and unmanned space flight.

There is no disagreement that work in all of these areas should continue and move forward on a progressive basis, with appropriate decisions and coordination of the specific projects and areas of effort. The policy issue relates to the scale of effort and relative priority of the work.

There are essentially two alternatives, indicated by the following figures:

Other NASA Programs
(Exclusive of amounts supporting manned lunar landing program)
1964 NOA - in billions

	NASA proposal	Illustrative alternate
Scientific investigations in space	$.6	$.4
Applications (communications & meteorology)	.2	.2
Future capabilities & supporting research & development	.8	.7
Total	1.6	1.3

NASA takes the view that the importance of maintaining the proposed general level of effort in the "other" areas is so great that if any reduction were to be made in the $6.2 billion budget request, it should be applied at least in part to the manned lunar landing program, in order to maintain a "balanced" total program. The Administrator and his principal assistants are fearful that the appeal and priority of the manned lunar landing program may turn NASA into a "one program agency" with loss of leadership and standing in the scientific community at home and abroad, and with inadequate provision [9] for moving ahead with developments required for future capabilities in apace. They point to the fact that to some extent the MLL and other programs are mutually supporting in a technical sense, although all scientific investigations and supporting research directly required for the manned lunar program have been identified and provided for in that program.

While recognizing the force of these arguments, it seems to me that (1) as in other research and development programs, the level of effort to be carried forward is, within limits, essentially a matter of degree, and (2) the decision to proceed with the manned lunar landing program as a matter of high urgency has been a unique sort of national decision which does not automatically endow other space objectives and programs with a special degree of urgency. Rather, it seems to me the appropriate national policy is to attempt to maintain a reasonable degree of balance between the very costly space programs, and research and development programs in other fields. Under the policies being applied to the 1964 budget, this would mean that the estimates for NASA programs other than the manned lunar landing should be treated on their merits in the same restrictive fashion as other programs. I feel that a restrictive approach is especially appropriate in 1964 in view of the tremendous peaking in 1964 fund requirements that will occur if alternative 1 is approved for the manned lunar landing program.

The practical effect on the 1964 budget of this policy difference is about $300 million in NOA and about $150 million in expenditures. While these amounts seem small compared to the totals for the space program, they are large compared to most of the other possibilities of adjustment in the 1964 [10] budget. The difference is not greater because NASA's proposals had already deferred to 1965 or later years initiation of most of the major new development projects under consideration, largely for reasons of technical feasibility, partly in recognition of the major effort required in 1964 on the manned lunar landing program. Our recommendation should not be equated with a "no new starts" policy, since even under the restrictive approach we feel would be appropriate, the program would include initiation of additional satellites of types currently available, new types of experiments, and some new development projects, as well as continuation of work already underway.

Defense and other space programs

The space programs of Defense, AEC, and the Weather Bureau do not present policy issues requiring resolution in advance of the final 1964 budget decisions. In the case of Defense, the Secretary and his assistants have taken a restrictive approach in their reviews, based on the conclusion that there are no valid new military requirements which justify at this time a major expansion in the military space programs. Special attention is being given in the budget reviews to the necessity for proceeding with the Titan III and Dynasoar projects, and to the approach that should be taken in the development of communications satellite systems. The communications satellite problem is complex, involving NASA, Defense, and prospectively the new corporation authorized at the last session of Congress. The alternatives and our recommendations on this matter will be presented to you at a later date.

[11] **Financial summary**

The financial effect on the 1964 budget of the policy alternatives that appear most pertinent on the basis of the foregoing discussion are summarized below. It should be recognized that all estimates shown are subject to further adjustment when the regular budget review is completed.

Fiscal Year 1964 - in billions

	Current agency estimates	Current BOB estimates
New Obligational Authority		
Manned lunar landing	$4.6	$4.6
Other NASA	1.6	1.3
Total NASA		
Defense space programs	1.6	1.6
AEC and Weather Bureau	.4	.4
Total NOA	8.2	7.9
Expenditures		
Manned lunar banding	3.4	3.4
Other NASA	1.2	1.0
Total NASA	4.6	4.4
Total Defense and other	1.9	1.9
Total expenditures	6.5	6.3

In closing, I should point out that under any alternative we will be faced with a large built-in further increase in expenditures in 1965 which we now tentatively estimate at about $1.3 billion.

Director

Attachment
[12]

SPACE ACTIVITIES OF THE U. S. GOVERNMENT
Based on agency estimates as of November 9, 1962 - Subject to change as budget reviews proceed

	New Obligational Authority - in millions				
	1963	1964	1965	1966	1967
National Aeronautics and Space Administration					
Manned Lunar Landing Program					
Spacecraft Development and Operations (Mercury, Gemini, Apollo, etc.)	$703	$1,536	$1,101	$978	$666
Launch Vehicle and Engine Development (Saturn, Advanced Saturn, and their engines)	660	1,028	796	579	361
Engineering Support (Systems engineering, integration, and checkout; aerospace medicine; launch operations)	72	244	207	173	165
Supporting Scientific Investigations in Space (Unmanned lunar exploration, orbiting solar observatories, radiation and bioscience satellites, etc.)	291	411	356	299	216
Other Support (Supporting research and development; tracking networks; NASA personnel and operation of installations)	397	609	569	517	316
Construction (Launch, ground test, laboratory, and support fac.)	586	785	343	91	51
Total, MLL Program	2,709	4,613	3,372	2,637	1,775
Other NASA Programs					
Other space sciences programs (Geophysical and astronomical satellites and unmanned exploration of Venus and Mars)	353	590	629	655	522
Applications programs (Development of meteorological and communications satellites)	129	186	144	108	102
Developments required for advanced manned space flight (Advanced engine development, nuclear rocket project, and studies of advanced manned space vehicles)	299	485	685	913	982
Other supporting research (General space technology, aeronautical research, and research grants and facilities for universities)	203	343	359	394	430
Provision for unspecified new programs	-	-	811	1,293	2,189
Total, Other NASA Programs	984	1,604	2,628	3,363	4,225
Total, NASA	3 693	6,217	6,000	6,000	6,000
Department of Defense					
Navigation satellite development and operation	45	35	*	*	*
Communications satellite development	95	76	*	*	*
Dynasoar manned space flight experiments	130	125	*	*	*
Dynasoar support at Vela nuclear weapons test detection experiments	26	26	*	*	*
Discoverer program	130	79	*	*	*
Titan III launch vehicle development	261	330	153	29	3
Large solid rocket development	40	34	*	*	*
Atlantic Missile Range (portion estimated as applicable to space activities)	80	88	*	*	*
Space tracking & detection systems	33	57	*	*	*
Minor projects, supporting research & development, laboratory operations, and miscellaneous	651	706	*	*	*
Total, Defense space activities	1,631	1,646	1,600	1,600	1,600

Atomic Energy Commission

Nuclear rocket development (Rover)	105	170	172	180	170
Space nuclear power development	95	128	187	214	204
Supporting activities	12	21	24	29	29
Total, AEC apace activities	212	319	383	423	403

Weather Bureau

Operational meteorological satellite system & related meteorological research	43	41	60	60	60
TOTAL, all space activities	5,579	8,223	8,043	8,083	8,063

* Current estimates not yet projected for all items by Defense; total shown is illustrative only.

Document III-14

Document title: James E. Webb, Administrator, NASA, to the President, November 30, 1962.

Source: NASA Historical Reference Collection, History Office, NASA Headquarters, Washington, D.C.

In November 1962, a controversy had arisen between NASA Administrator James Webb and the man he had selected to manage the Apollo program, Associate Administrator for Manned Space Flight R. Brainard Holmes, over the priority to be assigned to the Apollo program. Holmes argued that Apollo should be carried out on a crash basis and, if necessary, should be funded at the expense of other NASA programs. Webb, in contrast, believed that Apollo was just a part, albeit a very important part, of a balanced space effort aimed at across-the-board preeminence in space. In a November 21 meeting with President Kennedy and in this follow-up letter, Webb forcefully argued his position. Kennedy accepted the argument, and soon after Holmes left NASA. This letter presents a comprehensive overview of James Webb's concept of the space program that he was attempting to execute.

---

[1] At the close of our meeting on November 21, concerning possible acceleration of the manned lunar landing program, you requested that I describe for you the priority of this program in our overall civilian space effort. This letter has been prepared by Dr. Dryden, Dr. Seamans, and myself to express our views on this vital question.

The objective of our national space program is to become preeminent in all important aspects of this endeavor and to conduct the program in such a manner that our emerging scientific, technological, and operational competence in space is clearly evident.

To be preeminent in space, we must conduct scientific investigations on a broad front. We must concurrently investigate geophysical phenomena about the earth, analyze the sun's radiation and its effect on earth, explore the moon and the planets, make measurements in interplanetary space, and conduct astronomical measurements.

To be preeminent in space, we must also have an advancing technology that permits increasingly large payloads to orbit the earth and to travel to the moon and the planets. We must substantially improve our propulsion capabilities, must provide methods for delivering large amounts of internal power, must develop instruments and life support systems that operate for extended periods, and must learn to transmit large quantities of data over long distances.

To be preeminent in operations in space, we must be able to launch our vehicles at prescribed times. We must develop the capability to place payloads in exact orbits. We must maneuver in space and rendezvous with cooperative spacecraft and, for knowledge of the military potential with uncooperative spacecraft. We must develop techniques for landing on the moon and the planets, and for re-entry into the earth's atmosphere at increasingly high velocities. Finally, we must learn the process of fabrication, inspection, assembly, and check-out that will provide vehicles with life expectancies in space measured in years rather than months. Improved reliability is required for astronaut safety, long duration scientific measurements, and for economical meteorological and communications systems.

[2] In order to carry out this program, we must continually up-rate the competence of Government research and flight centers, industry, and universities, to implement their special assignments and to work together effectively toward common goals. We also must have effective working relationships with many foreign countries in order to track and acquire data from our space vehicles and to carry out research projects of mutual interest and to utilize satellites for weather forecasting and world-wide communications.

Manned Lunar Landing Program

NASA has many flight missions, each directed toward an important aspect of our national objective. The manned lunar landing program requires for its successful completion many, though not all, of these flight missions. Consequently, the manned lunar landing program provides currently a natural focus for the development of national capability in space and, in addition, will provide a clear demonstration to the world of our accomplishments in space. The program is the largest single effort within NASA, constituting three-fourths of our budget, and is being executed with the utmost urgency. All major activities of NASA, both in headquarters and in the field, are involved in this effort, either partially or full time.

In order to reach the moon, we are developing a launch vehicle with a payload capability 85 times that of the present Atlas booster. We are developing flexible manned spacecraft capable of sustaining a crew of three for periods up to 14 days. Technology is being advanced in the areas of guidance and navigation, re-entry, life support, and structures—in short, almost all elements of booster and spacecraft technology.

The lunar program is an extrapolation of our Mercury experience. The Gemini spacecraft will provide the answers to many important technological problems before the first Apollo flights. The Apollo program will commence with earth orbital maneuvers and culminate with the one-week trip to and from the lunar surface. For the next five to six years there will be many significant events by which the world will judge the competence of the United States in space.

The many diverse elements of the program are now being scheduled in the proper sequence to achieve this objective and to emphasize the major milestones as we pass them. For the years ahead, each of these tasks must be carried out on a priority basis.

[3] Although the manned lunar landing requires major scientific and technological effort, it does not encompass all space science and technology, nor does it provide funds to support direct applications in meteorological and communications systems. Also, university research and many of our international projects are not phased with the manned lunar program, although they are extremely important to our future competence and posture in the world community.

Space Science

As already indicated, space science includes the following distinct areas: geophysical, solar physics, lunar and planetary science, interplanetary science, astronomy, and space biosciences.

At present, by comparison with the published information from the Soviet Union, the United States clearly leads in geophysics, solar physics, and interplanetary science. Even here, however, it must be recognized that the Russians have within the past year launched a major series of geophysical satellites, the results of which could materially alter the balance. In astronomy, we are in a period of preparation for significant advances, using the Orbiting Astronomical Observatory which is now under development. It is not known how far the Russian plans have progressed in this important area. In space biosciences and lunar and planetary science, the Russians enjoy a definite lead at the present time. It is therefore essential that we push forward with our own programs in each of these important scientific areas in order to retrieve or maintain our lead, and to be able to identify those areas, unknown at this time, where an added push can make a significant breakthrough.

A broad-based space science program provides necessary support to the achievement of manned space flight leading to lunar landing. The successful launch and recovery of manned orbiting spacecraft in Project Mercury depended on knowledge of the pressure, temperature, density, and composition of the high atmosphere obtained from the nation's previous scientific rocket and satellite program. Considerably more space science data are required for the Gemini and Apollo projects. At higher altitudes than Mercury, the spacecraft will approach the radiation belt through which man will travel to reach the moon. Intense radiation in this belt is a major hazard to the crew. Information on the radiation belt will determine the shielding requirements and the parking orbit that must be used on the way to the moon.

[4] Once outside the radiation belt, on a flight to the soon, a manned spacecraft will be exposed to bursts of high speed protons released from time to time from flares on the sun. These bursts do not penetrate below the radiation belt because they are deflected by the earth's magnetic field, but they are highly dangerous to man in interplanetary space.

The approach and safe landing of manned spacecraft on the moon will depend on more precise information on lunar gravity and topography. In addition, knowledge of the bearing strength and roughness of the landing site is of crucial importance, lest the landing module topple or sink into the lunar surface.

Many of the data required for support of the manned lunar landing effort have already been obtained, but as indicated above there are many crucial pieces of information still unknown. It is unfortunate that the scientific program of the past decade was not sufficiently broad and vigorous to have provided us with most of these data. We can learn a lesson from this situation, however, and proceed now with a vigorous and broad scientific program not only to provide vital support to the manned lunar landing, but also to cover our future requirements for the continued development of manned flight in space, for the further exploration of space, and for future applications of space knowledge and technology to practical uses.

Advanced Research and Technology

The history of modern technology has clearly shown that preeminence in a given field of endeavor requires a balance between major projects which apply the technology, on the one hand, and research which sustains it on the other. The major projects owe their support and continuing progress to the intellectual activities of the sustaining research. These intellectual activities in turn derive fresh vigor and motivation from the projects. The philosophy of providing for an intellectual activity of research and an interlocking cycle of application must be a cornerstone of our National Space Program.

The research and technology information which was established by the NASA and its predecessor, the NACA, has formed the foundation for this nation's preeminence in aeronautics, as exemplified by our military weapons systems, our world market in civil jet airliners, and the unmatched manned flight within the atmosphere represented by the X-15.

[5] More recently, research effort of this type has brought the TFX concept to fruition and similar work will lead to a supersonic transport which will enter a highly competitive world market. The concept and design of these vehicles and their related propulsion, controls, and structures were based on basic and applied research accomplished years ahead. Government research laboratories, universities, and industrial research organizations were necessarily brought to bear over a period of many years prior to the appearance before the public of actual devices or equipment.

These same research and technological manpower and laboratory resources of the nation have formed a basis for the U.S. thrust toward pre-eminence in space during the last four years. The launch vehicles, spacecraft, and associated systems including rocket engines, reaction control systems, onboard power generation, instrumentation and equipment for communications, television and the measurement of the space environment itself have been possible in this time period only because of past research and technological effort. Project Mercury could not have moved as rapidly or as successfully without the information provided by years of NACA and later NASA research in providing a base of technology for safe re-entry heat shields, practical control mechanisms, and life support systems.

It is clear that a preeminence in space in the future is dependent upon an advanced research and technology program which harnesses the nation's intellectual and inventive genius and directs it along selective paths. It is clear that we cannot afford to develop hardware for every approach but rather that we must select approaches that show the greatest promise of payoff toward the objectives of our nation's space goals. Our research on environmental effects is strongly focused on the meteoroid problem in order to provide information for the design of structures that will insure their integrity through space missions. Our research program on materials must concentrate on those materials that not only provide meteoroid protection but also may withstand the extremely high temperatures which exist during re-entry as well as the extremely low temperatures of cryogenic fuels within the vehicle structure. Our research program in propulsion must explore the concepts of nuclear propulsion for early 1970 applications and the even more advanced electrical propulsion systems that may become operational in the mid 1970's. A high degree of selectivity must be and is exercised in all areas of research and advanced technology to ensure that we are working on the major items that contribute to the nation's goals that make up an over-all preeminence in space exploration. Research and technology must precede and pace these established goals or a stagnation of progress in space will inevitably result.

[6] **Space Applications**

The manned lunar landing program does not include our satellite applications activities. There are two such program areas under way and supported separately: meteorological satellites and communications satellites. The meteorological satellite program has developed the TIROS system, which has already successfully orbited six spacecraft and which has provided the foundation for the joint NASA-Weather Bureau planning for the national operational meteorological satellite system. This system will center on the use of the Nimbus satellite which is presently under development, with an initial research and development flight expected at the end of 1963. The meteorological satellite developments have formed an important position for this nation in international discussions of peaceful uses of space technology for world benefits.

NASA has under way a research and development effort directed toward the early realization of a practical communication satellite system. In this area, NASA is working with the Department of Defense on the Syncom (stationary, 24-hour orbit, communications satellite) project in which the Department of Defense is providing ground station support for NASA's spacecraft development; and with commercial interests, for example, AT&T on the Telstar project. The recent "Communications Satellite Act of 1962" makes NASA responsible for advice to and cooperation with the new Communications Satellite

Corporation, as well as for launching operations for the research end/or operational needs of the Corporation. The details of such procedures will have to be defined after the establishment of the Corporation. It is clear, however, that this tremendously important application of space technology will be dependent on NASA's support for early development and implementation.

University Participation

In our space program, the university is the principal institution devoted to and designed for the production, extension, and communication of new scientific and technical knowledge. In doing its job, the university intimately relates the training of people to the knowledge acquisition process of research. Further, they are the only institutions which produce more trained people. Thus, not only do they yield fundamental knowledge, but they are the sources of the scientific and technical manpower needed generally for NASA to meet its program objectives.

In addition to the direct support of the space program and the training of new technical and scientific personnel, the university is uniquely qualified to bring to bear the thinking of multidisciplinary [7] groups on the present-day problems of economic, political, and social growth. In this regard, NASA is encouraging the universities to work with local industrial, labor, and governmental leaders to develop ways and means through which the tools developed in the space program can also be utilized by the local leaders in working on their own growth problems. This program is in its infancy, but offers great promise in the working out of new ways through which economic growth can be generated by the spin-off from our space and related research and technology.

International Activity

The National Space Program also serves as the base for international projects of significant technical and political value. The peaceful purposes of these projects have been of importance in opening the way for overseas tracking and data acquisition sites necessary for manned flight and other programs which, in many cases, would otherwise have been unobtainable. Geographic areas of special scientific significance have been opened to cooperative sounding rocket ventures of immediate technical value. These programs have opened channels for the introduction of new instrumentation and experiments reflecting the special competence and talent of foreign scientists. The cooperation of other countries – indispensable to the ultimate achievement of communication satellite systems and the allocation of needed radio frequencies—has been obtained in the form of overseas ground terminals contributed by those countries. International exploitation and enhancement of the meteorological experiments through the synchronized participation of some 35 foreign nations represent another by-product of the applications program and one of particular interest to the less developed nations, including the neutrals, and even certain of the Soviet bloc satellite nations.

These international activities do not in most cases require special funding; indeed, they have brought participation resulting in modest savings. Nevertheless, this program of technical and political value can be maintained only as an extension of the underlying on-going programs, many of which are not considered part of the manned lunar landing program, but of importance to space science and direct applications.

Summary and Conclusion

In summarizing the views which are held by Dr. Dryden, Dr. Seamans, and myself, and which have guided our joint efforts to develop the National Space Program, I would emphasize that the manned lunar landing [8] program, although of highest national priority, will not by itself create the preeminent position we seek. The present interest of the United States in terms of our scientific posture and increasing prestige, and our future interest in terms of having an adequate scientific and technological base for space activities beyond the manned lunar landing, demand that we pursue an adequate, well-balanced space program in all areas, including those not directly related to the manned lunar landing. We strongly believe that the United States will gain tangible benefits from such a total accumulation of basic scientific and technological data as well as from the greatly increased strength of our educational institutions. For these reasons, we believe it would not be in the nation's long-range interest to cancel or drastically curtail on-going space science and technology development programs in order to increase the funding of the manned lunar landing program in fiscal year 1963.

The fiscal year 1963 budget for major hardware development and flight missions not part of the manned lunar landing program, as well as the university program, totals $400 million. This is the amount which the manned space flight program is short. Cancellation of this effort would eliminate all nuclear developments, our international sounding rocket projects, the joint U.S.-Italian San Marcos project recently signed by Vice President Johnson, all of our planetary and astronomical flights, and the communication and meteorological satellites. It should be realized that savings to the Government from this cancellation would be a small fraction of this total since considerable effort has already been expended in fiscal year 1963. However, even if the full amount could be realized, we would strongly recommend against this action.

In aeronautical and space research, we now have a program under way that will insure that we are covering the essential areas of the "unknown." Perhaps of one thing only can we be certain; that the ability to go into space and return at will increases the likelihood of new basic knowledge on the order of the theory that led to nuclear fission.

Finally, we believe that a supplemental appropriation for fiscal year 1963 is not nearly so important as to obtain for fiscal year 1964 the funds needed for the continued vigorous prosecution of the manned lunar landing program $4.6 billion) and for the continuing development of our program in space science ($670 million), advanced research and technology ($263 million), space application ($185 million), and advanced manned flight including nuclear propulsion ($485 million).

[9] The funds already appropriated permit us to maintain a driving, vigorous program in the manned space flight area aimed at a target date of late 1967 for the lunar landing. We are concerned that the efforts required to pass a supplemental bill through the Congress, coupled with Congressional reaction to the practice of deficiency spending, could adversely affect our appropriations for fiscal year 1964 and subsequent years, and permit critics to focus on such items as charges that "overruns stem from poor management instead of on the tremendous progress we have made and are making.

As you know, we have supplied the Bureau of the Budget complete information on the work that can be accomplished at various budgetary levels running from $5.2 billion to $6.6 billion for fiscal year 1964. We have also supplied the Bureau of the Budget with carefully worked out schedules showing that approval by you and the Congress of a 1964 level of funding of $6.2 billion together with careful husbanding and management of the $3.7 billion appropriated for 1963 would permit maintenance of the target dates necessary for the various milestones required for a final target date for the lunar landing of late 1967. The jump from $3.7 billion for 1963 to $6.2 billion for 1964 is undoubtedly going to raise more questions than the previous year jump from $1.8 billion to $3.7 billion.

If your budget for 1964 supports our request for $6.2 billion for NASA, we feel reasonably confident we can work with the committees and leaders of Congress in such a way as to secure their endorsement of your recommendation and the incident appropriations. To have moved in two years from President Eisenhower's appropriation request for 1962 of $1.1 billion to the approval of your own request for $1.8 billion, then for $3.7 billion for 1963 and on to $6.2 billion for 1964 would represent a great accomplishment for your administration. We see a risk that this will be lost sight of in charges that the costs are skyrocketing, the program is not under control, and so forth, if we request a supplemental in fiscal year 1963.

However, if it is your feeling that additional funds should be provided through a supplemental appropriation request for 1963 rather than to make the main fight for the level of support of the program on the basis of the $6.2 billion request for 1964, we will give our best effort to an effective presentation and effective use of any funds provided to speed up the manned lunar program.

> With much respect, believe me
> Sincerely yours,
> James E. Webb
> Administrator

Document III-15

Document title: John F. Kennedy, Memorandum for the Vice President, April 9, 1963.

Source: NASA Historical Reference Collection, History Office, NASA Headquarters, Washington, D.C.

Criticism of the priority assigned to the space program, and particularly Project Apollo, increased in 1963. As he had the previous year, President Kennedy asked for a careful review of the program. This time, however, unlike in 1962, Kennedy asked the vice president and the Space Council, rather than the Bureau of the Budget, to carry out the review. This increased the likelihood of an assessment generally favorable to the program as it then stood.

---

[1] In view of recent discussions, I feel the need to obtain a clearer understanding of a number of factual and policy issues relating to the National Space Program which seem to arise repeatedly in public and other contexts. With this objective in mind, I would appreciate having the Space Council give consideration, and replies, to the following questions:

1. What are the salient differences, as to planned technical and scientific accomplishments, between the NASA program as projected on January 1, 1961 for the years 1962 through 1970, and the NASA programs as redefined by the present Administration? What are the differences in the intended levels of funding, year by year, for the two projections? The costs of which new major projects, or what reestimates in the costs of projects appearing in both projections, are responsible for the year-by year differences in funding between them? What would be the differences in accomplishments assuming the two programs were successful?

2. What specifically are the principal benefits to the national economy we can expect to accrue from the present, greatly augmented program in the following areas: scientific knowledge; industrial productivity; education, at the various levels beginning with high school; and military technology? Please estimate in dollar terms the portion of the annual expenditure that will result in contributions in each of these areas, from the present NASA program, and from its predecessor.

[2] 3. What are some of the major problems likely to result from continuation of the national space program as now projected in the fields of industry, government and education? In particular, will research and development in the industrial and consumer products segments of the national economy suffer because of diversion of technical manpower away from it, and/or from increasing costs of such research and development?

4. To what extent could the program be reduced, beginning with FY 1964, in areas not directly affecting the Apollo program (and therefore not compromising the timetable for the first manned lunar landing)? What are these areas, and the dollar amounts

involved? Specifically, what would be the consequences in science, industrial productivity, education and in other areas, if such reductions were imposed? Conversely, where would you judge that the present projection merits expansion and, specifically, what kind of benefits, in what areas, would ensue from such increases?

5. Are we taking sufficient measures to insure the maximum degree of coordination and cooperation between NASA and the Defense Department in the areas of space vehicles development and facility utilization?

I would appreciate receiving your report concerning the above by May 15, 1963.

I have sent a copy of this memorandum to Mr. Webb, with the request that he assist in preparing the material needed for this review.

John F. Kennedy

Document III-16

Document title: Lyndon B. Johnson, Vice President, to the President, May 13, 1963, with attached report.

Source: NASA Historical Reference Collection, History Office, NASA Headquarters, Washington, D.C.

The vice president transmitted with this letter the results of the review requested by President Kennedy on April 9. The review compared NASA's 1962 plans with the 1960 NASA Long Range Plan. The report noted that the program accelerations that President Kennedy had announced in his May 1961 speech would require a budget over $30 billion greater during the 1960s than had been anticipated at the end of the Eisenhower administration.

[1] Dear Mr. President:

Your memorandum of April 9 to me asked that the Space Council give consideration and responses to five groups of questions regarding the National Space Program. Since most of the questions were directed toward NASA's participation in that program, major attention of the responses is also pointed in that direction.

In the process of preparing the attached reply, consultations were held with and inputs received from all agencies whose executive heads are members of the Council. Because of the detail involved, the written contributions from NASA and Defense accompany this report as reference Appendices A and B, respectively. Staff papers from State and AEC appear in Appendix C.

To assist in orderly and brief response, the five groups of questions in your memorandum were identified and numbered as follows: (1) Comparisons of current NASA program with that of previous Administration; (2) Benefits to national economy from NASA program; (3) Problems resulting from Space Program; (4) Reductions and expansions in NASA program, without affecting lunar project; (5) Coordination and cooperation between NASA and Defense.

All members of the Council, as well as the Executive Secretary, have concurred in this report.

Sincerely,
Lyndon B. Johnson...

[1 of Report]

I. Comparison of NASA Program with That of Previous Administration

Current and Projected NASA Budgets
(millions of dollars)

FY	1961	1962	1963	1964	1965	1966
Long-range Plan of Previous Administration (Tentative)	922	1159	1577	1674	1825	1931
Long-range Plan (1962) as Suggested by NASA	964	1822	3688	5712	5900	6000

FY	1967	1968	1969	1970	Total for Decade
Long-range Plan of Previous Administration (Tentative)	2056	2258	2239	2276	17,917
Long-range Plan (1962) as suggested by NASA	6000	6000	6000	6000	48,086

1. The major distinctions between the present NASA program and that of the previous Administration involve:
    a. The lunar project.
    b. Breadth of scientific endeavor.
    c. Expansion of space applications.
    d. Sense of urgency.

2. The operational plan of the previous Administration terminated with the 3-orbit MERCURY flight, and there was no continuation beyond that with the exception of studies, unsupported by any commitment to the necessary hardware and facilities, pointing to a landing on the Moon at an unspecified time after 1970, The plan of the present Administration is marshalling resources required for a round trip to the Moon in the 1967-68 period.

3. The previous plan proposed, but did not have Presidential approval, to commit $4.76 billion to manned space flight. The current plan would add $15.88 billion to this, bringing the cost of the manned and unmanned aspects of the lunar project to approximately $20 billion.

4. The extra money would buy major capital investments in facilities at Houston; Michoud, Mississippi Test Center; Cape Canaveral; and the [2] worldwide tracking network in direct support of the lunar project. In addition, the money is:
    a. Buying a strengthened support program in unmanned survey of the Moon, study of bioscience, and investigation of solar phenomena and space radiation.
    b. Giving us more support in advance technology as backup for APOLLO and a lead time for future missions to keep this country in front, such as the nuclear rockets.

c. Aiding the universities to add at least as many scientists and engineers to the national supply as the space program will draw from the pool.

5. The long-range plan of the previous Administration covering all NASA programs, amounted to $17.9 billion for the period 1961-70. It soon had to be re-estimated at $22.2 billion. The long-range plan of the present Administration came to $42.5 billion. NASA judgments, coupled with a projection of no annual budget in excess of $6 billion, would bring the total to $48 billion for the decade.

6. Over the 10-year period, the NASA program as now conceived:
a. Will run fairly consistently at $2\text{-}1/2$ times that of the 1961 tentative long-range plan.
b. Would run about 4 times more in manned space flight costs.
c. Would run about double in applications and tracking costs.
d. Would run more than $1\text{-}1/2$ times greater in space sciences and research and technology.

7. In over-all terms, the basic difference between the two programs is that the plan of the previous Administration represented an effort for a second place runner and the program of the present Administration is designed to make this country the assured leader before the end of the decade. This is not taking the narrow view that the project is just a race to reach the Moon first. Instead, it is an over-all coordinated program designed to return benefits to the economy and to the national security on a broad front. [3]

II. Benefits to National Economy from NASA Space Programs

1. It cannot be questioned that billions of dollars directed into research and development in an orderly and thoughtful manner will have a significant effect upon our national economy. No formula has been found which attributes specific dollar values to each of the areas of anticipated developments, However, the "multiplier" of space research and development will augment our economic strength, our peaceful posture, and our standard of living.

2. Even though specific dollar values cannot be set for these benefits, a mere listing of the fields which will be affected is convincing evidence that the benefits will be substantial. The benefits include:
a. Additional knowledge about the earth and the Sun's influence on the earth, the nature of interplanetary space environment, and the origin of the solar system as well as of life itself.
b. Increased ability and experience in managing major research and development efforts, expansion of capital facilities, encouragement of higher standards of quality production.
c. Accelerated use of liquid oxygen in steelmaking, coatings for temperature control of housing, efficient transfer of chemical energy into electrical energy, and wide-range advances in electronics.
d. Development of effective filters against detergents; increased accuracy (and therefore reduced costs) in measuring hot steel rods; improved medical equipment in human care; stimulation of the use of fiberglass refractory welding tape, high energy metal forming processes; development of new coatings for plywood and furniture; use of frangible tube energy absorption systems that can be adapted to absorbing shocks of failing elevators and emergency aircraft landings.
e. Improved communications, improved weather forecasting, improved forest fire detection, and improved navigation.
f. Development of high temperature gas-cooled graphite moderated reactors and liquid metal cooled reactors; development of radio-isotope power sources for both military and civilian uses; development of [4] instruments for monitoring degrees of

radiation; and application of thermoelectric and thermionic conversion of heat to electric energy.

 g. Improvements in metals, alloys, and ceramics.

 h. An augmentation of the supply of highly trained technical manpower.

 i. Greater strength for the educational system both through direct grants, facilities and scholarships and through setting goals that will encourage young people.

 j. An expansion of the base for peaceful cooperation among nations.

 k. Military competence. (It is estimated that between $600 and $675 million of NASA's FY 1964 budget would be needed for military space projects and would be budgeted by the Defense Department, if they were not already provided for in the NASA budget.)

III. Problems Resulting from the Space Program

1. The introduction of a vital new element into an economy always creates new problems but, otherwise, the nation's space program creates no major complications. The program has, to a lesser magnitude, the same problems which Defense budgets and programs have been creating for several years.

2. Despite claims to the contrary, there is no solid evidence that research and development in industry is suffering significantly from a diversion of technical manpower to the space program. NASA estimates that:

 a. The nation's pool of scientists and engineers was 1,400,000 as of January 1, 1963.

 b. NASA programs employed 42,000 of these scientists and engineers—only 9,000 directly on NASA payrolls.

 c. On this basis, the NASA space program currently draws upon only 3% of the national pool of scientists and engineers. [5]

 d. Taking into account anticipated expansion, NASA programs are not expected to absorb more than 7% of our country's total supply of scientists and engineers.

3. The majority of the technical people working for NASA fall in the category of engineering. However, NASA's education programs are designed to help the universities train additions to the nation's technical manpower needs.

4. NASA has undertaken to support the annual graduate training of 1000 Ph.D.'s $^1/_4$ of the estimated overall shortage of 4,000 per year. This program would more than replace those drawn upon by the agency.

5. In overall terms, NASA finds that diversion of manpower and resources is not a major problem arising from the space program. A major problem, however, is the need to minimize waste and inefficiency. To help meet this challenge, turnover of top level Government talent should be reduced and compensation more in line with responsibilities would contribute to this objective.

IV. Reductions and Expansions in the NASA Program Without Affecting the Lunar Project

1. The fiscal 1964 NASA budget is divided between $4.4 billion for the manned lunar landing program and $1.3 billion for a multi-project scientific, research, and technology development and applications effort. Therefore, only 23% of the total budget is unrelated to the manned lunar landing program.

2. There are approximately 60 programs, projects and activities within this 23% of the budget. Examples include geodesy, orbiting observatories, planetary and interplanetary probes, international satellites, university program, advanced propulsion, and communications and meteorological developments.

3. It is pertinent under this heading to recall that the NASA budget requests for fiscal 1964 were reduced from $6.2 billion to $5.7 billion before the presentation to Congress. Further reductions would:
    a. Lessen the quantity and quality of benefits to the economy.
    b. Give additional ammunition to those who criticize the major funding weight given to the lunar program on the grounds that it diverts money from other programs. [6]
    c. Disrupt manpower teams, delay the realization of goals, and ultimately lead to increased costs to the stretchout process.

4. Growth of the present favorable international attitude toward our space programs would be inhibited if the lunar program were favored through a reduction or elimination of projects which promote international cooperation or promise actual or potential benefits to foreign governments.

5. In light of current conditions, it is not considered practicable to increase the size of the program. However, in considering future budgets, attention should be directed toward such developments as:
    a. NASA/DOD space station competence.
    b. Nuclear rocket propulsion and auxiliary power.
    c. Interplanetary exploration.

V. Coordination and Cooperation Between NASA and Defense

1. The difficulties of assuring a single National space program have been recognized from the beginning, NASA and the Department of Defense carry the major burdens, but the program touches widely divergent agencies of government. In order to assist in coordination and in avoiding duplication, the following steps have been taken:
    a. The National Aeronautics and Space Council has been authorized and activated to advise and assist the President in coordinating the entire program.
    b. The Aeronautics and Astronautics Coordinating Board (and its six panels) has been organized for high-level managerial coordination to integrate major Space projects in the early stages of their development.
    c. Within the agencies, a number of coordinating arrangements operate at various levels.
    d. More than 50 joint written agreements have been worked out between NASA and DOD spelling out lines of action in such fields as development of launch vehicles and spacecraft, administration of range facilities, and planning for communication satellites.

[7] 2. However, it is inevitable that controversies will continue to arise in any field as new, as wide ranging, and as technically complicated as space. Areas in which cooperation could be further improved are:
    a. Coordination in joint planning of advanced projects to insure that divergent views are not prematurely curtailed and that unwarranted duplication between NASA and DOD is avoided in the initial development of these projects.
    b. Further strengthening of cross-fertilization in the areas of research and technology to insure that NASA research is available for the solution of critical military problems and that military research is available for the solution of NASA problems.

3. It must be kept in mind that no mechanical application of a formula will insure maximum cooperation and coordination and a minimum of duplication and waste. Continuous monitoring at a high level is essential at every stage of the development of the space program. The Space Council will continue to function on the premise that no relaxation of its efforts is possible.

Conclusion

There is one further point to be borne in mind. The space program is not solely a question of prestige, of advancing scientific knowledge, of economic benefit or of military development, although all of these factors are involved. Basically, a much more fundamental issue is at stake—whether a dimension that can well dominate history for the next few centuries will be devoted to the social system of freedom or controlled by the social system of communism.

The United States has made clear that it does not seek to "dominate" space and, in fact, has led the way in securing international cooperation in this field. But we cannot close our eyes as to what would happen if we permitted totalitarian systems to dominate the environment of the earth itself. For this reason our space program has an overriding urgency that cannot be calculated solely in terms of industrial, scientific, or military development. The future of society is at stake.

Document III-17

Document title: NASA, *Summary Report: Future Programs Task Group*, January 1965.

Source: NASA Historical Reference Collection, History Office, NASA Headquarters, Washington, D.C.

In the last year of the Kennedy administration, the Bureau of the Budget and the White House Office of Science and Technology shifted focus from whether or not to go forward with Apollo to what programs NASA was likely to propose to follow the lunar landing program. After Kennedy's assassination, President Johnson asked NASA Administrator Webb on January 30, 1964, to identify future objectives for the civilian space program. Webb was quite reluctant to identify NASA's goals and priorities in advance of any expression of political support, a position he had taken even during the debates preceding the decision to go to the Moon; he preferred that NASA identify a variety of paths it *could* take, and then have top policymakers choose the option they wished to pursue.

This is the approach NASA took in response to the president's request. Webb appointed a Future Programs Task Group, headed by Frank Smith of the Langley Research Center, to prepare an overview of the capabilities that NASA was developing during the 1960s and the uses to which they might be applied. The group's summary report, completed in January 1965 (several months late) and not released publicly until April, set no priorities and made no recommendation, except to continue a "balanced" program in all areas of space activity. Some figures have been omitted from the excerpt printed here.

[ii]
Foreword

This summary report of the Future Programs Task Group, directed by Francis B. Smith of Langley Research Center, presents the results of studies made during 1964 to answer inquiries made by President Johnson as to criteria and priorities for space missions to follow those now approved for the decade of the 1960's.

The President's request was for a review of space objectives in relation both to what we have learned from our space efforts and to the most important emerging concepts of missions needed for scientific purposes and for advances in technology. The President requested that our evaluation be made in relation to estimates of time and funds required to complete programs already approved and under way.

The Future Programs Task Group was established to develop materials to meet the President's request. It has studied: (1) the capability being created in the present aeronautical and space effort; (2) next-step or intermediate space missions that could use or extend this capability; and (3) a number of long-range missions which deserve serious attention. This summary report, resulting from these studies, provides a source of information on accomplishments to date and indicates the general time periods within which we can assume or forecast the availability of further scientific and technical knowledge. It is, in addition to providing a review for the President, a timely and valuable working document for use within NASA and other agencies as a foundation for further analysis and discussion looking toward decisions that can be based on a broad consensus as to values and timing.

A major concern of the Task Group has been to identify the areas and levels of technology required to accomplish the most likely future missions and to provide a basis for informed decisions relating to the allocation of resources and timing for those which may be approved. Considerable attention has been given to steps we need to take to insure that these areas and levels or technology are available as needed.

The long range developments section of this report contains a discussion of the technology development programs which are under way in NASA and a number which should be given careful consideration in making future plans. Many of these programs are broadly based, but are also essential to provide optional means to accomplish the minimum under study and also provide a strong basis for judgments bearing on the value, time and cost elements.

James E. Webb
Administrator...

[1] ## Summary Report: Future Programs Task Group

I. Introduction

The successful flight of Sputnik I, in its most fundamental aspect, meant that man had taken the first step toward the exploration of a new environment by means of a new technology. It also meant that in the USSR, which accomplished this first step, new horizons were opened and there was a surge of national pride and accomplishment. An internal drive was created that changed the posture of Soviet society and lifted it above many of the frictions and tensions of the existing status. Horizons were widened. Internationally, the leadership and image of the Soviet Union were vastly enhanced. The flights of Gagarin and other Soviet Cosmonauts added impetus to a marked degree.

In the United States and in the Free World, as we all know, the immediate effects were quite the opposite. However, since then, we have made tremendous progress under a broad based and balanced program aimed at achieving preeminence in aeronautics and space.

Down through the course of history, the mastery of a new environment, or of a major new technology, or of the combination of the two as we now see in space, has had profound effects on the future of nations; on their relative strength and security; on the relations with one another; on their internal economic, social and political affairs; and on the concepts of reality held by their people. From the elements of each such new situation which history records, all or most of the developments listed in Figure 1 have materialized.

The long-range effects of man's entry into space, in person and by instruments and machines, can be best forecast in terms of these considerations. As a new environment, space may well become as important to national security and national development as the land, the oceans and the atmosphere; rockets and spacecraft as important as automobiles, trucks, trains, ships, submarines and aircraft. The foreseeable returns from scientific advances, technical advances, and practical uses compare favorably with the returns yielded

Fig. 1

Developments Which Generally Follow a Nation's Mastery of a New Environment and Technology

I. An increase in power and position through:
   A. The prestige of being first in new accomplishments; of being in control or sharing control of the new environment; and of possessing new knowledge and technology.
   B. The establishment of strategic international positions for traffic, communications, and trade.
   C. Wide use of new resources.
   D. Military capability through use of the new environment and technology.
   E. An increase in initiative, pride and drive toward accomplishment in all walks of life.
II. The appearance of new developments in the relations among nations: by negotiation, by cooperation, or by conflict.
III. Changes within national societies as a result of the forces listed above; of actions taken to compete in the new environment and to develop and use the new technology; and of the interplay of new knowledge, new thought, increased resources, and changed social relations.

Fig. 2

Basic NASA Aeronautical and Space Objectives

I. The scientific measurement and understanding of the space environment.

II. The development of a broad-based national capability for manned and unmanned operations in space and close cooperation with the Department of Defense and other agencies having current or potential needs related to such capabilities.

III. The development of the practical uses of space.

IV. Continued advancement in all areas of aeronautics in order to maintain world leadership in this field.

V. An adequate level of research and development to support other government agencies with needs or interests in aeronautics and space.

VI. The bringing together of government, industry, and university capabilities into an effective national system for meeting the needs of space exploration and use.

VII. The maintenance of a technological base in aeronautics and space adequate to meet all non-military needs.

VIII. The strengthening and efficient utilization of the nation's aeronautical and space-related resources in science, engineering and technology.

IX. The maximum utilization of the scientific and technical results of the space effort for non-space purposes.

X. The use of space for further international cooperation and understanding and for the good of all mankind.

by the most vigorous past periods of exploration of newly opened segments of man's expanding frontier.

[2] If these larger considerations of the space effort are to be adequately dealt with in terms of national policy, they must be translated into brood objectives in order that particular programs and missions can be defined and evaluated. For the United States,

these objectives relate aeronautics to space and are contained in the Space Act of 1958. They are outlined in Figure 2.

Under the Space Act, NASA bears the general responsibility for continuously providing an adequate underlying aeronautical and space capability and cooperating with the military services and other agencies which have, or anticipate, specific missions and uses. In 1958, and again in 1961, two major periods of wide debate and assessment brought decisions to undertake missions and programs which accelerated our progress toward the achievement of these objectives. The capability which has been created through the work thus begun and now under way will be the basis for this analysis.

First, however, we need to understand that we face certain conditions and constraints....

[7]   **III. Major Capabilities Existing and Under Development**

The broad categories of capabilities which have been developed during the past 6 years, or are to be developed in current programs, are shown in Figure 4. The major categories are Aeronautics, Satellite Applications, Unmanned Exploration, Manned Operations, Launch Vehicles, and Technology.

Fig. 4

Major Capabilities Existing or Under Development

Aeronautics
  R&D hypersonic airplanes
  Operational supersonic military airplanes
  Commercial supersonic airplanes
  Improved subsonic aircraft including V/STOL

Satellite Applications
  Satellite pictures of Earth weather
  Intercontinental communications
    (including TV)

Unmanned Exploration
  Near-Earth exploration
  Solar effects
  Planetary and interplanetary probes
  Lunar probes and landers
  Biosatellites

Manned Operations
  Man in Earth orbit (1 to 2 weeks)
  Maneuver & rendezvous
  Lunar orbiting, landing and return

Launch Vehicles
  Up to 125 tons in Earth orbit
  Up to 47.5 tons to escape

Technology
  Power supplies of increased power and lifetime and decreased weight
  More accurate guidance and control
  Increased communications capability
  More accurate stabilization
  Life support for long periods
  Improved landing control systems
  Increased reliability

Fig. 5

FY 1965 In-House Aeronautical Research Effort
505 Professional Man Years

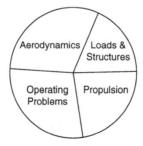

Aeronautics

The aeronautical program of NASA represents a continuation of a pattern of research activity developed by the National Advisory Committee for Aeronautics (NACA) over a period of more than 40 years. The in-house effort of this program is primarily basic and applied research activity at four NACA centers which were in existence in 1958 (the Langley Research Center, the Lewis Research Center, the Ames Research Center, and the Flight Research Center). This consists of in house and contracted-out work aimed at practical solutions of advanced problems of flight, but excludes the development of complete aircraft. Over many years the latter has been the responsibility of industry and other branches of the government with whom the NACA and now the NASA has developed effective working relations of collaboration and support.

For a number of years, the major portion of the NASA aeronautics effort has been in two areas. First, a basic research program in atmospheric flight. Three significant areas have received increasing attention—major increases in maximum flight speed, major decreases in minimum flight speed, and major increases in operational flexibility. Second, a continuing research and technology program in support of military and other government agencies and industry has pointed to continued evolutionary improvement of existing aircraft types.

Figure 5 shows the aeronautical research effort of NASA classified by broad areas and gives an idea of the size and distribution of the effort for Fiscal Year 1965. In-house effort is carried on by about 500 research professionals supported by 1500 additional in-house personnel. In addition to the costs of these personnel, $35.2 million of R&D funding for contracted-out research and development supports this effort....

Satellite Applications

[9] Significant successes have been achieved in NASA's applications satellite program during the past 6 years and the results are now being used to establish operational systems.

In the area of meteorology, nine TIROS satellites, launched since 1960, and one Nimbus satellite, launched in August 1964, have demonstrated the feasibility and value of Earth weather research and observation from satellites in orbit. One of the primary uses of the pictures returned by the TIROS satellite has been the identification and tracking of weather storms, including some 70 hurricanes and typhoons.

Based on NASA research and development success with TIROS, the United States Weather Bureau has adopted a modified TIROS system for its operational satellite system. This is expected to be ready during the winter of 1965-1966.

In the meantime, while we work toward operational systems, data from NASA's experimental TIROS weather satellites are used routinely by the Weather Bureau in the preparation of daily forecasts as well as for analysis in the area of climatology. The DOD also uses these data in the preparation of local forecasts in remote areas.

The Nimbus research and development weather (meteorological) satellite shown in Figure 9 is a significant advance over TIROS. Through three-axis stabilization and Earth orientation, continuous data on the Earth's weather is provided throughout its orbit. Its solar cells provide over 400 watts of power (10 times that of TIROS), and it can carry numerous types of meteorological experiments now emerging from our research program.

In addition to its television transmission system, the first Nimbus carried an Automatic Picture Transmission (APT) system and a High Resolution Infra-Red (HRIR) system. The APT system permits read-out of local cloud cover pictures by inexpensive ground stations, as shown by Figure 10, and will provide weather information to Department of Defense installations and the numerous foreign countries that have purchased or built, or plan to build, ground stations.

[10] A most significant achievement of Nimbus was demonstration of the capability and value of the HRIR system, illustrated in Figure 11. This system enables us to obtain for

the first time, in pictorial format and on a real time basis, cloud cover information from the dark side of the Earth. Cloud cover pictures such as these are reconstructed from measurements taken at night, and give an indication of cloud height as well as area coverage.

As illustrated in Figure 12, in the field of communications, the Echo passive satellite, and the Telstar, Relay and Syncom active satellites have experimentally proved out the technology for reliable, long-range point-to-point transmission of radio, television, telephone, teletype and facsimile via satellite. As a result, the Communications Satellite Corporation is now undertaking an international communications satellite system whose initial "Early Bird" satellite is based on NASA's Syncom.

Recently Syncom III, shown in Figure 13, transmitted real-time television of the Olympic Games from Japan to the United States. The precision achieved in the Syncom launch and positioning operations is indicated by the fact that Syncom III's period is almost synchronous and the inclination of the orbit is less than one-tenth of a degree from equatorial. This means that it moves north and south with respect to the Earth's equator less than 6 miles a day.

In addition to its primary communications function, Syncom has proved useful for scientific measurements used in better defining the shape of the Earth at the equator.

Fig. 13

SYNCOM Spacecraft
Active Synchronous Communication Satellite

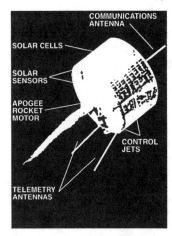

Launch Weight   146 lbs.
(including Apogee motor)
Control Systems   Gas jets
Propellant   Solid
Stabilization   Spin

Status
   SYNCOM I launched Feb. 14, 1963
     Achieved near synchronous orbit
   SYNCOM II launched July 26, 1963
     On station Aug. 16, 1963, at 55°W
     Communications "on" time
     over 4500 hours.
   SYNCOM III launched Aug. 19, 1964,
     into synchronous equatorial orbit
     over the Pacific Ocean.

Unmanned Exploration

The basic objective of this NASA program is to acquire fundamental knowledge of the space environment and of those phenomena which can be studied best from spacecraft, for both scientific and practical purposes....

[11] The unmanned space exploration program was initiated in 1958 when instruments carried by the first of the Explorer series of satellites revealed the existence of the Van Allen Belts encircling the Earth. Since that date, 26 Explorer and Monitor satellites... have been launched. These relatively small satellites, which have been placed in orbit by Jupiter C, Scout and Delta launch vehicles, are usually designed to make measurements of specific phenomena in space such as the distribution and energies of the particles trapped in the Earth's magnetic field (as with Explorers XIV and XV), ionospheric measurements

to determine electron densities and their variation bath diurnally and with changes in solar activity (as were made by Explorer VIII and the United Kingdom's Ariel 1), or other phenomena such as micrometeoroid flux and atmospheric structure.

As an example of some of the findings of one of the newer Explorers (Explorer XVI, launched November 27, 1963).... A transition region was found between the steady solar wind of interplanetary space and the magnetosphere, where the solar wind was turbulent and the magnetic field unsteady. Of extreme interest, also, was the discovery, on the fifth orbit, that the Moon as it moves through the solar wind apparently generates a wake that extends for a distance of at least 120,000 miles on the side away from the Sun.

This program of launching relatively inexpensive Explorer class satellites to make measurements of specific phenomena will be continued. A new series of Explorer satellites will carry payloads developed by universities and will also be used to continue international cooperation projects such as the U.K. Ariel II, the Canadian Alouette, and the Italian San Marco.

[12] The sounding rocket program is an important element of near-Earth exploration. It opens to investigation that vast region of the Earth's atmosphere that is too high for balloons to reach and too low for satellites. It also provides in-flight development testing of instruments and other equipment intended for later use in satellites. Further, new experimenters from universities, industry and foreign organizations are provided a logical and inexpensive way of gaining experience in space science techniques.

Shown in Figure 17 is the second generation of scientific satellites, the orbiting observatories. These larger satellites ore designed to make mare precise, more complex and better coordinated measurements of stellar, solar and geophysical phenomena.

The first Orbiting Solar Observatory, (OSO), was launched in March 1962, and successfully reported data on solar phenomena for well over a year. The second OSO, launched in February 1965, will continue making solar measurements during the present quiet period of the solar cycle.

The first of the Orbiting Geophysical Observatories (OGO) was launched in September 1964 into a highly elliptical orbit. Although 2 of the long booms shown did not deploy properly and the satellite was not stabilized as intended, 18 of the 20 experiments are operating and many of the objectives will be accomplished. The OGOs are designed to carry 20 to 50 experiments and will allow correlated measurements of Earth-related phenomena at a single point in space.

The first Orbiting Astronomical Observatory (OAO) will be launched in late 1965 or early 1966 and will allow the first extended observations of stars and planets from above the Earth's atmosphere. An eventual goal in this series of satellites to produce the capability of pointing a 36-inch telescope at a star to within plus or minus one-tenth of a second of arc, and one of its early experiments will be the mapping of the heavens in ultraviolet cave lengths.

Capabilities for interplanetary and planetary exploration were successfully demonstrated by Pioneer V, launched in 1960, and by the Venus probe, Mariner II, shown in Figure 18, which was launched in August 1962. Pioneer V set a record for that time by communicating to Earth from a distance of 22,000,000 miles, and returned new data on the interplanetary environment. Mariner II, 109 days after it was launched, passed close to the surface of Venus and [13] transmitted to Earth man's first close-up information about another planet. Although the data transmission capacity was limited, Mariner II gave us information on the surface temperature, magnetic fields, dust environment and radiation belts of Venus.

Mariner II also demonstrated the value of the mid-course maneuver capability on which we have standardized for guiding a spacecraft to a desired destination–in this case to within approximately 20,000 miles of Venus when it was 35,000,000 miles from the Earth.

The Mars probe, Mariner IV, was launched during the November 1964 opportunity. During its 8-month trip to the planet, this 575-pound spacecraft is making interplanetary measurements of the magnetic fields and solar winds. On arriving at the vicinity of the planet, the spacecraft, if operating properly, will make measurements of the Martian magnetic fields and radiation belts, collect some data on the Martian atmosphere, and will transmit to Forth about 20 television pictures of the planet's surface.

The Moon probe, Ranger VII, illustrated in Figure 19, gave man his first close-up look at the surface of Earth's nearest neighbor in space by transmitting approximately 4,300 television pictures to Earth in the last 15 minutes before it impacted on the lunar surface. Ranger also demonstrated our increased competence in mid-course maneuver capability, in this case to carry television cameras to within 10 miles of a preselected spot on the Moon's surface. It also demonstrated a communications capability for transmitting wide-band information over the quarter-million-mile Earth-Moon distance.

As illustrated in Figure 20, NASA is also developing the Lunar Orbiter and the Surveyor spacecraft for unmanned exploration of the Moon. Initial Lunar Orbiters, scheduled for launch in 1966, are designed to obtain topographic information by photographing an area of about 15,000 square miles with a resolution of 25 feet and of about 3,000 square miles with a resolution of 3 feet. Furthermore, the mass distribution and shape of the Moon can be determined from perturbations in the spacecraft's orbit. Later Lunar Orbiters will carry scientific instruments, as Well, that Will increase our knowledge of the lunar environment and of the surface and subsurface characteristics. The Surveyor spacecraft is designed to land on the Moon and make measurements of the bearing strength and composition of the lunar surface, to take close-up panoramic TV pictures of the lunar surface, to measure seismic activity, and to determine the flux of primary [14] and secondary particles impinging on the surface. The Surveyor and Lunar Orbiter will serve as a team to survey and select suitable sites for manned landings. The biosatellite program consists of orbital flights up to 30 days of recoverable capsules, which contain various biological experiments, illustrated conceptually in Figure 21. The experiments carried will range from studies of the effects of weightlessness and radiation on elemental cell functions to investigations of heart and nerve functions in primates immobilized for prolonged periods in a weightless condition. This program will use thrust-augmented Delta launch vehicles and take advantage of the recovery techniques developed by the Air Force in the Discoverer program.

Manned Operations

As illustrated in Figure 22, the current manned operations program provides an orderly progression of operational capabilities from the 2,900 pound Mercury spacecraft to the 7,000-pound Gemini, to the 95,000-pound Apollo-LEM system.

Figure 23 illustrates the progression of manned launch vehicles from the 368,000 pound thrust Atlas which launched the Mercury spacecraft to the 7 $^1/_2$ million-pound thrust Saturn V which will launch the Apollo.

The Mercury spacecraft, launched by the Atlas, provided this country's first capability for manned Earth-orbital flight and was used in the 3-orbit mission by John Glenn in 1962. The Mercury-Atlas system capability was later extended to accomplish Gordon Cooper's 22-orbit, 34-hour flight in 1963.

The Gemini two-man spacecraft, with its Titan II launch vehicle, will make possible missions of up to 14 days in Earth orbit, beginning in 1965. New equipment will permit orbit change, rendezvous and docking, and will enable the astronauts to venture out- [15] side the spacecraft into free space. Dual launches of the Gemini by Titan and the Agena by Atlas will place an unmanned Agena target into orbit and enable the Gemini astronauts to perfect the rendezvous and docking systems. These missions will verify the operations and techniques to be used later in the more ambitious Apollo missions.

In Project Gemini, NASA is providing a flexible, experimental space tool with which to flight test equipment, conduct scientific experiments, and develop techniques and provide training for Project Apollo. The Department of Defense Manned Orbital Laboratory will also make use of Gemini for the launch and return to Earth of the astronauts who will work in the laboratory.

As illustrated in Figure 24, the Gemini spacecraft consists of two major elements, the reentry module and the adapter, with a combined weight of 7,000 pounds. The reentry module provides life support and control equipment for the two crewmen, contains most of the experiments and also contains the rendezvous and recovery systems. The adapter element provides the link between the Titan II launch vehicle and the reentry module, and is composed of two sections, an equipment section to provide augmented life support, stabilization equipment and expendables for flight durations of up to 14 days, and a retrograde section to slow down the spacecraft from its orbital velocity.

Several of the Gemini missions are designated primarily as rendezvous missions to explore the feasibility of various modes of accomplishing rendezvous utilizing different levels of automation in the sensing and control equipment.

In a typical mission, an Agena engine will first be launched into a 160-nautical-mile circular orbit. The manned Gemini spacecraft will then be placed into a lower circular orbit at 130 nautical miles. The different periods of the two spacecraft in these concentric orbits will cause a continuing change in the relative position of the Gemini with respect to the Agena. When the relative positions are proper, the Gemini spacecraft will be accelerated in a transfer ellipse to the higher orbit where the rendezvous and docking will be accomplished. These missions will be short-lived because of the weight requirement for fuel which reduces the expendables that can be carried for life support, power supply, and stabilization.

[16] On the Gemini long-duration missions, primary emphasis will be placed an biomedical and behavioral aspects of man in a weightless condition; however, scientific and technical experiments are being planned for all missions. Specific experiments range all the way from visual definition experiments requiring no equipment and astronomical observations made with a 2-pound ultra-violet camera to radiometric or astronaut maneuvering experiments using equipment weighing as much as 200 pounds. The experiment program is tailored to the available weight, volume, and power in the spacecraft on each mission, as well as to the participation and accessibility which can be provided for the astronauts. Although the volume available for experiments within the pressurized cabin is limited, extra-vehicular operations are planned for the astronauts to permit free-space experiments, maneuvering, and other external operations such as the testing of manual dexterity and the use of specialized tools for spacecraft repair functions.

The Gemini spacecraft is already undergoing flight test. The successful launch of the first and second unmanned Gemini's in April 1964 and January 1965 will be followed soon by the first manned orbital flight. The easy access to the crew compartment is emphasized in Figure 25, which shows the Gemini spacecraft mockup.

The larger goals of the presently planned manned space flight program will be attained by the three-man Apollo-LEM system to be launched by the Saturn IB beginning in 1966 and by the Saturn V beginning in 1967. The Apollo Command and Service Modules, with fuel partially removed, will be launched first by the Saturn IB for Earth-orbital missions of up to 10 days duration. A number of such flights will be made in which rendezvous, docking, maneuvering, and other operations will be conducted.

Later, the total 47.5-ton Apollo-LEM spacecraft will be qualified in Earth orbit and eventually propelled to the Moon by the Saturn V, thus extending the area of space in which man can operate from near-Earth orbit to as far out as the Moon, and including the lunar surface.

The size of the Apollo-LEM spacecraft, shown previously in Figure 22, as compared with either the Mercury or Gemini spacecraft, is in part due to the longer duration of the Apollo missions, the larger heat-shield, and the increased crew; however, the major increase is due to the requirements for a large propulsion capability for maneuvering in space. The Service Module provides a propulsion capability for mid-course correction, lunar-orbit braking, and lunar-orbit escape, while the Lunar Excursion Module provides the capability for lunar landing and lunar takeoff.

The very large capability of the Apollo space exploration system (illustrated in Figure 26) will open a new era to manned space flight. Earth-orbital missions reaching out to synchronous orbit distances and of 14 days duration can be conducted, and lunar and other missions out to lunar distances will be possible, including one-day stays on the lunar surface for two men, or [17] 4 days stay in lunar orbit for three men.

On Earth-orbital, lunar-orbital, and lunar surface missions, provision is being made for the conduct of an extensive experimental program. Because of the increased size and the presence of man, these experiments will, in general, be more complex and extensive than those performed in unmanned vehicles. In the Command Module, volume has been provided for about 3 cubic feet of experimental equipment and with a return-to-Earth capability of about 80 pounds of instruments or lunar samples. In the Service Module, a complete empty bay provides an available volume of 250 cubic feet for the mounting of instruments; the weight available would depend upon the particular mission and the amount of fuel or other expendables required.

In the Lunar Excursion Module, 2 cubic feet of experimental equipment, weighing up to 80 pounds, can be installed within the existing ascent stage, and 15 cubic feet of instruments, weighing up to 250 pounds, an the descent stage in an area accessible to the astronauts while standing on the lunar surface. On all missions, however, the permissible weight of experiments must be evaluated against the comparable weight of expendables for fuel and life support, in order to extend the maneuver capability or duration of the mission.

A better impression of the room provided for experiments, as well as the progress that is being made in finalizing the design concept of the Apollo-LEM system, can be gained from the mockups of the major spacecraft elements shown in Figure 27. The area in the exposed bay in the Service Module could be utilized for installation of instruments that do not need direct monitoring by the astronauts. This space might also be used for carrying complete unmanned spacecraft in a piggy-back fashion for later deployment on unmanned space missions or for lunar surface probes.

Launch Vehicles

Figure 28 shows the boosters now included in the National Launch Vehicle Program which range from the Scout vehicle, capable of placing about 325 pounds in a 100-mile Earth orbit, to the Saturn V which will place about 250,000 pounds in the same orbit. The Thor/Delta vehicle, which will place about 930 pounds in a 100-nautical-mile Earth orbit or propel a 105 pound payload to escape velocity, has been the most successful of U. S. launch [18] vehicles, placing 26 payloads in Earth orbit out of 29 attempts. The capacity of the Thor/Delta has been improved recently by the addition of three 33-inch diameter strap-on solid rocket motors, giving about 25 per cent increase in Earth orbital payload capability. The thrust-augmented Thor/Agena will be capable of placing about 1,800 pounds in Earth orbit, when launched from the Western Test Range.

The two Atlas-based vehicles are the Atlas Agena and the Atlas/Centaur. The Atlas Agena can place up to 6,300 pounds in Earth orbit. It has been used successfully to launch the 750- to 800-pound Ranger probes to the Moon; and, on November 28, launched the 575-pound Mariner IV to Mars. The Atlas/Centaur, nearing completion of development, will accelerate 2,300 pounds to escape velocity or 9,700 pounds to Earth orbital velocity. It will be used as the launch vehicle for the Surveyor spacecraft designed to achieve a soft landing on the Moon.

The Centaur was the first rocket stage to use hydrogen and oxygen as fuel, a combination which gives an increase in specific impulse from about 300 seconds, available with standard fuels, to more than 400 seconds. This is an improvement of particular importance to missions requiring velocities equal to, or higher than, that far Earth escape.

The Titan series of launch vehicles is under development by the Department of Defense. The Titan II is used by NASA to launch the 7,000-pound, 10-foot diameter Gemini spacecraft. The Titan IIIA (not illustrated) consists of the Titan II to which an additional stage, called the trans-stage, has been added. The addition of two 120-inch solids to the IIIA produces the IIIC (illustrated in Figure 28), which will place about 25,000 pounds in low Earth orbit or propel about 5,000 pounds to escape velocities.

In the Saturn family of vehicles, the Saturn IB is capable of placing 35,000 pounds in Earth orbit; and the Saturn V, 250,000 pounds. The Saturn V will also accelerate 95,000 pounds to Earth escape velocity.

The Saturn program indicates the use of standardized rocket engines. The first stage of the Saturn IB is made up of eight LOX-RP-1 fueled H-1 engines – an up-rated version of the S-3 engines that were developed for the Atlas booster. The upper stage of the Saturn IB uses one hydrogen-oxygen S-2 engine which will also be used in a cluster of five to power the second stage of the Saturn V.

In the Saturn V, five 1,500,000-pound thrust F-1 engines power the first stage. The second stage will use five J-2 engines. The third stage uses one J-2 engine and is almost identical to the second stage of the Saturn IB.

These vehicles thus provide a wide range of launch capabilities based on a minimum number of [19] engine types. However, it is important to note that there are wide gaps in escape payload capability between the 950 pounds of the Atlas/Agena, the 2,300 pounds of the Atlas/Centaur, the 5,100 pounds of the Titan IIIC, the 13,000 pounds of the Saturn IB/Centaur, and the 95,000 pounds of the Saturn V.

Technology

The technological base which supports the development of the mission capabilities described has been made possible by the experience gained by our military services in the ballistic missile program and the broad research and technology development programs carried on by industry and by NASA. When the space age began in 1957, the reserve of technology which could be tapped to meet the immediate needs of the United States proved insufficient. The reliability and thrust of the launch vehicles, for example, were far short of that required to meet the challenge of the Soviet space program. However, due to the foresight exercised at that time in undertaking, without specific end uses in sight, the development of the 1-$^1/_2$ million-pound thrust F-1 engine, and other important projects, this country was able to make sound technical decisions when it became necessary to expand its space program in 1961. This expanded program is designed to assure United States leadership in space and to be ready to respond when national needs or objectives require new aeronautical or space systems. With respect to such a large, complex, and unknown environment as space, and the still not precisely defined characteristics of the Earth's atmosphere, this Nation would be oblivious to the lessons of history if it required that all its exploratory research and development efforts be matched to completely defined missions. It is clear from the 1958 Aeronautics and Space Act that NASA was established to make sure we would develop the capability which was clearly lacking at that time, and to develop the kind of policies and priorities that would do the job needed. Where there is reasonable promise of success in the development of such things as new materials, propulsion systems, or techniques, it is NASA policy to pursue these directions even though a specific use is not clearly defined. We have found that we can organize these efforts so as to point at broad classes of possible uses, giving the necessary technical base for options as to missions and the best ways to accomplish them. In his testimony before the Committee on Science and Astronautics on February 4, 1964, Dr. Hugh Dryden recalled how the United States, despite initial positions of advantage, failed to carry forward work of which it was capable in aeronautics, in jet propulsion, in ballistic missiles, and in the launch vehicles and spacecraft necessary for space exploration. The result in each case was that other nations moved ahead, placing the United States at a disadvantage and requiring an enormous effort to catch up. Our present relative position in space leaves no room for complacency. As Dr. Dryden said, "We must not delude ourselves or the nation with any thought that leadership in this fast moving age can be maintained with anything less than determined, whole-hearted, sustained effort."

It is on this basis that NASA is continuing to carry out a broad, long-range program in research and technology development. This program is aimed at the establishment of future mission capability, and it can be expected that new advances in technology will be made and will provide a better basis of judgment than we have had before as to the value of missions and projects and as to when they can be undertaken at reasonable costs and risks.

In the next two sections of this report, dealing with intermediate and long-range missions, we shall attempt to identify, when possible, the technological advances that are required for their accomplishment. These research and technology development programs will be discussed in [20] detail following the section on long-range missions.

Same examples of the capabilities which have been, or are being, developed to date are:

a. Solar cell power supplies capable of producing 650 watts.

b. Guidance and control capabilities for placing a spacecraft within a few thousand miles of a distant planet, or within a few miles of a given point on the Moon.

c. Communications technologies which provide almost continuous communications with manned spacecraft in Earth orbit, the transmission of about five television pictures per second from the Moon, or radio reception from a spacecraft over a 100 million miles in space.

d. Spacecraft stabilization technology which will enable precision instruments to be pointed, in some instances, to within $1/10$ second of arc.

e. Life support systems which will enable three men to remain in space for as long as 14 days and to venture as far as 250,000 miles from the Earth.

f. The reliability of both spacecraft and launch vehicles. Spacecraft reliability has been improved to the paint where many unmanned spacecraft now have lifetimes of well over one year, and manned spacecraft will be capable of dependable operation for 14 days or longer. Reliability of launch vehicles has been improved to such an extent that the per cent of successful vehicle launches has risen from about 60 per cent in 1960-61 to over 90 per cent at the present time.

g. A world-wide tracking, data acquisition, and communications system to support manned flight, scientific and application satellites, injection, monitoring, and deep space probes, as illustrated in Figure 29.

Along with the missions which are being undertaken and the capabilities which are being developed, the first years of the Nation's effort in space have produced a broad scientific and industrial base, and the facilities and management systems needed far carrying out an effective program of space exploration.

A major part of our present space capability is found in the expansion of NASA since 1961. About $2\text{-}1/3$ billion dollars have been invested in strengthening the industrial facilities and government laboratories associated with aeronautical and space research and in adding new installations. These include the Goddard Space Flight Center in Maryland, the Michoud Plant at New Orleans, the Mississippi Test Facility in southern Mississippi, the Manned Spacecraft Center at Houston, Texas, and the New Merritt Island Launch Facility at Cape Kennedy, of which the Saturn/Apollo Vehicle [21] Assembly Building is shown in a cutaway view in Figure 30. The exacting demands of space systems in the electronics Field have also required a new Electronics Research Center which will conduct and supervise research in this vital field from its location in Cambridge, Massachusetts.

NASA has contracted out more than 90 per cent of its research and development work. Over 1,600 manufacturing firms have held prime contracts of over $25,000 and about 20,000 firms have worked under prime or sub-contracts. Surveys made by 12 more prime contractors disclosed 3,000 subcontracts of over $10,000 to sub-contractors located in all 50 states. During slightly over 6 years of operation, NASA contracts have totaled more than $13 billion, adding great strength to the country's industrial base.

The conduct of space research and development, involving the design and manufacture of the most complex systems ever attempted, is demanding major improvements in methods of conducting large-scale organized effort. Included are new methods of production control, systems integration and checkout, and reliability and quality control.

The substantial expense involved in launching space vehicles, and the intricacy of the devices involved, have imposed unusual requirements of precision manufacture and quality assurance. As a consequence, increasing reliance is being placed upon incentive contracts, and new ways of encouraging improved government personnel and contractor performance are being developed.

The space program is indeed a large and varied research and development effort. The harsh environment of space requires major advances in all areas of technology, in materials, in electronics, in propulsion, in guidance and control, in power sources, in lubricants and coolants, in communications, in the integration of systems and the establishment of high levels of reliability, and in the maintenance of human life in space. It is already clear that our balanced and broadly-based space effort is producing important scientific and technological advances that are not limited to space use.

Experiments or pilot model efforts, through which these advances constituting major National resources for both security and economic growth can be made available quickly and efficiently for non-space use, are being carried out. NASA has established a program in technology utilization, with headquarters in Washington and with offices in each of the NASA centers. Innovations are being identified and described in appropriate publications; these are disseminated widely. Regional dissemination centers have been established on a trial basis at a number of universities. At each of these, NASA material is put on computer tapes and access provided to industrial concerns who support these centers through user fees and contributions. This system makes this material available within about 6 weeks of it's reporting date to Headquarters from both NASA centers and contractors, and makes it available on a selective basis conforming to the interests of the users. The system also provides a method by which the user can secure [22] complete, in-depth information on any advance in an area of particular interest. The objective of this program is to spread the advanced technical industrial capabilities developing in the space program within and beyond the government contractor population to the maximum extent practicable, and particularly to bring about the identification and practical utilization by American industry of new processes and products developing in the space program.

Scholars in the Nation's universities conduct much of the basic research, and prepare many of the experiments required for advances in space science. The breadth of the program has produced, at many universities, new requirements for interdisciplinary cooperation and participation among the scientific specialties, and between science and engineering.

NASA-supported research effort in universities involves both project-related research, as illustrated in Figure 31, and a Sustaining University Program comprised of training, research and facilities grants. Under the training program, 142 universities have received grants to support a total of 3,132 candidates for predoctoral training fellowships in space-related fields. Research grants under the Sustaining University Program have been made to 53 educational institutions, most of them involving interdisciplinary effort, and many of them "seed grants" aimed at strengthening research activity at universities capable of expanding their research programs. A total of 27 facilities grants is shown in Figure 32) have been made to universities to provide additional laboratory space required in the performance of space-related research.

A significant element in the overall NASA university effort, particularly in the case of those universities receiving facility grants, is the encouragement of a closer working relationship between the university research activities and those businesses and industries with which the university already has close relations. This aims to facilitate the transfer of space research results to practical, industrial application. Memoranda of understanding accompanying the facility grants provide that the university will, in an organized and interdisciplinary manner, seek ways in which such transfers can be achieved, and strive for closer relationships with the business community.

[23] Participation in the U. S. space program has not been limited to this country. Individuals or agencies in 69 nations throughout the world have joined the United States in space projects, including the establishment of tracking and data acquisition stations, as illustrated in Figure 33. In all of these projects, cooperation has been literal and substan-

tive, requiring significant contributions from both sides, without financial exchange, and meeting the test of scientific value.

NASA has launched four satellites in cooperation with Great Britain, Canada, and Italy and has existing agreements to launch others with all of these countries as well as with France and the European Space Research Organization. The present practice in these projects is that the cooperating country conceives and engineers the complete satellite, using its own resources.

Individual experiments proposed by foreign scientists, sponsored by their governments and selected on their merits, are also accommodated in NASA satellites. One British experiment flew on Explorer 1, and 12 other British, Dutch, and French experiments are scheduled for inclusion on NASA satellites which will be launched over the next few years.

NASA has participated in cooperative sounding rocket projects with 14 countries, involving more than 100 cooperative launchings, and currently has agreements for launching nearly 50 more in such projects. The multitude of foreign sites established for this program and the extent of capability stimulated by it vastly increase the possibilities for synoptic research, while reducing its cost.

A wide variety of ground-based cooperative projects involving foreign scientists has been organized to produce observations or measurements enhancing, and sometimes even necessary to, NASA's orbiting experiments. Thus, 42 countries have collected local meteorological information for correlation with TIROS observations, and 11 countries have already built, or will soon complete, ground terminals necessary for test transmissions in connection with our communications satellite programs.

Under international scientific and technical personnel exchanges, 103 gifted foreign Research Associates have contributed their talents to work in NASA centers, 84 International Graduate Fellows have trained in U. S. universities, and 180 foreign technicians have trained at NASA centers in support of cooperative projects and ground facility operations.

This completes the review of the capabilities which have been developed during the first 6 years of space exploration, or which will be developed within this decade. . . .

[61] **VI. Summary**

Our study of future programs has covered three major categories as illustrated earlier in Figures 4, 34, and 54, repeated here for ease of reference. These have covered:

    a. A review of the capabilities being developed by current programs;
    b. Intermediate missions which would support National objectives in space and afford steady progress toward longer-range goals, and at the some time make most effective use of capabilities developed thus far; and
    c. Long-range missions which may comprise the Nation's space exploration goals in the decades ahead.

In the areas of aeronautics, satellite applications, unmanned and manned space exploration, launch vehicles, and research and technology development, it is possible to trace horizontally the development path from 1958 to a decade or further into the future. It is obvious that there is increased uncertainty as the plans are projected into the future.

The details of these new missions, such as specific spacecraft designs and exact mission plans will, of course, be the subject of continued study by Headquarters and Field Centers of NASA, by interested government agencies, by universities, and by industry. Continued [62] space exploration will be an evolutionary process in which the next step is based largely on what was learned from the experience of preceding research and flight missions. The pace at which these new programs will be carried out will necessarily depend upon many other factors, such as the allocation of budgetary and manpower resources and the changing National needs of the future.

This study has not revealed any single area of space development which appears to require an overriding emphasis or a crash effort. Rather, it appears that a continued balanced program, steadily pursuing continued advancement in aeronautics, space sciences, manned space flight, and lunar and planetary exploration, adequately supported by a broad basic research and technology development program, still represents the wisest course. Further, it is believed that such a balanced program will not impose unreasonably large demands upon the Nation's resources and that such a program will lead to a preeminent role in aeronautics and space.

Fig. 34

Intermediate Missions—Extensions of Present Capabilities

Aeronautics
Supersonic transport
Hypersonic engine development
V/STOL

Satellite Applications
Applications technology satellites
Direct broadcast FM
Communications/navigation satellites
Meteorological observation technology

Unmanned Exploration
Observatories, Pioneers and Explorers continued
Planetary fly by, orbiters, and landers

Manned Operations
Earth orbit application (1 to 2 months)
Equatorial orbits
Polar orbits
Synchronous orbits
Rendezvous, inspection, repair, and rescue
Lunar mapping
Extended stay in Lunar space (3 to 14 days)

Launch Vehicles
Saturn 1B Centaur
7000 lb. high energy stage

Technology
Isotope power supplies (1 to 2 kW)
Guidance and control (within miles of point on Mars)
Communications (3000 bits/second from Mars)
Stabilization
Propellant storage
Life support (3 men, 1 to 2 months)
Sterilization
Reliability

Fig. 50

Extended Apollo Mission Capabilities

Mission	Configuration	Orbit N M Incl	Duration Days	Payload Lbs	ΔV fps	Equivalent Plane Change deg
Earth Orbit	Sat IB/XCSM/LEM AS	200/30°	30	5,000 0	0 2,000	— 4
Earth Orbit	Sat V/XCSM/LEM ASP Initial	200/30°	30	210,000 0	0 26,000	— 60
			90	200,000	0	—
Earth Orbit	Sat V/XCSM/LEM AS	200/Polar	30	17,500 0	0 4,000	— 9
Earth Orbit	Sat V/XCSM/LEM AS	200/Polar	60	12,500 2,000	0 2,000	— 4
Earth Orbit	Sat V/XCSM/LEM AS	Synchr/0°	30	9,500 0	0 1,500	— 9
Earth Orbit	Sat V/XCSM/LEM AS	Synchr/28°	60	10,000	—	—
Lunar Mapping	Sat V/XCSM/LEM ASP	80√/Polar	28	9,500	—	—
Lunar Surface Exploration	Sat V/XCSM/LEM Dual Launch	Lunar Surface	14	2,500	—	—

Notes XCSM—Apollo command and service module with additional subsystems and expendables
LEM AS—LEM ascent stage without subsystems, dependent on XCMS
LEM ASP—LEM AS plus LEM descent stage propulsion

Fig. 51

Manned Earth Orbit Experiments

I Space Sciences
   Biosciences
   Physical Sciences
   Astronomy/Astrophysics
II Earth Oriented Applications
   Atmospheric Science and Technology
   Earth Resources Survey and Inventory
   Communications
III Support for Space Operations
   Advanced Technology and Subsystems
   Operations Techniques/Subsystems
   Biomedical/Behavioral

Fig. 54

Long-Term Development

Aeronautics
   Hypersonic transports
   Recoverable orbital transport
   Commercial V/STOL aircraft

Satellite Applications
   Direct TV broadcast
   Navigation & traffic control
   Continuous global weather observation

Unmanned Exploration
   Probes & landers to distant planets
   Solar probes
   Galactic probes

Manned Space Exploration
   Conventional take off and landing
      of space vehicles
   Flexible Earth orbital operations
   Large permanent space laboratory
   Roving Lunar vehicles and Lunar bases
   Planetary exploration

Launch Vehicles and Propulsion
   1 million pounds in Earth orbit
   Recoverable boosters
   Nuclear engines
   Electric propulsion

Technology
   Nuclear & isotope power supplies (megawatt)
   Guidance & control (controlled landings
      at desired locations on other planets)
   Communications (wide band communications
      with planetary vehicles)
   Stabilization
   Permanent life support systems
   Reliability

Fig. 68

Lunar Landed Payloads

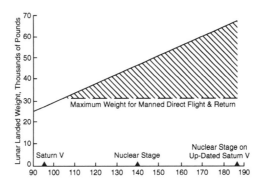

Fig. 69

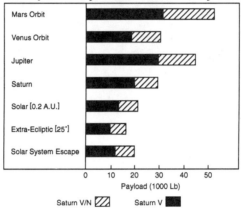

Interplanetary Mission Summary

Document III-18

Document title: James E. Webb, Administrator, NASA, to the President, August 26, 1966, with attached: James E. Webb, Administrator, NASA, to Honorable Everett Dirksen, U.S. Senate, August 9, 1966.

Source: NASA Historical Reference Collection, History Office, NASA Headquarters, Washington, D.C.

By 1966, the NASA budget had peaked, and the agency's future, once Apollo had been completed, was unclear. NASA Administrator Webb was becoming increasingly frustrated by the unwillingness of the White House and the Democratic-led Congress to support the budget for the space program that he thought was needed to continue a productive effort. The Bureau of the Budget had reduced NASA's fiscal year 1967 budget request by $712 million, and in August Senator William Proxmire (D-WI) proposed that the Congress reduce the budget by another $1 billion. To counter the Proxmire proposal, Webb had to seek the support of the Republican leader in the Senate, Everett Dirksen (R-IL).

Webb attached the letter he had sent to Dirksen, which sought his help in defeating the Proxmire amendment, in this August 26 letter to President Johnson in which he expressed his growing unhappiness with NASA's outlook. In particular he protested the guidelines for the fiscal year 1968 budget that had been given to the agency by the Bureau of the Budget, which to Webb's thinking were inconsistent with a plan for the president to give a speech setting out "a ringing challenge for the next half century in space."

---

[1] Dear Mr. President:

Almost six years ago when you urged me to accept the responsibilities which devolve upon the Administrator of the National Aeronautics and Space Administration, I asked if my task would be to carry out a preconceived program or to figure out what needed to be done and do it. You said, "the latter," and, of course, this was on the basis that President Kennedy would have to approve whatever you and I worked out. You will remember that in the sessions you had in 1961 with your advisers and Congressional leaders, I was quite reluctant to undertake the responsibility of building a transportation system to the moon and that you had to almost drive me to make the recommendation which you sent on to President Kennedy.

As to my discharge of this responsibility after the decision was made, and of the other responsibilities inherent in the aeronautical and space science activities of NASA, you are in position to judge. I believe the record justifies your continued support. There are few, if any, enterprises of such size and inherent difficulty that have yielded more total value in proportion to resources invested.

In presenting your 1967 budget to the Senate Subcommittee on Appropriations, I used this language and furnished you a marked copy:

"This budget has been carefully drawn by the President to reflect total national requirements. For NASA this is a particularly stringent budget. We are midway through a ten-year effort to achieve preeminence in all fields of aeronautics and space. This budget is less than we need to carry out this effort with greatest efficiency and minimum risk. Every expenditure that can be deferred until 1968 without causing gaps in our activity has been deferred. This budget provides for continuation of our ongoing efforts and a [2] few long lead-time items for the post-Apollo period. It provides no alternate or backup vehicles."

"The program we began presenting to you in 1961, and have elaborated in each succeeding year, was intended to meet fierce competition and end up ahead. It was also intended to give us a number of options in space from which we could choose those offering the greatest advantages at the least cost. The competition is still fierce, and we are not yet able to feel assurance that we will end up ahead in the option areas where the Russians are developing their strongest potential. A $5 billion budget level in the years ahead will not be adequate to develop and utilize the options we are now in the final stages of developing. Many of these show clear indications of usefulness far beyond their cost.

"In my view the main question which this committee must consider as it takes up the 1967 budget is whether we can or will continue to meet the challenges and pursue the opportunities opening up in space."

"Along with austerity, the NASA authorization request reflects the President's determination to provide sufficient resources to hold open for another year and not to foreclose the major decisions on future programs where failure to apply resources this year would make it impossible to act effectively next year. Most of these relate to whether to make use in 1970 and beyond of the space operational systems, space know-how, and facilities we have worked so hard to build up, or to begin their liquidation."

The combined effect of the action taken by the Senate and House Conferees on our appropriation, which puts us below the above-mentioned $5 billion figure for FY 1967, and the guidelines furnished us by the Bureau of the Budget for the 1968 submission leave no choice but to accelerate the rate at which we are carrying on the liquidation of some of the capabilities which we have built up. Important options which we have been holding open will be foreclosed. Further, the actions we must take will bring into play [3] forces of doubt and uncertainty in the minds of many whose competence, skill and courage have kept us above that thin line that divides success from failure.

There has not been a single important new space project started since you became President. Under the 1968 guidelines very little looking to the future can be done next year.

Struggle as I have to try to put myself in your place and see this from your point of view, I cannot avoid a strong feeling that this is not in the best interests of the country.

I know the heavy total responsibility which you bear, Mr. President, and believe firmly in the actions you are taking to make clear to the Communists that you have on call a large measure of power based on the kind of technology NASA is developing and that you are prepared to employ it to make sure they sustain a loss instead of a profit when they undertake excursions such as that in Vietnam. I have no desire to add to your burdens and have had serious doubts that I should involve you in a protest of your 1968 guidelines. However, when Mr. Moyers telephoned that you wanted to make a speech on space that would chart a course that would constitute a ringing challenge for the next half century, and include where we have come, where we have to go, and the benefits from the program, I decided I should let you know my feelings. They are set forth in the enclosed letter which I sent Senator Dirksen the day we had to collect up the votes to beat Senator Proxmire. I hope you can find time to read it. It is never an easy thing to decide the time has come to ask for help from the minority leader. The senior member of our committee, Senator Russell, voted to support Proxmire, as did Senator Robert Kennedy. Without the effective work of Senators Anderson, Magnuson, and Smith, with considerable help from Allott, we would now be facing a catastrophic emasculation of what we have labored so hard to build up.

If it is your purpose to enunciate a ringing challenge for the next half century in space, Bob Seamans and I will be right with you, but we cannot deliver the kind of successes we have had with the thin budgetary margins of the past three years.

With warm and affectionate personal regards, highest respect, and deep appreciation of your many acts of friendship and support, believe me

Sincerely yours,
James E. Webb
Administrator

[1] August 9, 1966

Honorable Everett Dirksen
United States Senate
Washington, D.C. 20510

Dear Senator Dirksen:

In accordance with your request for information on the effect of the Proxmire proposals to cut up to $1 billion from the space budget for Fiscal 1967, the best I can do in the short time between now and your deadline of noon is to state the following:

1. Through NASA, the nation is in the process of investing approximately $40 billion in the scientific measurement, development of engines, machines and the know-how to operate them, and in the use of this scientific knowledge, technical capability, know-how and machines to make use of both the air and the space region around the earth for practical, economic, and international purposes. Another factor is to make sure we do not wake up some day and find others in possession of the power to deny us the use of space. Beginning in 1958, the various non-military agencies of government were brought together to retrieve the position of leadership in space which we had lost to the Russians. A program looking toward the expenditure of from $22 to $25 billion over a period up to 1975 was initiated by President Eisenhower. With the dramatic capability demonstrated by the Gagarin flight in 1961, this was augmented and speeded up under the leadership of President Kennedy and Vice President Johnson with strong bi-partisan support. Over the past five years, the Congress has appropriated about $22 billion to carry out this effort and a dramatic build-up has taken place as demonstrated recently by the successful Gemini flights and the Surveyor landing on the moon. Right behind these tremendous efforts and these clear demonstrations of the correctness of our engineering approach, our knowledge of the environmental conditions to be met, and the validity of our system of management which allocates over 90 percent of the doing of this job to American industry, we now find ourselves facing an even greater requirement. The end result of an investment involving between $15 and $16 billion in advanced equipment that can far exceed anything we have seen demonstrated yet is now flowing toward our installations for test and on to Cape Kennedy for launch.

[2] 2. While the above five-year record has been achieved within the estimates of cost provided to the Congress at the beginning, we find that reductions made by the Congress in Presidential requests have been largely responsible for slowing up the program by two years and adding more than twice the amount of these reductions to the cost for doing the same amount of work. The reductions made by Congress over five years have amounted to $1,100,000,000 and the increase in cost will amount to $2.7 billion. The enclosed summary of this situation supplied to the Senate Committee on Aeronautical and Space Sciences at the request of Senator Margaret Chase Smith further explains what has happened.

3. For Fiscal Year 1967, the President reduced our request for funds by $712 million, with the result that under the most favorable circumstances the work force in the factories and laboratories of some 20,000 industrial companies, financed by NASA, will be reduced by from 60 to 80 thousand workers. This drastic reduction will have to be made at a time when the Civil Service personnel in NASA centers must take the responsibility for the final test and launch of the end results of the large investments referred to above. The Proxmire proposals would require a further cut of about 100,000 workers and the momentum and effectiveness of the program would, in my opinion, be utterly destroyed. These proposals can only be based on a complete lack of understanding of what it takes to build up a work force of over 450,000 people, proceed rapidly but without the waste of a crash program to develop advanced equipment that can operate with men out to the moon and with automated equipment out to Mars and Venus and then utilize this capability to increase the power of our nation to have on effective voice at the time the largest decisions as to the future of the development of the human race on this planet will be made. Those of us who have had to take the responsibility for what I am describing have little doubt that the balance of technological power among nations is rapidly becoming one of the most important determinants of national economic, social, and political viability as well as leadership in international affairs.

4. Over the past five years, NASA has invested about $22 billion in facilities to permit us, for from 25 to 50 years, to keep a constant challenge before the Russians or any other nation in the utilization of advanced aeronautical and space systems. American [3] industry has invested another $630 to $650 million in capital items, such as test stands, vacuum chambers, etc. The 1967 NASA budget includes $95 million to round out and complete this very large investment and make all of it worth more to the country.

5. We have already sent men into space 22 times before the eyes of the world and brought them back. One failure would have hurt our nation. Within the next three months, we should complete the 12 flights of the Gemini program and move on to the Apollo flights. This will involve the use of the very large Saturn boosters which concentrate in one machine the rough equivalent power of a small atomic bomb. Because of the danger, we must fuel and launch these machines automatically with no human being within miles of the launch pad except the three astronauts on the nose of the rocket. This has never been done before. The burden of doing the final perfection, correcting faults, proving reliability and launching these very large systems with the entire rocket and payload in place on every launch, even the first one, takes high competence, the availability at the launch site of every item required for success, and a good deal of self-confidence and guts. We have built the organization to see this job through, but we cannot hold it together on an up-and-down basis.

6. As to the period beyond 1970, the production lines for our nation's only really big boosters are going to grind to a halt unless we can buy the long lead-time items required to support them. Even if production is continued, these boosters are going to have nothing like the value they could have for our future if we cannot use the scientific and technical knowledge we are now buying at such great cost to do the necessary planning and testing of the payloads, earth sensing equipment, and requirements for operating over long peri-

ods in space which these boosters now open up to us. Senator Proxmire's proposals will, in my view, shortly put us back to the kind of frustration and inability to meet the USSR space challenge that we felt in 1958 with Sputnik and in 1961 with the Gagarin flight.

7. The capability to use our very large rocket engines, advanced electronics, and ability to marry these capabilities with those of the human being, as shown in our Gemini flights, has significance far beyond landing on the moon. What we have done [4] in space shows a can-do nation building strength in science, technology, engineering and management, teaming up its scientific, industrial and governmental institutions to meet the requirements for operating in the new and unlimited environment of space and developing the kind of national capability that will ensure that we are present when the big decisions affecting our future and that of hundreds of millions of people are made.

8. There is no doubt in my mind that cuts made in this program now will have to be restored and multiplied within the next year or two as the Russians begin to use the capability they are in process of developing for flying very large payloads. Beginning in the 1950's we saw them step over what we could do with our Atlas and Titan boosters with the Vostok and Voskhod systems. They clearly have the capability with the booster that has flown the Proton series to step over the capability of the Saturn 1B and get up to some 50,000 pounds in orbit. I believe they are now rapidly building the capability to leap-frog over the kind of payloads the Saturn V can boost into space.

9. In the years since 1958 NASA has shown the ability to get a great deal for every dollar of the investments made in aeronautics and space. We urgently need your support in order that some of the most important matters affecting the future of this nation are not put in jeopardy by on ill-considered action. The committees of Congress charged with the responsibility have officially approved the President's budget, and I would hope their judgment could be confirmed. Many of the statements being made in support of large cuts in the NASA budget simply will not stand up on close examination.

I appreciate your desire to understand this situation, and hope I have differentiated the NASA program from some of those you have characterized as "non-essential spending."

Sincerely yours,
James E. Webb
Administrator

Document III-19

Document title: James E. Webb, Administrator, Memorandum to Associate Administrator for Manned Space Flight, "Termination of the Contract for Procurement of Long Lead Time Items for Vehicles 516 and 517," August 1, 1968.

Source: NASA Historical Reference Collection, History Office, NASA Headquarters, Washington, D.C.

To ensure that there were enough heavy-lift boosters to complete the Apollo program, NASA had contracted for the elements of fifteen Saturn V vehicles. George Mueller, Associate Administrator for Manned Spaceflight, hoped to keep open the various production lines involved in the Saturn V program, anticipating that there would be other uses for the giant vehicle—extended lunar exploration and launching a space station, for example—that would require a heavy-lift capability during the 1970s. The first step in ensuring that this could be done was to contract for those components of the vehicle's S-1C first stage that required the longest time to manufacture. In mid-1968, Mueller requested authorization from James Webb to enter into such contracts.

Webb's answer was negative—no uses for Saturn Vs beyond the original fifteen had been approved, and the budget outlook for such approval was gloomy. This memorandum was thus the first step in a process that led to a 1970 decision to terminate the Saturn V program.

Memorandum to Associate Administrator for Manned Space Flight

SUBJECT: Termination of the Contract for Procurement of Long lead Time Items for Vehicles 516 and 517

REFERENCE: N memorandum to the Administrator, dated June 2, 1968, same subject
D memorandum to the Administrator, dated July 31, 1968
AD memorandum to M dated July 13, 1967

After reviewing the referenced documentation and in consideration of the FY 1969 budget situation, your request to expend additional funds for the procurement of long lead time items for the S-IC stages of the 516 and 517 vehicles is disapproved. This decision, in effect, limits at this time the production effort on Saturn through vehicle 515. No further work should be authorized for the development and fabrication of vehicles 516 and 517.

James E. Webb
Administrator

Document III-20

Document title: Bureau of the Budget, "National Aeronautics and Space Administration: Highlight Summary," October 30, 1968.

Source: NASA Historical Reference Collection, History Office, NASA Headquarters, Washington, D.C.

The career staff of the Bureau of the Budget (renamed the Office of Management and Budget in 1970) remains in position as administrations change, and it is an important contributor to continuity in government policies and programs. This summary, prepared during the last months of the Johnson administration but intended for whomever would enter the White House the following January, identifies the significant space policy issues that would have to be addressed by the new president. While Lyndon Johnson had remained committed to completing the Apollo program, the twin crises of the conflict in Southeast Asia and urban unrest in the United States had not allowed him to allocate resources to any major post-Apollo space objectives. As the first lunar landing approached, the space program was clearly at a crossroads.

[1] # National Aeronautics and Space Administration Highlight Summary

I. Program and Policy Issues

This paper discusses the major aspects of National Aeronautics and Space operations which warrant attention at an early point in 1969.

A. *Space Program Among Other National Priorities*

The resource requirements of the Viet Nam war and of pressing domestic needs, coupled with an apparent acceptance of the Soviet presence in space, have tended to push the civil space program down the scale of national priorities. As funding requirements for on-going programs have declined, it has been very difficult to obtain funds for new starts. Major space activities require large sums of money, and the development of equipment requires 3 to 8 years from go-ahead to flight. Therefore planning for space programs, and even annual budget decision, is very uncertain unless some general levels of funding commitment in future years can be assumed. In a period in which space enjoyed high priority, total programs were planned and budgeted around the expectation that $5-6 billion would be available in future years. Now future planning estimates range between $3 and $4B, and at the lower end of this scale our ability to undertake significant manned flight becomes marginal.

It appears that a two-fold major policy study should be undertaken to identify (1) the national needs served by space flight, and (2) the priority to be accorded the space program over the next several years in relation to other national priorities.

B. *Post Apollo Manned Space Flight*

Major decisions must be made in the 1970 and 1971 budgets. Funding variations of $\pm$ $2 billion from the present $2 billion per year base are involved. The Manned Lunar Landing is very likely to occur in late CY 1969, thereby ending what is generally considered the major cause of urgency in the progress of manned space flight.

As many as eight Saturn V launch vehicles with Apollo spacecraft will remain unused, as will 7 to 9 Saturn IB's. Budget decisions were made in 1969 to close down all these production lines on completion of Apollo program production. A short term Apollo Applications program has been defined to use the Saturn IB's in low earth orbit, but that program will pass its funding peak in FY 1970 and end in CY 1972. [2]

In the circumstances, pressure is mounting to budget significant sums for follow-on manned space flight activities, which forces the question of whether there should be a program of manned space flight after Apollo.

Termination, or even lengthy postponement, poses problems of abandoning expensive inventories, of local economic disorientation, and allegations of leaving all of outer space, including the Moon, to the U.S.S.R.

Continuation poses problems of funding, program rationale, program definition, and assignment of principal roles between NASA and Defense (see IV-A, below).

By landing a man on the Moon in 1969 we will have proven that we possess an engineering and technological capability to master the basic problems of very large scale manned space flight operations for periods of several days. The Gemini program proved our ability to keep men in orbit for periods of two weeks, and the Department of Defense MOL program is based on the assumption that man can function effectively in orbit for 30 days.

It is difficult to conceive of any use short of a manned planetary expedition that would require men to operate in orbit for more than 30 days. Most scientific endeavors that require the collection of data by means of space flight can be accomplished by unmanned systems at considerably less expense than the manned space flight systems.

The U.S.S.R. is continuing to develop a large rocket that can place payloads in orbit equivalent in size to those lifted by our Saturn V (285,000 pounds.) Only the Saturn V is in this weight lifting class, and no other combination of rocket stages currently existing in the U.S. can compete.

Our manned flight program was established, and expanded to include a manned lunar landing, by policy decisions in response to "technological challenge" from the U.S.S.R. An alternative to the policy of competition would be a policy of cooperation with U.S.S.R. in large manned flight endeavors.

Reasons for proceeding other than competition include enhancing the national prestige, advancing the general technology, or simply faith that manned space flight will ultimately return benefits to mankind in ways now unknown and unforeseen. None of these secondary arguments can be quantified and most are difficult to support.

The case for continuation of a manned space flight effort after Apollo is one of continuing to advance our capability to operate in space on a larger scale, for longer durations, for ultimate purposes that are unclear.

[3] C. *Unmanned Planetary Programs*

Pressures are strong from the scientific community to increase our pace of unmanned exploration of the planets. The National Academy of Sciences in its report, "Planetary Exploration; 1968-1975" urged NASA to begin an ambitious program of unmanned interplanetary flights, and recommended that a substantially increased fraction of the total NASA budget be devoted to unmanned planetary exploration. "This is an area in which the U.S.S.R. is competing strongly and one of those in which accomplishments have scientific as well as technological significance. Planetary investigations are basic research, however, and as such have no return in an economic sense. Even as a field of basic research planetary studies may have less long term social benefit than biosciences. Planetary programs require long lead time and firm commitment due to the limited planetary flight opportunities. Funding increases of $100M or more per year above the present $85M base are involved in these programs.

D. *Aeronautical R&D*

The growth of air transportation, the decline of emphasis on military aircraft, and the creation of the Department of Transportation have made commercial applications of key importance in determining the course of aeronautics research. NASA's aeronautics program should be considered within the context of overall government goals and objectives. See separate memorandum on this subject.

E. *Economic Applications*

Clientele groups, both within the Government and outside, are pressing NASA to increase its level of activity in development of satellites for communications, meteorology, navigation, and surveys of earth resources. Funding in this area runs about $100M per year and could easily double in the next two years. Though this is one of the few programs in NASA that shows promise of generating clear near-term benefits, in several areas, notably meteorology and earth resources, the basic cost/benefit ratio questions remain to be critically analyzed. Major management questions, possible reassignments of activities between agencies, and large increases in modest ongoing budgets are raised by the technical possibility of using satellites to serve the needs of several agencies. Interior, Agriculture, ESSA, DOT and Navy are among the clientele agencies in this area, as is the ComSat Corporation and other communications users.

[4] F. *Nuclear Rocket*

This joint AEC/NASA project, started in 1956, has established feasibility of a nuclear reactor-powered rocket engine. Over $1B has been spent to date and an additional $1.5-$2B would be required to develop a useable nuclear rocket stage. However, the advantages of nuclear propulsion do not begin to approximate the costs for missions short of a manned Mars landing. No national commitment has been made to undertake this mission which would cost $40-$100B. (see "B" above) Nevertheless, pressures are strong in NASA, industry and Congress to undertake the development of the nuclear rocket. See separate memorandum on this subject.

II. Budgetary Trends and Issues

NASA's funding level has declined from a high of $5.3B in appropriations in 1965 to the current 1969 appropriation of $4.0B. The 1969 operating level is $3.85B. The manned space flight activities account for over $2.0B in the current year.

The budget issues are those associated with each of the above items, plus the need to reassess the need to support the elaborate ground complex of Government, industry, and universities if the rate of space development activity should continue to decline. The cost of this ground complex is more than $1.5B per year (see IV-B, below).

III. Organization and Management Issues

A. *Use of support service contracts at NASA field centers*

NASA currently employs about 25,000 contractor personnel located in their laboratories in direct support of their 32,000 civil service employees. This is a problem from political, cost, and management standpoints. NASA is faced with a CSC ruling that several of those contracts are illegal. Others may not be administered within the Civil Service laws. At the same time, the agency is operating under the federal personnel ceiling constraints which make conversion to Civil Service difficult, and a future program level uncertainty which threatens the justification for keeping such large numbers of personnel.

B. *Scope of capability base for future space activities*

NASA currently spends between $1.5 and $2.0B per year to maintain a Government/ University/Industry basic capability to engage in space flight activity. This capability consists of the technical and management talent in the NASA laboratories, the world-wide satellite tracking and control networks, scientists and their research teams in universities, research and engineering teams in industry, and specialized ground test and launch facilities scattered around the U.S.

This basic capability complex was established on the assumption that the NASA budget would be about $5.5 B to $6 B per year. As the budget has declined to $3.85B, the flight program development activity has borne the brunt of the reductions and the support complex has been only slightly reduced. The question is whether to assume the possibility of increased funding levels and preserve the base, or to phase down on a long term basis on the assumption that lower funding levels will remain for the foreseeable future. [3]

C. *NASA advanced research and technology centers*

NASA has not yet developed a means to focus their in-house research on long range mission goals. The research program, costing in total around $400M in contract funds for space technology and aircraft technology, and in-house laboratory effort, is therefore diffused and general. It is difficult to judge how varying levels of funding in these areas relate to advancing the nation's ability to meet long-term space goals.

The laboratories do contain high calibre engineering and technical talent which could be used to serve other national needs besides aeronautical and space flight. Research and technology advancement in surface mass transportation, ocean engineering and other complex technological areas could well be done by NASA laboratories. [5]

IV. Inter-agency Relations

A. *NASA relationship to DOD space programs*

The NASA operates a space program for non-military purposes which consists of flight programs for collection of science data and for test and demonstration of new space-related technology, and of ground-based applied research and technology. The DOD operates a space program consisting of satellite flights contributing to defense operations and of ground based applied research and technology applicable to Defense oriented space flight. There are joint agency studies under way to review the two agency programs.

Certain economies may be achieved by reassigning and consolidating activities in such areas as standardization of equipment, ground based tracking networks, and technology programs.

A major policy problem concerns the future of earth orbital manned space flight in which DOD now has the Manned Orbiting Laboratory and NASA has the Apollo Applications program. In future, should we plan on two manned programs, a single program jointly run, or should a single agency be assigned responsibility for all manned space flight activities? [6]

B. As mentioned earlier in the areas of aeronautics and economic applications, there is a need to relate NASA's effort to these programs to the requirements established by the Departments of Transportation, Interior, Agriculture, Commerce, and others. The Government-wide goals, objectives and programs in the area of transportation and applications need to be established, and agency missions and roles delineated.

C. *Total space program funding*
Attached is a table showing the funding for space programs of all agencies 1958-1969.

NATIONAL AERONAUTICS AND SPACE ADMINISTRATION
Summary Budget Trends
(In Millions)

	Budget Authority	Outlays	Employment June 30 Permanent	Total, excl. Summer Youth
1963 actual	3,673	2,552	27,904	29,934
1968 actual	4,587	4,721	32,469	33,968
1969 current BOB estimate (tentative)	3,879	4,250	31,186	32,706

Document III-21

Document title: Charles Townes, et al., "Report of the Task Force on Space," January 8, 1969.

Source: NASA Historical Reference Collection, History Office, NASA Headquarters, Washington, D.C.

Richard M. Nixon was elected president in November 1968. Like the incoming Kennedy administration in 1960, Nixon appointed a number of blue-ribbon transition teams to advise the new government. Nixon's thirteen-person transition Task Force on Space was chaired by Nobelist Charles Townes of the University of California at Berkeley. Unlike the "Wiesner Report" prepared for the Kennedy transition, this report was not released to the public or the press.

[1] # Report of the Task Force on Space
January 8, 1969

Preamble

Development of space sciences and technology, exploitation of their uses, and exploration of the solar system inspire and attract human endeavor for many reasons. Since much effort and expense are involved, and plans for major moves in these fields must be made years in advance, it is prudent for any nation to consider carefully what, in the course of years and decades, is the likely importance and cost of such efforts to the nation and to humanity. Yet no one can assess with precision or surety the ultimate human value of our space program, and indeed we can expect that some of the more striking values are not yet visualized. However, what can now be foreseen, and historical experience in development of other areas of science and technology, make a convincing case that space exploration and utilization will have a tremendous impact on human thought, activity, and welfare. The space program has many facets, and the values of each cannot always be measured on the same scale. The more important aspects, not in order of priority, are:

1. *Exploration and Discovery.* Man's escape from the earth's surface, his exploration of the moon and planets and further penetration, at least by instruments, of space beyond the solar system represent one of the most exciting and appealing frontiers for human exploration of all time. Linked so closely with exploration as to not be really separable is a second aspect—

2. *Science.* The space program has provided new tools and unique capabilities for examination of some of the most challenging and basic scientific questions. For example, space observatories will have an important influence on our understanding of the history of the universe and yield enormous advances in astronomy, newly possible lunar and planetary [2] investigations should answer questions on the formation of the solar system and greatly increase our knowledge of geophysics, and exobiology may revolutionize our view of life.

3. *Use of Spacecraft and Associated Techniques for Civil or Commercial Benefit.* Some applications of space operations, such as communications satellites, already seem to be economical in terms of direct benefits to civilian life. Others, like weather-observing satellites, coupled with new sensing systems, offer realistic prospects of great advances in weather research and its applications. In such diverse areas as mineral and water resource development, forest and agricultural surveillance, and ocean monitoring, for example, substantial advances seem imminent and warrant vigorous research and development. In all of these cases, space technologies open up entirely new opportunities for achieving the global perspectives that are essential to the effective use of world resources and to the preservation or improvement of the quality of the human environment.

4. *World Cooperation and Stability.* Many aspects of space work stimulate and offer new opportunities to promote world unity and cooperation. Important among these are the fulfillment of common human aspirations in extending man's purview beyond the earth itself, the physical and logical impossibility of dividing space or satellite orbits along national lines, and the naturalness of global utilization of space operations. While capitalization on these aspects in the interest of world unity and stability will require care and subtlety, they do present new and potent opportunities for progress in this direction.

5. *National Security.* If one omits consideration of ballistic missiles, as we shall, there are still a large number of important direct applications [3] of space technology to military effectiveness. The DOD budget of about $2 billion for space work is an indication, and we think a reasonable reflection, of the present importance of military applications of space. Furthermore, the probability of additional unappreciated effects of space technology on military affairs and the rapidity of change in military technology give considerable importance to a high level of U.S. competence in all major areas of space technology and operations. Closely related to some aspects of national security is the question of—

6. ***Prestige.*** Prestige comprises a variety of real and sensible effects on the attitudes and responses of the U.S. citizenry, as well as other peoples. They are important to the confidence and well-being of our own citizens as well as to our international actions and national security. Prestige associated with space must in the long run be based, of course, on real values rather than on the appearance of accomplishment, and its effects need to be carefully judged by those versed in politics and social psychology, who at the same time are well informed about the technical and operational possibilities of the space program.

7. ***Technological Development.*** A successful space program gives not only the appearance of technological and organizational leadership, its stringent requirements demand and develop them. There are other conceivable technological programs, mostly less highly visible, which can give similar benefits for the general development of technology. However, the existence of a vigorous space program does provide an important stimulus to technology, and helps give U.S. industry a favorable competitive position in world markets.

[4]   **Summary of Issues and Conclusions**

Major issues and considerations in the present direction of the nation's space program are as follows:

1. ***Should the U.S. compete with the USSR in space activity?*** We believe it should not do so in detail, but that the U.S. effort must be as strong over-all as that of the Soviet Union. A decision to compete on this broad scale plays an important role in the budgetary level of space work, fixing it at something like the present level.

2. *Is any significant change required in thrust or content of the present space program?* A new look is required at the balance between the manned and unmanned segments of NASA space program, in order to ensure that the purposes and relative usefulness of each is properly assessed and fully exploited. Expanded research and development in use of unmanned devices for scientific investigation, and in a wide variety of useful applications, including communications, weather and earth resources surveys, seems strongly indicated.

3. *What should be the objectives and scope of the **manned program?*** While this issue is complex, and the function of man in space not yet clear, a considerable majority of the task force believes there is a substantial role for man in the long term, and that a continued manned flight program, including lunar exploration, is justified at present.

4. *What are the program items and their urgency for the immediate future?* Various items needing special consideration are

   a. A manned space station. We are against any present commitment to the construction of a large space station, but believe study of the possible purposes and design of such a station should be continued.

   [5] b. Apollo Applications Program. This program should proceed as a way of testing man's role in space, of allowing a healthy continuing manned space program, and for the biomedical and scientific information it will yield.

   c. Lunar exploration. Lunar exploration after the first Apollo landing will be exciting and valuable. But additional work needs to be initiated this year to provide for its full exploitation by means of an adequate mobility and extended stay on the lunar surface.

   d. Planetary exploration. The U.S. program for planetary exploration by instrumented probes needs to be strengthened and funds for such probes increased appreciably. However, the great majority of the task force is not in favor of a commitment at present to a manned planetary lander or orbiter.

   e. Astronomy and other sciences. The space program is important to a number of sciences, and can be of enormous benefit to astronomy. This potential should be continuously developed through sound and stable programs.

   f. Applications of spacecraft and associated techniques for civil and commercial benefit. We believe research and development of such applications should be supported strongly and increased in pace. Furthermore, the new administration should give considerable attention to their use in promoting international cooperation.

5. *The significance of space work to national security.* The space program is of great importance to national security, not only because of present direct military applications, and its effect on our posture, but also [6] to have available the necessary technology and skill to make or counter new military uses of space. Recommendations are discussed in a classified appendix.

6. *Cost reduction, and "low cost" boosters.* The unit costs of boosting payloads into space can be substantially reduced, but this requires an increased number of flights, or such an increase coupled with an expensive development program. We do not recommend initiation of such a development, but study of the technical possibilities and rewards. Some cost reductions in the space program can probably be made simply through experience and stabilization of the level of effort, and through coordination of future NASA and DOD programs.

7. *International affairs.* Space operations put in a new light many international questions and also lead naturally towards some areas on international cooperation. We believe these offer opportunities for initiatives and some progress towards world cooperation and stability, and the U.S. should exploit these opportunities with both care and vigor.

8. *Are organizational arrangements appropriate for the future space effort?* We believe the separation of nonmilitary from military space work which has been effectively produced by the creation of NASA, and the continuance of a strong, largely unclassified, space program without any direct military aspects is very important.

Organizational programs which need action or study include:
a. The DOD/NASA interface, where it is recommended that the new heads of the two organizations develop a plan for optimizing coordination.
b. The NASA organizational structure. Sometime after the first lunar landing, NASA should be reorganized on a more [7] functional basis rather than on a basis of use of manned or unmanned techniques, and in addition an out-standing scientist should be brought into its top administrative ranks.
c. The Space Council has not been very effective. We recommend changes.

The appropriate over-all budgetary level and rate of the space effort cannot be made precise without detailed examination. However, three considerations dominate in the general budgetary level required for the space program. One is the needed development and application of space technology directly for military problems. We have not examined the DOD budget of $2 billion for these purposes, but such a figure seems appropriate. A second is the need we see for a continued manned space flight program. For a successful, safe, and continuing manned program in NASA an annual budget of about $2 billion directly for this purpose is needed for fiscal 1970. Additional funds are of course necessary for many other parts of the NASA program, including some expansion on unmanned exploratory work. The third large and very pervasive factor affecting the budget is the need to maintain a generally competitive position with respect to the Soviet Union. We believe that approximately the present level of expenditure, $6 billion for the total space program, and about $4 billion for NASA, is needed for this purpose. This total amount, about $3/4$ of one per cent of the GNP, does not seem excessive in view of importance of the space developments to the nation.

A $4 billion budget represents a rather frugal amount to carry out NASA's many important tasks. But we believe it is adequate for the programs recommended here. In subsequent years some changes may be appropriate, [8] but we do not expect that any large fractional change will be desirable soon without a concomitant substantial change in the role of NASA or in the international situation.

The most reasonable way of effecting a large budget reduction in the future would be to postpone any development of new manned systems. Since most of the development and hardware purchases for Apollo have now been made and considerable number of boosters and space vehicles will remain after the first lunar landing, it is possible to have an active and successful manned program for several years while at the same time steadily decreasing the level of funding for manned space flight to perhaps $1.25 billion by fiscal 1972. This would be based on use of hardware already procured, which would permit continued manned space operation until 1975. An option representing a severely constrained manned program would be continuation of manned flight following 1975 with Saturn V equipment. Procurement lead time would require a decision about 1972, and annual acquisition and operational cost for a minimal program of two launches per year would level out at about $1.2 billion. Such a program would be based on extended use of present technology and not allow any new development of equipment for manned flight during this time. Such a plan is not recommended, since we believe a continued vigorous manned program beyond this period will be important.

Competition with the USSR

The Apollo commitment had its origin in a crisis of confidence in the technological superiority of the U.S., with implications concerning our national security. While this situation has changed radically and we believe that the nation can plan its space program with considerable confidence and [9] detachment, our plans must reflect the concurrent Soviet activity.

The USSR continues to expand its investments in nonmilitary and military space operations. It seems to be actively preparing for a long-term program of manned space flight activity, including both manned lunar flights and extended manned flights in earth orbit. In addition, the Russians are in a particularly strong position to compete in unmanned planetary exploration—for which they have a well-tested rocket more suitable than ours—and they are steadily strengthening their nonmilitary applications programs.

Our response to Soviet space activity must insure that we do not abdicate unilateral capabilities to the USSR whose potential impact on our security cannot be readily assessed. Nor should we permit ourselves to be completely dependent on Soviet sources for major areas of important scientific information. In applications areas the U.S. should insure the strength of its commercial and national security positions and take the initiative in international space cooperation.

The task force also believes that continuation of a vigorous program of space exploration, involving man's participation, is desirable in order that the U.S. shall remain competitive in this most visible area of space activity, although we recognize this as more a political than a technical question.

These views have the following consequence in policy:
1. We should remain competitive in each of the following areas under the principles given above:
 a. Manned and automated exploration of the solar system
 b. Military and civilian space applications
 c. Space science
 d. Technology relevant to the above

[10] 2. There is no need for our space goals to mirror those of the USSR in detail; we can and should design a program to meet our needs.

3. Continued efforts should be devoted to the ultimate goal of cooperation with the Soviet Union in manned exploration of the solar system, in the order that this area of prestige competition might be reduced in cost and become a force for political accommodation.

4. Current NASA budget levels are sufficient to support an adequately competitive space effort.

Objectives and Scope of the Manned Space Program

The remarkable success of the Apollo 8 mission has provided renewed insight to the dramatic public appeal of manned space flight and bolsters our confidence that the manned lunar landing may be accomplished as early as July 1969. With this convincing demonstration of our strength and capability in space technology we must examine and redefine the future role and objectives of manned space activity in our national space program. A decision regarding this role may be the most critical choice facing the new administration in regard to the space program.

The broad objectives of the space program, and particularly of its manned component, must be viewed realistically and objectively in two parts. The first part relates to the satisfaction of man's aspirations to explore his universe and extend his purview, coupled with the continued exercising of our national scientific, technological and industrial skill in a way that is dramatically appealing to the world public—a "show of constructive force," as it were. We will be measured, and we will measure ourselves against the Soviet Union by the quality and value of our space [11] activities, and thereby contribute to the over-all assessment of our relative strength and influence in the community of nations. Our accomplishments may further serve to provide an important domestic focus of national purpose and pride, a unifying and inspirational force of some consequence in the midst of difficult and divisive social problems.

The second class of objectives relates to man in space as a useful part of a scientific activity or a space applications operation. There are substantial differences of view among technically well-informed people about the future evolution of space technology, the role that man-in-space will play in it and how soon extensive practical use of man-in-space might come. Given a shirt-sleeve environment in which to work, men can probably work in the weightless state with an effectiveness nearly equivalent to their performance on the earth's surface. Doubts about the role of man in space arise in part from the rapid evolution of technology on earth toward the removal of man's intimate involvement in complex equipment and substitution of computer and other remote control systems. In part these doubts result from concern about risks to life that can never be reduced to zero. But primarily such doubts come from the great cost of placing man in orbit and sustaining him there with the necessary tools, propulsion, and other capabilities to be truly useful in a control or engineering role.

Whether these costs will be justified by the reductions in capital cost of space systems that manned operation, or manned repair and modernization in space might bring, and the value of man's dexterity in assisting with the assembly of complex systems in orbit will be to a large extent dependent on the total scale of space operations in the future and the reduction in costs of transport to orbit that new launch systems might bring. By this criterion [12] man cannot be said to "pay his way" in space today. There is a good reason to hope that in the long run man in earth orbit will be valuable in providing operational and engineering support to large-scale space operations and scientific experiments. Therefore, plans for future manned programs must recognize the fact that we do not know precisely what may be the proper or most useful functions of man in space, but it should be precisely our objective to find out.

It would be undesirable to define at this time a new goal that is both very ambitious in scope and highly restrictive in schedule, for example a manned landing on Mars before 1985, even though such a goal might be achievable. Such a commitment, adopted now, might inhabit our ability to establish a proper balance between the manned space program and the scientific and applications programs. On the other hand, there is probably some threshold budgetary level required to maintain a manned space flight capability in being, which may be between $1.2 and $2.0 billion per year. Some part of this manned space flight activity can be directed to the continued exploration of man's possible usefulness in space. The proposed Apollo Application Program, including the workshop experiments, will contribute to this end in the 1971-1972 period. Other than this program, the major focus of manned space flight during the coming half decade should be manned lunar exploration. It is inconceivable that we should terminate human exploration of the moon after one or two landings, with no activity beyond simply standing on the surface. However, continued manned exploration should be a thoughtfully integrated part of a total program of lunar exploration, utilizing unmanned landings and remotely controlled exploratory devices when they are advantageous. The manned landing should be infrequent, but planned to extend progressively man's roll in the exploration.

[13] It should also be noted that achievement of prestige through space achievements may require some shift of emphasis from manned to unmanned activity. With our apparent momentary lead in manned flight, it is likely that Soviet programs will emphasize strongly a massive commanded and automated exploration of the planets. These two aspects of space prestige must be considered carefully.

In the continued investigation of man's proper roll in either space science or space applications, it is desirable to avoid undue polarization along manned versus unmanned flight and instead to focus on the search for the most appropriate roles for the human being in the entire system, on the ground as well as in space. The objective should be to devise the most efficient means of conducting the entire activity, with the human intelligence operating in the most effective location. The focus should be on the mission itself, and the mission-oriented plan should include, where appropriate, the determination of an optimum combination of manned and unmanned flights. The present organization of NASA is not at all adapted to this approach.

Programs and Priorities—Space Stations and Apollo Applications Program

The Apollo Applications Program should contribute to our understanding of man's utility in space, but needs a much closer connection than has been achieved so far with the space science and the space applications programs and a sounder foundation in biomedical research. For this, management must put strong emphasis on the missions to be accomplished. The "manned space station" concept, proposed as a program for the later 1970's, is on much more doubtful ground. It is much too ambitious to be consistent with the present clear needs for continued exploration of man's usefulness in space. [14] On the other hand, it is not obviously an effective way of continuing to demonstrate for prestige purposes our manned space capability. Perhaps the most unique function of a space station would be to test man's ability for an extended space flight over times of a year or more, so that the practicality of a manned planetary mission could be examined. Such a test would be needed by the mid-70's if a manned Mars mission by the early 80's were planned. However, the desirability of such a mission is not yet clear, and the Apollo Applications Program may be able to give useful partial answers to the possibility of very long-durations space flights. It therefore seems premature to make any firm program decision regarding the proposed manned space station.

Programs and Priorities—Lunar Exploration

The primary goal of manned space flight in the 1970's which should be planned now is the scientific exploration of the moon, by both equipment and occasional manned landings using upgraded versions of the present Apollo system. Alternatives for this choice are:
  a. A commitment next year to a manned landing on Mars, which some of us believe could be carried out in the early or middle 1980's, if sufficient effort were made;
  b. An earth orbital space station to house perhaps six to nine men who would make occasional trips to and from earth.

A great majority of the task force opposes a commitment to a manned Mars landing at this time. It believes that the space program in this second decade should not be built around a single monolithic goal on a fixed timetable. The task force also recognizes that a Mars landing in the early or middle [15] 1980's would require a substantial expansion of the NASA budget in the next few years. It proposes that the space station receive further study without a binding commitment until its design and purposes are more clearly delineated and the possibilities of a radical reduction in the future of costs of transportation to orbit are more firmly established. It appears that the AAP program for manned flight, also scheduled for the 70's, might serve many of the purposes of a space station.

Mixed manned and unmanned lunar exploration has the following advantages as a primary goal of manned spaceflight in the next 5-8 years:

1. Exploration of the moon may reveal surprises which our studies of the earth did not lead us to suspect. The resulting new concepts about the evolution of planetary systems may have far-reaching impact on our understanding of earth resources, earthquakes, and other matters of great importance to mankind.

2. Building on the capability provided by Apollo, it provides the best opportunity in the next ten years for utilizing man's unique capabilities in space exploration, having a high potential for sustained scientific and public interest.

3. Lunar exploration makes best use of the already contracted inventory of Apollo Saturn V launch vehicles, of which there are sufficient to carry such a program from about 1973 through 1976.

4. It exploits our current "lead" over the Russians, although we can expect manned landing on the moon by the USSR before we can prepare the needed lunar exploration capability, about 1973.

5. A combined manned and unmanned approach is not only one of minimum cost for the maximum return in scientific knowledge, it examines [16] both the competition and the synergism between systems in which the man is either at hand or in a remote location. In that sense we suspect it may be the forerunner of the space technology of the distant future.

This program will require adding vehicles for mobility on the lunar surface and also provision of longer stay time.

Programs and Priorities—
Use of Spacecraft and Associated Techniques for Civil or Commercial Benefit

Satellites give new and uniquely valuable capabilities. These capabilities can be exploited for the benefit of all society and for specific practical applications. For such exploitation an expanded program of research and development, using both ground-based and space techniques is needed.

Because of the high level of technical development and diversification in booster launch use, guidance and control, and durability of electronic equipment in space, recent technological developments have so increased the long life potential of satellites that their operational cost is greatly reduced and leverage for future great cost reduction is large. All this has laid the ground work for application of satellites in the fields of

1. Communications as a radio relay or repeater with high information capacity. Specific applications include public and commercial communications, for example, telephoning, T.V., data collection and transmission and navigation aids. These have only begun, with greater expansion expected when questions of national and international policy and of public and private interests are resolved.

[17] 2. Observation using the electromagnetic spectrum of reading the earth's resources and environment. Users, present and possible, include those in the fields of meteorology, agriculture, forestry, water resources, navigation and traffic control, geodesy and cartography and oceanography.

The opportunities for application in communication and observation are of such social impact on man and have such unrealized economic benefits that their support by NASA should be an immediate *major* program. NASA with the support of other government agencies should have a strong satellite applications program within government and which also encourages the private sector for development and investment.

Programs and Priorities—Planetary Exploration

We consider that unmanned planetary exploration should be a major component of the future space program of the Unites States.

There are nine major planets and an uncountable number of smaller, planet-like objects in the solar system, each of whose motion is dominated by the gravitational attraction of the sun. Each of the planets is a "new world." Each has its own special properties and no two are alike. The origin of the entire system and the separate histories of each planet form one of the most engaging puzzles of astronomical science. Much has been learned and much more can be learned by the use of ground-based optical and radar telescopes. But truly definitive study of the planets must await on-the-spot observations by fly-by, orbiting, and landed spacecraft. The pioneering Venus and Mars missions, Mariners II, IV and V of the United States and mission Venus IV of the USSR, have demonstrated the effectiveness of automated equipment for detailed investigations of the planets and have already yielded substantial advances in knowledge.

[18] The United States now possesses the technological capability and the scientific sophistication to send powerful automated spacecraft to Mars, Venus, Mercury, and the giant outer planet Jupiter within the next five years and to the most distant outer planets Saturn, Uranus, Neptune, and Pluto within the following decade. The first objectives will be to learn the physical properties of the planets—the composition, structure and temperature of their atmospheres; and nature and temperature of their surfaces; their precise shapes, masses and magnetic characteristics; and their internal structures as inferred from such evidence.

Following rapidly behind such physical investigations will be attempts to establish the existence or absence of extraterrestrial life. The discovery of any form of life on another planet would be an event of outstanding scientific importance and of profound cultural and philosophical significance.

It is our opinion that a vigorous program of direct planetary exploration by automated spacecraft is readily encompassed by our national resources and will greatly increase the scope and depth of human knowledge and perceptive.

Programs and Priorities—Astronomy and Other Sciences

Curiosity as to the origin, the fundamental nature, and the form of our physical universe is a subject of profound interest to all civilized man. The present prospects of carrying out experiments and making observations from the environment of space provides an opportunity for studying the nature of the universe in ways heretofore impossible.

Until the advent of flight above the earth's atmosphere we were able to view the physical universe with blurred vision and in only two narrow wavelengths regions out of broad system of radiation by which the [19] processes of the stars and galaxies manifest themselves. A satisfactory start in exploiting the clear seeing beyond the earth's atmosphere has been made; we can point with pride to the success of the Solar Observatories, the Astronomical Observatory, and a host of cosmic and X-ray experiments, where these early observations have revolutionized our picture of processes occurring in our Galaxy. These experiments are only the pioneer steps in space science and there are clear-cut, long-range goals in several areas which must be borne in mind in planning the continued science program from space.

In the area of galactic and extragalactic astronomy, it is important to provide means to see the universe in all available wavelengths and with the highest possible angular resolution. This requires the ultimate construction of high sensitivity, high directivity X-ray and gamma-ray facilities, radio telescope arrays of diameters of miles and a sophisticated optical telescope of diffraction limited performance comparable in size and versatility to the largest now existing on the earth. All of these goals are within the capabilities of our program and can be achieved within the next decade by a vigorous program of progressively more refined experiments, each scientifically justified in itself.

The closest star, our sun, reveals new phenomena and interactions with interplanetary medium as it is studied with increasing spatial and energy resolution and we must work toward more sophisticated observations of this object. The interaction of this source of energy with the material between the planets and the earth, and the manner in which the cosmic rays are modulated as they enter the solar system is a subject of particular relevance to the astronomical and planetary programs of NASA.

[20] Furthermore, rather than looking at the moon and planets from a distance, we now have opportunities to view them at close range, and conduct experiments on their surfaces. The possibilities of studying the moon and planetary objects at first hand will vastly increase our understanding of geophysics and of the history of the earth and solar system.

Significance of the National Space Program to National Security

The national space program, taken as a whole, has been and will continue to be vital to national security. Certain parts of the program contribute very directly and with extremely high leverage to national security, while other parts make only an indirect and smaller contribution.

(The primary part of this section, which considers High Leverage Direct Contributions, is in a special classified appendix.)

Indirect Contributions of the Space Program to National Security

Indirect contributions of the space program to national security are important, but their naturally rather diffuse character makes it impractical to give more than a brief list of them here.

1. The national security, including in particular its diplomatic aspects, is substantially influenced by our apparent posture resulting from performance in the highly conspicuous areas of space science and technology, and space operations. Prestige factors are commented on in other sections of this report.

[21] 2. The space program and space-borne platforms have some unique potentials to help in a general way break down restrictions on free communications across borders and also to build healthy connections with other governments.

3. The space program provides challenging goals and severe tests of advanced technology and management techniques with are important to the nation's military effectiveness and economic success. Direct military programs and also some other civilian programs can provide a similar stimulus. However, the considerable human interest and the variety of new problems connected with the space program are notably effective in developing knowledge and trained personnel of importance to high technology and an adaptive military capability.

Reduction of Unit Costs of Space Operations

Much attention has been directed, particularly during the past six months, to the problem of achieving significantly lower costs in large boosters, without decrease in the reliability of the launching and boosting operations. It now seems clear that several different ways of achieving significantly lower launching and booster costs can be devised by taking full advantage of experience to date, and by applying current technology specifically to the purpose of reducing costs. The launching and booster costs per pound in low altitude orbit could be reduced by about a factor of 10—from $700-$1000 per pound to less then $100 per pound. (Enthusiasts suggest a reduction by a factor of 50.) This difference in cost could total many billions of dollars over a ten-year period. The exact savings depends, of course, on the number of launches one assumes. Each of the different [*ways*] [22] (recover both states, greatly simplified liquid propellant stages, and solid propellant stages) of accomplishing this reduction has its own vigorous proponents.

It does not appear necessary or desirable to initiate a major new program to achieve this cost reduction by any of the alternate approaches at this time. However, it is clear that continued priority should be given to the studies that are already under way, and that these studies should be augmented to provide a more complete understanding of the technical alternatives, and to make more complete economic comparisons for several different future levels of launching activity, projected over the next fifteen years. This work should be focussed and coordinated by DOD and NASA so as to provide by about 1 November 1969 information upon which a joint DOD-NASA program decision could be made.

International Cooperation

The space program provides many opportunities, a variety of stimuli, and some necessity for new initiatives in international cooperation. Space beyond the earth's atmosphere, including the heavenly bodies, has generally been recognized as common to the human race. Satellites must of necessity cross national boundaries and the tracking or retrieving of them is likely to extend past national frontiers. Furthermore, they are generally much more efficient when used on a global scale. And the exploration of space, like human knowledge, is naturally an inspiration and an enterprise best shared by all men.

We believe that the present technological position and national interest in the U.S. make it desirable to take vigorous initiatives towards international cooperation in space work, and to continually make clear our earnest desire [23] for such cooperation.

In general our policy and programs for international space cooperation has so far been important but modest. Much broader cooperation with selected nations or groups of nations would be valuable and is strongly recommended. Cooperation in scientific experiments with Italy, Canada, France, West Germany, Great Britain, and several other countries has been so successful that it seems profitable to increase these types of projects with the hope that the cooperating countries would gain in competence and play a larger role. We do have agreements with the Soviet Union for exchanging meteorological and magnetic data. Satellite communications are rapidly moving into the area of international agreement (Intelsat).

Active study should be begun to seek and to analyze initiatives and policies for the U.S. which would further international cooperation in peace work. The first lunar landing may offer a particular occasion for useful and arresting moves.

Consideration should be given to the merits of an international laboratory financed and staffed by all participating nations in proportion to their interest and devoted to intensive study of world-wide systems such as global weather prediction, or an earth resources satellite system, or both. The U.S. would participate in, but not finance, this laboratory. The relationship of such laboratories to ESRO, WMO, and to the U.N. would require careful study. One could also consider regional laboratories such as in Latin America or Africa where the individual countries cooperating in the program could read out data from earth applications satellites and work up these data for their own areas.

[24] It is suggested that space cooperation with the Soviet Union in the near future take the form of planning and scientific collaboration rather than join conduct of space activities. The most promising area might be in unmanned planetary exploration—one in which Soviet competence matches our own, and with obvious savings to both countries. In the future this might be extended to lunar exploration.

Organizational Issues—Importance of a Civilian Organization

Separation of the space program into a part directed towards military applications in the DOD and a largely unclassified part without strong military coloring in NASA has, we believe, been an eminently wise policy. It is especially important to easy cooperation of foreign nationals and governments with NASA, and to the very friendly attitude towards NASA, its bases and operations, which characteristically occurs abroad. We recommend careful efforts to see that this part of the space program continues to be clearly separated from military applications.

Organizational Issues—DOD/NASA Interface

In considering the relationship between military and nonmilitary space programs, the first question which arises is: are the roles and missions of NASA and DOD in the national space program correctly established. We believe the answer to this question is generally "yes." That is, the DOD should continue its responsibility for all space programs directly supporting military missions and NASA should continue its responsibility for more general space and aeronautical science and technology, for applications of space technology to nonmilitary purposes (such as [25] civil communications, civil navigation and traffic control, weather prediction, earth resources surveys), and for general exploratory programs in the near earth, lunar, and planetary regimes. Whether programs are "manned" or "unmanned" it is not really fundamental to the division of missions between these agencies, and both NASA and DOD should employ men or not employ men in space as suited to their basic missions, and as influenced by the projects involved.

The next question which must be considered are: Are improvements necessary in coordination and mutual support between NASA and DOD programs, and are there significant opportunities for cost-savings in stronger central management of supporting capabilities, such as booster vehicles, launching vehicles, rangers, tracking and communications networks, recovery forces and operational centers? We believe the answer to both of these questions is also "yes." That is, significant steps *should* be taken to provide stronger policies on coordination and mutual support between NASA and DOD programs, stronger central management and control of major new program planning and initiation, and stronger and more cost-sensitive management of supporting capabilities. These improvements are particularly needed relative to potential new manned space flight programs with potentially large budgetary impact.

This problem is complicated and requires mature, thorough, and objective study. Certainly no major changes in responsibilities or organizational reporting relationships should be introduced in the Apollo program prior to the lunar landing. In general, transfers of major organizational units and facilities between agencies may not be needed, if strong machinery or central coordination and management on a national basis can be effected, with suitable directed mutual support.

[26] We suggest that the new Secretary of Defense and the new Administrator of NASA be directed by the President to present specific recommendations aimed at these objectives.

Organizational Issues—Internal Organization of NASA

The present internal organization of NASA is oriented toward the achievement of a manned lunar landing by 1970, with manned and unmanned operations administratively divided. The organization is complex, often with no clear distinction between line and staff functions, and is considered inappropriate for the problems of the post-Apollo space program. While the present structure should not be seriously disturbed before the first lunar landing to avoid any possibility of interfering with this operation, after Apollo the administrative organization of NASA should be changed to correspond to program objectives rather than means of accomplishing them.

In the area of applications, NASA should be encouraged to continue its technical and scientific program leadership. This should continue beyond the initial research and engineering development stages into pilot operations. NASA should continue responsibility for total space flight experimental systems—that is, satellites, sensors, ground stations, test sites, and data processing. User agencies should participate actively in planning and in evaluating results, and in the establishment of budgetary controls. NASA should be organized to work in close cooperation with potential users, especially at the administrative and middle management level. Only thus, through shared responsibility, can the potential benefit of future operations be understood by those concerned, and programs designed for maximum efficiency and benefit.

[27] We believe that future scientific and applications returns from NASA investments can be substantially improved by strengthened policy and managerial direction. There are a number of ways this might be achieved, but one possibility is the return to a feature of the management structure under President Eisenhower: policy leadership is provided by an Administrator and his Deputy, one of whom should be an experienced executive with the primary political responsibility, the other a distinguished and internationally recognized scientist. Policy would be executed by a General Manager.

Legislation

New legislation relating to the space program may be required or appropriate during the next session of Congress in the following areas:

1. National Aeronautics and Space Council

The National Aeronautics and Space Council was established by the organic Act of Congress which created NASA (the National Aeronautics and Space Act of 1958), as a permanent mechanism for resolving policy differences and coordinating operations, primarily between the civilian and military space programs. The original Act provided for the President to be Chairman of the Council. This provision was later amended so as to substitute the Vice President (then Lyndon Johnson) as Chairman.

Although the new President will have the option of asking Congress to abolish the Council, or of not calling any meetings, we believe that as long as the Council exists and is used it should be made effective. For that purpose, there should be a strong staff and the President should be the Chairman. The later will require new legislation.

[28] 2. Communications

The capabilities and current use of satellites for communications purposes point to major imminent changes requiring legislation. For example, satellites can be used for communications within the United States. A proposal has been made to the Johnson Administration, by a task force on communications, to consolidate all U.S. international telecommunications into a single organization. We recommend early study of what legislation is needed in this area.

3. Rights to Inventions

The patent provisions of the National Aeronautics and Space Act are modeled on those of the Atomic Energy Act, and therefore differ radically from other laws governing rights to inventions made under similar circumstances. As the Act is now administered, title to inventions made under defined circumstances is vested in the Government unless the Administrator affirmatively determines that it should be vested in the inventor. We recommend that this emphasis be reversed.

Spencer M. Beresford
Lewis M. Branscomb
Francis H. Clauser
Harry H. Hess
Norman H. Horowitz
Samuel Lenher
Ruben F. Mettler
Charles R. O'Dell
Allen E. Puckett
Walter O. Roberts
Robert Seamans
Charles H. Townes (Chairman)
James A. Van Allen

January 10, 1969
Dr. Charles Townes, Chairman
Task Force on Space

Dear Charles:

In this letter I should like to present a view about the future of the U.S. space program that is somewhat divergent from the report of our task force. The space program is now rising to the climax of placing men on the moon. The world is acclaiming this an event which may herald a dawn of an age of exploration.

Mr. Nixon faces the task of planning the nation's future in space. He needs from us an assessment of the technological development that is possible. In our report I believe we have painted a picture which underestimates the potentialities of the future. In predicting progress today's problems loom large and tend to overshadow the inevitability of future development. My experience would indicate that the ingenuity of mankind can be relied upon to overcome today's obstacles and to carry us upward at an ever-accelerating rate.

Instinctively I feel if Mr. Nixon were to chart a bold program for us to explore the solar system and to push ahead with space science and applications, U.S. technology would be able to meet such a challenge. I think our rate of development can be considerably more rapid than presented in the task force report. For example, I believe we can place men on Mars before 1980. At the same time we can develop economical space transportation which will permit extensive exploration of the moon and in an even shorter time we can place large telescopes in orbit.

Whether we embark on such a space program is a decision that Mr. Nixon and the American people must make, balancing cost against historical perspective. I simply take this opportunity to record my views that as a nation we are capable of carrying through on such a challenge.

Cordially yours,
Francis H. Clauser

Professor Clauser has asked that the above view be submitted with the Task Force Report.

I associate myself with this minority view in believing the tone of the report does not reflect very well the real technical potentials of the longer range, nor the imperatives of that peculiar species, man. However, I endorse the report's conclusions and recommended present actions.

Charles H. Townes

Document III-22

Document title: Richard Nixon, Memorandum for the Vice President, the Secretary of Defense, the Acting Administrator, NASA, and the Science Adviser, February 13, 1969.

Source: NASA Historical Reference Collection, History Office, NASA Headquarters, Washington, D.C.

As the Nixon administration took office on January 20, 1969, it was clear to all that decisions with respect to the goals and pace of the space program after the first lunar landing needed to be made, and that some sort of review would be the first step toward such decisions. The new science adviser, Lee DuBridge, at first attempted to have the review carried out under his direction. DuBridge, who as president of the California Institute of Technology during the 1960s had clashed with NASA Administrator Webb, was thought to share the scientific community's skepticism regarding the value of human spaceflight. Thus NASA let it be known to the White House that it was opposed to DuBridge as the chair of the proposed space review.

By this memorandum, President Nixon established a Space Task Group, chaired by Vice President Spiro T. Agnew, to conduct the review. Agnew was chosen because he was by law the chairman of the National Aeronautics and Space Council. That council had fallen into disuse during the latter years of the Johnson administration, and the White House chose to assign the responsibility of staff support for the review to DuBridge and his staff in the Office of Science and Technology.

Memorandum for

The Vice President
The Secretary of Defense
The Acting Administrator, National Aeronautics and Space Administration
The Science Adviser

It is necessary for me to have in the near future definitive recommendation on the direction which the U. S. space program should take in the post Apollo period. I, therefore, ask the Secretary of Defense, the Acting Administrator of NASA, and the Science Adviser each to develop proposed plans and to meet together as a task group, with the Vice President in the chair, to prepare for me a coordinated program and budget proposal. In developing your proposed plans, you may wish to seek advice from the scientific, engineering, and industrial communities, from The Congress and the public. You will wish also to consult the Department of State (on international implications and cooperation) and other interested agencies, as appropriate, such as the Departments of Interior, Commerce, and Agriculture; the Atomic Energy Commission, and the National Science Foundation. I am asking the Science Adviser also to serve as staff officer for this task group and as coordinator of the staff studies.

I would like to receive the coordinated proposal by September 1, 1969.

Richard Nixon

Document III-23

Document title: T.O. Paine, Acting Administrator, NASA, Memorandum for the President, "Problems and Opportunities in Manned Space Flight," February 26, 1969.

Source: NASA Historical Reference Collection, History Office, NASA Headquarters, Washington, D.C.

The creation of a Space Task Group external to NASA as the means for reaching post-Apollo decisions was not totally welcomed by NASA. Unlike his predecessor James Webb, NASA Acting Administrator Thomas Paine preferred that the space agency decide internally what its priorities were and then seek support for them from the White House and Congress. NASA Headquarters had begun a long-range planning process in early 1968, and the various NASA field centers, particularly the Manned Spacecraft Center in Houston, had also been thinking about future programs as they worked on current programs.

By early 1969, NASA had identified a large, permanently occupied space station as its top-priority post-Apollo objective. Hoping to bypass the deliberations of the Space Task

Group and to get an early endorsement of such an undertaking, Thomas Paine went directly to the president with a carefully crafted case for such an action. The White House quickly rebuffed Paine's initiative, telling NASA that any decisions on future programs would await the recommendations of the Space Task Group.

---

[1] This memorandum is the first of several that I am preparing in response to your request of February 17, 1969, that I give you my views on the principal policy problems in space and aeronautics which now face your Administration, point out some of the opportunities for leadership initiatives now open to you, and give you my recommendations on the new directions which your Administration should set for the nation in space and aeronautics. These memoranda will also serve to indicate the alternative approaches NASA is examining in developing plans and proposals for the post-Apollo period as requested in your memorandum of February 13, 1969, and the basis for my recent recommendations to the Director of the Budget on amendments to the NASA FY 1970 Budget. Copies are being sent to the Vice President, the Secretary of Defense, and your Science Advisor as you requested, with additional copies to the Director of the Budget and Mr. Robert Ellsworth.

This memorandum outlines the problems, opportunities, and principal factors to be considered in *Manned Space Flight*, the area in our space program where NASA and your Administration are faced with the most urgent need for high-level decisions.

1. **Introduction** — NASA now has no approved plans or programs for manned space flight programs beyond the first Apollo manned lunar landings and the limited Apollo Applications earth orbital program now approved and underway. Sharply reduced space budgets over the past three years and the failure of the previous Administration to make the required decisions and provide the necessary resources for future programs have built in a period of low accomplishment which will become apparent during your Administration, and have left the program without a clear sense of future direction for the post-Apollo period. Positive and timely action must be taken by your Administration now to prevent the nation's programs in manned space flight from slowing to a halt in 1972.

The Apollo program served the nation well in providing a clear focus for the initial development and demonstration of manned space flight capabilities and technology. What is needed now, however, is a more balanced program for the next decade which will focus not on a single event but on sustained development and use of manned space [2] flight over a period of years. As discussed below, there are two principal program opportunities: one is a long-term carefully-planned program of manned exploration of the moon, the other is a wide range of activities involved in the progressive development and operation of a permanent manned station in earth orbit. I believe that (a) manned lunar exploration should be continued at an economical rate to the point where a sound decision on the future course the nation should follow with respect to the moon can be made on the basis of knowledge and experience gained from a series of manned missions, and (b) the nation should, in any case, focus our manned space flight program for the next decade on the development and operation of a permanent space station—a National Research Center in earth orbit—accessible at reasonable cost to experts in many disciplines who can conduct investigations and operations in space which cannot be effectively carried out on earth.

2. **Status of U.S. Programs and Plans** — If our Apollo flights continue to be successful we will achieve the first manned lunar landing later this year, possibly as early as this summer. We will then carry out three additional landings at different locations on the moon, but the improved equipment required for moving beyond this with a scientifically significant lunar exploration plan is restricted to the study stage. We will have a number of Saturn V boosters and Apollo spacecraft for future lunar missions left over from the Apollo program.

In earth orbit, the next major U.S. milestone is manned space flight is the Saturn I Workshop, which is now scheduled for launch in late 1971. This first step toward a space station will use existing Saturn IB rockets left over from the Apollo Program. Flight operations, including revisit and experimental Apollo telescope operations, will be completed in 1972. The military missions of the Air Force's smaller and more specialized Manned Orbiting Laboratory (MOL) are expected to take place about the same time.

There are no approved plans and no provision in the FY 1970 Budget for continued U.S. development or utilization of manned space flight beyond the Apollo moon flights, the single set of Saturn I Workshop and Apollo telescope missions, and the Air Force MOL program as currently planned. For the future of manned space flight beyond 1972 the present FY 1970 NASA Budget provides only small sums limited to studies of advanced manned lunar exploration and earth orbital space stations.

3. **USSR Prospects** — Recent USSR manned space flight activities substantiate previous indications that they are continuing strong programs pointed both at manned operations to the moon and at space station operations in earth orbit. Beyond this, they talk openly of future manned trips to the planets. While we now expect to land American astronauts on the moon before the Russians get there, the prospects are that during the period of our lunar flights in 1969-1970 the Soviets will, in addition to their manned lunar program, follow up their Soyuz 4-5 [3] success by pushing toward a dominant position in large-scale long-duration space station operations in earth orbit. They will have the required heavy-lift launch capability. A multi-man, multi-purpose USSR space station operating in orbit before the U.S. could match it would give the USSR a strong advantage in space research and operations. Their moving clearly ahead of the U.S. in this field would have a continuing impact on the rest of the world, particularly if the U.S. program did not include a strong program in the earth orbital space station area.

4. **Opportunity for Leadership** — The fact that the previous Administration deferred to you the setting of the nation's goals in manned space flight creates a problem, but it also gives you a unique opportunity for leadership that will clearly identify your Administration with the establishment of the nation's major goals in manned space flight for the next decade. The impact and positive image of your leadership would be seriously downgraded in the eyes of the nation, the Congress, and the public, in my view, if the U.S. were once again placed in the position of reacting to Soviet initiatives in space. For this reason, I believe that you should consider the advisability of initiating a general directive to define the future goals of manned space flight in the next few months, prior to your final decisions on the plans that will be recommended to you on September 1 by the members of the Task Group you have established. For example, a major thrust this summer by the USSR in the earth orbital space station field is a distinct possibility that would take the edge off your announcement of a similar U.S. objective in the fall. For the reasons given below, I believe that the case that a space station should be a major future U.S. goal is now strong enough to justify at least a general statement on your part that this will be one of our goals, with the understanding (which could be reaffirmed in your statement) that the scope, pace, specific uses, and detailed plans of the space station will be determined on the basis of the planning studies you have requested.

5. **Basic National Policy** — There is, I believe, almost unanimous agreement on the part of responsible leaders in your Administration, the Congress, industry, the scientific community, and the general public that the U.S. must continue manned space flight activities. The concerns and criticisms that have been expressed do not question the continuation of a manned space flight program but relate principally to (a) the cost of the program, (b) the value of specific goals, and (c) questions of priorities, within the space program or between the space program and other scientific fields or other national needs. However, virtually no responsible and thoughtful person, to my knowledge, advocates or is prepared to accept the prospect of the United States abandoning manned space flight to the Soviets to develop and exploit as they see fit.

It is very important that all concerned with planning the [4] future of our space programs recognize this basic question of national policy. Acceptance of the fact that as a matter of policy the nation must and will continue in manned space flight leads to the following four points which should be considered in our planning:

    a. Studies of our alternatives in future space programs should focus on *pace, objectives, and content* of the manned space flight program, not on whether the U.S. should have a manned program. Alternatives which have the effect of not supporting a continuing effective U.S. manned flight program are not acceptable. A balanced total space program must include a significant continuing manned space flight program as one of its key elements.

    b. The U.S. must be prepared to pay the annual cost of an advancing, effective manned space flight program, high though it may seem. An important early objective, however, must be to reduce the cost of manned space flight, without sacrificing safety, reliability, or accomplishment.

    c. An advancing, effective manned space flight program cannot at this stage be limited to repetitive flights of missions already flown but must provide for the continuous evolutionary development of new capabilities, new missions, new experiments, and new applications.

    d. Decisions and selections of future programs must be made on a continuing timely basis several years *before* current objectives are achieved; otherwise the long lead-times inherent in the space program will force dangerous and expensive breaks in continuity that will undermine the success of the program.

    5. **Effects of Decisions in the Previous Administration** — The failure, during the past three years, to make timely decisions and to take necessary future-oriented actions has placed our manned space flight program in a serious and difficult position for the early 1970's. The production of both Saturn IB and V launch vehicles has been terminated. The Saturn V vehicles now on order must either be launched on schedules stretched out to clearly uneconomical rates, rates which may be below the minimum acceptable for reliability and safety, or flown with experimental payloads that repeat previous missions without significant advances. The failure to develop and approve future goals and objectives has forced the program into expensive and unproductive "holding" operations in some areas and made it more difficult to focus sharply on the planning and preliminary development efforts which must precede future programs. The watchwords of budgetary actions for the past several years have [5] been "delay," "stretch-out," "defer," and "hold the options open." The results are that for the next several years the nation will be getting a smaller return on its great investment in manned space flight capability, and that the long-deferred decisions on future goals must be taken now at an earlier time than your Administration would otherwise prefer.

    6. **Recommended Approach** — I believe that your Administration should now speak out boldly about the nation's future in space. Instead of continuing to stretch out and *minimize* the manned space flight program at the risk of reducing it beyond the point where it can be effective, your Administration should (a) point out the fact that the nation must continue to move forward in manned space flight, (b) while seeking every economy, accept the costs that this entails, and (c) plan, announce, and support a new ten-year space program—including a strong program of manned space flight—of which this nation and the world will be proud. Your Administration's decisions in the next few months will determine the nation's direction and progress in space for many years.

    7. **Study of Future Directions** — The process established in your memorandum of February 13, 1969, provides a useful framework for the development of specific goals and plans for the future of our space program. It will, among other things, enable NASA to communicate to the other agencies involved the thinking and planning that we have had underway for some time, and help assure NASA that its planning is properly coordinated with future aerospace planning in DOD, DOT, and other departments.

    However, unless adequate provision is made in the FY 1970 Budget in time for Congressional action in the FY 1970 authorization and appropriation cycle, the implementation of plans decided upon next fall as a result of the Task Group recommendations will have to await the FY 1971 cycle. This would mean the loss of an entire year and the foreclosure of your option to move ahead promptly with a strong manned space flight program if that should be your decision.

For this reason, I believe that it is essential that the FY 1970 Budget be amended now to include the manned space flight funds—specifically deleted by the previous Administration—required to support moving ahead in lunar exploration and space station development. I can appreciate that you may be reluctant to decide now to amend the FY 1970 Budget, thus appearing to prejudge the recommendations to be made in September, but postponement will foreclose what may well be your most attractive option and will perpetuate and aggravate an already unsatisfactory situation.

8. **Future Directions and Goals** — As stated above, two major directions have been identified for the manned space flight [6] program in the next decade. One is the further exploration of the moon, with possibly the eventual goal of establishing a U.S. Lunar base; the other is the further development of manned flight in earth orbit, with the goal of establishing a permanent manned space station in earth orbit that will be accessible and useful for a wide range of scientific, engineering, and application purposes. An important part of the space station goal is the development of a low cost logistics system for shuttling people and equipment to and from the space station.

These goals have in common the fact that they are not focused on a single dramatic achievement to be accomplished by a certain date, as was the case in the Apollo program. However, they can provide in the second decade of space, as Apollo did in the first, the focus for continuing advances in U.S. space capabilities and technology which will be available to support future defense and civilian requirements and to sustain our long-term national technical and economic vitality.

9. **Lunar Exploration** — In lunar exploration, our immediate problem is to assure that we have adequate scientific and operational equipment to allow us to follow up the first few lunar landings with an effective initial program of exploration that will permit sound judgments on the potential value of more advanced future missions and the eventual establishment of a lunar base. If, as we now expect, we have early success in achieving the first manned landing on the moon, we will have Apollo hardware—launch vehicles and spacecraft—for as many as nine additional lunar missions, but we lack scientific and improved operational equipment for more than three of these. In order to proceed with these missions at an economical rate, we are preparing a budget amendment that will permit prompt initiation of procurement of additional scientific and operational equipment early in FY 1970. Your approval of this budget amendment now will not constitute a commitment to lunar exploration beyond that possible with the Saturn-Apollo hardware procured for the Apollo program. Decisions on an advanced program of lunar exploration requiring major redesign of the Apollo Lunar Module, the development of shelters and vehicles for use on the lunar surface, and the question of the ultimate goal of establishing a lunar base can and should be made in your review of the plans and proposals to be submitted next September.

10. **Space Station** — With respect to future manned earth orbital flight, the immediate problem is to assure that sufficient funds are available in FY 1970 to permit detailed planning and design studies to proceed, and to develop critical long lead-time subsystems that will be required in any future manned space flight program. Funds for these purposes were specifically excluded from the present FY 1970 Budget, except for a small amount for studies, and we are therefore preparing an appropriate amendment to the FY 1970 Budget. This budget [7] amendment can be approved now without a commitment on your part to a permanent space station as a major national goal. However, as stated in paragraph 4 above, we believe that it is in the national interest for you to endorse this as a general U.S. objective at this time. One possibility would be for you to give NASA and the Task Group a specific instruction at the time you approve the budget amendment that their recommendations to you in September should include proposals on the optimum program for establishing and utilizing a permanent U.S. space station.

11. **Space Station Concept** — The space station discussed here should become a central point for many activities in space and would be designed to carry on these activities

in an effective and economic manner. It would be located in the most advantageous position to conduct investigations and operations in the space environment, many important aspects of which cannot be duplicated in an earth-based environment. The best place to study space is in space. We have in mind a system consisting of general and special-purpose modules with a low-cost logistic support system that will permit ready access and return by many users and their equipment and supplies. The space station would not be launched as a single unit, but would evolve over a period of years by adding to a core new modules as they are required and developed. One of the key objectives is to develop the system in cooperation with the Department of Defense so that it can be adaptable for future military research as well as for a variety of non-military scientific, engineering, and other application purposes.

There are many potential valuable uses of such a space station, and new ones will be found as experts in many fields become familiar with the possibilities and are able to visit and actually use it. However, we believe strongly that the justification for proceeding now with this major project as a national goal does not, and should not be made to depend on the specific contributions that can be foreseen today in particular scientific fields like astronomy or high energy physics, in particular economic applications, such as earth resources surveys, or in specific defense needs. Rather, the justification for the space station is that it is clearly the next major evolutionary step in man's experimentation, conquest, and use of space. The development of man's capability to live and work economically and effectively in space for long periods of time is an essential prerequisite not only for operations in earth orbit, but for long stay times on the moon and in the distant future, manned travel to the planets. It is for these reasons that I believe that space station development should become one of your Administration's principal working goals for the action over the next decade.

12. **Saturn V Production** — Under NASA's reduced 1969 operating plan and its present FY 1970 Budget, the production of [8] Saturn V, the nation's largest launch vehicle, has been discontinued. The long-term future of the manned space flight program, as outlined above, will clearly require additional Saturn V launch vehicles, and we are therefore proposing a FY 1970 Budget amendment which will permit production to be resumed, at a very low rate, before "start up" costs become excessive. This amendment will not preclude other future decisions on large launch vehicles that might be made next fall, but it will assure that funds are available to provide the launch vehicles that will be needed. It will also get the U.S. out of what I believe to be a current untenable position of having discontinued production of our largest space booster at a time when the Soviets are expected to unveil a booster of this class or larger. For the reasons stated in paragraph 4 above, I recommend that you now take the initiative and announce this decision before the Russians launch their first booster in this class, so that your announcement will not be viewed as a reaction to the Soviet development.

13. **Cost** — In planning the space program careful consideration must, of course, be given each year, and especially at the time new major programs are undertaken, to the future budget levels required. Our national budget system wisely and necessarily provides for a review at least annually of both on-going and new programs, but long-term enterprises like major space programs require a policy commitment to follow through with the resources required over a period of many years. For these reasons, it is important that your Administration be prepared to accept the total budget levels required by the programs you determine to be in the national interest. NASA on its part has the obligation continually to search out the least costly ways of carrying out the approved programs and to make every effort to use the possibilities of new technology to reduce future costs. But most important of all, neither NASA nor the Administration should, in the name of economy, underestimate the resources that can realistically be expected to be required. We must meet our commitments.

Our present projections indicate that a balanced total NASA program that includes the recommended strong manned space flight program can be carried out with annual budgets over the next five years which will not rise above the $4.5 to $5.5 billion range.

More precise projections will depend on the nature of the future lunar exploration and space station programs decided upon and on future decisions in areas other than manned space flight. By the time we submit the planning proposals to you in September we will be able to state with considerable confidence the projected future estimated costs of alternative total programs.

A total annual program level of $4.5 - $5.5 billion compares to program and expenditure levels in the $5.0 - $6.0 billion range reached in the 1964-1967 period, which in the past two years has been reduced to $3.9 billion in our FY 1969 operating [9] plan and the present FY 1970 Budget. As we have informed the Director of the Budget, the FY 1970 NASA Budget amendments we are proposing in manned space flight amount to about $200 million and would bring our total 1970 Budget (including authority carried forward from FY 1969) to slightly under $4.1 billion. Even with this proposed amendment, however, NASA's outlays (expenditures) in FY 1970 will still decline $200 million from the $4.25 estimated for FY 1969.

This memorandum has given you my recommendation on the position your Administration should take with respect to the critical and urgent situation in manned space flight; other NASA problems and opportunities can be treated appropriately in the Task Group framework for your consideration in September. For the reasons stated above, and with the possibility of an initial lunar landing in July, I believe you should not defer initial consideration of the manned space flight problem. I therefore specifically recommend that you ask the members of the Task Group established in your memorandum of February 13, 1969, to meet within the next month and to consider as their first order of business the matters identified in this memorandum as requiring your early decision. They should then present their recommendations to you by the end of March. In anticipation of such a meeting, NASA will prepare and make available to the other members of the Task Group (a) detailed materials on the alternatives available, and (b) suggestions on how the recommended early decisions can be related to an effective process for developing overall space plans and alternatives for your consideration in September. I hope that this proposal will meet with your approval, and would, of course, be happy to discuss this matter further with you at your convenience.

T.O. Paine
Acting Administrator

Document III-24

Document title: Robert C. Seamans, Jr., Secretary of the Air Force, to Honorable Spiro T. Agnew, Vice President, August 4, 1969.

Source: NASA Historical Reference Collection, History Office, NASA Headquarters, Washington, D.C.

Robert Seamans had been NASA Associate Administrator and then Deputy Administrator during most of the 1960s, and returned to Washington to become Richard Nixon's Secretary of the Air Force. Given his background and the central role of the Air Force in military space, Secretary of Defense Melvin Laird asked Seamans to serve in his stead as the Department of Defense representative on the Space Task Group.

In the days following the July 20 landing of the Apollo 11 mission on the Moon, NASA decided to propose to the Space Task Group an ambitious program for the future, oriented to early human missions to Mars. Vice President Agnew supported such an initiative, and NASA scheduled an elaborate presentation of its proposals for August 4. Troubled by the direction that the Space Task Group deliberations were taking, Seamans came to the August 4 meeting with this letter, expressing a much more measured outlook for the next steps in space.

[1]
Dear Mr. Vice President:

The Department of Defense has carried out a comprehensive study of the various opportunities for using space technology to enhance national security. Options for increased space activity have been carefully reviewed by the Services, the Joint Chiefs, and the offices of the Secretary of Defense, and are the basis for a report that is being transmitted to you by Secretary Laird. As a member of your Space Task Group, I am writing this letter to give you certain of my own personal views.

Rocketry and advanced electronics have permitted us to accomplish unique missions in this decade. The landing of the Apollo 11 astronauts on the moon and their safe return to earth is the crowning achievement. However, NASA and DoD have accomplished many other highly significant missions that are important for scientific, technical and operational reasons. As a result of unmanned and manned space flight we know a great deal more about the sun, the moon, the earth, and our sister planets. We are developing a better understanding of meteorology, and are using satellites for communications, navigation, weather forecasting, mapping and surveillance. With this as background, let me outline a space program that I believe is relevant to our national needs. This program can provide focus in the next decade similar to that of Apollo in this decade, but with several rather than a single objective.

1. **Direct Service to Mankind**

We should capitalize on NASA's great scientific and technical capability to the maximum extent possible. By this I mean that NASA should wherever possible carry out work of direct relevance to man here on earth. ESSA of the Department of Commerce needs assistance to understand and predict the weather more accurately for longer periods of time. The Department of Commerce, Interior, and Agriculture need support that can be supplied by satellites if they are to carry out their responsibilities in such fields as oceanography, hydrology, [2] agriculture, ecology, etc. However, I am not only thinking of further satellite developments, but also the use of NASA's capability wherever pertinent to current national problems.

NASA should put increased emphasis in aeronautics. We, in the Department of Defense, have need for greater effort by NASA to support us in the development of military aircraft. The Department of Transportation needs major support if they are to implement a new air traffic control system.

The extent to which NASA can support HUD and HEW has not been determined, but it should be noted that advances in space are dependent upon extensive data processing. Data processing is required in cities for a wide variety of purposes, including traffic control, crime detection, communications and administration. Space exploration also requires in-depth investigation of waste management, fire prevention, materials development, construction of highly reliable equipment, all of vital importance to municipalities.

The medical and biological investigations of man in space have led to improved biological instrumentation, and a better understanding of physiology.

The applications of the NASA program are far reaching and considerably more effort should be expended to make the results available for the benefit of mankind. To accomplish this objective, program priorities will have to be revised and organizational changes will have to be made both internal to NASA and between NASA and other agencies.

2. **National Security**

As stressed in the Department of Defense report, space is an environment that provides many opportunities to improve and support military operations. These opportunities include improvements in communications, weather forecasting, navigation, surveillance, mapping and many others areas, all of which are discussed at length in the DoD report. However, from the standpoint of the Department of Defense, space exploration and the development of space technology are not ends in themselves, but rather provide means for accomplishing functions in support [3] of existing forces. *Each military space mission must be approved on a case-by-case basis and weighed carefully against other means for doing the same job.* In the DoD report, a program is defined that maintains the annual outlay at about its present level. However, options exist that would require an increase in annual outlay by about 50%. In any event major effort is required to improve DoD space capabilities by development of advanced systems with greater sensitivity, invulnerability, and longer on-orbit life time.

3. Extended Lunar Exploration Using Apollo

The objective of Apollo was to provide a manned transportation system from earth to the moon and return. From Apollo 11 we will derive significant scientific information, particularly by analysis of the lunar samples. Now that the lunar transportation system exists, we should use it for a continuing series of missions with the principal objective being to derive maximum scientific data from the moon. This will entail landing on different areas of the moon in order to bring back different materials and in order to implant a wide variety of instruments. To do this effectively, some additional mobility will be required for the astronauts on the lunar surface. *In a continuing manned lunar landing program it is important to proceed on a careful step-by-step basis reviewing scientific information from one flight before going to the next and using unmanned spacecraft where appropriate.*

4. Applications of Apollo to Earth Orbital Missions

The Apollo hardware can also be used for further exploration in earth orbit. The objectives here include the involvement of man for extended periods in earth orbit both to better understand man's performance in extended flight, and also to use man to make a wide variety of measurements both of the earth, of the immediate space environment, and of the sun and the stars. *The present Apollo Applications Program including relatively few missions should be expanded to include longer duration flights and wider variety of orbits.*

5. Space Transportation

The extent to which space is used either for exploration or national security depends upon the cost per pound in [4] orbit. Today's flights are expensive because most of the hardware is lost and even that which is returned cannot be readily and inexpensively repaired for additional missions. *I recommend that we embark on a program to study by experimental means including orbital tests the possibility of a Space Transportation System that would permit the cost per pound in orbit to be reduced by a substantial factor (ten times or more).* Although preliminary studies have been conducted by both the Department of Defense and NASA on new types of space transportation, it is not yet clear that we have the technology to make such a major improvement. Consequently, I believe we should not put a rigid time constraint on this objective, but rather embark on a flexible program where various alternatives are investigated. Then at a later date, if the decision is made to proceed with an operational system, it can be made with technical, funding, and schedule confidence.

6. Manned Space Station

Multi-manned space stations have been studied and evaluated for many years. The specific objectives for a space station are still not clear even though a large number of interesting possibilities have been suggested. I believe that ultimately a space station will be needed where man can live and work for long periods of time (a year or more), and where no special astronaut type training is required prior to a mission. This will permit scientific personnel to concentrate entirely on their specialties and to carry out projects where a space environment is required.

Even though the development of a large manned space station appears to be a logical step leading to further use and understanding of the space environment, I do not believe we should commit ourselves to the development of such a space station at this time. I believe that we should wait until we have had further experience with the Apollo Applications Program and a Space Transportation System before we embark on this mission. Knowledge derived from Apollo

Applications will give us a much greater understanding of the role that man can perform in space and only after a thorough investigation of a Space Transportation System can we predict with reasonable accuracy our ability to resupply such a station economically and to ferry astronauts, scientists, and engineers back and forth from earth to the space platform.

[5] 7. **Planetary Missions**

We are rapidly acquiring knowledge of Mars and Venus, the two closest planets, using unmanned spacecraft. *The unmanned planetary program should be expanded to include more thorough investigation of Mars and Venus, as well as exploration of the more distant planets.*

I don't believe we should commit this Nation to a manned planetary mission, at least until the feasibility and need are more firmly established. Experience must be gained in an orbiting space station before manned planetary missions can be planned. At this time we do not know the effect of placing a man in space for the one to two years required for a planetary trip. There may be serious physiological and psychological effects that must be understood and dealt with before such a trip can be considered a possibility.

A decision to travel to Mars, the only accessible planet, would require new launch vehicle stages and spacecraft modules and a greatly increased annual outlay that would compete with the resources needed to provide immediate benefits from NASA's capability.

In summary, the Department of Defense must use space to enhance its capability as appropriate to its various responsibilities. The NASA program, on the other hand, should continue to explore space with both unmanned and manned spacecraft, but with the solution of problems directly affecting man here on earth as its immediate objective. In the past, there has been much discussion as to whether NASA had the right balance between unmanned and manned flights and between science and technology. I believe that in the future the balance of NASA's activities should be shifted not from considerations of manned vs. unmanned flights nor from considerations of science vs. technology, but from considerations of how this highly trained, highly motivated agency can make this country and the world a more hopeful, and healthier place. Let us take the initiative and use the good will, the momentum and the skills demonstrated in Apollo to help solve many of our problems at home and abroad. But let us not give up exploration, rather let us also continue our exploration while validating its benefit to all mankind.

Respectfully,
Robert C. Seamans, Jr.

Document III-25

Document title: Space Task Group, *The Post-Apollo Space Program: Directions for the Future,* **September 1969.**

Source: NASA Historical Reference Collection, History Office, NASA Headquarters, Washington, D.C.

The report of the Space Task Group was presented to President Nixon on September 15, 1969. The report largely endorsed the hardware elements of the program that NASA had developed in the weeks following Apollo 11, including the development of a space station and a reusable transportation system. The Space Task Group as a whole did not recommend any particular option, though both the vice president and NASA Administrator Paine within a few days urged the president to approve Option II, which called for station and shuttle development by 1977 and an initial Mars expedition by 1986. In the days immediately preceding delivery of the report to the president, senior White House staff had demanded that the language of the report be changed so that the president not be put in the position of having only ambitious options from which to choose; thus the tone of the "Conclusions and Recommendations" section of the report is not consistent with the content of most of the rest of the document.

[i] **Conclusions and Recommendations**

The Space Task Group in its study of future directions in space, with recognition of the many achievements culminating in the successful flight of Apollo 11, views these achievements as only a beginning to the long-term exploration and use of space by man. We see a major role for this Nation in proceeding from the initial opening of this frontier to its exploitation for the benefit of mankind, and ultimately to the opening of new regions of space to access by man.

[ii] We have found increasing interest in the exploitation of our demonstrated space expertise and technology for the direct benefit of mankind in such areas as earth resources, communications, navigation, national security, science and technology, and international participation. We have concluded that the space program for the future must include increased emphasis upon space applications.

We have also found strong and wide-spread personal identification with the manned flight program, and with the outstanding men who have participated as astronauts in this program. We have concluded that a forward-looking space program for the future for this Nation should include continuation of manned space flight activity. Space will continue to provide new challenges to satisfy the innate desire of man to explore the limits of his reach.

We have surveyed the important national resource of skilled program managers, scientists, engineers, and workmen who have contributed so much to the success the space program has enjoyed. This resource together with industrial capabilities, government, and private facilities and growing expertise in space operations are the foundation upon which we can build.

We have found that this broad foundation has provided us with a wide variety of new and challenging opportunities from which to select our future directions. We have concluded that the Nation should seize these new opportunities, particularly to advance science and engineering, international relations, and enhance the prospects for peace.

We have found questions about national priorities, about the expense of manned flight operations, about new goals in space which could be interpreted as a "crash program." Principal concern in this area relates to decisions about a manned mission to Mars. We conclude that NASA has the demonstrated organizational competence and technology base, by virtue of the Apollo success and other achievements, to carry out a successful program to land man on Mars within 15 years. There are a number of precursor activities necessary before such a mission can be attempted. These activities can proceed without developments specific to a Manned Mars Mission—but for optimum benefit should be carried out with the Mars mission in mind. We conclude that a manned Mars mission should be accepted as a long-range goal for the space program. Acceptance of this goal would not give the manned Mars mission overriding priority relative to other program objectives, since options for decision on its specific date are inherent in a balanced program. Continuity of other unmanned exploration and applications efforts during periods of unusual budget constraints should be supported in all future plans.

We believe the Nation's future space program possesses potential for the following significant returns:

- new operational space applications to improve the quality of life on Earth
- non-provocative enhancement of our national security
- scientific and technological returns from space investments of the past decade and expansion of our understanding of the universe

- low-cost, flexible, long-lived, highly reliable, operational space systems with a high degree of commonality and reusability
  - international involvement and participation on a broad basis

[iii] Therefore, *we recommend*—

That this Nation accept the basic goal of a balanced manned and unmanned space program conducted for the benefit of all mankind.

To achieve this goal, the United States should emphasize the following *program objectives:*

- increase utilization of space capabilities for services to man, through an expanded space applications program
- enhance the defense posture of the United States and thereby support the broader objective of peace and security for the world through a program which exploits space techniques for accomplishment of military missions
- increase man's knowledge of the universe by conduct of a continuing strong program of lunar and planetary exploration, astronomy, physics, the earth and life sciences
- develop new systems and technology for space operations with emphasis upon the critical factors of: (1) commonality, (2) reusability, and (3) economy, through a program directed initially toward development of a new space transportation capability and space station modules which utilize this new capability
- promote a sense of world community through a program which provides opportunity for broad international participation and cooperation

As a focus for the development of new capability, *we recommend* the United States accept the long-range option or goal of manned planetary exploration with a manned Mars mission before the end of this century as the first target.

[iv] In proceeding towards this goal, three phases of activities can be identified:

- *initially*, activity should concentrate upon the dual theme of exploitation of existing capability and development of new capability, maintaining program balance within available resources.
- *second*, an operational phase in which new capability and new systems would be utilized in earth-moon space with groups of men living and working in this environment for extended periods of time. Continued exploitation of science and applications would be emphasized, making greater use of man or man-attendance as a result of anticipated lowered costs for these operations.
- *finally*, manned exploration missions out of earth-moon space, building upon the experience of the earlier two phases.

Schedule and budgetary implications associated with these three phases are subject to Presidential choice and decision at this time with detailed program elements to be determined in a normal annual budget and program review process. Should it be decided to develop concurrently the space transportation system and the modular space station, a rise of annual expenditures to approximately $6 billion in 1976 is required. A lower level of approximately $4-5 billion could be met if the space station and the transportation system were developed in series rather than in parallel.

For the Department of Defense, the space activities should be subject to continuing review relative to the Nation's needs for national security. Such review and decision processes are well established. However, the planned expansion of the DoD space technology effort and its documented interest in the Space Transportation System demands continued authoritative coordination through the Aeronautics and Astronautics Coordinating Board to assure that the national interests are met.

[v] The Space Task Group has had the opportunity to review the national space program at a particularly significant point in its evolution. We believe that the new directions we have identified can be both exciting and rewarding for this Nation. The environment in which the space program is viewed is a vibrant, changing one and the new opportunities that tomorrow will bring cannot be predicted with certainty. Our planning for the future should recognize this rapidly changing nature of opportunities in space.

We recommend that the National Aeronautics and Space Council be utilized as a mechanism for continuing reassessment of the character and pace of the space program.

[1] # The Post-Apollo Space Program: Directions for the Future

I. INTRODUCTION

With the successful flight of Apollo 11, man took his first step on a heavenly body beyond his own planet. As we look into the distant future it seems clear that this is a milestone—a beginning—and not an end to the exploration and use of space.

Success of the Apollo program has been the capstone to a series of significant accomplishments for the United States in space in a broad spectrum of manned and unmanned exploration missions and in the application of space techniques for the benefit of man. In the short span of twelve years man has suddenly opened an entirely new dimension for his activity.

In addition, the national space program has made significant contributions to our national security, has been a political instrument of international value, has produced new science and technology, and has given us not only a national pride of accomplishment, but has offered a challenge and example for other national endeavors.

The Nation now has the demonstrated capability to move on to new goals and new achievements in space in all of the areas pioneered during the decade of the sixties. In each area of space exploration what seemed impossible yesterday has become today's accomplishment. Our horizons and our competence have expanded to the point that we can consider unmanned missions to any region in our solar system; manned bases in earth orbit, lunar orbit or on the surface of the Moon; manned missions to Mars; space transportation systems that carry their payloads into orbit and then return and land as a conventional jet aircraft; reusable nuclear-powered rockets for space operations; remotely controlled roving science vehicles on the Moon or on Mars; and application of space capability to a variety of services of benefit to man here on earth.

Our opportunities are great and we have a broad spectrum of choices available to us. It remains only to chart the course and to set the pace of progress in this new dimension for man.

The Space Task Group, established under the chairmanship and direction of the Vice President (Appendices A and B), has examined the spectrum of new opportunities available in space, values and benefits from space activities, casts and resource implications of future options, and international aspects of the space program. A great wealth of data has been made available to the Task Group, including reports from the National Aeronautics and Space Administration and the Department of Defense reflecting very extensive planning and review activities, a detailed report from the President's Science Advisory Committee, views from [2] members of Congress, the National Academy of Sciences Space Science Board, and the American Institute of Aeronautics and Astronautics. In addition, a series of individual reports from a special group of distinguished citizens who were asked for their personal recommendations on the future course of the space program were of considerable value to the Task Group. This broad range of material was considered and evaluated as part of the Task Group deliberations. This report presents in summary form the views of the Space Task Group on the Nation's future directions in space.

[3]   **II. BACKGROUND**

Twelve years ago, when the first artificial Earth satellite was placed into orbit, most of the world's population was surprised and stunned by an achievement so new and foreign to human experience. Today people of all nations are familiar with satellites, orbits, the concept of zero 'g', manned operations in space, and a host of other aspects characteristic of this new age—the age of space exploration.

The United States has carried out a diversified program during these early years in space, requiring innovation in many fields of science, technology, and the human and social sciences. The Nation's effort has been interdisciplinary, drawing successfully upon a synergistic combination of human knowledge, management experience, and production know-how to bring this Nation to a position of leadership in space.

Space activities have become a part of our national agenda.

We now have the benefit of twelve years of space activity and our leadership position as background for our examination of future directions in space.

National Priorities

By its very nature, the exploration and exploitation of space is a costly undertaking and must compete for funds with other national or individual enterprises. Now that the national goal of manned lunar landing has been achieved, discussion of future space goals has produced increasing pressures for reexamination of, and possible changes in, our national priorities.

Many believe that funds spent for the space program contribute less to our national economic growth and social well-being than funds allocated for other programs such as health, education, urban affairs, or revenue sharing. Others believe that funds spent for space exploration will ultimately return great economic and social benefits not now foreseen. These divergent views will persist and must be recognized in making decisions on future space activities.

The Space Task Group has not attempted to reconcile these differences. Neither have we attempted to classify the space program in a hierarchy of national priorities. The Space Task Group has identified major technical and scientific challenges in space in the belief that returns will accrue to the society that takes up those challenges.

Values and Benefits

The magnitude of predicted great economic and social benefits from space activities cannot be precisely determined. Nevertheless, there should be a recognition that significant direct benefits have been realized as a result of space investments, particularly from applications programs, as a long-term result of space science activities, DOD space activities, and advancing technology. These direct benefits are only part of the total set of benefits from the space program, many of which are very difficult to quantify and therefore are not often given adequate consideration when costs and benefits from space activities are weighed or assessed in relation to other national programs.

[4] Benefits accrue in each of the following areas:

economic—directly through applications of space systems to services for man, and indirectly through potential for increased productivity resulting from advancing technology; improvements in reliability, quality control techniques, application of solid state electronics, and computer technology resulting from demands of space systems; advances in understanding and use of exotic new materials and devices with broad applicability; refinement of systems engineering and management techniques for extremely complex developments.

national security—directly through DOD space activities, and indirectly through enhancement of the national spirit and self-esteem; reinforcement of the image of the United States as a leader in advanced technology; strengthening of our international posture through demonstration that a free and democratic society can achieve a challenging, technologically sophisticated, long-term objective; maintenance of a broad base of highly skilled aerospace workers applicable to defense needs; and advancement of technology that may have relevance to defense use.

science—directly through support for ground and space research programs, indirectly through ability to open to observation new portions of the electromagnetic spectrum; opportunity to search for life on other planets, to make measurements in situ at the planets or in other regions of space, and to utilize the unique environment of space (high vacuum, zero "g") for experimental programs in the life sciences, physical sciences and engineering.

exploration—the opening of new opportunities to investigate and acquire knowledge about man's environment—which now has expanded to include not only the Earth, but potentially the entire solar system.

social—providing educational services through enhanced communications which improve treatment of social problems.

international relations—providing opportunities for cooperation; the identification of foreign interests with U.S. space objectives and programs, and their results.

What is the value to be placed upon these benefits, and how should the space program be constituted to provide the greatest return in each of these areas for a selected level of public investment?

The answers to these questions cannot be stated in absolute terms—there is no dollar value associated with national self-esteem or with many of the other benefits listed above, and there is no fixed program of missions without which these benefits will not accrue. As with many programs, there is, however, a lower limit of activity below which the viability of the program is threatened and a reasonable upper limit which is imposed by technological capability and rate of growth of the program.

These limits are a key consideration in the options discussed later in this report.

[5] **National Resource**

In the eleven years since its creation, NASA has provided the Nation with a broad capability for a wide variety of space activity, and has successfully completed a series of challenging tasks culminating in the first manned lunar landing. These accomplishments have involved rapid increases to peak annual expenditures of almost $6 billion and a peak civil service and contractor work force of 420,000 people. Expenditures for NASA have subsequently dropped over the last three years from this peak to the present level of about $4 billion and supporting manpower has dropped to about 190,000 people.

In addition to NASA space activity, the DOD has developed and operated space systems satisfying unique military requirements. Spending for military space grew rapidly in the early sixties and has increased gradually during the past few years to approximately $2 billion per year.

The Nation's space program has fostered the growth of a valuable reservoir of highly trained, competent engineers, managers, skilled workmen and scientists within government, industry and universities. The climactic achievement of Apollo 11 is tribute to their capability.

This resource together with supporting facilities, technology and organizational entities capable of complex management tasks grew and matured during the 1960's largely in response to the stimulation of Apollo, and if it is to be maintained, needs a new focus for its future.

Manned Space Flight

There has been universal personal identification with the astronauts and a high degree of interest in manned space activities which reached a peak both nationally and internationally with Apollo. The manned flight program permits vicarious participation by the man-in-the-street in exciting, challenging, and dangerous activity. Sustained high interest, judged in the light of current experience, however, is related to availability of new tasks and new mission activity—new challenges for man in space. The presence of man in space, in addition to its effect upon public interest in space activity, can also contribute to mission success by enabling man to exercise his unique capabilities, and thereby enhance mission reliability, flexibility, ability to react to unpredicted conditions, and potential for exploration.

While accomplishments related to man in space have prompted the greatest acclaim for our Nation's space activities, there has been increasing public reaction over the large investments required to conduct the manned flight program. Scientists have been particularly vocal about these high costs and problems encountered in performing science experiments as part of Apollo, a highly engineering oriented program in its early phases.

Much of the negative reaction to manned space flight, therefore, will diminish if costs for placing and maintaining man in space are reduced and opportunities for challenging new missions with greater emphasis upon science return are provided.

[6] ### Science and Applications

Although high public interest has resided with manned space flight, the Nation has also enjoyed a successful and highly productive science and applications program.

The list of more achievements in space science is great, ranging from our first exploratory orbital flights resulting in discoveries about the Earth and its environment to the most recent Mariner missions to the vicinity of Mars producing new data about our neighbor planet.

Both optical and radio astronomy have been stimulated by the opening of new regions of the electromagnetic spectrum and new fields of interest have been uncovered—notably in the high energy X-ray and gamma-ray regions. Astronomy is advancing rapidly at present, partly with the aid of observations from space, and a deeper understanding of the nature and structure of the universe is emerging. In planetary exploration, we have a unique opportunity to pursue a number of the major questions man has asked about his relation to the universe. What is the history of the formation and evolution of the solar system? Are there clues to the origin of life? Does life exist elsewhere in the solar system?

In the life sciences, questions about the effect of zero "g" upon living systems, demands of long-duration space flight upon our understanding of man and his interaction or response to his environment, both physiologically and psychologically, promise new insights into the understanding of complex living systems.

These are only a few of the disciplines that have profited from the program of research in space. Space science is not divorced from science on the ground, but is rather an extension of science which builds and depends vitally upon a strong ground-based foundation.

Building upon the basic science on the ground and in space, and upon the growing capability in the design, construction and launch of satellites, the United States pioneered in the development of space applications—notably communications, meteorology and navigation. Operational systems have been placed into service in each of these areas, and the potential for the future appears bright—not only in these areas but also in new fields such as earth resource surveying and oceanography.

International Aspects

Achievement of the Apollo goal resulted in a new feeling of "oneness" among men everywhere. It inspired a common sense of victory that can provide the basis for new initiatives for international cooperation.

The U.S. and the USSR have widely been portrayed as in a "race to the Moon" or as vying over leadership in space. In a sense, this has been on accurate reflection of one of the several strong motivations for U.S. space program decisions over the previous decade.

[7] Now with the successes of Apollo, of the Mariner 6 and 7 Mars flybys, of communications and meteorology applications, the U.S. is at the peak of its prestige and accomplishments in space. For the short term, the race with the Soviets has been won. In reaching our present position, one of the great strengths of the U.S. space program has been its open nature, and the broad front of solid achievement in science and applications that has accompanied the highly successful manned flight program.

The attitude of the American people has gradually been changing and public frustration over Soviet accomplishments in space, an important force in support of the Nation's acceptance of the lunar landing in 1961, is not now present. Today, new Soviet achievements are not likely to have the effect of those in the post. Nevertheless, the Soviets have continued development of capability for future achievements and dramatic missions of high political impact are possible. There is no sign of retrenchment or withdrawal by the Soviets from the public arena of space activity despite launch vehicle and spacecraft failures and the preemptive effect of Apollo 11.

The landing on the Moon has captured the imagination of the world. It is now abundantly clear to the man in the street, as well as to the political leaders of the world, that mankind now has at his service a new technological capability, an important characteristic of which is that its applicability transcends national boundaries. If we retain the identification of the world with our space program, we have on opportunity for significant political effects on nations and peoples and on their relationships to each other, which in the long run may be quite profound.

[9] ## III. GOALS AND OBJECTIVES

Goals

An important aspect in both popular acceptance of the space program and in the spirit, dedication and performance of those who are directly involved in space activity is the conviction that such activity is worthwhile and contributes to the quality of life on Earth.

Public support for the space program can be related to understanding of the values derived from space activity and to understanding and acceptance of long-term goals and objectives which establish the framework for the program.

In the National Aeronautics and Space Act of 1958, the Congress declared "...it is the policy of the United States that activities in space should be devoted to peaceful purposes for the benefit of all mankind." This policy statement, which served effectively as a guide to the first decade in space, must now be translated into clearly enunciated new long-range goals and program objectives for the post-Apollo space program.

We view the challenge of setting new goals, of providing a focus for our future space activities, of expanding the limits of man's reach and thereby demonstrating America's leadership in scientific and technological undertakings while maintaining the confidence of the people in the strength and purpose of our Nation, as the key to continued space leadership by the United States.

Facing this challenge, some would urge that our efforts should be restricted to exploitation of existing capability, pointing out, quite correctly, that exciting and challenging missions remain to be accomplished which can utilize the existing base. But such a course would risk loss of the foundation for future achievements—a foundation which depends largely on providing a new capability which challenges our technology.

One of the values of the lunar landing goal was that it carried a definite time for its accomplishment, which stressed our technology and served as basis for planning and for budget support. It was a national commitment, a demonstration of the will and determination of the American people and of our technological competence at a time when these attributes were being questioned by many.

The need for an expression of our strength and determination as a Nation has changed considerably since that time. Today the need is for guidance—for direction—to set before the people a vision of where we are going.

[10] Such a vision for the future should have a number of important qualities:

- it should have substantive values that are easily characterized and understood.
- it should have a long-term goal, a beacon, an aim for our activities to act as a guide to both short-term and longer range decisions.
- it should be sufficiently long-range to ensure that adequate opportunity exists for solid progress in a step-by-step fashion towards that long-term goal yet sufficiently within reach that each step draws measurably closer to that goal.
- it should be challenging both for man's spirit of adventure and of exploration and for man's technological capability.
- it should foster the simultaneous utilization of space capabilities for the welfare, security, and enlightenment of all people.

The Space Task Group has concluded that a balanced space program that exploits the great potential for automated and remotely-controlled spacecraft and at the same time maintains a vigorous manned flight program, can provide such a vision.

This balanced program would be based upon a framework in which the United States would:

- Accept, for the long term, the challenge of exploring the solar system, using both manned and unmanned expeditions.
- Develop on integrated and efficient space capability that will make Earth-Moon space easily and economically accessible for manned and unmanned systems.
- Maintain a steady return on space investments in applications, science, and technology.
- Use our space capability not only to extend the benefits of space to the rest of the world, but also to increase direct participation by the world community in both manned and unmanned exploration and use of space.

The balanced program for the future envisioned by the Task Group would possess several important characteristics:

flexibility. The ability to see clearly the opportunities that lie ahead in this new field is limited at best. Some opportunities will fade as we approach them while others, not even discernible at this time, will blossom to the first magnitude. This program will permit the course and time scale to be flexible, to adjust to variations in funding, to shifting national and international conditions, while preserving a guidepost for the future.

challenge. The space program has flourished under a set of goals that has demanded the highest standards of performance, and an incentive for excellence that has become characteristic of our space efforts. A balanced program of both challenging near-term objectives and long-range goals will enhance and preserve these attributes in the future. [11] **opportunity**. The Nation has in being significant capability for space activity. Abundant opportunities exist for further exploitation of this capability. A balanced program will permit adequate attention to applications and science while also creating new opportunities through development of new capability

In its deliberations, the Space Task Group considered a number of challenging new mission goals which were judged both technically feasible and achievable within a reasonable time, including establishment of a lunar orbit or surface base, a large 50-100 man earth-orbiting space base, and manned exploration of the planets. The Space Task Group believes that manned exploration of the planets is the most challenging and most comprehensive of the many long-range goals available to the Nation at this time, with manned exploration of Mars as the next step toward this goal. Manned planetary exploration would be a goal, not an immediate program commitment; it would constitute an understanding that within the context of a balanced space program, we will plan and move forward as a Nation towards the objective of a manned Mars landing before the end of this century. Mars is chosen because it is most earth-like, is in fairly close proximity to the Earth, and has the highest probability of supporting extraterrestrial life of all of the other planets in the solar system.

What are the implications of accepting this long-range goal or option on the character of the space program in the immediate future?

In a technical sense, the selection of manned exploration of the planets as a long-term option for the United States space program would act to focus a wide range of precursor activities and would be reflected in many decisions, large and small, where potential future applicability to long-lived manned planetary systems design will have relevance. In a broader sense such a selection would tend to reinforce and reaffirm the basic commitment to a long-term continued leadership position by the United States in space.

The Space Task Group sees acceptance of the long-term goal of manned planetary exploration as an important part of the future agenda for this Nation in space. The time for decisions on the development of equipment peculiar to manned mission to Mars will depend upon the level of support, in a budget sense, that is committed to the space program.

NASA has outlined plans that would include a manned Mars mission in 1981 with the development decision on a Mars Excursion Module in FY 1974, if the Nation were to accept this commitment. Such a program would result in maximum stimulation of our technology and creation of new capability. There are many precursor activities that will be required before a manned Mars mission is attempted, such as detailed study of biomedical aspects, both physiological and psychological, of flights lasting 500-600 days, unmanned reconnaissance of the planets, creation of highly reliable life support systems, power supplies, and propulsion capability adequate for the rigors of such a voyage and reliable enough to support man. Decision to proceed with a 1981 mission would require early attention to these precursor activities.

While launch of a manned Mars exploration mission appears achievable as early as 1981, it can also be accomplished at any one of the roughly biennial launch opportunities following this date, provided essential precursor activities have been carried out.

[12] Thus, the understanding that we are ultimately going to explore the planets with man provides a shaping function for the post-Apollo space program. However, in a balanced program containing other goals and objectives, this focus should not assume over-riding priority and cause sacrifice of other important activity in times of severe budget constraints. Flexibility in program content and options for decision on the specific date for a manned Mars mission are inherent in this understanding.

The Space Task Group, in response to the President's request for a "Coordinated program and budget proposal," has therefore chosen this balanced program as that plan best calculated to meet the Nation's needs for direction of its future space activity. In reaching this conclusion we have considered international and domestic influences, weighed and placed in perspective science and engineering development, exploration and application of space, manned and unmanned approaches to space missions, and have appraised interagency influences. Discussion of the principal objectives which describe this balanced program follows.

Program Objectives

Elements of the balanced program recommended by the Space Task Group can be identified within the following set of program objectives which define major emphases for future space activity:

- Application of space technology to the direct benefit of mankind
- Operation of military space systems to enhance national defense
- Exploration of the solar system and beyond
- Development of new capabilities for operating in space
- International participation and cooperation

1. **Application of space technology to the direct benefit of mankind.**
Focus: To increase utilization of space capabilities for services to man. Programs directed toward the application of the Nation's space capabilities to a wide range of services, such as air and ocean traffic control, world-wide navigation systems, environmental monitoring and prediction (weather, pollution), earth resource survey (crops, water resources, geological structures, oceanography) and communications have great potential for improving the quality of life on this planet Earth. Significant direct economic and social benefits from such applications have been forecast. Major contributions to management of domestic problems and greater opportunities for international cooperation could result from an expanded space applications program.

2. **Operation of military space systems to enhance national defense**
Focus: Enhance the defense posture of the United States and thereby support the broader objective of peace and security for the world.
[13] The Department of Defense is presently using space capabilities in the support of communications, weather forecasting, navigation, surveillance and mapping, and for other functions. Such space activity has been not an end in itself, but a means for accomplishing functions in support of existing forces and missions. Military uses of space have proven effective and space systems are now contenders for specific applications and missions. Each military space mission should continue to be decided on a case-by-case basis in competition with ground, sea, and airborne systems and should reflect priority given to national defense with consideration of arms limitation agreements, and other U.S. policy reactions. Exploitation of the unique characteristics of space systems by the Department of Defense can provide increased confidence in the ability of this Nation to defend itself from any aggressor and assurance that space will be used for peaceful purposes by all nations.

3. **Exploration of the solar system and beyond.**
Focus: Increase man's knowledge of the universe.
Exploration of the solar system and observations beyond the solar system should be important continuing broad objectives of the Nation's space program. Many unanswered scientific questions remain about the planets, the interplanetary medium, the sun—both as a type of star and as a source of the earth's energy—and about a variety of celestial objects, such as pulsars, quasars, X-ray and gamma ray sources. Both ground-and space-based experiments and observational programs will contribute to the quest for answers to these questions. Space platforms provide several unique advantages—such as ability to observe across the range of wave lengths of the electromagnetic spectrum (rather than only through specific atmospheric "windows," which is the case from the ground); freedom from local environmental conditions; potential for continuous observations (no day-night cycle); ability

to approach, orbit and land on extraterrestrial bodies—and also disadvantages—high cost, inaccessibility for easy repair and servicing, and long lead times for experiment modification. For these reasons a careful balance between investments in space and ground experiments should be maintained.

The major elements of such a program should be:

• **Planetary Exploration**—Unmanned planetary exploration missions continuing throughout the decade, both for science returns and, in the case of Mars and Venus, as precursors to later manned missions. The program should include progressively more sophisticated missions to the near planets as well as multiple-planet flyby missions to the outer planets taking advantage of the favorable relative positions of the outer planets in the late 1970's. Early missions to the asteroid belt and to the vicinity of a comet should be planned.

• **Astronomy, Physics, the Earth and Life Sciences**—In each of these disciplines, extension of existing or planned unmanned programs promises continued high science return. There are additional significant opportunities for experiments in connection with manned Earth orbital programs which should be exploited. Work in astronomy, physics and the life sciences, as well as work in the earth sciences and remote sensing, will form an essential part of the foundation for future applications benefits and will contribute to the broadening horizons of man as he acquires knowledge not only of his own planet but also about the rest of the universe.

[14] • **Lunar Exploration**—Apollo-type manned missions to continue exploration of the Moon should proceed. The launch rate should permit maximum responsiveness to new discoveries while maintaining mission safety and efficient utilization of support personnel. Early upgrading of lunar exploration capability beyond the basic Apollo level including enhanced mobility capability, and lunar rovers, is important to safe and efficient realization of significant returns over the longer term. An orbiting lunar station, followed by a surface-base, building upon Earth orbital space station and space transportation system developments, could be deployed as early as the latter half of the decade. Extension of manned lunar activity beyond upgraded Apollo capability should include consideration of these options.

4. **Development of new capabilities for operating in space.**

Focus: Develop new systems for space operations with emphasis upon the critical factors of: commonality, (2) reusability, and (3) economy.

Exploration and exploitation of space is costly with our current generation of expendable launch vehicles and spacecraft systems. This is particularly true for the manned flight program. Recovery and launch costs will become an even more significant factor when multiple re-visit and resupply missions to an Earth orbiting space station are contemplated. Future developments should emphasize:

• Commonality—the use of a few major systems for a wide variety of missions.
• Reusability—the use of the same system over a long period for a number of missions.
• Economy—for example, the reduction in the number of "throw away" elements in any mission; the reduction in the number of new developments required; the development of new program principles that capitalize on such capabilities as man-tending of space facilities; and the commitment to simplification of space hardware.

An integrated set of major new elements which satisfy these criteria are:

a. **A space station module** that would be the basic element of future manned activities in Earth orbit, of continued manned exploration of the Moon, and of manned expeditions to the planets. The space station will be a permanent structure, operating continuously to support 6-12 occupants who could be replaced at regular intervals. Initially, the space station would be in a low altitude, inclined orbit; later stations would be established in polar and synchronous orbits. The same space station module would also provide a permanent manned station in lunar orbit from which expeditions could be sent to the surface.

By joining together space station modules, a space base could be created. Occupied by 50-100 men, this base would be a laboratory in space where a broad range of physical and biological experiments would be performed.

Finally, the space station module would be the prototype of a mission module for manned expeditions to the planets.

[15] Such an array of space station modules would be designed to utilize the space transportation system described below.

    b. **A space transportation system that will:**
- Provide a major improvement over the present way of doing business in terms of cost and operational capability.
- Carry passengers, supplies, rocket fuel, other spacecraft, equipment, or additional rocket stages to and from orbit on a routine aircraft-like basis.
- Be directed toward supporting a spectrum of both DoD and NASA missions.

Although the concept of such a space transportation capability is not new, advances in rocket engine technology, additional experience in design for reentry conditions, and improved guidance, navigation and automated check-out systems now permit initiation of an experimental effort for a Space Transportation System with technical, operational, and economic characteristics satisfying the needs of both NASA and DoD. An orderly, phased, step-by-step development program could then be implemented including as potential components:

- A reusable chemically fueled **shuttle** operating between the surface of the Earth and low-earth orbit in an airline-type mode.
- A chemically fueled reusable **space tug** or vehicle for moving men and equipment to different earth orbits. This same tug could also be used as a transfer vehicle between the lunar-orbit base and the lunar surface.
- A reusable **nuclear stage** for transporting men, spacecraft and supplies between Earth orbit and lunar orbit and between low Earth orbit and geosynchronous orbit and for other deep space activities. The NERVA nuclear engine development program, presently underway and included in all of the options discussed later, provides the basis for this stage and represents a major advance in propulsion capability.

    c. **Advanced Technology Development**—In addition to the major vehicle developments listed above, a continuing program of investigation and exploration of new technology that can serve as the foundation for next generation systems is an essential component of the DoD, NASA, and other agency programs. A broad and aggressive program to advance our capabilities to operate in space during the next decade and to set the stage for the decade to follow is needed.

We foresee future requirements for larger and more efficient power supplies utilizing a range of energy sources, particularly nuclear systems, for continuing propulsion system improvements—both in performance and reliability, for improved understanding of the complex interface between man and machine, for advances in technology and systems design that result in lower cost development of new spacecraft, and for achievement of new levels of reliability. In the advanced technology program, we should emphasize biomedical research, space power and propulsion technology, both nuclear and non-nuclear, remotely controlled teleoperators, data management, multi-spectral sensors, communication and navigation technology, and experimental evaluation and demonstration of new concepts.

[16] 5. **International participation and cooperation.**

Focus: To promote a sense of world community; to optimize international scientific, technical, and economic participation; to apply space technology to mankind's needs; and to share the benefit and cost of space research and exploration.

To these ends, our international interests will be served best by (1) projects which afford maximum opportunities for direct foreign participation, (2) projects which yield economic and social benefits for other countries as well as ourselves, and (3) activities in which further international agreement and coordination might usefully be employed.

The past decade has demonstrated that programs like Project Apollo are virtually unrivaled in their capacity to catch the world's imagination and interest, win extensive admiration and respect for American achievements, and generate a common human experience. The decade has demonstrated also that effective ways can be found to share the practical benefits of space with people everywhere, as in space meteorology and communications. Modest but significant levels of direct participation in space flight research and exploration have also been successfully achieved through cooperative projects. Future program plans must seek to continue and substantially extend this experience.

We should also devote special effort to meliorate, between the space powers and others, the increasing gap in technological capability and the gap in awareness and understanding of new opportunities and responsibilities evolving in the space age.

If international participation and cooperation are to be expanded in an important way, there will have to be (1) a substantial raising of sights, interest and investment in space activity by the other nations able to do so in order to establish a base for major contributions by them; and (2) creation of attractive international institutional arrangements to take full advantage of new technologies and new applications for peoples in developing as well as advanced countries.

The most dramatic form of foreign participation in our program will be the inclusion of foreign astronauts. This should be approached in the context of substantive foreign contributions to the programs involved.

The form of cooperation most sought after by advanced countries will be technical assistance to enable them to develop their own capabilities. We should move toward a liberalization of our policies affecting cooperation in space activities, should stand ready to provide launch services and share technology wherever possible, and should make arrangements to involve foreign experts in the detailed definition of future United States space programs and in the conceptual and design studies required to achieve them. We should consider three further steps:

- The establishment of an international arrangement through which countries may be assured of launch services without being solely and directly dependent upon the United States.
- A division of labor between ourselves and other advanced countries or regional space organizations permitting assumptions of primary or joint responsibility for certain scientific or applications tasks in space.
- International sponsorship and support for planetary exploration such as that which was associated with the International Geophysical Year.

[17] The developing countries will be most attracted to (1) applications of space technology which serve their economic and social needs, and (2) the development of international institutional arrangements in which they can participate along with the advanced countries. Some examples are:

- Environmental studies and earth resource surveying via satellites;
- Direct broadcast via satellites of TV instructional and educational programs;
- Expanding arrangements to acquire and use meteorological data;
- Training opportunities in space applications and space-related disciplines.

To the extent that future practical space applications are achieved, there should be no significant technical obstacles to ensuring the sharing of benefits on a global basis. There will, however, be economic and political issues which require recognition and effective anticipation.

In the case of the USSR, experience over the past ten years makes clear that the central problem in developing space cooperation is political rather than technical or economic. Numerous specific technical opportunities for cooperation with the Soviet Union have been identified and are available. Indeed, many of them have been put to the Soviet Union in various forms through the years with little success. For example, we could formulate a series of graduated steps leading toward major cooperation. They would range from full and frank exchange of detailed space project results, at the lowest level, to prearranged complementary activities at the next level (e.g., mutual support of tracking requirements, coordinated satellite missions for specific tasks in space), and ultimately to fully integrated projects in which sub-systems could be provided by each side to carry out a total space mission of agreed character. The following possibilities merit serious consideration:

- In space research—earth orbital investigation of atmospheric dynamics and Earth's magnetic field; astronomical observations from earth satellites or lunar stations; satellite observation of solar phenomena, and lunar and planetary exploration.
- In practical applications—coordination of a continuing network of satellites to provide data for world-wide weather prediction and early warning of natural disasters; the development of capabilities for earth resource surveying via satellites.
- In manned flight—bio-medical research, space rescue, coordination of experiments and flight parameters for Earth orbiting space stations, lunar exploration, and exchange of astronauts.
- In tracking—to supplement each other's networks.

In view of the heavy commitment of the Soviets, planetary exploration appears to offer unusual opportunities for complementary activities.

[19] **IV. PROGRAM AND BUDGET OPTIONS**

The Space Task Group was asked to provide "definitive recommendation on the direction the U. S. space program should take in the post-Apollo period," through preparation of a "coordinated program and budget proposal." In the Section "Goals and Objectives," the Space Task Group has outlined the elements of this coordinated program.

We have also pointed out that there are upper and lower bounds to the Funding which will support a viable, productive and well disciplined program. Between these bounds there are many options both in program content and in total funding required. In this section we will explore the range of these options and their resource implications.

Clearly, there are a number of factors outside the space program and the intrinsic merit of it; goals and objectives that must be considered in determining the allocation of resources to the program. Demands of other domestic programs, international conditions, and state of economic health of our Nation are only a few of the major influences upon the specific budget for space in a given fiscal year.

Despite the highly variable nature of these influences, which produces a corresponding increasing uncertainty in projections of resource availability, it is important for planning purposes to look into the future and forecast the general nature of funding required to support decisions on content and pace of the program. Two basic questions arise. Is the Nation to exploit its existing capabilities, to expand those capabilities or reduce its participation in space activity? Is funding for space generally to remain at present levels, to increase dramatically or to decrease significantly below present levels?

We stand at a crossroads, with many sets of missions and new developments open to us and with three main avenues for funding to pursue these opportunities.

To assist in answering these questions and to provide a basis for Task Group analyses, NASA and DOD were each requested to prepare a set of alternative proposals or options that would cover a range of future resource levels and be consistent with the goals and objectives recommended by the Task Group.

NASA Options

The range of resource levels considered by the Task Group for NASA is shown in Figure 1.

[20] These include: (1) an upper bound, defined by a program conducted at a maximum pace—limited, not by funds, but by technology; (2) options I, II, and III which illustrate programs consistent with the Task Group recommendations, but conducted under varying degrees of funding restraints; and (3) a low level program constructed with an increased unmanned science and applications effort consistent with the Task Group recommendations but, because of the significantly lower budget levels, without a manned flight program after completion of Apollo and Apollo Applications.

A comparison of the timing of major mission accomplishments under the various programs is indicated in Table 1.

Although the program represented by the upper bound appears technically achievable, would provide maximum stimulation to our over-all capabilities, and is fully consistent with the Task Group recommendations, it represents on initial rate of growth of resources which cannot be realized because such budgetary requirements would substantially exceed predicted funding capabilities. This has therefore been rejected by the Space Task Group, and is presented only to demonstrate the upper bound of technological achievement.

We have therefore developed a set of options which falls within these limits to illustrate programs conducted at budget levels which appear possible during the next decade.

Option I is illustrative of a decision to increase funding dramatically and results in early accomplishment of the major manned and unmanned mission opportunities, including launch of a manned mission to Mars in the mid-1980's, establishment of an orbiting lunar station, a 50 man earth-orbit space base and a lunar surface base. Funding would rise from the present $4 billion level to $8-10 billion in 1980. Decision to proceed with

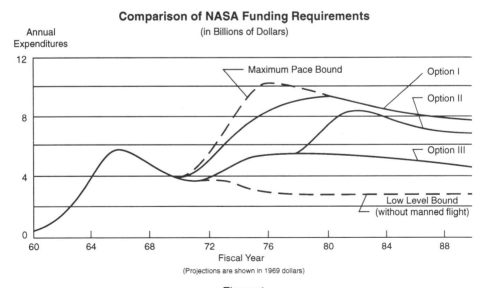

Figure 1

Comparative Program Accomplishments

Milestones	Maximum Pace	Program I	II, III	Low Level
Manned Systems				
Space Station (Earth Orbit)	1975	1976	1977	—
50-Man Space Base (Earth Orbit)	1980	1980	1984	—
100-Man Space Base (Earth Orbit)	1985	1985	1989	—
Lunar Orbiting Station	1976	1978	1981	—
Lunar Surface Base	1978	1980	1983	—
Initial Mars Expedition	1981	1983	II–1986 III–Open	—
Space Transportation System				
Earth-to-Orbit	1975	1976	1977	—
Nuclear Orbit Transfer Stage	1978	1978	1981	—
Space Tug	1976	1978	1981	—
Scientific				
Large Orbiting Observatory	1979	1979	1980	—
High Energy Astron. Capability	1973	1973	1981	1973
Out-of-Ecliptic Survey	1975	1975	1978	1975
Mars—High Resolution Mapping	1977	1977	1981	1977
Venus—Atmospheric Probes	1976	1976	Mid-80s	1976
Multiple Outer Planet "Tours"	1977–79	1977–79	1977–79	1977–79
Asteroid Belt Survey	1975	1975	1981	1975
Applications				
Earliest Oper. Earth Resource System	1975	1975	1976	1975
Demonstration of Direct Broadcast	1978	1978	Mid-80s	1978
Demonstration of Navigation Traffic Control	1974	1974	1976	1974

Table 1

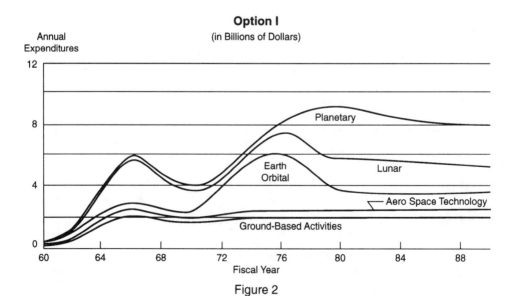

Figure 2

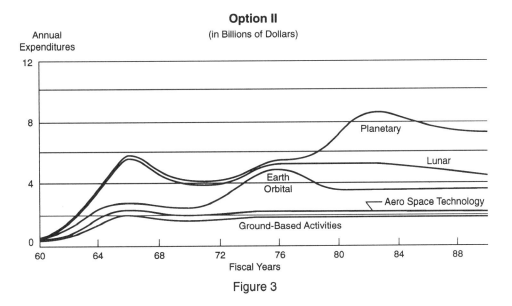

Figure 3

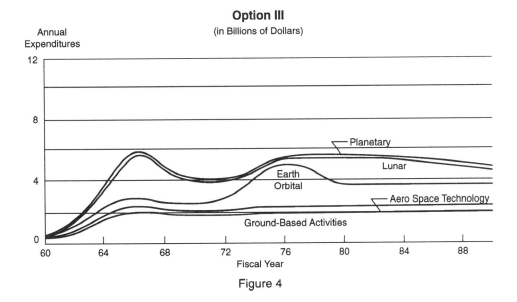

Figure 4

development of the space station, earth-to-orbit shuttle and the space tug would be required in FY 1971. Firm decisions [21] on other major systems or missions would not be needed until later years; for example, a decision to develop the Mars excursion module for an initial manned Mars expedition would not be required before FY 1974.

Options II and III illustrate a decision to maintain funding initially at recent levels and then gradually increasing. These options are identical with the exception that Option II includes a later decision to launch a manned planetary mission in 1986 and in Option III this decision is deferred. Both options demonstrate the effect of simultaneous development of the Space Transportation System and earth orbital space station module, each of

**Funding for NASA Program
Options I, II, and III**

Total Expenditures (Millions of Dollars)

	FY 70	71	72	73	74	75	76	77	78	79	80
Option I	3900	4250	4850	5850	6800	7700	8250	8750	9100	9350	9400
Option II	3900	3950	4050	4250	5000	5450	5500	5500	5650	6600	7650
Option III	3900	3950	4050	4250	5000	5450	5500	5500	5500	5500	5500

(Projections In 1969 Dollars)

Table 2

which is expected to require peak expenditure rates of the order of $1 billion per year, and both options include a substantial increase in unmanned science and applications from present levels but less than that in Option I. Maintaining the unmanned program at the Option I levels would require several hundred million dollars in additional funding. Decision to develop both space station and earth-to-orbit shuttle would be in about FY 1972, resulting in initial availability of these systems in 1977. Similarly, other major milestones would occur later, with decision on the Mars Excursion Module estimated for FY 1978. Funding for both options would remain approximately level at $4 billion for the next two fiscal years and then would rise to a peak of $5.7 billion in 1976—this increase reflecting simultaneous peak resource requirements of space station and space shuttle developments. If these developments were conducted in series, lower funding levels ($4-5 billion) could be achieved. Option II would have a later peak of nearly $8 billion in the early 1980's resulting from the manned Mars landing program.

Details of funding requirements for each of the program options are shown in Figure 2 through 4 and Table 2.

[23] The lower bound chosen by the Space Task Group illustrates a program conducted at significantly reduced funding levels. It is our judgment that, in order to achieve these significantly reduced NASA budgets, it would be necessary to reduce manned space flight operations below a viable minimum level. Therefore, this program has been constructed assuming a hiatus in manned flight following completion of Apollo applications and follow-on Apollo lunar missions. It thus sacrifices, for the period of such reduced budgets, program objectives relating to development of new capability, and the contribution of continuing manned space flight to several of the other program objectives recommended by the Task Group. It does, however, include a vigorous and expanded unmanned program of solar system exploration, astronomy, space applications for the benefit of man and potential for international cooperation. Funding for such a program would reduce gradually to a sustaining level of $2-3 billion depending upon the depth of change assumed for the supporting NASA facilities and manpower base.

The Space Task Group is convinced that a decision to phase out manned space flight operations, although painful, is the only way to achieve significant reductions in NASA budgets over the long term. At any level of mission activity, a continuing program of manned space flight, following use of launch vehicles and spacecraft purchased as part of Apollo, would require continued production of hardware, continued operation of extensive test, launch support and mission control facilities, and the maintenance of highly skilled teams of engineers, technicians, managers, and support personnel. Stretch-out of mission or production schedules, which can initially reduce total annual costs, would result in higher unit costs. More importantly, very low-level operations are highly wasteful of the skilled manpower required to carry out these operations and would risk deterioration of safety and reliability throughout the manned program. At some low level of activity, the viability at the program is in question. It is our belief that the interests of this Nation would not be served by a manned space flight program conducted at such levels.

DOD Options

A similar set of DOD Options, A through C, was constructed to illustrate three basically different levels of military space activity.

Three options are presented, not only to provide funding and program options, but also to characterize the band of choices within which a rational program of military space activities will evolve. Options A and C are considered to be the upper and lower boundaries of probable military space activity, with Option B being an example of an intermediate level.

Option A presumes a future in which the threat to national security could evolve in an increasingly hostile manner, thereby leading to increased priorities for national defense and military space activities. This option also provides for contingency efforts designed to accommodate a high degree of uncertainty in future international conditions. Cost effectiveness, technology availability, growth rate of resource application, and national policy constraints were considered in establishing this upper option for a full military space capability.

[24] Option B includes those efforts necessary to counter the known and generally accepted projections of the threat. In addition, it provides limited developmental activities toward those capabilities needed if the threat increases. Option B is a prototype program which recognizes the need to minimize cost increases over the next few years, but reflects the expectation that military space activity will increase to provide the necessary support to our military forces and posture. This option is consistent with national and DOD policies and with Force Structure planning.

Option C is directly responsive to current national economic constraints, and assumes that a lessening of world tensions will result in reduced emphasis on national defense. It, therefore, includes a lower level of system deployment than the other two options. It still includes, however, the technology and support effort necessary for contingency planning, together with those programs now considered to be reasonable and predictable requirements. Option C is the lower boundary of military space activity that will meet existing national defense needs, although implied in this option is a higher degree of risk than that inherent in Options A and B.

Annual resource requirements for the DOD options are shown in Figure 5.

Program Flexibility

In the option; submitted by NASA and DOD, resource requirements have been projected which represent a large number of decisions to be made in sequence over a number of years. Thus, the resource projections represent the upper envelope or sum of funds required to support these decisions. Many of these decisions are relatively independent—that is, an earth orbit space station module can be developed independently, without commitment to placing such a station in orbit around the moon, or sending such a module on a mission to Mars. In both of these examples, however, development of the space station module would [25] be the normal first step in achieving the lunar orbit station or Mars mission capability. An example of the set of major program elements and hence decision points inherent in the options described, based upon NASA Option II, is included as Figure 6. A diversity of specific programs with varying emphasis can be constructed by delaying or shifting initiation of funding for these major elements relative to other new developments.

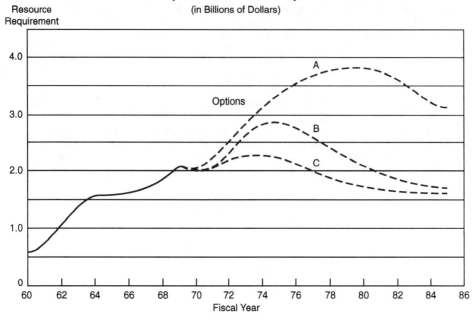

Figure 5

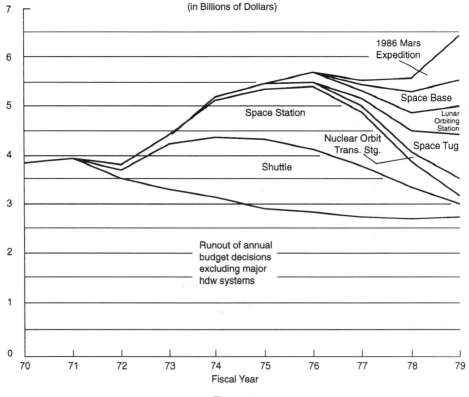

Figure 6

There is, therefore, a great amount of flexibility inherent in each of these options and adjustments to funding constraints may be made on a yearly basis as part of the normal budget process. Of course, once initiated, a specific major system development profits from continuity in funding—stretchout or major fluctuations in funding for a particular project generally increase the total costs associated with it.

The levels of activity for the NASA and the DOD programs are essentially independent, that is, selection of Options I or II for NASA could be consistent with an Option A, B, or C level of activity for DOD, since the DOD space activity will continue to be responsive to national defense needs and will be determined on a case-by-case basis under the budget and program established annually for the Defense Department. It is important, however, that continued coordination of the NASA and DOD programs and the effect of each agency's activity on a common industrial and facility base receive authoritative attention....

[29] Appendix B

SPACE TASK GROUP MEMBERSHIP

CHAIRMAN

THE HONORABLE SPIRO T. AGNEW
Vice President of the United States

MEMBERS

THE HONORABLE ROBERT C. SEAMANS
Secretary of the Air Force

THE HONORABLE THOMAS O. PAINE
Administrator
National Aeronautics and Space Administration

THE HONORABLE LEE A. DUBRIDGE
Science Adviser to the President

OBSERVERS

THE HONORABLE U. ALEXIS JOHNSON
Under Secretary of State
For Political Affairs

THE HONORABLE GLENN T. SEABORG
Chairman
Atomic Energy Commission

THE HONORABLE ROBERT P. MAYO
Director
Bureau of the Budget

Document III-26

Document title: **Robert P. Mayo, Director, Bureau of the Budget, Memorandum for the President, "Space Task Group Report," September 25, 1969.**

Source: NASA Historical Reference Collection, History Office, NASA Headquarters, Washington, D.C.

Bureau of the Budget Director Robert Mayo had first been critical of NASA's desire for a continued high budget when dealing with Thomas Paine's attempt to get an early presidential commitment to a space station and an increase in the NASA budget when the new Nixon administration had asked its agency heads to find ways of reducing their budgets. Mayo had been an observer in the Space Task Group, and he had made occasional comments about the difference between what the task group seemed to want to recommend and the budget outlook in coming years. This memorandum, written ten days after the president received the Space Task Group report, foreshadowed a bitter battle between NASA and the White House over the fiscal year 1971 budget level. NASA in October 1969 requested a fiscal year 1971 budget of $4.497 billion; by the time the President finally approved the budget the following January, it had been cut almost 25 percent, to $3.33 billion. Although Mayo left his position after the fiscal year 1971 budget was submitted, the newly renamed Office of Management and Budget continued to attempt to reduce NASA budget levels in subsequent years.

[1] # Memorandum for the President

Subject: Space Task Group Report

This memorandum presents a summary of my views on the Space Task Group Report and my recommendations as to the next steps in the decision process. I was an observer on the Space Task Group and, as such, participated in its discussions on the future of the space program, reserving the right to present to you my independent judgment as your Budget Director.

The report sets forth an excellent catalog of technical possibilities for the future. However, standing by itself, it has several shortcomings. In my view, these shortcomings impair its completeness as a vehicle for your *final* decision.

1. The report does not clearly differentiate between the values of the manned space flight program versus a much less costly unmanned program with its greater emphasis on scientific achievement and potential economic returns.

2. The Space Task Group could not, nor did it try to, assess the relative standing of the space program in our full range of national priorities. In order to do this, you might wish to have the report reviewed by the Cabinet—and perhaps the Security Council as well.

3. The Group could not address the future economic context within which the recommended space expenditure would have to be considered.

4. The report is written in such a way that your endorsement of any of the recommended program options implies endorsement of major new long-term development projects, which are included in *all three* of the program options. Therefore, in a practical sense, the report gives you little flexibility except as to timing (and therefore annual costs). The impact of this is only slightly softened by the assertion that the rate of progress toward the goals would be subject to annual budget decision. This reservation has very practical limits. All the defined options involve significant budget increases over current levels.

5. The Bureau of the Budget has not had the opportunity to review in detail the estimates set forth on page 22 of the report, but they vary sufficiently from other estimates

which have been used recently so that we believe they are significantly underestimated. Furthermore, these figures are presented in terms of 1969 dollars and are therefore further underestimated by reason of the inflation that has already taken place.

[2] Of course, there is no reflection on price increases that are almost certain to come in the years ahead.

The other decision factors that most concern me are related specifically to the 1971 budget, now under preparation, and to the budgets that you will be preparing during the remainder of your first term.

The 1971 problem is severe because of:

1. The inflation we are still trying to bring under control.
2. The need to assume continuation of the Vietnam conflict for budget preparation purposes.
3. The commitments we have already made in such areas as domestic welfare, manpower training, social security benefits, revenue sharing, airports/airways, mass transit, and supersonic transport development among others. Every one of these commitments requires outlay increases in 1971.
4. Uncontrollable items such as interest on the national debt.
5. Revenue losses associated with the tax bill—even with proposed Treasury amendments.

In light of these circumstances, I gave NASA an official budget planning target of $3.5 billion for 1971 ($350 million below 1970). This target was based on the assumption that after the manned lunar landing, some reduction in NASA's current budget levels could be made to ease our overall budget problem, without stopping the manned space program. All three options set forth in the report require 1971 budgets of at least $100 million plus price increases above the current NASA funding levels and further increases in following years. These increases will have to come from programs of other agencies.

Because the Space Task Group report has now been published, your endorsement now of any specific option will commit us to annual budget increases of at least the magnitudes specified in the report. Therefore, you could lose effective fiscal control of the program.

I am convinced that a forward-looking manned space program can be developed for you that does not involve commitments to significant near-term budget increases.

Such a program would involve a slower rate of manned Apollo flights than NASA now considers desirable. It would also involve consecutive rather than simultaneous development of a space transportation system and space station, which are necessary steps toward a manned Mars mission. I intend to explore such a program in some detail with Dr. Paine during the FY 1971 budget decision process. Such a program could be accelerated in the future if conditions permit.

[3] I believe this course would be preferable to announcing ambitious long-range plans now and then having to cut back in the future due to economic constraints.

In this circumstance, I recommend:

1. That you withhold announcement of your space program decision until after you have reviewed the report recommendations specifically in the context of the total 1971 budget problem.
2. That you ask the Cabinet and perhaps the NSC to consider the Space Task Group report during October or November and advise you of their views on its recommendations, so that you will have those views in mind during your budget decisions.
3. That you consider meeting with Tom Paine and me after I have had an

opportunity to discuss with him the lower cost program option I have described above. Your meeting could be planned for December, and could serve as the final step in your decision process on the NASA 1971 budget. At that time, it is essential that you specify program content as well as budget guidance in order to help maintain effective fiscal control of the program.

4. That your space program decisions be announced in the State of the Union address, the budget message, or a special message to the Congress in the spring of 1970.

Robert P. Mayo
Director

Document III-27

Document title: Peter M. Flanigan, Memorandum for the President, December 6, 1969.

Source: NASA Historical Reference Collection, History Office, NASA Headquarters, Washington, D.C.

It was not only a desire to reduce the federal budget that led the White House in late 1969 to seek a lower level of spending on space. There was also a belief among many of Richard Nixon's political and policy advisers that there was little political benefit to the president from major post-Apollo space initiatives. This memorandum from Assistant to the President Peter Flanigan, who had been asked by the president and his top policy adviser John Erlichman to be the White House link to NASA, is an example of the political input being provided to the president.

Memorandum for the President

The October 6 issue of *Newsweek* took a poll of 1,321 Americans with household incomes ranging from $5,000 to $15,000 a year. This represents 61% of the white population of the United States and is obviously the heart of your constituency. Of this group, 56% think the government should be spending less money on space exploration, and only 10% think the government should be spending more money.

Peter M. Flanigan

Document III-28

Document title: Caspar W. Weinberger, Deputy Director, Office of Management and Budget, via George P. Shultz, Memorandum for the President, "Future of NASA," August 12, 1971.

Source: NASA Historical Reference Collection, History Office, NASA Headquarters, Washington, D.C.

In the summer of 1971, NASA and the White House were once again locked in a bitter battle over the agency's future programs and budgets. At issue was whether NASA would receive approval to continue the Apollo program through the Apollo 16 and 17 missions and to begin developing a reusable space transportation system, the space shuttle. Observing the interactions between NASA and the Office of Management and Budget (OMB) staff and other White House elements, OMB Deputy Director Caspar Weinberger decided that budget cutting was going too far. His August 12, 1971, memorandum is the "smoking gun" with respect to the White House decision to approve the shuttle. President Nixon read the memorandum, wrote "OK" next to Weinberger's proposal to find money from elsewhere in the budget to fund NASA at a level adequate to support shuttle development, and wrote "I agree with Cap" next to OMB Director George Shultz's name on the first page of the memorandum. White House staffer Jon Huntsman reported the president's decision to Budget Director Schultz in a cover memorandum.

This exchange between Weinberger and the president was not reported to lower OMB staff, who continued through the rest of 1971 to oppose the shuttle and to propose further reductions in the NASA budget.

---

[1] **Memorandum for the President**

From: Caspar W. Weinberger
Via: George P. Shultz
Subject: Future of NASA

Present tentative plans call for major reductions or change in NASA, by eliminating the last two Apollo flights (16 and 17), and eliminating or sharply reducing the balance of the Manned Space Program (Skylab and Space Shuttle) and many remaining NASA programs.

I believe this would be a mistake.

1) The real reason for sharp reductions in the NASA budget is that NASA is entirely in the 28% of the budget that is controllable. In short we cut it because it is cuttable, not because it is doing a bad job or an unnecessary one.

2) We are being driven, by the uncontrollable items, to spend more and more on programs that offer no real hope for the future: Model Cities, OEO, Welfare, interest on National Debt, unemployment compensation, Medicare, etc. Of course, some of these have to be continued, in one form or another, but essentially they are program, not of our choice, designed to repair mistakes of the past, not of our making.

3) We do need to reduce the budget, in my opinion, but we should not make all our reduction decisions on the basis of what is reducible, rather than on the merits of individual programs.

[2] 4) There is real merit to the future of NASA, and to its proposed programs. The Space Shuttle and NERVA particularly offer the opportunity, among other things, to secure substantial scientific fall-out for the civilian economy at the same time that large numbers of valuable (and hard-to-employ-elsewhere) scientists and technicians are kept at work on projects that increase our knowledge of space, our ability to develop for lower cost space exploration, travel, and to secure, through NERVA, twice the existing propulsion efficiency of our rockets.

It is very difficult to re-assemble the NASA teams should it be decided later, after major stoppages, to re-start some of the long-range programs.

5) Recent Apollo flights have been very successful from all points of view. Most important is the fact that they give the American people a much needed lift in spirit, (and the people of the world an equally needed look at American superiority). Announcement now, or very shortly, that we were cancelling Apollo 16 and 17 (an announcement we would have to make very soon if any real savings are to be realized) would have a very bad effect, coming so soon after Apollo 15's triumph. It would be confirming in some respects, a belief that I fear is gaining credence at home and abroad: That our best years are behind us, that we are turning inward, reducing our defense commitments, and voluntarily starting to give up our super-power status, and our desire to maintain world superiority.

America should be able to afford something besides increased welfare, programs to repair our cities, or Appalachian relief and the like.

6) I do not propose that we necessarily fund all NASA seeks—only that we couple any announcement to that effect with announcements that we *are* going to fund space shuttles, NERVA, or other major, future NASA activities....

Document III-29

Document title: J.C.F., Administrator, Memorandum to Dr. Low, "Meeting with Ed David," August 24, 1971.

Source: NASA Historical Reference Collection, History Office, NASA Headquarters, Washington, D.C.

This memorandum catches the character of the NASA-White House interactions during 1971 as seen from NASA's perspective. Edward David was President Nixon's Science Adviser, and Russell Drew his top staff person on space issues. The Flax Committee, named after its chairman Dr. Alexander Flax, was an *ad hoc* panel of the President's Science Advisory Committee set up to advise the White House on NASA's shuttle proposal and possible alternatives. One of the panel's members, Dr. Eugene Fubini, was particularly active in the review. Richard McCurdy was the NASA official in charge of institutional management; the Office of Management and Budget was suggesting that the post-Apollo NASA had too many facilities and too many employees for the program envisaged for the 1970s.

---

[1] As we discussed, I met with Ed David, ostensibly to talk about the possibility of a $3.2 billion constant budget throughout the 70's and to get a feeling from him as to how much we can trust OMB and others with regard to holding the line at this figure if we came in with a "bare bones" budget of this magnitude, which included a minimum shuttle. Ed's feeling is that the Flax Committee (with Fubini leading the pack) is going to come in with some interesting options which I would judge to be consistent with the $3.2 billion budget and, perhaps, would include a shuttle of about $5 billion total investment running about 1 billion per year. I indicated that this might be in the same ball park, but we would have to examine any such ideas in depth before we could commit ourselves to any such programs; also, that we were thinking along similar lines but so far had not discussed them in any detail with the Flax Committee. Surprisingly enough, he felt this was the wise thing to do from our point of view and he would hope that we would continue to keep such studies confined to a small group in NASA until the time came to discuss them. I received a very definite impression that he would like to take credit for coming up with a reduced cost shuttle, and I didn't discourage him from this idea.

When it came to discussing tactics, he did agree that the two of us ought to sit down after the Flax Committee results were in and plan out a program together. However, his initial [2] thought was that he should propose the low-cost shuttle to OMB himself, but that we should try to resist in order to argue from a better bargaining position. I am not sure that this is a good way to proceed but his suggestion was based on the fact that we already recognize that OMB can't entirely be trusted to commit to any kind of program and that if we agreed too easily to the low-cost shuttle, they might try to work us down to a smaller budget yet. Basically, the strategy and tactics remain unresolved, except Ed did agree to chat further with us on the subject when the Flax Committee results were available.

I was personally a little discouraged by the conversation in the sense that he didn't feel there was anyone in OMB who could be completely trusted—not that they were dishonest, but that their sole function was to put a ratchet on the budget and couldn't make a commitment to hold the line on anything.

I tried out your ideas regarding the Space Council and, at first, he was quite defensive, indicating that OST perhaps served the function that we had in mind for the Space Council, particularly when the business of earth resources policy came up. However, after some discussion we agreed that the idea was worth considering, but he wanted to mull it over first. I think his thought was that perhaps he could chair the Space Council in the absence of the Vice President instead of "yours truly." I am afraid we are going to have some difficulty on this one, but I am willing to pursue it further if we still think it is a good idea.

Incidentally, I brought up another subject and that is his own views and those of OMB's toward our "operating base." He feels that OMB is unconvinced that there isn't considerable fat in the program because of the large operating base and although he knows Dick McCurdy's position on it, he is not himself convinced and is sure that OMB is not convinced. I indicated that we had just started our program with OMB [3] and, hopefully, Dick would be successful in a period of weeks to expose enough of OMB to the report that, at least, they would understand the problem better. Ed didn't volunteer to listen to Dick's analysis, but I think we ought to try to set up with both Ed David and Russ Drew at an appropriate time, even though they are lukewarm to the idea.

J.C.F.

Document III-30

Document title: Klaus P. Heiss and Oskar Morgenstern, Memorandum for Dr. James C. Fletcher, Administrator, NASA, "Factors for a Decision on a New Reusable Space Transportation System," October 28, 1971.

Source: NASA Historical Reference Collection, History Office, NASA Headquarters, Washington, D.C.

In 1970, the Office of Management and Budget had forced NASA to hire an external contractor to analyze the economic rationale for replacing existing or new expendable launch vehicles with a reusable space transportation system. The contractor chosen was Mathematica, Inc., a Princeton, New Jersey, firm headed by the distinguished economist Oskar Morgenstern. At Mathematica, the individual with primary responsibility for the study was a brash young Austrian-born economist, Klaus Heiss.

Mathematica submitted its analysis of the economic worth of a fully reusable space transportation system in May 1971, just as NASA was deciding that budget constraints would only allow the development of a partially reusable system. NASA asked Mathematica to examine the economics of a variety of possible designs for such a system.

Working closely with several aerospace contractors, Heiss came to the conclusion that a particular design, which he called a Thrust Assisted Orbiter Shuttle (TAOS), was the preferred alternative. Recognizing that the total study would not be completed in time to influence decisions on the shuttle program (the study was submitted in May 1972), Heiss and Morgenstern prepared this memorandum and circulated it among those involved in the space shuttle decision process.

---

[1]
(1) REUSABLE SPACE TRANSPORTATION SYSTEM IS ECONOMICALLY FEASIBLE, ASSUMING THAT THE LEVEL OF UNMANNED U.S. SPACE ACTIVITY WILL NOT BE LESS THAN IT HAS BEEN ON THE AVERAGE OVER THE LAST EIGHT YEARS.

(2) AMONG THE MANY SPACE SHUTTLE CONFIGURATIONS SO FAR INVESTIGATED, AND WHICH ARE DEEMED TO BE TECHNOLOGICALLY FEASIBLE, *A THRUST ASSISTED ORBITER SHUTTLE (TAOS)* WITH EXTERNAL HYDROGEN/OXYGEN TANKS EMERGES AT PRESENT AS THE *ECONOMICALLY PREFERRED* CHOICE. EXAMPLES OF SUCH CONCEPTS ARE RATO OF MCDONNELL DOUGLAS AND TAHO OF GRUMMAN-BOEING.

(3) *THE DEMAND FOR SPACE TRANSPORTATION* IN THE 1980's AND BEYOND BY THE NATIONAL AERONAUTICS AND SPACE ADMINISTRATION, THE DEPARTMENT OF DEFENSE, BUT PARTICULARLY BY COMMERCIAL AND OTHER USERS IS THE BASIS FOR THE ECONOMIC JUSTIFICATION FOR THE TAOS PROGRAM. SUBSTANTIAL FURTHER EFFORT IN THIS AREA IS NEEDED TO DETERMINE THESE EXPECTED NEEDS.

The following sets forth briefly, in a summary manner, the principal considerations which lead to conclusions (1) and (2). The following arguments, which in their entirety support the recommendation (2), contribute significantly to alleviating the doubts voiced by the Congress, the public and several branches of the Executive concerning the need for a new Space Transportation System. Such doubts have been raised because of the magnitude of the investment involved and the comparative technological difficulty of the proposed undertaking. [2]

I. Major Conclusions

1. In the May 31, 1971 report by MATHEMATICA, *Economic Analysis of New Space Transportation Systems*, the overall economic worth of a reusable space transportation system was examined. The study was based on the two-stage fully reusable concept then under investigation by Phase B contractors and NASA. That report has demonstrated how an economic justification of a space shuttle system, including a space tug, with an IOC date of 1978 has to be made. The report was *not* concerned with identifying the most economic choice among alternative space shuttle configurations to be considered.

2. The Baseline, fully reusable, space transportation system had attached to it a non-recurring cost of between $10 and $14 billion when the costs of all systems were included. This large investment outlay would be largely independent of the time span within which these funds are expended. These high non-recurring costs coupled with a relatively high risk led to the study of many alternate configurations. Among the many other approaches studied by NASA and industry, our calculations show the emergence of an *economical and acceptable solution to the question of the best strategy for NASA to achieve a reusable space transportation system for the 1980's at acceptable costs.*

3. Over 200 space programs were analyzed by MATHEMATICA, comparing (a) the Baseline two-stage fully reusable system, (b) the Baseline, external hydrogen tank system, (c) the Mark I-Mark II (reusable S1C) system, (d) the RATO system of McDonnell Douglas, (e) the TAHO system of Grumman-Boeing, (f) the Stage and One-Half of Lockheed Corporation, and (g) the Identical Vehicle Concept of McDonnell Douglas. The *Thrust Assisted Orbiter Shuttle concepts (TAOS) which include concepts like* [3] *RATO and TAHO, emerge as the most preferred systems within the space programs so far analyzed,* using the economic methodology as exemplified in the May 31, 1971 report. The common feature of TAOS concepts is a single orbiter with external hydrogen/oxygen tanks and rocket assists in the form of solid rocket motors or high pressure fed unmanned boosters. This eliminates the need to develop a large manned, reusable booster.

[4]   **II. Objectives of a Reusable Space Transportation System**

The *principal* objectives of a space shuttle system are considered to be:
1. A new capability of meeting all foreseeable space missions in NASA, DoD and elsewhere, *including* manned space flight capabilities.
2. Reduction of *space program costs* (manned, unmanned, NASA, DoD, commercial users) over the present expendable space transportation costs through reuse, refurbishment, maintenance, and updating of payloads.
3. Reduction of *space transportation costs* for all missions (low energy, high energy, manned).
4. *Option* of later transition to a fully reusable system.

Additional objectives supporting the major objectives are:
5. A low *non-recurring cost* to meet funding constraints.
6. Assurance of a *low cost per launch* of below $10 million—and if possible $5 million —justifiable when payload costs and effects are considered.

The work assigned to us and reported in the May 11 report showed clearly the *economic* justification of a fully reusable space transportation system and outlined some key questions that remain to be answered in order to assure an overall purpose to the space shuttle decision. Not yet analyzed in that report was the question of *the most economic shuttle configuration,* to meet the major objectives of NASA. Any decision on an economic new Space Transportation System will have to reconcile major constraints with those objectives.

[5] **III. Constraints of Decision on Configuration**

The key constraints that any decision on a particular configuration is confronted with are:

1. **Technological:** The technical feasibility of the alternative configurations studied by NASA and industry is assumed. However, for each alternative configuration the time-and-cost uncertainties were analyzed as far as presently possible. This still assumes that the concepts studied are indeed technically viable.

2. **Economic:**

(a) **Total cost and components.** Different configurations have very different costs associated with them as outlined, for example, by the Baseline, Two Stage Fully Reusable system, the Baseline Two Stage External Hydrogen Tank system, the Mark I-Mark II (reusable SIC) system, the Stage and One-Half system and the various Thrust Assisted Orbiter Shuttles (TAOS). In addition to total costs in research and development, investment, and operations (including the cost per launch), elements of uncertainty of various degrees are associated with individual subsystems of these configurations. NASA, industry contractors, and others are trying to analyze in part the cost component as well as the risk component in these different configurations. All possible different configurations, but certainly TAOS, have to be analyzed as to the advantages and disadvantages in cost, risk, and uncertainty that these configurations promise when compared to the two stage fully reusable original Baseline system of NASA.

[6] (b) **Timing of the space shuttle development and its systems.** In part the choice of the current Mark I-Mark II approach was forced by a peak funding requirement for space shuttle development of, say, $1 billion per year. In this approach, however, several important parts of the system would be postponed in some configurations *while* other configurations with the same total funding requirement assure *an early IOC date not only of the space shuttle alone, but also of the space tug.* The meeting of the funding requirements, the mission capabilities of the new system and the IOC dates of interim as well as final configurations have to be further studied in detail.

(c) **Timing of the Space Tug** should be such that its IOC date comes closely after the IOC date of the Space Shuttle. If European countries undertake the tug development —after assurance that NASA will have a Space Shuttle System!—then tug funding becomes a problem outside the NASA budget and these expenditures should not affect the shuttle decision itself. They were, however, fully allowed for in our analysis.

Within the above constraints, the *trade offs* of alternative configurations have to be studied with a number of alternative mission models. In particular, very close attention has to be paid to the very different *capabilities* that are given in terms of the overall system. In connection with the TAOS concept, tank costs are yet a major uncertainty that are included in our present studies and which may come down substantially with further technical change. Such a development would further favor the TAOS concept.

The key question raised in the May 31, 1971 report is: *Does there exist a precise and detailed NASA and national space program for the 1980's?* [7] We did receive detailed *mission* models of OSSA, OMSF, the DoD, non-NASA applications and others. Yet these continue to change substantially. A space program consists of individual missions which must be specified and integrated into an overall plan of not negligible firmness, though some flexibility must also be allowed for.

To allow the space shuttle decision on the basis of the Two Stage Shuttle funding requirements, many of the important missions were postponed recently by NASA to fit the shuttle development into the expected funding limitation. A far more sophisticated analysis needs to be done that allows the *scheduling* of types of payloads. The *importance* of payloads, the *interdependence* among payloads within missions and between missions, as well as an analysis of resupply, updating, maintenance, and reliability. Utilizing programming tools that are available today in operations research, substantial work can be performed, some of which is incorporated in the present ongoing work by our group.

Thus, within these constraints an acceptable Space Shuttle development program is indeed difficult: budget limitation by year, total program costs, the timing of different components of the system, the need for a Space Tug and an early full operational capability, and comprehensive and justified national space program alternatives for the 1980's.

[8]   **IV. Conclusions**

Among the 200 and more space programs analyzed, and comparing (a) the two stage, fully reusable system, (b) the Baseline, external hydrogen tank system, (c) the Mark I-Mark II (reusable SlC) system, (d) the RATO system of McDonnell Douglas, (e) the TAHO system of Grumman-Boeing, (f) the Stage and One-Half of Lockheed Corporation, and (g) the Identical Vehicle Concepts of McDonnell Douglas, the *Thrust Assisted Orbiter (TAOS) concepts emerge as the most economic systems within the space programs analyzed.* TAOS with external hydrogen and oxygen tanks, a 60 x 15 payload bay, and a 40,000 pound polar orbit capability, if possible by 1979, clearly dominates any other configuration.

The TAOS concept foregoes the development of a Two Stage Shuttle System. With the use of thrust assists of either solid rocket motors or high pressure fed systems—which can be made in part reusable for low staging velocities—the TAOS concepts promise a reduction of the non-recurring costs (RDT&E and initial fleet investment) from about $9 billion or more (two stage systems, including reusable SlC) to about $6 billion or less, with a minimal operating cost increase, if any, in the operating phase of the TAOS system.

The detailed economic justifications of the TAOS concept—when compared to any two stage reusable system are:

1. The non-recurring costs of TAOS are estimated by industry to be $6 billion or less over the period to 1979 or to 1984-85, depending on the objectives and choices of NASA.

2. The risks in the TAOS development are in balance lower but still substantial. Intact abort with external hydrogen and oxygen tanks is feasible; lagging performance in *the* engine area can be made up by added [9] external tank capability. A large reusable manned booster is not needed.

3. The TAOS's that were analyzed promise the *same capabilities* as the original two stage shuttle, including a 40,000 pound lift capability into polar orbit and a 60 x 15 feet payload bay.

4. The TAOS can carry the *Space Tug* and capture high energy missions from 1979 on.

5. The most economic TAOS would use the *advanced orbiter* engines immediately. Our calculations indicate that among the alternative TAOS configurations an early full operational capability (i. e., high performance engines on the orbiter) is economically most advantageous, and feasible, within budget constraints of $1 billion peak funding.

6. The TAOS *can* use J2S engines on the orbiter for an interim period.

7. The TAOS *abolishes* completely the immediate need to decide on a *reusable booster* and allows postponement of that decision without blocking later transition to that system if still desired. Thereby, TAOS eliminates or lowers the risk and potential cost overruns in booster development.

8. The TAOS can use *"parallel burn"* concepts, which, if feasible, may change the reusable booster decision.

9. Technological progress may make *tank costs,* and thrust assisted rocket costs less expensive, thus further aiding TAOS concepts when compared to two stage concepts.

10. TAOS assures NASA an *early program definition,* and a purpose to the agency. An agreement on TAOS will allow NASA Headquarters a quick and clear reorganization of major NASA centers to meet the TAOS development requirements economically.

11. The TAOS funding schedule makes an early Space Tug [10] development possible. The Space Tug is an important part of the Space Shuttle System. A 1979 Space Tug should recover its complete development costs before 1985 even with the stretched build up of Space Shuttle missions from 1979 to 1985.

12. A clear policy on TAOS development will give an incentive to European countries to undertake and fund the Space Tug development—thereby possibly even eliminating Space Tug funding from NASA budget considerations.

13. The cost per launch of TAOS can be as low as $6 million or a even less on an *incremental cost* basis, with reuse of parts of the thrust assist rockets (either SRM or pressure-fed). With Point 9 realized, the costs of TAOS would practically match the *costs per launch of two stage fully reusable systems.*

14. TAOS practically assures NASA of a reusable space transportation system with *major objectives achieved.* [11]

V. The Principal Open Problem: The Demand for Space Transportation

Within the analysis of payloads and their effects on a space shuttle system. The most important problem remains of what will *the demand for* space transportation be in the 1980's (1) *with* a space shuttle, (2) *without* a space shuttle. The demand for space transportation is an important function of the *costs* of doing space applications, the *reliability* of the space transportation system and the assured functioning of payloads as well as the *frequency* of launches over time. The present space programs analyzed by our group mainly concentrate on NASA and DoD applications. However, *it is our strong belief that the major portion of space transportation demand in the 1980's will come from economic applications of space technology to meet the growing needs of the U.S. and other developed and developing countries.*

The potential in this area is of major significance and will lead to completely new ways of looking at space and of evaluating expenditures of space programs like the space shuttle system. This needs to be further documented in detailed work. A major portion of the demand for space transportation will still come, of course, from NASA and DoD as exemplified by the mission models given to us. On commercial applications. However, an inter-agency, inter-industry, and international effort should be organized to study in detail the economic problems that can be alleviated or solved by using space technology. Communication and navigation systems are but one important component in this area. Earth resources applications, early warning systems, management information systems, television systems, along with completely new applications must he studied. The number of satellites and the different needs of many countries arc likely to be of such a scale that they will contribute substantially to the demand for space shuttle flights once this system is available. The effort that should be put [12] forth in this area is such that it be best undertaken by a group like the Space Task Group effort of 1969. However it should now be oriented towards the demand for space transportation, and within this, particularly to the economic applications area of space. It is also true that with regard to space shuttle missions further work needs to be done, particularly in the area of DoD missions in support of a decision like the space shuttle. The military implications of the additional capabilities offered by the Space Shuttle System beyond those of expendable space transportation systems have to be analyzed fully. If this were done, further strong support for the space shuttle decision should become available.

The point is sometimes made by NASA that the *technological* possibility of a space shuttle program suffices by itself to justify its construction, independent of the economic analysis. The economic analysis is not a challenge against the importance of technical efforts of a program like the space shuttle, but rather assures that a decision for the space shuttle is not based solely on technology. It is difficult to see how a program like the Space Shuttle can be undertaken, without a complete economic justification.

Last but not least, it seems to us, with all the potential promised by space and space applications, that *NASA has been very limited in the past in fulfilling its potential role and in realizing finally the importance of its function within the nation.* In this NASA is severely limited by its charter. In the present mood and the present state of the economy a program like the shuttle and its decision has to be *user* oriented, not in terms of who will build the shuttle and benefit by these expenditures, but rather in terms of who is going to use the space shuttle system and why the different agencies, corporations, and foreign countries will do so. NASA ought to adjust to such a reorientation of emphasis. [13]

Within the user area, there appear to be the following major needs:

1. **Military.** The military uses of space are in the present DoD mission model. These uses will presumably continue at this level into the 1980's. However, the missions analyzed so far in the context of the space shuttle system have to be supplemented by a comprehensive analysis of the military significance of the *added* capabilities that the Space Shuttle System offers in the 1980's. NASA will need the support of DoD and in getting this support, will have to initiate a complete re-evaluation of the additional capabilities that the space shuttle will give to DoD in addition to the "expendable mode" applications now provided for in the DoD mission model.

2. **Scientific.** The mission models of OSSA activities in the 1980's seem to cover the expected activity levels in the 1980's for scientific applications in the U.S. As scientific developments occur over the next decade, demands for new and additional missions may arise. At present NASA has fully worked out plans, schedules, and priorities. What remains is to explain further to the public and the Congress the great scientific value and ultimately practical importance of these activities. In doing so, however, this expenditure will have to be seen in the context of other national scientific and R&D efforts.

3. **Communications.** Applications are now fully developing and the encouragement of other countries becomes increasingly important. A new idea in this area would be to begin with a program by NASA to help *developing nations* in setting up communication systems within the *foreign* aid program of the U.S., through the United Nations and the International Bank for Reconstruction and Development. This could ultimately lead to a foreign aid program of *"Zero Dollar Outflow"* since the U.S. [14] contribution, as well as launch and maintenance services, of the space based systems would occur completely within the U.S.

4. **Earth Observations.** The need to assess the resources of the earth worldwide as well as for the U.S. is apparent. NASA is undertaking major demonstration programs in ERTS A and ERTS B. Yet when looking ahead to the 1980's it becomes apparent that given the many different uses of earth observation satellites, the different regions to be covered, as well as the programs of different nations, the demand for space based earth resource satellites will be much larger and specific to each field than is now contemplated. The function of a space shuttle system within such an application program would be tremendous and has not been analyzed. In this area alone one can foresee enough traffic to justify support for the development of a new space transportation system.

5. **Navigation.** Different navigation satellite systems are being used at present but mainly in DoD. A demand for a *reliable* system at low cost is apparent in the aviation and shipping industries, as well as in defense. A world wide system, again covering all industrial users and different regional applications will lead to a substantial increase in the number of satellites. New demands will be made concerning reliability and maintainability of these systems.

6. **Other Applications.** Several other possible world wide applications of space can be foreseen to which the space shuttle system would contribute significantly. Among these are the use of space based *production processes* which for safety considerations, gravity, environment or other technical reasons are either too expensive or not possible when earth based. Other areas concern the generation and transmission of *energy* [15] using potentially completely new sources (for example *solar* energy, fusion energy) for either large scale space based users, or in the more distant future, for use on earth.

We conclude this memorandum with the observation—though by now trivial and obvious, but nevertheless fundamental—that any expenditure of public funds must be justified, precisely as expenditure of private and business funds, by the *aims* and *purposes* of the expenditure. Technological possibilities alone carry no conviction, though they often

Document III-31

Document title: James C. Fletcher, "The Space Shuttle," November 22, 1971.

Source: NASA Historical Reference Collection, History Office, NASA Headquarters, Washington, DC.

The economic benefits of the shuttle were only one, and to the NASA leadership not the most important, reason for going ahead with the program. As the NASA-White House discussions about shuttle approval reached their climax, NASA prepared a "best case" paper reflecting its arguments for approving shuttle development.

[1]
The Space Shuttle

Summary

This paper outlines NASA's case for proceeding with the space shuttle. The principal points are as follows:

1. The U.S. cannot forego manned space flight.

2. The space shuttle is the only meaningful new manned space program that can be accomplished on a modest budget.

3. The space shuttle is a necessary next step for the practical use of space. It will help

— space science,

— civilian space applications,

— military space applications, and

— the U.S. position in international competition and cooperation in space.

4. The cost and complexity of today's shuttle is *one-half* of what it was six months ago.

5. Starting the shuttle now will have significant positive effect on aerospace employment. Not starting would be a serious blow to both the morale and health of the Aerospace Industry.

[2] ### The U.S. Cannot Forego Manned Space Flight

Man has worked hard to achieve—and has indeed achieved—the freedom of mobility on land, the freedom of sailing on his oceans, and the freedom of flying in the atmosphere.

And now, within the last dozen years, man has discovered that he can also have the freedom of space. Russians and Americans, at almost the same time, first took tentative small steps beyond the earth's atmosphere, and soon learned to operate, to maneuver, and to rendezvous and dock in near-earth space. Americans went on to set foot on the moon, while the Russians have continued to expand their capabilities in near-earth space.

Man has learned to fly in space, and man will continue to fly in space. This is fact. And, given this fact, the United States cannot forego its responsibility—to itself and to the free world—to have a part in manned space flight. Space is not all remote. Men in near-earth orbit can be less than 100 miles from any point on earth—no farther from the U.S. than Cuba. For the U.S. not to be in space, while others do have men in space, is unthinkable, and a position which America cannot accept.

[3] **Why the Space Shuttle?**

There are three reasons why the space shuttle is the right next step in manned space flight and the U.S. space program:

First, the shuttle is the only meaningful new manned space program which can be accomplished on a modest budget. Somewhat less expensive "space acrobatics" programs can be imagined but would accomplish little and be dead-ended. Additional Apollo or Skylab flights would be very costly, especially as left-over Apollo components run out, and would give diminishing returns. Meaningful alternatives, such as a space laboratory or a revisit to the moon to establish semi-permanent bases are *much* more expensive, and a visit to Mars, although exciting and interesting, is completely beyond our means at the present time.

Second, the space shuttle is needed to make space operations less complex and less costly. Today we have to mount an enormous effort every time we launch a manned vehicle, or even a large unmanned mission. The reusable space shuttle gives us a way to avoid this. This airplane-like spacecraft will make a launch into orbit an almost routine event—at a cost $1/_{10}$th of today's cost of space operations. How is this possible? Simply by not [4] throwing everything away after we have used it just once—just as we don't throw away an airplane after its first trip from Washington to Los Angeles.

The shuttle even looks like an airplane, but it has rocket engines instead of jet engines. It is launched vertically, flies into orbit under its own power, stays there as long as it is needed, then glides back into the atmosphere and lands on a runway, ready for its next use. And it will do this so economically that, if necessary, it can provide transportation to and from space each week, at an annual operating cost that is equivalent to only 15 percent of today's total NASA budget, or about the total cost of a single Apollo flight. Space operations would indeed become *routine*.

Third, the space shuttle is needed to do useful things. The long term need is clear. In the 1980's and beyond, the low cost to orbit the shuttle gives is essential for all the dramatic and practical future programs we can conceive. One example is a space station. Such a system would allow many men to spend long periods engaged in scientific, military, or even commercial activities in a more or less permanent station which could be visited cheaply and frequently and refurbished, [5] by means of a shuttle. Another interesting example is revisits to the moon to establish bases there; the shuttle would take the systems needed to earth orbit for assembly.

But what will the shuttle do before then? Why are routine operations so important? There is no single answer to these questions as there are many areas—in science, in civilian applications, and in military applications—where we can see now that the shuttle is needed; and there will be many more by the time routine shuttle services are actually available.

Take, for example, *space science.* Today it takes two to five years to get a new experiment ready for space flight, simply because operations in space are so costly that extreme care is taken to make everything just right. And because it takes so long, many investigations that should be carried out—to get fundamental knowledge about the sun, the stars, the universe, and, therefore, about ourselves on earth—are just not undertaken. At the same time, we have already demonstrated, by taking scientists and their instruments up in a Convair 990 airplane, that space science can be done in a much more straight-forward way with a much smaller investment in time and money, and with an ability to react quickly to new discoveries, because airplane operations are *routine*. This is what the shuttle will do for space science.

[6] Or take *civilian space applications*. Today new experiments in space communications, or in earth resources, are difficult and expensive for the same reasons as discussed under science. But with routine space operations instruments could quickly be adjusted until the optimum combination is found for any given application—a process that today involves several satellites, several years of time, and great expense.

One can also imagine new applications that would only be feasible with the routine operation of the space shuttle. For example, it may prove possible (with an economical space transportation system, such as the shuttle) to place into orbit huge fields of solar batteries—and then beam the collected energy down to earth. This would be a truly pollution-free power source that does not require the earth's latent energy sources. Or perhaps one could develop a global environmental monitoring system, international in scope, that could help control the mess man has made of our environment. These are just two examples of what might be done with *routine* space shuttle operations.

What about *military space applications?* It is true that our military planning has not yet defined a specific need for man in space for military purposes. But will this always be [7] the case? Have the Russians made the same decision? If not, the shuttle will be there to provide, quickly and routinely, for military operations in space, whatever they may be. It will give us a quick reaction time and the ability to fly ad hoc military missions whenever they are necessary. In any event, even without *new* military needs, the shuttle will provide the transportation for today's rocket-launched military spacecraft at substantially reduced cost.

Finally, the shuttle helps our *international* position—both our *competitive* position with the Soviets and our prospects of *cooperation* with them and with other nations.

Without the shuttle, when our present manned space program ends in 1973 we will surrender center stage in space to the only other nation that has the determination and capability to occupy it. The United States and the whole free world would then face a decade or more in which Soviet supremacy in space would be unchallenged. With the shuttle, the United States will have a clear space superiority over the rest of the world because of the low cost to orbit and the inherent flexibility and quick reaction capability of a reusable system. The rest of the world—the free world at least—would depend on the United States for launch of most of their payloads.

[8] On the side of cooperation, the shuttle would encourage far greater international participation in space flight. Scientists—as well as astronauts—of many nations could be taken along, with their own experiments, because shuttle operations will be routine. We are already discussing compatible docking systems with the Soviets, so that their spacecraft and ours can join in space. Perhaps ultimately men of all nations will work together in space—in joint environmental monitoring, international disarmament inspections, or perhaps even in joint commercial enterprises—and through these activities help humanity work together better on its planet earth. Is there a more hopeful way?

The Cost of the Shuttle Has Been Cut in Half

Six months ago NASA's plan for the shuttle was one involving heavy investment—$10 billion before the first manned orbital flight—in order to achieve a very low subsequent cost per flight—less than $5 million. But since then the design has been refined, and a trade-off has been made between investment cost and operational cost per flight. The result: a shuttle that can be developed for an investment of $4.5-$5 billion over a period of six years that will still only cost [9] around $10 million or less per flight. (This means 30 flights per year at an annual cost for space transportation of 10 percent of today's NASA total budget, or one flight per week for 15 percent.)

This reduction in investment cost was partly the result of a trade-off just mentioned, and partly due to a series of technical changes. The orbiter has been drastically reduced in size—from a length of 206 feet down to 110 feet. But the payload carrying capability has not been reduced: it is still 40,000 lbs. in polar orbit, or 65,000 lbs. in an easterly orbit, in a payload compartment that measures 15x60 feet.

The reduction in investment cost is highly significant. It means that the peak funding requirements, in any one year, can be kept down to a level that, even in a highly constrained NASA budget, will still allow for major advances in space science and applications, as well as in aeronautics.

The Shuttle and the Aerospace Industry

The shuttle is a technological challenge requiring the kind of capability that exists today in the aerospace industry. An accelerated start on the shuttle would lead to a direct employment of 8,800 by the end of 1972, and 24,000 by the end of 1973. This cannot compensate for the 270,000 laid off by NASA cutbacks since the peak of the Apollo program but would take [10] up the slack of further layoffs from Skylab and the remainder of the Apollo programs.

Conclusions

Given the fact that manned space flight is part of our lives, and that the U.S. must take part in it, it is essential to reduce drastically the complexity and cost of manned space operations. Only the space shuttle will do this. It will provide both *routine* and *quick reaction* space operations for space science and for civilian and military applications. The shuttle will do this at an investment cost that fits well within a highly constrained NASA budget. It will have low operating costs, and allow 30 to 50 space flights per year at a transportation cost equivalent to 10-15 percent of today's total NASA budget.

Document III-32

Document title: George M. Low, Deputy Administrator, NASA, Memorandum for the Record, "Meeting with the President on January 5, 1972," January 12, 1972.

Source: NASA Historical Reference Collection, History Office, NASA Headquarters, Washington, D.C.

NASA leaders James Fletcher and George Low were told in a January 3, 1972, meeting in the office of OMB Director George Shultz that the White House had made a decision to approve the development of a partially reusable shuttle with a 15-foot by 60-foot payload bay. The question of whether solid-fueled or liquid-fueled strap-on boosters would be used was left open for additional study. The next day, Low and Fletcher flew to California to meet on January 5 with President Richard Nixon, who was at the Western White House in San Clemente, for a discussion of the shuttle project. This memorandum records George Low's version of the meeting. After that meeting with the president, the White House announced approval of the shuttle to the press, and Fletcher and Low answered questions about the project.

---

[1] Jim Fletcher and I met with the President and John Ehrlichman for approximately 40 minutes to discuss the space shuttle. During the course of the discussion, the President either made or agreed with the following points:

1. **The Space Shuttle.** The President stated that we should stress civilian applications but not to the exclusion of military applications. We should not hesitate to mention the military applications as well. He was interested in the possibility of routine operations and quick reaction times, particularly as these would apply to problems of natural disasters, such as earthquakes or floods. When Dr. Fletcher mentioned a future possibility of collect-

ing solar power in orbit and beaming it down to earth, the President indicated that these kinds of things tend to happen much more quickly than we now expect and that we should not hesitate to talk about them now. He was also interested in the nuclear waste disposal possibilities. The President liked the fact that ordinary people would be able to fly in the shuttle, and that the only requirement for a flight would be that there is a mission to be performed. He also reiterated his concern for preserving the skills of the people in the aerospace industry.

In summary, the President said that even though we now know of many things that the shuttle will be able to do, we should realize that it will open up entirely new fields when we actually have the capability that the shuttle will provide. The President wanted to know if we [2] thought the shuttle was a good investment and, upon receiving our affirmative reply, requested that we stress the fact that the shuttle is not a "$7 billion toy," that it is indeed useful, and that it is a good investment in that it will cut operations costs by a factor of 10. But he indicated that even if it were not a good investment, we would have to do it anyway, because space flight is here to stay. Men are flying in space now and will continue to fly in space, and we'd best be part of it.

2. **International Cooperation.** The President said that he is most interested in making the space program a truly international program and that he had previously expressed that interest. He wanted us to stress international cooperation and participation for *all* nations. He said that he was disappointed that we had been unable to fly foreign astronauts on Apollo, but understood the reasons for our inability to do so. He understood that foreign astronauts of all nations could fly in the shuttle and appeared to be particularly interested in Eastern European participation in the flight program. However, in connection with international cooperation, he is not only interested in flying foreign astronauts, but also in other types of meaningful participation, both in experiments and even in space hardware development.

3. **USSR Cooperation.** The president was interested in our joint activities with the USSR in connection with the probes now in orbit around Mars. We also described to him the real possibility of conducting a joint docking experiment in the 1975 time period. The prospect of having Americans and Russians meet in space in this time period appeared to have great appeal to the President. He indicated that this should be considered as a possible item for early policy level discussions with the USSR.

The president asked John Ehrlichman to mention both the international aspects of the shuttle and the USSR docking possibilities to Henry Kissinger.

George H. Low
cc: A/Dr. Fletcher

Document III-33

Document title: Nick MacNeil, Carter-Mondale Transition Planning Group, to Stuart Eizenstat, Al Stern, David Rubenstein, Barry Blechman, and Dick Steadman, "NASA Recommendations," January 31, 1977.

Source: NASA Historical Reference Collection, History Office, NASA Headquarters, Washington, D.C.

Unlike Presidents-elect Kennedy and Nixon, Jimmy Carter did not appoint a blue ribbon group on space during his post-election transition. Instead, the NASA transition paper was prepared by one individual who took a generally skeptical view of NASA and most of its programs. Unlike earlier space transition reports, this document was completed after President Carter entered the White House.

[no pagination]   **Summary**

1. NASA's priorities are on the development end of R & D, not the basic research end. NASA directs our R & D resources toward centralized big technology, maintaining the defense R & D orientation of the aerospace industry.

2. The Shuttle has become the end, rather than the means, because NASA space policy has been shaped by the Office of (Manned) Space Flight. The Offices of Space Science, Applications, and Aeronautics Technology get the funds that are left over.

3. Alternative directions for space technology may be neglected because

(a) the Administrator's power to hire and fire top management inhibits effective dissent

(b) important NASA managers are from Defense and the aerospace industry

(c) NASA's budget is supported and approved by a space constituency....

1. Budget History

Perhaps the agency's growth, retraction, and resiliency can best be seen in its level of employment since 1962.

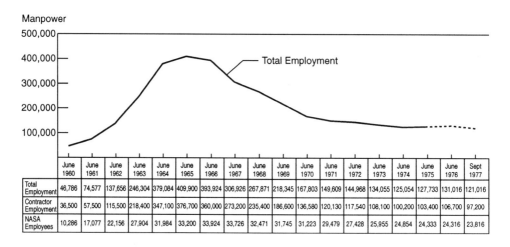

	June 1960	June 1961	June 1962	June 1963	June 1964	June 1965	June 1966	June 1967	June 1968	June 1969	June 1970	June 1971	June 1972	June 1973	June 1974	June 1975	June 1976	Sept 1977
Total Employment	46,786	74,577	137,656	246,304	379,084	409,900	393,924	306,926	267,871	218,345	167,803	149,609	144,968	134,055	125,054	127,733	131,016	121,016
Contractor Employment	36,500	57,500	115,500	218,400	347,100	376,700	360,000	273,200	235,400	186,600	136,580	120,130	117,540	108,100	100,200	103,400	106,700	97,200
NASA Employees	10,286	17,077	22,156	27,904	31,984	33,200	33,924	33,726	32,471	31,745	31,223	29,479	27,428	25,955	24,854	24,333	24,316	23,816

In real year dollars NASA funding is 70% [of what] it was in its peak year, and increasing....

3. Funding Justifications Unconvincing

a. NASA Mission Unclear

Much apprehension and uneasiness about the NASA budget would disappear if the civilian apace program, like its military counterpart, had clear objectives related to national goals.

DOD, with 38% of the space budget, would deny that its space efforts constitute a program; Defense programs are not ends but rather the means of accomplishing certain military missions, the purpose of which is to defend the nation and its allies from attack. Space programs have to compete with other means of accomplishing the same mission.

The entire NASA budget, on the other hand, is considered R & D. According to the National Science Foundation, "R & D is not an end in itself but is a means whereby national goals can be achieved more effectively and efficiently...."

What are these goals? NASA has more difficulty than most agencies in describing

EXPLORING THE UNKNOWN   561

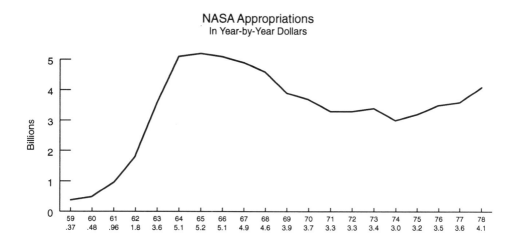

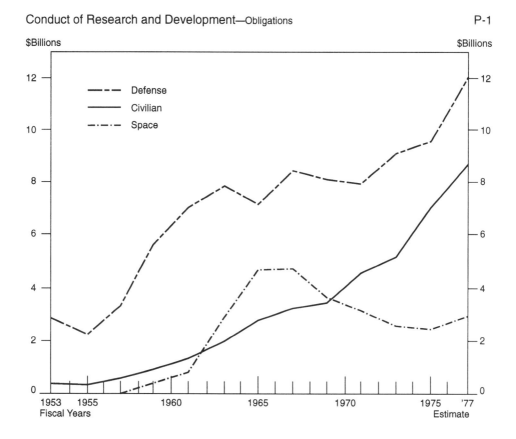

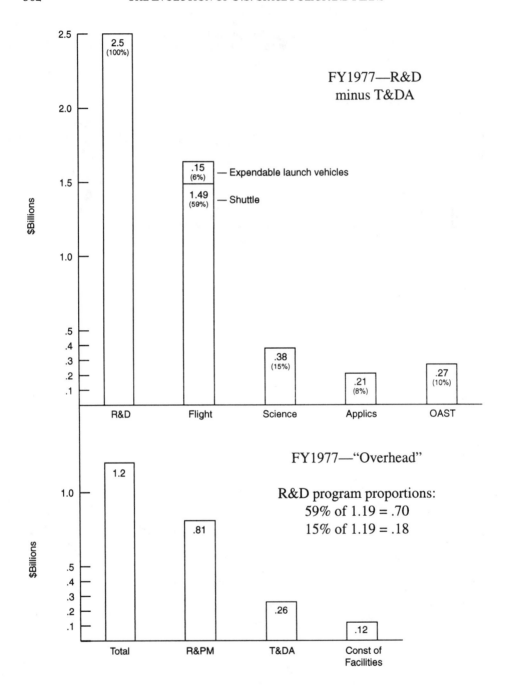

Construction of Facilities
and Research and Program Management

FY 77 Estimate
(Millions of Dollars)

Program Activities	C of F	R & PM	Function	R & PM
Space Flight	39.8	348.1	Personnel	612.4
Science	8.7	114.2	Travel & Transp.	19.7
Applications	—	87.1	Rent	61.7
Space Research	.7	75.3	Supplies	13.9
Aero Research	28.9	146.2	Equipment	2.5
Support 45.8	43.1		Other	103.9
	124.0	814.0		814.0

national goals in such a way that its programs relate to them. The law establishing NASA is no help in this regard. The National Aeronautics and Space Act of 1958 declares that the general welfare and security of the United States require "adequate provision" for aeronautical and space activities. But then it states that NASA must contribute to one or more of eight objectives, several of which go far beyond the usual understanding of welfare and security. Are we called as a nation to something greater than our welfare and security? There is no guide in law as to what provision is "adequate" for NASA's programs.

b. **The Budgeting Process**

Budgeting decisions are made in a framework provided by space scientists and engineers. This term is short-hand for those employed by NASA, by the aerospace industry, and by the universities. They decide what NASA's mission in space is . . . , they tell us the value of space activities, and they largely determine the share of available funds each program receives . . .

The club seems to achieve a Consensus in-house, by rallying around those programs with enough political appeal to have a spill-over or logjam-breaking effect for the most members. Thus seldom will scientists or engineers openly criticize programs that they consider ill-advised. Budget requests are made to OMB and the public with as little open dissent and as much gravity and consensus as possible. This behavior is the result of a shared outlook. It is aggravated by the ease with which most professional groups accept the "responsible" consensus.

It is true that independent budget evaluations are attempted by OMB, the Appropriations and Budget Committees, and the GAO. But as long as there is a general consensus within the club, and as long as evaluations are based on NASA-commissioned studies, these economy-oriented critiques will not be effectual. Indeed, not all these authorities are economy-oriented. As staffers become familiar with space activities they become interested in them. If pressures build to stimulate the economy, what better place than in one's favorite R & D program?

c. **Unconvincing Arguments**

Most agencies have a wide range of arguments to back up budget requests but they usually use these arguments informally. At budget hearings an agency will try to keep it simple. Informal arguments might lose some of their appeal to individual interests if they were listed together, and exposed to criticism.

Critics of a particular program would do a service if they took issue not only with the program's formal justification but with all the other claims that are made in support of it. However, the critic runs the risk of strengthening his case logically and weakening it here

National Aeronautics and Space Administration

FY 1978 Budget Estimates
($Millions)

Budget Authority	FY 1976	T. P.	FY 1978 Program Runout					
			FY 1977	FY 1978	FY 1979	FY 1980	FY 1981	FY 1982
Research & Development								
Space Shuttle	1,206.0	321.0	1,288.1	1,302.7	1,115.4	680.8	343.9	135.9
Space Flight Operations	188.7	48.4	202.2	297.6	360.4	508.7	594.0	592.1
Expendable Launch Vehicles	165.9	37.1	151.4	138.5	95.4	45.2	25.6	20.8
Suborbital Flight	1,560.6	406.5	1,641.7	1,738.8	1,571.2	1,234.7	963.5	748.8
Physics and Astronomy	159.3	43.5	166.3	234.1	270.2	266.9	264.0	235.7
Lunar & Planetary Expl	254.2	67.5	191.9	170.3	216.2	225.9	152.1	84.4
Life Sciences	20.6	5.4	22.1	36.4	51.1	58.5	63.8	67.9
Subtotal Science	434.1	116.4	380.3	440.8	537.5	551.3	479.9	388.0
Space Applications	178.2	47.7	198.2	224.8	242.8	266.4	163.0	135.5
Multi-Mission Modular S/C	-0-	-0-	-0-	25.0	40.0	21.0	2.5	-0-
Space Research & Tech.	74.9	19.3	82.0	115.0	114.7	112.9	~110.4	110.2
Aeronautical Res. & Tech.	175.4	43.8	190.1	245.6	302.1	311.6	264.4	'198.5
Subtotal OAST	250.3	63.1	272.1	360.6	416.8	424.5	374.8	308.7
Tracking & Data Acquisition	240.8	63.4	255.0	284.3	312.8	384.7	376.0	374.8
Technology Utilization	7.5	2.0	8.1	10.0	10.0	10.0	10.0	10.0
Energy Technology Applic.	5.9	1.5	6.0	8.5	10.5	5.0	5.0	5.0
Subtotal R&D	2,677.4	700.6	2,761.4	3,092.8	3,141.6	2,857.6	2,374.7	1,970.8
Construction of Facilities	82.1	10.7	118.1	19 5.6	200.O	161.0	125.O	110.0
Research & Program Management	792.3	220.8	813.0	818.5	818.5	818.5	818.5	818.5
Total NASA	3551.8	932.1	3,692.5	4,106.9	4,160.1	3,837.1	3,318.2	2,899.3
Additional Requirement								
Procurement of Fourth and Fifth Shuttle Orbiter				46.5	141.4	213.3	278.4	291.2
Grand Total	3,551.8	932.1	3,692.5	4,153.4	4301.5	4,050.4	3,596.6	3,190.5

National Aeronautics and Space Administration
New Starts in FY 1978 Budget
($Millions)

Research and Development	FY 1978	FY 1979	FY 1980	FY 1981	FY 1982	Balance	Total
Space Flight Operations	**15.0**						**15.0**
Space Industrialization	15.0						15.0
Physics and Astronomy	**36.0**	**79.4**	**92.0**	**95.7**	**66.8**		**435.0**
Space Telescope	36.0	79.4	92.0	95.7	66.8	65.1	435.0
Lunar and Planetary Exp.	**47.8**	**122.6**	**139.4**	**75.3**	**21.6**		**406.7**
Jupiter Orbiter Probe	20.7	78.7	102.0	61.4	18.9		281.7
Lunar Polar Orbiter	7.1	43.9	37.4	13.9	2.7		105.0
Mars Follow-on	20.0						20.0
Applications	**14.0**	**60.0**	**72.0**	**34.0**	**15.0**	**18.0**	**213.0**
Landsat D	14.0	60.0	72.0	34.0	15.0	18.0	213.0
Multi-Mission Modular Spacecraft	**25.0**	**40.0**	**21.0**	**2.5**			**88.5**
Aeronautics	**4.2**	**10.5**	**19.6**	**17.2**	**5.5**		**57.0**
Lift Cruise Fan Research Aircraft	4.2	10.5	19.6	17.2	5.5		57.0
Expendable Launch Vehicles	**.4**	**17.3**	**6.5**				
Landsat D		11.0	4.9				
Lunar Polar Orbiter	.4	6.3	1.6				
Tracking & Data Acquisition Support	**2.6**	**4.9**	**9.9**	**7.1**	**10.2**		
Total New Starts	145.0	334.7	360.4	231.8	119.1		

	1975 Actual ($Thousands)	1977 Budget Estimate ($Thousands)
Space Shuttle	797,500	1,288,100
Space Flight Operations	298,800	205,200
Expendable Launch Vehicles	139,500	151,400
Total	1,235,800	1,644,700

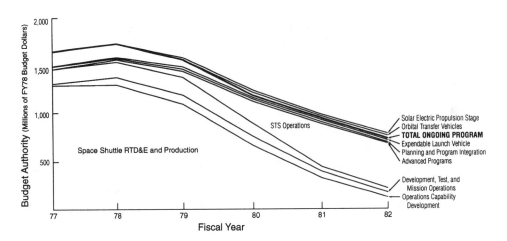

and there politically. Inaccurate claims can usually be asserted more quickly than they can be refuted.

Unconvincing arguments tend to weaken the aura of scientific invincibility and suggest a bureaucratic tendency to keep trying a multitude of arguments to weaken people's resistance, or to provide that particular argument which one group can accept. This list is by no means complete.

(1) The "Critical Threshold" Argument

NASA will maintain that funding must be kept at a certain level to preserve the necessary scientific and engineering base in people and facilities.

There is no one threshold, but a series of thresholds depending on the level and the purpose of R & D. The concept itself is suspect: if a base could be created when needed, it can be recreated. The costs of starting it up must be balanced against the costs of an entrenchment process that diverts the government's attention and funds from new problems, or new approaches to old problems.

(2) NASA's Stimulative Effect on the Economy

It is claimed that NASA expenditures are highly labor intensive, have a high multiplier effect, are not inflationary, and return the investment many times over due to the advanced technology involved.

Aside from the fact that these are the findings of studies commissioned by NASA (see following section on vested experts), the point is not how stimulative NASA spending is in absolute terms, but how stimulative it is compared to equivalent spending by some other agency in some other sector, or by different fiscal and monetary policies.

(3) The Level Budget "Commitment" of January 1972

NASA often refers to OMB assurances that it would have a funding floor in constant dollars to build the shuttle. Actually the "commitment" was made by NASA, not by OMB. The political process does not permit long-term commitments to controversial programs, yet claims of a "commitment" are still heard.

(4) The "Cutting Edge" of Technology

In simplest form this argument holds that what makes America preeminent is advanced technology, and that we depend on it for our defense and foreign exchange earnings. The "cutting edge" is never far from nuclear energy and the aerospace industry, and in these areas the high quality of research brings the highest return on our R & D dollars.

This argument confuses the value of R & D with subjective judgments on the value of different types of R & D. The issue should not be whether aircraft sales are a major earner of foreign exchange, but whether some other industry would have produced greater social and economic benefits if an equivalent amount had been invested in it. As to quality of research, talent follows money.

Our military and space efforts might well benefit from cheaper, more numerous and more expendable units. See Annex D,

(5) Individual Science Programs Vital

This tactic is to evaluate individual science programs in isolation from basic research policy. The stress is on the worthy objective and not on whether the program is cost effective, or whether data are related to results from recent or concurrent programs, or whether technology offers the possibility of leap-frogging to a more advanced stage.

The Space Telescope is a case in point. If observations are vastly improved outside the earth's atmosphere, why have observatories been built or upgraded recently in Chile, Mexico, Hawaii, Puerto Rico, and Arizona? Is there duplication from military space programs?

(6) National Security, or A Race with the Russians

The space club is not averse to taking a page out of DOD's book. When pressed, NASA will disclaim competition, but say the Russians are ahead.

DR. FLETCHER. We don't regard ourselves as being in a race with the Soviet Union. We do feel that we cannot fall too far behind in technology.

Some proponents will say that NASA programs have profound security implications. These claims suggest that DOD does not recognize certain defense needs, or that NASA should pay for a certain part of national defense.

(7) International Prestige

Akin to national defense is the notion that to keep our political and cultural values in high esteem, here and abroad, we must periodically give a display of technological virtuosity. Perhaps a winning team in sports or technology helps Americans feel less threatened by foreign developments beyond our control. We transfer vigor and Number 1 status in a particular field, to the nation as a whole. Selling international prestige on this basis panders to people's insecurities.

(8) The Call of Adventure

Adventure covers a variety of appeals to our emotions and imaginations.

– Vicarious space travel: e.g. the Shuttle will have hygienic facilities for both men and women and that "average" people—non-astronauts—will be placed in orbit, to obtain the "liberating perspectives" of space

– Creativity: e.g. the space program fills the same human need as cathedral-building in the Middle Ages.

– An Alternative to War: e.g. World War I might have been avoided if European nations could have vented their aggressiveness on space operations rather than armaments.

– A New Start for Mankind: e.g. artists' conceptions of space colonies, space factories.

– America's Destiny: e.g. the United States is the only country on this planet that can answer the riddle of man.

– Spectator Sport: e.g. Astronauts—technological sports figures—may do more to heighten this sense of adventure than to justify the added expense of manned over un-manned space missions. Perhaps they can be likened to a strong football team, that provides the gate receipts to support other athletic programs.

As with the international prestige appeal, there is a touch of "Madison Avenue" to this—space is more than R & D—it is patriotism, "gee-whiz" technology, entertainment, creativity, our national destiny. But the very success of these appeals to our emotions and imaginations shows that welfare and security are not the total of human aspiration. We enter a decision-making area full of risk for public policy which imposes certain responsibilities on government officials. Programs funded emotionally often lead to waste, empty psychological gratifications, and inflation. Ancient and recent history offer examples of peoples who have asserted their values and spirit in unprecedented, uneconomic programs that drained them, sometimes fatally, of their vitality and resources. The display of power was as important as the end it was put to. See Annex, Shuttle Justifications, 2g.

But non-economic or "irrational" motivations do exist, and they carry the potential for great creativity as well as great waste. Adventurous social programs and R & D programs have given us new knowledge, new powers and perhaps a new identity. Thus it is essential to argue over what kind of adventure we are getting into, and the costs. This is almost impossible when budget requests are made entirely on economic grounds, and the appeal to non-economic motivations is under the table. (See Recommendations.)

(9) Fait Accompli Statement

"The debate over manned vs. unmanned space flight was settled by the decision to build the Shuttle." This ploy can be used for most programs. It was a favorite for continuing the Vietnam war.

d. Expert vs. Popular Opinion

Related to the consensus of scientists and engineers with regard to budget requests is the absence of an outside vantage point that the layman could turn to for a professional but fresh perspective. The problem goes beyond the natural similarity of viewpoint of persons in the same field. As then Senator Mondale asked on May 9, 1972:

How can Congress and the public approve massive spending on new technology programs without the benefit of independent evaluations of such programs?

NASA's contractors are not likely to offer opinions which have not been checked with NASA. At times estimates suggest a form of blackmail:

NASA said that if the expendable alternate were selected, a further analysis might increase the development cost of the new expendable (launch vehicles) by about 1 billion dollars.[1]

On the one hand there must be a taxpayer counterweight to vested expert opinion. On the other hand there must be disinterested expert opinion to dampen public enthusiasm for space programs based on psychic gratifications rather than economic or scientific returns. Those who find entertainment or the solution to war in space may ultimately push space expenditures higher than space scientists and engineers. The object of both counterweights is to use national resources wisely.

4. Recommendations

a. **Outline National Goals—for example—**
  (1) The President's Economic Goals:
      – 4 ½% unemployment by 1981
      – inflation under x%
      – a balanced budget, amounting to 21% of GNP
      – a relatively favorable balance of trade
  (2) Defense Against Military Threat
  (3) Pollution at Acceptable Levels
  (4) International Collaboration, Project Humanitarian Values
  (5) Scientific Discovery
  (6) A program to Express National Values and Energy (?)

b. **Outline Corresponding Space Programs—for example—**
  (1) Defense Satellites
  (2) Scientific Probes, Experiments
  (3) Economic Application Satellites (crop and weather forecasting, resource management)
  (4) Pollution Detection Devices
  (5) Public Service Satellites (education, search and rescue)
  (6) Solar Energy Platform
  (7) Reimbursable Projects (communications satellites, space manufacturing)
  (8) International Cooperative Ventures (To train foreign scientists, share information, share the expense, use and seek superior talent.) To make these ventures effective the U.S. should avoid paternalism, or the notion that our resources give us a Manifest Destiny in space.

---

[1] Note that there is no comparison of *total* development costs of expendable and reusable launch systems.

(9) **Experimental Civilian R & D** Develop technology that applies to the way people live now, in this country and abroad.
See Annex D, NASA's R & D Direction.

c. **Accurate Labelling**

Avoid the scientific mystique. Justify programs in terms of all other activity being carried out to achieve the same broad objective. Set forth all the arguments used to support the program, strong or weak, point by point, if the program is based partly on non-economic considerations, such as curiosity or adventure, make that part of the appeal explicit, so that the rest of us can recognize the trade-offs and judge for ourselves whether the adventure will strengthen or weaken us in the long run.

d. **Downgrade Economic Objectives**

Economic stimulation should take a back seat when R & D programs are funded, because these programs invest in personnel and facilities that are far more specialized and influential, and multiply more rapidly, than the constituencies of non-R & D programs. Multiplying the supply of program administrators multiplies the demand for more of the same. This skews the economy more than it stimulates it. See Annex D, NASA's R & D Direction, Constituencies.

e. **Curb Budget Expansion**

Through Executive Order establish an obstacle course of hearings, studies and consultations for budget increases over, say, 5%. Once a benchmark budget has been set, vary the size of the slices, not the pie.... When priorities change, resources must be shifted, not added on. Scientists and engineers should be encouraged to blunt their spears on each other rather than the Administration.

f. **Use a Science/R&D Jury to Recommend R & D Priorities to the President**

Appoint a Science/R&D Council, headed by the Vice President, made up of distinguished laymen, to recommend allocation of R & D funding as to function and agency....

This Council would not resemble the President's new Committee on Science and Technology. It would present the president with a proposed R & D budget. Its members would represent labor, business, education, consumers, the press and other sectors without being weighted 2 to 1 in favor of engineers, scientists and bureaucrats. The members would serve full-time, for a year, without staff.

The Council would hear expert testimony from scientists, engineers, and those most knowledgeable about R & D. Its recommended budget would include military as well as civilian R & D in the space field, for example, the members would have security clearances adequate to allow them to try to fund military and space programs from the same "pie," minimizing duplication and maximizing multiple missions.

Discussion:

In seeking impartiality for decision-makers it would seem logical to assign laymen to determine *the over-all size* of the Science/R&D budget, and scientists and engineers to decide *how the R & D pie will be divided*. But more impartiality can be achieved by reversing the roles.

At the level of deciding between the nation's R & D and other non-defense goods and services (assuming this model is accepted, laymen are not disinterested, and may be too shortsighted to see the value of R & D, whereas the parochialism of scientific and engineering opinion would be less at the overall R & D level than at the level of funding individual R & D programs. At the program level, experts seek national commitments to their own programs, thus tending to jack up overall R & D on political considerations. Expert opinion at the *overall* R & D level, however, might dampen this effect. A compromise would be to set R & D within a narrow percentage range of general spending (not GNP).

R & D priorities are as political as they are scientific. A full debate is necessary. Without it we will be less likely to achieve mid-range budgetary stability and more importantly the lead-time necessary for contractors and scientists to prepare themselves for new problems and priorities.

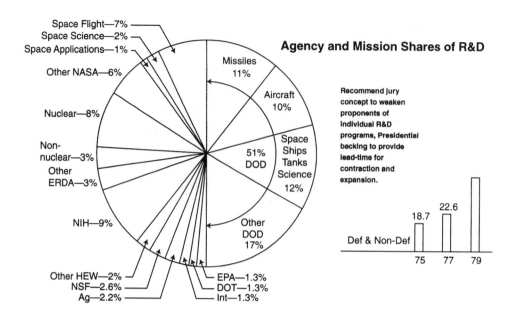
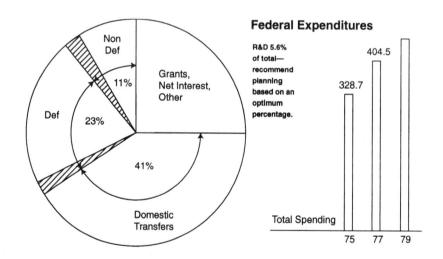

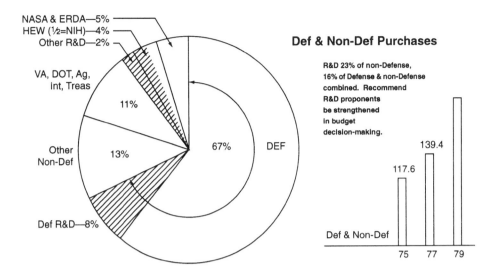

g. **Enforce ONE Circular A-109; Decentralize**

Depending on how one defines a need, circular A-109 could have prevented the Shuttle controversy. The circular states:

"When analysis of an agency's mission shows that a need for a new major system exists, such a need should not be defined in equipment terms, but should be defined in terms of the mission, purpose, capability, agency components involved, schedule and cost objectives, and operating constraints."

The present arrangement allows Space Flight to turn to Space Science and Space Applications and say "Here is your equipment, the Shuttle. Make use of it." Manned Space Flight will then find a new project. When it can no longer carry the expense of the Spacelab, or Space Industrialization, it will turn these half-started programs over to Science or Applications, the offices which should have controlled R & D from the beginning.

To take mission-orientation further, overhead could be funded out of the end-result offices (Science, Applications, and OAST). The NASA Comptroller would be split in three, and those three offices would draw up budget requests for C of F and R & PM. Facilities would bill those 3 offices for services rendered. (OMB and the GAO would have to ensure that billings represent the full cost of government facilities and personnel.) In effect all work would be contracted out, to either private or government contractors, whichever program management preferred.

Some of the advantages of decentralized budgeting are the following:

– it would weaken the agency's hierarchy, its institutional values, its growth as a bureaucracy.

– it would force economies on laboratories and facilities of marginal usefulness.

– it would increase the practical applications of independent (unstructured) R & D.

– it would make programs available to facilities, and facilities available to programs, across the board. Facilities and laboratories affected would be subject to a wider range of ideas and work opportunities.

– it would require ways of making the Civil Service more responsive to public needs.

h. **Reorient NASA Leadership**

Section 203 (b) (2) of the 1958 Aeronautics and Space Act allows the ASH Administrator to hire up to 425 executives, and set their salaries to the top Civil Service grades. This high number of excepted positions tends to unify top management. Unity is more beneficial to the implementation of policy than to the formation of it.

This system naturally lends itself to the notion of a network, and a perception that when RIFs come the Civil Service takes a disproportionate share. The system may also be related to NASA's poor Equal Employment Opportunity (EEO) record, discussed in Annex E.

Disturbing also is the number of former military personnel and former NASA contractors within the excepted positions. They cannot help but affect relations between NASA, Defense, and industry, and the kinds of work that NASA undertakes. Likewise a survey should be made of where NASA scientists have done their work. There may be a certain parochialism among the prestige institutions. This too may affect the kinds of work NASA does, who does it, and where.

If the thrust of this memorandum is followed, a new Administrator will have to come from outside the space club. He or she will have to be willing and able to use his authority to remove NASA veterans from excepted positions, and replace them with younger professionals. The purpose of these changes would be:
– to make NASA's personnel system more responsive to need, not less.
– implement the spirit of EEO.
– offset the steady increase in the average age of NASA employees.
– encourage disciplined dissent.

i. **Postpone the Appointment of a Science Adviser (OSTP) and a NASA Administrator Until These Issues Nave Been Discussed**

Do not approve new starts at NASA until the budget decision-making has been studied. Do not be rushed. If an attempt is made to challenge the experts who choose our options, appoint science and R & D officials who will support the new approach and make it work.

5. Options

The three options listed probably bear little relation to OMB options, which reflect expert opinion. My options suggest that we explore new directions for R & D, that we not commit ourselves to Shuttle operations, regardless of "cost-effectiveness," and that we give laymen a share in setting R & D priorities. To sum up, the options are based on keeping control of the agency.[2]

The options also reflect a bias toward Space Applications. Admittedly there are no options as to how Applications could use additional resources, but current NASA emphasis suggests that money (and talent) thrown at this area could bring significant results.

Option 1 - Appoint "jury" to recommend all R & D program priorities.

Budget effect - Unlikely to change level of space funding, but might favor Applications over Flight and Science.

Discussion

OMB states that R & D funding "is not a separately program[m]ed or budgeted activity of the Federal Government. Its funding must therefore be considered primarily in light of the potential contributions of science and technology to meeting agency or national goals and not as an end in itself."

Realizing that "therefore" belongs to the first sentence, not the second, the crucial point is that agency or national goals are slurred together. There is often a time-lag between *agency goals* and new perceptions of how *national goals* can be achieved. Since R & D needs more lead-time it is important that agency R & D decisions be subject to modification by a group with a totally national perspective.

---

[2] OMB may not see this as a problem. In discussing NASA's FY 1979 budget request, an OMB report states: "Substantial flexibility exists for reducing future year funding based on long-range policy and budget decisions in future budgets"—as if a program's constituency did not grow and gain a wider hearing, as if our investment does not bind us tighter to a program, with each passing year.

Advantages
1. Less overlap between military and civilian space programs.
2. Build broader consensus for longer-range planning, more lead-time for contractors.
3. A form of Executive oversight over Defense R & D.
4. More attention to national goals than agency goals.

Disadvantages
1. "Jury" unqualified to grasp issues involved.

2. "Jury" will become the captive of a particular R & D faction.

Option 2 - Build only three Shuttles. Use Shuttle for R & D and as required by individual missions.
Budget effect - Gradual reduction instead of sharp increase in Shuttle expenditure. FY 1978 is build-up year.
Discussion
Using the Shuttle as an R & D program for launch and payload reusability, while improving expendable systems, will provide greater flexibility. Some resources can be shifted to Space Applications. Publicize DOD distrust, and Mondale, Proxmire and GAO objections. OMB notes "widely divergent views."

Advantages
1. Change the big-program legacy of NASA; redirect R & D from "producers" to "consumers."
2. Take advantage of new broom; use press and public concern over inflation and bureaucracy.
3. Decision to put "Carter imprint" on Applications, give shuttle contractors an advantage in seeking Applications contracts.
4. Catch up in expendable vehicle technology, building Fords instead of Cadillacs.
5. More Science and Applications value per dollar spent, less drama.

Disadvantages
1. Political repercussions from areas surrounding affected facilities.

2. Wide currency of "cost-effectiveness" argument.

Option 3 - Expand the NASA charter to provide limited funding for specified technological breakthroughs.
Budget Effect - None.
Discussion
NASA coordinates with other agencies, industry and academia. It has capabilities in energy research, materials development, and across the spectrum of advanced technology. It put a man on the moon. It thinks more about the future than other agencies.

Why not challenge NASA to find technological breakthroughs to problems here on earth? NASA would serve as a gadfly, to weaken monopolization of R & D fields by other agencies. Congress and NASA would draw up a list of problems most susceptible to new technology, and NASA would in effect bid for a contract. New automobiles, insulation, and housing modules come to mind. See Annex U, NASA's R & D Direction, section 3.

Advantages	Disadvantages
1. Encourage new interdisciplinary approaches to old problems....	1. Maintain unneeded personnel and facilities on harebrained schemes.

Document III-34

Document title: Presidential Directive/NSC-37, "National Space Policy," May 11, 1978.

Source: NASA Historical Reference Collection, History Office, NASA Headquarters, Washington, D.C.

This directive resulted from a comprehensive review of U.S. space policy and programs undertaken during the early months of the Carter administration. It dealt primarily with the relationships among the civilian and national security portions of the national space program; its policy guidance with respect to the national security aspects of the effort was highly classified. The review was carried out under the auspices of the National Security Council, and it established a National Security Council Policy Review Committee chaired by the Director of the White House Office of Science and Technology Policy, Frank Press, as the mechanism for space policy formulation.

[1]
Presidential Directive/NSC-37
May 11, 1978

This directive establishes national policies which shall guide the conduct of United States activities in and related to the space programs and activities discussed below. The objectives of these policies are (1) to advance the interests of the United States through the exploration and use of space and (2) to cooperate with other nations in maintaining the freedom of space for all activities which enhance the security and welfare of mankind.

1. The United States space program shall be conducted in accordance with the following basic principles.
[2] a. [paragraph deleted during declassification review]
b. The exploration and use of outer space in support of the national well-being and policies of the United States.
c. Rejection of any claims to sovereignty over outer space or over celestial bodies, or any portion thereof, and rejection of any limitations on the fundamental right to acquire data from space.
d. The space systems of any nation are national property and have the right of passage through and operations in space without interference. Purposeful interference with operational space systems shall be viewed as an infringement upon sovereign rights.
e. The United States will pursue Activities in space in support of its right of self-defense.
f. [paragraph deleted during declassification review]
g. The United States will pursue space activities to increase scientific knowledge, develop useful civil applications of space technology, and maintain United States leadership in space.

h. The United States will conduct international cooperative space-related activities that are beneficial to the United States scientifically, politically, economically, and/or militarily.
i. [paragraph deleted during declassification review]
j. [paragraph deleted during declassification review]
[3] k. Close coordination, cooperation, and information exchange will be maintained among the space sectors to avoid unnecessary duplication and to allow maximum cross-utilization, in compliance with security and policy guidance, of all capabilities.

2. [remainder of page deleted during declassification review]
[4] 3. [paragraph deleted during declassification review]
4. The United States shall conduct civil space programs to increase the body of scientific knowledge about the earth and the universe; to develop and operate civil applications of space technology; to maintain United States leadership in space science, applications, and technology; and to further United States domestic and foreign policy objectives. The following policies shall govern the conduct of the civil space program.
a. The United States shall encourage domestic commercial exploitation of space capabilities and systems for economic benefit and to promote the technological position of the United States, except that all United States earth-oriented remote sensing satellites will require United States Government authorization and supervision of regulation.
b. [paragraph deleted during declassification review]
c. Data and results from the civil space programs will be provided the widest practical dissemination, except where specific exceptions defined by legislation, Executive Order, or directive apply.
d. [paragraph deleted during declassification review]
[5] e. [paragraph deleted during declassification review]
f. [paragraph deleted during declassification review]
5. The NSC Policy Review Committee shall meet when appropriate to provide a forum to all federal agencies for their policy views; to review and advise on proposed changes to national space policy; to resolve issues referred to the Committee; and to provide for orderly and rapid referral of open issues to the President for decision as necessary. The PRC will meet at the call of the Chairman for these purposes, and when so convened, will be chaired by the Director, Office of Science and Technology Policy.

Jimmy Carter

Document III-35

Document title: Zbigniew Brzezinski, Presidential Directive/NSC-42, "Civil and Further National Space Policy," October 10, 1978.

Source: NASA Historical Reference Collection, History Office, NASA Headquarters, Washington, D.C.

An initial assignment of the Policy Review Committee (Space) established by Presidential Directive/NSC-37 was to carry out a detailed review of civilian space policy and several other outstanding issues. NASA and its allies, recognizing that shuttle development was only a few years from completion, were beginning to lobby the White House for a new large-scale space initiative, and the president in this directive took a position on such a possibility. Other portions of the directive dealt with shuttle utilization for both civilian and national security missions.

[1] # Presidential Directive/NSC-42
October 10, 1978

This directive establishes national policies based on Presidential review of space policy issues submitted by the Policy Review Committee (Space). The President has approved civil and further national space policies which shall guide the conduct of United States space programs and activities discussed below. These policies are consistent with and; augment decisions reached in PD/NSC-37–National Space Policy.

ADMINISTRATION CIVIL SPACE POLICY. The United States' overarching civil space policy will be composed of three basic components.

First: Space activities will be pursued because they can be uniquely or more efficiently accomplished in space. Our space policy will become more evolutionary rather than centering around a single, massive engineering feat. Pluralistic objectives and needs of our society will set the course for future space efforts.

[2] Second: Our space policy will reflect a balanced strategy of applications, science, and technology development containing essential key elements that will:
– Emphasize applications that will bring important benefits to our understanding of earth resources, climate, weather, pollution, and agriculture.
– Emphasize space science and exploration in a manner that permits the nation to retain the vitality of its space technology base, yet provides short-term flexibility to impose fiscal constraints when conditions warrant.
– Take advantage of the flexibility of the Space Shuttle to reduce operating costs over the next two decades.
– Increase benefits by increasing efficiency through better integration and technology transfer among the national programs and through more joint projects.
– Assure US scientific and technological leadership for the security and welfare of the nation and to continue R&D necessary to provide the basis for later programmatic decisions.
– Provide for the private sector to take an increasing responsibility in remote sensing and other applications.
– Demonstrate advanced technological capabilities in open and imaginative ways having benefit for developing as well as developed countries.
– Foster space cooperation with nations by conducting joint programs.
– Confirm our support for the continued development of a legal regime for space that will assure its safe and peaceful use for the benefit of all mankind.

Third: It is neither feasible nor necessary at this time to commit the US to a high-challenge, highly-visible space engineering initiative comparable to Apollo. As the resources and manpower requirements for Shuttle development phase down, we will have the flexibility to give greater attention to new space applications and exploration, continue programs at present levels, or contract them. An adequate Federal budget commitment will be made to meet the objectives outlined above.

[3] **SPACE APPLICATIONS.** The President has approved the following:
Government Role in Remote Sensing
1. **Land Programs.** Experimentation and demonstrations will continue with LANDSAT as a developmental program. Operational uses of data from the experimental system will continue to be made by public and private users prepared to do so. Strategies for the future of our civil remote sensing efforts are to be addressed in the FY 1980 budget-review. This review should examine approaches to permit flexibility to best meet the appropriate technology mix, organizational arrangements, and potential to involve the private sector.

2. **Integrated Remote Sensing System.** NASA will chair an interagency task force to examine options for integrating current and future potential systems into an integrated

national system. This review will cover technical, programmatic, private sector, and institutional arrangements. Emphasis will be placed on user requirements; as such, agency participation will include Commerce, Agriculture, Interior, Energy, State, appropriate Executive Office participation, as well as Defense, the DCI, and others as appropriate. This task force will submit recommendations to the Policy Review Committee (Space) by August 1, 1979, for forwarding to the President prior to the FY 1981 budget review.

3. **Weather Programs.** In the FY 1980 budget review, OMB–in cooperation with Defense, the DCI, NASA, and NOAA–will conduct a cross-cut review of meteorological satellite programs to determine the potential for future budgetary savings and program efficiency. Based on this cross-cut, the Policy Review Committee (Space) will assess the feasibility and policy implications of program consolidation by April 1, 1979.

4. **Ocean Programs.** Any proposed FY 1980 new start for initial development of a National Oceanic Satellite System (NOSS) will be reviewed based on a ZBB priority ranking. The Policy Review Committee (Space) will assess the policy implications of combining civil and military programs as part of this process.

[4] 5. **Private Sector Involvement.** Under the joint chairmanship of Commerce and NASA, along with other appropriate agencies, a plan of action will be prepared by February 1, 1979, on how to encourage private investment and direct participation in the establishment and operations of civil remote sensing systems. NASA and Commerce jointly will be the contacts for the private sector on this matter and will analyze proposals received before submitting to the Policy Review Committee (Space) for consideration and action.

Communications Satellite R&D. NASA will undertake carefully selected communications technology R&D. The emphasis will be to provide better frequency and orbit utilization approaches. Specific projects selected will compete with other activities in the budget process.

Communications Satellite Services. Commerce's National Telecommunications and Information Administration (NTIA) will formulate policy to assist in market aggregation, technology transfer, and possible development of domestic and international public satellite services. This policy direction is intended to stimulate the aggregation of the public service market and for advanced research and development of technology for low-cost services. Under NTIA this effort will include: (a) an identified 4-year core budget for Commerce to establish a management structure–competitive against other budgetary priorities in Commerce–to purchase bulk services for domestic and international use; (b) support for advanced R&D on technologies to serve users with low-volume traffic requirements subject to its competitiveness against other applications expenditures; and (c) AID and Interior coordination with NTIA in translating domestic experience in emerging public service programs into potential programs for lesser-developed countries and remote territories. (U)

Long-term Economic Activity. It is too early to make a commitment to the development of a satellite solar power station or space manufacturing facility. There are very useful intermediate steps that would allow the development and testing of [5] key technologies and experience in space industrial operations without committing to full-scale projects. We will pursue an evolutionary program to stress science and basic technology-integrated with a complementary ground program–and will continue to evaluate the relative costs and benefits of proposed space activities compared to earth-based activities.

SPACE SCIENCE AND EXPLORATION GOALS

Priorities at any given time will depend upon the promise of the science, the availability of particular technology, and the budget situation in support of the following Presidentially approved goals:

– We will maintain US leadership in space science and planetary exploration and progress.

– The US will continue a vigorous program of planetary exploration to understand the origin and evolution of the solar system. Our goal is to continue the reconnaissance of the outer planets and to conduct more detailed exploration of Saturn, its moons, and its rings; to continue comparative studies of the neighboring planets, Venus and Mars; and to conduct reconnaissance of comets and asteroids.

– To utilize the space-telescope and free-flying satellites to usher in a new era of astronomy, as we explore interstellar molecules, quasars, pulsars, and black holes to expand our understanding of the universe and to complete the first all sky survey across the electromagnetic spectrum.

– To develop a better understanding of the sun and its interaction with the terrestrial environment. Space probes will journey towards the sun. Earth orbiting satellites will measure the variation in solar output and determine the resultant response of the earth's atmosphere.

– To use the Space Shuttle and Spacelab, in cooperation with the Western Europeans, to conduct basic research that complements earth-based life science investigations and human physiology research.

– Our policy in international space cooperation should include three primary elements: (1) support the best science available regardless of national origin, but expand our international planning and coordinating effort; (2) seek [6] supplemental foreign support only for selected experiments on spacecraft which have been chosen on the basis of sound scientific criteria; and (3) avoid lowering cooperative activities below the threshold where our science and international cooperative efforts would suffer.

STEPS TO INCREASE BENEFITS FOR RESOURCES EXPENDED. The President has approved the following:

Strategy to Utilize the Shuttle
1. [Paragraph deleted during declassification review]
2. [Paragraph deleted during declassification review]
3. Incremental improvements in the Shuttle transportation system will be made as they become necessary and will be examined in the context of emerging space policy goals. An interagency task force will make recommendations on what future capabilities are needed. Representation will include NASA, Defense, the DCI, Commerce, Interior, Agriculture, OMB, NSC, OSTP, State, and others as appropriate. This task force will submit the findings to the Policy Review Committee (Space) for transmittal to the President by August 1, 1979.
4. [Paragraph deleted during declassification review]

[7] **Technology Sharing**. The existing Program Review Board (PBS) will take steps to enhance technology transfer between the sectors. The objective will be, as directed in PD/NSC-37, to maximize efficient utilization of the sectors while maintaining necessary security and current management relationships among the sectors. The PBS will submit an implementation plan to the Policy Review Committee (Space) by May 15, 1979. In addition, the PBS will submit subsequent annual progress reports.

Zbigniew Brzezinski

Document III-36

Document title: George M. Low, Team Leader, NASA Transition Team, to Mr. Richard Fairbanks, Director, Transition Resources and Development Group, December 19, 1980, with attached: "Report of the Transition Team, National Aeronautics and Space Administration."

Source: NASA Historical Reference Collection, History Office, NASA Headquarters, Washington, D.C.

The transition team assembled to advise President-elect Ronald Reagan on space issues consisted of individuals with long experience in the field, both within and outside of NASA. It was chaired by George Low, who had left NASA in 1976 after a long career to become president of the Rensselaer Polytechnic Institute. The team's report provided a detailed set of recommendations and actions for the incoming administration.

December 19, 1980

Mr. Richard Fairbanks, Director
Transition Resources and Development Group
1726 M Street, NW
Washington, DC 20270

Dear Mr. Fairbanks:

I am pleased to submit the report of the transition team for the National Aeronautics and Space Administration (NASA). We hope you will find that it presents a balanced view of the status of the agency, its problems, strengths, and potentials. Team members received full cooperation from NASA officials. Our group worked together well, with frequent unanimity on identification and resolution of issues.

Recognizing that many members have been involved in the past with space programs, the team was particularly sensitive to its appearance of a pro-space bias. Members worked hard to prepare an objective report, with minimal personal advocacy. Team members have asked, however, that in this letter I emphasize our view that NASA and its civil space program represent an opportunity for positive accomplishment by the Reagan administration. In contrast with many government agencies that are mired in seemingly insoluble controversy, NASA can be many things in the future—the best in American accomplishment and inspiration for citizens.

We are pleased to have had the opportunity to aid the new administration and trust that our report will serve you and the next NASA Administrator well. The members of the team and I will be happy to provide additional consultation should it be needed.

Sincerely,
George M. Low
Team Leader
NASA Transition Team...

[1] **I. INTRODUCTION**

A. **Overview**

In 1958 the people of the United States set out to lead the world in space. By 1970 they had achieved their goal. Men walked on the moon, scientific satellites opened new windows to the universe, and communications satellites and new technologies brought economic return. With these came new knowledge and ideas, a sense of pride, and national prestige.

In 1980, by contrast, United States leadership and preeminence are seriously threatened and measurably eroded. The Soviet Union has established an essentially permanent manned presence in space, and is using this presence to meet economic, military, and foreign policy goals. Japan is broadcasting directly from space to individual homes and business, and France is moving ahead of the United States in preparing to reap the economic benefits of satellite resource observation. Ironically, U.S. commercial enterprises are turning to France to launch their satellites. In space science, the United States has decided to forego the rare opportunity to visit Halley's comet in 1986, yet the Soviet Union, the European Space Agency, and Japan are all planning such a venture.

Technically, it is within our means to reestablish U.S. preeminence in space. The civil space program and the National Aeronautics and Space Administration offer a number of options to carry out the purpose and direction of U.S. aeronautics and space activities. These options are examined in this report in full recognition of the need for fiscal restraint in the immediate future.

B. **The U.S. Aeronautics and Space Program in 1980**

The National Aeronautics and Space Administration (NASA) was created in 1958 by the National Aeronautics and Space Act (PL 85-568), largely as a response to the launch of Sputnik by the Soviet Union.

The Act declared that it is the policy of the United States that activities in space be developed to peaceful purposes for the benefit of all mankind, and that these activities (except those primarily associated with the defense of the United States) should be the responsibility of a civilian agency. [2] This agency—NASA—was chartered to carry out significant programs in aeronautics, space science, space technology and applications, and manned space flight.

In 1961, the President challenged the nation to land men on the Moon by the end of that decade. The Apollo project not only made the United States preeminent in space technology, but also instilled a sense of pride in the American people. Apollo's success was due to a long term commitment; adequate and stable financial support; a technological partnership among government, industry and universities; and disciplined managers drawn from within and outside the government.

Also in the past two decades, automated spacecraft explored Mercury, Venus, Mars, Jupiter and Saturn, while telescopes above the earth's atmosphere gave us new eyes to learn about our universe—the strange world of pulsars, quasars and black holes. The result was a new understanding of the past, present, and future of our total environment.

In the meantime, communications satellites have spawned an entire new industry, weather satellites can warn us of storms, and remote sensing satellites offer tremendous economic potential from assessing and managing the earth's resources.

At the end of 1980 we are on the eve of the launch of *Columbia*, the first Space Shuttle, and its promise to provide a multiplicity of benefits—in science, in exploration, in terrestrial applications, and in the security of our nation—from easy access to this new ocean of space.

C. **Aeronautics**

Since 1915 NASA (and its predecessor, the National Advisory Committee for Aeronautics) has been the world leader in aeronautical research. At NASA's laboratories are many of the national facilities and technical experts necessary to continue progress in the rapidly advancing field of aviation. NASA is also at the focal point of a unique partnership among industry, universities, the Department of Defense, and NASA itself that has been responsible for U.S. preeminence in aeronautics.

Built on the foundation of this research and technology base, the U.S. aviation industry employs about 1,000,000 Americans, ranks second largest among U.S. manufacturing employers, contributes more than any other manufacturing industry to the U.S. balance of trade, and has replaced agriculture as first in net trade contribution.

Continued advancements in research and technology are essential if the U.S. aviation industry is to remain a viable competitor in the world market.

D. The Space Program and U.S. Policy

In recent years the United States has lost its competitive edge in the world, militarily, commercially, and economically, [3] and our competition with the Soviet Union has taken on a new dimension.

The Soviet Union recognizes that science and technology are major factors in that competition. The nation that is strong in science and technology has the foundation to be strong in all other areas and will be perceived as a world leader.

Aeronautics and space can be major factors in our technological strength. They demand the very best in engineering, because the consequences of mistakes are great: the crash of an aircraft, or the complete failure of a spaceship.

A viable aviation industry and a strong space program are important visible elements in our international competition. Beyond these fundamental points, the United States civil space program, unlike many other government programs and agencies, has significant actual and potential impact on U.S. policy. Although some elements of the program have been so utilized, their potential in U.S. policy remains largely unrecognized and unrealized. The major factors are as follows:

1. National Pride and Prestige

National pride is how we view ourselves. Without a national sense of purpose and identity, national pride ebbs and flows in accordance with short-term events. The Iranian hostage situation and the abortive rescue mission have done harm to our national pride quite out of proportion to our true abilities as a nation. On the other hand, the recent Voyager visit to Saturn, reported by an enthusiastic press, made a significant contribution to our sense of self-worth. The space program has characteristics of American historic self-image: a sense of purpose; a pioneering spirit of exploration, discovery, and adventure; a challenge of frontiers and goals; a recognition of individual contributions and team efforts; and a firm sense of innovation and leadership.

National prestige is how others view us, the global perception of this country's intellectual, scientific, technological, and organizational capabilities. In recent history, the space program has been the unique positive factor in this regard. The Apollo exploration of the Moon restored our image in the post-Sputnik years, and the Voyager exploration of Saturn was a bright spot in an otherwise gloomy period of dwindling world recognition. With space programs we are a nation of the present and the future, while in the eyes of the world we become outward and forward looking.

2. Economics and Space Technology

A vigorous space program has provided many technological challenges to our nation. Efforts such as Apollo, Voyager, and the Space Shuttle have involved challenges and risks far more significant than those of short-term technological needs.

Meeting these challenges has resulted in a "technological push" to American industry, fostering significant innovation in [4] a wide range of high technology fields such as electronics, computers, science, aviation, communications and biomedicine. The return on the space investment is higher productivity, and greater competitiveness in the world market.

The space program also returns direct dividends, as in the field of satellite communications. The potential economic returns from satellite exploration for earth's resources are great.

3. Scientific Knowledge and Inspiration for the Nation's Youth

U.S. leadership in the scientific exploration of space has provided new knowledge about the earth and the universe, thus forming the basis for applied research and development–a significant factor in our society and economy.

The exploration of space has provided an inspirational focus for large numbers of young people who have become students of engineering and science. At a time when there is a shortage of technically trained people, when the U.S. productive vitality depends on the application of science, the space program could help attract young people into these fields.

4. Relation to U.S. Foreign Policy

Aspects of the civil space program can serve as instruments to develop and further U.S. foreign policy objectives. Not only can the space program contribute to how this country is viewed in the eyes of the world, but cooperative space activities, such as the U.S.-U.S.S.R. Apollo-Soyuz mission and European Space Agency payloads on the Space Shuttle, are important to other countries. Technology associated with the space program has resulted in strong economic and technological interaction with developed countries, as well as in important aid to underdeveloped countries, particularly in the areas of communications and resource exploration.

E. Observations

At the end of 1980 the U.S. civil space program stands at a crossroads. The United States has invested in a great capability for space exploration and applications, a capability that provides benefits in national pride and prestige, in science and technology, in the inspiration of young people, in foreign policy, and in economic gain.

Now this capability is waning. NASA and the space program are without clear purpose or direction....

[39]   **VI. SUMMARY OF RECOMMENDATIOMS**

NASA represents an important investment by the United States in aeronautics and space. The agency's programs have provided, and continue to offer, benefits in science and technology, in national pride and prestige, in foreign policy, and in economic gain. However, in recent years the agency has been underfunded, without purpose or direction. The new administration finds NASA at a crossroads, with possible moves toward either retrenchment or growth. The transition team has examined ten major areas and various options for dealing with them. For each issue, the team has made recommendations as follows:

A. **Presidential statement of purpose of the U.S. civil space program** (pages 5-7)

It is recommended:

1. That the President recognize the importance of the U.S. space program at an early date (e.g., the inaugural address) without yet making a commitment.

2. That the purpose and direction of the U.S. space effort be defined, and that a commitment to a viable space program be articulated by the President at a timely opportunity, such as the first flight of the Shuttle in the spring of 1981.

(N.B. A viable space program could be smaller than, equal to, or larger than the present one, but it must have purpose and direction.)

B. **NASA as an organization** (pages 8-11)

1. The NASA Administrator

It is recommended that the President select a politically experienced and strong manager as NASA Administrator, that he reestablish the Administrator's role as that of principal advisor on civil space matters, and that he be accessible to the Administrator as necessary.

2. Management capability

It is recommended that the Administrator, working either within the agency or with an outside group, assess NASA's vitality and discipline in management of complex projects, and make changes necessary to effect improvements.

3. Staffing

It is recommended that the dual problem of bringing experienced people from industry into governnent, and of [40] attracting bright young engineers and scientists into government service be addressed immediately, for the government as a whole and for NASA in particular.

4. The size of NASA

It is recommended that the question of whether or not NASA needs all its field centers be addressed as soon as the purpose of the aeronautics and space program is defined.

C. **Space policy and conflict resolution** (pages 12-16)

It is recommended that space policy development and conflict resolution be assigned to the NASA Administrator or special ad hoc groups as the need arises; and that consideration be given to a permanent space policy board for this purpose.

D. **The civil space program and national policy** (pages 14-15)

It is recommended that the administration develop an unequivocal statement of national space policy and an organizational framework that promotes economic exploitation of our capabilities and uses space to further our international goals.

In the area of remote sensing, the administration should undertake the development of an integrated civil program.

In foreign policy, the administration should develop procedures for the Department of State and other government agencies, together with industry, to employ space technology to further foreign policy objectives.

E. **Space Shuttle flight readiness** (pages 16-17)

It is recommended that

1. The NASA Administrator schedule immediate briefings and reviews, with NASA and contractor personnel, to become familiar with the Shuttle and its problems.

2. The Administrator obtain a formal assessment of Shuttle readiness from the Aerospace Safety Advisory Panel.

3. The Administrator seek the advice (outside the regular review process), of the knowledgeable outside experts.

4. The Administrator and/or Deputy participate in scheduled reviews and make specified Flight Readiness Firing and Launch decisions.

[41] F. **U.S. space launch capability** (pages 18-19)

It is recommended:

1. That existing plans for initial Shuttle operations, retention of expendable launch vehicles for the time being, and transfer of payloads to the Shuttle, be allowed to stand.

2. That at an appropriate time after the first flight (or flights) of the Shuttle, the President direct the Administrator of NASA to address the issues of Shuttle enhancement, continued Shuttle production, and expendable launch vehicle production; and to resolve them in the best interest of the United States, taking into account all users–commercial, civilian, government, DOD, and foreign.

G. **The transfer from research and development to operations** (pages 20-22)

It is recommended:

1. That the question of operational management of remote sensing satellite systems be addressed on an urgent basis (see section on "The Civil Space Program and National Policy").

2. That consideration be given to turning the operation of expendable launch vehicles over to a government agency other than NASA or to a private commercial organization in the next year.

3. That long term Space Shuttle operations be addressed only after some flight experience with the Shuttle is in hand.

H. **Aeronautics** (pages 23-24)

It is recommended that NASA's traditional role of research and technology support to civil and military aviation be reaffirmed, and perhaps even strengthened, to help stem the loss of U.S. leadership in aviation.

I. **NASA's role in areas other than aeronautics and space** (pages 25-26)

It is recommended that NASA's future role in non-aeronautics and non-space activities be confined to assistance to other agencies as requested for limited periods of time only, using cost reimbursements as possible, and that current long term commitments in other areas be eventually moved from NASA.

[42] J. **Personnel** (pages 27-30)

It is recommended that the new NASA Administrator review the situation of reemployed annuitants at an early date with the view of terminating the employment of many of them....

Requested New Starts
($Millions)

	Funding Requested	Funding Obtained
FY 1981		
Solar Electric Propulsion	20	7.5
Power Extension Package	17	—
Gamma Ray Observatory	19.1	19.1
Venus Orbiting Imaging Radar	30	—
National Oceanic Satellite System	15	5.8
Upper Atmospheric Research Satellite	10	—
Numerical Aerospace Simulator	3	—
FY 81 Total	114.1	32.4
FY 1982		
Venus Orbiting Imaging Radar	40	40
Halley Flyby	4	—
Halley Watch	1	1
Upper Atmospheric Research Sat. (Instrument only)	20	20
Geological Application	21.3	—
Numerical Aerodynamic Simulator	16	16
Large Composite Primary Structures	8	4
Energy Efficient Transport Technology Development	7	—
General Aviation Propulsion Technology	5.5	—
High Temperature Engine Core	6	—
Advanced Rotorcraft Technology	5	—
Cooperative Auto Research Program	6.5	—
Solar Power Systems	5	—
Research & Technology Base Augmentation	25	9
Solar Electric Propulsion	28	18
Shuttle Performance Augmentation	28	—
Shuttle Performance Augmentation (Study)	5	5
Power Extension Package	27	—
Advanced Space Transportation System Capability	4.5	—
FY 82 Total	262.8	113
Grand Total	376.9	137.9

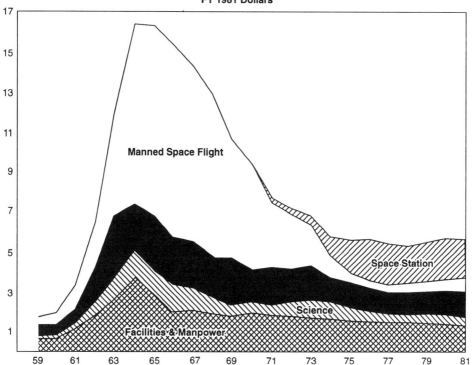

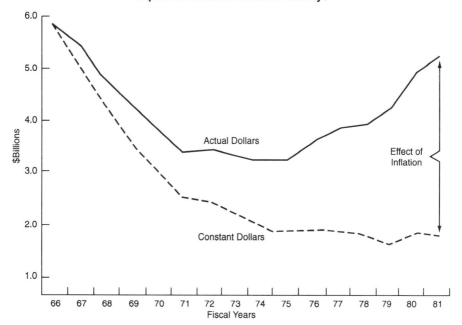

Total Employment on NASA Programs

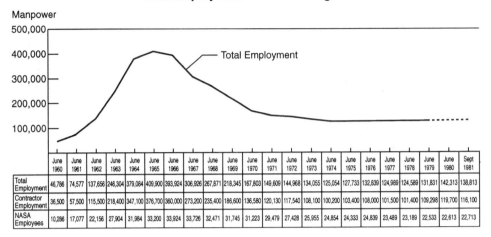

	June 1960	June 1961	June 1962	June 1963	June 1964	June 1965	June 1966	June 1967	June 1968	June 1969	June 1970	June 1971	June 1972	June 1973	June 1974	June 1975	June 1976	June 1977	June 1978	June 1979	June 1980	Sept 1981
Total Employment	46,786	74,577	137,656	246,304	379,084	409,900	393,924	306,926	267,871	218,345	167,803	149,609	144,968	134,055	125,054	127,733	132,839	124,989	124,589	131,831	142,313	138,813
Contractor Employment	36,500	57,500	115,500	218,400	347,100	376,700	360,000	273,200	235,400	186,600	136,580	120,130	117,540	108,100	100,200	103,400	108,000	101,500	101,400	109,298	119,700	116,100
NASA Employees	10,286	17,077	22,156	27,904	31,984	33,200	33,924	33,726	32,471	31,745	31,223	29,479	27,428	25,955	24,854	24,333	24,839	23,489	23,189	22,533	22,613	22,713

12/11/80

Budget History
(Millions of Dollars)

Subfunction Code		Annual Authorization				Appropriation	
		FY '77	FY '78	FY '79	FY '80	FY '81	FY '81**
	Research and Development	2856.4	3041.5	3522.6	4123.5	4436.8	4396.2
253	Space Shuttle	1383.1	1354.2	1628.3	1871.0	1873.0	1873.0
253	Space Flight Operations	202.7	267.8	315.9	463.3	779.5	769.5
253	Expendable Launch Vehicles	151.4	134.5	74.0	70.7	55.7	55.7
254	Physics & Astronomy	166.3	228.2	285.5	337.5	352.7	352.7
254	Planetary Exploration	192.1	153.2	187.1	220.2	179.6	179.6
254	Life Sciences	22.1	33.3	42.6	43.9	45.2	43.2
254	Space Applications	198.2	239.8	280.3	338.3	378.7	361.5
254	Technology Utilization	8.1	9.1	12.1	12.1	12.6	12.1
402	Aeronautical Research & Technology	191.1	234.0	275.1	309.3	289.8	282.55
254	Space Research & Technology	82.8	99.7	111.3	119.4	115.2	113.2
254	Energy Technology	3.5	7.5	5.0	5.0	4.0	4.0
255	Tracking & Data Acquisition	255.0	280.0	305.4	332.8	349.7	349.0
	Construction of Facilities	120.3	160.9	150.0	157.6	118.0	115.0
	Research & Program Management	845.1	892.8	940.4	1001.2	1033.2	1030.0
	Total	3821.8	4095.2	4612.6	5282.3	5587.9	5541.2
	Appropriation	3819.1	4063.7	4558.8	5243.4	5541.2	

**Does not include proposed 2% cut

Document III-37

Document title: Hans Mark and Milton Silveira, "Notes on Long Range Planning," August 1981.

Source: NASA Historical Reference Collection, History Office, NASA Headquarters, Washington, D.C.

Soon after taking office as NASA's Deputy Administrator in July 1981, Hans Mark and his assistant Milton Silveira prepared this document in response to a call from new Administrator James Beggs for ideas on the future of the space agency. The document was widely circulated within NASA, and soon adopted by Beggs and Mark as the basis for their initial actions as they assumed control.

[1]
Notes on Long Range Planning

The development of long range planning for NASA should be based on Section 5 in the 1958 Space Act requiring "the preservation of the role of the United States as a leader in aeronautical and space science and technology. . . ." This may be a difficult thing to do in view of limited funds that will be available to NASA in the coming years but the intent of the statement in the law is crystal clear and NASA must act accordingly.

1) FACILITIES

Fundamental to all that NASA does are the facilities that exist at NASA's research and development centers. It is not always recognized but the NASA aeronautical facilities are vital, not only to aircraft design, but also to the development of our space technology. For example, the 40' x 80' wind tunnel at the Ames Research Center which is justified solely as an aeronautical facility was used for testing the flying qualities of the Space Shuttle during the critical approach and landing phase. A one-third scale model was tested many hundreds of hours in the wind tunnel to insure performance, stability, and control characteristics. There are many other examples where wind tunnel and high temperature facilities are used to insure safe flight of a spacecraft as it passes through the atmospheric portion of its flight.

Broadly speaking, NASA's facilities fall into five separate categories:
1. Wind Tunnels
2. Flight and Operations Simulators
3. Propulsion Test Facilities
4. Experimental Airplanes
5. Computational Facilities

[2] Recently, heavy investments have been made in required wind tunnel facilities. Approximately $250M have been spent, improving the Ames 40' x 80' wind tunnel and building at Langley the High Reynolds Number Cryogenic Tunnel. Large investments have also been made in flight simulators, although more needs to be done in developing and building simulators to overcome current deficiencies. There is a need to develop more facilities for the simulation of operations and construction in space with a zero "g" environment and under demanding thermal conditions. The major aeronautical propulsion facility in the country is being developed by the United States Air Force at the Arnold Engineering Development Center. NASA must take advantage of this facility as best it can. NASA must also develop a policy toward the development of propulsion facilities at the Lewis Research Center. Particularly, NASA must also see to it that the rocket propulsion

test stands are adequate for programs in launch vehicles that may be initiated following the completion of the Space Shuttle program. Experimental aircraft tend to be more specialized toward specific flight configurations. However, there are some programs such as the F-8 fly-by-wire aircraft and the Boeing 737 control configured vehicle in which the aircraft are used more-or-less as general purpose simulation facilities. Computers are not usually regarded as facilities but they should be viewed as such. The Numerical Aerodynamic Simulator now being proposed is particularly important in this regard since it may overcome certain limitations in the simulations of the other facilities now operated by NASA (wind tunnels, propulsion facilities and flight simulators) if the promise of computational methods in aerodynamics, chemically reacting flows, and dynamic structures can be realized. The maintenance and development of the necessary facilities to accomplish the mission stated in the law must, therefore, have the highest institutional priority in NASA.

[3]   2) AERONAUTICS

Work in aeronautics by NASA, and the NACA prior to 1958, has traditionally been oriented toward the support of military and civil aviation. Future interest in the military is likely to be centered on the development of a new long range combat aircraft (LRCA) by the United States Air Force having low radar, infrared, and visible observables (i.e., stealth technology), the creation of a new family of V/STOL aircraft for the Navy, and the continuing enhancement of the performance of rotor craft for the Army. To maintain a lead in civil aircraft sales, continual improvements must be made for greater economy. The technology suitable for third level carriers (i.e. commuter airlines) is likely to be the major civil requirement. The latter is especially important in view of the inroads being made by foreign competition in that field. Right now the Dehavilland Twin Otter, the DHC-7, and the Shorts Skyvan dominate that field in the United States. In addition to all of these things, a strong basic research program in fluid mechanics, materials and other topics related to aeronautics and space vehicles must be maintained.

3) THE SPACE SHUTTLE

The major technological development carried out by NASA in the last decade is the Space Shuttle vehicle. That basic development is now nearly complete and the next step is to turn it into an operational system. This effort must have the highest programmatic priority in NASA for the coming years to realize a return for this large investment. It should take about three years to make the space Shuttle an operational transportation system. It is necessary to arrive at an agreed-upon definition of what is meant by "operational" and to determine whether NASA should be the agency that operates the Shuttle or whether some other institutional mechanism needs to be provided for that purpose. The organizational structure needs to be developed for [4] Shuttle Operations. No matter how the matter of Shuttle Operations is finally decided, the Johnson Space Center should phase out of the operational mission during the next three years. It is very unlikely that it will be possible to control costs of operation if the developmental attitudes that prevail at the Johnson Space Center dominate after the Space Shuttle becomes operational. The operations of the Space Shuttle, both launch as well as mission control, should be handled by the Kennedy Space Center and by Vandenberg Air Force Base once the West Coast launch facility is complete.

4) THE SPACE STATION

While the Space Shuttle becomes operational, a project to establish a permanent presence in space (i.e., a Space Station) should be initiated. This should become the major new goal of NASA and, some time during the next two years, the President should be persuaded to issue a statement proclaiming a national commitment to that effect. The necessary arguments that justify this step must be carefully developed now, and these arguments range from national security (i.e., arms control verification, military surveillance) to the improvement of space operations (i.e., satellite maintenance on orbit and other things of this kind). The necessary committees of the National Academy of Engineering, the National Academy of Sciences, and other bodies of this kind should be established to set up now the technical baselines for this new enterprise.

5) UNMANNED LAUNCH VEHICLES

The Shuttle program has led to the creation of a new propulsion technology which should be further exploited. It is now generally agreed that unmanned launch vehicles will not be phased out completely once the Shuttle is operational. They will always be necessary to supplement the Shuttle launch capability. The current launch vehicles, (Atlas, Titan, Delta) are based on technology that is now thirty [5] years old and should be replaced by more efficient and economical vehicles. New unmanned launch vehicles based on the Shuttle technology using solid rocket boosters and the Shuttle's main engine system should be developed. The solid rocket booster itself is an excellent rocket with a sea level thrust of the order of 2.5M lbs. Several solid rocket boosters strapped together could provide a formidable launch vehicle in terms of payload capacities. Such a vehicle with three solid rocket boosters could put into low earth orbit a payload weighing something like 100,000 lbs. and perhaps up to 20,000 lbs. into geosynchronous orbit with an appropriate upper stage. An important feature of the solid rocket booster is that they are recoverable which means that the cost advantages inherent in that property could be important. This new generation of launch vehicles would not be "expendable" although it would be unmanned.

6) SCIENTIFIC EXPLORATION

NASA has a fundamental responsibility to continue with the scientific exploration of objects in space and conditions in space. In the coming decade, scientific investigations conducted in earth orbit will be the most important because these take the best advantage of the unique properties of the Shuttle. Specifically, this means that astronomy, experiments involving certain cosmological things such as general relativity and experiments in zero gravity using Spacelab will be the dominant trend in scientific space research. An extremely important aspect of this are the medical and biological experiments to be done using the Shuttle to establish what must be done to permit people, animals, and plants to live in zero gravity conditions for lengthy periods. It is probable that exploration will be deemphasized somewhat until we have a Space Station that can serve as a base for the launching of a new generation of planetary exploration spacecraft. It is apparent that the return of samples from various bodies in the solar system will be given highest priority once that time arrives.

[6]
7) SPACE APPLICATIONS

The applications program should emphasize the scientific part of earth observations, specifically oceanography, geodesy, and things of this kind. In view of the Administration's policies with respect to technical demonstration programs, NASA should de-emphasize efforts to commercialize various applications projects. The applications program should also emphasize technology development and should cooperate closely with the national security community in these efforts. It is likely that the nation's surveillance satellites will move to geosynchronous orbit in the next two decades. This means that large space structures will be required, mirrors, antennas, and other systems of this type. NASA should be extremely active in the development of this technology and should establish the closest possible support of the national security community in achieving these objectives.

A few thoughts regarding future directions for NASA have been outlined in this paper. Obviously, much more detail needs to be done to develop some of these ideas. It is very important to begin now by setting up the proper procedural methods within NASA as well as the NASA advisory structure to make certain that these ideas are properly considered and developed into a coherent long range plan for the nation's aeronautical and space programs.

Hans Mark
Milton Silveira
August 1981

Document III-38

Document title: National Security Decision Directive Number 42, "National Space Policy," July 4, 1982.

Source: NASA Historical Reference Collection, History Office, NASA Headquarters, Washington, D.C.

In 1981, its first year in office, the Reagan administration issued a National Security Decision Directive (NSDD-8, November 13, 1981) that reiterated the central role of the Space Transportation System in U.S. space activities. The White House then initiated a comprehensive space policy review under the direction of new Science Adviser George Keyworth II. The results of that review were contained in NSDD-42, issued on July 4, 1982. This directive replaced NSDD-8 and the three Carter administration space policy statements, NSDD-37, 42, and 54. It also established as the primary forum for space policy formulation the National Security Council Senior Interagency Group (Space)—SIG (Space)—chaired by the assistant to the president for national security affairs. SIG (Space) was the locus of policymaking throughout the two terms that Ronald Reagan was president.

National Security Decision
Directive Number 42

[1] July 4, 1982

National Space Policy

I. INTRODUCTION AND PRINCIPLES

This directive establishes national policy to guide the conduct of United States space program and related activities; it supersedes Presidential Directives 37, 42, and 54, as well as National Security Decision Directive 8. This directive is consistent with and augments the guidance contained in existing directives, executive orders, and law. The decisions outlined in this directive provide the broad framework and the basis for the commitments necessary for the conduct of United States space programs.

The Space Shuttle is to be a major factor in the future evolution of United States space programs. It will continue to foster cooperation between the national security and civil efforts to ensure efficient and effective use of national resources. Specifically, routine use of the manned Space Shuttle will provide the opportunity to understand better and evaluate the role of man in space, to increase the utility of space programs, and to expand knowledge of the space environment.

The basic goals of United States space policy are to: (a) strengthen the security of the United States; (b) maintain United States space leadership; (c) obtain economic and scientific benefits through the exploitation of space; (d) expand United States private-sector investment and involvement in civil space and space-related activities; (e) promote international cooperative activities that are in the national interest; and (f) cooperate with other nations in maintaining the freedom of space for all activities that enhance the security and welfare of mankind.

[2] The United States space program shall be conducted in accordance with the following basic principles:

A. The United States is committed to the exploration and use of outer space by all nations for peaceful purposes and for the benefit of all mankind. [sentence deleted during declassification review]

B. The United States rejects any claims to sovereignty by any nation over outer space or celestial bodies, or any portion thereof, and rejects any limitations on the fundamental right to acquire data from space.

C. The United States considers the space systems of any nation to be national property with the right of passage through and operations in space without interference. Purposeful interference with space systems shall be viewed as infringement upon sovereign rights.

D. The United States encourages domestic commercial exploration of space capabilities, technology, and systems for national economic benefit. These activities must be consistent with national security concerns, treaties, and international agreements.

E. The United States will conduct international cooperative space-related activities that achieve sufficient scientific, political, economic, or national security benefits for the nation.

F. [paragraph deleted during declassification review]

G. The United States Space Transportation System (STS) is the primary space launch system for both national security and civil government missions. STS capabilities and capacities shall be developed to meet appropriate national needs and shall be available to authorized users—domestic and foreign, commercial, and governmental.

[3] H. The United States will pursue activities in space in support of its right of self-defense.

I. The United States will continue to study space arms control options. The United States will consider verifiable and equitable arms control measures that would ban or otherwise limit testing and deployment of specific weapons systems should those measures be compatible with United States national security. The United States will oppose arms control concepts or legal regimes that seek general prohibitions on the military or intelligence use of space. [declassified]

II. SPACE TRANSPORTATION SYSTEM

The Space Transportation System (STS) is composed of the Space Shuttle, associated upper stages, and related facilities. The following policies shall govern the development and operation of the STS:

A. The STS is a vital element of the United States space program and is the primary space launch system for both United States national security and civil government missions. The STS will be afforded the degree of survivability and security protection required for a critical national space resource.

B. The first priority of the STS program is to make the system fully operational and cost-effective in providing routine access to space.

C. The United States is fully committed to maintaining world leadership in space transportation with an STS capacity sufficient to meet appropriate national needs. The STS program requires sustained commitments by all affected departments and agencies. The United States will continue to develop the STS through the National Aeronautics and Space Administration (NASA) in cooperation with the Department of Defense (DoD). Enhancements of STS operational capability, upper stages, and efficient methods of deploying and retrieving payloads should be pursued as national requirements are defined.

D. United States Government spacecraft should be designed to take advantage of the unique capabilities of the STS. The completion of transition to the Shuttle should occur as expeditiously as practical.

[4] E. [paragraph deleted during declassification review]

F. Expandable launch vehicle operations shall be continued by the United States Government until the capabilities of the STS are sufficient to meet its needs and obligations. Unique national security considerations may dictate developing special-purpose launch capabilities.

G. For the near-term, the STS will continue to be managed and operated in an institutional arrangement consistent with the current NASA/DoD Memoranda of Understanding. Responsibility will remain in NASA for operational control of the STS for civil missions and in the DoD for operational control of the STS for national security missions. Mission management is the responsibility of the mission agency. As the STS operations mature, options will be considered for possible transition to a different institutional structure.

H. Major changes to STS program capabilities will require Presidential approval.

III. CIVIL SPACE PROGRAM (U)

The United States shall conduct civil space programs to expand knowledge of the Earth, its environment, the solar system, and the universe; to develop and promote selected civil applications of space technology; to preserve the United States leadership in critical aspects of space science, applications, and technology; and to further United States domestic and foreign policy objectives. Consistent with the National Aeronautics and Space Act, the following policies shall govern the conduct of the civil space program.

A. **Science, Applications, and Technology:** United States Government civil programs shall continue a balanced strategy of research, development, operations, and exploration for science, applications, and technology. The key objectives of these programs are to:

(1) Preserve the United States preeminence in critical major space activities to enable continued exploitation and exploration of space.

[5] (2) Conduct research and experimentation to expand understanding of: (a) astrophysical phenomena and the origin and evolution of the universe, through long-term astrophysical observation; (b) the Earth, its environment, and its dynamic relation with the Sun; (c) the origin and evolution of the solar system, through solar, planetary, and lunar sciences and exploration; and (d) the space environment and technology required to advance knowledge in the biological sciences.

(3) Continue to explore the requirements, operational concepts, and technology associated with permanent space facilities.

(4) Conduct appropriate research and experimentation in advanced technology and systems to provide a basis for future civil space applications.

B. **Private Sector Participation:** The United States Government will provide a climate conducive to expanded private sector investment and involvement in civil space activities, with due regard to public safety and national security. Private sector space activities will be authorized and supervised or regulated by the government to the extent required by treaty and national security.

C. **International Cooperation:** United States cooperation in international civil activities will:

(1) Support the public, nondiscriminatory direct readout of data from Federal civil systems to foreign ground stations and provision of data to foreign users under specified conditions.

(2) Continue cooperation with other nations by conducting joint scientific and research programs that yield sufficient benefits to the United States in areas such as access to foreign scientific and technological expertise, and access to foreign research and development facilities, and that serve other national goals. All international space ventures must be consistent with United States technology-transfer policy. [declassified]

D. **Civil Operational Remote Sensing:** Management of Federal civil operational remote sensing is the responsibility of the Department of Commerce. The Department of

Commerce will: (a) aggregate Federal needs for civil operational remote sensing to be met by either the private sector or the Federal government; (b) identify needed civil operational system research and development objectives; and (c) in coordination with other departments or agencies, provide for regulation of private-sector operational remote sensing systems.

[6] [page deleted during declassification review]

[7] [page deleted during declassification review]

[8] [paragraph deleted during declassification review]

(1) The fact that the United States conducts satellite photo-reconnaissance for peaceful purposes, including intelligence collection and the monitoring of arms control agreements, is unclassified. The fact that such photo-reconnaissance includes a near-real-time capability and is used to provide defense related information for indications and warning is also unclassified. All other details, facts and products concerning the national foreign intelligence space program are subject to appropriate classification and security controls.

(2) [paragraph deleted in declassification review]

VI. INTER-SECTOR RESPONSIBILITIES

[paragraphs A-F deleted during declassification review]
[9] G. The United States Government will maintain and coordinate separate national security and civil operational space systems when differing needs of the sectors dictate.

VII. IMPLEMENTATION

Normal interagency coordinating mechanisms will be employed to the maximum extent possible to implement the policies enunciated in this directive. To provide a forum to all Federal agencies for their policy views, to review and advise on proposed changes to national space policy, and to provide for orderly and rapid referral of space policy issues to the President for decisions as necessary, a Senior Interagency Group (SIG) on Space shall be established. The SIG (Space) will be chaired by the Assistant to the President for National Security Affairs and will include the Deputy or Under Secretary of State, Deputy or Under Secretary of Defense, Deputy or Under Secretary of Commerce, Director of Central Intelligence, Chairman of the Joint Chiefs of Staff, Director of the Arms Control and Disarmament Agency, and the [10] Administrator of the National Aeronautics and Space Administration. Representatives of the Office of Management and Budget and the Office of Science and Technology Policy will be include as observers. Other agencies or departments will participate based on the subjects to be addressed.

Document III-39

Document Title: National Security Decision Directive 5-83, "Space Station," April 11, 1983.

Source: NASA Historical Reference Collection, History Office, NASA Headquarters, Washington, D.C.

During 1982, NASA decided to push for presidential approval of a space station during 1983. To establish the basis for such a decision, SIG (Space) requested a study of NASA's station proposal and alternatives to it. President Reagan was briefed on the concept of a space station on April 7, 1983, and a few days later signed this directive establishing the terms of reference for the needed study. Ordinarily, Assistant to the President for National Security Affairs William Clark would have signed the directive as chairman of SIG (Space); the White House decided to have the president himself sign the document as an indication of the study's significance. Because the various agencies participating in the study mandated by the directive could not reach a consensus on a recommendation to the president, that study was never completed, and other paths were followed as the basis for President Reagan's decision to approve the station program.

Space Station

[1] April 11, 1983

OBJECTIVE

A study will be conducted to establish the basis for an Administration decision on whether or not to proceed with the NASA development of a permanently based, manned Space Station. This NSSD establishes the Terms of Reference for this study.

GUIDELINES

The specific policy issues to be addressed are the following (responsible agencies are indicated in parenthesis):

– How will a manned Space Station contribute to the maintenance of U.S. space leadership and to the other goals contained in our National Space Policy? (NASA)

– How will a manned Space Station best fulfill national and international requirements versus other means of satisfying them? (NASA/State for national and international civil space requirements; DOD/DCI for national security needs.)

– What are the national security implications of a manned Space Station? (DOD/DCI)

– What are the foreign policy implications, including arms control implications, of a manned Space Station? (State/NASA/ACDA)

– What is the overall economic and social impact of a manned Space Station? (NASA/Commerce/State)

These five policy issues will be addressed for each of the four scenarios outlined below.

In order to assess the policy issues in a balanced fashion, NASA will provide a background paper outlining four example scenarios that represent possible approaches for the continuation of this nation's manned space program. These example scenarios are:

[2] – Space Shuttle and Unmanned Satellites
– Space Shuttle and Unmanned Platforms
– Space Shuttle and an Evolutionary/Incrementally Developed Space Station
– Space Shuttle and a Fully Functional Space Station

A separate, unrelated, generic space requirements paper will be produced for use in addressing the national policy issues. The representative set of requirements for each space sector will be provided by DOD/DCI for national security and NASA/DOC for civil programs. A drafting group consisting of representatives of the DCI, DOD, DOC and NASA will coalesce the requirements into a single document. It will represent currently identifiable official agency statements of requirements for a Space Station. Long-term agency requirements and objectives should also be included.

IMPLEMENTATION

A Working Group under the Senior Interagency Group for Space has been established to conduct this study. The Working Group is chaired by NASA and includes representatives from DOD, DOC, DCI, DOS, and ACDA. The Working Group will produce a summary paper that assesses the issues and identifies policy options. Results of the study will be presented to the SIG (Space) not later than September 1983 prior to presentation to the President. Papers produced by the Working Group will not be distributed outside the Executive Branch without the approval of the SIG (Space). The SIG (Space) may issue more detailed Terms of Reference to implement this study.

Document III-40

Document title: "Revised Talking Points for the Space Station Presentation to the President and the Cabinet Council," November 30, 1983, with attached: "Presentation on Space Station," December 1, 1983, no pagination.

Source: NASA Historical Reference Collection, History Office, NASA Headquarters, Washington, D.C.

After having failed to get the support of a majority of the members of SIG (Space) for a recommendation to the president that he approve a space station program, NASA and its allies in the White House sought another path to get the issue before the president. NASA Administrator James Beggs was confident that the president would approve the program if only it was presented to him for decision. Cabinet Secretary Craig Fuller arranged for the station to be discussed at a meeting of the Cabinet Council for Commerce and Trade, a group not dominated by those opposed to the station. NASA Administrator Beggs made his presentation to the council, speaking from a set of staff-prepared "talking points" reproduced here.

FIRST VIEWGRAPH (IKE'S QUOTE ETC.)

- President Kennedy's decision to go to the moon chartered a course that resulted in leadership in space for the United States
- Incidentally, the Kennedy quotation is from a press conference in which the President is asked why he doesn't stop the Apollo program in light of budget concerns and other pressing needs
- President Nixon, against the wishes of many, continued America's commitment to leadership in space by approving the Space Shuttle

Link to next viewgraph

- This focus on leadership in space was reaffirmed in your Space Policy announced a year ago last July

NATIONAL SPACE POLICY VIEWGRAPH

- This policy sets forth goals and objectives that will keep America preeminent in space

Link to next viewgraph

- Today, the Space Shuttle makes us the leading Nation in space
- Tomorrow, America's preeminence in space can be achieved through a space station, manned and in permanent orbit around the earth

SHUTTLE VIEWGRAPH

- The Space Shuttle flies beautifully, and is something every American can see and be proud of
- Shuttle is operational, a ten year development program is over. By 1991, the earliest a space station could be in orbit, Shuttle will have been flying a full decade
- We brought the Shuttle in within 30% of the original budget projection:
  - declining NASA budget (24% in constant $, 1972-1981)
  - difficult technical hurdles
  - declining civil service work force
  - 20% between 1972 and 1981
  - 28,382 to 22,736

- Might be appropriate to mention here that a space station would cost $8 billion
  - describe what the $20 billion station is
  - launch by 1991

- The Shuttle has captured worldwide attention:
  - reassures our Allies of America's technological strength
  - concerns our enemies
  - impresses the fencesitters

- At a time when many in Europe and elsewhere focus on what divides us, the Shuttle has focused attention on what unites us (refer to Spacelab flight)
- Shuttle's impact is extraordinary. It exerts an influence over ordinary people and heads of government:
  - millions of Europeans turned out to see Enterprise
  - West German leadership
  - Mitterand

Link to next viewgraph

- Shuttle is routine transportation to Earth orbit. What's needed now, what was originally envisioned, is a place to Shuttle to...

THE WHAT IS A SPACE STATION VIEWGRAPH

- It's the logical extension of our past activities in space
- A United States Space Station would:
  - dominate the space environment for twenty years
  - stimulate commercial endeavors in space (recall the President's meeting with aerospace executives who emphasized commercial potential of space)
  - place in orbit an American outpost in space. With a space station, there would always be Americans working in space
  - be a national technology laboratory in space
  - check out and launch rockets to higher orbit
  - open up, for the first time, the possibility of assembling large satellites in space
  - stake out some options for the future, enabling a President in the years to come, to embark the United States upon missions that transcend the boundaries of earth:
    - back to a moon

- to an asteroid
- to the surface of Mars
- implement the overriding theme of your space policy: United States leadership in space

Link to next viewgraph

- In the 1990's, leadership in space will have a new dimension, something perhaps that Presidents Nixon and Kennedy could not foresee when they committed America to leadership in space…

A SPACE STATION WOULD STIMULATE VIEWGRAPH

- The new dimension will be the presence of the private sector in space
- The space program is going to change over the next 20 years
- no longer the monopoly of government
- no longer driven solely by motives of exploration and prestige

- Space is going to become a place of business:
- new products
- new services
- new benefits
- let me give you just one concrete example:
- McDonnell Douglas electrophoresis

- The government has a role to play in the commercialization of space
- sponsor K & D
- encourage entrepreneurs
- provide some essential support services

- This is where space station comes in:
- the station is a laboratory where pre-production research will need to be carried out
- the station is a servicing base where repairs to commercial equipment can be carried out. Modifications to equipment can be made, and spares can be stored for timely deployment

- In the future, there will be commercial enterprises making products in space, and a space station is going to make that possible
- Note importance of space program to scientific and technical education in the U.S, and to development of new technologies, both strong thrusts of this Administration

Link to next viewgraph

- These new commercial enterprises will involve the presence of man. Just as factories on the ground that have robots require men and women working in the plant to fix the robots and do things that machines can't, so will the space-based factories

MAN IN SPACE VIEWGRAPH

- Some people say you can do it all in space with robots. In fact, you must have man. He–and she–are the essential ingredient
- The presence of man is the key to leadership in space. And there are technical reasons for having man as well

- A principal reason for the excitement and attention the current Shuttle mission is having in Europe is the simple fact that there is a European scientist onboard the Columbia
- Compare Apollo 16 and Luna 16:
  - both were missions to the moon
  - both took place early in the 1970s
  - both brought lunar samples back to the Earth
  - but one, Apollo, was manned and captured the world's attention
  - while the other, an unmanned Russian rover, made no impression other than on the surface of the moon

- Man can repair and maintain spacecraft, and there are a number of satellites that have broken down where the capability to repair and maintain would be enormously beneficial:
  - Seasat
  - Solar Max
  - Landsat 4
  - IRAS

Link to next viewgraph

- The Soviets understand the importance of man in space

THE SOVIET VIEWGRAPHS

- The Soviets have an active, expanding space program in which cosmonauts play a central role
- The Soviet program:
  - is technically proficient: at present, two sophisticated Soviet scientific spacecraft are orbiting Venus. The Soviets were the first to take pictures from the surface of the moon and the first to transmit data from the surface of Venus
  - is an instrument of Soviet propaganda, particularly the cosmonauts (10 foreigners have flown with the Soviets)

- The centerpiece of the Soviet program is the Salyut Space Station:
  - about 49 feet long, and 42,000 lbs
  - usually a crew of two, sometimes four (five with Chratien)
  - automatic refueling capability
  - civil and military missions
  - overflies the U.S. 5-6 times a day
- What worries me is what the Soviets are up to. What are they planning to fly in the late 1980s and the 1990s? Will they be successful in their plans to dominate space?
  - CIA says they are expanding their level of activity and the CIA analysts expect qualitative improvements
    - the National Intelligence estimate indicates they are building:
    - heavy lift launch vehicle
    - a reusable Space Shuttle to Sly in '86 or '87
    - and they repeatedly have said and we have some evidence to support it that they intend to fly a large and permanent space station with up to 20 cosmonauts on board. I have no doubt that they can and will

- The Soviets are clearly taking their space program very seriously; they appear committed to a large space station–and I'm very concerned about it

Link to next viewgraph

- What are our alternatives?

SIG OPTIONS VIEWGRAPH

- The SIG study group outlined several options:
  - commit to a space station
  - build an Extended Duration Orbiter and unmanned space platforms, or
  - simply defer a decision on space station

- An Extended Duration Orbiter is an upgraded Shuttle that could stay up beyond the current limit of about nine days
- An unmanned platform is a satellite that would provide basic services such as power and data management to a bunch of scientific instruments on board the platform
- We have studied both of these for some time and have a good understanding of their capabilities

- The platform would let you do some good science, but:
  - it's not a staging base for transportation to higher orbit
  - it's not going to lead to the commercialization of space
  - it's not going to let you assembly large space systems
  - it's not going to impress many people, except for some scientists

- The Extended Duration Orbiter is something we have been looking at:
  - it would be nice to extend the orbiter's stay time
  - and we could devise some useful things to do
  - but the cost would be high for marginal improvements
  - $1.5 billion for orbiter mods
  - $1.6 billion for dedicated orbiter
  - plus $1.5 for a platform
  - and you still would not have continuous operations
  - just 40 days

- These may indeed be worthwhile projects, but they are hardly America's next step in space, and no one seems to be pushing them very hard

SIG OPTIONS VIEWGRAPH (CONTINUED)

- A decision to defer
  - details new commercial endeavors in space
  - simply means developing a station later, for a station is crucial to future operations in space
  - sends a signal to the American people that their space program is going to rest on its laurels, for the Shuttle will have flown for 10 years by the earliest time we could have a station ready to go
  - sends a signal to the Soviets that we are going to stand still in space

Link to next viewgraph

- A decision to defer or to build an extended duration orbiter in lieu of a space station really means that, in the years ahead, that we are going to forfeit our hard won leadership in space

LAST VIEWGRAPH

- First 25 years in space have been years of accomplishment for the United States. We have shown the world–and ourselves–what a Free People can do

- In 1958 the United States accepted the challenge of Sputnik and chartered a course from which I believe we can not now retreat
  - President Kennedy sent men to the moon, and in doing so sent a message to the world about America
  - we landed a scientific laboratory on the surface of Mars, in one of the most impressive scientific expeditions of all time
  - we began the exploration of the solar system and got a close-up look at the outer planets including Saturn with her intriguing rings
  - and we developed the most sophisticated flying machine the world has ever seen, one that routinely takes us into orbit around the Earth

- In doing all this, we developed new technologies and expanded the world of knowledge. And life here in the United States is better because of it. Our leadership in space these past 25 years told the world that America was strong and that America accepted the challenge of space, and that she was equal to the responsibilities of leadership.
- Now, today, here in this room, we must look forward to the next twenty-five years. The time to start a space station is now:
  - Shuttle development is over
  - technology is at hand
  - requirements have been analyzed
  - industry is ready
  - lead times are long
  - the stakes are enormous: leadership in space for the next 25 years

Document III-41

Document title: Caspar Weinberger, Secretary of Defense, to James M. Beggs, Administrator, NASA, January 16, 1984.

Source: NASA Historical Reference Collection, History Office, NASA Headquarters, Washington, D.C.

Throughout 1982 and much of 1983, NASA attempted to gain Department of Defense support for the space station program. That attempt was not successful, and in the fall of 1983, NASA decided to seek approval of the station program as solely a civilian program. In this letter, written after the decision to proceed with the station was made, Secretary of Defense Caspar "Cap" Weinberger spells out for the record his reservations about going ahead with the station.

---

[1] Dear Jim:

In your discussions and your correspondence after the 10 August Senior Interagency Group (Space) meeting and before the 1 December Cabinet Council for Commerce and Trade meeting you solicited my support for a space station commitment. Since this Department has been unable to identify any national security requirements that can be uniquely satisfied or capabilities that could be significantly enhanced by a manned space station, you have proposed that it proceed as a civil program.

My reservations about your proposal relate to cost and impact on the Space Transportation System.

The $8 billion estimate represents only a fraction of the actual costs required to achieve the initial capabilities you desire from a space station. Modules to make it operationally useful, and an extensive complement of instruments to support scientific missions, would inevitably multiply the total cost several times. In today's constrained fiscal environment, unprogrammed cost growths can only be funded at the expense of other programs. I have continuing concerns about the ability of the nation to support and sustain major commitments to defense programs, as well as new proposals like the President's Strategic Defense. You would not wish to cancel any of your approved civil programs to meet increased funding requirements for space station any more than we in Defense would like to see our national security budget jeopardized.

We remain firmly committed to the Space Transportation System. We have reconfigured all our payloads to be Shuttle compatible and have invested a considerable portion of our space related funding in Shuttle related projects. Our development of the west coast Shuttle launch facilities is a prime example of our commitment to the Space Transportation System. I believe that a major new start of this magnitude would inevitably divert NASA managerial talent and resources from the priority task of making the Space Transportation System fully operational and cost effective. With all our national security space programs committed to the Shuttle and dependent on it for their sole access to space, I am sure that you can appreciate my concern in this area.

[2] I regret not being able to endorse the modified thrust of the proposed space station, but the national security implications are too extensive and are not mitigated by calling it a civil program.

I will be pleased to discuss these issues with you further at your convenience.

Sincerely,
Cap

Document III-42

Document title: Office of the Press Secretary, "Fact Sheet: Presidential Directive on National Space Policy," February 11, 1988.

Source: NASA Historical Reference Collection, History Office, NASA Headquarters, Washington, D.C.

Between the issuance of the first Reagan administration space policy statement in July 1982 and 1987, there were a number of significant changes, including the *Challenger* accident, increased emphasis on the commercial uses of space, and the report of the blue ribbon National Commission on Space. A five-month SIG (Space) review during the second half of 1987 resulted in a new statement of national space policy reflecting these and other changes. President Reagan approved the new policy statement on January 5, but he withheld its release until a parallel review of commercial space policy initiatives being conducted by the Economic Policy Council was completed. The policy statement itself was classified; this unclassified summary was all that was publicly released.

[1]
Fact Sheet
Presidential Directive on National Space Policy

The President approved on January 5, 1988, a revised national space policy that will set the direction of U.S. efforts in space for the future. The policy is the result of a five-month interagency review which included a thorough analysis of previous Presidential decisions, the National Commission on Space report, and the implications of the Space Shuttle and expendable launch vehicle accidents. The primary objective of this review was to consolidate and update Presidential guidance on U.S. space activities well into the future.

The resulting Presidential Directive reaffirms the national commitment to the exploration and use of space in support of our national well being. It acknowledges that United States space activities are conducted by three separate and distinct sectors: two strongly interacting governmental sectors (Civil, and National Security) and a separate, non-governmental Commercial Sector. Close coordination, cooperation, and technology and information exchange will be maintained among sectors to avoid unnecessary duplication and promote attainment of United States space goals.

GOALS AND PRINCIPLES

The directive states that a fundamental objective guiding United States space activities has been, and continues to be, space leadership. Leadership in an increasingly competitive international environment does not require United States preeminence in all areas and disciplines of space enterprise. It does require United States preeminence in key areas of space activity critical to achieving our national security, scientific, technical, economic, and foreign policy goals.

- The overall goals of United States space activities are: (1) to strengthen the security of the United States; (2) to obtain scientific, technological, and economic benefits for the general population and to improve the quality of life on Earth through space-related activities; (3) to encourage continuing United States private-sector investment in space and related activities; (4) to promote international cooperative activities taking into account United States national security, foreign policy, scientific, and economic interests; (5) to cooperate with other nations in maintaining the freedom of space for all activities that enhance the security and welfare of mankind; and, as a long-range goal, (6) to expand human presence and activity beyond Earth orbit into the solar system.

- The directive states that United States space activities shall be conducted in accordance with the following principles:

– The United States is committed to the exploration and use of outer space by all nations for peaceful purposes and for the benefit of all mankind. "Peaceful purposes" allow for activities in pursuit of national security goals.

[2] – The United States will pursue activities in space in support of its inherent right of self-defense and its defense commitments to its allies.

– The Unites States rejects any claims to sovereignty by any nation over outer space or celestial bodies, or any portion thereof, and rejects any limitations on the fundamental right of sovereign nations to acquire data from space.

– The United States considers the space systems of any nation to be national property with the right of passage through and operations in space without interference. Purposeful interference with space systems shall be viewed as an infringement on sovereign rights.

– The United States shall encourage and not preclude the commercial use and exploitation of space technologies and systems for national economic benefit without direct Federal subsidy. These commercial activities must be consistent with national security interests, and international and domestic legal obligations.

– The United States shall encourage other countries to engage in free and fair trade in commercial space goods and services.

– The United States will conduct international cooperative space-related activities that are expected to achieve sufficient scientific, political, economic, or national security benefits for the nation. The United States will seek mutually beneficial international participation in its space and space-related programs.

CIVIL SPACE POLICY

The directive states that:

- The United States civil space sector activities shall contribute significantly to enhancing the Nation's science, technology, economy, pride, sense of well-being and direction, as well as United States world prestige and leadership. Civil sector activities shall comprise a balanced strategy of research, development, operations, and technology for science, exploration, and appropriate applications.
- The objectives of the United States civil space activities shall be (1) to expand knowledge of the Earth, its environment, the solar system, and the universe; (2) to create new opportunities for use of the space environment through the conduct of appropriate research and experimentation in advanced technology and systems (3) to develop space technology for civil applications and, wherever appropriate, make such technology available to the commercial sector; (4) to preserve the United States preeminence in critical aspects of space science, applications, technology, and manned space flight; (5) to establish a permanently manned presence in space; and (6) to engage in international cooperative efforts that further United States space goals.

COMMERCIAL SPACE POLICY

The directive states that the United States government shall not preclude or deter the continuing development of a separate, non-governmental Commercial Space Sector. Expanding private sector investment in space by the market-driven Commercial Sector generates economic benefits for the Nation and supports governmental Space Sectors with an increasing range of space goods and services. Governmental Space Sectors shall purchase commercially available space goods and services to the fullest extent feasible and shall not conduct [3] activities with potential commercial applications that preclude or deter Commercial Sector space activities except for national security or public safety reasons. Commercial Sector space activities shall be supervised or regulated only to the extent required by law, national security, international obligations, and public safety.

NATIONAL SECURITY SPACE POLICY

The directive further states that the United States will conduct those activities in space that are necessary to national defense. Space activities will contribute to national security objectives by 1) deterring, or if necessary defending against enemy attack; 2) assuring that forces of hostile nations cannot prevent our own use of space; 3) negating, if necessary, hostile space systems; and 4) enhancing operations of United States and Allied forces. Consistent with treaty obligations, the national security space program shall support such functions as command and control, communications, navigation, environmental monitoring, warning, and surveillance (including research and development programs which support these functions).

INTER-SECTOR POLICIES

This section contains policies applicable to, and binding on, the national security and civil space sectors:
- The United States Government will maintain and coordinate separate national security and civil operational space systems where differing needs of the sectors dictate.
- Survivability and endurance of national security space systems, including all necessary system elements, will be pursued commensurate with their planned use in crisis and conflict, with the threat, and with the availability of other assets to perform the mission.
- Government sectors shall encourage, to the maximum extent feasible, the

development and use of United States private sector space capabilities without direct Federal subsidy.

- The directive states that the United States Government will: (1) encourage the development of commercial systems which image the Earth from space competitive with or superior to foreign-operated civil or commercial systems; (2) discuss remote sensing issues and activities with foreign governments operating or regulating the private operation of remote sensing systems; and (3) continue a research and development effort for future advanced, remote sensing technologies. Commercial applications of such technologies will not involve direct Federal subsidy.

- The directive further states that assured access to space, sufficient to achieve all United States space goals, is a key element of national space policy. United States space transportations systems, must provide a balanced, robust, and flexible capability with sufficient resiliency to allow continued operations despite failures in any single system. The goals of United States space transportation policy are: (1) to achieve and maintain safe and reliable access to transportation in, and return from, space; (2) to exploit the unique attributes of manned and unmanned launch and recovery systems; (3) to encourage to the maximum extent feasible, the development and use of United States private sector space transportation capabilities without direct Federal subsidy; and (4) to reduce the costs of space transportation and related services.

- The directive also states that communications advancements are critical to all United States space sectors. To ensure necessary capabilities exist, the directive states [4] that the United States Government will continue research and development efforts for future advanced space communications technologies. These technologies, when utilized for commercial purposes, will be without direct Federal subsidy.

- The directive states that it is the policy of the United States to control or prohibit, as appropriate, exports of equipment and/or technology that would make an significant contribution to a foreign country's strategic military missile programs. Certain United States friends and allies will be exempted from this policy, subject to appropriate non-transfer and end-use assurances.

- The directive also states that the United States will consider and, as appropriate, formulate policy positions on arms control measures governing activities in space, and will conduct negotiations on such measures only if they are equitable, effectively verifiable, and enhance the security of the United States and its allies.

- The directive further states that all space sectors will seek to minimize the creation of space debris. Design and operations of space tests, experiments and systems will strive to minimize or reduce accumulation of space debris consistent with mission requirements and cost effectiveness.

IMPLEMENTING PROCEDURES

The directive states that normal interagency procedures will be employed wherever possible to coordinate the policies enunciated in this directive. To provide a forum to all Federal agencies for their policy views, to review and advise on proposed changes to national space policy issues to the President for decisions as necessary, a Senior Interagency Group (SIG) on Space shall continue to meet. The SIG (Space) will be chaired by a member of the National Security Council staff and will include appropriate representatives of the Department of State, Department of Defense (DOD), Department of Commerce (DOC), Department of Transportations (DOT), Director of Central Intelligence (DCI), Organization of the Joint Chiefs of Staff, United States Arms Control and Disarmament Agency, the National Aeronautics and Space Administration (NASA), Office of Management and Budget, and the Office of Science and Technology Policy. Other Executive agencies or departments will participate as the agenda of meeting shall dictate.

* *

POLICY GUIDELINES AND IMPLEMENTING ACTIONS

The directive also enumerates Policy Guidelines and Implementing Actions to provide a framework through which the policies in the directive shall be carried out. Agencies are directed to use this section as guidance on priorities, including preparation, review, and execution of budgets for space activities, within the overall resource and policy guidance provided by the President. Within 120 days of the date of this directive, affected Government agencies are directed to review their current policies for consistency with the directive and, where necessary, establish policies to implement the practices contained therein.

CIVIL SPACE SECTOR GUIDELINES

- The directive specifies that in conjunction with other agencies: NASA will continue the lead role within the Federal Government for advancing space science, exploration, and appropriate applications through the conduct of activities for research, technology, development and related operations; the National Oceanic and Atmospheric Administration will gather data, conduct research, and make predictions about the [5] Earth's environment; DOT will license and promote commercial launch operations which support civil sector operations.
- Space Science. NASA, with the collaboration of other appropriate agencies, will conduct a balanced program to support scientific research, exploration, and experimentation to expand understanding of: (1) astrophysical phenomena and the origin and evolution of the universe; (2) the Earth, its environment and its dynamic relationship with the Sun; (3) the origin and evolution of the solar system; (4) fundamental physical, chemical, and biological processes; (5) the effects of the space environment on human beings; and (6) the factors governing the origin and spread of life in the universe.
- Space Exploration. In order to investigate phenomena and objects both within and beyond the solar system, the directive states that NASA will conduct a balanced program of manned and unmanned exploration.
- Human Exploration. To implement the long-range goal of expanding human presence and activity beyond Earth orbit into the solar system the policy directs NASA to begin the systematic development of technologies necessary to enable and support a range of future manned missions. This technology program (Pathfinder) will be oriented toward a Presidential decision on a focused program of manned exploration of the solar system.
- Unmanned Exploration. The policy further directs NASA to continue to pursue a program of unmanned exploration where such exploration can most efficiently and effectively satisfy national space objectives by among other things: achieving scientific objectives where human presence is undesirable or unnecessary; exploring realms where the risk or costs of life support are unacceptable; and providing data vital to support future manned missions.
- Permanent Manned Presence. The directive states that NASA will develop the Space Station to achieve permanently manned operational capability by the mid-1990s. The directive further states that the Space Station will: (1) Contribute to United States preeminence in critical aspects of manned spaceflight; (2) provide support and stability to scientific and technological investigations; (3) provide early benefits, particularly in the materials of life sciences; (4) promote private sector experimentation preparatory to independent commercial activity; (5) allow evolution in keeping with the needs of Station users and the long-term goals of the United States; (6) provide opportunities for commercial sector participation; and (7) contribute to the longer term goal of expanding human presence and activity beyond Earth orbit into the solar system.
- Manned Spaceflight Preeminence. The directive specifies that approved programs such as efforts to improve the Space Transportation System (STS) and return it to safe flight and to develop, deploy and use the Space Station, are intended to ensure United

States preeminence in critical aspects of manned spaceflight.

 - Space Applications. The policy directs NASA and other agencies to pursue the identification and development of appropriate applications flowing from their activities. Agencies will seek to promote private sector development and implementation of applications. The policy also states that:

    – Such applications will create new capabilities, or improve the quality or efficiency of continuing activities, including long-term scientific observations.

    – NASA will seek to ensure its capability to conduct selected critical missions through an appropriate mix of assured access to space, on-orbit sparing, advanced [6] automation techniques, redundancy, and other suitable measures.

    – Agencies may enter cooperative research and development agreements on space applications with firms seeking to advance the relevant state-of-the-art consistent with United States Government space objectives.

    – Management of Federal civil operational remote sensing is the responsibility of the Department of Commerce. The Department of Commerce will: (1) consolidate Federal needs for civil operational remote sensing products to be met either by the private sector or the Federal government; (2) identify needed civil operational system research and development objectives; and (3) in coordination with other departments or agencies, provide for the regulation of private sector operational remote sensing systems.

 - Civil Government Space Transportation. The policy states the unique Space Transportation System (STS) capability to provide manned access to space will be exploited in those areas that offer the greatest national return, including contributing to United States preeminence in critical aspects of manned spaceflight. The STS fleet will maintain the Nation's capability and will be used to support critical programs requiring manned presence and other unique STS capabilities. In support of national space transportation goals, NASA will establish sustainable STS flight rates to provide for planning and budgeting of Government space programs. NASA will pursue appropriate enhancements to STS operational capabilities, upper stages, and systems for deploying, servicing, and retrieving spacecraft as national and user requirements are defined.

 - International Cooperation. The policy guidelines state that the United States will foster increased international cooperation in civil space activities by seeking mutually beneficial international participation in its civil space and space-related programs. The SIG (Space) Working Group on Space Science Cooperation with the U.S.S.R. shall be responsible for oversight of civil space cooperation with the Soviet Union. No such cooperative activity shall be initiated until an interagency review has been completed. The directive provides that United States cooperation in international civil space activities will:

    – Be consistent with United States technology transfer laws, regulations, Executive Orders and presidential directives.

    – Support the public, nondiscriminatory direct readout of data from Federal civil systems to foreign ground stations and the provision of data to foreign users under specified conditions.

    – Be conducted in such a way as to protect the commercial value of intellectual property developed with Federal support. Such cooperation will not preclude or deter commercial space activities by the United States private sector, except as required by national security or public safety.

COMMERCIAL SPACE SECTOR GUIDELINES

 - The directive states that NASA, and the Departments of Commerce, Defense, and Transportation will work cooperatively to develop and implement specific measures to foster the growth of private sector commercial use of space. A high-level focus for commercial space issues has been created through establishment of a Commercial Space Working Group of the Economic Policy Council. SIG (Space) will continue to coordinate the development and implementation of national space policy.

[7] - To stimulate private sector investment, ownership, and operation of space assets, the directive provides that the United States Government will facilitate private sector access to appropriate U.S. space-related hardware and facilities, and encourage the private sector to undertake commercial space ventures. The directive states that Governmental Space Sectors shall, without providing direct Federal subsidies:

– Utilize commercially available goods and services to the fullest extent feasible, and avoid actions that may preclude or deter commercial space sector activities except as required by national security or public safety. A space good or service is "commercially available" if it is currently offered commercially, or if it could be supplied commercially in response to a government service procurement request. "Feasible" means that such goods or services meet mission requirements in a cost-effective manner.

– Enter into appropriate cooperative agreements to encourage and advance private sector basic research, development, and operations while protecting the commercial value of the intellectual property developed;

– Provide for the use of appropriate Government facilities on a reimbursable basis;

– Identify, and eliminate or propose for elimination, applicable portions of United States laws and regulations that unnecessarily impede commercial space sector activities;

– Encourage free trade in commercial space activities. The United States Trade Representative will consult, or, as appropriate, negotiate with other countries to encourage free trade in commercial space activities. In entering into space-related technology development and transfer agreements with other countries, Executive Departments and agencies will take into consideration whether such countries practice and encourage free and fair trade in commercial space activities.

– Provide for the timely transfer of Government-developed space technology to the private sector in such a manner as to protect its commercial value, consistent with national security.

– Price Government-provided goods and services consistent with OMB Circular A-25.

– The directive also states that the Department of Commerce (DOC) will commission a study to provide information for future policy and program decisions on options for a commercial advanced earth remote sensing system. This study, to be conducted in the private sector under DOC direction with input from Federal Agencies, will consist of assessments of the following elements: (1) domestic and international markets for remote sensing data; (2) financing options, such as cooperative opportunities between government and industry in which the private sector contributes substantial financing to the venture, participation by other government agencies, and international cooperative partnerships; (3) sensor and data processing technology and; (4) spacecraft technology and launch options. The results of this study will include an action plan on the best alternatives identified during the study.

NATIONAL SECURITY SPACE SECTOR GUIDELINES

– General. The directive states that:

– The Department of Defense (DOD) will develop, operate, and maintain an assured mission capability through an appropriate mix of robust satellite control, assured access to [8] space, on-orbit sparing, proliferation, reconstitution or other means.

– The national security space program, including dissemination of data, shall be conducted in accordance with Executive Orders and applicable directives for the protection of national security information and commensurate with both the missions performed and the security measures necessary to protect related space activities.

– DOD will ensure that the military space program incorporates the support requirements of the Strategic Defense Initiative.

– Space Support. The directive states that:

– The national security space sector may use both manned and unmanned launch systems as determined by specific mission requirements. Payloads will be distributed among launch systems and launch sites to minimize the impact of loss of any single launch system or launch site on mission performance. The DOD will procure unmanned launch vehicles or services and maintain launch capability on both the East and West coasts. DOD will also continue to enhance the robustness of its satellite control capability through an appropriate mix of satellite autonomy and survivable command and control, processing, and data dissemination systems.

- DOD will study concepts and technologies which would support future contingency launch capabilities.
- Force Enhancement. The directive states that the national security space sector will develop, operate, and maintain space systems and develop plans and architectures to meet the requirements of operational land, sea, and air forces through all levels of conflict commensurate with their intended use.
- Space Control. The directive also states that:
- The DOD will develop, operate, and maintain enduring space systems to ensure its freedom of action in space. This requires an integrated combination of antisatellite, survivability, and surveillance capabilities.
- Antisatellite (ASAT) Capability. DOD will develop and deploy a robust and comprehensive ASAT capability with programs as required and with initial operational capability at the earliest possible date.
- DOD space programs will pursue a survivability enhancement program with long-term planning for future requirements. The DOD must provide for the survivability of selected, critical national security space assets (including associated terrestrial components) to a degree commensurate with the value and utility of the support they provide to national-level decision functions, and military operational forces across the spectrum of conflict.
- The United States will develop and maintain an integrated attack warning, notification, verification, and contingency reaction capability which can effectively detect and react to threats to United States space systems.
- Force Application. The directive states that the DOD will, consistent with treaty obligations, conduct research, development, and planning to be prepared to acquire and deploy space weapons systems for strategic defense should national security conditions dictate.

INTER-SECTOR GUIDELINES

The directive states that the following paragraphs identify selected, high priority cross-sector efforts and [9] responsibilities to implement plans supporting major United States space policy objectives:
- Space Transportation Guidelines.
- The United States national space transportation capability will be based on a mix of vehicles, consisting of the Space Transportation System (STS), unmanned launch vehicles (ULVs), and in-space transportation systems. The elements of this mix will be defined to support the mission needs of national security and civil government sectors of United States space activities in the most cost effective manner.
- As determined by specific mission requirements, the national security space sector will use the STS and ULVs. In coordination with NASA, the DOD will assure the Shuttle's utility to national defense and will integrate missions into the Shuttle system. Launch priority will be provided for national security missions as implemented by NASA-DOD agreements. Launches necessary to preserve and protect human life in space shall have the highest priority except in times of national security emergency.
- The STS will continue to be managed and operated in an institutional arrangement consistent with the current NASA/DOD Memorandum of Understanding. Responsibility will remain in NASA for operational control of the STS for civil missions, and in the DOD for operational control of the STS for national security missions. Mission management is the responsibility of the mission agency.

- United States commercial launch operations are an integral element of a robust national space launch capability. NASA will not maintain an expendable launch vehicle (ELV) adjunct to the STS. NASA will provide launch services for commercial and foreign payloads only where those payloads must be man-tended, require the unique capabilities of the STS, or it is determined that launching the payloads on the STS is important for national security or foreign policy purposes. Commercial and foreign payloads will not be launched on government owned or operated ELV systems except for national security or foreign policy reasons.

- Civil Government agencies will encourage, to the maximum extent feasible, a domestic commercial launch industry by contracting for necessary ELV launch services directly from the private sector or with DOD.

- NASA and the DOD will continue to cooperate in the development and use of military and civil space transportation systems and avoid unnecessary duplication of activities. They will pursue new launch and launch support concepts aimed at improving cost-effectiveness, responsiveness, capability, reliability, avaliability, maintainability and flexibility. Such cooperation between the national security and civil sectors will ensure efficient and effective use of national resources.

- The directive lists guidelines for the federal encouragement of commercial unmanned launch vehicles (ULVs):

- The United States Government fully endorses and will facilitate the commercialization of United States unmanned launch vehicles (ULVs).

- The Department of Transportation (DOT) is the lead agency within the Federal Government for developing, coordinating, and articulating Federal policy and regulatory guidance pertaining to United States commercial launch activities in consultation with DOD, State, NASA, and other concerned agencies. All Executive departments and agencies shall assist the DOT in carrying out its responsibilities as [10] set forth in the Commercial Space Launch Act and Executive Order 12465.

- The United States Government encourages the use of its launch and launch-related facilities for United States commercial launch operations.

- The United States Government will have priority use of Government facilities and support services to meet national security and critical mission requirements. The United States Government will make all reasonable efforts to minimize impacts on commercial operations.

- The United States Government will not subsidize the commercialization of ULVs, but will price the use of its facilities, equipment, and services with the goal of encouraging viable commercial ULV activities in accordance with the Commercial Space Launch Act.

- The United States Government will encourage free market competition within the United States private sector. The United States Government will provide equitable treatment for all commercial launch operators for the sale or lease of Government equipment and facilities consistent with its economic, foreign policy, and national security interests.

- NASA and DOD, for those unclassified and releasable capabilities for which they have responsibility shall, to the maximum extent feasible:

— Use best efforts to provide commercial launch firms with access, on a reimbursable basis, to national launch and launch-related facilities, equipment, tooling, and services to support commercial launch operations;

— Develop, in consultation with the DOT, contractual arrangements covering access by commercial launch firms to national launch and launch-related property and services they request in support of their operations;

— Provide technical advice and assistance to commercial launch firms on a reimbursable basis, consistent with the pricing guidelines herein; and

— Conduct, in coordination with DOT appropriate environmental analyses necessary to ensure that commercial launch operations conducted at Federal launch facilities are in compliance with the National Environmental Policy Act.

- The directive lists government ULV Pricing Guidelines. The price charged for the use of United States Government facilities, equipment, and service, will be based on the following principles:
  – Price all services (including those associated with production and launch of commercial ULVs) based on the direct costs incurred by the United States Government. Reimbursement shall be credited to the appropriation from which the cost of providing such property or service was paid.
  – The United States Government will not seek to recover ULV design and development costs or investments associated with any existing facilities or new facilities required to meet United States Government needs to which the U.S. Government retains title;
  – Tooling, equipment, and residual ULV hardware on hand at the completion of the United States Government's program will be priced on a basis that is in the best overall interest of the United States Government, taking into consideration that these sales will not constitute a subsidy to the private sector operator.

[11] - The directive also states that commercial launch firms shall:
  – Maintain all facilities and equipment leased from the United States Government to a level of readiness and repair specified by the United States Government;
  – Comply with all requirements of the Commercial Space Launch Act, all regulations issued under the Act, and all terms, conditions or restrictions of any license issued or transferred by the Secretary of Transportation under the Act.

- The directive establishes the following technology transfer guidelines:
  – The United States will work to stem the flow of advanced western space technology to unauthorized destinations. Executive departments and agencies will be fully responsible for protecting against adverse technology transfer in the conduct of their programs.
  – Sales of United States space hardware, software, and related technologies for use in foreign space projects will be consistent with relevant international and bilateral agreements and arrangements.

- The directive states that all Sectors shall recognize the importance of appropriate investments in the facilities and human resources necessary to support United States space objectives and maintain investments that are consistent with such objectives. A task force of the Commercial Space Working Group, in cooperation with OSTP, will conduct a feasibility study of alternate methods for encouraging, without direct Federal subsidy, private sector capital funding of United States space infrastructure such as ground facilities, launcher developments, and orbital assembly and test facilities. Coordinated terms of reference for this study shall be presented to the EPC and SIG (Space).

- The directive notes that the primary forum for negotiations on nuclear and space arms is the Nuclear and Space Talks (NST) with the Soviet Union in Geneva. The instructions to the United States Delegation will be consistent with this National Space Policy directive, established legal obligations, and additional guidance by the President. The United States will continue to consult with its Allies on these negotiations and ensure that any resulting agreements enhance the security of the United States and its Allies. Any discussions on arms control relating to activities in space in fora other than NST must be consistent with, and subordinate to, the foregoing activities and objectives.

- Finally the directive states that using NSC staff approved terms of reference, an IG (Space) working group will provide recommendations on the implementation of the Space Debris Policy contained in the Policy section of this directive.

Chapter Four

Organizing for Exploration

by Sylvia K. Kraemer

The Eisenhower administration's calculated policy of "open skies" and "peaceful uses of space" to enable satellite overflights of other nations virtually assured that the U.S. non-defense space program would be lodged in a civilian agency.[1] Eisenhower's uneasiness over an emerging military-industrial complex, expressed in his Farewell Address,[2] no doubt also contributed to his view that all non-defense-related space activities should be assigned to a new civilian organization. Scientists—who recognized that scientific exploration of space would fare better intertwined with a "peaceful," or nonmilitary, space program—agreed with Eisenhower. The president's own Science Advisory Committee, chaired by James R. Killian, Jr., of the Massachusetts Institute of Technology, favored creating a civilian national space agency out of the nucleus of the National Advisory Committee for Aeronautics (NACA).[3] In a March 1958 memorandum to President Eisenhower, Killian joined forces with Bureau of the Budget Director Percival Brundage and Nelson A. Rockefeller, chairman of the President's Advisory Committee on Government Organization, to make a lucid case for choosing NACA over the proposed alternatives, the most prominent of which were the Department of Defense (DOD), the Atomic Energy Commission, a private contractor, or a new Department of Science and Technology, to lead "the civil space effort." [IV-1, IV-2, IV-3, IV-4, IV-5]

A New Organization

Created by the National Aeronautics and Space Act (PL 85-568) [II-17], the National Aeronautics and Space Administration (NASA) opened for business on October 1, 1958, with a complement of nearly 8,000 employees transferred from the old NACA research laboratories: Langley Aeronautical Laboratory at Hampton, Virginia (est. 1917); Ames Aeronautical Laboratory at Moffett Field, California (est. 1939); the Flight Research Center at nearby Muroc Dry Lake (est. 1946), now known as the Dryden Flight Research Center; and the Lewis Flight Propulsion Laboratory in Cleveland, Ohio (est. 1940). By the end of 1960, NASA personnel rolls had nearly doubled to over 16,000. The principal increases were a result of the tripling of NASA Headquarters personnel and the addition of portions of the Army Ballistic Missile Agency (ABMA), renamed the George C. Marshall Space Flight Center, and the new Goddard Space Flight Center in Beltsville, Maryland. Most of Goddard's personnel had been transferred from the Naval Research Laboratory (NRL) and the Naval Ordnance Laboratory (NOL). The Jet Propulsion Laboratory of the California Institute of Technology, a contractor-owned and -operated facility involved in rocket

---

1. See R. Cargill Hall, "Origins of U.S. Space Policy: Eisenhower, Open Skies, and Freedom of Space," Chapter Two of this volume.
2. "Farewell Radio and Television Address to the American People," January 17, 1961, *Public Papers of the Presidents of the United States: Dwight D. Eisenhower, 1960-61* (Washington, DC: U.S. Government Printing Office, 1962), pp. 1035-40. Quote from pp. 1038-39; "military-industrial complex" phrase on p. 1038.
3. The political and legislative origins of the National Aeronautics and Space Administration are described in Walter A. McDougall, *...The Heavens and The Earth: A Political History of the Space Age* (New York: Basic Books, 1985), Chapter 7, "The Birth of NASA," and Enid Curtis Bok Schoettle, "The Establishment of NASA," in Sanford A. Lakoff, ed., *Knowledge and Power: Essays on Science and Government* (New York: Free Press, 1966).

research since 1936, was also transferred from the U.S. Army to NASA. The Manned Spacecraft Center in Houston and the Kennedy Space Center at Cape Canaveral were added within the next three years.[4]

Because of the way NASA was initially assembled, a little over 80 percent of NASA's technical core during the 1960s and 1970s—its engineers and scientists—held within its corporate memory the experience of working with NACA, ABMA, and the Navy organizations from which the Goddard Space Flight Center had drawn much of its personnel. Each group would bring its own institutional culture. Predominating among NASA's initial cadre, the scientists and research engineers of NACA (est. 1915) had based their careers in an institution that had conducted research in aerodynamics and aircraft structures and propulsion systems for both industrial and military clients. Informally structured, NACA had been overseen by its Main Committee and various technical subcommittees, and its work in aeronautical engineering was done largely by civil servants. Aside from its work in aeronautics, what distinguished NACA as an institution was the ethos that permeated its laboratories. With its emphasis on technical competence, evaluation of one's work by technical peers, and a collegial in-house research environment conducive to engineering innovation, the NACA centers were not well equipped for the sweeping institutional growth and change that would complicate their life after 1958.[5]

To the technical core of NACA were added, during NASA's first two years, the complementary Naval Research Laboratory habits of in-house engineering research and science and, by the ABMA group, the emphasis on in-house technical development characteristic of the Army's arsenal system. The presence at ABMA of a contingent of German rocket engineers reinforced its emphasis on in-house technical mastery and control. What these various components shared was a common culture that placed technical judgment above political competence. They undoubtedly also shared the conviction that they were embarked upon an exploratory venture unrivaled in the annals of mankind.[6]

The new agency's charter, the "Space Act of 1958," had given it broad latitude to contribute "to the general welfare and security of the United States" by expanding "human knowledge of phenomena in the atmosphere and space" and preserving "the role of the United States as a leader in aeronautical and space science and technology and in the application thereof to the conduct of peaceful activities within and outside the atmosphere." Within three years—not much time given the pace of policy evolution in most popularly elected governments—John F. Kennedy provided NASA a specific mission so compelling that debate over just how NASA's broad charter was to be carried out was effectively quieted.

The Cold War, most notably in the "Sputnik Crisis," and then the flight of Yuri Gagarin in 1961 stimulated not only the creation of NASA in 1958 but its tremendous expansion in the early 1960s to carry out the Apollo program.[7] After President John F. Kennedy issued

    4. By the end of 1960, the old NACA laboratories and Marshall Space Flight Center accounted for 49 percent and 33 percent, respectively, of NASA's employees. (The Manned Spacecraft Center in Houston, Texas, was added in 1961. The U.S. Army's Missile Firing Laboratory at Cape Canaveral, Florida, was added to Marshall Space Flight Center's organization in 1960 and was renamed the John F. Kennedy Space Center in 1963.) The 157 personnel who had been working on the Navy's Project Vanguard, which became the nucleus of the Goddard Space Flight Center (est. 1959), were transferred to NASA in 1958 from one of the Navy's own in-house research laboratories, the Naval Research Laboratory. They were soon joined by 63 more who had been working for the Naval Research Laboratory's Space Sciences and Theoretical Divisions. The next large group to transfer to NASA was the 5,367 civil servants from the Army Ballistic Missile Agency (ABMA) at Redstone Arsenal, in Huntsville, Alabama. ABMA had been essentially an in-house operation. The youngest NASA installations, the Manned Spacecraft Center (renamed Johnson Space Center in 1973) and the Kennedy Space Center, were initially staffed by personnel from Langley Research Center and the ABMA.
    5. Alex Roland, *Model Research: The National Advisory Committee for Aeronautics, 1915-1958* (Washington, DC: NASA SP-4103, 1985); Nancy Jane Petrovic, "Design for Decline: Executive Management and the Eclipse of NASA," Ph.D. Dissertation, University of Maryland, 1982 (Ann Arbor, MI: University Microfilms International, 1982).
    6. On NASA's culture, see Howard E. McCurdy, *Inside NASA: High Technology and Organizational Culture in the U.S. Space Program* (Baltimore, MD: Johns Hopkins University Press, 1993).
    7. Thanks to the GI Bill and its Korean War counterpart, the military services' reserve officers' training programs, cooperative work-education programs, and draft exemptions for those in engineering school or working for the government in engineering fields, NASA and its contractors were able to mobilize unprecedented numbers of engineers and scientists.

his challenge to the nation in May 1961 to send a man to the Moon and return him safely within the decade—a challenge framed within the Cold War contest between the Communist Bloc nations and the "free world"—NASA undertook a mobilization comparable, in relative scale, to that undertaken by the United States to fight World War II. The agency's civil service personnel rolls increased by a factor of three, while the men and women employed on NASA contracts increased by a factor of ten. Likewise, NASA's annual budget increased an order of magnitude between 1960 and 1965, from roughly $500 million to $5.2 billion.

Table 1: Dimensions of the Apollo Mobilization[8]

Overall Budget (billions)

	Amount	Percentage Increase
FY 61	$0.964	—
FY 62	$1.825	89%
FY 63	$3.674	101%
FY 64	$5.100	38%
FY 65	$5.250	2%

Construction of Facilities Budget (millions)

	Amount	Percentage Increase
FY 61	$98.2	—
FY 62	$217.1	121%
FY 63	$569.8	162%
FY 64	$546.6	-4%
FY 65	$522.2	-4%

Personnel

	In-House NASA	Contractor	Ratio
1958 (9/30)	8,040	—	—
1960	10,200	36,500	1:3.5
1961	17,500	57,000	1:3.3
1962	23,700	115,500	1:4.9
1963	29,900	218,400	1:7.3
1964	32,500	347,100	1:10.7
1965	34,300	376,700	1:11

Contracting Out

The private sector provided even more scientists, engineers, technicians, and supporting personnel for Apollo than did NASA. Throughout its history, between roughly 80 percent and 90 percent of NASA's budget has gone into goods, services, and development procured from the private sector through contracts. The notion of relying on private industry and universities did not originate with NASA's Apollo-era Administrator James E. Webb (1961-1968)—though both necessity and good politics made him a natural

8. Jane Van Nimmen and Leonard C. Bruno with Robert L. Rosholt, *NASA Historical Data Book, Vol. I, NASA Resources, 1958-1968*, (Washington, DC: NASA SP-4012, 1988), pp. 137-141, 134, 63-119.

champion of contracting out as the best way of getting the agency's work done. NACA had supplemented its in-house research with contracts to Stanford, the Massachusetts Institute of Technology, and other universities. To NASA's first administrator, T. Keith Glennan, and his ideologically sympathetic boss, President Eisenhower, reliance on the private sector came naturally.[9] [IV-7] Indeed, the practice had its roots deep in U.S. history. [IV-8]

Since the beginning of the republic, U.S. citizens have shared a widespread mistrust of large government establishments. Coupled with this mistrust has been a public faith in private enterprise that, through the mechanism of a free market, was thought the best guarantor of economic growth and a free society. On this usually bi-partisan ideological foundation, and partly in reaction to the alleged excesses of the New Deal, federal policy encouraged government agencies to acquire their goods and services from the private sector.

The military services had been acquiring equipment and logistics support from the private sector since the early nineteenth century; they were well schooled in government procurement. More recently, it was the U.S. Army and U.S. Air Force, created out of the U.S. Army Air Forces under the Defense Reorganization Act of 1947 that established the Department of Defense, that had the most experience with contracting to the private sector. As a result of the Army's Manhattan Project and the ballistic missile programs managed by the Air Force's Research and Development Command, both services came to rely on private contractors for advanced engineering and development work and, in some cases, to assist in the technical direction of other development contractors—the Air Force going so far as to create the Rand and Aerospace Corporations.

Contracting out by NASA also had great practical merit. Because most of the experience in the country to date in related missile and high-performance aircraft development centered in industry, which had worked as contractors to the military, the resources of industry could be marshalled more effectively *by* the government than reproduced *within* the government. NASA would be able to harness talent and institutional resources already in existence in the emerging aerospace industry and the country's leading research universities.[10] In 1959 the General Services Administration authorized NASA's use of the Armed Service Procurement Regulations of 1947, which contained important exemptions, suited to research and development work, from the principle of making awards to the "lowest responsible bidder." Contracting out promised the additional political advantage of dispersing Federal funds around the country and, as a consequence, creating within Congress a political constituency with a material interest in the health—and management—of the space program. The attempt to meld different institutional cultures into a single organization was not without its problems. For example, when the California Institute of Technology's Jet Propulsion Laboratory (JPL) became affiliated with NASA on January 1, 1959, its managers believed the lab would be called upon to play the dominant role in determining America's space exploration agenda. NASA had a much more limited role in mind for JPL, however, and the resulting conflict between these divergent expectations laid a foundation for lingering animosity between the two institutions.[11] [IV-6]

By 1961 the federal government had been contracting to the private sector for much of its research and development work for two decades, since World War II. Enough questions had been raised about the wisdom of that policy to prompt President John F. Kennedy to ask the director of the Bureau of the Budget to review it. Budget Director David E. Bell was joined in this task by the secretary of defense (Robert S. McNamara), the administrator of NASA (James E. Webb), the chairman of the Civil Service Commission (John W. Macy, Jr.), the chairman of the Atomic Energy Commission (Glenn T. Seaborg), the director of

---

9. On Glennan, see J.D. Hunley, ed., *The Birth of NASA: The Diary of T. Keith Glennan* (Washington, DC: NASA SP-4105, 1993).

10. One NASA installation, the Jet Propulsion Laboratory of the California Institute of Technology in Pasadena, California, would remain wholly a contractor operation. For an excellent and brief discussion of the NASA acquisition process, see Arnold S. Levine, *Managing NASA in the Apollo Era* (Washington, DC: NASA SP-4102, 1982), Chapter 4.

11. Clayton L. Koppes, *JPL and the American Space Program* (New Haven: Yale University Press, 1982).

the National Science Foundation (Alan T. Waterman), and the special assistant to the president for science and technology (Jerome B. Wiesner). The report—which came to be known as the "Bell Report"—constituted a detailed and comprehensive review of federal contracting for research and development. [IV-9]

The Bell Report affirmed the federal government's policy of relying "heavily on contracts with non-federal institutions to accomplish scientific and technical work needed for public purposes." At the same time, it cautioned that "the management and control of such programs must be firmly in the hands of full-time Government officials clearly responsible to the President and the Congress. With programs of the size and complexity now common," it continued,

> . . . the Government [must] have on its staff exceptionally strong and able executives, scientists, and engineers, fully qualified to weigh the views and advice of technical specialists, to make policy decisions concerning the types of work to be undertaken, when, by whom, and at what cost, to supervise the execution of work undertaken, and to evaluate the results.

This requirement, according to the Bell group, was not being met: "In recent years there has been a serious trend toward eroding the competence of the Government's research and development establishments—in part owing to the keen competition for scarce talent which has come from Government contractors." The solution, advised the budget director and heads of federal research and development agencies, was not "setting artificial or arbitrary limits on Government contractors" but creating a working environment and offering salaries that would better enable the government to compete with the private sector for top scientific and engineering talent. However wise and well-intentioned the Bell Report's recommendations may have been, they do not seem to have had great effect. "Contracting out" continues to this day to be a primary issue among NASA managers, scientists, and engineers.

Program Management

Not only were NASA's procurement procedures based on those of the military establishment, but NASA made extensive use of the military's experience in program management as well. The ratio of military detailees to civilians working in NASA increased steadily between 1960 and 1968.[12] Many of the detailees were Air Force or Navy career officers assigned to program or operations management positions. For example, 103 of the roughly 180 military detailees in NASA at the beginning of 1963 were career Navy or Air Force officers.[13] [IV-10] Though fewer in number, program managers who had honed their skills in private industry also helped to manage the NASA enterprise. For example, NASA's Office of Manned Space Flight was led during much of the 1960s and 1970s by men who had come from industry, such as George E. Mueller (Space Technology Laboratories), Dale D. Myers (North American Rockwell), and John F. Yardley (McDonnell Douglas Astronautics).

The epitome of the proven military program manager at NASA was U.S. Air Force Major General Samuel C. Phillips. Schooled in Air Force research and development

---

12. Nimmen and Bruno with Rosholt, *NASA Historical Data Book, Vol. I,* pp. 80-81, 98-99.
13. Albert F. Siepert, Memorandum to James E. Webb, February 8, 1963, NASA Historical Reference Collection, NASA Headquarters, Washington, DC. A list of positions "requiring USAF officers" forwarded by NASA to the Department of the Air Force in 1964 included: director, program control, Apollo; director, program control, Saturn V; deputy director for program management, Apollo spacecraft; assistant to director for program management, Saturn V; chief, configuration management, Apollo spacecraft; configuration management officer, Saturn V; chief, configuration management, Saturn I-IB; configuration management officer, Gemini; configuration management officer, Apollo launch site; assistant deputy director for program management, Apollo program office; configuration management officer; and chief, mission requirements, Apollo. Attachment to Eugene M. Zuckert, Secretary of the Air Force, Memorandum to Hugh L. Dryden, Deputy Administrator of NASA, May 27, 1964, NASA Historical Reference Collection.

program management at Wright-Patterson Air Force Base, Ohio, Phillips was assigned in 1959 to manage the development of the "Minuteman" intercontinental ballistic missile. Phillips was convinced that the development of a new technology system required that the program head have centralized authority over engineering, configuration management, procurement, testing, construction, manufacturing, logistics, and training. Phillips' success with the Minuteman program won the admiration of NASA's Associate Administrator for Manned Flight George Mueller, who brought Phillips to NASA where he served as deputy director and then program director for the Apollo program.[14]

What this conglomeration of assorted talents drawn from NASA and the military wrought was not simply the historic feat of placing Americans on the Moon and bringing them back safely. Less visible but no less important was their catalytic role in the emerging ability of U.S. industry to develop, manufacture, and operate large, complex, and sophisticated technical systems. In 1968, *Science* magazine, the publication of the American Association for the Advancement of Science, observed:

In terms of numbers of dollars or of men, NASA has not been our largest national undertaking, but in terms of complexity, rate of growth, and technological sophistication it has been unique.... It may turn out that [the space program's] most valuable spin-off of all will be human rather than technological: better knowledge of how to plan, coordinate, and monitor the multitudinous and varied activities of the organizations required to accomplish great social undertakings.[15]

NASA and military managers responsible for developing new aerospace technologies stimulated the government's contractors in U.S. industry to adopt program management and systems engineering strategies that would promote their survival in a market dominated by a few large federal customers.

The forces that have influenced the management strategies characteristic of U.S. industries at any given time have varied both with the nature of contemporary economic trends and with the nature of the goods being produced. For example, in the United States during the 1880s and 1890s, in an era before the triumph of mass media consumer advertising, companies sought to control markets by controlling production and/or prices. Firms producing relatively undifferentiated commodities (e.g., whiskey, salt, coal, tobacco, sugar, and kerosene) attempted to combine financial as well as management structures to achieve more effective market control within an industry. Toward the end of the century, such combinations were increasingly subject to state and federal anti-trust legislation. Successful prosecutions under the Sherman Anti-Trust Act of 1890 brought about the dissolution of such "horizontally integrated" firms as the Standard Oil Company of New Jersey and the American Tobacco Company.

Meanwhile, U.S. firms that began to produce increasingly complex manufactured items sought to achieve economies of scale in an expanding market through mass production and volume retailing (e.g., sewing machines, automobiles, and typewriters). By integrating vertically—controlling as many steps in the production of an item as possible, from raw material through manufacture and even marketing—firms (e.g., Carnegie Steel) combined to create even larger companies better able to withstand the economic oscillations of the period between the end of the Civil War and 1896.

The new large enterprises could no longer be administered informally, with control of markets the principal preoccupation of management. Creative managers of some of these enterprises (in, for example, the tobacco, meat-packing, and agricultural power machinery industries) developed the centralized, functionally departmentalized organizational structure. After 1900, a new wave of expansion occurred in industries exploiting new technologies such as electrification and the gasoline engine. Product diversification became a common strategy for expansion in firms that could exploit systematic research

---

14. That Phillips enjoyed continuing esteem long after Apollo was reflected in NASA's request that he head a comprehensive post-*Challenger* accident study of NASA's management practices.
15. Dale Wolfe, Executive Officer, American Association for the Advancement of Science, editorial for *Science*, November 15, 1968.

and development—firms in the chemical, rubber, automobile, and electrical industries. Product diversification, in turn, required a different organizational approach to management. The strategy of diversification was followed by decentralization in these firms' organizational structures.

Decentralization, however, posed its own administrative problems for these firms. How was authority to be distributed among headquarters and field activities? The most common solution was that developed by managers of the railroads nearly a half-century before: the multi-divisional line-and-staff organization, by which authority was delegated from headquarters to plant managers in the field (who could not otherwise be held accountable for the performance of their units), while managers of centrally located auxiliary or service functions set standards and procedures.[16]

In post-World War II America, several new forces began to make themselves felt on U.S. industry and, as a consequence, gave rise to new management strategies. Among these was the entrance of the public sector—primarily the federal government—into the marketplace as a significant buyer. Another was the emergence of a substantial market, and a responding productive capacity, for goods and services having highly sophisticated technological ("hardware," "software," and "services") components.

The importance of technological sophistication as a driving force in this new market cannot be overestimated. The largest public sector buyer, the military establishment, seeking out ever-improved weapons systems, funded industrial research and development both indirectly as a buyer of newer and more advanced systems and directly as the largest single investor in research and development.[17] How much the U.S. economy has been affected by these two factors—the federal government as buyer and that buyer's interest in new technologies—is reflected in the top five industries (measured by sales) in the United States in 1988. Heading the list are two U.S. industries well-established before World War II: petroleum refining ($284.3 billion) and motor vehicles and parts ($273.1 billion). Third, fourth, and fifth are industries that were initially stimulated by the federal government's post-World War II appetite for technologically sophisticated systems and its ability to find ways to pay for them: electronics ($115.3 billion), aerospace ($112.8 billion), and computers and office equipment ($112.6 billion).[18] The sales and capital represented by these figures grew on a foundation built of successfully managed government research and development programs.

To appreciate the complexity of the technical management and quality controls, not to mention coordination and accounting, that government and industrial managers faced in assuring the success of one major NASA program, consider the prime contracts awarded to industry to design, build, test, and certify the principal components of the Saturn V alone: Boeing Co., S-IC, first stage (powered by five F-1 engines); North American Aviation, S-II, second stage (powered by five J-2 engines); Douglas Aircraft Company, S-IVB, third stage (powered by a single J-2 engine); Rocketdyne Div. of North American Aviation, J-2 and F-1 engines; and International Business Machines (IBM), Saturn instrument unit.[19]

Were this the extent of industrial contractor involvement in the program, that would have been management challenge enough. In addition, a *partial* listing of the *subcontracts* these contractors awarded to other firms that "played a major role in the development and

---

16. Alfred D. Chandler, Jr., *Strategy and Structure: Chapters in the History of American Industrial Enterprise* (Cambridge, MA: M.I.T. Press, 1962), Chapters 1, 2, *passim.*

17. Ross M. Robertson, *History of the American Economy*, 2nd ed. (New York: Harcourt, Brace & World, Inc., 1964), p. 555. In 1946-47 the Federal government paid 24 percent, and industry paid 72 percent, of the dollars (est. $2.1 billion) spent on industrial research and development during that period. By 1969 the federal government's share of the total (est. $28 billion) had increased to 40 percent and industry's share declined to 58 percent. Which sector (private or public) actually spent the rapidly increasing number of dollars devoted to research and development during 1946-1969 underwent a comparable change: industry spent 62 percent of the nation's research and development dollars in 1946-47 and 76 percent in 1969

18. *The World Almanac and Book of Facts* (New York, 1990), p. 86. Data from *Fortune* magazine.

19. North American Aviation was bought by Rockwell and was known as North American Rockwell Corp. after September 1967. In 1967 Douglas Aircraft Co. and the McDonnell Corp. merged, becoming the McDonnell Douglas Corporation. The former Douglas division in California became the McDonnell Douglas Astronautics Co. (MDAC).

production of the Saturn V launch vehicle" would have to include the 50 subcontractors to Boeing, 91 subcontractors to Douglas Aircraft, 54 subcontractors to IBM, 28 subcontractors to North American Space Division, and 51 subcontractors to North American Rocketdyne.[20] These well over 250 firms provided innumerable parts and components, ranging from hydraulic hoses to analog computers, all of which had to meet exacting specifications for integrated fit and performance. "I wish to emphasize," remarked a Marshall Space Flight Center procurement officer during the bidding for the S-II stage contract, "that the important product that NASA will buy in this procurement is the efficient management of a stage system."[21]

So impressive was the management undertaking involved in developing and fabricating the Apollo/Saturn systems that even before the historic Apollo 11 mission left the launch pad on the morning of July 16, 1969, the Committee on Science and Astronautics of the U.S. House of Representatives asked key industry Apollo/Saturn contractors and NASA program managers to review their program management practices. [IV-11] Their published responses make tedious reading, littered as they are with charts and acronyms and general ineloquence, but they have an important story to tell. Unlike the industrial firms of earlier periods of U.S. history, the firms that supplied the aerospace programs of NASA and the military were engaged in the low-volume production of items that were complex, novel, and relatively unique; thousands of "end items" produced by dozens of different suppliers and manufacturers had to fit and function together, and be produced on schedule and at the levels of reliability called for by manned missions. Thus the efficiency-seeking attributes of the traditional "American system of manufacture" (use of standardized interchangeable parts and continuous process manufacture) no longer applied.

The "efficiency"-inspired organizational structure of functionally distinguished units (e.g., finance, accounting, marketing, research, facilities, engineering, testing, manufacture, logistics, etc.), adequate for the production of essentially undifferentiated products, would not suffice. "Early in the development phase of the Apollo/Saturn effort," recalled Rocketdyne's vice president of management planning and controls, "Rocketdyne management recognized that the traditional functional organizational alignment was not adequate to direct the effort of the various engine programs effectively. To ensure the necessary concentration of effort, it was decided to establish separate product organizations with responsibility for the development of specific types of engines."[22] Not all companies had been organized like Rocketdyne; Boeing's management was "basically decentralized and organized around product line responsibilities," one in which "the functional executive provides a unifying force which crosses the boundaries of the various line organizations. . . ." Nonetheless, at Boeing the "line organization managers" had the "ultimate authority and responsibility for carrying out The Boeing Co.'s contractual and related commitments to its customers."[23]

The novelty and relative uniqueness of the aerospace industry's products necessarily meant that little would be "standard"; the ability to respond intelligently and quickly to failures would become a critical management responsibility. That responsibility was felt especially acutely among government (NASA) managers responsible for the Saturn program's success:

. . . such [Apollo/Saturn program management] features as actions for early problem detection, actions and process for problem solving, and action and processes for recovery from anomalies and failures are basic features. . .[24] *. . . the system must provide visibility and flexibility. You need the visibility to identify nonproductive tasks and you need the flexibility to redirect the effort. Otherwise,*

---

20. Roger E. Bilstein, *Stages to Saturn: A Technological History of the Apollo/Saturn Launch Vehicles* (Washington, DC: NASA SP-4206, 1980), *passim*, and Appendix E.
21. Manned Space Flight Center, "Minutes of the Phase II Pre-Proposal Conference for Stage S-II Procurement on June 21, 1961," JSC files. Quoted in Bilstein, *Stages to Saturn*, p. 211.
22. "Apollo Program Management: Staff Study for the Subcommittee on NASA Oversight, U.S. House of Representatives, Committee on Science and Astronautics, 91st Cong., 1st sess. (Washington, DC: U.S. Government Printing Office, 1969), p. 122.
23. H.H. Gunning (Boeing Co.), *ibid.*, pp. 15-16.
24. Eberhard F.M. Rees (NASA Marshall Space Flight Center), *ibid.*, p. 9.

you would be using up limited resources on tasks that were no good. Visibility and flexibility imply a knowledgeable decision point close to the work.[25]

The project manager, the program manager, and their staff became the "knowledgeable decision" points "close to the work" that government and industry created to manage the development and production of specialized technological systems. "The heart of the Program Management System," explained one NASA program manager,

is the Project Manager who is responsible for the design, fabrication, test, delivery, and successful performance of a major piece of hardware, a product best exemplified by a stage of the launch vehicle. To achieve his goal, the Project Manager has clear lines of authority and responsibility as well as clear channels of coordination with supporting entities. These have been committed to clear, concise documented agreements. . . . In addition to management by product, such as the S-II Stage, the Program and Project managers also manage, to an extent, by function. These functional management elements. . .permeate the entire program. . . . These elements insure, within their disciplines, a continuous coordination between the functional elements [among other NASA organizations]. . .enabling many things to be handled at the working level. . . .[26]

Critical communication and coordination between government "customers" and industrial contractor organizations required of the latter that they develop management systems that paralleled, or mirrored, those NASA established. One Rocketdyne manager described NASA's (and DOD's) impact on the aerospace industry this way:

During the past seven years NASA has had a significant and favorable influence in the development of advanced management systems within Rocketdyne. Program planning and control requirements specified by both DoD and NASA have stimulated such management systems activity as development and implementation of the Rocketdyne Cost Management System, the Mechanized Production Control System, the Mechanized Inventory Control System coupled with the Required Inventory Control System, the Mechanized Quality Performance System, and the Mechanized Time-keeping System, to name a few. New concepts such as the well-defined program organization operating in a program/functional matrix relationship, the assignment of specific individuals to manage all activity on product-oriented elements of program work breakdown structures, and the application of the multiple accountability technique also saw their genesis during this period.[27]

Similar managerial adaptations occurred throughout the aerospace industry.

The government's and the aerospace industry's strategy for managing the design and development of large, complex, and relatively unique technical systems—or program management—had an important political dimension as well. The project (the development of a single entity or system) and the program (a cluster of interrelated projects) soon became, in effect, products and product lines marketed by the military and NASA to Congress and the White House. NASA learned, as the military had learned, that Congress, relatively stingy with funds for abstract and indefinite activities such as fundamental research, could be persuaded to open the public purse for clearly defined packages of concrete "end items" with specific missions. Concrete end items meant actual hardware contracts that might benefit particular congressional constituencies. The Apollo program, like the Manhattan Project before it, was just such a package. A *program* thus became a bureaucratic and budgetary device for framing and executing projects to explore space and advance aeronautical technology.[28] The design and execution of a successful project became the measure of success, as many of NASA's people got caught up in the annual

---

25. R.L. Brown (NASA Marshall Space Flight Center), *ibid.*, p. 13.
26. Edmund F. O'Connor, "General Program Management," *ibid.*, p. 247-48.
27. *Ibid.*, p. 126.
28. NACA's more modest aeronautical research role—the "service" it provided the military and aviation industry—was rapidly replaced by NASA's need to direct its research and development know-how to specific programs, in particular, the manned spaceflight sequence known as Mercury, Gemini, and Apollo. Conceptually and administratively, the NASA program became the umbrella under which projects were justified and planned, congressional authorization and appropriations obtained, private sector sources solicited and evaluated, contract awards made, and those contracts administered.

need to market the agency's projects and programs to Congress to obtain the appropriations necessary to sustain their work.

In an early (1961) reorganization, NASA sought to discourage internecine competition for resources that developed when an agency organized itself around hardware programs by identifying its own programs with broadly framed goals instead. The Apollo program represented one such goal.[29] The ultimate effectiveness of this approach, however, depended somewhat on the nature of the goal used—on the variety of realistic hardware approaches that could be used to achieve it. For example, the goal of "Space Sciences" was fairly diffuse; many hardware projects could be embraced by it. This was less true for costly projects. After Apollo, only the shuttle and the hoped-for space station—each a very specific hardware program that would require relatively large portions of the agency's total budget—emerged to satisfy the goal of manned space exploration. To appreciate the emergence and effect over time of the "program" both as a managerial and as a political device, note its absence in Hugh L. Dryden's speech on the fledgling space program, given when NASA was only a few months old.

A Culture at Risk

It would be difficult to exaggerate the significance of the policy of "contracting out" in terms of the way NASA went about its daily work. Virtually every aspect of the agency's business was ensnared in the dense forest of regulations and procedures of federal acquisitions policy. The number of procurement actions processed by NASA quadrupled from roughly 44,000 in 1960 to almost 190,000 in 1963; by 1965, NASA was processing almost 300,000 actions, or almost seven times the actions the agency was managing only five years before. The dollar value of the average NASA contract more than doubled as well. However, during the same period (1960–1965), NASA's personnel increased by only a factor of three, and only a fraction of them was qualified to manage or monitor contractors. Thus, the burden of implementing the government's "contract out" policy was borne increasingly by NASA's technical people. Engineers who had come to NASA (or earlier to NACA) to do engineering found themselves increasingly cast in the role of overburdened contract monitors, ever more remote from the "hands-on" work that had attracted them in the first place. [IV-19]

Originally an aggregate of essentially independent, in-house research organizations, NASA also struggled with the centralized controls inherent in large-scale program management. As NASA faced, after 1966, tighter budgets, competition among the former NACA laboratories, new NASA centers, and Headquarters intensified. Because the centers managed the contractors, and because the centers housed NASA's technical expertise, they acquired the power of fiefdoms—and were often so called. Nonetheless, NASA sought to retain the discipline orientation of NACA's decentralized laboratories—further accentuating a tension between aspirations of various research disciplines and program organization that would persist through much of NASA's institutional life in the next thirty years.

The agency's inherited culture struggled against centralization at the government-wide level as well. When NACA was transformed into NASA in 1958, the committee structure by which it had been administered was abandoned for a hierarchical and centralized management structure. Centralized federal administrative controls that evolved during the 1940s and 1950s—controls such as standardized personnel management, budgeting, procurement, and operating procedures—were imposed on NASA by the Bureau of the Budget (after 1970 the Office of Management and Budget, OMB), the Civil Service Commission (after 1979 the Office of Personnel Management, OPM), and ultimately, of course, the U.S. Congress.

The proportion of NASA's total in-house permanent workforce consisting of scientists and engineers gradually increased from one-third in 1958 to slightly less than one-half

---

29. Levine, *Managing NASA in the Apollo Era*, p. 5.

in 1970. At the same time, the ratio of NASA's contractor employees to civil service employees increased from roughly 3 to 1 in 1960 to 11 to 1 in 1965 (see Table 1). After the post-fiscal year 1966 downward slide in NASA's funding, that ratio declined. Assuming an increase in externally imposed, and thus difficult to change, administrative burdens on NASA from 1960 forward, those burdens had to be carried increasingly by the agency's civil service scientists and engineers.[30]

Among externally imposed management controls, the federal personnel system has proven as critical to NASA as federal acquisitions policy. NASA's predecessor, NACA, had struggled against civil service pay scales and hiring/promotion procedures and ceilings, which, NACA insisted, made it difficult to recruit and retain good engineers. NASA was able to obtain 525 "excepted" positions[31] to hire the talent it needed to carry out the Apollo program. However these were indeed exceptions—exceptions to a long-term, systemic disregard by the federal personnel system of its impact on the agency's culture of technical competence. That system was and remains strongly biased toward seniority and generic functions; it assumes that increases in rank and salary should be directly related to increasing supervisory or managerial responsibilities.

Compounding this systemic barrier to "advancement" for engineers has been a cultural prejudice that goes back to Greek and Roman antiquity, the notion that those who work with ideas have greater social value than those who work with their hands—or "things." For typical managers, the hierarchical and centralized structure of power in most organizations (not excepting NASA) reinforces their increasing remoteness, as they moved "up the ladder," from practical, day-to-day concerns and "hands-on" work. More than four-fifths of the NASA engineers recruited during NASA's first decade "advanced" into management positions, and among the older engineers who were employed with NASA or NACA before 1960, over 90 percent ended their careers in management positions. Occasionally, a NASA engineer has risen to the level of GS-16 without moving into management, but the widespread perception within the agency has been that the dual-career ladder works only for the very exceptional few. Thus many NASA engineers' occupations diverged increasingly from their vocations as they began to spend more of their days doing work for which they had not been trained and may have had little natural inclination. On the other hand, some NASA engineers, fearing obsolescence in engineering careers, considered management a legitimate and productive alternative for individuals with some understanding of how technical programs work.[32] Engineers turned managers could then leverage their knowledge and experience through the projects for which they were responsible.

Looking for a Mission

The Apollo program was unarguably an enormous achievement. Nevertheless, the transient motives behind the program, and the rapid mobilization of funds and personnel that made success possible, impeded the gradual evolution of a stable and broad public consensus about the nation's purpose in space. As more than 13,000 NASA engineers worked at their daily routines during the mid-1960s, pursuing the adventure to which President Kennedy had summoned them, the solid ground of common national purpose had already begun to shift ominously under their feet. By 1965, John F. Kennedy lay buried, and three years later he would be joined by Robert Kennedy, who, along with Martin Luther King, would be victims of violence. Violence in the United States, as race-related riots spread from urban ghetto to urban ghetto, was matched by U.S. violence abroad.

---

30. Sylvia Doughty Fries, "Apollo: A Pioneering Generation," International Astronautical Federation, 37th Congress (October 9, 1986), Ref. No. IAA-86-495.

31. Appointments are exempt from standard federal civil service classifications and salary ranges.

32. The information in this section is drawn from Sylvia Doughty Fries, *NASA Engineers and the Age of Apollo* (Washington, DC: NASA SP-4104, 1992).

Television, which had been acquired by 94 percent of all U.S. households by the mid-1960s, rendered these scenes of violence commonplace and provided a world stage for an outpouring of public protest against U.S. military involvement in Vietnam.[33] In March 1968, that champion of space exploration, President Lyndon B. Johnson—so tough in the battle against the North Vietnamese, so tough in the battle against poverty and race discrimination—formally abandoned any hope of reelection. Raising the specter of runaway inflation as costs for the war in Vietnam and the social programs of the "Great Society" mounted, Johnson's economic advisers had persuaded the president in 1965 that the budget for the space program would have to be contained. There was diminishing enthusiasm outside NASA for an ambitious space program to follow the Apollo adventure. In fiscal year 1966, NASA's budget began its downward slide (though actual outlays for 1966 were the highest of the decade).[34] [III-20]

The political consensus that had produced the visionary National Aeronautics and Space Act of 1958 began to dissipate before the first few Apollo missions were flown.[35] NASA's fiscal year 1971 budget took a battering from the Bureau of the Budget in 1969, forcing the cancellation of Apollo missions 18 through 20 and leading Webb's successor Thomas O. Paine to complain that the Bureau of the Budget had ignored the ambitious recommendations of the White House's own Space Task Group, chaired by Vice President Spiro T. Agnew. [III-25] A staunch supporter of a vigorous manned space program (and hence further Apollo manned expeditions to the Moon), Paine was willing to cease the continued production of the Saturn launch vehicle and to defer the Viking project to launch an unmanned spacecraft to land on the planet Mars to pay for further manned lunar missions. Viking survived, as did a proto-space station (Skylab) fashioned from Apollo-Saturn hardware and flown during 1973; but the mighty Saturn did not. NASA was able to persuade the Nixon administration that a new Space Transportation System (STS) featuring a reusable orbiter spacecraft and solid propellant rocket boosters, flying thirty or more times a year, would be an economical alternative to the use of large "throw away" launchers such as the Saturn.

The fortunes of NASA's authorizing legislation, the "Space Act," reflects a similar diminished priority for a great national adventure in space as successive amendments stripped the statute of its originally well-focused declaration of purpose. In 1964 NASA's ten top executives lost their special pay status. In 1973 the National Aeronautics and Space Council, which could have served as a vehicle by which the executive branch crafted an interagency consensus around a well-defined program, was abolished. From 1974 onward, NASA's authorizing statute became burdened with numerous charges to the agency, occasionally having only the most tangential relation to NASA's original purpose. At the same time, the addition of these new statutory directives reflected admiration for the agency's technical and managerial know-how. After all, "if NASA could send men to the Moon, why couldn't they also. . . ?" NASA was directed to develop and demonstrate "solar heating and cooling technologies" in 1974, to monitor and investigate the "chemical and physical integrity of the Earth's upper atmosphere" in 1975, to develop "more energy efficient and petroleum conserving and environment preserving ground propulsion systems" in 1976, to develop and demonstrate "electric and hybrid [ground] vehicle" technologies in 1976, and to develop advanced automobile propulsion systems *and* to assist "in bioengineering research, development, and demonstration programs designed to alleviate and minimize the effects of disability" in 1978. In the early 1980s NASA lost its privileged position as the U.S. arbiter of non-military space activity, as the agency was denied authority to

---

33. For one view of the decade, see Allen J. Matusow, *The Unraveling of America: A History of Liberalism in the 1960's* (New York: Harper & Row, 1984).

34. Robert A. Divine, "Lyndon B. Johnson and the Politics of Space," in Robert A. Divine, ed., *The Johnson Years: Vietnam, the Environment, and Science*, Vol. II (Lawrence: University Press of Kansas, 1987), pp. 217-53.

35. The last Apollo mission was the Apollo-Soyuz Test Project jointly conducted with the Soviet Union. An Apollo command and service module, equipped with a specially adapted docking module, joined with a Soyuz spacecraft in July 1975. The spacecraft spent two days docked together in orbit while American astronauts and Soviet cosmonauts ate and visited together and performed joint scientific investigations.

promulgate regulations for the granting of licenses for NASA patents and, in 1984, as the agency acquired statutory direction to "seek and encourage to the maximum extent possible the fullest commercial use of space." By 1988 NASA found itself required to contract with industry for Expendable Launch Vehicle (ELV) services.[36]

As public support for the civilian space program remained soft,[37] the number of government employees NASA was able to support declined to about two-thirds (in 1988) of the almost 36,000 people on the NASA payroll in 1966.[38] Faced with deteriorating support, NASA executives had a legitimate desire to protect the field centers, whose most skilled technical employees were essential to the agency's ability to go about its work. By designating "roles and missions" for each of the centers, NASA attempted to avoid duplication and assure each installation essential functions related to the particular project work assigned to it.[39] [IV-14] The elaborate institutional machinery developed to carry out Apollo could not be so easily disassembled, however, given the interlocking interests it had created among NASA's installations, contractors, and geographic regions and their representatives in Washington.

And so the organization that built America's civil space program in the high-noon of the Cold War groped about for a marketable mission. In 1971 Deputy Administrator George M. Low even contemplated recasting NASA as a national technology agency, responsible not only for aeronautics and space research and development, but also for a wide range of "technological solutions" for national problems such as alternative power and energy sources, environmental pollution, improved transportation systems, health care systems, productivity of services, education, and housing.[40] [IV-12] That others were thinking in this vein as well is apparent from the non-aerospace responsibilities added to NASA's authorizing legislation during the 1970s.

NASA's civil servants and various advisory groups carried out periodic studies during subsequent years to define NASA's goals, or to articulate a vision, for the civil space program. [IV-15, IV-16] There were, of course, those visionaries within the agency who had worked with NASA for decades and believed that if they tried harder the public could be persuaded not only to recognize the promise of an ambitious space program, but to pay for it. Such visionaries combined with bureaucratic entrepreneurs a decade later to persuade President Ronald Reagan in 1984 to pronounce his blessing on a program to design, build, and operate a true space station—an orbiting U.S. outpost in space that had been a NASA dream since the agency was first established.[41]

---

36. *National Aeronautics and Space Act of 1958, as Amended.* Printed for the use of the National Aeronautics and Space Administration (January 1990).

37. As measured by NASA appropriations, which have not recovered their 1965 level in constant dollars. See also "Towards a New Era in Space: Realigning Policies to New Realities," Committee on Space Policy, National Academy of Sciences and National Academy of Engineering (Washington, DC: National Academy Press, 1988).

38. NASA contractor employees outnumbered civil servants 3 to 1 in the early 1960s, ballooned to 10 to 1 in 1966, and subsided to about 2 to 1 in the 1980s. Nimmen and Bruno with Rosholt, *NASA Historical Data Book, Vol. I*, p. 118; *NASA Pocket Statistics* (Washington, DC: NASA, 1986), p. C-27. Numbers of current contractor employees can only be estimated.

39. Associate Director for Center Operations, on "Catalog of NASA Center Roles," April 16, 1976. Part of the intent of the "roles and missions" concept may have been to reduce inter-center rivalry, but institutional specialization has apparently done little to relieve institutional particularism.

40. George M. Low, Deputy Administrator, NASA, Memorandum for the Administrator, "NASA as a Technology Agency," May 25, 1971.

41. Sylvia Doughty Fries, "2001 to 1994: Political Environment and the Design of NASA's Space Station System," *Technology and Culture* 29 (July 1988): 568-93.

A Space Transportation System

Meanwhile, during the 1970s the more pragmatically minded bowed to the budgetary pressures that had come to dominate Washington's political climate. In 1971 NASA persuaded the Nixon White House that the proposed shuttle program[42] would "take the astronomical costs out of astronautics."[43] The agency had contracted with an economic research firm to investigate the economics of the proposed shuttle system. The economists reported in 1971—on the basis of figures and formulas that had to have been somewhat speculative—that such a system would be economical *assuming* a flight rate of "between 300 and 360 shuttle flights in the 1979-1990 period, or about 25 to 30 space shuttle flights per year."[44] [III-30] Even more portentous was what such a flight rate, in turn, assumed that NASA—its organizational strength rooted in its history as an advanced technology research and development organization—would be just as successful as the operator of a routine transportation system.

NASA Deputy Administrator George M. Low acknowledged that the agency would have to change to operate a cost-effective space transportation system, though whether he grasped just how fundamental a change was involved is not clear. [IV-13] The cost of "doing business in space, coupled with limited and essentially fixed resources available for space exploration," observed Low to his senior management team, "places severe limitations on the amount of productive work that NASA can do, unless we can develop means to lower the unit cost of space operations." Low correctly attributed that "high cost" to the "great sophistication" with which most space systems are designed in order to "operate acceptably with low allowable weight" and to the fact that "most systems are individually tailored for their mission, used once or twice, and then never used again. Thus the economies of producing a number of like systems are never attained." NASA would now, asserted Low, have to abandon the strategy of developing "individually tailored technologies" and, instead, "focus on *multiple-use, standardized* systems" (emphasis added).[45] In 1983, with the shuttle's series of flight tests completed, Congress added to the statutory activities in which NASA was authorized to engage "the *operation of a space transportation system. . .*" (emphasis added).

Although Low may not have thought of it in these terms, he was, in effect, asking the NASA organization to turn back the clock to a time when U.S. manufacturers evolved management strategies to achieve the efficiencies of standardized, volume production to exploit an expanding market. It was a bold risk that he was taking. To the extent that the nation's civil space program hinged on the success of the shuttle program, NASA would have to undertake the most profound reversal in its organizational culture that any organization could be asked to make. Would it succeed? Could the agency and its industrial partners unlearn the management strategies and habits they had had to learn in order to design and produce the complex and reliable aerospace systems that carried men to the Moon? Would NASA's inherited research culture be able to respond to the administrative and logistical demands of routine operational efficiency? And would an expanding market for space transportation support the need to divert scarce resources into the routine operation of "multiple-use, standardized systems"?

A partial answer came in the form of the report issued by the Presidential Commission on the space shuttle *Challenger* accident that had occurred January 28, 1986. [IV-17] Chaired by former Secretary of State William P. Rogers, the commission concluded that the fiery end of the STS-51L mission was caused by "the failure of the pressure seal in the aft field joint of the right Solid Rocket Motor. The failure was due to a faulty design unacceptably sensitive to a number of factors. These factors were the effects of temperature,

---

42. Properly referred to as the "Space Transportation System" (i.e., the Shuttle Orbiter, External Tank (non-recoverable), and twin Solid Rocket Boosters).

43. Statement by the President, the White House, January 5, 1972.

44. Mathematica, Inc., "Economic Analysis of the Space Shuttle System," National Aeronautics and Space Administration Contract NASW-2081 (January 1972).

45. George M. Low, Deputy Administrator, NASA, Memorandum to Addressees, "Space Vehicle Cost Improvement," May 16, 1972.

physical dimensions, the character of materials, the effects of reusability, processing, and the reaction of the joint to dynamic loading."[46] That was the *technical* cause. The commission was also impressed by other proximate causes of the accident to which it ultimately gave great weight: a top-level decision to launch that had been inadequately informed about the sensitivity of the O-rings on the Solid Rocket Boosters' aft field joints to the inordinately cold temperatures prevailing at the time of the launch, a "silent" safety, reliability, and quality assurance program, *and* an organizational failure to adapt to the requirements of a truly operational transportation system. These included lack of schedule discipline and inadequate logistics to support the flight rate that would enable the agency to deliver the economies promised when President Ronald Reagan announced in 1982 that "the first priority of the STS program is to make the system fully operational and cost-effective in providing routine access to space."[47] [III-38]

For the next two and a half years NASA redesigned known weaknesses in the shuttle's systems, elevated the status of the safety, reliability, and quality assurance organization, and tightened decision-making channels between its centers and headquarters. The result was a successful "return to flight" in September 1988. Wags remarked that the flight of STS-26 was probably the safest shuttle mission imaginable. Underlying management issues—especially whether NASA could, or even should, attempt to transform itself into an operations organization—proved more stubborn. When the agency undertook an assessment of its "management practices and...the effectiveness of the NASA organization," it turned for help to one of its most respected *program managers*, General Phillips.

Not surprisingly, the Phillips group, which reported back to NASA in December 1986 [IV-18], recommended (among other things) stronger program management, to be achieved through "strong headquarters program direction for each major NASA program, with clear assignment of responsibilities to the NASA centers involved," and improved "discipline and responsiveness to problems of the program management system." At the same time, the group insisted that "NASA must accept that it will be responsible for spaceflight operations for the foreseeable future." That NASA had not, to that point, fully accepted its operational responsibility was suggested by the fact that the agency's "present structure of organization and management does not assure adequate attention to operations requirements in system design or in the planning and conduct of operations and logistic support in the era of frequent shuttle flights and long-term operation of the space station."

To buttress the agency's ability to meet the operational needs of the shuttle program, the Phillips group called for the creation of a new associate administrator for operations, whose organization would include space tracking network and data systems and—eventually—the Kennedy Space Center. Two years later NASA did create an associate administrator-level Office of Space Operations, but it was not clear whether the new organization was merely old wine (the former Office of Space Tracking and Data Systems) in a new bottle. The competing demands of operations and research and development continued to trouble the agency whenever (as in 1990 and early 1991) its heightened safety procedures detected problems with shuttle hardware, requiring protracted "stand downs" of one or more shuttle spacecraft.

Compromise

Underscoring the uncertainty of NASA's mission and its standing within the constellation of Federal programs, President George H. Bush reestablished in April 1989 an interagency policy council for the nation's space activities when he created the National Space

---

46. *Report of the Presidential Commission on the Space Shuttle Challenger Accident, Vol. I* (Washington, DC: U.S. Government Printing Office, June 6, 1986), p. 72.
47. Quoted in *Report of the Presidential Commission on the Space Shuttle Challenger Accident*, p. 164.

Council, chaired by the vice president. Through the Advisory Committee on the Future of the U.S. Space Program, established in 1990 under the auspices of NASA and the National Space Council, a consensus emerged that NASA's primary business should continue to be what it had been in the 1960s—the scientific exploration of space and aerospace research and development. [IV-20] Asserting that "perfection" should be the single most important aim for NASA's organizational culture, the "Augustine Committee," informally named for its chairman, Norman R. Augustine, chairman and CEO of the Martin Marietta Corporation, explained:

... perfection can most closely be approached in an organization whose ethos is one of excellence and where this ethos permeates everything it does. . . . It must be clear to all that, in this culture, excellence is more important than schedule and more important than cost—even though these too are important—and that management at all levels can be reliably counted upon to act with this as its set of values (emphasis author's).[48]

At the same time, the committee recognized that, so long as NASA was responsible for the shuttle, the agency would have to adapt to the demands of a successful operating organization. The comments of many who spoke with the committee "frequently referred to the consuming effect this [flight operations] responsibility can have on NASA's senior management, limiting the time available for the planning and direction of leading-edge technological developments." Committee witnesses also expressed the belief that "the merging of operations into a largely developmental organization does not foster the building of a professional operations cadre which can best manage this vital responsibility."[49]

The committee added a refinement to the issue that had been provided by a 1988 National Academy of Public Administration (NAPA) study, also led by Phillips, of NASA Headquarters management. The NAPA study did not fault NASA for its weaknesses in operations management. Rather, it argued,

the term "operational" as applied to commercial aircraft, to ships, or to mass-produced articles of defense will most likely never apply to space systems in that same context. What we do see, however, are large, complex space systems such as the Shuttle and the Space Station that are or will be largely driven by operational issues—turnaround time between flights, manifesting, retrofitting of design changes for safety, cost or payload capability purposes, logistics, training of basic and science crew members, and so on. These are not the basic work of research and development leading to new concepts and ideas for future space systems, nor for expanding knowledge of the universe and discerning the implications of that knowledge for life on this planet or elsewhere.[50]

The NAPA report supported the earlier Phillips report recommendations and what the Augustine committee would recommend: "an organizational separation, from the top of the agency down, on the two matters of space flight operations and space system development." A new associate administrator position for space flight operations should be established, whose responsibilities should include space shuttle operations, ELV (expendable launch vehicle) operations, and the tracking and data systems organization. This individual should then be given the formidable task of "injecting operational requirements into new programs to assure that they can be effectively operated over their lifetimes at reasonable cost."[51] Just what leverage this individual would have at budget time over the prevailing research and development culture of the agency, the committee did not say. Shuttle operations themselves, however, might be less likely to receive short shrift, added the committee, if responsibility for the space shuttle was "eventually moved from a development oriented center [viz., Johnson Space Center] to the operationally oriented Kennedy

---

48. *Report of the Advisory Committee on the Future of the U.S. Space Program* (Washington, DC: U.S. Government Printing Office, December 1990), p. 16.
49. *Ibid.*, p. 38.
50. National Academy of Public Administration, Samuel C. Phillips, Chairman, *Effectiveness of NASA Headquarters: A Report for the National Aeronautics and Space Administration*, February 1988. Quoted in *ibid.*, p. 38.
51. *Ibid.*, p. 38.

Space Center." What NASA should strive for, urged the committee, is "safe operation [of the Shuttle], performed as efficiently and routinely as its complexity permits, and not burdened by excessive layers of management that are the legacy of the development era and recovery from the Challenger accident."[52]

And so, a compromise was struck. NASA should retain its identity and role as a research and development organization, the identity with which most of its people were comfortable and upon which its self-esteem depended, and it would not have to lose its most visible achievement—the shuttle—to do so. Suggestions that space shuttle operations be transferred to some other, and perhaps especially created, government entity, or to the private sector, had been rejected. But some significant portion of the organization would have to learn how to operate a transportation system. Whether Congress, or NASA's internal budgetary politics, would yield the wherewithal to do so remained to be seen.

How effectively an organization imbued with the values and habits of a research and development mission could adapt to the requirements of the efficient and cost-effective operation of a space transportation system was (setting aside perennial funding issues) one of the two principal issues facing the NASA organization at the beginning of the 1990s. The other was an old issue, one that could be traced back to the 1950s: the wisdom and consequences of the federal government's policy of "contracting out" for the bulk of its research and development work as well as for supplies and services.

In the spring of 1990, NASA's administrator asked the National Academy of Public Administration to revisit that question for NASA. The NAPA study, completed in January 1991, found still valid the 1962 Bell Report's guideline for what, and what should not, be contracted out. The government should *not* contract out. . .

decisions on what work is to be done, what objectives are to be set for the work, what time period and what costs are to be associated with the work, what the results are expected to be, and the evaluation, and the responsibilities for knowing whether the work has gone as it was supposed to go, and if it has not, what went wrong, and why, and how can it be corrected. . . .[53]

Having surveyed, with interviews and questionnaires, over 2,000 NASA scientists and engineers, the NAPA study team concluded that contracting out had indeed led to an erosion of strength among NASA's civil service scientists and engineers. Critics argued that that was a predictable conclusion, given the persons surveyed. It then proceeded to develop recommendations, most of which called on NASA's top management to provide better scrutiny of, and clearer guidelines for, the kinds of activities being contracted to the private sector. The context for these recommendations was the NAPA group's finding that "hands-on science and engineering work experience is essential to developing scientists and engineers with a level of knowledge that provides a sixth sense for spotting problems early, for being a smart buyer of technical products and services, and for being astute overseers of the work of technical contractors" and that NASA was not providing enough opportunities for this kind of work.[54]

The Augustine Commission, for its part, agreed that "an appropriate balance between in-house and external activity also should be developed." But this group saw the balance differently. In the more than three decades that had passed since NASA was created, there had developed a solid basis of space technology skills in both industry and academia; it was no longer necessary for NASA to match every development being contracted with comparable in-house laboratory skills. Citing the recent experience of national security aerospace research and development procurement, the committee argued that NASA could "buy smart" with fewer civil service project and program personnel. "NASA should concentrate its 'hands-on' expertise," the committee recommended, "in those areas unique to its mission, and avoid the excessive diversion of technical or mission

---

52. *Ibid.*, p. 40.
53. Quoted in National Academy of Public Administration, *Maintaining the Program Balance: The Distribution of NASA Science and Engineering Work Between NASA and Contractors and the Effect on NASA's In-House Technical Capability*, 2 vols. (Washington, DC: U.S. Government Printing Office, January 1991), Vol. I, p. 6.
54. *Ibid.*, Vol. I, p. x.

specialists to functions which could be performed elsewhere. Contract monitoring is best accomplished by a cadre of professional systems managers with appropriate experience. Increased use of performance requirements, rather than design specifications, will further increase the effectiveness of this approach."[55]

The Augustine Commission also called for more competitive government salaries for scientists and engineers, "pay for performance," and full use of existing flexibility within the government's personnel system. NASA should be a "pathfinding" agency for the development of an "advanced" federal personnel system that would reward excellence and special skills over seniority and generic tasks. Should NASA fail to persuade the Office of Personnel Management to allow the agency to revamp its personnel system, NASA might convert additional centers to federally funded research and development centers affiliated with major universities.[56] Whether NASA would succeed remains to be seen. Even if NASA were able to increase the number of high-caliber scientists and engineers within its ranks, would the practice of contracting out most of the agency's research and development work—leaving its own people to function as contract monitors—undermine its gains?

Conclusion

NASA's ongoing struggle to maintain its organizational momentum in the face of seemingly insuperable obstacles—public uncertainty, as well as its own, as to its overarching purpose; the constraining tendencies of federal regulations designed to keep political, bureaucratic, and technical power in check; and the need, time after time, to plead for funds and justify itself—is worth understanding not only because of what the agency does, but for what it represents. One obstacle NASA could not escape was the need to develop a large organization to carry out its work. That organization would perforce become a federal bureaucracy.

A creative bureaucracy seems to most a contradiction in terms. We rightly understand that the essence of a bureaucracy is depersonalized routine. Indeed, bureaucracies came into being so that the execution of laws and regulations in emerging nation-states might become less arbitrary, less capricious, and more accountable than it had been under personalized monarchical rule. No modern society with any aspiration to democracy would countenance surrendering its resources and destiny to a handful of solitary dreamers, however enticing the dream. Thus "organizing for exploration" was and remains the challenge facing the United States if it would venture across the frontier of outer space. The fact that managing the organization created to conduct that journey has proven difficult is less a sign of the failings of the travelers—though being human they have had failings enough—than a sign of the enormity of their task.

Document IV-1

Document title: J.R. Killian, Jr., "Memorandum on Organizational Alternatives for Space Research and Development," December 30, 1957.

Source: Dwight D. Eisenhower Papers, Eisenhower Library, Abilene, Kansas.

In the wake of Sputnik I and II, there was a wholesale reexamination of the U.S. organization for space-related activities. In 1955, when a scientific satellite program was initiated, it was given a low priority in comparison to other military efforts. At the time there was concern that even a small civilian space program, if given too many resources, could adversely affect critical ballistic missile programs. The issue was not so much one of cost, but the scarcity of human resources and development and test facilities. However, the political firestorm set off by the Soviet satellite brought into question the relatively low priority given the scientific space program. From the time the first Sputnik was launched

---

55. *Report of the Advisory Committee*, pp. 40-41.
56. *Ibid.*, pp. 40-42.

until NASA was established, almost all elements of the government were engaged in the debate on how best to redress the situation and reestablish the prestige of the United States. The failure of the first Vanguard launch on December 6, 1957, only intensified the calls for change. Sputnik also created the necessary impetus in the White House for the creation of the position of presidential science advisor. On November 7, James R. Killian, president of MIT, was appointed to this position. One of Killian's first duties was to address the issue of alternatives for space research organization. Some of his thoughts in this early memorandum eventually formed the basis of the administration's future policy toward the creation of a space agency.

---

[1] December 30, 1957

Memorandum on Organizational Alternatives for Space Research and Development

This memorandum is based upon the following assumptions:

A. That the Department of Defense proceeds with its announced plan for a Special Projects Division, reporting directly to the Secretary and including, as one of its major responsibilities, space research and development for the DOD.

B. That there is a broad area of non-military basic research relating to space which will command the interest and participation of scientists and engineers in a variety of non-government and government institutions.

With these assumptions in mind, we can proceed to a discussion of how the Government's sponsorship of space research and development can be handled and how the military and non-military programs can be related.

There have been proposals for a new Government agency analogous to either NACA or the AEC to handle all space research and development. In appraising this approach, the following considerations are of importance:

A. The DOD is committed to a space program and is in process of setting one up, although the nature of the program has not been clearly defined.

[2] B. Those aspects of space research and development which relate to the use of missile engines, and the testing and launching of vehicles must be closely associated with DOD missile programs. The necessity of such close association may dictate the placing of responsibility in the DOD for the development, testing, and use of rocketry for putting up space vehicles. It would seem unwise for a new agency, independent of the DOD, to have to create and use test facilities other than those built by DOD.

It seems of greatest importance that the DOD's own space program be very closely related to its missile program or for the two programs at some time to be merged.

These considerations seem to indicate clearly that the DOD must play a major role in space research and development if we are to use the nation's manpower and facilities in this area to the greatest advantage.

The DOD will, of course, be primarily concerned with those aspects of space research and development which will have military value. It is hard at this stage, however, to separate out of space R&D those elements, however basic and purely scientific, which would not contribute to military objectives. It seems entirely feasible for DOD to be the major sponsor and entrepreneur of space research and development, both military and "non-military."

There are many scientists and others, however, who are opposed to the centralization of all space R&D under the DOD. There are deeply felt convictions that the more purely scientific and non-military aspects of space research should not be under the control of the military. In the first place, [3] such an arrangement might improperly limit the program to narrowly concerned military objectives. In the second place, it would tag our basic space research as military and place the United States in the unfortunate position before the world of apparently tailoring all space research to military ends.

The problem of planning our non-military basic space research, then, becomes one of devising the means for non-military basic space research while at the same time taking advantage of the immense resources of the military missile and recon satellite programs, there are several possible ways of doing this:

A. The D.O.D. as a part of its program would establish a central space laboratory with a very broad charter which would permit the conduct of the most basic sort of research as well as R and D, having obvious military objectives. We see the pattern for this is such a Laboratory as the Los Alamos Scientific Laboratory of the A.E.C. Such a laboratory might also have the authority to sponsor research in civilian institutions.

B. The Department of Defense might confine itself to its military mission and some other agency or agencies external to the D.O.D. might engage in basic research. One obvious way of doing this would be to encourage N.A.C.A. to extend its space research and to provide it with the necessary funds to do so. A second [4] method (and this one might be handled along with an N.A.C.A. program) would be to provide funds either through the Department of Defense or otherwise to the National Research Counsel, the Council in turn sponsoring a series of projects in universities and industrial laboratories. The N.A.C.A. itself might do sub-contracting as indeed it does now to a limited extent. The problem here would be not to burden the N.A.C.A. with so large a program that the nature of N.A.C.A. would be changed. In its present form, it has been very successful but an undue enlargement of its program might reduce its effectiveness.

If either the N.A.C.A. or N.R.C. methods or both were followed, it would be necessary to carefully work out a cooperative arrangement with the D.O.D., for the D.O.D. would have to be an active partner with these agencies.

Such combination of sponsorship and programs would probably be the most advantageous way of carrying on space research for meeting both military and non-military objectives.

In considering these various alternatives and means, it is important to keep in mind existing resources available in the D.O.D., the Army's ABMA has a highly competent group for space research. The Air Force's BMC has important resources, including a going program for the development of a recon satellite. Cal Tech's Jet Propulsion Laboratory has advantages and resources for space research—a laboratory which has been closely associated with the Army. In the interest of conserving [5] man power and utilizing skill and experience already in being, these agencies must be considered in planning a new program. Some one or combination of these might well be made the nucleus of an extended program.

There should be some mechanism, however, which gives coherence to the broad program and which avoids a program encouraging inter service rivalries.

The overall plan must permit and provide for bold, imaginative research and planning. It must recognize the importance of providing the means and incentives for pure scientists to move effectively into space research without regard to practical applications. We must realize that in addition to such obvious objectives as space travel and reconnaissance, there are extraordinary opportunities to extend our knowledge of the earth and its environment and enormously to extend astronomical observations. It may well be that these kinds of pure, non-practical research objectives may prove to be the most important and in the end the most practical.

The overall plan, then, must keep steadily in view the need for those means and programs which will command the interest and participation of our best scientists. We must have far more than a program which appeals to the "space cadets." It must invoke, in the deepest sense, the attention of our best scientific minds if we as a nation are to become a leader in this field. If we do not achieve this, then other nations will continue to hold the leadership.

December 29, 1957   J.R. Killian, Jr.

Document IV-2

Document title: L.A. Minnich, Jr., "Legislative Leadership Meeting, Supplementary Notes," February 4, 1958.

Source: Dwight D. Eisenhower Papers, Eisenhower Library, Abilene, Kansas.

The Soviets had orbited *Sputnik I* four months prior to the meeting recorded by Minnich. By this time, it was all but certain that a new space agency would be created; however, its responsibilities, form, and location were still undecided. The question of the military or civilian character of a new agency was discussed in a regularly scheduled meeting among the president, vice president, other White House officials, and Republican leaders in Congress. The issue was raised in response to the impending reorganization of the Department of Defense, which was necessitated in part by the increasing sophistication and cost of weapons systems. Missiles and other space-related hardware were responsible for a significant portion of the technological revolution sweeping the military services at the time. At this time (February 1958), President Eisenhower had apparently not yet decided that most of the U.S. space program should be carried out under civilian auspices.

---

[1] ...**Outer Space Program** - A question was raised as to whether a new Space Agency should be set up within Department of Defense (as provided in the pending Defense appropriation bill), or be set up as an independent agency. The President's feeling was essentially a desire to avoid duplication, and priority for the present would seem to rest with Defense because of paramountcy of defense aspects. However, the President thought that in regard to non-military aspects, Defense could be the operational agent, taking orders from some non-military scientific group. The National Science Foundation, for instance, should not be restricted in any way in its peaceful research.

Dr. Killian had some reservations as to the relative interest and activity of military vs. peaceful aspects, as did the Vice President who thought our posture before the world would be better if non-military research in outer space were carried forward by an agency entirely separate from the military.

There was some discussion of the prospect of a lunar probe. Dr. Killian thought this might be next on the list of Russian efforts. He had some doubt as to whether the United States should at this late date attempt to press a lunar probe, but the question would be fully canvassed by the Science Advisory Committee in the broad survey it had under way.
[2] Dr. Killian thought the United States might do a lunar probe in 1960, or perhaps get to it on a crash program by 1959. Sen. Saltonstall had heard, however, that it might even be accomplished in 1958, if pressed hard enough.

[2] Dr. Killian outlined for the Leadership the various phases of future development (along the lines of the subsequent press release listing projects in the "soon," "later," and "much later" categories).

Sen. Knowland complained about having to get his information about Space research from the Democratic Senator from Washington (Jackson)—which was just as bad as having to learn from Mr. Symington anything there was to know about the Air Force.

The President was firmly of the opinion that a rule of reason had to be applied to these Space projects—that we couldn't pour unlimited funds into these costly projects where there was nothing of early value to the Nation's security. He recalled the great effort he had made for the Atomic Peace Ship but Congress would not authorize it, even though in his opinion it would have been a very worthwhile project.

And in the present situation, the President mused, he would rather have a good Redstone than be able to hit the moon, for we didn't have any enemies on the moon.

Sen. Knowland pressed the question of hurrying along with a lunar probe, because of the psychological factor. He recalled the great impact of Sputnik, which seemed to negate the impact of our large mutual security program. If we are close enough to doing a probe, he said, we should press it. The President thought it might be OK to go ahead with it if it could be accomplished with some missile already developed or nearly ready, but he didn't want to just rush into an all-out effort on each one of these possible glamor performances without a full appreciation of their great cost. Also, there would have to be a clear determination of what agency would have the responsibility.

The Vice President reverted to the idea of setting up a separate agency for "peaceful" research projects, for the military would be deterred from things that had no military value in sight. The President thought Defense would inevitably be involved since it presently had all the hardware, and he did not want further duplication. He did not preclude having eventually a great Department of Space....

Document IV-3

Document title: S. Paul Johnston, Memorandum for Dr. J. R. Killian, Jr., "Activities," February 21, 1958, with attached: Memorandum for Dr. J. R. Killian, Jr., "Preliminary Observations on the Organization for the Exploitation of Outer Space," February 21, 1958.

Source: NASA Historical Reference Collection, NASA History Office, NASA Headquarters, Washington, D.C.

On February 4, 1958, President Eisenhower announced that science advisor James R. Killian had appointed a panel to recommend the outlines of a space program and the organization to manage it. The so-called "Purcell Panel" (General James H. Doolittle, Chairman, NACA; Edwin Land, President, Polaroid Corporation; Herbert York, Director, Livermore Laboratory; and Edward Purcell, Professor of Physics, Harvard University), augmented by William Finan of the Bureau of the Budget and the staff support of S. Paul Johnston, Director of the Institute for Aeronautical Sciences, assessed organizational alternatives for the proposed agency. The task of inventing an organization to manage a space program was a difficult one. The number and strength of the claimants for the right to direct the space program had peaked in the wake of Sputnik. Several bills were already pending before Congress, which gave responsibility for space programs to the Department of Defense or to the Atomic Energy Commission. Johnston's thoughts on the subject eventually found their way into the March 5, 1958, memorandum [IV-4] to the president containing the formal proposal that NACA be reconstituted and given the responsibility for managing the nation's space program.

Memorandum for Dr. J. R. Killian, Jr.

FROM: S. Paul JOHNSTON
SUBJECT: Activities

1. During the past week, in accordance with your suggestion, I have conferred on the problem of organization and its legal implications with the following:

James A. Perkins, Vice President, Carnegie Corporation; John Cobb Cooper, Legal Consultant, Professor, International Air Law, McGill University; Dr. James Fisk, Vice President, Bell Telephone Laboratories; John J. Corson, McKinsey & Company; Don K. Price, Vice President, Ford Foundation; Dr. Edward Mason, Harvard University; Dean David Cavers, Harvard Law School

The above are in addition to the people we have talked to in the Bureau of the Budget at the meeting which you attended on Monday.

2. As a result of the above conferences I have prepared the attached memorandum which summarizes the various views which have been expressed on the organizational problem and which makes a recommendation which is my own by which appears to be consistent with the discussions of the past week. To date this has been discussed only with Dr. James Fisk.

S. P. Johnston

Attachment

[1] THE WHITE HOUSE
WASHINGTON
February 21, 1958

MEMORANDUM FOR DR. J. R. KILLIAN, JR.

Preliminary Observations on the Organization for the Exploitation of Outer Space

The exploitation of any unknown areas involves two distinct objectives, - one, *exploration* and two, *control*. The first is largely a scientific operation and the second largely military.

At the present time plans for the exploitation of outer space fall more nearly into civilian-scientific areas rather than into military areas. The "take" from the probing of outer space by rockets, satellites and interplanetary vehicles will be of more direct interest to the scientist than to the strategist. We can discount at this point most of the "Buck Rogers" type of thinking which anticipates hordes of little men in space helmets firing disintegrators into each other from flying saucers. Certainly, ICBM's will transit portions of outer space in performing their missions, but for the moment the chief military interest lies in better methods of surveillance, communications and long-range weather forecasting.

The potential space explorations in the immediate future are well outlined in a paper dated 14 February 1957 titled "Basic Objectives of a Continuing Program of Scientific Research in Outer Space" by Hugh Odishaw, Executive Director of the U.S. National Committee for the IGY of the National Academy of Sciences. A good layman's summary of the same subject appeared in a recent issue of LIFE magazine by Dr. Van Allen.

The *control* of outer space, basically a military matter, involves many troublesome questions of international law. The problem of the vertical extent of national sovereignty has yet to be determined. It appears to depend on the capability of any nation to deny access to space above its territory by physical means. No body of international law yet exists covering [2] the use of outer space. As a matter of fact, no acceptable definition has yet been evolved as to where "air-space" and "outer-space" begin and end. Maritime law has no such problem because, under most conditions, one is either afloat or ashore. The limits of the "high seas" have been determined by international agreement on the basis of very easily made physical measurements. With respect to outer space, however, such questions are wide open (a discussion of these problems is to be found in our files in papers on the subject by Professor John Cobb Cooper and others).

The control of radio-communications in our upper atmosphere and in space is another problem which must be settled by international agreement if a completely chaotic condition is to be avoided. Within the next ten years the probabilities are that dozens, if not hundreds, of objects will be in orbit around the earth. Apart from the question of sorting scientific intelligence from this "celestial junkyard" it will be highly important from a military point of view to be able to distinguish an incoming ICBM from other less lethal objects.

By any standards of comparison, the problems involved are tremendous and the programs which must be undertaken in their solution will be lengthy and costly. The technical feasibility studies and the forecasts that have been made by Doctors Purcell, York and others, anticipate the development of such items as booster rockets of one million to five million pounds of thrust in a period of 15 to 25 years. It is estimated that such development programs, quite apart from the missile requirement of the military, may cost anywhere from 500 million to a billion a year. We are, therefore, considering something of the general order of magnitude of the AEC. Obviously the Bureau of the Budget will exert an important influence in deciding whether the national economy can stand such a drain for such purposes.

General Organizational Requirements

In considering the proper organization to handle a project of such magnitude two factors must be taken into consideration—first, how to get the program off the ground immediately, i.e., how to get something started *now* with the facilities that are presently available and, second, how to gear-up for a long-range program to take care of the 5-10-25 year development. This leads to the thought that some sort of *Ad Hoc* organization could be set up in a very short time, possibly by Executive Order of the President, to take care of the immediate requirements. Such a group would [3] not only act as a temporary operating organization but would also initiate studies that would lead toward a more permanent organization on some basis that could be agreed upon by all departments of government and for which the necessary enabling legislature could be obtained.

Whatever plan is adopted, either for the short or for the long-range period, it would appear that certain basic characteristics should be incorporated. First of all, for reasons stated above, it should be a *civilian managed* organization both at the policy and at the operating levels. It must have wide contractual powers, and it must be free from the limitations of the Civil Service in hiring personnel. It must have access to, and be able to draw upon, all existing scientific talent in the country, both within government, and without, and it must be able to utilize the physical facilities that already exist in industry, universities, government laboratories and military installations. It must be able to purchase whatever hardware, systems or components it needs from all available sources. It must have its own physical facilities for testing completed vehicles and it must also be empowered to operate airborne and space vehicles.

Possible Organizational Patterns

To date four specific proposals have been made as to possible organizations to accomplish these ends. These include:
  1. the formation of an entirely new agency of government;
  2. assignment of the project to the AEC;
  3. establishment of the NACA as the controlling agency, with assistance from National Science Foundation, National Academy of Sciences, the military services, etc.;
  4. assignment of the project to the Advanced Research Project Agency of the Department of Defense (ARPA).

In the following paragraphs some of the advantages and disadvantages of the above suggestions will be briefly noted.

1. New Agency

The establishment of a wholly new agency may prove to be the eventual solution to the problem. Such an agency should report directly to the Executive Office of the President. It should be empowered by law to perform all the functions stated above and be given the necessary funds to accomplish them.

[4] The major difficulty would be in the time required to establish such an agency. New legislation would be required which might involve a very long time to debate and to formulate. It would need a new staff both on the management and on the scientific sides. This would take a long time to recruit, and in view of the overall shortage of scientific personnel in the country, would draw off key people from other necessary jobs. This procedure would also take a long time.

It would also need new facilities, with the inevitable delays in reaching decisions as to what was needed and where new laboratories should be located, before the planning and construction phases could begin.

In summary, the establishment of a new agency would require a very great legislative effort and a very long time to get into operation.

2. Atomic Energy Commission

Strong Congressional support is in evidence for assigning the mission to the AEC.

There is no question but that the AEC is organizationally sound and is a going concern. It already has the necessary authorization to contract for anything it needs and also is free from civil service restraint in hiring people. Its scope could very easily be expanded so that it could legally perform any additional assignment.

On the other hand, the technology of flight both in and out of the atmosphere is not a part of the normal AEC area of competence. Although it is true that nuclear propulsion for aerial and space vehicles comes within its field, consensus seems to be that practical utilization of such propulsion is 5 to 10 years away. AEC, therefore, has an interest in a very small part of the space exploitation picture but it has had little experience in such matters as high-speed aerodynamics, control, guidance, structures, telecommunications, etc.

Furthermore, the AEC is already engaged in a huge operation of great national importance. If it were asked to undertake an additional program of the magnitude contemplated for space exploration, its efforts [5] in each one might be so diluted that long delays in the production of end items would be inevitable and its overall effectiveness seriously impaired.

Although the AEC has unquestionably adequate management and all the authority it would need, it would be required to expand both its facilities and its staff into wholly new technical areas if it were given the space exploitation job.

3. National Advisory Committee for Aeronautics

Persuasive arguments can be made for assigning the responsibility for space exploration to the NACA. The Committee itself has suggested that with the support of the National Science Foundation and the National Academy of Sciences it could undertake the job by expanding its facilities.

The NACA is basically a civilian-operated, independent government agency. It has a long history of accomplishment. Its relations with the Congress and with the Executive Departments are good and it has an international standing for competence in scientific fields.

The NACA has been in the space exploitation field for a long time. Most of the work that has been done in extremely high altitude and high-speed aerodynamics on which the design of missiles and rockets has been based has been done in its laboratories. It has already made great progress in research in some of the very sophisticated propulsion systems required for space flight. It has recently established a special subcommittee in space flight technology made up of outstanding scientists in the field. Extending its interests into space technology would appear to be a logical evolutionary step from its research activities of the past 40-odd years.

The NACA budget for the coming year is of the order of 80 million and it has been authorized to expand its present personnel of 8,000 to 9,000. Its three laboratories (Langley, Ames and Lewis) and its missile firing range at Wallop's Island represent an aggregate investment of about 350 million dollars.

It has been argued that the difference between the size of current NACA operation and the proposed operation is so great that the result [6] would be, in effect, the establishment of a wholly new agency to which the NACA would be attached. There is no reason to believe, however, given proper authority and adequate funds, that the NACA could not expand its management functions to handle the larger assignment effectively as it did in 1942 to meet the comparably tremendous demands of World War II.

A moderate amount of legislation would be needed to assign the job to NACA. Its contractual authorization would have to be expanded, and the present civil service limitations on personnel would have to be relieved.

4. ARPA - Department of Defense

A strong case can be made for integration of the space program into the Department of Defense under ARPA on the grounds of immediate action. A great deal of hardware is already available, essential facilities (e.g., JPL, ABMA) exist. The facilities are well staffed and the experience level is high.

It has been suggested that whatever form of organization is agreed upon to initiate the space exploration program it should be attached temporarily to ARPA. If this were done it would appear to be important that some provision be made so that the entire outfit could be detached and assigned to some other agency in the future if it subsequently appeared desirable. It might happen that military interests might outweigh the purely scientific and civil aspects to the detriment of the latter. It would be difficult to avoid security restrictions, and participation in international programs of a purely scientific nature might thereby be hampered.

Under its present directive it seems that ARPA could take on the job with a minimum of additional legislation.

Suggested Compromise Program

Of the four proposals discussed above, No. 2, - i.e., assigning the project to the AEC, seems the least practical. As an example of appropriate organization and good management, it deserves careful study, but the problems under discussion here seem somewhat outside its main fields of interest.

None of the other proposals would satisfy *all* the requirements in themselves. A possible compromise suggests itself which might satisfy the requirement for immediate action and also lay the groundwork both as the organization and legislation for future action.

[7] This consists, in effect, of the immediate establishment of a provisional Space Exploration Control Group headed by a special assistant to the President and composed of the operating heads of the several government agencies who are already involved in research, development or operation of space vehicles. Several outstanding individuals from non-government organizations might also be included, but the total group should not be large. Their main function would be the implementation of national space policy as determined by the President and Congress, utilizing all assets and facilities which already exist in established government agencies and in industry. Their secondary function should be the determination of the kind of agency which should be established to put space exploitation on a permanent basis to handle the requests of the foreseeable future.

The suggested procedure might be outlined as follows:

[8] A. Short Range - By Executive Order for Immediate Action

1. *Appoint a Special Assistant to the President for Space Exploration* (This should be the *Chairman of the NACA* - See Footnote)

2. Appoint a *Provisional Board of Regents for Space Exploration* consisting of:
   a. Special Asst. to President for S.E. (Chairman)
   b. Scientific Advisor to President
   c. Director, AEC
   d. Director, NACA
   e. President, NSF
   f. Director, NAS
   g. Director, ARPA
   h. Two civilians, possibly from industry or science
3. Empower above to
   a. Establish immediate space objectives
   b. Establish program priorities
   c. Coordinate programs of associated agencies toward meeting established objectives
   d. Utilize funds already appropriated to the associated agencies to implement immediate objectives.
4. Instruct Special Assistant for Space Exploration to make immediate plans for the establishment of a Permanent Space Exploration Agency and to prepare the necessary legislation.

[9] B. *Long Range - By Legislation for Continuing Action*
1. Organize a permanent *Space Exploitation Agency*
2. Authorize the Agency to:
   a. establish, maintain and operate its own testing and operational facilities
   b. enter into whatever contractual arrangements may be necessary with government and civilian agencies
   c. hire personnel without regard to Civil Service restrictions
   d. operate air/space Vehicles

Document IV-4

Document title: James R. Killian, Jr., Special Assistant for Science and Technology; Percival Brundage, Director, Bureau of the Budget; Nelson A. Rockefeller, Chairman, President's Advisory Committee on Government Organization, Memorandum for the President, "Organization for Civil Space Programs," March 5, 1958, with attached: "Summary of Advantages and Disadvantages of Alternative Organizational Arrangements."

Source: Dwight D. Eisenhower Papers, Eisenhower Library, Abilene, Kansas.

As the preceding documents have shown [IV-1, IV-2, IV-3], there was substantial attention given within the Executive Office of the President during the December 1957-March 1958 period to how best to organize the nation's space effort. This memorandum was the culmination of that attention and laid the basis for President Eisenhower's decision to create a new civilian space agency.

---

[1] # Memorandum for the President

SUBJECT: Organization for Civil Space Programs

The Problem

As you know, there will soon be presented for your consideration civil space programs for the United States which will entail increased expenditures and the employment of important numbers of scientists, engineers and technicians.[1]

This Committee, in conjunction with the Director of the Bureau of the Budget and your Special Assistant for Science and Technology, have given consideration to the manner in which the executive branch should be organized to conduct the new program. This memorandum contains our joint findings and recommendations. The memorandum (1) discusses some of the factors which should be taken into account in establishing the government's organization for these civil space programs, (2) recommends a pattern of organization, and (3) indicates certain interim actions which will be necessary. Also attached is a summary of the advantages and disadvantages of certain alternative organizational arrangements.

[2] Discussions to date suggest that an aggressive space program will produce important civilian gains in the form of advances in general scientific knowledge and protection of the international prestige of the United States. These benefits will be in addition to such military uses of outer space as may prove feasible.

Establishing a Long Term Organization

Because of the importance of the civil interest in space exploration, the long term organization for Federal programs in this area should be under civilian control. Such civilian domination is also suggested by public and foreign relations considerations. However, civilian control does not envisage taking out from military control projects relating to missiles, anti-missile defense, reconnaissance satellites, military communications, and other space technology relating to weapons systems or direct military requirements.

[3] We have considered a number of different approaches to civil space organization. It is our conclusion that one of these alternatives provides a workable solution to the problem. The other principal alternatives have serious shortcomings which argue against their selection as a basis for space organization.

Recommendation No. 1. We recommend that leadership of the civil space effort be lodged in a strengthened and redesignated National Advisory Committee for Aeronautics.

The National Advisory Committee for Aeronautics (NACA), in a resolution adopted on January 16, 1958, has proposed that the national space program be implemented by the cooperative effort of the Department of Defense, the NACA, the National Academy of Sciences and the National Science Foundation, together with the universities, research institutions, and industrial companies of the nation. NACA further recommended that the development of space vehicles and the operations required for scientific research in space phenomena and space technology be conducted by the NACA when within its capabilities. NACA is now formulating a program which is expected to propose expansion of existing programs and the addition of supplementary research facilities.

[4]   **Factors Favoring NACA as the Principal Civil Space Agency**

1. NACA is a going Federal research agency with a large scientific and engineering staff (approximately 2,000 of its 7,500 employees are in these categories) and a large plant ($300,000,000 in laboratories and test facilities). It can expand its research program and increase its emphasis on space matters with a minimum of delay and can provide a functioning institutional setting for this activity.

2. NACA's aeronautical research has been progressively involving it in technical problems associated with space flight and its current facilities construction program is designed to be useful in space research. It has done research in rocket engines (including advanced

---

1. These programs do not include those projects relating to space vehicles and exploration which will be carried out in the Department of Defense under the direction of the Advanced Research Projects Agency (ARPA).

chemical propellants), it has developed materials and designs to withstand the thermal effects of high speeds in or on entering the earth's atmosphere, it conducts multi-stage rocket launchings, and in the X-15 project it has taken the leadership (in cooperation with the Navy and Air Force) in developing a manned vehicle capable of flights beyond the earth's atmosphere.

[5] 3. If NACA is not given the leading responsibility for the civil space program, its future research role will be limited to aircraft and missiles. Some of its present activities would have to be curtailed, and the logical paths of progress in much of its current work would be closed. It would, under such circumstances, be difficult for NACA to attract and retain the most imaginative and competent scientific and engineering personnel, and all aspects of its mission could suffer. Moreover, it is questionable whether it would be possible to define practicable boundaries between the missile and high performance aircraft research now performed by NACA and the space vehicle projects.

4. NACA has a long history of close and cordial cooperation with the military departments. This cooperation has taken place under a variety of arrangements, usually with little in the way of formalized agreements. Although new relationship problems are bound to arise from an augmented NACA role in space programs, the tradition of comity and civil-military accommodation which has been built up over the years will be a great asset in minimizing friction between the civil space agency and the Department of Defense.

[6] 5. Although much of its work has been done for the military departments, NACA is a civilian agency and widely recognized as such. A civilian setting for space programs is desirable and NACA satisfies this requirement.

6. Some of the principal problems in using NACA, as listed below, can be overcome by relatively limited accommodations to existing law and by appropriate administrative action. These measures are described in later paragraphs.

Problems in Using NACA as the Agency with Primary Responsibility for Civil Space Programs

1. NACA has in the past been concerned chiefly with research involving air breathing aircraft and missiles. NACA's competence in certain fields related to space flight (such as electronics and space medicine) will need to be augmented. NACA has also had little experience in the direct administration of large scale developmental contracts.

2. Many of the scientists who have done the most work on rocket engines and space vehicles are now employed by Defense Department agencies and by private contractors of the military services. Some means of utilizing such experienced personnel will have to be found which does not unduly impair the capacity of the Department of Defense to continue defense related aspects of missile and space activity.

[7] 3. The NACA is not in a position to push ahead with the immediate demonstration projects which may be necessary to protect the nation's world prestige. Therefore the military services may have to be relied on for such demonstrations while NACA is equipping itself for the full performance of the space job.

4. NACA suffers from some of the limitations imposed on civil service agencies, and some scientists are known to favor reliance on private research organizations operating under government contracts. Ceilings and numerical restrictions on the salaries of top scientific staff and the general lag in Classification Act salaries are among the obstacles to administration through government laboratories which pose problems in utilizing NACA.

5. NACA now spends around $100,000,000 per year. A civil space program may eventually entail additional annual expenditures substantially in excess of this amount. It is obvious that important changes in NACA will be required by such an expansion, and the agency may have some difficulty in assimilating the additional staff and functions.

[8] **Recommendation No. 2. We recommend that NACA's basic law be amended to give NACA the authority and flexibility to overcome or mitigate the problems noted above so that NACA can carry out its total program effectively.**

Specifically the amendments should:

a. Rename the NACA the National Aeronautical and Space Agency to get away from the limited connotations of the term "aeronautics" when used alone and to recognize that NACA has long since ceased to be an "advisory committee" as the term is customarily used.

b. Retain a board for top policy direction. Some changes in the composition of the present NACA board may be appropriate.

c. Provide for the appointment of a Director by the President by and with the advice and consent of the Senate.

d. Provide a system for the fixing of compensation of employees which, under appropriate Presidential controls, will permit the agency to pay rates which are reasonably competitive with the rates paid by non-Federal employers for comparable work. (This amendment will ease the salary limitations under the Classification Act of 1949 which have caused so much concern in and out of NACA.)

[9] Certain additional miscellaneous powers may also have to be given NACA if further investigation reveals that they are not already available and confirms that they will be of material assistance to the agency.

The above powers would give NACA as much flexibility as can reasonably be achieved by contract laboratories and would at the same time permit retention of the traditional NACA practice of conducting such research and testing through its own government employee staffed facilities as it determines to be desirable in carrying out a space program.

There will remain the need to refine relationships with the Department of Defense in space matters and to draw upon and utilize staff and experience now lodged in the laboratories of the military services and their contractors, but the reorganized NACA would be equipped to work out these problems in a flexible manner. Some Presidential intervention may prove necessary to bring about or implement agreements between the space agency and Defense, and it may also be desirable for the President to be given the specific authority to transfer to NACA space activities directly related to the civil program which are now being performed by other agencies.

[10] Overlapping between NACA's civil space program and the work of Defense on military projects should be kept to a minimum. This can be done if Defense, in a manner analogous to the practice followed on developing aircraft and missiles, makes appropriate use of NACA for supporting research and development on military space vehicles. An arrangement of this kind could reduce duplication without undermining the basic Defense Department responsibility for developing weapons systems and other military equipment.

Interim Measures

Recommendation No. 3. If you approve our recommended approach to space organization, we further recommend that a number of interim and short-term measures be given immediate attention.

Specifically, we propose:

a. An all-out attempt should be made to draft needed legislation within the next few weeks so that there will be some chance of final action during the current session of the Congress. At the same time decisions should be made with respect to the supplemental appropriations which will be required for NACA to get its part of the space program under way. If congressional [11] action can be secured on both matters before adjournment, the full civil space program under arrangements designed to serve long term needs can be launched this year.

If it proves impossible to obtain the enactment of the comprehensive legislation strengthening NACA during the current session, the passage of the general Classification Act revisions now pending, the authorization of addition super-grade and Public Law 313 positions, and the securing of supplemental appropriations would still enable NACA to get under way with a space program.

b. While awaiting congressional action we suggest that the President advise the NACA's top committee that it is being charged with the responsibility for developing and arranging for the execution of the civil space program. NACA will at first have to rely heavily upon the Department of Defense and its instrumentalities for interim development and demonstration projects. However, the problems created by such arrangements will be minimized once the President gives NACA the clear-cut authority required for it to select and monitor the advanced space projects entrusted to the Department of Defense during the transitional period.

[12] c. None of the immediate measures is more essential and fundamental than defining as clearly as possible just what the nation plans to do in the space field. At the same time an effort must be made to estimate with reasonable exactness the annual additional costs of the civil space program.

Immediate Action

If you concur in the recommendations set forth above, the director of the Bureau of Budget will proceed, in cooperation with this Committee, your Special Assistant for Science and Technology and other departments and agencies concerned, to develop for your consideration specific proposals for legislative and executive action.

> James B. Killian, Jr.,
> Special Assistant for
> Science and Technology
>
> Percival Brundage,
> Director, Bureau
> of the Budget
>
> Nelson A. Rockefeller,
> Chairman

[1]**Attachment**

Summary of Advantages and Disadvantages of Alternative Organizational Arrangements

1. Use of a private contractor to carry out the civil space program under supervision of NACA.

A variation of our recommended organizational approach is to select NACA as the civilian agency to supervise contracts with a private laboratory charged with developing and testing space vehicles. This is the pattern followed by the Atomic Energy Commission in much of its research. This approach has also been used to some extent by the military services in developing missiles.

Advantages

Contract operation is preferred by some scientific personnel as a means of circumventing government salary and administrative controls. It would retain NACA in a supervisory capacity while making use of selected private research organizations.

Disadvantages

This approach is in conflict with the traditional NACA practice of carrying out research largely through its own government-employee staffed laboratories: there is no assurance that a private research laboratory can be found to do the work on a sufficiently urgent schedule; and such greater flexibility as private laboratories may enjoy can also be provided NACA through the changes in law previously described.

[2] **Conclusion**

No real gains would flow from this alternative which could not be achieved under the preferred organization. It would be better to permit NACA to make its own decisions as to the extent to which it would use contracting authority in executing the space research program. It is assumed, of course, that NACA will, in fact, make fairly extensive use of research contracts, but on a selective basis.

2. Utilization of the Department of Defense

The recent Supplemental Military Construction Authorization Act authorizes the Secretary of Defense, for a period of one year, to carry on such space projects as may be designated by the President. It confers permanent authority for the Secretary or his designee to proceed with Missile and other space projects directly related to weapons systems and military requirements.

Advantages

The Department of Defense is now doing most of the current missile and satellite work: it has the bulk of the scientists and engineers active in these fields in its employ or on the rolls of its contractors; it will have to continue work on space vehicles on an interim basis for demonstration purposes; it is experienced in working with and utilizing the facilities of NACA; and it may be possible for a civilian agency of the Department to carry out the program.

[3] **Disadvantages**

The Department of Defense is a military agency in law and in the eyes of the world and placing the space program under it would be interpreted as emphasizing military goals: the space program is expected to produce benefits largely unrelated to the central mission of the Department of Defense; there is some danger that the non-military phases of space activity would be neglected; the Department is already so overloaded with its central military responsibilities that care should be taken to avoid charging it with additional civil functions; cooperation with other nations in international civil space matters could be made more difficult; and adequate civil-military cooperation can be achieved under the recommended organization without assigning inappropriate functions to Defense.

Conclusion

Since the space program has a relatively limited military significance, at least for the foreseeable future, and since the general scientific objectives should not be subordinated to military priorities, it is essential that the arrangements for space organization provide for leadership by a civilian agency.

[4] **3. Utilization of the Atomic Energy Commission**

There are now pending before the Congress bills which would authorize the Atomic Energy Commission to proceed with the development of vehicles for the exploration of outer space. Among these bills are S. 3117 (introduced by Senator Anderson) and S. 3000 (introduced by Senator Gore). The justification for these proposals is the role already being played by the Atomic Energy Commission in developing nuclear propelled jet and rocket engines.

Advantages

The Atomic Energy Commission is a civilian agency with competence in directing scientific research and development projects: it has had experience in managing research contracts and in working with the military agencies; and it is now charged with developing a nuclear rocket engine which may eventually be used to propel space vehicles.

Disadvantages

The Atomic Energy Commission is concerned chiefly with the use of a single form of energy and it is expected that chemical propellants, not atomic energy, will be the chief power source for space vehicles for years to come. Moreover, the Commission has virtually no experience or competence in most aspects of the design, construction and testing of space vehicles.

[5] Conclusion

The Atomic Energy Commission has a contribution to make in the space field. However, it should limit its work to the aspects of the space problem in which nuclear energy may have practical applications. An administration position along these lines has already been conveyed to the Chairman of the Atomic Energy Commission.

4. Creation of a Department of Science and Technology

Senators Humphrey, McClellan and Yarborough recently introduced S. 3126, a bill to create a Department of Science and Technology. The bill calls for the establishment of a new executive department which at the outset would contain or be given the functions of the National Science Foundation, the Patent Office, the Office of Technical Services of the Department of Commerce, the National Bureau of Standards, the Atomic Energy Commission and certain divisions of the Smithsonian Institution. The Secretary would also be authorized to establish institutes for basic research.

Advantages

The proposed department would provide a civilian setting for the administration of space programs, and it would give this and other scientific activities the prestige and accessibility to the President associated with departmental status.

[6] Disadvantages

The proposed department will be highly controversial, and there is no assurance that it can be established in time to assume the responsibility for civil space programs. It is also unlikely that science, of itself, will provide a sound basis for organizing an executive department.

Conclusion

There would be little prospect of getting such a reorganization approved and functioning in the near future. Even if the department could be created, it might not provide as good a setting for a high priority space program as that proposed under the preferred organization.

Document IV-5

Document title: Maurice H. Stans, Director, Bureau of the Budget, Memorandum for the President, "Responsibility for 'space' programs," May 13, 1958.

Source: Dwight D. Eisenhower Papers, Eisenhower Library, Abilene, Kansas.

Prior to the creation of NASA in 1958, the nation's space efforts were housed in various branches of the military services. The Naval Research Laboratory managed the

VANGUARD project; the Army was responsible for Explorer I, which led to the discovery of the Van Allen radiation belts; and the Army, Air Force, and Navy were responsible for a variety of booster programs. In addition, the Advanced Research Projects Agency (ARPA) had been created in early 1958 to house military space activities not closely linked to specific service missions. Although program content was generally dominated by military requirements, substantial portions of the military space program had no immediate relationship to established defense needs. However, many of these projects eventually proved essential to the human and scientific exploration of space. By March 1958, it had already been decided that a new civilian agency would best serve the scientific and prestige needs of the nation. What was absent was a plan for allocating responsibility for the various facets of the nation's space program. Even so, the desire to avoid duplication of effort and establish a new space agency in the shortest possible time necessitated the transfer of whole programs and in some cases military facilities to NASA. At the time no other alternative was tenable to those making the decisions. But, as the memorandum suggests, elements of the Department of Defense attempted to resist the transfer of their programs to the new civilian agency.

[1] # Memorandum for the President

Subject: Responsibility for "space" programs

In your letters of April 2, 1958, you directed the Secretary of Defense and the Chairman of the National Advisory Committee for Aeronautics to review and report to you which of the "space" programs currently underway or planned by Defense should be placed under the direction of *the new civilian space agency proposed* in your message to Congress. These instructions specifically stated that "the new Agency will be given responsibility for *all* programs except those peculiar to or primarily associated with military weapons systems or military operations."

It now appears that the two agencies have reached an agreement contemplating that certain space programs having no clear or immediate military applications would remain the responsibility of the Department of Defense. This agreement would be directly contrary to your instructions and to the concept underlying the legislation the administration has submitted to Congress.

The agreement is primarily the result of the determination of the Defense representatives not to relinquish control of programs in areas which they feel might some day have military significance. The NACA representatives apparently have felt obliged to accept an agreement on the best terms acceptable to Defense.

Specifically, Defense does not wish to turn over to the new agency all projects related to placing "man in space" and certain major component projects such as the proposed million pound thrust engine development. The review by your Scientific Advisory Committee did not see any immediate military applications of these projects.

The effect of the proposed agreement would be to divide responsibility for programs primarily of scientific interest between the two agencies. This would be an undesirable and unnecessary division of responsibility and would be highly impractical. There would not be any clear dividing line, and unnecessary overlap and duplication would be likely. The Bureau of the Budget would have an almost hopeless task in trying to keep the two parts of the program in balance, and problems on specific projects would constantly have to come to you for resolution. The net result of the proposed arrangement would be a less effective program at higher total cost.

[2] On the other hand, it will be relatively simple to work out practical working arrangements under which responsibility and control of the programs in question would clearly be assigned to the new agency as contemplated in your instructions, and the military interest would be recognized by the participation of the Department of Defense in the

planning and, where appropriate, the conduct of the programs.

In the circumstances, it is recommended that you direct that the two agencies consult with the Bureau of the Budget and Dr. Killian's office to be sure that any agreement reached is in accordance with the intent of your previous instructions. It is especially important that announcement of the agreement now being proposed be avoided at this stage of the consideration by the Congress of legislation to establish the new space agency.

If you approve this recommendation, there are *attached memoranda* to return to the Secretary of Defense and the Chairman of the National Advisory Committee for Aeronautics for your signature.

(Handled orally per President's instructions. AJG) 5/13/58

Signed by Maurice H. Stans
Director
5/13/58

5/14/58
I notified the Secretary of Defense (General Randall) and Dr. Dryden.

AJG

Document IV-6

Document title: W.H. Pickering, Director, Jet Propulsion Laboratory, to Dr. T. Keith Glennan, NASA, March 24, 1959.

Source: NASA Historical Reference Collection, NASA History Office, NASA Headquarters, Washington, D.C.

NASA was not created from whole cloth; rather, it was an amalgamation of a number of pre-existing programs and facilities. However, the attempt to meld different institutional cultures into a single organization was not without its problems. Management methods as well as institutional expectations often differed. The lingering animosity that was to plague the relationship between NASA Headquarters and the Jet Propulsion Laboratory (JPL) was present from the start of the relationship, as this letter demonstrates.

The Vega project mentioned in Pickering's letter was a JPL-favored three-stage booster for launching lunar missions. It was based on a modified Atlas ICBM, developed by Convair as the first stage, and a JPL-developed third stage. The project was canceled in December 1959 in favor of the Air Force Centaur. The Centaur was similar to the Vega, but its second stage used liquid hydrogen as a fuel. This stage was eventually developed, but not under JPL management, and used with the Atlas and Titan missiles to launch a variety of spacecraft.

---

[1] March 24, 1959
Dr. T Keith Glennan
National Aeronautics and Space Administration
1520 H Street, N.W.
Washington 25, D.C.

Dear Keith:

I was very glad to have had the opportunity to discuss some of our problems with you on Saturday, but I feel that I would like to enlarge on some of these topics with you. In so doing, I hope that you will regard this as a private communication between ourselves. My motives are essentially an attempt to clarify my position on some of these matters.

I. The Importance of the Vega Project to the U.S. and to NASA.

I believe that the Vega project is one of the most important actions which NASA must take this year. Vega is the first vehicle which NASA will build under its own direction for scientific and civilian purposes. If Vega is not started almost immediately then;

A. The Mars 1960 date will not be obtained with consequent loss of prestige to both the U.S. and NASA.

B. Vega will be scheduled only a few months ahead of Centaur and, therefore, a powerful argument will be presented that Vega is unnecessary. If this prevails, then the NASA space program will be subordinate to the Air Force military requirements for Centaur.

II. Why has not the Vega Project been Started Already?

As this letter is being written it appears that there is every hope that the project will, in fact, be finalized within the next week, but I think it is desirable to review some of the causes for the delay in getting it underway. I believe these have been, first, the reluctance of Headquarters to initiate authorization and funds for Vega activities to JPL, or to initiate a letter order with Convair. Second, the reluctance of Headquarters to delegate negotiating responsibility to JPL to work with Convair to develop the best possible program within the [2] JPL-Headquarters agreed upon objectives and money.

III. Problem Areas Which Exist Between Headquarters and JPL.

I am listing a number of bald statements without amplification or justification, but I think these are significant.

A. Headquarters is not yet willing to treat JPL like one of its own Laboratories, or at least in the manner in which JPL believes Headquarters treats its own Laboratories.

B. Headquarters appears to be too concerned with technical details of projects.

C. JPL does not seem to understand Headquarter's operating principles. JPL tries to cooperate when asked to do something by Headquarters, but is all too frequently frustrated when the expected results are not forthcoming.

D. JPL is concerned that Headquarters has not apparently established a Vega program office which will, in fact, be a focal point for all Vega program activities.

E. JPL would like to feel that Headquarters trusts us enough to ask for help or advice when needed, and also to invite us to pertinent meetings in which we are concerned.

F. JPL is prone to compare the Headquarters negotiations on the Vega program with the similar type of negotiations necessary when the Sergeant program was set up with the Army. The result is not flattering to Headquarters.

IV. What are the Problem Areas which JPL Believes Face NASA?

A. The flight program is too diversified and contains too many shots. Even the reduced schedule of approximately one shot per month is a very heavy program.

B. I believe that NASA faces some difficult problems with the Air Force, particularly in Vega and Centaur.

C. NASA should actively work to obtain its own launch and hanger facilities, both at AMR and PMR.

[3] D. NASA must clarify the relationship between Vega and Centaur when both are under NASA sponsorship. I believe that in spite of the obvious problems, the best answer is for JPL to be technically responsible for both. But, if this is planned, then JPL needs to be made cognizant of the present Centaur contractual and technical progress.

E. I think it is essential for NASA to move into the Goldstone area with at least a token propulsion activity. NASA needs to demonstrate an imaginative program to the public. A start on an advanced propulsion facility should be an excellent example.

V. What Needs to be Done?

It appears to me that the most urgent thing in the JPL-NASA picture is to get the Vega program going. It should be established as a program even if the objectives are not all as desired. The present contract is a down payment on a continuing program, not a final contract. It should be written as flexibly as possible. For example, an initial buy of six rounds is probably adequate for now. A firing rate of perhaps six per year could well be required so that it will be a continuing purchase of Vega vehicles.

In writing these comments to you, I hope you will not regard them as just petty criticisms. I have tried to put down some of the things which are bothering my people, and I believe it is fair to say that we at the Laboratory are trying to take a NASA view of the problem rather than a Laboratory view. I think that we all expected that we would encounter problems with our NASA operations, and I do not know that these problems have been any more serious than we had expected. However, I do feel the importance of trying to establish the best possible relationship between the Laboratory and Headquarters so that the Space Program will progress as effectively as possible.

Sincerely,
W.H. Pickering
Director

Document IV-7

Document title: T. Keith Glennan, *The Birth of NASA: The Diary of T. Keith Glennan* (Washington, DC: NASA Special Publication-4105, 1993), pp. 1-6.

The Eisenhower administration chose as first NASA Administrator T. Keith Glennan, who had been president of Case Institute of Technology in Cleveland, Ohio, since 1947. Up until almost the time that Glennan was asked to take the job, the leading candidate had been NACA Director Hugh Dryden. But Dryden was a career civil servant, a Democrat, and thought by some, particularly in Congress, to be insufficiently bold in his approach to the emerging space program. Glennan was an engineer who prior to World War II had worked primarily in the motion picture industry. He had one previous tour of duty in Washington, as a member of the Atomic Energy Commission from 1950 to 1952. In this excerpt from his diary, Glennan discusses how he came to NASA and the philosophy he brought to the position.

---

[1] In spite of my membership on the Board of the National Science Foundation, the agency providing the funding for the Vanguard Project, I had taken no more than casual interest in the efforts of this nation to develop a space program following the successful orbiting of Sputnik I by the Russians on 4 October 1957. The aftermath was marked by a continuous chorus of lament over the fact that the Soviet Union had stolen a march on the United States in fields that had seemed to be the special province of our own country. In reaction, President Eisenhower appointed Jim Killian of MIT as his Science Advisor. I thought this a most excellent appointment and sent a telegram to Jim congratulating him and stating that I would be happy to assist in any possible way.

In April 1958, the president sent to the Congress a bill calling for the establishment of an agency to develop and manage a national space program. Quite naturally, there was much debate about the actual management of this program—should it be handled by the military departments or by a civilian agency? The proponents of the civilian management won out, and the bill was passed and signed into law on 29 July by President Eisenhower. It called for the creation of the National Aeronautics and Space Administration using the then existing National Advisory Committee for Aeronautics as its foundation. That distinguished 43-year-old agency employed some 8,000 people, with major laboratories in Cleveland; Langley, Virginia; and Moffett Field, California. There were smaller field stations at Edwards Air [2] Force Base in California and at Wallops Island, Virginia. Its budget for the 1959 fiscal year had been set at $101 million as I recall.

The policy statement in the preface of the Act called for the establishment and prosecution of a program aimed at the development of useful knowledge of the space environment and the exploration and exploitation of that environment for peaceful purposes and for the benefit of all mankind. In recognition that space might well be used for military purposes, the law provided that any activities concerned principally with the defense of the nation were the responsibility of the Department of Defense.

As already stated, I paid about as much attention to all of these events as the ordinary citizen—not much more. Imagine my surprise when on 7 August 1958 I received a call from Jim Killian asking me to come immediately to Washington. I flew down on that same day and met with him at his apartment that evening. He said his purpose was to ask me, on behalf of President Eisenhower, to consider becoming administrator of the new agency, which of course was the National Aeronautics and Space Administration (NASA). He handed me a copy of the bill, which I had not previously seen. I read it through rather hurriedly and pointed out immediately the built-in conflict that seemed to me to be present whereby the Defense Department most certainly would dispute the claim of the civilian agency to important elements of any program that might be initiated. After some considerable discussion, I agreed to meet with the president the next morning.

The meeting with President Eisenhower was brief and very much to the point. He said he wanted to develop a program that would be sensibly paced and vigorously prosecuted. He made no mention of concern over accomplishments of the Soviet Union although it was clear he was concerned about the nature and quality of scientific and technological progress in this country. He seemed to rely on the advice of Jim Killian. I agreed that I would give the matter consideration and would give him a reply within a few days.

Discussions with Killian were followed by a visit to Don Quarles, Deputy Secretary of Defense. I had known Quarles for years, since my stay in New London during World War II. It was apparent that few people had been asked to recommend a candidate for the NASA job, and I gained the impression that Quarles had only heard about the proposal that I be offered the post. He urged me to take it but expressed some unhappiness over the fact that he had not acted more promptly on a matter troubling him—head of the research and development activity in the Office of the Secretary of Defense. He stated that he had intended to offer that job to me. Although flattered, I assured him that I would not have been able to accept because of my conviction that only a scientist should handle that job.

[3] Returning to Cleveland, I discussed these matters with my wife Ruth, several of my associates on the campus, and members of the board of trustees. Frederick C. Crawford [chairman of the board of trustees at Case Institute of Technology] urged me to take the post, and after two or three days of soul searching I called Killian to say I would accept—but only if Hugh Dryden, the director of the NACA, would endorse the appointment and would agree to serve as my deputy. Events began to move rapidly. Fred and I agreed that it would be desirable to ask Kent Smith to serve as acting president during my absence since John Hrones [Case's vice president for academic affairs] had been with us only a year and was not acquainted with all facets of the campus. Fred and I talked with Kent and Thelma and, in spite of the fact that they had planned a year abroad, Kent agreed to take on the job.

The swearing-in was set for 19 August in Washington. Ruth, Polly, and Sally drove down with me and Ruthie and Jack Packard attended the ceremony in the executive offices of the White House. A crowd of friends attended the brief ceremony, and the family had a chance to speak with President Eisenhower who [4] presided and handed us our Commissions, Hugh Dryden having been sworn in at the same time. Together with the Packards, we had lunch at LaSalle du Bois with everyone a bit punch drunk over events of the day. Ruth Packard and my Ruth immediately started a search for an apartment, and I returned to the NACA offices to become acquainted with members of the staff of that organization, soon to be absorbed by NASA.

Although my visit had been billed as casual, I found myself thrust into the problems of the new agency. Dryden called in Abe Silverstein and some of the top operating people who wanted to discuss budget. I will not try to describe the budget cycle in Washington agencies; suffice it to say that we were attempting to put together a budget that should have been initiated months before. Staff members were seeking my approval of a figure toward which they might work on the budget, which had to be submitted within weeks. Imagine my consternation when they proposed that we seek $615 million. The Case budget at that time was in the neighborhood of $6 or $7 million and I doubt that I had much feel for $615 million. Members of the staff made the point that when NASA was to be declared "ready for operations" we would be taking over from the Defense Department projects, together with manpower and funds already appropriated. It appeared that we would have about $300 million for FY 1959 (July 1958-June 1959). Their arguments must have been convincing, for I approved a budget for FY 1960 using the guideline figure of $615 million. This, then, was my introduction to what was to become one of the major activities of the federal government.

In accepting the appointment I had stipulated that I would take a vacation before reporting for duty and had set the reporting date at 9 September. In addition to taking a vacation, I had to complete my annual report for Case. We were able to find a cottage on Martha's Vineyard and after depositing the children in Cleveland we drove immediately to Wood's Hole and took the ferry to the Vineyard. This was a delightful spot, and I was able to complete my report even though I spent time on the telephone counseling with Dryden and others about additional top personnel.

I want to record my first brush with the inflexibility of bureaucratic procedure. The Case trustees had voted to continue my salary throughout the balance of 1958, paying me in a lump sum determined to be a legal procedure. I did not want to accept a check from the federal government until I was on the job, so I asked our financial officer at NASA to determine how this could be managed since my salary was supposed to begin when I was sworn in. He shook his head but agreed to make the attempt. When I returned to Washington on 9 September, I called him in. He stated that the only possible way to manage this affair was for me to accept the payment and to return it to the federal government as a gift. I would have to pay income tax. Since there seemed no way of circumventing these regulations, I decided to keep the salary although I suppose I could have paid the tax and returned the balance of the salary less the tax. The whole procedure seemed so unbusinesslike that I guess I acted as much in pique as from any sense of conviction.

Now my work began in earnest. Ruth was engaged, with the help of Mr. Bacome, a Cleveland decorator, in making the apartment livable. [5] We bought drapes, a bookcase and a room divider, a daybed, and a rug and shipped furniture from Cleveland. As I look back over my appointment schedules for those days, I wonder how I kept anything straight. I was concerned with acquiring a number of good men to fill top positions in the agency and I seem to have spent a good bit of my time on this task. Hardly a day passed without a visit from the representatives of some industrial concern—usually the president—and meetings with top people in the Department of Defense and some of the other agencies with which we would be dealing.

Although NACA contained many fine technical people, it had been an agency protected from the usual in-fighting found on the Washington scene. Its staff, composed of able people, had little depth and little experience in the management of large projects. Considerable thought had been given by the staff to the organization that might develop, and these plans served to get us underway. It became apparent almost immediately that further studies would be needed and that some good people would have to be hired.

Let me discuss the philosophy with which I approached this job—a philosophy about which I had thought while vacationing at Martha's Vineyard. First, having the conviction that our government operations were growing too large, I determined to avoid excessive additions to the federal payroll. Since our organizational structure was to be erected on the NACA staff, and their operation had been conducted almost wholly "in-house," I knew I would face demands on the part of our technical staff to add to in-house capacity. Indeed, approval had been given in the budget to initiate construction of a so-called "space control center" laboratory at Beltsville, Maryland, an action I approved. But I was convinced that the major portion of our funds must be spent with industry, education, and other institutions.

Second, it seemed to me that we were starting virtually from scratch and with little in the way of rocket-propelled launching systems. Thus it seemed to me that we should mount an aggressive program that would build on the advancing state of the art as we came to understand more about technologies with which we were dealing. Third, it seemed clear that we should not lose sight of the propaganda values residing in successful launches—yet we had to be aware of the limitations imposed upon us by the lack of availability of proven launch vehicle systems. This was because the military missile program was just reaching the testing stage and these same rocket-propelled units were going to have to serve as "booster systems" or, as we came to call them, launch vehicle systems for our space shots.

Fourth, in the nature of things it seemed necessary that we structure our program in accord with our own ideas of fields to be explored and the pace at which progress could and should be made. This meant we must avoid the undertaking of particular shots, the purpose of which would be propagandistic rather than directed toward solid accomplishment. Fifth, we faced the prospect of carrying to completion the projects [6] started by the Advanced Research Projects Agency of the Defense Department, called into being by Secretary Neil H. McElroy during the period between 4 October 1957 and the operational beginnings of NASA. At the same time we must be planning our own broadly-based program of science and technology and organizing to accomplish all these tasks.

Document IV-8

Document title: Anonymous, "Ballad of Charlie McCoffus," n.d.

Source: Johnson Space Center, Historical Documents Collection, Houston, Texas.

The doggerel that collects in people's desk drawers or on hallway bulletin boards sometimes tells more about an organization's culture than the dry, cautious language of bureaucratic prose. The "field centers" of the National Aeronautics and Space Administration shared little besides NASA's budget and a suspicious contempt for "headquarters" efforts to impose some coherent administrative order on the lot of them. Resentment of "headquarters" was strongest among former NACA aeronautical research laboratories. The oldest was Langley Memorial Aeronautical Laboratory, dedicated in 1920 and named for Smithsonian Institution Secretary Samuel P. Langley.

Ballad of Charlie McCoffus*

A young field engineer named Charlie McCoffus
Worked all day in the field and at night in the office,
Preparing reports and estimates too
To be picked all to bits by the Washington crew.

For the boys in D. C. and their double lensed specs,
Their sallow complexions and fried collar necks,
Care not for the time or trouble they make
If a comma is missing, or a carbon misplaced.

They fire it back with ill conceived jeers
To harass the poor hardworking field engineers.

---

* A ballad used by Langley in the early days!

To get back to Charlie he struggled along
Till an ache in his head told him something was wrong.
He went to the doctor and "Doctor" says he,
"There's a buzz in my brain. What's the matter with me?"

Well the medico thumped as the medicos do
and tested his pulse and his reflexes too,
And his head and his heart and his eyes and each lung
And Charlie said "Ah" and stuck out his tongue.

Then the doctor said "Well what a narrow escape!
But a brief operation will put you in shape.
I must take out your brain for a complete overhauling,
In the interim take a respite from your calling."

The weeks passed by and Charlie McCoffus
Never called for his brain at the medico's office.
The doctor got worried and gave Charlie a ring,
"You'd better come over and get the damn thing."

"Thanks Doc, I don't need it" said Charlie McCoffus.
"I have been transferred to the Washington Office."

So Charlie now wears a fried collar to work,
And hides in the lairs where the auditors lurk,
And his letters bring tremors of anger and fear
To the heart of each hardworking field engineer.

And the pride and the joy of the Washington Office
Is the brainless predacious young Charlie McCoffus.

Document IV-9

Document title: *Report to the President on Government Contracting for Research and Development*, Bureau of the Budget, U.S. Senate, Committee on Government Operations, 87th Cong., 2d sess. (Washington, DC: U.S. Government Printing Office, 1962), pp. vii-xiii, 1-24.

Public debates over the proper spheres of government and private enterprise, which became more intense with the growth of the federal establishment during and after World War II, drew a distinction between the "public sector" and "private sector" that was often more fiction than fact. By the end of the 1950s, there were entire industries that depended heavily on federal contracts, and there were federal agencies, such as the Atomic Energy Commission, whose facilities were operated almost entirely by private contractors. Was the government improperly transferring its authority and responsibilities to private industry? As early as 1961, the White House created a task force to examine this question. The group's report, issued in April 1962, became informally known as the "Bell Report," after its chairman, David Bell, Director of the Bureau of the Budget. This excerpt contains the body of the report's analysis and conclusions.

Report to the President on Government Contracting for Research and Development

[vii] Executive Office of the President,
Bureau of the Budget,
Washington, D.C., April 30, 1962

Dear Mr. President: As requested by your letter of July 31, 1961,[1] we have reviewed the experience of the Government in using contracts with private institutions and enterprises to obtain research and development work needed for public purposes.

The attached report presents our findings and conclusions. Without attempting to summarize the complete report, we include in this letter a few of our most significant conclusions, as follows:

1. Federally financed research and development work has been increasing at a phenomenal rate—from $100 million per year in the late 1930's to over $10 billion per year at present, with the bulk of the increase coming since 1950. Over 80 percent of such work is conducted today through non-Federal institutions rather than through direct Federal operations. The growth and size of this work, and the heavy reliance on non-Federal organizations to carry it out, have had a striking impact on the Nation's universities and its industries, and have given rise to the establishment of new kinds of professional and technical organizations. At present, the system for conducting Federal research and development work can best be described as a highly complex partnership among various kinds of public and private agencies, related in large part by contractual arrangements.

While many improvements are needed in the conduct of research and development work, and in the contracting systems used, it is our fundamental conclusion that it is in the national interest for the Government to continue to rely heavily on contracts with non-Federal institutions to accomplish scientific and technical work needed for public purposes. A partnership among public and private agencies is the best way in our society to enlist the Nation's resources and achieve the most rapid progress.

2. The basic purposes to be served by Federal research and development programs are public purposes, considered by the President and the Congress to be of sufficient national importance to warrant the expenditure of public funds. The management and control of such programs must be firmly in the hands of full-time Government officials clearly responsible to the President and the Congress. With programs of the size and complexity now common, this requires that the Government have on its staff exceptionally strong and able executives, scientists, and engineers, fully qualified to weigh the views and advice of technical specialists, to make policy decisions concerning the types of work to be undertaken, when, by whom, and at what cost, to supervise the execution of work undertaken, and to evaluate the results.

[viii] At the present time we consider that one of the most serious obstacles to the recruitment and retention of first-class scientists, administrators, and engineers in the Government service is the serious disparity between governmental and private compensation for comparable work. We cannot stress too strongly the importance of rectifying this situation, through congressional enactment of civilian pay reform legislation as you have recommended.

3. Given proper arrangements to maintain management control in the hands of Government officials, federally financed research and development work can be accomplished through several different means: Direct governmental operations of laboratories and other installations; operation of Government-owned facilities by contractors; grants and contracts with universities; contracts with not-for-profit corporations or with profit corporations. Choices among these means should be made on the basis of relative efficiency and effectiveness in accomplishing the desired work, with due regard to the need to

---

1. Ed. note.-see Annex 1, p. 25, for complete text of letter.

maintain and enlarge the long-term strength of the Nation's scientific resources, both public and private.

In addition, the rapid expansion of the use of Government contracts, in a field where 25 years ago they were relatively rare, has brought to the fore a number of different types of possible conflicts of interests, and these should be avoided in assigning research and development work. Clear-cut standards exist with respect to some of these potential conflict-of-interest situations—as is the case with respect to persons in private life acting as advisers and consultants to Government, which was covered in your memorandum of February 9, 1962. Some other standards are now widely accepted—for example, the undesirability of permitting a firm which holds a contract for technical advisory services to seek a contract to develop or to supply any major item with respect to which the firm has advised the Government. Still other standards are needed, and we recommend that you request the head of each department and agency which does a significant amount of contracting for research and development to develop, in consultation with the Attorney General, clear-cut codes of conduct, to provide standards and criteria to guide the public officials and private persons and organizations engaged in research and development activities.

4. We have identified a number of ways in which the contracting system can and should be improved, including:

Providing more incentives for reducing costs and improving performance;

Improving our ability to evaluate the quality of research and development work;

Giving more attention to feasibility studies and the development of specifications prior to inviting private proposals for major systems development, thus reducing "brochuremanship" with its heavy waste of scarce talent.

We have carefully considered the question whether standards should be applied to salaries and related benefits paid by research and development contractors doing work for the Government. We believe it is desirable to do so in those cases in which the system of letting contracts does not result in cost control through competition. We believe the basic standard to be applied should be essentially the same as the standard you recently recommended to the Congress with [ix] respect to Federal employees—namely, comparability with salaries and related benefits paid to persons doing similar work in the private economy. Insofar as a comparability standard cannot be applied—as would be the case with respect to the very top jobs in an organization, for example—we would make it the personal responsibility of the head of the contracting agency to make sure that reasonable limits are applied.

5. Finally, we considered that in recent years there has been a serious trend toward eroding the competence of the Government's research and development establishments—in part owing to the keen competition for scarce talent which has come from Government contractors. We believe it to be highly important to improve this situation—not by setting artificial or arbitrary limits on Government contractors but by sharply improving the working environment within the Government, in order to attract and hold first-class scientists and technicians. In our judgment, the most important improvements that are needed within Government are:

To ensure that governmental research and development establishments are assigned significant and challenging work;

To simplify management controls, eliminate unnecessary echelons of review and supervision, and give to laboratory directors more authority to command resources and make administrative decisions; and

To raise salaries, particularly in higher grades, in order to provide greater comparability with salaries available in private activities.

Action is under way along the first two lines—some of it begun as the result of our review. Only the Congress can act on the third aspect of the problem, and we strongly hope it will do so promptly.

In preparing this report, we have benefited from comments and suggestions by the Attorney General, the Secretaries of Agriculture, Commerce, Labor, and Health, Education, and Welfare, and the Administrator, Federal Aviation Agency, and they concur in general with our findings and conclusions.

>   Robert S. McNamara,
>   Secretary of Defense.
>
>   James E. Webb,
>   Administrator, National
>   Aeronautics and
>   Space Administration.
>
>   John W. Macy, Jr.,
>   Chairman, Atomic Energy Commission.
>
>   Dr. Alan T. Waterman,
>   Director, National Science Foundation.
>
>   Jerome B. Wiesner,
>   Special Assistant to the President
>   for Science and Technology.
>
>   David E. Bell,
>   Director, Bureau of the Budget....

[xi] **FOREWORD**

This report has been prepared in response to the President's letter of July 31, 1961, to the Director of the Bureau of the Budget, asking for a review of the use of Government contracts with private institutions and enterprises to obtain scientific and technical work needed for public purposes.

Such contracts have been used extensively since the end of World War II to provide for the operation and management of research and development facilities and programs, for analytical studies and advisory services, and for technical supervision of complex systems, as well as for the conduct of research and development projects.

As the President noted in his letter, there is a consensus that the use of contracts is appropriate in many cases. At the same time, a number of important issues have been raised, including the appropriate extent of reliance on contractors, the comparative salaries paid by contractors and the Government, the effect of extensive contracting on the Government's own research and development capabilities, and the extent to which contracts may have been used to avoid limitations which exist on direct Federal operations.

Accordingly, the President asked that the review focus on-

Criteria that should be used in determining whether to perform a function through a contractor or through direct Federal operations;

Actions needed to increase the Government's ability to review contractor operations and to perform scientific and technical work; and

Policies which should be followed by the Government in obtaining maximum efficiency from contractor operations and in reviewing contractor performance and costs (including standards for salaries, fees, and other items).

The President requested the following officials to participate in the study: The Secretary of Defense, the Chairman of the Atomic Energy Commission, the Chairman of the Civil Service Commission, the Administrator of the National Aeronautics and Space Administration, and the Special Assistant to the President for Science and Technology. The Director of the National Science Foundation was also invited to participate.

In making the review requested by the President, a great deal of material was available from hearings and reports of the Senate and House Committees on Appropriations, Armed Services, Judiciary, and Government Operations, the House Committees on Post Office and Civil Service and on Science and Astronautics, the second Hoover Commission, and various governmental and private studies. In addition, information was obtained:

By questionnaires to which 10 Federal agencies and 71 Government field installations, universities, and contract establishments respond; and

[xii] By interviews conducted at 28 Government field installations and non-Federal establishments, and with a number of agency headquarters officials.

These data were obtained and analyzed with respect to major policy implications by an interdepartmental staff group, which included representatives of each of the officials whom the President asked to participate in the review.

This report presents a summary analysis and recommendations growing out of this review. . . .

[1] **PART 1**
STATEMENT OF MAJOR ISSUES

Policy questions relating to Government contracting for research and development[2] must be considered in the perspective of the phenomenal growth, diversity, and change in Federal activities in this field.

FEDERAL RESEARCH AND DEVELOPMENT ACTIVITIES AND THEIR IMPACT

Prior to World War II, the total Federal research and development program is estimated to have cost annually about $100 million. In the fiscal year 1950, total Federal research and development expenditures were about $1.1 billion. In the fiscal year 1963, the total is expected to reach $12.4 billion.

The fundamental reason for this growth in expenditures has been the importance of scientific and technical work to the achievement of major public purpose. Since World War II the national defense effort has rested more and more on the search for new technology. Our military posture has come to depend less on production capacity in being and more on the race for shorter lead times in the development and deployment of new weapons systems and of countermeasures against similar systems in the hands of potential enemies. The Defense Department alone is expected to spend $7.1 billion on research and development in fiscal 1963, and the Atomic Energy Commission another $1.4 billion.

Aside from the national defense, science and technology are of increasing significance to many other Federal programs. The Nation's effort in nonmilitary space exploration—which is virtually entirely a research and development effort—is growing extremely rapidly; the National Aeronautics and Space Administration is expected to spend $2.4 billion in fiscal 1963, and additional sums related to the national space program will be spent by the Department of Commerce and other agencies. Moreover, scientific and technological efforts are of major significance in agriculture, health, natural resources, and many other federal programs.

The end of this period of rapid growth is not in sight. Public purposes will continue to require larger and larger scientific and technological efforts for as far ahead as we can see.

The increase in Federal expenditures for research and development has had an enormous impact on the Nation's scientific and technical resources. It is not too much to say

---

2. Note on terminology: The term "research and development" is used in this report in the sense in which it is used in the Federal budget—that is, it means the conduct of activities intended to obtain new knowledge or to apply existing knowledge to new uses. The Department of Defense uses the term "research, development, test, and evaluation," which is a somewhat fuller but more cumbersome term for the same concept. In this report the shorter term is used for convenience. For a summary of all Federal activities of this type, see Annex 3. "Federal Research and Development Programs," reprinted from "The Budget of the U.S. Government for fiscal year 1963."

that the major initiative and responsibility for promoting and financing research and development have in many important areas been shifted from private enterprise (including academic as well as business institutions) to the Federal [2] Government. Prior to World War II, the great bulk of the Nation's research achievements occurred with little support from Federal funds, although there were notable exceptions, such as in the field of agriculture. Today it is estimated by the National Science Foundation that the Federal budget finances about 65 percent of the total national expenditure for research and development. Moreover, the Federal share is rising.

Federal financing, however, does not necessarily imply Federal operation. As the Federal research and development effort has risen, there has been a steady reduction in the proportion conducted through direct Federal operations. Today about 80 percent of Federal expenditures for research and development are made through non-Federal institutions. Furthermore, while a major finding of this report is that the Government's capabilities for direct operations in research and development need to be substantially strengthened, there is no doubt that the Government must continue to rely on the private sector for the major share of the scientific and technical work which it requires.[3]

The effects of the extraordinary increase in Federal expenditures for research and development, and the increasing reliance on the private sector to perform such work, have been very far reaching.

The impact on private industry has been striking. In the past, the Government utilized profitmaking industry mainly for production engineering and the manufacture of final products—not for research and development. Industries with which it dealt in securing the bulk of its equipment were primarily the traditional large manufacturers for the civilian economy—such as the automotive, machinery, shipbuilding, steel, and oil industries, which relied on the Government for only a portion, usually a minority, of their sales and revenues. In the current scientific age, the older industries have declined in prominence in the advanced equipment area and newer research and development-oriented industries have come to the fore—such as those dealing in aircraft, rockets, electronics, and atomic energy.

There are significant differences among these newer industries and others. While the older industries were organized along mass-production principles, and used large numbers of production workers, the newer ones show roughly a 1-to-1 ratio between production workers and scientist-engineers. Moreover, the proportion of production workers is steadily declining. Between 1954 and 1959, production workers in the aircraft industry declined 17 percent while engineers and scientists increased 96 percent. Also, while the average ratio of research and development expenditures to sales in all industry is about 3 percent, the advanced weapons industry averages about 20 percent and the aerospace industry averages about 31 percent.

But the most striking difference is the reliance of the newer industries almost entirely on Government sales for their business. In 1958, a reasonably representative year, in an older industry, the automotive industry, military sales ranged from 5 percent for General Motors to 15 percent for Chrysler. In the same year in the aircraft industry, military sales ranged from a low of 67 percent for Beech Aircraft to a high of 99.2 percent for the Martin Co.

[3] The present situation, therefore, is one in which a large group of economically significant and technologically advanced industries depend for their existence and growth not on the open competitive market of traditional economic theory, but on the sales only to the U.S. Government. And, moreover, companies in these industries have the strongest incentives to seek contracts for research and development work, which will give them both the know-how and the preferred position to seek later follow-on production contracts.

The rapid increase in Federal research and development expenditures has had

---

3. Annex 4 provides data, supplied by the National Science Foundation, on the sources of funds for the national research and development effort and on the distribution of work between the various types of performing installations-direct Federal operations, industry, universities, and not-for-profit establishments.

striking effects on other institutions in our society apart from private industry.

There has been a major impact on the universities. The Nation has always depended largely on the universities for carrying out fundamental research. As such work has become more important to Government and more expensive, an increasing share—particularly in the physical and life sciences and engineering—has been supported by Federal funds. The total impact on a university can be sizable. Well over half of the research budgets of such universities as Harvard, Brown, Columbia, Massachusetts Institute of Technology, Stanford, California Institute of Technology, University of Illinois, New York University, and Princeton, for illustration, is supported by Federal funds.

New institutional arrangements have been established in many cases, related to but organized separately from the universities, in order to respond to the needs of the Federal Government. Thus, the Lincoln Laboratory of the Massachusetts Institute of Technology was established by contract with the Air Force to supply research and development services and to establish systems concepts for the continental air defense, and similarly the Jet Propulsion Laboratory was established at the California Institute of Technology to conduct research on rocket propulsion for the Department of the Army and later to supply space craft design and systems engineering services to the National Aeronautics and Space Administration. In addition, other research institutions—such as the Stanford Research Institute—which were established to conduct research on contract for private or public customers, now do a major share of their business with the Federal Government.

In addition to altering the traditional patterns of organization of private industry and the universities, the rise in Federal research and development expenditures has resulted in the creation of entirely new kinds of organizations.

One kind of organization is typified by the Rand Corp., established immediately after World War II, to provide operations research and other analytical services by contract to the Air Force. A number of similar organizations have been established since, more or less modeled on Rand, to provide similar services to other governmental agencies.

A second new kind of organization is the private corporation, generally not-for-profit but sometimes profit, created to furnish the Government with "systems engineering and technical direction" and other professional services. The Aerospace Corp., the MITRE Corp., the Systems Development Corp., and the Planning Research Corporation are illustrations.

[4] A third new organizational arrangement was pioneered by the Office of Scientific Research and Development during World War II and used by the Atomic Energy Commission, which took over the wartime atomic energy laboratories and added others—all consisting of facilities and equipment owned by the Government but operated under contract by private organizations, either industrial companies or universities.

Apart from their impact on the institutions of our society, Federal needs in research and development are placing critical demands on the national pool of scientific and engineering talent. The National Science Foundation points out that the country's supply of scientists and engineers is increasing at the fairly stable rate of 6 percent annually, while the number engaged in research and development activities is growing at about 10 percent each year. Accordingly, the task of developing our manpower resources in sufficient quality and quantity to keep pace with the expanding research and development effort is a matter of great urgency. The competition for scientists and engineers is becoming keener all the time and requires urgent attention to the expansion of education and training, and to the efficient use of the scientific and technical personnel we have now.

QUESTION AND ISSUES CONSIDERED IN THIS REPORT

The dynamic character of the Nation's research and development efforts, as summarized in the preceding paragraphs, has given rise to a number of criticisms and points of concern. For example, concern has been expressed that the Government's ability to perform essential management functions has diminished because of an increasing dependence on contractors to determine policies of a technical nature and to exercise the type of management functions which Government itself should perform. Some have criticized the new not-for-profit contractors, performing systems engineering and technical direction work for the Government, on the grounds that they are intruding on traditional functions performed by competitive industry. Some concern has been expressed that universities are undertaking research and development programs of a nature and size which may interfere with their traditional educational functions. The cost-reimbursement type of contracts the Government uses, particularly with respect to research and development work on weapons and space systems, have been criticized as providing insufficient incentives to keep costs down and ensure effective performance. Criticism has been leveled against relying so heavily on contractors to perform research and development work as simply a device for circumventing civil service rules and regulations.

Finally, the developments of recent years have inevitably blurred the traditional dividing lines between the public and private sectors of our Nation. A number of profound questions affecting the structure of our society are raised by our inability to apply the classical distinctions between what is public and what is private. For example, should a corporation created to provide services to Government and receiving 100 percent of its financial support from Government be considered a "public" or a "private" agency? In what sense is a business corporation doing nearly 100 percent of its business with the Government engaged in "free enterprise"?

[5] In light of these criticisms and concerns, an appraisal of the experience in using contracts to accomplish the Government's research and development purposes is evidently timely. We have not, however, in the course of the present review attempted to treat the fundamental philosophical issues indicated in the preceding paragraph. We accept as desirable the present high degree of interdependence and collaboration between Government and private institutions. We believe the present intermingling of the public and private sectors is in the national interest because it affords the largest opportunity for initiative and the competition of ideas from all elements of the technical community. Consequently, it is our judgment that the present complex partnership between Government and private institutions should continue.

On these assumptions, the present report is intended to deal with the practical question: What should the Government do to make the partnership work better in the public interest and with maximum effectiveness and economy?

We deal principally with three aspects of this main question.

There is first the question, What aspects of the research and development effort should be contracted out? This question falls into two parts. One part relates to those crucial powers to manage and control governmental activities which must be retained in the hands of public officials directly answerable to the President and Congress. Are we in danger of contracting out such powers to private organizations? If so, what should be done about it?

The other part of this question relates to activities which do not have to be carried out by Government officials, but on which there is an option: they may be accomplished either by direct Government operations or by contract with non-Federal institutions. What are the criteria that should guide this choice? And if a private institution is chosen, what are the criteria for choice as among universities, not-for-profit corporations, profit corporations, or other possible contractors?

The second question we deal with is what standards and criteria should govern contract terms in cases where research and development is contracted out. For example, to what extent is competition effective in ensuring efficient performance at low cost, and when—if at all—must special rules be established to control fees, salaries paid, and other elements of contractor cost?

The third question we deal with is how we can maintain strong research and development institutions as direct Governments operations. How can we prevent the best of the Government's research scientists, engineering, and administrators from being drained off to private institutions as a result of higher private salaries and superior private working environments, and how can we attract an adequate number of the most talented new college graduates to a career in Government service?

These questions are treated in the sections which follow.

[Blank page]

[7] **PART 2**
CONSIDERATIONS IN DECIDING WHETHER TO CONTRACT OUT RESEARCH AND DEVELOPMENT WORK

Generalizations about criteria for contracting out research and development work must be reached with caution, in view of the wide variety of different circumstances which must be covered.

A great many Government agencies are involved. The Department of Defense, the National Aeronautics and Space Administration, and the Atomic Energy Commission provide the bulk of Federal financing, but a dozen or more agencies also play significant roles.

Most Federal research and development work is closely related to the specific purpose of the agency concerned—to the creation of new weapons systems for the Department of Defense, for example, or the exploration of new types of atomic power reactors for the Atomic Energy Commission. But a significant portion of the research financed by the Federal Government is aimed at more general targets: to enlarge the national supply of highly trained scientists, for example, as is the case with some programs of the National Science Foundation. And even the most "mission oriented" agencies have often found it desirable to make funds available for basic research to advance the fundamental state of knowledge in fields that are relevant to their missions. Both the Department of Defense and the AEC, for example, make substantial funds available for fundamental research, not related to any specific item of equipment or other end product.

A great many different kinds of activity are involved, which have been classified by some under five headings:

(1) Fundamental research.
(2) Supporting research or exploratory development.
(3) Feasibility studies, operations analysis, and technical advice.
(4) Development and engineering of products, processes, or systems.
(5) Test and evaluation activities.

The lines between many of the activities listed are necessarily uncertain. Nevertheless, it is clear that "research and development" is a phrase that covers a considerable number of different kinds of activity.

Finally, there have been distinct historical developments affecting the different Government agencies. Some agencies, for example, have a tradition of relying primarily on direct Government operations of laboratories—others have precisely the opposite tradition of relying primarily on contracting for the operation of such installations.

Against this background of diversity in several dimensions, we have asked, what criteria should be used in deciding whether or not to contract out any given research and development task? In outline, our judgment on this question runs as follows:

[8] There are certain functions which should under no circumstances be contracted out. The management and control of the Federal research and development effort must be firmly in the hands of full-time Government officials clearly responsible to the President and the Congress.

Subject to this principle, many kinds of arrangements—including both direct Federal operations and the various patterns of contracting now in use—can and should be used to mobilize the talent and facilities needed to carry out the Federal research and development effort. Not all arrangements, however, are equally suitable for all purposes and under all circumstances, and discriminating choices must be made among them by the Government agencies having research and development responsibilities. These choices should be based primarily on two considerations:

(1) Getting the job done effectively and efficiently, with due regard to the long-term strength of the Nation's scientific and technical resources; and

(2) Avoiding assignments of work which would create inherent conflicts of interest.

Each of these judgments is elaborated below:

STRENGTHENING THE ABILITY OF THE GOVERNMENT TO MANAGE AND CONTROL RESEARCH AND DEVELOPMENT PROGRAMS

We regard it as axiomatic that policy decisions respecting the Government's research and development programs—decisions concerning the types of work to be undertaken when, by whom, and what cost—must be made by full-time Government officials clearly responsible to the President and to the Congress. Furthermore, such officials must be in a position to supervise the execution of work undertaken, and to evaluate the results. These are basic functions of management which cannot be transferred to any contractor if we are to have proper accountability for the performance of public functions and for the use of public funds.

To say this does not imply that detailed administration of each research and development task must be kept in the hands of top public officials. Indeed, quite the contrary is true, and an appropriate delegation of responsibility—either to subordinate public officials or by contract to private persons or organizations—for the detailed administration of research and development work is essential to its efficient execution.

It is not always easy to draw the line distinguishing essential management and control responsibilities which should not be delegated to private contractors (or, indeed, to governmental research organizations such as laboratories) from those which can and should be so assigned. Recognizing this difficulty, it nevertheless seems to be the case that in recent years there have been instances—particularly in the Department of Defense—where we have come dangerously close to permitting contract employees to exercise functions which belong with top Government management officials. Insofar as this has been true, we believe it is being rectified. Government agencies are now keenly aware of this problem and have taken steps to retain functions essential to the performance of their responsibility under the law.

[9] It is not enough, of course, to recognize that governmental managers must retain top management functions and not contract them out. In order to perform those functions effectively, they must be themselves competent to make the required management decisions and, in addition, have access to all necessary technical advice. Three conclusions follow:

First, where management decisions are based substantially on technical judgments, qualified executives, who can properly utilize the advice of technical consultants, from both inside and outside the Government, are needed to perform them. There must be sufficient technical competence within the Government so that outside technical advice does not become de facto technical decisionmaking. In many instances the executives making the decisions can and should have strong scientific backgrounds. In others, it is possible to have nonscientists so long as they are capable of understanding the technical issues involved and have otherwise appropriate administrative experience.

By and large, we believe it is necessary for the agencies concerned to give increased stress to the need to bring into governmental service as administrators men with scientific or engineering understanding, and during the development of Government career executives, to give many of them the opportunity, through appropriate training and experience, to strengthen their appreciation and understanding of scientific and technical matters. Correspondingly, scientists and engineers should be encouraged and guided to obtain, through appropriate training and experience, a broader understanding of management and public policy matters. The average governmental administrator in the years to come will be dealing with issues having larger and larger scientific and technical content, and his training and experience, both before he enters Government service and after he has joined, should reflect this fact.

At the present time, we are strongly persuaded that one of the most serious obstacles to acquiring and maintaining the managerial competence which the Government needs for its research and development programs is the discrepancy between governmental and private compensation for comparable work. This obstacle has been growing increasingly serious in recent years as increases in Federal pay have been concentrated primarily at the lower end of the pay scale—resulting in the anomalous situation that many officials of Government responsible for administering major elements of Federal research and development programs are paid substantially smaller salaries than personnel of universities, or business corporations, or of not-for-profit organizations who carry out subordinate aspects of those research and development programs.[4] We cannot stress too strongly the importance of rectifying this situation, and hope the Congress will take at this session the action which the President has recommended to reform Federal civilian pay scales.

Second, it is necessary for even the best qualified governmental managers to obtain technical advice from specialists. Such technical advice can be obtained from men within the Government or those outside. When it is obtained from persons outside of Government, special problems of potential conflict of interest are raised which were [10] dealt with in the President's recent memorandum entitled "Preventing Conflicts of Interest on the Part of Advisers and Consultants to the Government."

We believe it highly important for the Government to be able to turn to technical advice from its own establishment as well as from outside sources. One major source of this technical knowledge is the Government-operated laboratory or research installation and, as is made clear later in this report, we believe major improvements are needed at the present time in the management and staffing of these installations. A strong base of technical knowledge should be continually maintained within the Government service and available for advice to top management.

Third, we need to be particularly sensitive to the cumulative effects of contracting out Government work. A series of actions to contract out important activities, each wholly justified when considered on its own merits, may when taken together begin to erode the Government's ability to manage its research and development programs. There must be a high degree of awareness of this danger on the part of all governmental officials concerned. Particular attention must be given to strengthening the Government's ability to provide effective technical supervision in the letting and carrying out of contracts, and to developing more adequate measures for performance evaluation.

DETERMINING THE ASSIGNMENT OF RESEARCH AND DEVELOPMENT WORK

As indicated above, we considered it necessary and desirable to use a variety of arrangements to obtain the scientific and technical services needed to accomplish public purposes. Such arrangements include: direct governmental operations through laboratories or other installations; operation of Government-owned facilities by contractors; grants and contracts with universities and entities associated with universities; contracts with not-for profit corporations wholly or largely devoted to performing work for Government; and contracts with private business corporations. We also feel that innovation is still needed in these matters, and each agency should be encourage to seek new and better arrangements to accomplish its purposes. Choices among available arrangements should be based primarily on two factors:

Relative effectiveness and efficiency, and
Avoidance of conflicts of interest.

Relative effectiveness and efficiency

In selecting recipients, whether public or private, for research and development assignments, the basic rule (apart from the conflict-of-interest problem) should be to assign the job where it can be done most effectively and efficiently, with due regard to the strengthening of institutional resources as well as to the immediate execution of projects. This

---

4. Annex 5 summarizes information obtained during the present review regarding salaries and related benefits.

criterion does not, in our judgment lead to a conclusion that certain kinds of work should be assigned only to certain kinds of institutions. Too much depends on individual competence, historical evolution, and other special circumstances to permit any such simple rule to hold. However, it seems clear that some types of facilities have natural advantages which should be made use of.

Thus:

[11] Direct Federal operations, such as the governmental laboratory, enjoy a close and continuing relationship to the agency they serve, which permits maximum responsiveness to the needs of that agency and a maximum sense of sharing the mission of the agency. Such operations accordingly have a natural advantage in conducting research, feasibility studies, developmental and analytical work, user tests, and evaluations which directly support the management functions of the agency. Furthermore, an agency-operated research and development installation may provide a useful source of technical management personnel for its sponsor.

At the present time, we consider that the laboratories and other facilities available to Government are operating under certain important handicaps which should be removed if these facilities are to support properly the Federal research and development effort. These matters are discussed at some length in part 4 of this report.

Colleges and universities have a long tradition in basic research. The processes of graduate education and basic research have long been closely associated, and reinforce each other in many ways. This unique intellectual environment has proven to be highly conducive to successful undirected and creative research by highly skilled specialists. Such research is not amenable to management control by adherence to firm schedules, well-defined objectives, or predetermined methods of work. In the colleges and universities, graduate education and basic research constitute an effective means of introducing future research workers to their fields in direct association with experienced people in those fields, and in an atmosphere of active research work. Applied research appropriate to the universities is that which broadly advances the state of the art.

University-associated research centers are well suited to basic or applied research for which the facilities are so large and expensive that the research acquires the character of a major program best carried out in an entity apart from the regular academic organization. Research in such centers often benefits from the active participation of university scientists. At the same time, the sponsoring university (and sometimes other, cooperating universities) benefits from increased opportunities for research by its faculties and graduate students.

Not-for-profit organizations (other than universities and contractor-operated Government facilities), if strongly led, can provide a degree of independence, both from Government and from the commercial market, which may make them particularly useful as a source of objective analytical advice and technical services. These organizations have on occasion provided an important means for establishing a competent research organization for a particular task more rapidly than could have been possible within the less flexible administrative requirements of the Government.

Contractor-operated Government facilities appear to be effective, in some instances, in securing competent scientific and technical personnel to perform research and development work where very complex and costly facilities are required and the Government desires to maintain control of these facilities. Under such arrangements, it has been possible for the Government to retain most of the controls inherent in direct Federal operations, while at the same time gaining many of the advantages of flexibility with respect to staffing, organization, and management, which are inherent in university and industrial operations.

[12] Operations in the profit sector of the economy have special advantages when large and complex arrays of resources needed for advanced development and preproduction work must be marshaled quickly. If the contracting system is such as to provide appropriate incentives, operations for profit can have advantages in spurring efficiency, reducing costs, and speeding accomplishments. (It is plain that not all operations in this sector have

resulted in low costs or rapid and efficient performance; we regard this as a major problem for the contracting system and discuss it further in pt. 3 of this report.) Contractors in the profit sector may have the advantage of drawing on resources developed to satisfy commercial as well as governmental customers, which adds to the flexibility of procurement, and may permit resources to be phased in and out of Government work on demand.

The preceding paragraphs have stressed the advantages of these different types of organization. There are disadvantages relating to each type, which must also be taken into account. Universities, for example, are not ordinarily qualified—nor would they wish—to undertake major systems engineering contracts.

We repeat that the advantages—and disadvantages—noted above do not mean that these different types of arrangements should be given areas of monopoly on different kinds of work. There are, by common agreement, considerable advantages derived from the present diversity of operations. It permits great flexibility in establishing and directing different kinds of facilities and units, and in meeting the need for managing different kinds of jobs. Comparison of operations among these various types of organizations helps provide yardsticks for evaluating performance.

Moreover, this diversity helps provide many sources of ideas and of the critical analysis of ideas, on which scientific and technical progress depends. Indeed, we believe that some research (in contrast to development) should be undertaken by most types of organizations. Basic and applied research activities related to the mission of the organization help to provide a better intellectual environment in which to carry out development work. They also assist greatly in recruiting high-quality research staff.

In addition to the desirability of making use of the natural areas of advantage within this diversity of arrangements, there is one additional point we would stress. Activities closely related to governmental managerial decisions (such as those in support of contractor selection), or to activities inherently governmental (such as regulatory functions, or technical activities directly bound up with military operations), are likely to call for a direct Federal capability and to be less successfully handled by contract.

Conflicts of interest

There are at least three aspects of the conflict-of-interest problem which arise in connection with governmental research and development work.

First, there are problems relating to private individuals who serve simultaneously as governmental consultants and as officers, directors, or employees of private organizations with which the Government has a contractual relationship. Many of these individuals are among the Nation's most capable people in the research and development field and can be of very great assistance to Government agencies.

[13] The problems arising in their case with respect to potential conflicts of interest have been dealt with in the President's memorandum of February 9, referred to earlier in this report. The essential standard set out in that memorandum was that no individual serving as an adviser or consultant should render advice on an issue whose outcome would have a direct and predictable effect on the interests of the private organization which he serves. To this end, the President asked that arrangements be made whereby each adviser and consultant would disclose the full extent of his private interests, and the responsible Government officials would undertake to make sure that conflict-of-interest situations are avoided.

Second, there is a significant tendency to have on the boards of trustees and directors of the major universities, not-for-profit and profit establishments engaged in Federal research and development work, representatives of other institutions involved in such work. Such interlocking directorships may serve to reinforce and strengthen the overall management of private organizations which are heavily financed by the Government. Certainly it is in the public interest that organizations on whom so much reliance is placed for accomplishing public purposes, should be controlled by the most responsible, mature, and knowledgeable men available in the Nation. However, we see the clear possibility of conflict-of-interest situations developing through such common directorships that might be harmful to the public interest. Members of governing boards of private business enterprises, universities, or other organizations which advise the Government with respect to research and development activities are often simultaneously members of governing boards or organizations which receive or may receive contracts or grants from the Government for research, development, or production work. Unless these board members also serve as consultants to the Government, present conflict-of-interest laws do not apply. The spirit, if not the letter, of the standards of conduct for Government advisers set forth in the President's memorandum, in our judgment, can and should provide guidance to boards and their members with respect to the interrelationships among universities, not-for-profit organizations, and business corporations where Government business is involved. Some boards of trustees and directors have already taken action along these lines.

Beyond this, however, there is a third type of problem which requires consideration: This might be described as potential conflicts of interest relating to organizations rather than to individuals. It arises in several forms—not all of which by any means are yet fully understood. Indeed, in this area of potential conflicts of interest relating to individuals and organizations in the research and development field, we are in an early stage of developing accepted standards for conduct—unlike other fields, such as the law or medicine, where there are long-established standards of conduct.

One form of organizational conflict of interest relates to the distinction between organizations providing professional services (e.g., technical advice) and those providing manufactured products. A conflict of interest could arise, for example, if a private corporation received a contract to provide technical advice and guidance with respect to a weapons system for which that same private corporation later sought a development or production contract, or for which it sought to develop or supply a key sub-system or component. It is [14] clear that such conflict-of-interest situations can arise whether or not the profit motive is present. The managers of the not-for-profit institutions have necessarily a strong interest in the continuation and success of such institutions, and it is part good management of Federal research and development programs to avoid placing any contractor—whether profit or nonprofit—in a position where a conflict of interest could clearly exist.

Another kind of issue is raised by the question whether an organization which has been established to provide services to a Government agency should be permitted to seek contracts with other Government agencies—or with non-Government customers. The question has arisen particularly with respect to not-for-profit organizations established to provide professional services.

There is not a clear consensus on this question among Government officials and officers of the organizations in question. We have considered the question far enough to have the following tentative views:

In the case of organizations in the area of operations and policy research (such, for example, as the Rand Corp.), the principal advantages they have to offer are the detached quality and objectivity of their work. Here, too close control by any Government agency may tend to limit objectivity. Organizations of this kind should not be discouraged from dealing with a variety of clients, both in and out of Government.

On the other hand, a number of the organizations which have been established to provide systems engineering and technical direction (such, for example, as Aerospace Corp.) are at least for the time being of value principally as they act as agents of a single client. In time, as programs change and new requirements arise, it may be possible and desirable for such organizations also to achieve a fully independent financial basis, resting on multiple clients, but this would seem more likely to be a later rather than an earlier development.

Enough has been said to indicate that this general area of conflict of interest with respect to research and development work is turning up new kinds of questions and all the answers have not yet been found. We believe it important to continue to work toward setting forth standards of conduct, as was done by the President in his February memorandum. We recommend that the President instruct each department and agency head, in consultation with the Attorney General, to proceed to develop as much of a code of conduct for individuals and organizations in the research and development field as circumstances now permit.

Finally, we would note that beyond any formal standards, we cannot escape the necessity of relying on the sensitive conscience of officials in the Government and in private organizations to make sure that appropriate standards are continually maintained.

[15] **PART 3**
PROPOSAL FOR IMPROVING POLICIES AND PRACTICES APPLYING TO RESEARCH AND DEVELOPMENT CONTRACTING

During the course of this review, a number of suggestions arose which we believe to indicate desirable improvements in the Government's policies and practices applying to research and development contracting.

IMPROVING THE GOVERNMENT'S COMPETENCE AS A "SOPHISTICATED BUYER"

In order for the contracting system to work effectively, the first requirement is for the Government to be a sophisticated buyer—that is, to know what it wants and how to get it. Mention has already been made of the requirements this placed on governmental management officials. At this point four additional suggestions are made.

1. In the case of many large systems development projects, it has been the practice to invite private corporations to submit proposals to undertake research and development work—relating to a new missile system, for example, or a new aircraft system. Such proposals are often invited before usable and realistic specifications of the system have been worked out in sufficient detail. As a consequence, highly elaborate, independent, and expensive studies are often undertaken by the would-be contractors in the course of submitting their proposals. This is a very costly method of obtaining competitive proposals, and it unnecessarily consumes large amounts of the best creative talent this country possesses, both on the preparation of the proposals and their evaluation. Delivery time pressures may necessitate inviting proposals before specifications are completed, but we believe that practice can and should be substantially curtailed.

This would mean, in many instances, improving the Government's ability to accomplish feasibility studies, or letting special contracts for that purpose, before inviting proposals. In either event, it would require the acceptance of a greater degree of responsibility by Government managers for making preliminary decisions prior to inviting private proposals. We believe the gains from such a change would be substantial in the avoidance of unnecessary and wasteful use of scarce scientific and technical personnel as well as heavy costs to the private contractors concerned—costs which in most cases are passed on to the Government.

2. We believe there is a great deal of work to be done to improve the Government's ability to supervise and to evaluate the conduct of research and development efforts—whether undertaken through public or private facilities. We do not have nearly enough understanding as yet of how to know whether we are getting a good product for our money, whether research and development work is being [16] competently managed, or how to select the more competent from the less competent as between research and development establishments.

When inadequate technical criteria exist, there is a tendency to substitute conformity with administrative and fiscal procedures for evaluation of substantive performance. What is required is more exchange of information between agencies on their practices in contractor evaluation and on their experience with these practices. A continuing forum should be provided for such exchange. It is possible also that some central and fairly formal means of reporting methods and experience and recording them permanently should be established. We recommend that the Director of the new Office of Science and Technology, when established, be asked to study the possibility of establishing such a forum and the best means for providing information regarding evaluation practices.

3. With the tremendous proliferation of research and development operations and associated facilities in recent years, it has become difficult for the Government officials who arrange for such work to be done to be aware of all the facilities and manpower that are available. To maintain a complete and continuous roster of manpower, equipment, and organizations, sensitive to month-by-month changes, would undoubtedly be too costly in terms of its value.

Nevertheless, we believe that an organized attempt should be made to improve the current inventory of information on the scientific and technical resources of the country. We recommend that the National Science Foundation consider ways and means of improving the availability of such information for use by all concerned in public and private activities.

4. In addition, the expansion of the Nations's research and development effort has multiplied the difficulties of communication among researchers engaged on related projects at separate facilities, both public and private. It is clear that additional steps should be taken to further efforts to improve the system for the exchange of information in the field of science and technology.

At present, a panel on scientific information of the President's Science Advisory Committee is at work on this subject. We expect that its report will be followed by full-scale planning for the establishment of a more effective technical information exchange system, to support the needs of the operating scientist and the engineer.

IMPROVING ARRANGEMENTS WITH THE PRIVATE SECTOR TYPES OF CONTRACTS

The principal type of contract for research and development work which is made with private industry is the cost-plus-fixed-fee contract. Such contracts have been used in this area because of the inherent difficulty of establishing precise objectives for the work to be done and of making costs estimates ahead of time.

At the same time, this type of contract has well-known disadvantages. It provides little or no incentive for private managers to reduce costs or otherwise increase efficiency. Indeed, the cost-plus-fixed-fee contract, in combination with strong pressures from governmental managers to accomplish work on a rapid time schedule, probably provides incentives for raising rather than for reducing costs. If a corporation is judged in terms of whether it accomplishes a result by [17] a given deadline rather than by whether it accomplishes that result at minimum cost, it will naturally pay less attention to costs and more attention to speed of accomplishment. On the other hand, where there is no given deadline, the cost-plus-fixed-fee contract may serve to prolong the research and development work and induce the contractor to delay completion.

Consequently, we believe it to be desirable to replace cost-plus-fixed-fee contracting with fixed-price contracting wherever that is feasible—as it should be in the procurement of some late-stage development, test work, and services. Where it is judged that cost reimbursement must be retained as the contracting principle, it should be possible in many instances to include an incentive arrangement under which the fee would not be fixed, but would vary according to a pre-determined standard which would relate larger fees to lower costs, superior performance, and shorter delivery times. There is ample evidence to prove that if adequate incentives are given by rewards for outstanding performance, both time and money can be saved. Where the nature of the task permits, it may be desirable to include in the contract penalty provisions for inadequate performance.

Finally, if neither fixed-price nor incentive-type contracts are possible, it is still necessary for Government managers to insist on consideration being given to lower cost, as well as better products and shorter delivery times—and to include previous performance as one element in evaluating different contractors and the desirability of awarding them subsequent contracts.

Contract administration

The written contract itself, however well done, is only one aspect of the situation. The administration of a contract requires as much care and effort as the preparation of the contract itself. This is particularly important with respect to changes in system characteristics, for these changes often become the mechanism for justifying cost overruns. Other factors of importance in contract administration are fixing authority and responsibility in both Government and industry, excessive reporting requirements, and all-too-frequent lack of prearranged milestones for auditing purposes.

Reimbursable costs

Concern has been expressed because of significant differences among the various agencies in policies regarding which costs are eligible for reimbursement—notably with respect to some of the indirect costs. These differences are now being reviewed by the Bureau of the Budget with the cooperation of the Department of Defense, the National Aeronautics and Space Administration, the Atomic Energy Commission, and the General Services Administration.

Arrangements with universities

With respect to universities, Government agencies share responsibility for seeing that research and development financed at universities does not weaken these institutions or distort their functions which are so vital to the national interest.

Government agencies use both grants and contracts in financing research at universities, but in our judgment the grant has proved to be a simpler and more desirable device for Federal financing of fundamental research, where it is in the interest of the Government [18] not to exercise close control over the objectives and direction of research. Since all relevant Government agencies are now empowered to use grants instead of contracts in supporting basic research, the wider use of this authority should be encouraged.

Apart from this matter, three others seem worthy of comment. One arises from the extensive use of contracts (or grants) for specific and precisely identified projects. Often there is a tendency to believe that in providing support for a single specific project the chance of finding a solution to a problem is being maximized. In reality, however, less specific support often would permit more effective research in broad areas of science, or in interdisciplinary fields, and provide greater freedom in drawing in more scientists to participate in the work that is undertaken. Universities, too, often find project support cumbersome and awkward. A particular professor may be working on several projects financed by several Government agencies and must make arbitrary decisions in allocating expenses to a particular project. It thus appears both possible and desirable to move in the direction of using grants to support broader programs, or to support the more general activities of an institution, rather than to tie each allocation of funds to a specific project. A number of Government agencies have been moving in this direction and it would be desirable to expand the use of such forms of support as experience warrants.

At the same time, it would not, in our judgment, be appropriate to place major reliance on the institutional grant, since the major purpose of making grants in most cases is to assure that the university personnel and facilities concerned will be devoted to pursuing specific courses of inquiry.

A second problem associated with the support of research at universities is whether the Government should pay all costs, including indirect expenses or "overhead," associated with work financed by the Government. We believe this matter involves two related but distinct questions, which should be separated in considering the appropriate policy to be followed.

1. We believe there is no question that, in those cases in which it is desirable for the Government to pay the entire cost of work done at a university, the Government should pay for allowable indirect as well as direct costs. To do otherwise would be discriminatory against universities in comparison with other kinds of institutions. For purposes of financial and accounting simplicity, in those cases where grants are used, and it is desirable for the Government to pay all allowable costs, it may be possible to work out a uniform or average percentage figure which could be regarded as covering indirect costs.

2. We believe there are many cases in which it is neither necessary nor desirable for the Government to pay all the costs of the work to be done. In many fields of research, a university may gain a great deal from having the research in question done on its campus, with the participation of its faculty and students, and may be able and willing to share in the costs, either through its regular funds or through raising additional funds from foundations, alumni, or by other means. The extent and degree of cost-sharing can and should vary among different agencies and programs, and we are not prepared at this time to suggest any uniform standards—except the negative one that it would be plainly illogical to require that the university uniformly provide its share through the payment of all or a part of the indirect [19] costs. Only in the exceptional case would this turn out to be the best basis for determining the appropriate sharing of costs.

A third problem relates to the means for furnishing major capital assets for research at universities (such as a major building or a major piece of equipment such as a linear accelerator, synchrotron, or large computer. In most cases, it will be preferable to finance such facilities by a separate grant (or contract), which will ensure that careful attention is given to the long-term value of the asset and to the establishment of appropriate arrangements for managing and maintaining it.

Arrangements with respect to not-for-profit organizations other than universities

It has been the practice in contracting for research and development work with such organizations to cover all allowable costs and, in addition, to provide what is commonly called a "fee." The reason for paying a "fee" to not-for-profit organizations is quite different from the reason for paying a fee to profit-making contractors and therefore the term "fee" is misleading. The profitmaking contractor is engaged in business for profit. His profit and the return to his shareholders or investors can only come from the fee. In the case of the not-for-profit organizations, there are no shareholders, but there are two sound reasons to justify payment of a "development" or "general support" allowance to such organizations.

One is that such allowances provide some degree of operational stability and flexibility to organizations which otherwise would be very tightly bound to the precise limitations of cost financing of specific tasks; the allowances can be used to even out variations in the income of the organization resulting from variations in the level of contract work. A second justification is that most not-for-profit organizations must conduct some independent, self-initiated research if they are to obtain and hold highly competent scientists and engineers. Such staff members, it is argued, will only be attracted if they can share, to some extent, in independently directed research efforts.

We considered that both of these arguments have merit and, in consequence, support the continuation of these payments. Both arguments represent incentives to maintain the cohesiveness and the quality of the organization, which is in the interest of the Government. They should underlie the thinking of the Government representatives who negotiate contracts with not-for-profit organizations. But the amount of the "fee" or allowance in each instance must still be determined by bargaining between Government and contractor, in accordance with the independent relationship that is essential to successful contracting.

An important question relating to not-for-profit organizations, other than universities, concerns facilities and equipment. In our judgment, the normal rule should be that where facilities and equipment are required to perform research and development work desired by the Government, the Government should either provide the facilities and equipment, or cover their cost as part of the contract. This is the rule relating to profit organizations and would hold in general for not-for-profit organizations—but there are two special problems with respect to the latter.

[20] First, we believe it is generally not desirable to furnish funds through "fees" for the purpose of enabling a contractor to acquire major capital assets. On the other hand, the Government should not attempt to dictate what a contractor does with his "fee," provided it has been established on a sound and equitable basis, and if a contractor chooses to use part of his "fee" to acquire facilities for use in his self-initiated research we would see

no objection.

Second, we would think it equitable, where the Government has provided facilities, funds to obtain facilities, substantial working capital, or other resources to a contractor, it should, upon dissolution of the organization, be entitled to a first claim upon such resources. This would seem to be a matter which should be governed, insofar as possible, by the term of the contract—or, in the case of any newly established organizations, should be provided in the provisions of its charter.

Salaries and related benefits

In addition to the question of fees and allowances, there has been a great deal of concern over the salaries and related benefits received by persons employed on federally financed research and development work in private institutions, particularly persons employed in not-for-profit establishments doing work exclusively for the Government. Controls have been suggested or urged by congressional committees and others to make sure that there is no excessive expenditure of public funds and to minimize the undesirable competitive effect on the Federal career service.

We agree that where the contracting system does not provide built-in controls (for example, through competitive bidding), attention should be paid to the reasonableness of contractors' salaries and related benefits, and contractors should be reimbursed only for reasonable compensation costs.

The key question is how to decide what is reasonable and appropriate compensation. We believe the basic standard for reimbursement of salaries and related benefits should be one of comparability to compensation of persons doing similar work in the private economy. The President recently proposed to the Congress that the pay for Federal civilian employees should be based on the concept of reasonable comparability with employees doing similar work in the private economy. We believe this to be a sound principle which can be applied in the present circumstances as well.

Application of this comparability principle may require some special compensation surveys (perhaps made by the Bureau of Labor Statistics), which can and should be arranged for as necessary. Furthermore, there will undoubtedly be cases in which comparable data are difficult to obtain—as, for example, with respect to top management jobs. In such cases the specific approval of the head of the Government contracting agency or his designee should be required.

In view of the inherent complexity and sensitivity of this subject, we suggest that special administrative arrangements should be established in each agency. Contract policies respecting salaries and related benefits in each contracting agency should be controlled by an official reporting directly to the head of the agency (in the Department of Defense, to assure uniformity of treatment, by an official reporting directly to the Secretary of Defense), and salaries above a certain level—say $25,000—should require the personal approval of that official.

[21] **PART 4**
PROPOSALS FOR IMPROVING THE GOVERNMENT'S ABILITY TO CARRY OUT RESEARCH AND DEVELOPMENT ACTIVITIES DIRECTLY

Based on the evidence acquired in the course of this review, we believe there is no doubt that the effects of the substantial increase in contracting out Federal research and development work on the Government's own ability to execute research and development work have been deleterious.

The effects of the sharp rise in contracting out have included the following. First, contractors have often been able to provide a superior working environment for their scientists and engineers—better salaries, better facilities, better administrative support—making contracting operations attractive alternatives to Federal work. Second, it has often seemed that contractors have been given the more significant and more interesting work assignments, leaving Government research and development establishments with routine missions and static programs which do not attract the best talent. Third, additional burdens have often been placed on Government research establishments to assist in evaluating the work of increasing numbers of contractors and to train and educate less skilled contractor personnel—without adding to the total staff and thus detracting from the direct research work which appeals to the most competent personnel. Fourth, scientists in contracting institutions have often had freedom to move "outside of channels" in the Government hierarchy and to participate in program determination and technical advice at the highest levels—freedom frequently not available to the Government's own scientists. Finally, one of the most serious aspects of the contracting out process has been that it has provided an alternative to correcting the deficiencies in the Government's own operations.

In consequence, for some time there has been a serious trend toward the reduction of the competence of Government research and development establishments. Recently a number of significant actions have been started which are intended to reverse this trend. We point particularly to the strong leadership being given within the Defense Department by the Director of Defense Research and Engineering, in striving to raise the capabilities of the Department's laboratories and other research and development facilities.

Nevertheless, we believe the situation is still serious and that major efforts are required.

We consider it a most important objective for the Government to maintain first-class facilities and equipment of its own to carry out research and development work. This observation applies not only to the newer research and development agencies but equally to the older agencies such as Commerce, Interior, and Agriculture.

No matter how heavily the Government relies on private contracting, it should never lose a strong internal competence in research and development. By maintaining such competence it can be sure of [22] being able to make the difficult but extraordinarily important program decisions which rest on scientific and technical judgments. Moreover, the Government's research facilities are a significant source of management personnel.

Major steps seem to us to be necessary in the following matters:

1. It is generally recognized that having significant and challenging work to do is the most important element in establishing a successful research and development organization. It is suggested that responsibility should be assigned in each department and agency to the Assistant Secretary for Research and Development or his equivalent to make sure that assignments to governmental research facilities are such as to attract and hold first-class men. Furthermore, arrangements should be made to call on Government laboratory and development center personnel to a larger extent for technical advice and participation in broad program and management decisions—in contrast to the predominant use of outside advisers.

2. The evidence is compelling that managerial arrangements for many Government-operated research and development facilities are cumbersome and awkward. Several improvements are needed in many instances, including—

Delegating to research laboratory directors more authority to make program and personnel decisions, to control funds, and otherwise to command the resources which are necessary to carry out the mission of the installation;

Providing the research laboratory director a discretionary allotment of funds, to be available for projects of his choosing, and for the results of which he is to be responsible;

Eliminating where possible excess layers of echelons of supervisory management and ensuring that technical, administrative, and fiscal reviews be conducted concurrently and in coordinated fashion; and

Making laboratory research assignments in the form of a few major items with a reasonable degree of continuity rather than a multiplicity of small narrowly specific tasks; this will put responsibility for detailed definition of the work to be done at the laboratory level where it belongs.

To carry out these improvements will require careful and detailed analysis of the different situations in different agencies. Above all, it will require the energetic direction of top officials in each agency.

Plans have already been developed for joint teams of Civil Service Commission and Department of Defense research and manpower personnel to visit nine defense laboratories during April and May 1962, in order to analyze precisely what administrative restrictions exist that hamper research effectiveness. In this fashion, those unwarranted limitations that can be eliminated by executive action can be identified as distinguished from those that may require legislative change.

3. Salary limitations, as already mentioned, in our opinion play a major role in preventing the Government from obtaining or retaining highly competent men and women. Largely because of the lack of comparable salaries, the Government is not now and has not for at least the past 10 years been able to attract or retain its share of such critically necessary people as: recently graduated, highly recommended Ph.D.'s in mathematics and physics, recent B.S.-M.S. scientific and [23] engineering graduates in the upper 25 percent of their classes at top-ranked universities; good, experienced weapons system engineers and missile, space, and electronic specialists at intermediate and senior levels; and senior-level laboratory directors, scientific managers, and administrators. This obstacle will be substantially overcome if the Congress approves the President's recommendation to establish a standard of comparability with private pay levels for higher professional and technical jobs in the Federal service.

4. A special problem in the Defense Department is the relationship between uniformed and civilian personnel. This is a difficult and sensitive problem of which the Department of Defense is well aware. We do not attempt in this report to propose detailed solutions, but we do suggest that certain principles are becoming evident as a result of the experience of recent years.

It seems clear, for example, that the military services will have increasing need for substantial numbers of officers who have extensive scientific and technical training and experience. Such officers bring firsthand knowledge of operational conditions and requirements to research and development installations and, in turn, learn about the state of the art and the feasible applications of technology to military operations. The military officer is needed to communicate the needs of the user, to prepare the operational forces for new equipment, to plan for the use of developing equipment, and later to install it and supervise its use.

All of the above roles suggest that when military personnel are used in research and development activities, they should perform as "technical men" rather than "military men" except when there is a need for their military skills. Military command and direction become important only as one moves from the research end of the spectrum into the area where operational considerations predominate. Both at middle management and policy levels, a well-balanced mixture of military and civilian personnel may be most advantageous in programs designed to meet military needs.

In research, there are many instances in which the existence of military supervision, and the decreased opportunities for advancement because of military occupancy of top jobs, are among the principal reasons why the Defense Department has had difficulty in attracting outstanding civilian scientists and engineers. On the other hand, there are examples within the Department of cause in which enlightened policies of civil-military relationships have drawn on the strengths of each and produced excellent results. In such instances, the military head of the laboratory has usually concentrated on administrative problems and the civilian technical director has had complete control of technical programs.

Military officers should not be substituted for civilians in the direction and management of research and development unless they are technically qualified and their military background is directly needed and applicable.

In the course of the next year, the Department of Defense intends to give consideration to the delineation of those research and development installations in which operational considerations are predominant and those installations in which scientific and technical considerations are predominant. Having done so, the assignment of military officers [24] to head the former type of installation, and civilians (or equally qualified military officers) to head the latter will be encouraged. Furthermore, when military personnel are assigned to work in civilian-directed installations on the basis of their technical abilities, it is intended that they should be free of the usual rotation-of-duty requirements and not have separate lines of reporting.

5. In addition to the recommendations above, we have given consideration to the possible establishment, which might be called a Government institute. Such an institute would provide a means for reproducing within the Government structure some of the more positive attributes of the nonprofit corporation. Each institute would be created pursuant to authority granted by the Congress and be subject to the supervision of a Cabinet officer or agency head. It would, however, as a separate corporate entity directly managed by its own board of regents, enjoy a considerable degree of independence in the conduct of its internal affairs. An institute would have authority to operate its own career merit system, as the Tennessee Valley Authority does, would be able to establish a compensation system based on the comparability principle, and would have broad authority to use funds and to acquire and dispose of property.

The objective of establishing such an instrumentality would be to achieve in the administration of certain research and development programs the kind of flexibility which has been obtained by Government corporations while retaining, as was done with the Government corporation, effective public accountability and control.

We regard this idea as promising and recommend that the Bureau of the Budget study it further, in cooperation with some of the agencies having major research and development programs. It may well prove to be a useful additional means for carrying out governmental research and development efforts.

6. It would seem, based on the results of this review, that it would be possible and desirable to make more use of existing governmental facilities and avoid the creation of duplicate facilities. This is not as easy a problem as it might seem. It is ordinarily necessary for a laboratory, if it is to provide strong and competent facilities, to have a major mission and a major source of funding. This will limit the extent to which it is possible to make such facilities available for the work of other agencies. Nevertheless, in some cases and to some extent it is clearly possible to do this, and a continuing scrutiny is necessary in order to make sure that the facilities which the Government has are used to their fullest extent.

7. Finally, together with the better use of existing facilities, the Government must also make better use of its existing scientific and engineering personnel. This implies not only a careful watch over work assignments, but also a continual upgrading of the capabilities of Federal personnel through education and training. At the present time, technology is changing so rapidly that on-the-job scientists and engineers find themselves out of date after a decade or so out of the university. To remedy this, the Government must strengthen its educational program for its own personnel, to the extent of sending them back to the university for about an academic year every decade. This program, necessary as it is, will only become attractive if the employee is ensured of job security on his return from school and if his parent organization is allowed to carry him on its personnel roster....

Document IV-10

Document title: Albert F. Siepert to James E. Webb, Administrator, "Length of Tours of Certain Military Detailees," February 8, 1963.

Source: NASA Historical Reference Collection, NASA History Office, NASA Headquarters, Washington, D.C.

From its inception NASA administrators made a practice of accomplishing goals by marshalling outside resources rather than reproducing them within the agency. Although NASA was a civilian agency, in its early days it made extensive use of military expertise. In fact, the number of military personnel working for NASA increased steadily between 1960 and 1968, and military officers played key roles in the Apollo program. This 1963 memorandum emphasized the need for NASA to obtain the services of military detailees for the extended time needed for them to carry out the responsibilities assigned to them by NASA.

[1] ## Length of Tours of Certain Military Detailees

On April 13, 1959, the President approved an agreement between the Departments of Defense, Army, Navy, Air Force, and the National Aeronautics and Space Administration which provides for the detailing of military personnel for service with NASA. This agreement has made it possible for NASA to obtain from the military the services of many fine officers with skills and experience not obtainable from other sources. This cooperation on the part of the Department of Defense has contributed materially to the success of NASA's efforts.

The agreement provides that the normal tour of duty with NASA for military personnel on active duty will be three years. It also provides that NASA ". . . send a timely request to the military department concerned for any desired extension." Normally, the three-year tour is satisfactory. There are exceptions, however, and problems occasionally arise in obtaining extensions of more than one-year when NASA management happens to place a career officer with rather unique skills in a key program or management position. Even with the most careful planning, NASA's rapid growth has thus far made it impossible to plan ahead for an adequate understudy to take over these unusual assignments at the expiration of an automatic three- or four-year period. Jobs to which career officers have been assigned have in many instances grown considerably in terms of scope, responsibility, and urgency. Of greater significance, the ability of the officer himself to assume greater and greater responsibility and thereby become more critical to our needs makes it even more important for us to seek a more liberal interpretation of the provision in the basic agreement on extending tours of duty of career officers detailed to NASA when they occupy positions critical to our operations.

In two instances, by dealing with the military service concerned, we have been able to secure extensions greater than one year. The astronauts were granted a three-year extension because it was determined that such an extension was mutually beneficial to the officers, the military services, and NASA. Cdr. Albert J. Kelly was granted a three-year extension by the Navy because of his assignment as Director, Electronics and Control, Office of Advanced Research and Technology, a key executive position which was of considerable benefit to his career development. A one-year extension would not have provided sufficient time for NASA to secure and indoctrinate an acceptable replacement. Major Victor Hammond, Chief, National Range Support, Office of Tracking and Data Acquisition, is presently serving on a one year extension and no action has been taken to replace or to [2] obtain another extension. On the other hand, NASA has not attempted to obtain extended tours of duty of career officers when such an extension might stand in the way of the long-term career objectives of the officer. General Ostrander, General Roadman (returning to military assignment soon), Col. Heaton, and Col. Seaberg are good examples.

The Launch Operations Center at Cape Canaveral, more than any other NASA installation, has relied upon career officers of the military to staff key positions. Col. Asa Gibbs has occupied the position of liaison with the Atlantic Missile Range, thereby providing a focal point for all NASA range requirements. A request to extend his tour of duty of denied. Lt. Col. Ray Clark, presently serving as Special Assistant to Kurt Debus, has secured orders returning him to duty in a military assignment. A request for an extended tour of duty was disapproved. Major Rocco Petrone, presently serving as, Chief Heavy Space Vehicle Systems Office, is nearing the end of his three-year detail, and a request for extension of his tour of duty has been filed with the Department of the Army.

The loss of Lt. Col. Clark, and very possibly Major Petrone, will be a severe blow to the LOC organization. While the military objective of fulfilling the requirements of career development are recognized and understood, we feel we must do everything we can to obtain extended tours of duty for both Lt. Col. Clark and Major Petrone.

As background, the following table provides information on military details and extended tours of duty under terms of the agreement:

Career Military Officers Assigned to NASA

	Army	Navy	Air Force	Total
Presently Assigned	12	30	52	94
Previously Assigned	3	5	16	24
Assigned				118

Requests for Extended Tours of Duty

	Army	Navy	Air Force	Total
Requests	3	7	8	18
Approved	1	7 *	6 **	14
Disapproved	1	1	1	
Pending	1		1	

* 1 Navy + 2 Air Force extended for 2 years
** 4 Navy + 3 Air Force extended for 3 years

[3] The few times we have sought extensions of tours of duty clearly indicates that we have been most considerate of military objectives in furthering the career development of its officer personnel. Also, it is clearly evident that there is better cooperation on the part of the Navy and Air Force than there is by the Army.

Lt. Col. Clark and Major Petrone are key figures in the LOC organization which is now at the very beginning of a tremendous expansion. It is my opinion that every effort should be made to obtain two-year extensions of their tours of duty with NASA.

Document IV-11

Document title: U.S. Congress, House, Committee on Science and Astronautics, Subcommittee on NASA Oversight, Staff Study, "Apollo Program Management," 91st Cong., 1st sess. (Washington, DC: U.S. Government Printing Office, July 1969), pp. 59-74.

Olin ("Tiger") E. Teague (D-Texas), one of the space program's staunchest supporters in the House of Representatives and the Chairman of the Subcommittee on NASA Oversight of the House Committee on Science and Astronautics, struggled to protect NASA's budgets from cuts that began to occur after the agency's peak of $5.2 billion for FY 1965. In 1968 Teague wrote the presidents and chief executive officers of the Boeing Company, the McDonnell Douglas Astronautics Company, North American Rockwell Corporation, Grumman Aircraft Engineering Corporation, and the Space Division of the Chrysler Corporation, seeking their cooperation in a committee study of "those key management systems which have been adapted and developed in the Apollo program and which may have the potential for making a contribution to other large and complex technological programs." McDonnell Douglas, whose submission is reprinted here, designed and developed under contract to NASA the upper stage (S-IV) of the Saturn I launch vehicle and the third stage (S-IVB) of the Saturn IB and the Saturn V.

[59] # Apollo Program Management

PRESENTATION TO THE STAFF OF THE SUBCOMMITTEE ON NASA OVERSIGHT...

[61] **INTRODUCTION**

In May 1960 the McDonnell Douglas Astronautics Co.-Western Division (MDAC-WD) then known as the Douglas Aircraft Co., was awarded a NASA contract to design and develop the Saturn S-IV, the upper stage of the Saturn I launch vehicle, and the first of a family of three giant launch vehicles whose ultimate mission is manned exploration of the moon. In April 1962, this organization was also awarded a second NASA contract, to design and develop the Saturn S-IVB, the uppermost stage of the two other members of the launch vehicle family, the Saturn IB and Saturn V.

At the time of the first award, the organizational structure of that entity of the Company responsible for discharge of the newly contracted obligations was one which had been formed for the efficient and simultaneous development of various guided missile weapon systems for military forces. The advent of the Saturn/Apollo system, the greatest engineering task in history, required a much more intimate integration of Government and industry resources than had previously been the case. The necessity of this close integration of resources and effort was not immediately visible to either industry or Government, at least to the degree eventually required. Accordingly, over a period of time, the MDAC-WD found it necessary to realign its organizational structure and adjust its management techniques to accommodate the unique requirements of this great, joint, government-industry venture.

The report discusses how and why the MDAC-WD Saturn organization and management methods evolved to meet this challenge. It presents in chronological order how the organization configuration changed from an integrated functional form to a project form, to a divisional status, and then to a matrix form. It reviews the creation of new management tools to efficiently handle the requirements of precise configuration control, exacting quality standards, extensive contract change traffic, and even fundamental revision in the type of contract. The effectiveness of these management systems is then finally demonstrated by presenting the performances on cost control, schedule compliance, and flight program success.

SECTION 1
MANAGEMENT

1.1 Concepts

To manage the Saturn program, the Company devised a logical, systematic framework for the task with the ingredient of flexibility to accommodate program growth and change. Management concepts were influenced by the nature of the Company, Customer, work to be performed, and legal and regulatory requirements.

[62] **1.1.1 Company**

The chain of events involving organization and reorganization of the Company's missile and space systems efforts, covered elsewhere in this report, has influenced the Saturn/Apollo program. Two management principles were used: (1) provide autonomy and freedom to Company personnel to interface directly with the Customer's managers, and (2) provide top management with the means to evaluate program status and support the program manager's needs for resources. (This was made possible by the projectized program organization placed in a matrixed division framework.)

1.1.2 Customer
Interfacing effectively and expeditiously with Customer program managers at all levels became an overriding necessity. To achieve this, appropriate parallels were established between key organizations such as in the structuring of the NASA Apollo Program Office and (then titled) Douglas Saturn/Apollo Program Office.

1.1.3 Work to be Performed
Large scientific and technical staffs were established during the design and development phases, and the results of their efforts implemented by manufacturing and test groups. Physical requirements imposed by geography and logistics had to be met. The stage was to be manufactured in Southern California, tested in Northern California (NASA test facilities are in Alabama), with checkout and launch taking place in Florida. Stages were to be transported between these areas and facilities managed at each.

1.1.4 Legal and Regulatory Requirements
Finally, NASA requirements had to be expressed in contractual terms. MDAC-WD had to conform to NASA's procurement regulations and other federal regulations.

Increasingly precise requirements were embodied in the contract, and they exerted powerful influences on the management philosophies.

1.2 Organization
For best control, MDAC-WD consists of organizations structured vertically according to function and horizontally by product line. Saturn/Apollo, one of the 13 subdivisions in this matrix framework, is projectized for single-point management control. Thus, the vital functions of Development Engineering, Financial Management, Reliability and Launch Operations (R. & L.O.), and Operations are made directly available to Saturn by organizational structure....

Support given by these subdivisions to Saturn is (1) Huntington Beach Development Engineering–Engineering design and test, material research and production methods, standardization, etc., (2) Financial Management–Controller, contracts (including Work Order Authorizations), Operations Control (costs, pricing, budgets, schedules, program tracking, etc.), and financial forecasts and analysis, (3), R. & L.O.—Launch support services for Saturn/Apollo launches at KSC, including mission variation modifications, prelaunch preparations and stage testing, operating launch consoles, and participating in [63] countdown, and (4) Operations—Manufacturing Operations, Manufacturing Engineering, Reliability Assurance, Procurement, and Facilities.

1.3 Saturn/Apollo Program Subdivision
This subdivision manages the Company's Saturn/Apollo programs and coordinates support from other subdivisions. In certain areas cutting across the functional activities of several subdivisions, the Director, Assistant General Manager, MDAC—WD for the Saturn/Apollo program has established his own directorates, i.e., Director of Saturn Program Product Assurance, Director of Saturn Program Production, and others.

1.3.1 Director/Assistant General Manager
Mr. H. E. Bauer, Director/Assistant General Manager, has complete authority to plan, direct, and control the MDAC—WD resources applied to Saturn work. Mr. Bauer represents and acts for the Vice- President-General Manager of MDAC-WD in all matters concerning Saturn at all Division locations.

1.3.2 Directors and Staff

The directorates and staff elements ... comprise the essential links of a project organization, autonomous but supportable through a matrix structure by other Subdivisions. No attempt is made to describe functions and responsibilities in this report. ...

[64] In addition, support is given Saturn by the following subdivisions:

1.3.2.1 Information Systems

Developing and implementing integrated, management information system; the data from checkout, static firing, and flight operations for engineering analysis and evaluation, etc.

1.3.2.2 Advance Systems and Technology

Analysis of new product areas; conducting CRAD and IRAD programs, including the Saturn/Apollo.

1.4 Program Management Objectives

From its inception, the Saturn/Apollo effort was managed in a manner little recognized in textual theory on aerospace program management and management information control systems. Successes to date are attributable to the unique direct management techniques which were and are being used to meet program objectives.

1.4.1 Zero Flight Failures

The overriding objective was to avoid flight failures. MDAC-WD devised techniques and methods to produce articles of such quality and reliability that there would be no failures in flight. Administrative controls and requirements which had no direct bearing on flight success were subordinated to a secondary role, for later development in satisfying auditing agencies, internal and external.

1.4.2 Maximum Direct Communications

MDAC-WD and Customer personnel at middle management levels and higher were given freedom of direct communication for the sake of expediency and to avoid undue paperwork channels which might hinder progress on the day-by-day management of the program.

[65] 1.4.3 Flexibility With Short Reaction Time

Built into the management systems was the flexibility to respond to Customer direction, redirection, and changes in the program. These techniques are covered in subsequent sections of this report.

1.4.4 Change Management, Not Change Inhibition

This objective was achieved.

1.4.5 Schedule Compliance Ahead of Schedule Capability

Early in the program, management struggled to move on time from engineering release, through first hardware test, and first article acceptance firing and delivery to the Cape. They resolved to first meet, then get ahead of the schedule.

1.4.6 Outstanding Technical Capability

The record of achievements attests to the success of this objective. The S-IV program established an outstanding technical capability, which bore fruits in the S-IVB and related efforts.

1.4.7 Avoid Cost Overruns

This very important objective and the techniques used to overcome an actual overrun will be explained later in the report. It is noteworthy that, in so doing, management implemented numerous cost reduction activities that resulted in an on-target or underrun condition.

SECTION 2
KEY MANAGEMENT EVENTS AND ACTIONS

2.1 Reorganization

In 1966, the program organization underwent several significant changes. The projectized Program Control was abandoned in favor of the Division concept. The Sacramento Test Center now reported to the program director, instead of Saturn Engineering. Also, an effective Configuration Change Control Board had been created. The success of being able to communicate the role and relationship of program managers to the remainder of the Division's organizations, and the providing of sound management systems with which he can carry out his responsibilities, has reduced the need for projectized organizations. The reduction in projectized organizations in favor of recentralization of functional resources contributes to flexibility in the shifting of resources between programs, with the inherent benefits of better utilization of manpower, improved performance, and lower costs.

2.2 All-Up Test Concept

The Marshall Space Flight Center set the tone for the space program. Their philosophy was to drive the first stage with dummy upper stages and fly the needed number of development rounds with two-stage vehicles. MDAC-WD history through such programs as Nike/Ajax Nike/Hercules, Nike/Zeus, also led to the concept of progressive developments. Since they were unmanned vehicles, the company employed an incremental development methodology. Upon entering the S-IV and S-IVB programs, we were ready to accept the all-up test concept.

[66] Flights occurred with near-to-operational configurations on the very first launch, as opposed to flying with partially complete or alternative configurations, such as programmers instead of full guidance. This meant a most comprehensive ground test program, which prevented the revelation of hardware weaknesses in flight.

This brought the S-IVB into the flight stage far ahead of schedule. Originally, several decisions were made to fly men on the Saturn I with the S-IV. Economics dictated against a decision to have a Saturn I, an IB, and a V all going at once. In 1962, the decision was made to cancel. The S-IV was actually canceled before it ever flew.

The obvious consequence of the all-up test concept was the imposition of configuration control disciplines on the program, on the technical staff, on suppliers, and on the entire community. Rigorous management of configuration changes avoided a near chaotic condition which would have resulted from the inclusion of results from various test analyses in the hardware. This would have caused much difficulty in establishing the proper configuration for each acceptance firing or launch.

Ground rules were laid down that connected the all-up philosophy with configuration control. It was considered a law that anything that flew on 205 had to be flown on 204, and anything on 503 had to be on 502. That the rules were well conceived is attested to by the program's degree of success.

2.3 Saturn I Performance and the LOR Decision

Performance of the Saturn I program and the Saturn S-IV stage was technically outstanding. All six vehicles were launched with complete success, providing MDAC-WD with the technical capability and the baseline necessary to proceed into the S-IVB. The lunar orbital rendezvous decision was then made, which led to the requirement of the S-IVB stage and the S-II Stage using the J-2 liquid-oxygen/liquid/hydrogen engine. With that decision the Company entered into the contract definition phase for the S-IVB. Successful development of hydrogen technology had a tremendous effect on Apollo. In 1961, Pratt & Whitney experienced several accidents with their RL-10 engine, and there followed a wave of adverse sentiment against hydrogen. If work had not progressed to show that hydrogen was an easy material to work with, and that multiengined capability was not impossible to achieve, the entire Saturn/Apollo program might have been shaped quite differently.

2.4 Controls

Management awareness and control of Saturn required a continuous monitoring analysis, and evaluation of all program aspects in terms of cost, schedule, and technical performance. This in-depth surveillance of the program was made feasible through the operation of four high-level action boards chartered to review program progress. These boards were not only to review, but also carry out a wide range of overall management functions. Their authority cut across all program activities, and they were the principal apparatus by which management formulated policy and provided program direction. By name they are: the Configuration Change Control Board, Change Analysis Board, Senior Management Action Board, and the Senior Financial Management Review Board. They are real-time, decisionmaking [67] boards with the capability to record and establish their findings and convert them to firm contract language in the form of change orders and program adjustments.

2.4.1 Configuration Change Control Board (CCCB)

In one sense, the management process often begins with this board, which was established to coordinate the activities of the Saturn program. It is here that proposed contractual changes in the program are formally brought to the attention of Saturn top management and the initial decisions made for putting them into effect. The board is chaired by the Program Director.

The Configuration Change Control Board meets three times a week. It examines all contract change orders, supplemental agreements, proposed ECP's that adjust the contract, and all work effort that requires responsiveness across the Division. A NASA representative also attends, and the products of the staff work are brought before all of the directors. This community of directorates, located at Huntington Beach, sits in that meeting as formal members of the team. The community expands, depending on the particulars of the agenda. Members acquire a thorough understanding of what is contemplated, and may object, agree, or propose a change. It is not, however, a voting board.

Saturn management is, perhaps, unique in the depth of detail into which it goes—the turning on of any change order or supplemental agreement, of board items down to very small items. The program director has all of his decisionmakers immediately available—often in one room—and they have an opportunity to look at every important piece of work to be authorized, including details that many would consider completely unnecessary when related to the stature of the board.

2.4.2 Change Analysis Board (CAB)

All Company-initialed requests for changes are first presented to this board before being sent to the CCCB. Those changes deemed advisable that do not require a formal change in the contract can be approved or rejected on the spot by this board. For those changes requiring formal contractual change, an Engineering Change Proposal is prepared and submitted to the CCCB. Authority for the operation of this board rests with its chairman, who reports directly to the program director.

2.4.3 Senior Management Action Board

This biweekly board is one of the principal tools by which Saturn management controls the progress of the program. The meeting, chaired by the program director, is attended by senior management and key supervisory personnel. Characterized by incisive question-and-answer sessions, the meetings have come to be known as the "Black Tuesday" reviews.

2.4.4 Senior Financial Management Review

This review is presented each month by the Financial Management Subdivision to the program director and his staff. Target costs and expenditures of each of the operating departments are examined in detail, with particular emphasis given to estimates for the future. Budget adjustments to correct deviations and resolve potential problems are made on the basis of much of the information coming from this financial review process.

[68] ### 2.4.5 Other Reviews

Still other techniques used on the program embody the concept of review and re-review. The various levels of management or disciplines review the program progress. Communities are established to simply review the program on a short-term basis. In progress reviews with MSFC, however, the 30-45 review (meet for 30 minutes every 45 days) is presented only to division management.

In the launch mission reviews, any major tests may be reviewed, not only by the program but by an independent agency such as the Reliability and Launch Operations organization (R. & L.O.). These reviews are conducted by the Vehicle Flight Readiness Review Committee, which addresses itself to a specific test, such as an acceptance firing or a launch operation. They draw, from the division, appropriately skilled people to be the auditors. The program presents the status of the hardware, the configurations, significant failure and rejection reports, and open supplementary failure analysis documentation.

Another side benefit is the record of all of the examinations, findings, and actions taken to respond to those findings. These findings are presented to the director of the program. He must respond to all of those action items, in writing and to the satisfaction of the Reliability and Launch Operations organization—which is empowered to stop the test. It is then placed in the record for reference, should it become necessary.

The concept of review and re-review by different communities has been an important ingredient to ensure technical success.

2.5 Configuration Management Disciplines

Also in development during reorganization of the Saturn Program were the initial NASA/Contractor agreements involving application of the configuration management disciplines on program changes. These agreements were made contractual in early 1966 and represent a milestone in the achievement of program change control. Worthy of note is the fact that the agreement exceeded systems for control of the hardware only and provided methods by which all changes to the contract are defined and documented. Implementing these agreements significantly improved the control of the program.

2.6 Implementing Decisions

Given an organizational structure and an effective decisionmaking process, management decisions, once reached, still have to be put into effect. There are three such principal tools which have been adapted for use on the Saturn Program: a set of "management manuals" for issuing general operating directives; a work-management system for authorizing and implementing decisions; and a contract management procedure for identifying NASA requirements, negotiating contract provisions, and authorizing the work necessary to meet requirements.

2.7 Informal Communications

Superimposed upon the formal systems are the informal systems of communication through face-to-face contact. These are judged to be equally key to the success of the program. The management of the Saturn Program at MDAC-WD has not attempted to sit in an office examining status reports to reach significant management decisions. To the contrary, the program management's visibility is substantially [69] improved by daily personal contacts between Company and Customer personnel, and decisions are guided by information and facts which thus come to light.

2.8 Incentive Fees

NASA took positive steps to assure Apollo mission success by emphatically expressing to contractors the things most important to the Apollo Program. They did this in the way most meaningful to contractors: by increasing their profits for superlative performance and reducing profits for poor performance. The factors thus emphasized by NASA were: cost, schedule, and operational success.

At the time—in 1965 and early 1966—when the contract was converted from cost-plus-fixed-fee to cost-plus-incentive-fee, target costs and the cost-sharing incentive fees were agreed upon for the work then under contract. Provisional payment of incentive fees was to be made thereafter upon demonstration of specific evidences of superior performance.

The features have become a very important tool in controlling the technical performance, the schedules and the cost. It had been difficult to determine what would be paid for and what would not be paid for. Also, the many facets of the customer organization and MDAC-WD complicated contractual negotiations. Communications between MDAC-WD and the customer crystallized about these incentive features. The tradeoffs became crystal clear. Within 9 months to a year after signing, the confusion and difficulties in the program vanished because objectives were clearly defined. It now appears that the optimum time to incentivize is after the definition is well along.

Taken alone, a cost incentive could work to the detriment of other valuable considerations, but, combined with schedule and technical performance incentives, meaningful tradeoffs can be made to the benefit of the program. The greatest effect of these tradeoff considerations is to create in the whole organization—from top management right down to the man on the bench—an awareness of the importance for each of them to evaluate the effectiveness of every action that affects performance, cost, and schedule.

Meeting technical performance goals is the real make-or-break factor for an incentive contractor. Without this operational success, meeting cost and schedule goals is meaningless. So specific measures of technical excellence have been identified for comparing success, and fees are adjusted upward or downward in accordance with the results.

Performance is evaluated in terms of how flight missions are accomplished, the payload capability demonstrated, and telemetry responses shown. The performance requirements are based on technical requirements contained in CEI specifications. Flight-test plans are prepared for each flight by the Company and are followed afterward by final flight reports that set forth performance achievements. Certificates of performance achievement are submitted to NASA, and NASA's position is stated in return.

Schedule incentives are based on meeting three important milestones leading to delivery of completed S-IVB stages. The milestones here selected to provide NASA with the opportunity to review the effectiveness of the Company's work at regular, preplanned intervals. This is accomplished by the administrative technique of requiring from the [70] Company certificates stating the degree of completion of specific schedule-oriented actions and requiring from NASA prompt response—concurring or differing—so that program status is continually known by both parties.

In this way, the effect of all actions on delivery of contract end items becomes a part of total Saturn program activity, but is the special concern of the "Black Tuesday" reviews held biweekly and the primary concern of the Director for Saturn Program Production.

As a result of the incentive feature, MDAC-WD developed an improved reporting system in schedules and cost performance.

SECTION 3
ADDITIONAL MANAGEMENT ACTIONS

3.1 Introduction

The Saturn program introduced a new range of challenges for MDAC-WD. Management techniques geared to the production of aircraft in volume had to be slanted toward the complexity and state-of-the-art nature of the Saturn program. MDAC-WD relied heavily upon the quality and effectiveness of management practices that had been evolving steadily on prior programs but made them sufficiently flexible to be responsive to Saturn requirements. To do this, they established clear, detailed requirements, and provided for precise command and control through total program visibility. The success of this approach lies in the management record of the Saturn program.

3.2 Administrative

3.2.1 Interface

The number of agencies involved with Saturn, both within the Company and externally (Customer, associates, subcontractors, vendors and suppliers), is enormous. Communicating effectively with each, therefore, was a significant challenge. Both formal and informal lines of communication were established as were the means of transmitting information and documents to conduct daily business with NASA.

3.2.2 Evolution of the Role of the Program Manager

The role of the program manager in the newly formalized program office was not clearly understood by all levels of management. Further, a proper set of tools with which to carry out his responsibility was not available to the program manager. His capability to control his program depended somewhat upon his personal forcefulness and his success at inserting himself or members of his staff into the then existing Division's work authorization systems.

Customer organizations with responsibility to oversee the program were sympathetic to the frustrations of the program manager and began to express concern that their programs would not receive sufficient management attention.

To alleviate this situation, Division management responded with two courses of action: (1) Large programs were permitted significant projectization, especially in financial management and engineering; (2) A substantial effort was mounted to better define and publicize the responsibility and authority of the program manager. Position guides were carefully rewritten to assure that they carried a strong message [71] on that role. Division management directives were revised to define in operating directives the contribution and participation of the program manager. Probably the most significant change in the Division's management systems was the establishment of a Task Authorization Notice (TAN), the program manager's tool to authorize the release or cancellation of program plans and requirements to the Division's functional departments.

3.2.3 Company Standard Practice Bulletins

MDAC-WD sought to meet specific requirements of the Saturn program by expressly tailoring its Standard Practice Bulletins to the program and furnishing these documents to the Customer. This highly unusual amount of Customer orientation is somewhat reflected in that during 1964, over 507 SPB's were revised to improve MDAC-WD management systems. The Division sustains a concerted effort to continually revise, refine, and upgrade these management directives.

3.2.4 VIP Program

In 1964, a Value in Performance program (VIP) was implemented to produce superior product quality and personal excellence in work performance. The program emphasizes the importance of people, and enhances the feeling of each that he is a very important part of the Company. The program (1) motivates each person to take an increased interest in his job, (2) improves the quality of products and services, and (3) reduces costs and improves schedules.

The backbone of VIP accomplishment has been the establishment of meaningful measurable goals, and the subsequent attainment and improvement of these goals. Over 200 specific performance goals were established in 1967 of which 91 percent were achieved. This year Saturn/Apollo's VIP program has adopted the theme, "Management by Objectives." The objectives are goals which have become more specific and demand a high order of performance attainment. For MDAC-WD, the VIP program has been a factor in the dramatic increase in validated cost reductions, increasing from $16 million in 1964 to over $93 million in 1967.

In 1966 the Company received the U.S. Air Force's coveted Zero Defects honor, The Craftmanship Award. Of 3,500 competing companies, Douglas was one of the two to win this award. In October 1968, McDonnell Douglas was notified that they had achieved the Second or Sustained Craftsmanship Award for accomplishments in the field of motivation for the preceding year.

3.2.5 Supplier Motivation Program

NASA and MDAC-WD initiated a Supplier Motivation Program in early 1967. The intent was to advise MDAC-WD suppliers management of specific applications for the items they were manufacturing and thus motivate them to produce more reliable hardware.

The Company brought all suppliers of critical components to Huntington Beach to make them aware of the consequences of a failure in the critical component they were providing to the program. They were briefed thoroughly on failure-mode specification analysis of their individual piece of hardware, and the president of each Company completely understood what would happen should his component fail. They were shown their hardware on the stage, how it was handled, and [72] then asked to go back and examine the method by which they were providing this hardware. They were to determine whether they could detect anything that should be brought to MDAC-WD's attention, or anything they felt they should do internally in the preparation of their hardware. Sixty suppliers participated in 4 half-day sessions. A number of suppliers conducted awareness programs for employees and for their own suppliers. Their recommendations included design changes, and reverification of conformance to design requirements. The program benefited from these meetings before AS-5O1 was committed to launch at the Cape.

SECTION 4
SELECTED KEY PROBLEMS

4.1 Introduction

Managing Saturn has been almost as complicated and demanding a task as overcoming attendant technical difficulties. While geared to take on the management of this immense and complex program by valuable experience gained with Thor, Nike, and other families of missiles and space systems, no previous program compared with Saturn for scope, size and complexity. In retrospect, it can be seen that significant strides were made in learning how to control a major program of the size and magnitude of the Saturn project.

This section highlights some of the key management problems encountered by MDAC-WD with Saturn and how they were solved.

4.2 Effective Communication

On a program the size of Saturn/Apollo, the problem of communicating effectively impinges on all transactions, from the simplest, vis-a-vis, contact to major program negotiations. Throughout the program, at all levels, heavy emphasis was laid on the personal encounter. This basic philosophy was strengthened by firm and precisely defined requirements to document and record decisions made on the spot and under the duress of program schedules and requirements. The net effect of the decision to run the program on this basis, although intrinsically not measurable, was to expedite management and production decisions and raise morale.

A corollary of this decision lay in the necessity to so aline counterparts within the Company (as well as between those people and all external organizations) that each individual would be talking to others at precisely the right levels and in equally correct areas.

The Marshall Space Flight Center (MSFC) provided both in-depth technical and nontechnical control. Communications with the Industrial Operations Office, the Stage

Manager at MSFC, the laboratories at MSFC, and the technical communities within our Development Engineering organization, and supporting in-house technical activities had to be face to face. To realign the Saturn organization so that technical counterparts could be identified on a one-to-one basis, a technically oriented directorate was established, which could communicate to the Industrial Operations office, the Stage Manager at MSFC, and his corollary—the contracting officer. Saturn System Development supported that combination, so that the products that came out of technical interchanges and program requirements were crystallized into specific documents and became the contract end-item specifications.

[73] This was the beginning of effective control over the products of the technical working groups and the face-to-face interchange between technical counterparts. Real-time decisionmaking was implemented and authorized both by MSFC and Division management. The S-IVB, Stage Manager and the S-IVB Program Director made the principal program decisions, and all members of the program community accepted them.

Another area of communication, now formal, was the generation Change Orders that ultimately developed into contract requirements. To facilitate Change Order processing, the program director strengthened Saturn Systems Development and the manager of Saturn contracts. The Director of Development Engineering developed a supporting capability within his organization to assist in preparation of a formal response to Change Order direction. In concert, these organizations could quickly translate Change Order direction into work authorization.

4.3 Avoiding Cost Overruns

The number one priority was to achieve technical performance of the highest caliber. The second was to get the program on and ahead of schedule. The lowest priority was to avoid cost overruns (which should have been achieved had the program schedule remained intact). At present, the program is in an on-target position and in the process of realignment as a consequence of the schedule stretchout from the launch activity.

4.4 Program Schedule

The program is on schedule. Upon emerging from engineering release, and at the beginning of ground research programs, all contractors involved found themselves quite nervous about meeting schedule obligations. Several years ago, the goal was established of getting ahead and staying ahead of contract schedule. A vigorous program was initiated to obtain a complete set of hardware ahead of the contract schedule. This was probably the fundamental decision which permitted a get-ahead and stay-ahead-of-schedule capability. The procedure involved substantial risk, but resulted in avoidance of actual cost vulnerabilities inherent in major overtime panic situations generated in trying to meet contract schedules. Premium prices were sometimes paid to get these supplies into the system, but use of overtime and premium time was weighed very carefully by Saturn management and the NASA Resident Manager.

4.5 Information Retrieval

Saturn management does not maintain a program control room, with charts, graphs and schedule status on the walls. By themselves, such charts are considered out of date by anywhere from an hour to a month, depending on how responsive the system is. Instead, management developed a recording technique which retained the real-time decisions of those responsible and converted them properly to contract language. That was the essence of the unique feature of the Saturn/Apollo program management.

4.6 Capability Retention

A key consideration in the retention of a high-level of technical competence is that, for all practical purposes, Saturn has a fleet of S-IB, S-IVB's and Saturn V, S-IVB's. Five were launched on the IB program and two on the Saturn V Program, which means that [74] some 20-odd stages in inventory have yet to be flown. A technical and supporting staff must be maintained, capable of handling any problems which could come out relative to new mission assignments or anomalies.

4.7 Mission Failure Avoidance

To avoid mission failures, management went into a very comprehensive, in-depth, system, subsystem, and component development. The object was early exposure of weaknesses through repetitive forced exposures. The underlying and most fundamental activities are the ground test program, development tests, qualification tests, formal qualification tests, repeat qualification tests, and reliability verification tests, which are essentially component and subsystem oriented. In the system area are the factory checkout at Huntington Beach, the preacceptance firing checkout at Sacramento, the acceptance firings at Sacramento, and the postacceptance firing checkout at Sacramento. At KSC, prior to launch, there are a very elaborate set of validation, subsystem, and systems tests. Major opportunities for reducing costs on a program such as this (in area of debate) are to reduce or delete acceptance firings or repetitive subsystem and system tests at KSC.

The basic formal qualification activity on this program will soon diminish. A group of reliability verification tests will be eliminated entirely. Qualification tests on selective items will be repeated for some time to provide an opportunity for forced exposure to weaknesses inherent either in the design, in the manufacturing technique, or the production acceptance testing technique. Each one of these areas, although it is an opportunity to reduce costs on the program, also must be weighed as another opportunity of forcing an exposure of something that has escaped through the reliability assurance and quality programs. Large cost returns may be realized by deleting some of these activities. They may or may not be cost effective. The major tradeoff becomes nontechnical and political in nature, very rapidly.

SECTION 5
CONCLUSIONS

However economical their cost performance, however timely their schedule performance, the management systems described herein cannot be said to have justified the customer's investment unless the technical performance of the products assures him that his overall program objective can be achieved. In recognition of this, each management system element has actively participated in technical operations to assure success in technical performance. Unquestionably, the most significant technical operation assuring the customer of the effectiveness of these management systems is the performance of the product in acceptance testing and in-flight operations. Recapitulating, 23 flight vehicles have experienced successful acceptance testing. Thirteen flight vehicles have been successfully launched in either developmental or operational flight test configurations. While these acceptance and flight test programs have not been flawless (almost by definition developmental programs cannot be) the high incidence of success, MDAC-WD believes, bespeaks the effectiveness of the management systems it has devised and operated to control the S-IV and S-IVB programs. Its measure lies in the fact that current planning for the next flight is directed toward manned circumnavigation of the moon.

Document IV-12

Document title: George M. Low, Deputy Administrator, NASA, Memorandum for the Administrator, "NASA as a Technology Agency," May 25, 1971.

Source: James C. Fletcher Papers, Special Collections, Marriott Library, University of Utah, Salt Lake City.

The political consensus on the importance of space that produced the National Aeronautics and Space Act of 1958 and the Apollo program began to dissipate even before the first few Apollo missions were completed. As public support deteriorated, NASA executives found it more difficult to protect not just their programs, but their mission and institution as well. Maintaining NASA's infrastructure depended in part on the identification of marketable missions that the agency could pursue. A 1971 White House review of how government-funded technology could be applied to the nation's problems stimulated NASA's deputy administrator to reassess the agency's future role. His May 25, 1971, memorandum to the administrator printed here describes his position on the subject.

---

[1] SUBJECT: NASA as a Technology Agency

These are some thoughts as to why it might make sense to assign to NASA the government-wide responsibility for the application of technology to national needs.

There are many national problems that require, at least in part, technology solutions and often, at the same time, require a systems management approach. These problems can be found, for example, in the areas of power and energy, pollution, transportation system, health care systems, productivity of services, education, and housing.

NASA has demonstrated a capability to solve difficult technological problems and to apply systems management and know-how in the solution of these problems. In these efforts, NASA has established a working relationship with the aerospace industry that would be difficult for other agencies to duplicate. At the same time, the aerospace community has a surplus of talent that could be applied to these problems, if properly controlled and managed. It, therefore, appears to be logical that NASA should be the agency to undertake the newly needed technological tasks.

There are two alternative ways in which this could be done. First, NASA could provide its services to other agencies; second, NASA could do these things in its own right as part of an expanded NASA mission.

If the first alternative were to be followed, NASA could apply some of its inhouse personnel resources (say, up to ten percent or 3,000) for any direct inhouse efforts and get [2] funding for out-of-house efforts by transfers of funds from other agencies. These other agencies would provide for the funds in their own budgets. This alternative could be done today without any change in existing laws.

Were we to take the second alternative (to do these tasks as part of an expanded NASA mission), then the job would be assigned directly to NASA and budgeted for by NASA. This, however, would require a change in the Space Act. The major disadvantage of the second alternative would be that other agencies would be reluctant to let go of jobs that they now consider to be their own. However, the second alternative, I believe, would be much more likely to succeed.

A word about the kinds of jobs that NASA could undertake. First, I believe that they should be in the general area of applied technology. It is in this area that NASA has the talent and the demonstrated capability. Also, the jobs must be doable, and they must be adequately supported. Finally, they should be tasks that are not now clearly assigned and capably carried out by other agencies.

If it were desired to change NASA's name, I would vote for something like "Aeronautics, Space and Applied Technology Administration."

Should we be asked to undertake a job like this, the first step would be to form a task team, reporting to NASA, to define the charter for the new agency, and to formulate the required government reorganization legislation.

George M. Low
Deputy Administrator

Document IV-13

Document title: George M. Low, Deputy Administrator, NASA, Memorandum to Addressees, "Space Vehicle Cost Improvement," May 16, 1972.

Source: NASA Historical Reference Collection, NASA History Office, NASA Headuqarters, Washington, D.C.

NASA's dwindling budgets and its aspirations for an aggressive space program in early 1970s were incompatible. However, the NASA leadership did not perceive themselves as wholly at the mercy of the political environment in which the agency existed. Although politicians established spending levels, NASA's top administrators, as engineers and scientists, believed it was within the organization's ability to reduce the cost of doing business in space. Doing so would not only allow them to live within externally imposed budgets, but also to pursue aggressive institutional and programmatic goals as well. This memorandum of May 16, 1972, from George M. Low, the NASA Deputy Administrator, to several senior NASA officials emphasizes the importance of reducing costs and creating greater efficiencies inside the agency.

---

[1] SUBJECT: Space Vehicle Cost Improvement

The high cost of doing business in space, coupled with limited and essentially fixed resources available for space exploration, places severe limitations on the amount of productive work that NASA can do, unless we can develop means to lower the unit cost of space operations. It therefore becomes an item of first-order business for each of us to find ways to drastically reduce the costs of all elements of space missions.

A fundamental reason for high costs has been the fact that most space systems are designed with great sophistication so as to operate acceptably with low allowable weight. However, as the cost of space transportation is decreased (especially with the shuttle, but even with some existing launch vehicles) a great many designs should be optimized for high reliability and low cost—in general with weight being a secondary consideration.

Another reason for high cost has been that most systems are individually tailored for their mission, used once or twice, and then never used again. Thus the economies of producing a number of like systems are never attained. Now that we have acquired a considerable background of experience as to the kinds and needs of space missions, we can better plan for multiple-use types of equipment.

I am convinced that major cost improvements can be realized, and that this matter should become a first order item of business for all of us. A basic approach to lowering the costs of space systems should include the following:

[2] 1. A detailed understanding of exactly where we spend our money. We need to identify those areas where a substantial cost improvement would be worthwhile in that it would have a major impact on the cost of the end product. In other words, we need to define the things with the greatest potential pay-off for cost improvement.

2. The determination of range of requirements (for the systems or subsystems with the highest potential payoff) for our spacecraft of the future. (So that we can develop a few "standard" systems, instead of individually tailored systems for each requirement.)

3. The development of "standard" systems or subsystems, designed for low cost and high reliability. (We need a catalog, ultimately, of available preferred parts.)

4. A method for assuring that as a rule only the "standard" systems are used.

I consider this effort of such high importance and priority that I am prepared to devote whatever resources are required, both in-house and on contract, to achieve significant results.

To begin with, I am hereby establishing a task force, chaired by Del Tischler, to carry out steps l and 2, above, and to develop a plan, goals and objectives for steps 3 and 4. I want each of the addressees to provide the necessary support to the task force, especially in terms of experienced people.

My plan is to have task force members named within one week, and to be in business in two weeks. Thereafter I intend to meet with the task force on a biweekly basis, and to have its final report in six months.

The task force is authorized to place requirements on the various line organizations to accomplish its objectives.

George M. Low

Document IV-14

Document title: E.S. Groo, Associate Administrator for Center Operations, NASA, to Center Directors, "Catalog of NASA Center Roles," April 16, 1976.

Source: NASA Historical Reference Collection, NASA History Office, NASA Headuqarters, Washington, D.C.

The elaborate institutional machinery inherited by NASA from the National Advisory Committee for Aeronautics (NACA), supplemented by that developed to carry out Apollo, could not be easily disassembled, nor demobilized after the completion of the Apollo program, given the interlocking interests it had created among NASA's installations, contractors, and geographic regions and their representatives in Washington. By designating "roles and missions" for each of its field centers, NASA Headquarters attempted to avoid duplication, reduce intercenter rivalry, and assure each installation adequate work to utilize its special capabilities and facilities.

---

[1] SUBJECT: Catalog of NASA Center Roles

Enclosed is a copy of the catalog of NASA Center Roles, dated April 1976, developed on the basis of decisions reached during the Institutional Assessment conducted earlier this year. The primary purposes of the catalog are to describe in a consistent way the programmatic responsibilities of the Centers and to serve as a guide in the assignment of work to the Centers.

The catalog has been reviewed by the Program Associate Administrators and reflects the changes they have proposed. As we discussed at the last Center Directors' meeting, the document is now forwarded to you for your comments. Dr. Naugle and I will entertain specific proposals which would further clarify the document within the context of the Institutional Assessment decisions.

While it is possible that refinements to the catalog should be made based upon your suggestions, we should, in the meantime, assume this document to be the definitive statement of the roles and missions of the Centers on which new program assignments will be based. There will, or course, be changes from time to time in the catalog as roles and missions evolve and all changes will be issued in writing and signed jointly by Dr. Naugle and me and, where appropriate, by Mr. Yardley.

We are now developing a procedure for the review and approval of major work assignments to the Centers. This [2] procedure, which we expect to issue in about one month, will recognize the catalog as the baseline document in the assignment of work.

E. S. Groo
Enclosure...

[1 of Enclosure]
CENTER ROLES
Introduction and Rationale

Assignment of specific responsibilities to NASA Field Centers is one of the keystones in the process by which the Nation's goals in Aeronautics and Space are met. Field Center responsibilities relate, in their broadest context, to these major goals. These goals are:

• gaining new fundamental knowledge about the earth, solar system and universe through maintaining a strong program in *space science and exploration*;
• bringing the benefits of space and space technology to bear for the direct and immediate benefit of man on earth through cooperation in *applications* oriented activities with a wide range of users and non-space mission oriented agencies;
• facilitating improvements in aircraft design and operations through the provision of an on-going *aeronautics research and technology* base;
• maintaining a strong base of *space research and technology* as a national resource which can serve to evolve and/or support new initiatives in space exploration or applications; and
• making space more accessible to both domestic and foreign users through development and operation of economical *space transportation* and the operation of efficient *tracking and data acquisition* systems.

These goals translate into a set of broad program areas to which the roles in this document are related.

Within these broad program areas, a Center is assigned and carries out both *principal* roles and *supporting* roles.

Principal: Roles of fundamental importance in supporting the Agency's overall goals. They serve as a basis for deploying resources to Centers over the longer term. They also represent areas of Center excellence and expertise that [2] is clearly discernible within NASA and recognized as a national capability.

Supporting: Roles of more limited scope or tentative nature supporting the Agency's overall goals. Such roles can also support principal roles for which other Centers or government agencies generally have the lead. They may also support a Center's own principal role or roles, or are discrete roles assigned to a Center because of a specific expertise a Center can provide in a particular discipline.*

Each NASA Field Center represents particular areas of special capability which, when considered on an Agency-wide basis, form the core of our national capability in aeronautics and space. The special capabilities highlighted herein consist of areas of technical excellence and facilities of superior merit – technical facilities which may be of unique or almost unique character and constitute, in themselves, a national resource. Consideration of such special capabilities is integral to the process of assigning Field Center responsibilities within the Agency's overall program.

Summarized on the following pages are highlights of Center capabilities and statements of role responsibilities current as of April 1976. Roles are grouped according to overall emergency goals by broad program areas – so that "Applications," for instance, has a broader context than just Office of Applications programs and includes Technology Utilization and Energy programs as well. The same can be said of "Space Research and Technology" vs. the Office of Aeronautics and Space Technology programs.

For supporting roles, any other Centers having related responsibilities – either principal or supporting – are noted in parentheses following each supporting role description. There are a few cases where a Center shares principal responsibility with another Center

---

* Many of these roles were previously identified in the institutional assessment as "limited roles." Others may have been identified as "broad roles" but are now judged to be supportive to principal roles assigned to the Center.

or acts as an alternate to another Center in the development of space hardware. Such roles are so identified where appropriate.

[3] AMES RESEARCH CENTER
Special Capabilities

- Areas of Technical Excellence
  - Biology
  - Human factors and man-machine interactions
  - Fluid dynamics and heat transfer
  - Aerodynamics and flight dynamics
  - Flight stability and control
  - Technical project management

- Facilities of Superior Merit
  - 40 X 80 ft. Wind Tunnel
  - Flight Simulator for Advanced Aircraft
  - Illiac IV
  - C-141 Airborne Infrared Observatory
  - High-Enthalpy Arc Jets
  - Unitary Wind Tunnel Complex
  - 3.5 ft. Hypersonic Wind Tunnel
  - Biological Containment Facility
  - Vertical Gun

[4] AMES RESEARCH CENTER

PROGRAM AREAS:
Aeronautics
Space Science and Exploration
Space Research and Technology
Applications
Space Transportation

Principal and Supporting Roles

- Aeronautics
- Principal

    • **Short-haul aircraft technology** - developing a technology base for facilitating incorporation of short-haul aircraft into overall air transportation systems.
    • **Helicopter technology*** - developing a technology base for improving efficiency and flexibility for both civil and military use.
    • **Computational fluid mechanics** - furthering the state-of-the-art through the definition of new systems, both hardware and software, for application to aeronautical and other related areas such as weather and climate, etc.
    • **Fluid simulation** - improving the state-of-the-art to permit more effective use of simulators in aircraft design and validation of flight simulation.
    • **Human-vehicle interactions** - furthering the state-off-the-art through the study of man-machine and other human factor interactions and considerations involved in aircraft operations.

---

* Under study

• **Fundamental aerodynamics** - advancing the general state-of-the-art, both theoretical and experimental. (Shared principal responsibility with *LaRC;* supporting responsibilities: DFRC, JPL)

• **Fire resistant materials** - developing a technology base for internal application in aircraft. (Supporting responsibilities: JPL, JSC)

[5] - Supporting

• **Aviation system studies** - conducted to help define technical and system requirements. (Shared supporting responsibility with LaRC)

• **Aircraft structures** - improving predictive capability for structural lifetimes in degrading chemical environments, unsteady aerodynamic loads and aeroelasticity, and high temperature fuel tank sealants. (Principal responsibility: *LaRC;* supporting responsibility: DFRC)

• **Acoustics noise reduction** - using ARC unique fullscale low speed wind tunnel to study airframe noise and forward velocity effects. (Principal responsibility: *LaRC;* supporting responsibilities: DFRC, LeRC)

• **Aviation safety** - contributing to advances through joint efforts with the FAA and other appropriate agencies. Advanced tire materials, and wake vortex studies. (Supporting responsibilities: DFRC, JPL, LaRC, LeRC, MSFC, WFC)

• **Wind tunnel support** - provision of facility support to industry and other government agencies. (Other Centers having unique or outstanding facilities may provide similar support.)

• **Military support** - provision of military aviation systems technology support. (Other Centers providing military aeronautics support: DFRC, LaRC, LeRC)

• **General aviation aircraft technology** - developing a technology base for improving agricultural aircraft. (Principal responsibility: *LaRC*)

• Space Science and Exploration
- Principal

• **Extraterrestrial life detection** - developing and applying the analytical basis for life detection in space, including experiment design and management.

[6] • **Biological experiments** - developing and implementing experiments for determining effects of space flight environment on (non-human) living organisms.

• **Level IV life sciences integration** - developing, integrating and operating space flight hardware to conduct in-flight biomedical experiments and experiments on non-human living organisms.

• **Airborne research operations** - operating instrumented jet aircraft for the purpose of conducting airborne science experiments.

• **Planetary probes** - developing thermoprotection systems required for planetary atmosphere entry probes and managing probe development.

• **Pioneer** - completing the currently approved series, including associated flight operations. Phase out to be concluded after Pioneer Venus.[*]

- Supporting

• **Planetary science analysis techniques** - developing and applying techniques for analysis of planetary atmosphere and mass. To be completed in early 1980. (Principal responsibility: *JPL*; supporting responsibility: GSFC)

---

[*] Future pioneer spacecraft will be managed by JPL

- **Astronomical observation techniques** - focus on airborne science and the development of IR techniques and supporting systems for use in Spacelab payloads. (Principal responsibility: *GSFC;* alternate responsibility: JPL)
- **Upper atmospheric research** - providing aircraft based sampling and contributing to model development. (Principal responsibilities: *JPL, GSFC;* supporting responsibilities: LaRC, LeRC)
- **Spacelab bioresearch** - supporting development of Spacelab life science research capability through common operating research equipment development. (Principal responsibility: *JSC*)

- Space Research and Technology
  - Principal

[7] • **Planetary entry technology*** - advancing thermal heat protection technology for planetary entry. (Supporting responsibility: LaRC)
- **Biomedical support systems** - developing advanced technology for development of long duration life support systems.

  - Supporting

- **Fundamental research** - focus on quantum and surface states in solids. (Supporting responsibilities: JPL, LaRC, LeRC)
- **Space vehicle structures and materials technology** - focus on prediction of dynamic loading parameters related to space vehicles. (Principal responsibilities: *MSFC, LaRC;* supporting responsibilities: GSFC, JPL)
- **Space energy processes and systems technology** -furthering state-of-the-art in key areas such as heat pipes for thermal control and high power gas dynamic laser technology. (Principal responsibility: *LeRC;* supporting responsibilities: GSFC, JPL)
- **Technology experiments in space** - definition and development of experiments in areas consistent with ARC's other Space Research and Technology roles. (Principal responsibilities: *JSC, LaRC;* supporting responsibilities: DFRC, GSFC, JPL, LeRC, MSFC)
- **Shuttle technology** - Shuttle vehicle technology development and ground facility testing in the areas of thermal protection systems, dynamics and aeroelasticity. (Shared supporting activity with LaRC)
- **Space technology studies** - conducted to help define technology and systems requirements. (Supporting responsibilities: GSFC, JPL)
- **Medical research** - utilizing non-human specimens to derive information and develop countermeasures needed to solve space medicine problems. (Principal responsibility: *JSC*)

[8] • Applications
  - Principal

- **Airborne instrumentation research** - providing aircraft platform support for applications oriented sensor research and development.
- **Technology transfer**
  - **Technology utilization** - conducting projects to establish applicability of NASA technology in health care, participate in technology transfer to industry, identify and report new technology, and document results of secondary applications of NASA technology.
  - **Specialized applications tasks** - draw upon unique Center capabilities inclued under other Center roles to advance the application of space related techniques. Current emphasis is on space processing and video compression techniques.

* Under study

- **Regional applications transfer** - current effort is a joint demonstration activity with USGS/EROS, the Pacific Northwest Regional Commission, and State agencies with Idaho, Washington and Oregon.

- Supporting

• **Energy technology** - conducting energy related materials investigations. (Principal responsibility: *LeRC*; supporting responsibilities: JPL, JSC, MSFC)

• Space Transportation
- Supporting

• **Passenger selection criteria** - establishment of medical criteria for non-crew passenger selection.

[9] DRYDEN FLIGHT RESEARCH CENTER
Special Capabilities

- Areas of Technical Excellence
  - Flight research instrumentation
  - Flight dynamics and controls
  - Flight research operations

- Facilities of Superior Merit
  - High Temperature Loads Facility
  - 600-Mile Instrumented Range
  - Remote Piloted Research Facility
  - Airborne Launch Aircraft
  - General Purpose Airborne Simulator

[10] DRYDEN FLIGHT RESEARCH CENTER

PROGRAM AREAS:
Aeronautics
Space Transportation
Applications
Space Research and Technology
Tracking and Data Acquisition*

Principal and Supporting Roles

• Aeronautics
- Principal

• **Aeronautical flight research** - providing a broad-based flight research and test capability including tracking and data acquisition for the Agency in support of aeronautics and other programs as required. (This principal role represents the composite of the supporting roles given below.)

• **Remotely piloted vehicle research** - development of research aircraft, and management/operation of flight experiments.

---

* Included within other program areas as indicated

- Supporting

• **Fundamental aerodynamics** - contributing to state- of-the-art advancement through flight testing of aerodynamics concepts. (Principal responsibilities: *ARC, LaRC*; supporting responsibility: JPL)
• **Aircraft structures** - contributing to technology base with focus on flight loads measurements. (Principal responsibility: *LaRC*; supporting responsibility: ARC)
• **Acoustics and aircraft noise reduction** - focus on flight measurements of airframe noise. (Principal responsibility: *LaRC*; supporting responsibilities: ARC, LeRC)
• **Short-haul aircraft technology** - support of *ARC* role through participation in flight testing of short-haul aircraft and systems.
• **Long-haul aircraft systems** - support of *LaRC* role through flight testing of long-haul aircraft and [11] systems, with focus on digital fly-by-wire experiments and active controls aircraft flight experiments.
• **Aviation safety** - contributing to advances through flight testing of devices/systems for wake vortex marking and minimization; definition of atmospheric conditions for supersonic acceleration and cruise. (Supporting responsibilities: ARC, JPL, LaRC, LeRC, MSFC, WFC)
• **Military support** - provide flight research support to the DOD. (Other Centers providing military aeronautics support: ARC, LaRC, LeRC)

• Space Transportation
- Supporting

• **Shuttle orbiter development** - conducting approach and landing tests in support of JSC. Provide landing and recovery capability during OFT and contingency recovery capability after OFT. (Principal responsibility: *JSC*)

• Applications
- Supporting

• **Technology transfer**
- **Technology transfer** - identify and report new technology, participate in technology transfer to public service and private organizations, and document results of secondary applications of NASA technology.

• Space Research and Technology
- Supporting

• **Space vehicle configurations technology** - analysis and study of the effect of operational considerations on the design and test program of manned research vehicles. (Principal responsibility: *LaRC*; supporting responsibility: MSFC)
• **Technology experiments in space** - definition and development of experiments consistent with DFRC's [12] other Space Research and Technology roles. (Principal responsibilities: *JSC, LaRC*; supporting responsibilities: ARC, GSFC, JPL, LeRC, MSFC)

[13] GODDARD SPACE FLIGHT CENTER
Special Capabilities

• Areas of Technical Excellence
  • Space and earth sciences
  • Data systems and analysis
  • Sensors and instrument systems
  • Flight systems - automated
  • Tracking and data acquisition and communications
  • Technical project management

- Space flight operations
- Mission operations control and information processing

- Facilities of Superior Merit
  - Spaceflight Tracking and Data Network
  - Spacecraft Magnetic Test Facility
  - Optical Systems Laboratories
  - Optical Tracking and Communications Facility
  - Thermal-Vacuum Simulation and Test Facilities
  - Dynamic Test Chamber
  - Remote Sensing Information Processing Facilities
  - Operations Communications Network
  - Mission Operations Control Centers

[14] GODDARD SPACE FLIGHT CENTER

PROGRAM AREAS:
Space Science and Exploration
Space Research and Technology
Applications
Tracking and Data Acquisition
Space Transportation

Principal and Supporting Roles

- Space Science and Exploration
- Principal

• **Earth orbit spacecraft development** - for science, including spacecraft propulsion systems. Emphasis on automated, standard spacecraft system and free flyers, including experiment integration.

• **Earth orbit flight operations** - planning and conducting flight operations for earth orbit science spacecraft.

• **Physics and astronomy** - developing the technical discipline base, developing and implementing flight experiments. (Includes planetary astronomy and the transfer of AMPS.)

• **Upper atmospheric research** - developing the technical discipline base, developing and implementing flight experiments.

• **Sounding rocket development, procurement and operations** - developing and procuring sounding rockets, and carrying out all phases of operations from mission/flight planning to landing and recovery. Payload carrier development, development and management of experiments, experiment management support to other institutions, launch operations and tracking and data acquisition are included. (Most GSFC sounding rocket activities involve the higher performance, more complex vehicle support systems. Most activities involving lower performance vehicle systems are assigned to WFC.)

[15] • **Spacelab payloads** - development, integration, and data processing for Spacelab payloads in astrophysics, solar terrestrial physics, and astronomy.

- Supporting

• **Planetary science** - developing and applying techniques for the analysis of planetary atmospheres. (Principal responsibility: *JPL*; supporting responsibility: ARC)

• **Lunar science** - phase out by FY 79. Continuation of unique computer programs for processing lunar and planetary remote sensing data currently being used for lunar and Venera data. Those unique computer programs that cannot be economically transferred will continue.

- Space Research and Technology
- Principal

• **Information systems technology** - developing and maintaining technical discipline base. (Supporting responsibilities: JPL, JSC, LaRC)

- Supporting

• **Space vehicle structures and materials technology** - contributing to technology base with focus on reducing cost of structural evaluation and reliability demonstration for space flight hardware. (Principal responsibilities: *LaRC, MSFC*; supporting responsibilities: ARC, JPL)
• **Guidance and control technology** - contributing to technology base with focus on magnetic suspension systems. (Principal responsibility: *JPL*; supporting responsibilities: LaRC, MSFC)
• **Space energy processes and systems technology** -contributing to space technology base - space power system component test and evaluation. (Principal responsibility: *LeRC*; supporting responsibilities: ARC, JPL)
• **Sensor and data acquisition technology** - focus on CCD astronomical sensor work for application in [16] space astronomy. (Principal responsibility: *LaRC*; supporting responsibilities: JPL, MSFC)
• **Space technology studies** - focus on earth applications spacecraft technology requirements. (Supporting responsibilities: ARC, JPL)
• **Technology experiments in space** - definition and development of experiments in areas consistent with GSFC's other Space Research and Technology roles. (Principal responsibilities: *JSC, LaRC*; supporting responsibilities: ARC, DFRC, JPL, LeRC, MSFC)

- Applications
- Principal

• **Earth orbital spacecraft development** - for applications, including spacecraft propulsion systems. Emphasis on automated, standard spacecraft system and free flyers, including experiment integration.
• **Earth orbit flight operations** - planning and conducting flight operations for earth orbit applications spacecraft.
• **Technology transfer**

- **Applications system verifications test** - acquire, process, and disseminate LANDSAT coverage to JSC.
- **Technology utilization** - conducting projects to establish applicability of NASA technology in public service activities, participate in technology transfer to industry, identify and report new technology, and document results of secondary applications of NASA technology.

• **Applications R&D** - developing the technical discipline base, developing and implementing experiments in the following Applications disciplines:
  - **weather and climate**
  - **earth and ocean dynamics** (JPL shares principal responsibility)
  [17] - **communications**

- Supporting

• Contributing to the discipline base, developing and implementing experiments in:
  - **environmental monitoring** (Principal responsibility: *LaRC*; supporting responsibility: JPL)
  - **earth resources** (Principal responsibility: *JSC*)

- Tracking and Data Acquisition
- Principal

  • **Tracking and data acquisition support operations** -planning and conducting support for earth orbit spacecraft. Includes flight control, tracking, data acquisition, communications, and information processing. (Tracking and data acquisition responsibilities include orbital phase of all mission types, such as manned, deep space, etc.)
  • **Tracking and data acquisition systems** - planning, development, and implementation of network, data processing, communications, and mission control systems and facilities for earth orbit spacecraft.

- Space Transportation
- Principal

  • **Launch vehicle procurement** - for science/applications oriented missions. Current focus on sounding rockets and Delta (includes procurement for Delta).

- Supporting

  • **Flight operations** - network planning and implementation support for Shuttle including ALT and OFT.

[18] **JET PROPULSION LABORATORY**
Special Capabilities

- Areas of Technical Excellence
  - Space sciences
  - Space flight mechanics and flight systems
  - Space guidance and control
  - Tracking and data acquisition
  - Sensors and instrument systems
  - Technical project management
  - Deep space flight operations

- Facilities of Superior Merit
  - Space Flight Operations Facility
  - Deep Space Network
  - Rocket Propulsion Test Facilities
  - Solid Propellant Processing Laboratory
  - Table Mountain Solar Test Facilities
  - Electric Propulsion Laboratories
  - Radio Telescope Facility

[19] **JET PROPULSION LABORATORY**

PROGRAM AREAS:
Space Science and Exploration Applications
Space Research and Technology Aeronautics
Tracking and Data Acquisition

Principal and Supporting Roles

- Space Science and Exploration
- Principal

- **Planetary spacecraft development** - development of automated spacecraft for deep space exploration. Includes experiment integration and all aspects of spacecraft systems technology, with special emphasis on guidance and control, space power systems and the procurement of spacecraft propulsion systems.
- **Space flight operations** - conduct of flight operations for *deep space* missions involving automated spacecraft. Includes mission/flight planning, and flight command and control. (ARC retains flight control of current Pioneer series.)
- **Lunar/planetary science** - development of discipline base in lunar and planetary sciences, including developing and applying techniques for analysis of planetary characteristics (except *geosciences* for which *JSC* has principal responsibility, along with returned sample handling and analysis).
- **Upper atmospheric research** - developing and testing advanced instrumentation for atmospheric constituent analysis; conducting diffusion studies and contributing to model development. (Principal responsibility: *GSFC*; supporting responsibilities: ARC, LaRC, LeRC)
- **Science/Applications spacecraft development** - serves as alternate center to GSFC for earth orbital spacecraft development. Current focus is SEASAT. (Principal responsibility: *GSFC*)

[20] - Supporting

- **Space physics** - contributing to discipline base with focus on particles and fields; development of space physics experiments for planetary missions. Cometary physics work will continue under "planetary science" designation. (Principal responsibility: *GSFC*)
- **Space astronomy** - contributing to discipline base with focus on ground based radio astronomy, relatively and celestial mechanics, IR astronomy, laboratory and high-energy astrophysics. (Principal responsibility: *GSFC*; supporting responsibility: ARC)
- **Lunar/planetary geoscience** - conducting earth based observations, theoretical studies, analog studies, and developing science experiment concepts. (Principal responsibility: *JSC*)

- Applications
- Principal

- **Technology transfer**
  - **Specialized applications tasks** - utilizing the unique capability associated with other roles to meet discrete needs. Current emphasis is on communications and space processing.
  - **Technology utilization** - conducting projects to establish applicability of NASA technology in biomedicine and other fields, participate in technology transfer to industry, identify and report new technology, and document results of secondary applications of NASA technology.
- **Science/Applications spacecraft development** - serves as alternate center to GSFC for earth orbital spacecraft development. Current focus is SEASAT. (Principal responsibility: *GSFC*)
- **Earth and ocean dynamics** - focus on contributing to discipline base, data analysis and investigation management, and spacecraft payload/experiment development. Current emphasis on ocean sensor experiments related to SEASAT.

[21] - Supporting

- **Weather and climate** - focus on sensor development for solar radiation measurement, definition of weather and climate-related experiments. (Principal responsibility: *GSFC*; supporting responsibility: LaRC)

• **Environmental monitoring** - focus on development of advanced instrumentation. (Principal responsibility: *LaRC*; supporting responsibility: GSFC)
• **Energy technology** - conducting energy R&D, primarily on a reimbursable basis, with principal focus on photovóltaics and advanced coal energy extraction technology. (Principal responsibility: *LeRC*; supporting responsibilities: ARC, JSC, MSFC)

• Space Research and Technology
- Principal

• **Teleoperator technology** - focus on teleoperator/robot technology and communication delayed control techniques for exploration.
• **Guidance and control technology** - developing and maintaining a broad technology base in guidance and control systems. (Supporting responsibilities: GSFC, LaRC, MSFC)

- Supporting

• **Sensor and data acquisition technology** - focus on planetary imaging, failure modeling and prediction. (Principal responsibility: *LaRC*; supporting responsibilities: MSFC, GSFC)
• **Information systems technology** - focus on planetary data processing and transfer systems. (Principal responsibility: *GSFC*; supporting responsibilities: JSC, LaRC)
• **Space vehicle structures and materials technology** - focus on planetary expandable structures and dynamic response. (Principal responsibilities: *LaRC, MSFC*; supporting responsibilities: ARC, GSFC)
[22] • **Space propulsion systems technology** - focus on planetary spacecraft propulsion and low-cost solids. (Principal responsibility: *LeRC*; supporting responsibility: MSFC)
• **Energy processes and systems technology** - focus on long life, high energy density power systems for planetary spacecraft. (Principal responsibility: *LeRC*; supporting responsibilities: ARC, GSFC)
• **Space technology studies** - focus on planetary spacecraft technology requirements. (Supporting responsibilities: ARC, GSFC)
• **Fundamental research** - focus on photon-matter interactions and energy transformation research. (Supporting responsibilities: ARC, LaRC, LeRC)
• **Technology experiments in space** - definition and development of experiments in areas consistent with JPL's other Space Research and Technology roles. (Principal responsibilities: *JSC, LaRC*; supporting responsibilities: ARC, DFRC, GSFC, LeRC, MSFC)

• Aeronautics
- Supporting

• **Fire resistant materials** - focus on fire resistant polymers and anti-misting fuels. (Principal responsibility: *ARC*; supporting responsibility: JSC)
• **Propulsion systems** - focus on hydrogen enrichment of piston engine fuels and reducing oxides of nitrogen via unconventional combustor design. (Principal responsibility: *LeRC*; supporting responsibilities: DFRC, LaRC)
• **Fundamental aerodynamics** - focus on fundamental fluid mechanics, non-linear wave interactions. (Principal responsibilities: *ARC, LaRC*; supporting responsibility: DFRC)
• **Aviation safety** - focus on wake vortex marking techniques. (Supporting responsibilities: ARC, DFRC, LaRC, LeRC, MSFC, WFC)

[23] • Tracking and Data Acquisition
- Principal

• **Tracking and data acquisition support operations** - planning and conducting tracking, command, and data acquisition support for planetary spacecraft and radio science.

• **Tracking and data acquisition systems** -planning, development, and implementation of network systems and facilities for planetary spacecraft and radio science.

[24] JOHNSON SPACE CENTER
Special Capabilities

- Areas of Technical Excellence
  - Biotechnology and space medicine
  - Extraterrestrial materials analysis
  - Space flight mechanics - manned vehicles
  - Data systems and analysis
  - Sensors and instrument systems
  - Space flight systems - manned vehicles
  - Flight crew training and mission simulation
  - Mission operations - manned vehicles
  - Technical project management

- Facilities of Superior Merit
  - Space Environment Simulation Laboratories
  - Docking Test Facility

- Simulation and Training Facility
  - Mockup and Integration Laboratory
  - Mapping Sciences Laboratory
  - Geology and Geochemistry Laboratory
  - Test and Evaluation Laboratories for All Major Spacecraft Systems and Subsystems
  - Earth Resources Laboratory (Slidell, LA)
  - Mission Control Center
  - Lunar Curatorial Facility

[25] JOHNSON SPACE CENTER

PROGRAM AREAS:
Space Transportation
Space Science and Exploration Applications
Aeronautics
Space Research and Technology

Principal and Supporting Roles

- Space Transportation
- Principal

  • **Manned vehicles** - development of manned space vehicles and
  - **Shuttle** - development of the orbiter and lead Center for management of the Shuttle system.
  - **Advanced missions** - focus is on space station, advanced transportation systems and construction of a satellite space power station-definition activities (MSFC and JSC have co-equal roles through definition, development responsibilities are not yet assigned).
  - **Environmental and crew support systems** -develop and demonstrate EC/LSS and EVA systems suitable for the space transportation systems and other advanced needs.
  - **Advanced developments** - development of prototypes, long lead time systems, and new procedures and software for advanced systems.

• **Operations** - operational planning, crew selection and training, space transportation system flight control, experiment/payload flight control for Spacelab and STS utilization planning/payload accommodation studies.
  • **STS sustaining engineering** - providing sustaining engineering and logistical support for STS hardware. Includes Shuttle configuration management, Shuttle sustaining engineering and orbiter operational procurement. (To be phased over to KSC at a future point, yet to be identified.)

[26] • Space Science and Exploration
  - Principal

   • **Lunar and planetary geosciences** - developing and maintaining the technical discipline base for lunar/planetary geosciences and extraterrestrial sample handling techniques.
   • **Space medicine** - defining and developing in-flight biomedical experiments to assess human physiological response to space flight environments. (Supporting responsibility: ARC)
   • **Spacelab bioresearch** - development of Spacelab life science research capability through Common Operating Research Equipment development. (Supporting responsibility: ARC)

  - Supporting

   • **Physics and astronomy** - phase out by FY 79 of all science activity including payload definition activity.
   • **Upper atmospheric research** - phase out by FY 79 of modeling and measurement activities.

• Applications
  - Principal

   • **Earth resources** - provide a discipline base for earth resources applications including airborne instrumentation research, data interpretative techniques, and space-based flight sensors.
   • **Technology transfer**
    - **Application systems verification tests** -conducting interagency operational tests to demonstrate automated natural resources inventory systems. Current emphasis includes the Large Area Crop Inventory Experiment and the Louisiana Environmental Information System.
    - **Specialized applications tasks** - drawing on unique capabilities associated with other roles to meet discrete needs. Current emphasis involves life sciences space processing.
   [27] - **Technology utilization** - conducting projects to establish applicability of NASA technology in health care, participate in technology transfer to industry, identify and report new technology, and document results of secondary applications of NASA technology.

  - Supporting

   • **Energy technology** - complete assigned energy efficient utility systems program. (Principal responsibility: *LeRC;* supporting responsibilities: ARC, JPL, MSFC)

• Aeronautics
  - Supporting

• **Fire resistant materials** - performing evaluation tests of fire resistant materials for use in aircraft. (Principal responsibility: *ARC*; supporting responsibility: JPL)

• Space Research and Technology
- Principal

• **Technology experiments in space** - management of Orbiter Experiments program. Definition and development of experiments in areas consistent with JSC's other Space Research and Technology roles. (Principal responsibilities: *JSC, LaRC*; supporting responsibilities: ARC, DFRC, GSFC, JPL, LeRC, MSFC)
• **Medical research** - establishing human baseline data and developing countermeasures to solve space medicine problems.
• **Food systems technology** - developing nutritional requirements and food processing systems in support of human space flight.

[28] - Supporting

• **Information systems technology** - contributing to technical discipline base, with focus on advanced software for manned spacecraft data systems. (Principal responsibility: *GSFC*; supporting responsibilities: JPL, LaRC)

[29] **KENNEDY SPACE CENTER**
Special Capabilities

- Areas of Technical Excellence
  - Flight systems testing
  - Facility and equipment operations
  - Launch operations
  - Technical management

- Facilities of Superior Merit
  - Launch Complexes
  - Operations and Checkout Facilities
  - Central Instrumentation Facility
  - Fluid Test Area
  - Landing Strip for Shuttle

[30] **KENNEDY SPACE CENTER**

PROGRAM AREAS:
Space Transportation
Applications

Principal and Supporting Roles

• Space Transportation
- Principal

• **Launch systems development** - provide launch systems support for all Agency flight programs.
• **Unmanned launch operations** - includes launch preparations and checkout for current inventory of launch vehicles.
• **STS ground operations** - includes launch operations, STS turnaround, Levels I and II integration, Spacelab Level III integration, integrated logistics and transportation and post-landing operations, and flight line medical and biomedical support.

• **STS sustaining engineering** - includes configuration management, operational hardware accommodations and mods. (This responsibility will be phased over from JSC at a future point, yet to be identified.)

• Applications
- Principal

• **Technology transfer**
- **Regional applications transfer** - remote sensing applications involving studies of thermal pollution and methods of sensing crop freeze exposure over large areas.
- **Specialized applications tasks** - support to NSTL and studies of changes in requirements, procedures and techniques for processing space applications type payloads for Spacelab.
- **Technology utilization** - conducting projects to establish applicability of NASA technology in public safety and other fields, participate [31] in technology transfer to industry, identify and report new technology, and document results of secondary applications of NASA technology.

- Supporting

• **Bicentennial exhibition** - support major science and technology exhibition of national scope in conjunction with Bicentennial celebration.

[32] **LANGLEY RESEARCH CENTER**
Special Capabilities

- Areas of Technical Excellence
  - Structures and aerostructural dynamics
  - Flight mechanics and configurations
  - Flight stability, control, and performance
  - Sensors and instrument systems
  - Avionics
  - Flight acoustics
  - Aerothermodynamics
  - Technical project management

- Facilities of Superior Merit
  - 8 ft. Transonic Pressure Tunnel
  - Transonic Dynamics Tunnel
  - 16 ft. Transonic Tunnel
  - V/STOL Tunnel
  - Unitary Wind Tunnel
  - 8 ft. High Temperatures Structures Tunnel
  - Fatigue Laboratory
  - Aircraft Noise Reduction Facility
  - Scramjet Test Facility
  - Real Gas/Viscous Effects Entry Simulation Facilities
  - Differential Maneuvering Simulator

[33] **LANGLEY RESEARCH CENTER**

PROGRAM AREAS:
Aeronautics
Space Research and Technology
Applications
Space Transportation
Space Science and Exploration

Principal and Supporting Roles

- Aeronautics
  - Principal

    - **Long-haul aircraft technology** - developing a technology base for improving long-haul aircraft as cost effective, safe and environmentally compatible transportation modalities.
    - **General aviation aircraft technology** - developing and maintaining an engineering technology base related to improving general aviation aircraft.
    - **Acoustics and noise reduction** - conducting research and development of technology related to reducing aircraft noise.
    - **Aircraft structures** - development of technology base for facilitating structural advances.
    - **Helicopter technology***- developing a technology base for improving efficiency and flexibility for both civil and military use.
    - **Fundamental aerodynamics** - advancing the general state-of-the-art, both theoretical and experimental.

  - Supporting

    - **Avionics technology** - developing a technology base related to improving avionics.
    - **Computational fluid mechanics** - contributing to technology base, with emphasis on the prediction of aerodynamic characteristics of 3-D aerodynamic configurations. (Principal responsibility: *ARC*)
    - [34] • **Propulsion systems** - contributing to technology base of air breathing propulsion systems by advancing the state-of-the-art hypersonic propulsion. (Principal responsibility: *LeRC*; supporting responsibilities: DFRC, JPL)
    - **Remotely piloted vehicle research** - contributing to the technology base of highly maneuvering aircraft through analytical studies, experimental studies in wind tunnels and test evaluations on the differential maneuvering simulator. (Principal responsibility: *DFRC*)
    - **Aviation safety** - contributing to safety advances with focus on wake vortex minimization. (Supporting responsibilities: ARC, DFRC, JPL, LeRC, MSFC, WFC)
    - **Aviation systems studies** - with focus on foreign technology assessment. (Shared support responsibility with ARC)
    - **Military support** - supporting military aviation advances through work for DOD. (ARC, DFRC and LeRC also provide military aeronautics support.)
    - **Wind tunnel support** - provision of wind tunnel support to industry and other government agencies. (Other Centers having unique or outstanding facilities may provide similar support.)

- Space Research and Technology
  - Principal

    - **Space vehicle structures and materials** - developing technology base to facilitate advances.
    - **Space vehicle configurations technology** - developing technology base related to advanced configuration including advanced space transportation concepts.
    - **Technology experiments in space** - development and management of the Long Duration Exposure Facility and Advanced Technology Laboratory. Definition and development of experiments in areas consistent with LaRC's other Space Research and Technology roles.

* Under study

[35] • **Sensor and data acquisition technology** - contributing to the technology base of sensors and devices. (Supporting responsibilities: GSFC, JPL, MSFC)

- Supporting

• **Guidance and control technology** - contributing to technology base, with focus on multi-purpose stabilization systems. (Principal responsibility: *JPL*; supporting responsibilities: GSFC, MSFC)
• **Information systems technology** - contributing to technology base, with focus on solid state data storage. (Principal responsibility: *GSFC*; supporting responsibilities: JPL, JSC)
• **Fundamental research** - focus on high density plasma phenomena. (Supporting responsibilities: ARC, JPL, LeRC)
• **Planetary entry technology*** - provide planetary and earth entry aerothermodynamics experimental and analytical data. (Principal responsibility: *ARC*)
• **Shuttle technology** - Shuttle vehicle technology development and ground facility testing in the areas of thermal protection systems, dynamics and aeroelasticity. (Shared supporting role with ARC)

• Applications
- Principal

• **Environmental quality monitoring technology** - developing improved techniques for environmental monitoring. Includes maintenance of discipline base, experiment development/management, data analysis and investigator management and specialized ground/aircraft investigations. Also includes development of Shuttle payloads related to environmental monitoring.
• **Technology transfer**
- **Specialized applications tasks** - drawing on unique competence related to other roles to perform discrete tasks. Current emphasis [36] involves earth resources and space processing studies.
- **Technology utilization** - conducting projects to establish applicability of NASA technology in environmental fields, participate in technology transfer to industry, identify and report new technology, and document results of secondary applications of NASA technology.

- Supporting

• **Weather and climate** - contributing to discipline base. Emphasis on earth radiation budget. (Principal responsibility: *GSFC*; supporting responsibility: JPL)
• **Earth and ocean dynamics** - contributing to discipline base. Emphasis on wave modeling and ocean sensor experiments. (Principal responsibilities: *GSFC, JPL*)

• Space Transportation
- Supporting

• **Launch vehicle development** - development and procurement for science/applications missions, includes scout and meteorological sounding rockets. (Principal responsibilities: *GSFC, LeRC*)

• Space Science and Exploration
- Principal

---

* Under study

- **Viking** - completion of the Viking project including extended Viking mission management.

  - Supporting

  - **Physics and astronomy** - phase out of all physics and astronomy science and spacecraft development management.
  - **Upper atmospheric research** - conduct stratospheric emissions research relative to Shuttle operations.
  - **Planetary/lunar science** - phase out of all activities.

[37] LEWIS RESEARCH CENTER

- Areas of Technical Excellence
  - Acoustics
  - Materials
  - Space propulsion systems
  - Energy processes and systems
  - Internal flow dynamics
  - Heat transfer
  - Instrument and control systems
  - Technical project management

- Facilities of Superior Merit
  - Engine Research Building
  - Turbine Combustor Facility
  - Engine Fan and Jet Noise Facility
  - Zero Gravity Facility
  - Icing Research Tunnel
  - 8 x 6 and 10 x 10 ft. Wind Tunnels

[38] LEWIS RESEARCH CENTER

PROGRAM AREAS:
Aeronautics
Space Transportation
Space Research and Technology
Applications
Space Science and Exploration

Principal and Supporting Roles

- Aeronautics
- Principal

  - **Propulsion systems** - development of advanced aeronautical propulsion systems (except hypersonic). Focus on efficiency and environmental compatibility.
  - Supporting

  - **Wind tunnel support** - testing and facility operations support to DOD, other government agencies and industry. (Other Centers having unique or outstanding facilities may provide similar support.)
  - **Aviation safety** - contributing to advances, with focus on lightning hazards, rotor burst protection and high-energy brakes. (Supporting responsibilities: ARC, DFRC, JPL, LaRC, MSFC, WFC)

• **Acoustics and noise reduction** - focus on internal engine noise reduction. (Principal responsibility: *LaRC*; supporting responsibilities: ARC, DFRC)
• **Military support** - provision of propulsion systems technology support to DOD. (ARC, DFRC and LaRC also provide military aeronautics support.)

• Space Transportation
- Principal

• **Centaur** - development and procurement of Centaur launch vehicle system.

[39] • Space Research and Technology
- Principal

• **Space propulsion systems technology** - development and maintenance of space propulsion systems technology base.
• **Space energy processes and systems technology** - development and maintenance of technology base.

- Supporting

• **Fundamental research** - contributing to basic knowledge of metals and ceramics at atomic/molecular level. (Supporting responsibilities: ARC, JPL, LaRC)
• **Technology experiments in space** - definition and development of experiments in areas consistent with LeRC's other Space Research and Technology roles. (Principal responsibilities: *JSC, LaRC*; supporting responsibilities: ARC, DFRC, GSFC, JPL, MSFC)

• Applications
- Principal

• **Energy technology** - conducting energy related R&D, primarily on a reimbursable basis, with broad emphasis on solar, gas turbine, ground propulsion and other appropriate terrestrial energy systems.
• **Technology transfer**
- **Application systems verification tests** - demonstrate through exploratory tests the use of remote sensing techniques to improve current operational techniques. Current emphasis is on the use of satellite data to enhance ocean navigation, particularly shipping operations in Arctic areas.
[40] - **Regional applications transfer** - utilizing remote sensing techniques to monitor pollution, water quality, and land reclamation potential in cooperation with various neighboring governments.
- **Technology utilization** - conducting projects to establish applicability of NASA technology in public service activities, participate in technology transfer to industry, identify and report new technology, and document results of secondary applications of NASA technology.

- Supporting

• **Communications** - development of high-power communications technology oriented toward satellite-based applications. Includes experiment development and management. (Principal responsibility: *GSFC*)

• Space Science and Exploration
- Supporting

• **Upper atmospheric research** - contributing to discipline base, with emphasis on support of Global Atmospheric Sampling Program. (Principal responsibilities: *GSFC, JPL*; supporting responsibilities: ARC, LaRC)

[41] MARSHALL SPACE FLIGHT CENTER
Special Capabilities

- Areas of Technical Excellence
  - Launch vehicle flight mechanics and control
  - Structures and aerostructural dynamics
  - Materials
  - Propulsion systems
  - Space vehicle flight systems
  - Data systems and analysis
  - Technical project management

- Facilities of Superior Merit
  - Neutral Buoyancy Facility
  - X-Ray Telescope Facility
  - Acoustic Model Engineering Test Facility
  - External Tank Structural Test Facility
  - Dynamics Test Facility
  - Solid Rocket Booster Structural Test Facility
  - Structures and Materials Laboratory

[42] MARSHALL SPACE FLIGHT CENTER

PROGRAM AREAS:
Space Transportation
Space Science and Exploration
Applications
Space Research and Technology
Aeronautics

Principal and Supporting Roles

- Space Transportation
- Principal

• **Propulsion systems** - design, development and procurement of major propulsion-oriented systems and subsystems. Current focus on Shuttle-related systems, including Shuttle main engine, solid rocket booster, external tanks and interim upper stage in cooperation with the Air Force. Advanced development effort includes TUG and solar electric propulsion systems.
• **Manned vehicle** - design, development and procurement of manned vehicle systems on "as assigned" basis.
- **Spacelab** - focus on systems engineering management, development interface with European Space Agency and procurement.
- **Advanced missions** - focus is on space station, advanced transportation systems and construction of a satellite space power station - definition activities (MSFC and JSC have co-equal roles through definition; development responsibilities are not yet assigned).
- **Advanced development** - technology advances focused on advanced missions identified above within those disciplines assigned. Termination and transfer of all biotechnology efforts.
• **STS sustaining engineering** - providing sustaining engineering for assigned STS hardware.

[43] • Space Science and Exploration
  - Principal

    • **Spacelab mission management** - management of Spacelab I and II missions.
    • **Specialized automated spacecraft** - design and development of large, complex and/or specialized automated spacecraft as assigned. Current focus on spacecraft and systems/experiment integration for STS, HEAO, and Gravity Probe B spacecraft development. (Principal responsibility: *GSFC*; alternate responsibility: JPL)

  - Supporting

    • **Physics and astronomy science** - phase out by FY 79 of science discipline base with retention of a minimal science capability to fulfill such scientific interfaces as are required to support space science mission and spacecraft management roles.

• Applications
  - Principal

    • **Space processing** - developing space processing discipline base, developing and managing space processing experiments for Spacelab.
    • **Data management** - development of applications oriented data management discipline base. Contributing overall data management systems expertise in support of advanced high data rate systems development.
    • **Technology transfer**
    - **Regional applications transfer** - transfer of aerospace technology to State and local agencies in the Southeastern United States with particular emphasis on applications of earth resources data from satellites.
    - **Specialized applications tasks** - drawing on capability related to other roles provides [44] discrete support in such as is related to laser applications in earth and ocean dynamics.
    - **Technology utilization** - conducting projects to establish applicability of NASA technology in transportation, manufacturing, and other fields; participate in technology transfer to industry; identify and report new technology; and document results of secondary applications of NASA technology.

  - Supporting

    • **Spacelab payload definition** - definition of requirements for an Atmospheric Cloud Physics Laboratory for flight as a partial payload of the Spacelab. (Principal responsibility: *GSFC*; supporting responsibilities: JPL, LeRC)
    • **Energy technology** - conducting energy related systems studies for reimbursable activity with primary focus on solar heating and cooling and advanced coal extraction technology. (Principal responsibility: *LeRC*; supporting responsibilities: ARC, JPL, JSC)

• Space Research and Technology
  - Principal

    • **Space vehicle structures and materials** - contributing to large complex space vehicle structures and materials technology base. (Shared responsibility with *LaRC*)

  - Supporting

    • **Space propulsion systems technology** - contributing to space propulsion systems technology base, with focus on launch vehicle propulsion, solar electric propulsion, system performance and technology assessment, and contamination control. (Principal responsibility: *LeRC*; supporting responsibility: JPL)

• **Space vehicle configuration technology** - contributing to technology base for advanced space vehicle configuration. (Principal responsibility: *LaRC*; supporting responsibility: DFRC)

[45] • **Guidance and control technology** - contributing to guidance and control technology base. Focus on inertial components. (Principal responsibility: *JPL*; supporting responsibilities: GSFC, LaRC)

• **Sensor and data acquisition technology** - contributing to fundamental electronics technology base, with focus on long-life reliable circuits. (Principal responsibility: *LaRC*; supporting responsibilities: GSFC, JPL)

• **Information systems technology** - contributing to technology base, with focus on high capacity data systems for applications use. (Principal responsibility: *GSFC*; supporting responsibilities: JPL, JSC, LaRC)

• **Technology experiments in space** - definition and development of experiments in areas consistent with MSFC's other Space Research and Technology roles. (Principal responsibilities: *JSC, LaRC*; supporting responsibilities: ARC, DFRC, GSFC, JPL, LeRC)

- Aeronautics
- Supporting

• **Aviation safety** - contributing to advances in aviation safety through improved understanding of turbulence phenomena. (Supporting responsibilities: ARC, DFRC, JPL, LaRC, LeRC, WFC)

[46] WALLOPS FLIGHT CENTER
Special Capabilities

- Areas of Technical Excellence
  - Operations support
  - Tracking and data acquisition
  - Small project management

- Facilities of Superior Merit
  - Sounding Rocket Range
  - World-Wide Mobile Launch Tracking and Telemetry Capability
  - Research Airport

[47] WALLOPS FLIGHT CENTER

PROGRAM AREAS:
Space Science and Exploration
Applications
Aeronautics
Tracking and Data Acquisition*

Principal and Supporting Roles

- Space Science and Exploration
- Principal

• **Sounding rocket development, procurement, and operations** - developing and procuring sounding rockets and carrying out all phases of operations, from mission/flight planning to landing and recovery. Payload carrier development, development and management of experiments, experiment management support to other institutions, launch operations and tracking and data acquisition are included. (Most WFC sounding rocket

---
* Included within other program areas

activities involve lower performance vehicle support systems. Most activities involving higher performance systems are assigned to GSFC.)
 • **Balloon program** - Managing, Monitoring, scheduling, and technical analysis of OSS funded balloon activities conducted by other agencies (NRL and NSF at the present time).

- Applications
- Principal

   • **Technology transfer**
   - **Regional applications transfer** - identify, demonstrate and evaluate specific practical applications of remote sensing technology with emphasis on those of particular concern to the Chesapeake Bay regional resource managers.
 [48] - **Specialized applications tasks** - undertaking desirable tasks in areas related to other roles. Current emphasis includes pollution monitoring and atmospheric measurement techniques.
   - **Technology utilization** - identify and report new technology, participate in technology transfer to public service and private organizations, and document results of secondary applications of NASA technology.

   - Supporting

   • **Sounding rocket payload carrier development and experiment management support** - provided in the following applications disciplines:
     - **Weather and climate** - (Principal responsibility: *GSFC*; supporting responsibilities: JPL, LaRC)
     - **Space processing** - (Principal responsibility: *MSFC*)
     - **Earth and ocean dynamics** - (Principal responsibilities: *GSFC, JPL*; supporting responsibility: LaRC)

- Aeronautics
- Supporting

   • **Aviation safety** - contributing to advances in aviation operations through improved instrumentation and procedures in critical phases such as approach and landing. (Supporting responsibilities: ARC, DFRC, LaRC, LeRC, MSFC, JPL)

Document IV-15

Document title: James C. Fletcher, Administrator, NASA, Memorandum to Bob Frosch, "Problems and Opportunities at NASA," May 9, 1977.

Source: NASA Historical Reference Collection, NASA History Office, NASA Headquarters, Washington, D.C.

 In the aftermath of the completion of the lunar landing phase of the Apollo program in December 1972, NASA as a post-Apollo, transitional institution was very much in an uncertain and potentially unstable situation. In this memorandum, James Fletcher, who headed NASA from May 1971 to March 1977, reflects for his successor Robert Frosch on the major institutional and programmatic issues facing the agency. Of particular interest are Fletcher's observations on keeping the NASA institutional base intact or at least ensuring a flow of new people into the agency. The "Al" referred to by Dr. Fletcher is Alan Lovelace, NASA Deputy Administrator under Frosch. The project called LACIE (Large

Area Crop Inventory Experiment) was an Earth observation project using Landsat satellites. The Jupiter Orbiter Project (JOP) was later renamed Galileo.

[1] May 9, 1977
(Dictated May 6)

SUBJECT: Problems and Opportunities at NASA

Continuing our discussion in writing on some of the things that are less sensitive, let me raise some issues not in any particular order but simply for the record. Please feel free to share these with Al if you feel you would like to do so. He is already aware of most of them.

1. **Applications Program**. In my view, the Applications Program is the "wave of the future" as far as NASA's public image is concerned. It is the most popular program (other than aeronautics) in the Congress and as you begin to visit with community leaders, you will understand it is clearly the most popular program with them as well. The Application Program consists mainly of communications satellites, weather satellites, LACIE, and earthquake research. There are problems in each of these areas:

a. **Communication Satellites**. We temporarily phased out of this program in 1973 due to a severe budget cut. At the time, it seemed like industry was picking it up most rapidly and was something they could do without much help from NASA. I had serious misgivings when this decision was made since I realized that it was the part of the Applications Program which had the greatest public visibility and was the most obvious example of transfer to industry. We were able to keep a skeleton group aboard to support OTP and FCC and, to a limited extent, the existing ATS/CTS satellites. However, at this point in time, I believe we need to get back into the business one way or another. The search and rescue satellite was a small attempt in this direction. Also, the work we are doing with NOAA and the Coast Guard to monitor fishing vessels within the 200-mile limit (they install the transponders) is also a small step in that direction. The National Academy study prepared under the chairmanship of [2] Bill Davenport was a good one, and I think it is time we started following along the tracks that they recommended. I'm afraid, however, that OMB is going to give us problems.

b. **Weather Satellites**. To many, weather satellites are mostly talk and not much show. I had been at NASA four years before I realized that NOAA was not using weather satellites at all in their weather forecasting but rather used them as backup for their forecasters and occasionally for monitoring severe storms such as tornadoes and hurricanes. Weather satellites, however, have been used extensively by the Navy and by the Air Force for overseas forecasting, I think very effectively, and just recently NOAA's Numerical Weather Service in Suitland has begun making global weather forecasts for overseas construction and a variety of military uses.

The real potential, however, of weather satellites lies in the possibility of 5-day (possibly up to 2-week) forecasts and it has only been clear in the last year or two what the technical problems really are in making such forecasts. Bob Cooper is very much aware of the problem, as is Bob Jastrow of GISS, so I won't try to elaborate further on it except to say that what is really needed is some broad-gauge scientific talent to be involved rather than the specialized, narrow scope meteorologists who have been working the problem at NOAA (and for that matter at NASA also).

c. **LACIE**. The LACIE program is not going well and OMB is very much aware of this. If this program fails, it is going to reflect on NASA's credibility in the Applications area. What is needed here also is a new approach to the problem either organizationally or by using people of different technical background. The people now involved in the program at Houston are not the most talented, and they have been doing the same thing for too many years. It has not had high-level attention at Houston because, of course, the Shuttle is their main future. It may be that the program can be handled better by simply shifting the focus from Houston to Goddard. (I recommended this two years ago but got less than an enthusiastic reception from the Office of Applications.)

[3] d. **Earthquake Research**. I'm afraid we have no program in earthquake research, but we were able to get funds from the Congress and OMB by labeling some of our "tectonic plate motion" investigations improperly. As near as I can tell, what we are doing is scientific research only and this does not relate directly to predicting earthquakes, although admittedly it might add to the scientific base on which future earthquake prediction techniques might be predicated.

There are a lot of opportunities in Applications that we may be missing which may or may not be related directly to the programs in which we are now engaged. Electronic mail, wired suburbs, the Cooper/Augenstein Global Information System and, of course, a leadership responsibility for a national climate program are all things that Al is aware we could move into; however, it does take aggressive leadership to pursue these opportunities. We don't have that in the Applications Office itself. In fact, to pursue these new programs, it might be wisest to set up a separate office outside of Applications and leave the marketing of current programs (a, b, c, and d above) to the Applications Office.

In addition to opportunities and problems, we have personnel problems in the Office of Applications, which I'd be glad to discuss with you sometime.

2. **The MSFC Institutional Problem**. As I indicated to you in our discussion, NASA has an overall institutional problem which arises from the fact that we had to trim out civil service staff by almost a factor of two since its peak during the Apollo days. This has caused a number of problems that go with aging institutions generally, but our problems were accelerated because of the rapid RIFing that went on in the late 60's and early 70's. We still have a large number of competent people at NASA, but we are not bringing in new blood either at the younger age group or at the middle age level. There are three principal reasons for this. One is that some of the glamour has worn off from the Apollo days; second, there are other interesting fields in which scientists and engineers can become involved (in my judgment none of them compete with [4] what goes on at NASA but, of course, I'm prejudiced); and, third, new employees feel insecure knowing that the last ones hired are usually the first to leave in case of a RIF. This would be a dilemma for any agency in such a situation and even though we try diligently to protect our best people, we are still in danger of approaching mediocrity.

This is especially severe at Marshall where some of the largest cuts were made. Some time in the 1980-81 period, we face severe manpower cuts at this Center. An obvious solution would be to close the Center unless some new program came along that would keep the staff fully occupied. Because of the urgent need for the talent that Marshall has for the Space Shuttle development, we have tried to put new programs there (such as space telescope, HEAO, etc.) and have allowed them to do a considerable amount of in-house work on the Shuttle to make good use of their personnel. Closing Marshall has been on OMB's agenda ever since I came to NASA, although from time-to-time they have also suggested JPL, Ames, and Lewis. We have always resisted this very strongly on the basis that (a) the initial cost of replacing the facility would be very high, and (b) we couldn't afford to risk the Space Shuttle program. The real reason, however, is that there is no guarantee that by closing a Center we would be allowed to build back to the institutional base we had before the closing, and we might find ourselves in the same RIF situation but be one Center smaller. The only possible solution that I can see is to get a commitment from the President himself (the OMB Director's commitment can always be overturned) that if we do close the Center we will be allowed to build back substantially in order to bring in new personnel. Most people in Headquarters would laugh at this suggestion but I think that it is one that ought to be considered early on in your tenure. My own bias, of course, would be to try to find work to put into MSFC and use the Center as a national resource, which it indeed is, but so far efforts along these lines have not been successful.

3. **Space Science Program**. As you have already undoubtedly picked up, we are in a dilemma on space science at NASA because it seems to be strongly supported by the White House (President, Science Adviser, OMB, etc.) but poorly supported by the Congress. Congress seems to go for the Applications [5] Program, the Aeronautics Program, and the so-called "space spectaculars" such as Apollo, Skylab, ASTP, Viking, etc. Space solar power is an excellent example of such a spectacular and is a case in point. Apparently the reason for this dilemma is that OMB feels the Applications Program should more properly be left to the user agencies or to industry, which are always slow to support new satellite programs, whereas, Congress, especially the Space Committees, doesn't care about the user agencies because that's not their responsibility. In science, however, we have a clear mandate since we are our own user but somehow Congress recognizes that science of any kind is not popular among the general public (ask Herb Rowe for polls on that subject) and although low-profile science can get through Congress fairly easily, large bites such as the space telescope, JOP and Viking Follow-on (perhaps) seem to have difficult times. I have no pat solution for this dilemma except to continue to work the problem as we have been doing.

4. **Senate Power Base**. I used to raise my eyebrows when Jim Webb talked about a "power base" in Congress. Having just come from academia, this seemed a crude way of operating; however, after being here six years, I'm beginning to see what he meant. In the Senate especially, but to some degree also in the House, there are individuals who seem to sway the rest of the body. In the House this is less clear but certainly Tip O'Neill, George Mahon and to a lesser extent, Jim Wright, could be put into this category. In past years, Wilbur Mills and Eddie Hebert served that function, but I'm not sure their successors have quite moved into such strong positions.

Incidentally, Tiger Teague has a great deal of respect in the House, and when he is willing and able to acknowledge this respect, he can be very helpful; however, in recent years his health has been a definite handicap and this is one of our problems at the moment in the House Appropriations Subcommittee. Tiger says he is not going to run again, but when he gets his lightweight leg and his spirits improve, I wouldn't bet against his running.

In the Senate, however, the situation has changed drastically. When I first heard that Senator Proxmire was going to be [6] Chairman of our Appropriations Subcommittee, it looked like "the end of the world" until we began to work the problem. It began to be clear that the ex officio votes on the Proxmire Subcommittee by the Senate Space Committee Chairman and Majority Leader were enough to swing the rest with no problem at all. The votes typically were 8 or 9 to 1, with Senator Proxmire's being the only negative vote. The loss of Senator Moss was considerable even though he did not have the leverage that some of the other Senators had. He was the Chairman of the Democratic Caucus and the #3 democrat in the Senate and on occasion could swing a fair number of votes. Senator Goldwater, of course, was the undisputed conservative leader in the Senate and consequently both sides of the house could be swayed by him. So it was not only the loss of Senator Moss but the loss of those ex officio votes that caused us to lose leverage in the Senate. Senator Stevenson is just learning the business but I think in time he, along with his strong staff members, should be great support especially if they are able to involve Senator Magnuson in helping him to influence some of their colleagues. Meanwhile, I'm afraid we are forced into falling back on the Proxmire Subcommittee itself.

Although we have strong support in Senator Stennis and, I believe, Senator Sasser on the Democratic side and I think all four on the Republican side, this is not enough to be considered strong support in the sense of adding in programs that the House may have taken out. This latter situation occurred many times in the past through the help of Senators Moss and Goldwater, but this year we simply can't count on it. On the other hand, I think the support is strong enough so that they are not likely to make further cuts.

The one redeeming feature in the Senate reorganization is the position of Senator Cranston as Majority Whip. He is a strong space supporter in his own right but, being a California Senator, has vested interests as well. Senator McClellan also has a great deal of influence as does Senator Stennis but those are primarily on the conservative side of the House and the number of conservative democrats is becoming fewer each year. Senator Jackson, of course, [7] is powerful as in earlier years but so far has not had any impact on NASA's programs. Senator Cannon has moved up in stature since his recent reelection, having been one of the few western democratic Senators to be returned to the Senate. I had hoped that he would end up in one way or another as Chairman of one of our committees but that was not to be. I think becoming better acquainted with Senator Cannon can be a great help both by influencing votes in the Appropriations Committee and, of course, in the Commerce Committee itself.

These are all things that must be tracked very carefully, and I'm afraid roles are changing so rapidly that I can only alert you to the problems. Pete Crow and Joe Allen, I think, understand the situation pretty well and should be able to help make the appropriate contacts. Judy Cole, if she stays, is excellent on the G-2 and has a very good working relationship with the staff of the Senate Budget Committee. Although run by Senator Muskie, it is not yet clear how much impact it will have.

5. **Aeronautics**. I won't dwell on this subject since Al Lovelace is very familiar with the problem, but simply mention that we need to revive the fundamental work that the old NACA used to do. (The Aeronautical Centers should be at least as good as NRL is to the Navy, but so far not a single member of the NASA organization has been elected to the National Academy of Science as has Herb Friedman of NRL.) It is not clear how to do this but, of course, it is related to the institutional problem of bringing in stronger scientific and creative new talent.

6. **The Shuttle Launch Phase**. Undoubtedly Al must have mentioned to you that my biggest concern on the Shuttle at the moment (aside from operational costs) is the technical difficulties involved in the launch phase. As you know, Houston is the lead Center for Shuttle development and performed very well on the Apollo spacecraft and the LM, and also carrying out operations in space. They had very little to do with the development of the Saturn launch vehicle, which was done out of Huntsville. Wernher von Braun and the people [8] he brought with him both from Germany and from within the United States had an in-house capability second to none in the world.* As a result, if you look back in the records, you will find very few difficulties with the Saturn itself and, in fact, the extra weight-carrying margin of the Saturn saved the Apollo program more than once. Incidentally, neither George Low nor John Yardley has had this launch vehicle background, and so with the loss of Rocco Petrone, we have never really had anybody in Headquarters who had much experience in this area.

I guess the question is, why do I consider this different from space problems generally? It comes down to something like the following: With the spacecraft itself during its flight in space and its landing and its attitude control system and its life support systems, etc., we had the capability to build highly redundant systems. So if we ran into a problem, there was usually time to find a "workaround" and, in fact, in every case except the fire on the ground, we managed a workaround good enough to bring the astronauts back. On the other hand, look at all of the things that went wrong during the Apollo/Skylab series. If we hadn't had this redundancy, we would have lost essentially every mission. In the case of the booster, however, there was no time for any significant workarounds on the ground. There is some redundancy built in but not an excessive amount. Therefore, testing analyses and engineering intuition have been the backbone of the launch vehicle business from the days of the V-2. Clearly the combined vehicle consisting of the two solid rockets, the external tank, and the Shuttle is the most complicated launch vehicle ever built. My big concern is whether or not analysis and testing on the ground are sufficient to ensure the reliability of this phase of the flight profile or whether engineering judgment and experience which were the hallmark of the von Braun group aren't still necessary for a guaranteed success. So far I have discussed this with Al Lovelace and Walt Williams only. Walt, of

---

*[handwritten note] Korolov played the same role in the U.S.S.R. When he died, the Soviets were not able to make a single new launch vehicle work.

course, has had extensive experience with Air Force launch vehicles but again nothing like the experience of the Huntsville group. [9] Perhaps I'm overly concerned about this problem, but when you consider the value of the payload even on the first flight and the consequences of a failure, I'd have to put it as one of my high-priority items for the near term.

7. **Pet Projects.** There are a number of things which I have tried to keep going because I believed in them, but most had a low level, which you may want to discontinue:

a. **Hypersonic Transport.** I have always felt, aside from environmental problems, that it would be a rather straightforward development to build a commercial vehicle for long-distance travel (say from New York to Delhi or from New York to Bahrein), but I'm not sure that the airplane is the best way to do it. I have the uncomfortable feeling that it might be simpler to remove the energy from a returning space vehicle by means of high-drag devices such as parachutes, blunt bodies, retrorockets, etc., rather than with wings. This is heresy at NASA but you must understand that I came up through the rocket route, not the airplane route. I did, however, go to the trouble of bringing in the parachute people to see whether indeed parachutes could be built that would allow large transports to be dropped through the atmosphere in much the same way as the Apollo capsule, and it always seems to be technically feasible but on the surface more optimal than the Shuttle itself. Needless to say, I didn't want to emphasize this in the middle of the Shuttle program.

b. **Heavy Lift Launch Vehicle.** For putting large quantities of payloads in space, it seems there are better ways of doing this than the Shuttle itself since the missions are all one way and all you need to recover are pieces of the launch vehicle itself. This can be done easily with parachutes. We might easily gain a factor of 10 to 1 over the cost per pound now required by the Shuttle.

c. **Solar Sailing**. I am sure you remember with some ambivalence Dick Garwin partly for his abrasive tone but also for his tremendous creativity. In 1972, he strongly urged me to look into the possibility of using lightweight [10] materials to "sail" around the solar system. It took four years for the NASA "system" to respond. Bruce Murray picked it up and is now running with it. In my opinion this will be the way we will move out to Mars and other planets in the future even when we decide to go there with manned missions.

d. **Personal Communication Systems**. I really do believe that some day we will want to have person-to-person communications systems, not necessarily for the wristwatch variety but at least of the pocket calculator variety in which any person can dial any other person long distance from his car, from the golf course or wherever. This, I think, is a straightforward use of a high-powered, highly directional stationary satellite. I don't believe cost tradeoffs of this system have been made, and I'm not sure how much a person would pay for such a convenience.

e. **Technology Transfer**. The early studies made on the relationship between high technology and national productivity were very exciting indeed. The whole problem is not very well understood by economists and, I do not believe, other people in government. People seem to equate high technology with new inventions or new products instead of with productivity, and the picture gets all out of proportion. Paul Kochanowski, a former Brookings Fellow at NASA, understood the problem very well and I learned what little I know from him. It does seem to me that the impact on our economy of the technology such as NASA develops and as portions of DoD develop is absolutely enormous. I therefore have encouraged further economic studies of this process but you may wish to discontinue it.

f. **Broadening NASA's Responsibility**. As Al has probably indicated to you, I've always felt that NASA's managerial talents as well as some of its technical talents have been underutilized, and we ought to move into areas that are now the responsibilities of other agencies. This is a severe bureaucratic problem, and I'm not sure you'll want to get into that but if you do, the best time to do it [11] is during a change of Administration, as you well know, before the bureaucracy becomes firmly entrenched.

8. **Public Affairs**. During the Apollo days and before, NASA provided an excellent public information service to the media and, generally speaking, the public was well in-

formed about the so-called "space spectaculars." At this point in time though we need to move to a public *relations* program; that is, an aggressive program to inform the public as to how their money is being spent and what they get for it. This is a much different problem and I have asked Bob Newman and Herb Rowe to put together a program plan which presumably has been done by now but is awaiting guidance. The last session we had brought out the fact that the focus on this aggressive program ought to be on applications and spinoffs, but we really hadn't come down to the heart of the matter and that is how to have one or two simple themes which describe NASA's contributions to the nation. My own feeling is that we need outside expertise on this one and although we brought in Burson-Marsteller, a first-rate Chicago outfit, I value the advice of Jim Mortensen of Young and Rubicam much more highly. Jim is a broad, thoughtful person interested in the space program and is willing to contribute his service freely when he has the time available. Todd Groo's experience in this area is also helpful. All of these latter are more creative than Herb and I have indicated to him that I wanted all of these other men to be heavily involved in any program plans for the future. You may wish to change that.

There are other items that I could mention here and still more that I will think of before I leave, but I expect I have covered 90 percent of the biggest issues.

James C. Fletcher

Document IV-16

Document title: Task Force for the Study of the Mission of NASA, NASA Advisory Council, "Study of the Mission of NASA," October 12, 1983, pp. 1-9.

Source: NASA Historical Reference Collection, NASA History Office, NASA Headquarters, Washington, D.C.

The NASA Advisory Council and its standing committees are descendants of the National Advisory Committee for Aeronautics (NACA), which, through its technical committees, oversaw the research conducted by NACA's Langley, Lewis, and Ames research laboratories. Lacking the statutory authority of its predecessor, the NASA Advisory Council acts as an informal "board of directors" to the NASA administrator. The council's 1983 "Study of the Mission of NASA," chaired by Daniel J. Fink, was a detailed and comprehensive effort to chart a course for the U.S. civil space agency in a changed political and technical environment. It reviewed in a comprehensive manner the overall mission of the agency and recommended alterations for its activities for the next 20 to 40 years in both aeronautics and space. While the task force members said they unanimously agreed with the elements of the mission statement that emerged, they admitted some disagreement as to whether the mission area of "Exploration of the Solar System" should be viewed as an overarching theme to guide the forward technological thrusts of the agency. While some strongly endorsed this as a central focus for NASA's future space activity, the majority were concerned that such a specific identification would result in the diminution of the other important missions. A special area of concern was the space shuttle, and the report recommended that a new NASA organization, with resources "fenced" from those of the rest of the agency, be established to manage the shuttle program.

---

[1] # Study of the Mission of NASA

EXECUTIVE SUMMARY

A task force of the NASA Advisory Council was authorized by the NASA Administrator in July 1982 to study the long-range missions of NASA and present recommendations for the future course and direction of the Agency over a period of the next 20 to 40 years. The Task Force membership consisted of selected members of the NASA Advisory Council, augmented by additional participants representing the requisite areas of expertise.

1. The Mission of NASA

The Task Force recommends this "Mission of NASA" that rests on current statute and policy and provides a framework for NASA's activities for the next 20 to 40 years.

Mankind has acquired the ability to move within and beyond the confines of the surface and the atmosphere of Earth, creating apparently limitless opportunities for beneficial human activity. In this regard, NASA has a dual mission—in space and the atmosphere—portions of which are overlapping.

NASA's space mission is to conduct activities on behalf of the people of the United States in collaboration with other nations, to:
- explore the solar system and study its planetary processes, including, as appropriate, those governing the Earth, for the benefit of humankind,
- pursue a program of fundamental scientific research in space to expand human knowledge,
- plan and implement space technology programs and research into the use of the environment of space in order to provide for the continued advance of the national space capability and its exploitation for public and commercial purposes,
- and, to achieve these ends, create the capability for an expanded human presence in space and develop and assure the operation of launch and space vehicles.

NASA's mission in aeronautics is to maintain, and augment as appropriate, an aeronautics research and technology program which contributes materially to the U.S. leadership in civil and military aviation.

[2] The Task Force believes that it is inevitable that human habitation will eventually extend beyond the confines of the Earth in many ways and on a scale far larger than is currently envisioned. Although it may not now be productive to debate the specific nature or the timing of this most dramatic of all human ventures, it is appropriate to use such a venture as a distant goal to guide our search for an understanding of the solar system and to stimulate the further advance of humankind.

2. Key Missions Considered

In arriving at this broad statement of the Mission of NASA, the Task Force considered a wide range of specific mission areas:
- Exploration of the Solar System, Including the Planet Earth, for Human Benefit
- Fundamental Space Science
- Space Technology
- Space Applications
- STS Operations
- Aeronautics
- Human Spaceflight Research
- International Relationships.

a. Exploration of the Solar System, Including Planet Earth, for Human Benefit

NASA's mission in space science can be considered in two distinct parts: one is space science conducted to contribute directly to human welfare and national need, and the other is fundamental space science conducted to expand human knowledge. The first is currently focused on solar system exploration, including the planet Earth, while the second covers the full range of space science fields, from astronomy to space physics.

NASA's mission in the exploration of the solar system itself contains two related parts, which together provide a major objective for the next half century. The first involves exploration of the solar system with the eventual goal of utilizing the resources and knowledge of space for human benefit. The second brings a planetary perspective to our own planet and leads to the goal of understanding those processes involved in global surface and atmosphere change especially important to living systems.

To accomplish this long-range mission, the Task Force recommends that:

• The United States, in collaboration with other interested nations, vigorously pursue a program of exploration and understanding of the solar system for ultimate human benefit.
[3] • A more aggressive pursuit of the technologies of robotics, teleoperation, and machine intelligence be undertaken to maintain a proper lead and balance in the design of NASA programs.
• NASA accept a leadership role in major aspects of the study of the planet Earth, in particular in those areas which concern global changes of importance to the support of human life.
• Adequate funding be provided for ground-based, laboratory, and theoretical work as well as for space systems.
• NASA seek a lasting national commitment in order to achieve these goals, which will require new approaches to project procedures and budgetary policies including steady funding levels and program continuity.

b. Fundamental Space Science

In fundamental science, NASA shares a responsibility with the National Science Foundation, with the break in responsibility occurring at the construction of spacecraft and space instruments. Working relations are generally satisfactory, but policy interactions have been limited. This division of responsibility has left a gray area of ground-based observations and theoretical and laboratory work, all essential for realizing the full advantages of space missions. Problems also exist in the provision of adequate funding to maintain effectiveness in research capabilities built laboriously over decades. The Task Force sees no simple solutions to these problems, but is convinced that the effort to find solutions must continue.

The Task Force recommends that:

• Space science must remain a principal part of NASA's mission because research in space science has become a central element of scientific research in the United States and the prospects for future major advances in this field remain bright.
• Adequate and stable levels of funding for ground-based, laboratory, and theoretical research, a key part of NASA's scientific program, be provided independently of the fluctuating needs for spacecraft and instrument construction.
• NASA take a lead role in assuring that the activities sponsored by NASA and NSF are properly coordinated.

[4] **c. Space Technology**

The Task Force considered whether NASA should support the space technology needs of other government agencies, both military and civil, and the private sector. A recent study by the Aeronautics and Space Engineering Board of the National Research Council recommended that NASA's program be redirected to the needs of the broad national space constituency, and endorsed the concept of a role in space technology analogous to NASA's traditional role in aeronautics.

The developing capability to conduct space experimentation on or with the support of the Shuttle provides NASA with a unique resource for space R&T and one that is critical to the technology needs of many space users. A space station could provide an even greater laboratory in space—a "field center"—that would be truly unique in this regard. The Task Force concurs with the ASEB in its recommendation that NASA's mission in space R&T be supportive of total national space requirements.

The Task Force recommends that:

• NASA's mission in space research and technology be supportive of total national requirements, considering future needs of the civil, military, and commercial sectors. In terms comparable to those for NASA's mission in aeronautics, NASA should have the mission in space R&T to:
  - Fund, direct, and implement space research, technology, and demonstration programs in support of its own and other civil space activities; and support DOD technology needs where the results have broad application and are not duplicative of other government-funded effort.
  - Encourage and facilitate, together with other appropriate agencies, the transfer of space R&T results to and within U.S. industry.
  - Manage, maintain, and operate space research, development, test, and evaluation facilities. Use of the Space Shuttle and eventually a space station as a laboratory facility in space should be exploited for development of new space technologies.
• Funding for NASA space R&T be increased selectively to permit implementing these recommendations. The following criteria are suggested for identification of technologies to be accorded high priority:
[5] - The technology is in the national interest and will fill a reasonably established future requirement.
  - The technology offers payoff significantly greater than that presently available at a reasonable cost and is at least comparable in payoff and cost to alternative approaches to the same end.
  - The technology program is not likely to be undertaken, on a timely basis, by others, either public or private.
  - The technology is in an area in which NASA already has a demonstrated capability, or if in a new area, is one in which NASA can readily build the capability and expertise without duplicating an equivalent capability outside.
• Funding augmentations in high-priority technologies be provided only after reasonable assurance that ongoing technology developments of little potential value are being phased down and that a base level of more basic research is being maintained across the full spectrum of disciplines to assure that new technology opportunities applicable to future missions are not missed.

d. Space Applications

There appears to be little question that NASA should perform research and technology development in major space applications areas such as telecommunications, meteorology, Earth resources, and materials processing in space. However, there is much concern about how far NASA should go to provide utility demonstration and early operation of space applications systems.

NASA's role in space applications should be compatible with the overall mission of NASA. The present assumption is that NASA's primary focus should be on space R&D with involvement in operations only if necessary. The other working assumption is that NASA's primary emphasis should be on the civil side, although support of military and other defense interests is not excluded.

The Task Force recommends that the NASA mission include the study of space and space technology for civil applications, by:

• Continuing the identification of possible civil applications of potential value and by conducting preliminary studies of potential benefits, users, and markets. This includes taking the leadership in systematic reviews of existing and possible civil applications of space

[6] • Conducting or supporting research and technology development on essential components and subsystems of space and ground systems for civil space applications.

• Conducting, with suitable cooperative arrangements with the private or public agencies, tests, demonstrations, and experimental user operations of new types of spacecraft, spaceborne systems, and ground systems for civil space applications in cases where:
  - Potential advantages, uses, and users have been identified.
  - The private sector or other agencies cannot reasonably be expected to pay the full cost.

Administration recognition of this mission of NASA should be sought.

• Arranging for the transfer to private or public agencies, as appropriate, of useful applications systems employing space technologies, unless it is in the national interest for NASA to become the operating agency.

e. STS Operations

Realization of the maximum operational efficiency for Shuttle operations is an immediate task which requires a major amount of attention from senior NASA management. A continuation of the current STS management approach might create a serious diversion of NASA resources and attention from its more traditional roles of aeronautics and space R&D and space exploration and experimentation. The view also has been expressed that NASA is not the proper organization to achieve the operating efficiencies, levels of customer satisfaction, or degree of market development desired. Therefore, resolution of the technical and management issues involved in Shuttle operations can have a major impact on future NASA missions.

Various options for the management of STS operations were examined by the Task Force, including commercial management by an industrial organization, full operation by DOD, creation of a new Federal agency for Shuttle operations management, establishment of a quasi-government corporation, creation of a new organization within NASA, and continuation of the present management arrangements in NASA.

The study group concluded that the present small size of the Shuttle fleet, its lack of maturity and high costs of operation, and the lack of DOD interest in assuming full operating responsibility, all militate against shifting Shuttle operations outside [7] of NASA in the near future. The potential for future diversion of management attention and resources from the STS to other NASA programs suggests the advisability of further segregating the Shuttle operating management organization from the rest of the NASA organization, at least as an evolutionary, timely step.

The Task Force recommends that:

• A new NASA organization be created at the appropriate time within NASA to focus on Shuttle operations and utilization, including marketing activities and sustaining engineering support. This organization, which should be headed by a Deputy Administrator, should have fenced manpower, finances, and facilities

• The Shuttle Operations organization continue to enhance customer services and market development activities through the application of resources both internal and external to NASA as appropriate. Consideration should be given to contracting out the market development activities

• The Deputy Administrator for Shuttle Operations be charged with proposing the evolutionary steps for future management of Shuttle operations

- The Shuttle operations organization not undertake future STS development activities, such as a follow-on Shuttle or new launch capability. It should, however, help define future requirements for such major improvements
- The value of the STS as a national resource due to its unique capabilities to provide for manned spaceflight, defense missions, space science support, and research and technology development in the space environment, as well as its special capabilities for satellite retrieval and spacecraft servicing, be recognized. The total costs associated with this national resource value should not be charged to the Shuttle's launch service users.
- As markets for the STS develop, NASA consider shifting pricing of Shuttle services towards an incremental (i.e., Institutional) cost basis. This is the usual practice in NASA research facilities and, given the national resource character of the Shuttle for civil and military Government programs, is more appropriate than pricing on the basis of total recovery of recurring costs (industrial funding).

[8] **f. Aeronautics**

When the Task Force was established, the issue of NASA's mission in aeronautics was thought to be one that would require a great deal of attention as a result of questions raised in recent years regarding the role of the Federal government vis-a-vis that of the private sector. However, a broad policy statement on the Federal mission in aeronautical R&T has been issued by the Office of Science and Technology Policy, which resolved most of the issues. The Task Force addressed two derivative issues: demonstration of technology for civil aviation, and effective relations with the DOD.

The Task Force recommends that:

[9] • NASA's mission in aeronautics, subject to specific approval, include support to the industry in civil aviation technology demonstration programs provided that they are in the national interest, the industry cannot effectively conduct the programs without that support, and the support fits naturally within NASA's capabilities.
- NASA continue to assist, communicate with, and cooperate with all branches of the DOD in all matters relating to aeronautics. The first priority should be in basic aeronautics research and technologies, and should include providing access to and utilization of both human resources and laboratories and other physical test facilities as appropriate. The second priority should be in mission-oriented systems work, as requested or when special expertise or facilities are available.

g. Human Spaceflight Research

While the effects of spaceflight have proved to be manageable in flights of the durations experienced up to now, there are additional concerns when prolonged duration spaceflight, as in permanent space stations or eventually in interplanetary flight, is considered. Five areas which require intensive research include: prolonged exposure to zero gravity; provision of oxygen, food, and water; provision of an adequate social and organizational environment; exposure to ionizing radiation; and extra-vehicular activities.

The Task Force believes that it is inevitable that people will seek to explore the solar system, not only by remote sensing or even by automated acquisition of samples, but by being there, and thus ultimately extend the domain of human life beyond the confines of the Earth. The requirement for mission durations ranging from months to years is implicit. Given the significance of the issues and the lengthy interactive R&D process required, the Task Force recommends that:

[9] NASA give high priority to the continuing program needed for the development of the capability to keep people healthy, effective, and well motivated over the long periods required for manned exploration of the solar system. The development of effective countermeasures for the disturbances associated with zero gravity requires research ex-

tending over many years, and must be addressed now to avoid later constraints on manned exploration missions.

h. International Relationships

Bearing in mind that one of the basic goals of the National Space Policy is to promote international cooperative activities in the national interest, the Task Force examined the international aspects of NASA's roles and missions. NASA's international role is very important to the maintenance of the image of the United States as the technological leader of the free world. There is need for more binding commitments in our cooperative activities in space ventures with other nations. Further, the U.S. policy on technology transfer was observed to be counter-productive because it limits other nations in their ability to procure U.S. space products rather than develop such products domestically.

The Task Force recommends that:

- NASA take the steps necessary to ensure greater awareness within the U.S. government of its value and that of its aerospace programs as an instrument of U.S. foreign policy.
- Cooperative agreements between NASA and foreign or international agencies be developed and maintained consistent with long-term foreign policy objectives as well as with scientific and technological objectives to achieve a greater degree of constancy and stability.

Document IV-17

Document title: *Report of the Presidential Commission on the Space Shuttle Challenger Accident, Vol. I* (Washington, DC: U.S. Government Printing Office, June 6, 1986), pp. 164-77.

The January 28, 1986, explosion of the Space Shuttle *Challenger* and the ensuing investigation invited comparison with the events that followed the launch-pad fire of the Apollo 204 spacecraft almost 19 years before that resulted in the deaths of three astronauts. During the earlier accident a politically strong administrator was at the helm of NASA; James E. Webb persuaded the White House to allow NASA to take the lead in the accident investigation. That investigation was largely technical, and it was sufficiently rigorous and critical to be seen as credible. It resulted primarily in engineering changes; what managerial changes Webb made as a result were surgical in nature, lest the agency's entire management corps be cast into confusion. In contrast, after the *Challenger* accident NASA's internal investigation took a back seat to the work of a White House-appointed commission, chaired by former Secretary of State William P. Rogers. NASA was unable to seize the initiative because, among other factors, its own top management was in disarray. The agency had been without a permanent administrator for two months, and Acting Administrator William Graham was an "outsider" not widely trusted within the agency. The report of the "Rogers Commission" was deliberate and thorough and, as this excerpt suggests, gave as much emphasis to the accident's managerial as to its technical origins.

---

[164] **Pressures on the System**
With the 1982 completion of the orbital flight test series, NASA began a planned acceleration of the Space Shuttle launch schedule. One early plan contemplated an eventual rate of a mission a week, but realism forced several downward revisions. In 1985, NASA published a projection-calling for an annual rate of 24 flights by 1990. Long before the Challenger accident, however, it was becoming obvious that even the modified goal of two flights a month was overambitious.

In establishing the schedule, NASA had not provided adequate resources for its attainment. As a result, the capabilities of the system were strained by the modest nine-mission rate of 1985, and the evidence suggests that NASA would not have been able to accomplish the 15 flights scheduled for 1986. These are the major conclusions of a Commission examination of the pressures and problems attendant upon the accelerated launch schedule.

On the same day that the initial orbital tests concluded—July 4, 1982—President Reagan announced a national policy to set the direction of the U.S. space program during the following decade. As part of that policy, the President stated that:

"The United States Space Transportation System (STS) is the primary space launch systems for both national security and civil government missions."

Additionally, he said:

"The first priority of the STS program is to make the system fully operational and cost-effective in providing routine access to space."

From the inception of the Shuttle, NASA had been advertising a vehicle that would make space operations "routine and economical." The greater the annual number of flights, the greater the degree of routinization and economy, so heavy emphasis was placed on the schedule. However, the attempt to build up to 24 missions a year brought a number of difficulties, among them the compression of training schedules, the lack of spare parts, and the focusing of resources on near-term problems.

One effect of NASA's accelerated flight rate and the agency's determination to meet it was the dilution of human and material resources that could be applied to any particular flight.

The part of the system responsible for turning the mission requirements and objectives into flight software, flight trajectory information and crew training materials was struggling to keep up with the flight rate in late 1985, and forecasts showed it would be unable to meet its milestones for 1986. It was falling behind because its resources were strained to the limit, strained by the flight rate itself and by the constant changes it was forced to respond to within that accelerating schedule. Compounding the problem was the fact that NASA had difficulty evolving from its single-flight focus to a system that could efficiently support the projected flight rate. It was slow in developing a hardware maintenance plan for its reusable fleet and slow in developing the capabilities that would allow it to handle the higher volume of work and training associated with increased flight frequency.

[165] Pressures developed because of the need to meet customer commitments, which translated into a requirement to launch a certain number of flights per year and to launch them on time. Such considerations may occasionally have obscured engineering concerns. Managers may have forgotten—partly because of past success, partly because of their own well-nurtured image of the program—that the Shuttle was still in a research and development phase. In his testimony before a U.S. Senate Appropriations subcommittee on May 5, 1982, following the third flight of the Space Shuttle, James Beggs, then the NASA Administrator, expressed NASA's commitment

"The highest priority we have set for NASA is to complete development of the Shuttle and turn it into an operational system. Safety and reliability of flight and the control of operational costs are primary objectives as we move forward with the Shuttle program."

Sixteen months later, arguing in support of the Space Station, Mr. Beggs said, "We can start anytime.... There's no compelling reason [why] it has to be 1985 rather than '86 or '87. The point that we have made is that the Shuttle is now operational." The prevalent attitude in the program appeared to be that the Shuttle should be ready to emerge from the developmental stage, and managers we determined to prove it "operational."

Various aspects of the mission design and development process were directly affected by that determination. The sections that follow will discuss the pressures exerted on the system by the flight rate, the reluctance to relax the optimistic schedule, and the attempt to assume an operational status.

Planning of a Mission

The planning and preparation for a Space Shuttle flight require close coordination among those making the flight manifest, those designing the flight and the customers contracting NASA's services. The goals are to establish the manifest; define the objectives, constraints and capabilities of the mission; and translate those into hardware, software and flight procedures.

There are major program decision points in the development of every Shuttle flight. At each of these points, sometimes called freeze points, decisions are made that form the basis for further engineering and product development. The disciplines affected by these freeze points include integration hardware, engineering, crew timeline, flight design and crew training.

The first major freeze point is at launch minus 15 months. At that time the flight is officially defined: the launch date, Orbiter and major payloads are all specified, and initial design and engineering are begun based on this information.

The second major freeze point is at launch minus 7.7 months, the cargo integration hardware design. Orbiter vehicle configuration, flight design and software requirements are agreed to and specified. Further design and engineering can then proceed.

Another major freeze point is the flight planning and stowage review at launch minus five months. At that time, the crew activity timeline and the crew compartment configuration, which includes middeck payloads and payload specialist assignments, are established. Final design, engineering and training are based on these products.

Development of Flight Products

The "production process" begins by collecting all mission objectives, requirements and constraints specified by the payload and Space Shuttle communities at the milestones described above. That information is interpreted and assimilated as various groups generate products required for a Space Shuttle flight: trajectory data, consumables requirements, Orbiter flight software, Mission Control Center software and the crew activity plan, to name just a few.

Some of these activities can be done in parallel, but many are serial. Once a particular process has started, if a substantial change is made to the flight, not only does that process have to be started again, but the process that preceded it and supplied its date may also need to be repeated. If one group fails to meet its due date, the group that is next in the chain will start late. The delay then cascades through the system.

Were the elements of the system meeting their schedules? Although each group believed it had an adequate amount of time allotted to perform its function, the system as a whole was falling [166] behind. An assessment of the system's overall performance is best made by studying the process at the end of the production chain: crew training. Analysis of training schedules for previous flights and projected training schedules for flights in the spring and summer of 1986 reveals a clear trend: less and less time was going to be available for crew members to accomplish their required training....

The production system was disrupted by several factors including increased flight rate, lack of efficient production processing and manifest changes.

Changes in the Manifest

Each process in the production cycle is based on information agreed upon at one of the freeze points. If that information is later changed, the process may have to be repeated. The change could be a change in manifest or a change to the Orbiter hardware or software. The hardware and software changes in 1985 usually were mandatory changes; perhaps some of the manifest changes were not.

The changes in the manifest were caused by factors that fall into four general categories: hardware problems, customer requests, operational [167] constraints and external factors. The significant changes made in 1985 are shown in the accompanying table. The following examples illustrate that a single proposed change can have extensive impact, not because the change itself is particularly difficult to accommodate (though it may be), but because each change necessitates four or five other changes. The cumulative effect can be substantial....

When a change occurs, the program must choose a response and accept the consequences of that response. The options are usually either to maximize the benefit to the customer or to minimize the adverse impact on Space Shuttle operations. If the first option is selected, the consequences will include short-term and/or long-term effects....

1985 Changes in the Manifest
Hardware Problems
Tracking and Data Relay Satellite (canceled 51-E, added 61-M).
Synchronous Communication Satellite (added to 61-C).
Synchronous Communication Satellite (removed from 61-C).
OV-102 late delivery from Palmdale (changed to 51-G, 51-I, and 61-A).
Customer Requests
HS-376 (removed from 51-I).
G-Star (removed from 61-C).
Satellite Television Corporation-Direct Broadcast Satellite (removed from 61-E).
Westar (removed from 61-C).
Satellite Television Corporation-Direct Broadcast Satellite (removed from 61-H).
Electrophoresis Operations in Space (removed from 61-B).
Electrophoresis Operations in Space (removed from 61-H).
Hubble Space Telescope (swap with Earth Observation Mission).
Operational Constraints
No launch window for Skynet/Indian Satellite Combination (61-H).
Unacceptable structural loads for Tracking and Data Relay Satellite/Indian Satellite (61-H).
Landing weight above allowable limits for each of the following missions: 61-A, 61-E, 71-A, 61-K.
External Factors
Late addition of Senator Jake Garn (R-Utah) (51-D).
Late addition of Representative Bill Nelson (D-Florida) (61-C).
Late addition of Physical Vapor Transport Organic Solid experiment (51-I).

[168] Operational constraints (for example, a constraint on the total cargo weight) are imposed to ensure that the combination of payloads does not exceed the Orbiter's capabilities. An example involving the Earth Observation Mission Spacelab flight is presented in the NASA Mission Planning and Operations Team Report in Appendix J. That case illustrates that changes resulting from a single instance of a weight constraint violation can cascade through the entire schedule.

External factors have been the cause of a number of changes in the manifest as well. The changes discussed above involve major payloads, but changes to other payloads or to payload specialists can create problems as well. One small change does not come alone; it generates several others. A payload specialist was added to mission 61-C only two months before its scheduled lift off. Because there were already seven crew members assigned to the flight, one had to be removed. The Hughes payload specialist was moved from 61-C to 51-L just three months before 51-L was scheduled to launch. His experiments were also added to 51-L. Two middeck experiments were deleted from 51-L as a result, and the deleted experiments would have reappeared on later flights. [169] Again, a "single" late change affected at least two flights very late in the planning and preparation cycles.

The effects of such changes in terms of budget, cost and manpower can be significant. In some cases, the allocation of additional resources allows the change to be accommodated with little or no impact to the overall schedule. In those cases, steps that need to be re-done can still be accomplished before their deadlines. The amount of additional resources required depends, of course, on the magnitude of the change and when the change occurs: early changes, those before the cargo integration review, have only a minimal impact; changes at launch minus five months (two months after the cargo integration review) can carry a major impact, increasing the required resources by approximately 30 percent. In the missions from 41-C to 51-L, only 60 percent of the major changes occurred before the cargo integration review. More than 20 percent occurred after launch minus five months and caused disruptive budget and manpower impacts.

Engineering flight products are generated under a contract that allows for increased expenditures to meet occasional high workloads. [170] Even with this built-in flexibility, however, the requested changes occasionally saturate facilities and personnel capabilities. The strain on resources can be tremendous. For short periods of two to three months in mid-1985 and early 1986, facilities and personnel were being required to perform at roughly twice the budgeted flight rate.

If a change occurs late enough, it will have an impact on the serial processes. In these cases, additional resources will not alleviate the problem, and the effect of the change is absorbed by all downstream processes, and ultimately by the last element in the chain. In the case of the flight design and software reconfiguration process, that last element is crew training. In January, 1986, the forecasts indicated that crews on flights after 51-L would have significantly less time than desired to train for their flights....

"Operational" Capabilities

For a long time during Shuttle development, the program focused on a single flight, the first Space Shuttle mission. When the program became "operational," flights came more frequently, and the same resources that had been applied to one flight had to be applied to several flights concurrently. Accomplishing the more pressing immediate requirements diverted attention from what was happening to the system as a whole. That appears to be one of the many telling differences between a "research and development" program and an "operational program." Some of the differences are philosophical, some are attitudinal and some are practical.

Elements within the Shuttle program tried to adapt their philosophy, their attitude and their requirements to the "operational era." But that era came suddenly, and in some cases, there had not been enough preparation for what "operational" might entail. For example, routine and regular post-flight maintenance and inspections are critical in an operational program; spare parts are critical to flight readiness in an operational fleet; and the software tools and training facilities developed during a test program may not be suitable for the high volume of work required in an operational environment. In many respects, the system was not prepared to meet an "operational" schedule.

As the Space Shuttle system matured, with numerous changes and compromises, a comprehensive set of requirements was developed to ensure the success of a mission. What evolved was a system in which the preflight processing, flight planning, flight control and flight training were accomplished with extreme care applied to every detail. This process checked and rechecked everything, and though it was both labor- and time-intensive, it was appropriate and necessary for a system still in the developmental phase. This process, however, was not capable of meeting the flight rate goals.

After the first series of flights, the system developed plans to accomplish what was required to support the fight rate. The challenge was to streamline the processes through automation, standardization, and centralized management, and to convert from the developmental phase to the mature system without a compromise in quality. It required that experts carefully analyze their areas to determine what could be standardized and automated, then take the time to do it.

But the increasing flight rate had priority—quality products had to be ready on time. Further, schedules and budgets for developing the needed facility improvements were not adequate. Only the time and resources left after supporting the flight schedule could be directed toward efforts to streamline and standardize. In 1985, NASA was attempting to develop the capabilities of a production system. But it was forced to do that while responding—with the same personnel—to a higher flight rate.

At the same time the flight rate was increasing, a variety of factors reduced the number of skilled personnel available to deal with it. Theses included retirements, hiring freezes, transfers to other programs like the Space Station and transitioning to a single contractor for operations support.

[171] The flight rate did not appear to be based on assessment of available resources and capabilities and was not reduced to accommodate the capacity of the work force. For example, on January 1, 1986, a new contract took effect at Johnson that consolidated the entire contractor work force under a single company. This transition was another disturbance at a time when the work force needed to be performing at full capacity to meet the 1986 flight rate. In some important areas, a significant fraction of workers elected not to change contractors. This reduced the work force and its capabilities, and necessitated intensive training programs to qualify the new personnel. According to projections, the work force would not have been back to full capacity until the summer of 1986. This drain on a critical part of the system came just as NASA was beginning the most challenging phase of its flight schedule.

Similarly, at Kennedy the capabilities of the Shuttle processing and facilities support work force became increasingly strained as the Orbiter turnaround time decreased to accommodate the accelerated launch schedule. This factor has resulted in overtime percentages of almost 28 percent in some directorates. Numerous contract employees have worked 72 hours per week or longer and frequent 12-hours shifts. The potential implications of such overtime for safety were made apparent during the attempted launch of mission 61-C on January 6, 1986, when fatigue and shiftwork were cited as major contributing factors to a serious incident involving a liquid oxygen depletion that occurred less than five minutes before scheduled lift off....

Responding to Challenges and Changes

Another obstacle in the path toward accommodation of a higher flight rate is NASA's legendary "can-do" attitude. The attitude that enabled the agency to put men on the moon and to build the Space Shuttle will not allow it to pass up an exciting challenge—even though accepting the challenge may drain resources from the more mundane (but necessary) aspects of the program.

A recent example is NASA's decision to perform a spectacular retrieval of two communications satellites whose upper-stage motors had failed to raise them to the proper geosynchronous orbit. NASA itself then proposed to the insurance companies who owned the failed satellites that the agency design a mission to rendezvous with them in turn and that an astronaut in a jet backpack fly over to escort the satellites into the Shuttle's payload bay for a return to Earth.

The mission generated considerable excitement within NASA and required a substantial effort to develop the necessary techniques, hardware and procedures. The mission was conceived, created, designed and accomplished within 10 months. The result, mission 51-A (November, 1984), was a resounding success, as both failed satellites were successfully returned to Earth. The retrieval mission vividly demonstrated the service that astronauts and the Space Shuttle can perform.

Ten months after the first retrieval mission, NASA launched a mission to repair another communications satellite that had failed in low-Earth orbit. Again, the mission was developed and executed on relatively short notice and was resoundingly successful for both NASA and the satellite insurance industry.

The satellite retrieval missions were not isolated occurrences. Extraordinary efforts on NASA's part in developing and accomplishing missions will, and should, continue, but such efforts will be a substantial additional drain on resources. NASA cannot both accept the relatively spur-of-[172] the-moment missions that its "can-do" attitude tends to generate and also maintain the planning and scheduling discipline required to operate as a "space truck" on a routine and cost-effective basis. As the flight rate increases, the cost in resources and the accompanying impact on future operations must be considered when infrequent but extraordinary efforts are undertaken. The system is still not sufficiently developed as a "production line" process in terms of planning or implementation procedures. It cannot routinely or even periodically accept major disruptions without considerable cost. NASA's attitude historically has reflected the position that "We can do anything," and while that may essentially be true, NASA's optimism must be tempered by the realization that it cannot do everything.

NASA has always taken a positive approach to problem solving and has not evolved to the point where its officials are willing to say they no longer have the resources to respond to proposed changes....

[173] It is important to determine how many flights can be accommodated, and accommodated safely. NASA must establish a realistic level of expectation, then approach it carefully. Mission schedules should be based on a realistic assessment of what NASA can do safely and well, not on what is possible with maximum effort. The ground rules must be established firmly, and then enforced.

The attitude is important, and the word operational can mislead. "Operational" should not imply any less commitment to quality or safety, nor a dilution of resources. The attitude should be, "We are going to fly high risk flights this year; every one is going to be a challenge, and every one is going to involve some risk, so we had better be careful in our approach to each."...

[176] **Findings**

1. The capabilities of the system were stretched to the limit to support the flight rate in winter 1985/1986. Projections into the spring and summer of 1986 showed a clear trend; the system, as it existed, would have been unable to deliver crew training software for scheduled flights by the designated dates. The result would have been an unacceptable compression of the time available for the crews to accomplish their required training.

2. Spare parts are in critically short supply. The Shuttle program made a conscious decision to postpone spare parts procurements in favor of budget items of perceived higher priority. Lack of spare parts would likely have limited flight operations in 1986.

3. Stated manifesting policies are not enforced. Numerous late manifest changes (after the cargo integration review) have been made to both major payloads and minor payloads throughout the Shuttle program.

• Late changes to major payloads or program requirements can require extensive resources (money, manpower, facilities) to implement.

• If many late changes to "minor" payloads occur, resources are quickly absorbed.

• Payload specialists frequently were added to a flight well after announced deadlines.

• Late changes to a mission adversely affect the training and development of procedures for subsequent missions.

[177] 4. The scheduled flight rate did not accurately reflect the capabilities and resources.

• The flight rate was not reduced to accommodate periods of adjustment in the capacity of the work force. There was no margin in the system to accommodate unforeseen hardware problems.

• Resources were primarily directed toward supporting the flights and thus not enough were available to improve and expand facilities needed to support a higher flight rate.

5. Training simulators may be the limiting factor on the flight rate: the two current simulators cannot train crews for more than 12-15 flights per year.

6. When flights come in rapid succession, current requirements do not ensure that critical anomalies occurring during one flight are identified and addressed appropriately before the next flight.

Document IV-18

Document title: Samuel C. Phillips, NASA Management Study Group, "Recommendations to the Administrator," December 30, 1986.

Source: NASA Historical Reference Collection, NASA History Office, NASA Headquarters, Washington, D.C.

In the wake of the *Challenger* accident, former NASA Administrator James C. Fletcher was asked by President Reagan to return to the agency. One of Fletcher's early actions was to seek the advice of the National Academy of Public Administration on the management issues facing him. The academy's response was to organize a NASA Management Study Group, headed by retired General Samuel C. Phillips, who had been program manager for Apollo. The group's final report provided an overview of NASA's management problems in the post-*Challenger*, Space Station era.

---

[1] # Summary Report of the NASA Management Study Group Recommendations

I. INTRODUCTION

The NASA Management Study Group (NMSG) was established under the auspices of the National Academy of Public Administration at the request of the Administrator of NASA to assess NASA's Management practices and to evaluate the effectiveness of the NASA organization. The NMSG addresses first the organization and management of the space station program, then the restructuring of the space shuttle program, and finally NASA's overall organization and management.

Recommendations of the NMSG on the space station program were made in the form of oral briefings to the Administrator and other officials of NASA on June 26, 1986, and have subsequently been largely implemented. With respect to the space shuttle program the NMSG contributed to and reviewed the study led by Astronaut Robert Crippen and participated in the discussions that led to the Administrator's decisions announced on November 5, 1986, with which the NMSG has concurred.

This report summarizes the conclusions and recommendations of the NMSG on the overall management and organization of NASA. Detailed findings and draft recommendations were presented and discussed on several occasions during the course of the study in oral briefings to the Administrator and to the Advisory Panel of the National Academy of Public Administration. A presentation was made to the entire team of NASA top headquarters officials and center directors at an all day meeting on November 25, 1986. A final report, in the form of a revised oral briefing taking account of the comments of the Advisory Panel and the NASA officials after the November 25 meeting, was presented to Administrator and Deputy Administrator on December 16, 1986.

[2] ### II. GENERAL OBSERVATIONS

The NMSG study has concentrated on identifying issues in need of special attention by NASA management at this time. As a result, our recommendations focus on areas where changes or improvement may be required.

We must emphasize, at the outset, therefore, that a principal finding of our study is that NASA is fundamentally a sound institution, with many outstanding people with strong dedication to the success of NASA and its programs. We also recognize that many positive steps have been taken in recent months to strengthen the organization, management, and practices of NASA, and that some NMSG recommendations were adopted during the course of our study. The conclusions and recommendations set forth below should be viewed in this context.

The NMSG recognizes that NASA management is conditioned to a significant degree by factors in the external environment over which NASA has only limited control. NASA must conform to Administration policies, budgetary restrictions, Congressional guidance, and the increasingly complex web of legal and regulatory constraints affecting procurement, personnel, and other areas. As a result of the Challenger accident, NASA faces increased critical scrutiny by Congress and the media, a long hiatus in space flights, and some unrealistic public expectations of risk-free space flight. On the other hand, NASA and its program have the President's personal interest and support, and there is, we believe, strong public and Congressional support as well.

In this situation, NASA has the challenge of coping with its external environment and managing its affairs in a way that earns the respect and continued support of the Administration, Congress, and the public. To reestablish NASA's leadership [3] position in space and aeronautics, management excellence is as essential as technical excellence. Our recommendations are intended as suggestions to help NASA achieve the level of excellence it must have.

III. PRINCIPAL RECOMMENDATIONS

The principal recommendations of the NMSG can be summarized as follow:

1. Establish strong headquarters program direction for each major NASA program, with clear assignment of responsibilities to the NASA centers involved.

2. Improve the discipline and responsiveness to problems of the program management system.

3. Place shuttle and space station programs under a single Associate Administrator when the Administrator is satisfied that recovery of the shuttle will not thereby be compromised.

4. Increase management emphasis on space flight operations.

5. Place special management emphasis on establishing NASA world-class leadership in advanced technology in selected areas of both space and aeronautical technology.

6. Establish a formal planning process within NASA to enunciate long-range goals and lay out program, institutional, and financial plans for meeting them.

7. Strengthen agency-wide leadership in developing and managing people, facilities, equipment, and other institutional resources.

[4] 8. Improve management of NASA's external relations.

9. Strengthen the Office of the Administrator and ease the workload of the Administrator and Deputy Administrator.

These and other NMSG recommendations are discussed briefly in the following sections for each of the areas covered by the NMSG study.

IV. PROGRAM MANAGEMENT

Effective management of its technical program is NASA's central task. Five of the principal NMSG recommendations and many subsidiary recommendations are in this area.

1. **Establish strong headquarters program direction for each major NASA program, with clear assignment of responsibilities to the NASA center involved.**
　　a. Large multi-center spaceflight programs should be managed by a strong program director at headquarters supported by a competent program office in the Washington area. The functions of the headquarters program office should include systems engineering (a support contractor may be needed); program planning and control; management of operations and interfaces with users; safety, reliability, and quality assurance;

and other functions as appropriate. Program managers at each center should have clearly defined responsibilities and accountability to the headquarters program director. The NMSG has concurred in the actions now being taken to structure the shuttle and space station programs in this way. The NMSG also believes that the Technical Management Information System (TMIS) proposed for the space station program should be initiated but should be subject to periodic [5] review by non-advocates and outside experts to ensure that the expected utility is being achieved.

    b. Single center spaceflight programs or projects should also have a program director at headquarters with the overall program control functions of establishing requirements, reviewing progress, and approving changes as necessary. A central program control staff at the Program AA level could support the directors of several smaller programs. The program or project manager at the center should be responsible for planning and implementing the program (including systems engineering), for keeping the program director regularly informed of status and problems, and for requesting his approval of major changes that may be necessary.

    c. NASA should avoid organizing major programs so that large tasks are assigned to more than one center unless technical demands or the scale of the program clearly require substantial contribution from more than one center.

    d. A highly qualified independent office of safety, reliability, and quality assurance is an essential requirement for assuring safety and success in NASA programs. The NMSG has reviewed the goals, organization, priorities, and general plans of the new office recently established in NASA and agrees with the actions already taken and now planned.

    e. NASA Headquarters and each center should assess their procurement practices to seek to minimize the long lead times in placing contracts and to assure that proper emphasis is placed on contract structure, contractor selection, and contract administration.

    [6] 2. **Improving the discipline and responsiveness of the program management system.**

    a. Reinstitute the former system of Program Approval Documents (PAD's) as the basic agreement between the Administrator and the Program Associate Administrator responsible for the program. The PAD should contain the official statement of the program objectives and scope, how the program is to be performed, the responsibilities of the participating organizations, the total resources required (dollars, people, facilities and support from other organizations), and cost and schedule baselines against which progress can be measured. Program control documents at successive lower levels of management should be integrated into a system consistent with and supporting the PADS.

    b. Revitalize regular status reviews at each successive level of management at which progress is measured against the approved baselines, current and potential problems are fully discussed and actions assigned. The Administrator or Deputy Administrator should conduct periodic reviews of all major NASA programs.

    c. Strengthen the agency's independent cost estimating and program assessment capabilities at headquarters and at the centers.

    3. **Place shuttle and space station programs under a single Associate Administrator when the Administrator is satisfied that recovery of the shuttle will not thereby be compromised.** Although now in very different stages of development, the shuttle and space station programs should be unified to ensure proper attention to compatibility of space station design and operational planning with the shuttle and its capabilities, [7] operational availability, and requirements for logistic support. Nevertheless, the programs should not be combined until it is clear that the NASA's top priority task of returning the shuttle to flight status will not thereby be adversely affected. Until the programs are combined under one AA, the offices of Space Station and Space Flight should jointly prepare plans for the Administrator's approval which clearly define their responsibilities and relationships.

4. **Increase management emphasis on spaceflight operations.** NASA must accept that it will be responsible for spaceflight operations for the foreseeable future - shuttle, space station, man-tended and free-flying spacecraft, deep space probes, etc. The present structure of organization and management does not assure adequate attention to operations requirements in system design or in the planning and conduct of operations and logistic support in the era of frequent shuttle flights and long-term operations of the space station. A better delineation between development and operations activities is needed even before the shuttle or space station become operational. It is also important that steps be taken to accommodate users more efficiently without compromising safety. At the same time, the shuttle recovery program must not be placed at risk. Therefore, NASA should:

    a. Strengthen management of operations in the space shuttle program at headquarters and the NASA centers. Steps to do this are now under way.

    b. Ensure responsiveness to operational and user requirements in the design and development of the space station. The Offices of Space Station (OSS), Space Science and Applications, and Aeronautics and Space Technology should jointly prepare plans for the Administrator's approval which clearly define their responsibilities and relationships. OSS should ensure [8] that its organization and procedures provide adequate linkages with all major user constituencies.

    c. Establish a new Associate Administrator for Operations to develop a comprehensive plan for managing NASA spaceflight operations, to be implemented when shuttle recovery is complete. Initial priority should be given to planning for the future management of manned, mantended, and related operations. The present Offices of Space Tracking and Data Systems should become a division in the new Office of Spaceflight Operations. The NMSG anticipates that at some point in the future, the Kennedy Space Center would also be placed under the Office of Spaceflight Operations.

5. **Place special management emphasis on establishing NASA world-class leadership in advanced technology in selected areas of both space and aeronautical technology.** The NMSG believes that NASA's efforts to develop advanced technologies beyond the requirements of current spaceflight programs, on which the U.S. future in space and aeronautics will depend, need more emphasis and a clearer sense of direction. Specifically, NASA should:

    a. Strengthen capabilities for advanced research and technology development at all NASA centers.

    b. Limit spaceflight program management activities at NASA OAST research centers. This should permit a stronger focus on advanced research and technology.

    c. Seek to establish stronger linkages between the NASA research centers and industry in space technology, comparable to those that now exist in aeronautics.

[9] V. PLANNING

6. **Establish a formal planning process within NASA to enunciate long-range goals and lay out program, institutional, and financial plans for meeting them.** NMSG believes that a formal iterative planning process that involves direct participation of the entire NASA line organization at headquarters and the center would materially assist NASA by giving a clearer sense of direction and better focus to its programs.

    a. A biennial planning process should be instituted to develop detailed program, institutional, and financial plans for the next five years and skeletal plans for the ten years beyond.

    b. The plans should be developed by the line organization, based on goals and guidelines enunciated by the Administrator after taking account of the views and recommendations of the NASA program offices, congressional reports, scientific and other advisory groups, and other constituencies.

    c. The present Strategic Planning Council should be retained and its role broadened to include an annual evaluation of progress against plans.

d. A small planning support staff should be established in a new Policy and Planning Support Office reporting to the Administrator, to analyze and integrate planning within the agency and to publish and update agency plans. (See also VII-9-c.)

[10] **VI. INSTITUTIONAL MANAGEMENT**

7. **Strengthen agency-wide leadership in developing and managing people, facilities, equipment, and other institutional resources.** The NMSG believes that more attention needs to be given at headquarters and the centers to improving the management of NASA as an institution, both to make current agency operations more efficient and to assure the future strength of NASA capabilities. The NMSG recommends that NASA:

a. Appoint an Associate Deputy Administrator-Institution to provide a focus on institutional management in the Office of the Administrator. This official would assist, and when appropriate act for the Administrator and Deputy Administrator on institutional matters generally, including determination of requirements and distribution of resources for manpower, facilities, and institutional funding.

b. Strengthen the institutional management capabilities of the Program Associate Administrators, who should continue to be responsible for supervising NASA field centers as at present. Each Program Associate Administrator having supervision of a field center must assure the center's responsiveness to the requirements of programs assigned by other Associate Administrators.

c. Establish institutional planning as an integral part of the NASA planning process, to include planning for personnel, facilities, major equipment and support service contractor requirements and for the evolution of the assigned roles and missions of NASA Centers. A small staff focused on institutional planning should be included in the new Policy and Planning Support office recommended below (VII-9-c).

[11] d. Place a new special management emphasis on human resources in NASA to enhance efforts to acquire, retain, and make full utilization of the best possible people to conduct and manage NASA's work. Where necessary to meet its special needs, NASA should seek administrative or legislative relief from general government requirements that impede effective human resources management.

VII. EXECUTIVE MANAGEMENT

8. **Improve management of NASA's external relations.**
   a. Give special management attention to ensuring that NASA:
      (1) Keeps Congress informed on a timely basis of matters of importance or special interest.
      (2) Is effectively represented in dealings with other agencies, other governments, and industry.
      (3) Maintains the NASA tradition of openness in its relations with the media and the public.
   b. Consolidate under the Associate Administrator for External Relations the functions of public, international, and industry affairs, with either de facto or actual responsibility for legislative affairs.
   c. Reaffirm to all headquarters offices and field centers the requirement for consistent agency policies and actions in external affairs under the functional management leadership of the headquarters staff offices.

[12] 9. **Strengthen the Office of the Administrator and ease the workload of the Administrator and Deputy Administrator.** The NMSG believes that these needs can best be addressed by appointing two new senior officials within the Office of the Administrator and the establishment of a small policy and planning support staff unit. Specifically, NASA should:

a. Appoint an Associate Deputy Administrator-Policy to assist, and where appropriate act for the Administrator, on policy, external affairs, and related matters. The Administrator continually faces problems in the policy and external affairs areas that are growing in number and complexity. Coping with these problems now requires major personal involvements of the Administrator, creating the risks of insufficient attention to policy matters, missed opportunities for leadership, and diversion from other important responsibilities. The Associate Deputy Administrator-Policy would share the Administrator's and Deputy Administrator's workload and help ensure effective use and participation of NASA staff and program offices in policy matters and external affairs.

  b. Appoint an Associate Deputy Administrator-Institution to assist, and where appropriate act for the Administrator or Deputy Administrator in the management of NASA as an institution (IV-7-a).

  c. Establish a small Policy and Planning Support Staff for policy analysis and to support the program and institutional planning processes. This staff would provide a resource for the Office of the Administrator to perform or coordinate selected policy studies and analysis as assigned, and to assist in the review of studies and analysis done elsewhere in the agency. It [13] would also provide support for the program and institutional planning processes as previously recommended.

Document IV-19

Document title: NASA, "The Hubble Space Telescope Optical Systems Failure Report," November 1990, pp. iii-v, 9-1 to 9-4, 10-1 to 10-4.

Source: NASA Historical Reference Collection, NASA History Office, NASA Headquarters, Washington, D.C.

  The Hubble Space Telescope was launched in May 1990 with an aberration in its primary mirror that made the telescope unable to carry out significant aspects of its planned observations. The fault in the mirror was introduced during its manufacture in the early 1980s and not detected in the testing program that preceded assembly and launch of the telescope. NASA in July 1990 established a Board of Investigation to identify reasons for the fault in the Hubble mirror and for the failure to detect the fault prior to launch. The board was chaired by Jet Propulsion Laboratory Director Lew Allen.

  These excerpts from the board's report reflect the shortfalls in NASA technical management and quality assurance that contributed, along with the performance of the mirror manufacturer Perkin-Elmer, to the problems with the Hubble mirror.

[iii] EXECUTIVE SUMMARY

  The Hubble Space Telescope (HST) was launched aboard the Space Shuttle Discovery on April 24, 1990. During checkout on orbit, it was discovered that the telescope could not be properly focused because of a flaw in the optics. The HST Project Manager announced this failure on June 21, 1990. Both of the high-resolution imaging cameras (the Wide Field/Planetary Camera and the Faint Object Camera) showed the same characteristic distortion, called spherical aberration, that must have originated in the primary mirror, the secondary mirror, or both.

  The National Aeronautics and Space Administration (NASA) Associate Administrator for the Office of Space Science and Applications then formed the Hubble Space Telescope Optical Systems Board of Investigation on July 2, 1990, to determine the cause of the flaw in the telescope, how it occurred, and why it was not detected before launch. The Board conducted its investigation to include interviews with personnel involved in the fabrication and test of the telescope, review of documentation, and analysis and test of the equipment used in the fabrication of the telescope's mirrors. The information in this report is based exclusively on the analysis and tests requested by the Board, the testimony given to the Board, and the documentation found during this investigation.

Continued analysis of images transmitted from the telescope indicated that most, if not all, of the problem lies in the primary mirror. The Board's investigation of the manufacture of the mirror proved that the mirror was made in the wrong shape, being too much flattened away from the mirror's center (a 0.4-wave rms wavefront error at 632.8 nm). The error is ten times larger than the specific tolerance.

The primary mirror is a disc of glass 2.4 m in diameter, whose polished front surface is coated with a very thin layer of aluminum. When glass is polished, small amounts of material are worn away, so by selectively polishing different parts of a mirror, the shape is altered. During the manufacture of all telescope mirrors there are many repetitive cycles in which the surface is tested by reflecting light from it; the surface is then selectively polished to correct any errors in its shape. The error in the HST's mirror occurred because the optical test used in this process was not set up correctly; thus the surface was polished into the wrong shape.

The primary mirror was manufactured by the Perkin-Elmer Corporation, now Hughes Danbury Optical Systems, Inc., which was the contractor for the Optical Telescope Assembly. The critical optics used as a template in shaping the mirror, the reflective null corrector (RNC), consisted of two small mirrors and a lens. The [iv] RNC was designed and built by the Perkin-Elmer Corporation for the HST Project. This unit had been preserved by the manufacturer exactly as it was during the manufacture of the mirror. When the Board measured the RNC, the lens was incorrectly spaced from the mirrors. Calculations of the effect of such displacement on the primary mirror show that the measured amount, 1.3 mm, accounts in detail for the amount and character of the observed image blurring.

No verification of the reflective null corrector's dimensions was carried out by Perkin-Elmer after the original assembly. There were, however, clear indications of the problem from auxiliary optical tests made at the time, the results of which have been studied by the Board. A special optical unit called an inverse null corrector, designed to mimic the reflection from a perfect primary mirror, was built and used to align the apparatus; when so used, it clearly showed the error in the reflective null corrector. A second null corrector, made only with lenses, was used to measure the vertex radius of the finished primary mirror. It, too, clearly showed the error in the primary mirror. Both indicators of error were discounted at the time as being themselves flawed.

The Perkin-Elmer plan for fabricating the primary mirror placed complete reliance on the reflective null corrector as the only test to be used in both manufacturing and verifying the mirror's surface with the required precision. NASA understood and accepted this plan. This methodology should have alerted NASA management to the fragility of the process and the possibility of gross error, that is, a mistake in the process, and the need for continued care and consideration of independent measurements.

The design of the telescope and the measuring instruments was performed well by skilled optical scientists. However, the fabrication was the responsibility of the Optical Operations Division at the Perkin-Elmer Corporation (P-E), which was insulated from review or technical supervision. The P-E design scientists, management, and Technical Advisory Group, as well as NASA management and NASA review activities, all failed to follow the fabrication process with reasonable diligence and, according to testimony, were unaware that discrepant data existed, although the data were of concern to some members of P-E's Optical Operations Division. Reliance on a single test method was a process which was clearly vulnerable to simple error. Such errors had been seen in other telescope programs, yet no independent tests were planned, although some simple tests to protect against major error were considered and rejected. During the critical time period, there was great concern about cost and schedule, which further inhibited consideration of independent tests.

The most unfortunate aspect of this HST optical system failure, however, is that the data revealing these errors were available from time to time in the fabrication [v] process, but were not recognized and fully investigated at the time. Reviews were inadequate, both internally and externally, and the engineers and scientists who were qualified to analyze the test data did not do so in sufficient detail. Competitive, organizational, cost, and schedule pressures were all factors in limiting full exposure of all the test information to qualified reviewers....

[9-1] **CHAPTER IX**

WHY THE ERROR WAS NOT DETECTED PRIOR TO FLIGHT

The explanations for why the HST error was not detected before launch can be separated into two categories: factual and judgmental. Based on the test plan that was in place at the time of the fabrication of the HST mirrors, the factual issues presented in this Chapter were events that should have warned the Project personnel of the existence of a problem. The judgmental issues that follow are conclusions based on the Board's own expertise.

A. FACTUAL STATEMENTS

1. Complete reliance was placed on the reflective null corrector (RNC) to determine the shape of the primary mirror. It was determined that the RNC would be certified only by accurate measurement of the elements and the spacings. Although test philosophy placed great emphasis on "certification" of the RNC, the Board could not find documentation that the RNC was certified. In spite of the total reliance on the RNC, no independent measurements were made of the optical-element spacings of the RNC to verify the values. Although the RNC was designed so that spacings could be rechecked without disassembly, the actual implementation did not permit such measurements, and no remeasurement of spacings was made after initial assembly.

2. The erroneous measurement of the spacing of the field lens of the RNC led to the need to install spacers to increase the separation of the field lens from the lower mirror. The bolts securing the field-lens basket were not staked, suggesting a lack of quality surveillance, since securing bolts was a common and easily observable inspection to conduct. These anomalies should have led to a Material Review Board (MRB) approval document and a thorough consideration of the cause. Although the NASA representative recalls approving such an MRB, no documentation was found.

3. After the RNC was assembled in the laboratory, an INC was set up below the RNC. The INC was intended to simulate a perfect mirror below the RNC so that any errors in the null corrector could be detected. The interferograms taken when using the INC to align the RNC/CORI indicated a spherical aberration pattern (see Figure D-3). The full RNC/CORI assembly was then moved to the top of the optical telescope assembly test chamber, and each time the primary mirror was tested the INC was used to check the alignment of the setup. As before, the same spherical aberration distortion was evident in the fringes. These aberration fringes [9-2] could not be aligned out and were incorrectly attributed to the spacing errors in the lens system of the INC. Perkin-Elmer's Optical Operation Division believed that the INC was not reliable when, in fact, it was quite accurate enough to detect the gross error, and indeed did so.

4. The vertex radius measurement taken by the refractive null corrector (RvNC) indicated the presence of spherical aberration (see Figure D-2). This information was dismissed, as it was in the case for the INC, because the RvNC was believed to be less precise than the RNC and therefore not reliable. It has been determined that the RvNC was easily accurate enough to detect the spherical aberration the existed, and its reliability should not have been discounted.

5. There were two other occasions when a careful analysis of the data might have revealed the problem:

  a. The primary mirror was ground and polished to an approximate shape, about 1 wavelength rms, using the RNC for the test. This took place at Perkin-Elmer's facility in Wilton, Connecticut. The mirror was then transferred to P-E's Danbury facility, where the RNC was the test instrument for final polishing. At the time of transfer, the interferograms obtained with the RvNC were compared with those obtained from the RNC, and the discrepancy could have been noted. However, the data and the circumstances of transfer are unclear, and the requirements for transfer appeared to be adequately met; therefore no concern was noted.

b. After the assembly of the OTA, tests were performed to assure proper focus position. Those tests were made with a 0.36-mn telescope (subaperture test), and careful analysis of the data might have revealed the problem. However, the data were complicated by gravity sag because the OTA was mounted horizontally, and only the focus position was verified.

6. A range of feasible tests to verify the shape of the primary mirror were considered, but not carried out. Finally, no end-to-end tests were planned or implemented to verify the performance of the OTA.

B. JUDGMENTAL STATEMENTS

The following judgements are offered with the recognition that there were many distraction and crises during this period-cost, schedule, threat of cancellation, mirror contamination, possibility of mirror distortion caused by [9-3] mount, etc. Nevertheless, the flaw occurred and, as can now be seen, these are factors that bear on that occurrence.

1. The proposal of P-E, accepted by NASA, to rely entirely on the RNC should have alerted knowledgeable people in P-E and NASA that special attention was required to certify the RNC; to the need for independent validation of the RNC and/or the primary mirror; and to the need to examine and review the test data for any indications of inconsistency. A project test plan that considered the various measurements, the possibilities of error in each, and the feasibility of independent checks should have been prepared by the implementing organization and externally reviewed.

2. The conclusion by P-E, accepted by NASA, that the RNC was the only device that would yield an accuracy of 0.01 wave rms at 632.8 nm led P-E to fail to consider any independent measurement which would yield less accuracy. In fact, such independent data were obtained incidental to other measurements and were rationalized away due to this mindset.

3. The HST development program was complex and challenging and there were many issues demanding management attention; the primary mirror was only one of these. Although the telescope was recognized as a particular challenge, with a primary mirror requiring unprecedented performance, there was a surprising lack of participation by optical experts with experience in the manufacture of large telescopes during the fabrication phase. The NASA Project management did not have the necessary expertise to critically monitor the optical activities of the program and to probe deeply enough into the adequacy and competence of the review process that was established to guard against technical errors. The record of reviews reveals no sensitively to in-process data and no questioning of the test method.

4. The NASA Scientific Advisory Group did not have the depth of experience and skill to critically monitor the fabrication and test results of a large aspheric mirror. However, this Group should have recognized the criticality of the figure of the primary mirror and the fragility of the metrology approach, and these concerns should have impelled them to penetrate the process and ask for validation.

5. A highly competitive environment existed between Perkin-Elmer and the Eastman Kodak subcontractor. Although the manufacturing process and the method of measurement for the backup primary mirror were reviewed and approved by P-E, there was limited additional technical exchange of experience. NASA did not utilize the opportunity offered by this directed subcontract to validate, and gain confidence in, the P-E approach to the primary mirror manufacture.

[9-4] 6. Perkin-Elmer line management did not review or supervise their Optical Operations Division adequately. In fact, the management structure provided a strong block against communication between the people actually doing the job and higher level experts both within and outside of P-E.

7. The P-E Technical Advisory Group did not probe at all deeply into the optical manufacturing processes and, although they recognized the fragility of the measuring approach, they did not adequately assert their concerns or follow up with data reviews. This is particularly surprising since the members were aware of the history of manufacture of other Ritchey-Chretien telescopes, where spherical aberration was known to be a common problem.

8. The most capable optical scientists at P-E were involved closely with the production of the 1.5-m demonstration mirror and the design of the HST mirror and the test apparatus. However, fabrication of the HST mirror was the responsibility of the Optical Operations Division of P-E, which did not include optical design scientists and which did not use the skills external to the Division which were available at Perkin-Elmer.

9. The Optical Operations Division at P-E operated in a "closed-door" environment which permitted discrepant data to be discounted without review. During the testimony, it was indicated that some technical personnel in the Optical Operations Division were deeply concerned at the time that the discrepant optical data might indicate a flaw. There are no indications that these concerns were formally expressed outside this Division.

10. The quality assurance people at P-E, NASA, and DCAS (Defense Contract Administration Services, now Defense Contract Management Command) were not optical experts and, therefore, were not able to distinguish the presence of inconsistent data results from the optical tests. The DCAS people concentrated mainly on safety issues.

11. The basic product assurance requirements and formal review processes were procedurally adequate to raise critical issues in most safety, material, and handling matters, but not in optical matters.

12. The inability of P-E to provide the Board with vital archival data on the design and manufacture of the primary mirror is an indication of inadequate documentation practices, which hampered the Board in determining the source of the primary mirror error.

[10-1] **CHAPTER X**

LESSONS LEARNED

A. IDENTIFY AND MITIGATE RISK

The Project Manager must make a deliberate effort to identify those aspects of the project where there is a risk of error with serious consequences for the mission. Upon recognizing the risks the manager must consider those actions which mitigate that risk.

[10-2] In this case, the primary mirror fabrication task was identified as particularly challenging due to the stringent performance requirements. The contractor clearly specified in the proposal that total reliance would be placed on a single test instrument and that no optical performance tests would be made at higher levels of assembly. Therefore, OTA performance would be determined by component tests and great care in precision assembly. Although NASA accepted this proposal, the methodology should have alerted NASA management to the fragility of the process, the possibility of gross error (that is, mistake in the process), and the need for continued care and consideration of independent tests.

The history of spherical aberration in the primary mirrors of Ritchey-Chretien telescopes was known to some of the optical scientists involved, but did not lead to specific recommendations early in the Project. Late in the Project an advisory group did call out the risk of gross error and suggested simple tests to check for such errors. This recommendation was not seriously considered, primarily due to total lack of concern that such a risk was reasonable, but also in view of cost and schedule problems.

Several methods of detecting the flaw were inherent in the testing, but Project management did not recognize the value of or need for independent tests. Project management was concerned about the performance specifications and directed a subcontract to Eastman Kodak Company for an alternate primary mirror. The Eastman Kodak mirror was fabricated and tested using quite different techniques. The mirror or the instrumentation could also have served as cross-checks for gross error. Such error checks were not made, again due to total lack of concern about the possibility of gross error. Project management failed to identify a significant risk and therefore failed to consider mitigating actions. A formal discipline such as fault-tree analysis might have assisted the manager in directing his attention to this risk.

[10-3] B. MAINTAIN GOOD COMMUNICATION WITHIN THE PROJECT

While proper delegation of responsibility and authority is important, this delegation must not restrict communication such that problems are not subject to review. In this case, the Optical Operations Division of P-E was allowed to operated in an artisan, closed-door mode. The impermeability of this Division seems astounding. The optical designers at P-E did not learn how their designs were being implemented; e.g., if the designer of the null correctors had been following their use, the data from the INC and the RNC likely would not have been discounted. The data indicating the flaw was of great concern to some members of the division. Testimony indicates that their concerns were addressed at the level of the head of metrology and the division manager, but were not satisfied by the decision to rely only on the RNC data and remained deeply concerned. Their concerns and the data which caused them did not seem to come to the attention of anyone external to the division. P-E management should have been sensitive and open to these concerns. The P-E Technical Advisory Group should have found out what was going on in the Division and insisted on reviewing in-process data. NASA Project management should have been aware that communications were failing with the Optical Operations Division.

Contributing to poor communications was an apparent philosophy at MSFC at the time to resolve issues at the lowest possible level and to consider problems that surfaced at reviews to be indications of bad management.

A culture must be developed in any project which encourages concerns to be expressed and which ensures that those concerns which deal with a potential risk to the mission cannot be disposed without appropriate review, a review which includes NASA project management.

C. UNDERSTAND ACCURACY OF CRITICAL MEASUREMENTS

The project manager must understand the accuracy of critical measurements. P-E concluded, based on design considerations, that the RNC was the only test device which could achieve the required precision. They stated that its performance could not be determined by optical test but would be determined by component and assembly measurements which could be made in situ. P-E engineers regarded the RNC as "certified" and the INC and RvNC as "uncertified." The terms were not defined, and "certification" was not documented. P-E discounted evidence of spherical aberration from INC and RNC measurements on the basis of "uncertified" status. In fact, the Board reviewed a recent as-built error analysis of both devices. The review showed the RNC to be [10-3] accurate to 0.02 wave rms and the INC to 0.14 wave rms. This indicates that the INC is a factor of three more accurate than the error observed in the INC/RNC interferograms. While in-process data were not subject to external review, which is another lesson, the methodology of test instrument use was reviewed by P-E and NASA management. This review could and should have questioned the judgment not to use the INC or the RNC as independent checks of the *accuracy* of the RvNC even though the *precision* was not to specification. Project management must understand critical tests and measurement.

In addition, the project management must seriously consider the classification of test equipment that directly impacts the flight hardware. The RNC was classified as standard test equipment, which means that the RNC was not subject to the rigorous documentation and review requirements demanded of items classified as flight hardware equipment. Under the contract, there were no Government regulations requiring that records for the RNC be maintained. Considering the importance placed on the RNC in the test program, management should have upgraded the level of classification of this equipment.

Key decisions, test results, and changes in plans and procedures must be adequately documented. In preparing such documentation, individuals are forced to review and explain inconsistencies in the test data. This also provides a communication link to those individuals who are responsible for overseeing the project.

D. ENSURE CLEAR ASSIGNMENT OF RESPONSIBILITY

Project managers must ensure clear assignment of responsibility to QA and Engineering. NASA QA personnel were not optical system experts. The Project relied upon P-E Engineering to establish test and fabrication procedures, and P-E or NASA QA generally verified that Engineering approved and certified accomplishment of procedures. However, at times, NASA management seemed to rely on QA to verify the adequacy of procedures and the fact that they were satisfactorily accomplished. This lack of clarity apparently led to incomplete documentation and may have contributed to faulty procedures. The project manager must know what QA can and cannot do, and when it is necessary to rely on engineering for verifying its own procedures, management should be alert to the need for independent checks.

Quality assurance, to be truly effective, must have an independent reporting path to top management.

[10-4] ### E. REMEMBER THE MISSION DURING CRISIS

There will be a period of crisis in cost or schedule during most challenging projects. The project manager must be especially careful during such periods that the project does not become distracted and fail to give proper consideration to prudent action. At one point in the fabrication cycle of the primary mirror, an urgent recommendation for independent tests to check for gross error entered the system, but was apparently not acted upon. Again, at the completion of mirror polishing, the final review of data for a final report was abandoned and the team reassigned as a cost-cutting measure.

F. MAINTAIN RIGOROUS DOCUMENTATION

The project manager should ensure that documentation covering design, development, fabrication, and testing is rigorously prepared, indexed, and maintained. Because quality, at a minimum, consists in meeting requirements, it is not possible to determine whether the necessary quality is being achieved if the requirements are not set forth in sufficient detail and maintained in retrievable archival form. Adequate documentation also helps maintain a disciplined approach to fabrication and testing processes, especially with so complicated a project as the HST.

Document IV-20

Document title: *Report of the Advisory Committee on the Future of the U.S. Space Program*, (Washington, DC: U.S. Government Printing Office, December 1990), pp. 47-48.

On July 25, 1990, the White House announced the creation of a blue-ribbon panel to make a comprehensive assessment of the status of the U.S. civilian space program. This announcement was the result of increasing dissatisfaction on the part of the National Space Council with both NASA's response to President Bush's 1989 call for what came to be known as the "Space Exploration Initiative" and NASA's technical and mangerial performance as evidenced by the grounding of the space shuttle in June 1990 because of problems in its fuel lines and the discovery that the Hubble Space Telescope had been launched with an improperly shaped primary mirror. The original intent was to have the panel report only to the vice president in his role as the chairman of the National Space Council, but NASA Administrator Richard Truly successfully argued that it should also report to him. The panel was chaired by Martin Marietta Corporation Chief Executive Officer Norman R. Augustine, and it was composed of a cross-section of individuals knowledgeable about all aspects of U.S. space efforts. The panel held a series of hearings and conducted fact-finding visits around the country between September and November 1990. It issued its report on December 17, 1990. Published here are the principal recommendations of the commission.

---

[47] **Principal Recommendations**

This report offers specific recommendations pertaining to civil space goals and program content as well as suggestions relating to internal NASA management. These are summarized below in four primary groupings. In order to implement fully these recommendations and suggestions, the support of both the Executive Branch and Legislative Branch will be needed, and of NASA itself.

Principal Recommendations Concerning Space Goals

It is recommended that the United States' future civil space program consist of a balanced set of five principal elements:

- a science program, which enjoys highest priority within the civil space program, and is maintained at or above the current fraction of the NASA budget (Recommendations 1 and 2);
- a Mission to Planet Earth (MTPE) focusing on environmental measurements (Recommendation 3);
- a Mission from Planet Earth (MFPE), with the long-term goal of human exploration of Mars, preceded by a modified Space Station which emphasizes life-sciences, an exploration base on the moon, and robotic precursors to Mars (Recommendations 4, 5, 6, and 7);
- a significantly expanded technology development activity, closely coupled to space mission objectives, with particular attention devoted to engines + a robust space transportation system (Recommendation 9).

Principal Recommendations Concerning Programs

With regard to program content, it is recommended that:

- the strategic plan for science currently under consideration be implemented (Recommendation 2);
- a revitalized technology plan be prepared with strong input from the mission offices, and that it be funded (Recommendation 8);
- Space Shuttle missions be phased over to a new unmanned (heavy-lift) launch vehicle except for mission where human involvement is essential or other critical national needs dictate (Recommendation 9);
- Space Station Freedom be revamped to emphasize life-sciences and human space operations, and include microgravity research as appropriate. It should be reconfigured to reduce cost and complexity; and the current 90-day time limit on redesign should be extended if a thorough reassessment is not possible in that period (Recommendation 6);
- a personnel module be provided, as planned, for emergency return from Space Station Freedom, and that initial provisions be made for two-way missions in the event of unavailability of the Space Shuttle (Recommendation 11).

Principal Recommendations Concerning Affordability

It is recommended that the NASA program be structured in scope so as not to exceed a funding profile containing approximately 10 percent real growth per year throughout the remainder of the decade and then remaining at that level, including but not limited to the following actions:

- redesign and reschedule the Space Station Freedom to reduce cost and complexity (Recommendation 6);

- defer or eliminate the planned purchase of another orbiter (Recommendation 10);
- place the Mission from Planet Earth on a "go-as-you-pay" basis, i.e., tailoring the schedule to match the availability of funds (Recommendation 5).

Principal Recommendations Concerning Management

With regard to management of the civil space program, it is recommended that:
- an Executive Committee of the Space Council be established which includes the Administrator of NASA (Recommendation 12);
- major reforms be made in the civil service regulations as they apply to specialty skills; or, if that is not possible, exemptions be granted to NASA for at least 10 percent of its employees to operate under a tailored personnel system; or, as a final [48] alternative, that NASA begin selectively converting at least some of its centers into university-affiliated Federally Funded Research and Development Centers (Recommendations 14 and 15);
- NASA management review the mission of each center to consolidate and refocus centers of excellence in currently relevant fields with minimum overlap among centers (Recommendation 13).

It is considered by the Committee that the *internal* organization of any institution should be the province of, and at the discretion of, those bearing ultimate responsibility for the performance of that institution....
- That the current headquarters structure be revamped, disestablishing the positions of certain existing Associate Administrators...
- an exceptionally well-qualified independent cost analysis group be attached to headquarters with ultimate responsibility for all top-level cost estimating including cost estimates provided outside of NASA;
- a systems concept and analysis group reporting to the Administrator of NASA be established as a Federally Funded Research and Development Center;
- multi-center projects be avoided wherever possible, but when this is not practical, a strong and independent project office reporting to headquarters be established near the center having the principle share of the work for that project; and that this project office have a systems engineering staff and full budget authority (ideally industrial funding—i.e., funding allocations related specifically to end-goals).

In summary, we recommend:

1) Establishing the science program as the highest priority element of the civil space program, to be maintained at or above the current fraction of the budget.

2) Obtaining exclusions for a portion of NASA's employees from existing civil service rules or, failing that, beginning a gradual conversion of selected centers to Federally Funded Research and Development Centers affiliated with universities, using as a model the Jet Propulsion Laboratory.

3) Redesigning the Space Station Freedom to lessen complexity and reduce cost, taking whatever time may be required to do this thoroughly and innovatively.

4) Pursuing a Mission from Planet Earth as a complement to the Mission to Planet Earth, with the former having Mars as its very long-term goal—but relieved of schedule pressures and progressing according to the availability of funding.

5) Reducing our dependence on the Space Shuttle by phasing over to a new unmanned heavy lift launch vehicle for all but missions requiring human presence.

The Committee would be pleased to meet again in perhaps six months should the NASA Administrator so desire, in order to assist on the implementation process. In the meantime, NASA may wish to seek the assistance of its regular outside advisory group, the NASA Advisory Council, to provide independent and ongoing advice for implementing these findings.

Each of the recommendations herein is supported unanimously by the members of the Advisory Committee on the Future of the U.S. Space Program.

Biographical Appendix

A

Spiro T. Agnew (1918-) was elected vice president of the United States in November 1968, serving under Richard M. Nixon. He served as chair of the 1969 Space Task Group that developed a long-range plan for a post-Apollo space effort. *The Post-Apollo Space Program: Directions for the Future* (Washington, DC: President's Science Advisory Council, September 1969) developed an expansive program that included building a space station, a space shuttle, a lunar base, and a mission to Mars (this last goal had been endorsed by Agnew at the time of the *Apollo 11* launch in July 1969). This plan was not accepted by Nixon, and only the Space Shuttle was approved for development. See Roger D. Launius, "NASA and the Decision to Build the Space Shuttle, 1969-72," *The Historian* 57 (Autumn 1994): 17-34.

Neil A. Armstrong (1930-) was the first American to set foot on the Moon, on July 20, 1969, as commander of *Apollo 11*. He had become an astronaut in 1962, after having served as a test pilot with the National Advisory Committee for Aeronautics (1955-1958) and NASA (1958-1962). He also flew as command pilot on *Gemini VIII* in March 1966. In 1970 and 1971 he was deputy associate administrator for the Office of Advanced Research and Technology at NASA headquarters. In 1971 he left NASA to become a professor of aerospace engineering at the University of Cincinnati and to undertake private consulting. See Neil A. Armstrong, *et al., First on the Moon: A Voyage with Neil Armstrong, Michael Collins and Edwin E. Aldrin, Jr.* (Boston: Little, Brown, 1970); Neil A. Armstrong, et al., *The First Lunar Landing: 20th Anniversary/as Told by the Astronauts, Neil Armstrong, Edwin Aldrin, Michael Collins* (Washington, DC: National Aeronautics and Space Administration EP-73, 1989).

Henry H. (Hap) Arnold (1886-1950) was commander of the Army Air Forces in World War II and was the only air commander ever to attain the five-star rank of general of the armies. He was especially interested in the development of sophisticated aerospace technology to give the United States an edge in the achievement of air superiority, and he fostered the development of such innovations as jet aircraft, rocketry, rocket-assisted take-off, and supersonic flight. After a lengthy career as an Army aviator and commander that spanned the two world wars, he retired from active service in 1945. See Henry H. Arnold, *Global Mission* (New York: Harper & Brothers, 1949); Flint O. DuPre, *Hap Arnold: Architect of American Air Power* (New York: Macmillan, 1972); Thomas M. Coffey, *Hap: The Story of the U.S. Air Force and the Man Who Built It* (New York: Viking, 1982).

Isaac Asimov (1920-1992) was born in Petrovichi, Russia, and came to the United States in 1923. He became a member of the faculty, in biochemistry, at Boston University but gained his greatest fame as a writer of extremely sophisticated science fiction. He is best known for the Foundation trilogy (1951-1953), as well as *I, Robot* (1950) and *Fantastic Voyage* (1966). In all, Asimov published more than 200 books during his life, many of them fiction but also some nonfiction. See "Isaac Asimov," biographical file, NASA Historical Reference Collection, NASA Headquarters, Washington, D.C.

Norman R. Augustine (1935-) was born in Denver, Colorado, and has long been a key person in the aerospace industry. He became chairman and CEO of the Martin Marietta Corporation in the 1980s. Previously, he had served as the Under Secretary of the Army, Assistant Secretary of the Army for Research and Development, and as an Assistant Director of Defense Research and Engineering in the Office of the Secretary of Defense. In 1990 he was appointed to head the Advisory Committee on the Future of the U.S. Space Program for the Bush administration. This panel produced the *Report of the Advisory Committee on the Future of the U.S. Space Program* (Washington, DC: Government Printing Office, December 1990). The study was enormously important in charting the course of the space program in the first half of the 1990s. See Norman R. Augustine, *Augustine's Laws* (Washington, DC: American Institute for Aeronautics and Astronautics, 1984); "Norman R. Augustine," biographical file, NASA Historical Reference Collection.

B

James E. Beggs (1926-) served as NASA Administrator from July 10, 1981, to December 4, 1985, when he took an indefinite leave of absence pending disposition of an indictment from the Justice Department for activities taking place prior to his tenure at NASA. This indictment was later dismissed, and the U.S. Attorney General apologized to Beggs for any embarrassment. His resignation from NASA was effective on February 25, 1986. Prior to NASA, Beggs had been executive vice president and a director of General Dynamics Corp. in St. Louis, Missouri. Previously, he had served with NASA in 1968-1969 as associate administrator for the Office of Advanced Research and Technology. From 1969 to 1973 he was under secretary at the Department of Transportation. He went to Summa Corp., Los Angeles, California, as managing director of operations and joined General Dynamics in January 1974. Before joining NASA, he also had been with Westinghouse Electric Corp., in Sharon, Pennsylvania, and in Baltimore, Maryland, for thirteen years. A 1947 graduate of the U.S. Naval Academy, he served with the Navy until 1954. In 1955 he received a master's degree from the Harvard Graduate School of Business Administration.

David E. Bell (1919-   ) was budget director for President Kennedy from 1961 to 1962. A Harvard University-trained economist, Bell had previously been a member of the staff of the Bureau of the Budget and special assistant to the president during the Truman administration before returning to the Harvard faculty during the late 1950s. Between 1962 and 1966 he served as head of the Agency for International Development, and then he was vice president of the Ford Foundation. While budget director, Bell was responsible for working with NASA in establishing a realistic financial outlook for Project Apollo.

Spencer M. Beresford (1918-1992) was a general counsel for NASA between 1963 and 1973. A Washington lawyer, he served as a naval officer in World War II and the Korean War, but in 1954 he became general counsel for the Foreign Operations Administration. In 1957 he joined the Legislative Reference Service of the Library of Congress, and in 1958 and 1959, he was special counsel to the House Select Committee on Astronautics and Space Exploration. He performed a similar duty for the House Committee on Science and Technology, 1959-1962. After completing his assignment at NASA in 1973, Beresford became general counsel for the Office of Technology Assessment. See "Spencer M. Beresford," biographical file, NASA Historical Reference Collection.

Chesley Bonestell (1888-1986) was a world-famous artist who was known as the creator of significant space-oriented artwork. From 1944 on, he mostly worked in space art and illustrated numerous books such as Willy Ley's *The Conquest of Space* and articles such as Wernher von Braun's articles for the *Collier's* magazine series on spaceflight in the 1950s. He also illustrated space sets for science fiction films such as *Destination Moon* (1950). See "Chesley Bonestell," *Ad Astra*, July/August 1991, p. 9.

Karel J. Bossart (1904-1975) was a pre-World War II immigrant from Belgium, who was involved early on in the development of rocket technology with the Convair Corporation. In the 1950s he was largely responsible for designing the Atlas ICBM booster with a very thin, internally pressurized fuselage instead of massive struts and a thick metal skin. See Richard E. Martin, *The Atlas and Centaur "Steel Balloon" Tanks: A Legacy of Karel Bossart* (San Diego: General Dynamics Corp., 1989); Robert L. Perry, "The Atlas, Thor, Titan, and Minuteman," in Eugene M. Emme, ed., *A History of Rocket Technology* (Detroit, MI: Wayne State University Press, 1964), pp. 143-55; John L. Sloop, *Liquid Hydrogen as a Propulsion Fuel, 1945-1959* (Washington, DC: NASA SP-4404, 1978), pp. 173-77.

Lewis M. Branscomb (1926-   ) is a Harvard University-trained physicist who served in a variety of university and public service posts before becoming the chief scientist of the IBM Corporation (*American Men and Women of Science, 1989-1990* [New York: R.R. Bowker, 1990], p. 692).

Wernher von Braun (1912-1977) was the leader of what has been called the "rocket team," which had developed the German V-2 ballistic missile in World War II. At the conclusion of the war, von Braun and some of his chief assistants—as part of a military operation called Project Paperclip—came to America and were installed at Fort Bliss in El Paso, Texas, to work on rocket development and use the V-2 for high altitude research. They used launch facilities at the nearby White Sands Proving Ground in New Mexico. Later, in 1950 von Braun's team moved to the Redstone Arsenal near Huntsville, Alabama, to concentrate on developing a new missile for the Army. They built the Army's Jupiter ballistic missile, and before that the Redstone, used by NASA to launch the first Mercury capsules. The story of von Braun and the "rocket team" has been told many times. See, as examples, David H. DeVorkin, *Science With a Vengeance: How the Military Created the US Space Sciences After World War II* (New York: Springer-Verlag, 1992); Frederick I. Ordway III and Mitchell R. Sharpe, *The Rocket Team* (New York: Thomas Y. Crowell, 1979); Erik Bergaust, *Wernher von Braun* (Washington, DC: National Space Institute, 1976).

Styles Bridges (1898-1961) (R-NH) served as governor of New Hampshire, 1935-1937, and was elected to the Senate in 1936. During the early years of NASA, he was the ranking Republican member of the Appropriations Committee, a member of the Armed Services Committee and its preparedness investigating subcommittee, as well as the Aeronautical and Space Sciences Committee. He was the leader of his party's conservative wing and a strong proponent of military preparedness. Bryce Harlow told Eisenhower in 1958 that Bridges was "a walking 25 votes in the Senate, the most skilled maneuverer on the Republican side" (quoted in Robert A. Divine, *The Sputnik Challenge: Eisenhower's Response to the Soviet Satellite* [New York: Oxford University Press, 1993], p. 140).

Bernard Brodie (1910-1977) was a well-known political scientist who specialized in studies of Cold War strategy, especially nuclear policy. With a Ph.D. from the University of Chicago, he was a member of Project Rand, later the Rand Corporation, and prepared numerous studies and books for public policy purposes.

Detlev W. Bronk (1897-1975) was president of the National Academy of Sciences, 1950-1962, and a member of the National Aeronautics and Space Council. A scientist, he was president of The Johns Hopkins University, 1949-1953, and Rockefeller University, 1953-1968.

Percival Brundage (1892-1981) was the first deputy director and then director of the Bureau of the Budget, 1954-1958. Thereafter, he went on to a series of business and financial positions.

Nikolai A. Bulganin was chairman of the Soviet Council of Ministers, and he was heavily involved in the negotiations over the freedom of space issue in terms of flying over territories.

George H.W. Bush (1924-) served as president of the United States between 1989 and 1993. Before that, he had been a diplomat, director of the Central Intelligence Agency (CIA), and vice president under Ronald Reagan (1981-1989).

Vannevar Bush (1890-1974) was one of the most powerful members of the scientific and technological elite to emerge during World War II. An aeronautical engineer on the faculty at the Massachusetts Institute of Technology, Bush lobbied to create and then headed the National Defense Research Committee in 1940 to oversee science and technology in the federal government. Later, its name was changed to the Office of Science Research and Development, and Bush used it as a means to build a powerful infrastructure for scientific research in support of the federal government. Although he went to the Carnegie Institution after the war, Bush remained a powerful force in shaping post-war science and technology by serving on numerous federal advisory committees and preparing several influential reports. See David Petechuk, "Vannevar Bush," in Emily J. McMurray, *et al.*, eds., *Notable Twentieth-Century Scientists* (New York: Gale Research Inc., 1995), pp. 285-88.

C

Howard W. Cannon (1912-) (D-NV) was first elected to the Senate in 1958 and served until 1983.

Jimmy Carter (1924-) was president of the United States between 1977 and 1981. Previously, he had been a naval officer and businessman before entering politics. He entered politics in the Georgia State Legislature (1962-1966) and later served as the governor of Georgia (1971-1975).

Benjamin Chidlaw (1900-) was a career U.S. Air Force officer. He entered the U.S. Army Air Corps in 1924 as a pilot and progressed through a series of ranks until he became chief of the Materiel Division at General Headquarters, Army Air Forces, in 1942. He was deputy commander of the Army Air Forces for the Mediterranean Theater, 1944-1945, and deputy at the Air Material Command, 1945-1949. He served as commander of several research and development organizations for the Air Force and retired as a four-star general in 1955.

William Clark was Ronald Reagan's assistant for National Security Affairs and chair of the Senior Interagency Group (Space) that worked on the decision to develop the Space Station.

Arthur C. Clarke (1917-), one of the most well-known science fiction authors, has also been an eloquent writer on behalf of the exploration of space. In 1945, before the invention of the transistor, Clarke wrote an article in *Extraterrestrial Relays* describing the possibility of geosynchronous orbit and the development of communication relays by satellite. He also wrote several novels, and the most well-known was *2001: A Space Odyssey*, based on a screenplay of the same name that he prepared for Stanley Kubrick. The movie is still one of the most realistic depictions of the rigors of spaceflight ever to be filmed.

Francis H. Clauser (1913-) was a leading research aerodynamicist in academia and the aerospace industry until the 1970s. He worked with the Douglas Aircraft Co., 1937-1946, and served as chair of aerospace studies at The Johns Hopkins University, 1946-1960. He then served in a variety of academic appointments; from 1969 until retirement in 1980, he was the Clark B. Millikin Professor of Aerospace Engineering at the California Institute of Technology.

Ansley Johnson Coale (1917-) received a Ph.D. from Princeton University in 1947 and worked in several capacities with the federal government in social science and population statistics. He became a professor of economics at Princeton in 1947 and also directed the Office of Population Research between 1959 and 1973. He was especially involved in research associated with population loss from a nuclear holocaust.

William Congreve (1772-1828) of Great Britain was an artillery officer and inventor, who was best known for his work on black powder rockets that could be used for bombardment of enemy fortifications. He based his rocketry work on the pioneering work of Indian prince Hyder Ali, who had successfully used them against the British in 1792 and 1799 at Seringapatam. Congreve's rockets were used in the Napoleonic Wars and in the War of 1812 (Frank H. Winter, *The First Golden Age of Rocketry: Congreve and Hale Rockets of the Nineteenth Century* [Washington, DC: Smithsonian Institution Press, 1990]).

Donald Clarence Cook (1909-1981) was a government official, lawyer, and businessman who held numerous posts from 1935 to 1945 in the Securities and Exchange Commission (SEC), as well as staff positions in other agencies and in Congress before being appointed SEC member in 1949. In 1952 he became chair of the SEC. He joined the American Electric Power Company in 1953, and he served as its president between 1962 and 1972 and as its chair from 1971 to 1976 ("Cook, Donald C[larence]," *Current Biography 1982*, p. 462).

Nicolaus Copernicus (1473-1543) of Poland symbolized the spirit of scientific inquiry that came to dominate the Renaissance. The son of a prosperous merchant, when his father died Copernicus was raised by his uncle, Lucas Watzelrode, the Bishop of Ermland. He was educated at the University of Cracow, where he excelled at mathematics, and at the University of Bologna in Italy, where he began to study astronomy. Copernicus developed complex models of movement for the Earth and other planets around the Sun. His "Heliocentric Solar System" concept gained acceptance slowly, but a century after his death was accepted as the norm for the scientific community (Edward Rosen, "Nicolaus Copernicus," *Dictionary of Scientific Biography* [New York: Charles Scribner's Sons, 1971], 3: 401-402).

John J. Corson (1905-1990) was a management consultant with McKinsey & Co. since 1951, remaining there until 1966. T. Keith Glennan contracted with McKinsey & Co. for a series of studies. These included: "Organizing Headquarters Functions," two volumes, December 1958; "Financial Management—NASA-JPL Relationships," February 1959; "Security and Safety—NASA-JPL Relationships," February 1959; "Facilities Construction—NASA-JPL Relationships," February 1959; "Procurement and Subcontracting—NASA-JPL Relationships," February 1959; "NASA-JPL Relationships and the Role of the Western Coordination Office," March 1959; "Providing Supporting Services for the Development Operations Division," January 1960, on the transfer of the Army Ballistic Missile Agency to NASA; "Report of the Advisory Committee on Organization," October 1960; and "An Evaluation of NASA's Contracting Politics, Organization, and Performance," October 1960. All are in "T. Keith Glennan," correspondence files, NASA Historical Reference Collection.

Alan Cranston (1914-   ) (D-CA) served in the U.S. Senate from 1969 to 1991.

Robert Cutler (1895-1974) was a lawyer and banking executive. He practiced law in Boston from 1922 to 1942 and then became president and director of the Old Colony Trust Co., 1946-1953, and its chairman for the next several years. He served as special assistant for security affairs for President Eisenhower, 1953-1960. From 1960 to 1962 he served as executive director of the Inter-American Development Bank.

Cyrano de Bergerac, Savinien (1619-1655) was a French writer whose works combined political satire and fantasy. As a young man, he joined the company of guards, but he was wounded at the siege of Arras in 1640 and retired from military life. He then studied under philosopher and mathematician Pierre Gassendi, whose influence was significant. His two best known written works were his two novels of spaceflight to the Moon. He has become famous in the twentieth century largely through the 1897 novel by Edmond Rostand, who described him as a gallant and brilliant, but ugly man with the large nose ("Cyrano de Bergerac, Savinien," *The New Encyclopedia Britannica* [Chicago: Encyclopedia Britannica, Inc., 1987 ed.], 3: 829).

D

Edward E. David, Jr. (1925-   ), served as science advisor to President Richard M. Nixon in 1970 and then as director of the Office of Science and Technology. Previously, he had served as executive director of research of Bell Telephone Laboratories, 1950-1970. For a discussion of the President's Science Advisory Committee, see Gregg Herken, *Cardinal Choices: Science Advice to the President from Hiroshima to SDI* (New York: Oxford University Press, 1992).

Merton E. Davies (1917-   ) was educated at Stanford University and worked at the Douglas Aircraft Co., 1940-1948, and at the Rand Corporation since 1948. He served as a member of the U.S. delegation to the Surprise Attack Conference in Geneva in 1958 and on the imaging teams of *Mariner 6* and *7* in 1969, *Mariner 9* in 1971, and *Voyager* in 1977.

Kurt H. Debus (1908-1983) earned a B.S. in mechanical engineering (1933) and an M.S. (1935) and Ph.D. (1939) in electrical engineering, all from the Technical University of Darmstadt in Germany. He became an assistant professor at the university after receiving his degree. During the course of World War II, he became an experimental engineer at the A-4 (V-2) test stand at Peenemünde (see entry for Wernher von Braun), rising to become superintendent of the test stand and test firing stand for the rocket. In 1945, he came to the United States with a group of engineers and scientists headed by von Braun. From 1945 to 1950 the group worked at Fort Bliss, Texas, and then moved to the Redstone Arsenal in Huntsville, Alabama. From 1952 to 1960 Debus was chief of the missile firing laboratory of the Army Ballistic Missile Agency. In this position, he was located at Cape Canaveral, Florida, where he supervised the launching of the first ballistic missile fired from there, an Army Redstone. When the Army Ballistic Missile Agency became part of NASA, Debus continued to supervise missile and space vehicle launchings, first as director of the Launch Operations Center and then of the Kennedy Space Center (as it was renamed in December 1963). He retired from that position in 1974 ("Kurt H. Debus," biographical file, NASA Historical Reference Collection).

Thomas Digges was an astronomer and mathematician who modified Dante's medieval conceptions of the universe in his *Description of the Caelestiall Orbes* (1576), adopting a Copernican view that placed the Sun in the center of the universe.

Everett Dirksen (1896-1969) (R-IL) served in the U.S. Senate from 1951 to 1969 and in the U.S. House of Representatives from 1933 to 1949. He served as the Republican leader in the Senate from 1959 until 1969 (*Current Biography 1969*, p. 465).

Walt Disney (1901-1966) was the creator of Mickey Mouse and several other animated characters. In 1955 his weekly television series aired the first of three programs related to spaceflight. The first of these, "Man in Space," premiered on March 9, 1955, with an estimated audience of 42 million. The second show, "Man and the Moon," also aired in 1955 and sported the powerful image of a wheel-like space station as a launching point for a mission to the Moon. The final show, "Mars and Beyond," premiered on December 4, 1957, after the launching of Sputnik I (obituary in *New York Times*, December 16, 1966, p. 1).

Allen Frances Donovan (1914-   ) was an accomplished aeronautical engineer who worked for several aeronautical firms between 1936 and 1946. He then headed the aeronautical mechanics department at Cornell University from 1946 to 1955. He later became a corporate executive with the Aerospace Corporation, serving as senior vice president, technical, 1960-1978. He also served on several government advisory boards, including the President's Science Advisory Committee, 1959-1968.

James H. Doolittle (1896-1993) was a longtime aviation promoter, air racer, Air Force officer, and aerospace research and development advocate. He had served with the U.S. Army Air Corps between 1917 and 1930, and then he was manager of the aviation section for Shell Oil Co. between 1930 and 1940. In World War II Doolittle won early fame for leading the April 1942 bombing of Tokyo, and then he was commander of a succession of air units in Africa, the Pacific, and Europe. He was promoted to the rank of lieutenant general in 1944. After the war he was a member of the Air Force's Scientific Advisory Board and the President's Scientific Advisory Committee. At the time of Sputnik, he was chair of the National Advisory Committee for Aeronautics and the Air Force Scientific Advisory Board. In 1985, the Senate approved his promotion in retirement to four-star general (General James H. (Jimmy) Doolittle with Carroll V. Glines, *I Could Never Be So Lucky Again: An Autobiography* [New York: Bantam Books, 1991]; Carroll V. Glines, *Jimmy Doolittle: Daredevil Aviator and Scientist* [New York: Macmillan, 1972]; "James H. Doolittle," biographical file, NASA Historical Reference Collection).

Walter Dornberger (1895-1980) was Wernher von Braun's military superior during the German rocket development program of World War II. He oversaw the effort at Peenemünde to build the V-2, fostering internal communication and successfully advocating the program to officials in the German army. He also assembled the team of highly talented engineers under von Braun's direction and provided the funding and staff organization necessary to complete the technology project. After World War II Dornberger came to the United States and assisted the Department of Defense with the development of ballistic missiles. He also worked for the Bell Aircraft Co. for several years, helping to develop hardware for Project BOMI, a rocket-powered spaceplane. See Walter R. Dornberger, *V-2*, trans. by James Cleugh and Geoffrey Halliday (New York: Viking, 1958); Gerald L. Borrowman, "Walter R. Dornberger," *Spaceflight* 23 (April 1981): 118-19.

Russell C. Drew (1931-   ) has been an influential physicist who served in the U.S. Navy from 1953 through 1973, and he spent much of his career working on nuclear submarine ballistic missile programs. He also served as assistant to the President's Science Advisor, 1966-1972, and director of the staff of the President's Space Task Group. His last assignment, as a naval captain, was as the head of the Office of Naval Research (London). Thereafter, he served as the director of the Science and Technology Policy Office of the National Science Foundation, 1973-1976, and in several capacities in the aerospace industry since 1976.

Hugh L. Dryden (1898-1965) was a career civil servant and an aerodynamicist by discipline who had also begun life as something of a child prodigy. He graduated at age 14 from high school and went on to earn an A.B. in three years from The Johns Hopkins (1916). Three years later (1919) he earned his Ph.D. in physics and mathematics from the same institution, even though he had been a full-time employee of the National Bureau of Standards since June 1918. His career at the Bureau of Standards, which lasted until 1947, was devoted to studying airflow, turbulence, and particularly the problems of the boundary layer—the thin layer of air next to an airfoil that causes drag. In 1920 he became chief of the aerodynamics section in the bureau. His work in the 1920s on measuring turbulence in wind tunnels facilitated research in NACA that produced the laminar flow wings used in the P-51 Mustang and other World War II aircraft. From the mid-1920s to 1947, his publications became essential reading for aerodynamicists around the world. During World War II his work on a glide bomb named the Bat won him a Presidential Certificate of Merit. He capped his career at the Bureau of Standards by becoming its assistant director and then associate director during his final two years there. He then served as director of NACA from 1947-1958, after which he became deputy administrator of NASA under T. Keith Glennan and James E. Webb (Richard K. Smith, *The Hugh L. Dryden Papers, 1898-1965* [Baltimore, MD: The Johns Hopkins University Library, 1974]).

Lee A. DuBridge (1901-   ), a physicist with a Ph.D. from the University of Wisconsin (1926), became director of the radiation laboratory at the Massachusetts Institute of Technology after an academic career capped to that point by a deanship at the University of Rochester, 1938-1941. He was president of the California Institute of Technology between 1946 and 1969, when he resigned to serve as science advisor to President Richard M. Nixon. He had been involved in several governmental science advisory organizations before taking up his formal White House duties in 1969 and serving in that capacity until 1970 ("Lee A. DuBridge," biographical file, NASA Historical Reference Collection).

Allen W. Dulles (1893-1969), younger brother of President Eisenhower's more famous secretary of state, served as director of the Central Intelligence Agency (CIA) from 1953 to 1961.

John Foster Dulles (1888-1959) served as secretary of state under President Eisenhower, 1953-1959.

John R. Dunning (1892-1975) was a physicist who conducted the early experiments in nuclear fission that helped lay the groundwork for developing the atomic bomb. He later became the dean of the School of Engineering at Columbia University (obituary in *New York Times,* August 28, 1975, p. 36).

Frederick C. Durant III (1916-   ) was heavily involved in rocketry in the United States between the end of World War II and the mid-1960s. He worked for several different aerospace organizations, including Bell Aircraft Corp., Everett Research Laboratory, the Naval Air Rocket Test Station, and the Maynard Ordnance Test Station. He later became the director of astronautics for the National Air and Space Museum, Smithsonian Institution. In addition, he was an officer in several spaceflight organizations, such as president of the American Rocket Society (1953), president of the International Astronautical Federation (1953-1956), and governor of the National Space Club (1961).

E

Dwight D. Eisenhower (1890-1969) was president of the United States between 1953 and 1961. Previously, he had been a career U.S. Army officer and was Supreme Allied Commander in Europe during World War II. As president, he was deeply interested in the use of space technology for national security purposes and directed that ballistic missiles and reconnaissance satellites be developed on a crash basis. On Eisenhower's space efforts, see Rip Bulkeley, *The Sputniks Crisis and Early United States Space Policy* (Bloomington: Indiana University Press, 1991); R. Cargill Hall, "The Eisenhower Administration and the Cold War: Framing American Astronautics to Serve National Security," *Prologue: Quarterly of the National Archives* 27 (Spring 1995): 59-72; Robert A. Divine, *The Sputnik Challenge: Eisenhower's Response to the Soviet Satellite* (New York: Oxford University Press, 1993).

John D. Erlichman was a senior assistant to the president during the Nixon administration. See John Erlichman, *Witness to Power: The Nixon Years* (New York: Simon and Schuster, 1982).

F

Philip J. Farley (1916-   ) earned a Ph.D. from the University of California, Berkeley, in 1941 and was on the faculty at Corpus Christi Junior College from 1941 to 1942 before entering government work—for the Atomic Energy Commission, 1947-1954, and for the State Department, 1954-1969. From 1957 to 1961 he was a special assistant to the secretary of state for disarmament and atomic energy, and from 1961 to 1962 his responsibilities shifted to atomic energy and outer space. After several years of assignment to the North Atlantic Treaty Organization (NATO), he returned to Washington and became deputy secretary of state for political-military affairs, 1967-1969. Then from 1969 to 1973 he became deputy director of the U.S. Arms Control and Disarmament Agency.

William Finan was a staff member with the Bureau of the Budget during the Eisenhower administration. He was a member of the Purcell Panel that assessed spaceflight capabilities for the U.S. government in 1957 and 1958.

Daniel J. Fink was chair of the NASA Advisory Council's Task Force that produced the 1983 "Study of the Mission of NASA."

James Brown Fisk (1910-1981) received his Ph.D. in physical science from the Massachusetts Institute of Technology in 1935 and served in a variety of educational and industry positions. Most importantly, he was heavily involved in work at Bell Telephone Labs, serving as president from 1959 (obituary in *New York Times*, August 13, 1981, p. D21).

Peter M. Flanigan (1923-) was an assistant to the president on the White House staff, 1969-1974. Previously, he had been involved in investment banking with Dillon, Read, and Co. He returned to business when he left government service. His position in the White House from 1969 to 1972 involved him in efforts to gain approval to build the Space Shuttle.

Alexander H. Flax (1921-) was an aeronautical engineer, with a Ph.D. in physics, who worked in several important positions in universities and industry. He worked for Curtiss-Wright, 1940-1944; the Piasecki Helicopter Corporation, 1944-1946; and Cornell University, 1946-55. He served in scientific positions with the U.S. Air Force, 1955-1969, and as assistant secretary of the Air Force for research and development, 1963-1969. Thereafter, he became vice president for research for the Institute for Defense Analysis.

James C. Fletcher (1919-1991) was born on June 5, 1919, in Millburn, New Jersey. He received an undergraduate degree in physics from Columbia University and a doctorate in physics from the California Institute of Technology. After holding research and teaching positions at Harvard and Princeton Universities, he joined Hughes Aircraft in 1948 and later worked for the Guided Missile Division of the Ramo-Wooldridge Corporation. In 1958 Fletcher co-founded the Space Electronics Corporation in Glendale, California, which after a merger became the Space General Corporation. He was later named systems vice president of the Aerojet General Corporation in Sacramento, California. In 1964 he became president of the University of Utah, a position he held until he was named NASA administrator in 1971. He served until 1977. He served as NASA administrator a second time, for nearly three years following the loss of the Space Shuttle *Challenger* in 1986 until 1989. During his first administration at NASA, Dr. Fletcher was responsible for beginning the Shuttle effort. During his second tenure he presided over the effort to recover from the *Challenger* accident. See "Fletcher, James C., Administrator's Files," NASA Historical Reference Collection.

Gerald R. Ford (1913-) (R-MI) was elected to the House of Representatives in 1948 and served there until he became vice president in 1973 following the resignation of Spiro T. Agnew. He then served as president, 1974-1977, following Richard M. Nixon's resignation in the wake of the Watergate break-in.

William C. Foster, later the head of the Arms Control and Disarmament Agency, was President Eisenhower's representative to the U.S./U.S.S.R. summit at Geneva, Switzerland, in 1955. One of his responsibilities was to obtain freedom of space for overflight by spacecraft.

Robert A. Frosch (1928-) was NASA administrator throughout the administration of President Jimmy Carter, 1977-1981. He earned undergraduate and graduate degrees in theoretical physics at Columbia University, and between September 1951 and August 1963 he worked as a research scientist and director of research programs for Hudson Laboratories of Columbia University. Until 1953 he worked on problems in underwater sound, sonar, oceanography, marine geology, and marine geophysics. Thereafter, Frosch was first associate and then director of the laboratories. In September 1963 Dr. Frosch came to Washington to work with the Advanced Research Projects Agency (ARPA), Department of Defense, serving as director for nuclear test detection (Project VELA) and then as deputy director of the Advanced Research Projects Agency. In July 1966 he became assistant secretary for research and development for the Navy, responsible for all Navy programs of research, development, engineering, test, and evaluation. From January 1973 to July 1975 he served as assistant executive director of the United Nations Environmental Program. While at NASA Frosch was responsible for overseeing the continuation of the development effort on the Space Shuttle. During his tenure, the project underwent testing of the first orbiter, *Enterprise*, at NASA's Dryden Flight Research Facility in southern California. The orbiter made its first free flight in the atmosphere on August 12, 1977. He left NASA with the change of administrations in January 1981 to become vice president for research at the General Motors Research Laboratories. See "Frosch, Robert A., Administrator's Files," NASA Historical Reference Collection.

Eugene G. Fubini (1913-) was a noted physicist. A native of Italy, he came to the United States in 1938 to work for CBS and was responsible for microwave and international broadcasting. He worked for the U.S. military in World War II and then in a succession of technical and scientific positions with the Department of Defense in the postwar era. Since 1969 he has served as a consultant with Texas Instruments and IBM.

Craig Fuller was President Ronald Reagan's Cabinet secretary in the early 1980s and arranged for NASA's space station proposal to be discussed at a meeting of the Cabinet Council for Commerce and Trade.

G

Yuri Gagarin (1934-1968) was the Soviet cosmonaut who became the first human in space with a one-orbit mission aboard the spacecraft *Vostok 1* on April 12, 1961. The great success of that feat made the gregarious Gagarin a global hero, and he was an effective spokesman for the Soviet Union until his death in an unfortunate aircraft accident.

Galileo Galilei (1564-1642) used the newly invented telescope to view the bodies of the universe and to develop to its most advanced state in the pre-nineteenth century the "Heliocentric System" of the Solar System. Galileo made four important observations that convinced him that Copernican cosmology was correct, as described by writers Lloyd Motz and Jefferson Hane Weaver: "(1) the moon's surface is cratered and highly irregular, thus negating the theory that celestial bodies are 'perfect'; (2) the phases of Venus and those of the moon are similar, proving that Venus revolves around the sun and not around the earth; (3) four moons (satellites) revolve around Jupiter, illustrating in miniature the Copernican model of the solar system; and (4) the Milky Way consists of numerous points of light, which Galileo correctly interpreted as very distant stars" (p. 42). Galileo ran afoul of ecclesiastical authorities because of his observations, but they quickly became the standard explanation for understanding the workings of the universe (Lloyd Motz and Jefferson Hane Weaver, *The Story of Physics* [New York: Avon Books, 1992]).

Dave Garroway (1913-1982) was a television and radio personality who hosted the "Today Show" for NBC between 1952 and 1961 (obituary in *New York Times*, July 22, 1982, p. D22).

S. Everett Gleason (1905-   ) was a longtime government official in the Department of State and for a time its official historian.

T. Keith Glennan (1905-1995) was the first NASA administrator. Born in 1905 in Enderlin, North Dakota, Glennan was educated at Yale University, and he then worked in the sound motion picture industry with the Electrical Research Products Company. He was also studio manager of Paramount Pictures, Inc., and Samuel Goldwyn Studios in the 1930s. Glennan joined Columbia University's Division of War Research in 1942, serving through the war, first as administrator and then as director of the U.S. Navy's Underwater Sound Laboratories at New London, Connecticut. In 1947 he became president of the Case Institute of Technology. During his administration, Case rose from a primarily local institution to rank with the top engineering schools in the nation. From October 1950 to November 1952 he served as a member of the Atomic Energy Commission. He also served as administrator of NASA while on leave from Case, between August 7, 1958, and January 20, 1961. Upon leaving NASA Glennan returned to the Case Institute of Technology, where he continued to serve as president until 1966. See J.D. Hunley, ed., *The Birth of NASA: The Diary of T. Keith Glennan* (Washington, DC: NASA SP-4105, 1993).

Robert H. Goddard (1882-1945) was one of the three most prominent pioneers of rocketry and spaceflight theory. He earned his Ph.D. in physics at Clark University in 1911 and went on to become head of the Clark physics department and director of its physical laboratories. He began to work seriously on rocket development in 1909 and is credited with launching the world's first liquid-propellant rocket in 1926. He continued his rocket development work with the assistance of a few technical assistants throughout the remainder of his life. Although he developed and patented many of the technologies later used on large rockets and missiles—including film cooling, gyroscopically controlled vanes, and a variable-thrust rocket motor—only the last of these contributed directly to the furtherance of rocketry in the United States. Goddard kept most of the technical details of his inventions a secret and thus missed the chance to have the kind of influence his real abilities promised. At the same time, he was not good at integrating his inventions into a workable system, so his own rockets failed to reach the high altitudes he sought. See Milton Lehman, *Robert H. Goddard: A Pioneer of Space Research* (New York: Da Capo, 1988); J.D. Hunley, "The Enigma of Robert H. Goddard," *Technology and Culture* 36 (April 1995—forthcoming).

Harry J. Goett (1910-   ) earned a degree in physics from Holy Cross College in 1931 and one in aeronautical engineering from New York University in 1933. After holding a number of engineering posts with private firms, he became a project engineer at Langley Aeronautical Laboratory in 1936. He later moved to Ames Aeronautical Laboratory, where he was chief of the full-scale and flight research division, 1948-1959. During the the last year at Ames he became director of the Goddard Space Flight Center, a post he held until July 1965, when he became a special assistant to NASA Administrator James E. Webb. Later that year he became director for plans and programs at Philco's Western Development Laboratories in California and ultimately retired from a position with Ford Aerospace and Communications ("Harry J. Goett," biographical file, NASA Historical Reference Collection).

Barry M. Goldwater (1909-   ) (R-AZ) was a U.S. senator from 1953 to 1965. In 1964 he ran unsuccessfully for president of the United States against Lyndon Johnson. He was an outspoken conservative and became the leader and later elder statesman for the right wing of the Republican party.

Andrew Jackson Goodpaster (1915-) was a career Army officer who served as defense liaison officer and secretary of the White House staff from 1954 to 1961, being promoted to brigadier general during that period. He later was deputy commander, U.S. forces in Vietnam, 1968-1969, and commander-in-chief, U.S. Forces in Europe, 1969-1974. He retired in 1974 as a four-star general but returned to active duty in 1977 and served as superintendent of the U.S. Military Academy, a post he held until his second retirement in 1981.

William R. Graham (1937-), with a Ph.D. in physics from the California Institute of Technology and a Ph.D. in electrical engineering from Stanford University, was a founder and executive of R&D Associates, Marina Del Rey, California, and became deputy administrator of NASA on November 25, 1985. In 1980 Graham served as an advisor to candidate Ronald Reagan and was a member of the president-elect's transition team. Graham had also served for three years prior to coming to NASA as chair of the General Advisory Committee on Arms Control and Disarmament. Graham left NASA in October 1986 to become director of the White House Office of Science and Technology Policy, a position he held until June 1989 when he left government service to join Jaycor, a high-technology company headquartered in San Diego, California. See "Graham, William R., Deputy Administrator Folders," NASA Historical Reference Collection.

Virgil I. "Gus" Grissom (1927-1967) was chosen with the first group of astronauts in 1959. He was the pilot for the 1961 Mercury-Redstone 4 (*Liberty Bell 7*) mission (a suborbital flight), command pilot for *Gemini III*, and backup command pilot for *Gemini VI*. He had been selected as commander of the first Apollo flight at the time of his death in the Apollo fire in January 1967. See Betty Grissom and Henry Still, *Starfall* (New York: Thomas Y. Crowell, 1974); The Astronauts Themselves, *We Seven* (New York: Simon and Schuster, 1962).

Aristid V. Grosse (1905-) was born in Riga, Russia, and trained in engineering at the Technische Hochschule in Berlin. He came to the United States in 1930 and was on the chemistry faculty at the University of Chicago, 1931-1940. He then went to Columbia University briefly before working on the Manhattan Project during the war years. In 1948 he became a faculty member at Temple University, presiding over the Research Institute (now Franklin Institute) through 1969.

H

Fritz Haber (1868-1934) was a German research chemist who received the Nobel Prize for developing nitrates from ammonia, which were put to numerous agricultural and industrial uses.

John P. Hagen (1908-1990) was director of the Vanguard program during the 1950s. He had been an astronomer at Wesleyan University, 1931-1935, before working for the Naval Research Laboratory, 1935-1958. With the creation of NASA, he became the assistant director of spaceflight development, 1958-1960, and in 1962 he returned to higher education, becoming a professor of astronomy at Pennsylvania State University (obituary in *New York Times*, September 1, 1990, p. 25).

James C. Hagerty (1909-1981) had been on the staff of the *New York Times* from 1934 to 1942, the last four years as legislative correspondent in the paper's Albany bureau. He served as executive assistant to New York Governor Thomas Dewey from 1943 to 1950 and then as Dewey's secretary for the next two years, before becoming press secretary for President Eisenhower from 1953 to 1961.

Edward Everett Hale (1822-1909) was a writer in the United States during the middle part of the nineteenth century. He was best known for his short story "The Man Without a Country," about a conspirator in the 1803 attempt of Aaron Burr to create a separate nation in the American West. He was widely regarded as one of the foremost literary figures of his time and was the primary speaker at the dedication of the Civil War cemetery in Gettysburg in 1863, at which Abraham Lincoln gave his famous address (Jean Holloway, *Edward Everett Hale: A Biography* [Austin: University of Texas Press, 1956]).

Donald H. Heaton was an Air Force officer who from 1951 to 1957 as a lieutenant colonel and colonel had served on various subcommittees of the NACA committee on power plants for aircraft as well as on the committee itself. Available information does not indicate just when he joined NASA headquarters, but the August 1959 telephone directory shows him working in the office of the assistant director of propulsion within the Office of Space Flight Development. He served in a variety of positions connected with launch vehicles, and in June 1961 Associate Administrator Robert Seamans appointed him chairman of an ad hoc task group to formulate plans and determine the resources necessary to carry out a manned lunar landing. His group submitted its summary report in August 1961. He appears to have left NASA headquarters sometime between June and October 1963. See "Donald H. Heaton," biographical file, NASA Historical Reference Collection and headquarters telephone directories for the period; on his committee's report, see especially Courtney G. Brooks, James M. Grimwood, and Loyd S. Swenson, Jr., *Chariots for Apollo: A History of Manned Lunar Spacecraft* (Washington, DC: NASA SP-4205, 1979), pp. 45, 70-72.

F. Edward Hebert (1901-1979) (D-LA) was elected to the U.S. House of Representatives in 1932 and came to Washington as part of the Democratic sweep that led to the "New Deal" legislation of 1933-1935. He retired from office in 1976 after being stripped of his chairmanship of the House Armed Services Committee (obituary in *New York Times*, December 31, 1979, p. A13).

Robert A. Heinlein (1907-1988) was a well-known science fiction author who began publishing stories before World War II and continued a celebrated career until his death. He published more than sixty books; among the best known were *Starship Troopers* (1952), *Stranger in a Strange Land* (1961), and *The Moon is a Harsh Mistress* (1966) (obituary in *New York Times*, May 10, 1988, p. D2).

Klaus P. Heiss (1942-) is an Austrian-born economist who prepared a major economic feasibility study for the Space Shuttle program in 1971. He later worked with Econ, Inc., and founded and headed Space Transportation Corp., in Princeton, New Jersey. See "Heiss, Klaus P.," biographical file, NASA Historical Reference Collection.

Christian A. Herter (1895-1966) was under secretary of state, 1957-1959, and then succeeded John Foster Dulles as secretary of state from 1959-1961. He never achieved the level of mutual understanding with President Eisenhower that Dulles had enjoyed, however, and thus failed to have the sort of influence in developing the administration's foreign policy that his predecessor had achieved (Chester A. Pach and Elmo Richardson, *The Presidency of Dwight D. Eisenhower* [Lawrence, KS: University Press of Kansas, 1987], p. 204).

Harry H. Hess (1906-1969) was one of the senior scientists involved in analyzing the lunar samples returned to Earth by Project Apollo. Blair Professor of Geology at Princeton University, he was chair of the Space Science Board of the National Academy of Sciences during the Apollo era.

William M. Holaday (1901-) was special assistant to the secretary of defense for guided missiles between 1957 and 1958. He was then Department of Defense director of guided missiles in 1958 and chairman of the civilian-military liaison committee, 1958-1960. Previously, Holaday had been associated with a variety of research and development activities, notably as director of research for the Socony-Mobil Oil Co., 1937-1944 ("William M. Holaday," biographical file, NASA Historical Reference Collection).

D. Brainard Holmes (1921-) was involved in the management of high-technology efforts in private industry and the federal government. He was on the staff of Bell Telephone Laboratories, 1945-1953, and at RCA, 1953-1961. He then became deputy associate administrator for manned spaceflight at NASA, 1961-1963. Thereafter, he assumed a series of increasingly senior positions with Raytheon Corp., and he served as chairman of Beech Aircraft since 1982. See "D. Brainard Holmes," biographical file, NASA Historical Reference Collection; "Holmes, D(yer) Brainard," *Current Biography 1963*, pp. 191-92.

Donald F. Hornig (1920-), a chemist, was a research associate at the Woods Hole Oceanographic Laboratory, 1943-1944, and a scientist and group leader at the Los Alamos Scientific Laboratory, 1944-1946. He taught chemistry at Brown University starting in 1946, rising to the directorship of Metcalf Research Laboratory, 1949-1957, and also serving as associate dean and acting dean of the graduate school from 1952 to 1954. He was Donner Professor of Science at Princeton from 1957 to 1964, as well as chairman of the chemistry department from 1958 to 1964. He was a special assistant to the U.S. president on science and technology from 1964 to 1969 and president of Brown University from 1970 to 1976. See Gregg Herken, *Cardinal Choices: Science Advice to the President from Hiroshima to SDI* (New York: Oxford University Press, 1992).

Norman H. Horowitz (1915-) was a biologist educated at the California Institute of Technology (Caltech), receiving a Ph.D. in 1939. He made a career as a scientist at both Caltech and the Jet Propulsion Laboratory in Pasadena, California. At the Jet Propulsion Laboratory, he worked as a scientist on the Viking Mars lander program.

Hubert H. Humphrey (1911-1978) (D-MN) served in the U.S. Senate, 1949-1964 and 1971-1978. As a senator, he pressed for the creation of a Cabinet-level Department of Science and Technology in early 1958, which was defeated by the president's proposal to establish NASA. He was vice president of the United States between 1965 and 1968 under Lyndon Johnson (obituary in *New York Times*, January 14, 1978, p. 1).

Jerome C. Hunsaker (1886-1984) was a senior aeronautical engineer at the Massachusetts Institute of Technology. He was heavily involved in the development of the science of flight in America for the first three-quarters of the twentieth century. See Roger D. Launius, "Jerome C. Hunsaker," in Emily J. McMurray, *et al.*, eds., *Notable Twentieth-Century Scientists* (New York: Gale Research Inc., 1995), pp. 980-81.

J

Henry M. ("Scoop") Jackson (1912-1983) (D-WA) was first elected to the House of Representatives in 1940 and to each succeeding Congress until 1952, when he was elected to the Senate, where he served until the mid-1980s. During the Eisenhower administration he was a leading advocate of greater attention to the development of the U.S. missile program.

Robert Jastrow (1925-) earned a Ph.D. in theoretical physics from Columbia in 1948 and pursued post-doctoral studies at Leiden, Princeton (Institute for Advanced Studies), and the University of California at Berkeley before becoming an assistant professor at Yale, 1953-1954. He then served on the staff at the Naval Research Laboratory from 1954 to 1958. In the last year he was appointed chief of the theoretical division of the Goddard Space Flight Center. He became director of the Goddard Institute of Space Studies in 1961 and stayed at its helm for twenty years before becoming professor of earth sciences at Dartmouth. He specialized in nuclear physics, plasma physics, geophysics, and the physics of the Moon and terrestrial planets ("Robert Jastrow," biographical file, NASA Historical Reference Collection).

Clarence L. (Kelly) Johnson (1910-1990) was one of the foremost aircraft designers in the United States. As the head of the Lockheed Aircraft Corporation's famous "Skunk Works" design center, he headed the effort to build the U-2 reconnaissance aircraft in the 1950s. He also worked on the F-80 "Shooting Star," which was the first U.S. jet aircraft, and the SR-71 "Blackbird" reconnaissance plane that still holds speed records. During World War II he was also responsible for the design of the P-38 twin-tailed fighter, "Lightning." He worked for Lockheed from 1933 until his retirement as senior vice president in 1975. See Clarence L. "Kelly" Johnson with Maggie Smith, *Kelly: More Than My Share of it All* (Washington, DC: Smithsonian Institution Press, 1985).

Louis A. Johnson (1891-1966) was the assistant Secretary of the U.S. Department of War (1937-1940) and then secretary of defense, 1949-1950. See obituary in *New York Times*, April 25, 1966, p. 31.

Lyndon B. Johnson (1908-1973) (D-TX) was elected to the House of Representatives in 1937 and served until 1949. He was a senator from 1949 to 1961, vice president of the United States under President Kennedy from 1960 to 1963, and president from then until 1969. Best known for the social legislation he passed during his presidency and for his escalation of the war in Vietnam, he was also highly instrumental in revising and passing the legislation that created NASA and in supporting the U.S. space program as chairman of the Committee on Aeronautical and Space Sciences and of the preparedness subcommittee of the Senate Armed Services Committee. He was later effective as chairman of the National Aeronautics and Space Council when he was vice president. (On his role in support of the space program, Robert A. Divine, "Lyndon B. Johnson and the Politics of Space," in Robert A. Divine, ed., *The Johnson Years: Vietnam, the Environment, and Science* [Lawrence: University of Kansas Press, 1987], pp. 217-53; Robert Dallek, "Johnson, Project Apollo, and the Politics of Space Program Planning," unpublished paper delivered at a symposium on "Presidential Leadership, Congress, and the U.S. Space Program," sponsored by NASA and American University, March 25, 1993.)

U. Alexis Johnson (1908-) was a longtime member of the U.S. Foreign Service and served in a number of embassies around the world. A specialist in Asian affairs, he was attached to the embassy in Tokyo, 1935-1938; consul general to Japan, 1947-1949; and ambassador to Japan, 1966-1969. He served on several international commissions and in numerous senior positions with the Department of State in Washington, D.C., most significantly as under secretary of state for political affairs beginning in 1969 until his retirement.

S. Paul Johnston was director of the Institute for Aeronautical Sciences. He was also a member of the 1957-1958 Purcell Panel that assessed spaceflight capabilities for the U.S. government.

K

Joseph Kaplan (1902-1991) was born in Tapolcza, Hungary, and came to the United States in 1910. He trained as a physicist at The Johns Hopkins University and worked on the faculty of the University of California at Berkeley from 1928 until his retirement in 1970. He directed the university's Institute of Geophysics, later the Institute of Geophysics and Planetary Physics, from the time of its creation in 1944. Kaplan was heavily involved in efforts in the 1950s to launch the first artificial Earth satellite, serving as the chair of the U.S. National Committee for the International Geophysical Year, 1953-1963. See "Kaplan, Joseph," biographical file, NASA Historical Reference Collection; Joseph Kaplan, "The Aeronomy Story: A Memoir," in R. Cargill Hall, ed., *Essays on the History of Rocketry and Astronautics: Proceedings of the Third Through the Sixth History Symposia of the International Academy of Astronautics* (Washington, DC: NASA Conference Publication 2014, 1977), 2: 423-27; Joseph Kaplan, "The IGY Program," *Proceedings of the IRE*, June 1956, pp. 741-43.

Theodore von Kármán (1881-1963) was a Hungarian aerodynamicist who founded the Aeronautical Institute at Aachen before World War I and achieved a world-class reputation in aeronautics through the 1920s. In 1930 Robert A. Millikan and his associates at the California Institute of Technology lured von Kármán from Aachen to become the director of the Guggenheim Aeronautical Laboratory at Caltech (GALCIT). There, he trained a generation of engineers in theoretical aerodynamics and fluid dynamics. With its eminence in physics, physical chemistry, and astrophysics as well as aeronautics, it proved to be an almost ideal site for the early development of U.S. ballistic rocketry. See Judith R. Goodstein, *Millikan's School: A History of California Institute of Technology* (New York: W.W. Norton, 1991); Clayton R. Koppes, *JPL and the American Space Program: A History of the Jet Propulsion Laboratory* (New Haven: Yale University Press, 1982); Michael H. Gorn, *The Universal Man: Theodore von Kármán's Life in Aeronautics* (Washington, DC: Smithsonian Institution Press, 1992).

Amron Harry Katz (1915-   ) was a physicist who worked with the Rand Corporation in Santa Monica, California, between 1954 and 1969. He was a specialist in aerospace reconnaissance.

William W. Kellogg (1917-   ) was a meteorologist with the Rand Corporation between 1947 and 1959. Thereafter, he held a senior position with the National Center for Atmospheric Research in Boulder, Colorado.

John F. Kennedy (1916-1963) was president of the United States, 1961-1963. In 1960 Kennedy, a senator from Massachusetts between 1953 and 1960, ran for president as the Democratic candidate, with party "wheelhorse" Lyndon B. Johnson as his running mate. Using the slogan, "Let's get this country moving again," Kennedy charged the Republican Eisenhower administration with doing nothing about the myriad social, economic, and international problems that festered in the 1950s. He was especially hard on Eisenhower's record in international relations, taking a "cold warrior" position on a supposed "missile gap" (which turned out not to be the case) wherein the United States lagged far behind the Soviet Union in ICBM technology. On May 25, 1961, President Kennedy announced to the nation a goal of sending an American to the Moon before the end of the decade. The human spaceflight imperative was a direct outgrowth of it; Projects Mercury (at least in its latter stages), Gemini, and Apollo were each designed to execute it. On this subject, see Walter A. McDougall, . . . *The Heavens and the Earth: A Political History of the Space Age* (New York: Basic Books, 1985); John M. Logsdon, *The Decision to Go to the Moon: Project Apollo and the National Interest* (Cambridge, MA: MIT Press, 1970).

Robert F. Kennedy (1925-1968) was attorney general during the administration of his brother, John F. Kennedy, and a candidate for the Democratic nomination for the presidency in 1968 at the time of his assassination. He was involved in the 1961 decision to go to the Moon as a senior advisor (as well as attorney general) in the Kennedy administration. On his career, see Arthur M. Schlesinger, Jr., *Robert Kennedy and His Times* (Boston: Houghton Mifflin, 1978).

Johann Kepler (1571-1630), a young German astronomer, began work with Tycho Brahe in Prague, Czechoslovakia, in 1599. When Brahe died in 1601, Kepler inherited his position and continued his observations for a method of mathematically solidifying the Copernican view of the universe. He developed his three laws of planetary motion, and he was interested in cosmology and dabbled in astrology. His last book, *Somnium*, was completed shortly before his death and related a fantastic story of space travel that was memorable for its exposition of the Copernican model to explain planetary motion (Owen Gingerich, "Johnnes Kepler," *Dictionary of Scientific Biography* [New York: Charles Scribner's Sons, 1970], 7: 289-90).

Robert S. Kerr (1896-1963) (D-OK) had been governor of Oklahoma from 1943-1947 and was elected to the Senate the following year. From 1961 until 1963 he chaired the Aeronautical and Space Sciences Committee. See Anne Hodges Morgan, *Robert S. Kerr: The Senate Years* (Norman: University of Oklahoma Press, 1977).

George A. Keyworth II (1939-   ) was director of the Office of Science and Technology Policy and science advisor to President Ronald Reagan between 1981 and 1986. Formerly the head of the Los Alamos Scientific Laboratory, Keyworth was a Ph.D. in nuclear physics from Duke University in 1968. He began work at Los Alamos after graduation and remained there until 1981. See "Keyworth, George A(lbert), 2d," *Current Biography Yearbook 1986*, pp. 265-68.

Nikita S. Khrushchev (1894-1971) was premier of the Soviet Union from 1958 to 1964 and first secretary of the Communist Party from 1953 to 1964. He was noted for an astonishing speech in 1956 denouncing the crimes and blunders of Joseph Stalin and for gestures of reconciliation with the West in 1959-1960, ending with the breakdown of a Paris summit with President Eisenhower and the leaders of France and Great Britain in the wake of Khrushchev's announcement that the Soviets had shot down an American U-2 reconnaissance aircraft over the Urals on 1 May 1960. Then in 1962 Khrushchev attempted to place Soviet medium-range missiles in Cuba. This led to an intense crisis in October, after which Khrushchev agreed to remove the missiles if the United States promised to make no more attempts to overthrow Cuba's Communist government. Although he could be charming at times, Khrushchev was also given to bluster (extending even to shoe-pounding at the U.N.) and was a tough negotiator, although he believed, unlike his predecessors, in the possibility of Communist victory over the West without war. For further information about him, see his *Khrushchev Remembers: The Last Testament* (Boston: Little, Brown, 1974); Edward Crankshaw, *Khrushchev: A Career* (New York: Viking, 1966); Michael R. Beschloss, *Mayday: Eisenhower, Khrushchev and the U-2 Affair* (New York: Harper and Row, 1986); Robert A. Divine, *Eisenhower and the Cold War* (New York: Oxford University Press, 1981).

James R. Killian, Jr. (1904-1988), was president of the Massachusetts Institute of Technology (MIT) between 1949 and 1959. He was on leave between November 1957 and July 1959 when he served as the first presidential science advisor. President Dwight D. Eisenhower established the President's Science Advisory Committee (PSAC), which Killian chaired, following the Sputnik crisis. After leaving the White House staff in 1959, Killian continued his work at MIT, but in 1965 he began working with the Corporation for Public Broadcasting to develop public television. Killian described his experiences as a presidential advisor in *Sputnik, Scientists, and Eisenhower: A Memoir of the First Special Assistant to the President for Science and Technology* (Cambridge, MA: MIT Press, 1977). For a discussion of the PSAC, see Gregg Herken, *Cardinal Choices: Science Advice to the President from Hiroshima to SDI* (New York: Oxford University Press, 1992).

Jeane J. Kirkpatrick (1926-) was U.S. Permanent Representative to the United Nations.

Henry Kissinger (1923-) was assistant to the president for national security affairs, 1969-1973, under President Richard Nixon and secretary of state thereafter until 1977 under Nixon and President Gerald Ford. In these positions he was especially involved in international aspects of spaceflight, particularly the joint Soviet/American flight, the Apollo-Soyuz Test Project, in 1975.

George B. Kistiakowsky (1900-1982) was a pioneering chemist at Harvard University, associated with the development of the atomic bomb, and later an advocate of banning nuclear weapons. He served as science advisor to President Eisenhower from July 1959 to the end of the administration. He later served on the advisory board to the Arms Control and Disarmament Agency from 1962 to 1969 (*New York Times*, December 9, 1982, p. B21; "George B. Kistiakowsky," biographical file, NASA Historical Reference Collection).

William F. Knowland (1908-1974) (R-CA) served in the Senate between 1945 and 1959 (*Washington Post*, October 5, 1959, p. C3; *Guide to Research Collections of Former United States Senators*, 1789-1982 [Washington, DC: Government Printing Office, 1983], p. 291).

Joseph J. Knopow was a young Lockheed engineer who helped develop an infrared radiometer and telescope to detect the hot exhaust gases emitted by long-range jet bombers and, more important, large rockets in the mid-1950s. This aircraft-tracker and missile-detection system became a standard method of targeting enemy air- and spacecraft.

L

Melvin Laird (1922-) was secretary of defense during the Nixon administration.

Edwin Land was president of the Polaroid Corporation, as well as a member of the 1957-1958 Purcell Panel that assessed spaceflight capabilities for the U.S. government.

Harold Lasswell (1902-1978) was a political scientist at Yale University. He was especially interested in pubic opinion polling, the uses of propaganda, and the democratic political process.

James S. Lay, Jr. (1911-1987), was a senior official in the National Security Council, first as assistant executive secretary, 1947-1950, and then as executive secretary, 1950-1961. He then served as deputy assistant to the director of the Central Intelligence Agency (CIA), 1961-1964, and the executive secretary of the Intelligence Board through 1971.

Tom Lehrer (1928-) was a satirist who wrote and recorded several folk songs in the 1960s that made light of current events. His last album, *That Was the Year That Was* (1965), contained the satirical song "Wernher von Braun," dealing with the relationship of science to ethics. See "Lehrer, Tom," *Current Biography 1982*, pp. 227-30.

Curtis E. LeMay (1906-1990) was a career Air Force officer who entered the Army Air Corps in the 1920s and rose through a series of increasingly responsible Army Air Forces commands in World War II. After the war, LeMay built the Strategic Air Command into the premier nuclear deterrent force in the early 1950s. He also served as deputy chief of staff, 1957-1961, and chief of staff, 1961-1965, of the U.S. Air Force. He retired as a four-star general in 1965, and he ran for vice president with independent candidate George C. Wallace in 1968. See Thomas M. Coffey, *Iron Eagle: The Turbulent Life of General Curtis LeMay* (New York: Crown Pub., 1986).

Samuel Lenher (1905-   ) was a chemical manufacturing executive with the Dupont Corporation in Wilmington, Delaware, from 1929 until his retirement.

Willy Ley (1906-1969) was an extremely effective popularizer of spaceflight, first in Germany and then after 1935 in the United States, to which he emigrated after Hitler's ascension to power. He helped found the large and significant German "Verein fur Raumschiffahrt" (Society for Spaceship Travel, or VfR) in 1927. He also wrote several books that dealt with the dream of spaceflight. One of the most important was *Rockets: The Future of Travel Beyond the Stratosphere*, first published in 1944. In it Ley labored to convince interested readers that rockets would soon be able to carry humans off the surface of the Earth. One of the earliest books on rocketry for the general public, this work became a reference source for future science fiction and reality writing. A revised edition appeared in 1947, titled *Rockets and Space Travel*, and another in 1952, *Rockets, Missiles, and Space Travel*. An obituary can be found in the *New York Times*, June 25, 1969, p. 47.

Charles A. Lindbergh (1902-1974) was an early aviator who gained fame as the first pilot to fly solo across the Atlantic in 1927. His public stature following this flight was such that he became an important voice on behalf of aerospace activities until his death. He served on a variety of national and international boards and committees, including the central committee of the National Advisory Committee for Aeronautics in the United States. He became an expatriate living in Europe, following the kidnapping and murder of his two-year-old son in 1932. In Europe during the rise of fascism, Lindbergh assisted American aviation authorities by providing them with information about European technological developments. After 1936 he was especially important in warning the United States of the rise of Nazi air power. He assisted with the war effort in the 1940s by serving as a consultant to aviation companies and the government, and after the war he lived quietly in Connecticut and then in Hawaii. See Walter S. Ross, *The Last Hero: Charles A. Lindbergh* (New York: Harper and Row, 1967).

James E. Lipp (1910-   ) earned a Ph.D. in aeronautical engineering from the California Institute of Technology in 1935, and he then worked for the Douglas Aircraft Co., 1935-1948. Thereafter, he went to work for the Rand Corporation and eventually headed its aerospace division.

Alan M. Lovelace (1929-   ) was born in St. Petersburg, Florida, and was educated at the University of Florida, Gainesville, receiving a B.S. in chemistry in 1951, an M.S. in organic chemistry in 1952, and a Ph.D. in organic chemistry in 1954. Shortly after the end of the Korean conflict, he served in the U.S. Air Force from 1954 to 1956. Thereafter, Dr. Lovelace began work as a government scientist at the Air Force Materials Laboratory at Wright-Patterson Air Force Base in Dayton, Ohio. In January 1964 he was appointed chief scientist of the Air Force Materials Laboratory. In 1967 he was named director of the Air Force Materials Laboratory, and in October 1972 he was named director of science and technology for the Air Force Systems Command at headquarters, Andrews Air Force Base, Maryland. In September 1973 he became the principal deputy to the assistant secretary of the Air Force for research and development. In September 1974 Dr. Lovelace left the Department of Defense to become the associate administrator of the NASA Office of Aeronautics and Space Technology. With the departure of George Low as NASA deputy administrator in June 1976, Dr. Lovelace became deputy administrator, serving until July 1981. He retired from NASA to accept a position as corporate vice president—science and engineering with the General Dynamics Corporation in St. Louis, Missouri. See "Lovelace, Alan M.," Deputy Administrator files, NASA Historical Reference Collection.

George M. Low (1926-1984), a native of Vienna, Austria, came to the United States in 1940 and received an aeronautical engineering degree from Rensselaer Polytechnic Institute (RPI) in 1948 and an M.S. in the same field from that school in 1950. He joined NACA in 1949, and at the Lewis Flight Propulsion Laboratory, he specialized in experimental and theoretical research in several fields. He became chief of manned spaceflight at NASA headquarters in 1958. In 1960 he chaired a special committee that formulated the original plans for the Apollo lunar landings. In 1964 he became deputy director of the Manned Spacecraft Center in Houston, the forerunner of the Johnson Space Center. He became deputy administrator of NASA in 1969 and served as acting administrator from 1970 to 1971. He retired from NASA in 1976 to become president of RPI, a position he still held at his death. In 1990 NASA renamed its quality and excellence award after him ("Low, George M.," Deputy Administrator files, NASA Historical Reference Collection).

Percival Lowell (1855-1916) was the U.S. astronomer who predicted the existence of the planet Pluto. A Boston Brahmin, Lowell was a gentleman scholar who was involved in literature, writing several books on his travels around the globe. He also served as counselor and foreign secretary to the Korean Special Mission to the United States. Lowell developed an interest in astronomy in middle age, and he founded an observatory in Flagstaff, Arizona, to study the Solar System, especially Mars. He was enamored with the prospect of life on the red planet and theorized that its "canals" were the product of intelligent life (William Graves Hoyt, *Lowell and Mars* [Tucson: University of Arizona Press, 1976]).

M

Richard C. McCurdy (1909-   ), an engineer specializing in petroleum, was associate administrator for organization and management at NASA headquarters, Washington, D.C., 1970-1973, and a consultant to the agency from 1973 to 1982.

Neil H. McElroy (1904-1972) became secretary of defense in 1957 and served through 1959. He had previously been president of Procter & Gamble and returned there in December 1959 to become chairman of the board. He served in that position until October 1972, a month before his death.

Walter A. MacNair (1901-   ) was an electrical engineer who worked with the Bell Telephone Laboratories, 1929-1952, and the Consolidated Electrodynamics Corporation thereafter.

Robert S. McNamara (1916-   ) was secretary of defense during the Kennedy and Johnson administrations, 1961-1968. Thereafter, he served as president of the World Bank, where he remained until retirement in 1981. As secretary of defense in 1961 McNamara was intimately involved in the process of approving Project Apollo by the Kennedy administration. See "McNamara, Robert S(trange)," *Current Biography Yearbook 1987*, pp. 408-13; John M. Logsdon, *The Decision to Go to the Moon: Project Apollo and the National Interest* (Cambridge, MA: MIT Press, 1970).

John W. Macy, Jr., was chair of the Civil Service Commission during the Kennedy administration. He served as a member of a study committee in 1961 to ascertain the viability of "contracting out" considerable functions in aerospace research and development. The 1961 study was known as the "Bell Report" because the chair of the committee was David E. Bell, director of the Bureau of the Budget.

Frank J. Malina (1912-1981) was a young Ph.D. student at the California Institute of Technology in the mid-1930s, when he began an aggressive rocket research program to design a high-altitude sounding rocket. Beginning in late 1936 Malina and his colleagues started the static testing of rocket engines in the canyons above the Rose Bowl, with mixed results, but a series of tests eventually led to the development of the WAC Corporal rocket during World War II. After the war Malina worked with the United Nations and eventually retired to Paris to pursue a career as an artist. See "Malina, Frank J.," biographical file, NASA Historical Reference Collection.

Gordon Manning was a journalist for several periodicals. He was a staff writer for *Collier's*, 1948-1949, and worked in a series of increasingly responsible positions for *Newsweek*, 1949-1964. Between 1961 and 1964 he was executive editor. Thereafter, he worked with television, first as vice president and director of news for CBS, 1964-1972, and then as executive producer of NBC News, 1975-1978.

Hans Mark (1929-   ) became NASA deputy administrator in July 1981. He had previously served as secretary of the Air Force from July 1979 until February 1981 and as under secretary of the Air Force since 1977. In February 1969 Mark became director of NASA's Ames Research Center in Mountain View, California, where he managed the center's research and applications efforts in aeronautics, space science, life science, and space technology. Born in Mannheim, Germany, he came to the United States in 1940, and he became a citizen in 1945. He received a Ph.D. in physics from the Massachusetts Institute of Technology in 1954. Upon leaving NASA he became Chancellor of the University of Texas at Austin. See "Mark, Hans," Deputy Administrator files, NASA Historical Reference Collection.

Robert P. Mayo (1916-   ) was an economist and President Richard Nixon's first director of the Bureau of the Budget. On July 1, 1970, when the Bureau of the Budget was replaced with the Office of Management and Budget, Mayo was shifted to the White House as a presidential assistant. In July 1970 he left Washington to assume the presidency of the Federal Reserve Bank of Chicago ("Mayo, Robert P(orter)," *Current Biography 1970*, pp. 282-84).

John B. Medaris (1902-1990) was a major general commanding the Army Ballistic Missile Agency when T. Keith Glennan tried to incorporate it into NASA in the late 1950s. He attempted to retain the organization as part of the Army, but with a series of Department of Defense agreements, the Air Force obtained primacy in space activities, and Medaris could not succeed in his effort. Medaris also worked with Wernher von Braun to launch *Explorer I* in early 1958. He retired from the Army in 1969 and became an Episcopal priest, later joining an even more conservative Anglican-Catholic church ("Medaris, John Bruce," biographical file, NASA Historical Reference Collection; John B. Medaris with Arthur Gordon, *Countdown for Decision* [New York: Putnam, 1960]).

Ruben F. Mettler (1924-   ) was an electronics and engineering company executive who worked for the Hughes Aircraft Co., 1949-1954; Ramo-Wooldridge Corp., 1955-1958; TRW Space Technology Laboratories, 1958-1965; and TRW Systems Group, 1965-1968. He became president and chief operating officer of TRW Inc., 1969-1977, and then TRW chairman of the board and CEO, 1977-1988.

Stuart Miller (1927-   ) was a research engineer in industry, working with the Chrysler Corporation, 1952-1953, and the General Electric Co., 1953-1977.

Robert A. Millikan (1868-1953) was a Nobel Prize-winning physicist at the California Institute of Technology (Caltech). Best known for his research on cosmic rays, he also built Caltech into a world-class educational and scientific institution, over which he presided until his retirement in 1946. For more information on Millikan, see Robert H. Kargon, *The Rise of Robert Millikan: Portrait of a Life in American Science* (Ithaca, NY: Cornell University Press, 1982); *The Autobiography of Robert A. Millikan* (New York: Prentice-Hall, 1950).

Wilbur D. Mills (1909-1992) (D-AR) was a member of the U.S. House of Representatives from 1939 to 1977. He served as chair of the powerful House Ways and Means Committee, 1957-1975 (obituary in *New York Times,* May 3, 1992, p. I53).

L. Arthur Minnich, Jr. (1918-   ), was assistant staff secretary in the White House, 1953-1960. A historian by training, he also served on the faculty of Lafayette College before 1953. After leaving the White House, he served as the executive secretary of UNESCO.

Oskar Morgenstern (1902-   ) was a German-born and -trained economist. He came to the United States in 1925 and worked at Princeton University after 1938. He founded and headed Mathematica, Inc., which provided economic analyses to government and industry.

Frank E. "Ted" Moss (1906-   ) (D-UT) was first elected to the Senate in 1958 and served until 1977. Between 1972 and 1977 he served as chair of the Senate Space Committee.

George E. Mueller (1918-   ) was associate administrator for the Office of Manned Space Flight at NASA headquarters, 1963-1969, where he responsible for overseeing the completion of Project Apollo and for beginning the development of the Space Shuttle. He moved to the General Dynamics Corporation, as senior vice president in 1969, and remained until 1971. He then became president of the Systems Development Corporation, 1971-1980, and then its chairman and chief executive officer, 1981-1983. See "Mueller, George E.," biographical file, NASA Historical Reference Collection.

Edmund Muskie (1914-   ) (D-ME) served in the U.S. Senate, 1959-1981.

Dale D. Myers (1922-   ) served as NASA deputy administrator from October 1986 until 1989. He had previously been under secretary of the U.S. Department of Energy from 1977 to 1979. From 1974 to 1977 he was vice president at Rockwell International and president of the North American Aircraft Group in El Segundo, California. He was also the associate administrator for the Office of Manned Space Flight at NASA from 1970 to 1974. From 1969 to 1970 Myers served as vice president/program manager of the Space Shuttle Program at Rockwell International. He was also vice president and program manager of the Apollo Command/Service Module Program at North American-Rockwell from 1964 to 1969. After leaving NASA in 1989 Myers returned to private industry. See "Myers, Dale D.," Deputy Administrator files, NASA Historical Reference Collection.

N

John E. Naugle (1923-   ) was trained as a physicist at the University of Minnesota and began his career studying cosmic rays by launching balloons to high altitudes. In 1959 he joined NASA's Goddard Space Flight Center in Greenbelt, Maryland, where he developed projects to study the magnetosphere. In 1960 he took charge of NASA's fields and particles research program. He also served as NASA's associate administrator for the Office of Space Science and as the agency's chief scientist before his retirement in 1981. See John E. Naugle, *First Among Equals: The Selection of NASA Space Science Experiments* (Washington, DC: NASA SP-4215, 1991).

Richard G. Neustadt (1919-   ) was a Harvard University-trained political scientist who made a career in public policy analysis. He served for a time (1946-1953) with the federal government in Washington and thereafter in academia at Columbia University (1954-1964) and Harvard University (since 1964). He was an informal advisor to presidents and their associates between the 1940s and the 1980s. See Richard E. Neustadt and Ernest R. May, *Thinking in Time: The Uses of History for Decision Makers* (New York: Free Press, 1986).

Isaac Newton (1642-1727) created a scientific explanation of the workings of the universe that held sway until the twentieth century. Based on the concept of gravity and three laws of motion that related to it, the Newtonian construct placed astronomy and physics on a firm mathematical foundation. Born in England, Newton was educated at Trinity College in Cambridge. As a relatively young man, by 1667 he had developed his ideas on universal gravitation and its consequences, the nature of white light, and the calculus. In the same year, he was elected a fellow of Trinity College and two years later succeeded to the chair of his mentor Isaac Barrow. In 1696 Newton was named warden of the mint and became its master in 1699. While still officially associated with Cambridge, his work at the mint effectively ended Newton's academic career (James R. Newman, ed., *The World of Mathematics* [New York: Simon and Schuster, 1956], pp. 256-78; Lloyd Motz and Jefferson Hane Weaver, *The Story of Physics* [New York: Avon Books, 1992]).

Kenneth D. Nichols (1907-) worked on the Manhattan Project in World War II and served in a variety of special weapons activities with the Department of Defense. In the early 1950s he was involved in directing the guided missile research and development effort for the secretary of defense. He also held posts with the Atomic Energy Commission and with industry.

Richard M. Nixon (1913-1994) was president of the United States between January 1969 and August 1974. Early in his presidency Nixon appointed a Space Task Group under the direction of Vice President Spiro T. Agnew to assess the future of spaceflight for the nation. Its report recommended a vigorous post-Apollo exploration program culminating in a human expedition to Mars. Nixon did not approve this plan, but he did decide in favor of building one element of it, the Space Shuttle, which was approved on January 5, 1972. See Roger D. Launius, "NASA and the Decision to Build the Space Shuttle, 1969-72," *The Historian* 57 (Autumn 1994): 17-34.

O

Hermann J. Oberth (1894-1989) is one of the three recognized fathers of spaceflight. A Transylvanian by birth but a German in his family heritage, he was educated at the Universities of Klausenburg, Munich, Gottingen, and Heidelberg. His doctoral dissertation being rejected because it did not fit into any established scientific discipline, he published it privately as *Die Rakete zu den Planetenräumen* (*The Rocket into Interplanetary Space*) in 1923. It and its expanded version, titled *Ways to Spaceflight* (1929), set forth the basic principles of spaceflight and directly inspired many subsequent spaceflight pioneers, including Wernher von Braun. See his "Hermann Oberth: From My Life," *Astronautics,* June 1959, pp. 38-39, 100-106; Frank Winter, *Rockets into Space* (Cambridge, MA: Harvard University Press, 1990), pp. 17-25; Helen B. Walters, *Hermann Oberth: Father of Space Travel* (New York: Macmillan, 1962).

Charles R. O'Dell (1937-) was trained as an astronomer at the University of Wisconsin and was project scientist for the Hubble Space Telescope project, 1972-1983, at the Marshall Space Flight Center. He has been on the astronomy faculty at several universities, including the University of Houston where he is Buchanan Professor of Astrophysics.

Hugh Odishaw (1916-1984) became assistant to the director of the National Bureau of Standards from 1946 to 1954, served as executive director of the U.S. National Committee for the International Geophysical Year from 1954 to 1965, and then became the executive secretary of the Division of Physical Sciences in the National Academy of Sciences from 1966 to 1972.

Thomas F. (Tip) O'Neill (1912-1994) (D-MA) served in the U.S. House of Representatives from 1953 until 1987. For much of his later service in the House, he was speaker.

Don Richard Ostrander (1914-1972) was a career Air Force officer who became a major general in 1958. He was deputy commander of the Advanced Research Projects Agency in 1959 and became director of NASA's launch vehicle programs in late 1959 as NASA began taking over responsibility for the Saturn program. He left NASA in 1961 and retired from the Air Force in 1965 as vice commander of the Ballistic Systems Division, Air Force Systems Command, to become vice president for planning of the Bell Aero Systems Corporation ("Don Richard Ostrander," biographical file, NASA Historical Reference Collection).

Carl F.J. Overhage (1910-) earned his Ph.D. in physics at the California Institute of Technology in 1937 and served as acting director of research for Technicolor Motion Picture Corp. until 1941, when he joined the staff of the radiation laboratory at the Massachusetts Institute of Technology (MIT) from 1942 to 1945. After a stint with Eastman Kodak from 1946 to 1954, he joined the Lincoln Laboratories of MIT, becoming its director from 1957 to 1964, after which he served as a professor of engineering.

P

Thomas O. Paine (1921-1992) was appointed deputy administrator of NASA on January 31, 1968. Upon the retirement of James E. Webb on October 8, 1968, he was named acting administrator of NASA. He was nominated as NASA's third administrator March 5, 1969, which was confirmed by the Senate on March 20, 1969. During his leadership the first seven Apollo manned missions were flown, in which twenty astronauts orbited the earth, fourteen traveled to the Moon, and four walked on its surface. Paine resigned from NASA on September 15, 1970, and he returned to the General Electric Co. in New York City as vice president and executive of the Power Generation Group, where he remained until 1976. In 1985 the White House chose Paine as chair of a National Commission on Space to prepare a report on the future of space exploration. Since leaving NASA fifteen years earlier Paine had been a tireless spokesperson for an expansive view of what should be done in space. The Paine Commission took most of a year to prepare its report, largely because it solicited public input in hearings throughout the United States. The commission's report, *Pioneering the Space Frontier,* was published in a lavishly illustrated, glossy format in May 1986. It espoused a "pioneering mission for 21st-century America"—"to lead the exploration and development of the space frontier, advancing science, technology, and enterprise, and building institutions and systems that make accessible vast new resources and support human settlements beyond Earth orbit, from the highlands of the Moon to the plains of Mars." The report also contained a "Declaration for Space," which included a rationale for exploring and settling the Solar System and outlined a long-range space program for the United States.

Richard S. Perkin (1906-) was co-founder and president of Perkin-Elmer Corp., 1937-1960, and then chairman of the board.

James A. Perkins (1911-) was vice president of the Carnegie Corporation from 1951 to 1963 and president of Cornell University from 1963 to 1969. He served on the Kimpton Committee of 1959 to assess the space effort.

Rocco Petrone (1926-) was heavily involved at NASA with the development of the Saturn V booster used to launch Apollo spacecraft to the Moon in the 1960s and early 1970s. He worked at the Marshall Space Flight Center and became its director in 1973. He left Marshall in 1974 for a position at NASA headquarters in Washington, D.C., in 1974, and he retired from the agency in 1975. He then became president and chief executive officer of the National Center for Resource Recovery.

Samuel C. Phillips (1921-1990) was trained as an electrical engineer at the University of Wyoming, but he also participated in the Civilian Pilot Training Program during World War II. Upon his graduation in 1942 Phillips entered the Army infantry but soon transferred to the air component. As a young pilot, he served with distinction in the Eighth Air Force in England—earning two distinguished flying crosses, eight air medals, and the French croix de guerre—but he quickly became interested in aeronautical research and development. He became involved in the development of the incredibly successful B-52 bomber in the early 1950s and headed the Minuteman intercontinental ballistic missile program in the latter part of the decade. In 1964, by this time an Air Force general, Phillips was lent to NASA to head the Apollo moon landing program, which, of course, was unique in its technological accomplishment. He went back to the Air Force in the 1970s and commanded the Air Force Systems Command prior to his retirement in 1975. See "Gen. Samuel C. Phillips of Wyoming," *Congressional Record*, August 3, 1973, S-15689; Rep. John Wold, "Sam Phillips: One Who Led Us to the Moon," *NASA Activities*, May/June 1990, pp. 18-19; obituary in *New York Times*, February 1, 1990, p. D1.

William H. Pickering (1910-) obtained his bachelor's and master's degrees in electrical engineering and then a Ph.D. in physics from the California Institute of Technology before becoming a professor of electrical engineering there in 1946. In 1944 he organized the electronics efforts at the Jet Propulsion Laboratory (JPL) to support guided missile research and development, becoming project manager for Corporal, the first operational missile that JPL developed. From 1954 to 1976 he was director of JPL, which developed the first U.S. satellite (*Explorer I*), the first successful U.S. cislunar space probe (*Pioneer IV*), the Mariner flights to Venus and Mars in the early to mid-1960s, the Ranger photographic missions to the Moon in 1964-65, and the Surveyor lunar landings of 1966-1967 ("William H. Pickering," biographical file, NASA Historical Reference Collection).

William Proxmire (1915-) (D-WI) served in the U.S. Senate between 1957 and 1989.

Claudius Ptolemy (fl. 127-145) of Alexandria, Egypt, was responsible for the development of the "Ptolemaic System" of understanding the universe. It placed the Earth at its center with the planets, Moon, Sun, and stars orbiting overhead. Ptolemy based his system on observations of celestial bodies and the application of mathematical models that adequately explained the movements he observed. He also catalogued 1,022 stars (Owen T. Gingerich, gen. ed., *The Cambridge General History of Astronomy*, Vol. 1 [New York: Cambridge University Press, 1984]).

Allen E. Puckett (1919-) earned his Ph.D. at the California Institute of Technology in 1949 and went to work for Hughes Aircraft Co. that year, becoming its executive vice president from 1965 to 1977 and its president thereafter. He served as a member of the Nixon transition team's Task Force on Space, which was led by Dr. Charles Townes, to make recommendations on the new administration's efforts in aerospace.

Edward M. Purcell (1912-) was a professor of physics at Harvard University and also served on the president's Scientific Advisory Committee from 1957 to 1960 and 1962 to 1965. He had been co-winner of the Nobel Prize in physics in 1952 (with Felix Bloch) for the discovery of nuclear magnetic resonance in solids.

Donald L. Putt (1905-1988) was a career U.S. Air Force officer who specialized in the management of aerospace research and development activities. Trained as an engineer, he entered the Army Air Corps in 1928 and worked in a series of increasingly responsible posts at the Air Materiel Command and general headquarters of the Air Force. From 1948 to 1952 he was director of research and development for the Air Force, and he was first vice commander and then commander of the Air Research and Development Command between 1952 and 1954. Thereafter until his retirement in 1958, he served as deputy chief of the development staff at Air Force headquarters.

Q

Donald A. Quarles (1894-1959) was a deputy secretary of defense between 1957 and 1959. Just after World War II he had been a vice president first at Western Electric Co. and later at Sandia National Laboratories, but in 1953 he accepted the position of assistant secretary of defense (research and development). He was also secretary of the Air Force between 1955 and 1957.

R

Ronald Reagan (1911-) was elected president of the United States in 1980 and assumed office in January 1981; he served until 1989. During his presidency the maiden flight of the Space Shuttle took place. In 1984 he mandated the construction of an orbital space station. Reagan declared that "America has always been greatest when we dared to be great. We can reach for greatness again. We can follow our dreams to distant stars, living and working in space for peaceful, economic, and scientific gain. Tonight I am directing NASA to develop a permanently manned space station and to do it within a decade." See Sylvia D. Fries, "2001 to 1994: Political Environment and the Design of NASA's Space Station System," *Technology and Culture* 29 (July 1988): 568-93.

Sally K. Ride (1951-) was the first American woman to fly in space. She was chosen as an astronaut in 1978 and served as a mission specialist for STS-7 (1983) and for STS-41G (1984). She was also a member of the Presidential Commission on the Space Shuttle *Challenger* Accident in 1986, and from 1986 to 1987 she chaired a NASA task force that prepared a report on the future of the civilian space program, titled *Leadership and America's Future in Space* (Washington, DC: U.S. Government Printing Office, 1987). Ride resigned from NASA in 1987 to join the Center for International Security and Arms Control at Stanford University. She left Stanford in 1989 to assume the directorship of the California Space Institute, part of the University of California at San Diego. See "Ride, Sally K.," biographical file, NASA Historical Reference Collection.

Louis N. Ridenour (1911-) received his Ph.D. in physics from the California Institute of Technology in 1936, and he began work at Princeton University. In 1938 he moved to the University of Pennsylvania, where he remained until 1947. He then went to the University of Illinois, but he left there in 1951 to become vice president of the International Telemeter Corp. He also served in several positions with scientific organizations in the federal government, most significantly as chief scientist with the U.S. Air Force in the early 1950s.

Walter O. Roberts (1915-1990) was an astronomer at the University of Colorado's High Altitude Observatory. He was also instrumental in the creation of the National Center for Atmospheric Research in 1960, and he directed the program on food, climate, and the world's future for the Aspen Institute for Humanistic Studies, 1974-1981. He was heavily involved in the debate over "nuclear winter" and the possibility of the "Greenhouse Effect" on the Earth in the 1980s. See "Roberts, Walter Orr," *Current Biography Yearbook 1990*, p. 660.

Nelson A. Rockefeller (1909-1979) was vice president of the United States from 1974 to 1977. He had previously been the Republican governor of New York, 1958-1973 (obituary in *New York Times*, January 26, 1979, p. 27).

William P. Rogers (1913-) was chair of the presidentially mandated blue ribbon commission investigating the *Challenger* accident in January 1986. It found that the failure had resulted from a poor engineering decision, an O-ring used to seal joints in the solid rocket booster that was susceptible to failure at low temperatures, introduced innocently enough years earlier. Rogers kept the commission's analysis on that technical level, and he documented the problems in exceptional detail. The commission, after some prodding by Nobel Prize-winning scientist Richard P. Feynman, did a credible job of grappling with the technologically difficult issues associated with the accident. See *Report of the Presidential Commission on the Space Shuttle Challenger Accident, Vol. I* (Washington, DC: U.S. Government Printing Office, June 6, 1986).

H.E. Ross was one of the leaders of the British Interplanetary Society from the time of its inception in 1933. Ross wrote a 1939 article in the society's journal that outlined a method of accomplishing a lunar mission. The effort leading to the article had begun in London in February 1937 when the British Interplanetary Society formed a technical committee to conduct feasibility studies.

Herbert J. Rowe (1924-) was NASA associate administrator for external affairs, 1975-1978. He also worked with several high-technology industrial firms, including the Aerovax Corporation.

Richard B. Russell, Jr. (1897-1971) (D-GA), was a U.S. Senator from 1933 until his death. He was an influential force in the Senate, and he served as chair of the Senate Armed Services Committee, 1951-1969.

Cornelius Ryan was an influential journalist who worked for *Collier's* magazine in the 1950s and was in large measure responsible for the issues of the magazine devoted to space that appeared between 1952 and 1955. He became best known for his World War II trilogy: *The Longest Day: June 6, 1944* (1959); *A Bridge Too Far* (1974); and *The Last Battle* (1966).

S

Robert M. Salter, Jr. (1920-), was a physicist who worked with North American Aviation, 1946-1948; the Rand Corporation, 1948-1954; Lockheed Aircraft Co., 1954-1959; Quantatron, Inc., 1960-1962; and Xerad, Inc., since 1962. He was responsible for much of the early thinking at Rand on the possibility of an artificial Earth-orbiting satellite.

Leverett Saltonstall (1892-1979) (R-MA) was governor of Massachusetts from 1939 to 1944, when he won election to the U.S. Senate. He served in the Senate from then until 1967 and became one of its Republican leaders.

Giovanni Schiaparelli (1835-1910) was an Italian astronomer and senator of the Kingdom of Italy. He studied astronomy in Berlin, beginning in 1854 under Johann F. Encke. Two years later he was appointed assistant observer at Pulkovo Observatory, Russia. In 1860 he returned to Italy as an observer at Brera Observatory in Milan. There he made controversial observations of Martian *canali*, or straight lines, that set off speculation about the possibility of intelligent life who had constructed them. He also discovered the asteroid Hesperia and correctly calculated the Perseid meteor showers (Frederick I. Ordway III, "The Legacy of Schiaparelli and Lowell," *Journal of the British Interplanetary Society,* January 1986, pp. 18-22).

Bernard A. Schriever (1910-) earned a B.S. in architectural engineering from Texas A&M in 1931 and was commissioned in the Army Air Corps Reserve in 1933 after completing pilot training. Following broken service, he received a regular commission in 1938. He earned an M.A. in aeronautical engineering from Stanford in 1942 and then flew sixty-three combat missions in B-17s with the 19th Bombardment Group in the Pacific Theater during World War II. In 1954 he became commander of the Western Development Division (soon renamed the Air Force Ballistic Missile Division), and from 1959 to 1966 he was commander of its parent organization, the Air Research and Development Command, renamed the Air Force Systems Command in 1961. As such, he presided over the development of the Atlas, Thor, and Titan missiles, which served not only as military weapon systems but also as boosters for NASA's space missions. In developing these missiles, Schriever instituted a systems approach, whereby the various components of the Atlas and succeeding missiles underwent simultaneous design and testing as part of an overall "weapons system." Schriever also introduced the notion of concurrency, which has been given various interpretations but essentially allowed the components of the missiles to enter production while still in the testing phase, thereby speeding up development. He retired as a general in 1966. See Jacob Neufeld, "Bernard A. Schriever: Challenging the Unknown," *Makers of the United States Air Force* (Washington, DC: Office of Air Force History, 1986), pp. 281-306; Robert L. Perry, "Atlas, Thor . . .," in Eugene M. Emme, ed., *A History of Rocket Technology* (Detroit, MI: Wayne State University Press, 1964), pp. 144-160; Robert A. Divine, *The Sputnik Challenge: Eisenhower's Response to the Soviet Satellite* (New York: Oxford University Press, 1993), p. 25.

Glenn T. Seaborg (1912-) earned a Ph.D. in physics from the University of California at Berkeley in 1937 and worked on the Manhattan Project in Chicago during World War II. Afterward, he became associate director of Berkeley's Lawrence Radiation Laboratory, where he and associates isolated several transuranic elements. For this work, Seaborg received the Nobel Prize in 1951. He also served as chair of the Atomic Energy Commission, 1961-1971, and thereafter returned to the faculty of the University of California at Berkeley. See David Petechuk, "Glenn T. Seaborg," in Emily J. McMurray, *et al.*, eds., *Notable Twentieth-Century Scientists* (New York: Gale Research Inc., 1995), pp. 1803-1806.

Robert C. Seamans, Jr. (1918-), had been involved in aerospace issues since he completed his Sc.D. degree at the Massachusetts Institute of Technology (MIT) in 1951. He was on the faculty at MIT's department of aeronautical engineering between 1949 and 1955, when he joined the Radio Corporation of America as manager of the Airborne Systems Laboratory. In 1958 he became the chief engineer of the Missile Electronics and Control Division and joined NASA in 1960 as associate administrator. In December 1965, he became NASA deputy administrator. He left NASA in 1968 and became secretary of the Air Force in 1969, serving until 1973. Seamans was president of the National Academy of Engineering from May 1973 to December 1974, when he became the first administrator of the new Energy Research and Development Administration. He returned to MIT in 1977, becoming dean of its School of Engineering in 1978. In 1981 he was elected chair of the board of trustees of Aerospace Corp. ("Robert C. Seamans, Jr.," biographical file, NASA Historical Reference Collection; Robert C. Seamans, Jr., *Aiming at Targets* [Beverly, MA: Memoirs Unlimited, 1994]).

Alan B. Shepard, Jr. (1923-), was a member of the first group of seven astronauts in 1959 chosen to participate in Project Mercury. He was the first American in space, piloting Mercury-Redstone 3 (*Freedom 7*) and was backup pilot for Mercury-Atlas 9. He was subsequently grounded because of an inner ear ailment until May 7, 1969 (during which time he served as chief of the Astronaut Office). Upon returning to flight status Shepard commanded Apollo 14, and in June 1971, he resumed duties as chief of the Astronaut Office. He retired from NASA and the U.S. Navy on August 1, 1974, to join the Marathon Construction Company of Houston, Texas, as partner and chairman. See Alan Shepard and Deke Slayton, *Moonshot: The Inside Story of America's Race to the Moon* (New York: Turner Publishing, Inc., 1994); The Astronauts Themselves, *We Seven* (New York: Simon and Schuster, 1962).

George P. Shultz (1920-) served as director of the Office of Management and Budget after 1970, during the Nixon administration. Before that he had been Nixon's secretary of labor. During the Reagan administration, 1981-1989, Shultz served as secretary of state ("Shultz, George P(ratt)," *Current Biography Yearbook 1988*, pp. 525-30).

Albert F. Siepert (1915-) was a longtime federal employee who entered federal service in 1937 and moved from being executive officer for the National Institutes of Health to NASA in 1958. In 1959 he was NASA's chief negotiator in the transfer of the Army Ballistic Missile Agency to the space agency from his position as director of business administration, and in 1963 he moved to the deputy director position at the Kennedy Space Center in Florida. In 1969 Siepert left NASA to become a program associate at the University of Michigan's Institute for Social Research ("Albert F. Siepert," biographical file, NASA Historical Reference Collection).

Milton A. Silveira (1929-) was a longtime NASA employee, who worked at the agency's Lewis Research Center, 1955-1963, and at the Manned Spacecraft Center in Houston, 1963-1967. He also served as deputy manager of the orbiter project at the Johnson Space Center, 1967-1981; assistant to the deputy administrator at NASA, 1981-1983; and NASA chief engineer, 1983-1986.

Abe Silverstein (1908-), who earned a B.S. in mechanical engineering (1929) and an M.E. (1934) from Rose Polytechnic Institute, was a longtime NACA manager. He had worked as an engineer at the Langley Aeronautical Laboratory between 1929 and 1943 and had moved to the Lewis Laboratory (later, Research Center) in a succession of management positions, the last (1961-1970) as director of the center. Interestingly, in 1958 Case Institute of Technology had awarded him an honorary doctorate. When T. Keith Glennan arrived at NASA, Silverstein was on a rotational assignment to the Washington headquarters as director of the Office of Space Flight Development (later, Office of Space Flight Programs) from the position of associate director at Lewis, which he had held since 1952. During his first tour at Lewis he had directed investigations leading to significant improvements in reciprocating and early turbojet engines. At NASA headquarters he helped create and direct the efforts leading to the spaceflights of Project Mercury and establish the technical basis for the Apollo program. As Lewis's director he oversaw a major expansion of the center and the development of the Centaur launch vehicle. He retired from NASA in 1970 to take a position with Republic Steel Corp. On the career of Silverstein, see Virginia P. Dawson, *Engines and Innovation: Lewis Laboratory and American Propulsion Technology* (Washington, DC: NASA SP-4306, 1991), passim; "Abe Silverstein," biographical file, NASA Historical Reference Collection.

S. Fred Singer (1924-), a physicist at the University of Maryland, proposed a Minimum Orbital Unmanned Satellite of the Earth (MOUSE) at the fourth Congress of the International Astronautics Federation in Zurich, Switzerland, in the summer of 1953. It had been based on two years of previous study conducted under the auspices of the British Interplanetary Society, which had built on the post-war research of the V-2 rocket. The Upper Atmosphere Rocket Research Panel at White Sands discussed Singer's plan in April 1954. In May Singer presented his MOUSE proposal at the Hayden Planetarium's fourth Space Travel Symposium. MOUSE was the first satellite proposal widely discussed in non-governmental engineering and scientific circles, although it never was adopted. See "Singer, S. Fred," biographical file, NASA Historical Reference Collection.

Maurice H. Stans (1908-   ) was a longtime Republican in Washington. He served in several positions with the Eisenhower administration, notably as deputy director of the Bureau of the Budget between 1957 and 1958 and then as its director from 1958 to 1961. In 1969 he was appointed secretary of commerce for the Nixon administration and served until 1972. He was finance director of the 1972 Nixon re-election campaign and pleaded guilty in 1975 to five misdemeanor charges of violating campaign laws ("Maurice H. Stans," biographical file, NASA Historical Reference Collection).

Frank Stanton (1908-   ) earned a Ph.D. from Ohio State University in 1935 and went on to become a business executive, serving most notably as president of CBS, Inc., from 1946 to 1971 and its vice chairman from 1971 to 1973.

Edward V. Stearns (1922-   ) was trained in physics at the University of California at Berkeley and worked in several research positions in industry and universities. He was a physicist with the Rand Corporation, 1949-1954, and assistant chief engineer with the Lockheed Missile and Space Co. after 1954.

John C. Stennis (1901-1995) (D-MS) was elected to the Senate in 1947 and served until 1989. He was a member of the Appropriations, Armed Services, and Aeronautical and Space Sciences Committees in the early 1960s. In 1988 NASA's National Space Technology Laboratories in Mississippi became the John C. Stennis Space Center in his honor (" John C. Stennis," biographical file, NASA Historical Reference Collection).

Ted Stevens (1923-   ) (D-AK) was elected to the U.S. Senate in 1968 and has served to the present.

Lewis L. Strauss (1915-1974) was chairman of the Atomic Energy Commission from 1953 to 1958 and was secretary of commerce from 1958 to 1959. He also held the rank of admiral in the U.S. Navy.

Stuart Symington (1901-1988) (D-MO) served in the Senate between 1953 and 1977. He entered government in 1945 when his fellow Missourian, Harry S. Truman, appointed him chair of the Surplus Property Board. He later served Truman as secretary of the Air Force and was an outspoken advocate of building a strong aerospace presence. As such, he repeatedly charged the Eisenhower administration with balancing the budget at the expense of national security and was one of its most vocal critics after the launch of Sputnik, predicting what proved to be a fallacious missile gap between the United States and the Soviet Union. He left the Senate in 1977 (*New York Times*, December 15, 1988, p. D26; Robert A. Divine, *The Sputnik Challenge: The U.S. Response to the Soviet Satellite* [New York: Oxford University Press, 1993], pp. 20, 43, 125, 178-183).

T

Olin ("Tiger") E. Teague (1910-1981) (D-TX) was first elected to the House of Representatives in 1946 and served in each succeeding Congress through the 95th (1977-1979). He was appointed to the new Science and Astronautics Committee in the 86th Congress (1959-1961).

Charles H. Townes (1915-   ) was trained in physics at Duke University and specialized in the development of laser and maser technology. He first worked for the Bell Telephone Laboratories, and in 1948 he joined the faculty of Columbia University, leaving there in 1961 to move to the Massachusetts Institute of Technology and on to the University of California. For his work on the maser, Townes received the Nobel Prize in 1964. See David E. Newton, "Charles H. Townes," in Emily J. McMurray, *et al.*, eds., *Notable Twentieth-Century Scientists* (New York: Gale Research Inc., 1995), pp. 2042-44.

Richard H. Truly (1937-   ) was a career naval aviator who split time between naval assignments and NASA in the 1960s. In 1965 he was selected to participate in the Air Force's Manned Orbiting Laboratory program and transferred to NASA as an astronaut in August 1969. He served as capsule communicator for all three Skylab missions in 1973 and the Apollo-Soyuz mission in 1975. He was also involved in the Space Shuttle flight test program, and he piloted *Columbia* (STS-2) in 1981 and *Challenger* (STS-8) in 1983. He became NASA's associate administrator for the Office of Space Flight on February 20, 1986, leading the effort to return to flight following the *Challenger* accident. He served as NASA administrator between 1989 and 1992, and he then became vice president and director of the Georgia Tech Research Institute, Georgia Institute of Technology, in Atlanta ("Truly, Richard H.," NASA Administrator Folders, NASA Historical Reference Collection).

H.S. Tsien (1909-   ) was a Chinese national who received a Ph.D. in aeronautics in 1939 from the California Institute of Technology (Caltech) and worked on the development of rocket technology at his alma mater through World War II. He was on the faculty of the Massachusetts Institute of Technology from 1946 to 1949, when he returned to Caltech. In the 1950s his loyalty to democratic institutions was questioned, and he was deported from the United States to the People's Republic of China. There, he was largely responsible for the development of ICBM rocket technology, especially the "Long March" launch vehicle.

Konstantin E. Tsiolkovskiy (1857-1935) is one of the three recognized pioneers of spaceflight. A schoolteacher in Kaluga, Russia, Tsiolovskiy theorized about the flight of rockets and spacecraft, calculated many of the equations required for the successful launch of rockets, and speculated on the development of space vehicles and permanent space colonies. See Arkady Kosmodemyansky, *Konstantin Tsiolkovskiy* (Moscow, USSR: Nauka, 1985).

Nathan F. Twining (1897-1982) was a career pilot in the Army and the Air Force, commanding the 13th Air Force in the Pacific, the 15th Air Force in Europe, and then the 20th Air Force again in the Pacific during World War II. He became chief of staff of the Air Force in 1953 and chairman of the Joint Chiefs of Staff from 1957 to 1960 (Donald J. Mrozek, "Nathan F. Twining: New Dimensions, a New Look," in John L. Frisbee, ed., *Makers of the United States Air Force* [Washington, DC: Office of Air Force History, 1987], pp. 257-80).

V

Max Valier (1893-1930) was an early advocate of the use of rockets for spaceflight. A German, he had been educated in engineering in Berlin, and as a young man in the 1920s he began experimenting with rockets with the "Verein fur Raumschiffahrt" (VfR), the Society for Spaceship Travel of which Wernher von Braun and Hermann Oberth were prominent members. He was also interested in using rockets for propelling ground vehicles, and he built a rocket-powered automobile. He died in a crash of this car in 1930 (I. Essers, *Max Valier: A Pioneer of Space Travel* [Washington, DC: NASA TT F-664, 1976]).

James A. Van Allen (1914-) was a pathbreaking astrophysicist best known for his work in magnetospheric physics. Van Allen's January 1958 *Explorer I* experiment established the existence of radiation belts—later named for the scientist—that encircled the Earth, representing the opening of a broad research field. Extending outward in the direction of the Sun approximately 40,000 miles, as well as stretching out with a trail away from the Sun to approximately 370,000 miles, the magnetosphere is the area dominated by Earth's strong magnetic field. See James A. Van Allen, *Origins of Magnetospheric Physics* (Washington, DC: Smithsonian Institution Press, 1983); David E. Newton, "James A. Van Allen," in Emily J. McMurray, *et al.*, eds., *Notable Twentieth-Century Scientists* (New York: Gale Research Inc., 1995), pp. 2070-72.

Cyrus R. Vance (1917-) had a long career as a senior government official in various Democratic administrations. He had been general counsel for the Department of Defense during the Kennedy administration of the early 1960s and was also secretary of the Army from 1962 to 1964. He was deputy secretary of defense from 1964 to 1967. He served as secretary of state for President Jimmy Carter in the latter half of the 1970s ("Vance, Cyrus R[oberts]," *Current Biography 1977*, pp. 408-11).

Jules Verne (1828-1905) was one of the leading writers of his time, as well as one of the founders of the literary genre of science fiction. He described in his novels the possibility of spaceflight, the use of submarines for travel beneath the ocean, and a variety of other visionary technologies that were realized in the twentieth century (I.O. Evans, *Jules Verne and His Work* [New York: Twayne, 1966]).

W

Alan T. Waterman (1892-1967) was the first director of the National Science Foundation (NSF), from its founding in 1951 until 1963. Waterman received his Ph.D. in physics from Princeton University in 1916; he then served with the Army's Science and Research Division in World War I, on the faculty of Yale University in the interwar years, with the War Department's Office of Scientific Research and Development in World War II, and with the Office of Naval Research between 1946 and 1951. He and NASA leaders contended over control of the scientific projects to be undertaken by the space agency, with Waterman's NSF being used as an advisory body in the selection of space experiments. See "Waterman, First NSF Head, Dies at 75," *Science* 158 (8 December 1967): 1293; Norriss S. Hetherington, "Winning the Initiative: NASA and the U.S. Space Science Program," *Prologue: The Journal of the National Archives* 7 (Summer 1975): 99-108; John E. Naugle, *First Among Equals: The Selection of NASA Space Science Experiments* (Washington, DC: NASA SP-4215, 1991).

James E. Webb (1906-1992) was NASA administrator between 1961 and 1968. Previously, he had been an aide to a congressman in New Deal Washington, an aide to Washington lawyer Max O. Gardner, and a business executive with the Sperry Corporation and the Kerr-McGee Oil Co. He had also been director of the Bureau of the Budget between 1946 and 1950 and under secretary of state from 1950 to 1952 (W. Henry Lambright, *Powering Apollo: James E. Webb of NASA* [Baltimore, MD: The Johns Hopkins University Press, 1995]).

R.S. Wehner (1915-) was a research scientist with the Radio Corporation of America, 1943-1945; the Airborne Instrument Laboratory, 1945-1948; the Rand Corporation, 1948-1951; and the Hughes Aircraft Co., 1951-1959.

Caspar W. Weinberger (1917-), a longtime Republican government official, was a senior member of the Nixon, Ford, and Reagan administrations. For Nixon he was deputy director (1970-1972) and director (1972-1976) of the Office of Management and Budget. In this capacity, had a leading role in shaping the direction of NASA's major effort of the 1970s, the development of a reusable Space Shuttle. For Reagan he served as secretary of defense, where he also oversaw the use of the Shuttle in the early 1980s for the launching of classified Department of Defense payloads into orbit. See "Weinberger, Caspar W(illard)," *Current Biography 1973*, pp. 428-30.

H.G. Wells (1866-1946) was a noted futurist and one of the founders of the literary genre of science fiction. His novels described a future filled with technology, some of it terrifying, and contact with extraterrestrial beings, much of it disastrous (Lovat Dickson, *H.G. Wells: His Turbulent Life* [New York: Atheneum, 1969]).

Edward C. Welsh (1909-) had a long career in various private and public enterprises. He had served as legislative assistant to Senator Stuart Symington (D-MO), 1953-1961, and was the executive secretary of the National Aeronautics and Space Council through the 1960s.

Fred L. Whipple (1906-) was a University of California at Berkeley Ph.D. in astronomy who served on the faculty of Harvard University. He was involved in efforts in the early 1950s to expand public interest in the possibility of spaceflight through a series of symposia at the Hayden Planetarium in New York City and articles in *Collier's* magazine. He was also heavily involved in planning for the International Geophysical Year, 1957-1958. As a pathbreaking astronomer he pioneered research on comets. See Raymond E. Bullock, "Fred Lawrence Whipple," in Emily J. McMurray, *et al.*, eds., *Notable Twentieth-Century Scientists* (New York: Gale Research Inc., 1995). pp. 2167-70.

Jerome B. Wiesner (1915-1994) was science advisor to President John F. Kennedy. He had been a faculty member of the Massachusetts Institute of Technology and had served on President Eisenhower's Science Advisory Committee. During the presidential campaign of 1960 Wiesner had advised Kennedy on science and technology issues and prepared a transition team report on the subject that questioned the value of human spaceflight. As Kennedy's science advisor he tussled with NASA over the lunar landing commitment and the method of conducting it. See Gregg Herken, *Cardinal Choices: Science Advice to the President from Hiroshima to SDI* (New York: Oxford University Press, 1992).

Walter C. Williams (1919-) earned a B.S. in aerospace engineering from Louisiana State University in 1939 and went to work for NACA in 1940, serving as a project engineer to improve the handling, maneuverability, and flight characteristics of World War II fighters. Following the war he went to what became Edwards Air Force Base to set up flight tests for the X-1, including the first human supersonic flight by Capt. Charles E. Yeager in October 1947. He became the founding director of the organization that became the Dryden Flight Research Facility. In September 1959 he assumed the associate directorship of the new NASA space task group at Langley that was created to carry out Project Mercury. He later became director of operations for the project and then associate director of NASA's Manned Spacecraft Center in Houston, subsequently renamed the Johnson Space Center. In 1963 Williams moved to NASA headquarters as deputy associate administrator of the Office of Manned Space Flight. From 1964 to 1975 he was a vice president for Aerospace Corporation. Then from 1975 to 1982 he served as chief engineer of NASA, retiring in 1982 ("Walter C. Williams," biographical file, NASA Historical Reference Collection).

Charles E. Wilson (1886-1972) was an industrialist with General Electric who worked with the Office of Defense Mobilization in the 1950s.

Y

John F. Yardley (1925-) was an aerospace engineer who worked with the McDonnell Aircraft Corporation on several NASA human spaceflight projects from the 1950s and into the 1970s. He also served as NASA associate administrator for the Office of Space Flight between 1974 and 1981. Thereafter, he returned to McDonnell Douglas as president, 1981-1988 ("Yardley, John F.," biographical file, NASA Historical Reference Collection).

Chuck Yeager (1923-) was the U.S. Air Force test pilot who piloted the X-1 research aircraft on the first supersonic powered flight in 1947. Thereafter, he served in several Air Force positions, retiring as a brigadier general. See Chuck Yeager, *Yeager* (New York: Bantam Books, 1982).

Herbert F. York (1923-) had been associated with scientific research in support of national defense since World War II. He was director of the Livermore Radiation Laboratory for the University of California before moving to the Department of Defense in March 1958 as chief scientist of the Advanced Research Projects Agency. He became the Department of Defense's director of research and engineering in December 1958 during a Department of

Defense reorganization; this was the third-ranking civilian office after the secretary and deputy secretary of defense. He served as director of defense research and engineering until 1961. He then moved to the University of California at San Diego, where he was chancellor and a professor of physics. He also served as a member of the President's Science Advisory Committee under both Eisenhower and Johnson and was later the chief negotiator for the comprehensive test ban during the Carter administration ("Dr. Herbert F. York," biographical file, NASA Historical Reference Collection; Herbert F. York, *Making Weapons, Talking Peace: A Physicist's Odyssey from Hiroshima to Geneva* [New York: Basic Books, 1987]).

Index

A

Abbot, Charles G., 136, 140
"Ad Hoc Panel on Man-in-Space, Report of," 378, 408-12
Adams, L.H., 295
Adams, Walter S., 136
Advanced Research Projects Agency (ARPA), DOD, 225, 226, 227, 634, 636-37, 643, 650
Advent project, 447
Aerobee Launch Vehicle, 14, 303, 310
Aerobee-Hi Launch Vehicle, 223
Aerojet Corp., 11, 14, 153, 154, 156, 157
Aeronautical Board, War Department, 214
Aeronautical Sciences, Journal of, 145-53
Aeronautics and Astronautics Coordinating Board, 472, 524
Aerophysics Development Corp., 278, 280
Aerospace Corp., 436, 614, 657, 664
Africa, 509
Agena Booster, 404, 406, 407, 425, 430, 441-42, 480-83
Agnew, Spiro T., and Space Task Group (1969), 383-85, 513, 519, 543, 622
Agriculture, Department of, 497, 499, 513, 520, 570-71, 577, 654, 670
Air Corps Jet Propulsion Research Project, 153, 154, 156, 157
Air Defense Command, 205, 209
Air Force Meteorological Research Center, Cambridge, Massachusetts, 203
Air Force Proving Ground, Cocoa, Florida, 182
Air Force Regulation 200-2, 210
Air Force, United States, 178, 207-11, 214, 215, 217, 222, 224, 236, 245, 274, 275, 359, 373, 384, 389, 436, 515, 587, 614-15, 639, 644-46, 657, 673, 683, 708, 716; and Ballistic Missile Command (BMC), 630
Air Intelligence Service Squadron, 210
Air Research and Development Command (ARDC), 221, 269, 278
Air Technical Intelligence Center, 202-206, 210
Allen, Joe, 715
Allen, Lew, 735
Allis-Chalmers, Inc., 218
Almagest, 1
Alouette project, and Canada, 479
Alverez, Luis W., 202, 205, 206, 241
American Association for the Advancement of Science (AAAS), 616
American Interplanetary Society, 9, 12, 281
American Institute of Aeronautics and Astronautics, 281
American Rocket Society, 12, 157, 274; and "Utility of an Artificial Unmanned Earth Satellite: A Proposal to the National Science Foundation, Prepared by the ARS Space Flight Committee, November 24, 1954, On the," 281-94
American Telephone and Telegraph (AT&T) Corp., 439, 464
Ames Research Center, California, 389, 477, 587, 611, 647, 717; and field center roles, 689-711
Anacostia, Maryland, 202
"Analysis of Reports of Unidentified Aerial Objects," 210
Anderson, Clinton P., 492, 642
Annapolis, Maryland, 12
Antwerp, Belgium, 13
Apollo Applications Program (AAP), 383-85, 499, 501, 504-05, 514, 521, 714
Apollo, Project, 9, 409, 441, 457, 460-63, 484, 469, 495-96, 502-03, 506, 513-15, 517, 522, 527-29, 532, 545, 547-48, 598, 612-13, 616, 620, 622, 686, 714, 716, 730; and Apollo 11, 384, 385, 392, 519, 521-23, 525, 529, 556, 580-81, 618, 672; and Command Module, 481-82; and decision for, 379-81; and Lunar Excursion Module (LEM), 481-82, 517, 715; review of, 381-82; and Saturn Launch Vehicle, 674-77, 681-82, 684; and Service Module, 481-82; and project, 582
Aquinas, Thomas, 22
Archer, Harry J., 283
Ariel projects, and United Kingdom, 479
Arms Control and Disarmament Agency, 228, 593-94, 604

Armstrong, Neil A., 392
Army Air Corps, United States, 10, 11
Army Air Forces, United States, 12, 15, 153, 213, 214, 614
Army Ballistic Missile Agency (ABMA) (also see Marshall Space Flight Center), 611-12, 630, 636
Army Map Service, 289
Army Ordnance Guided Missile Development Group/Division, 178, 189, 195, 274-81
Army, United States, 14-15, 179, 180, 214, 221-22, 227, 274, 275, 612, 614, 630, 644, 657, 673; and "Minimum Satellite Vehicle: Based on Components Available from Missile Developments of the Army Ordnance Corps," 274-81
Arnold, Henry H. (Hap), 10-11, 153, 213, 215
Asia, Southeast, 439
Asimov, Isaac, 16, 17
Associated Universities, Inc., 206
Astronautica Acta, 314-24
Astronomy, 1-3
Atlantic Missile Range, 399, 460, 673
Atlantic Monthly, 3-4, 23-55
Atomic Energy Act, 511
Atomic Energy Commission (AEC), 268, 341-42, 413, 436, 443, 449-50, 452, 455, 459, 461, 497, 513, 611, 614, 629-37, 642-43, 647, 654-55, 657, 659, 667
Atomic Peace Ship (also see Eisenhower, Dwight), 632
Atwood, Wallace W., Jr., 136, 140, 295
Atlas Launch Vehicle, 15-16, 219, 224, 310, 359, 373, 379, 388, 404, 406, 410, 419, 425, 430, 441-42, 449, 451, 462, 480, 483, 494, 589, 645
Auburn, Massachusetts, 7, 136
Augustine, Norman R., and "Augustine Commission," 626-28, 741-743
Autour de la Lune (Around the Moon), 4

B

Bahrein, 716
Baker, James, 217
Baker-Nunn, Inc., 227
Ballistic Missile Early Warning System (BMEWS), 227
Baltimore, Maryland, 5, 8
Barr, William J., 283
Battelle Memorial Institute, 203
Bauer, H.E., 676
"Bazooka," 8, 156
"Beacon Hill" Study, 217
Beech Aircraft Company, 656
Beggs, James M., 587, 724; and decision to build the space station, 390-92, 600; and named NASA administrator, 389
Bell Aircraft Co., 8
Bell, David, 614; and "Bell Report," 615, 627, 651-72
Bell Laboratories, 214, 217, 288, 439, 633
Bell Telephone Corp., 280, 633
Bellefountain, Ohio, 204
Bellfortis, 5
Beltsville, Maryland, 649
Bendix Aviation, Inc., 218
Beresford, Spencer M., 511
Berkner, Lloyd V., 202, 203, 206; and National Committee for the IGY's "Summary of the Eighth Meeting," 295-308
Berning, W., 299
Bing Crosby Enterprises, 272
"B.I.S. Space Ship, The," 140-45
Bison Bomber, 218
Blechman, Barry, 559
Bloemfontein, South Africa, 195
Blue Book, Project, 208
Blue Scout program, 451

Board of National Intelligence Estimates, CIA, 215-16
Boeing Corporation, 550, 552, 617-18, 674
Bollay, Eugene, and "Utility of an Artificial Unmanned Earth Satellite: A Proposal to the National Science Foundation, Prepared by the ARS Space Flight Committee, November 24, 1954, On the," 281-94
Bollay, William, 280
Bonestell, Chesley, 177, 179
Bossart, Karel J., 15
Boston American, 86
Boston, Massachusetts, 178
Bowen, Ira S., and "Utility of an Artificial Unmanned Earth Satellite: A Proposal to the National Science Foundation, Prepared by the ARS Space Flight Committee, November 24, 1954, On the," 281-94
Branscomb, Lewis M., 511
von Braun, Wernher, 6, 9, 12-15, 17-19, 84, 140, 178, 179, 267, 427, 433, 715; and Apollo decision, 379-81; and "Can We Get to Mars?," 176, 195-200; and *Collier's* magazine, 17-19, 176-77; and "Crossing the Last Frontier," 176, 179-88; and decision to build the space shuttle, 386-88; and "Man in the Moon: The Journey," 176, 189-94; and "Minimum Satellite Vehicle: Based on Components Available from Missile Developments of the Army Ordnance Corps," 274-81; and Nixon space policy, 385-86; and post-Apollo planning, 382-85; and Project Orbiter, 221-22, 308, 310; and V-2 (A-4) Rocket, 12-15, 17, 180-81, 237, 239, 267, 429; and Walt Disney, 19
Breckinridge, Henry, 136
"Brick Moon, The," 3-4, 24-55
Bridges, Styles, 380, 433-37
Britain, 509
British-Australian Guided Missile Range, Australia, 312
British Interplanetary Society, 9, 314; and "B.I.S. Space Ship, The," 140-45
British Interplanetary Society, Journal of the, 9; and "B.I.S. Space Ship, The," 140-45
Brodie, Bernard, 215
Brookhaven National Laboratories, 206
Bronk, Detlov W., and "Memorandum of Discussion at the 322d Meeting of the National Security Council," 324-28; and National Committee for the IGY's "Summary of the Eighth Meeting," 295-308
Brown, George, 427, 437-38
Brown University, 378, 408, 657
Brundage, Percival, 324-25, 611, 637, 641; and "Memorandum of Discussion at the 322d Meeting of the National Security Council," 324-28
Brzezinski, Zbigniew, 575
Bulganin, Nikolai, 223, 350
Bumper, Project, 14-15
Burbank, California, 222
Bureau of Aeronautics, U.S. Navy, 12, 156, 157, 214
Bureau of the Budget (BOB), 307, 408, 437, 439, 454-55, 466-67, 473, 490, 495, 499, 611, 614, 620, 622, 632, 637-38, 643-45; Apollo decision, 379-81; and "Bell Report," 651-72; and "Memorandum of Discussion at the 322d Meeting of the National Security Council," 324-28; and National Aeronautics and Space Act of 1958, 334, 377; and post-Apollo planning, 382-85; renamed Office of Management and Budget (OMB), 386; and review of Apollo, 381-82, 435; and Space Task Group (1969), 384-85, 544-45 (see also Office of Management of Budget)
Bureau of Labor Statistics, 669
Bureau of Standards, 643
Burr, Aaron, 23
Burston-Marstellar, 717
Bush, George H., 625, 741
Bush, Vannevar, 382

C

C-119 Aircraft, 216
California Institute of Technology (Caltech), 10-11, 145, 154, 156, 206, 219, 298, 512, 611, 614, 630, 657
California, University of, Berkeley, 206, 383
California, University of, Los Angeles (UCLA), 178
Cambridge, Massachusetts, 203
"Camel News Caravan," 18
"Can We Get to Mars?," 176, 195-200
Canada, 486, 507; and Alouette project, 479
Cannon, Joseph, 715
Cape Canaveral, Florida, 15, 429, 436, 469, 612, 673

Cape Kennedy, Florida, and Merritt Island Launch Facility, 484
Carnegie Corporation, 633
Carnegie Institution of Washington, 87, 136, 140
Carpenter Steel Co., 97
Carter, James E. (Jimmy), 388-89, 390, 559, 575; and "Civil and Further National Space Policy" (NSC 42), 389, 575
Case Institute of Technology (CIT), 377, 647-49
Cavers, Dean David, 633
Centaur Booster, 404, 406, 407, 410, 419, 425, 429, 441, 450-51, 482-83, 487, 645-46, 707
Central Intelligence Agency (CIA), 19-20, 201-206, 215-16, 217, 220, 225, 326, 329, 331, 373-75, 593-94, 598, 604
Challenger Space Shuttle, 390, 392, 624, 730-31; and "Rogers Commission Report," 723-29
Chicago, Illinois, 269
Chidlaw, Benjamin, 11
Chrysler Corporation, 656, 674
Churchill, Winston, 178
Cicero, 1-2
Chadwell, E. Marshall, 206
China, People's Republic of, 10
"Civil and Further National Space Policy," 389
Civil Service Commission, United States (also see Office of Personnel Management), 614, 620, 671
Civil War, 616
Clark College/University, Worcester, Massachusetts, 7, 86, 135, 136, 137-38, 140
Clark, Ralph, E., 206
Clark, Ray, 673-74
Clark, William, 594
Clarke, Arthur C., 16, 17
Clauser, Francis H., 215, 511-12
Cleveland, Ohio, 377, 647-49
Coale, Ansley, 215
Cocoa, Florida, 182, 312
Cole, Judy, 715
College Park, Maryland, 314
Collier's Magazine, 17-19, 176-77; and "Can We Get to Mars?," 176, 195-200; and "Crossing the Last Frontier," 176, 179-88; and "Is There Life on Mars?," 176, 194-95; and "Man in the Moon: The Journey," 176, 189-94; and "What Are We Waiting For?," 18, 176, 177-79
Columbia Broadcasting System (CBS), 18, 437
Columbia University, 267, 657
Columbus, Ohio, 203
Commerce, Department of, 227, 442, 449, 499, 513, 520, 577, 592-95, 604, 606-07, 654-55, 670; and Office of Technical Services, 643
Commercial Space Working Group, 610
Committee on Guided Missiles, DOD, 214
Committee on the Peaceful Uses of Outer Space, 229
Committee on Space Research (COSPAR), 368
Committee on Special Capabilities, DOD, 221-23
Communications Satellite Act of 1962, 464
Communications Satellite Corporation, and "Early Bird" satellite, 478
ComSat Corporation, 497
Congress, United States, 228, 307, 326, 328, 498, 511, 513, 515, 546, 563, 568, 618, 620, 627, 631-32, 636, 645, 647, 674, 713-14, 734; and Apollo decision, 379-81; and "Bell Report," 651-672; and decision to build the space shuttle, 386-88; and post-Apollo planning, 382-85; and National Aeronautics and Space Act of 1958, 334-45, 377, 397, 529; and Nixon space policy, 385-86; and Paine Commission, 392; and review of Apollo, 381-82; and Space Shuttle program, 550, 554
Congreve, William, 5
Conquest of Space, The, 176
Consolidate Vultee Aircraft (Convair) Division, 15, 645-46
Cook, Donald, 427, 437
Cooper, Gordon, 480
Cooper, John Cobb, 633
Cooper, William 712
Copernicus, Nicolas, 2, 22
Cornell Aeronautical Laboratories, 217

Corson, John J., 633
Courier Communications Satellite, 227, 442
Cranston, Alan, 714
Crawford, Frederick C., 648
Crippen, Robert, 730
Crow, Pete, 715
"Crossing the Last Frontier," 176, 179-88
Cuba, 556
Cutler, Robert, 219; and "Memorandum of Discussion at the 322d Meeting of the National Security Council," 324-28
Curtiss-Wright Aeronautical Co., 8
Curtiss-Wright XLR25-CW-1 Engine, 8
Cyrano de Bergerac, Savinion de, 3

D

Davenport, William 712
David, Edward, 548-49
Davies, E. Merton, 215
Day the Earth Stood Still, The, 17, 20
Defense, Department of (DOD), 8, 11, 15, 21, 181, 214, 216, 220, 221, 224, 225, 228, 266, 269, 306, 398, 400, 418, 423, 424, 428, 433, 436, 439-41, 449, 451, 454-55, 459-60, 464, 468, 471-72, 498, 500, 502, 509-10, 516, 518-20, 522, 524, 526-27, 532, 534, 537, 541-43, 560, 570-72, 577, 580, 604-09, 611, 619, 629-31, 638-44, 647-50, 654-55, 659, 667, 669-73, 716, 721-22; and Apollo decision, 379-81; and commercialization of satellite communications, 447; and decision to build the space shuttle, 386-88; and decision to build the space station, 390-92; and Defense Reorganization Act of 1947, 614; and Explorer program, 221-22, 228, 644; and freedom of space, 213–29, 230, 308; and International Geophysical Year (IGY), 19, 20, 200-201, 220-23, 224-25, 227, 228, 314; and Manned Orbital Laboratory (MOL), 480, 496, 499, 515; and *Meeting the Threat of Surprise Attack*, 219-20; and "Memorandum of Discussion at the 322d Meeting of the National Security Council," 324-28; and National Reconnaissance Office, 373-75; and National Security Council, 200, 218, 222, 229, 308-13, 324-28; and Nixon space policy, 382-86; and "Open Skies" doctrine, 213-29, 230; and "Policy on Outer Space, U.S." (NSC 5814), 345-59, 360; and "Policy on U.S. Scientific Satellite Program, Draft Statement of" (NSC 5520), 200, 308-13, 326, 328; and post-Apollo planning, 382-85; and "Problem of Space Programs," 413-15; and "Project Feed Back Summary Report," 269-74; and Project Vanguard, 14, 221-23, 224-25, 228, 395, 402, 644; and "Preliminary U.S. Policy on Outer Space" (NSC 5814/1), 360-61, 377; and reconnaissance programs, 219-20, 222-23, 227-28, 348, 354, 373-75; and "Report on the Present Status of the Satellite Problem," 267-69; and review of Apollo, 381-82; and satellite reconnaissance, 216-17, 221-24, 227-28, 269-74, 348, 354, 373-75; and Space Shuttle program, 549-54, 582, 591-93; and Space Station program, 594-95; and "Surprise Attack Conference," 227-28; and Technological Capabilities Panel ("Surprise Attack Panel"), 218-19, 221, 225, 309; and U-2 program, 219-20, 222-23, 227; and "U.S. Policy on Outer Space," 362-73; and "Wiesner Report," 379, 416-23
Dehavilland Twin Otter, 588
Delhi, India, 716
Delta Launch Vehicle, 388, 404, 478, 482, 589
Description of the Caelestiall Orbes, 22
Dessau, Germany, 9
Digges, Thomas, 22
Dirksen, Everett, 383, 490, 492
Discoverer program, 451, 460
Disney, Walt, 19
Distant Early Warning (DEW) Line, 218, 219, 227
Divine Comedy, The, 22
Donovan, Allen, 217
Doolittle, James A., 394, 632
Dorman, B.L., 283
Dornberger, Walter, 13
Douglas Aircraft Company, Inc., 236, 617-18, 675; and "Preliminary Design of an Experimental World-Circling Spaceship," 236-45
Dow, W.G., 299
Drew, Russ, 549
Droessler, Earl J., 295
Dryden, Hugh L., 394, 435-39, 461, 464, 483, 620, 645, 647-48
Dryden Flight Research Center, California, 477, 611; and field center roles, 689-711

DuBridge, Lee A., 219, 512-13, 543
Dulles, Allen W., 225, 329; and "Memorandum of Discussion at the 322d Meeting of the National Security Council," 324-28
Dulles, John Foster, 331
Dunning, John R., 266-67
Durant, Frederick C., III, 201-206, 274
Dyna-Soar, Project, 227, 426, 451, 459-60

E

"Early Bird" satellite system, and Communications Satellite Corporation, 478
Earth Resources Observation System (EROS), 693
Earth Resources Technology Satellite (ERTS), 554
Eastman Kodak, Inc., 217
Eaton, E.L., 299
Echo Communications Satellite, 442, 478
Economics Policy Council (EPC), 610
Edwards Air Force Base, California, 647
von Eichstadt, Konrad Kyser, 5
Einstein, Albert, 395
Eisenhower, Dwight D., 19, 218, 388, 466, 468, 492, 511, 614, 628, 631, 637, 643, 647-48; and Ad Hoc Panel on Man-in-Space, 378, 408-12; and establishes NASA, 226; and Atomic Peace Ship, 632; and Explorer program, 221-22, 228; and freedom of space, 213-29, 230, 308; and establishes National Reconnaissance Office, 373-75; and Geneva Summit, 222-23, 227-28; and International Geophysical Year (IGY), 19, 20, 200-201, 220-23, 224-25, 227, 314; and "Introduction to Outer Space," 332-34; and *Meeting the Threat of Surprise Attack*, 219-20; and "Memorandum of Discussion at the 322d Meeting of the National Security Council," 324-28; and National Security Council, 200, 218, 222, 229, 308-13, 324-28; and "Open Skies" doctrine, 213-29, 230, 611; and no more "Pearl Harbors," 215-16, 221, 222-23, 225, 227; and "Policy on Outer Space, U.S." (NSC 5814), 345-59, 360; and "Policy on U.S. Scientific Satellite Program, Draft Statement of" (NSC 5520), 200, 308-13, 326, 328; and "Preliminary U.S. Policy on Outer Space" (NSC 5814/1), 360-61, 377; and "Project Feed Back Summary Report," 269-74; and Project Vanguard, 14, 221-23, 224-25, 228, 395, 402; and reconnaissance programs, 219-20, 222-23, 227-28, 348, 354, 373-75; and "Report on the Present Status of the Satellite Problem," 267-69; and satellite reconnaissance, 216-17, 221-24, 227-28, 269-74, 348, 354, 373-75; and "Surprise Attack Conference," 227-28; and Technological Capabilities Panel ("Surprise Attack Panel"), 218-19, 221, 225, 309; and U-2 program, 219-20, 222-23, 227; and U.S. Policy on Outer Space," 362-73
Eizenstat, Stuart, 559
El Paso, Texas, 13
Electronics Research Center, Massachusetts, 484
Ellsworth, Robert, 514
Energy, Department of, 577
Energy Research and Development Administration (ERDA), 570
"Engineering Techniques in Relation to Human Travel at Upper Altitudes," 262-66
Environmental Protection Agency (EPA), United States, 570
Environmental Science Services Administration (ESSA), 520
Ehrlichman, John, 35, 546, 558-59
"Exploration of the Universe with Reaction Machines," 59-84
European Space Agency (ESA), 580, 582, 708
European Space Research Organization (ESRO), 486, 509
Explorer Project, 221-22, 228, 478-79, 486, 644
Extended Duration Orbiter (also see Shuttle, Space), 599

F

Fairbanks, Richard, 579
Farley, Clare, 438
Federal Aviation Administration (FAA), 654
Federal Communications Commission (FCC), 439
Feed Back, Project, 218, 221, 245, 269-74
Finan, William, 632
Fink, Daniel J., and "Study of the Mission of NASA," 717-23
First Men in the Moon, The, 4
Fisk, James, 633

Flagstaff, Arizona, 4, 55
Flanigan, Peter M., 546
Flax, Alexander, and "Flax Committee," 548
Fleming, John A., 136
Fletcher, James C., 549, 567, 685, 730; and decision to develop the space shuttle, 386-88, 555, 558-59; and *Leadership and America's Future in Space*, 392-93; and named NASA administrator, 386, 392; and "Problems and Opportunities at NASA," 711-17
"Flight Analysis of the Sounding Rocket," 145-53
Flight Research Center, Dryden, California, 477
Forbidden Planet, 17
Ford Foundation, 633
Forrestal, James, 178
Fort, Charles, 204
Fort Bliss, Texas, 13
Fort Devens, Massachusetts, 135, 136
Fort McHenry, Maryland, 5
Foster, William C., 228
Fournet, Dewey J., 202, 203
France, 509, 580
Frau im Mond (*The Girl in the Moon*), 85
Freeman, Fred, 177, 179
Friedman, Herb, 715
Frosch, Robert, 711
Fubini, Eugene, 548
Fuller, Craig, 595
Future Programs Task Group, 473-90

G

Gagarin, Yuri, 379, 423, 492, 494
Galilei, Galileo, 2, 3, 395
Gallup Poll, 16
Gamma Ray Observatory, 584
Gardner, Trevor, 218-19
Garland, William N., 203, 206
Garn, Jake, 726
Gemini program, 457, 460, 462-63, 480-82, 492-94, 496
General Services Administration (GSA), 614, 667
Garroway, Dave, 18
General Accounting Office, United States (GAO), 563, 571
General Motors Corporation, 656
General Tire and Rubber Co., 11
Geneva, Switzerland, 222, 227-28, 610
Geological Survey, United States (USGS), 693
Gerson, N.C., 298, 299
Gleason, S. Everett, and "Memorandum of Discussion at the 322d Meeting of the National Security Council," 324-28
Glenn, John, 480
Glennan, T. Keith, 377, 614, 645, 647
Global Atmospheric Sampling Program, 708
Goddard, Robert H., 1, 6-7, 10, 11, 12, 13, 16, 59, 133, 146, 157; and *A Method of Reaching Extreme Altitudes*, 7, 86-132; and *Liquid-Propellant Rocket Development*, 8, 134-40
Goddard Space Flight Center, Maryland, 378, 484, 611-12; and field center roles, 689-711
Goett, Harry J., 379
Goldwater, Barry, 714
Gore, Albert Sr., 642
Goudsmit, Samuel A., 202, 206
Graham, William, 723
Gravity Probe-B project, 709
Great Falls, Montana, 202, 204
Greenbelt, Maryland, 378
Griggs, David T., 205

Grissom, Virgil I. (Gus), 15
Groo, Elmer S. (Todd), 688, 717
Grosse, Aristid V., 266; "Report on the Present Status of the Satellite Problem," 267-69
Grumman Corporation, 550, 552, 674
Guggenheim Aeronautical Laboratory, California Institute of Technology (GALCIT), 10-11; and "A Review and Preliminary Analysis of Long-Range Rocket Projectiles," 155-76; and "Flight Analysis of the Sounding Rocket," 145-53; and "Memorandum on the Possibilities of Long-Range Rocket Projects," 153-55
Guggenheim, Daniel, 134, 135, 137, 140
Guggenheim, Florence, 134, 137, 140
Guggenheim Fund, 7-8, 134, 137,140

H

Haber, Fritz, 177
Haber, Heinz, 178
Hagen, John P., and "Memorandum of Discussion at the 322d Meeting of the National Security Council," 324-28
Hagerty, James C., 200-201, 331
Hale, Edward Everett, 3-4; "Brick Moon, The," 24-55
Haley, Andrew G., and "Utility of an Artificial Unmanned Earth Satellite: A Proposal to the National Science Foundation, Prepared by the ARS Space Flight Committee, November 24, 1954, On the," 281-94
Halley's comet, 580, 584
Hammond, Victor, 673
Haneda Air Force Base, Japan, 204
Hansen, 437-38
Harvard Observatory, 217, 243
Harvard University, 178, 189, 194, 217, 228, 633, 657
Haurwitz, B., 298, 299
Hayden Planetarium, New York City, 18, 176, 178, 314
Hayward, Thomas, 427
Health, Education, and Welfare, Department of, 520, 570-71, 654
Heaton, Donald, 673
Hebert, Eddie, 714
Heinlein, Robert A., 16
Heiss, Klaus P., 549
Heller, Gerhard, 280
Herter, Christian A., and "Memorandum of Discussion at the 322d Meeting of the National Security Council," 324-28
Hess, Harry H., 511
High Energy Astronomical Observatory (HEAO), 709
Hitchcock, James J., 220
Hodgkins Fund, 7, 86
Holaday, William M., and "Memorandum of Discussion at the 322d Meeting of the National Security Council," 324-28
Hollywood, California, 179
Holmes, D. Brainerd, 381, 454, 461
Hoover, Lt. Cdr., 275
Hoover, Herbert, Jr., 223
Hornig, Donald F., and Ad Hoc Panel on Man-in-Space, 378, 408-12
Horowitz, Norman H., 511
Housing and Urban Development, Department of, 520
Hrones, John, 648
Hubble Space Telescope (also see Space Telescope project), 726, 741; and "Hubble Space Telescope Optical Systems Failure Report," 735-741
Hughes Danbury Optical Systems, Inc. (also see Perkin Elmer Corp.), and Hubble Space Telescope, 736
Humphrey, Hubert, 643
Hunsaker, Jerome C., 10
Huntsman, Jon, 547
Huntsville, Alabama, 14, 15, 17, 179, 189, 195, 281, 429, 715-16
Hynek, J. Allen, 200-206, 207

I

IBM Corp., 209, 617-18
Illinois, University of, 657
Infrared Astronomy Satellite (IRAS), 598
Institute for Aeronautical Sciences, 632
Integrated National Space Plan, 440, 446
Intelligence Advisory Committee, CIA, 202
Intelsat, 509
Intercontinental Ballistic Missiles (ICBM), 10, 15-16, 217, 395, 432, 446, 448, 633, 645
Interior, Department of, 497, 499, 513, 520, 570-71, 577, 670
International Astronautical Federation, 274, 314
International Bank for Reconstruction and Development, 554
International Council of Scientific Unions (ICSU), 20-21, 295, 368
International Geophysical Year (IGY), 19, 20, 200-201, 220-23, 224-25, 227, 228, 274, 329, 366; and National Committee's "Summary of the Eighth Meeting," 295-308; and "Policy on Outer Space, U.S." (NSC 5814), 345-59, 360; and "Policy on U.S. Scientific Satellite Program, Draft Statement of" (NSC 5520), 200, 308-13, 326, 328; and "Preliminary U.S. Policy on Outer Space" (NSC 5814/1), 360-61, 377
International Polar Year, 21
International Scientific Radio Union, 314; and National Committee's "Summary of the Eighth Meeting," 295-308
International Union of Geodesy and Geophysics (IUGG), 314; and National Committee's "Summary of the Eighth Meeting," 295-308
"Introduction to Outer Space," 332-34
Inyokern, California, 312
"Is There Life on Mars?," 176, 194-95
Italy, 486, 509
Itek, Inc., 217

J

Jackson, Henry, 631, 715
Jacobs, Kenneth H., 283
Japan, 580
Jastrow, Robert, 712
Jet-Assisted Take-Off (JATO), 8, 10-11
Jet Propulsion, 281-94
Jet Propulsion Laboratory (JPL), Pasadena, California, 11, 14-15, 611, 630, 636, 645-47, 657, 691-92, 735, 743; and affiliation with NASA, 614; and field center roles, 689-711; and "Memorandum on the Possibilities of Long-Range Rocket Projects," 153-55; and "A Review and Preliminary Analysis of Long-Range Rocket Projectiles," 155-76; and sounding rockets, 145-53
Johns Hopkins University, 206
Johnson, Kelly, 222
Johnson, Louis, 217
Johnson, Lyndon B., 434, 468, 473, 490, 495, 511; and Apollo Applications Program (AAP), 383-85; and Apollo decision, 379-81; and "Great Society," 622; and National Aeronautics and Space Act of 1958, 334-45, 377; and post-Apollo planning, 382-83; and review of Apollo, 381-82, 424, 427, 429, 433; and U.S.-Italian San Marcos project, 466
Johnson Space Center, Texas, 588, 626, 715; and field center roles, 689-711
Johnson, U. Alexis, 543
Johnston, S. Paul, 632-33
Joiner, Col. W.H., 154
Joint Chiefs of Staff, 359-60
Journal of Aeronautical Sciences, 145-53
Journal of the British Interplanetary Society, 9, 140-45
Joyce, J. Wallace, 295
Juno Launch Vehicle, 406
Jupiter, 507, 580
Jupiter Launch Vehicle, 14-15, 219, 221, 395, 402, 419, 478, 712

K

Kahn, Genghis, 5
Kaluga, Soviet Union, 59
Kaplan, Joseph, 177, 178; and National Committee for the IGY's "Summary of the Eighth Meeting," 295-308
Kapsitsa, Peter, 308
von Kármán, Theodore, 10-11; and "Memorandum on the Possibilities of Long-Range Rocket Projects," 153-55
Karpenko, Anatoly, 308
"Katusha," 11
Katz, Amron, 215
Kellogg, William, 215
Kelly, Albert J., 673
Kendrick, J.B., 280
Kennedy, John F., 229, 383, 388, 467-68, 473, 491, 597, 614, 621; and Apollo as crash program, 461; and Apollo decision, 379-81, 595, 600; and Gagarin flight, 379, 423, 612; and launch vehicles, 414-15; and prestige in space, 414; and review of Apollo, 381-82, 424, 427; and "Problem of Space Programs," 413-15; and Alan Shepard ceremony, 440; and space policy of, 379-82; and speech before Congress on sending Americans to the Moon, 453-54; and "Wiesner Report," 379, 416-23, 499
Kennedy, Robert, 492, 621
Kennedy Space Center, Florida, 588, 612, 625-27, 685, 733; and field center roles, 689-711
Kepler, Johann, 3
Kerr, Robert S., 380, 433-35
Key, Francis Scott, 5
Keyworth, George A., II, 392, 590
Khrushchev, Nikita, 223
Killian, James R., 637, 641, 645, 647-48; and "Introduction to Outer Space," 332-34; and "Killian Committee," 397, 416-23, 628-31; and *Meeting the Threat of Surprise Attack*, 219-20; and Presidential Science Advisor, 225, 611; and "Purcell Panel," 632; and Technological Capabilities Panel ("Surprise Attack Panel"), 218-19, 222, 225, 309
King, Martin Luther, 621
Kirkpatrick, Jeanne, 392
Kissinger, Henry, 559
Kistiakowsky, George, 228, 241, 408
Klaatu, 17, 20
Klep, Rolf, 177, 179
Knopow, Joseph J., 224
Knowland, William, 631
Kochanowski, Paul, 716
Korolev, Sergei, 715

L

Labor, Department of, 654
Laird, Melvin, 384, 519-20
Land, Edwin, 217, 219, 632
Landsat program, 565, 598
Lang, Fritz, 85
Langley Research Center, Virginia, 477, 587, 611, 647, 650, 717; and field center roles, 689-711
Langley, Samuel P., 650
Large Area Crop Inventory Experiment (LACIE), 711-13
Lasswell, Harold, 215
Latin America, 509
Lawrence Livermore National Laboratory, 632
Lay, James S., 312-13
Leghorn, Richard, 217
Lehrer, Tom, 12-13
LeMay, Curtis E., 213-14, 236
Lenher, Samuel, 511
Lewis, C.S., 5
Lewis Research Center, Ohio, 477, 587, 611, 647, 717; and field center roles, 689-711
Ley, Willy, 18-19, 176, 179
Life, 176

Lincoln Laboratory, 657
Lincoln, Project, 217
Lindbergh, Charles A., 7, 134, 135, 136, 140
Lipp, James E., 215; and "Project Feed Back Summary Report," 269-74; and "The Utility of a Satellite Vehicle for Reconnaissance," 245-61, 269
Liquid-Propellant Rocket Development, 8, 134-40
Little, Arthur D., Inc., 206
Locke, Richard Adams, 23
Lockheed Corporation, 550, 552
Lockheed Missile Systems Division, 222, 223, 224
LOKI Boosters, 275-79
London, United Kingdom, 331, 350
Long Playing Rocket Project, 295; and National Committee for the IGY's "Summary of the Eighth Meeting," 295-308
Long Range Plan, NASA (1959), 377-38, 403-407
Los Alamos, New Mexico, 241
Los Alamos National Laboratory, 241, 630
Los Angeles, California, 178, 269
Lovelace, Alan, 711-17
Low, George M., 386-88, 389, 548, 558-59, 579, 623-24, 685-88, 715
Lowell Observatory, Flagstaff, Arizona, 4, 55, 195
Lowell, Percival, 4, 55-58, 195
Luna 16 project, 598
Lunar Excursion Module (LEM), 481, 488

M

McClellan, John, 643, 714
McCloskey, Chester M., 283
McCoffus, Charlie, Ballad of, 650-51
McDaniel, Keith K., 283
McElroy, Neil, 225, 650
McKinsey & Company, 633
McMillan, Dr., 241
MacNair, Walter, 214
McNamara, Robert S., 426, 439-40, 614, 654; and Apollo decision, 379-81; and post-Apollo planning, 382-85; and review of Apollo, 381-82, 424
MacNeil, Nick, 559
Macy, John W., 614, 654
Malina, Frank J., 10-11, 12, 154; and "A Review and Preliminary Analysis of Long-Range Rocket Projectiles," 155-76; and "Flight Analysis of the Sounding Rocket," 145-53
Magnus, Albert, 5
Magnuson, Warren, 492, 714
Mahon, George, 714
"Man and the Moon," 19
"Man in the Moon: The Journey," 176, 189-94
"Man in Space," 19
"Man Without a Country, The," 23
Manned Spacecraft Center, Texas, 484, 513, 612
Manhattan Project, 178, 241, 266, 614
Manned Orbital Laboratory (MOL), and Department of Defense, 480, 496, 499, 515
Manning, Gordon, 18, 176-77
Marble, John P., 295
Mark, Hans, 389, 587-89; and decision to build the space station, 390-92
Mariner project, 479, 507, 528
Mars, 55-58, 460, 493, 522, 580, 597, 600, 742-43; and manned landing, 497, 504, 523, 530, 537, 540, 545; and Mariner probe, 479, 507, 528; and Viking project, 622
"Mars and Beyond," 19
Mars and Its Canals, 4, 55
Mars Excursion Module, 531, 540
Martha's Vineyard, Massachusetts, 649

Mason, Edward, 633
Marshall Space Flight Center, 429-30, 611, 618, 678, 680, 683-84, 715-16; and field center roles, 689-711; and possible closing of, 713-14
Martin, Glenn L, Co., 14
Martin Marietta Corp., 626, 656, 741
Martz, E.P., Jr., 280
Marvin, C.F., 136
Maryland, University of, 314
Massachusetts Institute of Technology (MIT), 10, 217, 230, 379, 416, 611, 614, 647, 657
Mathematica, Inc., 549-50
Mayo, Robert P., 385, 543-46
McCurdy, Richard, 548-49
McDonnell Douglas Corporation, 550, 552, 615, 674-77, 683
McGill University, 633
Medaris, John B., 221
Medicare program, 547
Meeting the Threat of Surprise Attack, 219-20
Megiste Syntaxis, 1
"Memorandum of Discussion at the 322d Meeting of the National Security Council," 324-28
Mercury, 507, 580
Mercury, Project, 378, 379, 404, 408-409, 411, 419, 422, 426, 429-30, 441, 450, 460, 463-64, 469, 480-81
Merriam, John C., 136, 140
Merritt Island Launch Facility, Florida, and Cape Kennedy, 484
Method of Reaching Extreme Altitudes, A, 7, 86-132
Mettler, Ruben F., 511
Michoud Assembly Facility, Louisiana, 469, 484
Miller, Stuart, 217
Millikin, Robert A., 10, 136
Mills, Wilbur, 714
Milwaukee Sentinel, 86
"Minimum Orbital Unmanned Satellite of the Earth (MOUSE), Studies of a," 314-24
"Minimum Satellite Vehicle: Based on Components Available from Missile Developments of the Army Ordnance Corps," 274-81
Minimum Satellite Vehicle Project, 274-81
Minnesota Mining and Manufacturing Co., 272
Minnich, L.A., 631
Minuteman Launch Vehicle, 219, 432, 616
Mirabilus Mundi, De (On the Wonders of the World), 5
Mission from Planet Earth program (MFPE), 742-43
Mission to Planet Earth program (MTPE), 742-43
Missile Detection and Alarm System (MIDAS), 227, 451
Mississippi Test Center, 469, 484
Mitre Corporation, 657
Mitterand, Francois, 596
Model Cities program, 547
Moffett Field, California, 647
Mondale, Walter, 568
Monitor project, 478
Moon Hoax, 23
Moore, Gary, 18
Morgan Spring Co., 100
Morgenstern, Oskar, 549
Mortensen, Jim, 717
Moss, Frank, 714
Mount Wilson Observatory, 87, 283
Mueller, George, 494, 615-16
Munger, William P., 283
Murray, Bruce, 716
Museum of Natural History, New York City, 176
Muskie, Edmund, 715
Myers, Dale D., 615

N

National Academy of Engineering, 589
National Academy of Public Administration (NAPA), 626-27, 730
National Academy of Sciences, 10, 153, 221, 308, 589, 634, 637-38, 712, 715; and "Memorandum of Discussion at the 322d Meeting of the National Security Council," 324-28; and National Committee for the IGY's "Summary of the Eighth Meeting," 295-308; and "Planetary Exploration; 1968-1975" report", 497
National Advisory Committee for Aeronautics (NACA), 14, 226, 334, 340, 345, 359, 394, 449, 463-64, 477, 580, 611-12, 614, 620-21, 629-42, 644-45, 647-49, 688, 715, 717; and Special Committee on Space Technology, 394-403
National Aeronautics and Space Act of 1958, 226, 334-45, 377, 397, 476, 483, 529, 563, 571, 580, 587, 611-12, 622, 685-86
National Aeronautics and Space Administration (NASA), 6, 8, 14, 345, 640; and "Ad Hoc Panel on Man-in-Space, Report of," 378, 408-12; and administrator of, 337-38; and Advisory Committee on the Future of the U.S. Space Program ("Augustine Commission"), 626-28, 741-743; and Apollo, 9; and Apollo Applications Program (AAP), 383-85, 501, 514, 521; and Apollo decision, 379-81; and Armed Service Procurement Regulations of 1947; and Carter space policy, 388-89; and "Civil and Further National Space Policy," 389; and creation, 611; and decision to build the space shuttle, 386-88; and decision to build the space station, 390-92; and economic stimulation, 566, 569; and Equal Employment Opportunity (EEO) record, 572; and established, 226, 334-45; and international cooperation, 339, 354-55; and launch vehicles, 414-15, 419-20; and *Leadership and America's Future in Space*, 392-93; and Long Range Plan of (1959), 377-38, 403-407; and "Management Study Group Recommendations," 730-35; and Mercury project, 378, 379, 404, 408-409, 411, 419, 422, 430, 464, 469, 480-81; and missions of, 227, 335-36, 392-93; and National Commission on Space, 392; and Nixon space policy, 383-86; and *Pioneering the Space Frontier*, 392; and plans of, 1959-1960, 377-79; and post-Apollo planning, 382-85; and prestige in space, 414; and "Problems and Opportunities in Manned Space Flight," 384; and "Problem of Space Programs," 413-15; and property rights to inventions, 342-44; and Reagan space policy, 389-90, 623; and recommendations from NACA, 394-403; and relations with Department of Defense, 11, 339; and reports of, 339-40; and review of Apollo, 381-82; and Space Task Group (1969), 383-85, 392, 513, 515, 519, 530-31, 540, 543-45, 553; and Strategic Planning Council, 733; and "Study of the Mission of NASA," 717-23; and "Wiesner Report," 379, 416-23, 499
National Aeronautics and Space Council, 336-37, 415, 418, 433-34, 467-68, 472, 502, 511, 525, 548, 622; and Apollo decision, 379-81; and review of Apollo, 381-82, 424, 427; and post-Apollo planning, 382-85; and "Problem of Space Programs," 413-15; and "U.S. Policy on Outer Space," 362-73; and Space Task Group (1969), 383-85, 513; and "Wiesner Report," 379, 416-23
National Broadcasting Corp. (NBC), 18
National Commission on Space, 392
National Committee for the IGY, U.S., 221; and "Summary of the Eighth Meeting," 295-308
National Defense Research Committee, 153, 154, 156
National Environmental Policy Act, 609
National Geographic Society, 195
National Indications Center, CIA, 220, 228
National Institutes of Health (NIH), 570-71
National Oceanic and Atmospheric Administration (NOAA), 577, 605, 712
National Research Council, 298, 630
National Science Foundation (NSF), 221, 223, 225, 226, 308, 449, 513, 560, 570, 615, 631, 634, 637-38, 643, 647, 654, 656-57, 659, 666, 710; and "Memorandum of Discussion at the 322d Meeting of the National Security Council," 324-28; and National Committee for the IGY's "Summary of the Eighth Meeting," 295-308; and "Utility of an Artificial Unmanned Earth Satellite: A Proposal to the National Science Foundation, Prepared by the ARS Space Flight Committee, November 24, 1954, On the," 281-94
National Space Council, 625-26, 741, 743
National Security Act of 1947, 214, 373
National Security Council (NSC), 200, 218, 222, 229, 307, 437, 544-45, 604; and Civil and Further National Space Policy" (NSC 42), 575-78, 590-93; and "Memorandum of Discussion at the 322d Meeting of the National Security Council," 324-28; and "National Space Policy" (NSC 37), 574-76, 610; and "Policy on Outer Space, U.S." (NSC 5814), 345-59, 360; and "Policy on U.S. Scientific Satellite Program, Draft Statement of" (NSC 5520), 200, 308-13, 326, 328; and Policy Review Committee (PRC), 575, 577; and "Preliminary U.S. Policy on Outer Space" (NSC 5814/1), 360-61, 377; and "U.S. Policy on Outer Space" (Space Council, 1960), 362-73
National Security Decision Directives, 590
National Telecommunications and Information Administration (NTIA), 577
National Watch Committee, 220

Nauchnoye Obozreniye (Science Review), 6
Naugle, John E., 688
Navaho Launch Vehicle, 15, 217
Naval Engineering Experiment Station, 12
Naval Ordnance Test Station, California, 312, 611
Naval Research Laboratory (NRL), 14, 221, 223, 286, 298, 301, 611-12, 643, 710, 715
Naval Research, Office of, 274, 275
Naval School of Aviation Medicine, 284
"Navigation of Space, The," 6
Navy Air Missile Test Center, California, 312
Navy, United States, 12, 15, 156, 157, 202, 214, 221-22, 223, 224, 225, 227, 274, 275, 279, 359, 380, 436, 497, 612, 615, 639, 644, 673, 715
Neasham, R.S., 202
Nekrassoff, V.A., 329
Nelson, Bill, 726
Neptune, 507
Neustadt, Richard E., 413-415
New Deal, 614
Newell, Homer E., Jr., and National Committee for the IGY's "Summary of the Eighth Meeting," 295-308; and "Utility of an Artificial Unmanned Earth Satellite: A Proposal to the National Science Foundation, Prepared by the ARS Space Flight Committee, November 24, 1954, On the," 281-94
Newman, Bob, 717
Newspaper Publishers Association, 438
Newsweek, 546
Newton, Isaac, 2-3, 5, 180, 217, 395
New York City, 716
New York City, New York, 18, 86, 176, 178, 269
New York Times, 7, 86, 133
New York University, 657
Nichols, Kenneth D., 266
Nike Project, 683; and Ajax Booster, 678; and Hercules Booster, 678; and Zeus Booster, 432, 678
Nimbus communications satellite, 439, 449, 477
Nixon, Richard M., 345, 379, 388, 512-13, 519, 522, 597, 622; and decision to build the space shuttle, 86-88, 558, 624; and post-Apollo planning, 383-85; and space policy of, 385-86; and Space Task Group (1969), 383-85; and Task Force on Space, 499
Nobel Prize, 10
North Atlantic Treaty Organization (NATO), 228, 349
North American Aviation, Inc., 218; and Rocketdyne Division, 617-19
North American Rockwell Corp., 674
North American Weather Consultants, 289
Nova Launch Vehicle, 406, 407, 409-12, 441, 450
Nuclear Engine for Rocket Vehicle Application (NERVA), 547
Nyack, J. Allen, 202

O

Oberth, Hermann, 6, 9, 11, 13, 16, 23, 59, 84-86
O'Dell, Charles, R., 511
Oder, Frederick C.E., 202, 203
Odishaw, Hugh, and National Committee for the IGY's "Summary of the Eighth Meeting," 295-308, 633
Office of Defense Mobilization, 218
Office of Management and Budget (OMB), 386, 546-49, 563, 572, 577, 593, 604, 607, 620, 712, 714; and decision to develop the space shuttle, 386-88, 571 (see also Bureau of the Budget)
Office of Personnel Management (OPM), United States, 620, 628
Ohio State University, 206, 207
O'Keefe, John, and "Utility of an Artificial Unmanned Earth Satellite: A Proposal to the National Science Foundation, Prepared by the ARS Space Flight Committee, November 24, 1954, On the," 281-94
O'Neill, Tip, 714
"Open Skies" Doctrine, 213-29, 230
Orion, Project, 420
Orbiter, Project, 221-22, 308, 310
Orbiting Astronomical Observatory (OAO), 463, 479, 507

Orbiting Geophysical Observatory (OGO), 479
Orbiting Solar Observatory (OSO), 479, 507
Ostrander, Don R., 673
Out of the Silent Planet, 5
Overhage, Carl, 217

P

Pacific Missile Range, 399
Pacific Northwest Regional Commission, 693
Pacific Palisades, California, 278, 280
Packard, Jack, 648
Page, Thornton, 202, 206
Paine, Thomas O., 522, 622; and chairs National Commission on Space, 392; and named NASA administrator, 384; and *Pioneering the Space Frontier*, 392; and "Problems and Opportunities in Manned Space Flight," 384; and resigns as NASA administrator, 386; and Space Task Group (1969), 383-85, 513-14, 519, 543-44
Pal, George, 17
Palomar Observatory, 283
Paperclip, Project, 13
Pasadena, California, 10
Patent and Trademark Office, United States, 643
Pathfinder program, 605
Patrick Air Force Base, Florida, 312
Patterson, Robert, 213, 215
Patton, James R., Jr.r, 283
Pauli, Fritz K., 281
PBM3C Flying Boat, 12
Pearl Harbor, Hawaii, 215, 216, 225
Peavey, R.C., and National Committee for the IGY's "Summary of the Eighth Meeting," 295-308
Peenemünde Rocket Development Center, 13, 267, 331
People's Republic of China, 10
Perelandra, 5
Perkin, Richard, 217
Perkin-Elmer, Inc. (also see Hughes Danbury Optical Systems, Inc.), 217; and Hubble Space Telescope, 736
Perkins, James A., 633
Petrone, Rocco, 673-74, 715
Phillips, Samuel C., 615-16, 625; and "NASA Management Study Group Recommendations," 730-35
Photo Interpretation Laboratory, Anacostia, Maryland, 202
Photo Records and Services Division, U.S. Air Force, 218
Physics and Medicine of the Upper Atmosphere: A Study of the Aeropause, 262-66
Pickering, William H., 645-47, and National Committee for the IGY's "Summary of the Eighth Meeting," 295-308
Pierce, John R., and "Utility of an Artificial Unmanned Earth Satellite: A Proposal to the National Science Foundation, Prepared by the ARS Space Flight Committee, November 24, 1954, On the," 281-94
"Pilot Lights of the Apocalypse: A Playlet in One Act," 230-35
Pioneer project, 479
Pioneering the Space Frontier, 392
Planning Research Corporation, 657
Pluto, 507
Point Mugu, California, 312
Polaero, Hans R., 281
Polaris Missile, 15, 219
Polaroid, Inc., 217, 632
"Policy on U.S. Scientific Satellite Program, Draft Statement of" (NSC 5520), 200, 308-13, 326, 328
"Policy on Outer Space, U.S." (NSC 5814), 345-59, 360
Policy on Outer Space, U.S." (Space Council, 1960), 362-73
Pompton Plains, New Jersey, 12
Popular Science News, 6
Port Huron, Michigan, 204
Porter, Richard B., 283
Possony, Stephen, 202
Pravda, 329-30
"Preliminary Design of an Experimental World-Circling Spaceship," 236-45

"Preliminary U.S. Policy on Outer Space" (NSC 5814/1), 360-61, 377
Presidential Science Advisor, 225, 392; and "Introduction to Outer Space," 332-34; and "Killian Committee," 397, 416-23, 629
Presidential Science Advisory Committee (PSAC), 309, 548, 666; and Ad Hoc Panel on Man-in-Space, 378, 408-12; and Apollo decision, 379-81; and "Introduction to Outer Space," 332-34; and "Killian Committee," 397, 416-23, 629; and post-Apollo planning, 382-85; and review of Apollo, 281-82; and Space Task Group (1969), 383-85
President's Advisory Committee on Government Organization, 611, 637
Presque Isle, Maine, 204
Price, Don K., 633
Princeton University, Princeton, New Jersey, 86, 215, 657
"Problems and Opportunities in Manned Space Flight," 384
Proxmire, William, 383, 490, 492-94, 714-15
Ptolemy, 1-2
Puckett, Allen E., 511
Purcell, Edward, 217; and "Purcell Panel," 632, 634
Putt, Donald, 224

Q

Quarles, Donald A., 221-22, 223, 224, 225, 228, 329, 648; and "Report on the Present Status of the Satellite Problem," 267-69

R

Radcliff, J.D., 199
Radiation Laboratory, MIT, 230
Radio Corporation of America (RCA), 218, 227, 271-74
Die Rakete zu den Planetenraumen (*Rockets in Planetary Space*), 6, 84-86
Rand Corp., 213-15, 216, 217, 221, 224, 225, 230, 245, 614, 657, 664; and "Project Feed Back Summary Report," 269-74; and "The Utility of a Satellite Vehicle for Reconnaissance," 245-61, 269
Ranger project, 430, 442, 446, 480
Rayburn, Sam, 345
Reaction Motors, Inc., 12
Reagan, Ronald, 389-92, 579, 593-94, 625, 730; and decision to build the space station, 390-92, 623; and *Leadership and America's Future in Space*, 392-93; and National Commission on Space, 392; and *Pioneering the Space Frontier*, 392
Redstone Arsenal, Alabama, 14, 176, 189, 195, 274, 276, 280-81
Redstone Launch Vehicle, 15, 406, 632; and "Minimum Satellite Vehicle: Based on Components Available from Missile Developments of the Army Ordnance Corps," 274-81
Renssellaer Polytechnic Institute, 389, 579
Research and Development Committee/Board, War Department/DOD, 214, 215, 217
"Review and Preliminary Analysis of Long-Range Rocket Projectiles, A," 155-76
Ride, Sally K., 392-93; and *Leadership and America's Future in Space*, 392-93
Ridenour, Louis N., 215, 217, 241; and "Military Security and the Atomic Bomb," 230; and "Pilot Lights of the Apocalypse: A Playlet in One Act," 230-35
Roadman, Charles, 673
Roberts, E.B., 295
Roberts, Walter O., 511
Robertson, H.P., 202, 205, 206
Rockefeller, Nelson A., 312-13, 611, 637, 641
Rocketdyne Division (also see North American Aviation Co.), 617-19
Rocketry; *Collier's* and, 17-19; GALCIT and, 10-11; Robert H. Goddard and, 1, 6-7, 10, 11, 12, 13, 16, 59, 86-132, 133, 134-40; and ICBMs, 10, 15-16, 432; and modern war, 11-13; and post-war development, 13-15; Project Bumper and, 14-15; Reaction Motors, Inc., and, 12; rocket societies and, 8-9; science fiction and, 16-17; and technology of, 5, 10-15;
Rogers, William P., and Presidential Commission on Challenger Accident, 624, 723-729
Rome, Italy, 301, 305
Romick, Darrell C., 283
Root, Eugene, 215
Rose Bowl, Pasadena, California, 0
Rosen, Milton W., 295, 298; and National Committee for the IGY's "Summary of the Eighth Meeting," 295-308; and "Utility of an Artificial Unmanned Earth Satellite: A Proposal to the National Science Foundation, Prepared by the ARS Space Flight Committee, November 24, 1954, On the," 281-94

Ross, H.E., and "B.I.S. Space Ship, The," 140-45
Roswell, New Mexico, 7-8, 134
Rover, Project, 420, 443, 449-50, 454, 461
Rowe, Herb, 714, 717
Rubel, John H., 435-36
Rubenstein, David, 559
Rudolph, Walter N., 295
Rupelt, E.J., 202
Russell, Richard, 383, 492
Ryan, Cornelius, 18, 176-77

S

St. Elmo's Fire, 204-205
SAINT program, 451
Salter, Robert M., Jr., 215; and "Engineering Techniques in Relation to Human Travel at Upper Altitudes," 262-66; and "Project Feed Back Summary Report," 269-74; and "Utility of a Satellite Vehicle for Reconnaissance, The," 245-61, 269
Saltonstall, Leverett, 631
Samek, Michael J., 283
SAMOS program, 451
San Antonio, Texas, 177
San Clemente, California, 388
San Francisco Examiner, 86
San Marcos project, 466
Sänger, Eugen, 10, 146
Santa Monica, California, 215, 236, 269
Sasser, Jim, 714
Saturday Evening Post, 176
Saturn, 507, 580, 600
Saturn Launch Vehicle, 6, 383, 386, 404, 406, 407, 409-12, 413, 419-20, 425, 426, 429-30, 433, 441, 450, 457, 460, 481-84, 487, 489, 493-96, 503, 506, 514-18, 617-18, 622, 675-78, 680-84, 715
Schacter, Oscar, 179
Schade, Otto, 272
Schaefer, Hermann J., and "Utility of an Artificial Unmanned Earth Satellite: A Proposal to the National Science Foundation, Prepared by the ARS Space Flight Committee, November 24, 1954, On the," 281-94
Schiaparelli, Giovanni, 4, 195
Scientific Research and Development, Office of, 657
Schilling, G.F., and National Committee for the IGY's "Summary of the Eighth Meeting," 295-308
Schriever, Bernard A., 15-16, 427
Schultz, George, 387
Science Advisory Committee, Office of Defense Mobilization, 218, 219
Scientific Advisory Panel, CIA, 19-20
Science and Technology, proposed Department of, 611, 643
Science and Technology Policy, Office of (OSTP), 390, 473, 572, 575, 578, 593, 604, 610, 665, 722
"Scientific Satellite Program," 221
Scout Launch Vehicle, 14-15, 404, 406, 478, 482
Seaborg, Glenn T., 543, 614, 673
Seamans, Robert C., Jr., 384-85, 461, 464, 492, 511, 519, 522, 543
Seasat project, 598, 698
Seifert, Howard S., 283
Senior Interagency Group (SIG) for Space, 390, 604, 606, 610; and decision to build the space station, 390-92, 590-95, 599, 600
Sergeant Launch Vehicle, 14
Shapley, A. Harlow, 243; and National Committee for the IGY's "Summary of the Eighth Meeting," 295-308
Shepard, Alan B., Jr., 15, 380, 440, 453
Sherman Anti-Trust Act of 1890, 616
Shultz, George P., 546-47, 558
Shuttle, Space (also see Space Transportation System), 547, 549, 551-59, 564-68, 571-73, 576, 578, 580-99, 600-01, 608, 624-28, 702-03, 708-09, 715-16, 721-23, 742-43; and Carter space policy, 388-89; and decision to develop, 386-88; and "Rogers Commission Report," 723-30
Sidereus Nuncius, 2

Sidereal Messenger, 2
Siepert, Albert F., 672
Silveira, Milton, 389, 587-89
Silverstein, Abe, 648
Singer, S. Fred, "Minimum Orbital Unmanned Satellite of the Earth (MOUSE), Studies of a," 314-24; and National Committee for the IGY's "Summary of the Eighth Meeting," 295-308
Siple, Paul A., 295
Skyhook Project, 216
Skylab Project, 547, 556, 558, 622, 714-15
Skyvan, Shorts, 588
Slidell, Louisiana, 700
Smith, A.M.O., and "Flight Analysis of the Sounding Rocket," 145-53
Smith, Francis B. (Frank), and Future Programs Task Group 473
Smith, Harry B., 202
Smith, Kent, 648
Smith, Margaret Chase, 493
Smithsonian Institution, 7, 86, 134, 135, 137, 140, 643, 650
Smyth Report, 178
Snark Missile, 217
Society for Spaceship Travel, 8-9
Solar Max project, 598
Somnium (Dream), 3
Soviet Union, 6, 11, 12, 18, 177, 424, 427-430, 432, 434, 436, 438, 440, 444-49, 453-57, 462, 474, 483, 490, 492-97, 501-03, 506-07, 509, 515, 529, 536, 555-59, 567, 580-81, 598, 631, 647; and Ad Hoc Panel on Man-in-Space report, 378, 408-12; and Apollo decision, 379-81; and capabilities in space, 345-59, 362-73; and Cold War, 613; and decision to build the space shuttle, 386-88; and freedom of space, 213-29, 230, 309; and Geneva Summit, 222-23, 227-28; and International Geophysical Year (IGY), 19, 20, 200-201, 220-23, 224-25, 227, 228; and *Meeting the Threat of Surprise Attack,* 219-20; and Nixon space policy, 383-86; and Nuclear and Space Talks (NST), 610; and "Open Skies" doctrine, 213-29; "Policy on Outer Space, U.S." (NSC 5814), 345-59, 360; and "Policy on U.S. Scientific Satellite Program, Draft Statement of" (NSC 5520), 200, 308-13, 326, 328; and post-Apollo planning, 382-85; and "Preliminary U.S. Policy on Outer Space" (NSC 5814/1), 360-61, 377; and reconnaissance programs, 219-20, 222-23, 227-28, 348, 354, 373-75; and "Report on the Present Status of the Satellite Problem," 267-69; and review of Apollo, 381-82; and satellite reconnaissance, 216-17, 221-24, 227-28, 269-74, 348, 354, 373-75; and Soyuz, 515, 582; and "Surprise Attack Conference," 227-28; and Technological Capabilities Panel ("Surprise Attack Panel"), 218-19, 221, 225, 309; and U-2 program, 219-20, 222-23, 227; and "U.S. Policy on Outer Space" (Space Council, 1960), 362-73; and Yuri Gagarin flight, 379, 423, 474, 492, 494, 612
Soyuz project, 515, 582
Space Debris Policy (also see National Security Council, and National Space Policy), 610
Space Exploration Control Group, 636-37
Space Exploration Initiative (SEI), 741
Spaceflight; *Collier's* and, 17-19; and dreams of, 1, 3-5; and enthusiasm for, 1, 3-5; and escape velocity for, 7; GALCIT and, 10-11; and imagination for, 16-20; and progenitors of, 6-8; and rocket societies, 8-9
Spacelab, 596, 692, 701-02, 709
Space Task Group (1969), 383-85, 392, 515, 522-26, 530-31, 536-37, 540, 543-45, 553, 622
Space Telescope project, 565, 567, 578, 714
SPACETRACK program, 451
Space Transportation System (also see Space Shuttle), 521, 524, 534, 538-39, 545, 549-51, 557-59, 590-93, 600-01, 605-09, 622, 624-28, 702-03, 708-09, 721-23, 742-43; and "Rogers Commission Report," 723-30
Space Travel Symposium, 176, 178, 314
SPASUR program, 451
Special Committee for the International Geophysical Year (CSAGI), and National Committee for the IGY's "Summary of the Eighth Meeting," 295-308
Special Committee on Space Technology (NACA), 394-403
Spilhaus, A.F., and National Committee for the IGY's "Summary of the Eighth Meeting," 295-308
Sprattling, Willis, Jr., 283
Sputnik I, 16, 17, 19, 21, 220, 225, 228, 329-30, 331, 332, 394, 448, 474, 494, 580, 600, 612, 628-29, 632, 647
Sputnik II, 225, 228, 628-29
Stanford Research Institute, 657
Stanford University, 614, 657
Stans, Maurice, 643-45
Stanton, Frank, 427, 437-38

"Star Spangled Banner, The," 5
State, Department of, 220, 223, 268, 307, 326, 331, 438, 577, 594, 604
Station, Space, 390-92, 538-39, 545, 585, 588-89, 593-98, 605, 620, 623, 626, 724, 730-33, 742
Steadman, Dick, 559
Stearns, Edwards, 215
Stehling, Kurt R., 283
Stennis, John, 714-15
Stephenson, H.K., 295
Stern, Al, 559
Stevenson, Adlai, 714
Stevenson, David B., 202
STOCK, Project, 203
Strategic Defense Initiative, 607
Strauss, Louis L., 268, 327
Strong, Philip G., 202
Stroud, W., 299
Suez Crisis, 220
"Summary of the Eighth Meeting," National Committee for the IGY, 295-308
Sunnyvale, California, 223
"Surprise Attack Conference," 227-28
Surprise Attack Panel ("Technological Capabilities Panel"), 218-19, 222, 225, 309; and *Meeting the Threat of Surprise Attack*, 219-20
Surveyor program, 442, 446, 492
Symington, Stuart, 631
Syncom satellites project, 464, 478
System for Nuclear Auxiliary Power (SNAP), 449
Systems Development Corporation, 657
Swayze, John Cameron, 18

T

Task Force on Space, 499-500
Tass, 329-30
Teague, Olin E. (Tiger), 674, 714
Technological Capabilities Panel ("Surprise Attack Panel"), 218-19, 222, 225, 309; and *Meeting the Threat of Surprise Attack*, 219-20
Telstar project, 464, 478
Temple University, 266
Tennessee Valley Authority, 672
Terre à la Lune, De la (*From the Earth to the Moon*), 4
Texas, rural, story of electricity for, 439
That Hideous Strength, 5
Third Law of Motion, Newton's, 5, 180
Thor Launch Vehicle, 16, 219, 225, 359, 373, 404, 406, 419, 449, 451, 482, 683
Thrust Assisted Orbiter Shuttle design (TAOS), 549-53
Tikhonravov, M.K., 177
Tiros Communication Satellite, 227, 439, 442, 447, 464, 477, 486
Titan Launch Vehicle, 15, 16, 219, 388, 419, 425, 430, 441-42, 459-60, 480-82, 494, 589, 645
"Today" Show, 18
Toftoy, H.N., 275
Townes, Charles, 383; and Task Force on Space, 499-500, 511-12
Townsend, John W., Jr., and National Committee for the IGY's "Summary of the Eighth Meeting," 295-308
Tracking and Data Relay Satellite, 726
Trade Representative, United States, 607
TRANSIT program, 451
Transportation, Department of, 497, 499, 516, 570-71, 604, 606, 609
Treasury, Department of, 571
Tremonton, Utah, 202, 204
Truly, Richard, 741
Truman, Harry S., 214; and "Report on the Present Status of the Satellite Problem," 267-69

Tsien, H.S., 10, 154; and "A Review and Preliminary Analysis of Long-Range Rocket Projectiles," 155-76
Tsiolkovskiy, Konstantin Edwardovich, 6, 23, 309, 329, 330, 356; and "Exploration of the Universe with Reaction Machines," 59-84
Tuhy, Ivan E., 283
Tuve, Merle A., 295
Twining, Nathan F., 359-60

U

U-2 Aircraft, 219-20, 222-23, 227
Unidentified Flying Objects (UFO), 19-20, 201-11
United Kingdom, 486; and Ariel project, 479
United Nations, 10, 179, 227-28, 229, 349, 509, 554
United States Army, see "Army, United States"
United States Army Air Corps, see "Army Air Corps, United States"
United States Army Air Forces, see "Army Air Forces, United States"
United States Air Force, see "Air Force, United States"
United States Congress, see "Congress, United States"
United States Navy, see "Navy, United States"
Upper Atmosphere Research Panel, 277, 314
Uranus, 507
Utah, University of, 386
"Utility of an Artificial Unmanned Earth Satellite: A Proposal to the National Science Foundation, Prepared by the ARS Space Flight Committee, November 24, 1954, On the," 281-94
"Utility of a Satellite Vehicle for Reconnaissance, The," 245-61, 269

V

V-1 Rocket, 13
V-2 (A-4) Launch Vehicle, 13-15, 17, 180-81, 237, 239, 267, 303, 314, 429
Valier, Max, 9
Van Allen, James A., 177, 285, 511; and "Minimum Orbital Unmanned Satellite of the Earth (MOUSE), Studies of a," 314-24; and National Committee for the IGY's "Summary of the Eighth Meeting," 295-308, 633
Van Allen Belts, 478, 644
Vance, Cyrus, 373-75
Vandenberg Air Force Base, California, 588
Vanguard, Project, 14, 221-23, 224-25, 228, 267, 274, 329, 395, 402, 449, 629, 644, 647; and "Memorandum of Discussion at the 322d Meeting of the National Security Council," 324-28; and "Policy on Outer Space, U.S." (NSC 5814), 345-59, 360; and "Policy on U.S. Scientific Satellite Program, Draft Statement of" (NSC 5520), 200, 308-13, 326, 328; and "Preliminary U.S. Policy on Outer Space" (NSC 5814/1), 360-61, 377
Vega project, 645-47
Venus, 460, 493, 522; and Soviets' shot at, 430; and Mariner probe, 479, 507, 580, 598
Verein fur Raumschiffahrt (VfR or Society for Spaceship Travel), 8-9, 12-13, 18, 179
Verne, Jules, 4, 7, 16, 133
Vertical/Short Take off and Landing (V/STOL) project, 476, 487, 489, 587, 703
Vestine, E.H., 295
Veterans Administration, United States, 571
Vietnam, 491, 495-96, 622
Viking Launch Vehicle, 14, 221, 223, 303, 310
Viking Project (probe to Mars), 622, 714
Vitro, Inc., 218
Voskhod system, 494
Vostok system, 494
Voyage dans la Lune (The Voyage to the Moon), 3
Voyager project, 581

W

WAC Corporal Rocket, 12, 14-15, 180-82, 267
Walker, T.B., 299
Wallops Island Facility, Virginia, 399, 436, 636, 647; and field center roles, 689-711
War Department, 135, 213, 214
War of the Worlds, 4-5, 20

Washington, D.C., 11, 18, 203, 204, 209, 269, 275, 295, 297, 298, 301, 312
Washington Post, 308
Waterman, Alan T., 221, 223, 615, 654; and "Memorandum of Discussion at the 322d Meeting of the National Security Council," 324-28; and National Committee for the IGY's "Summary of the Eighth Meeting," 295-308
Weather Bureau, United States, 227, 449, 452, 455, 459, 461, 464, 477
Webb, James E., 379, 384, 389, 434, 457, 461, 467-68, 490, 492, 494-95, 512, 613-14, 622, 654, 672, 714; and Apollo decision, 379-81; and post-Apollo planning, 382-83; and *Challenger* accident investigation, 723; and Future Programs Task Group, 723-24; and review of Apollo, 381-82, 423, 433-440
Wege zur Raumschiffahrt (Ways to Spaceflight), 85
Wehner, R.S., and "The Utility of a Satellite Vehicle for Reconnaissance," 245-61, 269
Weinberger, Caspar W., 546-47; and decision to build the space shuttle, 386-87; and decision to build the space station, 390-92, 600-01
Wells, H.G., 4, 16, 20
Welsh, Edward C., 427, 437
West Germany, 509, 596
Western Test Range, 482
Westford program, 451
Westinghouse Electric Corp., 218
"What Are We Waiting For?," 18, 176, 177-79
Whipple, Fred L., 176-77, 178, 189; and "Is There Life on Mars?," 176, 194-95; and National Committee for the IGY's "Summary of the Eighth Meeting," 295-308
White Sands Proving Ground, New Mexico, 13-14, 180, 267, 278, 280, 312, 314
Wiesner, Jerome B., 615, 654; and lunar orbital rendezvous approach, 456; and named presidential science advisor, 379; and "Wiesner Report," 379, 416-23, 499; and review of Apollo program, 424
Williams, Walt, 715
Wilson, Charles E., 215, 225; and "Memorandum of Discussion at the 322d Meeting of the National Security Council," 324-28
Wise, Robert, 17
Woo, Harry, 202
Woods Hole, Massachusetts, 649
Woomera, Australia, 312
Worcester, Massachusetts, 7, 93, 100
World Meteorological Organization (WMO), 509
World War I, 8, 567
World War II, 8, 16, 230, 613-14, 617, 636, 647-48, 655-57; and GALCIT, 10-11; and Project Paperclip, 13; and Reaction Motors, Inc., 12; and rocketry development, 11-13; and scientific research, 213; and Wernher von Braun, 12-13
Wright Field, Ohio, 280
Wright, Jim, 714
Wright-Patterson Air Force Base, Ohio, 616
Wright, Cdr., 275
WS 117L Program, 221, 222, 223, 224, 308, 358-59, 373
WS 119L Program, 216-17
Wyckoff, P.H., 299

X

X-1 Aircraft, 12
X-2 Aircraft, 8
X-15 Aircraft, 359, 451, 463, 639

Y

Yaak, Montana, 204
Yale University, 215
Yarborough, Ralph W., 643
Yardley, John F., 615, 715
Yeager, Chuck, 392
Young and Rubicam, 717
York, Herbert, 632, 634

Z

Zurich, Switzerland, 314

The NASA History Series

Reference Works, NASA SP-4000:

Grimwood, James M. *Project Mercury: A Chronology.* (NASA SP-4001, 1963).

Grimwood, James M., and Hacker, Barton C., with Vorzimmer, Peter J. *Project Gemini Technology and Operations: A Chronology.* (NASA SP-4002, 1969).

Link, Mae Mills. *Space Medicine in Project Mercury.* (NASA SP-4003, 1965).

Astronautics and Aeronautics, 1963: Chronology of Science, Technology, and Policy. (NASA SP-4004, 1964).

Astronautics and Aeronautics, 1964: Chronology of Science, Technology, and Policy. (NASA SP-4005, 1965).

Astronautics and Aeronautics, 1965: Chronology of Science, Technology, and Policy. (NASA SP-4006, 1966).

Astronautics and Aeronautics, 1966: Chronology of Science, Technology, and Policy. (NASA SP-4007, 1967).

Astronautics and Aeronautics, 1967: Chronology of Science, Technology, and Policy. (NASA SP-4008, 1968).

Ertel, Ivan D., and Morse, Mary Louise. *The Apollo Spacecraft: A Chronology, Volume I, Through November 7, 1962.* (NASA SP-4009, 1969).

Morse, Mary Louise, and Bays, Jean Kernahan. *The Apollo Spacecraft: A Chronology, Volume II, November 8, 1962-September 30, 1964.* (NASA SP-4009, 1973).

Brooks, Courtney G., and Ertel, Ivan D. *The Apollo Spacecraft: A Chronology, Volume III, October 1, 1964-January 20, 1966.* (NASA SP-4009, 1973).

Ertel, Ivan D., and Newkirk, Roland W., with Brooks, Courtney G. *The Apollo Spacecraft: A Chronology, Volume IV, January 21, 1966-July 13, 1974.* (NASA SP-4009, 1978).

Astronautics and Aeronautics, 1968: Chronology of Science, Technology, and Policy. (NASA SP-4010, 1969).

Newkirk, Roland W., and Ertel, Ivan D., with Brooks, Courtney G. *Skylab: A Chronology.* (NASA SP-4011, 1977).

Van Nimmen, Jane, and Bruno, Leonard C., with Rosholt, Robert L. *NASA Historical Data Book, Vol. I: NASA Resources, 1958-1968.* (NASA SP-4012, 1976, rep. ed. 1988).

Ezell, Linda Neuman. *NASA Historical Data Book, Vol II: Programs and Projects, 1958-1968.* (NASA SP-4012, 1988).

Ezell, Linda Neuman. *NASA Historical Data Book, Vol. III: Programs and Projects, 1969-1978.* (NASA SP-4012, 1988).

Astronautics and Aeronautics, 1969: Chronology of Science, Technology, and Policy. (NASA SP-4014, 1970).

Astronautics and Aeronautics, 1970: Chronology of Science, Technology, and Policy. (NASA SP-4015, 1972).

Astronautics and Aeronautics, 1971: Chronology of Science, Technology, and Policy. (NASA SP-4016, 1972).

Astronautics and Aeronautics, 1972: Chronology of Science, Technology, and Policy. (NASA SP-4017, 1974).

Astronautics and Aeronautics, 1973: Chronology of Science, Technology, and Policy. (NASA SP-4018, 1975).

Astronautics and Aeronautics, 1974: Chronology of Science, Technology, and Policy. (NASA SP-4019, 1977).

Astronautics and Aeronautics, 1975: Chronology of Science, Technology, and Policy. (NASA SP-4020, 1979).

Astronautics and Aeronautics, 1976: Chronology of Science, Technology, and Policy. (NASA SP-4021, 1984).

Astronautics and Aeronautics, 1977: Chronology of Science, Technology, and Policy. (NASA SP-4022, 1986).

Astronautics and Aeronautics, 1978: Chronology of Science, Technology, and Policy. (NASA SP-4023, 1986).

Astronautics and Aeronautics, 1979-1984: Chronology of Science, Technology, and Policy. (NASA SP-4024, 1988).

Astronautics and Aeronautics, 1985: Chronology of Science, Technology, and Policy. (NASA SP-4025, 1990).

Gawdiak, Ihor Y. Compiler. *NASA Historical Data Book, Vol. IV: NASA Resources, 1969-1978.* (NASA SP-4012, 1994).

Noordung, Hermann. *The Problem of Space Travel: The Rocket Motor.* Ernst Stuhlinger, and J.D. Hunley, with Jennifer Garland. Editors. (NASA SP-4026, 1995).

Management Histories, NASA SP-4100:

Rosholt, Robert L. *An Administrative History of NASA, 1958-1963.* (NASA SP-4101, 1966).

Levine, Arnold S. *Managing NASA in the Apollo Era.* (NASA SP-4102, 1982).

Roland, Alex. *Model Research: The National Advisory Committee for Aeronautics, 1915-1958.* (NASA SP-4103, 1985).

Fries, Sylvia D. *NASA Engineers and the Age of Apollo.* (NASA SP-4104, 1992).

Glennan, T. Keith. *The Birth of NASA: The Diary of T. Keith Glennan,* edited by J.D. Hunley. (NASA SP-4105, 1993).

Project Histories, NASA SP-4200:

Swenson, Loyd S., Jr., Grimwood, James M., and Alexander, Charles C. *This New Ocean: A History of Project Mercury.* (NASA SP-4201, 1966).

Green, Constance McL., and Lomask, Milton. *Vanguard: A History.* (NASA SP-4202, 1970; rep. ed. Smithsonian Institution Press, 1971).

Hacker, Barton C., and Grimwood, James M. *On Shoulders of Titans: A History of Project Gemini.* (NASA SP-4203, 1977).

Benson, Charles D. and Faherty, William Barnaby. *Moonport: A History of Apollo Launch Facilities and Operations.* (NASA SP-4204, 1978).

Brooks, Courtney G., Grimwood, James M., and Swenson, Loyd S., Jr. *Chariots for Apollo: A History of Manned Lunar Spacecraft.* (NASA SP-4205, 1979).

Bilstein, Roger E. *Stages to Saturn: A Technological History of the Apollo/Saturn Launch Vehicles.* (NASA SP-4206, 1980).

Compton, W. David, and Benson, Charles D. *Living and Working in Space: A History of Skylab.* (NASA SP-4208, 1983).

Ezell, Edward Clinton, and Ezell, Linda Neuman. *The Partnership: A History of the Apollo-Soyuz Test Project.* (NASA SP-4209, 1978).

Hall, R. Cargill. *Lunar Impact: A History of Project Ranger.* (NASA SP-4210, 1977).

Newell, Homer E. *Beyond the Atmosphere: Early Years of Space Science.* (NASA SP-4211, 1980).

Ezell, Edward Clinton, and Ezell, Linda Neuman. *On Mars: Exploration of the Red Planet, 1958-1978.* (NASA SP-4212, 1984).

Pitts, John A. *The Human Factor: Biomedicine in the Manned Space Program to 1980.* (NASA SP-4213, 1985).

Compton, W. David. *Where No Man Has Gone Before: A History of Apollo Lunar Exploration Missions.* (NASA SP-4214, 1989).

Naugle, John E. *First Among Equals: The Selection of NASA Space Science Experiments.* (NASA SP-4215, 1991).

Wallace, Lane E. *Airborne Trailblazer: Two Decades with NASA Langley's Boeing 737 Flying Laboratory.* (NASA SP-4216, 1994).

Center Histories, NASA SP-4300:

Rosenthal, Alfred. *Venture into Space: Early Years of Goddard Space Flight Center.* (NASA SP-4301, 1985).

Hartman, Edwin, P. *Adventures in Research: A History of Ames Research Center, 1940-1965.* (NASA SP-4302, 1970).

Hallion, Richard P. *On the Frontier: Flight Research at Dryden, 1946-1981.* (NASA SP-4303, 1984).

Muenger, Elizabeth A. *Searching the Horizon: A History of Ames Research Center, 1940-1976.* (NASA SP-4304, 1985).

Hansen, James R. *Engineer in Charge: A History of the Langley Aeronautical Laboratory, 1917-1958.* (NASA SP-4305, 1987).

Dawson, Virginia P. *Engines and Innovation: Lewis Laboratory and American Propulsion Technology.* (NASA SP-4306, 1991).

Dethloff, Henry C. *"Suddenly Tomorrow Came . . .": A History of the Johnson Space Center, 1957-1990.* (NASA SP-4307, 1993).

Hansen, James R. *Spaceflight Revolution: NASA Langley Research Center from Sputnik to Apollo.* (NASA SP-4308, 1995).

General Histories, NASA SP-4400:

Corliss, William R. *NASA Sounding Rockets, 1958-1968: A Historical Summary.* (NASA SP-4401, 1971).

Wells, Helen T., Whiteley, Susan H., and Karegeannes, Carrie. *Origins of NASA Names.* (NASA SP-4402, 1976).

Anderson, Frank W., Jr., *Orders of Magnitude: A History of NACA and NASA, 1915-1980.* (NASA SP-4403, 1981).

Sloop, John L. *Liquid Hydrogen as a Propulsion Fuel, 1945-1959.* (NASA SP-4404, 1978).

Roland, Alex. *A Spacefaring People: Perspectives on Early Spaceflight.* (NASA SP-4405, 1985).

Bilstein, Roger E. *Orders of Magnitude: A History of the NACA and NASA, 1915-1990.* (NASA SP-4406, 1989).

Logsdon, John M. Logsdon, with Lear, Linda J., Warren-Findley, Jannelle, Williamson, Ray A., and Day, Dwayne A. *Exploring the Unknown: Selected Documents in the History of the U.S. Civil Space Program, Volume I: Organizing for Exploration.* (NASA SP-4407, 1995).

"New Series in NASA History," published by The Johns Hopkins University Press:

Cooper, Henry S. F., Jr. *Before Lift-Off: The Making of a Space Shuttle Crew.* (1987).

McCurdy, Howard E. *The Space Station Decision: Incremental Politics and Technological Choice.* (1990).

Hufbauer, Karl. *Exploring the Sun: Solar Science Since Galileo.* (1991).

McCurdy, Howard E. *Inside NASA: High Technology and Organizational Change in the U.S. Space Program.* (1993).

Lambright, W. Henry. *Powering Apollo: James E. Webb of NASA.* (1995).

MORGANTOWN PUBLIC LIBRARY
373 SPRUCE STREET
MORGANTOWN, WV 26505